Solutions Manual

Engineering Mechanics *Dynamics*

EIGHTH EDITION

R. C. Hibbeler

PRENTICE HALL, Upper Saddle River, NJ 07458

Associate Editor: Alice Dworkin
Production Editor: James Buckley
Supplement Cover Designer: Liz Nemeth
Special Projects Manager: Barbara A. Murray
Supplement Cover Manager: Paul Gourhan
Manufacturing Buyer: Julia Meehan
Editorial Assistant: Andrea Au

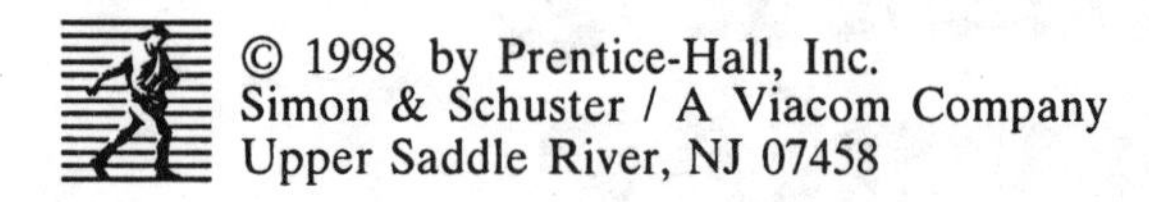

Simon & Schuster / A Viacom Company
Upper Saddle River, NJ 07458

Printed in the United States of America

10 9 8 7 6 5 4 3 2 1

ISBN 0-13-080259-X

Prentice-Hall International (UK) Limited, *London*
Prentice-Hall of Australia Pty. Limited, *Sydney*
Prentice-Hall Canada, Inc., *London*
Prentice-Hall Hispanoamericana, S.A., *Mexico*
Prentice-Hall of India Private Limited, *New Delhi*
Prentice-Hall of Japan, Inc., *Tokyo*
Simon & Schuster Asia Pte. Ltd., *Singapore*
Editora Prentice-Hall do Brazil, Ltda., *Rio de Janeiro*

CONTENTS

To the Instructor

This manual contains the solutions to all the problems in *Engineering Mechanics: Dynamics, Eighth Edition.* As stated in the preface of the text, the problems in every section are arranged in an approximate order of increasing difficulty. Be aware that answers to all but every fourth problem, which is indicated by an asterisk (*), are listed in the back of the book. Also, those problems indicated by a square (■) will require additional numerical work. The solutions may be readily determined using a numerical analysis method as outlined in Appendix B.

You may wish to use one of the lists of homework problems given on the following pages. Here you will find three lists for which the answers are in the back of the book, a fourth list for problems without answers, and a fifth sheet which can be used to develop your own personal syllabus. The prepared lists generally represent assigments with an easy, a moderate, and sometimes a more challenging problem.

If you have any questions regarding the solutions in this manual, I would greatly appreciate hearing from you.

R. C. Hibbeler

hibbeler@bell south.net

Section	Topic	Assignment List 1
12.1	Rectilinear Motion	12.9, 12.15, 12.21
12.2	Graphical Solutions	12.41, 12.47, 12.51
12.3-12.5	Curvilinear x,y,z Motion	12.69, 12.79, 12.89
12.6	Curvilienar n,t,b Motion	12.101, 12.111, 12.118
12.7	Curvilinear r,θ,z motion	12.139, 12.149, 12.167
12.8	Dependent Motion	12.173, 12.177, 12.183, 12.186
12.9	Independent Motion	12.193, 12.198, 12.202
13.1-13.4	Equation of Motion x,y,z	13.2, 13.11, 13.21, 13.33
13.5	Equation of Motion n,t,b	13.55, 13.63, 13.70
13.6	Equation of Motion r,θ,z	13.86, 13.91, 13.102
13.7	Space Mechanics	13.113, 13.117, 13.122
14.1-14.3	Principle of Work and Energy	14.2, 14.9, 14.17,14.25
14.4	Power and Efficiency	14.42, 14.50, 14.55
14.5-14.6	Conservation of Energy	14.70, 14.78, 14.83
15.1	Linear Momentum	15.2, 15.9, 15.21
15.2-15.3	Conservation of Linear Momentum	15.33, 15.38, 15.45
15.4	Impact	15.57, 15.65, 15.77, 15.83
15.5	Angular Momentum	15.90, 15.99, 15.105
*15.8	Steady Fluid Flow	15.111, 15.115, 15.117
*15.9	Variable Mass	15.123, 15.127, 15.131
16.1-16.3	Rotation About a Fixed Axis	16.2, 16.10, 16.18
16.4	Absolute Motion Analysis	16.33, 16.38, 16.45
16.5-16.6	Relative Velocity Analysis	16.51, 16.58, 16.66
16.7	Inst. Center of Zero Veloctiy	16.87, 16.91, 16.95
16.8-16.9	Relative Acceleration Analysis	16.106, 16.113, 16.121
16.10	Motion Using Rotating Axes	16.131, 16.135, 16.143
17.1	Moment of Inertia	17. 7, 17.13, 17.17
17.2-17.3	Eq. of Motion-Translation	17.25, 17.30, 17.37,17.49
17.4	Eq. of Motion-Fixed Axis Rotation	17.54, 17.61, 17.70
17.5	Eq. of Motion-Gen. Plane Motion	17.89, 17.98, 17.102
18.1-18.4	Principle of Work and Energy	18.2, 18.9, 18.17, 18.22
18.5	Conservation of Energy	18.42, 18.47, 18.54
19.1-19.2	Principle of Impulse and Momentum	19.5, 19.10, 19.17, 19.23
19.3	Cons. of Angular Momentum	19.33, 19.37, 19.41
19.4	Eccentric Impact	19.47, 19.51, 19.55
*20.1-*20.2	Rotation About a Fixed Point	20.1, 20.5, 20.9
*20.3	General Motion	20.15, 20.23, 20.27
*20-4	Relative Motion Analysis	20.33, 20.39, 20.43
*21.1	Moment of Inertia	21.2, 21.6, 21.13
*21.2-*21.3	Imp. and Mom., Work and Energy	21.23, 21.26, 21.35
*21.4	Euler Equations of Motion	21.42, 21.46, 21.51
*21.5-*21.6	Gyroscopic and Torque Free Motion	21.63, 21.67, 21.75, 21.79
22.1	Free Vibration	22.3, 22.10, 22.19
22.2	Energy Methods	22.30, 22.34, 22.38
*22.3-*22.4	Forced Vib., Damped Free Vib.	22.41, 22.45, 22.59
*22.5-*22.6	Damped Forced Vib., Elec. Analog	22.61, 22.65, 22.66

Section	Topic	Assignment List 2
12.1	Rectilinear Motion	12.7, 12.18, 12.22
12.2	Graphical Solutions	12.38, 12.46, 12.54
12.3-12.5	Curvilinear x,y,z Motion	12.71, 12.81, 12.90
12.6	Curvilienar n,t,b Motion	12.106, 12.113, 12.119
12.7	Curvilinear r,θ,z motion	12.141, 12.153, 12.169
12.8	Dependent Motion	12.174, 12.179, 12.185, 12.187
12.9	Independent Motion	12.195, 12.199, 12.203
13.1-13.4	Equation of Motion x,y,z	13.3, 13.14, 13.23, 13.35
13.5	Equation of Motion n,t,b	13.59, 13.65, 13.71
13.6	Equation of Motion r,θ,z	13.87, 13.93, 13.103
13.7	Space Mechanics	13.114, 13.118, 13.123
14.1-14.3	Principle of Work and Energy	14.5, 14.10, 14.19,14.27
14.4	Power and Efficiency	14.43, 14.51, 14.57
14.5-14.6	Conservation of Energy	14.71, 14.79, 14.85
15.1	Linear Momentum	15.5, 15.11, 15.22
15.2-15.3	Conservation of Linear Momentum	15.34, 15.39, 15.46
15.4	Impact	15.58, 15.67, 15.78, 15.81
15.5	Angular Momentum	15.91, 15.101, 15.106
*15.8	Steady Fluid Flow	15.113, 15.115, 15.118
*15.9	Variable Mass	15.125, 15.129, 15.133
16.1-16.3	Rotation About a Fixed Axis	16.3, 16.11, 16.21
16.4	Absolute Motion Analysis	16.34, 16.39, 16.47
16.5-16.6	Relative Velocity Analysis	16.53, 16.59, 16.69
16.7	Inst. Center of Zero Velocity	16.89, 16.93, 16.97
16.8-16.9	Relative Acceleration Analysis	16.107, 16.114, 16.122
16.10	Motion Using Rotating Axes	16.133, 16.137, 16.145
17.1	Moment of Inertia	17. 2, 17.14, 17.18
17.2-17.3	Eq. of Motion-Translation	17.26, 17.31, 17.38,17.50
17.4	Eq. of Motion-Fixed Axis Rotation	17.55, 17.62, 17.73
17.5	Eq. of Motion-Gen. Plane Motion	17.91, 17.99, 17.103
18.1-18.4	Principle of Work and Energy	18.3, 18.11, 18.19, 18.23
18.5	Conservation of Energy	18.43, 18.49, 18.57
19.1-19.2	Principle of Impulse and Momentum	19.6, 19.11, 19.18, 19.25
19.3	Cons. of Angular Momentum	19.34, 19.38, 19.42
19.4	Eccentric Impact	19.49, 19.53, 19.57
*20.1-*20.2	Rotation About a Fixed Point	20.2, 20.6, 20.10
*20.3	General Motion	20.17, 20.25, 20.29
*20-4	Relative Motion Analysis	20.34, 20.41, 20.45
*21.1	Moment of Inertia	21.3, 21.10, 21.14
*21.2-*21.3	Imp. and Mom., Work and Energy	21.25, 21.27, 21.37
*21.4	Euler Equations of Motion	21.43, 21.47, 21.53
*21.5-*21.6	Gyroscopic and Torque Free Motion	21.65, 21.69, 21.75, 21.77
22.1	Free Vibration	22.5, 22.14, 22.21
22.2	Energy Methods	22.31, 22.35, 22.39
*22.3-*22.4	Forced Vib., Damped Free Vib.	22.42, 22.47, 22.58
*22.5-*22.6	Damped Forced Vib., Elec. Analog	22.62, 22.66, 22.67

Section	Topic	Assignment List 3
12.1	Rectilinear Motion	12.11, 12.15, 12.23
12.2	Graphical Solutions	12.39, 12.43, 12.57
12.3-12.5	Curvilinear x,y,z Motion	12.75, 12.85, 12.91
12.6	Curvilienar n,t,b Motion	12.109, 12.114, 12.121
12.7	Curvilinear r,θ,z motion	12.145, 12.154, 12.162
12.8	Dependent Motion	12.175, 12.181, 12.182, 12.189
12.9	Independent Motion	12.197, 12.201, 12.205
13.1-13.4	Equation of Motion x,y,z	13.6, 13.18, 13.27, 13.37
13.5	Equation of Motion n,t,b	13.61, 13.66, 13.73
13.6	Equation of Motion r,θ,z	13.90, 13.94, 13.106
13.7	Space Mechanics	13.115, 13.119, 13.126
14.1-14.3	Principle of Work and Energy	14.6, 14.13, 14.23,14.29
14.4	Power and Efficiency	14.45, 14.53, 14.58
14.5-14.6	Conservation of Energy	14.73, 14.81, 14.86
15.1	Linear Momentum	15.6, 15.13, 15.23
15.2-15.3	Conservation of Linear Momentum	15.35, 15.41, 15.47
15.4	Impact	15.59, 15.69, 15.74, 15.79
15.5	Angular Momentum	15.93, 15.102, 15.107
*15.8	Steady Fluid Flow	15.114, 15.118, 15.119
*15.9	Variable Mass	15.126, 15.130, 15.134
16.1-16.3	Rotation About a Fixed Axis	16.5, 16.13, 16.25
16.4	Absolute Motion Analysis	16.35, 16.41, 16.46
16.5-16.6	Relative Velocity Analysis	16.54, 16.61, 16.71
16.7	Inst. Center of Zero Velocity	16.90, 16.94, 16.98
16.8-16.9	Relative Acceleration Analysis	16.109, 16.115, 16.123
16.10	Motion Using Rotating Axes	16.134, 16.138, 16.146
17.1	Moment of Inertia	17. 5, 17.15, 17.19
17.2-17.3	Eq. of Motion-Translation	17.27, 17.34, 17.39,17.51
17.4	Eq. of Motion-Fixed Axis Rotation	17.57, 17.63, 17.75
17.5	Eq. of Motion-Gen. Plane Motion	17.93, 17.101, 17.105
18.1-18.4	Principle of Work and Energy	18.5, 18.13, 18.21, 18.27
18.5	Conservation of Energy	18.45, 18.50, 18.58
19.1-19.2	Principle of Impulse and Momentum	19.7, 19.13, 19.22, 19.27
19.3	Cons. of Angular Momentum	19.35, 19.39, 19.43
19.4	Eccentric Impact	19.50, 19.54, 19.58
*20.1-*20.2	Rotation About a Fixed Point	20.3, 20.7, 20.11
*20.3	General Motion	20.18, 20.26, 20.30
*20-4	Relative Motion Analysis	20.35, 20.42, 20.47
*21.1	Moment of Inertia	21.5, 21.11, 21.15
*21.2-*21.3	Imp. and Mom., Work and Energy	21.26, 21.29, 21.38
*21.4	Euler Equations of Motion	21.45, 21.49, 21.54
*21.5-*21.6	Gyroscopic and Torque Free Motion	21.66, 21.70, 21.78, 21.79
22.1	Free Vibration	22.6, 22.15, 22.22
22.2	Energy Methods	22.33, 22.37, 22.39
*22.3-*22.4	Forced Vib., Damped Free Vib.	22.43, 22.49, 22.57
*22.5-*22.6	Damped Forced Vib., Elec. Analog	22.63, 22.65, 22.67

Section	Topic	Assignment Without Answers in Book
12.1	Rectilinear Motion	12.12, 12.16, 12.20
12.2	Graphical Solutions	12.40, 12.48, 12.60
12.3-12.5	Curvilinear x,y,z Motion	12.76, 12.84, 12.92
12.6	Curvilienar n,t,b Motion	12.108, 12.116, 12.120
12.7	Curvilinear r,θ,z motion	12.144, 12.156, 12.164
12.8	Dependent Motion	12.172, 12.180, 12.184, 12.188
12.9	Independent Motion	12.196, 12.200, 12.204
13.1-13.4	Equation of Motion x,y,z	13.8, 13.20, 13.32, 13.40
13.5	Equation of Motion n,t,b	13.60, 13.68, 13.80
13.6	Equation of Motion r,θ,z	13.88, 13.96, 13.108
13.7	Space Mechanics	13.112, 13.120, 13.124
14.1-14.3	Principle of Work and Energy	14.8, 14.16, 14.24,14.32
14.4	Power and Efficiency	14.44, 14.52, 14.60
14.5-14.6	Conservation of Energy	14.72, 14.80, 14.92
15.1	Linear Momentum	15.8, 15.16, 15.28
15.2-15.3	Conservation of Linear Momentum	15.36, 15.44, 15.52
15.4	Impact	15.60, 15.72, 15.76, 15.80
15.5	Angular Momentum	15.92, 15.100, 15.108
*15.8	Steady Fluid Flow	15.112, 15.116, 15.120
*15.9	Variable Mass	15.124, 15.128, 15.132
16.1-16.3	Rotation About a Fixed Axis	16.4, 16.12, 16.24
16.4	Absolute Motion Analysis	16.36, 16.40, 16.44
16.5-16.6	Relative Velocity Analysis	16.52, 16.60, 16.72
16.7	Inst. Center of Zero Velocity	16.88, 16.92, 16.100
16.8-16.9	Relative Acceleration Analysis	16.108, 16.116, 16.124
16.10	Motion Using Rotating Axes	16.132, 16.136, 16.144
17.1	Moment of Inertia	17. 4, 17.16, 17.20
17.2-17.3	Eq. of Motion-Translation	17.28, 17.32, 17.44,17.52
17.4	Eq. of Motion-Fixed Axis Rotation	17.56, 17.64, 17.80
17.5	Eq. of Motion-Gen. Plane Motion	17.96, 17.100, 17.108
18.1-18.4	Principle of Work and Energy	18.4, 18.16, 18.24, 18.32
18.5	Conservation of Energy	18.44, 18.52, 18.60
19.1-19.2	Principle of Impulse and Momentum	19.4, 19.12, 19.24, 19.28
19.3	Cons. of Angular Momentum	19.36, 19.40, 19.44
19.4	Eccentric Impact	19.48, 19.52, 19.56
*20.1-*20.2	Rotation About a Fixed Point	20.4, 20.8, 20.12
*20.3	General Motion	20.20, 20.24, 20.28
*20-4	Relative Motion Analysis	20.36, 20.40, 20.48
*21.1	Moment of Inertia	21.4, 21.12, 21.16
*21.2-*21.3	Imp. and Mom., Work and Energy	21.24, 21.32, 21.36
*21.4	Euler Equations of Motion	21.44, 21.48, 21.52
*21.5-*21.6	Gyroscopic and Torque Free Motion	21.64, 21.72, 21.76, 21.80
22.1	Free Vibration	22.4, 22.16, 22.24
22.2	Energy Methods	22.32, 22.36, 22.40
*22.3-*22.4	Forced Vib., Damped Free Vib.	22.44, 22.48, 22.52
*22.5-*22.6	Damped Forced Vib., Elec. Analog	22.60, 22.64, 22.68

Section	Topic	Assignment List
12.1	Rectilinear Motion	
12.2	Graphical Solutions	
12.3-12.5	Curvilinear x,y,z Motion	
12.6	Curvilinear n,t,b Motion	
12.7	Curvilinear r,θ,z motion	
12.8	Dependent Motion	
12.9	Independent Motion	
13.1-13.4	Equation of Motion x,y,z	
13.5	Equation of Motion n,t,b	
13.6	Equation of Motion r,θ,z	
13.7	Space Mechanics	
14.1-14.3	Principle of Work and Energy	
14.4	Power and Efficiency	
14.5-14.6	Conservation of Energy	
15.1	Linear Momentum	
15.2-15.3	Conservation of Linear Momentum	
15.4	Impact	
15.5	Angular Momentum	
*15.8	Steady Fluid Flow	
*15.9	Variable Mass	
16.1-16.3	Rotation About a Fixed Axis	
16.4	Absolute Motion Analysis	
16.5-16.6	Relative Velocity Analysis	
16.7	Inst. Center of Zero Velocity	
16.8-16.9	Relative Acceleration Analysis	
16.10	Motion Using Rotating Axes	
17.1	Moment of Inertia	
17.2-17.3	Eq. of Motion-Translation	
17.4	Eq. of Motion-Fixed Axis Rotation	
17.5	Eq. of Motion-Gen. Plane Motion	
18.1-18.4	Principle of Work and Energy	
18.5	Conservation of Energy	
19.1-19.2	Principle of Impulse and Momentum	
19.3	Cons. of Angular Momentum	
19.4	Eccentric Impact	
*20.1-*20.2	Rotation About a Fixed Point	
*20.3	General Motion	
*20-4	Relative Motion Analysis	
*21.1	Moment of Inertia	
*21.2-*21.3	Imp. and Mom., Work and Energy	
*21.4	Euler Equations of Motion	
*21.5-*21.6	Gyroscopic and Torque Free Motion	
22.1	Free Vibration	
22.2	Energy Methods	
*22.3-*22.4	Forced Vib., Damped Free Vib.	
*22.5-*22.6	Damped Forced Vib., Elec. Analog	

12-1. A ball is thrown downward from a 50-ft tower with an initial speed of 18 ft/s. Determine the speed at which it hits the ground and the time of travel.

$(+\downarrow)$ $\quad v^2 = v_0^2 + 2a_c(s - s_0)$

$v^2 = 18^2 + 2(32.2)(50 - 0)$

$v = 59.53 = 59.5$ ft/s **Ans**

$(+\downarrow)$ $\quad v = v_0 + a_c t$

$59.53 = 18 + 32.2t$

$t = 1.29$ s **Ans**

12-2. A car has an initial speed of 25 m/s and a constant deceleration of 3 m/s². Determine the velocity of the car when $t = 4$ s. What is the displacement of the car during the 4-s time interval? How much time is needed to stop the car?

$v = v_0 + a_c t$

$v = 25 + (-3)(4) = 13$ m/s **Ans**

$\Delta s = s - s_0 = v_0 t + \frac{1}{2} a_c t^2$

$\Delta s = s - 0 = 25(4) + \frac{1}{2}(-3)(4)^2 = 76$ m **Ans**

$v = v_0 + a_c t$

$0 = 25 + (-3)(t)$

$t = 8.33$ s **Ans**

12-3. If a particle has an initial velocity of $v_0 = 12$ ft/s to the right, at $s_0 = 0$, determine its position when $t = 10$ s, if $a = 2$ ft/s² to the left.

$(\overset{+}{\rightarrow})$ $\quad s = s_0 + v_0 t + \frac{1}{2} a_c t^2$

$= 0 + 12(10) + \frac{1}{2}(-2)(10)^2$

$= 20$ ft **Ans**

***12-4.** A particle travels along a straight line with a velocity $v = (12 - 3t^2)$ m/s, where t is in seconds. When $t = 1$ s, the particle is located 10 m to the left of the origin. Determine the acceleration when $t = 4$ s, the displacement from $t = 0$ to $t = 10$ s, and the distance the particle travels during this time period.

$v = 12 - 3t^2 \quad (1)$

$a = \frac{dv}{dt} = -6t\Big|_{t=4} = -24 \text{ m/s}^2$ **Ans**

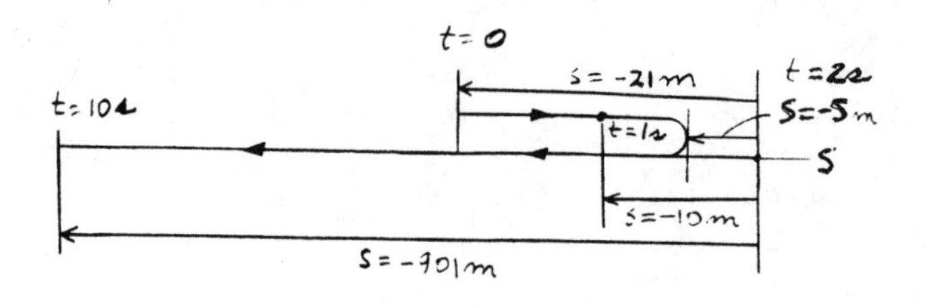

$\int_{-10}^{s} ds = \int_1^t v\,dt = \int_1^t (12 - 3t^2)\,dt$

$s + 10 = 12t - t^3 - 11$

$s = 12t - t^3 - 21$

$s|_{t=0} = -21$

$s|_{t=10} = -901$

$\Delta s = -901 - (-21) = -880$ m **Ans**

From Eq. (1) :

$v = 0$ when $t = 2s$

$s|_{t=2} = 12(2) - (2)^3 - 21 = -5$

$s_T = (21 - 5) + (901 - 5) = 912$ m **Ans**

12-5. Starting from rest, a particle moving in a straight line has an acceleration of $a = (2t - 6)$ m/s², where t is in seconds. What is the particle's velocity when $t = 6$ s, and what is its position when $t = 11$ s?

$a = 2t - 6$

$dv = a\,dt$

$\int_0^v dv = \int_0^t (2t - 6)\,dt$

$v = t^2 - 6t$

$ds = v\,dt$

$\int_0^s ds = \int_0^t (t^2 - 6t)\,dt$

$s = \frac{t^3}{3} - 3t^2$

When $t = 6$ s,

$v = 0$ **Ans**

When $t = 11$ s,

$s = 80.7$ m **Ans**

12-6. A particle moves along a straight line such that its position is defined by $s = (t^3 - 3t^2 + 2)$ m. Determine the average velocity, the average speed, and the acceleration of the particle when $t = 4$ s.

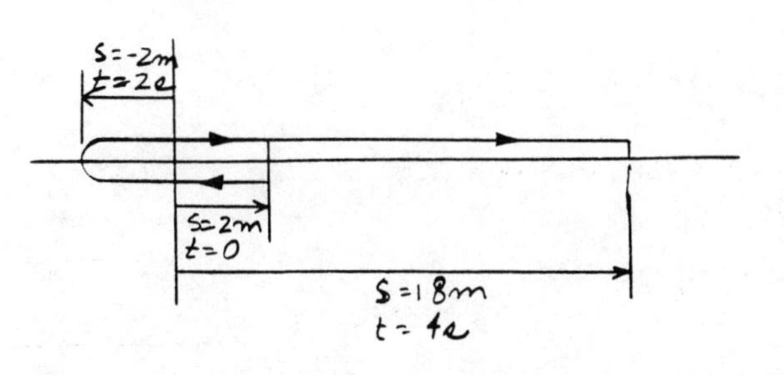

$$s = t^3 - 3t^2 + 2$$

$$v = \frac{ds}{dt} = 3t^2 - 6t$$

$$v = 0 \text{ at } t = 0, \; t = 2$$

$$a = \frac{dv}{dt} = 6t - 6$$

$$s|_{t=0} = 2$$

$$s|_{t=2} = -2$$

$$s|_{t=4} = 18$$

$$v_{avg} = \frac{\Delta s}{\Delta t} = \frac{18-2}{4-0} = 4 \text{ m/s} \quad \textbf{Ans}$$

$$(v_{sp})_{avg} = \frac{s_T}{\Delta t} = \frac{4+20}{4-0} = 6 \text{ m/s} \quad \textbf{Ans}$$

$$a|_{t=4} = 6(4) - 6 = 18 \text{ m/s}^2 \quad \textbf{Ans}$$

12-7. A particle moves along a straight line such that its position is defined by $s = (2t^3 + 3t^2 - 12t - 10)$ m. Determine the velocity, average velocity, and the average speed of the particle when $t = 3$ s.

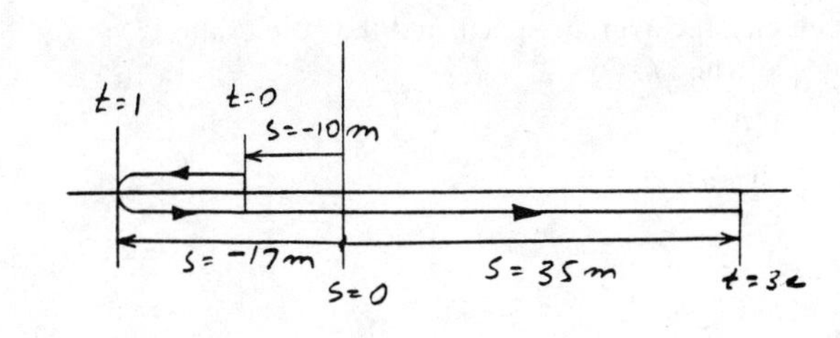

$$s = 2t^3 + 3t^2 - 12t - 10$$

$$v = \frac{ds}{dt} = 6t^2 + 6t - 12$$

$$v|_{t=3} = 6(3)^2 + 6(3) - 12 = 60 \text{ m/s} \quad \textbf{Ans}$$

$$v = 0 \text{ at } t = 1, \; t = -2$$

$$s|_{t=0} = -10$$

$$s|_{t=1} = -17$$

$$s|_{t=3} = 35$$

$$v_{avg} = \frac{\Delta s}{\Delta t} = \frac{35-(-10)}{3-0} = 15 \text{ m/s} \quad \textbf{Ans}$$

$$(v_{sp})_{avg} = \frac{s_T}{\Delta t} = \frac{(17-10)+(17+35)}{3-0} = 19.7 \text{ m/s} \quad \textbf{Ans}$$

***12-8.** The velocity of a particle traveling in a straight line is given by $v = (6t - 3t^2)$ m/s, where t is in seconds. If $s = 0$ when $t = 0$, determine the particle's deceleration and position when $t = 3$ s. How far has the particle traveled during the 3-s time interval, and what is its average speed?

$v = 6t - 3t^2$

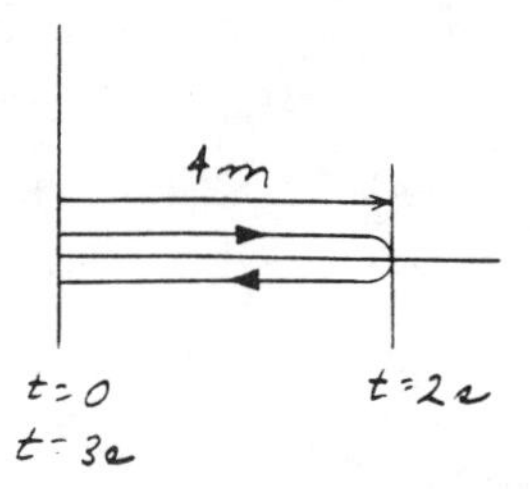

$a = \dfrac{dv}{dt} = 6 - 6t$

At $t = 3$ s

$a = -12 \text{ m/s}^2$ **Ans**

$ds = v\,dt$

$\int_0^s ds = \int_0^t (6t - 3t^2)\,dt$

$s = 3t^2 - t^3$

At $t = 3$ s

$s = 0$ **Ans**

Since $v = 0 = 6t - 3t^2$, when $t = 0$ and $t = 2$ s,

when $t = 2$ s, $s = 3(2)^2 - (2)^3 = 4$ m

$s_T = 4 + 4 = 8$ m **Ans**

$(v_{sp})_{avg} = \dfrac{s_t}{t} = \dfrac{8}{3} = 2.67$ m/s **Ans**

12-9. A particle moves along a straight line such that its position is defined by $s = (t^2 - 6t + 5)$ m. Determine the average velocity, the average speed, and the acceleration of the particle when $t = 6$ s.

$s = t^2 - 6t + 5$

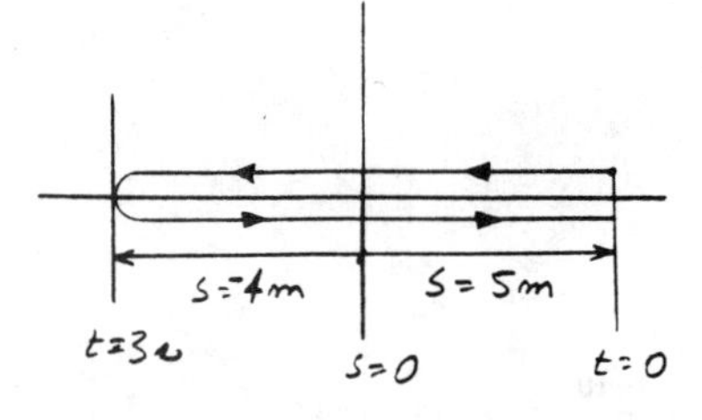

$v = \dfrac{ds}{dt} = 2t - 6$

$a = \dfrac{dv}{dt} = 2$

$v = 0$ when $t = 3$

$s|_{t=0} = 5$

$s|_{t=3} = -4$

$s|_{t=6} = 5$

$v_{avg} = \dfrac{\Delta s}{\Delta t} = \dfrac{0}{6} = 0$ **Ans**

$(v_{sp})_{avg} = \dfrac{s_T}{\Delta t} = \dfrac{9+9}{6} = 3$ m/s **Ans**

$a|_{t=6} = 2 \text{ m/s}^2$ **Ans**

12-10. A particle travels along a straight line such that in 2 s it moves from an initial position $s_A = +0.5$ m to a position $s_B = -1.5$ m. Then in another 4 s it moves from s_B to $s_C = +2.5$ m. Determine the particle's average velocity and average speed during the 6-s time interval.

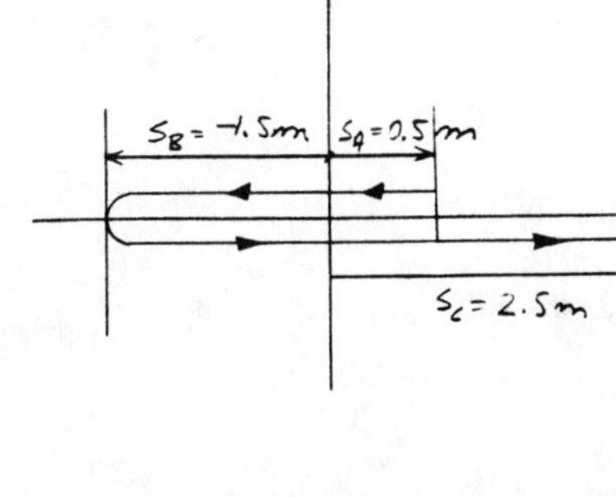

$$\Delta s = (s_C - s_A) = 2 \text{ m}$$

$$s_T = (0.5 + 1.5 + 1.5 + 2.5) = 6 \text{ m}$$

$$t = (2 + 4) = 6 \text{ s}$$

$$v_{avg} = \frac{\Delta s}{t} = \frac{2}{6} = 0.333 \text{ m/s} \quad \textbf{Ans}$$

$$(v_{sp})_{avg} = \frac{s_T}{t} = \frac{6}{6} = 1 \text{ m/s} \quad \textbf{Ans}$$

12-11. A particle is moving along a straight line such that its position is defined by $s = (10t^2 + 20)$ mm, where t is in seconds. Determine (a) the displacement of the particle during the time interval from $t = 1$ s to $t = 5$ s, (b) the average velocity of the particle during this time interval, and (c) the acceleration when $t = 1$ s.

$$s = 10t^2 + 20$$

(a) $s|_{1\,s} = 10(1)^2 + 20 = 30$ mm

$s|_{5\,s} = 10(5)^2 + 20 = 270$ mm

$\Delta s = 270 - 30 = 240$ mm **Ans**

(b) $\Delta t = 5 - 1 = 4$ s

$$v_{avg} = \frac{\Delta s}{\Delta t} = \frac{240}{4} = 60 \text{ mm/s} \quad \textbf{Ans}$$

(c) $a = \dfrac{d^2s}{dt^2} = 20 \text{ mm/s}^2$ (for all t) **Ans**

***12-12.** The acceleration of a particle as it moves along a straight line is given by $a = (2t - 1)$ m/s², where t is in seconds. If $s = 1$ m and $v = 2$ m/s when $t = 0$, determine the particle's velocity and position when $t = 6$ s. Also, determine the total distance the particle travels during this time period.

$t = 0$, d, $s = 1m$, $s = 67m$

$$a = 2t - 1$$

$$dv = a\,dt$$

$$\int_2^v dv = \int_0^t (2t - 1)\,dt$$

$$v = t^2 - t + 2$$

$$ds = v\,dt$$

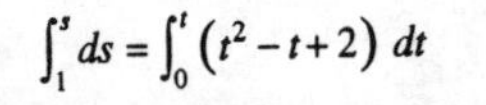

$$\int_1^s ds = \int_0^t (t^2 - t + 2)\,dt$$

$$s = \frac{1}{3}t^3 - \frac{1}{2}t^2 + 2t + 1$$

When $t = 6$ s

$v = 32$ m/s **Ans**

$s = 67$ m **Ans**

Since $v \neq 0$ for $0 \le t \le 6$ s, then

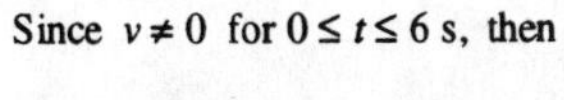

$d = 67 - 1 = 66$ m **Ans**

12-13. A car is to be hoisted by elevator to the fourth floor of a parking garage, which is 48 ft above the ground. If the elevator can accelerate at 0.6 ft/s², decelerate at 0.3 ft/s², and reach a maximum speed of 8 ft/s, determine the shortest time to make the lift, starting from rest and ending at rest.

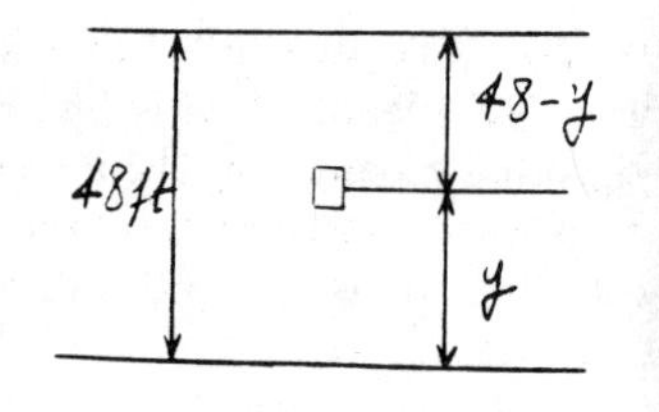

$+\uparrow\ v^2 = v_0^2 + 2\,a_c(s - s_0)$

$v_{max}^2 = 0 + 2(0.6)(y - 0)$

$0 = v_{max}^2 + 2(-0.3)(48 - y)$

$0 = 1.2\,y - 0.6(48 - y)$

$y = 16.0$ ft, $\quad v_{max} = 4.382$ ft/s < 8 ft/s

$+\uparrow\ v = v_0 + a_c t$

$4.382 = 0 + 0.6\,t_1$

$t_1 = 7.303$ s

$0 = 4.382 - 0.3\,t_2$

$t_2 = 14.61$ s

$t = t_1 + t_2 = 21.9$ s **Ans**

12-14. The position of a particle on a straight line is given by $s = (t^3 - 9t^2 + 15t)$ ft, where t is in seconds. Determine the position of the particle when $t = 6$ s and the total distance it travels during the 6-s time interval. *Hint:* Plot the path to determine the total distance traveled.

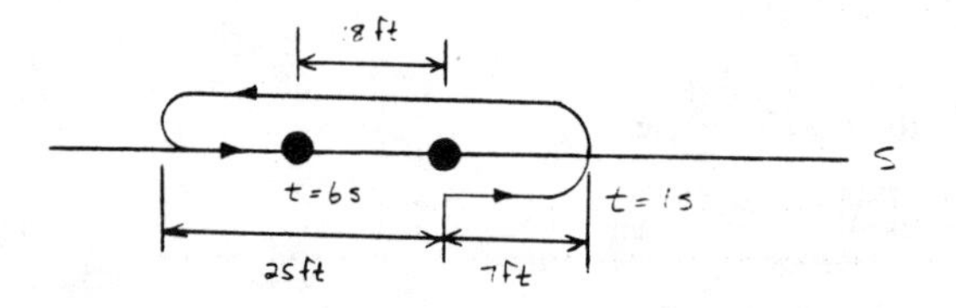

$s = t^3 - 9t^2 + 15t$

$v = \dfrac{ds}{dt} = 3t^2 - 18t + 15$

$v = 0$ when $t = 1$ s and $t = 5$ s

$t = 0,\ s = 0$

$t = 1$ s, $s = 7$ ft

$t = 5$ s, $s = -25$ ft

$t = 6$ s, $s = -18$ ft **Ans**

$s_T = 7 + 7 + 25 + (25 - 18) = 46$ ft **Ans**

12-15. A particle has an initial speed of 27 m/s. If it experiences a deceleration of $a = (-6t)$ m/s², where t is in seconds, determine the distance traveled before it stops.

$a = -6t$

$dv = a\,dt$

$\int_{27}^{v} dv = -\int_0^t 6t\,dt$

$v - 27 = -3t^2$

$v = 27 - 3t^2$

$v = 0$ at $t = 3$

$ds = v\,dt$

$\int_0^s ds = \int_0^t (27 - 3t^2)\,dt$

$s = 27t - t^3$

$s|_{t=3} = 27(3) - (3)^3 = 54$ m **Ans**

■*12-16. A particle has an initial speed of 27 m/s. If it experiences a deceleration of $a = (-6t)$ m/s^2, where t is in seconds, determine its velocity when it travels 10 m. How much time does this take?

$a = -6t$

$dv = a\,dt$

$\int_{27}^{v} dv = -\int_0^t 6t\,dt$

$v - 27 = -3t^2$

$v = 27 - 3t^2$

$v = 0$ when $t = 3$

$ds = v\,dt$

$\int_0^s ds = \int_0^t (27 - 3t^2)\,dt$

$s = 27t - t^3$

At $s = 10$ m $\qquad 10 = 27t - t^3 \qquad t^3 - 27t + 10 = 0$

Solving for the smallest positive root $t < 3$ s

$t = 0.3723 = 0.372$ s **Ans**

$v|_{t=0.3723} = 27 - 3(0.3723)^2 = 26.6$ m/s **Ans**

12-17. A car starts from rest and moves along a straight line with an acceleration of $a = (3s^{-1/3})$ m/s^2, where s is in meters. Determine the car's velocity and position when $t = 6$ s.

$a = 3s^{-\frac{1}{3}}$

$a\,ds = v\,dv$

$\int_0^s 3s^{-\frac{1}{3}}\,ds = \int_0^v v\,dv$

$\frac{3}{2}(3)s^{\frac{2}{3}} = \frac{1}{2}v^2$

$v = 3s^{\frac{1}{3}}$

$\frac{ds}{dt} = 3s^{\frac{1}{3}}$

$\int_0^s s^{-\frac{1}{3}}\,ds = \int_0^t 3\,dt$

$\frac{3}{2}s^{\frac{2}{3}} = 3t$

$s = (2t)^{\frac{3}{2}}$

$s|_{t=6} = (2(6))^{\frac{3}{2}} = 41.57 = 41.6$ m **Ans**

$v|_{t=6} = 3(41.57)^{\frac{1}{3}} = 10.4$ m/s **Ans**

12-18. A car starts from rest and moves along a straight line with an acceleration of $a = (3s^{-1/3})\ \text{m/s}^2$, where s is in meters. Determine the car's acceleration when $t = 4$ s.

$$a = 3s^{-\frac{1}{3}}$$

$$a\,ds = v\,dv$$

$$\int_0^s 3s^{-\frac{1}{3}}\,ds = \int_0^v v\,dv$$

$$\frac{3}{2}(3)s^{\frac{2}{3}} = \frac{1}{2}v^2$$

$$v = 3s^{\frac{1}{3}}$$

$$\frac{ds}{dt} = 3s^{\frac{1}{3}}$$

$$\int_0^s s^{-\frac{1}{3}}\,ds = \int_0^t 3\,dt$$

$$\frac{3}{2}s^{\frac{2}{3}} = 3t$$

$$s = (2t)^{\frac{3}{2}}$$

$$s|_{t=4} = (2(4))^{\frac{3}{2}} = 22.62 = 22.6 \text{ m}$$

$$a|_{t=4} = 3(22.62)^{-\frac{1}{3}} = 1.06 \text{ m/s}^2 \qquad \textbf{Ans}$$

12-19. Car B is traveling a distance d ahead of car A. Both cars are traveling at 60 ft/s when the driver of B suddenly applies the brakes, causing his car to decelerate at 12 ft/s². It takes the driver of car A 0.75 s to react (this is the normal reaction time for drivers). When he applies his brakes, he decelerates at 15 ft/s². Determine the minimum distance d between the cars so as to avoid a collision.

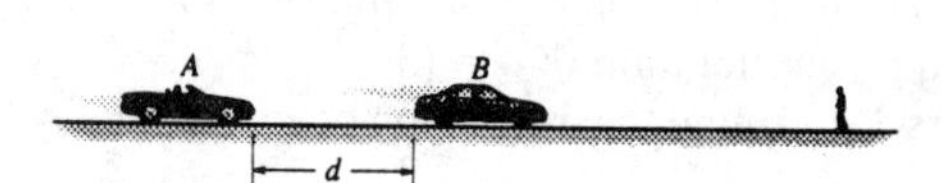

For B:

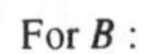

$(\overset{+}{\rightarrow})$ $\quad v = v_0 + a_c t$

$$v_B = 60 - 12t$$

$(\overset{+}{\rightarrow})$ $\quad s = s_0 + v_0 t + \frac{1}{2}a_c t^2$

$$s_B = d + 60t - \frac{1}{2}(12)t^2 \qquad (1)$$

For A:

$(\overset{+}{\rightarrow})$ $\quad v = v_0 + a_c t$

$$v_A = 60 - 15(t - 0.75), \quad \{t > 0.75\}$$

$(\overset{+}{\rightarrow})$ $\quad s = s_0 + v_0 t + \frac{1}{2}a_c t^2$

$$s_A = 60(0.75) + 60(t - 0.75) - \frac{1}{2}(15)(t - 0.75)^2, \quad \{t > 0.75\} \quad (2)$$

Require $v_A = v_B$ the moment of closest approach.

$$60 - 12t = 60 - 15(t - 0.75)$$

$$t = 3.75 \text{ s}$$

Worst case without collision would occur when $s_A = s_B$.

At $t = 3.75$ s, from Eqs. (1) and (2):

$$60(0.75) + 60(3.75 - 0.75) - 7.5(3.75 - 0.75)^2 = d + 60(3.75) - 6(3.75)^2$$

$$157.5 = d + 140.625$$

$$d = 16.9 \text{ ft} \qquad \textbf{Ans}$$

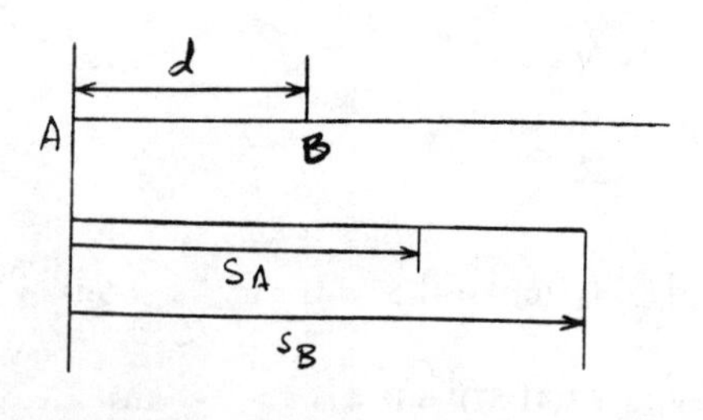

***12-20.** A particle is moving along a straight line such that its speed is defined as $v = (-4s^2)$ m/s, where s is in meters. If $s = 2$ m when $t = 0$, determine the velocity and acceleration as functions of time.

$$v = -4s^2$$

$$\frac{ds}{dt} = -4s^2$$

$$\int_2^s s^{-2}\, ds = \int_0^t -4\, dt$$

$$-s^{-1}\big|_2^s = -4t\big|_0^t$$

$$t = \frac{1}{4}\left(s^{-1} - 0.5\right)$$

$$s = \frac{2}{8t+1}$$

$$v = -4\left(\frac{2}{8t+1}\right)^2 = \left(-\frac{16}{(8t+1)^2}\right) \text{ m/s} \qquad \textbf{Ans}$$

$$a = \frac{dv}{dt} = \frac{16(2)(8t+1)(8)}{(8t+1)^4} = \left(\frac{256}{(8t+1)^3}\right) \text{ m/s}^2 \qquad \textbf{Ans}$$

12-21. A particle is moving along a straight line such that its acceleration is defined as $a = (4s^2)$ m/s^2, where s is in meters. If $v = -100$ m/s when $s = 10$ m and $t = 0$, determine the particle's velocity as a function of position.

$$a = 4s^2$$

$$v\, dv = a\, ds$$

$$\int_{-100}^{v} v\, dv = \int_{10}^{s} 4s^2\, ds$$

$$\frac{1}{2}v^2\Big|_{-100}^{v} = \frac{4}{3}s^3\Big|_{10}^{s}$$

$$\frac{1}{2}\left(v^2 - (-100)^2\right) = \frac{4}{3}\left(s^3 - (10)^3\right)$$

$$v = -\left(2.667s^3 + 7333.3\right)^{\frac{1}{2}} \text{ m/s} \qquad \textbf{Ans}$$

12-22. A particle is moving along a straight line such that its acceleration is defined as $a = (-2v)$ m/s^2, where v is in meters per second. If $v = 20$ m/s when $s = 0$ and $t = 0$, determine the particle's position, velocity, and acceleration as functions of time.

$$a = -2v$$

$$\frac{dv}{dt} = -2v$$

$$\int_{20}^{v} \frac{dv}{v} = \int_0^t -2\, dt$$

$$\ln\frac{v}{20} = -2t$$

$$v = \left(20e^{-2t}\right) \text{m/s} \qquad \textbf{Ans}$$

$$a = \frac{dv}{dt} = \left(-40e^{-2t}\right) \text{m/s}^2 \qquad \textbf{Ans}$$

$$\int_0^s ds = v\, dt = \int_0^t \left(20e^{-2t}\right) dt$$

$$s = -10e^{-2t}\big|_0^t = -10\left(e^{-2t} - 1\right)$$

$$s = 10\left(1 - e^{-2t}\right) \text{m} \qquad \textbf{Ans}$$

12-23. A particle is moving along a straight line such that its acceleration is defined as $a = (-2v)$ m/s^2, where v is in meters per second. If $v = 20$ m/s when $s = 0$ and $t = 0$, determine the particle's velocity as a function of position and the distance the particle moves before it stops.

$$a = -2v$$

$$a\,ds = v\,dv$$

$$-2v\,ds = v\,dv$$

$$-2\int_0^s ds = \int_{20}^v dv$$

$$-2\,s = v - 20$$

$v = (20 - 2s)$ m/s **Ans**

When $v = 0$,

$s = 10$ m **Ans**

Note : From the solution of Prob. 12 – 22 , $t \to \infty$ for this to occur.

***12-24.** A particle, initially at the origin, moves along a straight line through a fluid medium such that its velocity is defined as $v = 1.8(1 - e^{-0.3t})$ m/s, where t is in seconds. Determine the displacement of the particle during the first 3 s.

$$v = 1.8\left(1 - e^{-0.3t}\right)$$

$$ds = v\,dt$$

$$\Delta s = \int_0^3 1.8\left(1 - e^{-0.3t}\right)dt = 1.8\left(t + \frac{1}{0.3}e^{-0.3t}\right)\Big|_0^3$$

$$\Delta s = 1.8\left(3 + \frac{1}{0.3}e^{-0.3(3)}\right) - 1.8\left(0 + \frac{1}{0.3}\right)$$

$\Delta s = 1.84$ m **Ans**

■12-25. A particle moves along a straight line with an acceleration of $a = 5/(3s^{1/3} + s^{5/2})$ m/s^2, where s is in meters. Determine the particle's velocity when $s = 2$ m, if it starts from rest when $s = 1$ m. Use Simpson's rule to evaluate the integral.

$$a = \frac{5}{\left(3s^{\frac{1}{3}} + s^{\frac{5}{2}}\right)}$$

$$a\,ds = v\,dv$$

$$\int_1^2 \frac{5\,ds}{\left(3s^{\frac{1}{3}} + s^{\frac{5}{2}}\right)} = \int_0^v v\,dv$$

$$0.8351 = \frac{1}{2}v^2$$

$v = 1.29$ m/s **Ans**

12-26. A sandbag is dropped from a balloon which is ascending vertically at a constant speed of 6 m/s. If the bag is released with the same upward velocity of 6 m/s when $t = 0$ and hits the ground when $t = 8$ s, determine the speed of the bag as it hits the ground and the altitude of the balloon at this instant.

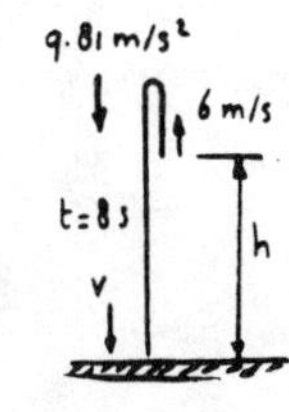

$(+\downarrow) \qquad s = s_0 + v_0 t + \frac{1}{2} a_c t^2$

$$h = 0 + (-6)(8) + \frac{1}{2}(9.81)(8)^2$$

$$= 265.92 \text{ m}$$

During $t = 8$ s, the balloon rises

$$h' = vt = 6(8) = 48 \text{ m}$$

Altitude $= h + h' = 265.92 + 48 = 314$ m **Ans**

$(+\downarrow) \qquad v = v_0 + a_c t$

$$v = -6 + 9.81(8) = 72.5 \text{ m/s}$$ **Ans**

12-27. A particle is moving along a straight line such that when it is at the origin it has a velocity of 4 m/s. If it begins to decelerate at the rate of $a = (-1.5v^{1/2})$ m/s^2, where v is in m/s, determine the distance it travels before it stops.

$$a = \frac{dv}{dt} = -1.5v^{\frac{1}{2}}$$

$$\int_4^v v^{-\frac{1}{2}}\, dv = \int_0^t -1.5\, dt$$

$$2v^{\frac{1}{2}}\Big|_4^v = -1.5t\Big|_0^t$$

$$2\left(v^{\frac{1}{2}} - 2\right) = -1.5t$$

$v = (2 - 0.75t)^2$ m/s (1)

$$\int_0^s ds = \int_0^t (2 - 0.75t)^2\, dt = \int_0^t \left(4 - 3t + 0.5625t^2\right) dt$$

$s = 4t - 1.5t^2 + 0.1875t^3$ (2)

From Eq. (1), the particle will stop when

$$0 = (2 - 0.75t)^2$$

$$t = 2.667 \text{ s}$$

$s|_{t=2.667} = 4(2.667) - 1.5(2.667)^2 + 0.1875(2.667)^3 = 3.56$ m **Ans**

***12-28.** A particle is moving along a straight line such that when it is at the origin it has a velocity of 4 m/s. If it begins to decelerate at the rate of $a = (-1.5v^{1/2})$ m/s^2, where v is in m/s, determine the particle's position and velocity when $t = 2$ s.

$$a = \frac{dv}{dt} = -1.5v^{\frac{1}{2}}$$

$$\int_4^v v^{-\frac{1}{2}}\,dv = \int_0^t -1.5\,dt$$

$$2v^{\frac{1}{2}}\Big|_4^v = -1.5t\Big|_0^t$$

$$2\left(v^{\frac{1}{2}} - 2\right) = -1.5t$$

$$v = (2 - 0.75t)^2 \text{ m/s}$$

$$v\big|_{t=2} = (2 - 0.75(2))^2 = 0.25 \text{ m/s} \qquad \textbf{Ans}$$

$$\int_0^s ds = \int_0^t (2 - 0.75t)^2\,dt = \int_0^t \left(4 - 3t + 0.5625t^2\right)dt$$

$$s = 4t - 1.5t^2 + 0.1875t^3$$

$$s\big|_{t=2} = 4(2) - 1.5(2)^2 + 0.1875(2)^3 = 3.5 \text{ m} \qquad \textbf{Ans}$$

12-29. A particle is moving with a velocity of v_0 when $s = 0$ and $t = 0$. If it is subjected to a deceleration of $a = -kv^3$, where k is a constant, determine its velocity and position as functions of time.

$$a = \frac{dv}{dt} = -kv^3$$

$$\int_{v_0}^v v^{-3}\,dv = \int_0^t -k\,dt$$

$$-\frac{1}{2}\left(v^{-2} - v_0^{-2}\right) = -kt$$

$$v = \left(2kt + \left(\frac{1}{v_0^2}\right)\right)^{-\frac{1}{2}} \qquad \textbf{Ans}$$

$$ds = v\,dt$$

$$\int_0^s ds = \int_0^t \frac{dt}{\left(2kt + \left(\frac{1}{v_0^2}\right)\right)^{\frac{1}{2}}}$$

$$s = \frac{2\left(2kt + \left(\frac{1}{v_0^2}\right)\right)^{\frac{1}{2}}}{2k}\Bigg|_0^t$$

$$s = \frac{1}{k}\left[\left(2kt + \left(\frac{1}{v_0^2}\right)\right)^{\frac{1}{2}} - \frac{1}{v_0}\right] \qquad \textbf{Ans}$$

12-30. A car can have an acceleration and a deceleration of 5 m/s². If it starts from rest, and can have a maximum speed of 60 m/s, determine the shortest time it can travel a distance of 1200 m when it stops.

Time and distance to reach 60 m/s :

$a = 5\ \text{m/s}^2$

$v = v_0 + a_c t$

$60 = 0 + 5t$

$t = 12\ \text{s}$

$s = s_0 + v_0 t + \frac{1}{2} a_c t^2$

$s_1 = 0 + 0 + \frac{1}{2}(5)(12)^2 = 360\ \text{m}$

This is the same time and distance to stop.

Time and distance to travel at 60 m/s :

$s_2 = 1200 - 2(360) = 480\ \text{m}$

$s = vt$

$480 = 60t$

$t = 8\ \text{s}$

Total time :

$t = 12 + 8 + 12 = 32\ \text{s}$ **Ans**

12-31. A ball A is thrown vertically upward from the top of a 30-m-high building with an initial velocity of 5 m/s. At the same instant another ball B is thrown upward from the ground with an initial velocity of 20 m/s. Determine the height from the ground and the time at which they pass.

Origin at roof :

Ball A :

$(+\uparrow)$ $s = s_0 + v_0 t + \frac{1}{2} a_c t^2$

$-s = 0 + 5t - \frac{1}{2}(9.81)t^2$

Ball B :

$(+\uparrow)$ $s = s_0 + v_0 t + \frac{1}{2} a_c t^2$

$-s = -30 + 20t - \frac{1}{2}(9.81)t^2$

Solving,

$t = 2\ \text{s}$ **Ans**

$s = 9.62\ \text{m}$

Distance from ground,

$d = (30 - 9.62) = 20.4\ \text{m}$ **Ans**

Also, origin at ground,

$s = s_0 + v_0 t + \frac{1}{2} a_c t^2$

$s_A = 30 + 5t + \frac{1}{2}(-9.81)t^2$

$s_B = 0 + 20t + \frac{1}{2}(-9.81)t^2$

Require

$s_A = s_B$

$30 + 5t + \frac{1}{2}(-9.81)t^2 = 20t + \frac{1}{2}(-9.81)t^2$

$t = 2\ \text{s}$ **Ans**

$s_B = 20.4\ \text{m}$ **Ans**

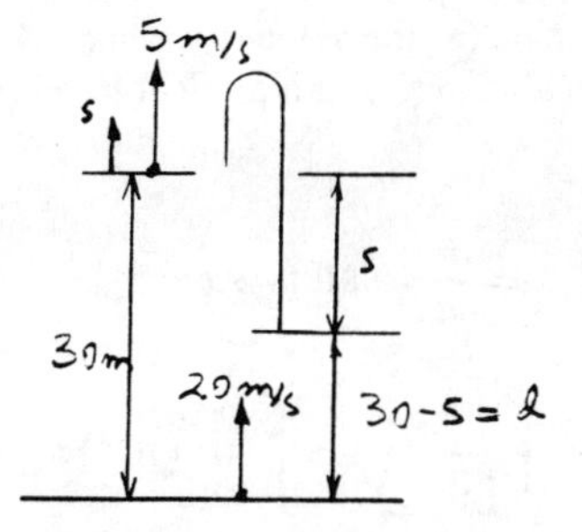

***12-32.** A motorcycle starts from rest at $t = 0$ and travels along a straight road with a constant acceleration of 6 ft/s^2 until it reaches a speed of 50 ft/s. Afterwards it maintains this speed. Also, when $t = 0$, a car located 6000 ft down the road is traveling toward the motorcycle at a constant speed of 30 ft/s. Determine the time and the distance traveled by the motorcycle when they pass each other.

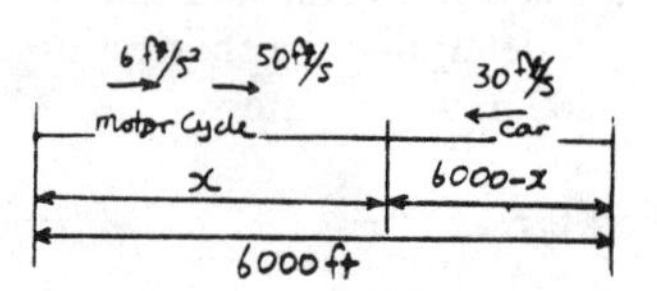

Motorcycle :

$(\overset{+}{\rightarrow})$ $\quad v = v_0 + a_c t'$

$50 = 0 + 6t'$

$t' = 8.33 \text{ s}$

$v^2 = v_0^2 + 2a_c(s - s_0)$

$(50)^2 = 0 + 2(6)(s' - 0)$

$s' = 208.33 \text{ ft}$

In $t' = 8.33$ s car travels

$s'' = v_0 t' = 30(8.33) = 250 \text{ ft}$

Distance between motorcycle and car :

$s = v_0 t; \quad 6000 - 250 - 208.33 = 5541.67 \text{ ft}$

When passing occurs for motorcycle :

$s = v_0 t; \quad x = 50(t'')$

For car :

$s = v_0 t; \quad 5541.67 - x = 30(t'')$

Solving,

$x = 3463.54 \text{ ft}$

$t'' = 69.27 \text{ s}$

Thus, for the motorcycle,

$t = 69.27 + 8.33 = 77.6 \text{ s}$ **Ans**

$s_m = 208.33 + 3463.54 = 3.67(10)^3 \text{ ft}$ **Ans**

t=0
50 ft/s
30 ft/s
t=0
208.33 ft
x
5541.67-x
250 ft
t=8.33s
t=8.33s

12-33. If the effects of atmospheric resistance are accounted for, a freely falling body has an acceleration defined by the equation $a = 9.81[1 - v^2(10^{-4})]$ m/s^2, where v is in m/s and the positive direction is downward. If the body is released from rest at a *very high altitude,* determine (a) the velocity when $t = 5$ s, and (b) the body's terminal or maximum attainable velocity (as $t \to \infty$).

(a) $\quad a = \dfrac{dv}{dt} = 9.81\left[1 - v^2\left(10^{-4}\right)\right]$

$$\int_0^v \frac{dv}{\left[10^4 - v^2\right]} = \int_0^t 9.81\left(10^{-4}\right)dt \qquad (1)$$

$$\frac{1}{100}\tanh^{-1}\left(\frac{v}{100}\right)\Big|_0^v = 9.81\left(10^{-4}\right)t$$

$$\tanh^{-1}\left(\frac{v}{100}\right) = \left(9.81\left(10^{-2}\right)t\right) \qquad (2)$$

$$v = 100\tanh\left(9.81\left(10^{-2}\right)(5)\right)$$

$= 100\tanh(0.4905) = 45.5 \text{ m/s}$ **Ans**

(b) From Eq. (2), with $t \to \infty$,

$v = 100\tanh\infty = 100 \text{ m/s}$ **Ans**

Also note that Eq. (1) can be written as

$$10^4\int_0^v \frac{dv}{\left[10^4 - v^2\right]} = 9.81t$$

$$10^4\left[\left(\frac{1}{2(10^2)}\right)\ln\left(\frac{10^2 + v}{10^2 - v}\right)\right]_0^v = 9.81t$$

$$50\left[\ln\left(\frac{100 + v}{100 - v}\right) - \ln 1\right] = 9.81t$$

When $t = 5$ s,

$$\frac{100 + v}{100 - v} = e^{0.981} = 2.667$$

$100 + v = 266.7 - 2.667v$

$v = \dfrac{166.7}{3.667} = 45.5 \text{ m/s}$ **Ans**

12-34. As a body is projected to a high altitude above the earth's *surface*, the variation of the acceleration of gravity with respect to altitude y must be taken into account. Neglecting air resistance, this acceleration is determined from the formula $a = -g_0[R^2/(R+y)^2]$, where g_0 is the constant gravitational acceleration at sea level, R is the radius of the earth, and the positive direction is measured upward. If $g_0 = 9.81 \text{ m/s}^2$ and $R = 6356$ km, determine the minimum initial velocity (escape velocity) at which a projectile should be shot vertically from the earth's surface so that it does not fall back to the earth. *Hint:* This requires that $v = 0$ as $y \to \infty$.

$$v\,dv = a\,dy$$

$$\int_v^0 v\,dv = -g_0 R^2 \int_0^\infty \frac{dy}{(R+y)^2}$$

$$\frac{v^2}{2}\Big|_v^0 = \frac{g_0 R^2}{R+y}\Big|_0^\infty$$

$$v = \sqrt{2 g_0 R}$$

$$= \sqrt{2(9.81)(6356)(10)^3}$$

$$= 11167 \text{ m/s} = 11.2 \text{ km/s} \qquad \textbf{Ans}$$

12-35. Accounting for the variation of gravitational acceleration a with respect to altitude y (see Prob. 12-34), derive an equation that relates the velocity of a freely falling particle to its altitude. Assume that the particle is released from rest at an altitude y_0 from the earth's surface. With what velocity does the particle strike the earth if it is released from rest at an altitude $y_0 = 500$ km? Use the numerical data in Prob. 12-34.

From Prob. 12–34,

$$(+\uparrow) \qquad g = -g_0 \frac{R^2}{(R+y)^2}$$

Since $g\,dy = v\,dv$

then

$$-g_0 R^2 \int_{y_0}^{y} \frac{dy}{(R+y)^2} = \int_0^v v\,dv$$

$$g_0 R^2 \left[\frac{1}{R+y}\right]_{y_0}^{y} = \frac{v^2}{2}$$

$$g_0 R^2 \left[\frac{1}{R+y} - \frac{1}{R+y_0}\right] = \frac{v^2}{2}$$

Thus

$$v = -R\sqrt{\frac{2g_0(y_0-y)}{(R+y)(R+y_0)}} \qquad \textbf{Ans}$$

When $y_0 = 500$ km, $y = 0$,

$$v = -6356(10^3)\sqrt{\frac{2(9.81)(500)(10^3)}{6356(6356+500)(10^6)}}$$

$$v = -3016 \text{ m/s} = 3.02 \text{ km/s} \downarrow \qquad \textbf{Ans}$$

*12-36. When a particle falls through the air, its initial acceleration $a = g$ diminishes until it is zero, and thereafter it falls at a constant or terminal velocity v_f. If this variation of the acceleration can be expressed as $a = (g/v_f^2)(v_f^2 - v^2)$, determine the time needed for the velocity to become $v < v_f$. Initially the particle falls from rest.

$$\frac{dv}{dt} = a = \left(\frac{g}{v_f^2}\right)(v_f^2 - v^2)$$

$$\int_0^v \frac{dv}{v_f^2 - v^2} = \frac{g}{v_f^2}\int_0^t dt$$

$$\frac{1}{2v_f}\ln\left(\frac{v_f + v}{v_f - v}\right)\Bigg|_0^v = \frac{g}{v_f^2}t$$

$$t = \frac{v_f}{2g}\ln\left(\frac{v_f + v}{v_f - v}\right)$$ **Ans**

12-37. The speed of a train during the first minute has been recorded as follows:

t (s)	0	20	40	60
v (m/s)	0	16	21	24

Plot the v–t graph, approximating the curve as straight-line segments between the given points. Determine the total distance traveled.

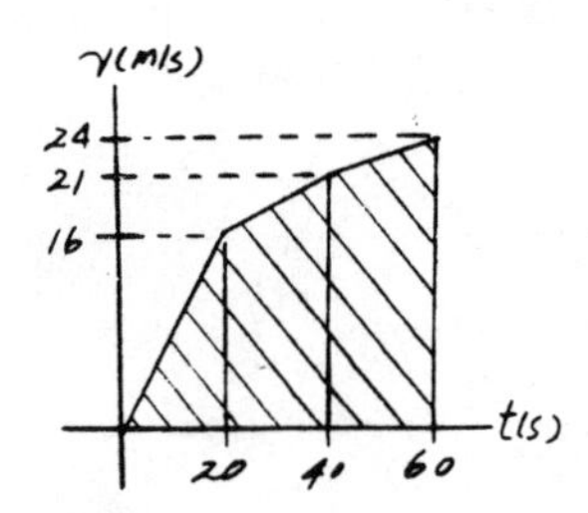

The total distance traveled is equal to the area under the graph.

$$s_T = \frac{1}{2}(20)(16) + \frac{1}{2}(40-20)(16+21) + \frac{1}{2}(60-40)(21+24) = 980\text{ m}$$ **Ans**

12-38. A two-stage missile is fired vertically from rest with the acceleration shown. In 15 s the first stage A burns out and the second stage B ignites. Plot the v–t and s–t graphs which describe the motion of the second stage for $0 \le t \le 20$ s.

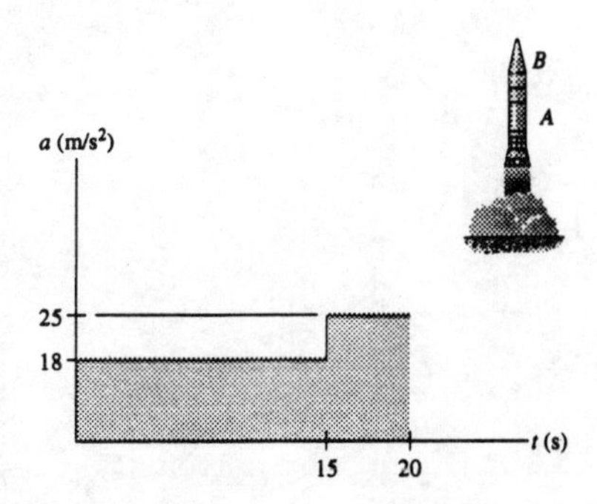

Since $v = \int a\,dt$, the constant lines of the $a-t$ graph become sloping lines for the $v-t$ graph.
The numerical values for each point are calculated from the total area under the $a-t$ graph to the point.

At $t = 15$ s, $\quad v = (18)(15) = 270$ m/s
At $t = 20$ s, $\quad v = 270 + (25)(20 - 15) = 395$ m/s

Since $s = \int v\,dt$, the sloping lines of the $v-t$ graph become parabolic curves for the $s-t$ graph.
The numerical values for each point are calculated from the total area under the $v-t$ graph to the point.

At $t = 15$ s, $\quad s = \frac{1}{2}(15)(270) = 2025$ m

At $t = 20$ s, $\quad s = 2025 + 270(20 - 15) + \frac{1}{2}(395 - 270)(20 - 15) = 3687.5$ m $= 3.69$ km

Also;

$0 \le t \le 15$:

$a = 18$

$v = v_0 + a_c t = 0 + 18t$

$s = s_0 + v_0 t + \frac{1}{2} a_c t^2 = 0 + 0 + 9t^2$

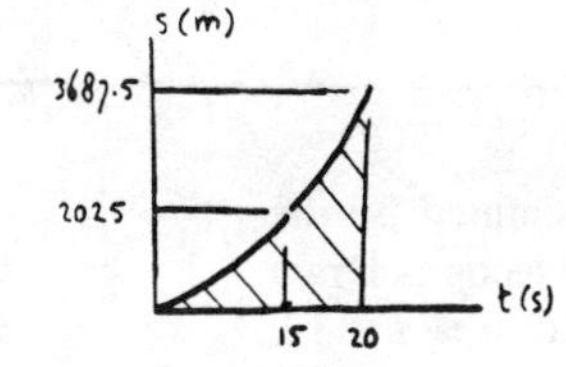

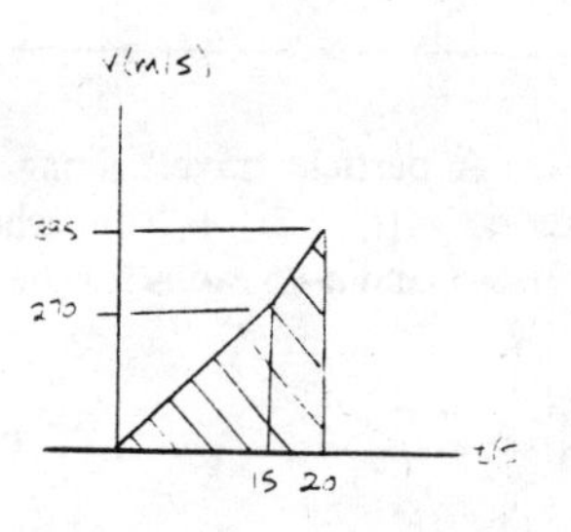

At $t = 15$:

$v = 18(15) = 270$

$s = 9(15)^2 = 2025$

$15 \le t \le 20$:

$a = 25$

$v = v_0 + a_c t = 270 + 25(t - 15)$

$s = s_0 + v_0 t + \frac{1}{2} a_c t^2 = 2025 + 270(t - 15) + \frac{1}{2}(25)(t - 15)^2$

When $t = 20$:

$v = 395$ m/s

$s = 3687.5$ m $= 3.69$ km

12-39. If the position of a particle is defined as $s = (5t - 3t^2)$ ft, where t is in seconds, construct the s–t, v–t, and a–t graphs for $0 \le t \le 10$ s.

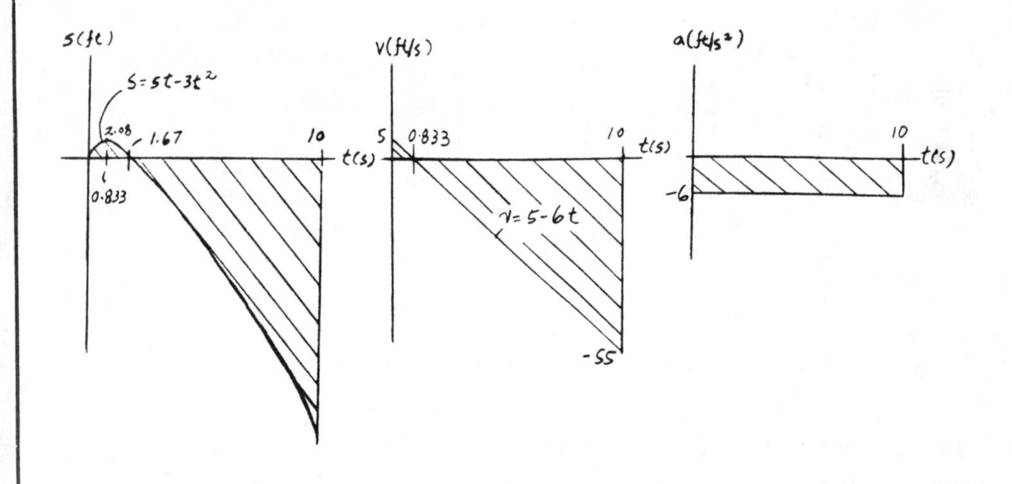

***12-40.** If the position of a particle is defined by $s = [2 \sin (\pi/5)t + 4]$ m, where t is in seconds, construct the s–t, v–t, and a–t graphs for $0 \le t \le 10$ s.

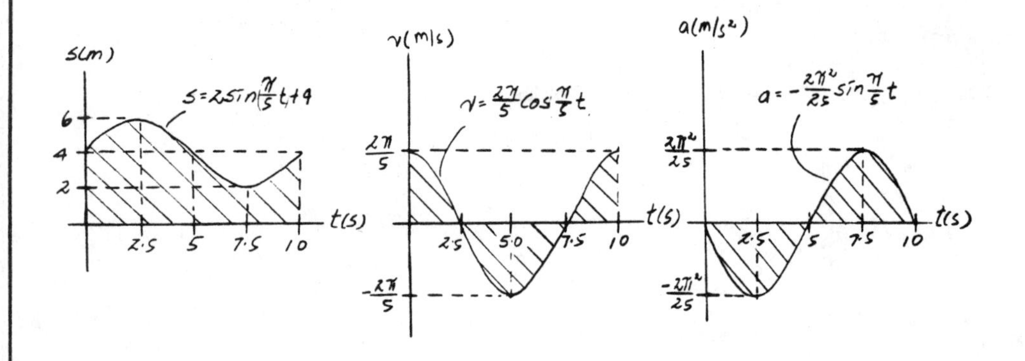

12-41. A particle travels along a curve defined by the equation $s = (t^3 - 3t^2 + 2t)$ m, where t is in seconds. Draw the s–t, v–t, and a–t graphs for the particle for $0 \le t \le 3$ s.

$s = t^3 - 3t^2 + 2t$

$v = \dfrac{ds}{dt} = 3t^2 - 6t + 2$

$a = \dfrac{dv}{dt} = 6t - 6$

$v = 0$ at $0 = 3t^2 - 6t + 2$

$t = 1.577$ s, and $t = 0.4226$ s,

$s|_{t=1.577} = -0.385$ m

$s|_{t=0.4226} = 0.385$ m

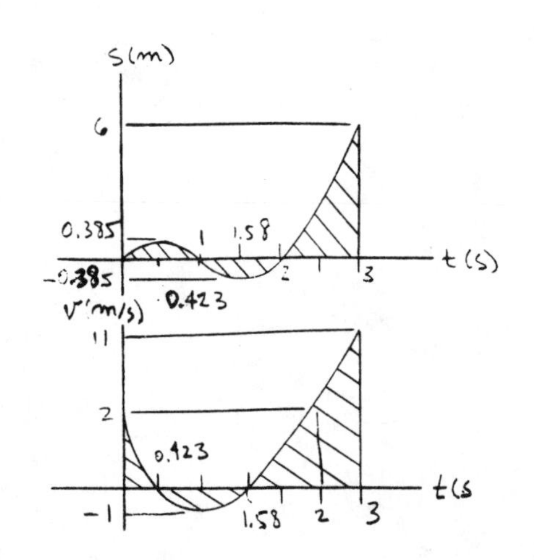

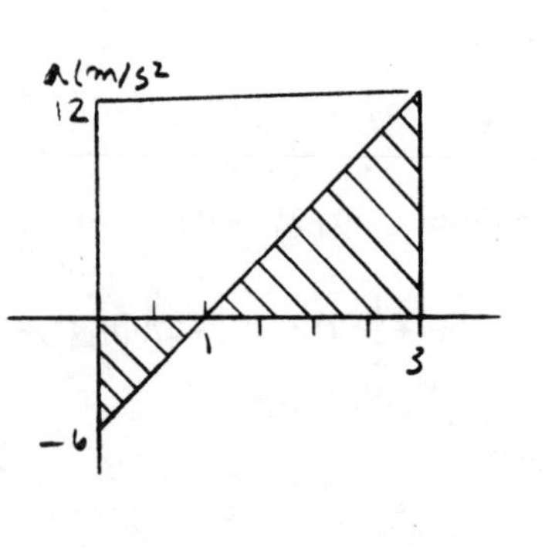

12-42. An airplane lands on the straight runway, originally traveling at 110 ft/s when $s = 0$. If it is subjected to the deceleration shown, determine the time t' needed to stop the plane and construct the s–t graph for the motion.

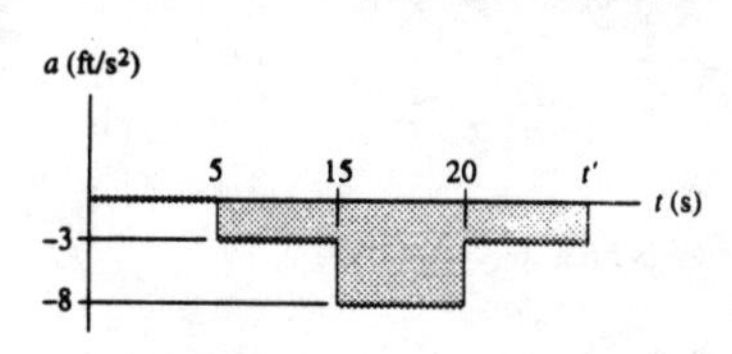

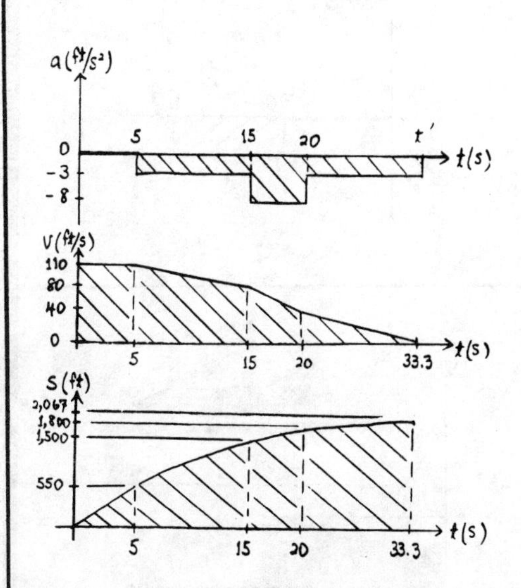

$v_0 = 110$ ft/s

$$\Delta v = \int a\,dt$$

$$0 - 110 = -3(15-5) - 8(20-15) - 3(t' - 20)$$

$t' = 33.3$ s **Ans**

12-43. A car starting from rest moves along a straight track with an acceleration as shown. Determine the time t for the car to reach a speed of 50 m/s and construct the v–t graph that describes the motion until the time t.

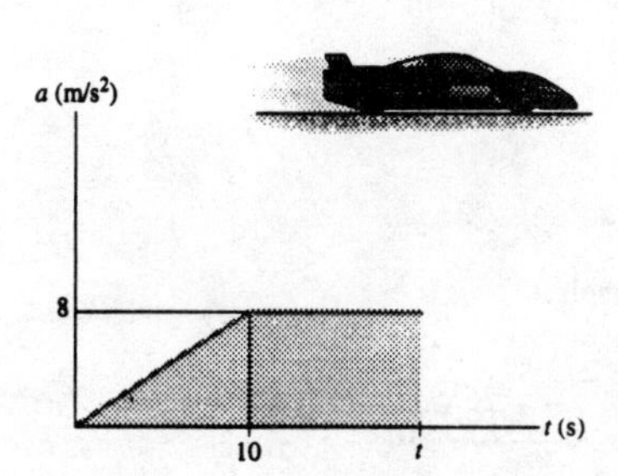

For $0 \le t \le 10$ s,

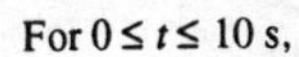

$$a = \frac{8}{10}t$$

$$dv = a\,dt$$

$$\int_0^v dv = \int_0^t \frac{8}{10}t\,dt$$

$$v = \frac{8}{20}t^2$$

At $t = 10$ s,

$$v = \frac{8}{20}(10)^2 = 40 \text{ m/s}$$

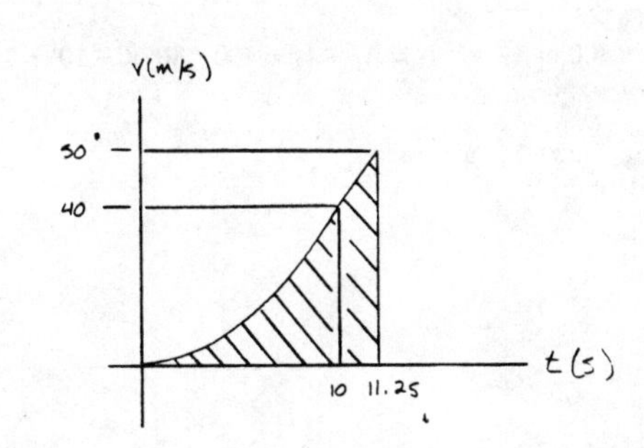

For $t > 10$ s,

$$a = 8$$

$$dv = a\,dt$$

$$\int_{40}^v dv = \int_{10}^t 8\,dt$$

$$v - 40 = 8t - 80$$

$$v = 8t - 40$$

When $v = 50$ m/s

$$t = \frac{50 + 40}{8} = 11.25 \text{ s}$$ **Ans**

***12-44.** A car travels up a hill with the speed shown. Determine the total distance the car moves until it stops ($t = 60$ s). Plot the a–t graph.

Distance traveled is area under $v-t$ graph.

$$s = (10)(30) + \frac{1}{2}(10)(30) = 450 \text{ m} \qquad \textbf{Ans}$$

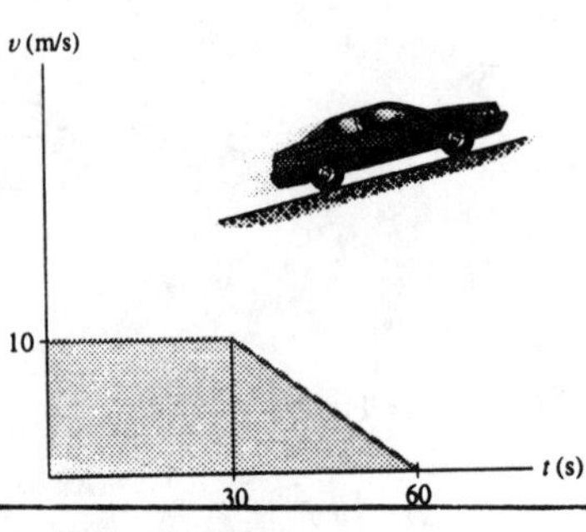

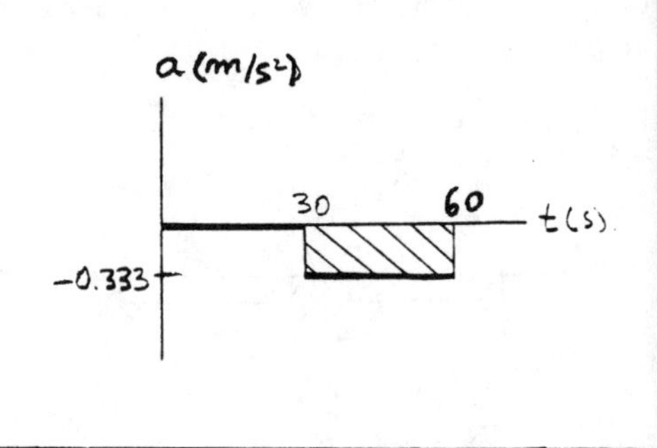

12-45. From experimental data, the motion of a jet plane while traveling along a runway is defined by the v–t graph shown. Construct the s–t and a–t graphs for the motion.

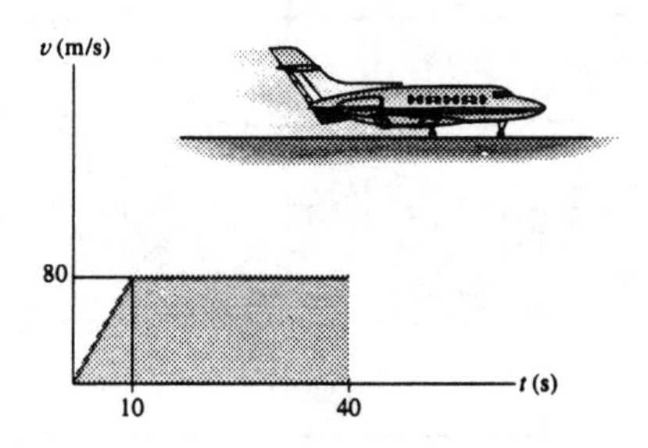

From the $v-t$ graph :

$0 \le t \le 10$ s $\qquad a = \dfrac{\Delta v}{\Delta t} = \dfrac{80}{10} = 8 \text{ m/s}^2$

$10 \le t \le 40$ s $\qquad a = \dfrac{\Delta v}{\Delta t} = \dfrac{(80-80)}{(40-10)} = 0$

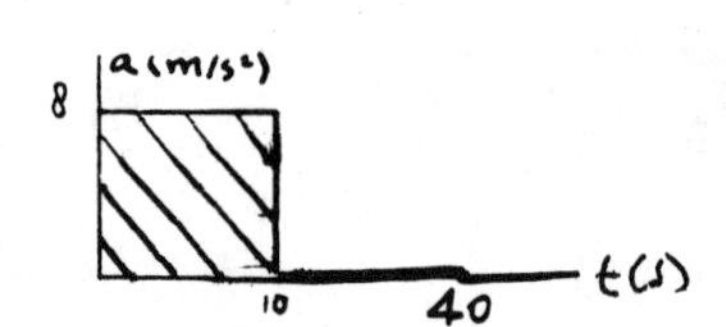

From the $v-t$ graph at $t_1 = 10$ s and $t_2 = 40$ s :

$s_1 = A_1 = \dfrac{1}{2}(10)(80) = 400 \text{ m} \qquad s_2 = A_1 + A_2 = 400 + 80(40-10) = 2800 \text{ m}$

The equations defining the $s-t$ graph are :

$0 \le t \le 10$:

$$ds = v\,dt$$

$$\int_0^s ds = \int_0^t 8t\,dt$$

$$s = 4t^2$$

$10 \le t \le 40$:

$$ds = v\,dt$$

$$\int_{400}^s ds = \int_{10}^t 80\,dt$$

$$s = 80t - 400$$

s (m)
2800
400
s = 80t − 400
t (s)
10
40
s = 4t²

12-46. A car travels along a straight road with the speed shown by the v–t graph. Determine the total distance the car travels until it stops when $t = 48$ s. Also plot the s–t and a–t graphs.

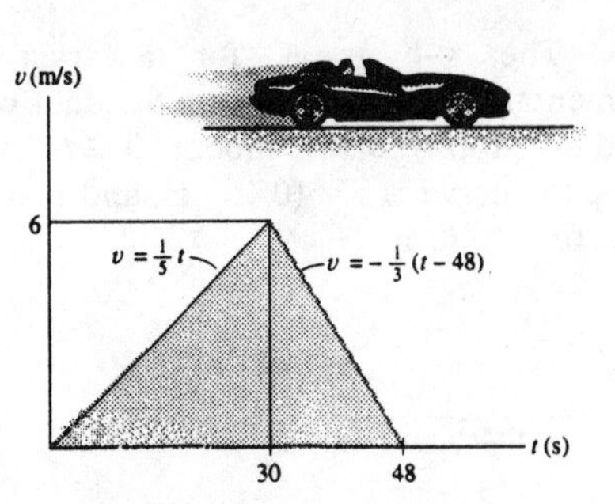

For $0 \le t \le 30$ s,

$$v = \frac{1}{5}t$$

$$a = \frac{dv}{dt} = \frac{1}{5}$$

$$ds = v\,dt$$

$$\int_0^s ds = \int_0^t \frac{1}{5}t\,dt$$

$$s = \frac{1}{10}t^2$$

When $t = 30$ s, $s = 90$ m,

$$v = -\frac{1}{3}(t-48)$$

$$a = \frac{dv}{dt} = -\frac{1}{3}$$

$$ds = v\,dt$$

$$\int_{90}^s ds = \int_{30}^t -\frac{1}{3}(t-48)\,dt$$

$$s = -\frac{1}{6}t^2 + 16t - 240$$

When $t = 48$ s,

$s = 144$ m **Ans**

Also, from the $v-t$ graph

$$\Delta s = \int v\,dt \qquad s - 0 = \frac{1}{2}(6)(48) = 144 \text{ m} \quad \textbf{Ans}$$

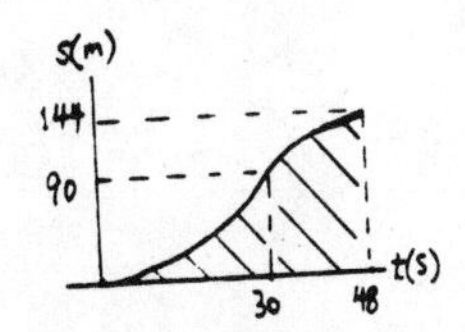

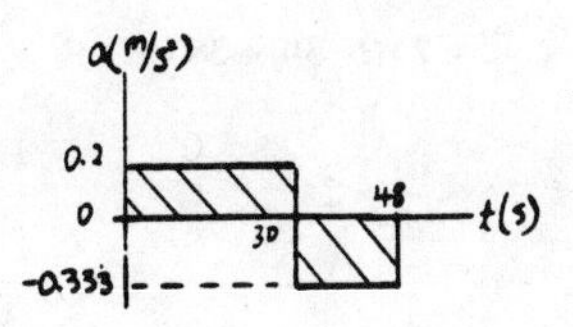

12-47. The v–t graph for the motion of a train as it moves from station A to station B is shown. Draw the a–t graph and determine the average speed and the distance between the stations.

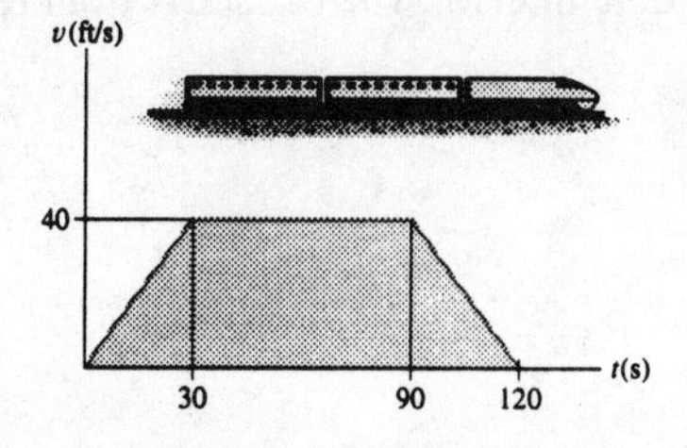

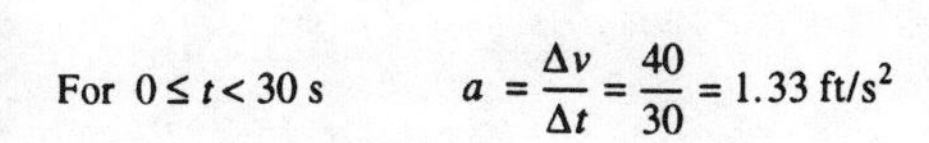

For $0 \le t < 30$ s $\qquad a = \dfrac{\Delta v}{\Delta t} = \dfrac{40}{30} = 1.33 \text{ ft/s}^2$

For $30 < t < 90$ s $\qquad a = \dfrac{\Delta v}{\Delta t} = 0$

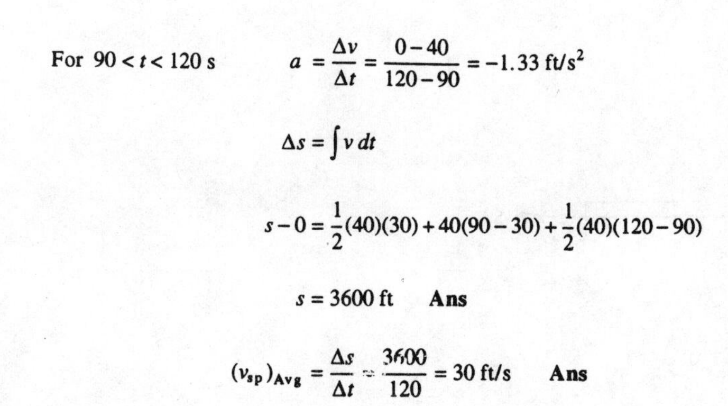

For $90 < t < 120$ s $\qquad a = \dfrac{\Delta v}{\Delta t} = \dfrac{0-40}{120-90} = -1.33 \text{ ft/s}^2$

$$\Delta s = \int v\,dt$$

$$s - 0 = \frac{1}{2}(40)(30) + 40(90-30) + \frac{1}{2}(40)(120-90)$$

$s = 3600$ ft **Ans**

$$(v_{sp})_{Avg} = \frac{\Delta s}{\Delta t} = \frac{3600}{120} = 30 \text{ ft/s} \quad \textbf{Ans}$$

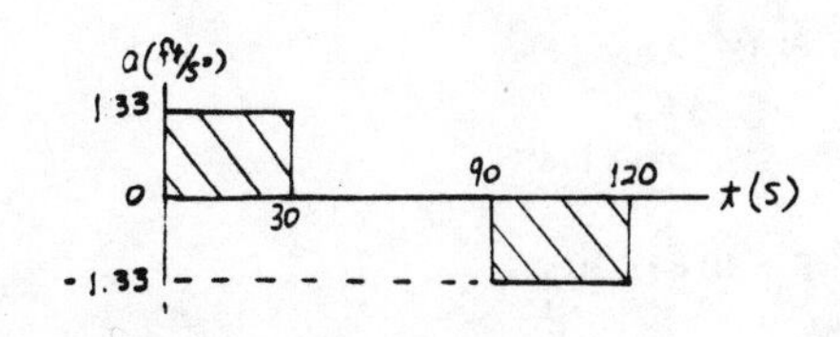

***12-48.** The s–t graph for a train has been experimentally determined. From the data, construct the v–t and a–t graphs for the motion; $0 \le t \le 40$ s. For $0 \le t \le 30$ s, the curve is $s = (0.4t^2)$ m, and then it becomes straight for $t \ge 30$ s.

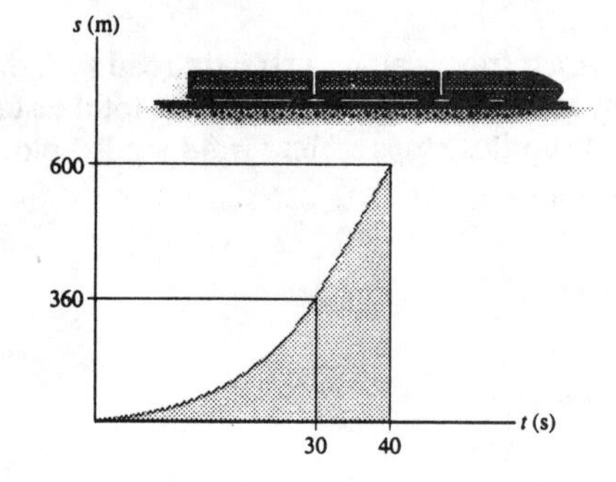

$0 \le t \le 30$:

$$s = 0.4t^2$$

$$v = \frac{ds}{dt} = 0.8t$$

$$a = \frac{dv}{dt} = 0.8$$

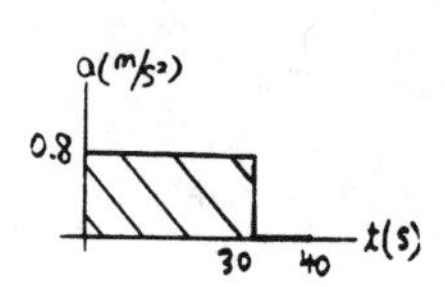

$30 \le t \le 40$:

$$s - 360 = \left(\frac{600-360}{40-30}\right)(t-30)$$

$$s = 24(t-30) + 360$$

$$v = \frac{ds}{dt} = 24$$

$$a = \frac{dv}{dt} = 0$$

12-49. The v–t graph for the motion of a car as it moves along a straight road is shown. Draw the a–t graph and determine the maximum acceleration during the 30-s time interval. The car starts from rest at $s = 0$.

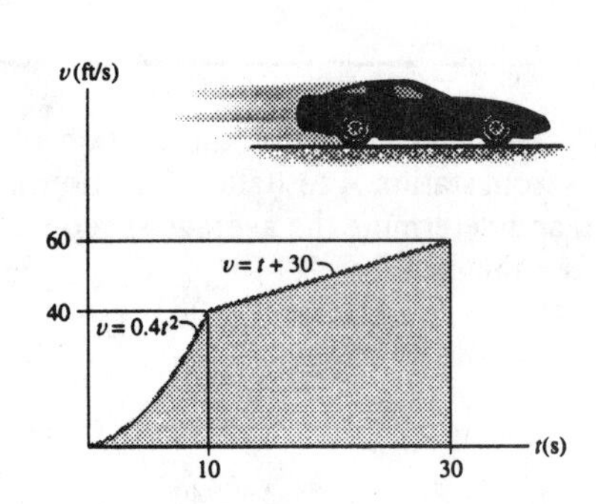

For $t < 10$ s :

$$v = 0.4t^2$$

$$a = \frac{dv}{dt} = 0.8t$$

At $t = 10$ s :

$$a = 8 \text{ ft/s}^2$$

For $10 < t \le 30$ s :

$$v = t + 30$$

$$a = \frac{dv}{dt} = 1$$

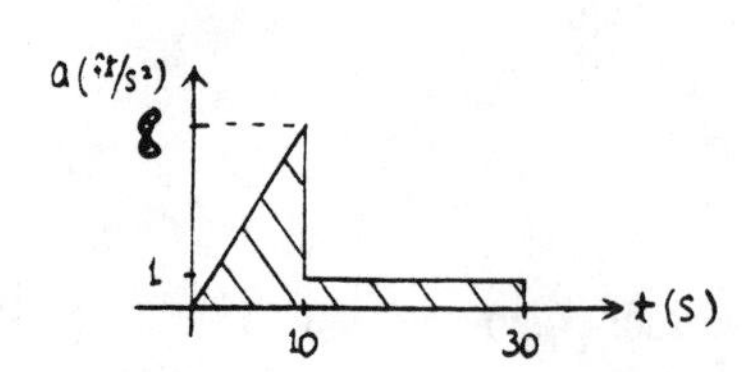

$a_{max} = 8 \text{ ft/s}^2$ **Ans**

12-50. The v–t graph for the motion of a car as it moves along a straight road is shown. Draw the s–t graph and determine the average speed and the distance traveled for the 30-s time interval. The car starts from rest at $s = 0$.

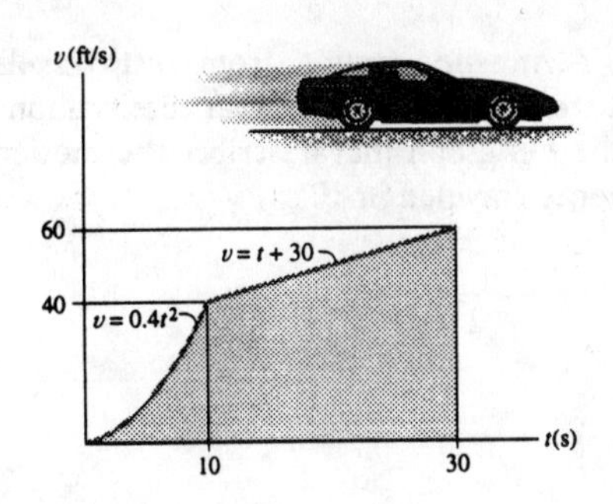

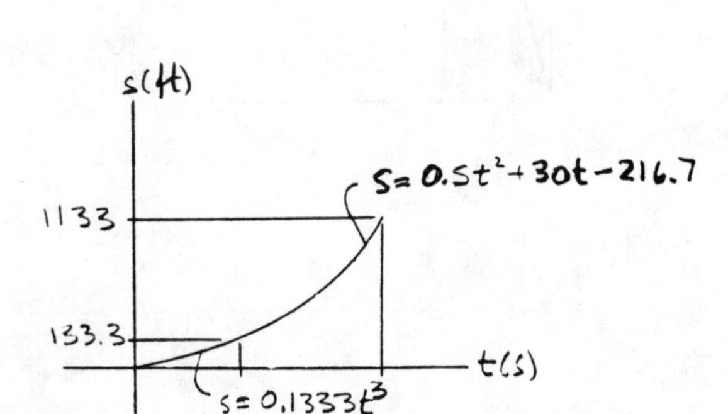

For $t < 10$ s,

$$v = 0.4t^2$$

$$ds = v\,dt$$

$$\int_0^s ds = \int_0^t 0.4t^2\,dt$$

$$s = 0.1333t^3$$

At $t = 10$ s,

$$s = 133.3 \text{ ft}$$

For $10 < t < 30$ s,

$$v = t + 30$$

$$ds = v\,dt$$

$$\int_{133.3}^s ds = \int_{10}^t (t + 30)\,dt$$

$$s = 0.5t^2 + 30t - 216.7$$

At $t = 30$ s,

$$s = 1133 \text{ ft}$$

$$(v_{sp})_{Avg} = \frac{\Delta s}{\Delta t} = \frac{1133}{30} = 37.8 \text{ ft/s} \quad \textbf{Ans}$$

$$s_T = 1133 \text{ ft} = 1.13(10^3) \text{ ft} \quad \textbf{Ans}$$

12-51. A missile starting from rest travels along a straight track and for 10 s has an acceleration as shown. Draw the v–t graph that describes the motion and find the distance traveled in 10 s.

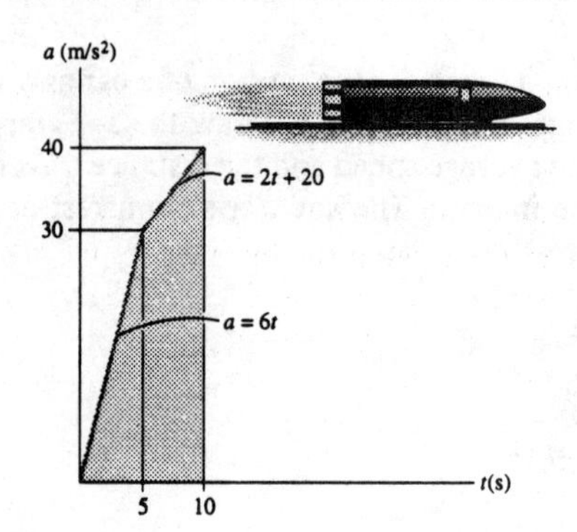

For $t \leq 5$ s,

$$a = 6t$$

$$dv = a\,dt$$

$$\int_0^v dv = \int_0^t 6t\,dt$$

$$v = 3t^2$$

When $t = 5$ s,

$$v = 75 \text{ m/s}$$

For $5 < t < 10$ s,

$$a = 2t + 20$$

$$dv = a\,dt$$

$$\int_{75}^v dv = \int_5^t (2t + 20)\,dt$$

$$v - 75 = t^2 + 20t - 125$$

$$v = t^2 + 20t - 50$$

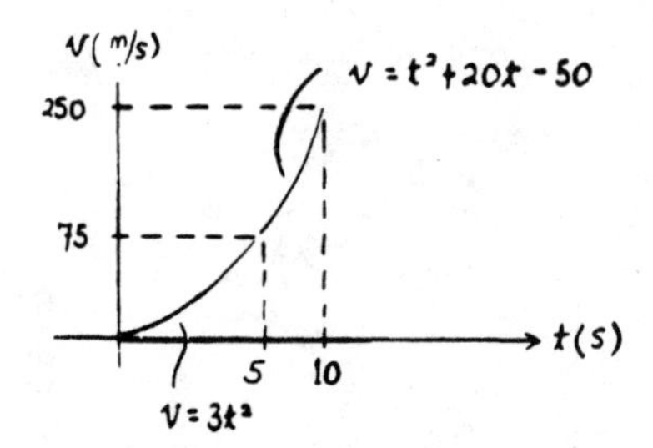

When $t = 10$ s,

$$v = 250 \text{ m/s}$$

Distance at $t = 5$ s :

$$ds = v\,dt$$

$$\int_0^s ds = \int_0^5 3t^2\,dt$$

$$s = (5)^3 = 125 \text{ m}$$

Distance at $t = 10$ s :

$$ds = v\,dv$$

$$\int_{125}^s ds = \int_5^{10} \left(t^2 + 20t - 50\right) dt$$

$$s - 125 = \frac{1}{3}t^3 + 10t^2 - 50t\Big]_5^{10}$$

$$s = 917 \text{ m} \quad \textbf{Ans}$$

***12-52.** A man riding upward in a freight elevator accidentally drops a package off the elevator when it is 100 ft from the ground. If the elevator maintains a constant upward speed of 4 ft/s, determine how high the elevator is from the ground the instant the package hits the ground. Draw the v–t curve for the package during the time it is in motion. Assume that the package was released with the same upward speed as the elevator.

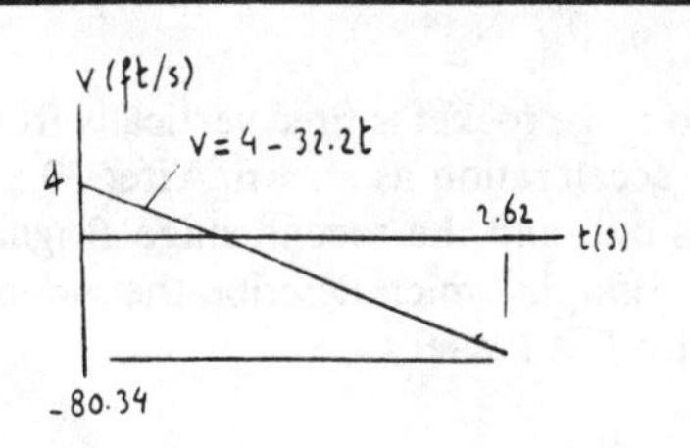

For package :

$(+\uparrow)\quad v^2 = v_0^2 + 2a_c(s_2 - s_0)$

$v^2 = (4)^2 + 2(-32.2)(-100-0)$

$v = 80.35 \text{ ft/s} \downarrow$

$(+\uparrow)\quad v = v_0 + a_c t$

$-80.35 = 4 + (-32.2)t$

$t = 2.620 \text{ s}$

For elevator :

$(+\uparrow)\quad s_2 = s_0 + vt$

$s = 100 + 4(2.620)$

$s = 110 \text{ ft}$ **Ans**

12-53. Two cars start from rest side by side and travel along a straight road. Car A accelerates at 4 m/s² for 10 s and then maintains a constant speed. Car B accelerates at 5 m/s² until reaching a constant speed of 25 m/s and then maintains this speed. Construct the a–t, v–t, and s–t graphs for each car until $t = 15$ s. What is the distance between the two cars when $t = 15$ s?

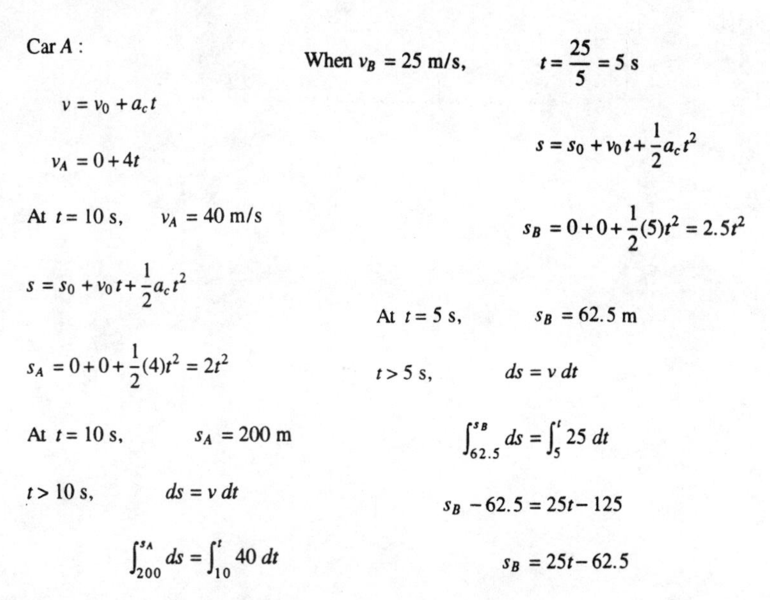

Car A :

$v = v_0 + a_c t$

$v_A = 0 + 4t$

At $t = 10$ s, $\quad v_A = 40$ m/s

$s = s_0 + v_0 t + \frac{1}{2}a_c t^2$

$s_A = 0 + 0 + \frac{1}{2}(4)t^2 = 2t^2$

At $t = 10$ s, $\quad s_A = 200$ m

$t > 10$ s, $\quad ds = v\,dt$

$\int_{200}^{s_A} ds = \int_{10}^{t} 40\,dt$

$s_A = 40t - 200$

At $t = 15$ s, $\quad s_A = 400$ m

Car B :

$v = v_0 + a_c t$

$v_B = 0 + 5t$

When $v_B = 25$ m/s, $\quad t = \frac{25}{5} = 5$ s

$s = s_0 + v_0 t + \frac{1}{2}a_c t^2$

$s_B = 0 + 0 + \frac{1}{2}(5)t^2 = 2.5t^2$

At $t = 5$ s, $\quad s_B = 62.5$ m

$t > 5$ s, $\quad ds = v\,dt$

$\int_{62.5}^{s_B} ds = \int_{5}^{t} 25\,dt$

$s_B - 62.5 = 25t - 125$

$s_B = 25t - 62.5$

When $t = 15$ s, $\quad s_B = 312.5$

Distance between the cars is

$\Delta s = s_A - s_B = 400 - 312.5 = 87.5 \text{ m}$ **Ans**

Car A is ahead of car B.

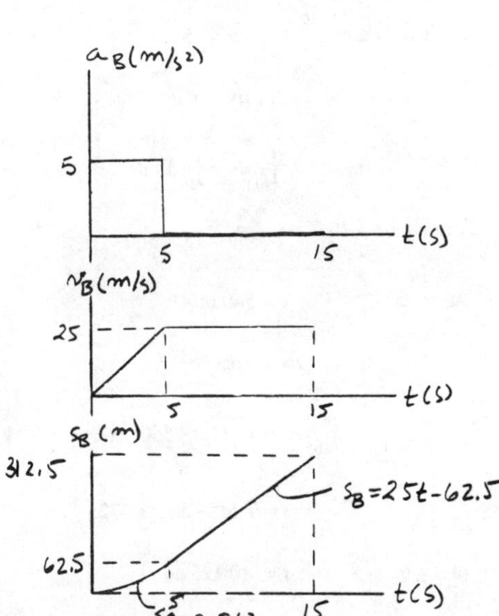

12-54. A two-stage rocket is fired vertically from rest at $s = 0$ with an acceleration as shown. After 30 s the first stage A burns out, and the second stage B ignites. Plot the v–t and s–t graphs which describe the motion of the second stage for $0 \le t \le 60$ s.

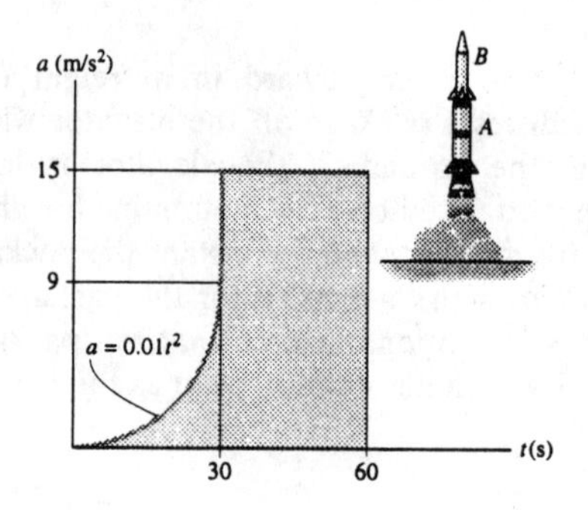

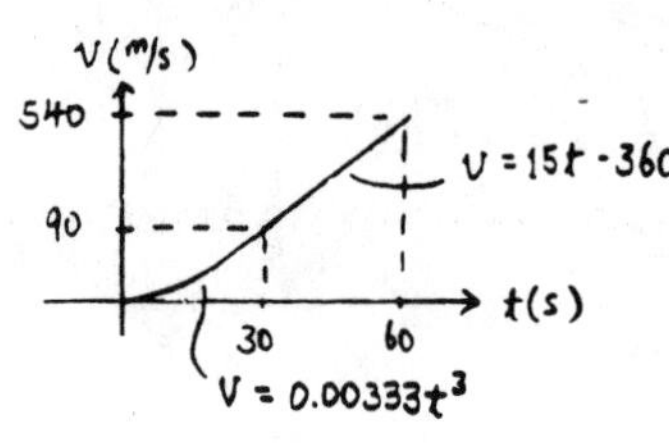

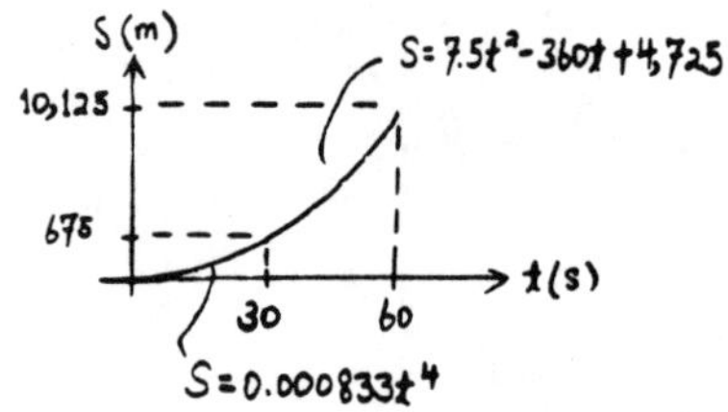

For $0 \le t < 30$ s $\quad a = 0.01t^2$

$$dv = a\,dt$$

$$\int_0^v dv = \int_0^t 0.01t^2\,dt$$

$$v = 0.00333t^3$$

At $t = 30$ s, $\quad v = 90$ m/s

$$ds = v\,dt$$

$$\int_0^s ds = \int_0^t 0.00333t^3\,dt$$

$$s = 0.000833t^4$$

At $t = 30$ s, $\quad s = 675$ m

For $30 < t \le 60$ s $\quad a = 15$

$$dv = a\,dt$$

$$\int_{90}^v dv = \int_{30}^t 15\,dt$$

$$v = 15t - 360$$

At $t = 60$ s, $\quad v = 540$ m/s

$$ds = v\,dt$$

$$\int_{675}^s ds = \int_{30}^t (15t - 360)\,dt$$

$$s = 7.5t^2 - 360t + 4725$$

At $t = 60$ s, $\quad s = 10\,125$ m

12-55. The a–t graph for a motorcycle traveling along a straight road has been estimated as shown. Determine the time needed for the motorcycle to reach a maximum speed of 100 ft/s and the distance traveled in this time. Draw the v–t and s–t graphs. The motorcycle starts from rest at $s = 0$.

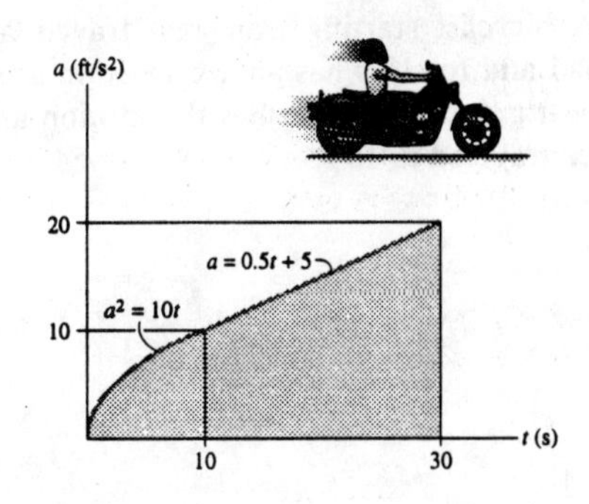

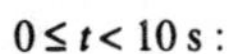

$0 \le t < 10$ s :

$$a = \sqrt{10}\, t^{\frac{1}{2}}$$

$$dv = a\, dt$$

$$\int_0^v dv = \int_0^t \sqrt{10}\, t^{\frac{1}{2}}\, dt$$

$$v = \frac{2}{3}\sqrt{10}\, t^{\frac{3}{2}}$$

When $t = 10$ s, $\quad v = \frac{2}{3}\sqrt{10}\,(10)^{\frac{3}{2}} = 66.67$ ft/s < 100 ft/s

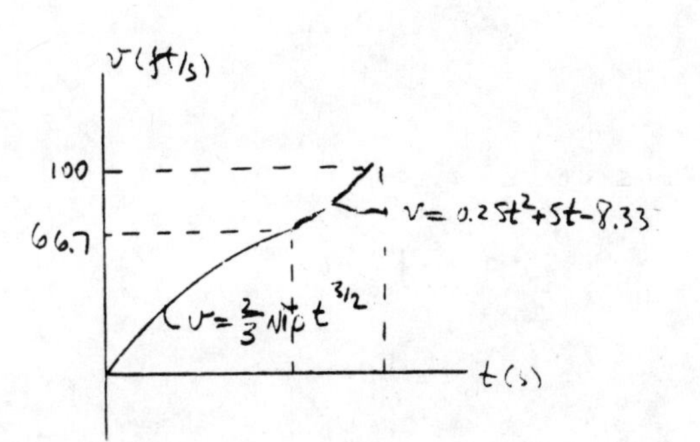

$$ds = v\, dt$$

$$\int_0^s ds = \int_0^t \frac{2}{3}\sqrt{10}\, t^{\frac{3}{2}}\, dt$$

$$s = \frac{4}{15}\sqrt{10}\, t^{\frac{5}{2}}$$

When $t = 10$ s, $\quad s = \frac{4}{15}\sqrt{10}\,(10)^{\frac{5}{2}} = 266.67$ ft

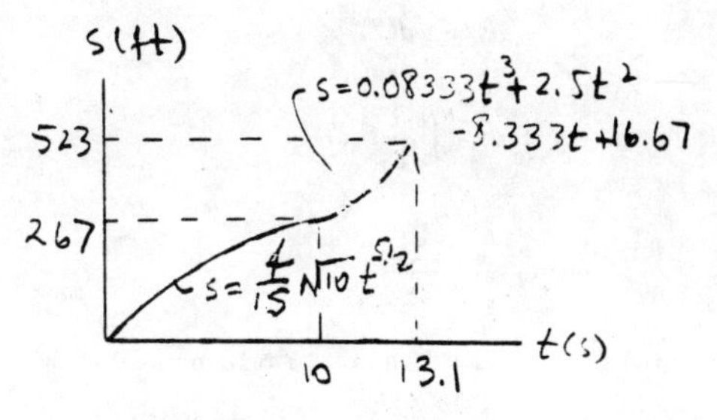

$t > 10$ s :

$$a = 0.5t + 5$$

$$dv = a\, dt$$

$$\int_{66.67}^v dv = \int_{10}^t (0.5t + 5)\, dt$$

$$v - 66.67 = 0.25t^2 + 5t - 75$$

$$v = 0.25t^2 + 5t - 8.333$$

When $v = 100$ ft/s,

$$0.25t^2 + 5t - 108.333 = 0$$

Solving for the positive root: $\quad t = 13.09\text{ s} = 13.1\text{ s}$ **Ans**

$$ds = v\, dt$$

$$\int_{266.67}^s ds = \int_{10}^t v\, dt = \int_{10}^t \left(0.25t^2 + 5t - 8.333\right) dt$$

$$s - 266.67 = 0.08333t^3 + 2.5t^2 - 8.333t - 250$$

$$s = 0.08333t^3 + 2.5t^2 - 8.333t + 16.67$$

$$s|_{t=13.09} = 0.08333(13.09)^3 + 2.5(13.09)^2 - 8.333(13.09) + 16.67 = 523\text{ ft}$$ **Ans**

***12-56.** A bicyclist starting from rest travels along a straight road and for 10 s has an acceleration as shown. Draw the v–t graph that describes the motion and find the distance traveled in 10 s.

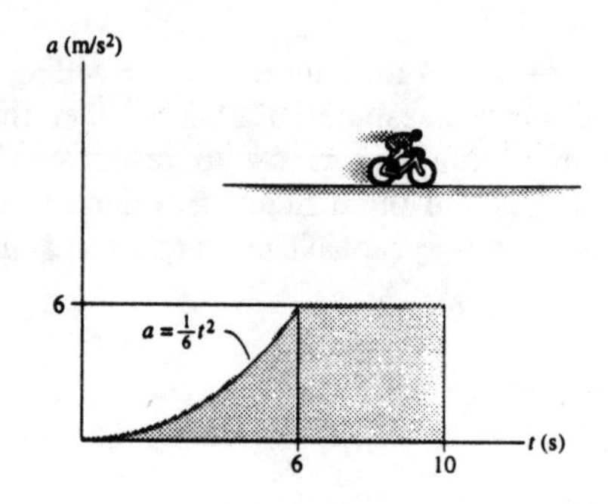

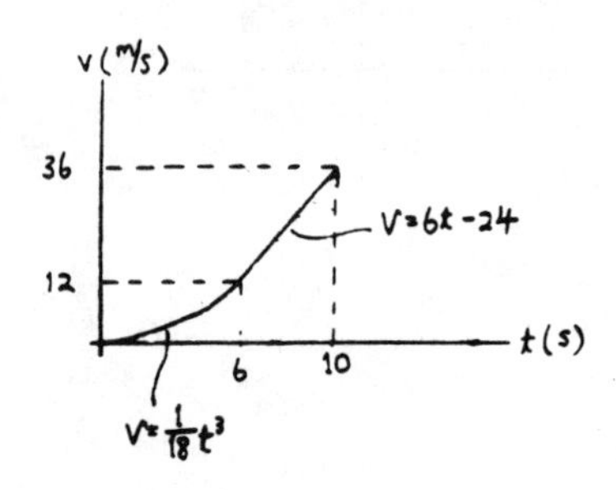

For $0 \le t < 6$ $\quad dv = a\,dt$

$$\int_0^v dv = \int_0^t \frac{1}{6}t^2\,dt$$

$$v = \frac{1}{18}t^3$$

$$ds = v\,dt$$

$$\int_0^s ds = \int_0^t \frac{1}{18}t^3\,dt$$

$$s = \frac{1}{72}t^4$$

When $t = 6$ s, $\quad v = 12$ m/s $\quad s = 18$ m

For $6 < t \le 10$ $\quad dv = a\,dt$

$$\int_{12}^v dv = \int_6^t 6\,dt$$

$$v = 6t - 24$$

$$ds = v\,dt$$

$$\int_{18}^s ds = \int_6^t (6t - 24)\,dt$$

$$s = 3t^2 - 24t + 54$$

When $t = 10$ s, $\quad v = 36$ m/s

$s = 114$ m **Ans**

12-57. The v–t graph of a car while traveling along a road is shown. Draw the s–t and a–t graphs for the motion.

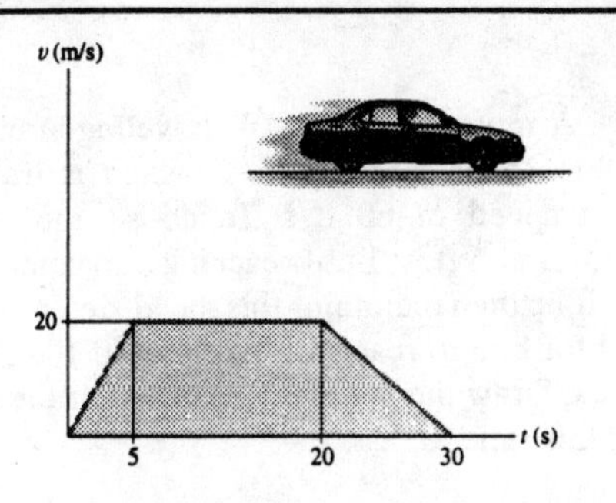

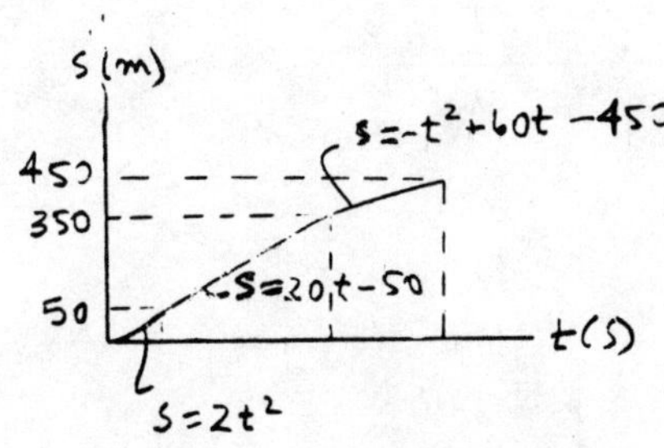

$0 \le t \le 5 \qquad a = \frac{\Delta v}{\Delta t} = \frac{20}{5} = 4 \text{ m/s}^2$

$5 \le t \le 20 \qquad a = \frac{\Delta v}{\Delta t} = \frac{20-20}{20-5} = 0 \text{ m/s}^2$

$20 \le t \le 30 \qquad a = \frac{\Delta v}{\Delta t} = \frac{0-20}{30-20} = -2 \text{ m/s}^2$

From the $v-t$ graph at $t_1 = 5$ s, $t_2 = 20$ s, and $t_3 = 30$ s,

$s_1 = A_1 = \frac{1}{2}(5)(20) = 50 \text{ m}$

$s_2 = A_1 + A_2 = 50 + 20(20-5) = 350 \text{ m}$

$s_3 = A_1 + A_2 + A_3 = 350 + \frac{1}{2}(30-20)(20) = 450 \text{ m}$

The equations defining the portions of the $s-t$ graph are

$0 \le t \le 5 \text{ s} \qquad v = 4t; \quad ds = v\,dt; \quad \int_0^s ds = \int_0^t 4t\,dt; \quad s = 2t^2$

$5 \le t \le 20 \text{ s} \qquad v = 20; \quad ds = v\,dt; \quad \int_{50}^s ds = \int_5^t 20\,dt; \quad s = 20t - 50$

$20 \le t \le 30 \text{ s} \qquad v = 2(30-t); \quad ds = v\,dt; \quad \int_{350}^s ds = \int_{20}^t 2(30-t)\,dt; \quad s = -t^2 + 60t - 450$

12-58. A motorcyclist at A is traveling at 60 ft/s when he wishes to pass the truck T which is traveling at a constant speed of 60 ft/s. To do so the motorcyclist accelerates at 6 ft/s^2 until reaching a maximum speed of 85 ft/s. If he then maintains this speed, determine the time needed for him to reach a point located 100 ft in front of the truck. Draw the v–t and s–t graphs for the motorcycle during this time.

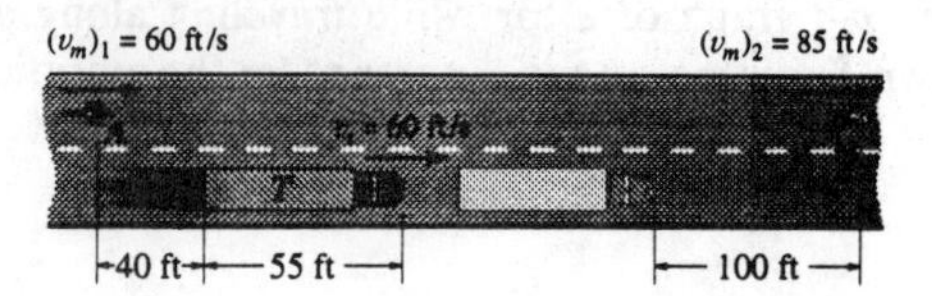

Motorcycle :

Time to reach 85 ft/s,

$$v = v_0 + a_c t$$

$$85 = 60 + 6t$$

$$t = 4.167 \text{ s}$$

$$v^2 = v_0^2 + 2a_c(s - s_0)$$

Distance traveled,

$$(85)^2 = (60)^2 + 2(6)(s_m - 0)$$

$$s_m = 302.08 \text{ ft}$$

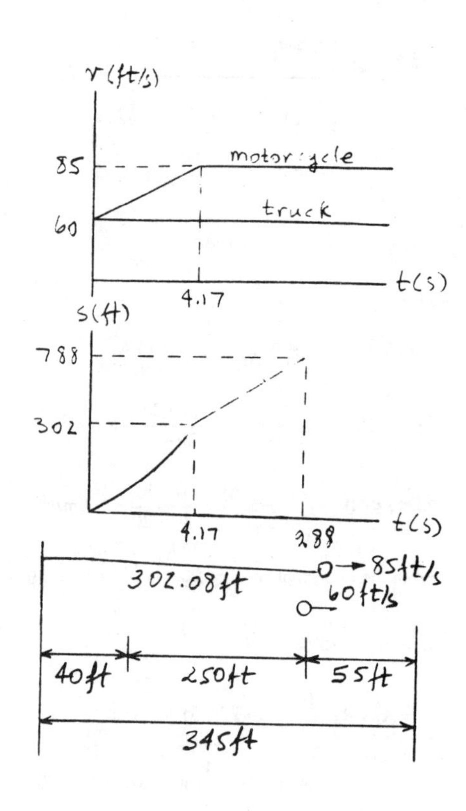

In $t = 4.167$ s, truck travels

$s_t = 60(4.167) = 250$ ft

Further distance for motorcycle to travel : $40 + 55 + 250 + 100 - 302.08 = 142.92$ ft

Motorcycle :

$$s = s_0 + v_0 t$$

$$(s + 142.92) = 0 + 85t'$$

Truck :

$$s = 0 + 60t'$$

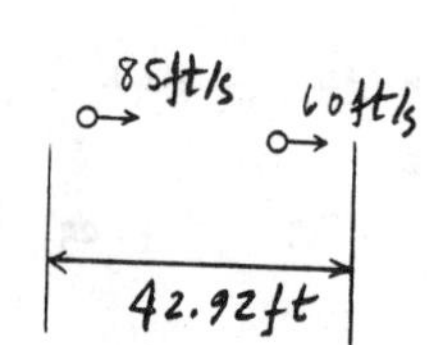

Thus $t' = 5.717$ s

$t = 4.167 + 5.717 = 9.88$ s **Ans**

Total distance motorcycle travels

$s_T = 302.08 + 85(5.717) = 788$ ft

12-59. The jet car is originally traveling at a speed of 20 m/s when it is subjected to the acceleration shown in the graph. Determine the car's maximum speed and the time t when it stops.

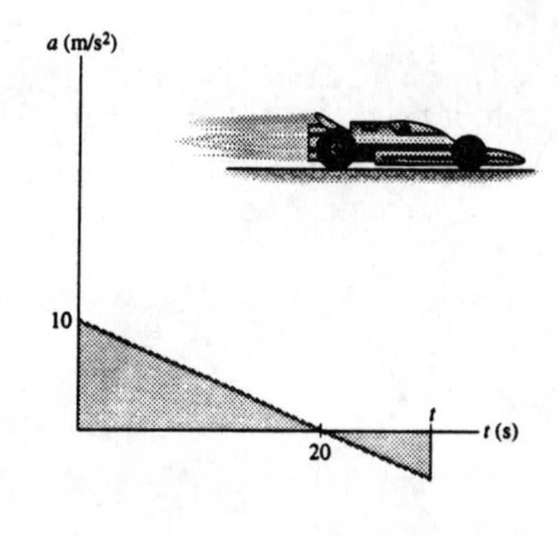

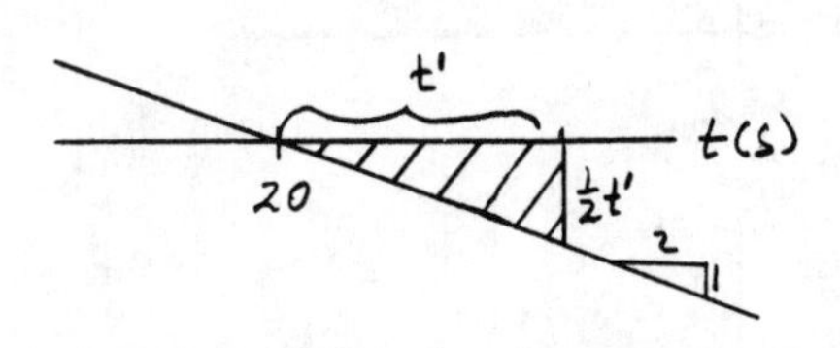

$$a = -0.5t + 10$$

$$dv = a\,dt$$

$$\int_{20}^{v} dv = \int_{0}^{t} (-0.5t + 10)\,dt$$

$$v = -0.25t^2 + 10t + 20$$

Maximum speed :

$$a = \frac{dv}{dt} = -0.5t + 10 = 0$$

$$t = 20 \text{ s}$$

$$v_{max} = -0.25(20)^2 + 10(20) + 20 = 120 \text{ m/s} \quad \textbf{Ans}$$

Also, find the area under the $a-t$ graph $0 \leq t \leq 20$ s,

$$\Delta v = v_{max} - 20 = \frac{1}{2}(10)(20) \qquad v_{max} = 120 \text{ m/s} \quad \textbf{Ans}$$

When the car stops, $v = 0$,

$$0 = -0.25t^2 + 10t + 20$$

Solve for the positive root,

$$t = 41.9 \text{ s} \quad \textbf{Ans}$$

Also, use $a-t$ graph, require negative area to equal 120.

$$120 = \frac{1}{2}(t')\left[\frac{1}{2}(t')\right]$$

$$t' = 21.9 \text{ s}$$

Thus,

$$t' + 20 = 41.9 \text{ s} \quad \textbf{Ans}$$

***12-60.** The a–t graph for a car is shown. Construct the v–t and s–t graphs if the car starts from rest at $t = 0$. At what time t' does the car stop?

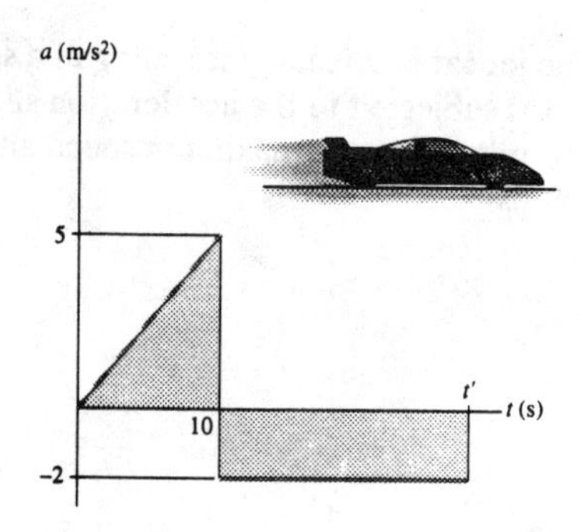

Velocity :

For $0 \le t \le 10$ s $\quad \dfrac{dv}{dt} = 0.5t \quad \int_0^v dv = \int_0^t 0.5t\,dt \quad v = 0.25t^2$

At $t = 10$ s $\quad v = 0.25(10)^2 = 25$ m/s

For $10\text{ s} \le t \le t'$ $\quad \dfrac{dv}{dt} = -2 \quad \int_{25}^{v} dv = \int_{10}^{t} -2dt \quad v = -2t + 45$

At $t = t'$, $v = 0$ $\quad v = -2t' + 45 = 0 \quad t' = 22.5$ s **Ans**

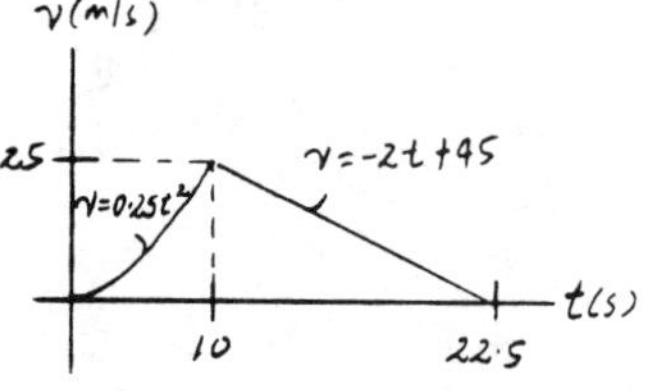

Position :

For $0 \le t \le 10$ s $\quad \dfrac{ds}{dt} = 0.25t^2 \quad \int_0^s ds = \int_0^t 0.25t^2 dt$

$s = 0.0833t^3$

At $t = 10$ s $\quad s = 0.0833(10)^3 = 83.3$ m

For $10\text{ s} \le t \le t' = 22.5$ s $\quad \dfrac{ds}{dt} = -2t + 45 \quad \int_{83.3}^{s} ds = \int_{10}^{t} (-2t + 45)dt$

$s = -t^2 + 45t - 267$

At $t = 22.5$ s $\quad s = -22.5^2 + 45(22.5) - 267 = 240$ m

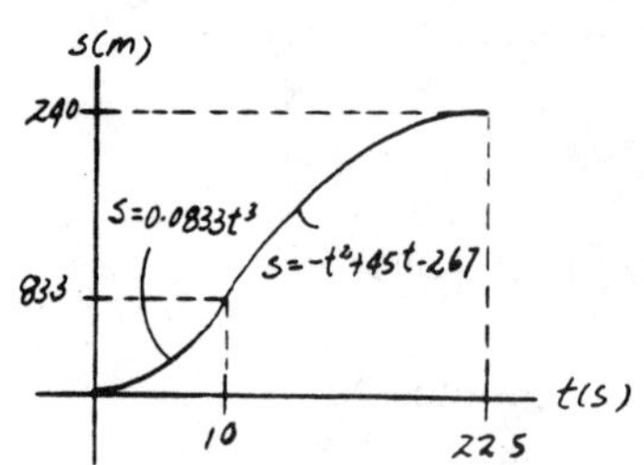

12-61. The a–s graph for a train traveling along a straight track is given for the first 400 m of its motion. Plot the v–s graph. $v = 0$ at $s = 0$.

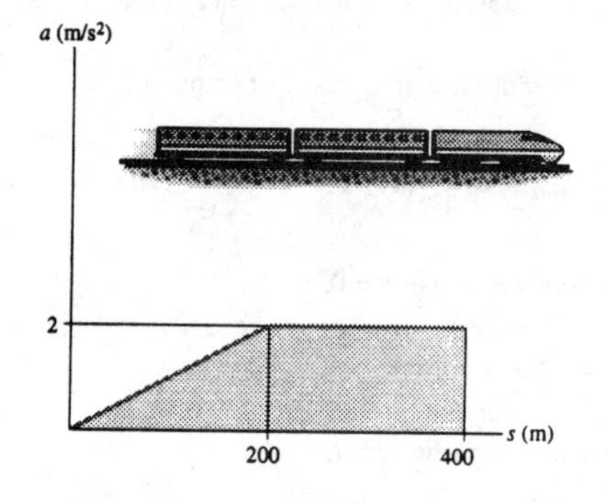

$0 \le s \le 200: \quad a = \dfrac{1}{100}s$

$$a\,ds = v\,dv$$

$$\int_0^s \frac{1}{100}s\,ds = \int_0^v v\,dv$$

$$\frac{1}{200}s^2 = \frac{1}{2}v^2$$

$$v = 0.1s$$

At $s = 200$, $\quad v = 20$ m/s

$200 \le s \le 400: \quad a = 2$

$$a\,ds = v\,dv$$

$$\int_{200}^{s} 2\,ds = \int_{20}^{v} v\,dv$$

$$2(s - 200) = \frac{1}{2}(v^2 - 400)$$

$$v^2 = 4s - 400$$

At $s = 400$ m, $\quad v = \sqrt{4(400) - 400} = 34.6$ m/s

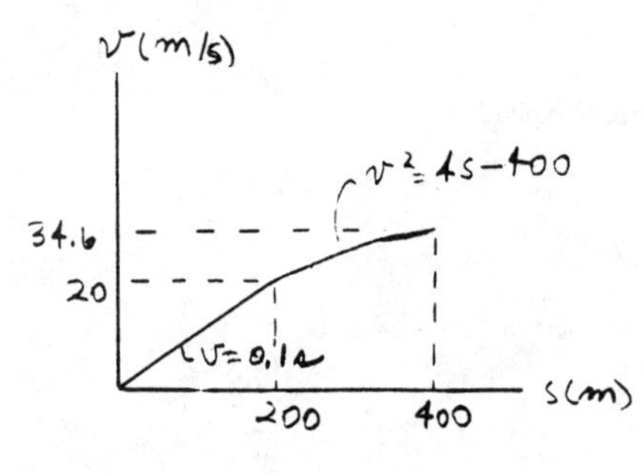

12-62. The v–s graph for a test vehicle is shown. Determine its acceleration when $s = 100$ m and when $s = 175$ m.

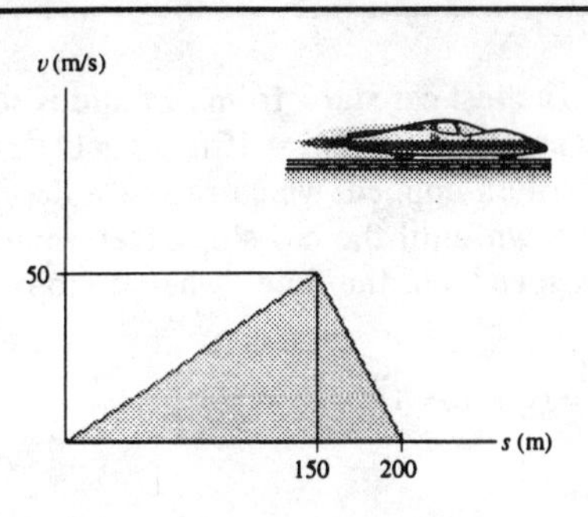

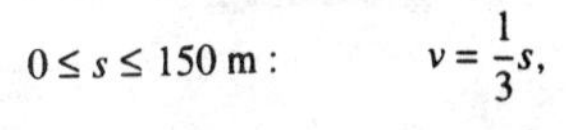

$0 \le s \le 150$ m: $\quad v = \frac{1}{3}s,$

$$dv = \frac{1}{3}ds$$

$$v\,dv = a\,ds$$

$$\frac{1}{3}s\left(\frac{1}{3}ds\right) = a\,ds$$

$$a = \frac{1}{9}s$$

At $s = 100$ m, $\quad a = \frac{1}{9}(100) = 11.1 \text{ m/s}^2$ **Ans**

$150 \le s \le 200$ m: $\quad v = 200 - s,$

$$dv = -ds$$

$$v\,dv = a\,ds$$

$$(200 - s)(-ds) = a\,ds$$

$$a = s - 200$$

At $s = 175$ m, $\quad a = 175 - 200 = -25 \text{ m/s}^2$ **Ans**

12-63. The rocket sled starts from rest at $s = 0$ and is subjected to an acceleration as shown by the a–s graph. Draw the v–s graph and determine the time needed to travel 500 ft.

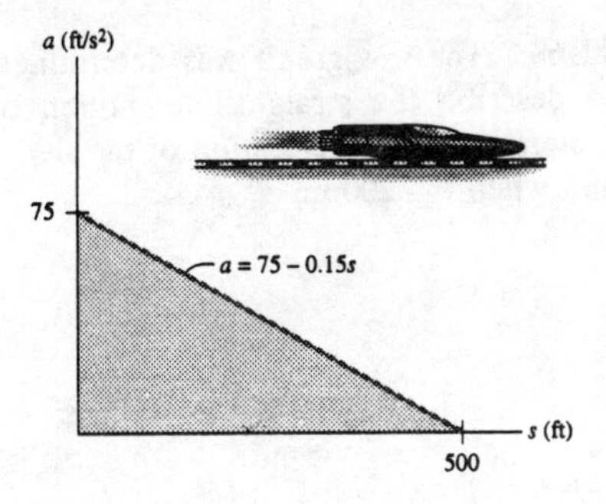

$$a = 75 - 0.15s$$

$$a\,ds = v\,dv$$

$$\int_0^s (75 - 0.15s)\,ds = \int_0^v v\,dv$$

$$v = \sqrt{150s - 0.15s^2}$$

$$v = \frac{ds}{dt}$$

$$\int_0^s \frac{ds}{\sqrt{150s - 0.15s^2}} = \int_0^t dt$$

$$t = -\frac{1}{\sqrt{0.15}}\left[\sin^{-1}\left(\frac{150 - 0.3\,s}{150}\right) - \frac{\pi}{2}\right]$$

At $s = 500$ ft,

$t = 4.06$ s **Ans**

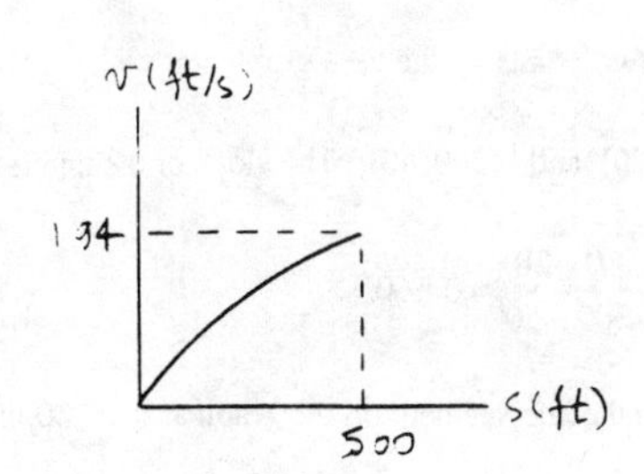

***12-64.** The test car starts from rest and is subjected to a constant acceleration of $a_c = 15\ \text{ft/s}^2$ for $0 \le t < 10$ s. The brakes are then applied, which causes a deceleration at the rate shown until the car stops. Determine the car's maximum speed and the time t when it stops.

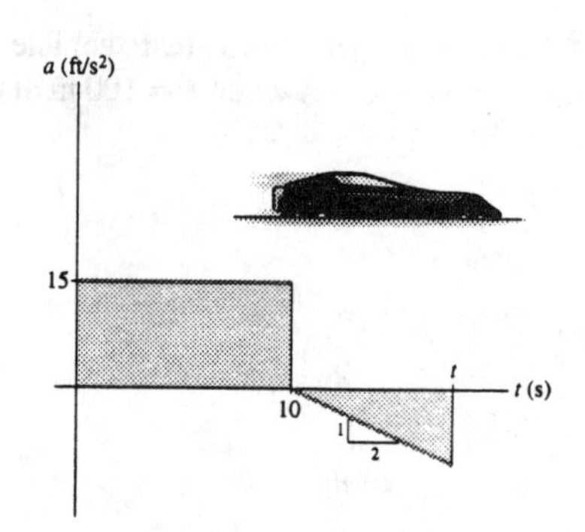

$v_{max} = A_1 = (15)(10) = 150$ ft/s **Ans**

From the graph, for $t > 10$ s, $\quad a = -\frac{1}{2}(t-10)$

$$dv = a\,dt$$

$$\int_{150}^{v} dv = \int_{10}^{t} -\frac{1}{2}(t-10)\,dt$$

$$v - 150 = -\frac{1}{2}\left[\frac{1}{2}t^2 - 10t\right]_{10}^{t}$$

$$v = 150 - \frac{1}{4}t^2 + 5t + \frac{1}{4}(10)^2 - \frac{1}{2}(10)^2$$

$$= -\frac{1}{4}t^2 + 5t + 125$$

When the car stops, $\quad v = 0 = -\frac{1}{4}t^2 + 5t + 125 \quad (1)$

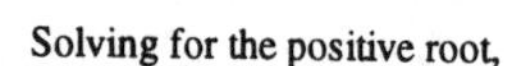

Solving for the positive root,

$t = 34.5$ s **Ans**

Using the $a - t$ graph, we can obtain the same result by requiring

$$A_1 + A_2 = (15)(10) + \frac{1}{2}(a)(t-10) = 0$$

$$150 + \frac{1}{2}\left[-\frac{1}{2}(t-10)\right](t-10) = 0$$

$$-\frac{1}{4}t^2 + 5t + 125 = 0$$

Which is the same as Eq. (1).

12-65. The v–s graph was determined experimentally to describe the straight-line motion of a rocket sled. Determine the acceleration of the sled when $s = 100$ m, and when $s = 200$ m.

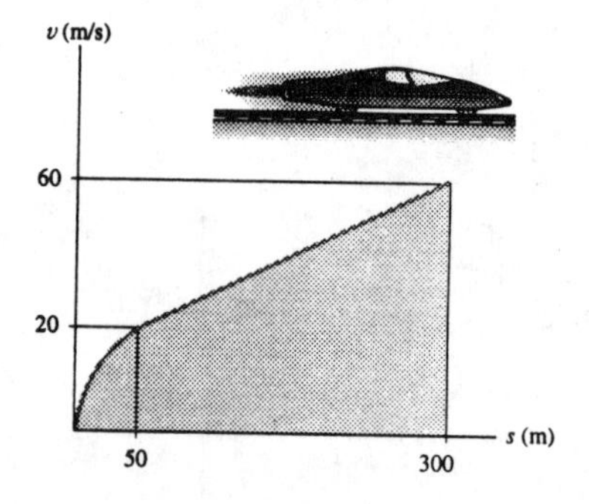

$$a = v\frac{dv}{ds}$$

The two points on the $v - s$ graph are

(50, 20) and (300, 60). The slope of the line is

$$\frac{dv}{ds} = \frac{60-20}{300-50} = 0.160$$

From the line segment on the graph at $s = 100$ m, $v = 28$ m/s,

$a = 28(0.160) = 4.48\ \text{m/s}^2$ **Ans**

In a similar manner, $s = 200$ m, $v = 44.0$ m/s, so that

$a = 44.0(0.160) = 7.04\ \text{m/s}^2$ **Ans**

12-66. If the velocity of a particle is defined as $\mathbf{v}(t) = \{0.8t^2\mathbf{i} + 12t^{1/2}\mathbf{j} + 5\mathbf{k}\}$ m/s, determine the magnitude and coordinate direction angles α, β, γ of the particle's acceleration when $t = 2$ s.

$$\mathbf{v}(t) = 0.8t^2\mathbf{i} + 12t^{\frac{1}{2}}\mathbf{j} + 5\mathbf{k}$$

$$\mathbf{a} = \frac{d\mathbf{v}}{dt} = 1.6t\mathbf{i} + 6t^{-\frac{1}{2}}\mathbf{j}$$

When $t = 2$ s, $\quad \mathbf{a} = 3.2\mathbf{i} + 4.243\mathbf{j}$

$$a = \sqrt{(3.2)^2 + (4.243)^2} = 5.31 \text{ m/s}^2 \qquad \textbf{Ans}$$

$$\mathbf{u}_a = \frac{\mathbf{a}}{a} = 0.6022\mathbf{i} + 0.7984\mathbf{j}$$

$$\alpha = \cos^{-1}(0.6022) = 53.0° \qquad \textbf{Ans}$$

$$\beta = \cos^{-1}(0.7984) = 37.0° \qquad \textbf{Ans}$$

$$\gamma = \cos^{-1}(0) = 90.0° \qquad \textbf{Ans}$$

12-67. A particle travels along a path such that its position is $\mathbf{r} = \{(2 \sin t)\mathbf{i} + 2(1 - \cos t)\mathbf{j}\}$ ft, where t is in seconds and the arguments for the sine and cosine are given in radians. Find the equation which describes the path in terms of x and y and show that the magnitudes of the particle's velocity and acceleration are constant. What is the magnitude of the particle's displacement from $t = 0$ to $t = 2$ s?

$$\mathbf{r} = 2\sin t\,\mathbf{i} + 2(1 - \cos t)\,\mathbf{j}$$

$$x = 2\sin t \qquad y = 2(1 - \cos t)$$

$$\cos t = \frac{\sqrt{4 - x^2}}{2}$$

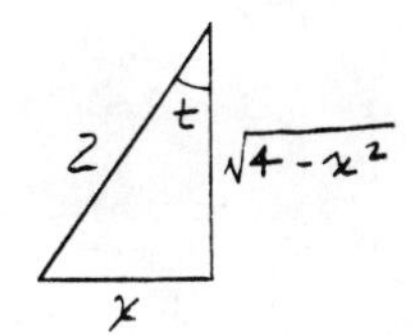

$$y = 2\left(1 - \frac{\sqrt{4 - x^2}}{2}\right) = 2 - \sqrt{4 - x^2}$$

$$4 - x^2 = (2 - y)^2$$

$$4 - x^2 = 4 - 4y + y^2$$

$$y^2 + x^2 - 4y = 0 \qquad \textbf{Ans}$$

$$\mathbf{v} = \frac{d\mathbf{r}}{dt} = 2\cos t\,\mathbf{i} + 2\sin t\,\mathbf{j}$$

$$v = \sqrt{(2)^2(\cos^2 t + \sin^2 t)} = 2 \text{ ft/s} \qquad \textbf{Ans}$$

$$\mathbf{a} = \frac{d\mathbf{v}}{dt} = -2\sin t\,\mathbf{i} + 2\cos t\,\mathbf{j}$$

$$a = \sqrt{(2)^2(\sin^2 t + \cos^2 t)} = 2 \text{ ft/s}^2 \qquad \textbf{Ans}$$

$$\mathbf{r}|_{t=0} = 0$$

$$\mathbf{r}|_{t=2} = 2\sin 2\mathbf{i} + 2(1 - \cos 2)\mathbf{j} = 1.819\mathbf{i} + 2.832\mathbf{j}$$

$$r = \sqrt{(1.819)^2 + (2.832)^2} = 3.37 \text{ ft} \qquad \textbf{Ans}$$

***12-68.** A particle is traveling with a velocity of $\mathbf{v} = \{3\sqrt{t}e^{-0.2t}\mathbf{i} + 4e^{-0.8t^2}\mathbf{j}\}$ m/s, where t is in seconds. Determine the magnitude of the particle's displacement from $t = 0$ to $t = 3$ s. Use Simpson's rule with $n = 100$ to evaluate the integrals. What is the magnitude of the particle's acceleration when $t = 2$ s?

$$ds = v\,dt$$

$$\Delta s_x = \int_0^3 3\sqrt{t}\,e^{-0.2t}\,dt = 7.341$$

$$\Delta s_y = \int_0^3 4\,e^{-0.8t^2}\,dt = 3.963$$

Thus,

$$\Delta s = \sqrt{(7.341)^2 + (3.963)^2} = 8.34 \text{ m} \qquad \textbf{Ans}$$

$$a_x = \dot{v}_x = 3\left(\frac{1}{2}\right)t^{-\frac{1}{2}}e^{-0.2t} + 3\sqrt{t}\,e^{-0.2t}(-0.2)\Big|_{t=2} = 0.1422$$

$$a_y = \dot{v}_y = 4\,e^{-0.8t^2}(-0.8)(2t)\Big|_{t=2} = -0.5218$$

$$a = \sqrt{(0.1422)^2 + (-0.5218)^2} = 0.541 \text{ m/s}^2 \qquad \textbf{Ans}$$

12-69. The position of particles A and B is described by $\mathbf{r}_A = \{2t\mathbf{i} + (t^2 - 1)\mathbf{j}\}$ ft and $\mathbf{r}_B = \{(t + 2)\mathbf{i} + (2t^2 - 5)\mathbf{j}\}$ ft, respectively, where t is in seconds. Determine the point where the particles collide and their speeds just before the collision.

When collision occurs $\mathbf{r}_A = \mathbf{r}_B$,

$$2t = t + 2$$

$$t^2 - 1 = 2t^2 - 5$$

Both equations are solved when $t = 2$ s. Then,

$$\mathbf{r}_A = 4\mathbf{i} + 3\mathbf{j} \quad \text{and} \quad \mathbf{r}_B = 4\mathbf{i} + 3\mathbf{j}$$

Collision occurs at point (4 ft, 3 ft) **Ans**

$$\mathbf{v}_A = \frac{d\mathbf{r}_A}{dt} = 2\mathbf{i} + 2t\mathbf{j}$$

$$v_A = \sqrt{2^2 + 4^2} = 4.47 \text{ ft/s} \qquad \textbf{Ans}$$

$$\mathbf{v}_B = \frac{d\mathbf{r}_B}{dt} = 1\mathbf{i} + 4t\mathbf{j}$$

$$v_B = \sqrt{1^2 + 8^2} = 8.06 \text{ ft/s} \qquad \textbf{Ans}$$

12-70. The car travels from A to B, and then from B to C, as shown in the figure. Determine the magnitude of the displacement of the car and the distance traveled.

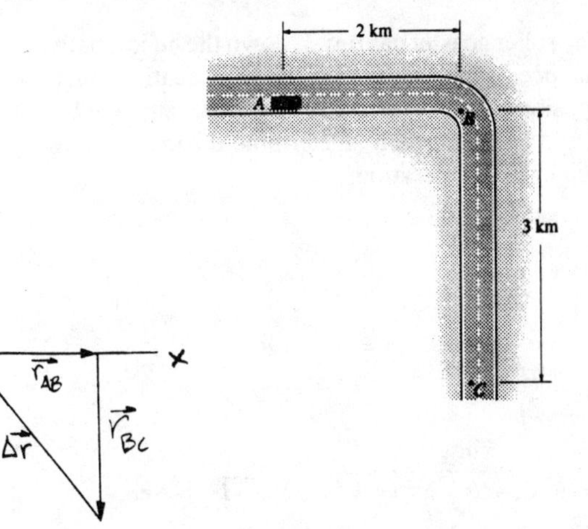

Displacement : $\Delta \mathbf{r} = \{2\mathbf{i} - 3\mathbf{j}\}$ km

$$\Delta r = \sqrt{2^2 + 3^2} = 3.61 \text{ km} \quad \textbf{Ans}$$

Distance traveled :

$$d = 2 + 3 = 5 \text{ km} \quad \textbf{Ans}$$

12-71. A particle travels along the curve from A to B in 1 s. If it takes 3 s for it to go from A to C, determine its *average velocity* when it goes from B to C.

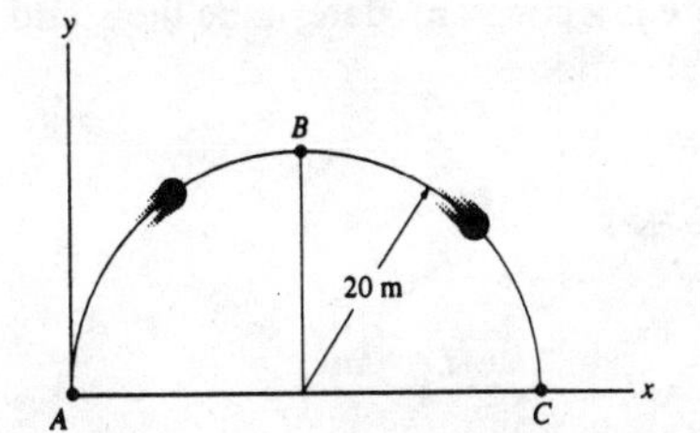

Time from B to C is $3 - 1 = 2$ s

$$\mathbf{v}_{avg} = \frac{\Delta \mathbf{r}}{\Delta t} = \frac{(\mathbf{r}_{AC} - \mathbf{r}_{AB})}{\Delta t} = \frac{40\mathbf{i} - (20\mathbf{i} + 20\mathbf{j})}{2} = \{10\mathbf{i} - 10\mathbf{j}\} \text{ m/s} \quad \textbf{Ans}$$

***12-72.** A particle moves with curvilinear motion in the x–y plane such that the y component of motion is described by the equation $y = (7t^3)$ m, where t is in seconds. If the particle starts from rest at the origin when $t = 0$, and maintains a *constant* acceleration in the x direction of 12 m/s^2, determine the particle's speed when $t = 2$ s.

$v = 0$ at $t = 0$

$a_x = \ddot{x} = 12$

$v_x = (v_x)_0 + a_c\, t$

$\quad = 0 + 12(2) = 24$ m/s

$y = 7t^3$

$$v_y = \frac{dy}{dt} = 21t^2$$

When $t = 2$ s,

$v_y = 21(2)^2 = 84$ m/s

$$v = \sqrt{(24)^2 + (84)^2} = 87.4 \text{ m/s} \quad \textbf{Ans}$$

12-73. The roller coaster car travels down the helical path at constant speed such that the parametric equations that define its position are $x = c \sin kt$, $y = c \cos kt$, $z = h - bt$, when c, h, and b are constants. Determine the magnitudes of its velocity and acceleration.

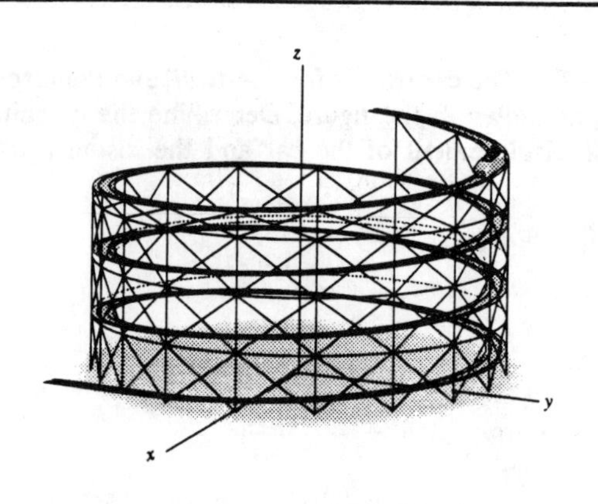

$x = c \sin kt \qquad \dot{x} = ck \cos kt \qquad \ddot{x} = -ck^2 \sin kt$

$y = c \cos kt \qquad \dot{y} = -ck \sin kt \qquad \ddot{y} = -ck^2 \cos kt$

$z = h - bt \qquad \dot{z} = -b \qquad \ddot{z} = 0$

$v = \sqrt{(ck \cos kt)^2 + (-ck \sin kt)^2 + (-b)^2} = \sqrt{c^2k^2 + b^2}$ **Ans**

$a = \sqrt{(-ck^2 \sin kt)^2 + (-ck^2 \cos kt)^2 + 0} = ck^2$ **Ans**

12-74. The path of a particle is defined by $x + y = 1$. If the component of velocity along the x axis is $v_x = \cos kt$, where k is a constant, determine the x and y components of acceleration.

$v_x = \cos kt$

$a_x = \dfrac{dv_x}{dt} = -k \sin kt$ **Ans**

$y = 1 - x \qquad v_y = -v_x \qquad a_y = -a_x$

$a_y = k \sin kt$ **Ans**

12-75. The path of a particle is defined by $y^2 = 4kx$, and the component of velocity along the y axis is $v_y = ct$, where both k and c are constants. Determine the x and y components of acceleration.

$y^2 = 4kx$

$2yv_y = 4kv_x$

$2v_y^2 + 2ya_y = 4ka_x$

$v_y = ct$

$a_y = c$ **Ans**

$2(ct)^2 + 2yc = 4ka_x$

$a_x = \dfrac{c}{2k}\left(y + ct^2\right)$ **Ans**

***12-76.** A particle travels along the circular path $x^2 + y^2 = r^2$, where r is the radius. If the component of velocity along the x axis is $v_x = t^2$, determine the x and y components of velocity and acceleration.

$$x^2 + y^2 = r^2$$

$$2xv_x + 2yv_y = 0$$

$$xv_x = -yv_y \qquad (1)$$

$$v_x^2 + xa_x = -v_y^2 - ya_y \qquad (2)$$

$$v_x = t^2 \qquad \textbf{Ans}$$

$$a_x = 2t \qquad \textbf{Ans}$$

From Eqs. (1) and (2) :

$$v_y = -\frac{x}{y}t^2 \qquad \textbf{Ans}$$

$$(t^2)^2 + x(2t) = -\left(-\frac{x}{y}t^2\right)^2 - ya_y$$

$$a_y = -\frac{t^4}{y} - \frac{2tx}{y} - \frac{x^2}{y^3}t^4$$

$$a_y = -\frac{2tx}{y} - \frac{t^4}{y^3}(y^2 + x^2)$$

$$a_y = -\frac{2tx}{y} - \frac{t^4 r^2}{y^3} \qquad \textbf{Ans}$$

12-77. The motorcycle travels with constant speed v_0 along the path that, for a short distance, takes the form of a sine curve. Determine the x and y components of its velocity at any instant on the curve.

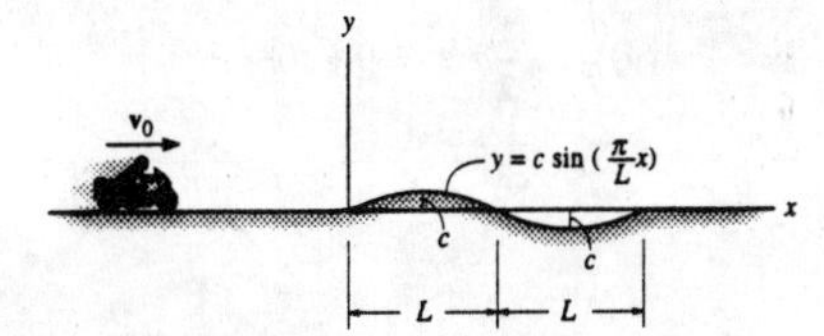

$$y = c\sin\left(\frac{\pi}{L}x\right)$$

$$y = \frac{\pi}{L}c\left(\cos\frac{\pi}{L}x\right)\dot{x}$$

$$v_y = \frac{\pi}{L}c\,v_x\left(\cos\frac{\pi}{L}x\right)$$

$$v_0^2 = v_y^2 + v_x^2$$

$$v_0^2 = v_x^2\left[1 + \left(\frac{\pi}{L}c\right)^2\cos^2\left(\frac{\pi}{L}x\right)\right]$$

$$v_x = v_0\left[1 + \left(\frac{\pi}{L}c\right)^2\cos^2\left(\frac{\pi}{L}x\right)\right]^{-\frac{1}{2}} \qquad \textbf{Ans}$$

$$v_y = \frac{v_0\pi c}{L}\left(\cos\frac{\pi}{L}x\right)\left[1 + \left(\frac{\pi}{L}c\right)^2\cos^2\left(\frac{\pi}{L}x\right)\right]^{-\frac{1}{2}} \qquad \textbf{Ans}$$

12-78. The flight path of the helicopter as it takes off from A is defined by the parametric equations $x = (2t^2)$ m and $y = (0.04t^3)$ m, where t is the time in seconds after takeoff. Determine the distance the helicopter is from point A and the magnitudes of its velocity and acceleration when $t = 10$ s.

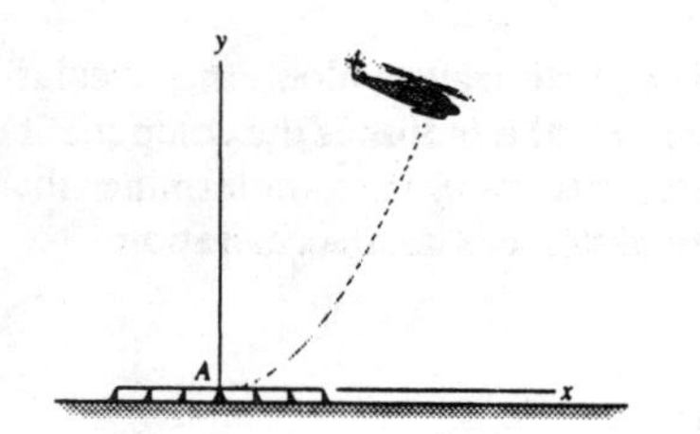

$x = 2t^2 \qquad y = 0.04t^3$

At $t = 10$ s, $\quad x = 200$ m $\quad y = 40$ m

$d = \sqrt{(200)^2 + (40)^2} = 204$ m **Ans**

$v_x = \dfrac{dx}{dt} = 4t$

$a_x = \dfrac{dv_x}{dt} = 4$

$v_y = \dfrac{dy}{dt} = 0.12t^2$

$a_y = \dfrac{dv_y}{dt} = 0.24t$

At $t = 10$ s,

$v = \sqrt{(40)^2 + (12)^2} = 41.8$ m/s **Ans**

$a = \sqrt{(4)^2 + (2.4)^2} = 4.66$ m/s^2 **Ans**

12-79. At the instant shown particle A is traveling to the right at 10 ft/s and has an acceleration of 2 ft/s^2. Determine the initial speed v_0 of particle B so that when it is fired at the same instant from the angle shown it strikes A. Also, at what speed does it strike A?

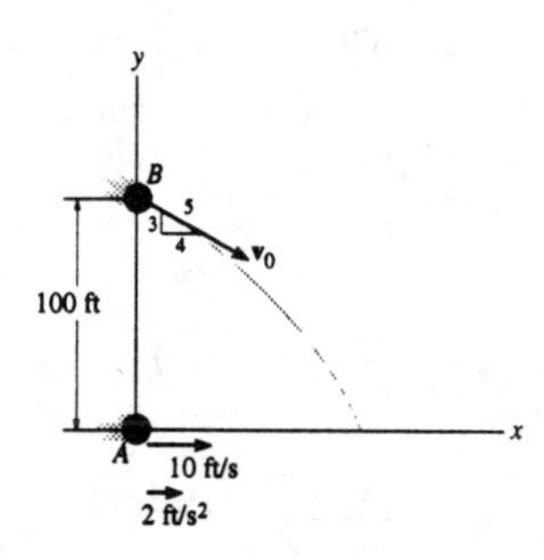

Particle A :

$a_x = 2$ ft/s^2

$v_x = v_0 + a_c t = 10 + 2t$

$x = x_0 + (v_x)_0 t + \frac{1}{2} a_c t^2 = 0 + 10t + t^2$

Particle B :

$v_x = (v_x)_0 = \frac{4}{5} v_0$

$x = (v_x)_0 t = \frac{4}{5} v_0 t$

$v_y = (v_y)_0 + a_c t = -\frac{3}{5} v_0 - 32.2t$

$y = y_0 + (v_y)_0 t + \frac{1}{2} a_c t^2 = 100 - \frac{3}{5} v_0 t - \frac{1}{2}(32.2)t^2$

Require :

$x_A = x_B \qquad 10t + t^2 = \frac{4}{5} v_0 t$

$v_0 = 1.25(10 + t)$

$y_A = y_B \qquad 0 = 100 - \frac{3}{5} v_0 t - 16.1t^2$

$0 = 100 - \frac{3}{5}(1.25)(10 + t)t - 16.1t^2$

$0 = 16.85t^2 + 7.5t - 100$

$t = 2.224$ s

$v_0 = 1.25(10 + 2.224) = 15.28 = 15.3$ ft/s **Ans**

$v_{B_x} = \frac{4}{5}(15.28) = 12.224$

$v_{B_y} = -\frac{3}{5}(15.28) - 32.2(2.224) = -80.77$

$v_B = \sqrt{(12.224)^2 + (-80.77)^2} = 81.7$ ft/s **Ans**

***12-80.** The pitcher throws the baseball horizontally with a speed of 140 ft/s from a height of 5 ft. If the batter is 60 ft away, determine the time needed for the ball to arrive at the batter and the height *h* at which it passes the batter.

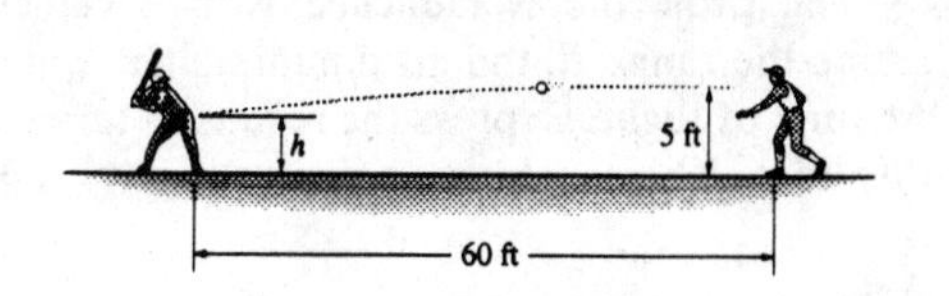

$(\overset{+}{\leftarrow})$ $\quad s = vt; \quad 60 = 140t$

$$t = 0.4286 = 0.429 \text{ s} \quad \textbf{Ans}$$

$(+\uparrow)$ $\quad s = s_0 + v_0 t + \frac{1}{2} a_c t^2$

$$h = 5 + 0 + \frac{1}{2}(-32.2)(0.4286)^2 = 2.04 \text{ ft} \quad \textbf{Ans}$$

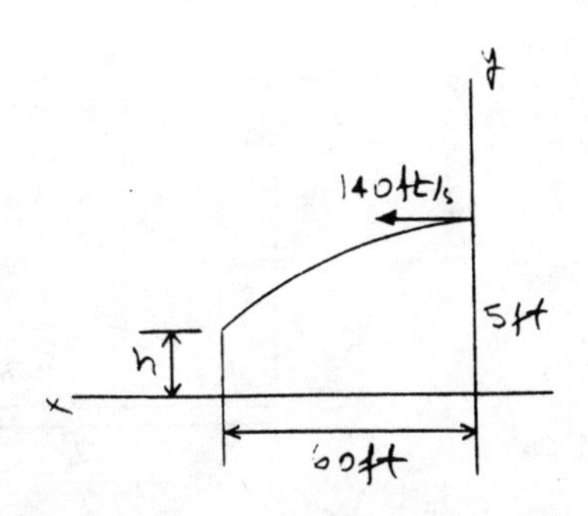

12-81. The nozzle of a garden hose discharges water at the rate of 15 m/s. If the nozzle is held at ground level and directed $\theta = 30°$ from the ground, determine the maximum height reached by the water and the horizontal distance from the nozzle to where the water strikes the ground.

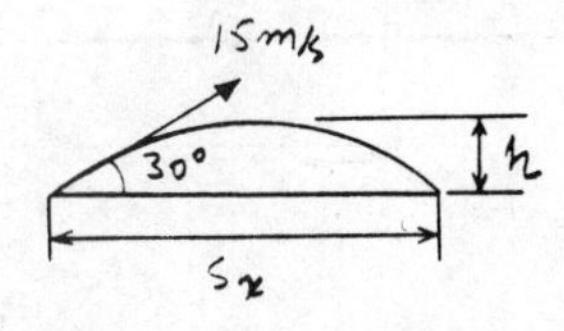

$$(v_0)_x = 15 \cos 30° = 12.99 \text{ m/s}$$

$$(v_0)_y = 15 \sin 30° = 7.5 \text{ m/s}$$

Maximum height :

$(+\uparrow)$ $\quad v^2 = v_0^2 + 2a_c(s - s_0)$

$$0 = (7.5)^2 + 2(-9.81)(h - 0)$$

$$h = 2.87 \text{ m} \quad \textbf{Ans}$$

Time of travel to top of path :

$(+\uparrow)$ $\quad v = v_0 + a_c t$

$$0 = 7.5 + (-9.81)t$$

$$t = 0.7645 \text{ s}$$

Total time along path

$$t = 2(0.7645) = 1.529 \text{ s}$$

Range

$$s_x = v_x t = (12.99)(1.529) = 19.9 \text{ m} \quad \textbf{Ans}$$

12-82. The projectile is launched with a velocity $\mathbf{v}_0$. Determine the range R, the maximum height h attained, and the time of flight. Express the results in terms of the angle θ and v_0. The acceleration due to gravity is g.

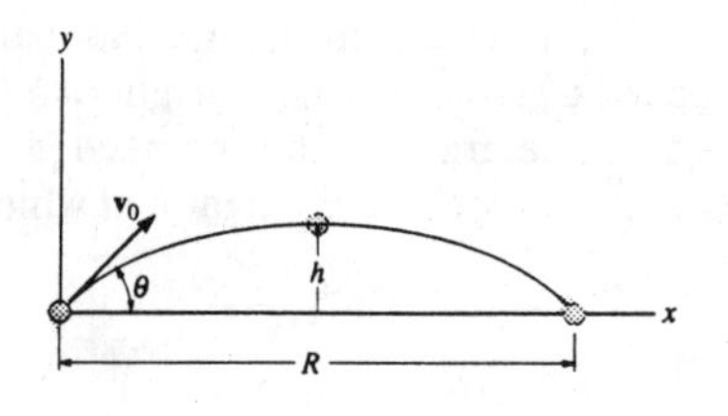

$(\overset{+}{\rightarrow})$ $\quad s = s_0 + v_0 t$

$$R = 0 + (v_0 \cos\theta)t$$

$(+\uparrow)$ $\quad s = s_0 + v_0 t + \frac{1}{2} a_c t^2$

$$0 = 0 + (v_0 \sin\theta)t + \frac{1}{2}(-g)t^2$$

$$0 = v_0 \sin\theta - \frac{1}{2}(g)\left(\frac{R}{v_0 \cos\theta}\right)$$

$$R = \frac{v_0^2}{g}\sin 2\theta \qquad \textbf{Ans}$$

$$t = \frac{R}{v_0 \cos\theta} = \frac{v_0^2(2\sin\theta\cos\theta)}{v_0 g\cos\theta}$$

$$= \frac{2v_0}{g}\sin\theta \qquad \textbf{Ans}$$

$(+\uparrow)$ $\quad v^2 = v_0^2 + 2a_c(s - s_0)$

$$0 = (v_0 \sin\theta)^2 + 2(-g)(h - 0)$$

$$h = \frac{v_0^2}{2g}\sin^2\theta \qquad \textbf{Ans}$$

12-83. Show that if a projectile is fired at an angle θ from the horizontal with an initial velocity $\mathbf{v}_0$, the *maximum* range the projectile can travel is given by $R_{\max} = v_0^2/g$, where g is the acceleration of gravity. What is the angle θ for this condition?

$(v_0)_x = v_0 \cos\theta$

$(v_0)_y = v_0 \sin\theta$

After time t,

$(\overset{+}{\rightarrow})$ $\quad s_x = (v_0)_x t; \qquad x = v_0 \cos\theta t \qquad (1)$

$(+\uparrow)$ $\quad s_y = (v_0)_y t + \frac{1}{2} a_c t^2; \qquad y = (v_0 \sin\theta)t - \frac{1}{2} g t^2 \qquad (2)$

Substituting Eq. (1) into Eq. (2), $\quad y = x\tan\theta - \dfrac{g x^2}{2v_0^2\cos^2\theta}$

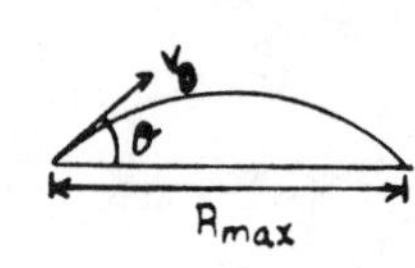

Set y = 0 to determine the range, $x = R$:

$$R = \frac{2v_0^2 \sin\theta\cos\theta}{g} = \frac{v_0^2 \sin 2\theta}{g}$$

R_{max} occurs when $\sin 2\theta = 1$ or,

$\theta = 45°$ $\quad$ **Ans**

This gives : $\quad R_{max} = \dfrac{v_0^2}{g}$ $\quad$ **Q.E.D**

***12-84.** The catapult is used to launch a ball such that it strikes the wall of the building at the maximum height of its trajectory. If it takes 1.5 s to travel from A to B, determine the velocity $\mathbf{v}_A$ at which it was launched, the angle of release θ, and the height h.

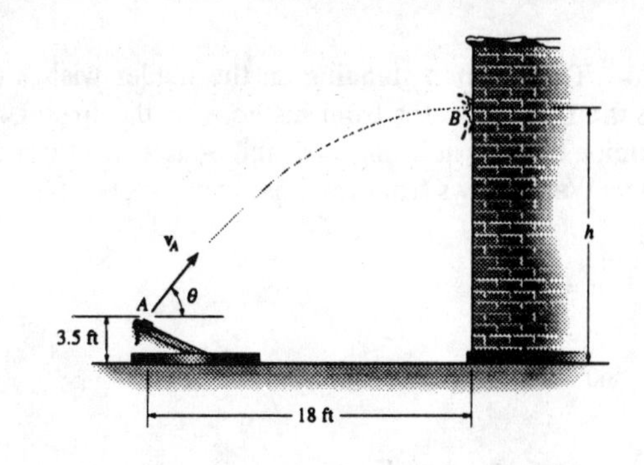

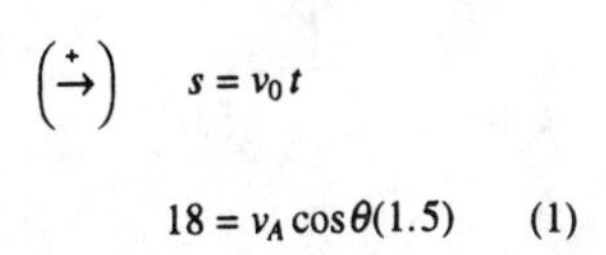

$(\overset{+}{\rightarrow})\quad s = v_0 t$

$$18 = v_A \cos\theta (1.5) \qquad (1)$$

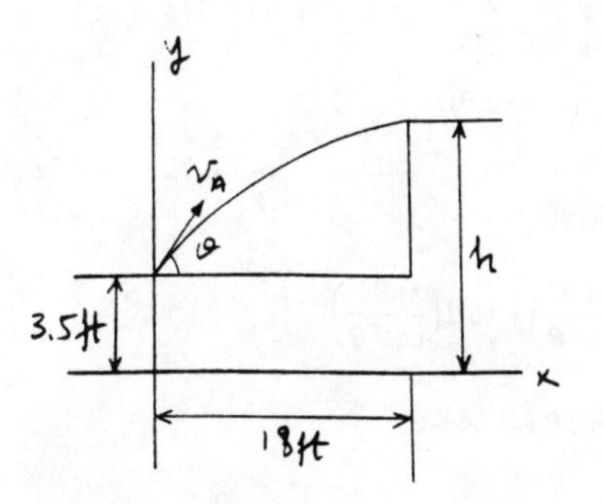

$(+\uparrow)\quad v^2 = v_0^2 + 2a_c(s - s_0)$

$$0 = (v_A \sin\theta)^2 + 2(-32.2)(h - 3.5)$$

$(+\uparrow)\quad v = v_0 + a_c t$

$$0 = v_A \sin\theta - 32.2(1.5) \qquad (2)$$

To solve, first divide Eq. (2) by Eq. (1), to get θ. Then

$\theta = 76.0°$ **Ans**

$v_A = 49.8$ ft/s **Ans**

$h = 39.7$ ft **Ans**

12-85. From a videotape, it was observed that a pro football player kicked a football 126 ft during a measured time of 3.6 seconds. Determine the initial speed of the ball and the angle θ at which it was kicked.

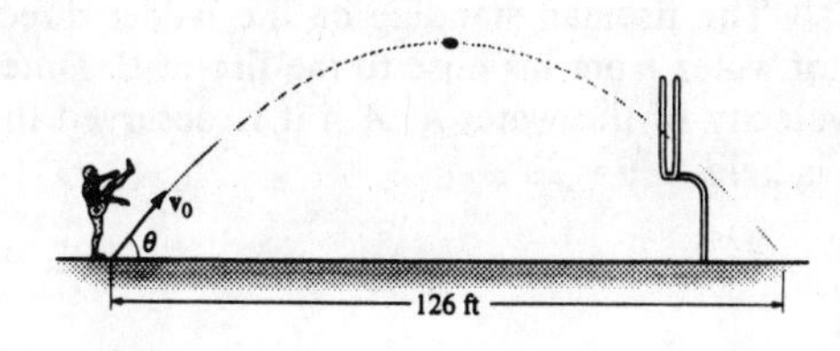

$(\overset{+}{\rightarrow})\quad s = s_0 + v_0 t$

$$126 = 0 + (v_0)_x (3.6)$$

$$(v_0)_x = 35 \text{ ft/s}$$

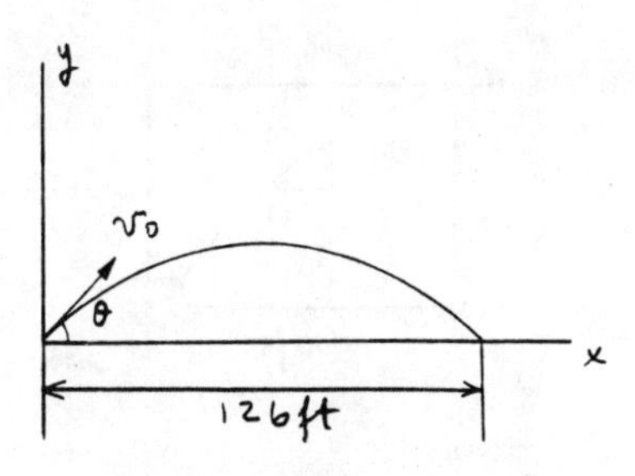

$(+\uparrow)\quad s = s_0 + v_0 t + \frac{1}{2} a_c t^2$

$$0 = 0 + (v_0)_y (3.6) + \frac{1}{2}(-32.2)(3.6)^2$$

$$(v_0)_y = 57.96 \text{ ft/s}$$

$v_0 = \sqrt{(35)^2 + (57.96)^2} = 67.7$ ft/s **Ans**

$\theta = \tan^{-1}\left(\frac{57.96}{35}\right) = 58.9°$ **Ans**

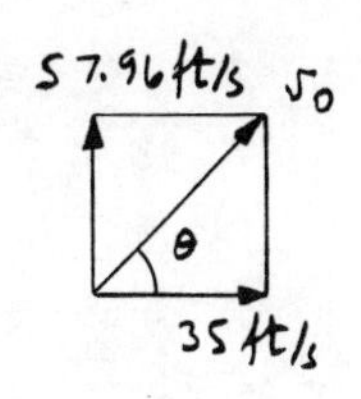

▪12-86. The fireman standing on the ladder wishes to direct the flow of water from his hose to the fire at B. Determine two possible angles θ_1 and θ_2 at which this can be done. Water flows from the hose at $v_A = 300$ ft/s.

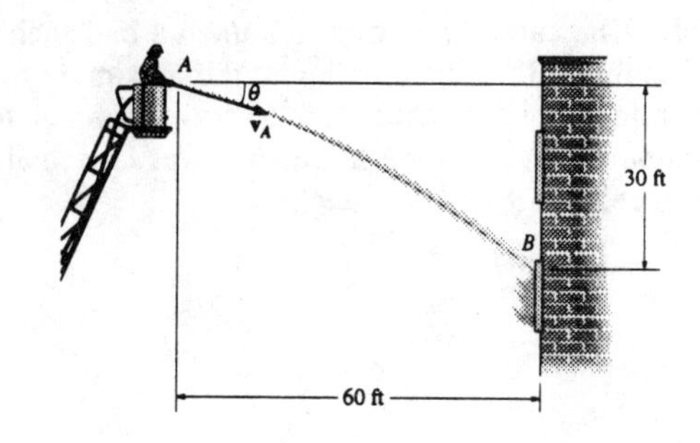

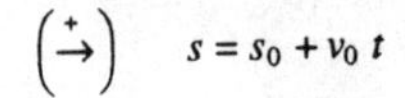

$(\overset{+}{\rightarrow})$ $\quad s = s_0 + v_0 t$

$$60 = 0 + (300 \cos\theta)t$$

$(+\uparrow)$ $\quad s = s_0 + v_0 t + \frac{1}{2} a_c t^2$

$$-30 = 0 - (300 \sin\theta)t + \frac{1}{2}(-32.2)t^2$$

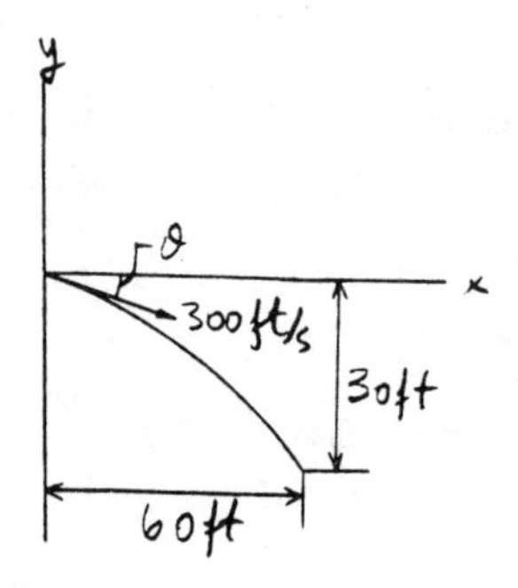

Thus,

$$30 = 300 \sin\theta\left(\frac{0.2}{\cos\theta}\right) + 16.1\left(\frac{0.04}{\cos^2\theta}\right)$$

$$30 \cos^2\theta = 30 \sin 2\theta + 0.644$$

Solving,

$\theta_1 = 26.0°$ (below horizontal) **Ans**

$\theta_2 = -89.4° = 89.4°$ (above horizontal) **Ans**

12-87. The fireman standing on the ladder directs the flow of water from his hose to the fire at B. Determine the velocity of the water at A if it is observed that the hose is held at $\theta = 20°$.

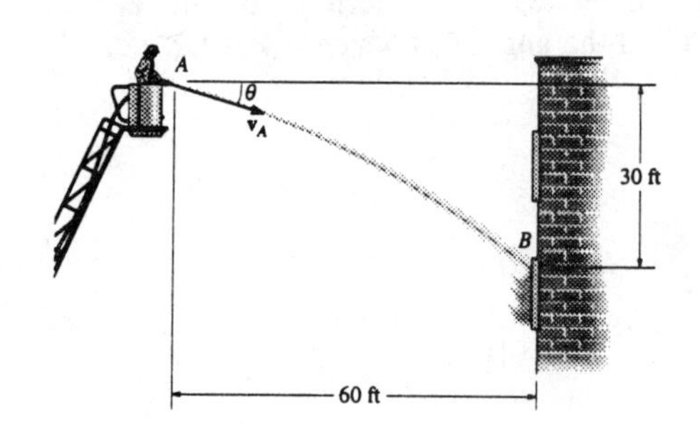

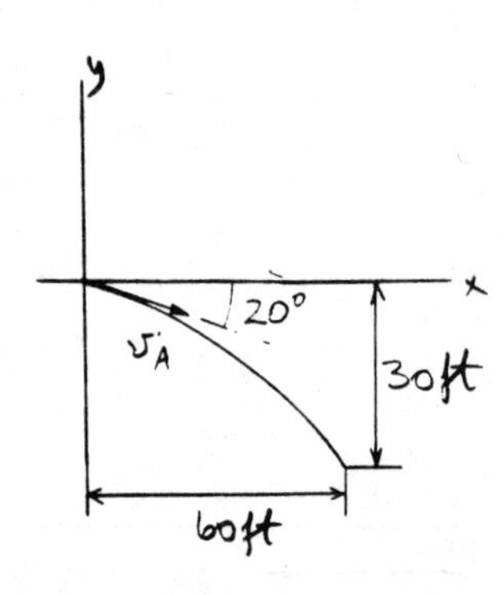

$(\overset{+}{\rightarrow})$ $\quad s = s_0 + v_0 t$

$$60 = 0 + (v_A \cos 20°)t$$

$$t = \frac{63.851}{v_0}$$

$(+\uparrow)$ $\quad s = s_0 + v_0 t + \frac{1}{2} a_c t^2$

$$-30 = 0 - v_A \sin 20°\left(\frac{63.851}{v_A}\right) + \frac{1}{2}(-32.2)\left(\frac{63.851}{v_A}\right)^2$$

$v_A = 89.7$ ft/s **Ans**

***12-88.** A projectile is given a velocity $\mathbf{v}_0$ at an angle ϕ above the horizontal. Determine the distance d to where it strikes the sloped ground. The acceleration due to gravity is g.

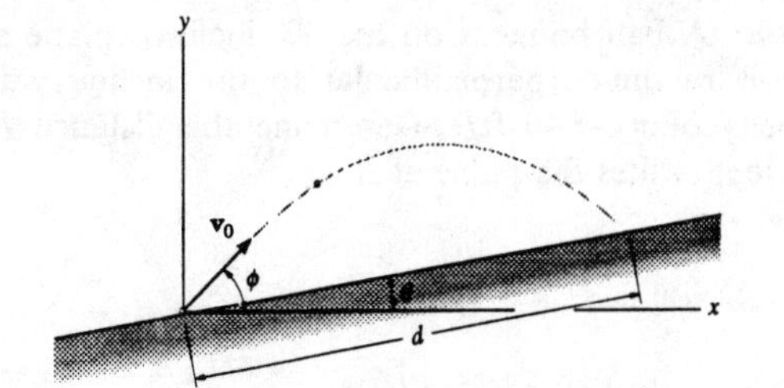

$(\overset{+}{\rightarrow})$ $\quad s = s_0 + v_0 t$

$$d\cos\theta = 0 + v_0(\cos\phi)t$$

$(+\uparrow)$ $\quad s = s_0 + v_0 t + \frac{1}{2}a_c t^2$

$$d\sin\theta = 0 + v_0(\sin\phi)t + \frac{1}{2}(-g)t^2$$

Thus,

$$d\sin\theta = v_0\sin\phi\left(\frac{d\cos\theta}{v_0\cos\phi}\right) - \frac{1}{2}g\left(\frac{d\cos\theta}{v_0\cos\phi}\right)^2$$

$$\sin\theta = \cos\theta\tan\phi - \frac{gd\cos^2\theta}{2v_0^2\cos^2\phi}$$

$$d = (\cos\theta\tan\phi - \sin\theta)\frac{2v_0^2\cos^2\phi}{g\cos^2\theta}$$

$$d = \frac{v_0^2}{g\cos\theta}(\sin 2\phi - 2\tan\theta\cos^2\phi) \quad \textbf{Ans}$$

12-89. A projectile is given a velocity $\mathbf{v}_0$. Determine the angle ϕ at which it should be launched so that d is a maximum. The acceleration due to gravity is g.

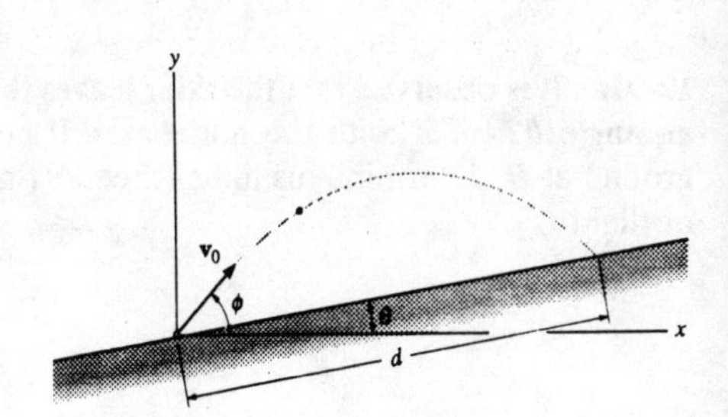

$(\overset{+}{\rightarrow})$ $\quad s_x = s_0 + v_0 t$

$$d\cos\theta = 0 + v_0(\cos\phi)t$$

$(+\uparrow)$ $\quad s_y = s_0 + v_0 t + \frac{1}{2}a_c t^2$

$$d\sin\theta = 0 + v_0(\sin\phi)t + \frac{1}{2}(-g)t^2$$

Thus,

$$d\sin\theta = v_0\sin\phi\left(\frac{d\cos\theta}{v_0\cos\phi}\right) - \frac{1}{2}g\left(\frac{d\cos\theta}{v_0\cos\phi}\right)^2$$

$$\sin\theta = \cos\theta\tan\phi - \frac{gd\cos^2\theta}{2v_0^2\cos^2\phi}$$

$$d = (\cos\theta\tan\phi - \sin\theta)\frac{2v_0^2\cos^2\phi}{g\cos^2\theta}$$

$$d = \frac{v_0^2}{g\cos\theta}(\sin 2\phi - 2\tan\theta\cos^2\phi) \quad \textbf{Ans}$$

Require :

$$\frac{d(d)}{d\phi} = \frac{v_0^2}{g\cos\theta}\left[\cos 2\phi(2) - 2\tan\theta(2\cos\phi)(-\sin\phi)\right] = 0$$

$$\cos 2\phi + \tan\theta\sin 2\phi = 0$$

$$\frac{\sin 2\phi}{\cos 2\phi}\tan\theta + 1 = 0$$

$$\tan 2\phi = -\text{ctn}\,\theta$$

$$\phi = \frac{1}{2}\tan^{-1}(-\text{ctn}\,\theta) \quad \textbf{Ans}$$

12-90. A ball bounces on the 30° inclined plane such that it rebounds perpendicular to the incline with a velocity of $v_A = 40$ ft/s. Determine the distance R to where it strikes the plane at B.

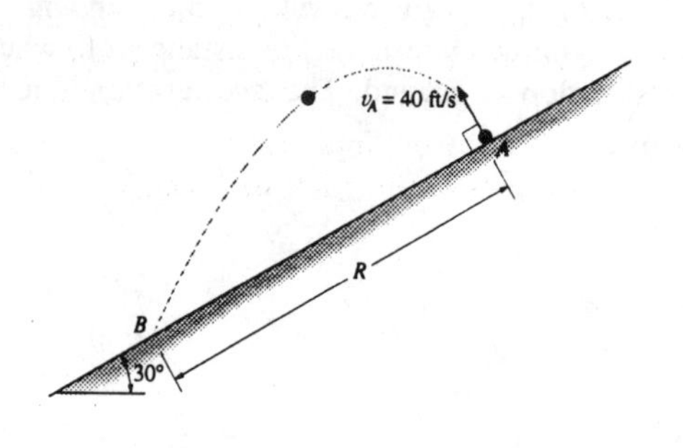

$(v_A)_x = 40\sin 30° = 20$ ft/s

$(v_A)_y = 40\cos 30° = 34.64$ ft/s

$(\overset{+}{\leftarrow})\qquad s = s_0 + v_0 t$

$R\cos 30° = 0 + 20t$

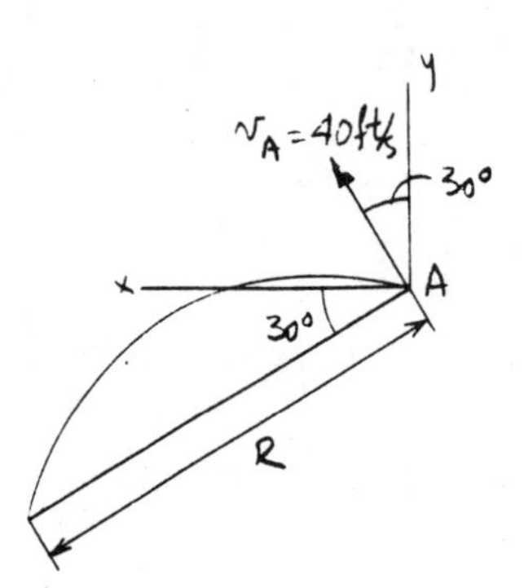

$(+\uparrow)\qquad s = s_0 + v_0 t + \frac{1}{2}a_c t^2$

$-R\sin 30° = 0 + 34.64t + \frac{1}{2}(-32.2)t^2$

Combining the equations,

$20t\tan 30° = -34.64t + 16.1t^2$

$46.19 = 16.1t$

$t = 2.87$ s

Thus,

$R = \dfrac{20(2.87)}{\cos 30°} = 66.3$ ft **Ans**

12-91. It is observed that the skier leaves the ramp A at an angle $\theta_A = 25°$ with the horizontal. If he strikes the ground at B, determine his initial speed v_A and the time of flight t_{AB}.

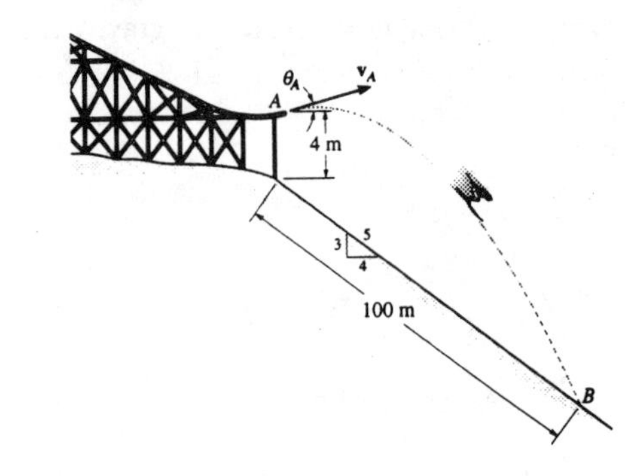

$(\overset{+}{\rightarrow})\qquad s = v_0 t$

$100\left(\frac{4}{5}\right) = v_A\cos 25° t_{AB}$

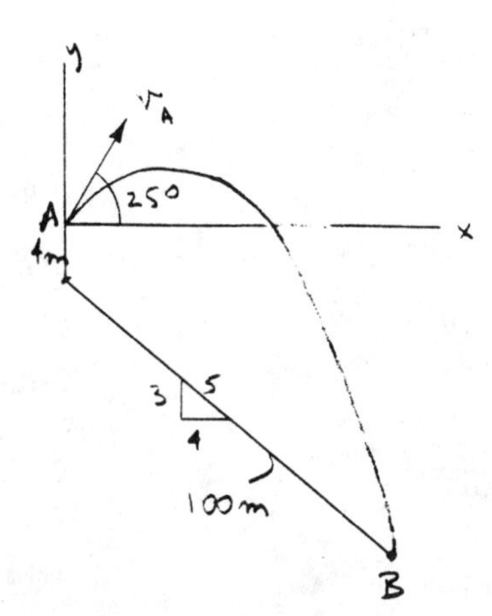

$(+\uparrow)\qquad s = s_0 + v_0 t + \frac{1}{2}a_c t^2$

$-4 - 100\left(\frac{3}{5}\right) = 0 + v_A\sin 25° t_{AB} + \frac{1}{2}(-9.81)t_{AB}^2$

Solving,

$v_A = 19.4$ m/s **Ans**

$t_{AB} = 4.54$ s **Ans**

***12-92.** The drinking fountain is designed such that the nozzle is located from the edge of the basin as shown. Determine the maximum and minimum speed at which water can be ejected from the nozzle so that it does not splash over the sides of the basin at B and C.

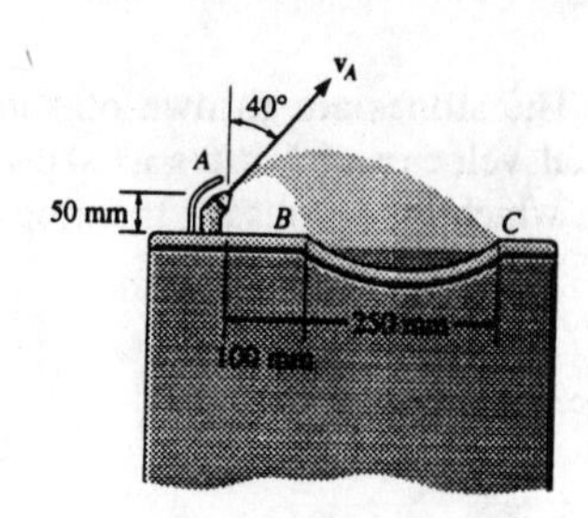

Horizontal Motion :

$$(\overset{+}{\rightarrow}) \qquad s = v_0 t$$

$$R = v_A \sin 40° t \qquad t = \frac{R}{v_A \sin 40°} \qquad (1)$$

Vertical Motion :

$$(+\uparrow) \qquad s = s_0 + v_0 t + \tfrac{1}{2} a_c t^2$$

$$-0.05 = 0 + v_A \cos 40° t + \tfrac{1}{2}(-9.81)t^2 \qquad (2)$$

Substituting Eq. (1) into (2) yields :

$$-0.05 = v_A \cos 40° \left(\frac{R}{v_A \sin 40°}\right) + \frac{1}{2}(-9.81)\left(\frac{R}{v_A \sin 40°}\right)^2$$

$$v_A = \sqrt{\frac{4.905R^2}{\sin 40° (R\cos 40° + 0.05 \sin 40°)}}$$

At point B, $R = 0.1$ m.

$$v_{min} = v_A = \sqrt{\frac{4.905(0.1)^2}{\sin 40° (0.1 \cos 40° + 0.05 \sin 40°)}} = 0.838 \text{ m/s} \qquad \textbf{Ans}$$

At point C, $R = 0.35$ m.

$$v_{max} = v_A = \sqrt{\frac{4.905(0.35)^2}{\sin 40° (0.35 \cos 40° + 0.05 \sin 40°)}} = 1.76 \text{ m/s} \qquad \textbf{Ans}$$

12-93. The stones are thrown off the conveyor with a horizontal velocity of 10 ft/s as shown. Determine the distance d down the slope to where the stones hit the ground at B.

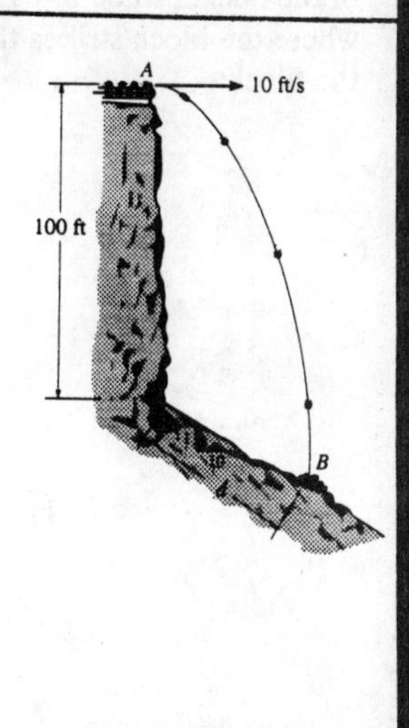

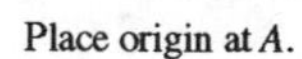

Place origin at A.

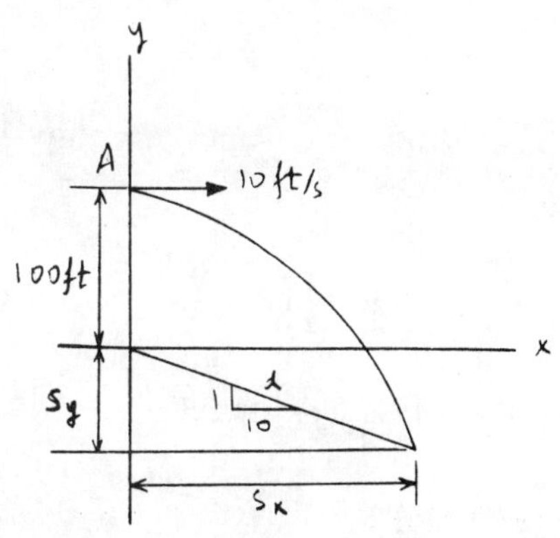

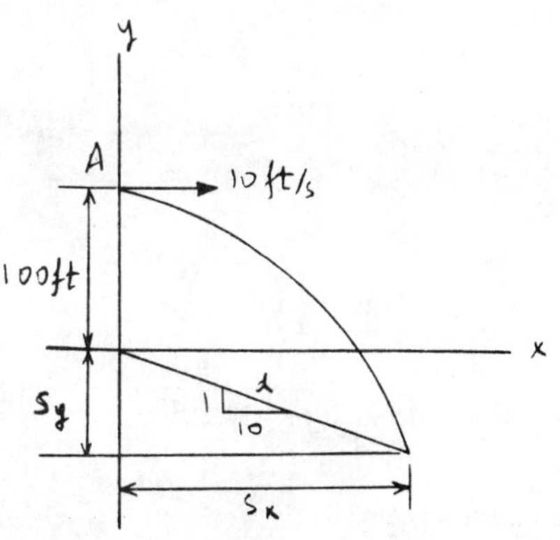

$$(\overset{+}{\rightarrow}) \qquad s = s_0 + v_0 t$$

$$s_x = 0 + 10t$$

$$(+\uparrow) \qquad s = s_0 + v_0 t + \frac{1}{2} a_c t^2$$

$$-s_y = 100 + 0 + \frac{1}{2}(-32.2)t^2$$

$$s_y = 16.1t^2 - 100$$

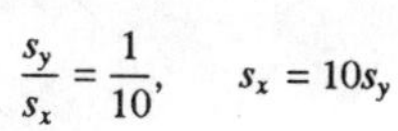

$$\frac{s_y}{s_x} = \frac{1}{10}, \qquad s_x = 10 s_y$$

$$s_y = \frac{10t}{10} = t$$

$$t = 16.1t^2 - 100$$

$$16.1t^2 - 1t - 100 = 0$$

Solving for the positive root, $t = 2.5235$ s

$$s_x = 10(2.5235) = 25.235 \text{ ft}$$

$$s_y = 16.1(2.5235)^2 - 100 = 2.5235 \text{ ft}$$

$$d = \sqrt{(25.235)^2 + (2.5235)^2} = 25.4 \text{ ft} \qquad \textbf{Ans}$$

12-94. The stones are thrown off the conveyor with a horizontal velocity of 10 ft/s as shown. Determine the speed at which the stones hit the ground at B.

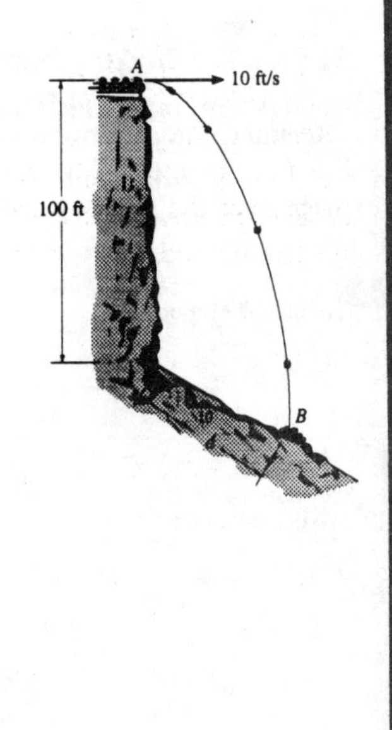

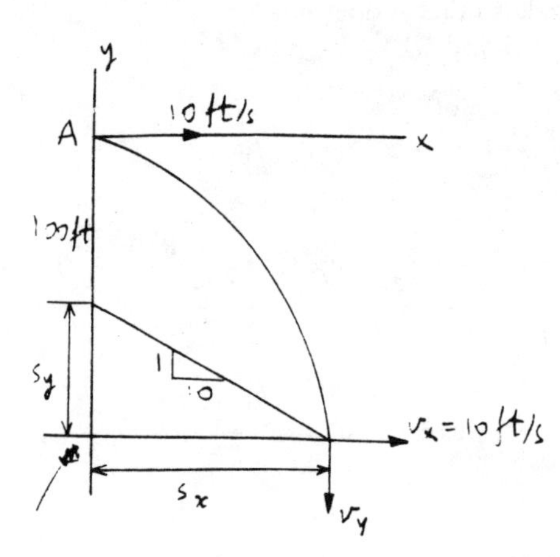

Place origin at A.

$(\overset{+}{\rightarrow})$ $s = s_0 + v_0 t$

$$s_x = 0 + 10t$$

$(+\uparrow)$ $s = s_0 + v_0 t + \frac{1}{2}a_c t^2$

$$-100 - s_y = 0 + \frac{1}{2}(-32.2)t^2$$

$$\frac{s_y}{s_x} = \frac{1}{10}, \quad s_x = 10s_y$$

$$10t = 10s_y, \quad s_y = t$$

$$-100 - t = -16.1t^2$$

$$16.1t^2 - 1t - 100 = 0$$

Solving for the positive root, $t = 2.5235$ s

$(+\uparrow)$ $v_y = (v_0)_y + a_c t$

$$v_y = 0 - 32.2(2.5235) = -81.256 \text{ ft/s}$$

$$v_x = 10 \text{ ft/s}$$

$$v = \sqrt{(10)^2 + (-81.256)^2} = 81.9 \text{ ft/s} \quad \textbf{Ans}$$

12-95. The buckets on the conveyor travel with a speed of 15 ft/s. Each bucket contains a block which falls out of the bucket when $\theta = 120°$. Determine the distance s to where the block strikes the conveyor. Neglect the size of the block.

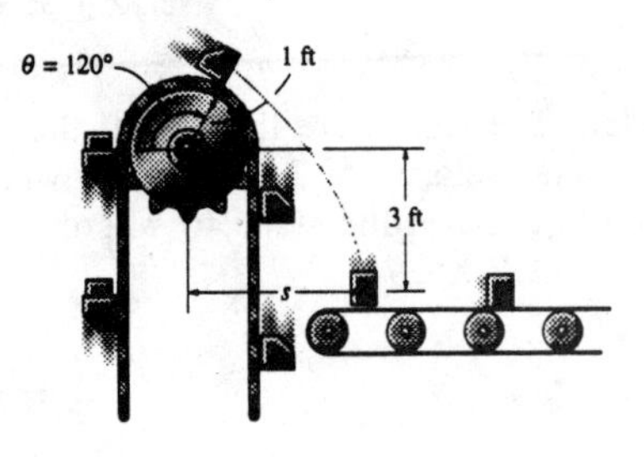

Vertical Motion :

$(+\downarrow)$ $s_y = (s_0)_y + v_y t + \frac{1}{2}a_c t^2$

$$3 + 1\cos 30° = 0 + 15\sin 30° t + \frac{1}{2}(32.2)t^2$$

Take the positive root $t = 0.3096$ s

Horizontal Motion :

$(\overset{+}{\rightarrow})$ $s_x = (s_0)_x + v_x t$

$$s - 1\sin 30° = 0 + 15\cos 30°(0.3096)$$

$$s = 4.52 \text{ ft} \quad \textbf{Ans}$$

***12-96.** The missile at A takes off from rest and rises vertically to B, where its fuel runs out in 8 s. If the acceleration varies with time as shown, determine the missile's height h_B and speed v_B. If by internal controls the missile is then suddenly pointed 45° as shown, and allowed to travel in free flight, determine the maximum height attained, h_C, and the range R to where it crashes at D.

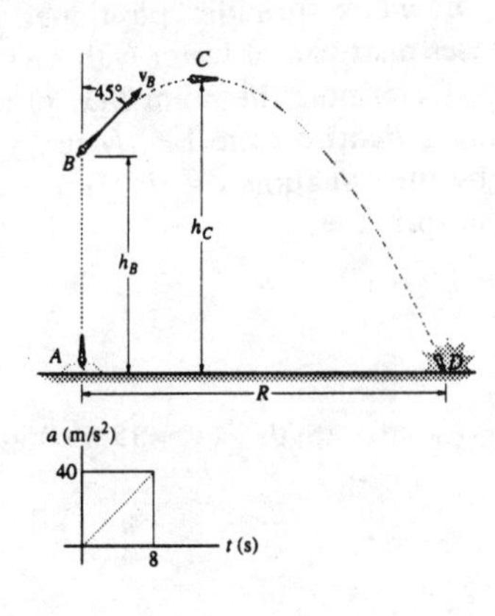

$$a = \frac{40}{8}t = 5t$$

$$dv = a\,dt$$

$$\int_0^v dv = \int_0^t 5t\,dt$$

$$v = 2.5t^2$$

When $t = 8$ s, $\quad v_B = 2.5(8)^2 = 160$ m/s **Ans**

$$ds = v\,dt$$

$$\int_0^s ds = \int_0^t 2.5t^2\,dt$$

$$s = \frac{2.5}{3}t^3$$

$$h_B = \frac{2.5}{3}(8)^3 = 426.67 = 427 \text{ m}$$ **Ans**

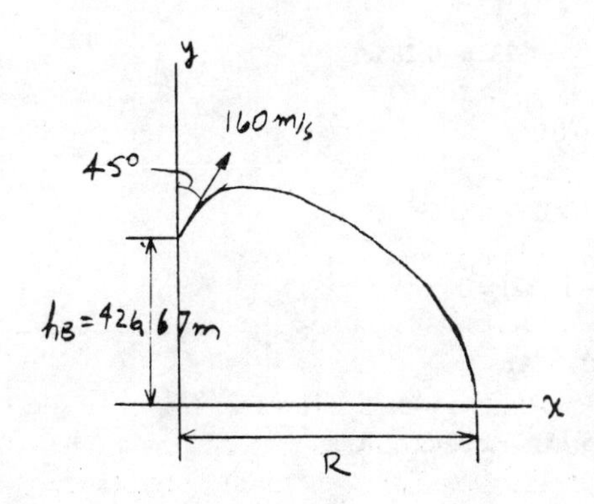

$$(v_B)_x = 160\sin 45° = 113.14 \text{ m/s}$$

$$(v_B)_y = 160\cos 45° = 113.14 \text{ m/s}$$

$(+\uparrow)$ $\quad v^2 = v_0^2 + 2a_c(s - s_0)$

$$0^2 = (113.14)^2 + 2(-9.81)(s_C - 426.67)$$

$$h_C = 1079.1 \text{ m} = 1.08 \text{ km}$$ **Ans**

$(\overset{+}{\rightarrow})$ $\quad s = s_0 + v_0 t$

$$R = 0 + 113.14t$$

$(+\uparrow)$ $\quad s = s_0 + v_0 t + \frac{1}{2}a_c t^2$

$$0 = 426.67 + 113.14t + \frac{1}{2}(-9.81)t^2$$

Solving for the positive root, $t = 26.36$ s

Then,

$R = 113.14(26.36) = 2983.0 = 2.98$ km **Ans**

12-97. The water sprinkler, positioned at the base of a hill, releases a stream of water with a velocity of 15 ft/s as shown. Determine the point $B(x, y)$ where the water strikes the ground on the hill. Assume that the hill is defined by the equation $y = (0.05x^2)$ ft and neglect the size of the sprinkler.

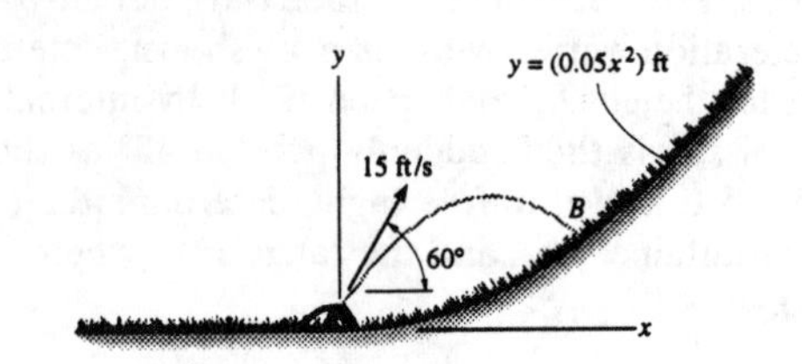

$v_x = 15\cos 60° = 7.5$ ft/s $\quad v_y = 15\sin 60° = 12.99$ ft/s

$(\overset{+}{\rightarrow})\quad s = v_0 t$

$x = 7.5t$

$(+\uparrow)\quad s = s_0 + v_0 t + \frac{1}{2}a_c t^2$

$y = 0 + 12.99t + \frac{1}{2}(-32.2)t^2$

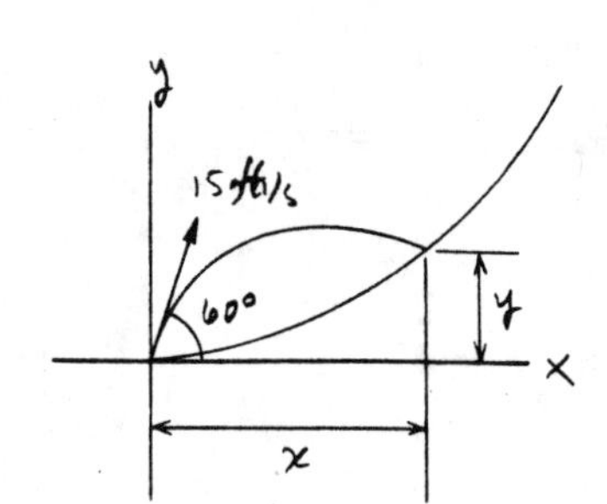

$y = 1.732x - 0.286x^2$

Since $y = 0.05x^2$,

$0.05x^2 = 1.732x - 0.286x^2$

$x(0.336x - 1.732) = 0$

$x = 5.15$ ft **Ans**

$y = 0.05(5.15)^2 = 1.33$ ft **Ans**

Also,

$(\overset{+}{\rightarrow})\quad s = v_0 t$

$x = 15\cos 60° t$

$(+\uparrow)\quad s = s_0 + v_0 t + \frac{1}{2}a_c t^2$

$y = 0 + 15\sin 60° t + \frac{1}{2}(-32.2)t^2$

Since $y = 0.05x^2$

$12.99t - 16.1t^2 = 2.8125t^2 \quad t = 0.6869$ s

So that,

$x = 15\cos 60°(0.6868) = 5.15$ ft **Ans**

$y = 0.05(5.15)^2 = 1.33$ ft **Ans**

12-98. The ball is thrown from the tower with a velocity of 20 ft/s as shown. Determine the x and y coordinates to where the ball strikes the slope. Also, determine the speed at which the ball hits the ground.

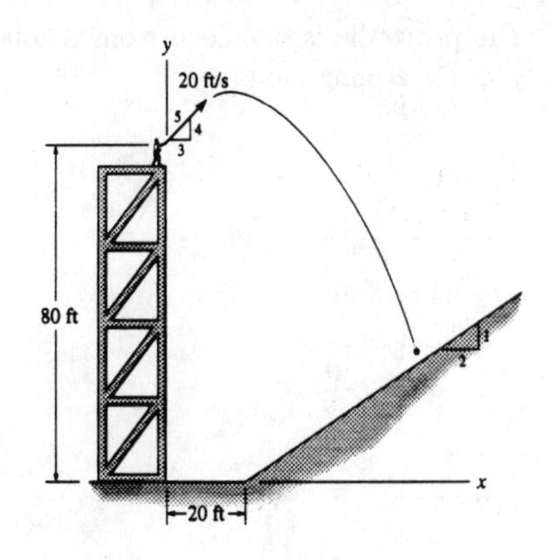

Assume ball hits slope.

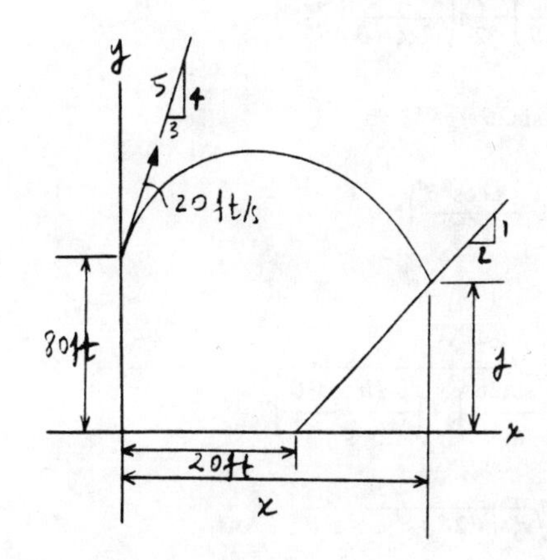

$(\overset{+}{\rightarrow})$ $\quad s = s_0 + v_0 t$

$$x = 0 + \frac{3}{5}(20)t = 12t$$

$(+\uparrow)$ $\quad s = s_0 + v_0 t + \frac{1}{2}a_c t^2$

$$y = 80 + \frac{4}{5}(20)t + \frac{1}{2}(-32.2)t^2 = 80 + 16t - 16.1t^2$$

Equation of slope : $\quad y - y_1 = m(x - x_1)$

$$y - 0 = \frac{1}{2}(x - 20)$$

$$y = 0.5x - 10$$

Thus,

$$80 + 16t - 16.1t^2 = 0.5(12t) - 10$$

$$16.1t^2 - 10t - 90 = 0$$

Choosing the positive root :

$t = 2.6952$ s

$x = 12(2.6952) = 32.3$ ft **Ans**

Since 32.3 ft > 20 ft , assumption is valid.

$y = 80 + 16(2.6952) - 16.1(2.6952)^2 = 6.17$ ft **Ans**

$(\overset{+}{\rightarrow})$ $\quad v_x = (v_0)_x = \frac{3}{5}(20) = 12$ ft/s

$(+\uparrow)$ $\quad v_y = (v_0)_y + a_c t = \frac{4}{5}(20) + (-32.2)(2.6952) = -70.785$ ft/s

$v = \sqrt{(12)^2 + (-70.785)^2} = 71.8$ ft/s **Ans**

12-99. The projectile is launched from a height h with a velocity $\mathbf{v}_0$. Determine the range R.

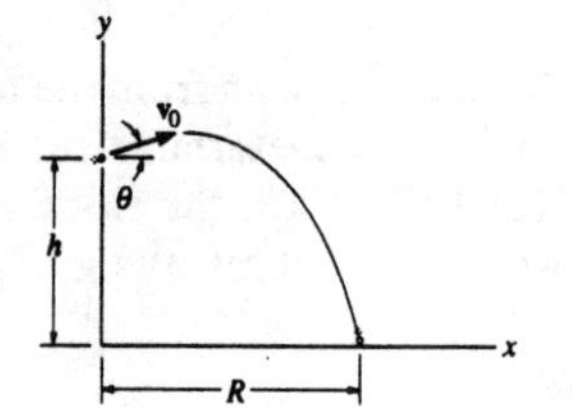

$(\overset{+}{\rightarrow})\qquad s = s_0 + v_0 t$

$$R = 0 + v_0 \cos\theta t$$

$(+\uparrow)\qquad s = s_0 + v_0 t + \frac{1}{2} a_c t^2$

$$0 = h + v_0 \sin\theta t + \frac{1}{2}(-g)t^2$$

Thus,

$$0 = h + v_0 \sin\theta\left(\frac{R}{v_0 \cos\theta}\right) - \frac{1}{2}g\left(\frac{R^2}{v_0^2 \cos^2\theta}\right)$$

$$0 = 2v_0^2 h \cos^2\theta + R v_0^2 \sin 2\theta - gR^2$$

$$R^2 - R\left(\frac{v_0^2 \sin 2\theta}{g}\right) - \left(\frac{2v_0^2 h \cos^2\theta}{g}\right) = 0$$

For positive R;

$$R = \frac{v_0^2 \sin 2\theta}{2g} + \frac{1}{2}\sqrt{\left(\frac{v_0^2 \sin 2\theta}{g}\right)^2 + \left(\frac{8v_0^2 h \cos^2\theta}{g}\right)}$$

$$R = \left(\frac{v_0^2 \sin 2\theta}{2g}\right) + \frac{v_0}{2g}\sqrt{v_0^2 \sin^2 2\theta + 8gh \cos^2\theta} \qquad \textbf{Ans}$$

Note: For $h = 0$, $R = \left(\frac{v_0^2 \sin 2\theta}{g}\right)$, and for $\theta = 90°$, $R = 0$.

***12-100.** A particle is moving along a curved path at a constant speed of 60 ft/s. The radii of curvature of the path at points P and P' are 20 and 50 ft, respectively. If it takes the particle 20 s to go from P to P', determine the acceleration of the particle at P and P'.

$$a_t = 0$$

$$a_P = (a_n)_P = \frac{v^2}{\rho_P} = \frac{60^2}{20} = 180 \text{ ft/s}^2 \qquad \textbf{Ans}$$

$$a_{P'} = (a_n)_{P'} = \frac{v^2}{\rho_{P'}} = \frac{60^2}{50} = 72 \text{ ft/s}^2 \qquad \textbf{Ans}$$

12-101. A car travels along a horizontal curved road that has a radius of 600 m. If the speed is uniformly increased at a rate of 2000 km/h^2, determine the magnitude of the acceleration at the instant the speed of the car is 60 km/h.

$$a_t = \left(\frac{2000\text{ km}}{\text{h}^2}\right)\left(\frac{1000\text{ m}}{1\text{ km}}\right)\left(\frac{1\text{ h}}{3600\text{ s}}\right)^2 = 0.1543\text{ m/s}^2$$

$$\upsilon = \left(\frac{60\text{ km}}{\text{h}}\right)\left(\frac{1000\text{ m}}{1\text{ km}}\right)\left(\frac{1\text{ h}}{3600\text{ s}}\right) = 16.67\text{ m/s}$$

$$a_n = \frac{\upsilon^2}{\rho} = \frac{16.67^2}{600} = 0.4630\text{ m/s}^2$$

$$a = \sqrt{a_t^2 + a_n^2} = \sqrt{0.1543^2 + 0.4630^2} = 0.488\text{ m/s}^2 \qquad \textbf{Ans}$$

12-102. The car travels along the curve having a radius of 300 m. If its speed is uniformly increased from 15 m/s to 27 m/s in 3 s, determine the magnitude of its acceleration at the instant its speed is 20 m/s.

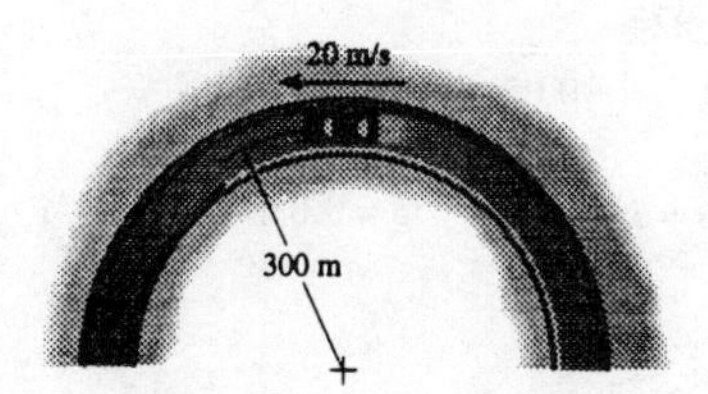

$$a_n = \frac{v^2}{\rho} = \frac{(20)^2}{300} = 1.33\text{ m/s}^2$$

$$v = v_0 + a_c t$$

$$27 = 15 + a_t(3)$$

$$a_t = 4\text{ m/s}^2$$

$$a = \sqrt{(1.33)^2 + 4^2} = 4.22\text{ m/s}^2 \qquad \textbf{Ans}$$

12-103. A boat is traveling along a circular curve having a radius of 100 ft. If its speed at $t = 0$ is 15 ft/s and is increasing at $\dot{\upsilon} = (0.8t)$ ft/s^2, determine the magnitude of its acceleration at the instant $t = 5$ s.

$$\int_{15}^{\upsilon} d\upsilon = \int_0^5 0.8t\,dt$$

$$\upsilon = 25\text{ ft/s}$$

$$a_n = \frac{\upsilon^2}{\rho} = \frac{25^2}{100} = 6.25\text{ ft/s}^2$$

At $t = 5$ s, $\qquad a_t = \dot{\upsilon} = 0.8(5) = 4\text{ ft/s}^2$

$$a = \sqrt{a_t^2 + a_n^2} = \sqrt{4^2 + 6.25^2} = 7.42\text{ ft/s}^2 \qquad \textbf{Ans}$$

***12-104.** A boat is traveling along a circular path having a radius of 20 m. Determine the magnitude of the boat's acceleration if at a given instant the boat's speed is $v = 5$ m/s and the rate of increase in the speed is $\dot{v} = 2$ m/s^2.

$$a_t = 2 \text{ m/s}^2$$

$$a_n = \frac{v^2}{\rho} = \frac{5^2}{20} = 1.25 \text{ m/s}^2$$

$$a = \sqrt{a_t^2 + a_n^2} = \sqrt{2^2 + 1.25^2} = 2.36 \text{ m/s}^2 \qquad \textbf{Ans}$$

***12-105.** Starting from rest, a bicyclist travels around a horizontal circular path, $\rho = 10$ m, at a speed of $v = (0.09t^2 + 0.1t)$ m/s, where t is in seconds. Determine the magnitudes of his velocity and acceleration when he has traveled $s = 3$ m.

$$\int_0^s ds = \int_0^t (0.09t^2 + 0.1t)\, dt$$

$$s = 0.03t^3 + 0.05t^2$$

When $s = 3$ m, $\qquad 3 = 0.03t^3 + 0.05t^2$

Solving,

$$t = 4.147 \text{ s}$$

$$v = \frac{ds}{dt} = 0.09t^2 + 0.1t$$

$$v = 0.09(4.147)^2 + 0.1(4.147) = 1.96 \text{ m/s} \qquad \textbf{Ans}$$

$$a_t = \frac{dv}{dt} = 0.18t + 0.1\Big|_{t=4.147\text{ s}} = 0.8465 \text{ m/s}^2$$

$$a_n = \frac{v^2}{\rho} = \frac{1.96^2}{10} = 0.3852 \text{ m/s}^2$$

$$a = \sqrt{a_t^2 + a_n^2} = \sqrt{(0.8465)^2 + (0.3852)^2} = 0.930 \text{ m/s}^2 \qquad \textbf{Ans}$$

12-106. The truck travels in a circular path having a radius of 50 m at a speed of 4 m/s. For a short distance from $s = 0$, its speed is increased by $\dot{v} = (0.05s)$ m/s^2, where s is in meters. Determine its speed and the magnitude of its acceleration when it has moved $s = 10$ m.

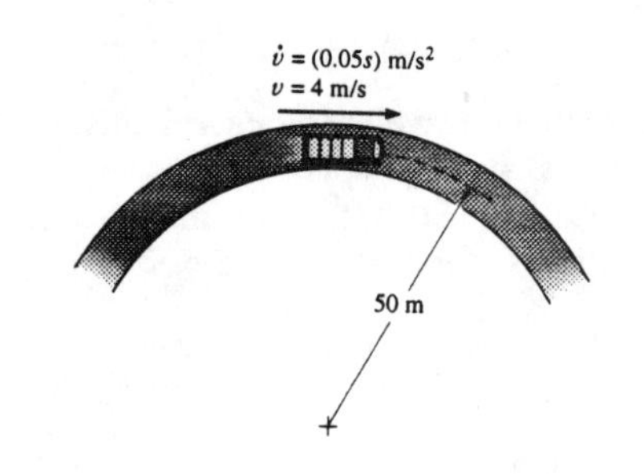

$$v\, dv = a_t\, ds$$

$$\int_4^v v\, dv = \int_0^{10} 0.05s\, ds$$

$$0.5v^2 - 8 = \frac{0.05}{2}(10)^2$$

$$v = 4.583 = 4.58 \text{ m/s} \qquad \textbf{Ans}$$

$$a_n = \frac{v^2}{\rho} = \frac{(4.583)^2}{50} = 0.420 \text{ m/s}^2$$

$$a_t = 0.05(10) = 0.5 \text{ m/s}^2$$

$$a = \sqrt{(0.420)^2 + (0.5)^2} = 0.653 \text{ m/s}^2 \qquad \textbf{Ans}$$

12-107. The satellite S travels around the earth in a circular path with a constant speed of 20 Mm/h. If the acceleration is 2.5 m/s^2, determine the altitude h. Assume the earth's diameter to be 12 713 km.

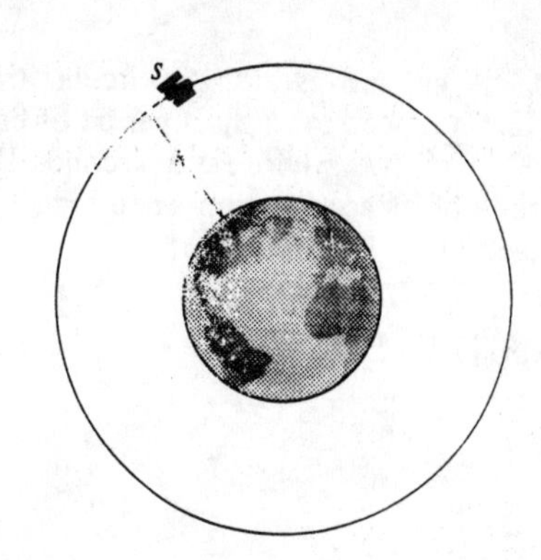

$$v = 20 \text{ Mm/h} = \frac{20(10^6)}{3600} = 5.56(10^3) \text{ m/s}$$

Since $a_t = \dfrac{dv}{dt} = 0$, then,

$$a = a_n = 2.5 = \frac{v^2}{\rho}$$

$$\rho = \frac{(5.56(10^3))^2}{2.5} = 12.35(10^6) \text{ m}$$

The radius of the earth is

$$\frac{12\,713(10^3)}{2} = 6.36(10^6) \text{ m}$$

Hence,

$$h = 12.35(10^6) - 6.36(10^6) = 5.99(10^6) \text{ m} = 5.99 \text{ Mm} \quad \textbf{Ans}$$

***12-108.** A particle P moves along the curve $y = (x^2 - 4)$ m with a constant speed of 5 m/s. Determine the point on the curve where the maximum magnitude of acceleration occurs and compute its value.

$$y = (x^2 - 4)$$

$$a_t = \frac{dv}{dt} = 0,$$

To obtain maximum $a = a_n$, ρ must be a minimum. This occurs at :

$$x = 0, \quad y = -4 \text{ m} \quad \textbf{Ans}$$

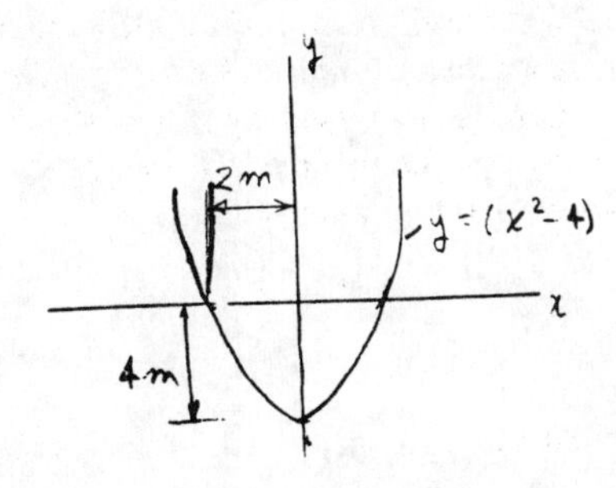

Hence,

$$\left.\frac{dy}{dx}\right|_{x=0} = 2x = 0; \quad \frac{d^2y}{dx^2} = 2$$

$$\rho_{min} = \frac{\left[1 + \left(\dfrac{dy}{dx}\right)^2\right]^{\frac{3}{2}}}{\left|\dfrac{d^2y}{dx^2}\right|} = \frac{[1+0]^{\frac{3}{2}}}{|2|} = \frac{1}{2}$$

$$(a)_{max} = (a_n)_{max} = \frac{v^2}{\rho_{min}} = \frac{5^2}{\frac{1}{2}} = 50 \text{ m/s}^2 \quad \textbf{Ans}$$

12-109. A car moves along a circular track of radius 250 ft, and its speed for a short period of time $0 \leq t \leq 2$ s is $v = 3(t + t^2)$ ft/s, where t is in seconds. Determine the magnitude of its acceleration when $t = 2$ s. How far has it traveled in $t = 2$ s?

$$v = 3(t + t^2)$$

$$a_t = \frac{dv}{dt} = 3 + 6t$$

When $t = 2$ s,

$$a_t = 3 + 6(2) = 15 \text{ ft/s}^2$$

$$a_n = \frac{v^2}{\rho} = \frac{\left[3(2+2^2)\right]^2}{250} = 1.296 \text{ ft/s}^2$$

$$a = \sqrt{(15)^2 + (1.296)^2} = 15.1 \text{ ft/s}^2 \qquad \textbf{Ans}$$

$$ds = v\,dt$$

$$\int ds = \int_0^2 3(t + t^2)\,dt$$

$$\Delta s = \frac{3}{2}t^2 + t^3\Big]_0^2$$

$$\Delta s = 14 \text{ ft} \qquad \textbf{Ans}$$

■12-110. The car travels along the curved path such that its speed is increased by $\dot{v} = (0.5e^t)$ m/s², where t is in seconds. Determine the magnitudes of its velocity and acceleration after the car has traveled $s = 18$ m starting from rest. Neglect the size of the car.

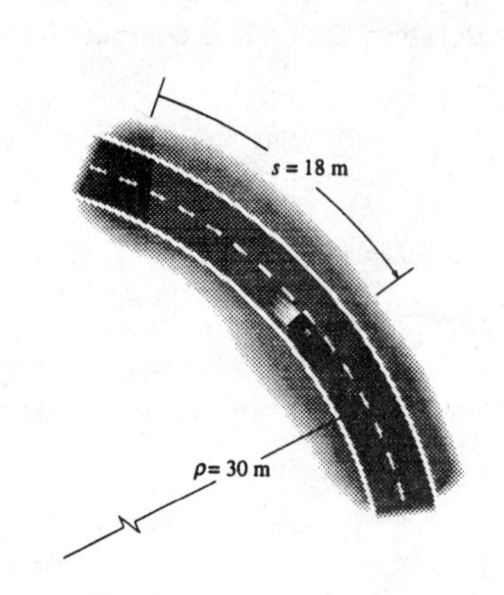

$$\int_0^v dv = \int_0^t 0.5e^t dt$$

$$v = 0.5(e^t - 1)$$

$$\int_0^{18} ds = 0.5\int_0^t (e^t - 1)\,dt$$

$$18 = 0.5(e^t - t - 1)$$

Solving, $\quad t = 3.7064$ s

$$v = 0.5(e^{3.7064} - 1) = 19.85 \text{ m/s} = 19.9 \text{ m/s} \qquad \textbf{Ans}$$

$$a_t = \dot{v} = 0.5e^t\big|_{t=3.7064\text{ s}} = 20.35 \text{ m/s}^2$$

$$a_n = \frac{v^2}{\rho} = \frac{19.85^2}{30} = 13.14 \text{ m/s}^2$$

$$a = \sqrt{a_t^2 + a_n^2} = \sqrt{20.35^2 + 13.14^2} = 24.2 \text{ m/s}^2 \qquad \textbf{Ans}$$

12-111. The Ferris wheel turns such that the speed of the passengers is increased by $\dot{v} = (4t)$ ft/s^2, where t is in seconds. If the wheel starts from rest when $\theta = 0°$, determine the magnitudes of the velocity and acceleration of the passengers when the wheel turns $\theta = 30°$.

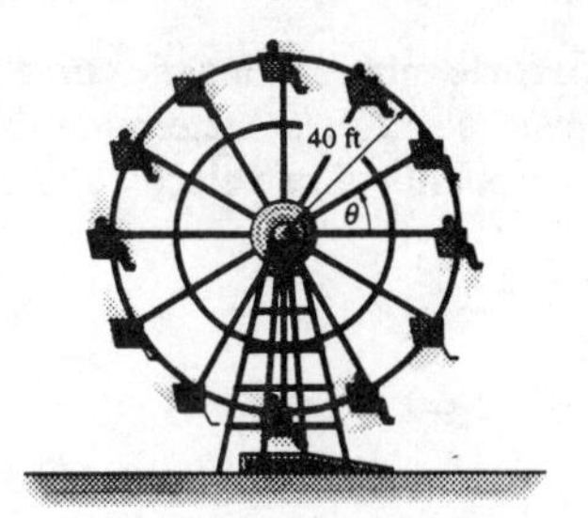

$$\int_0^v dv = \int_0^t 4t\,dt$$

$$v = 2t^2$$

$$\int_0^s ds = \int_0^t 2t^2\,dt$$

$$s = \frac{2}{3}t^3$$

When $s = \frac{\pi}{6}(40)$ ft, $\quad \frac{\pi}{6}(40) = \frac{2}{3}t^3 \quad t = 3.1554$ s

$v = 2(3.1554)^2 = 19.91$ ft/s $= 19.9$ ft/s **Ans**

$a_t = \dot{v} = 4t\,|_{t=3.1554\,s} = 12.62$ ft/s^2

$$a_n = \frac{v^2}{\rho} = \frac{19.91^2}{40} = 9.91 \text{ ft/s}^2$$

$a = \sqrt{a_t^2 + a_n^2} = \sqrt{12.62^2 + 9.91^2} = 16.0$ ft/s^2 **Ans**

***12-112.** A package is dropped from the plane which is flying with a constant horizontal velocity of $v_A = 150$ ft/s. Determine the normal and tangential components of acceleration and the radius of curvature of the path of motion (a) at the moment the package is released at A, where it has a horizontal velocity of $v_A = 150$ ft/s, and (b) *just before* it strikes the ground at B.

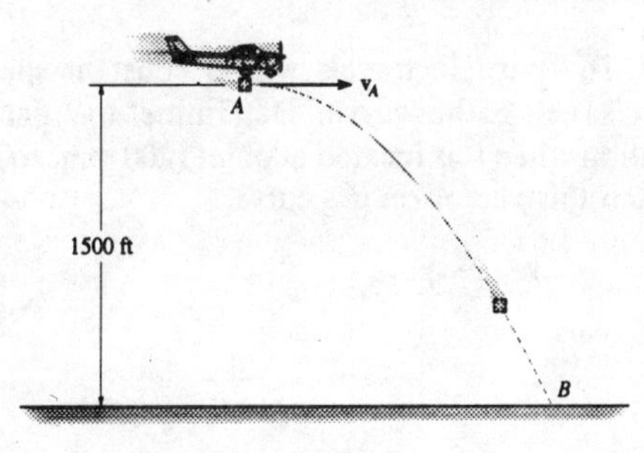

Initially (Point A):

$(a_n)_A = g = 32.2$ ft/s^2 **Ans**

$(a_t)_A = 0$ **Ans**

$$(a_n)_A = \frac{v_A^2}{\rho_A}; \quad 32.2 = \frac{(150)^2}{\rho_A}$$

$\rho_A = 698.8$ ft **Ans**

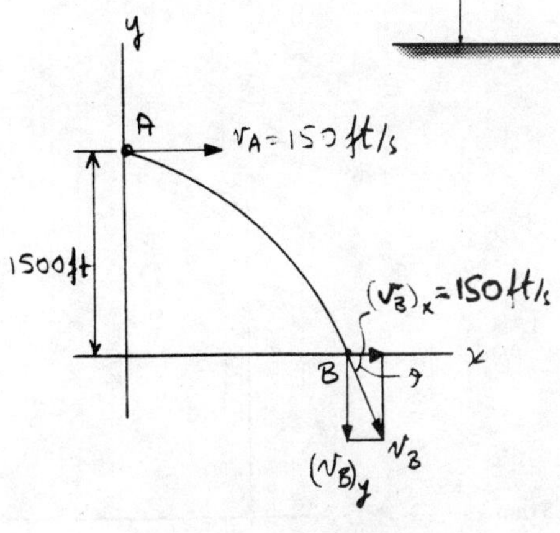

$(v_B)_x = (v_A)_x = 150$ ft/s

$(+\downarrow) \quad v^2 = v_0^2 + 2a_c(s - s_0)$

$(v_B)_y^2 = 0 + 2(32.2)(0 - 1500)$

$(v_B)_y = -310.8$ ft/s

$v_B = \sqrt{(150)^2 + (-310.8)^2} = 345.1$ ft/s

$$\theta = \tan^{-1}\left(\frac{v_{B_y}}{v_{B_x}}\right) = \tan^{-1}\left(\frac{310.8}{150}\right) = 64.23°$$

$(a_n)_B = g\cos\theta = 32.2\cos 64.24° = 14.0$ ft/s^2 **Ans**

$(a_t)_B = g\sin\theta = 32.2\sin 64.24° = 29.0$ ft/s^2 **Ans**

$$(a_n)_B = \frac{v_B^2}{\rho_B}; \quad 14.0 = \frac{(345.1)^2}{\rho_B}$$

$\rho_B = 8509.8$ ft $= 8.51(10^3)$ ft **Ans**

12-113. A particle moves along the curve $y = \sin x$ with a constant speed $v = 2$ m/s. Determine the normal and tangential components of its velocity and acceleration at any instant.

$$y = \sin x$$

$$\frac{dy}{dx} = \cos x$$

$$\frac{d^2y}{dx^2} = -\sin x$$

$$v_t = 2 \text{ m/s} \qquad \textbf{Ans}$$

$$v_n = 0 \qquad \textbf{Ans}$$

$$a_t = \frac{dv}{dt} = 0 \qquad \textbf{Ans}$$

$$a_n = \frac{v^2}{\rho} = \frac{2^2}{\rho}$$

$$\rho = \frac{[1 + (\frac{dy}{dx})^2]^{3/2}}{\left|\frac{d^2y}{dx^2}\right|} = \frac{(1 + \cos^2 x)^{3/2}}{|-\sin x|}$$

$$a_n = \frac{4 \sin x}{(1 + \cos^2 x)^{3/2}} \qquad \textbf{Ans}$$

12-114. The particle travels with a constant speed of 300 mm/s along the curve. Determine the particle's acceleration when it is located at point (200 mm, 100 mm) and sketch this vector on the curve.

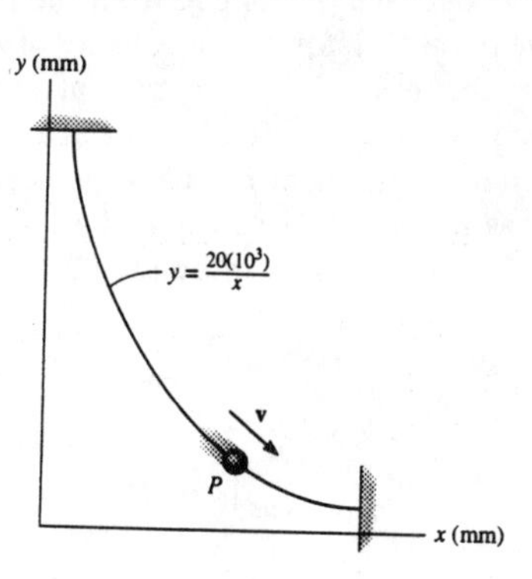

$$v = 300 \text{ mm/s}$$

$$a_t = \frac{dv}{dt} = 0$$

$$y = \frac{20(10^3)}{x}$$

$$\left.\frac{dy}{dx}\right|_{x=200} = -\frac{20(10^3)}{x^2} = -0.5$$

$$\left.\frac{d^2y}{dx^2}\right|_{x=200} = \frac{40(10^3)}{x^3} = 5\left(10^{-3}\right)$$

$$\rho = \frac{\left[1+\left(\frac{dy}{dx}\right)^2\right]^{\frac{3}{2}}}{\left|\frac{d^2y}{dx^2}\right|} = \frac{\left[1+(-0.5)^2\right]^{\frac{3}{2}}}{|5(10^{-3})|} = 279.5 \text{ mm}$$

$$a_n = \frac{v^2}{\rho} = \frac{(300)^2}{279.5} = 322 \text{ mm/s}^2$$

$$a = \sqrt{a_t^2 + a_n^2}$$

$$= \sqrt{(0)^2 + (322)^2} = 322 \text{ mm/s}^2 \qquad \textbf{Ans}$$

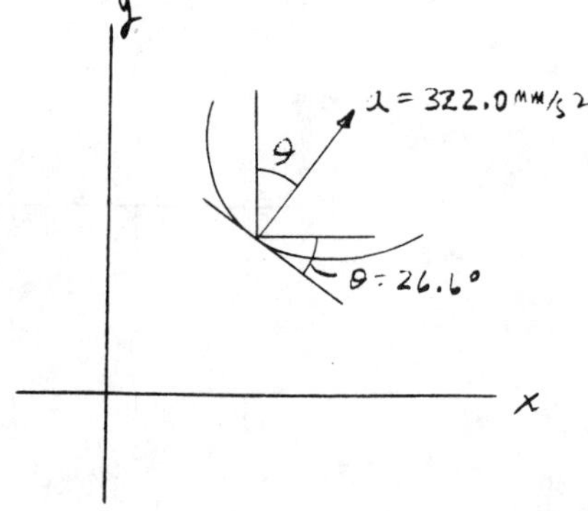

Since $\frac{dy}{dx} = -0.5$,

$\theta = \tan^{-1}(-0.5) = -26.6°$

***12-115.** The jet plane is traveling with a speed of 120 m/s which is decreasing at 40 m/s^2 when it reaches point A. Determine the magnitude of its acceleration when it is at this point. Also, specify the direction of flight, measured from the x axis.

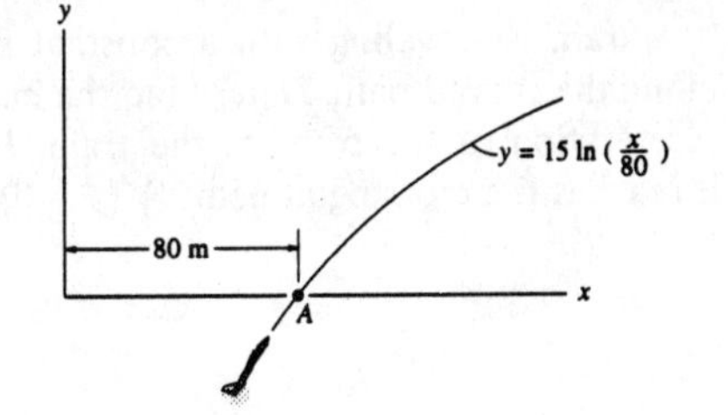

$$y = 15\ln\left(\frac{x}{80}\right)$$

$$\frac{dy}{dx} = \frac{15}{x}\bigg|_{x=80\text{ m}} = 0.1875$$

$$\frac{d^2y}{dx^2} = -\frac{15}{x^2}\bigg|_{x=80\text{ m}} = -0.002344$$

$$\rho\bigg|_{x=80\text{ m}} = \frac{\left[1+(\frac{dy}{dx})^2\right]^{3/2}}{\left|\frac{d^2y}{dx^2}\right|}\Bigg|_{x=80\text{ m}}$$

$$= \frac{\left[1+(0.1875)^2\right]^{3/2}}{|-0.002344|} = 449.4\text{ m}$$

$$a_n = \frac{v^2}{\rho} = \frac{(120)^2}{449.4} = 32.04\text{ m/s}^2$$

$$a_t = -40\text{ m/s}^2$$

$$a = \sqrt{(-40)^2 + (32.04)^2} = 51.3\text{ m/s}^2 \qquad \textbf{Ans}$$

Since

$$\frac{dy}{dx} = \tan\theta = 0.1875$$

$$\theta = 10.6° \qquad \textbf{Ans}$$

***12-116.** The jet plane is traveling with a constant speed of 110 m/s along the curved path. Determine the magnitude of the acceleration of the plane at the instant it reaches point A ($y = 0$).

$$y = 15\ln\left(\frac{x}{80}\right)$$

$$\frac{dy}{dx} = \frac{15}{x}\bigg|_{x=80\text{ m}} = 0.1875$$

$$\frac{d^2y}{dx^2} = -\frac{15}{x^2}\bigg|_{x=80\text{ m}} = -0.002344$$

$$\rho\bigg|_{x=80\text{ m}} = \frac{\left[1+(\frac{dy}{dx})^2\right]^{3/2}}{\left|\frac{d^2y}{dx^2}\right|}\Bigg|_{x=80\text{ m}}$$

$$= \frac{\left[1+(0.1875)^2\right]^{3/2}}{|-0.002344|} = 449.4\text{ m}$$

$$a_n = \frac{v^2}{\rho} = \frac{(110)^2}{449.4} = 26.9\text{ m/s}^2$$

Since the plane travels with a constant speed, $a_t = 0$. Hence

$$a = a_n = 26.9\text{ m/s}^2 \qquad \textbf{Ans}$$

12-117. A train is traveling with a constant speed of 14 m/s along the curved path. Determine the magnitude of the acceleration of the front of the train, B, at the instant it reaches the crossing at point A ($y = 0$).

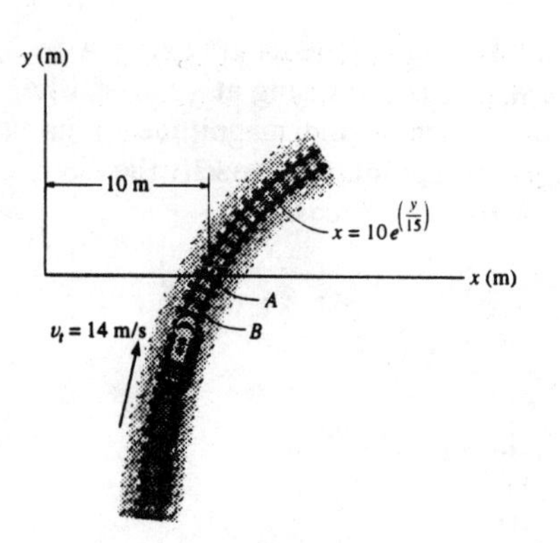

$$x = 10e^{\left(\frac{y}{15}\right)}$$

$$y = 15\ln\left(\frac{x}{10}\right)$$

$$\frac{dy}{dx} = 15\left(\frac{10}{x}\right)\left(\frac{1}{10}\right) = \frac{15}{x}$$

$$\frac{d^2y}{dx^2} = -\frac{15}{x^2}$$

At $x = 10$,

$$\rho = \frac{\left[1+\left(\frac{dy}{dx}\right)^2\right]^{\frac{3}{2}}}{\left|\frac{d^2y}{dx^2}\right|} = \frac{\left[1+(1.5)^2\right]^{\frac{3}{2}}}{|-0.15|} = 39.06 \text{ m}$$

$$a_t = \frac{dv}{dt} = 0$$

$$a_n = a = \frac{v^2}{\rho} = \frac{(14)^2}{39.06} = 5.02 \text{ m/s}^2 \qquad \textbf{Ans}$$

12-118. When the motorcyclist is at A, he increases his speed along the vertical circular path at the rate of $\dot{v} = (0.3t)$ ft/s^2, where t is in seconds. If he starts from rest at A, determine the magnitudes of his velocity and acceleration when he reaches B.

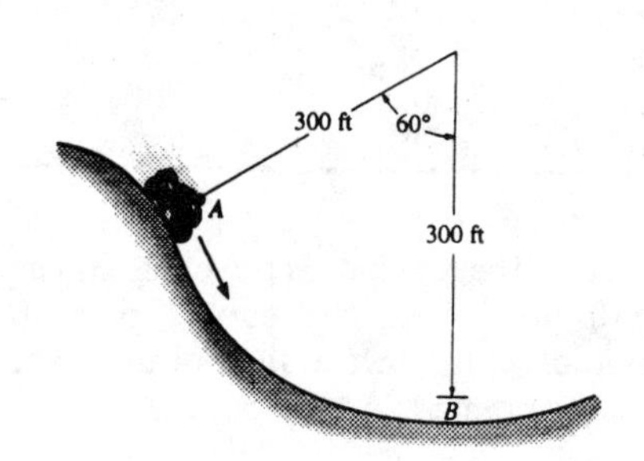

$$\int_0^v dv = \int_0^t 0.3t\,dt$$

$$v = 0.15t^2$$

$$\int_0^s ds = \int_0^t 0.15t^2\,dt$$

$$s = 0.05t^3$$

When $s = \frac{\pi}{3}(300)$ ft, $\quad \frac{\pi}{3}(300) = 0.05t^3 \quad t = 18.453$ s

$$v = 0.15(18.453)^2 = 51.08 \text{ ft/s} = 51.1 \text{ ft/s} \qquad \textbf{Ans}$$

$$a_t = \dot{v} = 0.3t\big|_{t=18.453\text{ s}} = 5.536 \text{ ft/s}^2$$

$$a_n = \frac{v^2}{\rho} = \frac{51.08^2}{300} = 8.696 \text{ ft/s}^2$$

$$a = \sqrt{a_t^2 + a_n^2} = \sqrt{(5.536)^2 + (8.696)^2} = 10.3 \text{ ft/s}^2 \qquad \textbf{Ans}$$

12-119. When the car reaches point A it has a speed of 4 m/s, which is increasing at a constant rate of 2 m/s^2. Determine the time required to reach point B and the magnitudes of its velocity and acceleration.

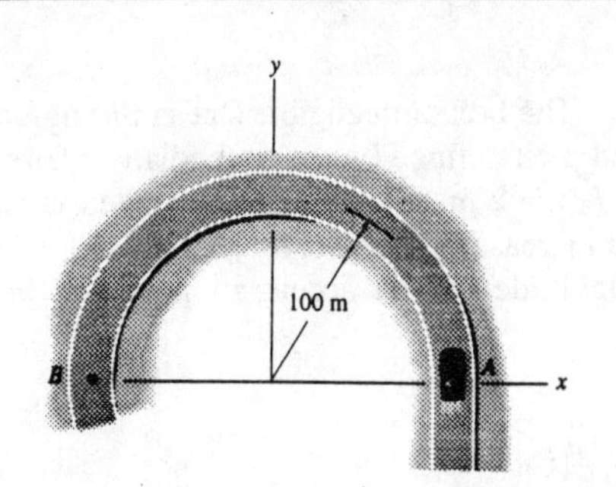

$$a_t = \dot{v} = 2 \text{ m/s}^2$$

$$s = 100\pi$$

$$s = s_0 + v_0 t + \frac{1}{2} a_c t^2$$

$$100\pi = 0 + 4t + \frac{1}{2}(2)t^2$$

$$t^2 + 4t - 314.159 = 0$$

Solving for the positive root

$t = 15.84 \text{ s} = 15.8 \text{ s}$ **Ans**

$$v = v_0 + a_c t$$

$= 4 + 2(15.84) = 35.67 = 35.7 \text{ m/s}$ **Ans**

$$a_n = \frac{v^2}{\rho} = \frac{(35.67)^2}{100} = 12.73 \text{ m/s}^2$$

$a = \sqrt{(2)^2 + (12.73)^2} = 12.9 \text{ m/s}^2$ **Ans**

***12-120.** The motorcyclist travels along the curve at a constant speed of 30 ft/s. Determine his acceleration when he is located at point A. Neglect the size of the motorcycle and rider for the calculation.

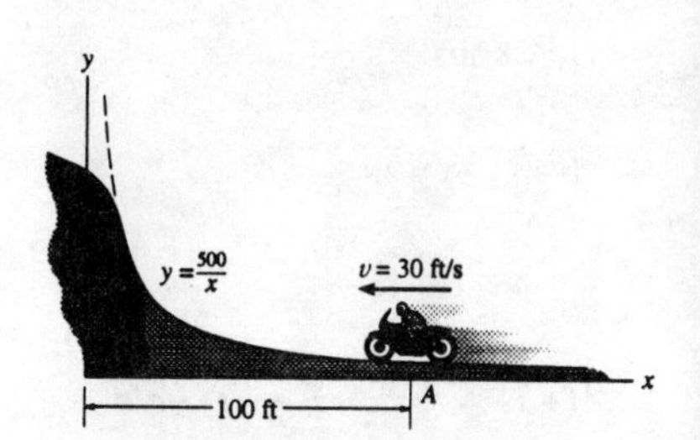

$$\frac{dy}{dx} = -\frac{500}{x^2}\bigg|_{x=100 \text{ ft}} = -0.05$$

$$\frac{d^2y}{dx^2} = \frac{1000}{x^3}\bigg|_{x=100 \text{ ft}} = 0.001$$

$$\rho\bigg|_{x=100 \text{ ft}} = \frac{\left[1 + \left(\frac{dy}{dx}\right)^2\right]^{3/2}}{\left|\frac{d^2y}{dx^2}\right|}\Bigg|_{x=100 \text{ ft}}$$

$$= \frac{\left[1 + (-0.05)^2\right]^{3/2}}{|0.001|} = 1003.8 \text{ ft}$$

$$a_n = \frac{v^2}{\rho} = \frac{30^2}{1003.8} = 0.897 \text{ ft/s}^2$$

Since the motorcyclist travels with a constant speed, $a_t = 0$. Hence

$a = a_n = 0.897 \text{ ft/s}^2$ **Ans**

12-121. The box of negligible size is sliding down along a curved path defined by the parabola $y = 0.4x^2$. When it is at A ($x_A = 2$ m, $y_A = 1.6$ m), the speed is $v_B = 8$ m/s and the increase in speed is $dv_B/dt = 4$ m/s^2. Determine the magnitude of the acceleration of the box at this instant.

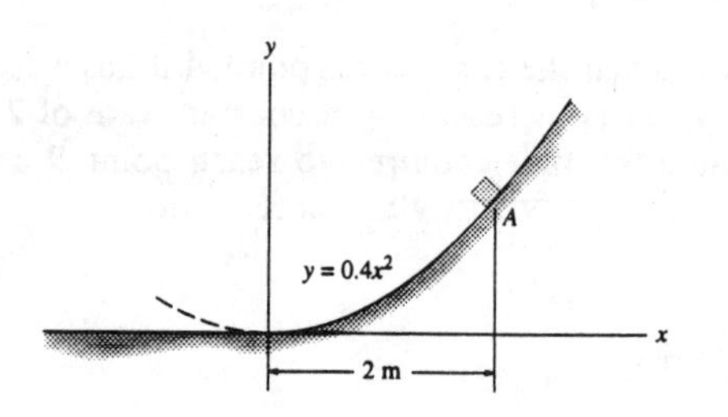

$$y = 0.4\,x^2$$

$$\left.\frac{dy}{dx}\right|_{x=2\text{ m}} = 0.8x\Big|_{x=2\text{ m}} = 1.6$$

$$\left.\frac{d^2y}{dx^2}\right|_{x=2\text{ m}} = 0.8$$

$$\rho = \left.\frac{\left[1+\left(\frac{dy}{dx}\right)^2\right]^{3/2}}{\left|\frac{d^2y}{dx^2}\right|}\right|_{x=2\text{ m}} = \frac{\left[1+(1.6)^2\right]^{3/2}}{|0.8|} = 8.396\text{ m}$$

$$a_n = \frac{v_B^{\,2}}{\rho} = \frac{8^2}{8.396} = 7.622\text{ m/s}^2$$

$$a = \sqrt{a_t^2 + a_n^2} = \sqrt{(4)^2 + (7.622)^2} = 8.61\text{ m/s}^2$$ **Ans**

12-122. The ball is ejected horizontally from the tube with a speed of 8 m/s. Find the equation of the path, $y = f(x)$, and then find the ball's velocity and the normal and tangential components of acceleration when $t = 0.25$ s.

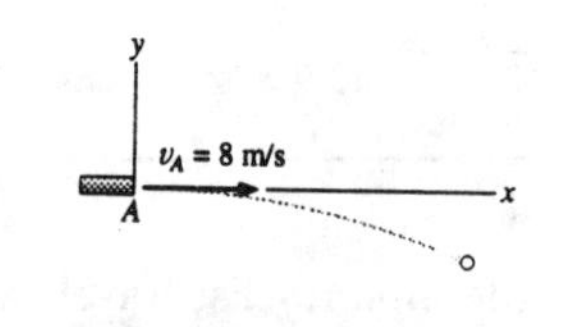

$v_x = 8$ m/s

$(\overset{+}{\rightarrow})$ $\quad s = v_0 t$

$x = 8t$

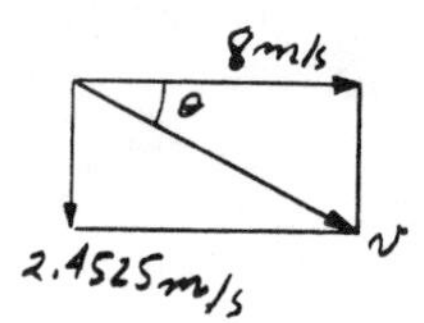

$(+\uparrow)$ $\quad s = s_0 + v_0 t + \frac{1}{2}a_c t^2$

$$y = 0 + 0 + \frac{1}{2}(-9.81)t^2$$

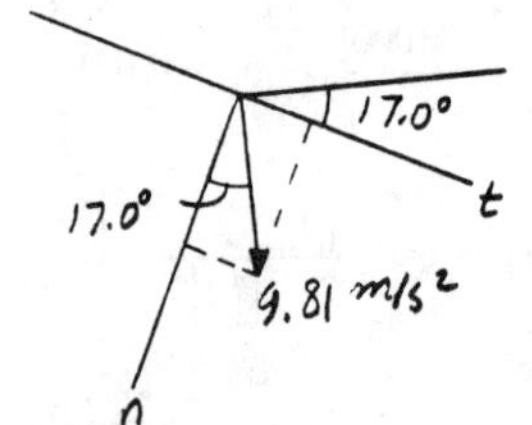

$$y = -4.905t^2$$

$$y = -4.905\left(\frac{x}{8}\right)^2$$

$$y = -0.0766x^2 \quad \text{(Parabola)}$$ **Ans**

$v = v_0 + a_c t$

$v_y = 0 - 9.81t$

When $t = 0.25$ s,

$v_y = -2.4525$ m/s

$v = \sqrt{(8)^2 + (2.4525)^2} = 8.37$ m/s **Ans**

$$\theta = \tan^{-1}\left(\frac{2.4525}{8}\right) = 17.04°$$

$a_x = 0 \quad a_y = 9.81$ m/s^2

$a_n = 9.81\cos 17.04° = 9.38$ m/s^2 **Ans**

$a_t = 9.81\sin 17.04° = 2.88$ m/s^2 **Ans**

12-123. Cars move around the "traffic circle" which is in the shape of an ellipse. If the speed limit is posted at 60 km/h, determine the maximum acceleration experienced by the passengers.

$$\frac{x^2}{a^2} + \frac{y^2}{b^2} = 1$$

$$b^2x^2 + a^2y^2 = a^2b^2$$

$$b^2(2x) + a^2(2y)\frac{dy}{dx} = 0$$

$$\frac{dy}{dx} = -\frac{b^2x}{a^2y}$$

$$\frac{dy}{dx}y = \frac{-b^2x}{a^2}$$

$$\frac{d^2y}{dx^2}y + \left(\frac{dy}{dx}\right)^2 = \frac{-b^2}{a^2}$$

$$\frac{d^2y}{dx^2}y = \frac{-b^2}{a^2} - \left(\frac{-b^2x}{a^2y}\right)^2$$

$$\frac{d^2y}{dx^2} = \frac{-b^4}{a^2y^3}$$

$$\rho = \frac{\left[1 + \left(\frac{-b^2x}{a^2y}\right)^2\right]^{3/2}}{\left|\frac{-b^4}{a^2y^3}\right|}$$

At $x = a,\ y = 0$,

$$\rho = \frac{b^2}{a}$$

Then

$$a_t = 0$$

$$a_{max} = a_n = \frac{v^2}{\rho} = \frac{v^2}{\frac{b^2}{a}} = \frac{v^2a}{b^2}$$

Set $a = 60$ m, $b = 40$ m, $v = \dfrac{60(10^3)}{3600} = 16.67$ m/s

$$a_{max} = \frac{(16.67)^2(60)}{(40)^2} = 10.4 \text{ m/s}^2 \quad \textbf{Ans}$$

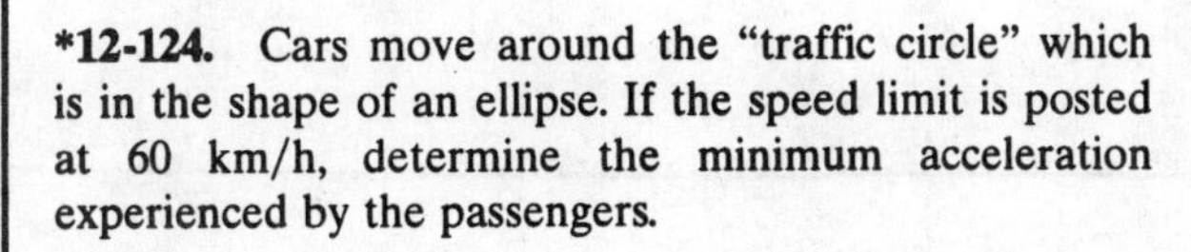

***12-124.** Cars move around the "traffic circle" which is in the shape of an ellipse. If the speed limit is posted at 60 km/h, determine the minimum acceleration experienced by the passengers.

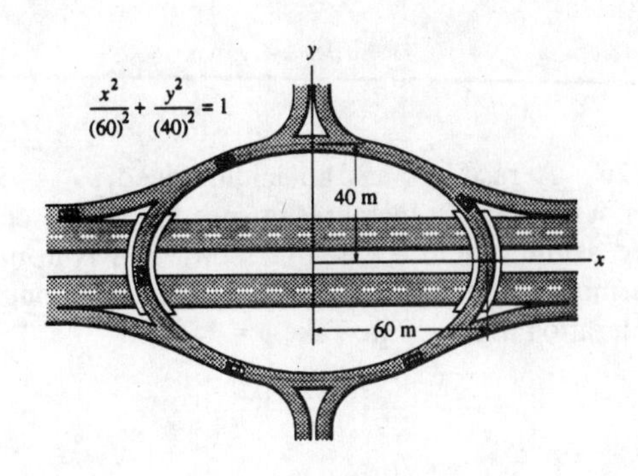

$$\frac{x^2}{a^2} + \frac{y^2}{b^2} = 1$$

$$b^2x^2 + a^2y^2 = a^2b^2$$

$$b^2(2x) + a^2(2y)\frac{dy}{dx} = 0$$

$$\frac{dy}{dx} = -\frac{b^2x}{a^2y}$$

$$\frac{dy}{dx}y = \frac{-b^2x}{a^2}$$

$$\frac{d^2y}{dx^2}y + \left(\frac{dy}{dx}\right)^2 = \frac{-b^2}{a^2}$$

$$\frac{d^2y}{dx^2}y = \frac{-b^2}{a^2} - \left(\frac{-b^2x}{a^2y}\right)^2$$

$$\frac{d^2y}{dx^2}y = \frac{-b^2}{a^2} - \left(\frac{b^4}{a^2y^2}\right)\left(\frac{x^2}{a^2}\right)$$

$$\frac{d^2y}{dx^2}y = \frac{-b^2}{a^2} - \frac{b^4}{a^2y^2}\left(1 - \frac{y^2}{b^2}\right)$$

$$\frac{d^2y}{dx^2}y = \frac{-b^2}{a^2} - \frac{b^4}{a^2y^2} + \frac{b^2}{a^2}$$

$$\frac{d^2y}{dx^2} = \frac{-b^4}{a^2y^3}$$

$$\rho = \frac{\left[1 + \left(\frac{-b^2x}{a^2y}\right)^2\right]^{3/2}}{\left|\frac{-b^4}{a^2y^3}\right|}$$

At $x = 0,\ y = b$,

$$\rho = \frac{a^2}{b}$$

Thus

$$a_t = 0$$

$$a_{min} = a_n = \frac{v^2}{\rho} = \frac{v^2}{\frac{a^2}{b}} = \frac{v^2b}{a^2}$$

Set $a = 60$ m, $b = 40$ m,

$$v = \frac{60(10)^3}{3600} = 16.67 \text{ m/s}$$

$$a_{min} = \frac{(16.67)^2(40)}{(60)^2} = 3.09 \text{ m/s}^2 \quad \textbf{Ans}$$

12-125. The race car travels around the circular track with a speed of 16 m/s. When it reaches point A it increases its speed at $\dot{v} = (\frac{4}{3}v^{1/4})$ m/s^2, where v is in m/s. Determine the velocity and acceleration of the car when it reaches point B. Also, how much time is required for it to travel from A to B?

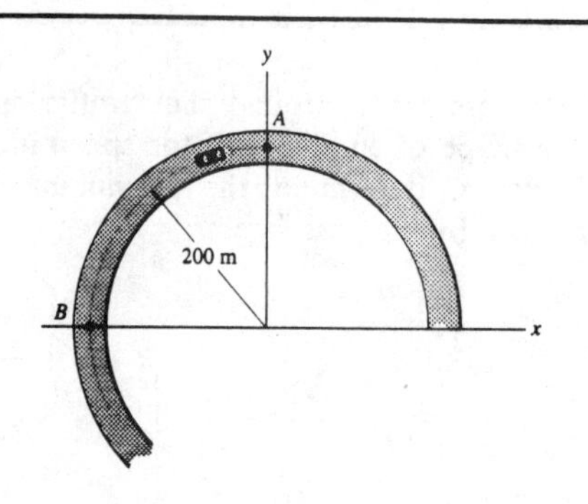

$$a_t = \frac{4}{3}v^{\frac{1}{4}}$$

$$dv = a_t\,dt$$

$$dv = \frac{4}{3}v^{\frac{1}{4}}\,dt$$

$$\int_{16}^{v} 0.75\frac{dv}{v^{\frac{1}{4}}} = \int_0^t dt$$

$$v^{\frac{3}{4}}\Big|_{16}^{v} = t$$

$$v^{\frac{3}{4}} - 8 = t$$

$$v = (t+8)^{\frac{4}{3}}$$

$$ds = v\,dt$$

$$\int_0^s ds = \int_0^t (t+8)^{\frac{4}{3}}\,dt$$

$$s = \frac{3}{7}(t+8)^{\frac{7}{3}}\Big|_0^t$$

$$s = \frac{3}{7}(t+8)^{\frac{7}{3}} - 54.86$$

For $s = \frac{\pi}{2}(200) = 100\pi = \frac{3}{7}(t+8)^{\frac{7}{3}} - 54.86$

$t = 10.108$ s $= 10.1$ s **Ans**

$v = (10.108+8)^{\frac{4}{3}} = 47.551 = 47.6$ m/s **Ans**

$$a_t = \frac{4}{3}(47.551)^{\frac{1}{4}} = 3.501 \text{ m/s}^2$$

$$a_n = \frac{v^2}{\rho} = \frac{(47.551)^2}{200} = 11.305 \text{ m/s}^2$$

$a = \sqrt{(3.501)^2 + (11.305)^2} = 11.8$ m/s^2 **Ans**

12-126. A race car has an initial speed $v_A = 15$ m/s when it is at A. If it increases its speed along the circular track, at the rate $a_t = (0.4s)$ m/s^2, where s is in meters, determine the car's normal and tangential components of acceleration at $s = 10$ m. Take $\rho = 150$ m.

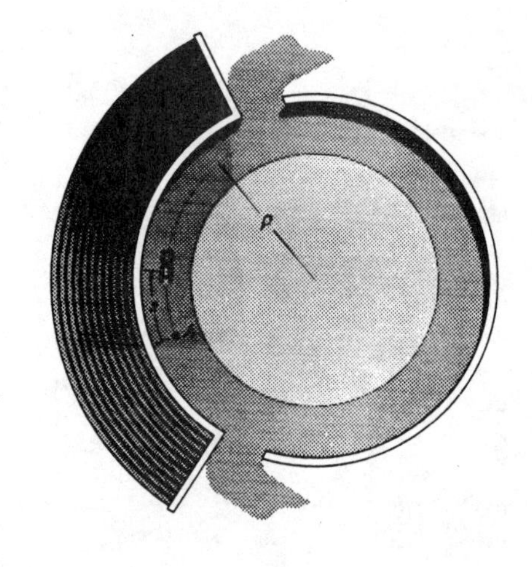

$$a_t = 0.4s$$

$$a\,ds = v\,dv$$

$$\int_0^s 0.4s\,ds = \int_{15}^{v} v\,dv$$

$$0.2\,s^2 = \frac{v^2}{2} - 112.5$$

$$v^2 = 0.4s^2 + 225$$

At $s = 10$ m, $v^2 = 0.4(10)^2 + 225 = 265$

$a_n = \frac{v^2}{\rho} = \frac{265}{150} = 1.77$ m/s^2 **Ans**

$a_t = 0.4(10) = 4$ m/s^2 **Ans**

12-127. The race car has an initial speed $v_A = 15$ m/s at A. If it increases its speed along the circular track at the rate $a_t = (0.4s)$ m/s^2, where s is in meters, determine the time needed for the car to travel 20 m. Take $\rho = 150$ m.

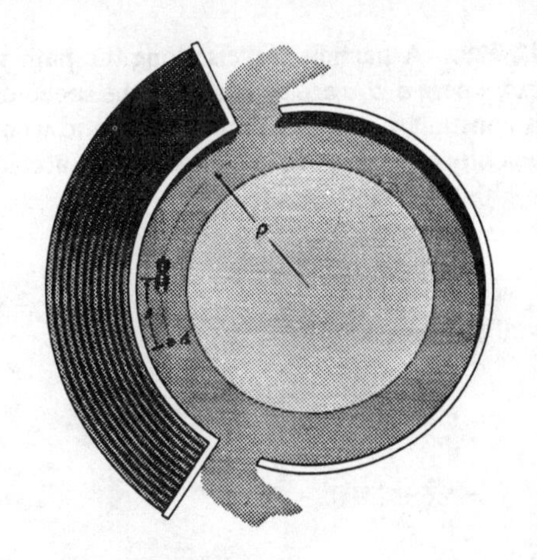

$$a_t = 0.4s = \frac{v\,dv}{ds}$$

$$a\,ds = v\,dv$$

$$\int_0^s 0.4s\,ds = \int_{15}^{v} v\,dv$$

$$\left.\frac{0.4s^2}{2}\right|_0^s = \left.\frac{v^2}{2}\right|_{15}^{v}$$

$$\frac{0.4s^2}{2} = \frac{v^2}{2} - \frac{225}{2}$$

$$v^2 = 0.4s^2 + 225$$

$$v = \frac{ds}{dt} = \sqrt{0.4s^2 + 225}$$

$$\int_0^s \frac{ds}{\sqrt{0.4s^2 + 225}} = \int_0^t dt$$

$$\int_0^s \frac{ds}{\sqrt{s^2 + 562.5}} = 0.632\,456t$$

$$\ln\left(s + \sqrt{s^2 + 562.5}\right)\Big|_0^s = 0.632\,456t$$

$$\ln\left(s + \sqrt{s^2 + 562.5}\right) - 3.166\,196 = 0.632\,456t$$

At $s = 20$ m,

$t = 1.21$ s **Ans**

***12-128.** A spiral transition curve is used on railroads to connect a straight portion of the track with a curved portion. If the spiral is defined by the equation $y = (10^{-6})x^3$, where x and y are in feet, determine the magnitude of the acceleration of a train engine moving with a constant speed of 40 ft/s when it is at point $x = 600$ ft.

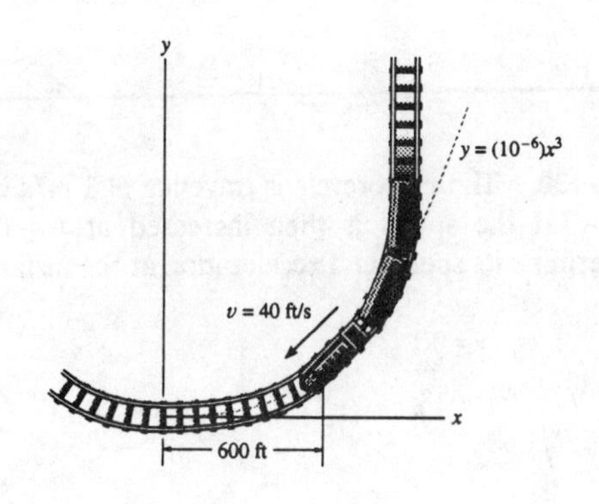

$$y = (10)^{-6}x^3$$

$$\left.\frac{dy}{dx}\right|_{x=600\text{ ft}} = 3(10)^{-6}x^2\Big|_{x=600\text{ ft}} = 1.08$$

$$\left.\frac{d^2y}{dx^2}\right|_{x=600\text{ ft}} = 6(10)^{-6}x\Big|_{x=600\text{ ft}} = 3.6(10)^{-3}$$

$$\rho\Big|_{x=600\text{ ft}} = \left.\frac{\left[1+\left(\frac{dy}{dx}\right)^2\right]^{3/2}}{\left|\frac{d^2y}{dx^2}\right|}\right|_{x=600\text{ ft}} = \frac{\left[1+(1.08)^2\right]^{3/2}}{|3.6(10)^{-3}|} = 885.7\text{ ft}$$

$$a_n = \frac{v^2}{\rho} = \frac{40^2}{885.7} = 1.81\text{ ft/s}^2$$

$$a = \sqrt{a_t^2 + a_n^2} = \sqrt{0 + (1.81)^2} = 1.81\text{ ft/s}^2$$ **Ans**

12-129. A particle travels along the path $y = a + bx + cx^2$, where a, b, c are constants. If the speed of the particle is constant, $v = v_0$, determine the x and y components of velocity and the normal component of acceleration when $x = 0$.

$$y = a + bx + cx^2$$

$$\dot{y} = b\dot{x} + 2cx\dot{x}$$

$$\ddot{y} = b\ddot{x} + 2c(\dot{x})^2 + 2cx\ddot{x}$$

When $x = 0$, $\quad \dot{y} = b\dot{x}$

$$v_0^2 = \dot{x}^2 + b^2\dot{x}^2$$

$$v_x = \dot{x} = \frac{v_0}{\sqrt{1+b^2}} \qquad \textbf{Ans}$$

$$v_y = \frac{v_0 b}{\sqrt{1+b^2}} \qquad \textbf{Ans}$$

$$a_n = \frac{v_0^2}{\rho}$$

$$\rho = \frac{\left[1 + \left(\frac{dy}{dx}\right)^2\right]^{\frac{3}{2}}}{\left|\frac{d^2y}{dx^2}\right|}$$

$$\frac{dy}{dx} = b + 2cx$$

$$\frac{d^2y}{dx^2} = 2c$$

At $x = 0$, $\quad \rho = \dfrac{(1+b^2)^{3/2}}{2c}$

$$a_n = \frac{2c\,v_0^2}{(1+b^2)^{3/2}} \qquad \textbf{Ans}$$

■12-130. The motorcycle is traveling at 1 m/s when it is at A. If the speed is then increased at $\dot{v} = 0.1$ m/s², determine its speed and acceleration at the instant $t = 5$ s.

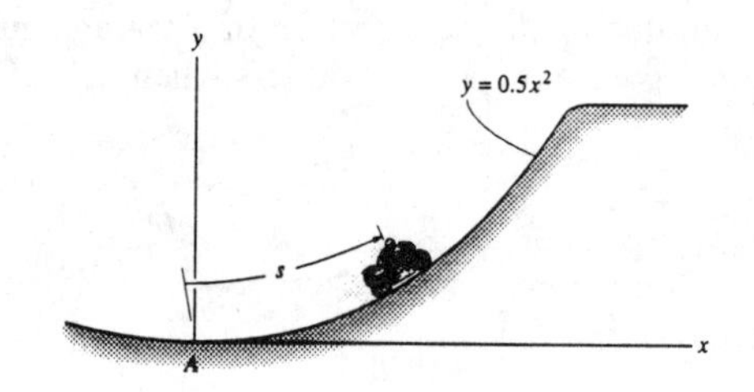

$$a_t = \dot{v} = 0.1$$

$$s = s_0 + v_0 t + \frac{1}{2}a_c t^2$$

$$s = 0 + 1(5) + \frac{1}{2}(0.1)(5)^2 = 6.25 \text{ m}$$

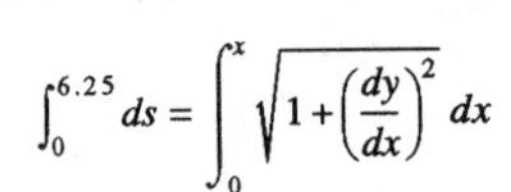

$$\int_0^{6.25} ds = \int_0^x \sqrt{1+\left(\frac{dy}{dx}\right)^2}\,dx$$

$$y = 0.5x^2$$

$$\frac{dy}{dx} = x$$

$$\frac{d^2y}{dx^2} = 1$$

$$6.25 = \int_0^x \sqrt{1+x^2}\,dx$$

$$6.25 = \frac{1}{2}\left[x\sqrt{1+x^2} + \ln\left(x+\sqrt{1+x^2}\right)\right]_0^x$$

$$x\sqrt{1+x^2} + \ln\left(x+\sqrt{1+x^2}\right) = 12.5$$

Solving,

$$x = 3.184 \text{ m}$$

$$\rho = \frac{\left[1+\left(\frac{dy}{dx}\right)^2\right]^{\frac{3}{2}}}{\left|\frac{d^2y}{dx^2}\right|} = \frac{\left[1+x^2\right]^{\frac{3}{2}}}{|1|}\Bigg|_{x=3.184} = 37.17 \text{ m}$$

$$v = v_0 + a_c t$$

$$= 1 + 0.1(5) = 1.5 \text{ m/s} \qquad \textbf{Ans}$$

$$a_n = \frac{v^2}{\rho} = \frac{(1.5)^2}{37.17} = 0.0605 \text{ m/s}^2$$

$$a = \sqrt{(0.1)^2 + (0.0605)^2} = 0.117 \text{ m/s}^2 \qquad \textbf{Ans}$$

■12-131. The car travels around the circular track having a radius of $r = 300$ m such that when it is at point A it has a velocity of 5 m/s, which is increasing at the rate of $\dot{v} = (0.06t)$ m/s^2, where t is in seconds. Determine the magnitudes of its velocity and acceleration when it has traveled one-third the way around the track.

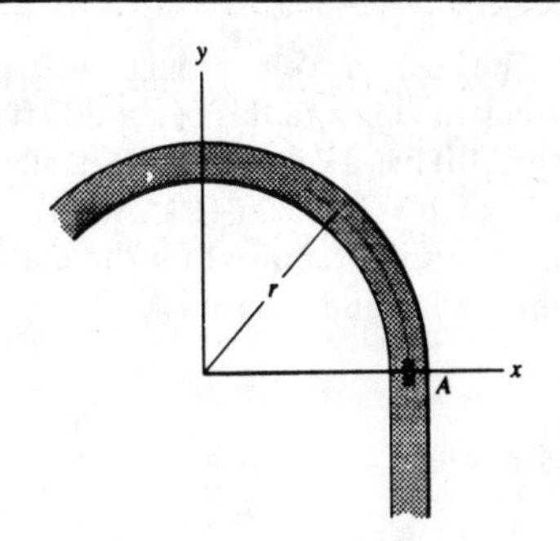

$a_t = v = 0.06t$

$dv = a_t\, dt$

$\int_5^v dv = \int_0^t 0.06t\, dt$

$v = 0.03t^2 + 5$

$ds = v\, dt$

$\int_0^s ds = \int_0^t \left(0.03t^2 + 5\right) dt$

$s = 0.01t^3 + 5t$

$s = \frac{1}{3}(2\pi(300)) = 628.3185$

$0.01t^3 + 5t - 628.3185 = 0$

Solve for the positive root,

$t = 35.58$ s

$v = 0.03(35.58)^2 + 5 = 42.978 \text{ m/s} = 43.0 \text{ m/s}$ **Ans**

$a_n = \frac{v^2}{\rho} = \frac{(42.978)^2}{300} = 6.157 \text{ m/s}^2$

$a_t = 0.06(35.58) = 2.135 \text{ m/s}^2$

$a = \sqrt{(6.157)^2 + (2.135)^2} = 6.52 \text{ m/s}^2$ **Ans**

***12-132.** The car travels around the portion of a circular track having a radius of $r = 500$ ft such that when it is at point A it has a velocity of 2 ft/s, which is increasing at the rate of $\dot{v} = (0.002s)$ ft/s^2, where t is in seconds. Determine the magnitudes of its velocity and acceleration when it has traveled three-fourths the way around the track.

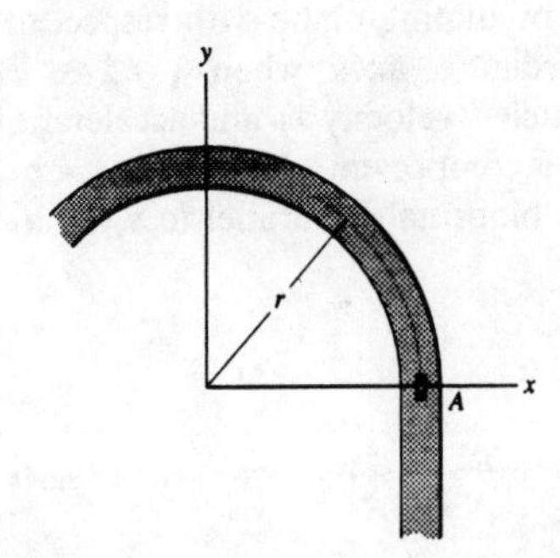

$a_t = 0.002\, s$

$a_t\, ds = v\, dv$

$\int_0^s 0.002s\, ds = \int_2^v v\, dv$

$0.001s^2 = \frac{1}{2}v^2 - \frac{1}{2}(2)^2$

$v^2 = 0.002s^2 + 4$

$s = \frac{3}{4}[2\pi(500)] = 2356.194$ ft

$v^2 = 0.002(2356.194)^2 + 4$

$v = 105.39 \text{ ft/s} = 105 \text{ ft/s}$ **Ans**

$a_n = \frac{v^2}{\rho} = \frac{(105.39)^2}{500} = 22.21 \text{ ft/s}^2$

$a_t = 0.002(2356.194) = 4.712 \text{ ft/s}^2$

$a = \sqrt{(22.21)^2 + (4.712)^2} = 22.7 \text{ ft/s}^2$ **Ans**

12-133. The car travels around the portion of the circular track having a radius of $r = 500$ ft such that when it is at point A it has a velocity of 2 ft/s, which is increasing at the rate of $\dot{v} = (0.002s)$ ft/s^2, where t is in seconds. Determine the time required for the car to travel three-fourths the way around the track.

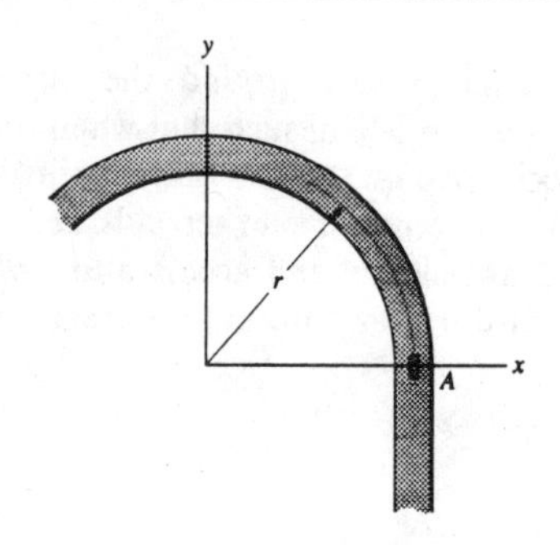

$$a_t = 0.002\, s$$

$$a_t\, ds = v\, dv$$

$$\int_0^s 0.002s\, ds = \int_2^v v\, dv$$

$$0.001s^2 = \frac{1}{2}v^2 - \frac{1}{2}(2)^2$$

$$v^2 = 0.002s^2 + 4$$

$$ds = \sqrt{0.002s^2 + 4}\, dt$$

$$\int_0^t dt = \int_0^s \frac{ds}{\sqrt{0.002s^2 + 4}} = 22.36 \int_0^s \frac{ds}{\sqrt{s^2 + 2000}}$$

$$t = 22.36 \ln\left(s + \sqrt{s^2 + 2000}\right)\Big|_0^s$$

$$t = 22.36 \ln\left(s + \sqrt{s^2 + 2000}\right) - 84.9781$$

$$s = \frac{3}{4}[2\pi(500)] = 2356.194$$

$$t = 22.36 \ln\left(2356.194 + \sqrt{(2356.194)^2 + 2000}\right) - 84.9781$$

$$t = 104 \text{ s} \qquad \textbf{Ans}$$

12-134. The motion of a particle along a fixed path is defined by the parametric equations $r = 8$ ft, $\theta = (4t)$ rad, and $z = (6t^2)$ ft, where t is in seconds. Determine the unit vector that specifies the direction of the binormal axis to the osculating plane with respect to a set of fixed x, y, z coordinate axes when $t = 2$ s. *Hint:* Formulate the particle's velocity $\mathbf{v}_P$ and acceleration $\mathbf{a}_P$ in terms of their **i, j, k** components. Note that $x = r \cos\theta$ and $y = r \sin\theta$. The binormal is parallel to $\mathbf{v}_P \times \mathbf{a}_P$. Why?

$$r = 8 \text{ ft} \qquad \theta = 4t \qquad z = 6t^2$$

$$x = r\cos\theta = 8\cos 4t \qquad y = r\sin\theta = 8\sin 4t \qquad z = 6t^2$$

Hence,

$$\mathbf{r}_P = 8\cos(4t)\mathbf{i} + 8\sin(4t)\mathbf{j} + 6t^2\mathbf{k}$$

$$\mathbf{v}_P = -32\sin(4t)\mathbf{i} + 32\cos(4t)\mathbf{j} + 12t\mathbf{k}$$

$$\mathbf{a}_P = -128\cos(4t)\mathbf{i} - 128\sin(4t)\mathbf{j} + 12\mathbf{k}$$

When $t = 2$ s,

$$\mathbf{v}_P = -32\sin(8 \text{ rad})\mathbf{i} + 32\cos(8 \text{ rad})\mathbf{j} + 24\mathbf{k} = -31.659\mathbf{i} - 4.6560\mathbf{j} + 24\mathbf{k}$$

$$\mathbf{a}_P = -128\cos(8 \text{ rad})\mathbf{i} - 128\sin(8 \text{ rad})\mathbf{j} + 12\mathbf{k} = 18.624\mathbf{i} - 126.64\mathbf{j} + 12\mathbf{k}$$

Since $\mathbf{a}_P$ and $\mathbf{v}_P$ are in the $n-t$ plane, and the binormal axis is perpendicular to this plane, then by definition of the vector cross product, we have

$$\mathbf{b} = \mathbf{v}_P \times \mathbf{a}_P = \begin{vmatrix} \mathbf{i} & \mathbf{j} & \mathbf{k} \\ -31.659 & -4.6560 & 24 \\ 18.624 & -126.64 & 12 \end{vmatrix} = 2983.44\mathbf{i} + 826.89\mathbf{j} + 4096\mathbf{k}$$

$$b = \sqrt{(2983.44)^2 + (826.89)^2 + (4096)^2} = 5134.38$$

$$\mathbf{u}_b = \frac{\mathbf{b}}{b} = 0.581\mathbf{i} + 0.161\mathbf{j} + 0.798\mathbf{k} \qquad \textbf{Ans}$$

Note : It is also possible to define the binormal axis using $\mathbf{a}_P \times \mathbf{v}_P$ for the calculation. For this case, $\mathbf{u}_{b'} = -\mathbf{u}_b = -0.581\mathbf{i} - 0.161\mathbf{j} - 0.798\mathbf{k}$.

12-135. A particle P travels along an elliptical spiral path such that its position vector $\mathbf{r}$ is defined by $\mathbf{r} = \{2\cos(0.1t)\mathbf{i} + 1.5\sin(0.1t)\mathbf{j} + (2t)\mathbf{k}\}$ m, where t is in seconds and the arguments for the sine and cosine are given in radians. When $t = 8$ s, determine the coordinate direction angles α, β, and γ, which the binormal axis to the osculating plane makes with the x, y, and z axes. *Hint:* Solve for the velocity $\mathbf{v}_P$ and acceleration $\mathbf{a}_P$ of the particle in terms of their $\mathbf{i}, \mathbf{j}, \mathbf{k}$ components. The binormal is parallel to $\mathbf{v}_P \times \mathbf{a}_P$. Why?

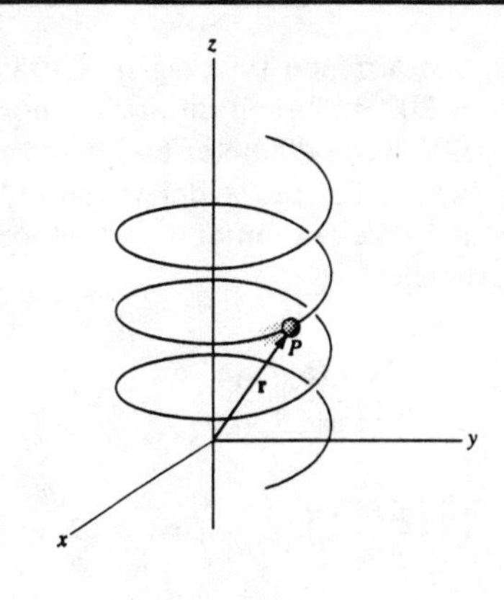

$$\mathbf{r}_P = 2\cos(0.1t)\mathbf{i} + 1.5\sin(0.1t)\mathbf{j} + 2t\mathbf{k}$$

$$\mathbf{v}_P = \dot{\mathbf{r}} = -0.2\sin(0.1t)\mathbf{i} + 0.15\cos(0.1t)\mathbf{j} + 2\mathbf{k}$$

$$\mathbf{a}_P = \ddot{\mathbf{r}} = -0.02\cos(0.1t)\mathbf{i} - 0.015\sin(0.1t)\mathbf{j}$$

When $t = 8$ s,

$$\mathbf{v}_P = -0.2\sin(0.8\text{rad})\mathbf{i} + 0.15\cos(0.8\text{rad})\mathbf{j} + 2\mathbf{k} = -0.143\,47\mathbf{i} + 0.104\,51\mathbf{j} + 2\mathbf{k}$$

$$\mathbf{a}_P = -0.02\cos(0.8\text{rad})\mathbf{i} - 0.015\sin(0.8\text{rad})\mathbf{j} = -0.013\,934\mathbf{i} - 0.010\,76\mathbf{j}$$

Since the binormal vector is perpendicular to the plane containing the $n-t$ axis, and $\mathbf{a}_P$ and $\mathbf{v}_P$ are in this plane, then by the definition of the cross product,

$$\mathbf{b} = \mathbf{v}_P \times \mathbf{a}_P = \begin{vmatrix} \mathbf{i} & \mathbf{j} & \mathbf{k} \\ -0.143\,47 & 0.104\,51 & 2 \\ -0.013\,934 & -0.010\,76 & 0 \end{vmatrix} = 0.021\,52\mathbf{i} - 0.027\,868\mathbf{j} + 0.003\mathbf{k}$$

$$b = \sqrt{(0.02152)^2 + (-0.027868)^2 + (0.003)^2} = 0.035\,338$$

$$\mathbf{u}_b = 0.608\,99\mathbf{i} - 0.788\,62\mathbf{j} + 0.085\mathbf{k}$$

$$\alpha = \cos^{-1}(0.608\,99) = 52.5^\circ \quad \textbf{Ans}$$

$$\beta = \cos^{-1}(-0.788\,62) = 142^\circ \quad \textbf{Ans}$$

$$\gamma = \cos^{-1}(0.085) = 85.1^\circ \quad \textbf{Ans}$$

Note: The direction of the binormal axis may also be specified by the unit vector $\mathbf{u}_{b'} = -\mathbf{u}_b$, which is obtained from $\mathbf{b}' = \mathbf{a}_P \times \mathbf{v}_P$.
For this case, $\alpha = 128^\circ$, $\beta = 37.9^\circ$, $\gamma = 94.9^\circ$ **Ans**

***12-136.** The time rate of change of acceleration is referred to as the *jerk*, which is often used as a means of measuring passenger discomfort. Calculate this vector, $\dot{\mathbf{a}}$, in terms of its cylindrical components, using Eq. 12–32.

$$\mathbf{a} = \left(\ddot{r} - r\dot{\theta}^2\right)\mathbf{u}_r + \left(r\ddot{\theta} + 2\dot{r}\dot{\theta}\right)\mathbf{u}_\theta + \ddot{z}\mathbf{u}_z$$

$$\dot{\mathbf{a}} = \left(\dddot{r} - \dot{r}\dot{\theta}^2 - 2r\dot{\theta}\ddot{\theta}\right)\mathbf{u}_r + \left(\ddot{r} - r\dot{\theta}^2\right)\dot{\mathbf{u}}_r + \left(\dot{r}\ddot{\theta} + r\dddot{\theta} + 2\ddot{r}\dot{\theta} + 2\dot{r}\ddot{\theta}\right)\mathbf{u}_\theta + \left(r\ddot{\theta} + 2\dot{r}\dot{\theta}\right)\dot{\mathbf{u}}_\theta + \dddot{z}\mathbf{u}_z + \ddot{z}\dot{\mathbf{u}}_z$$

But, $\dot{\mathbf{u}}_r = \dot{\theta}\mathbf{u}_\theta \quad \dot{\mathbf{u}}_\theta = -\dot{\theta}\mathbf{u}_r \quad \dot{\mathbf{u}}_z = 0$

Substituting and combining terms yields

$$\dot{\mathbf{a}} = \left(\dddot{r} - 3\dot{r}\dot{\theta}^2 - 3r\dot{\theta}\ddot{\theta}\right)\mathbf{u}_r + \left(3\dot{r}\ddot{\theta} + r\dddot{\theta} + 3\ddot{r}\dot{\theta} - r\dot{\theta}^3\right)\mathbf{u}_\theta + \left(\dddot{z}\right)\mathbf{u}_z \quad \textbf{Ans}$$

12-137. An airplane is flying in a straight line with a velocity of 200 mi/h and an acceleration of 3 mi/h^2. If the propeller has a diameter of 6 ft and is rotating at an angular rate of 120 rad/s, determine the magnitudes of velocity and acceleration of a particle located on the tip of the propeller.

$$v_{Pl} = \left(\frac{200\text{ mi}}{\text{h}}\right)\left(\frac{5280\text{ ft}}{1\text{ mi}}\right)\left(\frac{1\text{ h}}{3600\text{ s}}\right) = 293.3\text{ ft/s}$$

$$a_{Pl} = \left(\frac{3\text{ mi}}{\text{h}^2}\right)\left(\frac{5280\text{ ft}}{1\text{ mi}}\right)\left(\frac{1\text{ h}}{3600\text{ s}}\right)^2 = 0.001\ 22\text{ ft/s}^2$$

$$v_{Pr} = 120(3) = 360\text{ ft/s}$$

$$v = \sqrt{v_{Pl}^2 + v_{Pr}^2} = \sqrt{(293.3)^2 + (360)^2} = 464\text{ ft/s} \qquad \textbf{Ans}$$

$$a_{pr} = \frac{v_{Pr}^2}{\rho} = \frac{(360)^2}{3} = 43\ 200\text{ ft/s}^2$$

$$a = \sqrt{a_{Pl}^2 + a_{Pr}^2} = \sqrt{(0.001\ 22)^2 + (43\ 200)^2} = 43.2(10^3)\ \text{ft/s}^2 \qquad \textbf{Ans}$$

12-138. A particle is moving along a circular path having a radius of 4 in. such that its position as a function of time is given by $\theta = \cos 2t$, where θ is in radians and t is in seconds. Determine the magnitude of the acceleration of the particle when $\theta = 30°$.

When $\theta = \frac{\pi}{6}$ rad, $\quad \frac{\pi}{6} = \cos 2t \quad t = 0.5099$ s

$$\dot{\theta} = \frac{d\theta}{dt} = -2\sin 2t\Big|_{t=0.5099\text{ s}} = -1.7039\text{ rad/s}$$

$$\ddot{\theta} = \frac{d^2\theta}{dt^2} = -4\cos 2t\Big|_{t=0.5099\text{ s}} = -2.0944\text{ rad/s}^2$$

$r = 4 \qquad \dot{r} = 0 \qquad \ddot{r} = 0$

$$a_r = \ddot{r} - r\dot{\theta}^2 = 0 - 4(-1.7039)^2 = -11.6135\text{ in./s}^2$$

$$a_\theta = r\ddot{\theta} + 2\dot{r}\dot{\theta} = 4(-2.0944) + 0 = -8.3776\text{ in./s}^2$$

$$a = \sqrt{a_r^2 + a_\theta^2} = \sqrt{(-11.6135)^2 + (-8.3776)^2} = 14.3\text{ in./s}^2 \qquad \textbf{Ans}$$

12-139. A train is traveling along the circular curve of radius $r = 600$ ft. At the instant shown, its angular rate of rotation is $\dot{\theta} = 0.02$ rad/s, which is decreasing at $\ddot{\theta} = -0.001$ rad/s^2. Determine the magnitudes of the train's velocity and acceleration at this instant.

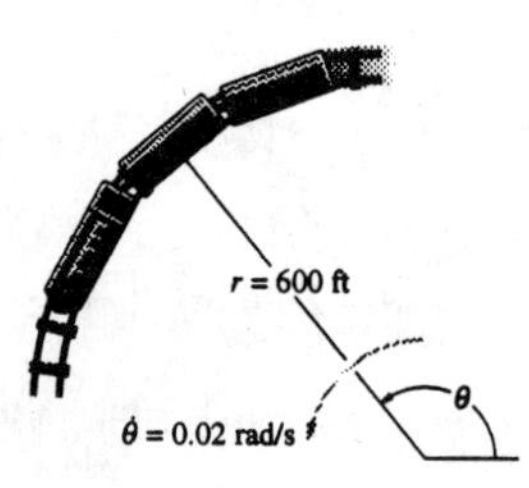

$r = 600 \qquad \dot{r} = 0 \qquad \ddot{r} = 0$

$$v_r = \dot{r} = 0$$

$$v = v_\theta = r\dot{\theta} = 600(0.02) = 12\text{ ft/s} \qquad \textbf{Ans}$$

$$a_r = \ddot{r} - r\dot{\theta}^2 = 0 - 600(0.02)^2 = -0.24\text{ ft/s}^2$$

$$a_\theta = r\ddot{\theta} + 2\dot{r}\dot{\theta} = 600(-0.001) + 0 = -0.6\text{ ft/s}^2$$

$$a = \sqrt{a_r^2 + a_\theta^2} = \sqrt{(-0.24)^2 + (-0.6)^2} = 0.646\text{ ft/s}^2 \qquad \textbf{Ans}$$

***12-140.** If a particle moves along a path such that $r = (2\cos t)$ ft and $\theta = (t/2)$ rad, where t is in seconds, plot the path $r = f(\theta)$ and determine the particle's radial and transverse components of velocity and acceleration.

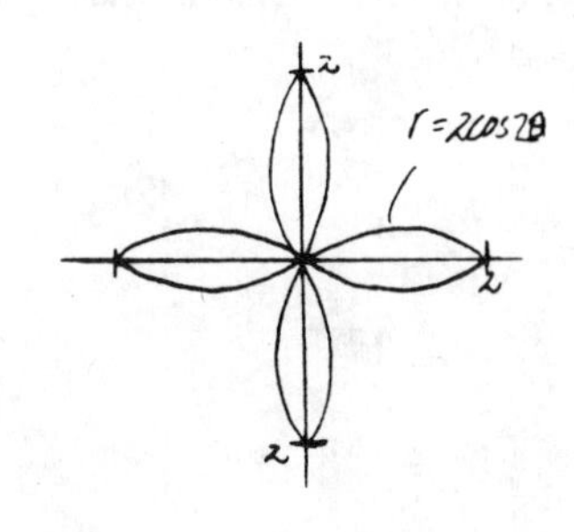

$r = 2\cos t \qquad \dot{r} = -2\sin t \qquad \ddot{r} = -2\cos t$

$\theta = \dfrac{t}{2} \qquad \dot{\theta} = \dfrac{1}{2} \qquad \ddot{\theta} = 0$

$v_r = \dot{r} = -2\sin t$ **Ans**

$v_\theta = r\dot{\theta} = 2\cos t\left(\dfrac{1}{2}\right) = \cos t$ **Ans**

$a_r = \ddot{r} - r\dot{\theta}^2 = -2\cos t - 2\cos t\left(\dfrac{1}{2}\right)^2 = -\dfrac{5}{2}\cos t$ **Ans**

$a_\theta = r\ddot{\theta} + 2\dot{r}\dot{\theta} = 2\cos t(0) + 2(-2\sin t)\left(\dfrac{1}{2}\right) = -2\sin t$ **Ans**

12-141. If a particle moves along a path such that $r = (2\sin t^2)$ m and $\theta = t^2$ rad, where t is in seconds, plot the path $r = f(\theta)$ and determine the particle's radial and transverse components of velocity and acceleration as functions of time.

$r = 2\sin t^2 \qquad \dot{r} = 4t\cos t^2 \qquad \ddot{r} = 4\cos t^2 - 8t^2\sin t^2$

$\theta = t^2 \qquad \dot{\theta} = 2t \qquad \ddot{\theta} = 2$

$v_r = \dot{r} = 4t\cos t^2$ **Ans**

$v_\theta = r\dot{\theta} = 2\sin t^2(2t) = 4t\sin t^2$ **Ans**

$a_r = \ddot{r} - r\dot{\theta}^2$

$= 4\cos t^2 - 8t^2\sin t^2 - 2\sin t^2(2t)^2$

$= 4\cos t^2 - 16t^2\sin t^2$ **Ans**

$a_\theta = r\ddot{\theta} + 2\dot{r}\dot{\theta}$

$= 2\sin t^2(2) + 2(4t\cos t^2)(2t)$

$= 4\sin t^2 + 16t^2\cos t^2$ **Ans**

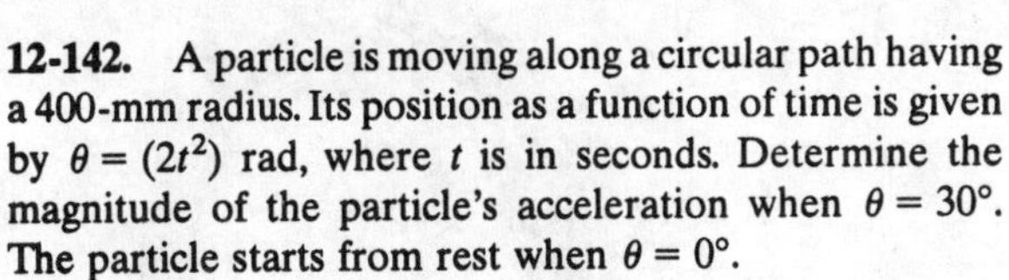

12-142. A particle is moving along a circular path having a 400-mm radius. Its position as a function of time is given by $\theta = (2t^2)$ rad, where t is in seconds. Determine the magnitude of the particle's acceleration when $\theta = 30°$. The particle starts from rest when $\theta = 0°$.

$r = 400$ mm $\qquad \dot{r} = 0 \qquad \ddot{r} = 0$

$\theta = 2t^2 \qquad \dot{\theta} = 4t \qquad \ddot{\theta} = 4$

$a_r = \ddot{r} - r\dot{\theta}^2 = 0 - 400(4t)^2 = -6400t^2$

$a_\theta = r\ddot{\theta} + 2\dot{r}\dot{\theta} = 400(4) + 0 = 1600$

When $\theta = 30° = 30°\left(\dfrac{\pi}{180°}\right) = 0.5236$ rad, then,

$0.5236 = 2t^2, \qquad t = 0.5117$ s

Hence,

$a = \sqrt{(-6400(0.5117)^2)^2 + (1600)^2} = 2316.76\ \text{mm/s}^2 = 2.32\ \text{m/s}^2$ **Ans**

12-143. If a particle moves along a path such that $r = (e^{at})$ m and $\theta = t$, where t is in seconds, plot the path $r = f(\theta)$, and determine the particle's radial and transverse components of velocity and acceleration.

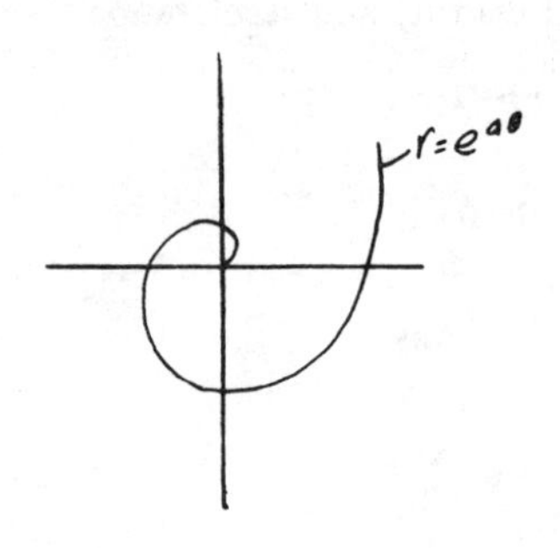

$r = e^{at}$ $\quad \dot{r} = ae^{at}$ $\quad \ddot{r} = a^2 e^{at}$

$\theta = t$ $\quad \dot{\theta} = 1$ $\quad \ddot{\theta} = 0$

$v_r = \dot{r} = ae^{at}$ **Ans**

$v_\theta = r\dot{\theta} = e^{at}(1) = e^{at}$ **Ans**

$a_r = \ddot{r} - r\dot{\theta}^2 = a^2e^{at} - e^{at}(1)^2 = e^{at}(a^2 - 1)$ **Ans**

$a_\theta = r\ddot{\theta} + 2\dot{r}\dot{\theta} = e^{at}(0) + 2(ae^{at})(1) = 2ae^{at}$ **Ans**

***12-144.** A car is traveling along the circular curve having a radius $r = 400$ ft. At the instant shown, its angular rate of rotation is $\dot{\theta} = 0.025$ rad/s, which is decreasing at the rate $\ddot{\theta} = -0.008$ rad/s^2. Determine the radial and transverse components of the car's velocity and acceleration at this instant and sketch these components on the curve.

$r = 400$ $\quad \dot{r} = 0$ $\quad \ddot{r} = 0$

$\dot{\theta} = 0.025$ $\quad \ddot{\theta} = -0.008$

$v_r = \dot{r} = 0$ **Ans**

$v_\theta = r\dot{\theta} = 400(0.025) = 10$ ft/s **Ans**

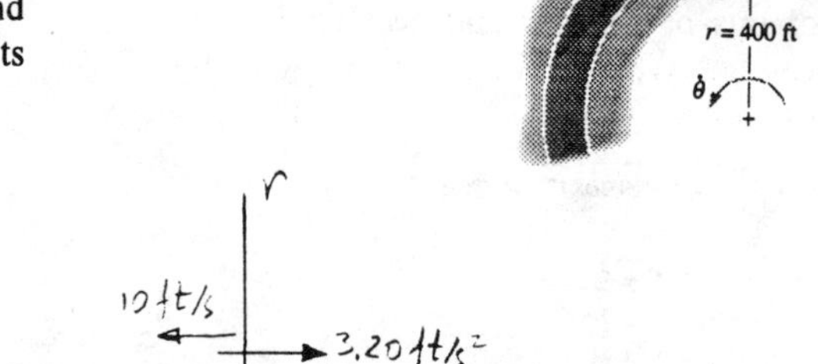

$a_r = \ddot{r} - r\dot{\theta}^2 = 0 - 400(0.025)^2 = -0.25$ ft/s^2 **Ans**

$a_\theta = r\ddot{\theta} + 2\dot{r}\dot{\theta} = 400(-0.008) + 0 = -3.20$ ft/s^2 **Ans**

12-145. A car is traveling along the circular curve of radius $r = 400$ ft with a constant speed of $v = 30$ ft/s. Determine the angular rate of rotation $\dot{\theta}$ of the radial line r and the magnitude of the car's acceleration.

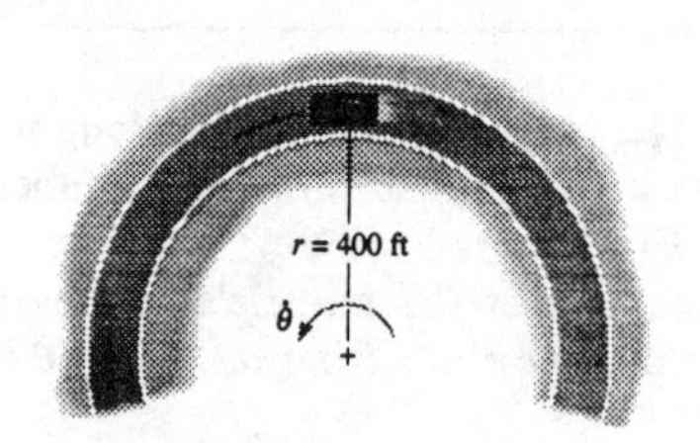

$r = 400$ ft $\quad \dot{r} = 0$ $\quad \ddot{r} = 0$

$v_r = \dot{r} = 0$ $\quad v_\theta = r\dot{\theta} = 400\left(\dot{\theta}\right)$

$$v = \sqrt{(0)^2 + \left(400\dot{\theta}\right)^2} = 30$$

$\dot{\theta} = 0.075$ rad/s **Ans**

$\ddot{\theta} = 0$

$a_r = \ddot{r} - r\dot{\theta}^2 = 0 - 400(0.075)^2 = -2.25$ ft/s^2

$a_\theta = r\ddot{\theta} + 2\dot{r}\dot{\theta} = 400(0) + 2(0)(0.075) = 0$

$a = \sqrt{(-2.25)^2 + (0)^2} = 2.25$ ft/s^2 **Ans**

12-146. A particle is moving along a circular path having a radius of 6 in. such that its position as a function of time is given by $\theta = \sin 3t$, where θ is in radians, the argument for the sine is in degrees, and t is in seconds. Determine the acceleration of the particle at $\theta = 30°$. The particle starts from rest at $\theta = 0°$.

$r = 6 \text{ in.}, \quad \dot{r} = 0, \quad \ddot{r} = 0$

$\theta = \sin 3t$

$\dot{\theta} = 3\cos 3t$

$\ddot{\theta} = -9\sin 3t$

At $\theta = 30°$,

$\frac{30°}{180°}\pi = \sin 3t$

$t = 10.525 \text{ s}$

Thus,

$\dot{\theta} = 2.5559 \text{ rad/s}$

$\ddot{\theta} = -4.7124 \text{ rad/s}^2$

$a_r = \ddot{r} - r\dot{\theta}^2 = 0 - 6(2.5559)^2 = -39.196$

$a_\theta = r\ddot{\theta} + 2\dot{r}\dot{\theta} = 6(-4.7124) + 0 = -28.274$

$a = \sqrt{(-39.196)^2 + (-28.274)^2} = 48.3 \text{ in./s}^2$ **Ans**

12-147. A particle travels around a lituus, defined by the equation $r^2\theta = a^2$, where a is a constant. Determine the particle's radial and transverse components of velocity and acceleration as a function of θ and its time derivatives.

$r^2\theta = a^2$

$r = a\theta^{-\frac{1}{2}}$

$\dot{r} = a\left(-\frac{1}{2}\right)\theta^{-\frac{3}{2}}\dot{\theta}$

$\ddot{r} = -\frac{1}{2}a\left(-\frac{3}{2}\theta^{-\frac{5}{2}}\dot{\theta}^2 + \theta^{-\frac{3}{2}}\ddot{\theta}\right)$

$v_r = \dot{r} = -\frac{1}{2}a\theta^{-\frac{3}{2}}\dot{\theta}$ **Ans**

$v_\theta = r\dot{\theta} = a\theta^{-\frac{1}{2}}\dot{\theta}$ **Ans**

$a_r = \ddot{r} - r\dot{\theta}^2 = -\frac{1}{2}a\left(-\frac{3}{2}\theta^{-\frac{5}{2}}\dot{\theta}^2 + \theta^{-\frac{3}{2}}\ddot{\theta}\right) - a\theta^{-\frac{1}{2}}\dot{\theta}^2$

$= a\left[\left(\frac{3}{4}\theta^{-2} - 1\right)\theta^{-\frac{1}{2}}\dot{\theta}^2 - \frac{1}{2}\theta^{-\frac{3}{2}}\ddot{\theta}\right]$ **Ans**

$a_\theta = r\ddot{\theta} + 2\dot{r}\dot{\theta} = a\theta^{-\frac{1}{2}}\ddot{\theta} + 2(a)\left(-\frac{1}{2}\right)\theta^{-\frac{3}{2}}\dot{\theta}\left(\dot{\theta}\right) = a\left[\ddot{\theta} - \frac{\dot{\theta}^2}{\theta}\right]\theta^{-\frac{1}{2}}$ **Ans**

***12-148.** A particle travels around a limaçon, defined by the equation $r = b - a\cos\theta$, where a and b are constants. Determine the particle's radial and transverse components of velocity and acceleration as a function of θ and its time derivatives.

$r = b - a\cos\theta$

$\dot{r} = a\sin\theta\dot{\theta}$

$\ddot{r} = a\cos\theta\dot{\theta}^2 + a\sin\theta\ddot{\theta}$

$v_r = \dot{r} = a\sin\theta\dot{\theta}$ **Ans**

$v_\theta = r\dot{\theta} = (b - a\cos\theta)\dot{\theta}$ **Ans**

$a_r = \ddot{r} - r\dot{\theta}^2 = a\cos\theta\dot{\theta}^2 + a\sin\theta\ddot{\theta} - (b - a\cos\theta)\dot{\theta}^2$

$= (2a\cos\theta - b)\dot{\theta}^2 + a\sin\theta\ddot{\theta}$ **Ans**

$a_\theta = r\ddot{\theta} + 2\dot{r}\dot{\theta} = (b - a\cos\theta)\ddot{\theta} + 2\left(a\sin\theta\dot{\theta}\right)\dot{\theta}$

$= (b - a\cos\theta)\ddot{\theta} + 2a\dot{\theta}^2\sin\theta$ **Ans**

12-149. A particle travels along a portion of the "four-leaf rose" defined by the equation $r = (5 \cos 2\theta)$ m. If the angular velocity of the radial coordinate line is $\dot{\theta} = (3t^2)$ rad/s, where t is in seconds, determine the radial and transverse components of the particle's velocity and acceleration at the instant $\theta = 30°$. When $t = 0$, $\theta = 0°$.

$\dot{\theta} = 3\,t^2$

$\ddot{\theta} = 6\,t$

$\int_0^{\theta} d\theta = \int_0^{t} 3\,t^2\,dt$

$\theta = t^3$

At $\theta = 30° = \dfrac{\pi}{6}$

$t = (\pi/6)^{1/3} = 0.806$

$\dot{\theta} = 1.95$

$\ddot{\theta} = 4.84$

$r = 5\cos 2\theta$

$\dot{r} = -10 \sin 2\theta\,\dot{\theta}$

$\ddot{r} = -10(2\cos 2\theta(\dot{\theta})^2 + \sin 2\theta\,\ddot{\theta})$

At $\theta = 30°$

$r = 2.5$

$\dot{r} = -16.88$ m/s

$\ddot{r} = -79.86$ m/s^2

$v_r = \dot{r} = -16.9$ m/s **Ans**

$v_\theta = r\,\dot{\theta} = 2.5(1.95) = 4.87$ m/s **Ans**

$a_r = \ddot{r} - r(\dot{\theta})^2 = -79.86 - 2.5(1.95)^2 = -89.4$ m/s^2 **Ans**

$a_\theta = r\,\ddot{\theta} + 2\,\dot{r}\,\dot{\theta} = 2.5(4.84) + 2(-16.88)(1.95) = -53.7$ m/s^2 **Ans**

12-150. A block moves outward along the slot in the platform with a speed of $\dot{r} = (4t)$ m/s, where t is in seconds. The platform rotates at a constant rate of 6 rad/s. If the block starts from rest at the center, determine the magnitudes of its velocity and acceleration when $t = 1$ s.

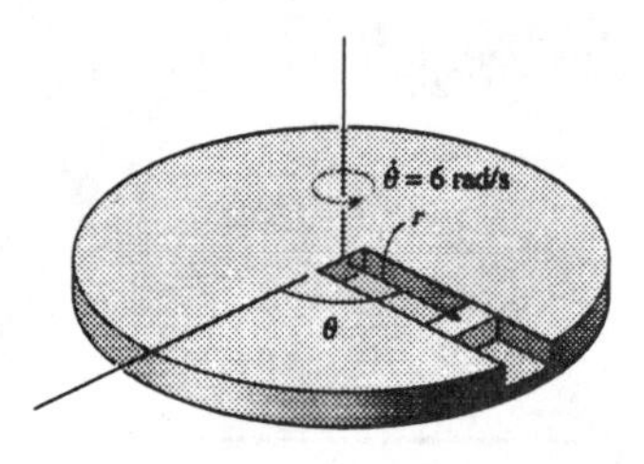

$\dot{r} = 4t|_{t=1} = 4 \qquad \ddot{r} = 4$

$\dot{\theta} = 6 \qquad \ddot{\theta} = 0$

$\int_0^{r} dr = \int_0^{1} 4t\,dt$

$r = 2t^2\Big]_0^1 = 2$ m

$v = \sqrt{(\dot{r})^2 + (r\dot{\theta})^2} = \sqrt{(4)^2 + [2(6)]^2} = 12.6$ m/s **Ans**

$a = \sqrt{(\ddot{r} - r\dot{\theta}^2)^2 + (r\ddot{\theta} + 2\dot{r}\dot{\theta})^2} = \sqrt{[4 - 2(6)^2]^2 + [0 + 2(4)(6)]^2} = 83.2$ m/s^2 **Ans**

12-151. The small washer is sliding down the cord OA. When it is halfway down the cord, its speed is 200 mm/s and its acceleration is 10 mm/s^2. Express the velocity and acceleration of the washer at this point in terms of its cylindrical components.

$$OA = \sqrt{(400)^2 + (300)^2 + (700)^2} = 860.23 \text{ mm}$$

$$OB = \sqrt{(400)^2 + (300)^2} = 500 \text{ mm}$$

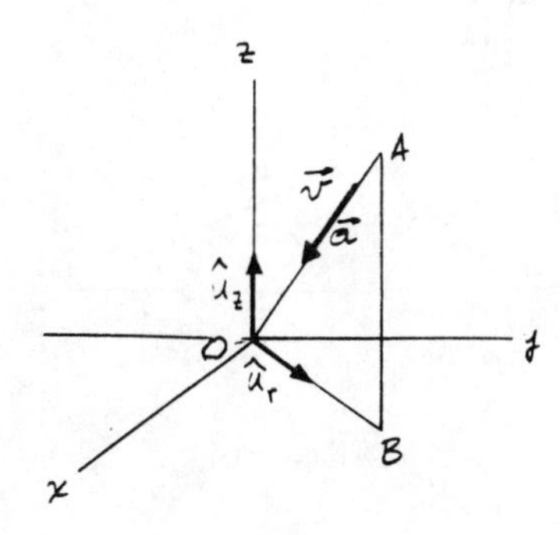

$$v_r = (200)\left(\frac{500}{860.23}\right) = 116 \text{ mm/s}$$

$$v_\theta = 0$$

$$v_z = (200)\left(\frac{700}{860.23}\right) = 163 \text{ mm/s}$$

Thus, $\mathbf{v} = \{-116\mathbf{u}_r - 163\mathbf{u}_z\}$ mm/s **Ans**

$$a_r = 10\left(\frac{500}{860.23}\right) = 5.81$$

$$a_\theta = 0$$

$$a_z = 10\left(\frac{700}{860.23}\right) = 8.14$$

Thus, $\mathbf{a} = \{-5.81\mathbf{u}_r - 8.14\mathbf{u}_z\}$ mm/s^2 **Ans**

***12-152.** The airplane on the amusement park ride moves along a path defined by the equations $r = 4$ m, $\theta = (0.2t)$ rad, and $z = (0.5 \cos \theta)$ m, where t is in seconds. Determine the cylindrical components of the velocity and acceleration of the airplane when $t = 6$ s.

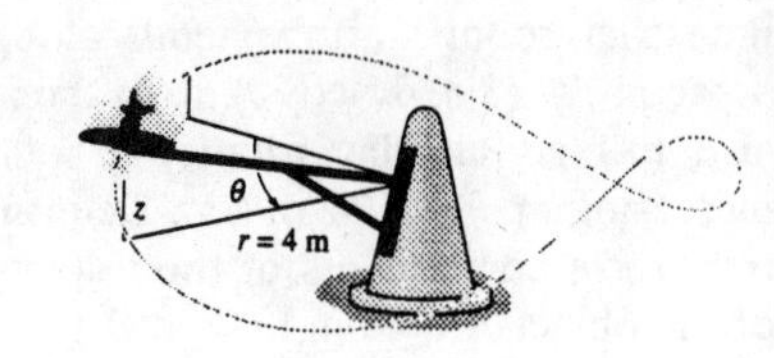

$r = 4$ m $\qquad \theta = 0.2t|_{t=6\text{ s}} = 1.2$ rad

$\dot{r} = 0 \qquad \dot{\theta} = 0.2$ rad/s

$\ddot{r} = 0 \qquad \ddot{\theta} = 0$

$z = 0.5\cos\theta \qquad \dot{z} = -0.5\sin\theta\dot{\theta}|_{\theta=1.2\text{ rad}} = -0.0932$ m/s

$$\ddot{z} = -0.5\left[\cos\theta\dot{\theta}^2 + \sin\theta\,\ddot{\theta}\right]\Big|_{\theta=1.2\text{ rad}} = -0.007247 \text{ m/s}^2$$

$v_r = \dot{r} = 0$ **Ans**

$v_\theta = r\dot{\theta} = 4(0.2) = 0.8$ m/s **Ans**

$v_z = \dot{z} = -0.0932$ m/s **Ans**

$a_r = \ddot{r} - r\dot{\theta}^2 = 0 - 4(0.2)^2 = -0.16$ m/s^2 **Ans**

$a_\theta = r\ddot{\theta} + 2\dot{r}\dot{\theta} = 4(0) + 2(0)(0.2) = 0$ **Ans**

$a_z = \ddot{z} = -0.00725$ m/s^2 **Ans**

12-153. The automobile is traveling from a parking deck down along a cylindrical spiral ramp at a constant speed of $v = 1.5$ m/s. If the ramp descends a distance of 12 m for every full revolution, $\theta = 2\pi$ rad, determine the magnitude of the car's acceleration as it moves along the ramp, $r = 10$ m. *Hint:* For part of the solution, note that the tangent to the ramp at any point is at an angle of $\phi = \tan^{-1}(12/[2\pi(10)]) = 10.81°$ from the horizontal. Use this to determine the velocity components v_θ and v_z, which in turn are used to determine $\dot{\theta}$ and $\dot{z}$.

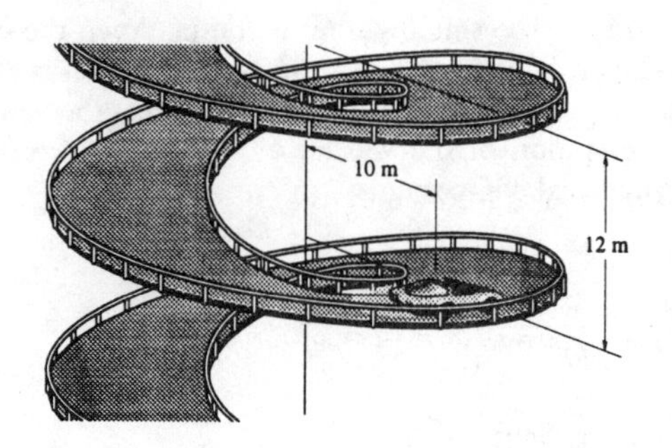

$$\phi = \tan^{-1}\left(\frac{12}{2\pi(10)}\right) = 10.81°$$

$v = 1.5$ m/s

$v_r = 0$

$v_\theta = 1.5\cos 10.81° = 1.473$ m/s

$v_z = -1.5\sin 10.81° = -0.2814$ m/s

Since

$r = 10 \quad \dot{r} = 0 \quad \ddot{r} = 0$

$v_\theta = r\dot{\theta} = 1.473 \qquad \dot{\theta} = \dfrac{1.473}{10} = 0.1473$

Since $\ddot{\theta} = 0$

$a_r = \ddot{r} - r\dot{\theta}^2 = 0 - 10(0.1473)^2 = -0.217$

$a_\theta = r\ddot{\theta} + 2\dot{r}\dot{\theta} = 10(0) + 2(0)(0.1473) = 0$

$a_z = \ddot{z} = 0$

$a = \sqrt{(-0.217)^2 + (0)^2 + (0)^2} = 0.217$ m/s² **Ans**

12-154. Because of telescopic action, the end of the industrial robotic arm extends along the path of the limaçon $r = (1 + 0.5\cos\theta)$ m. At the instant $\theta = \pi/4$, the arm has an angular rotation $\dot{\theta} = 0.6$ rad/s, which is increasing at $\ddot{\theta} = 0.25$ rad/s². Determine the radial and transverse components of the velocity and acceleration of the object A held in its grip at this instant.

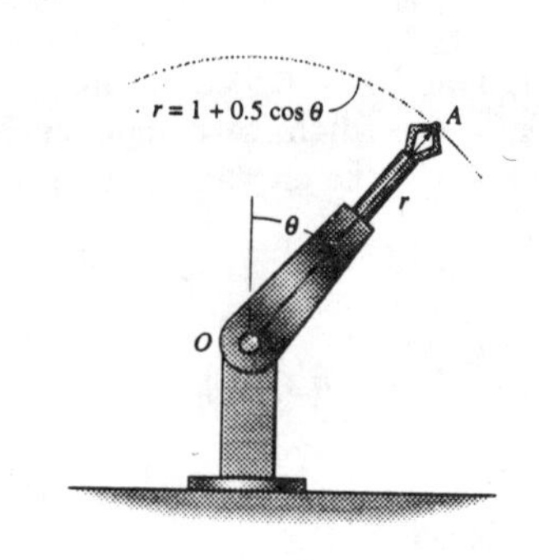

$\theta = \dfrac{\pi}{4} = 45°$

$\dot{\theta} = 0.6$

$\ddot{\theta} = 0.25$

$r = 1 + 0.5\cos\theta$

$\dot{r} = -0.5\sin\theta\dot{\theta}$

$\ddot{r} = -0.5\left(\cos\theta\dot{\theta}^2 + \sin\theta\ddot{\theta}\right)$

At $\theta = 45°$

$r = 1.354 \qquad \dot{r} = -0.2121 \qquad \ddot{r} = -0.2157$

$v_r = \dot{r} = -0.212$ m/s **Ans**

$v_\theta = r\dot{\theta} = 1.354(0.6) = 0.812$ m/s **Ans**

$a_r = \ddot{r} - r\dot{\theta}^2 = -0.2157 - (1.354)(0.6)^2 = -0.703$ m/s² **Ans**

$a_\theta = r\ddot{\theta} + 2\dot{r}\dot{\theta} = 1.354(0.25) + 2(-0.2121)(0.6) = 0.0838$ m/s² **Ans**

12-155. The boy slides down the slide at a constant speed of 2 m/s. If the slide is in the form of a helix, defined by the equations $r = 1.5$ m and $z = -\theta/\pi$, determine the boy's angular velocity about the z axis, $\dot{\theta}$, and the magnitude of his acceleration.

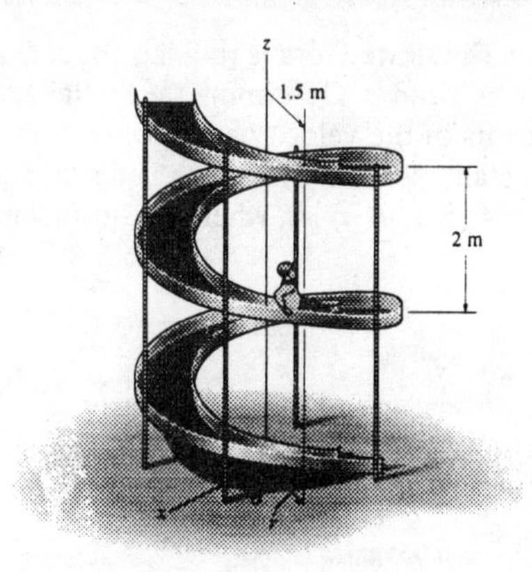

$$z = -\frac{\theta}{\pi} \qquad \dot{z} = -\frac{\dot{\theta}}{\pi} \qquad \ddot{z} = -\frac{\ddot{\theta}}{\pi}$$

$$r = 1.5 \qquad \dot{r} = 0 \qquad \ddot{r} = 0$$

$$v_r = \dot{r} = 0$$

$$v_\theta = r\dot{\theta} = 1.5\dot{\theta}$$

$$v_z = \dot{z} = -\frac{\dot{\theta}}{\pi}$$

$$2 = \sqrt{(0)^2 + \left(1.5\dot{\theta}\right)^2 + \left(-\dot{\theta}/\pi\right)^2}$$

$$4 = 2.25\dot{\theta}^2 + 0.1013\dot{\theta}^2$$

$$\dot{\theta} = 1.3043 = 1.30 \text{ rad/s} \qquad \textbf{Ans}$$

$$\ddot{\theta} = 0$$

$$a_r = \ddot{r} - r\dot{\theta}^2$$

$$= 0 - 1.5(1.3043)^2 = -2.552$$

$$a_\theta = r\ddot{\theta} + 2\dot{r}\dot{\theta}$$

$$= 1.5(0) + 0 = 0$$

$$a_z = \ddot{z} = -\frac{(0)}{\pi} = 0$$

$$a = \sqrt{(-2.552)^2 + (0)^2 + (0)^2} = 2.55 \text{ m/s}^2 \qquad \textbf{Ans}$$

***12-156.** The motion of particle B is controlled by the rotation of the grooved link OA. If the link is rotating at a constant angular rate of $\dot{\theta} = 6$ rad/s, determine the magnitudes of the velocity and acceleration of B at the instant $\theta = \pi/2$ rad. The spiral path is defined by the equation $r = (40\theta)$ mm, where θ is in radians.

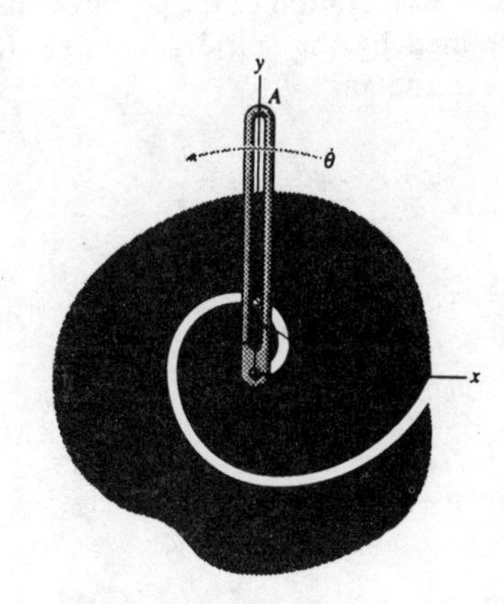

$$\dot{\theta} = 6 \qquad \ddot{\theta} = 0$$

$$r = 40\theta \qquad \dot{r} = 40\dot{\theta} = 240 \qquad \ddot{r} = 0$$

At $\theta = \dfrac{\pi}{2}$,

$$v_r = \dot{r} = 240$$

$$v_\theta = r\dot{\theta} = 40\left(\frac{\pi}{2}\right)(6) = 377.0$$

$$v = \sqrt{(240)^2 + (377.0)^2} = 447 \text{ mm/s} \qquad \textbf{Ans}$$

$$a_r = \ddot{r} - r\dot{\theta}^2 = 0 - 40\left(\frac{\pi}{2}\right)(6)^2 = -2261.9$$

$$a_\theta = r\ddot{\theta} + 2\dot{r}\dot{\theta} = 0 + 2(240)(6) = 2880$$

$$a = \sqrt{(-2261.9)^2 + (2880)^2} = 3662 \text{ mm/s}^2 \qquad \textbf{Ans}$$

12-157. The slotted fork is rotating about O at a constant rate of $\dot{\theta} = 3$ rad/s. Determine the radial and transverse components of the velocity and acceleration of the pin A at the instant $\theta = 360°$. The path is defined by the spiral groove $r = (5 + \theta/\pi)$ in., where θ is in radians.

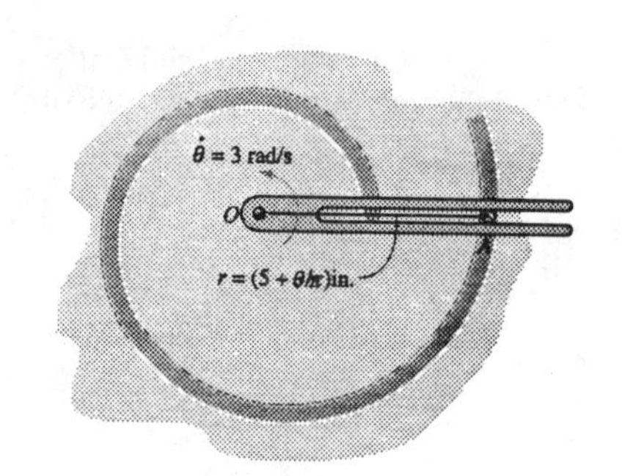

$$r = 5 + \frac{\theta}{\pi}\bigg|_{\theta = 2\pi} = 7 \qquad \dot{r} = \frac{\dot{\theta}}{\pi} \qquad \ddot{r} = \frac{\ddot{\theta}}{\pi}$$

$$\theta = 360° = 2\pi \qquad \dot{\theta} = 3 \qquad \ddot{\theta} = 0$$

$$v_r = \dot{r} = \frac{3}{\pi} = 0.955 \text{ in./s} \qquad \textbf{Ans}$$

$$v_\theta = r\dot{\theta} = 7(3) = 21 \text{ in./s} \qquad \textbf{Ans}$$

$$a_r = \ddot{r} - r\dot{\theta}^2 = 0 - 7(3)^2 = -63 \text{ in./s}^2 \qquad \textbf{Ans}$$

$$a_\theta = r\ddot{\theta} + 2\dot{r}\dot{\theta} = 0 + 2\left(\frac{3}{\pi}\right)(3) = 5.73 \text{ in./s}^2 \qquad \textbf{Ans}$$

12-158. The slotted fork is rotating about O at $\dot{\theta} = 3$ rad/s, which is increasing at $\ddot{\theta} = 2$ rad/s² when $\theta = 360°$. Determine the radial and transverse components of the velocity and acceleration of the pin A at this instant. The path is defined by the spiral groove $r = (5 + \theta/\pi)$ in., where θ is in radians.

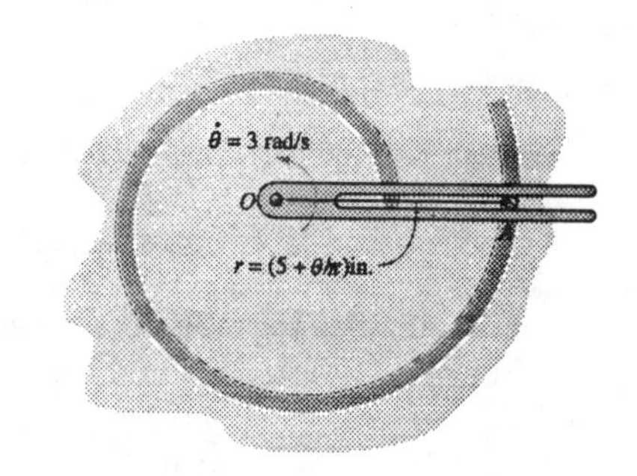

$$r = 5 + \frac{\theta}{\pi}\bigg|_{\theta = 2\pi} = 7 \qquad \dot{r} = \frac{\dot{\theta}}{\pi} \qquad \ddot{r} = \frac{\ddot{\theta}}{\pi}$$

$$\theta = 360° = 2\pi \qquad \dot{\theta} = 3 \qquad \ddot{\theta} = 2$$

$$v_r = \dot{r} = \frac{3}{\pi} = 0.955 \text{ in./s} \qquad \textbf{Ans}$$

$$v_\theta = r\dot{\theta} = 7(3) = 21 \text{ in./s} \qquad \textbf{Ans}$$

$$a_r = \ddot{r} - r\dot{\theta}^2 = \frac{2}{\pi} - 7(3)^2 = -62.4 \text{ in./s}^2 \qquad \textbf{Ans}$$

$$a_\theta = r\ddot{\theta} + 2\dot{r}\dot{\theta} = 7(2) + 2\left(\frac{3}{\pi}\right)(3) = 19.7 \text{ in./s}^2 \qquad \textbf{Ans}$$

12-159. The partial surface of the cam is that of a logarithmic spiral $r = (40e^{0.05\theta})$ mm, where θ is in radians. If the cam is rotating at a constant angular rate of $\dot{\theta} = 4$ rad/s, determine the magnitudes of the velocity and acceleration of the follower rod at the instant $\theta = 30°$.

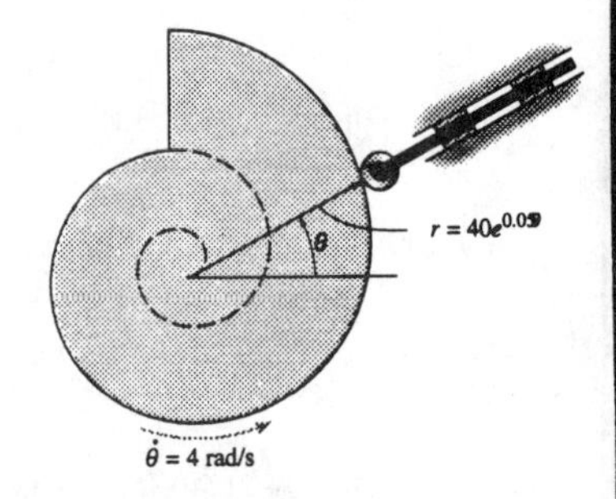

$$r = 40e^{0.05\theta} \qquad r = 40e^{0.05\left(\frac{\pi}{6}\right)} = 41.0610$$

$$\dot{r} = 2e^{0.05\theta}\dot{\theta} \qquad \dot{r} = 2e^{0.05\left(\frac{\pi}{6}\right)}(4) = 8.2122$$

$$\ddot{r} = 0.1e^{0.05\theta}\left(\dot{\theta}\right)^2 + 2e^{0.05\theta}\ddot{\theta} \qquad \ddot{r} = 0.1e^{0.05\left(\frac{\pi}{6}\right)}(4)^2 + 0 = 1.64244$$

$$\theta = \frac{\pi}{6} \qquad v_r = \dot{r} = 8.2122$$

$$\dot{\theta} = 4 \qquad v_\theta = r\dot{\theta} = 41.0610(4) = 164.24$$

$$\ddot{\theta} = 0 \qquad v = \sqrt{(8.2122)^2 + (164.24)^2} = 164 \text{ mm/s} \qquad \textbf{Ans}$$

$$a_r = \ddot{r} - r\dot{\theta}^2 = 1.642\,44 - 41.0610(4)^2 = -655.33$$

$$a_\theta = r\ddot{\theta} + 2\dot{r}\dot{\theta} = 0 + 2(8.2122)(4) = 65.6976$$

$$a = \sqrt{(-655.33)^2 + (65.6976)^2} = 659 \text{ mm/s}^2 \qquad \textbf{Ans}$$

***12-160.** Solve Prob. 12-159, if the cam has an angular acceleration of $\ddot{\theta} = 2\ \text{rad/s}^2$ when its angular velocity is $\dot{\theta} = 4$ rad/s at $\theta = 30°$.

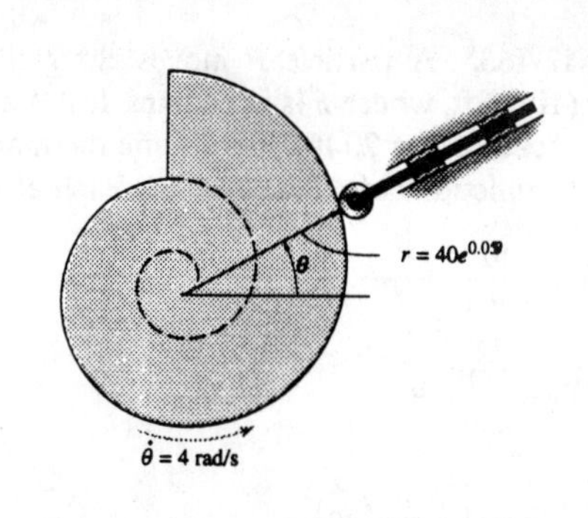

$r = 40e^{0.05\theta}$

$\dot{r} = 2e^{0.05\theta}\dot{\theta}$

$\ddot{r} = 0.1e^{0.05\theta}\left(\dot{\theta}\right)^2 + 2e^{0.05\theta}\ddot{\theta}$

$\theta = \dfrac{\pi}{6}$

$\dot{\theta} = 4$

$\ddot{\theta} = 2$

$r = 40e^{0.05\left(\frac{\pi}{6}\right)} = 41.0610$

$\dot{r} = 2e^{0.05\left(\frac{\pi}{6}\right)}(4) = 8.2122$

$\ddot{r} = 0.1e^{0.05\left(\frac{\pi}{6}\right)}(4)^2 + 2e^{0.05\left(\frac{\pi}{6}\right)}(2) = 5.749$

$v_r = \dot{r} = 8.2122$

$v_\theta = r\dot{\theta} = 41.0610(4) = 164.24$

$v = \sqrt{(8.2122)^2 + (164.24)^2} = 164\ \text{mm/s}$ **Ans**

$a_r = \ddot{r} - r\dot{\theta}^2 = 5.749 - 41.0610(4)^2 = -651.2$

$a_\theta = r\ddot{\theta} + 2\dot{r}\dot{\theta} = 41.0610(2) + 2(8.2122)(4) = 147.8197$

$a = \sqrt{(-651.2)^2 + (147.8197)^2} = 668\ \text{mm/s}^2$ **Ans**

12-161. The rod OA rotates counterclockwise with a constant angular velocity of $\dot{\theta} = 5$ rad/s. Two pin-connected slider blocks, located at B, move freely on OA and the curved rod whose shape is a limaçon described by the equation $r = 100(2 - \cos\theta)$ mm. Determine the speed of the slider blocks at the instant $\theta = 120°$.

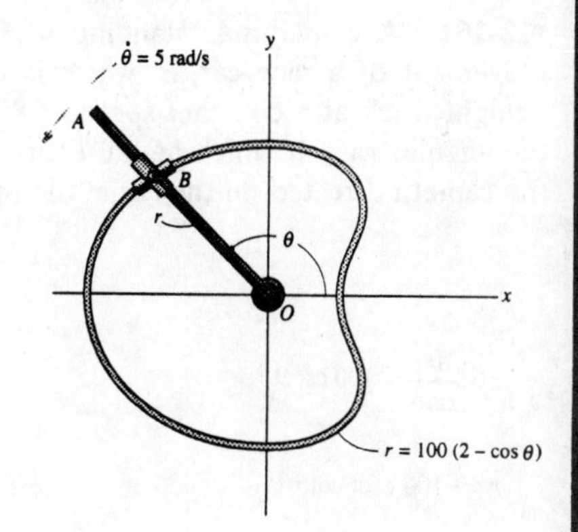

$\dot{\theta} = 5$

$r = 100(2 - \cos\theta)$

$\dot{r} = 100\sin\theta\dot{\theta} = 500\sin\theta$

$\ddot{r} = 500\cos\theta\dot{\theta} = 2500\cos\theta$

At $\theta = 120°$,

$v_r = \dot{r} = 500\sin 120° = 433.013$

$v_\theta = r\dot{\theta} = 100(2 - \cos 120°)(5) = 1250$

$v = \sqrt{(433.013)^2 + (1250)^2} = 1322.9\ \text{mm/s} = 1.32\ \text{m/s}$ **Ans**

12-162. Determine the magnitude of the acceleration of the slider blocks in Prob. 12-161 when $\theta = 120°$.

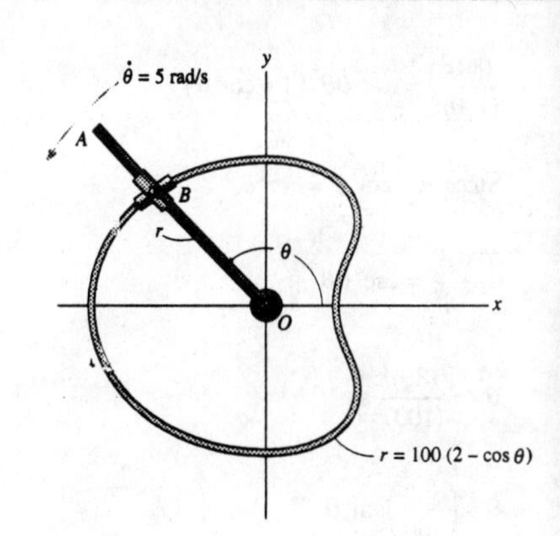

$\dot{\theta} = 5$

$\ddot{\theta} = 0$

$r = 100(2 - \cos\theta)$

$\dot{r} = 100\sin\theta\dot{\theta} = 500\sin\theta$

$\ddot{r} = 500\cos\theta\dot{\theta} = 2500\cos\theta$

$a_r = \ddot{r} - r\dot{\theta}^2 = 2500\cos\theta - 100(2 - \cos\theta)(5)^2 = 5000(\cos 120° - 1) = -7500\ \text{mm/s}^2$

$a_\theta = r\ddot{\theta} + 2\dot{r}\dot{\theta} = 0 + 2(500\sin\theta)(5) = 5000\sin 120° = 4330.1\ \text{mm/s}^2$

$a = \sqrt{(-7500)^2 + (4330.1)^2} = 8660.3\ \text{mm/s}^2 = 8.66\ \text{m/s}^2$ **Ans**

12-163. A particle P moves along the spiral path $r = (10/\theta)$ ft, where θ is in radians. If it maintains a constant speed of $v = 20$ ft/s, determine the magnitudes v_r and v_θ as functions of θ and evaluate each at $\theta = 1$ rad.

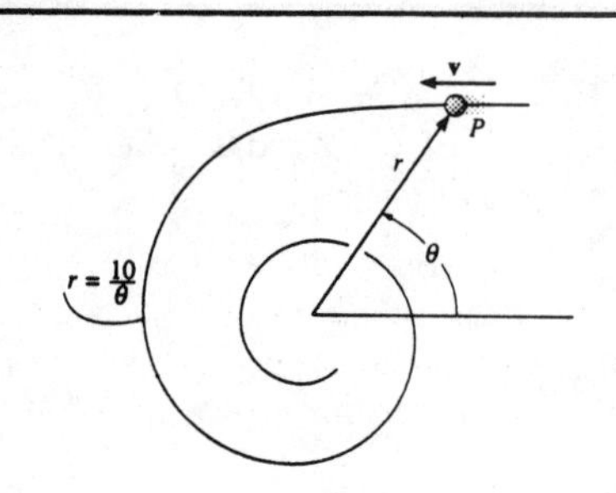

$$r = \frac{10}{\theta}$$

$$\dot{r} = -\left(\frac{10}{\theta^2}\right)\dot{\theta}$$

Since $v^2 = \dot{r}^2 + \left(r\dot{\theta}\right)^2$

$$(20)^2 = \left(\frac{10^2}{\theta^4}\right)\dot{\theta}^2 + \left(\frac{10^2}{\theta^2}\right)\dot{\theta}^2$$

$$(20)^2 = \left(\frac{10^2}{\theta^4}\right)(1+\theta^2)\dot{\theta}^2$$

Thus, $\quad \dot{\theta} = \dfrac{2\theta^2}{\sqrt{1+\theta^2}}$

$$v_r = \dot{r} = -\left(\frac{10}{\theta^2}\right)\left(\frac{2\theta^2}{\sqrt{1+\theta^2}}\right) = -\frac{20}{\sqrt{1+\theta^2}} \qquad \textbf{Ans}$$

$$v_\theta = r\dot{\theta} = \left(\frac{10}{\theta}\right)\left(\frac{2\theta^2}{\sqrt{1+\theta^2}}\right) = \frac{20\theta}{\sqrt{1+\theta^2}} \qquad \textbf{Ans}$$

When $\theta = 1$ rad,

$$v_r = \left(-\frac{20}{\sqrt{2}}\right) = -14.1 \text{ ft/s} \qquad \textbf{Ans}$$

$$v_\theta = \left(\frac{20}{\sqrt{2}}\right) = 14.1 \text{ ft/s} \qquad \textbf{Ans}$$

***12-164.** A cameraman standing at A is following the movement of a race car, B, which is traveling along a straight track at a constant speed of 80 ft/s. Determine the angular rate at which he must turn in order to keep the camera directed on the car at the instant $\theta = 60°$.

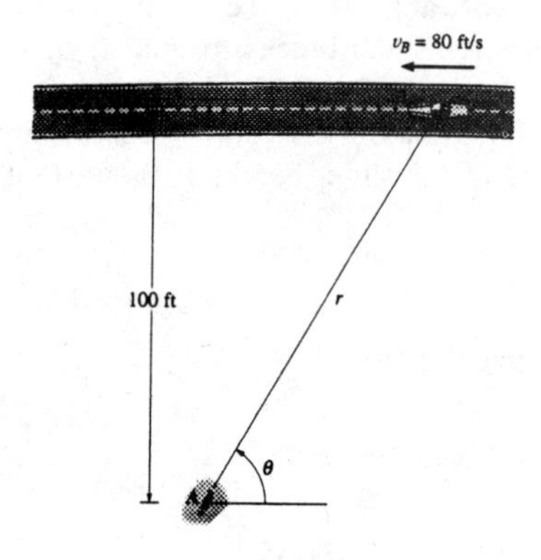

$$r = \frac{100}{\sin\theta} = 100 \csc\theta$$

$$\dot{r} = -100 \csc\theta\cot\theta\dot{\theta}$$

$$v^2 = \left(\dot{r}\right)^2 + \left(r\dot{\theta}\right)^2$$

$$(80)^2 = (-100)^2\csc^2\theta\cot^2\theta\dot{\theta}^2 + (100)^2\csc^2\theta\dot{\theta}^2$$

$$\frac{(80)^2}{(100)^2} = \csc^2\theta\dot{\theta}^2\left(1+\cot^2\theta\right)$$

Since $1+\cot^2\theta = \csc^2\theta$,

$$\frac{(80)^2}{(100)^2} = \csc^4\theta\dot{\theta}^2$$

$$\dot{\theta}^2 = \frac{(80)^2}{(100)^2}\sin^4\theta$$

$$\dot{\theta} = \left(\frac{80}{100}\right)\sin^2\theta$$

At $\theta = 60°$,

$$\dot{\theta} = \left(\frac{8}{10}\right)\sin^2 60° = 0.6 \text{ rad/s} \qquad \textbf{Ans}$$

***12-165.** The double collar C is pin connected such that one collar slides over a fixed rod and the other slides over a rotating rod AB. If the angular velocity of AB is given as $\dot{\theta} = (e^{0.5t^2})$ rad/s, where t is in seconds, and the path defined by the fixed rod is $r = |0.4 \sin\theta + 0.2|$ m, determine the radial and transverse components of the collar's velocity and acceleration when $t = 1$ s. When $t = 0$, $\theta = 0°$. Use Simpson's rule to determine θ at $t = 1$ s.

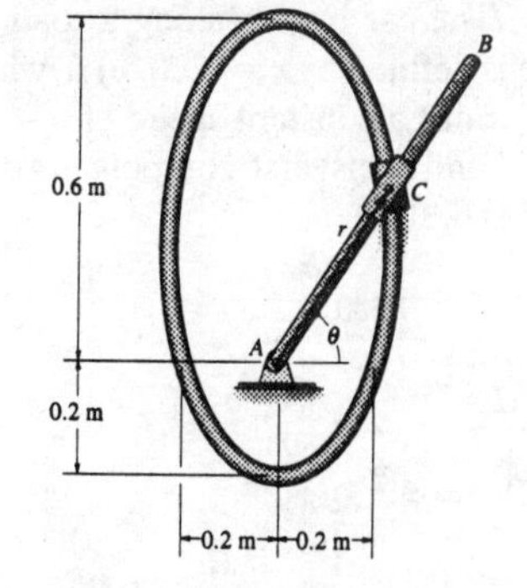

$\dot{\theta} = e^{0.5t^2}\big|_{t=1\text{ s}} = 1.649$ rad/s

$\ddot{\theta} = e^{0.5t^2}(t)\big|_{t=1\text{ s}} = 1.649$ rad/s^2

$\theta = \int_0^1 e^{0.5t^2}\,dt = 1.195 \text{ rad} = 68.47°$

$r = 0.4\sin\theta + 0.2$

$\dot{r} = 0.4\cos\theta\dot{\theta}$

$\ddot{r} = -0.4\sin\theta\dot{\theta}^2 + 0.4\cos\theta\ddot{\theta}$

At $t = 1$ s,

$r = 0.5721$

$\dot{r} = 0.2421$

$\ddot{r} = -0.7694$

$v_r = \dot{r} = 0.242$ m/s **Ans**

$v_\theta = r\dot{\theta} = 0.5721(1.649) = 0.943$ m/s **Ans**

$a_r = \ddot{r} - r\dot{\theta}^2 = -0.7694 - 0.5721(1.649)^2 = -2.32$ m/s^2 **Ans**

$a_\theta = r\ddot{\theta} + 2\dot{r}\dot{\theta} = 0.5721(1.649) + 2(0.2421)(1.649) = 1.74$ m/s^2 **Ans**

12-166. For a short time the bucket of the backhoe traces the path of the cardioid $r = 25(1 - \cos\theta)$ ft. Determine the magnitudes of the velocity and acceleration of the bucket when $\theta = 120°$ if the boom is rotating with an angular velocity of $\dot{\theta} = 2$ rad/s and an angular acceleration of $\ddot{\theta} = 0.2$ rad/s^2 at the instant shown.

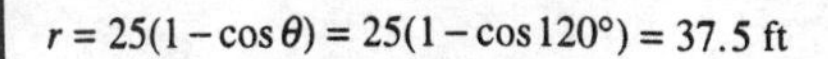

$r = 25(1 - \cos\theta) = 25(1 - \cos 120°) = 37.5$ ft

$\dot{r} = 25\sin\theta\dot{\theta} = 25\sin 120°(2) = 43.30$ ft/s

$\ddot{r} = 25\left[\cos\theta\dot{\theta}^2 + \sin\theta\ddot{\theta}\right] = 25\left[\cos 120°(2)^2 + \sin 120°(0.2)\right] = -45.67$ ft/s^2

$v_r = \dot{r} = 43.30$ ft/s

$v_\theta = r\dot{\theta} = 37.5(2) = 75$ ft/s

$v = \sqrt{v_r^2 + v_\theta^2} = \sqrt{43.30^2 + 75^2} = 86.6$ ft/s **Ans**

$a_r = \ddot{r} - r\dot{\theta}^2 = -45.67 - 37.5(2)^2 = -195.67$ ft/s^2

$a_\theta = r\ddot{\theta} + 2\dot{r}\dot{\theta} = 37.5(0.2) + 2(43.30)(2) = 180.71$ ft/s^2

$a = \sqrt{a_r^2 + a_\theta^2} = \sqrt{(-195.67)^2 + 180.71^2} = 266$ ft/s^2 **Ans**

12-167. The car travels along a road which for a short distance is defined by $r = (200/\theta)$ ft, where θ is in radians. If it maintains a constant speed of $v = 35$ ft/s, determine the radial and transverse components of its velocity when $\theta = \pi/3$ rad.

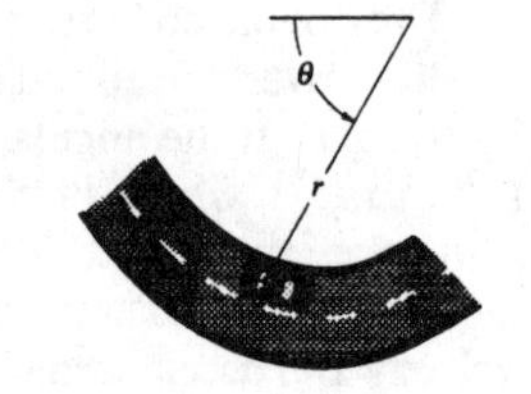

$$r = \frac{200}{\theta}\bigg|_{\theta=\pi/3\text{ rad}} = \frac{600}{\pi}\text{ ft}$$

$$\dot{r} = -\frac{200}{\theta^2}\dot{\theta}\bigg|_{\theta=\pi/3\text{ rad}} = -\frac{1800}{\pi^2}\dot{\theta}$$

$$v_r = \dot{r} = -\frac{1800}{\pi^2}\dot{\theta} \qquad v_\theta = r\dot{\theta} = \frac{600}{\pi}\dot{\theta}$$

$$v^2 = v_r^2 + v_\theta^2$$

$$35^2 = \left(-\frac{1800}{\pi^2}\dot{\theta}\right)^2 + \left(\frac{600}{\pi}\dot{\theta}\right)^2$$

$$\dot{\theta} = 0.1325\text{ rad/s}$$

$$v_r = -\frac{1800}{\pi^2}(0.1325) = -24.2\text{ ft/s}$$ **Ans**

$$v_\theta = \frac{600}{\pi}(0.1325) = 25.3\text{ ft/s}$$ **Ans**

***12-168.** The pin follows the path described by the equation $r = (0.2 + 0.15\cos\theta)$ m. At the instant $\theta = 30°$, $\dot{\theta} = 0.7$ rad/s and $\ddot{\theta} = 0.5$ rad/s². Determine the magnitudes of the pin's velocity and acceleration at this instant. Neglect the size of the pin.

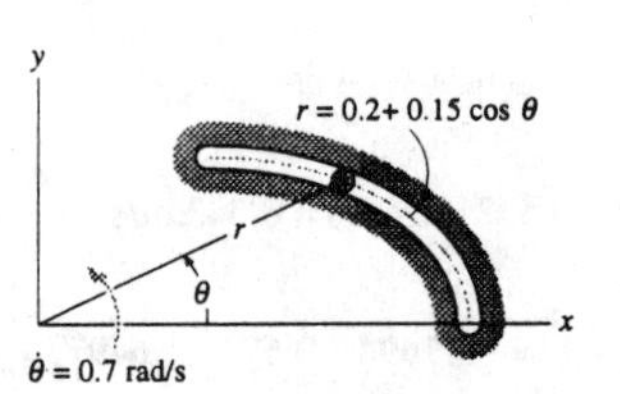

$$r = 0.2 + 0.15\cos\theta = 0.2 + 0.15\cos 30° = 0.3299\text{ m}$$

$$\dot{r} = -0.15\sin\theta\dot{\theta} = -0.15\sin 30°(0.7) = -0.0525\text{ m/s}$$

$$\ddot{r} = -0.15\left[\cos\theta\dot{\theta}^2 + \sin\theta\ddot{\theta}\right] = -0.15\left[\cos 30°(0.7)^2 + \sin 30°(0.5)\right] = -0.10115\text{ m/s}^2$$

$$v_r = \dot{r} = -0.0525\text{ m/s}$$

$$v_\theta = r\dot{\theta} = 0.3299(0.7) = 0.2309\text{ m/s}$$

$$v = \sqrt{v_r^2 + v_\theta^2} = \sqrt{(-0.0525)^2 + (0.2309)^2} = 0.237\text{ m/s}$$ **Ans**

$$a_r = \ddot{r} - r\dot{\theta}^2 = -0.10115 - 0.3299(0.7)^2 = -0.2628\text{ m/s}^2$$

$$a_\theta = r\ddot{\theta} + 2\dot{r}\dot{\theta} = 0.3299(0.5) + 2(-0.0525)(0.7) = 0.09145\text{ m/s}^2$$

$$a = \sqrt{a_r^2 + a_\theta^2} = \sqrt{(-0.2628)^2 + (0.09145)^2} = 0.278\text{ m/s}^2$$ **Ans**

12-169. The mechanism of a machine is constructed so that for a short time the roller at A follows the surface of the cam described by the equation $r = (0.3 + 0.2 \cos \theta)$ m. If $\dot{\theta} = 0.5$ rad/s and $\ddot{\theta} = 0$, determine the magnitudes of the roller's velocity and acceleration at the instant $\theta = 30°$. Neglect the size of the roller. Also determine the velocity components $(\mathbf{v}_A)_x$ and $(\mathbf{v}_A)_y$ of the roller at this instant. The rod to which the roller is attached remains vertical and can slide up or down along the guides while the guides move horizontally to the left.

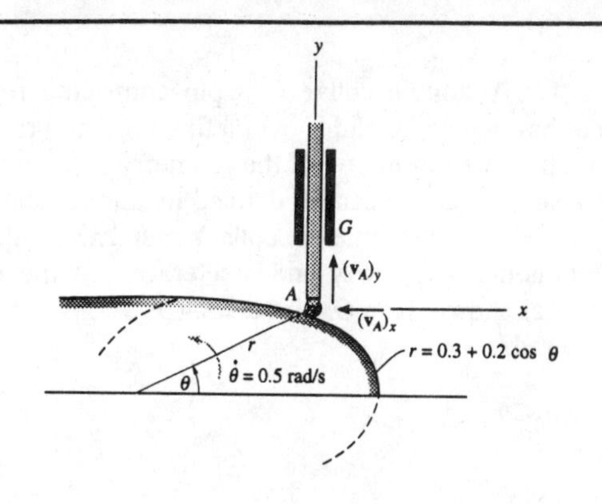

$\dot{\theta} = 0.5$

$\ddot{\theta} = 0$

$r = (0.3 + 0.2 \cos\theta)$

$\dot{r} = -0.2 \sin\theta\dot{\theta}$

$\ddot{r} = -0.2\left(\cos\theta\dot{\theta}^2 + \sin\theta\ddot{\theta}\right)$

At $\theta = 30°$,

$r = 0.4732$

$\dot{r} = -0.05$

$\ddot{r} = -0.04330$

$v_r = \dot{r} = -0.05$

$v_\theta = r\dot{\theta} = 0.473(0.5) = 0.2366$

$v = \sqrt{(-0.05)^2 + (0.2366)^2} = 0.242$ m/s **Ans**

$a_r = \ddot{r} - r\dot{\theta}^2 = -0.04330 - 0.4732(0.5)^2 = -0.1616$ m/s^2

$a_\theta = r\ddot{\theta} + 2\dot{r}\dot{\theta} = 0 + 2(-0.05)(0.5) = -0.05$ m/s^2

$a = \sqrt{(-0.1616)^2 + (-0.05)^2} = 0.169$ m/s^2 **Ans**

$(\overset{+}{\rightarrow})$ $(v_A)_x = -0.05 \cos30° - 0.2366 \sin30° = -0.162$ m/s **Ans**

$(+\uparrow)$ $(v_A)_y = -0.05 \sin30° + 0.2366 \cos30° = 0.180$ m/s **Ans**

12-170. The crate slides down the section of the spiral ramp such that $r = (0.5z)$ ft and $z = (100 - 0.1t^2)$ ft, where t is in seconds. If the rate of rotation about the z axis is $\dot{\theta} = 0.04\pi t$ rad/s, determine the magnitudes of the velocity and acceleration of the crate at the instant $z = 10$ ft.

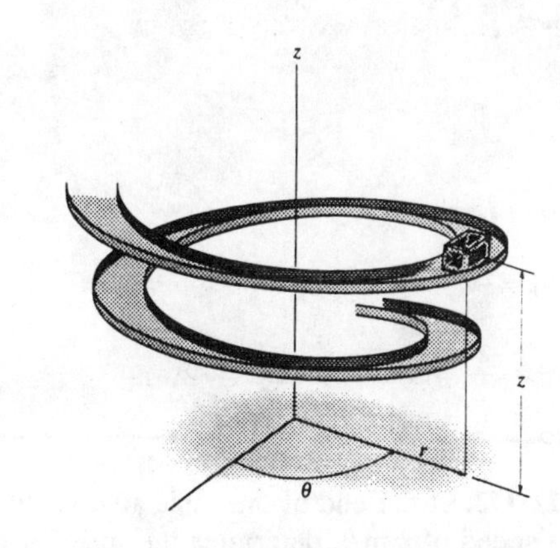

$r = 0.5z$

$z = 100 - 0.1t^2$

Thus,

$r = 50 - 0.05t^2$

$\dot{r} = -0.1t$

$\ddot{r} = -0.1$

$\dot{\theta} = 0.04\pi\, t$ rad/s $= 0.12566t$ rad/s

$\ddot{\theta} = 0.12566$

$\dot{z} = -0.2t$

$\ddot{z} = -0.2$

At $z = 10$ ft,

$10 = 100 - 0.1t^2$

$t = 30$ s

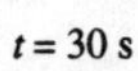

$r = 50 - 0.05(30)^2 = 5$

$\dot{r} = -0.1(30) = -3$

$\ddot{r} = -0.1$

$\dot{\theta} = 0.12566(30) = 3.76991$

$\ddot{\theta} = 0.12566$

$\dot{z} = -0.2(30) = -6$

$\ddot{z} = -0.2$

$v_r = \dot{r} = -3$

$v_\theta = r\dot{\theta} = 5(3.76991) = 18.850$

$v_z = \dot{z} = -6$

$v = \sqrt{(-3)^2 + (18.850)^2 + (-6)^2} = 20.0$ ft/s **Ans**

$a_r = \ddot{r} - r\dot{\theta}^2 = -0.1 - 5(3.76991)^2 = -71.16$

$a_\theta = r\ddot{\theta} + 2\dot{r}\dot{\theta} = 5(0.12566) + 2(-3)(3.76991) = -21.99$

$a_z = \ddot{z} = -0.2$

$a = \sqrt{(-71.16)^2 + (-21.99)^2 + (-0.2)^2} = 74.5$ ft/s^2 **Ans**

12-171. A double collar C is pin connected together such that one collar slides over a fixed rod and the other slides over a rotating rod. If the geometry of the fixed rod for a short distance can be defined by a lemniscate, $r^2 = (4\cos 2\theta)\ \text{ft}^2$, determine the collar's radial and transverse components of velocity and acceleration at the instant $\theta = 0°$ as shown. Rod OA is rotating at a constant rate of $\dot{\theta} = 6$ rad/s.

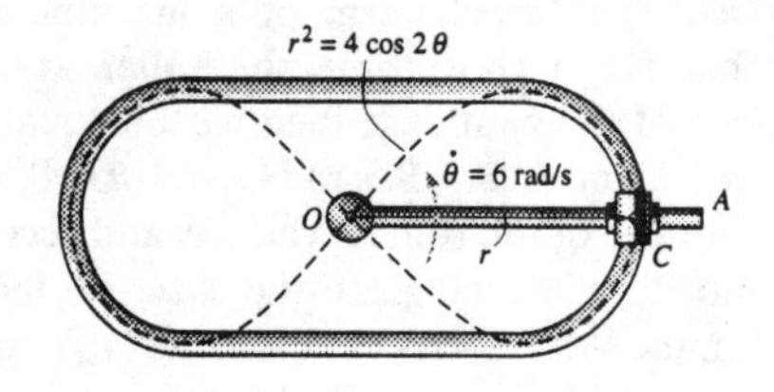

$$r^2 = 4\cos 2\theta$$

$$r\dot{r} = -4\sin 2\theta\dot{\theta}$$

$$r\ddot{r} + \dot{r}^2 = -4\sin 2\theta\ddot{\theta} - 8\cos 2\theta\dot{\theta}^2$$

When $\theta = 0,\ \dot{\theta} = 6,\ \ddot{\theta} = 0$

$$r = 2,\ \dot{r} = 0,\ \ddot{r} = -144$$

$$v_r = \dot{r} = 0 \qquad \textbf{Ans}$$

$$v_\theta = r\dot{\theta} = 2(6) = 12\ \text{ft/s} \qquad \textbf{Ans}$$

$$a_r = \ddot{r} - r\dot{\theta}^2 = -144 - 2(6)^2 = -216\ \text{ft/s}^2 \qquad \textbf{Ans}$$

$$a_\theta = r\ddot{\theta} + 2\dot{r}\dot{\theta} = 2(0) + 2(0)(6) = 0 \qquad \textbf{Ans}$$

***12-172.** If the end of the cable at A is pulled down with a speed of 2 m/s, determine the speed at which block B rises.

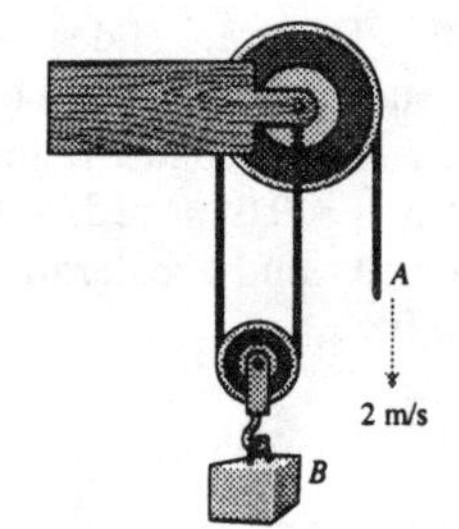

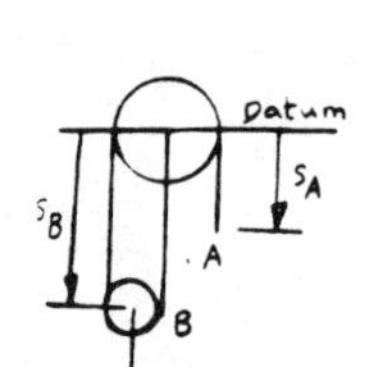

$$2s_B + s_A = l$$

$$2v_B = -v_A$$

$$v_B = \frac{-2}{2} = -1\ \text{m/s} = 1\ \text{m/s}\uparrow \qquad \textbf{Ans}$$

12-173. If the end of the cable at A is pulled down with a speed of 2 m/s, determine the speed at which block B rises.

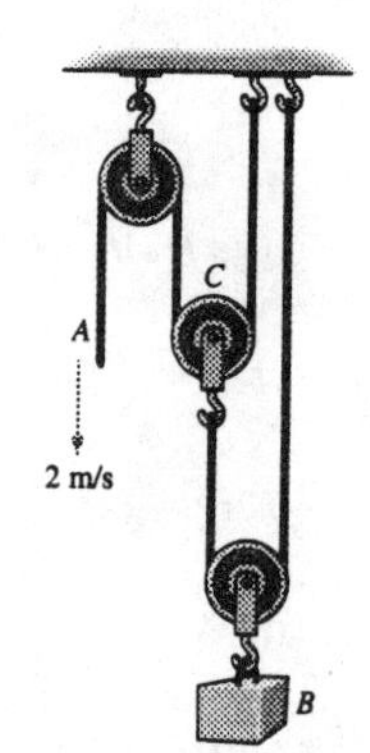

Two cords :

$$s_A + 2s_C = l$$

$$s_B + (s_B - s_C) = l'$$

Thus,

$$v_A = -2v_C$$

$$2v_B = v_C$$

Thus,

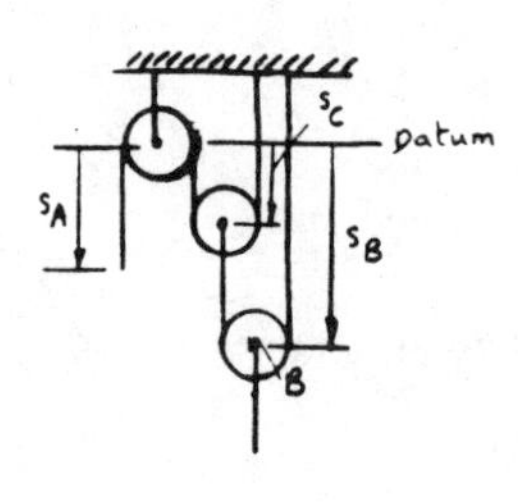

$$v_B = \frac{-v_A}{4} = \frac{-2}{4} = -0.5\ \text{m/s} = 0.5\ \text{m/s}\uparrow \qquad \textbf{Ans}$$

12-174. If the end of the cable at A is pulled down with a speed of 2 m/s, determine the speed at which block B rises.

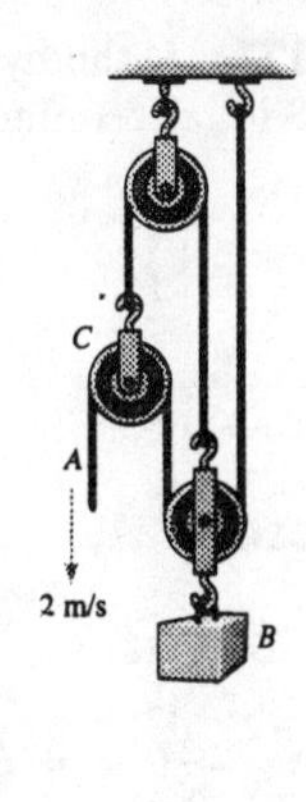

Two cords :

$s_C + s_B = l$

$s_B + (s_B - s_C) + (s_A - s_C) = l'$

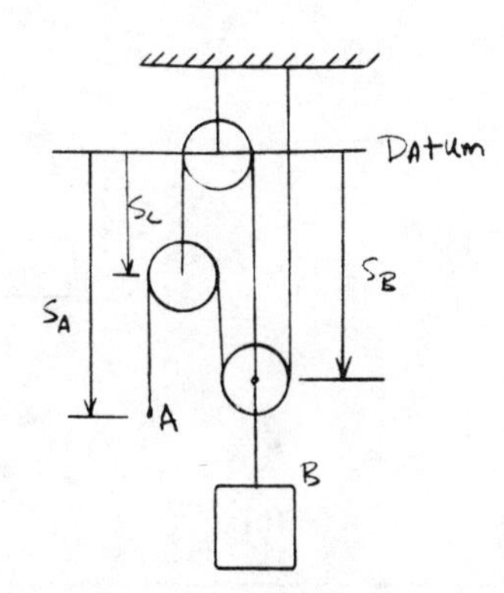

Thus,

$v_C = -v_B$

$2v_B - 2v_C + v_A = 0$

Thus,

$v_B = \frac{-v_A}{4} = \frac{-2}{4} = -0.5 \text{ m/s} = 0.5 \text{ m/s} \uparrow$ **Ans**

12-175. The mine car C is being pulled up the incline using the motor M and the rope-and-pulley arrangement shown. Determine the speed v_P at which a point P on the cable must be traveling toward the motor to move the car up the plane with a constant speed of $v = 2$ m/s.

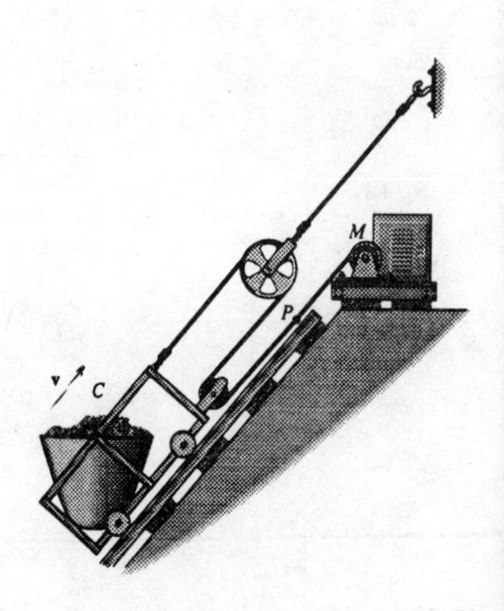

$2s_C + (s_C - s_P) = l$

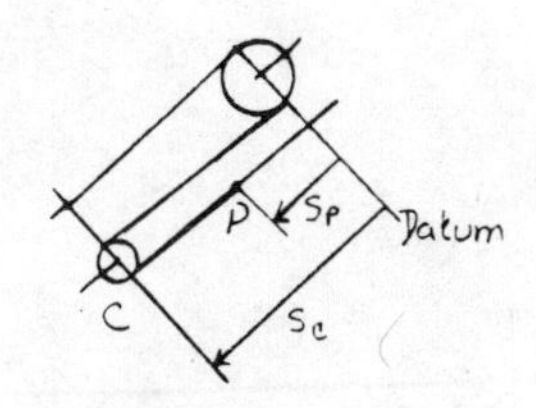

Thus,

$3v_C - v_P = 0$

Hence,

$v_P = 3(-2) = -6 \text{ m/s} = 6 \text{ m/s} \nearrow$ **Ans**

***12-176.** Determine the displacement of the log if the truck at C pulls the cable 4 ft to the right.

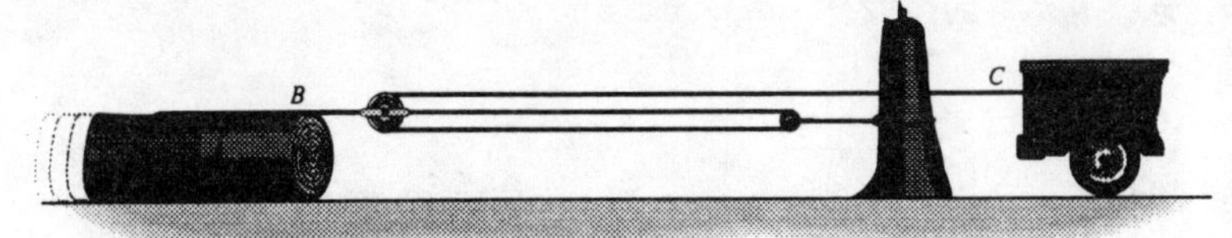

$2s_B + (s_B - s_C) = l$

$3s_B - s_C = l$

$3\Delta s_B - \Delta s_C = 0$

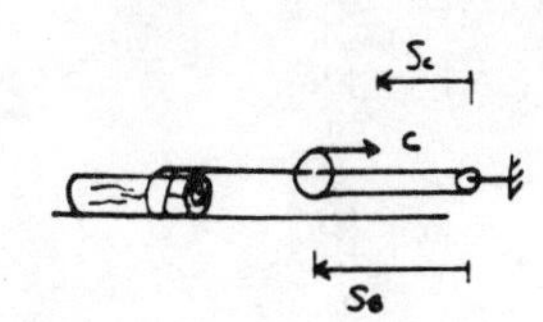

Since $\Delta s_C = -4$, then

$3\Delta s_B = -4$

$\Delta s_B = -1.33 \text{ ft} = 1.33 \text{ ft} \rightarrow$ **Ans**

12-177. If the hydraulic cylinder at H draws rod BC in by 8 in., determine how far the slider at A moves.

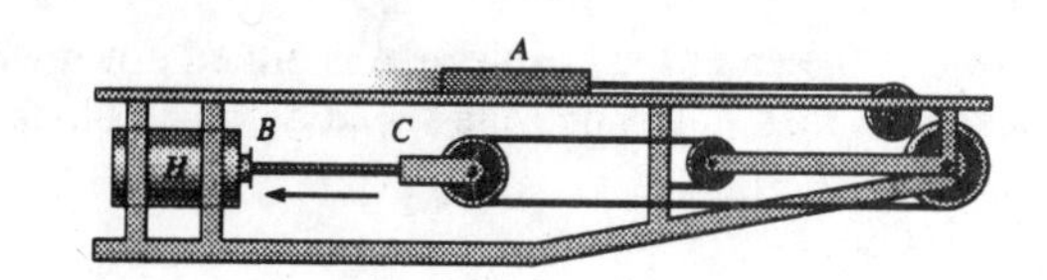

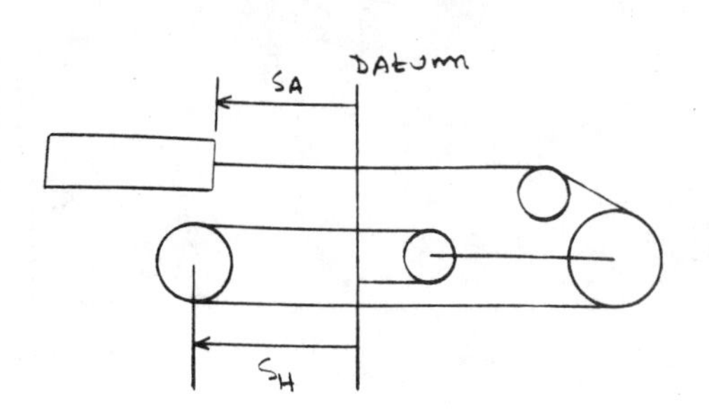

$$2s_H + s_A = l$$

$$2\Delta s_H = -\Delta s_A$$

$$-\Delta s_A = 2(8)$$

$\Delta s_A = -16 \text{ in.} = 16 \text{ in.} \rightarrow$ **Ans**

12-178. If the hydraulic cylinder at H draws in rod BC at 2 ft/s, determine the speed of the slider at A.

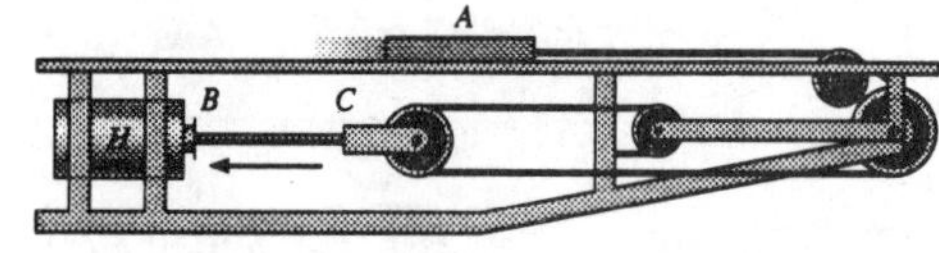

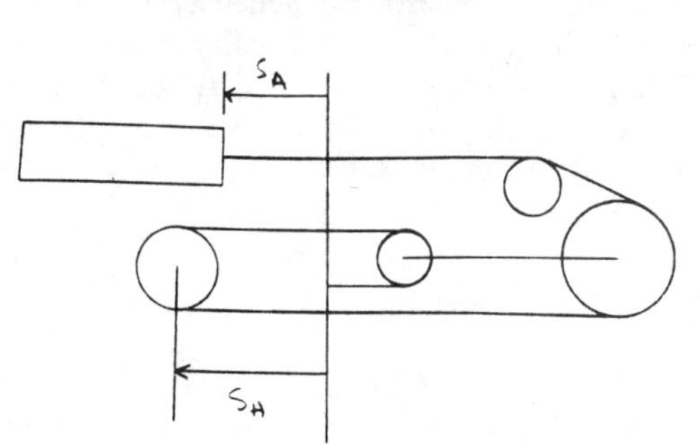

$$2s_H + s_A = l$$

$$2v_H = -v_A$$

$$2(2) = -v_A$$

$v_A = -4 \text{ ft/s} = 4 \text{ ft/s} \rightarrow$ **Ans**

12-179. The cable at B is pulled downwards at 4 ft/s, and is slowing at 2 ft/s². Determine the velocity and acceleration of block A at this instant.

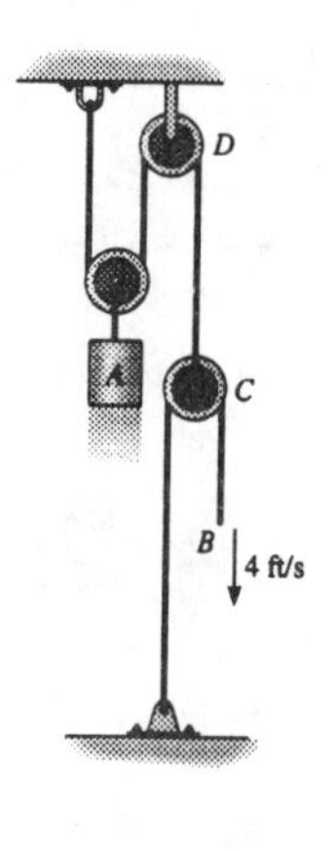

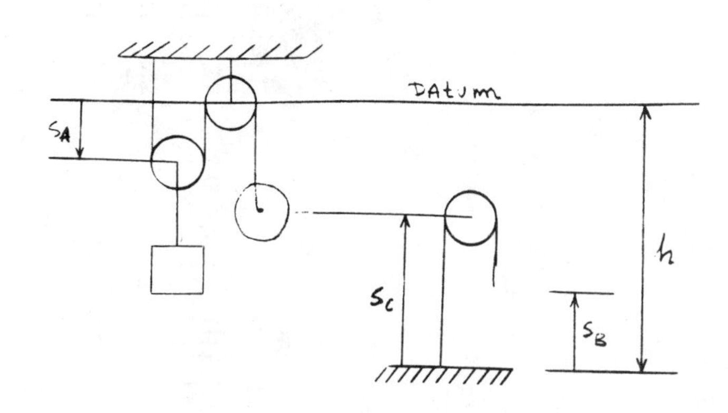

$$2s_A + (h - s_C) = l$$

$$2v_A = v_C$$

$$s_C + (s_C - s_B) = l$$

$$2v_C = v_B$$

$$v_B = 4v_A$$

$$a_B = 4a_A$$

Thus,

$$-4 = 4v_A$$

$v_A = -1 \text{ ft/s} = 1 \text{ ft/s} \uparrow$ **Ans**

$$2 = 4a_A$$

$a_A = 0.5 \text{ ft/s} = 0.5 \text{ ft/s}^2 \downarrow$ **Ans**

***12-180.** The hoist is used to lift the load at *D*. If the end *A* of the chain is traveling downward at $v_A = 5$ ft/s and the end *B* is traveling upward at $v_B = 2$ ft/s, determine the velocity of the load at *D*.

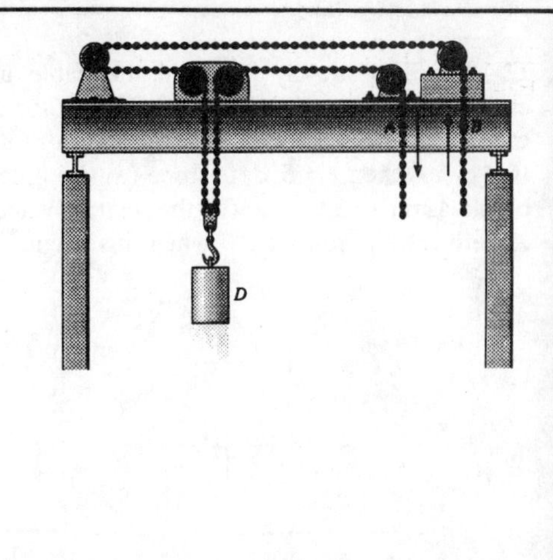

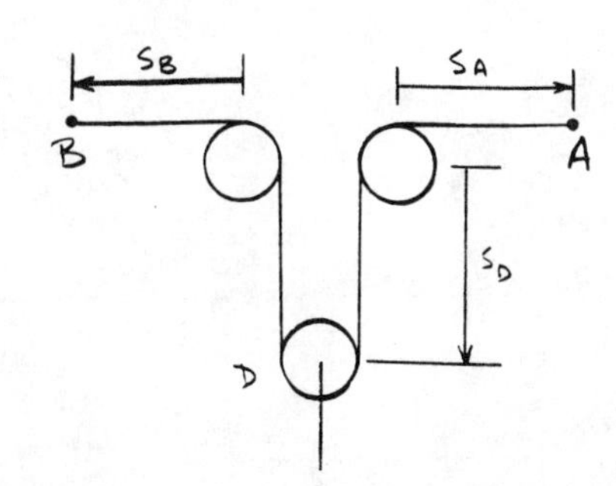

$$s_A + s_B + 2s_D = l$$

$$v_A + v_B + 2v_D = 0$$

$$5 - 2 + 2v_D = 0$$

$v_D = -1.5 \text{ ft/s} = 1.5 \text{ ft/s} \uparrow$ **Ans**

12-181. If block *A* is moving downward with a speed of 4 ft/s while *C* is moving up at 2 ft/s, determine the speed of block *B*.

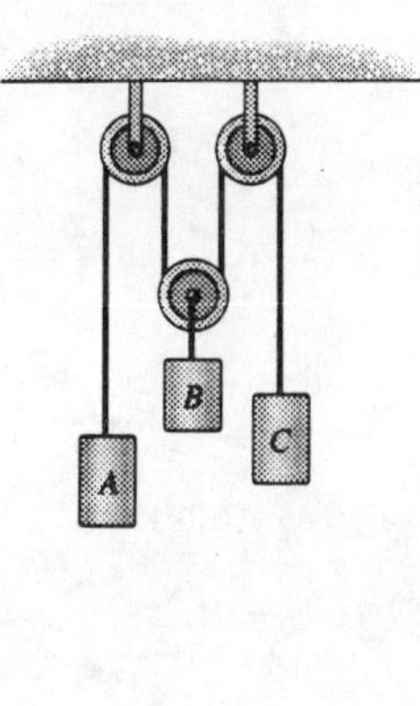

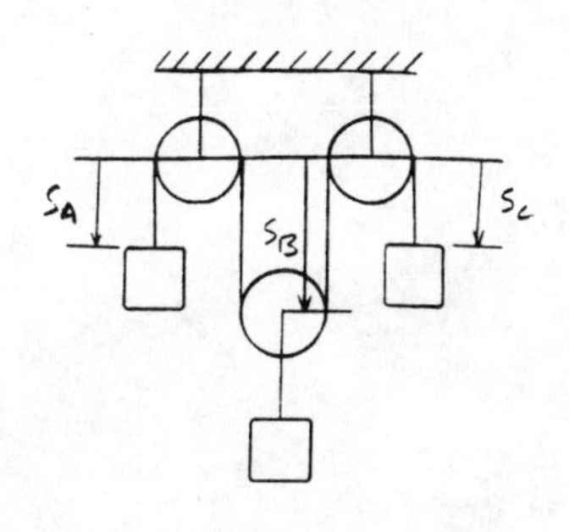

$$s_A + 2s_B + s_C = l$$

$$v_A + 2v_B + v_C = 0$$

$$4 + 2v_B - 2 = 0$$

$v_B = -1 \text{ ft/s} = 1 \text{ ft/s} \uparrow$ **Ans**

12-182. If block *A* is moving downward at 6 ft/s while block *C* is moving down at 18 ft/s, determine the relative velocity of block *B* with respect to *C*.

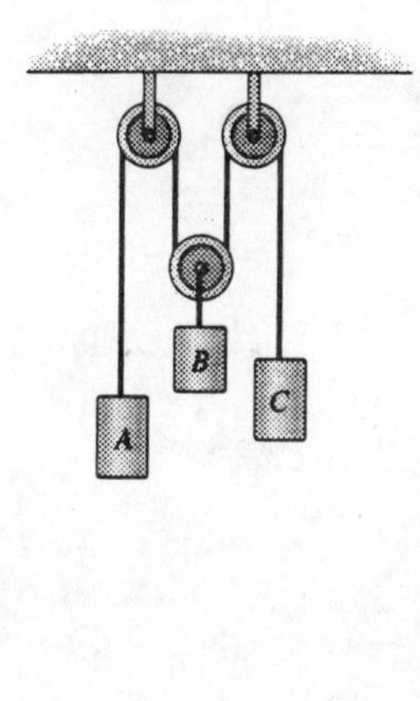

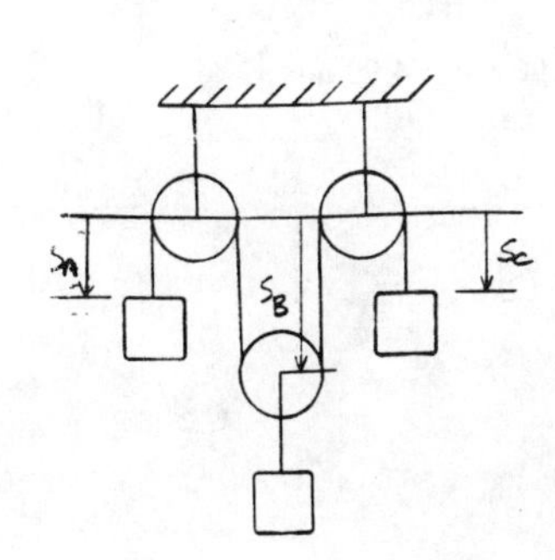

$$s_A + 2s_B + s_C = l$$

$$v_A + 2v_B + v_C = 0$$

$$6 + 2v_B + 18 = 0$$

$$v_B = -12 \text{ ft/s} = 12 \text{ ft/s} \uparrow$$

$$\mathbf{v}_B = \mathbf{v}_C + \mathbf{v}_{B/C}$$

$$[12 \uparrow] = [18 \downarrow] + [v_{B/C} \uparrow]$$

$v_{B/C} = 30 \text{ ft/s} \uparrow$ **Ans**

12-183. The motor draws in the cable at C with a constant velocity of $v_C = 4$ m/s. The motor draws in the cable at D with a constant acceleration of $a_D = 8$ m/s^2. If $v_D = 0$ when $t = 0$, determine (a) the time needed for block A to rise 3 m, and (b) the relative velocity of block A with respect to block B when this occurs.

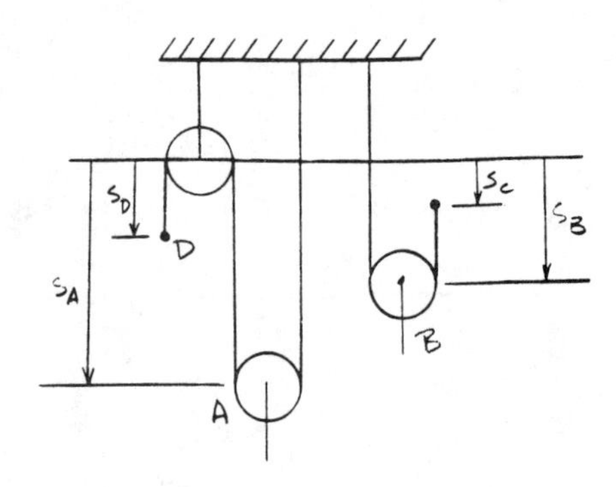

(a) $a_D = 8$ m/s^2

$v_D = 8\,t$

$s_D = 4\,t^2$

$s_D + 2s_A = l$

$\Delta s_D = -2\Delta s_A$ (1)

$\Delta s_A = -2\,t^2$

$-3 = -2\,t^2$

$t = 1.2247 = 1.22$ s **Ans**

(b) $v_A = \dot{s}_A = -4\,t = -4(1.2247) = -4.90$ m/s $= 4.90$ m/s $\uparrow$

$s_B + (s_B - s_C) = l'$

$2v_B = v_C = -4$

$v_B = -2$ m/s $= 2$ m/s $\uparrow$

$(+\downarrow)$ $\mathbf{v}_A = \mathbf{v}_B + \mathbf{v}_{A/B}$

$-4.90 = -2 + v_{A/B}$

$v_{A/B} = -2.90$ m/s $= 2.90$ m/s $\uparrow$ **Ans**

***12-184.** If motors at A and B draw in their attached cables with an acceleration of $a = (0.2t)$ m/s^2, where t is in seconds, determine the speed of the block when it reaches a height of $h = 4$ m, starting from rest. Also, how much time does it take to reach this height?

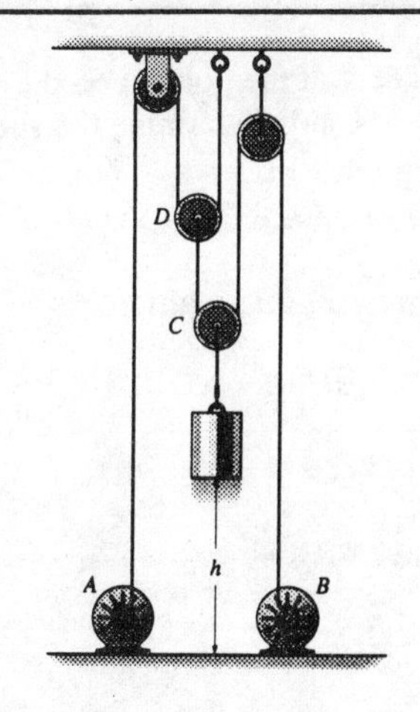

$s_A + 2s_D = l$

$(s_C - s_D) + s_C + s_B = l'$

$\Delta s_A = -2\Delta s_D \qquad 2\Delta s_C - \Delta s_D + \Delta s_B = 0$

If $\Delta s_C = -4$ and $\Delta s_A = \Delta s_B$, then,

$\Delta s_A = -2\Delta s_D \qquad 2(-4) - \Delta s_D + \Delta s_A = 0$

$\Delta s_D = -2.67$ m $\qquad \Delta s_A = \Delta s_B = 5.33$ m

Thus,

$v_A = -2v_D \qquad (1)$

$2v_C - v_D + v_B = 0 \qquad (2)$

$a = 0.2t$

$dv = a\,dt$

$\int_0^v dv = \int_0^t 0.2t\,dt$

$v = 0.1t^2$

$ds = v\,dt$

$\int_0^s ds = \int_0^t 0.1t^2\,dt$

$s = \frac{0.1}{3}t^3 = 5.33$

$t = 5.428$ s $= 5.43$ s **Ans**

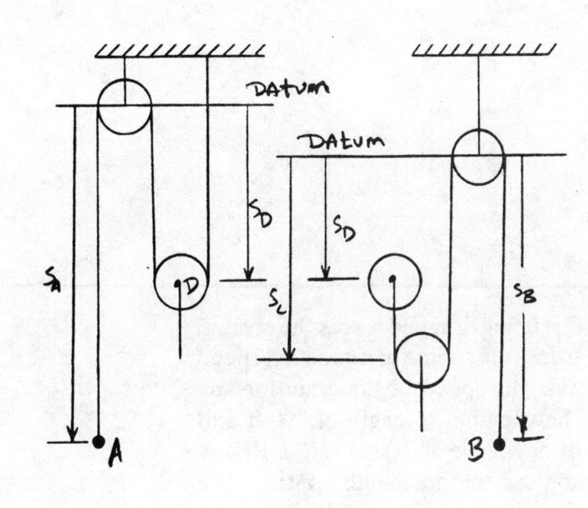

$v = 0.1(5.428)^2 = 2.947$ m/s

$v_A = v_B = 2.947$ m/s

Thus, from Eqs. (1) and (2) :

$2.947 = -2v_D$

$v_D = -1.474$

$2v_C - (-1.474) + 2.947 = 0$

$v_C = -2.21$ m/s $=$ 2.21 m/s $\uparrow$ **Ans**

12-185. If the point A on the cable is moving upward at $v_A = 14$ m/s, determine the speed of block B.

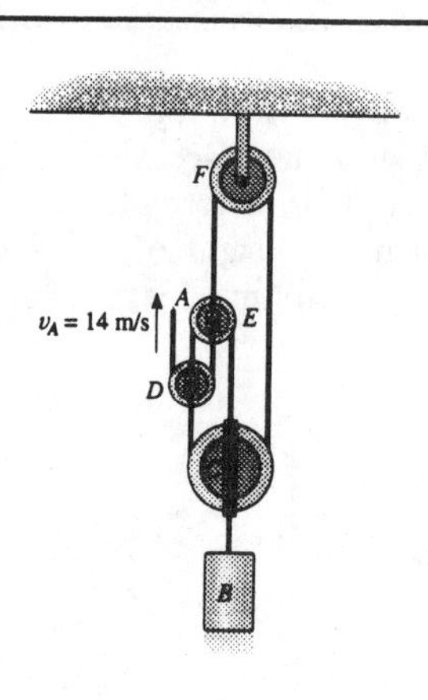

$s_E + s_C + (s_C - s_D) = l$

$(s_D - s_E) + (s_C - s_E) = l'$

$(s_D - s_E) + (s_D - s_A) = l''$

$v_E + 2v_C = v_D$

$2v_E - v_C = v_D$

$v_E + v_A = 2v_D$

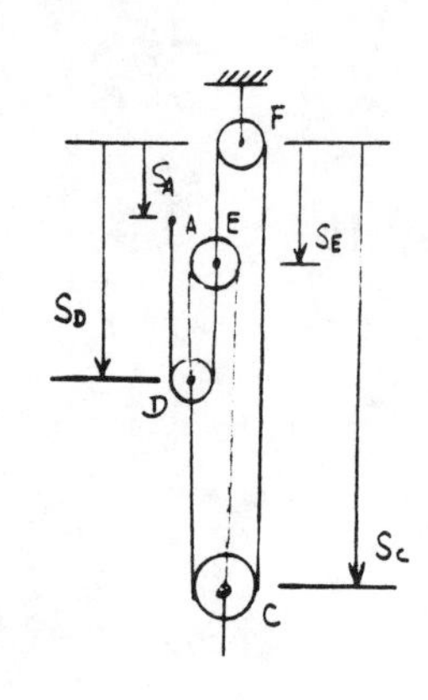

Thus

$v_E + 2v_C = 2v_E - v_C$

$3v_C = v_E$

$2v_E - v_C = \frac{v_E}{2} + \frac{v_A}{2}$

$3v_E = 2v_C + v_A$

$3(3v_C) = 2v_C + v_A$

$v_A = 7v_C$

If $v_A = -14$ m/s,

$v_B = v_C = -2 \text{ m/s} = 2 \text{ m/s} \uparrow$ **Ans**

***12-186.** The crate C is being dragged across the ground by the truck T. If the truck is traveling at a constant speed of $v_T = 6$ ft/s, determine the speed of the crate for any angle θ of the rope. The rope has a length of 100 ft and passes over a pulley of negligible size at A. *Hint:* Relate the coordinates x_T and x_C to the length of the rope and take the time derivative. Then substitute the trigonometric relation between x_C and θ.

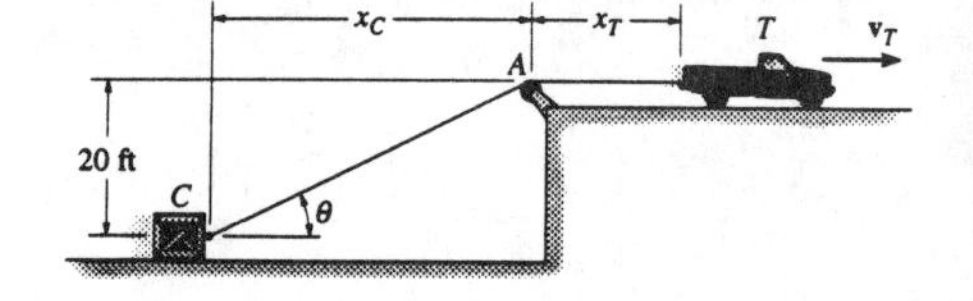

$\sqrt{(20)^2 + x_C^2} + x_T = l = 100$

$\frac{1}{2}\left((20)^2 + (x_C)^2\right)^{-\frac{1}{2}}\left(2x_C\dot{x}_C\right) + \dot{x}_T = 0$

Since $\dot{x}_T = v_T = 6$ ft/s, $v_C = \dot{x}_C$, and

$x_C = 20 \operatorname{ctn}\theta$

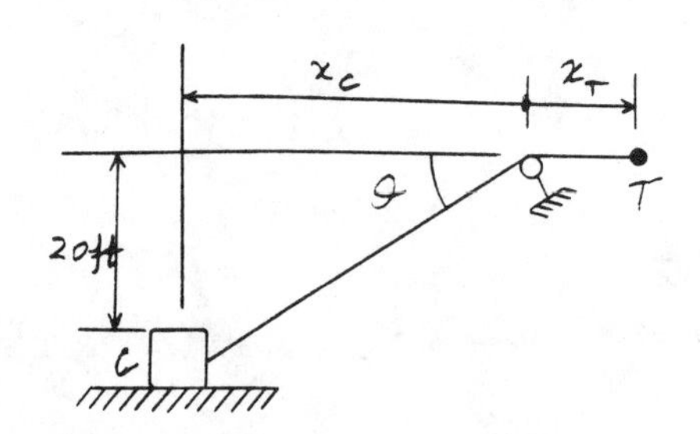

Then,

$\frac{(20 \operatorname{ctn}\theta) v_C}{(400 + 400 \operatorname{ctn}^2\theta)^{\frac{1}{2}}} = -6$

Since $1 + \operatorname{ctn}^2\theta = \csc^2\theta$,

$\left(\frac{\operatorname{ctn}\theta}{\csc\theta}\right) v_C = \cos\theta\, v_C = -6$

$v_C = -6 \sec\theta = (6 \sec\theta) \text{ ft/s} \rightarrow$ **Ans**

12-187. The motion of the collar at A is controlled by a motor at B such that when the collar is at $s_A = 3$ ft it is moving upwards at 2 ft/s and slowing down at 1 ft/s^2. Determine the velocity and acceleration of the cable as it is drawn into the motor at B at this instant.

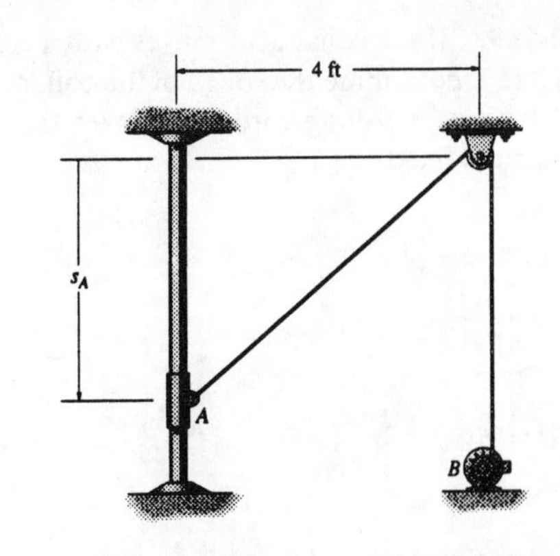

$$\sqrt{s_A^2 + 4^2} + s_B = l$$

$$\frac{1}{2}(s_A^2 + 16)^{-\frac{1}{2}}(2s_A)\dot{s}_A + \dot{s}_B = 0$$

$$\dot{s}_B = -s_A\dot{s}_A\left(s_A^2 + 16\right)^{-\frac{1}{2}}$$

$$\ddot{s}_B = -\left[\left(\dot{s}_A\right)^2\left(s_A^2 + 16\right)^{-\frac{1}{2}} + s_A\ddot{s}_A\left(s_A^2 + 16\right)^{-\frac{1}{2}} + s_A\dot{s}_A\left(-\frac{1}{2}\right)\left(s_A^2 + 16\right)^{-\frac{3}{2}}\left(2s_A\dot{s}_A\right)\right]$$

$$\ddot{s}_B = \frac{\left(s_A\dot{s}_A\right)^2}{\left(s_A^2 + 16\right)^{\frac{3}{2}}} - \frac{\left(\dot{s}_A\right)^2 + s_A\ddot{s}_A}{\left(s_A^2 + 16\right)^{\frac{1}{2}}}$$

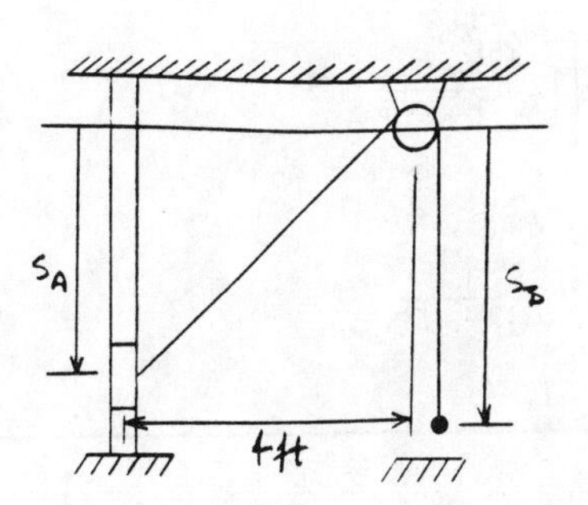

Evaluating these equations :

$$\dot{s}_B = -3(-2)\left((3)^2 + 16\right)^{-\frac{1}{2}} = 1.20 \text{ ft/s} \downarrow$$ **Ans**

$$\ddot{s}_B = \frac{((3)(-2))^2}{((3)^2 + 16)^{\frac{3}{2}}} - \frac{(-2)^2 + 3(1)}{((3)^2 + 16)^{\frac{1}{2}}} = -1.11 \text{ ft/s}^2 = 1.11 \text{ ft/s}^2 \uparrow$$ **Ans**

***12-188.** The roller at A is moving upward with a velocity of $v_A = 3$ ft/s and has an acceleration of $a_A = 4$ ft/s^2 when $s_A = 4$ ft. Determine the velocity and acceleration of block B at this instant.

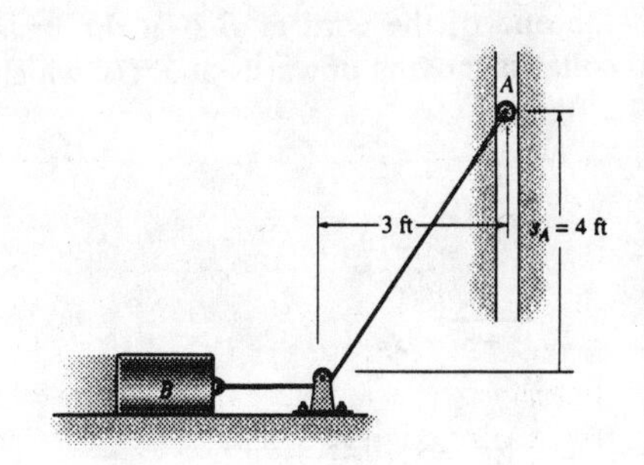

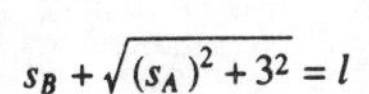

$$s_B + \sqrt{(s_A)^2 + 3^2} = l$$

$$\dot{s}_B + \frac{1}{2}\left[(s_A)^2 + 3^2\right]^{-\frac{1}{2}}(2s_A)\dot{s}_A = 0$$

$$\dot{s}_B + \left[s_A^2 + 9\right]^{-\frac{1}{2}}\left(s_A\dot{s}_A\right) = 0$$

$$\ddot{s}_B - \left[(s_A)^2 + 9\right]^{-\frac{3}{2}}\left(s_A^2\dot{s}_A^2\right) + \left[s_A^2 + 9\right]^{-\frac{1}{2}}\left(\dot{s}_A^2\right) + \left[s_A^2 + 9\right]^{-\frac{1}{2}}\left(s_A\ddot{s}_A\right) = 0$$

At $s_A = 4$ ft, $\dot{s}_A = 3$ ft/s, $\ddot{s}_A = 4$ ft/s^2

$$\dot{s}_B + \left(\frac{1}{5}\right)(4)(3) = 0$$

$v_B = -2.4$ ft/s $= 2.40$ ft/s $\rightarrow$ **Ans**

$$\ddot{s}_B - \left(\frac{1}{5}\right)^3(4)^2(3)^2 + \left(\frac{1}{5}\right)(3)^2 + \left(\frac{1}{5}\right)(4)(4) = 0$$

$a_B = -3.85$ ft/s^2 $= 3.85$ ft/s^2 $\rightarrow$ **Ans**

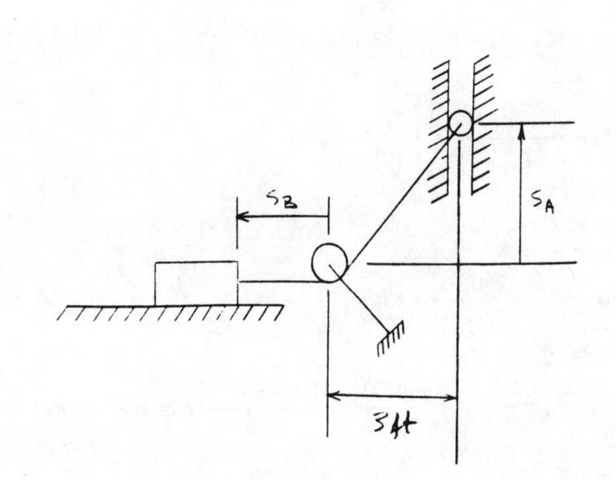

12-189. If the collar at A moves with a constant velocity of 2 ft/s, determine the speed of the collar at B when s_A = 12 ft. The 33-ft-long cord wraps over the small pulley at A and is fixed at C.

$$l_{AB} + l_{AC} = 33$$

$$\sqrt{s_A^2 + s_B^2} + \sqrt{s_A^2 + 5^2} = 33$$

$$\tfrac{1}{2}\left(s_A^2 + s_B^2\right)^{-\frac{1}{2}}\left(2s_A\dot{s}_A + 2s_B\dot{s}_B\right) + \tfrac{1}{2}\left(s_A^2 + 25\right)^{-\frac{1}{2}}\left(2s_A\dot{s}_A\right) = 0$$

$$-\frac{s_A\dot{s}_A}{\left(s_A^2 + s_B^2\right)^{\frac{1}{2}}} - \frac{s_B\dot{s}_B}{\left(s_A^2 + s_B^2\right)^{\frac{1}{2}}} = \frac{s_A\dot{s}_A}{\left(s_A^2 + 25\right)^{\frac{1}{2}}}$$

$$\dot{s}_B = -\frac{s_A\dot{s}_A}{s_B}\left[1 + \left(\frac{s_A^2 + s_B^2}{s_A^2 + 25}\right)^{\frac{1}{2}}\right]$$

When $s_A = 12$,

$$\sqrt{(12)^2 + s_B^2} + \sqrt{(12)^2 + 5^2} = 33$$

$$s_B = 16 \text{ ft}$$

$$\dot{s}_B = -\frac{12(2)}{16}\left[1 + \left(\frac{(12)^2 + (16)^2}{(12)^2 + 25}\right)^{\frac{1}{2}}\right] = -3.81 \text{ ft/s}$$

Thus,

$v_B = 3.81$ ft/s ↑ **Ans**

12-190. The cord is attached to the pin at C and passes over the two pulleys at A and D. The pulley at A is attached to the smooth collar that travels along the vertical rod. Determine the velocity and acceleration of the end of the cord at B if at the instant s_A = 4 ft the collar is moving upwards at 5 ft/s, which is decreasing at 2 ft/s².

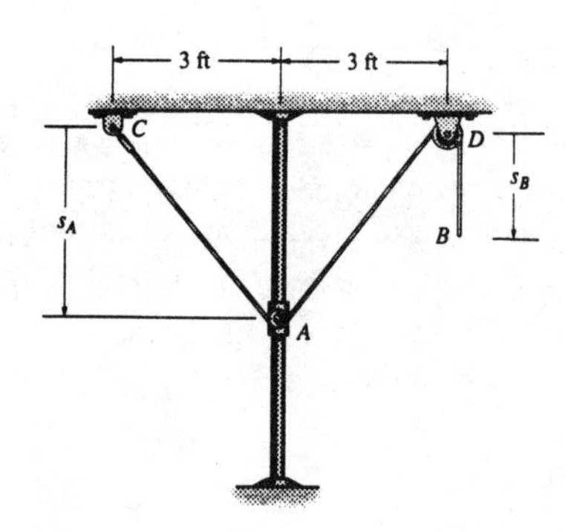

$$2\sqrt{s_A^2 + 3^2} + s_B = l$$

$$2\left(\frac{1}{2}\right)\left(s_A^2 + 9\right)^{-\frac{1}{2}}\left(2s_A\dot{s}_A\right) + \dot{s}_B = 0$$

$$\dot{s}_B = -\frac{2s_A\dot{s}_A}{\left(s_A^2 + 9\right)^{\frac{1}{2}}}$$

$$\ddot{s}_B = -2\dot{s}_A^2\left(s_A^2 + 9\right)^{-\frac{1}{2}} - \left(2s_A\ddot{s}_A\right)\left(s_A^2 + 9\right)^{-\frac{1}{2}} - \left(2s_A\dot{s}_A\right)\left[\left(-\frac{1}{2}\right)\left(s_A^2 + 9\right)^{-\frac{3}{2}}\left(2s_A\dot{s}_A\right)\right]$$

$$\ddot{s}_B = -\frac{2\left(\dot{s}_A^2 + s_A\ddot{s}_A\right)}{\left(s_A^2 + 9\right)^{\frac{1}{2}}} + \frac{2\left(s_A\dot{s}_A\right)^2}{\left(s_A^2 + 9\right)^{\frac{3}{2}}}$$

At $s_A = 4$ ft,

$$v_B = \dot{s}_B = -\frac{2(4)(-5)}{(4^2 + 9)^{\frac{1}{2}}} = 8 \text{ ft/s} \uparrow$$ **Ans**

$$a_B = \ddot{s}_B = -\frac{2\left[(-5)^2 + (4)(2)\right]}{(4^2 + 9)^{\frac{1}{2}}} + \frac{2[(4)(-5)]^2}{(4^2 + 9)^{\frac{3}{2}}} = -6.80 \text{ ft/s}^2 = 6.80 \text{ ft/s}^2 \uparrow$$ **Ans**

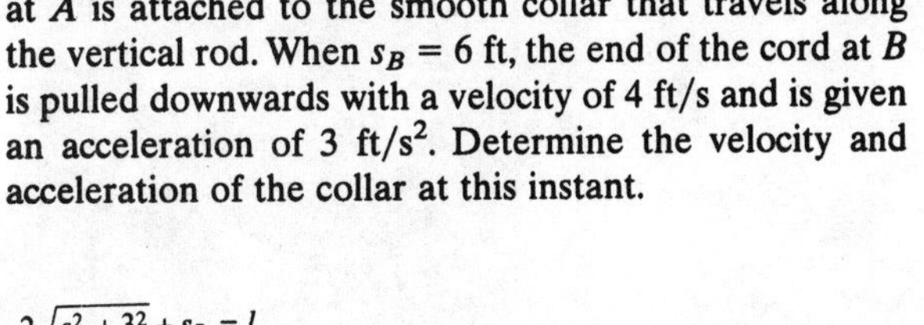
12-191. The 16-ft-long cord is attached to the pin at C and passes over the two pulleys at A and D. The pulley at A is attached to the smooth collar that travels along the vertical rod. When $s_B = 6$ ft, the end of the cord at B is pulled downwards with a velocity of 4 ft/s and is given an acceleration of 3 ft/s^2. Determine the velocity and acceleration of the collar at this instant.

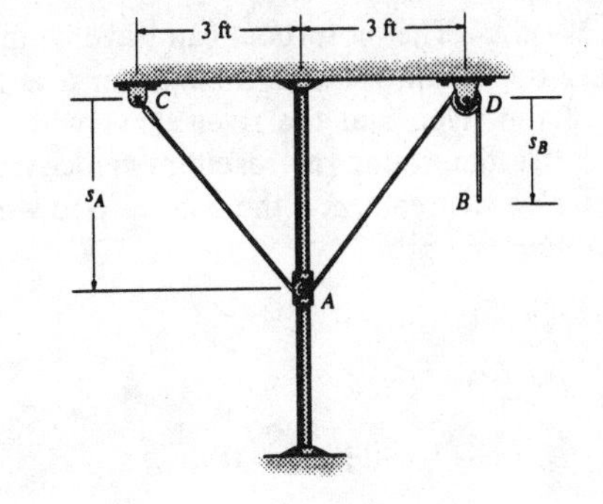

$$2\sqrt{s_A^2 + 3^2} + s_B = l$$

$$2\left(\frac{1}{2}\right)\left(s_A^2 + 9\right)^{-\frac{1}{2}}\left(2s_A\dot{s}_A\right) + \dot{s}_B = 0$$

$$\dot{s}_B = -\frac{2s_A\dot{s}_A}{\left(s_A^2 + 9\right)^{\frac{1}{2}}}$$

$$\ddot{s}_B = -2\dot{s}_A^2\left(s_A^2 + 9\right)^{-\frac{1}{2}} - \left(2s_A\ddot{s}_A\right)\left(s_A^2 + 9\right)^{-\frac{1}{2}} - \left(2s_A\dot{s}_A\right)\left[\left(-\frac{1}{2}\right)\left(s_A^2 + 9\right)^{-\frac{3}{2}}\left(2s_A\dot{s}_A\right)\right]$$

$$\ddot{s}_B = -\frac{2\left(\dot{s}_A^2 + s_A\ddot{s}_A\right)}{\left(s_A^2 + 9\right)^{\frac{1}{2}}} + \frac{2\left(s_A\dot{s}_A\right)^2}{\left(s_A^2 + 9\right)^{\frac{3}{2}}}$$

At $s_B = 6$ ft, $\dot{s}_B = 4$ ft/s, $\ddot{s}_B = 3$ ft/s^2

$$2\sqrt{s_A^2 + 3^2} + 6 = 16$$

$$s_A = 4 \text{ ft}$$

$$4 = -\frac{2(4)\left(\dot{s}_A\right)}{\left(4^2 + 9\right)^{\frac{1}{2}}}$$

$v_A = \dot{s}_A = -2.5$ ft/s $= 2.5$ ft/s $\uparrow$ **Ans**

$$3 = -\frac{2\left[(-2.5)^2 + 4\left(\ddot{s}_A\right)\right]}{\left(4^2 + 9\right)^{\frac{1}{2}}} + \frac{2\left[4(-2.5)\right]^2}{\left(4^2 + 9\right)^{\frac{3}{2}}}$$

$a_A = \ddot{s}_A = -2.4375 = 2.44$ ft/s^2 $\uparrow$ **Ans**

***12-192.** Collars A and B are connected to the cord that passes over the small pulley at C. When A is located at D, B is 24 ft to the left of D. If A moves at a constant speed of 2 ft/s to the right, determine the speed of B when A is 4 ft to the right of D.

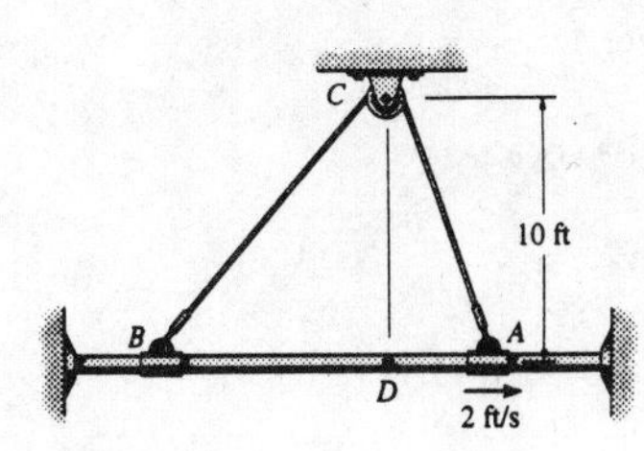

$$l = \sqrt{(24)^2 + (10)^2} + 10 = 36 \text{ ft}$$

$$\sqrt{(10)^2 + s_B^2} + \sqrt{(10)^2 + s_A^2} = 36$$

$$\frac{1}{2}\left(100 + s_B^2\right)^{-\frac{1}{2}}\left(2s_B\dot{s}_B\right) + \frac{1}{2}\left(100 + s_A^2\right)^{-\frac{1}{2}}\left(2s_A\dot{s}_A\right) = 0$$

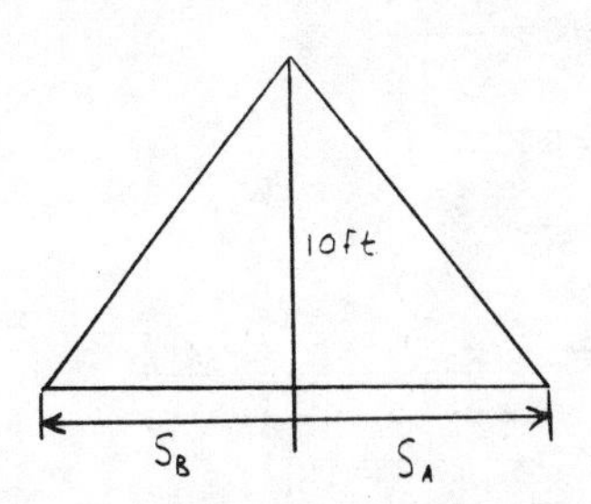

$$\dot{s}_B = -\left(\frac{s_A\dot{s}_A}{s_B}\right)\left(\frac{100 + s_B^2}{100 + s_A^2}\right)^{\frac{1}{2}}$$

At $s_A = 4$,

$$\sqrt{(10)^2 + s_B^2} + \sqrt{(10)^2 + (4)^2} = 36$$

$$s_B = 23.163 \text{ ft}$$

Thus,

$$\dot{s}_B = -\left(\frac{4(2)}{23.163}\right)\left(\frac{100 + (23.163)^2}{100 + 4^2}\right)^{\frac{1}{2}} = -0.809 \text{ ft/s} = 0.809 \text{ ft/s} \rightarrow$$ **Ans**

12-193. The motorboat can travel with a speed of $v_B = 12$ ft/s in still water. If it heads for C at the opposite side of the river, and the river flows with a speed of $v_R = 5$ ft/s, determine the resultant velocity of the boat. How far downstream, d, is the boat carried when it reaches the other side at D?

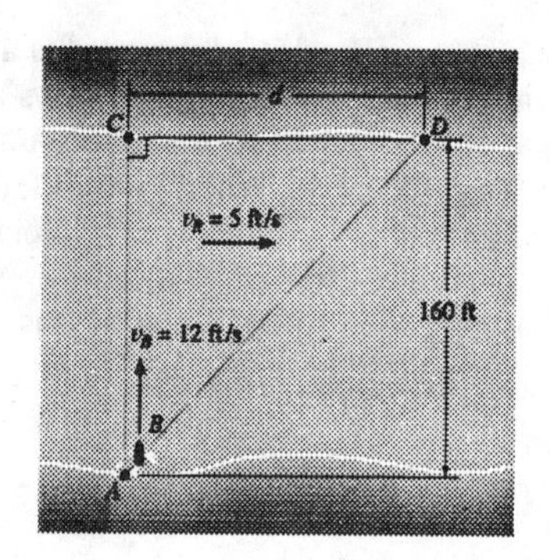

$\mathbf{v}_B = \mathbf{v}_R + \mathbf{v}_{B/R}$

$v_B \sin\theta \mathbf{i} + v_B \cos\theta \mathbf{j} = 5\mathbf{i} + 12\mathbf{j}$

$v_B \sin\theta = 5$

$v_B \cos\theta = 12$

Solving :

$v_B = 13$ ft/s **Ans**

$\theta = 22.6°$ **Ans**

$d = 160 \tan 22.6° = 66.7$ ft **Ans**

12-194. The motor at C pulls in the cable with an acceleration $a_C = (3t^2)$ m/s^2, where t is in seconds. The motor at D draws in its cable at $a_D = 5$ m/s^2. If both motors start at the same instant from rest when $d = 3$ m, determine (a) the time needed for $d = 0$, and (b) the relative velocity of block A with respect to block B when this occurs.

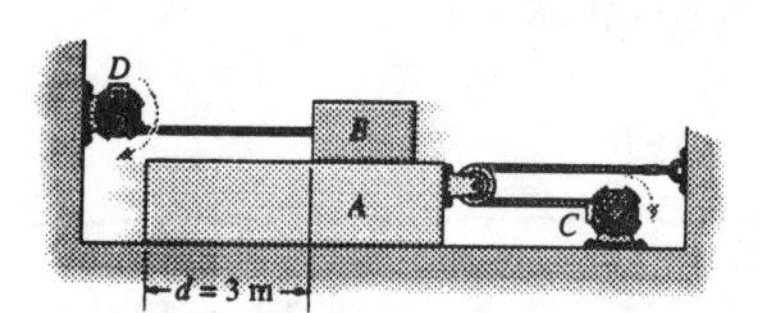

For A :

$s_A + (s_A - s_C) = l$

$2v_A = v_C$

$2a_A = a_C = -3t^2$

$a_A = -1.5t^2 = 1.5t^2 \rightarrow$

$v_A = 0.5t^3 \rightarrow$

$s_A = 0.125t^4 \rightarrow$

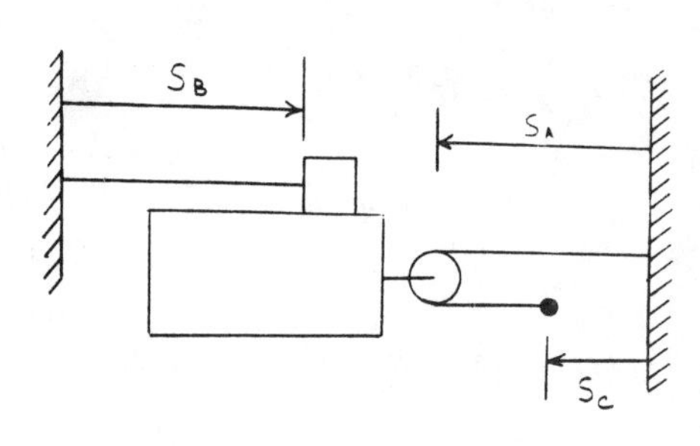

For B :

$a_B = 5$ m/s^2 $\leftarrow$

$v_B = 5t \leftarrow$

$s_B = 2.5t^2 \leftarrow$

Require $s_A + s_B = d$

$0.125t^4 + 2.5t^2 = 3$

Set $u = t^2$ $0.125u^2 + 2.5u = 3$

The positive root is $u = 1.1355$. Thus,

$t = 1.0656 = 1.07$ s **Ans**

$v_A = 0.5(1.0656)^3 = 0.6050$

$v_B = 5(1.0656) = 5.3281$ m/s

$\mathbf{v}_A = \mathbf{v}_B + \mathbf{v}_{A/B}$

$0.6050\mathbf{i} = -5.3281\mathbf{i} + v_{A/B}\mathbf{i}$

$v_{A/B} = 5.93$ m/s $\rightarrow$ **Ans**

12-195. Sand falls from rest 0.5 m vertically onto a chute. If the sand then slides with a velocity of $v_C = 2$ m/s down the chute, determine the relative velocity of the sand just falling on the chute at A with respect to the sand sliding down the chute. The chute is inclined at an angle of 40° with the horizontal.

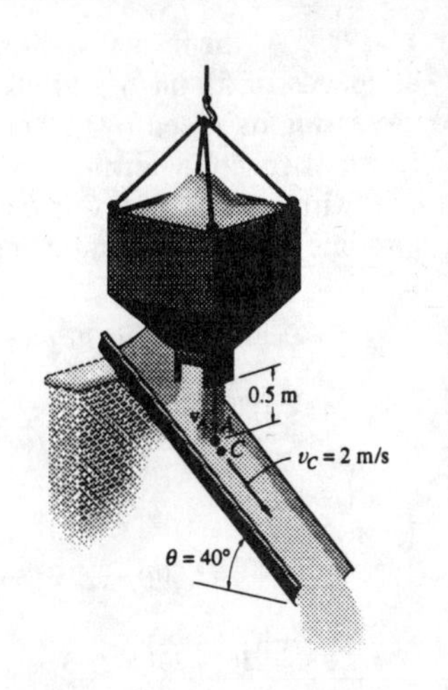

$$(+\uparrow) \quad v_A^2 = (v_A)_1^2 + 2a_c(s_A - s_{A1})$$

$$v_A^2 = 0 + 2(-9.81)(0.5 - 0)$$

$$v_A = -3.1321 \text{ m/s}$$

$$\mathbf{v}_A = \mathbf{v}_C + \mathbf{v}_{A/C}$$

$$-3.1321\mathbf{j} = 2\cos 40°\mathbf{i} - 2\sin 40°\mathbf{j} + (v_{A/C})_x\mathbf{i} + (v_{A/C})_y\mathbf{j}$$

$$0 = 2\cos 40° + (v_{A/C})_x$$

$$-3.1321 = -2\sin 40° + (v_{A/C})_y$$

Solving,

$$(v_{A/C})_x = -1.5321$$

$$(v_{A/C})_y = -1.8465$$

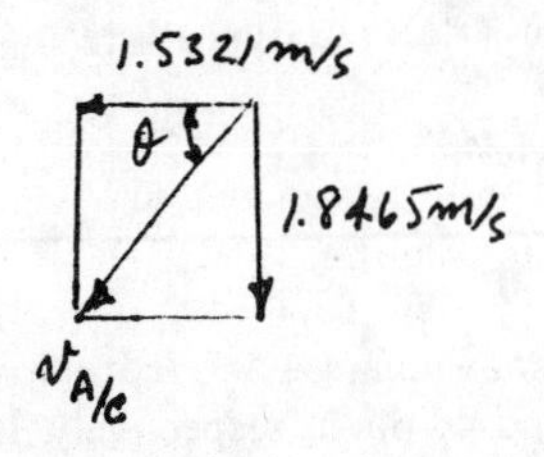

$$v_{A/C} = \sqrt{(-1.5321)^2 + (-1.8465)^2} = 2.40 \text{ m/s} \qquad \textbf{Ans}$$

$$\theta = \tan^{-1}\left(\frac{1.8465}{1.5321}\right) = 50.3° \swarrow \qquad \textbf{Ans}$$

***12-196.** Two planes, A and B, are flying at the same altitude. If their velocities are $v_A = 600$ km/h and $v_B = 700$ km/h, such that the angle between their straight-line courses is $\theta = 35°$, determine the velocity of plane B with respect to plane A.

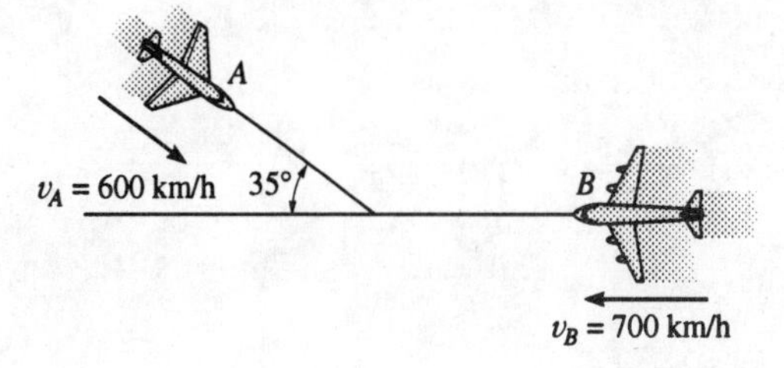

$$\mathbf{v}_A = 600\cos 35°\mathbf{i} - 600\sin 35°\mathbf{j} = \{491.5\mathbf{i} - 344.1\mathbf{j}\} \text{ km/h}$$

$$\mathbf{v}_B = \{-700\mathbf{i}\} \text{ km/h}$$

$$\mathbf{v}_{B/A} = \mathbf{v}_B - \mathbf{v}_A$$

$$= (-700\mathbf{i}) - (491.5\mathbf{i} - 344.1\mathbf{j}) = \{-1191\mathbf{i} + 344\mathbf{j}\} \text{ km/h}$$

$$v_{B/A} = \sqrt{(-1191)^2 + 344^2} = 1240 \text{ km/h} \qquad \textbf{Ans}$$

$$\theta = \tan^{-1}\frac{344}{1192} = 16.1° \qquad \textbf{Ans}$$

12-197. At the instant shown, cars A and B are traveling at speeds of 55 mi/h and 40 mi/h, respectively. If B is increasing its speed by 1200 mi/h^2, while A maintains a constant speed, determine the velocity and acceleration of B with respect to A. Car B moves along a curve having a radius of curvature of 0.5 mi.

$\mathbf{v}_B = -40\cos 30°\mathbf{i} + 40\sin 30°\mathbf{j} = \{-34.64\mathbf{i} + 20\mathbf{j}\}$ mi/h

$\mathbf{v}_A = \{-55\mathbf{i}\}$ mi/h

$\mathbf{v}_{B/A} = \mathbf{v}_B - \mathbf{v}_A$

$= (-34.64\mathbf{i} + 20\mathbf{j}) - (-55\mathbf{i}) = \{20.36\mathbf{i} + 20\mathbf{j}\}$ mi/h

$v_{B/A} = \sqrt{20.36^2 + 20^2} = 28.5$ mi/h **Ans**

$\theta = \tan^{-1}\dfrac{20}{20.36} = 44.5°$ **Ans**

$(a_B)_n = \dfrac{v_B^2}{\rho} = \dfrac{40^2}{0.5} = 3200$ mi/h^2 $\qquad (a_B)_t = 1200$ mi/h^2

$\mathbf{a}_B = (3200\cos 60° - 1200\cos 30°)\mathbf{i} + (3200\sin 60° + 1200\sin 30°)\mathbf{j}$

$= \{560.77\mathbf{i} + 3371.28\mathbf{j}\}$ mi/h^2

$\mathbf{a}_A = \mathbf{0}$

$\mathbf{a}_{B/A} = \mathbf{a}_B - \mathbf{a}_A$

$= \{560.77\mathbf{i} + 3371.28\mathbf{j}\} - \mathbf{0} = \{560.77\mathbf{i} + 3371.28\mathbf{j}\}$ mi/h^2

$a_{B/A} = \sqrt{(560.77)^2 + (3371.28)^2} = 3418$ mi/h^2 **Ans**

$\theta = \tan^{-1}\dfrac{3371.28}{560.77} = 80.6°$ **Ans**

12-198. At the instant shown, cars A and B are traveling at speeds of 55 mi/h and 40 mi/h, respectively. If B is decreasing its speed at 1500 mi/h^2 while A is increasing its speed at 800 mi/h^2, determine the acceleration of B with respect to A. Car B moves along a curve having a radius of curvature of 0.75 mi.

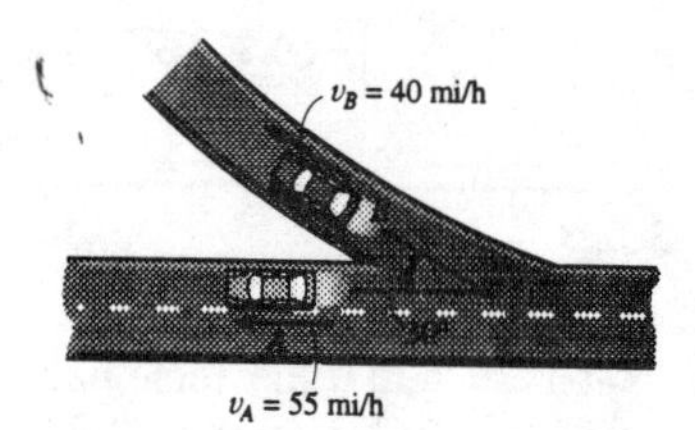

$(a_B)_n = \dfrac{(40)^2}{0.75} = 2133.33$ mi/h^2

$(a_B)_t = 1500$ mi/h^2

$\mathbf{a}_B = \mathbf{a}_A + \mathbf{a}_{B/A}$

$2133.33\sin 30°\mathbf{i} + 2133.33\cos 30°\mathbf{j} = -800\mathbf{i} + (a_{B/A})_x\mathbf{i} + (a_{B/A})_y\mathbf{j}$

$2133.33\sin 30° + 1500\cos 30° = -800 + (a_{B/A})_x$

$(a_{B/A})_x = 3165.705 \rightarrow$

$2133.33\cos 30° - 1500\sin 30° = (a_{B/A})_y$

$(a_{B/A})_y = 1097.521 \uparrow$

$(a_{B/A}) = \sqrt{(1097.521)^2 + (3165.705)^2}$

$a_{B/A} = 3351$ mi/h^2 **Ans**

$\theta = \tan^{-1}\left(\dfrac{1097.521}{3165.705}\right) = 19.1°$ **Ans**

12-199. At the instant shown, the car at A is traveling at 10 m/s around the curve while increasing its speed at 5 m/s^2. The car at B is traveling at 18.5 m/s along the straightaway and increasing its speed at 2 m/s^2. Determine the relative velocity and relative acceleration of A with respect to B at this instant.

$$\mathbf{v}_A = 10\cos 45°\mathbf{i} - 10\sin 45°\mathbf{j} = \{7.071\mathbf{i} - 7.071\mathbf{j}\} \text{ m/s}$$

$$\mathbf{v}_B = \{18.5\mathbf{i}\} \text{ m/s}$$

$$\mathbf{v}_{A/B} = \mathbf{v}_A - \mathbf{v}_B$$

$$= (7.071\mathbf{i} - 7.071\mathbf{j}) - 18.5\mathbf{i} = \{-11.429\mathbf{i} - 7.071\mathbf{j}\} \text{ m/s}$$

$$v_{A/B} = \sqrt{(-11.429)^2 + (-7.071)^2} = 13.4 \text{ m/s} \quad \textbf{Ans}$$

$$\theta = \tan^{-1}\frac{7.071}{11.429} = 31.7° \quad \textbf{Ans}$$

$$(a_A)_n = \frac{v_A^2}{\rho} = \frac{10^2}{100} = 1 \text{ m/s}^2 \qquad (a_A)_t = 5 \text{ m/s}^2$$

$$\mathbf{a}_A = (5\cos 45° - 1\cos 45°)\mathbf{i} + (-1\sin 45° - 5\sin 45°)\mathbf{j}$$

$$= \{2.828\mathbf{i} - 4.243\mathbf{j}\} \text{ m/s}^2$$

$$\mathbf{a}_B = \{2\mathbf{i}\} \text{ m/s}^2$$

$$\mathbf{a}_{A/B} = \mathbf{a}_A - \mathbf{a}_B$$

$$= (2.828\mathbf{i} - 4.243\mathbf{j}) - 2\mathbf{i} = \{0.828\mathbf{i} - 4.24\mathbf{j}\} \text{ m/s}^2$$

$$a_{A/B} = \sqrt{0.828^2 + (-4.243)^2} = 4.32 \text{ m/s}^2 \quad \textbf{Ans}$$

$$\theta = \tan^{-1}\frac{4.243}{0.828} = 79.0° \quad \textbf{Ans}$$

***12-200.** An aircraft carrier is traveling forward with a velocity of 50 km/h. At the instant shown, the plane at A has just taken off and has attained a forward horizontal air speed of 200 km/h, measured from still water. If the plane at B is traveling along the runway of the carrier at 175 km/h in the direction shown, determine the velocity of A with respect to B.

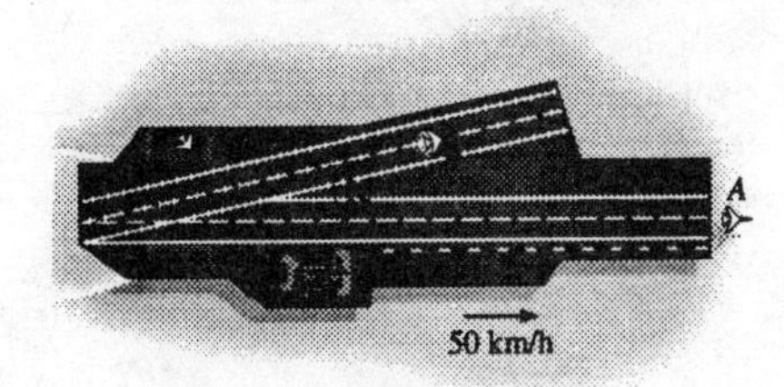

$$\mathbf{v}_B = \mathbf{v}_C + \mathbf{v}_{B/C}$$

$$\mathbf{v}_B = 50\mathbf{i} + 175\cos 15°\mathbf{i} + 175\sin 15°\mathbf{j} = 219.04\mathbf{i} + 45.293\mathbf{j}$$

$$\mathbf{v}_A = \mathbf{v}_B + \mathbf{v}_{A/B}$$

$$200\mathbf{i} = 219.04\mathbf{i} + 45.293\mathbf{j} + (v_{A/B})_x\mathbf{i} + (v_{A/B})_y\mathbf{j}$$

$$200 = 219.04 + (v_{A/B})_x$$

$$0 = 45.293 + (v_{A/B})_y$$

$$(v_{A/B})_x = -19.04$$

$$(v_{A/B})_y = -45.293$$

$$v_{A/B} = \sqrt{(-19.04)^2 + (-45.293)^2} = 49.1 \text{ km/h} \quad \textbf{Ans}$$

$$\theta = \tan^{-1}\left(\frac{45.293}{19.04}\right) = 67.2° \quad \textbf{Ans}$$

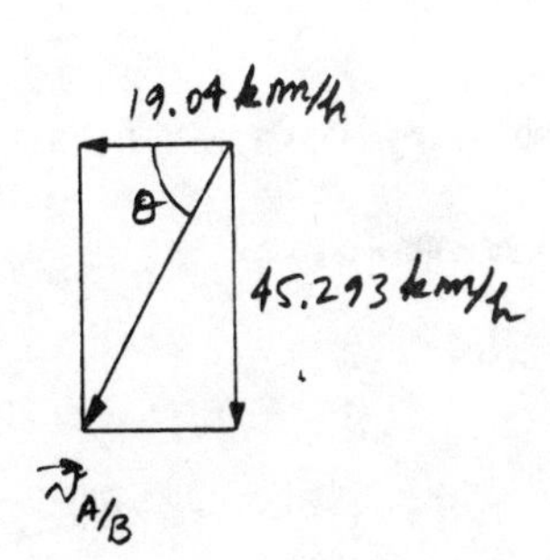

12-201. At the instant shown, cars A and B are traveling at speeds of 30 mi/h and 20 mi/h, respectively. If B is increasing its speed by 1200 mi/h^2, while A maintains a constant speed, determine the velocity and acceleration of B with respect to A.

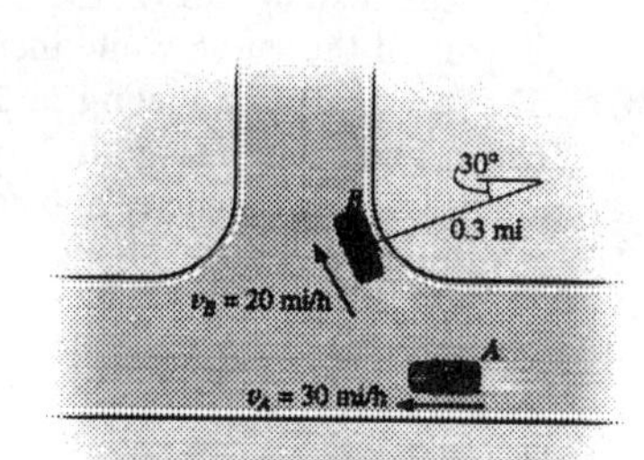

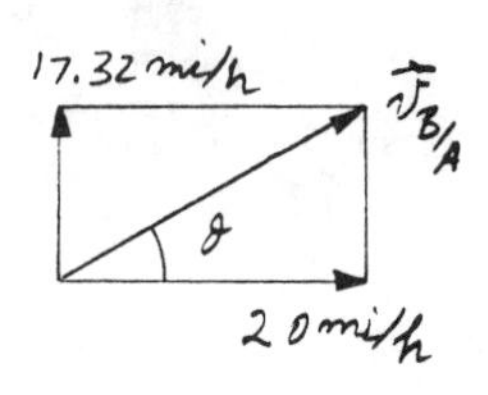

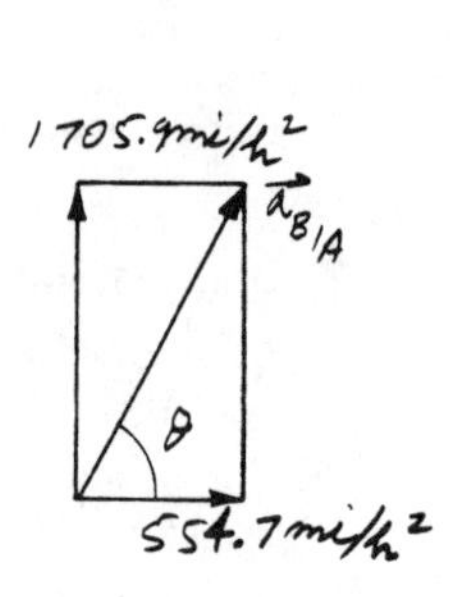

$\mathbf{v}_B = \mathbf{v}_A + \mathbf{v}_{B/A}$

$-20\sin 30°\mathbf{i} + 20\cos 30°\mathbf{j} = -30\mathbf{i} + (v_{B/A})_x\,\mathbf{i} + (v_{B/A})_y\,\mathbf{j}$

$-20\sin 30° = -30 + (v_{B/A})_x$

$20\cos 30° = (v_{B/A})_y$

Solving :

$(v_{B/A})_x = 20$

$(v_{B/A})_y = 17.32$

$v_{B/A} = \sqrt{(20)^2 + (17.32)^2} = 26.5$ mi/h **Ans**

$\theta = \tan^{-1}\left(\frac{17.32}{20}\right) = 40.9°$ ∠ **Ans**

$(a_B)_n = \frac{v^2}{\rho} = \frac{20^2}{0.3} = 1333.3$ mi/h

$\mathbf{a}_B = \mathbf{a}_A + \mathbf{a}_{B/A}$

$-1200\sin 30°\mathbf{i} + 1200\cos 30°\mathbf{j} + 1333.3\cos 30°\mathbf{i} + 1333.3\sin 30°\mathbf{j} = \mathbf{0} + (a_{B/A})_x\,\mathbf{i} + (a_{B/A})_y\,\mathbf{j}$

$-1200\sin 30° + 1333.3\cos 30° = (a_{B/A})_x$

$1200\cos 30° + 1333.3\sin 30° = (a_{B/A})_y$

Solving :

$(a_{B/A})_x = 554.7, \quad (a_{B/A})_y = 1705.9$

$a_{B/A} = \sqrt{(554.7)^2 + (1705.9)^2} = 1.79(10^3)$ mi/h^2 **Ans**

$\theta = \tan^{-1}\left(\frac{1705.9}{554.7}\right) = 72.0°$ ∠ **Ans**

12-202. At the instant shown, car A has a speed of 20 km/h, which is being increased at the rate of 300 km/h^2 as the car enters an expressway. At the same instant, car B is decelerating at 250 km/h^2 while traveling forward at 100 km/h. Determine the velocity and acceleration of car A with respect to car B.

$\mathbf{v}_A = \{-20\mathbf{j}\}$ km/h $\qquad \mathbf{v}_B = \{100\mathbf{j}\}$ km/h

$\mathbf{v}_{A/B} = \mathbf{v}_A - \mathbf{v}_B$

$= (-20\mathbf{j} - 100\mathbf{j}) = \{-120\mathbf{j}\}$ km/h

$v_{A/B} = 120$ km/h $\downarrow$ **Ans**

$$(a_A)_n = \frac{v_A^2}{\rho} = \frac{20^2}{0.1} = 4000 \text{ km/h}^2 \qquad (a_A)_t = 300 \text{ km/h}^2$$

$\mathbf{a}_A = -4000\mathbf{i} + (-300\mathbf{j})$

$= \{-4000\mathbf{i} - 300\mathbf{j}\}$ km/h^2

$\mathbf{a}_B = \{-250\mathbf{j}\}$ km/h^2

$\mathbf{a}_{A/B} = \mathbf{a}_A - \mathbf{a}_B$

$= (-4000\mathbf{i} - 300\mathbf{j}) - (-250\mathbf{j}) = \{-4000\mathbf{i} - 50\mathbf{j}\}$ km/h^2

$a_{A/B} = \sqrt{(-4000)^2 + (-50)^2} = 4000$ km/h^2 **Ans**

$\theta = \tan^{-1}\dfrac{50}{4000} = 0.716°$ **Ans**

12-203. At the instant shown car A is traveling with a velocity of 30 m/s and has an acceleration of 2 m/s^2 along the highway. At the same instant B is traveling on the trumpet interchange curve with a speed of 15 m/s, which is decreasing at 0.8 m/s^2. Determine the relative velocity and relative acceleration of B with respect to A at this instant.

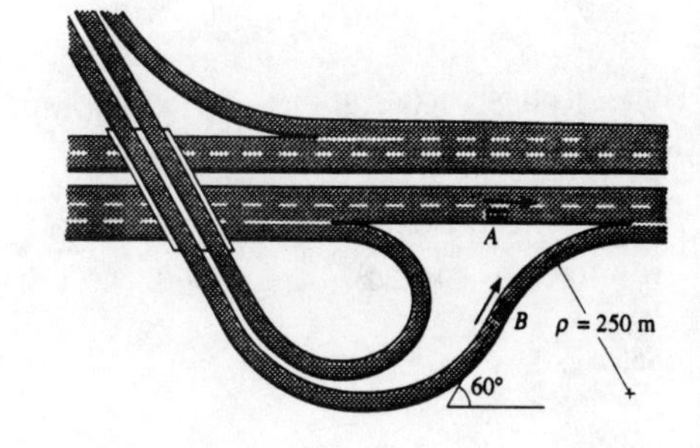

$\mathbf{v}_B = \mathbf{v}_A + \mathbf{v}_{B/A}$

$15\cos 60°\mathbf{i} + 15\sin 60°\mathbf{j} = 30\mathbf{i} + (v_{B/A})_x\mathbf{i} + (v_{B/A})_y\mathbf{j}$

$15\cos 60° = 30 + (v_{B/A})_x$

$15\sin 60° = 0 + (v_{B/A})_y$

$(v_{B/A})_x = -22.5 = 22.5$ m/s $\leftarrow$

$(v_{B/A})_y = 12.99$ m/s $\uparrow$

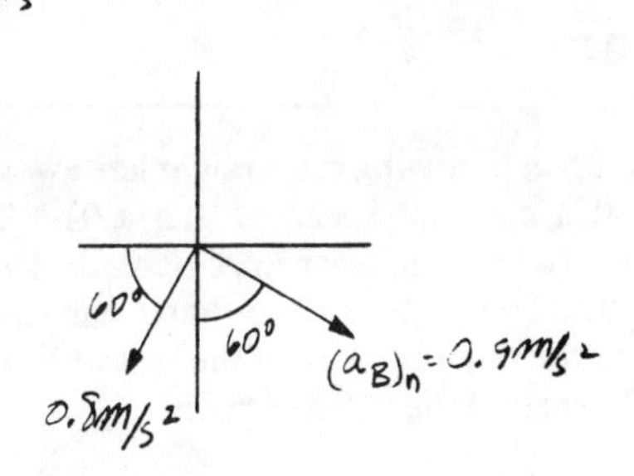

$v_{B/A} = \sqrt{(22.5)^2 + (12.99)^2} = 26.0$ m/s **Ans**

$\theta = \tan^{-1}\left(\dfrac{12.99}{22.5}\right) = 30°$ **Ans**

$$(a_B)_n = \frac{v^2}{\rho} = \frac{15^2}{250} = 0.9 \text{ m/s}^2$$

$\mathbf{a}_B = \mathbf{a}_A + \mathbf{a}_{B/A}$

$-0.8\cos 60°\mathbf{i} - 0.8\sin 60°\mathbf{j} + 0.9\sin 60°\mathbf{i} - 0.9\cos 60°\mathbf{j} = 2\mathbf{i} + (a_{B/A})_x\mathbf{i} + (a_{B/A})_y\mathbf{j}$

$-0.8\cos 60° + 0.9\sin 60° = 2 + (a_{B/A})_x$

$-0.8\sin 60° - 0.9\cos 60° = (a_{B/A})_y$

$(a_{B/A})_x = -1.6206$ ft/s^2 $= 1.6206$ m/s^2 $\leftarrow$

$(a_{B/A})_y = -1.1428$ ft/s^2 $= 1.1428$ m/s^2 $\downarrow$

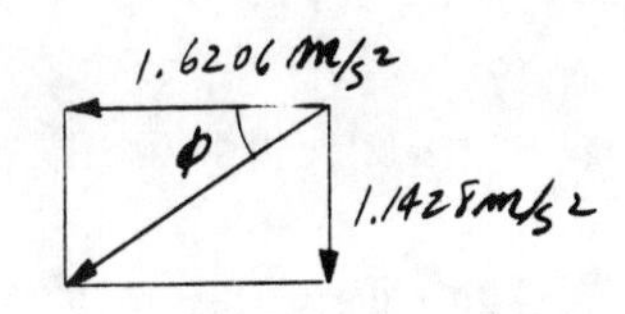

$a_{B/A} = \sqrt{(1.6206)^2 + (1.1428)^2} = 1.98$ m/s^2 **Ans**

$\phi = \tan^{-1}\left(\dfrac{1.1428}{1.6206}\right) = 35.2°$ **Ans**

***12-204.** The two cyclists A and B travel at the same constant speed v. Determine the speed of A with respect to B if A travels along the circular track, while B travels along the diameter of the circle.

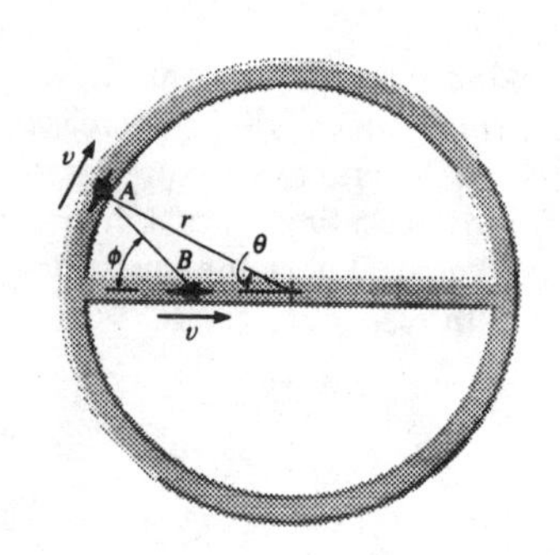

$\mathbf{v}_A = v\sin\theta\mathbf{i} + v\cos\theta\mathbf{j}$ $\qquad \mathbf{v}_B = v\,\mathbf{i}$

$\mathbf{v}_{A/B} = \mathbf{v}_A - \mathbf{v}_B$

$= (v\sin\theta\mathbf{i} + v\cos\theta\mathbf{j}) - v\,\mathbf{i}$

$= (v\sin\theta - v)\mathbf{i} + v\cos\theta\mathbf{j}$

$v_{A/B} = \sqrt{(v\sin\theta - v)^2 + (v\cos\theta)^2}$

$= \sqrt{2v^2 - 2v^2\sin\theta}$

$= v\sqrt{2(1 - \sin\theta)}$ **Ans**

12-205. The airplane has a speed relative to the wind of 100 mi/h. If the wind relative to the ground is 10 mi/h, determine the angle θ at which the plane must be directed in order to travel in the direction of the runway. Also, what is its speed relative to the runway?

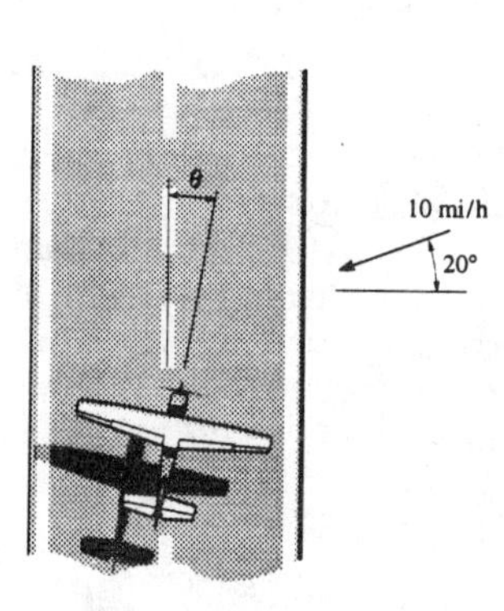

$\mathbf{v}_P = \mathbf{v}_{P/W} + \mathbf{v}_W$

$v_P\mathbf{j} = 100\sin\theta\mathbf{i} + 100\cos\theta\mathbf{j} - 10\cos 20°\mathbf{i} - 10\sin 20°\mathbf{j}$

$0 = 100\sin\theta - 10\cos 20°$

$v_P = 100\cos\theta - 10\sin 20°$

Solving,

$v_P = 96.1$ mi/h **Ans**

$\theta = 5.39°$ **Ans**

12-206. The boy A is moving in a straight line away from the building at a constant speed of 4 ft/s. The boy C throws the ball B horizontally when A is at $d = 10$ ft. At what speed must C throw the ball so that A can catch it? Also determine the relative speed of the ball with respect to boy A at the instant the catch is made.

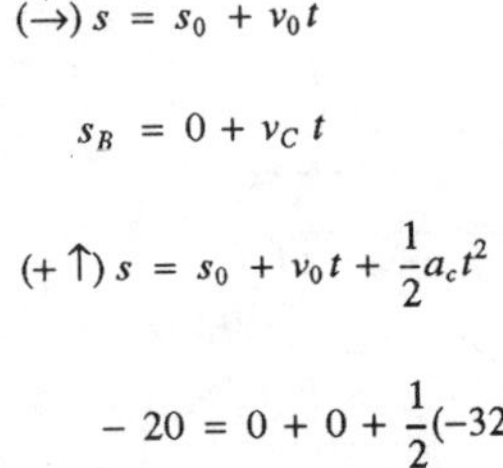

For B:

$(\overset{+}{\rightarrow})\ s = s_0 + v_0 t$

$s_B = 0 + v_C\, t$

$(+\uparrow)\ s = s_0 + v_0 t + \frac{1}{2}a_c t^2$

$-20 = 0 + 0 + \frac{1}{2}(-32.2)t^2$

$t = 1.1146$ s

For A:

$(\overset{+}{\rightarrow})\ s = s_0 + v_0 t$

$s_A = 10 + 4\,t$

Require $s_B = s_A$ at $t = 1.1146$ s

$v_C(1.1146) = 10 + 4(1.1146)$

$v_C = 12.97 = 13.0$ ft/s **Ans**

When ball is caught

$(v_{Bx})_2 = 12.97$ ft/s $\rightarrow$

$(+\uparrow)\quad v = v_0 + a_c t$

$(v_{By})_2 = 0 - 32.2(1.114) = -35.89$ ft/s $= 35.89$ ft/s $\downarrow$

Thus,

$\mathbf{v}_B = \mathbf{v}_A + \mathbf{v}_{B/A}$

$12.97\mathbf{i} - 35.89\mathbf{j} = 4\mathbf{i} + (v_{B/A})_x\mathbf{i} - (v_{B/A})_y\mathbf{j}$

$(v_{B/A})_x = 8.97$ ft/s $\rightarrow$

$(v_{B/A})_y = 35.89$ ft/s $\downarrow$

$v_{B/A} = \sqrt{(8.97)^2 + (35.89)^2} = 37.0$ ft/s **Ans**

12-207. The boy A is moving in a straight line away from the building at a constant speed of 4 ft/s. At what horizontal distance d must he be from C in order to make the catch if the ball is thrown with a horizontal velocity of $v_C = 10$ ft/s? Also determine the relative speed of the ball with respect to the boy A at the instant the catch is made.

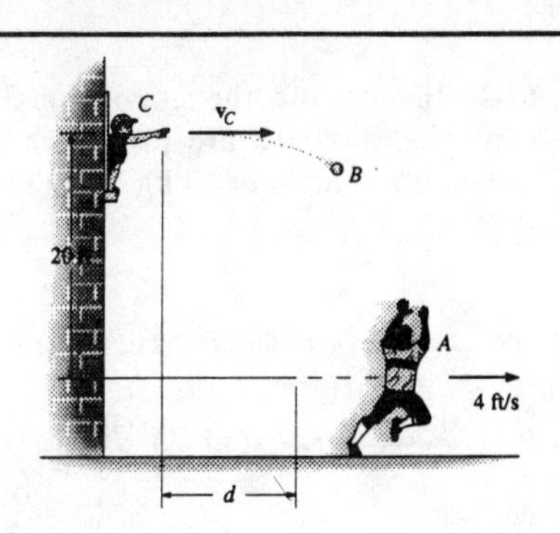

$(\overset{+}{\rightarrow})$ $\quad s = s_0 + v_0 t$

$$x = 0 + 10t$$

$(+\uparrow)$ $\quad s = s_0 + v_0 + \frac{1}{2} a_c t^2$

$$-20 = 0 + 0 + \frac{1}{2}(-32.2)t^2$$

$t = 1.1146$ s

$d = 11.146 \text{ ft} = 11.1 \text{ ft}$ **Ans**

$(\overset{+}{\rightarrow})$ $\quad (v_{Bx})_2 = 10 \text{ ft/s} \rightarrow$

$(+\uparrow)$ $\quad v = v_0 + a_c t$

$$(v_{By})_2 = 0 - 32.2(1.1146) = -35.89 \text{ ft/s} = 35.89 \text{ ft/s} \downarrow$$

$$\mathbf{v}_B = \mathbf{v}_A + \mathbf{v}_{B/A}$$

$$10\mathbf{i} - 35.89\mathbf{j} = 4\mathbf{i} + (v_{B/A})_x\mathbf{i} - (v_{B/A})_y\mathbf{j}$$

$(v_{B/A})_x = 6 \text{ ft/s} \rightarrow$

$(v_{B/A})_y = 35.89 \text{ ft/s} \downarrow$

$v_{B/A} = \sqrt{6^2 + (35.89)^2} = 36.4 \text{ ft/s}$ **Ans**

***12-208.** At a given instant, two particles A and B are moving with a speed of 8 m/s along the paths shown. If B is decelerating at 6 m/s^2 and the speed of A is increasing at 5 m/s^2, determine the acceleration of A with respect to B at this instant.

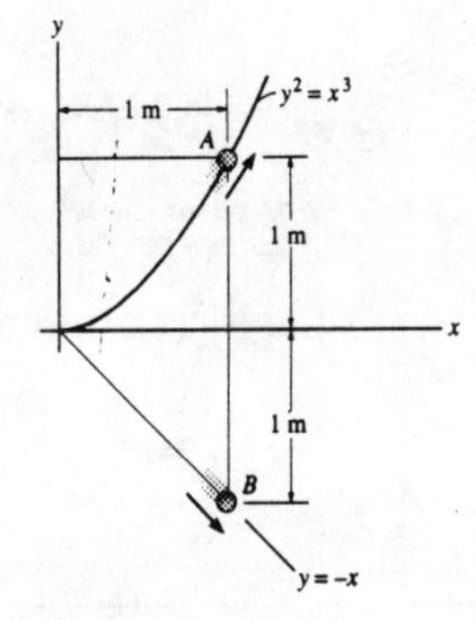

$$y^2 = x^3$$

$$2y\frac{dy}{dx} = 3x^2 \qquad \frac{dy}{dx} = \frac{3x^2}{2y}\bigg|_{x=1,\, y=1} = 1.5$$

$$\theta = \tan^{-1}(1.5) = 56.31°$$

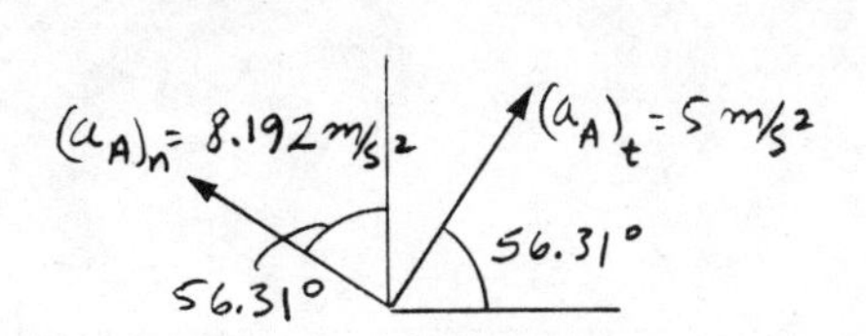

$$2\left(\frac{dy}{dx}\right)^2 + 2y\left(\frac{d^2y}{dx^2}\right) = 6x$$

At $x = 1,\ y = 1$

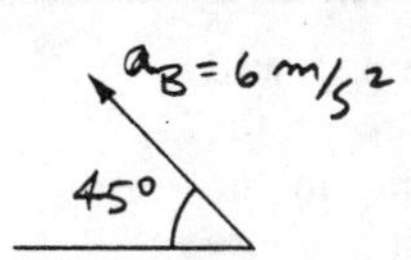

$$2(1.5)^2 + 2(1)\left(\frac{d^2y}{dx^2}\right) = 6(1) \qquad \frac{d^2y}{dx^2} = 0.75$$

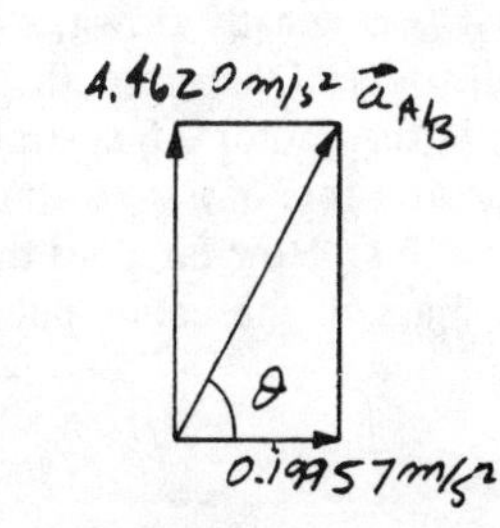

$$\rho = \frac{\left[1 + \left(\frac{dy}{dx}\right)^2\right]^{\frac{3}{2}}}{\left|\frac{d^2y}{dx^2}\right|} = \frac{\left[1 + (1.5)^2\right]^{\frac{3}{2}}}{0.75} = 7.812$$

$$(a_A)_n = \frac{v_A^2}{\rho} = \frac{8^2}{7.812} = 8.192$$

$$\mathbf{a}_A = \mathbf{a}_B + \mathbf{a}_{A/B}$$

$$-8.192 \sin 56.31°\mathbf{i} + 8.192 \cos 56.31°\mathbf{j} + 5 \cos 56.31°\mathbf{i} + 5 \sin 56.31°\mathbf{j}$$

$$= -6 \cos 45°\mathbf{i} + 6 \sin 45°\mathbf{j} + (a_{A/B})_x\mathbf{i} + (a_{A/B})_y\mathbf{j}$$

$$-8.192 \sin 56.31° + 5 \cos 56.31° = -6 \cos 45° + (a_{A/B})_x$$

$$8.192 \cos 56.31° + 5 \sin 56.31° = 6 \sin 45° + (a_{A/B})_y$$

$$(a_{A/B})_x = 0.199\,57 \qquad (a_{A/B})_y = 4.4620$$

$a_{A/B} = \sqrt{(0.199\,57)^2 + (4.4620)^2} = 4.47 \text{ m/s}^2$ **Ans**

$\theta = \tan^{-1}\left(\frac{4.4620}{0.1995}\right) = 87.4°$ ∠ **Ans**

13-1. Determine the gravitational attraction between two spheres which are just touching each other. Each sphere has a mass of 10 kg and a radius of 200 mm.

The distance between the centers of the spheres is $r = 400\text{ mm} = 0.4\text{ m}$.

$$F = G\frac{m_1 \cdot m_2}{r^2} = 66.73(10^{-12})\left(\frac{(10)(10)}{(0.4)^2}\right) = 41.7(10^{-9})\text{N} = 41.7\text{ nN} \quad \textbf{Ans}$$

13-2. The 10-lb block has an initial velocity of 10 ft/s on the smooth plane. If a force $F = (2.5t)$ lb, where t is in seconds, acts on the block for 3 s, determine the final velocity of the block and the distance the block travels during this time.

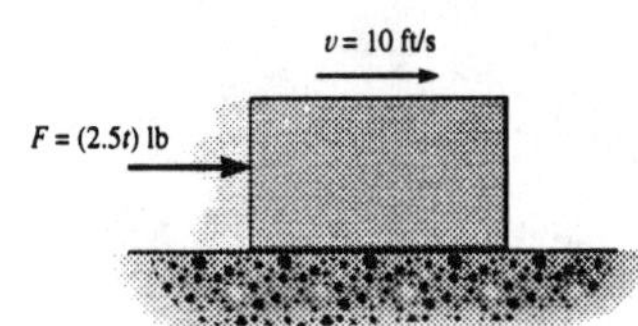

$$\xrightarrow{+} \Sigma F_x = ma_x; \qquad 2.5t = \left(\frac{10}{32.2}\right)a$$

$$a = 8.05t$$

$$dv = a\,dt$$

$$\int_{10}^{v} dv = \int_0^t 8.05t\,dt$$

$$v = 4.025t^2 + 10$$

When $t = 3$ s,

$$v = 46.2\text{ ft/s} \quad \textbf{Ans}$$

$$ds = v\,dt$$

$$\int_0^s ds = \int_0^t (4.025t^2 + 10)\,dt$$

$$s = 1.3417t^3 + 10t$$

When $t = 3$ s,

$$s = 66.2\text{ ft} \quad \textbf{Ans}$$

13-3. The 300-kg bar B, originally at rest, is being towed over a series of small rollers. Determine the force in the cable when $t = 5$ s, if the motor M is drawing in the cable for a short time at a rate of $v = (0.4t^2)$ m/s, where t is in seconds $(0 \leq t \leq 6\text{ s})$. How far does the bar move in 5 s? Neglect the mass of the cable, pulley, and the rollers.

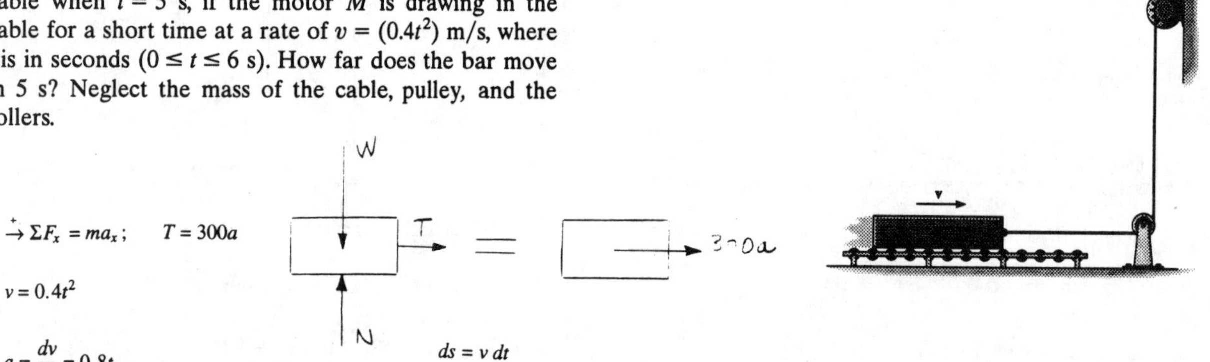

$$\xrightarrow{+} \Sigma F_x = ma_x; \qquad T = 300a$$

$$v = 0.4t^2$$

$$a = \frac{dv}{dt} = 0.8t$$

When $t = 5$ s, $a = 4\text{ m/s}^2$

$$T = 300(4) = 1200\text{ N} = 1.20\text{ kN} \quad \textbf{Ans}$$

$$ds = v\,dt$$

$$\int_0^s ds = \int_0^5 0.4t^2\,dt$$

$$s = \left(\frac{0.4}{3}\right)(5)^3 = 16.7\text{ m} \quad \textbf{Ans}$$

***13-4.** The van is traveling at 20 km/h when the coupling of the trailer at A fails. If the trailer has a mass of 250 kg and coasts 45 m before coming to rest, determine the constant horizontal force F created by rolling friction which causes the trailer to stop.

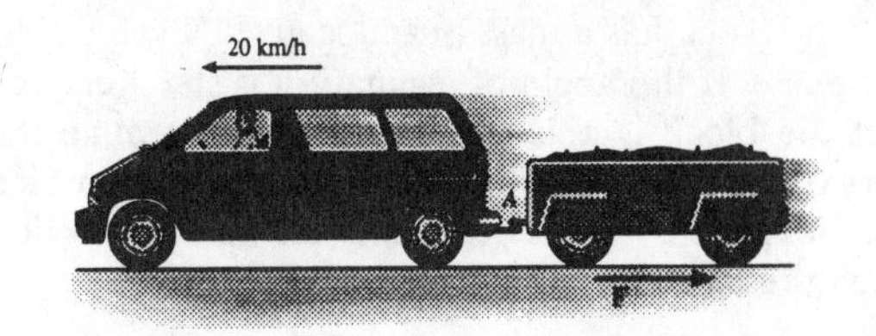

$$20\text{ km/h} = \frac{20(10^3)}{3600} = 5.556\text{ m/s}$$

$(\overset{+}{\leftarrow})$ $\quad v^2 = v_0^2 + 2a_c(s - s_0)$

$$0 = 5.556^2 + 2(a)(45 - 0)$$

$$a = -0.3429\text{ m/s}^2 = 0.3429\text{ m/s}^2 \rightarrow$$

$\overset{+}{\rightarrow} \Sigma F_x = ma_x;\quad F = 250(0.3429) = 85.7\text{ N}\qquad$ **Ans**

13-5. A crate having a mass of 60 kg falls horizontally off the back of a truck which is traveling at 80 km/h. Determine the coefficient of kinetic friction between the road and the crate if the crate slides 45 m on the ground with no tumbling along the road before coming to rest. Assume that the initial speed of the crate along the road is 80 km/h.

$$80\text{ km/h} = \frac{80(10^3)}{3600} = 22.22\text{ m/s}$$

$+\uparrow \Sigma F_y = ma_y;\quad N_C - 60(9.81) = 0\qquad N_C = 588.6\text{ N}$

$\overset{+}{\rightarrow} \Sigma F_x = ma_x;\quad \mu_k(588.6) = 60a\qquad a = 9.81\mu_k$

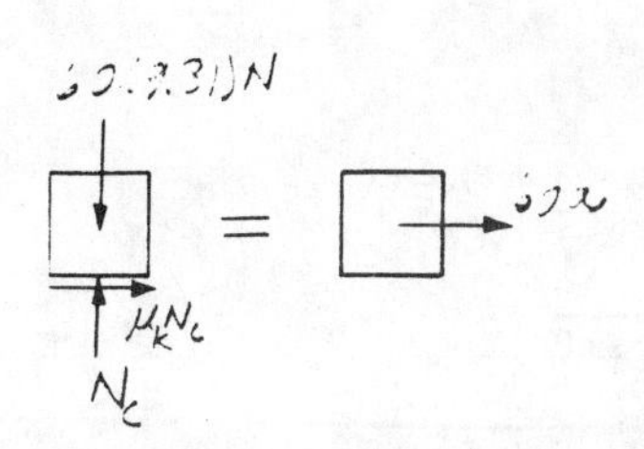

$$v^2 = v_0^2 + 2a_c(s - s_0)$$

$$0 = (22.22)^2 - 2a(45 - 0)$$

$$a = 5.487\text{ m/s}^2$$

Thus,

$\mu_k = \dfrac{5.487}{9.81} = 0.559\qquad$ **Ans**

13-6. The crane lifts the 700-kg bin with an initial acceleration of 3 m/s². Determine the force in each of the supporting cables due to this motion.

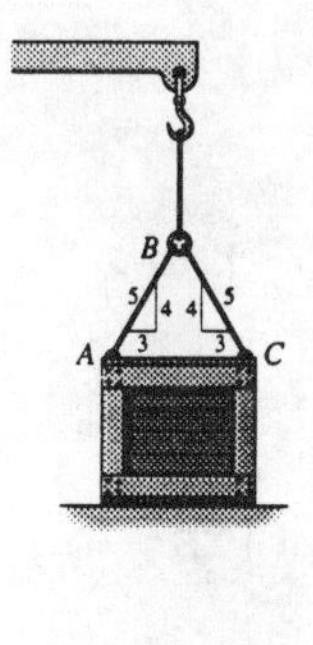

$+\uparrow \Sigma F_y = ma_y;\quad 2T\left(\frac{4}{5}\right) - 700(9.81) = 700(3)$

$T = 5.60\text{ kN}\qquad$ **Ans**

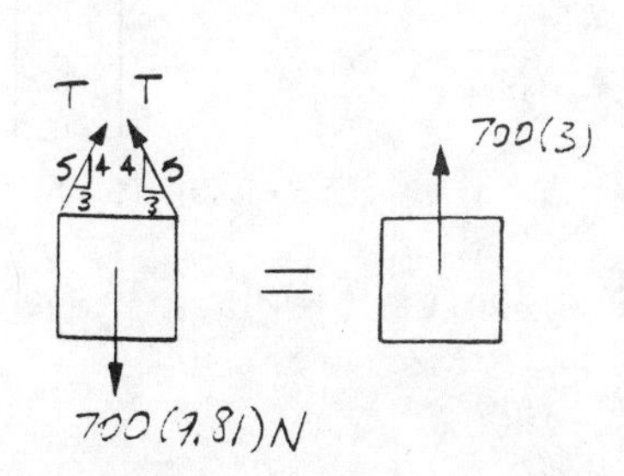

13-7. The block has a mass m and is given a velocity $\mathbf{v}_0$ up the plane. If the angle of inclination is steep enough so that the block is able to slide down, determine the velocity of the block when it returns to the point at which it was launched. The coefficient of kinetic friction between the block and the plane is μ_k.

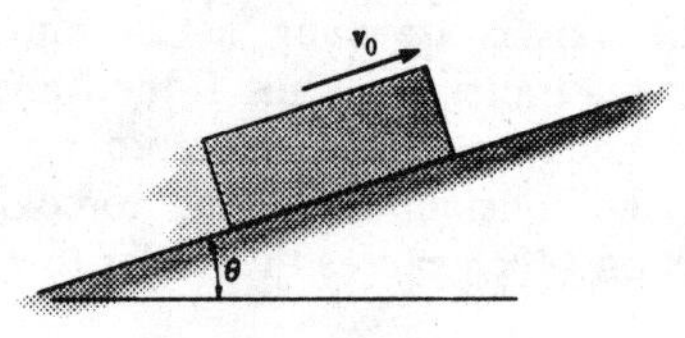
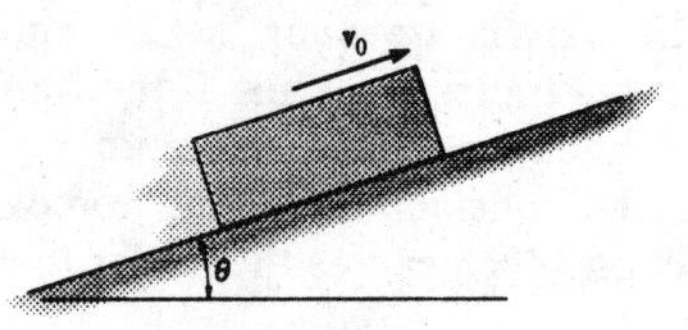

Up the plane :

$(+\nwarrow)\qquad N - mg\cos\theta = 0$

$N = mg\cos\theta$

$(+\nearrow)\qquad -\mu_k N - mg\sin\theta = ma$

$a = -g(\mu_k\cos\theta + \sin\theta)$

$(+\nearrow)\qquad v^2 = v_0^2 + 2a_c(s - s_0)$

$0 = v_0^2 + 2(-g)(\mu_k\cos\theta + \sin\theta)(s - 0)$

$$s = \frac{v_0^2}{2g(\mu_k\cos\theta + \sin\theta)}$$

Down the plane :

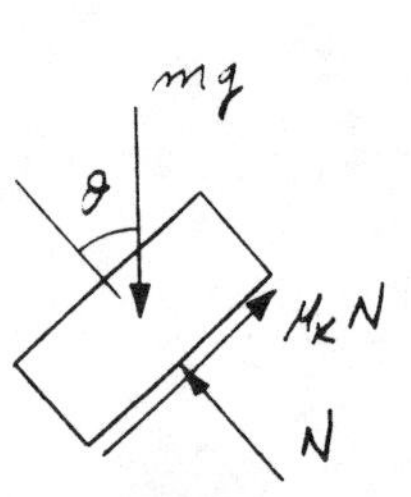

$(+\nwarrow)\qquad N - mg\cos\theta = 0$

$N = mg\cos\theta$

$(+\swarrow)\qquad -\mu_k N + mg\sin\theta = ma$

$a = g(-\mu_k\cos\theta + \sin\theta)$

$(+\swarrow)\qquad v^2 = v_0^2 + 2a_c(s - s_0)$

$v_1^2 = 0 + 2(g)(-\mu_k\cos\theta + \sin\theta)(s' - 0)$

$$s' = \frac{v_1^2}{2g(-\mu_k\cos\theta + \sin\theta)}$$

Require $s = s'$

$$\frac{v_1^2}{2g(-\mu_k\cos\theta + \sin\theta)} = \frac{v_0^2}{2g(\mu_k\cos\theta + \sin\theta)}$$

$$v_1^2 = v_0^2\left(\frac{\sin\theta - \mu_k\cos\theta}{\sin\theta + \mu_k\cos\theta}\right)$$

$$v_1 = v_0\left(\frac{1 - \mu_k\,\mathrm{ctn}\,\theta}{1 + \mu_k\,\mathrm{ctn}\,\theta}\right)^{\frac{1}{2}} \qquad \textbf{Ans}$$

***13-8.** The 200-kg crate is suspended from the cable of a crane. Determine the force in the cable when $t = 2$ s if the crate is moving upward with (a) a constant velocity of 2 m/s, and (b) a speed of $v = (0.2t^2 + 2)$ m/s, where t is in seconds.

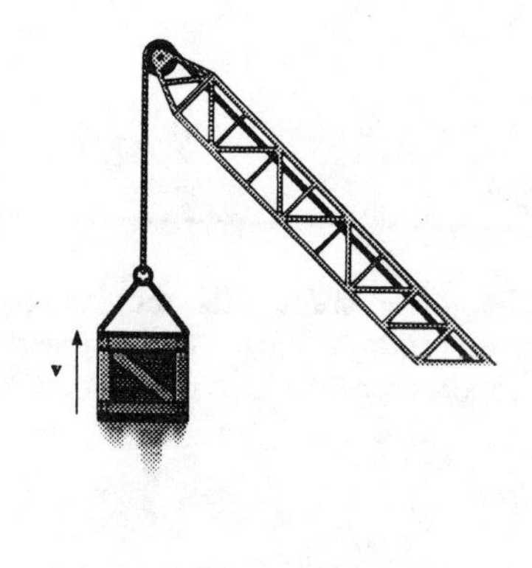

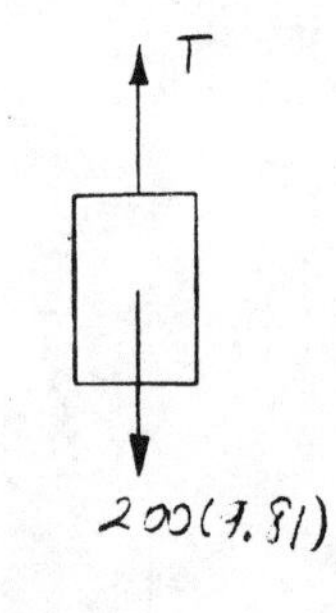

a) $\quad +\uparrow\Sigma F_y = ma_y;\qquad T - 200(9.81) = 0$

$T = 1.96$ kN $\quad$ **Ans**

b) $\quad v = 0.2t^2 + 2$

$$a = \frac{dv}{dt} = 0.4t\Big|_{t=2\text{ s}} = 0.8\text{ m/s}^2$$

$+\uparrow\Sigma F_y = ma_y;\qquad T - 200(9.81) = 200(0.8)$

$T = 2.12$ kN $\quad$ **Ans**

13-9. The elevator E has a mass of 500 kg, and the counterweight at A has a mass of 150 kg. If the motor supplies a constant force of 5 kN on the cable at B, determine the speed of the elevator when $t = 3$ s, starting from rest. Neglect the mass of the pulleys and cable.

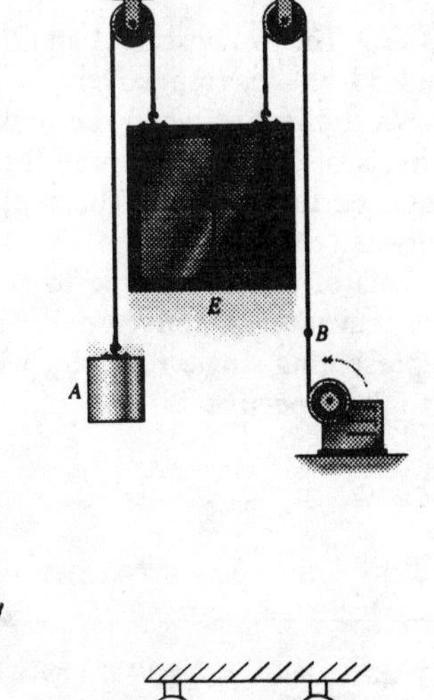

For A :

$+\downarrow \Sigma F_y = ma_y;\quad 150(9.81) - T = 150a_A \quad (1)$

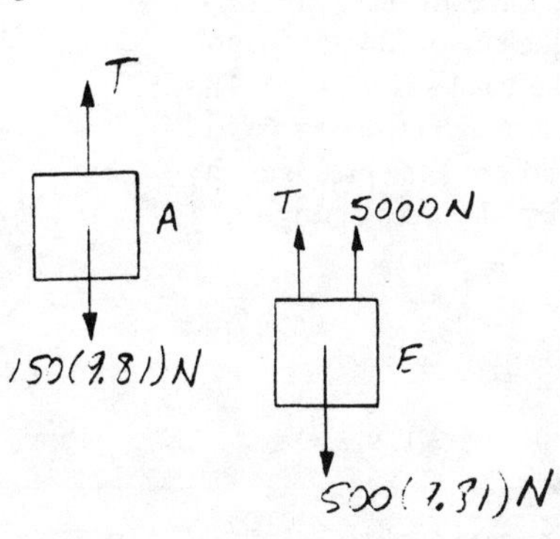

For E :

$+\downarrow \Sigma F_y = ma_y;\quad 500(9.81) - 5000 - T = 500a_E \quad (2)$

$s_A + s_E = l$

$a_A = -a_E \quad (3)$

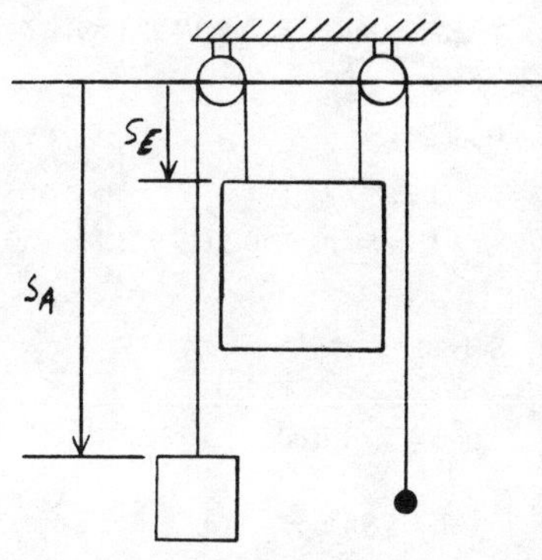

Solving :

$T = 1110$ N

$a_E = -2.410 \text{ m/s}^2 = 2.410 \text{ m/s}^2 \uparrow$

$(+\uparrow) \quad v = v_0 + a_c t$

$v_E = 0 + 2.410(3) = 7.23 \text{ m/s} \uparrow$ **Ans**

13-10. The elevator E has a mass of 500 kg and the counterweight at A has a mass of 150 kg. If the elevator attains a speed of 10 m/s after it rises 40 m, determine the constant force developed in the cable at B. Neglect the mass of the pulleys and cable.

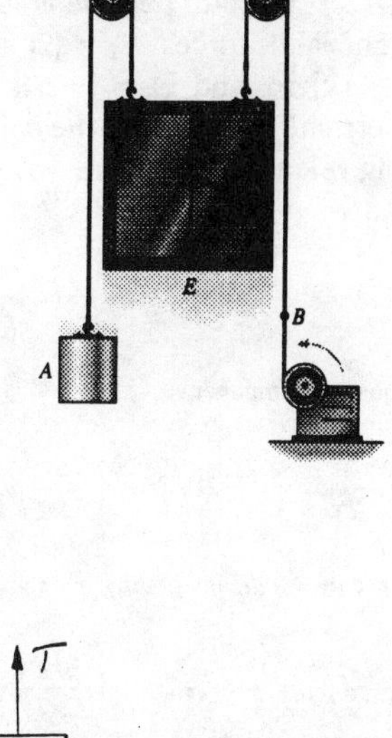

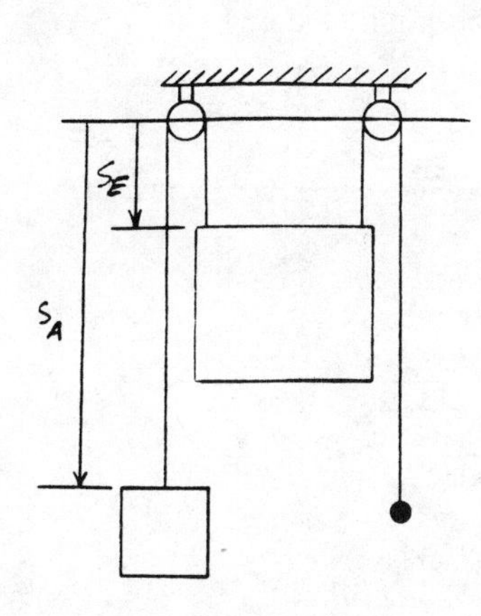

$(+\downarrow) \quad v^2 = v_0^2 + 2a_c(s - s_0)$

$(-10)^2 = (0)^2 + 2a_E(-40 - 0)$

$a_E = -1.25 \text{ m/s}^2 = 1.25 \text{ m/s}^2 \uparrow$

$s_A + s_E = l$

$a_A = -a_E$

For A :

$+\downarrow \Sigma F_y = ma_y;\quad 150(9.81) - T = 150(1.25) \quad T = 1.284 \text{ kN}$

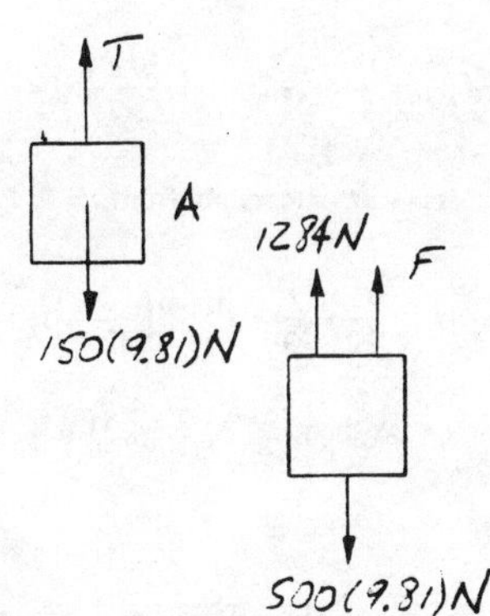

For E :

$+\downarrow \Sigma F_y = ma_y;\quad 500(9.81) - 1284 - F = 500(-1.25) \quad F = 4.25 \text{ kN}$ **Ans**

13-11. The two boxcars A and B have a weight of 20 000 lb and 30 000 lb, respectively. If they are freely coasting down the incline when the brakes are applied to all the wheels of car A, determine the force in the coupling C between the two cars. The coefficient of kinetic friction between the wheels of A and the tracks is $\mu_k = 0.5$. The wheels of car B are free to roll. Neglect their mass in the calculation. *Suggestion:* Solve the problem by representing single resultant normal forces acting on A and B, respectively.

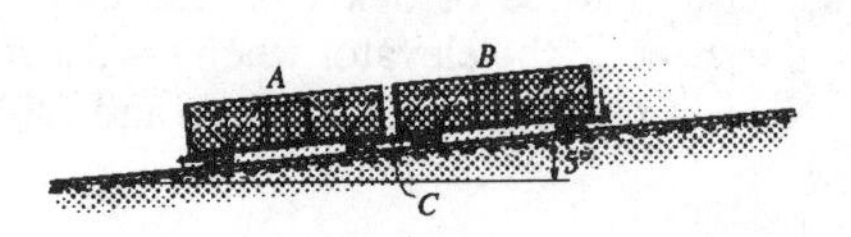

Car A :

$+\nwarrow \Sigma F_y = 0;\quad N_A - 20\,000\cos 5° = 0 \qquad N_A = 19\,923.89 \text{ lb}$

$+\nearrow \Sigma F_x = ma_x;\quad 0.5(19\,923.89) - T - 20\,000\sin 5° = \left(\frac{20\,000}{32.2}\right)a \qquad (1)$

Both cars :

$+\nearrow \Sigma F_x = ma_x;\quad 0.5(19\,923.89) - 50\,000\sin 5° = \left(\frac{50\,000}{32.2}\right)a$

Solving,

$a = 3.61 \text{ ft/s}^2$

$T = 5.98$ kip **Ans**

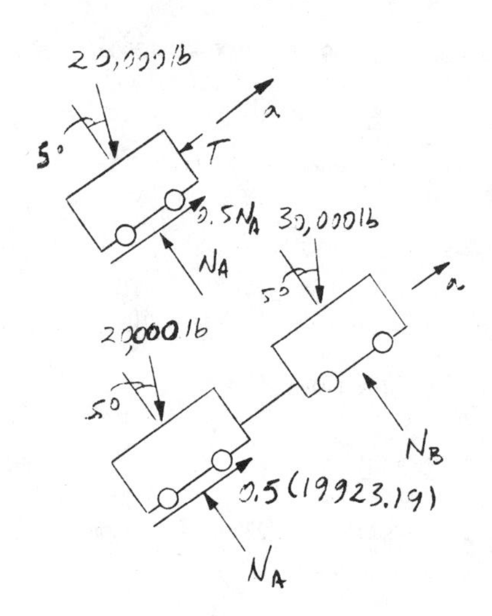

***13-12.** The 6-lb particle is subjected to the action of its weight and forces $\mathbf{F}_1 = \{2\mathbf{i} + 6\mathbf{j} - 2t\mathbf{k}\}$ lb, $\mathbf{F}_2 = \{t^2\mathbf{i} - 4t\mathbf{j} - 1\mathbf{k}\}$ lb, and $\mathbf{F}_3 = \{-2t\mathbf{i}\}$ lb, where t is in seconds. Determine the distance the ball is from the origin 2 s after being released from rest.

$\Sigma \mathbf{F} = m\mathbf{a};\quad (2\mathbf{i} + 6\mathbf{j} - 2t\mathbf{k}) + (t^2\mathbf{i} - 4t\mathbf{j} - 1\mathbf{k}) - 2t\mathbf{i} - 6\mathbf{k} = \left(\frac{6}{32.2}\right)(a_x\mathbf{i} + a_y\mathbf{j} + a_z\mathbf{k})$

Equating components :

$\left(\frac{6}{32.2}\right)a_x = t^2 - 2t + 2 \qquad \left(\frac{6}{32.2}\right)a_y = -4t + 6 \qquad \left(\frac{6}{32.2}\right)a_z = -2t - 7$

Since $dv = a\,dt$, integrating from $v = 0$, $t = 0$, yields

$\left(\frac{6}{32.2}\right)v_x = \frac{t^3}{3} - t^2 + 2t \qquad \left(\frac{6}{32.2}\right)v_y = -2t^2 + 6t \qquad \left(\frac{6}{32.2}\right)v_z = -t^2 - 7t$

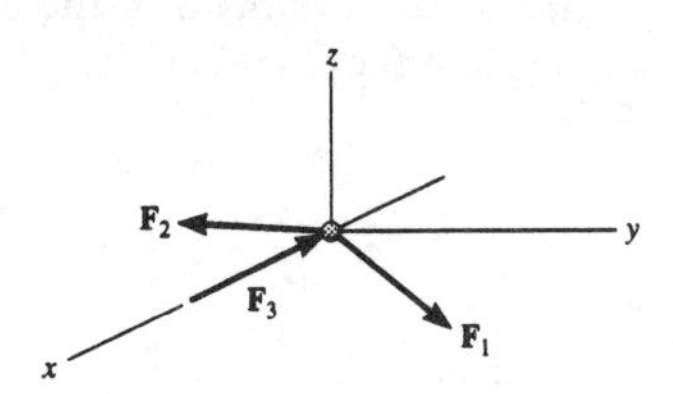

Since $ds = v\,dt$, integrating from $s = 0$, $t = 0$ yields

$\left(\frac{6}{32.2}\right)s_x = \frac{t^4}{12} - \frac{t^3}{3} + t^2 \qquad \left(\frac{6}{32.2}\right)s_y = -\frac{2t^3}{3} + 3t^2 \qquad \left(\frac{6}{32.2}\right)s_z = -\frac{t^3}{3} - \frac{7t^2}{2}$

When $t = 2$ s then, $\quad s_x = 14.31$ ft, $\quad s_y = 35.78$ ft $\quad s_z = -89.44$ ft

Thus,

$s = \sqrt{(14.31)^2 + (35.78)^2 + (-89.44)^2} = 97.4$ ft **Ans**

13-13. The 2-lb particle A is acted upon by the force system $\mathbf{F}_1 = \{2\mathbf{i} + 6\mathbf{j} - 2\mathbf{k}\}$ lb, $\mathbf{F}_2 = \{3\mathbf{i} - 1\mathbf{k}\}$ lb, and $\mathbf{F}_3 = \{1\mathbf{i} - t^2\mathbf{j} - 2\mathbf{k}\}$ lb, where t is in seconds. Determine the distance the particle is from the origin 3 s after it has been released from rest.

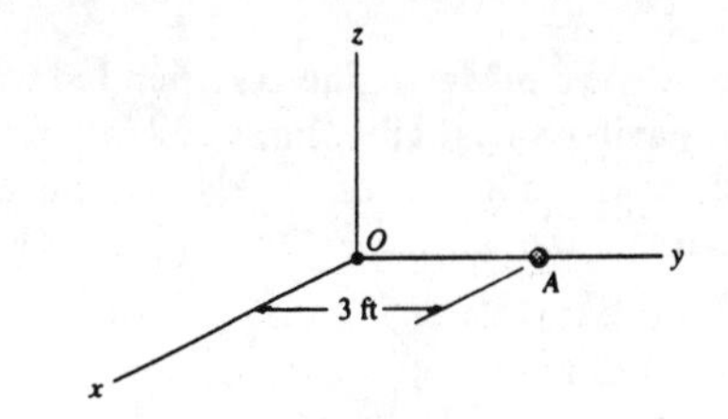

$\Sigma F_x = ma_x;\quad 2+3+1 = \left(\frac{2}{32.2}\right)a_x$

$\Sigma F_y = ma_y;\quad 6-t^2 = \left(\frac{2}{32.2}\right)a_y$

$\Sigma F_z = ma_z;\quad -2-1-2-2 = \left(\frac{2}{32.2}\right)a_z$

F3 F2 F1 2 lb = 2/32.2 a

Since $dv = a\,dt$, integrating from $v = 0$ when $t = 0$,

$\left(\frac{2}{32.2}\right)v_x = 6t \qquad \left(\frac{2}{32.2}\right)v_y = 6t - \frac{t^3}{3} \qquad \left(\frac{2}{32.2}\right)v_z = -7t$

Since $ds = v\,dt$, integrating from $s_x = s_z = 0$, $s_y = 3$ ft when $t = 0$,

$\left(\frac{2}{32.2}\right)s_x = 3t^2 \qquad \left(\frac{2}{32.2}\right)s_y = 3t^2 - \frac{t^4}{12} + 0.186\,34 \qquad \left(\frac{2}{32.2}\right)s_z = -\frac{7}{2}t^2$

When $t = 3$ s, $\quad s_x = 434.70$ ft $\quad s_y = 329.025$ ft $\quad s_z = -507.15$ ft

$s = \sqrt{(434.70)^2 + (329.25)^2 + (-507.15)^2} = 745$ ft **Ans**

13-14. The 3.5-Mg engine is suspended from a spreader beam AB having a negligible mass and is hoisted by a crane which gives it an acceleration of 4 m/s² when it has a velocity of 2 m/s. Determine the force in chains CA and CB during the lift.

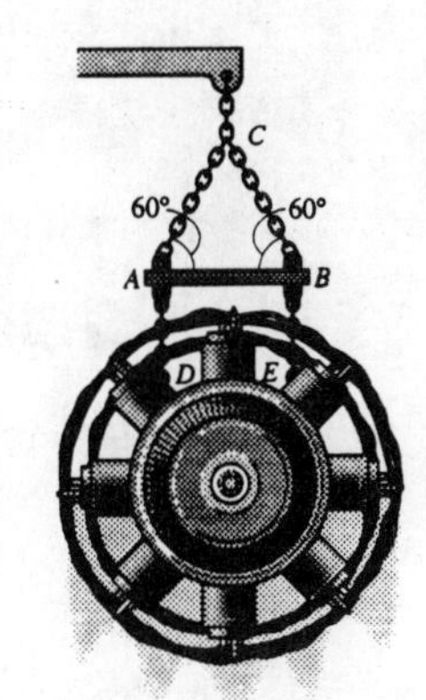

System :

$+\uparrow \Sigma F_y = ma_y;\quad T' - 3.5(10^3)(9.81) = 3.5(10^3)(4)$

$T' = 48.335$ kN

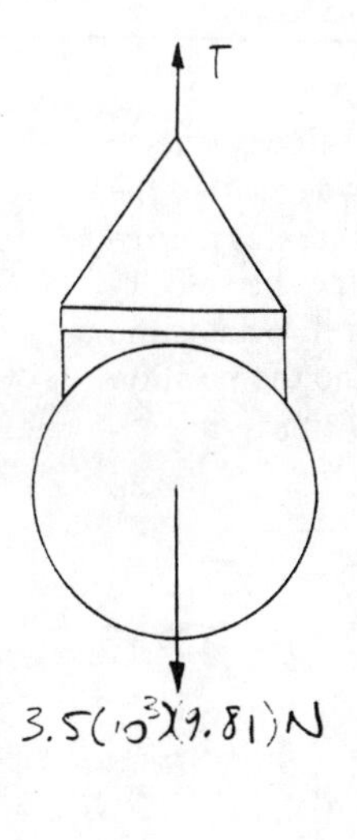

Joint C :

$+\uparrow \Sigma F_y = ma_y;\quad 48.335 - 2\,T\cos 30° = 0$

$T = T_{CA} = T_{CB} = 27.9$ kN **Ans**

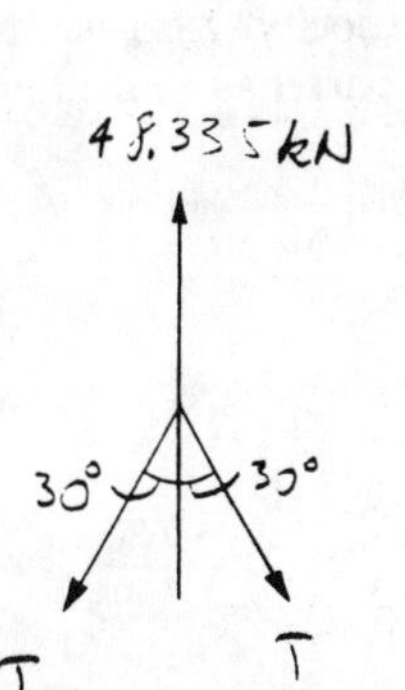

13-15. The 3.5-Mg engine is suspended from a spreader beam having a negligible mass and is hoisted by a crane which exerts a force of 40 kN on the hoisting cable. Determine the distance the engine is hoisted in 4 s, starting from rest.

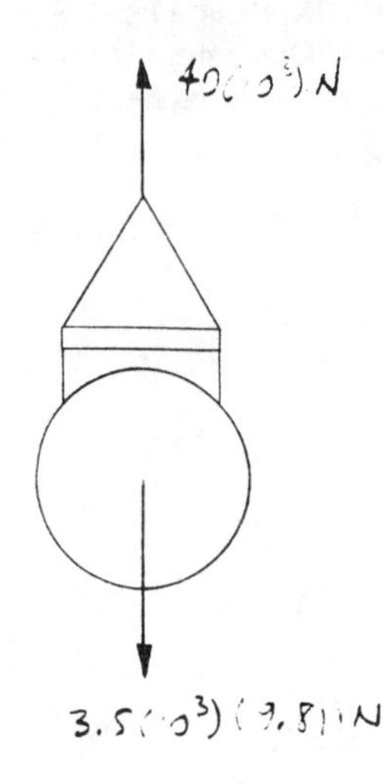

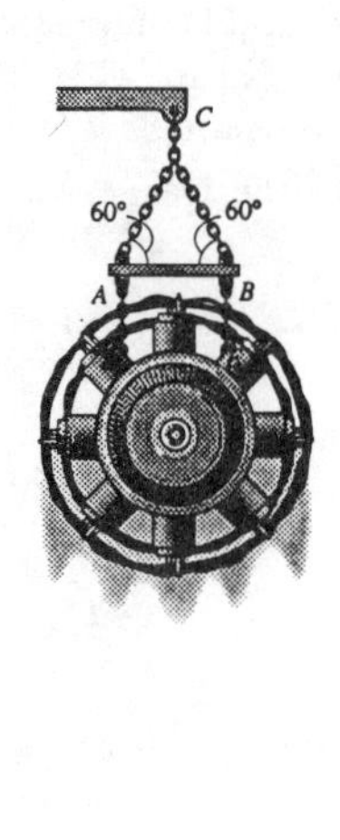

System :

$+\uparrow \Sigma F_y = ma_y; \quad 40(10^3) - 3.5(10^3)(9.81) = 3.5(10^3)a$

$a = 1.619 \text{ m/s}^2$

$(+\uparrow) \quad s = s_0 + v_0 t + \frac{1}{2}a_c t^2$

$s = 0 + 0 + \frac{1}{2}(1.619)(4)^2 = 12.9 \text{ m}$ **Ans**

***13-16.** The 3.5-Mg engine is suspended from a 500-kg spreader beam and hoisted by a crane which gives it an acceleration of 4 m/s^2 when it has a velocity of 2 m/s. Determine the force in chains AC and AD during the lift.

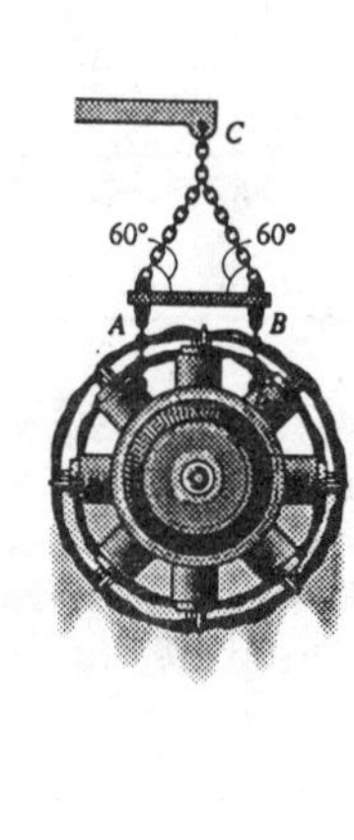

System :

$+\uparrow \Sigma F_y = ma_y; \quad 2T\sin 60° - 500(9.81) - 3.5(10^3)(9.81) = [3.5(10^3) + 500]4$

$T_{AC} = T = 31.9 \text{ kN}$ **Ans**

Engine :

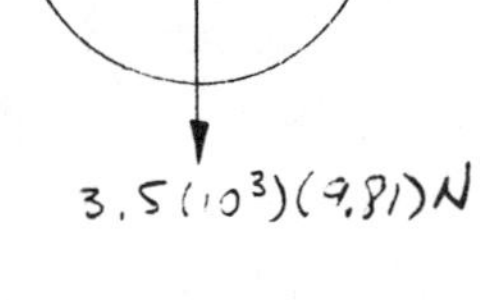

$+\uparrow \Sigma F_y = ma_y; \quad 2T' - 3.5(10^3)(9.81) = 3.5(10^3)(4)$

$T_{AD} = T' = 24.2 \text{ kN}$ **Ans**

13-17. The bullet of mass m is given a velocity due to gas pressure caused by the burning of powder within the barrel of the gun. Assuming this pressure creates a force of $F = F_0 \sin(\pi t/t_0)$ on the bullet, determine the velocity of the bullet at any instant it is in the barrel. What is the bullet's maximum velocity? Also, determine the position of the bullet in the barrel as a function of time.

$\xrightarrow{+} \Sigma F_x = ma_x; \quad F_0 \sin\left(\frac{\pi t}{t_0}\right) = ma$

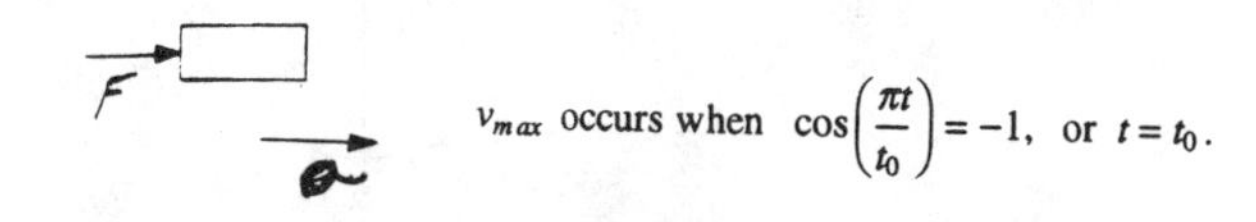

$a = \frac{dv}{dt} = \left(\frac{F_0}{m}\right)\sin\left(\frac{\pi t}{t_0}\right)$

$\int_0^v dv = \int_0^t \left(\frac{F_0}{m}\right)\sin\left(\frac{\pi t}{t_0}\right) dt \qquad v = -\left(\frac{F_0 t_0}{\pi m}\right)\cos\left(\frac{\pi t}{t_0}\right)\Big]_0^t$

$v = \left(\frac{F_0 t_0}{\pi m}\right)\left(1 - \cos\left(\frac{\pi t}{t_0}\right)\right)$ **Ans**

v_{max} occurs when $\cos\left(\frac{\pi t}{t_0}\right) = -1$, or $t = t_0$.

$v_{max} = \frac{2F_0 t_0}{\pi m}$ **Ans**

$\int_0^s ds = \int_0^t \left(\frac{F_0 t_0}{\pi m}\right)\left(1 - \cos\left(\frac{\pi t}{t_0}\right)\right)dt$

$s = \left(\frac{F_0 t_0}{\pi m}\right)\left[t - \frac{t_0}{\pi}\sin\left(\frac{\pi t}{t_0}\right)\right]_0^t$

$s = \left(\frac{F_0 t_0}{\pi m}\right)\left(t - \frac{t_0}{\pi}\sin\left(\frac{\pi t}{t_0}\right)\right)$ **Ans**

13-18. The man pushes on the 60-lb crate with a force **F**. The force is always directed down at 30° from the horizontal as shown, and its magnitude is increased until the crate begins to slide. Determine the crate's initial acceleration if the static coefficient of friction is $\mu_s = 0.6$ and the kinetic coefficient of friction is $\mu_k = 0.3$.

Force to produce motion :

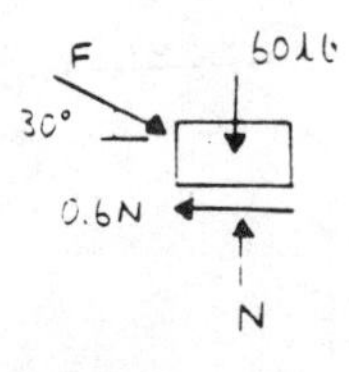

$\xrightarrow{+} \Sigma F_x = 0; \qquad F\cos 30° - 0.6N = 0$

$+\uparrow \Sigma F_y = 0; \qquad N - 60 - F\sin 30° = 0$

$N = 91.80 \text{ lb} \qquad F = 63.60 \text{ lb}$

Since $N = 91.80$ lb,

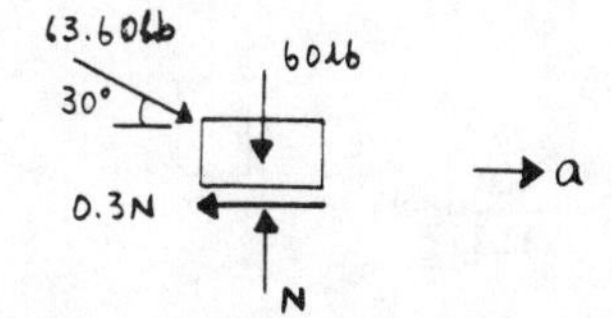

$\xrightarrow{+} \Sigma F_x = ma_x; \qquad 63.60 \cos 30° - 0.3(91.80) = \left(\frac{60}{32.2}\right)a$

$a = 14.8 \text{ ft/s}^2$ **Ans**

13-19. A force of $F = 15$ lb is applied to the cord. Determine how high the 30-lb block A rises in 2 s starting from rest. Neglect the weight of the pulleys and cord.

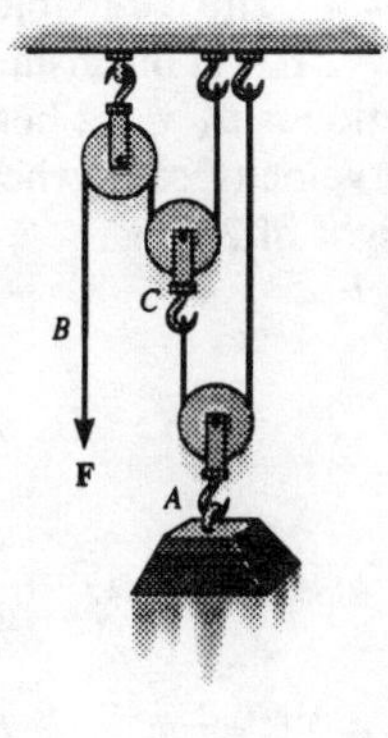

Block :

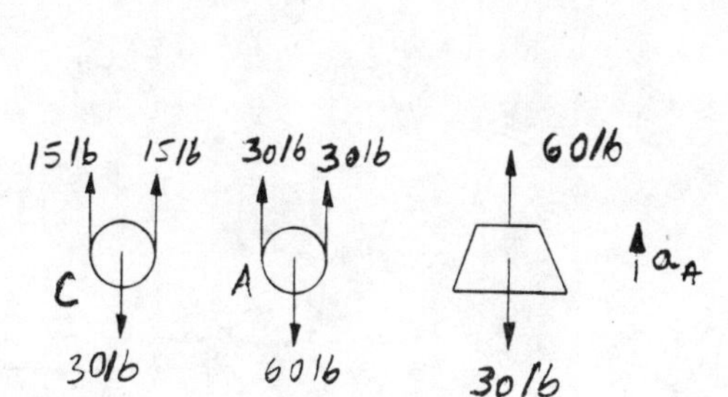

$+\uparrow \Sigma F_y = ma_y; \qquad -30 + 60 = \left(\frac{30}{32.2}\right)a_A$

$a_A = 32.2 \text{ ft/s}^2$

$(+\uparrow) \qquad s = s_0 + v_0 t + \frac{1}{2}a_c t^2$

$s = 0 + 0 + \frac{1}{2}(32.2)(2)^2$

$s = 64.4 \text{ ft}$ **Ans**

***13-20.** Determine the constant force F which must be applied to the cord in order to cause the 30-lb block A to have a speed of 12 ft/s when it has been displaced 3 ft upward starting from rest. Neglect the weight of the pulleys and cord.

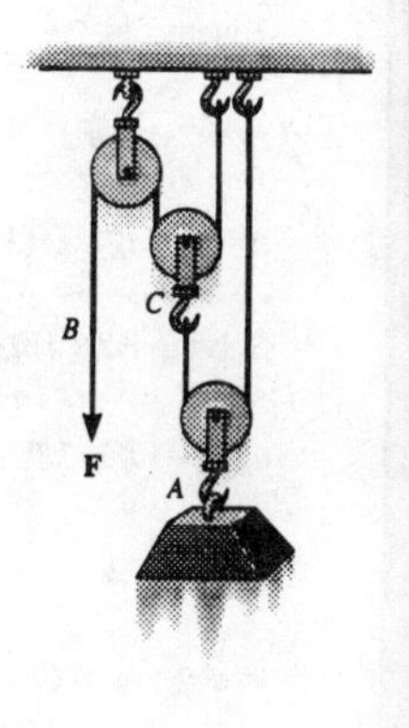

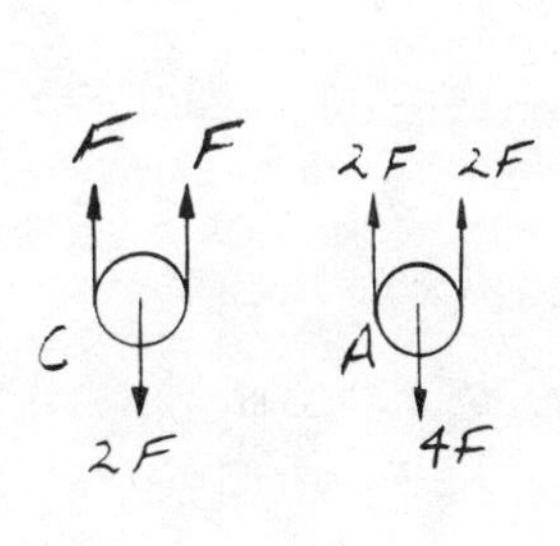

$(+\uparrow) \qquad v^2 = v_0^2 + 2a_c(s - s_0)$

$(12)^2 = 0 + 2(a)(3)$

$a = 24 \text{ ft/s}^2$

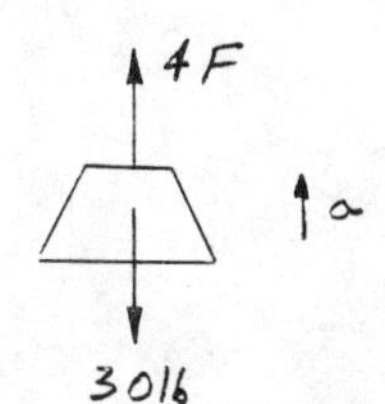

$+\uparrow \Sigma F_y = ma_y; \qquad -30 + 4F = \left(\frac{30}{32.2}\right)(24)$

$F = 13.1 \text{ lb}$ **Ans**

13-21. The 400-lb cylinder at A is hoisted using the motor and the pulley system shown. If the speed of point B on the cable is increased at a constant rate from zero to $v_B = 10$ ft/s in $t = 5$ s, determine the tension in the cable at B to cause the motion.

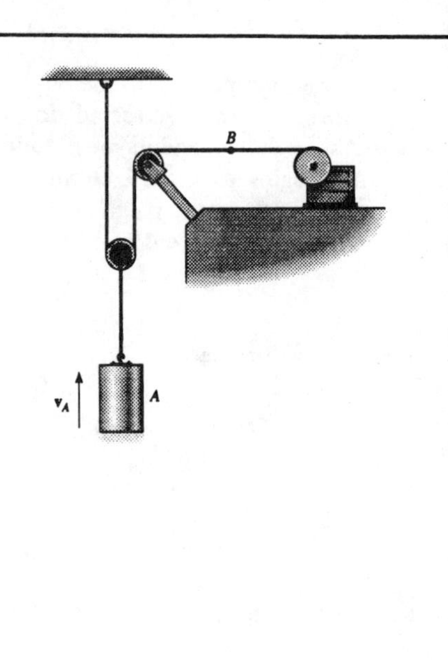

$$2s_A + s_B = l$$

$$2a_A = -a_B$$

$(\overset{+}{\rightarrow})$ $\quad v = v_0 + a_c t$

$$10 = 0 + a_B(5)$$

$$a_B = 2 \text{ ft/s}^2$$

$$a_A = -1 \text{ ft/s}^2$$

$+\downarrow \Sigma F_y = ma_y; \quad 400 - 2T = \left(\frac{400}{32.2}\right)(-1)$

Thus, $\quad T = 206$ lb $\quad$ **Ans**

13-22. The 10-lb block A is traveling to the right at $v_A = 2$ ft/s at the instant shown. If the coefficient of kinetic friction is $\mu_k = 0.2$ between the surface and A, determine the velocity of A when it has moved 4 ft. Block B has a weight of 20 lb.

Block A :

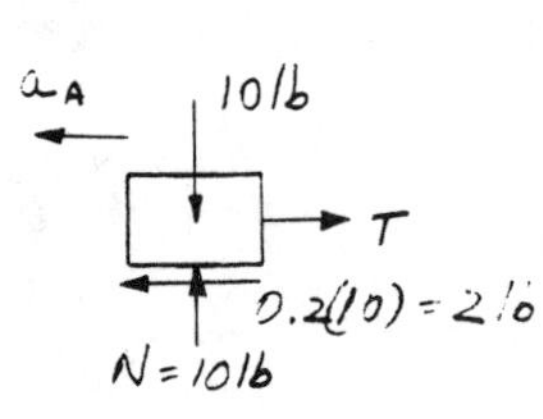

$\overset{+}{\leftarrow} \Sigma F_x = ma_x; \quad -T + 2 = \left(\frac{10}{32.2}\right)a_A \quad (1)$

Weight B :

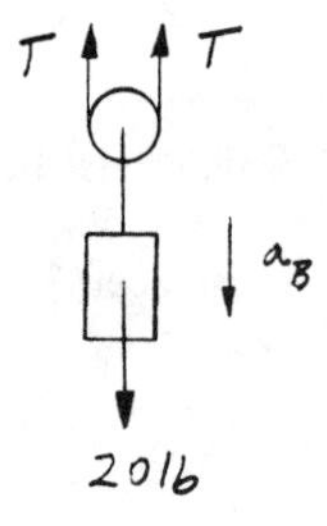

$+\downarrow \Sigma F_y = ma_y; \quad 20 - 2T = \left(\frac{20}{32.2}\right)a_B \quad (2)$

Kinematics :

$$s_A + 2s_B = l$$

$$a_A = -2a_B \quad (3)$$

Solving Eq.s (1) – (3) :

$a_A = -17.173$ ft/s^2 $\quad a_B = 8.587$ ft/s^2 $\quad T = 7.33$ lb

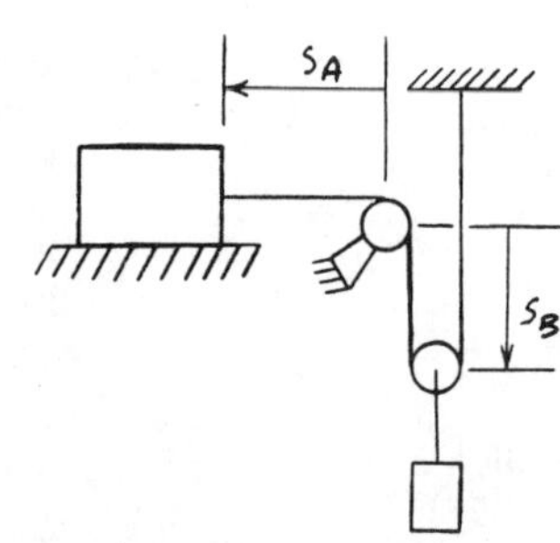

$$v^2 = v_0^2 + 2a_c(s - s_0)$$

$$v^2 = (2)^2 + 2(17.173)(4 - 0)$$

$v = 11.9$ ft/s $\quad$ **Ans**

13-23. The motor M pulls in its attached rope causing an acceleration of 6 m/s². Determine this towing force. The coefficient of kinetic friction between the 50-kg crate and the plane is $\mu_k = 0.3$. Neglect the mass of the pulleys and rope.

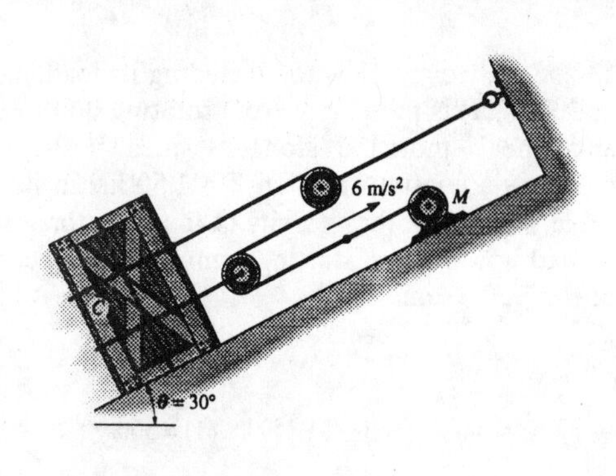

$+\nwarrow \Sigma F_y = ma_y;\quad N_C - 50(9.81)\cos 30° = 0 \quad N_C = 424.79 \text{ N}$

$+\swarrow \Sigma F_x = ma_x;\quad -3T + 0.3(424.79) + 50(9.81)\sin 30° = 50a_C \quad (1)$

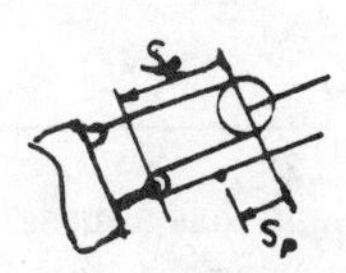

Kinematics :

$2s_C + (s_C - s_P) = l$

Taking two time derivatives, yields,

$3a_C = a_P$

Thus,

$a_C = \frac{-6}{3} = -2 \text{ m/s}^2$

Substituting into Eq. (1) and solving,

$T = 158$ N **Ans**

***13-24.** At a given instant the 5-lb weight A is moving downward with a speed of 4 ft/s. Determine its speed 2 s later. Block B has a weight of 6 lb, and the coefficient of kinetic friction between it and the horizontal plane is $\mu_k = 0.3$. Neglect the mass of the pulleys and cord.

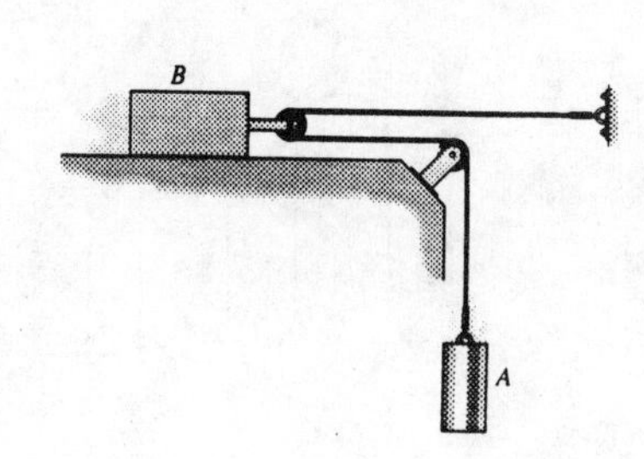

$\overset{+}{\leftarrow} \Sigma F_x = ma_x;\quad 1.8 - 2T = \left(\frac{6}{32.2}\right)a_B$

$+\downarrow \Sigma F_y = ma_y;\quad 5 - T = \left(\frac{5}{32.2}\right)a_A$

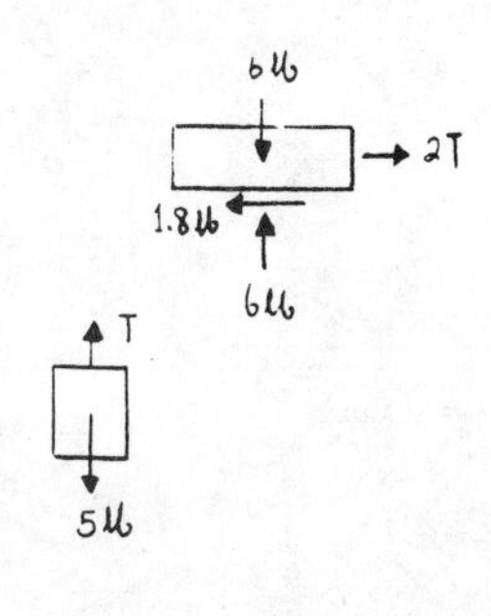

$s_A + 2s_B = l$

$a_A = -2a_B$

Solving,

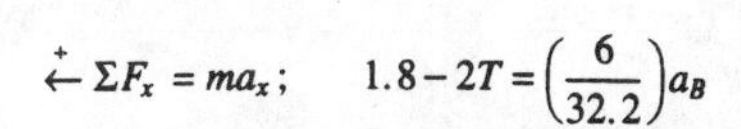

$T = 1.84$ lb

$a_A = 20.31 \text{ ft/s}^2$

$a_B = -10.16 \text{ ft/s}^2$

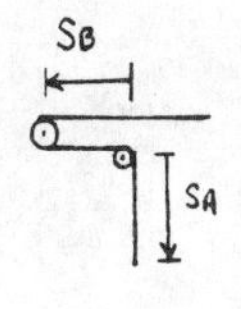

$(+\downarrow) \quad v = v_0 + a_c t$

$v_A = 4 + 20.31(2) = 44.6$ ft/s **Ans**

13-25. A freight elevator, including its load, has a mass of 500 kg. It is prevented from rotating due to the track and wheels mounted along its sides. If the motor M develops a constant tension $T = 1.50$ kN in its attached cable, determine the velocity of the elevator when it has moved upward 3 m starting from rest. Neglect the mass of the pulleys and cables.

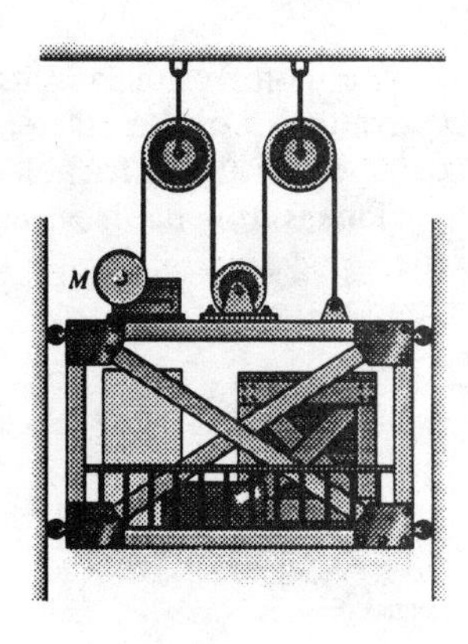

$+\uparrow \Sigma F_y = ma_y;\qquad 1500(4) - 500(9.81) = 500a$

$a = 2.19 \text{ m/s}^2$

1500(4) N

500(9.81) N

$(+\uparrow)\qquad v^2 = v_0^2 + 2a_c(s - s_0)$

$v^2 = 0 + 2(2.19)(3)$

$v = 3.62 \text{ m/s}$ **Ans**

13-26. A freight elevator, including its load, has a mass of 500 kg. It is prevented from rotating by using the track and wheels mounted along its sides. Starting from rest, in $t = 2$ s, the motor M draws in the cable with a speed of 6 m/s, *measured relative to the elevator.* Determine the constant acceleration of the elevator and the tension in the cable. Neglect the mass of the pulleys and cables.

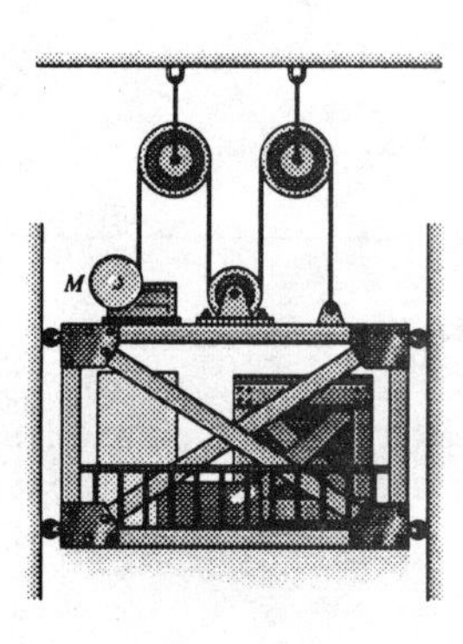

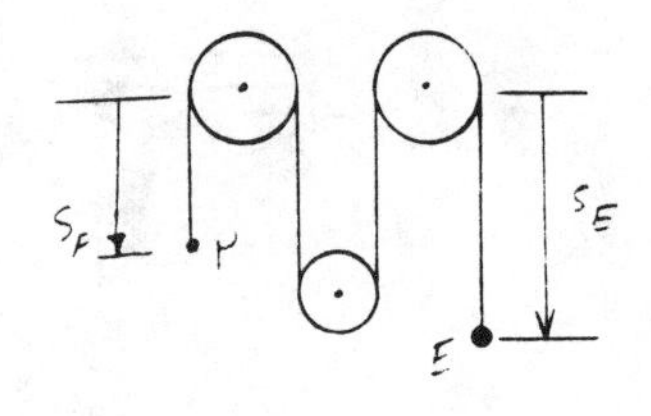

$3s_E + s_P = l$

$3v_E = -v_P$

$(+\downarrow)\qquad v_P = v_E + v_{P/E}$

$-3v_E = v_E + 6$

$v_E = -\frac{6}{4} = -1.5 \text{ m/s} = 1.5 \text{ m/s}\uparrow$

$(+\uparrow)\qquad v = v_0 + a_c t$

$1.5 = 0 + a_E(2)$

$a_E = 0.75 \text{ m/s}^2\uparrow$ **Ans**

$+\uparrow \Sigma F_y = ma_y;\qquad 4T - 500(9.81) = 500(0.75)$

$T = 1320 \text{ N} = 1.32 \text{ kN}$ **Ans**

13-27. At the instant shown the 100-lb block A is moving down the plane at 5 ft/s while being attached to the 50-lb block B. If the coefficient of kinetic friction is $\mu_k = 0.2$, determine the acceleration of A and the distance A slides before it stops. Neglect the mass of the pulleys and cables.

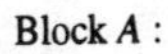

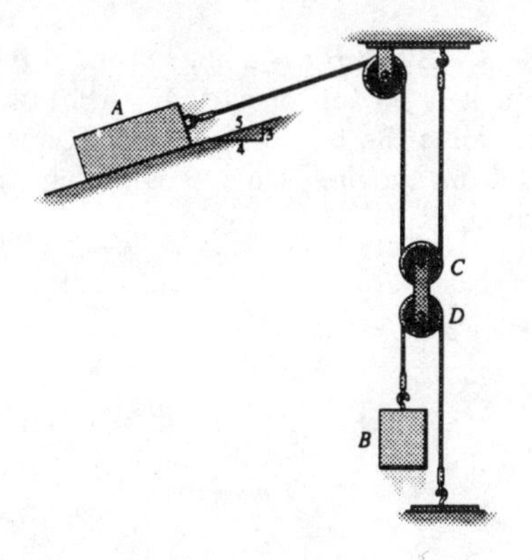

Block A :

$+\swarrow \Sigma F_x = ma_x;\qquad -T_A - 0.2N_A + 100\left(\frac{3}{5}\right) = \left(\frac{100}{32.2}\right)a_A$

$+\nwarrow \Sigma F_y = ma_y;\qquad N_A - 100\left(\frac{4}{5}\right) = 0$

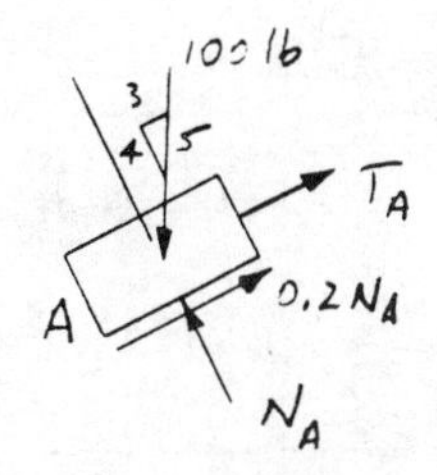

Thus,

$T_A - 44 = -3.1056a_A \qquad (1)$

Block B :

$+\uparrow \Sigma F_y = ma_y;\qquad T_B - 50 = \left(\frac{50}{32.2}\right)a_B$

$T_B - 50 = 1.553a_B \qquad (2)$

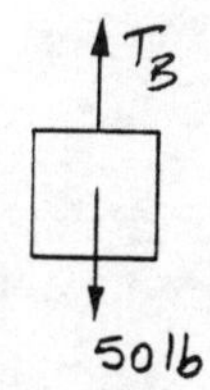

Pulleys at C and D :

$+\uparrow \Sigma F_y = 0;\qquad 2T_A - 2T_B = 0$

$T_A = T_B \qquad (3)$

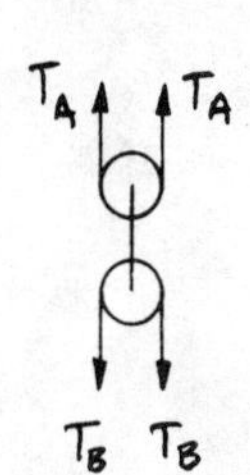

Kinematics :

$s_A + 2s_C = l$

$s_D + (s_D - s_B) = l'$

$s_C + d + s_D = d'$

Thus,

$a_A = -2a_C$

$2a_D = a_B$

$a_C = -a_D,$

so that $\quad a_A = a_B \qquad (4)$

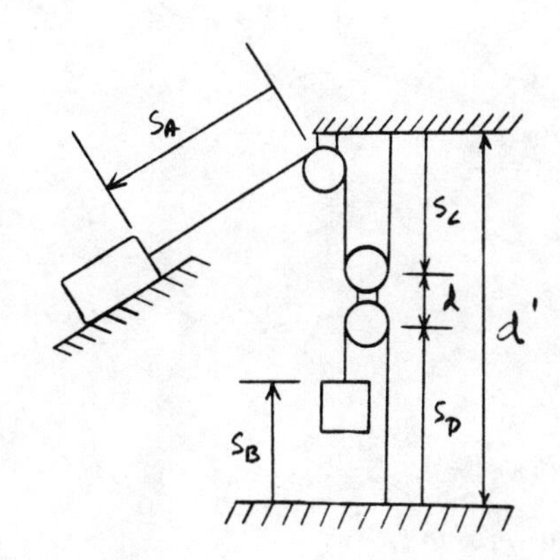

Solving Eqs. (1)–(4) :

$a_A = a_B = -1.288 \text{ ft/s}^2$

$T_A = T_B = 48.0 \text{ lb}$

Thus,

$a_A = 1.29 \text{ ft/s}^2$ **Ans**

$(+\swarrow)\qquad v^2 = v_0^2 + 2a_c(s - s_0)$

$0 = (5)^2 + 2(-1.288)(s - 0)$

$s = 9.70 \text{ ft}$ **Ans**

***13-28.** Block B has a mass m and is released from rest when it is on top of cart A, which has a mass of $3m$. Determine the tension in cord CD needed to hold the cart from moving while B is sliding down A. Neglect friction.

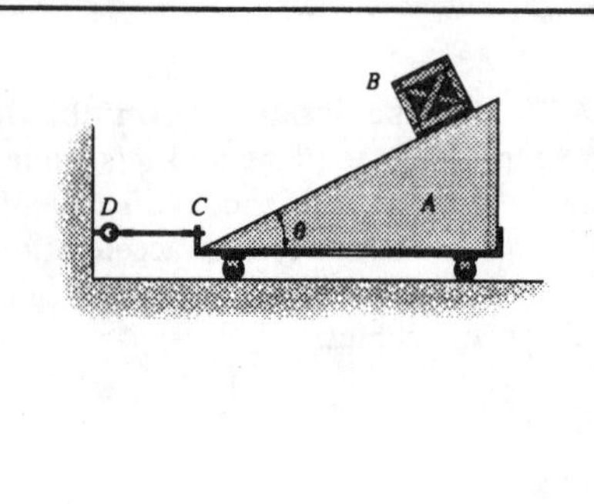

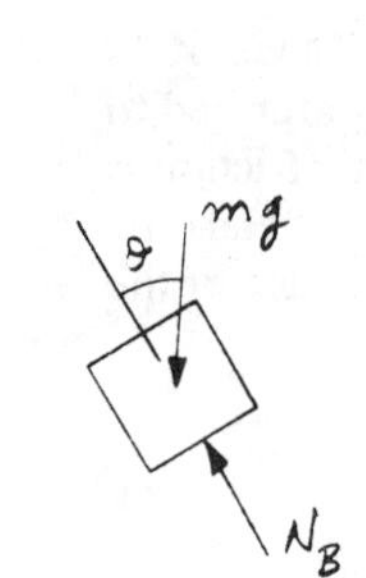

Block B :

$+\nwarrow \Sigma F_y = ma_y;$ $\quad N_B - mg\cos\theta = 0$

$$N_B = mg\cos\theta$$

Cart :

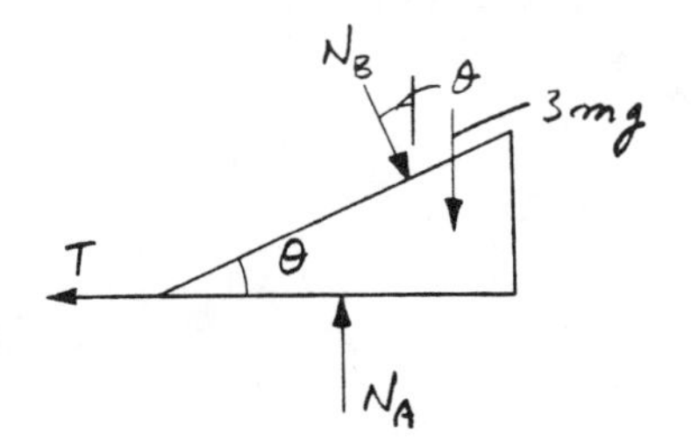

$\overset{+}{\rightarrow} \Sigma F_x = ma_x;$ $\quad -T + N_B \sin\theta = 0$

$$T = mg\sin\theta\cos\theta$$

$$T = \left(\frac{mg}{2}\right)\sin 2\theta \quad \textbf{Ans}$$

13-29. Block B has a mass m and is released from rest when it is on top of cart A, which has a mass of $3m$. Determine the tension in cord CD needed to hold the cart from moving while B is sliding down A. The coefficient of kinetic friction between A and B is μ_k.

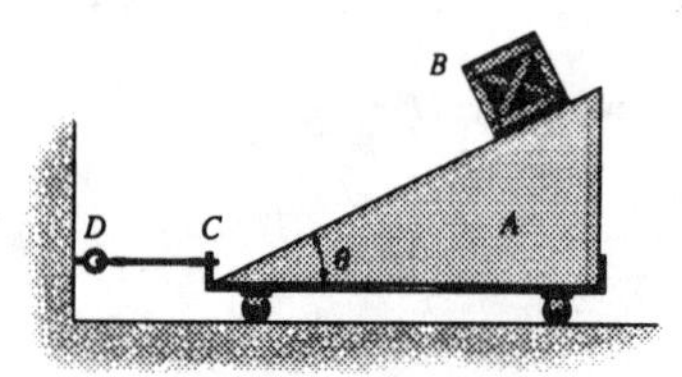

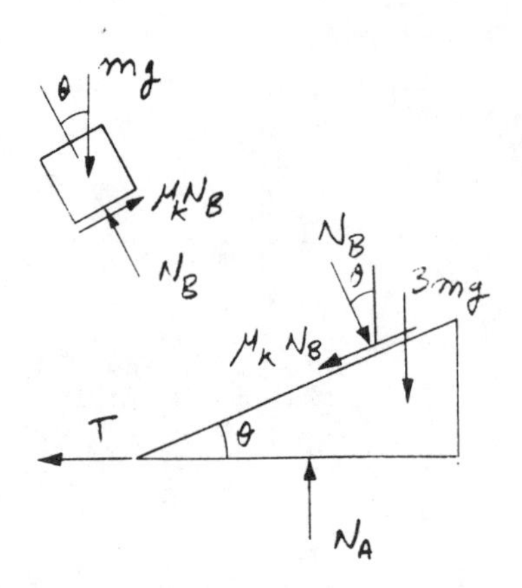

Block B :

$+\nwarrow \Sigma F_y = ma_y;$ $\quad N_B - mg\cos\theta = 0$

$$N_B = mg\cos\theta$$

Cart :

$\overset{+}{\rightarrow} \Sigma F_x = ma_x;$ $\quad -T + N_B \sin\theta - \mu_k N_B \cos\theta = 0$

$$T = mg\cos\theta(\sin\theta - \mu_k\cos\theta) \quad \textbf{Ans}$$

■**13-30.** The tanker has a weight of $800(10^6)$ lb and is traveling forward at $v_0 = 3$ ft/s in still water when the engines are shut off. If the drag resistance of the water is proportional to the speed of the tanker at any instant and can be approximated by $F_D = (400(10^3)v)$ lb, where v is in ft/s, determine the time needed for the tanker's speed to become 1.5 ft/s. Given the initial velocity of $v_0 = 3$ ft/s, through what distance must the tanker travel before it stops?

$$\overset{+}{\rightarrow} \Sigma F_x = ma_x; \qquad -\left(400\left(10^3\right) v\right) = \left(\frac{800(10^6)}{32.2}\right) a$$

$$a = \frac{dv}{dt}$$

$$\int_0^t -0.0161\, dt = \int_3^v \frac{dv}{v}$$

$$-0.0161t = \ln\left(\frac{v}{3}\right) \qquad (1)$$

When $v = 1.5$ ft/s,

$t = 43.1$ s **Ans**

$$v = \frac{ds}{dt} = 3e^{-0.0161t}$$

$$\int_0^{s_{max}} ds = \int_0^{\infty} 3e^{-0.0161t} dt$$

$$s_{max} = \left.\frac{-3e^{-0.0161t}}{0.0161}\right|_0^{\infty} = 186 \text{ ft} \qquad \textbf{Ans}$$

Note that from Eq. (1) it is seen that as $v \rightarrow 0$, $t \rightarrow \infty$. Hence it takes an infinite amount of time to stop the tanker. In reality, however, the drag equation $F_D = \left(400\left(10^3\right) v\right)$ lb changes as the tanker slows down, and hence the dependence of v on t also changes.

13-31. The 2-kg shaft CA passes through a smooth journal bearing at B. Initially, the springs, which are coiled loosely around the shaft, are unstretched when no force is applied to the shaft. In this position $s = s' = 250$ mm and the shaft is originally at rest. If a horizontal force of $F = 5$ kN is applied, determine the speed of the shaft at the instant $s = 50$ mm, $s' = 450$ mm. The ends of the springs are attached to the bearing at B and the caps at C and A.

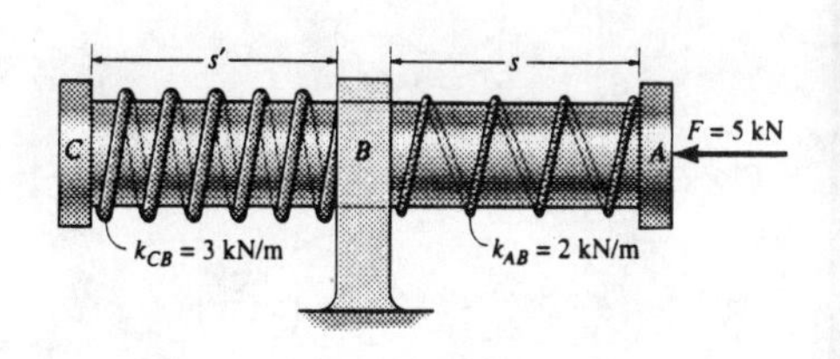

$$F_{CB} = k_{CB}x = 3000x \qquad F_{AB} = k_{AB}x = 2000x$$

$$\overset{+}{\leftarrow} \Sigma F_x = ma_x; \qquad 5000 - 3000x - 2000x = 2a$$

$$2500 - 2500x = a$$

$$a\, dx = v\, dv$$

$$\int_0^{0.2} (2500 - 2500x)\, dx = \int_0^v v\, dv$$

$$2500(0.2) - \left(\frac{2500(0.2)^2}{2}\right) = \frac{v^2}{2}$$

$v = 30$ m/s **Ans**

***13-32.** The spring mechanism is used as a shock absorber for railroad cars. Determine the maximum compression of spring HI if the fixed bumper R of a 5-Mg railroad car, rolling freely at 2 m/s, strikes the plate P. Bar AB slides along the guide paths CE and DF. The ends of all springs are attached to their respective members and are originally unstretched.

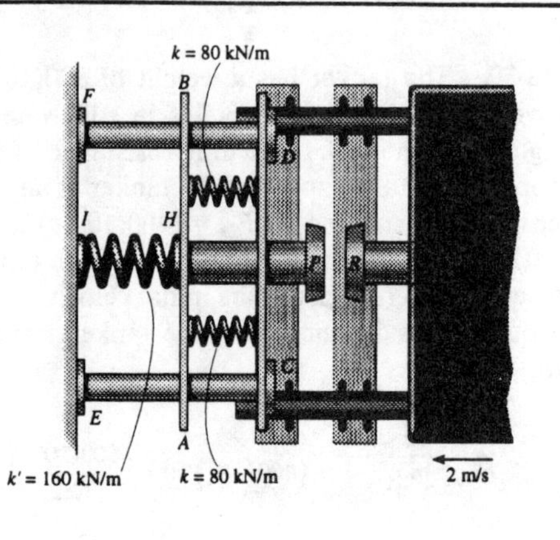

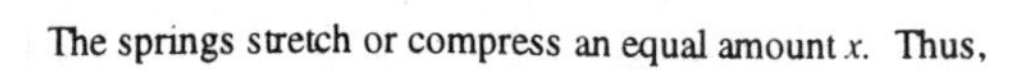

The springs stretch or compress an equal amount x. Thus,

$$F_s = (2k + k')x = \left[2(80)\left(10^3\right) + 160\left(10^3\right)\right]x = 320\,000x$$

$\overset{+}{\rightarrow} \Sigma F_x = ma_x;\quad 320\,000x = 5000a$

$$a = 64x$$

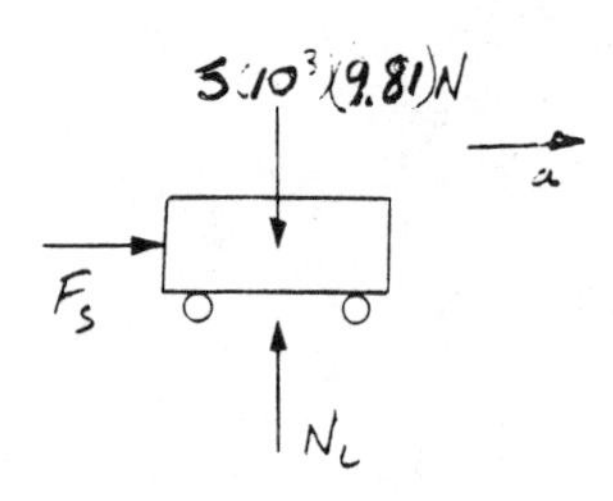

$(\overset{+}{\leftarrow})\quad a\,dx = v\,dv$

$$-\int_0^x 64x\,dx = \int_2^0 v\,dv$$

$$32x^2 = 2$$

$x = 0.25$ m $= 250$ mm **Ans**

13-33. The 10-lb block A and the 20-lb block B are initially at rest. If a force of $P = 20$ lb is applied to B as shown, determine the acceleration of each block. The coefficient of kinetic friction between any two surfaces is $\mu_k = 0.2$, and the coefficient of static friction is $\mu_s = 0.3$.

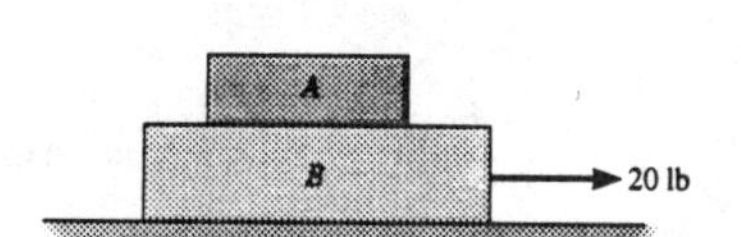

Assume A slips on B as B slips.

Block A :

$\overset{+}{\rightarrow} \Sigma F_x = ma_x;\quad 2 = \left(\frac{10}{32.2}\right)a_A$

$$a_A = 6.44 \text{ ft/s}^2$$

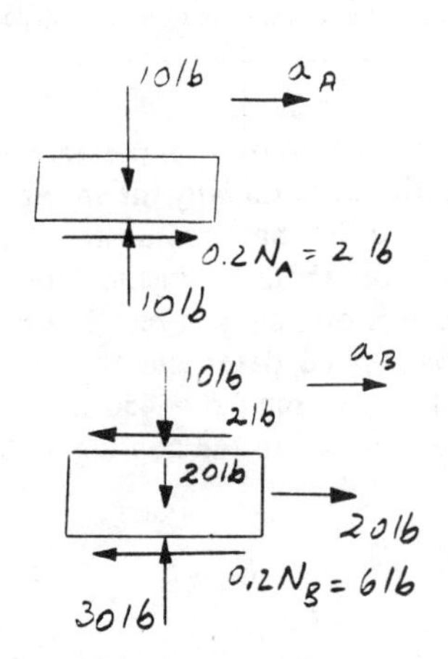

Block B :

$\overset{+}{\rightarrow} \Sigma F_x = ma_x;\quad 20 - 6 - 2 = \left(\frac{20}{32.2}\right)a_B$

$$a_B = 19.32 \text{ ft/s}^2$$

Note that for A, $F_{max} = 0.3N_A = 3$ lb and if A was not to slip on B then, for B,

$\overset{+}{\rightarrow} \Sigma F_x = ma_x;\quad 20 - 6 - 3 = \left(\frac{20}{32.2}\right)a_B$

$$a_B = 17.71 \text{ ft/s}^2$$

For A,

$\overset{+}{\rightarrow} \Sigma F_x = ma_x;\quad 3 = \left(\frac{10}{32.2}\right)a_B = 9.66 \text{ ft/s}^2 < 17.71 \text{ ft/s}^2$ OK A slips on B.

Also, at the ground,

$F_{max} = 0.3(30) = 9$ lb < 20 lb, so indeed motion occurs.

Thus,

$a_A = 6.44$ ft/s^2 **Ans**

$a_B = 19.3$ ft/s^2 **Ans**

13-34. The 10-kg block A rests on the 50-kg plate B in the position shown. Neglecting the mass of the rope and pulley, and using the coefficients of kinetic friction indicated, determine the time needed for block A to slide 0.5 m *on the plate* when the system is released from rest.

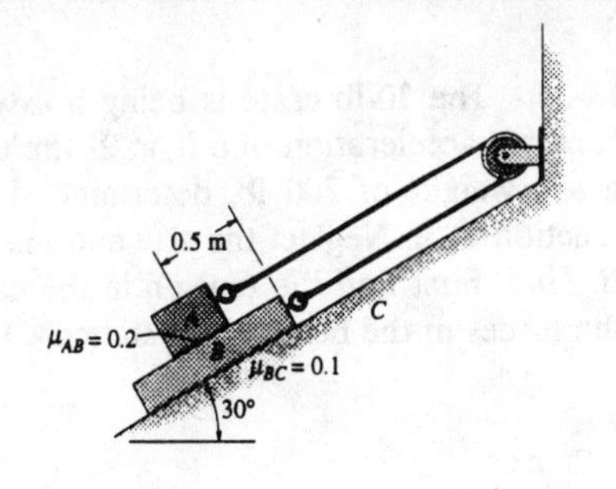

Block A :

$$+\nwarrow \Sigma F_y = ma_y; \quad N_A - 10(9.81)\cos 30° = 0 \quad N_A = 84.96 \text{ N}$$

$$+\swarrow \Sigma F_x = ma_x; \quad -T + 0.2(84.96) + 10(9.81)\sin 30°) = 10a_A$$

$$T - 66.04 = -10a_A \quad (1)$$

Block B :

$$+\nwarrow \Sigma F_y = ma_y; \quad N_B - 84.96 - 50(9.81)\cos 30° = 0$$

$$N_B = 509.7 \text{ N}$$

$$+\swarrow \Sigma F_x = ma_x; \quad -0.2(84.96) - 0.1(509.7) - T + 50(9.81\sin 30°) = 50a_B$$

$$177.28 - T = 50a_B \quad (2)$$

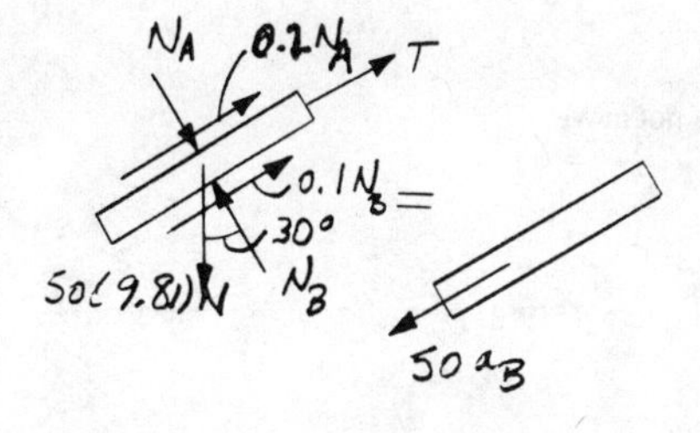

$$s_A + s_B = l$$

$$\Delta s_A = -\Delta s_B$$

$$a_A = -a_B \quad (3)$$

Solving Eqs. (1) − (3) :
$a_B = 1.854 \text{ m/s}^2$
$a_A = -1.854 \text{ m/s}^2 \quad T = 84.58 \text{ N}$

In order to slide 0.5 m along the plate the block must move 0.25 m. Thus,

$$(+\swarrow) \quad s_B = s_A + s_{B/A}$$

$$-\Delta s_A = \Delta s_A + 0.5$$

$$\Delta s_A = -0.25 \text{ m}$$

$$(+\swarrow) \quad s_A = s_0 + v_0 t + \frac{1}{2}a_A t^2$$

$$-0.25 = 0 + 0 + \frac{1}{2}(-1.854)t^2$$

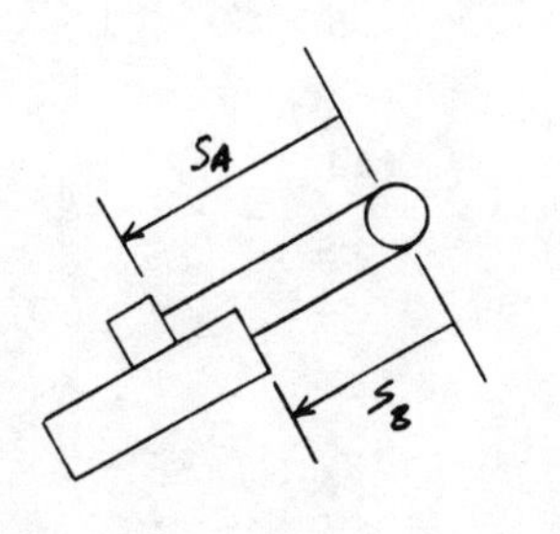

$t = 0.519$ s **Ans**

13-35. The 30-lb crate is being hoisted upward with a constant acceleration of 6 ft/s^2. If the uniform beam AB has a weight of 200 lb, determine the components of reaction at A. Neglect the size and mass of the pulley at B. *Hint:* First find the tension in the cable, then analyze the forces in the beam using statics.

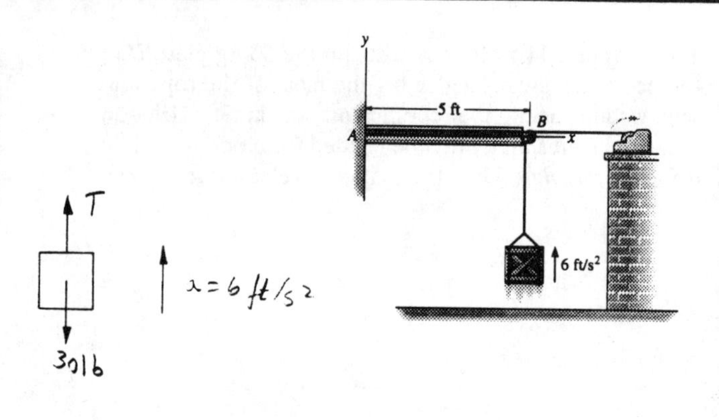

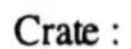

Crate :

$$+\uparrow \Sigma F_y = ma_y; \quad T - 30 = \left(\frac{30}{32.2}\right)(6) \quad T = 35.59 \text{ lb}$$

Beam :

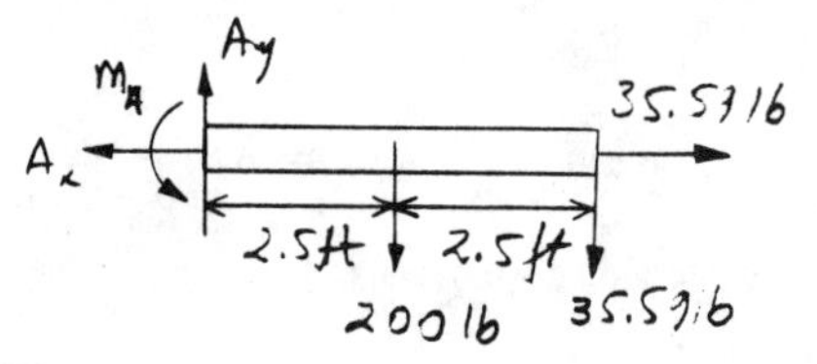

$$\xrightarrow{+} \Sigma F_x = 0; \quad -A_x + 35.59 = 0 \quad A_x = 35.6 \text{ lb} \quad \textbf{Ans}$$

$$+\uparrow \Sigma F_y = 0; \quad A_y - 200 - 35.59 = 0 \quad A_y = 236 \text{ lb} \quad \textbf{Ans}$$

$$\circlearrowleft +\Sigma M_A = 0; \quad M_A - 200(2.5) - (35.59)(5) = 0 \quad M_A = 678 \text{ lb} \cdot \text{ft} \quad \textbf{Ans}$$

***13-36.** If cylinders B and C have a mass of 15 kg and 10 kg, respectively, determine the required mass of A so that it does not move when all the cylinders are released. Neglect the mass of the pulleys and the cords.

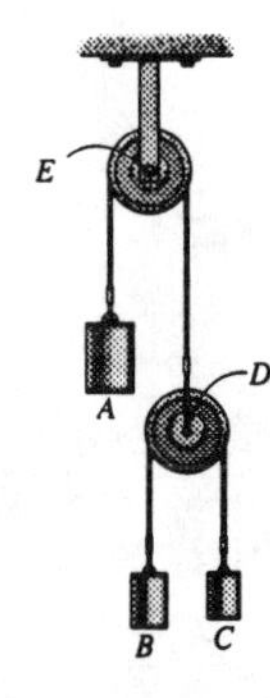

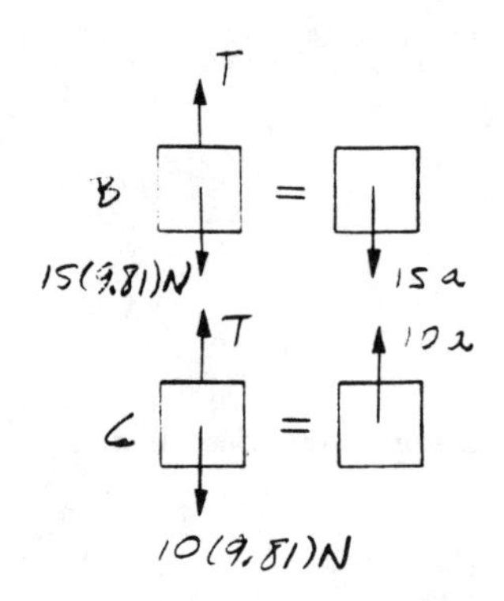

Point D does not move.

For B :

$$+\downarrow \Sigma F_y = ma_y; \quad 15(9.81) - T = 15a$$

For C :

$$+\uparrow \Sigma F_y = ma_y; \quad -10(9.81) + T = 10a$$

Solving,

$$a = 1.962 \text{ m/s}^2 \quad T = 117.72 \text{ N}$$

By statics :

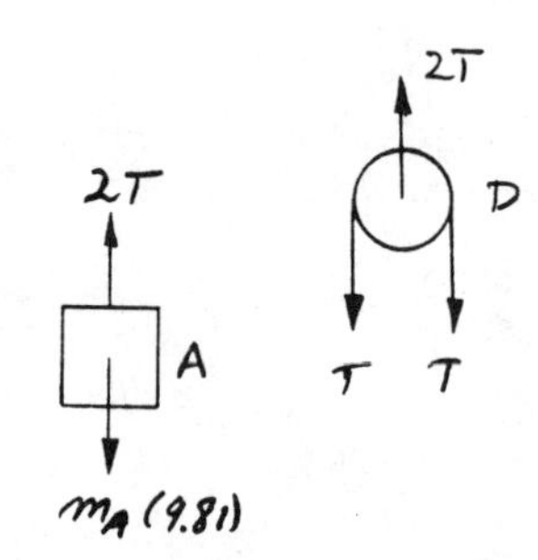

For A :

$$+\uparrow \Sigma F_y = ma_y; \quad 2T - m_A(9.81) = 0$$

$$m_A = \frac{2(117.72)}{9.81} = 24 \text{ kg} \quad \textbf{Ans}$$

13-37. Determine the acceleration of block A when the system is released. The coefficient of kinetic friction and the weight of each block are indicated. Neglect the mass of the pulleys and cord.

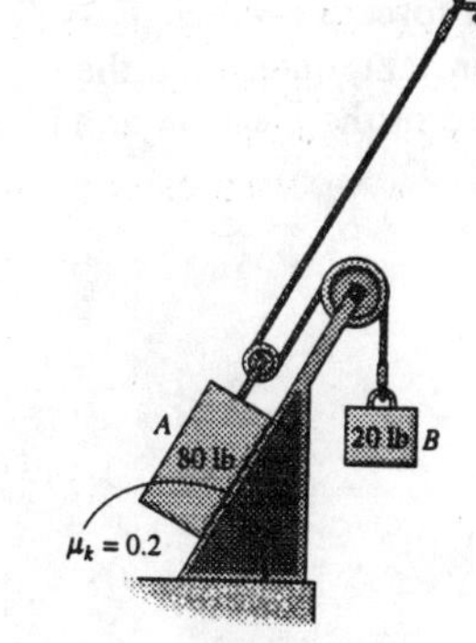

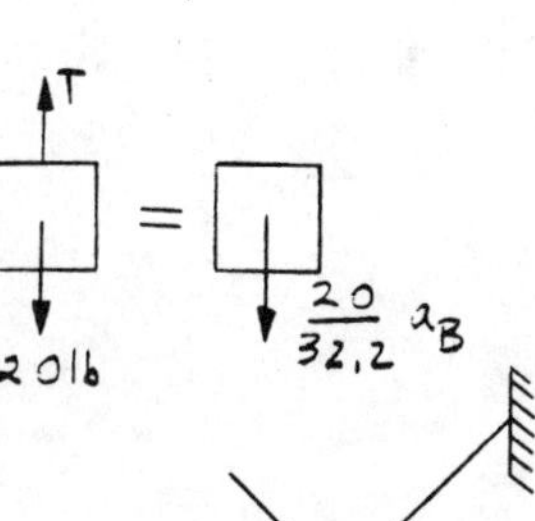

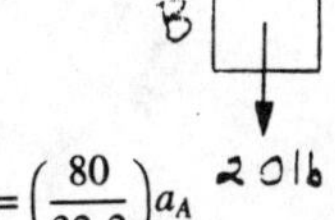

Block A :

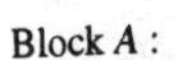

$+\nwarrow \Sigma F_y = ma_y;\quad N_A - 80\cos 60° = 0$

$+\swarrow \Sigma F_x = ma_x;\quad 80\sin 60° - 0.2N_A - 2T = \left(\frac{80}{32.2}\right)a_A$

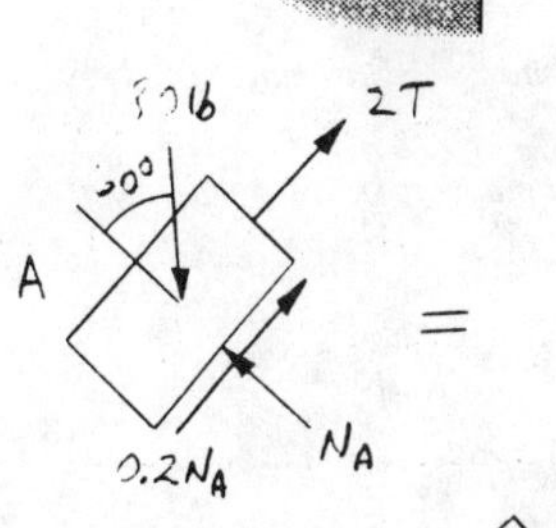

Block B :

$+\downarrow \Sigma F_y = ma_y;\quad -T + 20 = \left(\frac{20}{32.2}\right)a_B$

$2s_A + s_B = l$

$2a_A = -a_B$

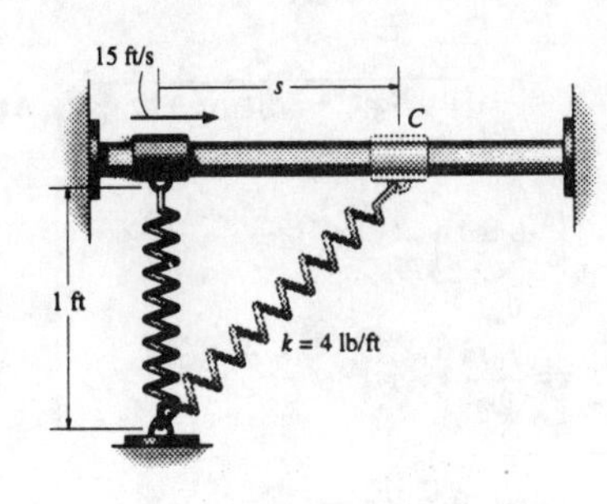

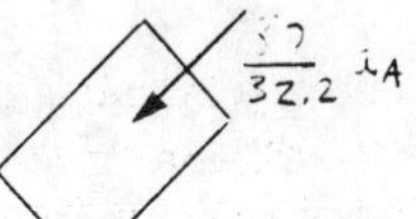

Solving,

$N_A = 40$ lb $\quad T = 25.32$ lb $\quad a_B = -8.57$ ft/s^2

$a_A = 4.28$ ft/s^2 **Ans**

13-38. The 2-lb collar C fits loosely on the smooth shaft. If the spring is unstretched when $s = 0$ and the collar is given a velocity of 15 ft/s, determine the velocity of the collar when $s = 1$ ft.

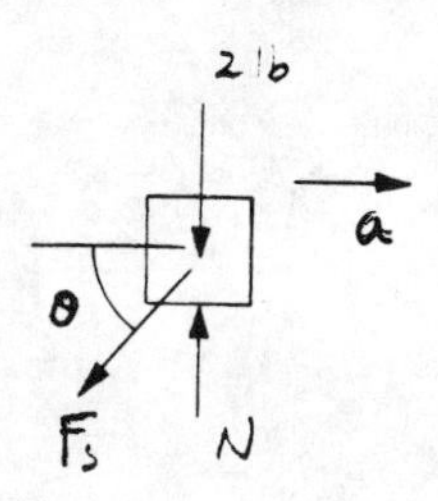

$F_s = kx;\quad F_s = 4\left(\sqrt{1+s^2} - 1\right)$

$\overset{+}{\rightarrow} \Sigma F_x = ma_x;\quad -4\left(\sqrt{1+s^2} - 1\right)\left(\frac{s}{\sqrt{1+s^2}}\right) = \left(\frac{2}{32.2}\right)\left(v\frac{dv}{ds}\right)$

$-\int_0^1 \left(4s\,ds - \frac{4s\,ds}{\sqrt{1+s^2}}\right) = \int_{15}^{v}\left(\frac{2}{32.2}\right)v\,dv$

$-\left[2s^2 - 4\sqrt{1+s^2}\right]_0^1 = \frac{1}{32.2}\left(v^2 - 15^2\right)$

$v = 14.6$ ft/s **Ans**

13-39. An electron of mass m is discharged with an initial horizontal velocity of v_0. If it is subjected to two fields of force for which $F_x = F_0$ and $F_y = 0.3F_0$, where F_0 is constant, determine the equation of the path, and the speed of the electron at any time t.

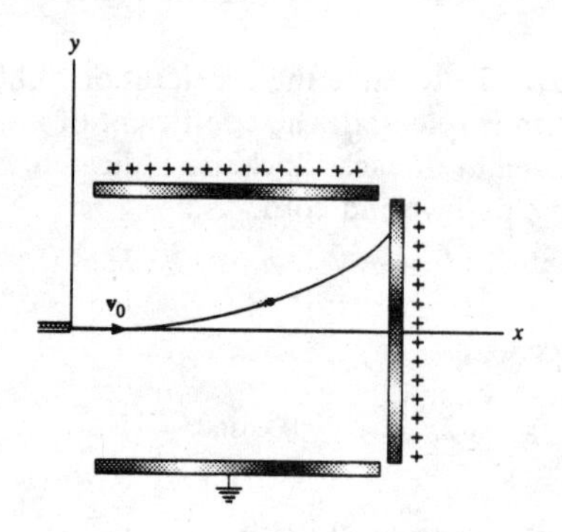

$\xrightarrow{+} \Sigma F_x = ma_x; \qquad F_0 = ma_x$

$+\uparrow \Sigma F_y = ma_y; \qquad 0.3F_0 = ma_y$

Thus,

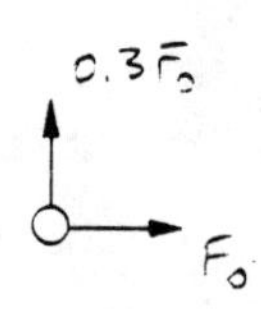

$$\int_{v_0}^{v_x} dv_x = \int_0^t \frac{F_0}{m}\, dt$$

$$v_x = \frac{F_0}{m}t + v_0$$

$$\int_0^{v_y} dv_y = \int_0^t \frac{0.3F_0}{m}\, dt \qquad v_y = \frac{0.3F_0}{m}t$$

$$v = \sqrt{\left(\frac{F_0}{m}t + v_0\right)^2 + \left(\frac{0.3F_0}{m}t\right)^2}$$

$$= \frac{1}{m}\sqrt{1.09F_0^2t^2 + 2F_0 tmv_0 + m^2v_0^2} \qquad \textbf{Ans}$$

$$\int_0^x dx = \int_0^t \left(\frac{F_0}{m}t + v_0\right) dt$$

$$x = \frac{F_0 t^2}{2m} + v_0 t$$

$$\int_0^y dy = \int_0^t \frac{0.3F_0}{m} t\, dt$$

$$y = \frac{0.3F_0 t^2}{2m}$$

$$t = \left(\sqrt{\frac{2m}{0.3F_0}}\right) y^{\frac{1}{2}}$$

$$x = \frac{F_0}{2m}\left(\frac{2m}{0.3F_0}\right)y + v_0\left(\sqrt{\frac{2m}{0.3F_0}}\right) y^{\frac{1}{2}}$$

$$x = \frac{y}{0.3} + v_0\left(\sqrt{\frac{2m}{0.3F_0}}\right) y^{\frac{1}{2}} \qquad \textbf{Ans}$$

***13-40.** In the cathode-ray tube, electrons having a mass m are emitted from a source point S and begin to travel horizontally with an initial velocity $\mathbf{v}_0$. While passing between the grid plates a distance l, they are subjected to a vertical force having a magnitude eV/w, where e is the charge of an electron, V the applied voltage acting across the plates, and w the distance between the plates. After passing clear of the plates, the electrons then travel in straight lines and strike the screen at A. Determine the deflection d of the electrons in terms of the dimensions of the voltage plate and tube. Neglect gravity and the slight vertical deflection which occurs between the plates.

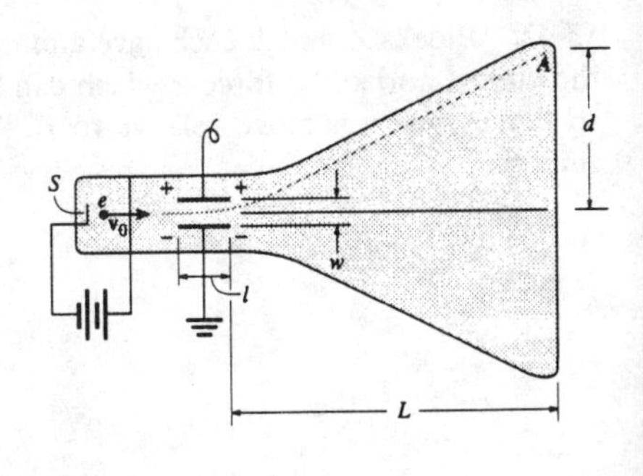

$v_x = v_0$

$t_1 = \dfrac{l}{v_0}$

t_1 is the time between plates.

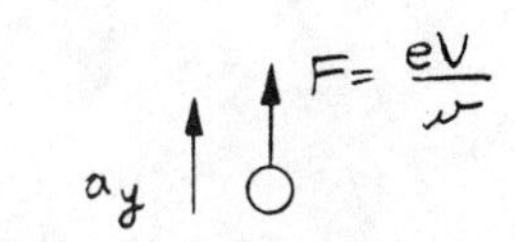

$t_2 = \dfrac{L}{v_0}$

t_2 is the time to reach screen.

$+\uparrow \Sigma F_y = ma_y; \qquad \dfrac{eV}{w} = ma_y$

$a_y = \dfrac{eV}{mw}$

During t_1 constant acceleration,

$(+\uparrow) \qquad v = v_0 + a_c t$

$v_y = a_y t_1 = \left(\dfrac{eV}{mw}\right)\left(\dfrac{l}{v_0}\right)$

During time t_2, $a_y = 0$

$d = v_y t_2 = \left(\dfrac{eVl}{mwv_0}\right)\left(\dfrac{L}{v_0}\right)$

$d = \dfrac{eVLl}{v_0^2 wm}$ **Ans**

13-41. If a horizontal force $P = 12$ lb is applied to block A determine the acceleration of block B. Neglect friction.

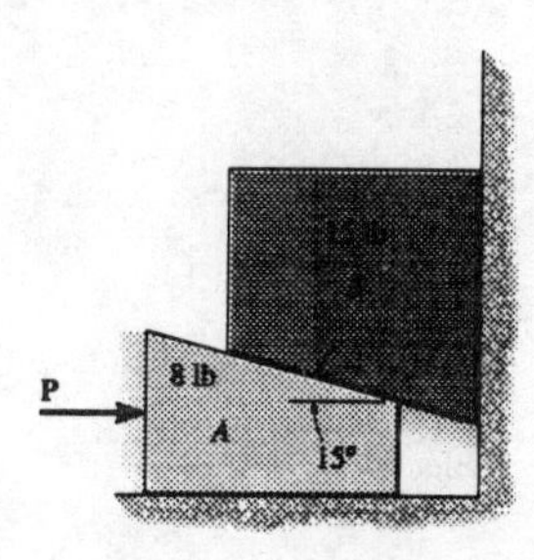

Block A :

$\xrightarrow{+} \Sigma F_x = ma_x; \qquad 12 - N_B \sin 15° = \left(\dfrac{8}{32.2}\right)a_A \qquad (1)$

Block B :

$+\uparrow \Sigma F_y = ma_y; \qquad N_B \cos 15° - 15 = \left(\dfrac{15}{32.2}\right)a_B \qquad (2)$

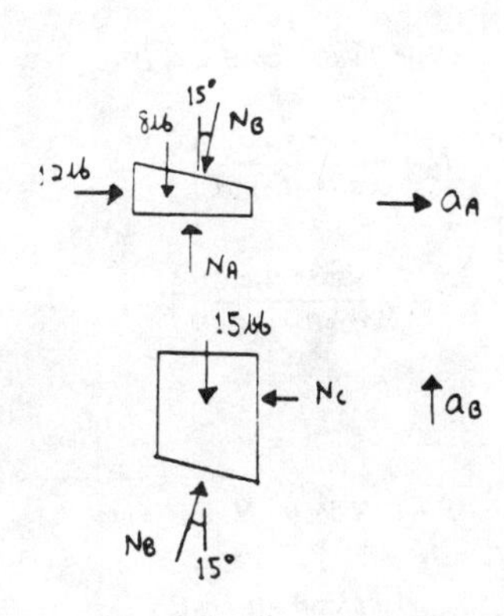

$s_B = s_A \tan 15°$

$a_B = a_A \tan 15° \qquad (3)$

Solving Eqs. (1) – (3),

$a_A = 28.3 \text{ ft/s}^2 \qquad N_B = 19.2 \text{ lb}$

$a_B = 7.59 \text{ ft/s}^2$ **Ans**

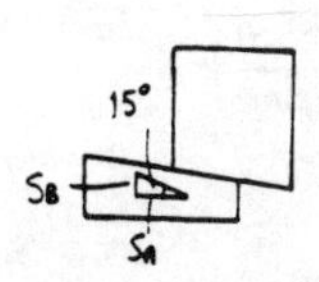

13-42. Blocks A and B each have a mass m. Determine the largest horizontal force P which can be applied to B so that A will not move relative to B. All surfaces are smooth.

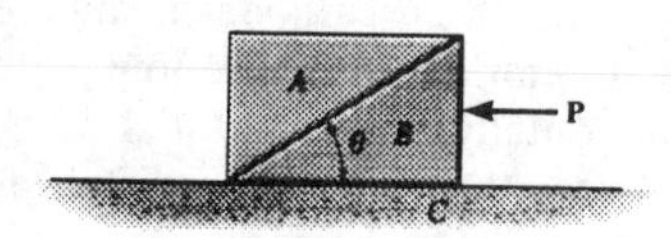

Require

$a_A = a_B = a$

Block A :

$+\uparrow \Sigma F_y = 0; \qquad N\cos\theta - mg = 0$

$\overset{+}{\leftarrow} \Sigma F_x = ma_x; \qquad N\sin\theta = ma$

$$a = g\tan\theta$$

Block B :

$\overset{+}{\leftarrow} \Sigma F_x = ma_x; \qquad P - N\sin\theta = ma$

$$P - mg\tan\theta = mg\tan\theta$$

$$P = 2mg\tan\theta \qquad \textbf{Ans}$$

13-43. Blocks A and B each have a mass m. Determine the largest horizontal force P which can be applied to B so that A will not slip up B. The coefficient of static friction between A and B is μ_s. Neglect any friction between B and C.

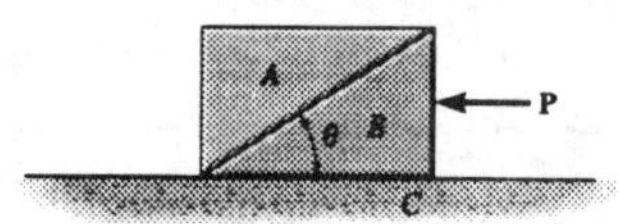

Require

$a_A = a_B = a$

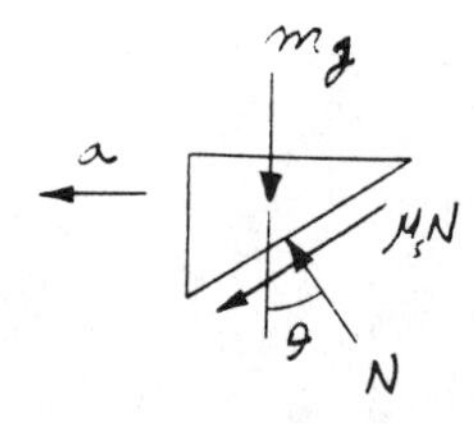

Block A :

$+\uparrow \Sigma F_y = 0; \qquad N\cos\theta - \mu_s N\sin\theta - mg = 0$

$\overset{+}{\leftarrow} \Sigma F_x = ma_x; \qquad N\sin\theta + \mu_s N\cos\theta = ma$

$$N = \frac{mg}{\cos\theta - \mu_s\sin\theta}$$

$$a = g\left(\frac{\sin\theta + \mu_s\cos\theta}{\cos\theta - \mu_s\sin\theta}\right)$$

Block B :

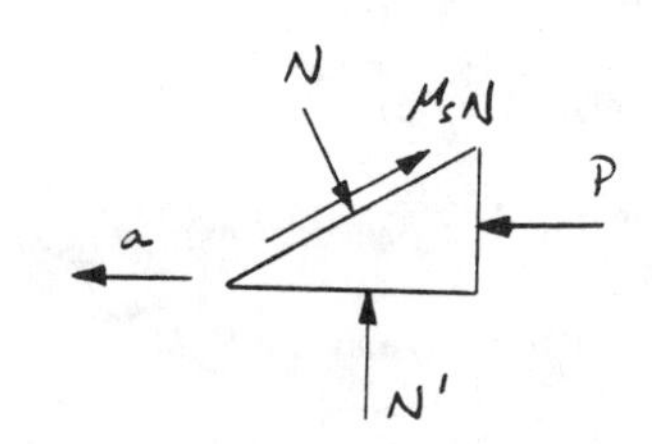

$\overset{+}{\leftarrow} \Sigma F_x = ma_x; \qquad P - \mu_s N\cos\theta - N\sin\theta = ma$

$$P - mg\left(\frac{\sin\theta + \mu_s\cos\theta}{\cos\theta - \mu_s\sin\theta}\right) = mg\left(\frac{\sin\theta + \mu_s\cos\theta}{\cos\theta - \mu_s\sin\theta}\right)$$

$$P = 2mg\left(\frac{\sin\theta + \mu_s\cos\theta}{\cos\theta - \mu_s\sin\theta}\right) \qquad \textbf{Ans}$$

***13-44.** Determine the force P that is needed to pull down the cord with an acceleration of 4 ft/s^2 at the instant $\theta = 30°$. The block has a weight of 5 lb and starts from rest. Neglect the mass of the pulleys and the cord.

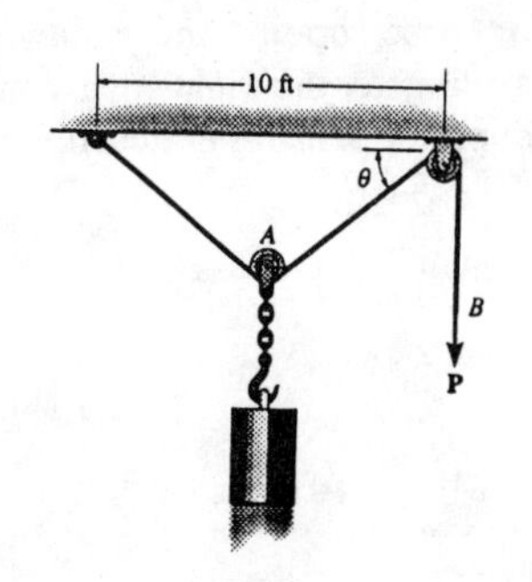

$+\downarrow \Sigma F_y = ma_y;\qquad -2P\sin 30° + 5 = \left(\frac{5}{32.2}\right)a_A \qquad (1)$

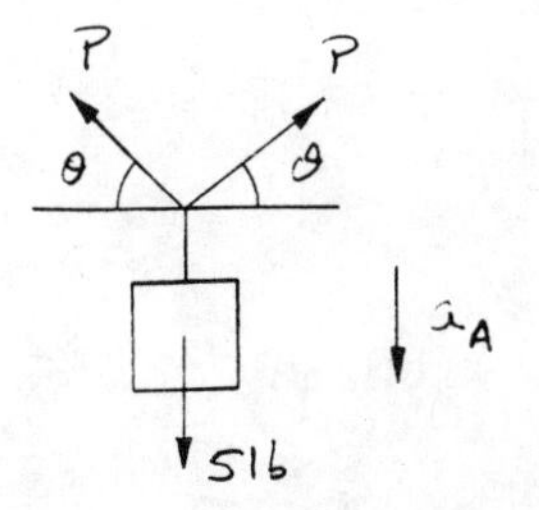

Kinematics :

$$s_B + 2\sqrt{s_A^2 + 5^2} = l$$

$$\dot{s}_B + 2\left(\frac{1}{2}\right)\left(s_A^2 + 25\right)^{-\frac{1}{2}}\left(2s_A\dot{s}_A\right) = 0 \qquad (2)$$

$$\ddot{s}_B - \frac{1}{2}\left(s_A^2 + 25\right)^{-\frac{3}{2}}\left(2s_A\dot{s}_A\right)^2 + \left(s_A^2 + 25\right)^{-\frac{1}{2}}\left(2\dot{s}_A^2 + 2s_A\ddot{s}_A\right) = 0 \qquad (3)$$

Since $\dot{s}_A = 0$, from Eq. (2), $\dot{s}_B = 0$. When $\theta = 30°$.

$s_A = 5\tan 30° = 2.88675$ ft

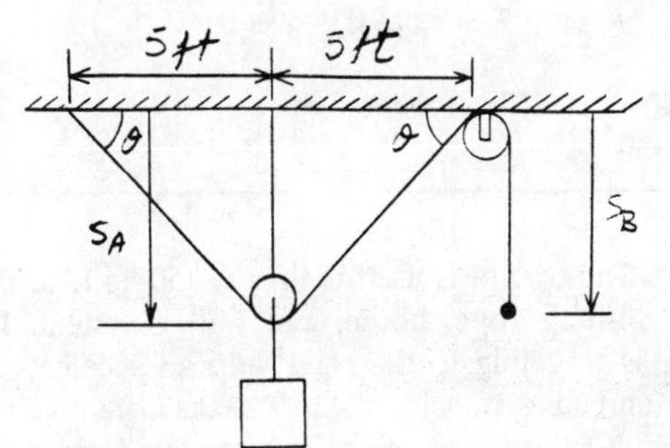

$\ddot{s}_B = 4$

Then from Eq. (3) :

$$4 - 0 + \left[(2.88675)^2 + 25\right]^{-\frac{1}{2}}(0 + 2(2.88675)a_A) = 0$$

$a_A = -4$ ft/s^2

From Eq. (1) :

$$-2P\sin 30° + 5 = \left(\frac{5}{32.2}\right)(-4)$$

$P = 5.62$ lb **Ans**

13-45. Block A has a mass of 10 kg and is hoisted using the rope and pulley system shown. If the collar is moving to the right at 10 m/s and has a deceleration of 5 m/s^2 at the instant shown, determine the tension in the cable connected to the collar at this instant.

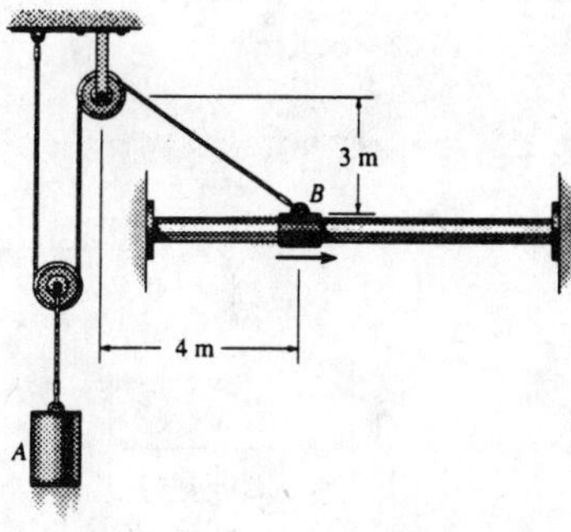

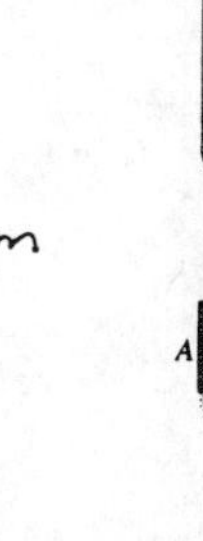

$$2s_A + \sqrt{s_B^2 + 3^2} = l$$

$$2\dot{s}_A + \frac{1}{2}\left(s_B^2 + 9\right)^{-\frac{1}{2}}\left(2s_B\dot{s}_B\right) = 0$$

$$\dot{s}_A = -\frac{s_B\dot{s}_B}{2\left(s_B^2 + 9\right)^{\frac{1}{2}}}$$

$$\ddot{s}_A = -\frac{1}{2}\left(s_B\dot{s}_B\right)\left(-\frac{1}{2}\right)\left(s_B^2 + 9\right)^{-\frac{3}{2}}\left(2s_B\dot{s}_B\right) - \frac{1}{2}\left(s_B^2 + 9\right)^{-\frac{1}{2}}\left(\dot{s}_B\right)^2 - \frac{1}{2}\left(s_B^2 + 9\right)^{-\frac{1}{2}}\left(\ddot{s}_B s_B\right) = 0$$

$$\ddot{s}_A = -\frac{1}{2}\left[\frac{\dot{s}_B^2 + s_B\ddot{s}_B}{\left(s_B^2 + 9\right)^{\frac{1}{2}}} - \frac{s_B^2\dot{s}_B^2}{\left(s_B^2 + 9\right)^{\frac{3}{2}}}\right]$$

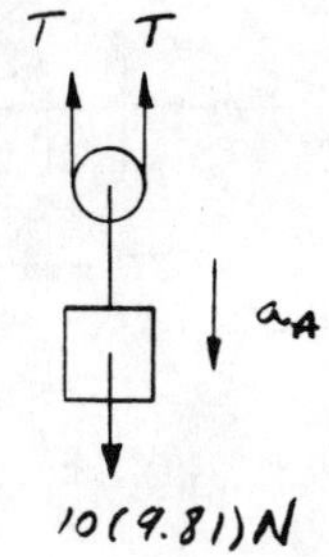

$+\downarrow \Sigma F_y = ma_y;\qquad -2T + 10(9.81) = -10(1.6)$

$$a_A = -\frac{1}{2}\left[\frac{(10)^2 + 4(-5)}{((4)^2 + 9)^{\frac{1}{2}}} - \frac{(4)^2(10)^2}{((4)^2 + 9)^{\frac{3}{2}}}\right] = -1.6\text{ m/s}^2 = 1.6\text{ m/s}^2 \uparrow$$

$T = 57.0$ N **Ans**

***13-46.** The tractor is used to lift the 150-kg load B with the 24-m-long rope, boom, and pulley system. If the tractor is traveling to the right at a constant speed of 4 m/s, determine the tension in the rope when $s_A = 5$ m. When $s_A = 0$, $s_B = 0$.

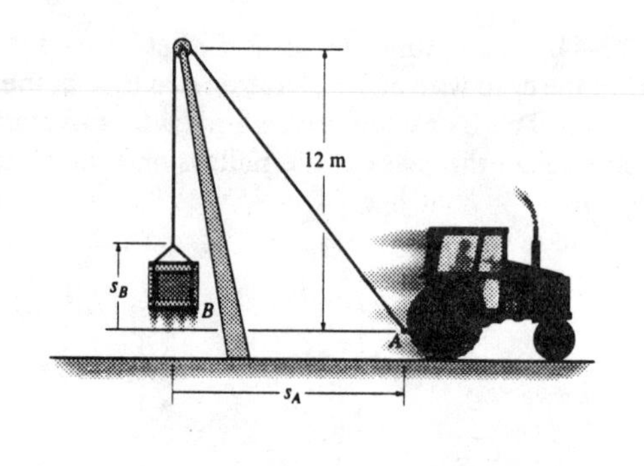

$$12 - s_B + \sqrt{s_A^2 + (12)^2} = 24$$

$$-\dot{s}_B + \left(s_A^2 + 144\right)^{-\frac{1}{2}}\left(s_A \dot{s}_A\right) = 0$$

$$-\ddot{s}_B - \left(s_A^2 + 144\right)^{-\frac{3}{2}}\left(s_A \dot{s}_A\right)^2 + \left(s_A^2 + 144\right)^{-\frac{1}{2}}\left(\dot{s}_A^2\right) + \left(s_A^2 + 144\right)^{-\frac{1}{2}}\left(s_A \ddot{s}_A\right) = 0$$

$$\ddot{s}_B = -\left[\frac{s_A^2 \dot{s}_A^2}{\left(s_A^2 + 144\right)^{\frac{3}{2}}} - \frac{\dot{s}_A^2 + s_A \ddot{s}_A}{\left(s_A^2 + 144\right)^{\frac{1}{2}}}\right]$$

$$a_B = -\left[\frac{(5)^2(4)^2}{((5)^2 + 144)^{\frac{3}{2}}} - \frac{(4)^2 + 0}{((5)^2 + 144)^{\frac{1}{2}}}\right] = 1.0487 \text{ m/s}^2$$

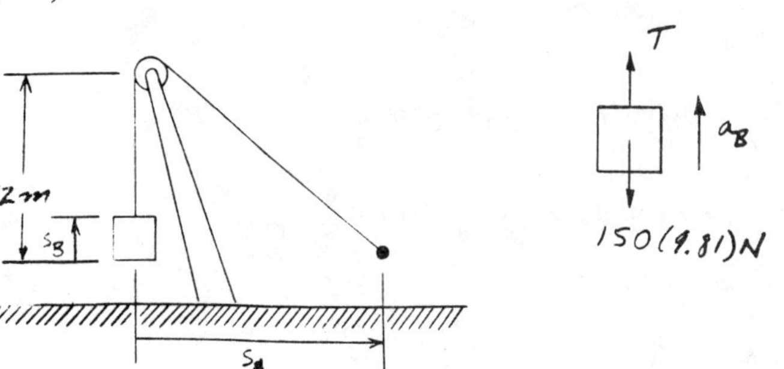

$+\uparrow \Sigma F_y = ma_y;\qquad T - 150(9.81) = 150(1.0487)$

$T = 1.63$ kN **Ans**

13-47. The tractor is used to lift the 150-kg load B with the 24-m-long rope, boom, and pulley system. If the tractor is traveling to the right with an acceleration of 3 m/s^2 and has a velocity of 4 m/s at the instant $s_A = 5$ m, determine the tension in the rope at this instant. When $s_A = 0$, $s_B = 0$.

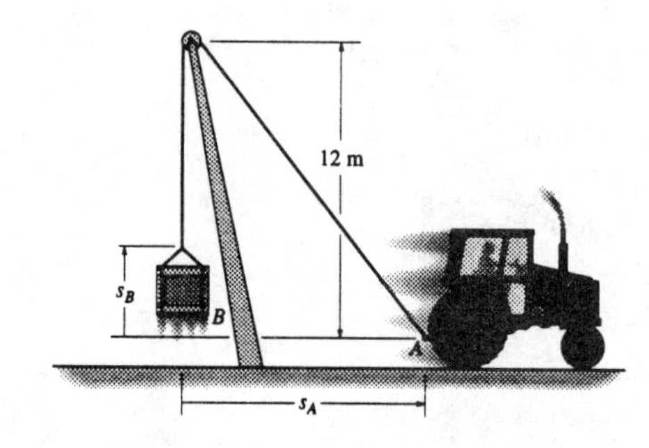

$$12 - s_B + \sqrt{s_A^2 + (12)^2} = 24$$

$$-\dot{s}_B + \frac{1}{2}\left(s_A^2 + 144\right)^{-\frac{1}{2}}\left(2 s_A \dot{s}_A\right) = 0$$

$$-\ddot{s}_B - \left(s_A^2 + 144\right)^{-\frac{3}{2}}\left(s_A \dot{s}_A\right)^2 + \left(s_A^2 + 144\right)^{-\frac{1}{2}}\left(\dot{s}_A^2\right) + \left(s_A^2 + 144\right)^{-\frac{1}{2}}\left(s_A \ddot{s}_A\right) = 0$$

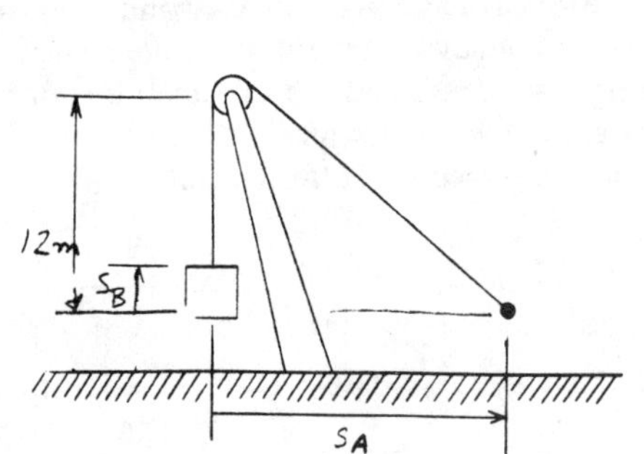

$$\ddot{s}_B = -\left[\frac{s_A^2 \dot{s}_A^2}{\left(s_A^2 + 144\right)^{\frac{3}{2}}} - \frac{\dot{s}_A^2 + s_A \ddot{s}_A}{\left(s_A^2 + 144\right)^{\frac{1}{2}}}\right]$$

$$a_B = -\left[\frac{(5)^2(4)^2}{((5)^2 + 144)^{\frac{3}{2}}} - \frac{(4)^2 + (5)(3)}{((5)^2 + 144)^{\frac{1}{2}}}\right] = 2.2025 \text{ m/s}^2$$

$+\uparrow \Sigma F_y = ma_y;\qquad T - 150(9.81) = 150(2.2025)$

$T = 1.80$ kN **Ans**

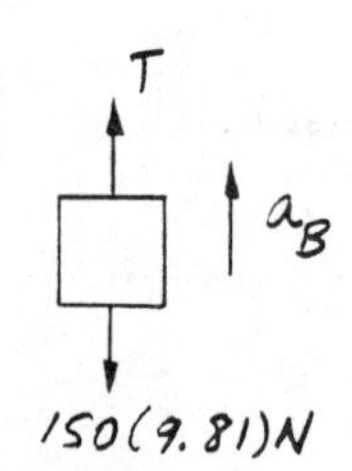

***13-48.** Cylinder B has a mass m and is hoisted using the cord and pulley system shown. Determine the magnitude of force $\mathbf{F}$ as a function of the block's vertical position y so that when $\mathbf{F}$ is applied the block rises with a constant acceleration $\mathbf{a}_B$. Neglect the mass of the cord, pulleys, hook and chain.

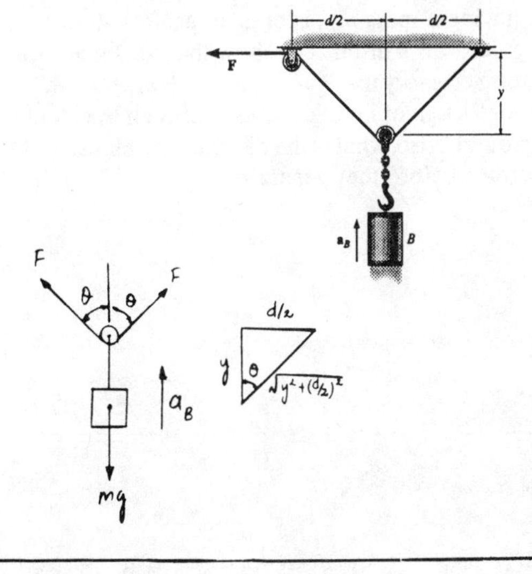

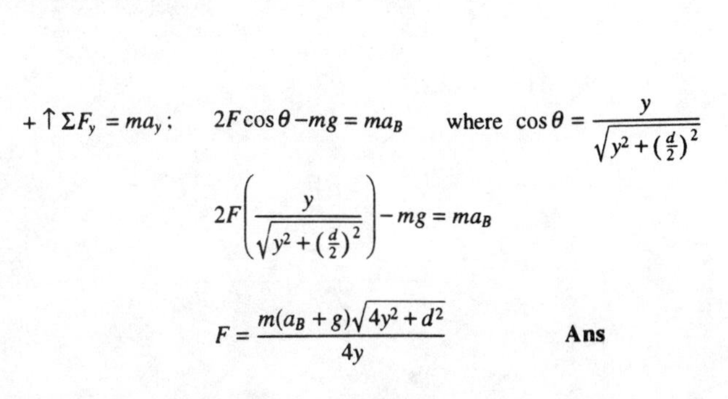

$+\uparrow \Sigma F_y = ma_y;\qquad 2F\cos\theta - mg = ma_B \qquad \text{where } \cos\theta = \dfrac{y}{\sqrt{y^2+(\frac{d}{2})^2}}$

$$2F\left(\frac{y}{\sqrt{y^2+(\frac{d}{2})^2}}\right) - mg = ma_B$$

$$F = \frac{m(a_B+g)\sqrt{4y^2+d^2}}{4y} \qquad \textbf{Ans}$$

13-49. Block A has a mass m_A and is attached to a spring having a stiffness k and unstretched length l_0. If another block B, having a mass m_B, is pressed against A so that the spring deforms a distance d, determine the distance both blocks slide on the smooth surface before they begin to separate. What is their velocity at this instant?

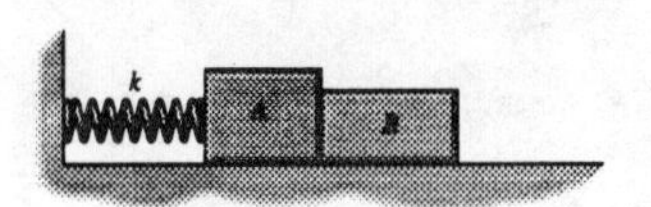

Block A :

$\overset{+}{\rightarrow} \Sigma F_x = ma_x;\qquad -k(x-d) - N = m_A a_A$

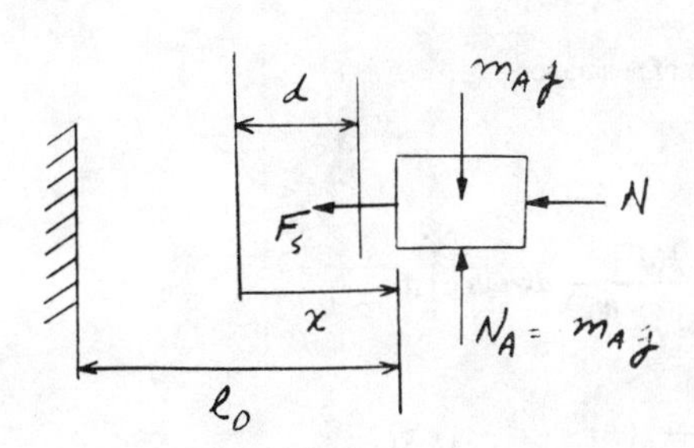

Block B :

$\overset{+}{\rightarrow} \Sigma F_x = ma_x;\qquad N = m_B a_B$

Since $a_A = a_B = a$,

$-k(x-d) - m_B a = m_A a$

$$a = \frac{k(d-x)}{(m_A+m_B)} \qquad N = \frac{km_B(d-x)}{(m_A+m_B)}$$

$N = 0$ when $d - x = 0$, or $\quad x = d \quad$ **Ans**

$v\,dv = a\,dx$

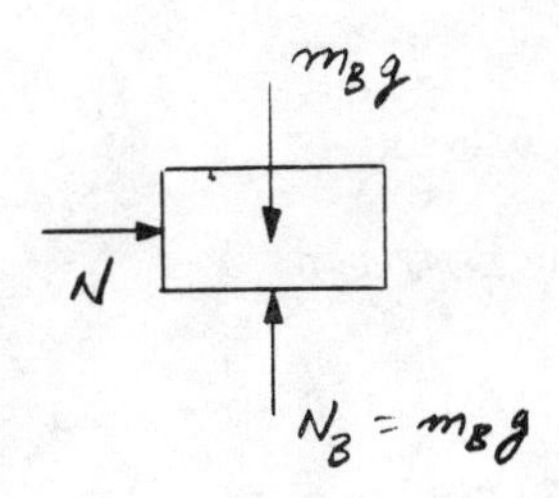

$$\int_0^v v\,dv = \int_0^d \frac{k(d-x)}{(m_A+m_B)}\,dx$$

$$\frac{1}{2}v^2 = \frac{k}{(m_A+m_B)}\left[(d)x - \frac{1}{2}x^2\right]_0^d = \frac{1}{2}\frac{kd^2}{(m_A+m_B)}$$

$$v = \sqrt{\frac{kd^2}{(m_A+m_B)}} \qquad \textbf{Ans}$$

13-50. Block A has a mass m_A and is attached to a spring having a stiffness k and unstretched length l_0. If another block B, having a mass m_B, is pressed against A so that the spring deforms a distance d, show that for separation to occur it is necessary that $d > 2\mu_k g(m_A + m_B)/k$, where μ_k is the coefficient of kinetic friction between the blocks and the ground. Also, what is the distance the blocks slide on the surface before they separate?

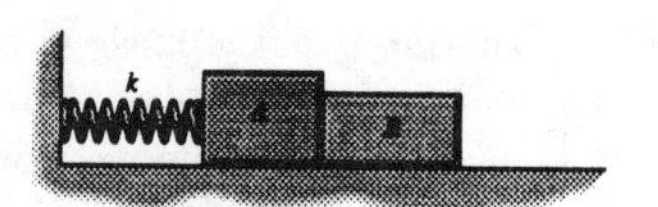

Block A :

$\xrightarrow{+} \Sigma F_x = ma_x;\qquad -k(x-d) - N - \mu_k m_A g = m_A a_A$

Block B :

$\xrightarrow{+} \Sigma F_x = ma_x;\qquad N - \mu_k m_B g = m_B a_B$

Since $a_A = a_B = a$,

$$a = \frac{k(d-x) - \mu_k g(m_A + m_B)}{(m_A + m_B)} = \frac{k(d-x)}{(m_A + m_B)} - \mu_k g$$

$$N = \frac{km_B(d-x)}{(m_A + m_B)}$$

$N = 0$, then $x = d$ for separation. **Ans**

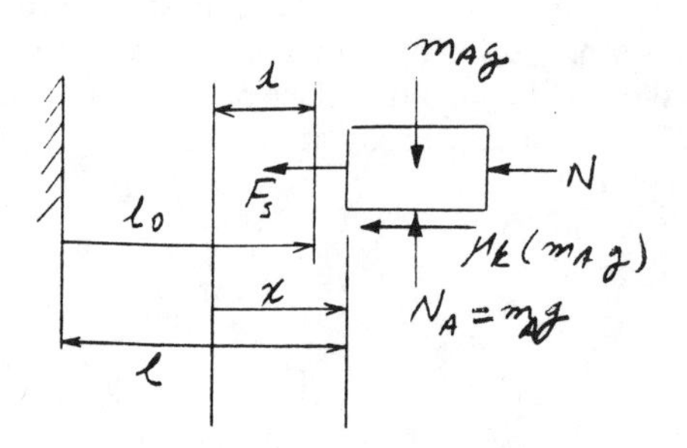

At the moment of separation :

$v\,dv = a\,dx$

$$\int_0^v v\,dv = \int_0^d \left[\frac{k(d-x)}{(m_A + m_B)}dx - \mu_k g\right] dx$$

$$\frac{1}{2}v^2 = \frac{k}{(m_A + m_B)}\left[(d)x - \frac{1}{2}x^2 - \mu_k g\,x\right]_0^d$$

$$v = \sqrt{\frac{kd^2 - 2\mu_k g(m_A + m_B)d}{(m_A + m_B)}}$$

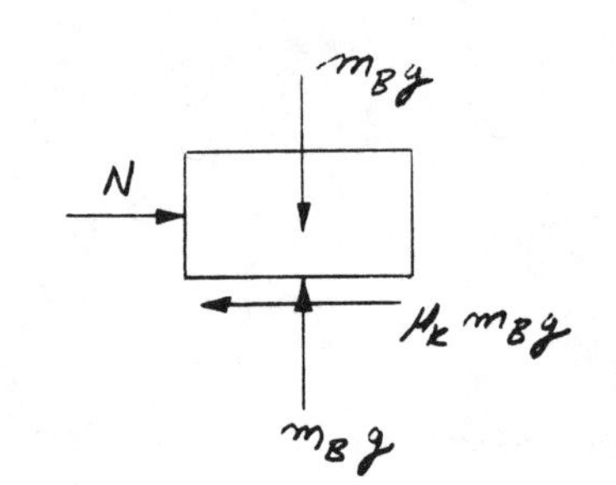

Require $v > 0$, so that

$$kd^2 - 2\mu_k g(m_A + m_B)d > 0$$

Thus,

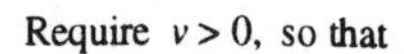

$$kd > 2\mu_k g(m_A + m_B)$$

$$d > \frac{2\mu_k g}{k}(m_A + m_B) \qquad \textbf{Q.E.D.}$$

13-51. The weight of a particle varies with altitude such that $W = m(gr_0^2)/r^2$, where r_0 is the radius of the earth and r is the distance from the earth's center. If the particle is fired vertically with a velocity v_0 from the earth's surface, determine its velocity as a function of position r. What is the smallest velocity v_0 required to escape the earth's gravitational field, what is r_{max}, and what is the time required to reach this altitude?

$+\uparrow \Sigma F_y = ma_y; \qquad -m\left(\frac{gr_0^2}{r^2}\right) = ma$

$$a = -\frac{gr_0^2}{r^2}$$

$m\left(\frac{gr_0^2}{r^2}\right)$

$v\,dv = a\,dr$

$$\int_{v_0}^{v} v\,dv = \int_{r_0}^{r} -gr_0^2\frac{dr}{r^2}$$

$$\frac{1}{2}(v^2 - v_0^2) = -gr_0^2\left[-\frac{1}{r}\right]_{r_0}^{r} = gr_0^2\left(\frac{1}{r} - \frac{1}{r_0}\right)$$

$$v = \sqrt{v_0^2 - 2gr_0\left(1 - \frac{r_0}{r}\right)} \qquad \textbf{Ans}$$

For minimum escape, require $v = 0$,

$$v_0^2 - 2gr_0\left(1 - \frac{r_0}{r}\right) = 0$$

$$r_{max} = \frac{2gr_0^2}{2gr_0 - v_0^2} \qquad \textbf{Ans} \qquad (1)$$

$r_{max} \to \infty$ when $v_0^2 \to 2gr_0$

Escape velocity is

$$v_{esc} = \sqrt{2gr_0} \qquad \textbf{Ans}$$

From Eq. (1), using the value for v_0 from Eq. (2),

$$v = \frac{dr}{dt} = \sqrt{\frac{2gr_0^2}{r}}$$

$$\int_{r_0}^{r}\frac{dr}{\sqrt{\frac{2gr_0^2}{r}}} = \int_0^t dt$$

$$\frac{1}{\sqrt{2gr_0^2}}\left[\frac{2}{3}r^{\frac{3}{2}}\right]_{r_0}^{r_{max}} = t$$

$$t = \frac{2}{3r_0\sqrt{2g}}\left(r_{max}^{\frac{3}{2}} - r_0^{\frac{3}{2}}\right) \qquad \textbf{Ans}$$

***13-52.** Determine the mass of the sun, knowing that the distance from the earth to the sun is 149.6(10⁶) km. *Hint:* Use Eq. 13–1 to represent the force of gravity acting on the earth.

$$\Sigma F_n = ma_n; \qquad G\frac{M_eM_s}{R^2} = M_e\frac{v^2}{R} \qquad M_s = \frac{v^2R}{G}$$

F_G = $m_e\frac{v^2}{R}$

$$v = \frac{s}{t} = \frac{2\pi(149.6)(10^9)}{365(24)(3600)} = 29.81(10^3)\text{ m/s}$$

$$M_s = \frac{[(29.81)(10^3)]^2(149.6)(10^9)}{66.73(10^{-12})} = 1.99(10^{30})\text{ kg} \qquad \textbf{Ans}$$

13-53. The sports car, having a mass of 1700 kg, is traveling horizontally along a 20° banked track which is circular and has a radius of curvature of $\rho = 100$ m. If the coefficient of static friction between the tires and the road is $\mu_s = 0.2$, determine the *maximum constant speed* at which the car can travel without sliding up the slope. Neglect the size of the car.

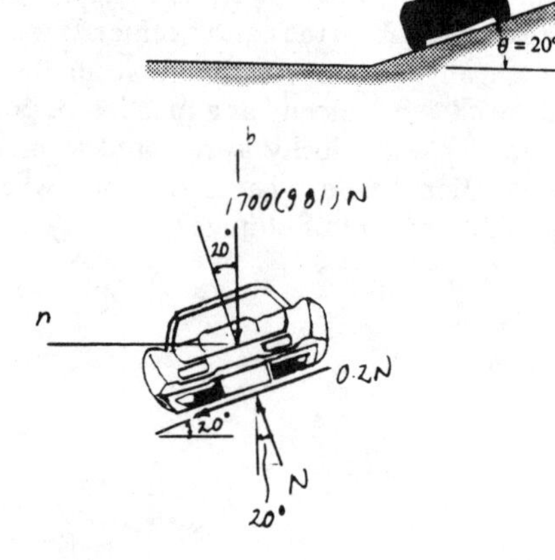

$+\uparrow \Sigma F_b = 0;\qquad N\cos 20° - 0.2N\sin 20° - 1700(9.81) = 0$

$N = 19\,140.6 \text{ N}$

$\overset{+}{\leftarrow} \Sigma F_n = ma_n;\qquad 19\,140.6\sin 20° + 0.2(19\,140.6)\cos 20° = 1700\left(\frac{v_{max}^2}{100}\right)$

$v_{max} = 24.4 \text{ m/s}$ **Ans**

13-54. Using the data in Prob. 13-53, determine the *minimum speed* at which the car can travel around the track without sliding down the slope.

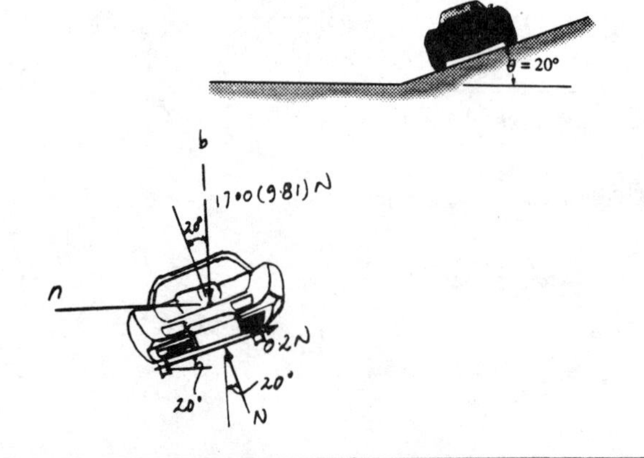

$+\uparrow \Sigma F_b = 0;\qquad N\cos 20° + 0.2N\sin 20° - 1700(9.81) = 0$

$N = 16543.1 \text{ N}$

$\overset{+}{\leftarrow} \Sigma F_n = ma_n;\qquad 16543.1\sin 20° - 0.2(16543.1)\cos 20° = 1700\left(\frac{v_{min}^2}{100}\right)$

$v_{min} = 12.2 \text{ m/s}$ **Ans**

13-55. The 1.40-Mg helicopter is traveling at a constant speed of 40 m/s along the horizontal curved path while banking at $\theta = 30°$. Determine the force acting normal to the blade, i.e., in the y' direction, and the radius of curvature of the path.

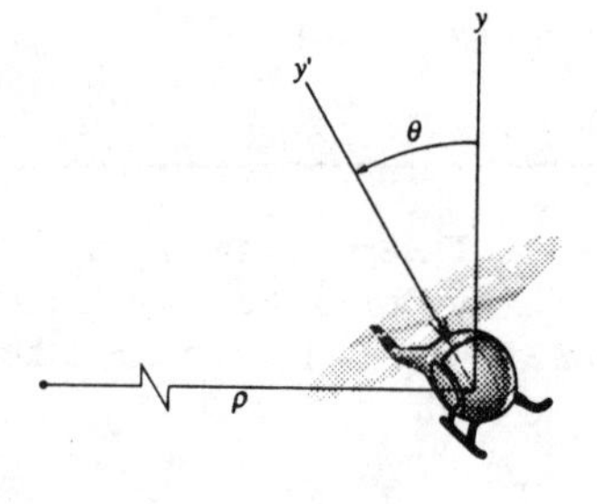

$+\uparrow \Sigma F_y = ma_y;\qquad N\cos 30° - 1400(9.81) = 0$

$N = 15\,859 \text{ N} = 15.9 \text{ kN}$ **Ans**

$\overset{+}{\leftarrow} \Sigma F_n = ma_n;\qquad 15\,859\sin 30° = 1400\left(\frac{(40)^2}{\rho}\right)$

$\rho = 282 \text{ m}$ **Ans**

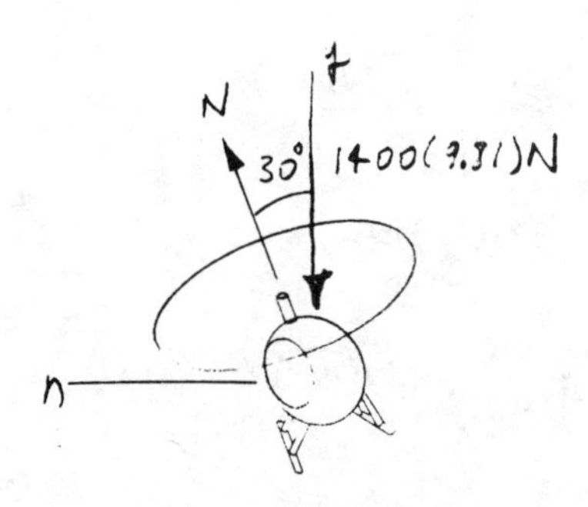

***13-56.** The 1.40-Mg helicopter is traveling at a constant speed of 33 m/s along the horizontal curved path having a radius of curvature of $\rho = 300$ m. Determine the force the blade exerts on the frame and the bank angle θ.

$+\uparrow \Sigma F_y = ma_y; \qquad F_y - 1400(9.81) = 0$

$F_y = 13\ 734$ N

$\overset{+}{\leftarrow} \Sigma F_n = ma_n; \qquad F_n = 1400\left(\frac{(33)^2}{300}\right) = 5082$ N

$F_R = \sqrt{(13\ 734)^2 + (5082)^2} = 14.6$ kN **Ans**

$\theta = \tan^{-1}\left(\frac{F_n}{F_y}\right) = \tan^{-1}\left(\frac{5082}{13\ 734}\right) = 20.3°$ **Ans**

13-57. Determine the constant speed of the satellite S so that it circles the earth with an orbit of radius $r = 20$ Mm. The mass of the earth is $5.976(10^{24})$ kg. *Hint:* Use Eq. 13–1.

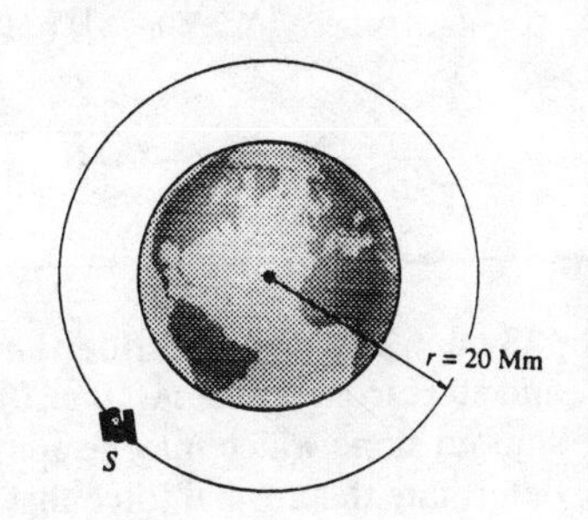

$\Sigma F_n = ma_n; \qquad G\frac{M_e M_s}{r^2} = M_s \frac{v^2}{r}$

$v = \sqrt{\frac{GM_e}{r}} = \sqrt{\frac{66.73(10^{-12})[5.976(10^{24})]}{20(10^6)}} = 4465.3$ m/s

$v = 16.1$ Mm/h **Ans**

13-58. Prove that if the block is released from rest at point B of a smooth path of *arbitrary shape,* the speed it attains when it reaches point A is equal to the speed it attains when it falls freely through a distance h; i.e., $v = \sqrt{2gh}$.

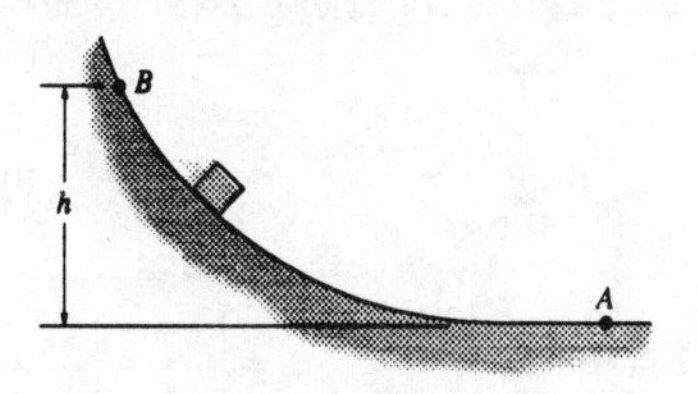

$+\searrow \Sigma F_t = ma_t; \qquad mg\sin\theta = ma_t \qquad a_t = g\sin\theta$

$v\,dv = a_t\,ds = g\sin\theta\,ds \qquad$ However $dy = ds\sin\theta$

$\int_0^v v\,dv = \int_0^h g\,dy$

$\frac{v^2}{2} = gh$

$v = \sqrt{2gh}$ **Q.E.D.**

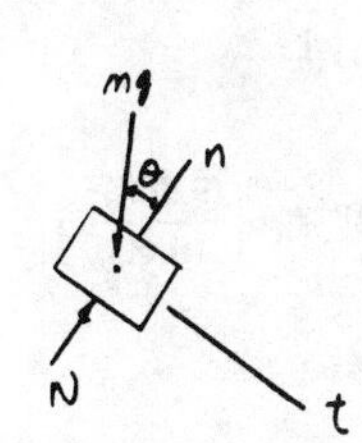

13-59. The sled and rider have a total mass of 80 kg and start from rest at $A(10\text{ m}, 0)$. If the sled descends the smooth slope, which may be approximated by a parabola, determine the normal force that the ground exerts on the sled at the instant it arrives at point B. Neglect the size of the sled and rider. *Hint:* Use the result of Prob. 13-58.

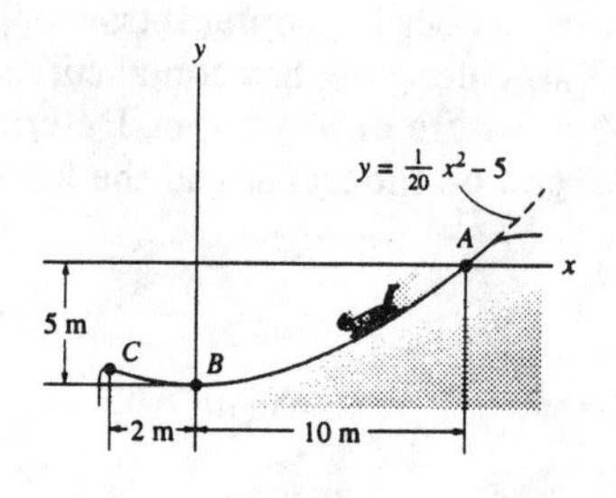

Velocity of the sled : $x = 0,\ \ h = -5$ m.

$$v = \sqrt{2gh} = \sqrt{2(9.81)(5)} = 9.9045\text{ m/s}$$

$$\frac{dy}{dx} = \frac{1}{10}x\bigg|_{x=0} = 0 \qquad \frac{d^2y}{dx^2} = \frac{1}{10}$$

$$\rho = \frac{\left[1+\left(\frac{dy}{dx}\right)^2\right]^{3/2}}{\left|\frac{d^2y}{dx^2}\right|}\Bigg|_{x=0} = \frac{(1+0^2)^{3/2}}{\left|\frac{1}{10}\right|} = 10\text{ m}$$

n

80(9.81) N

t

N_b

$$+\uparrow\Sigma F_n = ma_n; \qquad N_b - 80(9.81) = 80\left(\frac{(9.9045)^2}{10}\right)$$

$$N_b = 1.57\text{ kN} \qquad \textbf{Ans}$$

***13-60.** The sled and rider have a total mass of 80 kg and start from rest at $A(10\text{ m}, 0)$. If the sled descends the smooth slope which may be approximated by a parabola, determine the normal force that the ground exerts on the sled at the instant it arrives at point C. Neglect the size of the sled and rider. *Hint:* Use the result of Prob. 13-58.

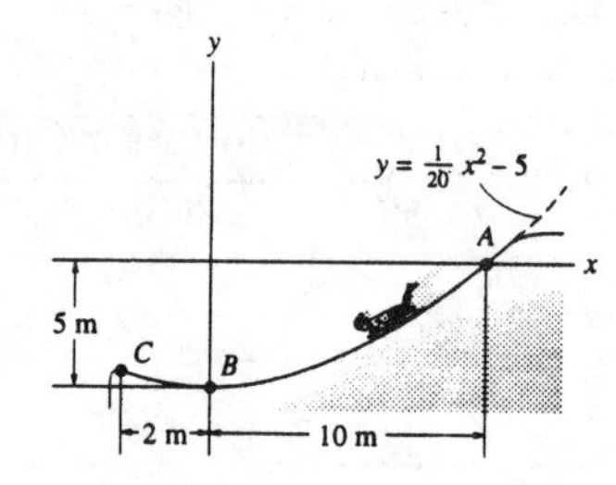

Velocity of the sled : $x = -2$ m, $h = -4.80$ m.

$$v = \sqrt{2gh} = \sqrt{2(9.81)(4.80)} = 9.704\text{ m/s}$$

$$\frac{dy}{dx} = \frac{1}{10}x\bigg|_{x=-2\text{ m}} = -0.2 \qquad \frac{d^2y}{dx^2} = \frac{1}{10}$$

$$\rho = \frac{\left[1+\left(\frac{dy}{dx}\right)^2\right]^{3/2}}{\left|\frac{d^2y}{dx^2}\right|}\Bigg|_{x=-2\text{ m}} = \frac{\left[1+(-0.2)^2\right]^{3/2}}{\left|\frac{1}{10}\right|} = 10.606\text{ m}$$

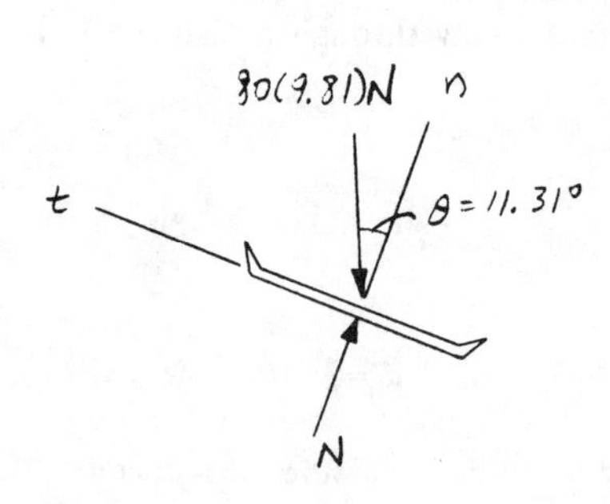

$$\tan\theta = \frac{dy}{dx}\bigg|_{x=-2\text{ m}} = -0.2 \qquad \theta = |11.31°|$$

$$\nearrow + \Sigma F_n = ma_n; \qquad N - 80(9.81)\cos 11.31° = 80\left(\frac{(9.704)^2}{10.606}\right)$$

$$N = 1.48\text{ kN} \qquad \textbf{Ans}$$

13-61. The plane is traveling at a constant speed of 800 ft/s along the curve $y = 20(10^{-6})x^2 + 5000$, where x and y are in feet. If the pilot has a weight of 180 lb, determine the normal and tangential components of the force the seat exerts on the pilot when the plane is at its lowest point.

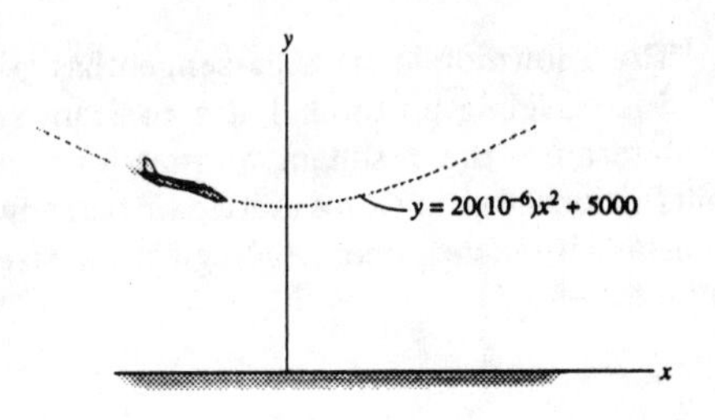

$$y = 20(10^{-6})x^2 + 5000$$

$$\frac{dy}{dx} = 40(10^{-6})x\Big|_{x=0} = 0$$

$$\frac{d^2y}{dx^2} = 40(10^{-6})$$

$$\rho = \frac{\left[1+\left(\frac{dy}{dx}\right)^2\right]^{\frac{3}{2}}}{\left|\frac{d^2y}{dx^2}\right|} = \frac{[1+0]^{\frac{3}{2}}}{|40(10)^{-6}|} = 25(10^3)$$

$+\uparrow \Sigma F_n = ma_n;\quad F_n - 180 = \left(\frac{180}{32.2}\right)\left(\frac{(800)^2}{25(10^3)}\right)$

$F_n = 323 \text{ lb}$ **Ans**

$\overset{+}{\rightarrow} \Sigma F_t = ma_t;\quad F_t = 0$ **Ans**

13-62. The jet plane is traveling at a constant speed of 1000 ft/s along the curve $y = 20(10^{-6})x^2 + 5000$, where x and y are in feet. If the pilot has a weight of 180 lb, determine the normal and tangential components of the force the seat exerts on the pilot when $y = 10\,000$ ft.

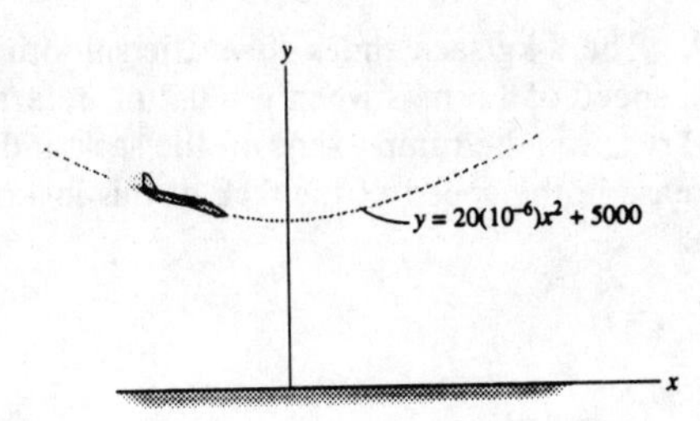

$$y = 20(10^{-6})x^2 + 5000$$

$$10\,000 = 20(10^{-6})x^2 + 5000$$

$$x = 15\,811 \text{ ft}$$

$$\frac{dy}{dx} = \tan\theta = 40(10^{-6})x\Big|_{x=15\,811} = 0.63246$$

$$\theta = 32.31^\circ$$

$$\frac{d^2y}{dx^2} = 40(10^{-6})$$

$$\rho = \frac{\left[1+\left(\frac{dy}{dx}\right)^2\right]^{\frac{3}{2}}}{\left|\frac{d^2y}{dx^2}\right|} = \frac{\left[1+(0.63246)^2\right]^{\frac{3}{2}}}{|40(10^{-6})|} = 41.413(10^3) \text{ ft}$$

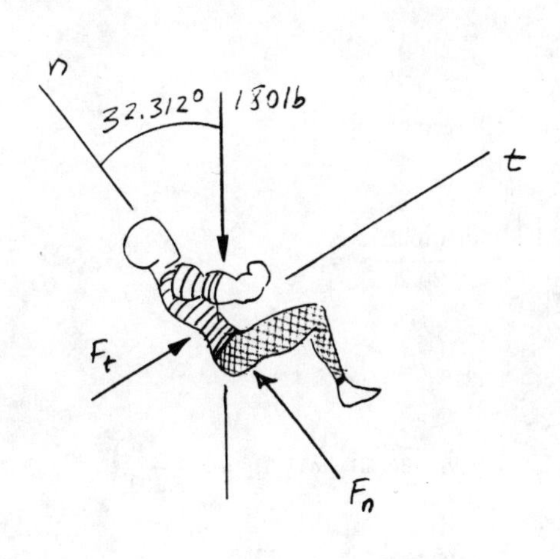

$+\nwarrow \Sigma F_n = ma_n;\quad F_n - 180\cos 32.31^\circ = \left(\frac{180}{32.2}\right)\left(\frac{(1000)^2}{41.413(10)^3}\right)$

$F_n = 287 \text{ lb}$ **Ans**

$+\nearrow \Sigma F_t = ma_t;\quad F_t - 180\sin 32.31^\circ = 0$

$F_t = 96.2 \text{ lb}$ **Ans**

13-63. The snowmobile with passenger has a 200-kg mass and is traveling up the hill at a constant speed of 6 m/s. Determine the resultant normal force and the resultant frictional traction force exerted on the snowmobile at the instant it reaches point A. Neglect the size of the snowmobile.

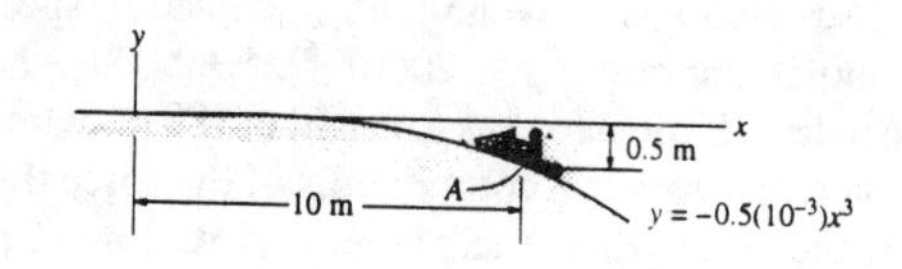

$$y = -0.5(10^{-3})x^3$$

$$\frac{dy}{dx} = -0.0015x^2\Big|_{x=10\text{ m}} = -0.15 \qquad \frac{d^2y}{dx^2} = -0.003x\Big|_{x=10\text{ m}} = -0.03$$

$$\rho = \frac{\left[1+\left(\frac{dy}{dx}\right)^2\right]^{3/2}}{\left|\frac{d^2y}{dx^2}\right|}\Bigg|_{x=10\text{ m}} = \frac{\left[1+(-0.15^2)\right]^{3/2}}{|0.03|} = 34.46\text{ m}$$

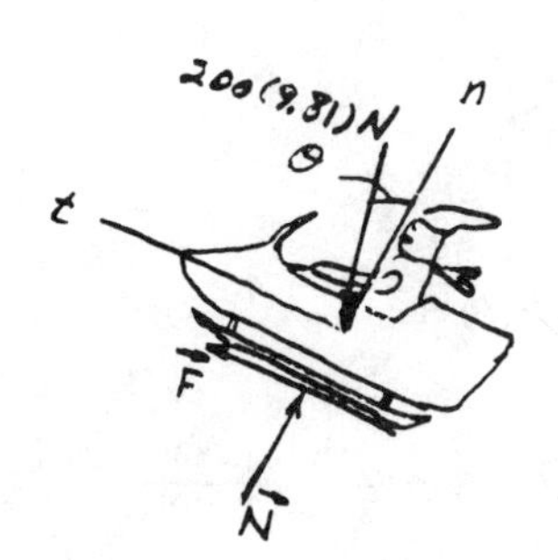

$$\tan\theta = \frac{dy}{dx}\Big|_{x=10\text{ m}} = -0.15 \qquad \theta = -8.531^\circ$$

$+\swarrow \Sigma F_n = ma_n;\qquad -N + 200(9.81)\cos 8.531^\circ = 200\left(\frac{6^2}{34.46}\right)$

$N = 1731.4\text{ N} = 1.73\text{ kN}$ **Ans**

$+\nwarrow \Sigma F_t = ma_t;\qquad F - 200(9.81)\sin 8.531^\circ = 0$

$F = 291\text{ N}$ **Ans**

***13-64.** The 8-kg sack slides down the smooth ramp. If it has a speed of 1.5 m/s when $y = 0.2$ m, determine the normal reaction the ramp exerts on the sack and the rate of increase in the speed of the sack at this instant.

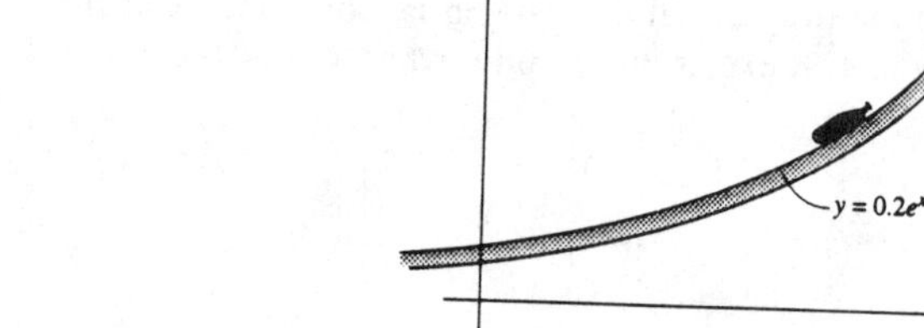

$$y = 0.2 \qquad x = 0$$

$$y = 0.2e^x$$

$$\frac{dy}{dx} = 0.2e^x\Big|_{x=0} = 0.2$$

$$\frac{d^2y}{dx^2} = 0.2e^x\Big|_{x=0} = 0.2$$

$$\rho = \frac{\left[1+\left(\frac{dy}{dx}\right)^2\right]^{\frac{3}{2}}}{\left|\frac{d^2y}{dx^2}\right|} = \frac{\left[1+(0.2)^2\right]^{\frac{3}{2}}}{|0.2|} = 5.303$$

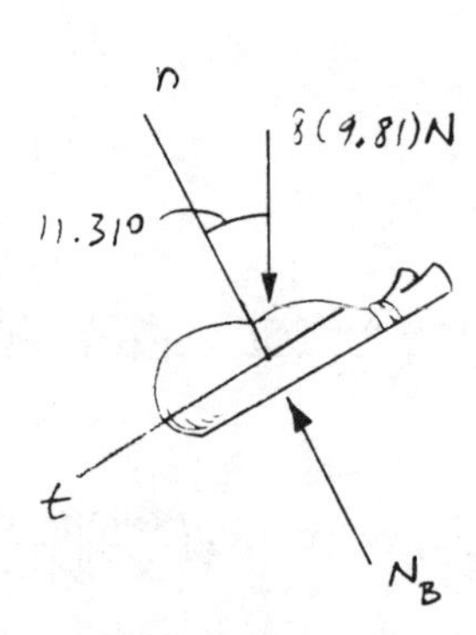

$$\theta = \tan^{-1}(0.2) = 11.31^\circ$$

$+\nwarrow \Sigma F_n = ma_n;\qquad N_B - 8(9.81)\cos 11.31^\circ = 8\left(\frac{(1.5)^2}{5.303}\right)$

$N_B = 80.4\text{ N}$ **Ans**

$+\swarrow \Sigma F_t = ma_t;\qquad 8(9.81)\sin 11.31^\circ = 8a_t$

$a_t = 1.92\text{ m/s}^2$ **Ans**

13-65. The 200-kg snowmobile with passenger is traveling down the hill at a constant speed of 6 m/s. Determine the resultant normal force and the resultant frictional force exerted on the tracks at the instant it reaches point A. Neglect the size of the snowmobile.

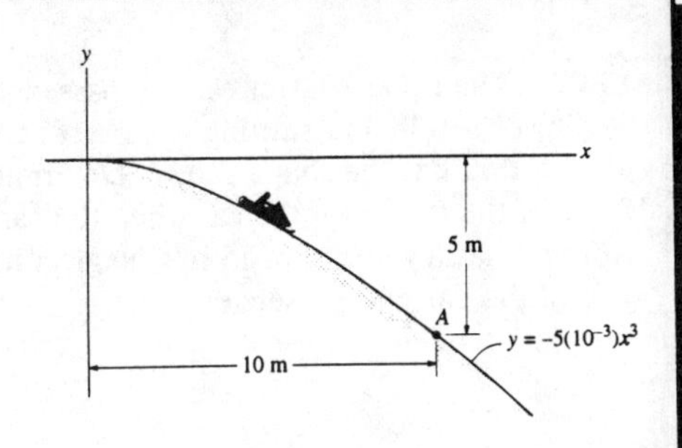

$$y = -5(10^{-3})x^3\big|_{x=10\text{ m}} = -5\text{ m}$$

$$\frac{dy}{dx} = -15(10^{-3})x^2\Big|_{x=10\text{ m}} = -1.5$$

$$\frac{d^2y}{dx^2} = -30(10^{-3})x\Big|_{x=10\text{ m}} = -0.3$$

$$\theta = \tan^{-1}(-1.5) = -56.31^\circ$$

$$\rho = \frac{\left[1+\left(\frac{dy}{dx}\right)^2\right]^{\frac{3}{2}}}{\left|\frac{d^2y}{dx^2}\right|} = \frac{\left[1+(-1.5)^2\right]^{\frac{3}{2}}}{|-0.3|} = 19.53\text{ m}$$

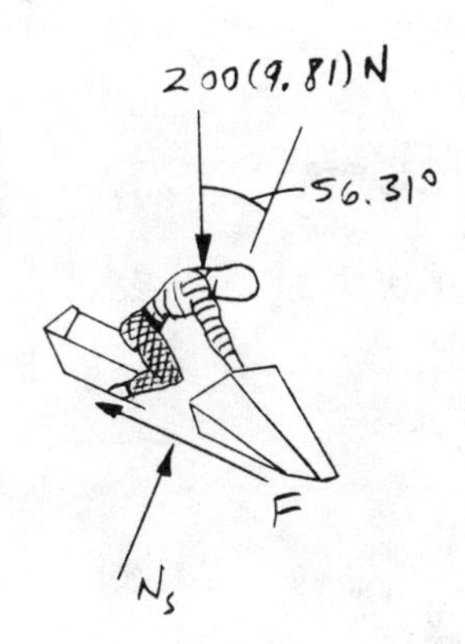

$+\searrow \Sigma F_t = ma_t;\qquad -F + 200(9.81)\sin 56.31^\circ = 0$

$F = 1632\text{ N} = 1.63\text{ kN}$ **Ans**

$+\swarrow \Sigma F_n = ma_n;\qquad -N_s + 200(9.81)\cos 56.31^\circ = 200\left(\frac{(6)^2}{19.53}\right)$

$N_s = 720\text{ N}$ **Ans**

13-66. The 200-kg snowmobile with passenger is traveling down the hill such that when it is at point A, it is traveling at 4 m/s and increasing its speed at 2 m/s^2. Determine the resultant normal force and the resultant frictional force exerted on the tracks at this instant. Neglect the size of the snowmobile.

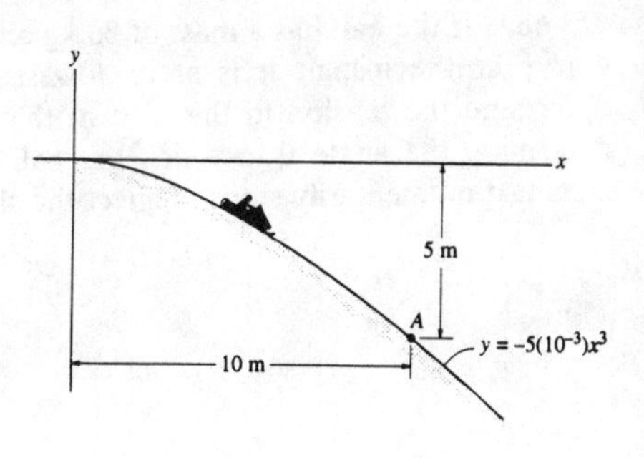

$$y = -5(10^{-3})x^3\big|_{x=10\text{ m}} = -5\text{ m}$$

$$\frac{dy}{dx} = -15(10^{-3})x^2\Big|_{x=10\text{ m}} = -1.5$$

$$\frac{d^2y}{dx^2} = -30(10^{-3})x\Big|_{x=10\text{ m}} = -0.3$$

$$\theta = \tan^{-1}(-1.5) = -56.31^\circ$$

$$\rho = \frac{\left[1+\left(\frac{dy}{dx}\right)^2\right]^{\frac{3}{2}}}{\left|\frac{d^2y}{dx^2}\right|} = \frac{\left[1+(-1.5)^2\right]^{\frac{3}{2}}}{|-0.3|} = 19.53\text{ m}$$

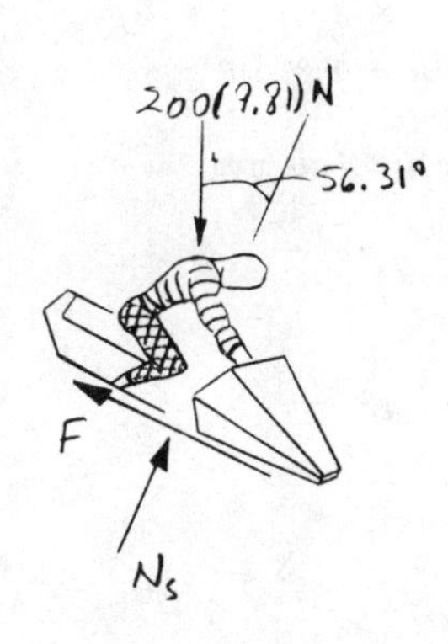

$+\searrow \Sigma F_t = ma_t;\qquad -F + 200(9.81)\sin 56.31^\circ = 200(2)$

$F = 1.23\text{ kN}$ **Ans**

$+\swarrow \Sigma F_n = ma_n;\qquad -N_s + 200(9.81)\cos 56.31^\circ = 200\left(\frac{(4)^2}{19.53}\right)$

$N_s = 924\text{ N}$ **Ans**

13-67. The roller coaster car and passenger have a total weight of 600 lb and starting from rest at A travel down a track that has the shape shown. Determine the normal force of the tracks on the car when the car is at point B, where it has a velocity of 15 ft/s. Neglect friction and the size of the car and passenger.

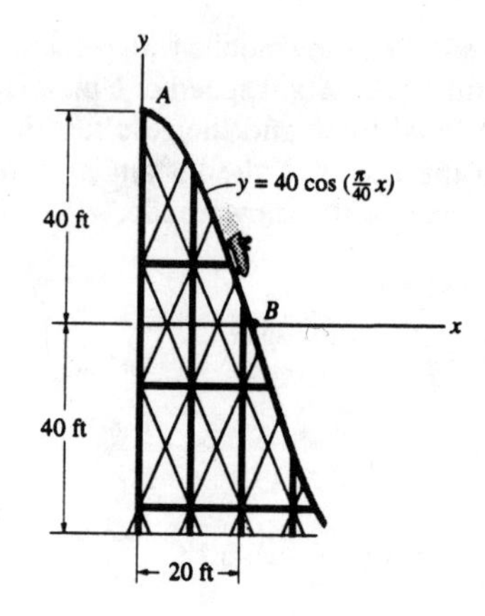

At point B :

$$y = 40\cos\left(\frac{\pi}{40}x\right)$$

$$\frac{dy}{dx} = -\pi\sin\left(\frac{\pi}{40}x\right)\Big|_{x=20} = -\pi$$

$$\theta = \tan^{-1}(-\pi) = -72.34°$$

$$\frac{d^2y}{dx^2} = -\frac{\pi^2}{40}\cos\left(\frac{\pi}{40}x\right)\Big|_{x=20} = 0$$

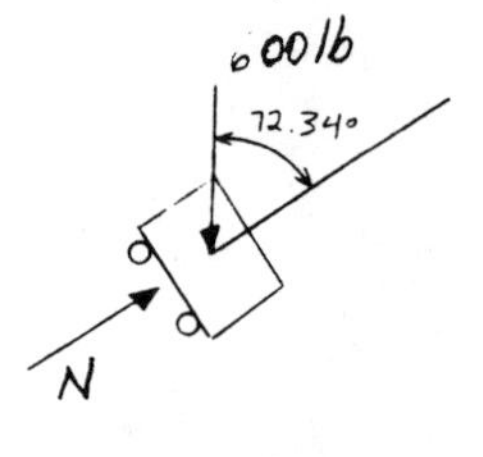

$$\rho = \frac{\left[1+\left(\frac{dy}{dx}\right)^2\right]^{\frac{3}{2}}}{\left|\frac{d^2y}{dx^2}\right|} = \frac{1}{0} \to \infty$$

$$+\nearrow\Sigma F_n = ma_n; \quad N_t - 600\cos 72.34° = \left(\frac{600}{32.2}\right)\left(\frac{(v_B)^2}{\infty}\right)$$

$N_t = 182$ lb **Ans**

***13-68.** If the ball has a mass of 30 kg and a speed v = 4 m/s at the instant it is at its lowest point, $\theta = 0°$, determine the tension in the cord at this instant. Also, determine the angle θ to which the ball swings at the instant it momentarily stops. Neglect the size of the ball.

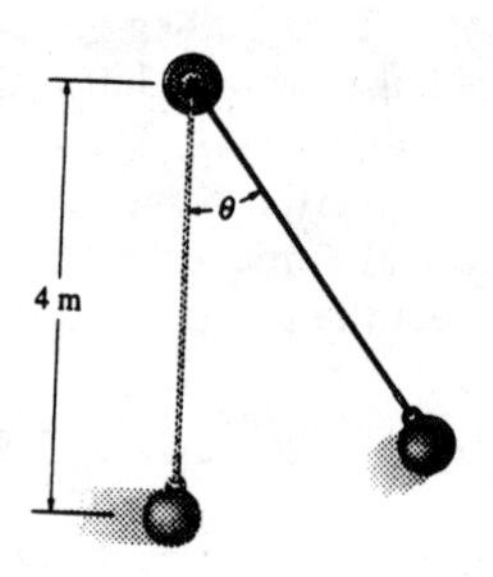

$$+\uparrow\Sigma F_n = ma_n; \quad T - 30(9.81) = 30\left(\frac{(4)^2}{4}\right)$$

$T = 414$ N **Ans**

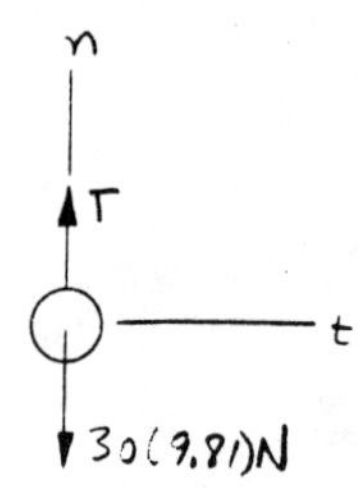

$$+\swarrow\Sigma F_t = ma_t; \quad -30(9.81)\sin\theta = 30a_t$$

$$a_t = -9.81\sin\theta$$

$a_t\,ds = v\,dv$ Since $ds = 4\,d\theta$, then

$$-9.81\int_0^{\theta}\sin\theta\,(4d\theta) = \int_4^0 v\,dv$$

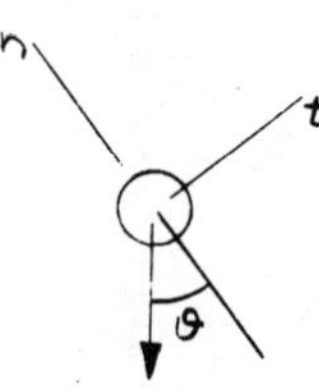

$$9.81(4)\cos\theta\Big]_0^{\theta} = -\frac{1}{2}(4)^2$$

$$39.24(\cos\theta - 1) = -8$$

$\theta = 37.2°$ **Ans**

13-69. The ball has a mass of 30 kg and a speed $v = 4\text{ m/s}$ at the instant it is at its lowest point, $\theta = 0°$. Determine the tension in the cord and the rate at which the ball's speed is decreasing at the instant $\theta = 20°$. Neglect the size of the ball.

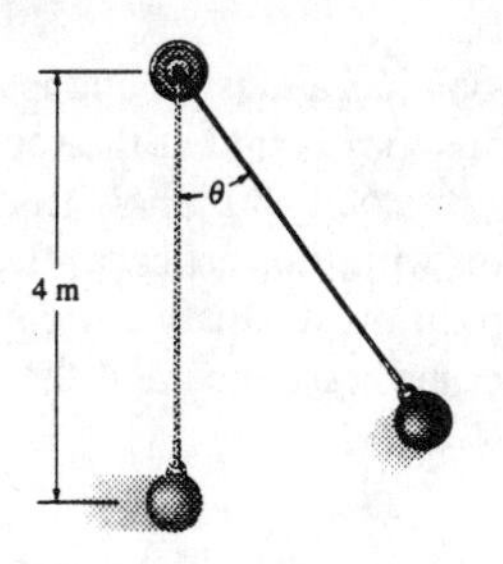

$+\nwarrow \Sigma F_n = ma_n;\quad T - 30(9.81)\cos\theta = 30\left(\frac{v^2}{4}\right)$

$+\nearrow \Sigma F_t = ma_t;\quad -30(9.81)\sin\theta = 30a_t$

$a_t = -9.81\sin\theta$

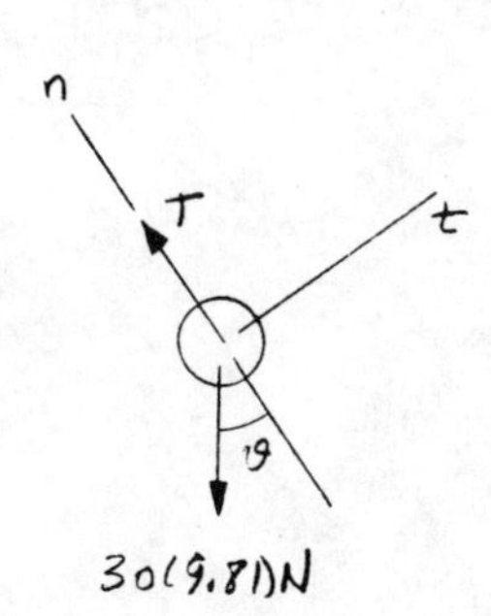

$a_t\,ds = v\,dv$ Since $ds = 4\,d\theta$, then

$-9.81\int_0^\theta \sin\theta\,(4\,d\theta) = \int_4^v v\,dv$

$9.81(4)\cos\theta\Big|_0^\theta = \frac{1}{2}(v)^2 - \frac{1}{2}(4)^2$

$39.24(\cos\theta - 1) + 8 = \frac{1}{2}v^2$

At $\theta = 20°$

$v = 3.357\text{ m/s}$

$a_t = -3.36\text{ m/s}^2 = 3.36\text{ m/s}^2 \swarrow$ **Ans**

$T = 361\text{ N}$ **Ans**

13-70. The 5-kg pendulum bob B is released from rest when $\theta = 0°$. Determine the initial tension in the cord and also at the instant the bob reaches point D, $\theta = 45°$. Neglect the size of the bob.

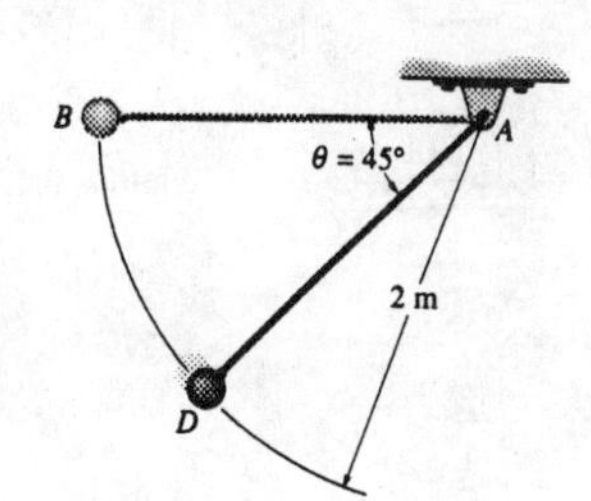

Initially, $v = 0$ so $a_n = 0$.

$\overset{+}{\rightarrow} \Sigma F_n = ma_n;\quad T = 0$ **Ans**

In the general position

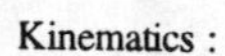

$+\searrow \Sigma F_t = ma_t;\quad 5(9.81)\cos\theta = 5a_t$

$a_t = 9.81\cos\theta$

$+\nearrow \Sigma F_n = ma_n;\quad T - 5(9.81)\sin\theta = 5\left(\frac{v^2}{2}\right)\quad (1)$

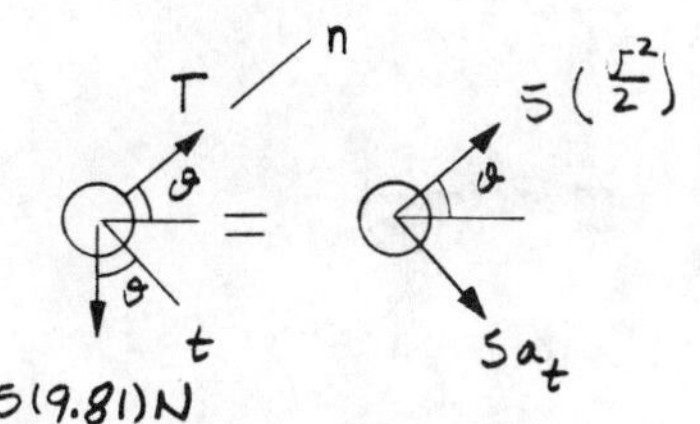

Kinematics :

$v\,dv = a_t\,ds$, where $ds = 2\,d\theta$

$\int_0^v v\,dv = 19.62\int_{0°}^{45°} \cos\theta\,d\theta$

$\frac{1}{2}v^2 = 19.62\left[\sin\theta\right]_{0°}^{45°} = 13.87$

$v = 5.268\text{ m/s}$

Substituting into Eq. (1), with $\theta = 45°$ yields

$T = 104\text{ N}$ **Ans**

13-71. The ball has a mass m and is attached to the cord of length l. The cord is tied at the top to a swivel and the ball is given a velocity $\mathbf{v}_0$. Show that the angle θ which the cord makes with the vertical as the ball travels around the circular path must satisfy the equation $\tan\theta \sin\theta = v_0^2/gl$. Neglect air resistance and the size of the ball.

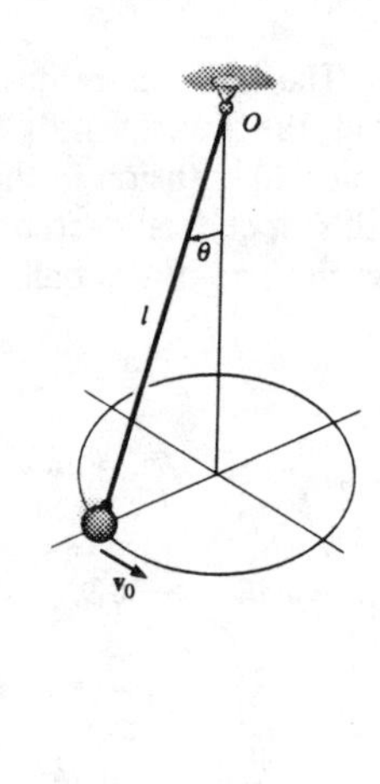

$\xrightarrow{+} \Sigma F_n = ma_n;\quad T\sin\theta = m\left(\frac{v_0^2}{r}\right)$

$+\uparrow \Sigma F_b = 0;\quad T\cos\theta - mg = 0$

Since $r = l\sin\theta \quad T = \dfrac{mv_0^2}{l\sin^2\theta}$

$$\left(\frac{mv_0^2}{l}\right)\left(\frac{\cos\theta}{\sin^2\theta}\right) = mg$$

$$\tan\theta\sin\theta = \frac{v_0^2}{gl}\qquad \textbf{Q.E.D.}$$

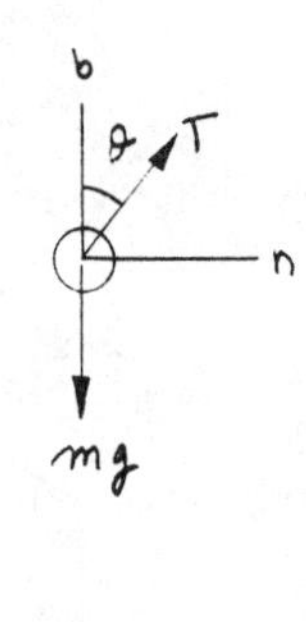

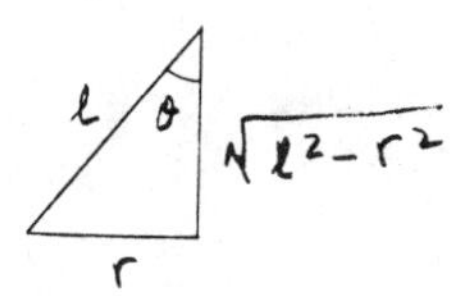

***13-72.** The smooth block B, having a mass of 0.2 kg, is attached to the vertex A of the right circular cone using a light cord. The cone is rotating at a constant angular rate about the z axis such that the block attains a speed of 0.5 m/s. At this speed, determine the tension in the cord and the reaction which the cone exerts on the block. Neglect the size of the block.

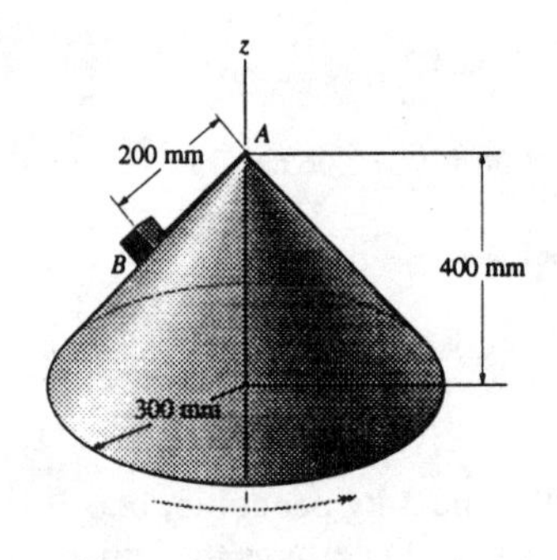

$\dfrac{\rho}{200} = \dfrac{300}{500};\quad \rho = 120\text{ mm} = 0.120\text{ m}$

$+\nearrow \Sigma F_y = ma_y;\quad T - 0.2(9.81)\left(\frac{4}{5}\right) = \left[0.2\left(\frac{(0.5)^2}{0.120}\right)\right]\left(\frac{3}{5}\right)$

$T = 1.82\text{ N}\qquad$ **Ans**

$+\searrow \Sigma F_x = ma_x;\quad N_B - 0.2(9.81)\left(\frac{3}{5}\right) = -\left[0.2\left(\frac{(0.5)^2}{0.120}\right)\right]\left(\frac{4}{5}\right)$

$N_B = 0.844\text{ N}\qquad$ **Ans**

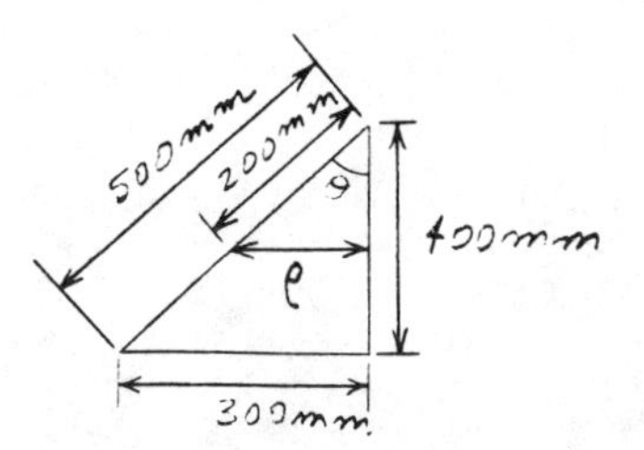

Also,

$\xrightarrow{+} \Sigma F_n = ma_n;\quad T\left(\frac{3}{5}\right) - N_B\left(\frac{4}{5}\right) = 0.2\left(\frac{(0.5)^2}{0.120}\right)$

$+\uparrow \Sigma F_b = 0;\quad T\left(\frac{4}{5}\right) + N_B\left(\frac{3}{5}\right) - 0.2(9.81) = 0$

$T = 1.82\text{ N}\qquad$ **Ans**

$N_B = 0.844\text{ N}\qquad$ **Ans**

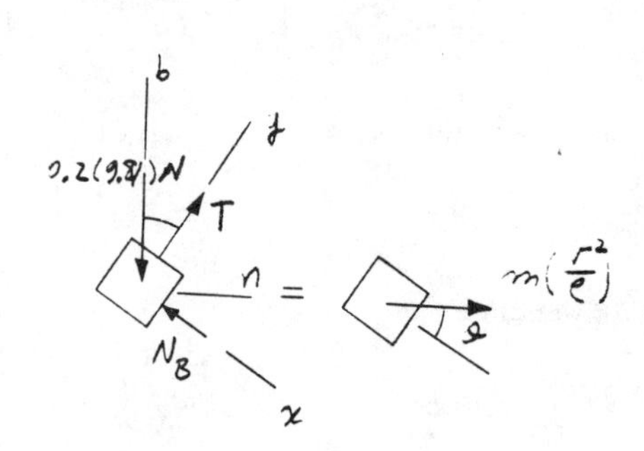

13-73. The rotational speed of the disk is controlled by a 30-g smooth contact arm AB which is spring-mounted on the disk. When the disk is *at rest,* the center of mass G of the arm is located 150 mm from the center O, and the preset compression in the spring is 20 mm. If the initial gap between B and the contact at C is 10 mm, determine the (controlling) speed v_G of the arm's mass center, G, which will close the gap. The disk rotates in the horizontal plane. The spring has a stiffness of $k = 50$ N/m, and its ends are attached to the contact arm at D and to the disk at E.

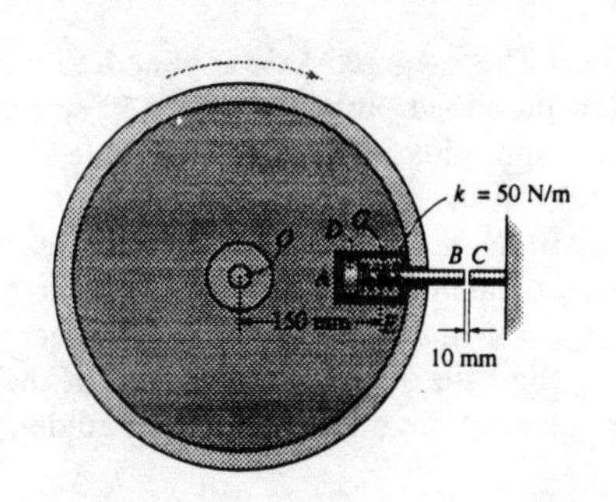

$\overset{+}{\leftarrow} \Sigma F_n = m a_n; \quad 50(0.03) = 0.03\left(\frac{v_G^2}{0.160}\right)$

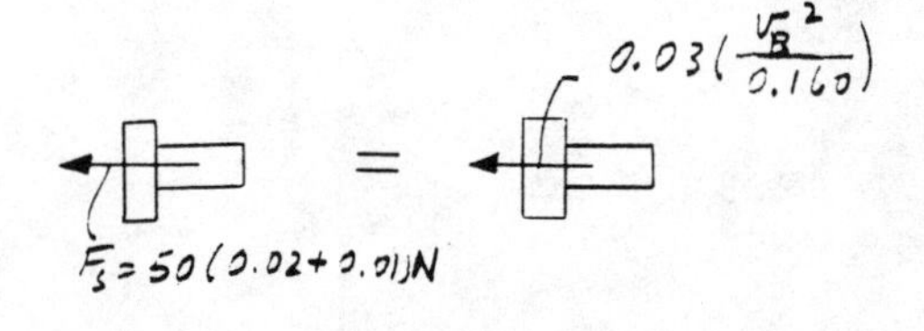

$v_G = 2.83$ m/s **Ans**

13-74. The collar A, having a mass of 0.75 kg, is attached to a spring having a stiffness of $k = 200$ N/m. When rod BC rotates about the vertical axis, the collar slides outward along the smooth rod DE. If the spring is unstretched when $s = 0$, determine the constant speed of the collar in order that $s = 100$ mm. Also, what is the normal force of the rod on the collar? Neglect the size of the collar.

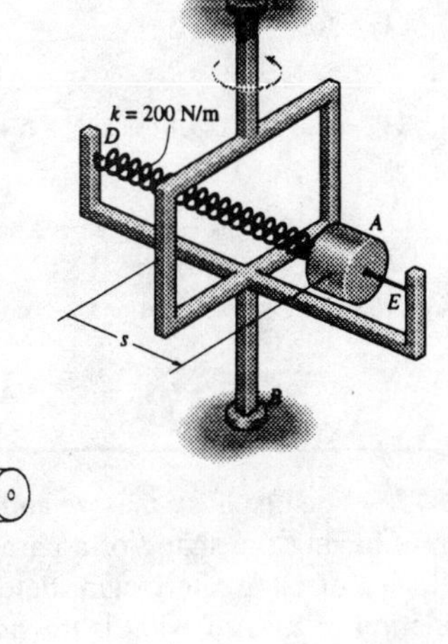

$\Sigma F_b = 0; \quad N_b - 0.75(9.81) = 0 \quad N_b = 7.36$

$\Sigma F_n = m a_n; \quad 200(0.1) = 0.75\left(\frac{v^2}{0.10}\right)$

$\Sigma F_t = m a_t; \quad N_t = 0$

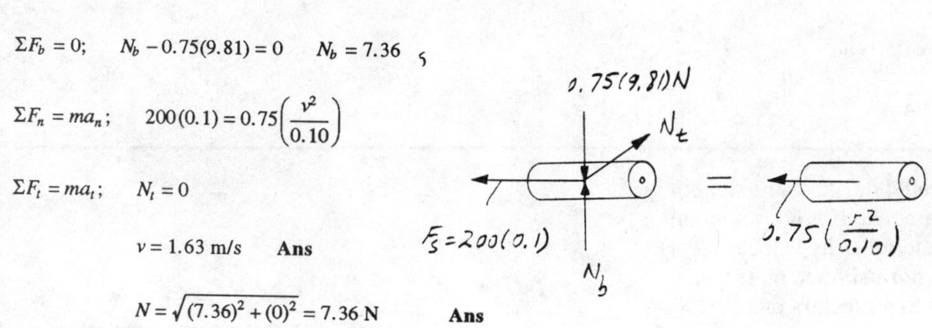

$v = 1.63$ m/s **Ans**

$N = \sqrt{(7.36)^2 + (0)^2} = 7.36$ N **Ans**

13-75. The 10-lb suitcase slides down the curved ramp for which the coefficient of kinetic friction is $\mu_k = 0.2$. If at the instant it reaches point A it has a speed of 5 ft/s, determine the normal force on the suitcase and the rate of increase of its speed.

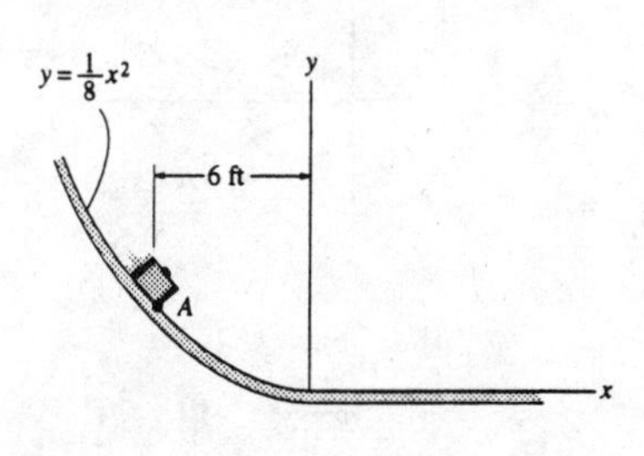

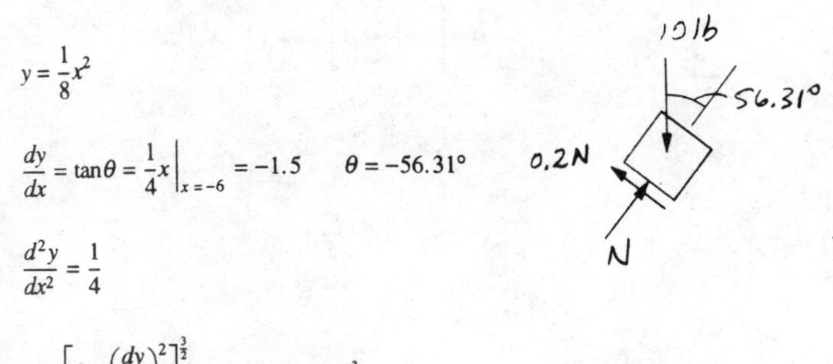

$$y = \frac{1}{8}x^2$$

$$\frac{dy}{dx} = \tan\theta = \frac{1}{4}x\bigg|_{x=-6} = -1.5 \qquad \theta = -56.31^\circ$$

$$\frac{d^2y}{dx^2} = \frac{1}{4}$$

$$\rho = \frac{\left[1+\left(\frac{dy}{dx}\right)^2\right]^{\frac{3}{2}}}{\left|\frac{d^2y}{dx^2}\right|} = \frac{\left[1+(-1.5)^2\right]^{\frac{3}{2}}}{\left|\frac{1}{4}\right|} = 23.436 \text{ ft}$$

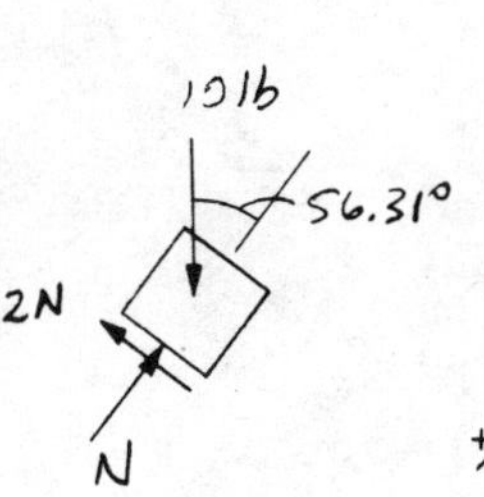

$+\nearrow \Sigma F_n = m a_n; \quad N - 10\cos 56.31^\circ = \left(\frac{10}{32.2}\right)\left(\frac{(5)^2}{23.436}\right)$

$N = 5.8783 = 5.88$ lb **Ans**

$+\searrow \Sigma F_t = m a_t; \quad -0.2(5.8783) + 10\sin 56.31^\circ = \left(\frac{10}{32.2}\right)a_t$

$a_t = 23.0$ ft/s^2 **Ans**

***13-76.** The 6-kg block is confined to move along the smooth parabolic path. The attached spring restricts the motion and, due to the roller guide, always remains horizontal as the block descends. If the spring has a stiffness of $k = 10$ N/m, and an unstretched length of 0.5 m, determine the normal force of the path on the block at the instant $x = 1$ m and the block has a speed of 4 m/s. Also, what is the rate of increase in speed of the block at this point? Neglect the mass of the roller guide and the spring.

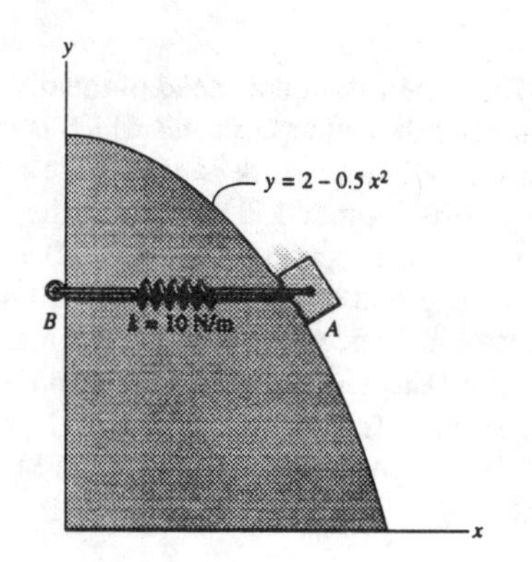

$$y = 2 - 0.5x^2$$

$$\frac{dy}{dx} = \tan\theta = -x\Big|_{x=1} = -1 \qquad \theta = -45^\circ$$

$$\frac{d^2y}{dx^2} = -1$$

$$\rho = \frac{\left[1 + \left(\frac{dy}{dx}\right)^2\right]^{\frac{3}{2}}}{\left|\frac{d^2y}{dx^2}\right|} = \frac{\left[1 + (-1)^2\right]^{\frac{3}{2}}}{|-1|} = 2.8284 \text{ m}$$

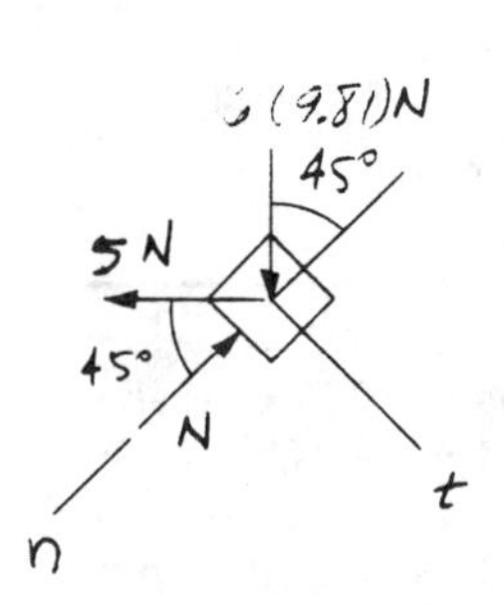

$$F_s = kx = 10(1 - 0.5) = 5 \text{ N}$$

$$+\swarrow \Sigma F_n = ma_n; \qquad 6(9.81)\cos 45^\circ - N + 5\cos 45^\circ = 6\left(\frac{(4)^2}{2.8284}\right)$$

$$N = 11.2 \text{ N} \qquad \textbf{Ans}$$

$$+\searrow \Sigma F_t = ma_t; \qquad 6(9.81)\sin 45^\circ - 5\sin 45^\circ = 6a_t$$

$$a_t = 6.35 \text{ m/s}^2 \qquad \textbf{Ans}$$

13-77. The box has a mass m and slides down the smooth chute having the shape of a parabola. If it has an initial velocity of v_0 at the origin, determine its velocity as a function of x. Also, what is the normal force on the box, and the tangential acceleration as a function of x?

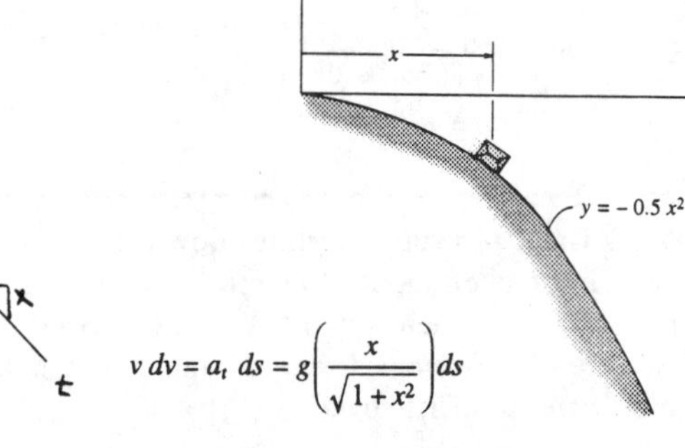

$$y = -\frac{1}{2}x^2$$

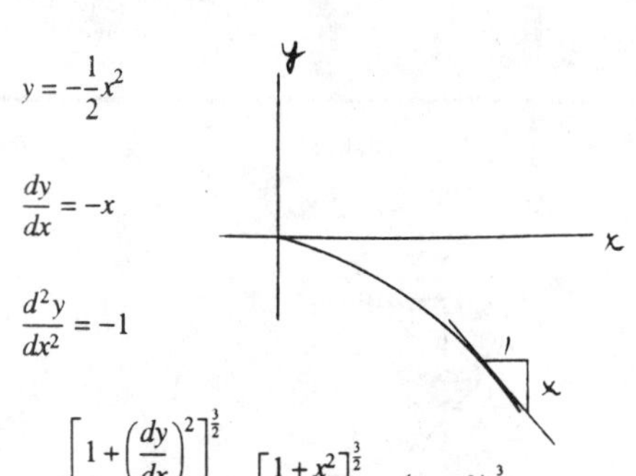

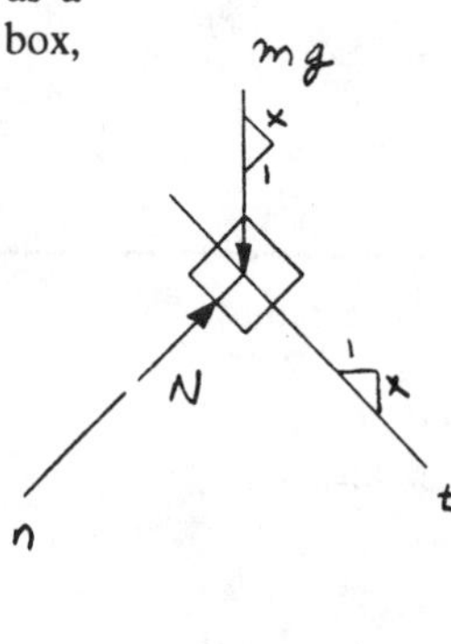

$$\frac{dy}{dx} = -x$$

$$\frac{d^2y}{dx^2} = -1$$

$$\rho = \frac{\left[1 + \left(\frac{dy}{dx}\right)^2\right]^{\frac{3}{2}}}{\left|\frac{d^2y}{dx^2}\right|} = \frac{\left[1 + x^2\right]^{\frac{3}{2}}}{|-1|} = (1 + x^2)^{\frac{3}{2}}$$

$$+\swarrow \Sigma F_n = ma_n; \qquad mg\left(\frac{1}{\sqrt{1+x^2}}\right) - N = m\left(\frac{v^2}{(1+x^2)^{\frac{3}{2}}}\right) \qquad (1)$$

$$+\searrow \Sigma F_t = ma_t; \qquad mg\left(\frac{x}{\sqrt{1+x^2}}\right) = ma_t$$

$$a_t = g\left(\frac{x}{\sqrt{1+x^2}}\right) \qquad \textbf{Ans}$$

$$v\,dv = a_t\,ds = g\left(\frac{x}{\sqrt{1+x^2}}\right)ds$$

$$ds = \left[1 + \left(\frac{dy}{dx}\right)^2\right]^{\frac{1}{2}} dx = (1 + x^2)^{\frac{1}{2}}\,dx$$

$$\int_{v_0}^{v} v\,dv = \int_0^x gx\,dx$$

$$\frac{1}{2}v^2 - \frac{1}{2}v_0^2 = g\left(\frac{x^2}{2}\right)$$

$$v = \sqrt{v_0^2 + gx^2} \qquad \textbf{Ans}$$

From Eq. (1):

$$N = \frac{m}{\sqrt{1+x^2}}\left[g - \frac{(v_0^2 + gx^2)}{(1+x^2)}\right] \qquad \textbf{Ans}$$

13-78. The 35-kg box has a speed of 2 m/s when it is at A on the smooth ramp. If the surface is in the shape of a parabola, determine the normal force on the box at the instant $x = 3$ m. Also, what is the rate of increase in its speed at this instant?

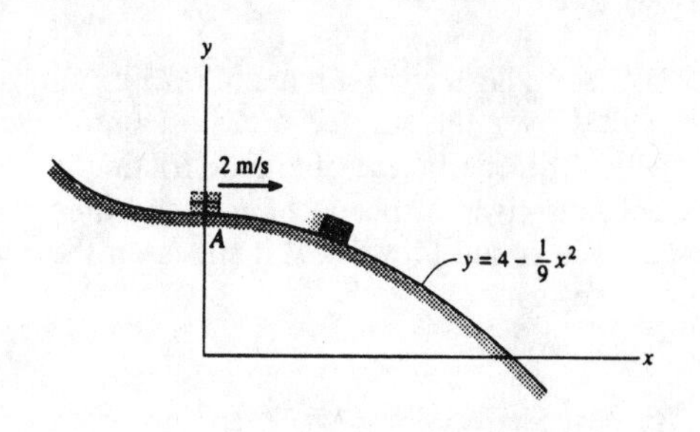

$$y = 4 - \frac{1}{9}x^2$$

$$\frac{dy}{dx} = \tan\theta = -\frac{2}{9}x\Big|_{x=3} = -0.6667 \qquad \theta = -33.69^\circ$$

$$\frac{d^2y}{dx^2} = -\frac{2}{9}$$

$$\rho = \frac{\left[1+\left(\frac{dy}{dx}\right)^2\right]^{\frac{3}{2}}}{\left|\frac{d^2y}{dx^2}\right|} = \frac{\left[1+\left(-\frac{2}{9}x\right)^2\right]^{\frac{3}{2}}}{\left|-\frac{2}{9}\right|} = 4.5\left(1+0.04938x^2\right)^{\frac{3}{2}}\Big|_{x=3} = 7.812 \text{ m}$$

$$+\swarrow\Sigma F_n = ma_n; \qquad 35(9.81)\cos\theta - N = 35\left(\frac{v^2}{7.812}\right) \qquad (1)$$

$$+\searrow\Sigma F_t = ma_t; \qquad 35(9.81)\sin\theta = 35a_t$$

$$a_t = 9.81\sin\theta \qquad (2)$$

$$v\,dv = a_t\,ds$$

$$v\,dv = 9.81\sin\theta\,ds$$

$$v\,dv = 9.81\left(-\frac{dy}{ds}\right)ds = -9.81\,dy$$

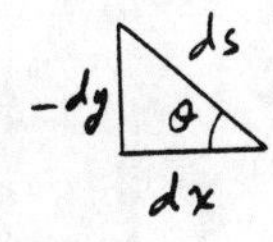

When $x = 0$, $y = 4$. When $x = 3$, $y = 4 - \frac{1}{9}(3)^2 = 3$. Thus,

$$\int_2^v v\,dv = -\int_4^3 9.81\,dy$$

$$\frac{1}{2}v^2 - \frac{1}{2}(2)^2 = -9.81(3-4)$$

$$v = 4.86 \text{ m/s}$$

From Eqs. (1) and (2) for $\theta = 33.69^\circ$

$$35(9.81)\cos 33.69^\circ - N = 35\left(\frac{(4.86)^2}{7.812}\right)$$

$N = 180$ N **Ans**

$a_t = 9.81\sin 33.69^\circ = 5.44 \text{ m/s}^2$ **Ans**

13-79. A collar having a mass of 0.75 kg and negligible size slides over the surface of a horizontal circular rod for which the coefficient of kinetic friction is $\mu_k = 0.3$. If the collar is given a speed of 4 m/s and then released at $\theta = 0°$, determine how far, s, it slides on the rod before coming to rest.

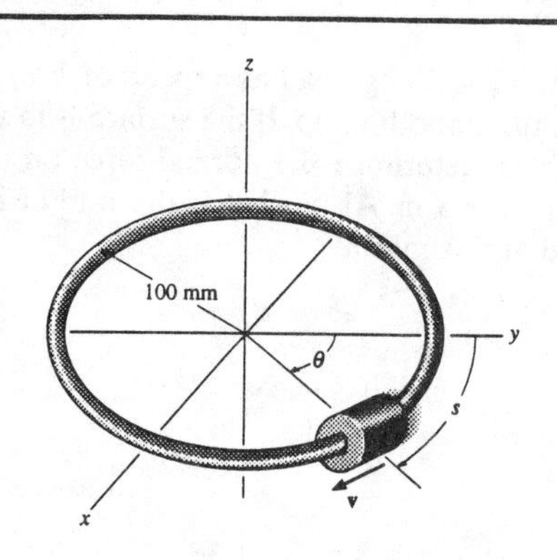

$$\Sigma F_t = ma_t; \qquad -0.3N_C = 0.75a_t \qquad (1)$$

$$\Sigma F_n = ma_n; \qquad (N_C)_n = 0.75\left(\frac{v^2}{0.1}\right)$$

$$\Sigma F_z = 0; \qquad (N_C)_z - 0.75(9.81) = 0 \qquad (N_C)_z = 7.3575$$

Hence, $\quad N_C = \sqrt{(7.3575)^2 + (7.5v^2)^2}$

Since $v\,dv = a_t\,ds$, then from Eq. (1),

$$v\,dv = -\frac{0.3}{0.75}\left(\sqrt{(7.3575)^2 + (7.5v^2)^2}\right)ds$$

$$\int_4^v \frac{v\,dv}{\left(\sqrt{(7.3575)^2 + (7.5v^2)^2}\right)} = \int_0^s -0.4\,ds$$

$$\int_4^v \frac{v\,dv}{\sqrt{(0.981)^2 + v^4}} = -3s$$

$$s = -\frac{1}{6}\ln\left(\frac{v^2 + \sqrt{(0.981)^2 + v^4}}{32.03}\right)$$

When $v = 0$,

$$s = -\frac{1}{6}\ln\left(\frac{0.981}{32.03}\right)$$

$$= 0.581 \text{ m} \qquad \textbf{Ans}$$

***13-80.** The 5-lb collar slides on the smooth rod, so that when it is at A it has a speed of 10 ft/s. If the spring to which it is attached has an unstretched length of 3 ft and a stiffness of $k = 10$ lb/ft, determine the normal force on the collar and the acceleration of the collar at this instant.

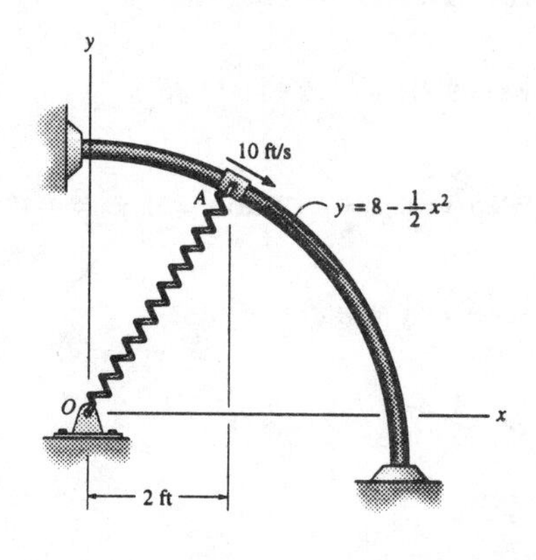

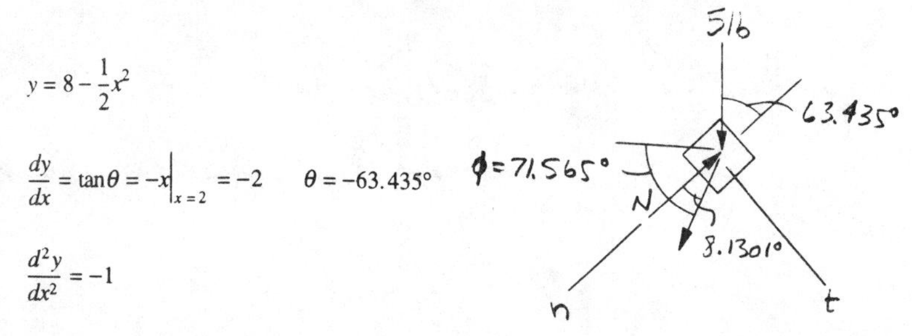

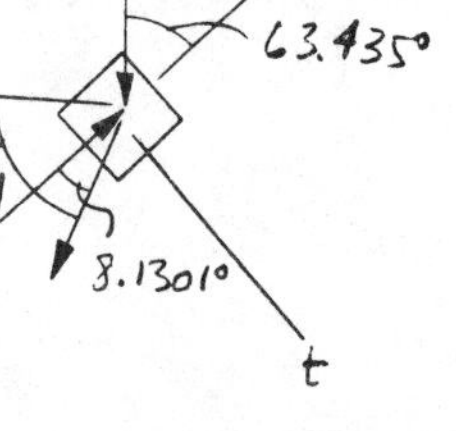

$$y = 8 - \frac{1}{2}x^2$$

$$\frac{dy}{dx} = \tan\theta = -x\Big|_{x=2} = -2 \qquad \theta = -63.435°$$

$$\frac{d^2y}{dx^2} = -1$$

$$\rho = \frac{\left[1 + \left(\frac{dy}{dx}\right)^2\right]^{\frac{3}{2}}}{\left|\frac{d^2y}{dx^2}\right|} = \frac{(1 + (-2)^2)^{\frac{3}{2}}}{|-1|} = 11.18 \text{ ft}$$

$$y = 8 - \frac{1}{2}(2)^2 = 6$$

$$OA = \sqrt{(2)^2 + (6)^2} = 6.3246$$

$$F_s = kx = 10(6.3246 - 3) = 33.246 \text{ lb}$$

$$\tan\phi = \frac{6}{2} = 71.565°$$

$$+\swarrow \Sigma F_n = ma_n; \qquad 5\cos 63.435° - N + 33.246\cos 8.1301° = \left(\frac{5}{32.2}\right)\left(\frac{(10)^2}{11.18}\right)$$

$$N = 33.8 \text{ lb} \qquad \textbf{Ans}$$

$$+\searrow \Sigma F_t = ma_t; \qquad 5\sin 63.435° + 33.246\sin 8.1301° = \left(\frac{5}{32.2}\right)a_t$$

$$a_t = 59.08 \text{ ft/s}^2$$

$$a_n = \frac{v^2}{\rho} = \frac{(10)^2}{11.18} = 8.9443 \text{ ft/s}^2$$

$$a = \sqrt{(59.08)^2 + (8.9443)^2} = 59.8 \text{ ft/s}^2 \qquad \textbf{Ans}$$

13-81. The 5-lb packages ride on the surface of the conveyor belt. If the belt starts from rest and increases to a constant speed of 2 ft/s in 2 s, determine the maximum angle θ so that none of the packages slip on the inclined surface AB of the belt. The coefficient of static friction between the belt and a package is $\mu_s = 0.3$. At what angle ϕ do the packages first begin to slip off the surface of the belt after the belt is moving at its constant speed of 2 ft/s? Neglect the size of the packages.

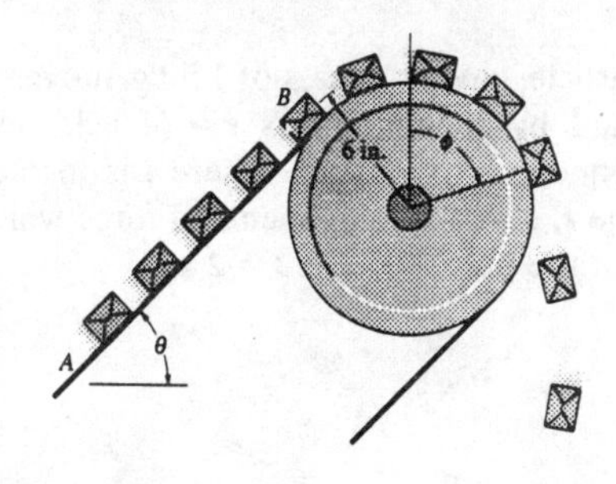

$v = v_1 + a_c t; \quad 2 = 0 + a_c(2); \quad a_c = 1 \text{ ft/s}^2$

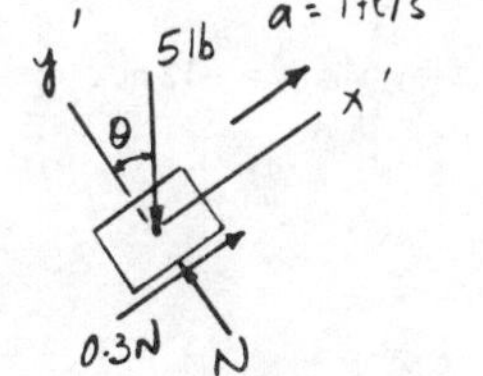

$+\nwarrow \Sigma F_{y'} = ma_{y'}; \qquad N - 5\cos\theta = 0 \qquad (1)$

$+\nearrow \Sigma F_{x'} = ma_{x'}; \qquad 0.3N - 5\sin\theta = \frac{5}{32.2}(1) \qquad (2)$

Solving Eqs. (1) and (2) yields :

$\theta = 15.0°$ **Ans**

$N = 4.83$ lb

For circular motion

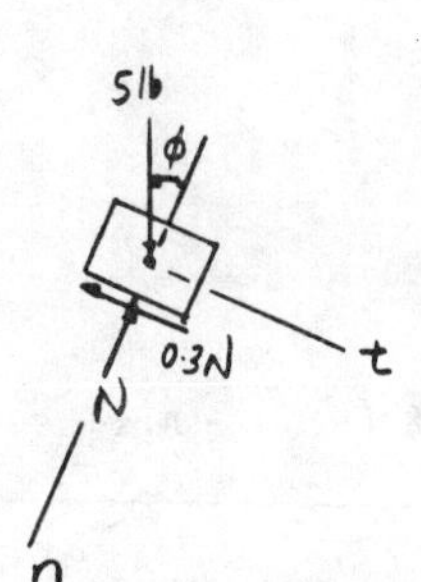

$+\swarrow \Sigma F_n = ma_n; \qquad 5\cos\phi - N = \frac{5}{32.2}\left(\frac{2^2}{0.5}\right) \qquad (3)$

$+\searrow \Sigma F_t = ma_t; \qquad 5\sin\phi - 0.3N = 0 \qquad (4)$

Solving Eqs. (3) and (4) yields :

$\phi = 12.6°$ **Ans**

$N = 3.64$ lb

13-82. A ball having a mass of 2 kg and negligible size moves within a smooth vertical circular slot. If it is released from rest when $\theta = 10°$, determine the force of the slot on the ball when the ball arrives at points A and B.

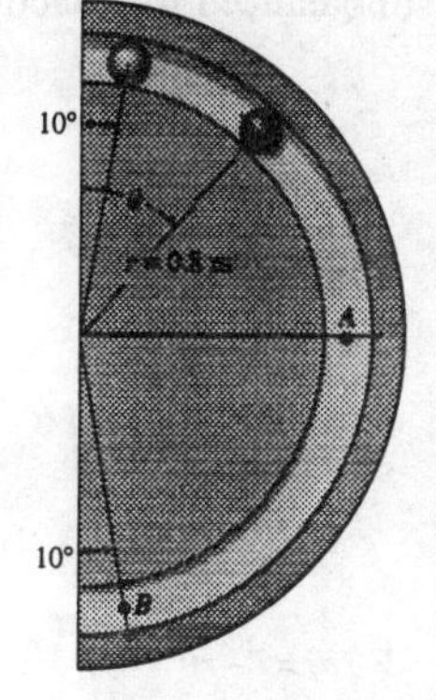

$+\searrow \Sigma F_t = ma_t; \qquad 2(9.81)\sin\theta = 2a_t$

$a_t = 9.81\sin\theta \qquad (1)$

$+\swarrow \Sigma F_n = ma_n; \qquad -N_s + 2(9.81)\cos\theta = 2\left(\frac{v^2}{0.8}\right) \qquad (2)$

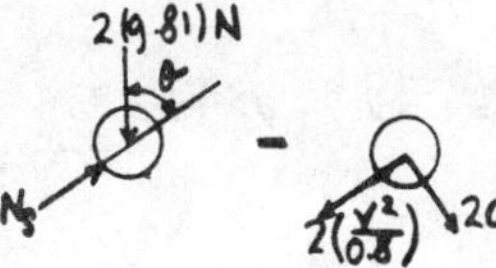
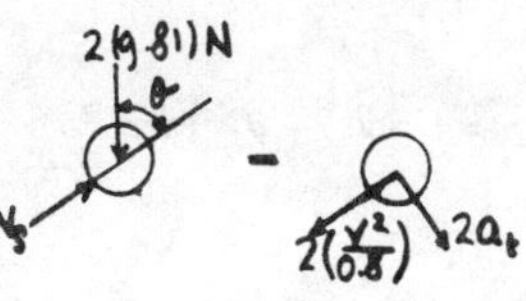

$v\,dv = a_t\,ds = a_t(0.8\,d\theta)$

$v\,dv = 9.81\sin\theta(0.8\,d\theta)$

At A : $\displaystyle\int_0^{v_A} v\,dv = 7.848\int_{10°}^{90°}\sin\theta\,d\theta$

$\frac{1}{2}v_A^2 = 7.848[-\cos 90° + \cos 10°]$

$v_A = 3.932$ m/s

From Eq (2) :

$N_s = 2(9.81)\cos 90° - 2\left(\frac{(3.932)^2}{0.8}\right) = -38.6$ N

$N_s = 38.6$ N **Ans**

At B : $\displaystyle\int_0^{v_B} v\,dv = 7.848\int_{10°}^{170°}\sin\theta\,d\theta$

$\frac{1}{2}v_B^2 = 7.848[-\cos 170° + \cos 10°]$

$v_B = 5.560$ m/s

$N_s = 2(9.81)\cos 170° - 2\left(\frac{(5.560)^2}{0.8}\right)$

$N_s = -96.6$ N

So that $N_s = 96.6$ N **Ans**

13-83. A particle, having a mass of 1.5 kg, moves along a path defined by the equations $r = (4 + 3t)$ m, $\theta = (t^2 + 2)$ rad, and $z = (6 - t^3)$ m, where t is in seconds. Determine the r, θ, and z components of force which the path exerts on the particle when $t = 2$ s.

$r = 4 + 3t|_{t=2\text{ s}} = 10$ m $\qquad \dot{r} = 3$ m/s $\qquad \ddot{r} = 0$

$\theta = t^2 + 2 \qquad \dot{\theta} = 2t|_{t=2\text{ s}} = 4$ rad/s $\qquad \ddot{\theta} = 2$ rad/s^2

$z = 6 - t^3 \qquad \dot{z} = -3t^2 \qquad \ddot{z} = -6t|_{t=2\text{ s}} = -12$ m/s^2

$a_r = \ddot{r} - r\dot{\theta}^2 = 0 - 10(4)^2 = -160$ m/s^2

$a_\theta = r\ddot{\theta} + 2\dot{r}\dot{\theta} = 10(2) + 2(3)(4) = 44$ m/s^2

$a_z = \ddot{z} = -12$ m/s^2

$\Sigma F_r = ma_r; \qquad F_r = 1.5(-160) = -240$ N $\qquad$ **Ans**

$\Sigma F_\theta = ma_\theta; \qquad F_\theta = 1.5(44) = 66$ N $\qquad$ **Ans**

$\Sigma F_z = ma_z; \qquad F_z - 1.5(9.81) = 1.5(-12) \qquad F_z = -3.28$ N $\qquad$ **Ans**

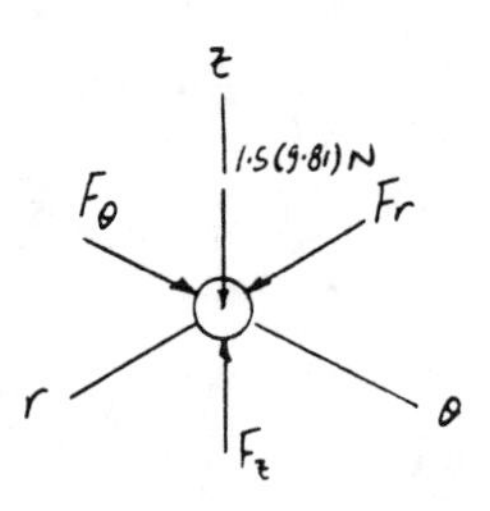

***13-84.** The path of motion of a 5-lb particle in the horizontal plane is described in terms of polar coordinates as $r = (2t + 1)$ ft and $\theta = (0.5t^2 - t)$ rad, where t is in seconds. Determine the magnitude of the unbalanced force acting on the particle when $t = 2$ s.

$r = 2t + 1|_{t=2\text{ s}} = 5$ ft $\qquad \dot{r} = 2$ ft/s $\qquad \ddot{r} = 0$

$\theta = 0.5t^2 - t|_{t=2\text{ s}} = 0$ rad $\qquad \dot{\theta} = t - 1|_{t=2\text{ s}} = 1$ rad/s $\qquad \ddot{\theta} = 1$ rad/s^2

$a_r = \ddot{r} - r\dot{\theta}^2 = 0 - 5(1)^2 = -5$ ft/s^2

$a_\theta = r\ddot{\theta} + 2\dot{r}\dot{\theta} = 5(1) + 2(2)(1) = 9$ ft/s^2

$\Sigma F_r = ma_r; \qquad F_r = \frac{5}{32.2}(-5) = -0.7764$ lb

$\Sigma F_\theta = ma_\theta; \qquad F_\theta = \frac{5}{32.2}(9) = 1.398$ lb

$F = \sqrt{F_r^2 + F_\theta^2} = \sqrt{(-0.7764)^2 + (1.398)^2} = 1.60$ lb $\qquad$ **Ans**

13-85. Determine the magnitude of the unbalanced force acting on a 5-kg particle at the instant $t = 2$ s, if the particle is moving along a horizontal path defined by the equations $r = (2t + 10)$ m and $\theta = (1.5t^2 - 6t)$ rad, where t is in seconds.

$r = 2t + 10|_{t=2\,s} = 14$

$\dot{r} = 2$

$\ddot{r} = 0$

$\theta = 1.5t^2 - 6t$

$\dot{\theta} = 3t - 6|_{t=2\,s} = 0$

$\ddot{\theta} = 3$

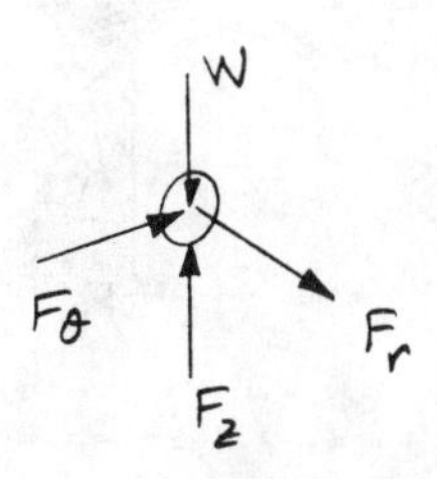

$a_r = \ddot{r} - r\dot{\theta}^2 = 0 - 0 = 0$

$a_\theta = r\ddot{\theta} + 2\dot{r}\dot{\theta} = 14(3) + 0 = 42$

Hence,

$\Sigma F_r = ma_r; \quad F_r = 5(0) = 0$

$\Sigma F_\theta = ma_\theta; \quad F_\theta = 5(42) = 210$ N

$F = \sqrt{(F_r)^2 + (F_\theta)^2} = 210$ N **Ans**

13-86. The 4-kg spool slides along the rotating rod. At the instant shown, the angular rate of rotation of the rod is $\dot{\theta} = 6$ rad/s, which is increasing at $\ddot{\theta} = 2$ rad/s^2. At this same instant, the spool is moving outward along the rod at 3 m/s, which is increasing at 1 m/s^2 when $r = 0.5$ m. Determine the radial frictional force and the normal force of the rod on the spool at this instant.

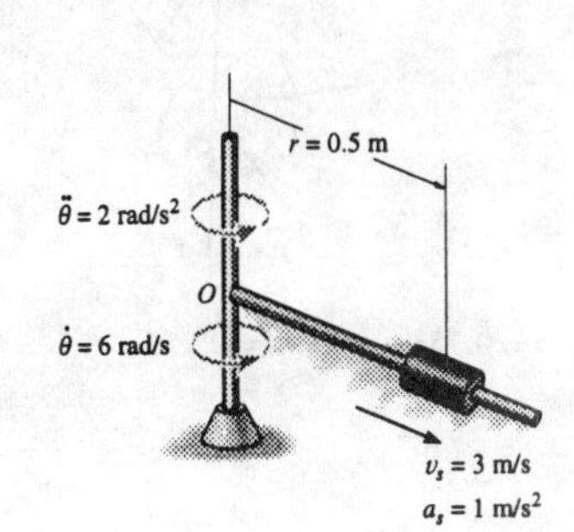

$r = 0.5 \quad \dot{r} = 3 \quad \ddot{r} = 1$

$\dot{\theta} = 6 \quad \ddot{\theta} = 2$

$a_r = \ddot{r} - r\dot{\theta}^2 = 1 - 0.5(6)^2 = -17$

$a_\theta = r\ddot{\theta} + 2\dot{r}\dot{\theta} = 0.5(2) + 2(3)(6) = 37$

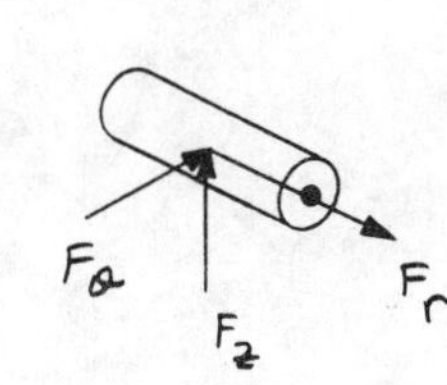

Hence,

$\Sigma F_r = ma_r; \quad F_r = 4(-17) = -68$ N

$\Sigma F_\theta = ma_\theta; \quad F_\theta = 4(37) = 148$ N

$\Sigma F_z = ma_z; \quad F_z - 4(9.81) = 0 \quad F_z = 39.24$ N

$F_{frict} = 68$ N **Ans**

Normal force is $N_C = \sqrt{(148)^2 + (39.24)^2} = 153$ N **Ans**

13-87. The girl has a mass of 50 kg. She is seated on the horse of the merry-go-round which undergoes constant rotational motion $\dot{\theta} = 1.5$ rad/s. If the path of the horse is defined by $r = 4$ m, $z = (0.5 \sin \theta)$ m, determine the maximum and minimum force F_z the horse exerts on her during the motion.

$\dot{\theta} = 1.5 \qquad \ddot{\theta} = 0$

$z = 0.5\sin\theta \qquad \dot{z} = 0.5\cos\theta\dot{\theta} \qquad \ddot{z} = -0.5\sin\theta\dot{\theta}^2 + 0.5\cos\theta\ddot{\theta}$

$+\uparrow \Sigma F_z = ma_z; \qquad F_z - 50(9.81) = 50\left[-0.5\sin\theta(1.5)^2 + 0\right]$

$F_z = 490.5 - 56.25\sin\theta$

Max. when $\sin\theta = -1$, $\qquad (F_z)_{max} = 547$ N **Ans**

Min. when $\sin\theta = 1$, $\qquad (F_z)_{min} = 434$ N **Ans**

***13-88.** A car of a roller coaster travels along a track which for a short distance is defined by a conical spiral, $r = \frac{3}{4}z$, $\theta = -1.5z$, where r and z are in meters and θ in radians. If the angular motion $\dot{\theta} = 1$ rad/s is always maintained, determine the r, θ, z components of reaction exerted on the car by the track at the instant $z = 6$ m. The car and passengers have a total mass of 200 kg.

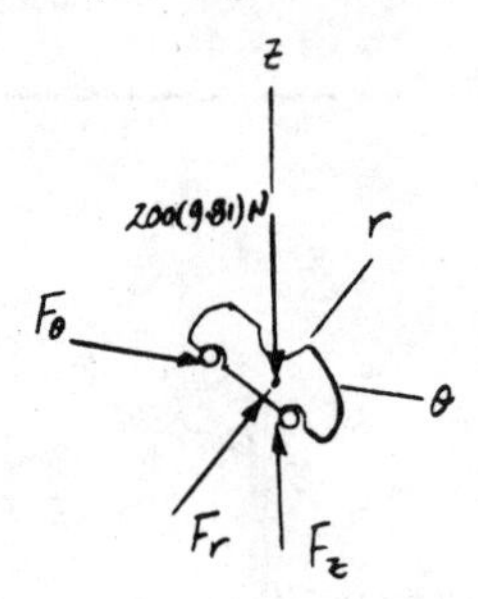

$r = 0.75z \qquad \dot{r} = 0.75\dot{z} \qquad \ddot{r} = 0.75\ddot{z}$

$\theta = -1.5z \qquad \dot{\theta} = -1.5\dot{z} \qquad \ddot{\theta} = -1.5\ddot{z}$

$\dot{\theta} = 1 = -1.5\dot{z} \qquad \dot{z} = -0.6667$ m/s $\qquad \ddot{z} = 0$

At $z = 6$ m,

$r = 0.75(6) = 4.5$ m $\qquad \dot{r} = 0.75(-0.6667) = -0.5$ m/s $\qquad \ddot{r} = 0.75(0) = 0 \qquad \ddot{\theta} = 0$

$a_r = \ddot{r} - r\dot{\theta}^2 = 0 - 4.5(1)^2 = -4.5$ m/s^2

$a_\theta = r\ddot{\theta} + 2\dot{r}\dot{\theta} = 4.5(0) + 2(-0.5)(1) = -1$ m/s^2

$a_z = \ddot{z} = 0$

$\Sigma F_r = ma_r; \qquad F_r = 200(-4.5) \qquad F_r = -900$ N **Ans**

$\Sigma F_\theta = ma_\theta; \qquad F_\theta = 200(-1) \qquad F_\theta = -200$ N **Ans**

$\Sigma F_z = ma_z; \qquad F_z - 200(9.81) = 0 \qquad F_z = 1962$ N $= 1.96$ kN **Ans**

13-89. Using a forked rod, a smooth cylinder C having a mass of 0.5 kg is forced to move along the *horizontal slotted* path $r = (0.5\theta)$ m, where θ is in radians. If the angular position of the arm is $\theta = (0.5t^2)$ rad, where t is in seconds, determine the force of the forked rod on the cylinder and the normal force of the slot on the cylinder at the instant $t = 2$ s. The cylinder is in contact with only *one edge* of the rod and slot at any instant.

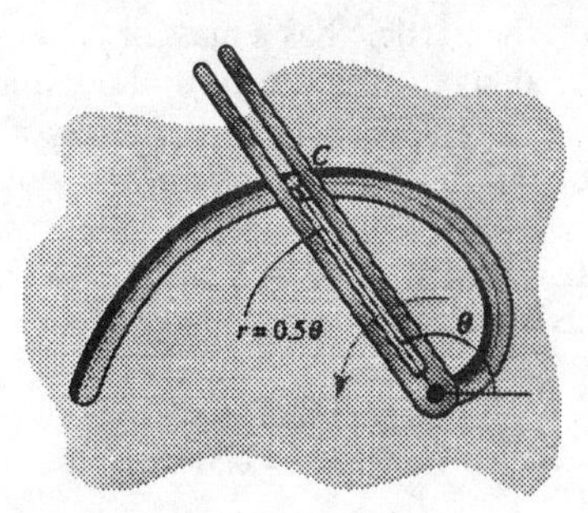

$r = 0.5\theta \qquad \dot{r} = 0.5\dot{\theta} \qquad \ddot{r} = 0.5\ddot{\theta}$

$\theta = 0.5t^2 \qquad \dot{\theta} = t \qquad \ddot{\theta} = 1$

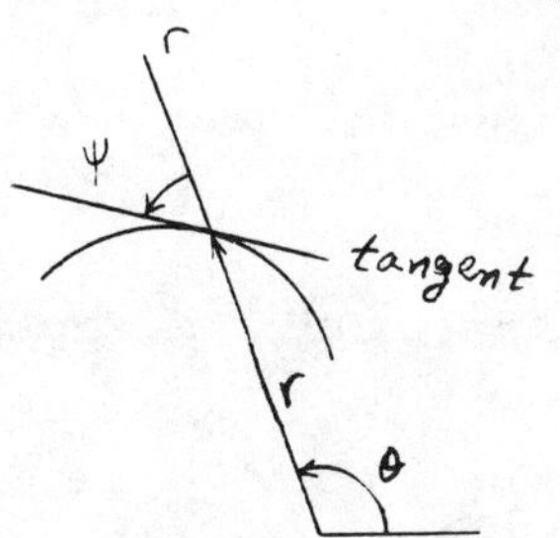

At $t = 2$ s,

$\theta = 2 \text{ rad} = 114.59° \qquad \dot{\theta} = 2 \text{ rad/s} \qquad \ddot{\theta} = 1 \text{ rad/s}^2$

$r = 1 \text{ m} \qquad \dot{r} = 1 \text{ m/s} \qquad \ddot{r} = 0.5 \text{ m/s}^2$

$\tan\psi = \dfrac{r}{dr/d\theta} = \dfrac{0.5(2)}{0.5} \qquad \psi = 63.43°$

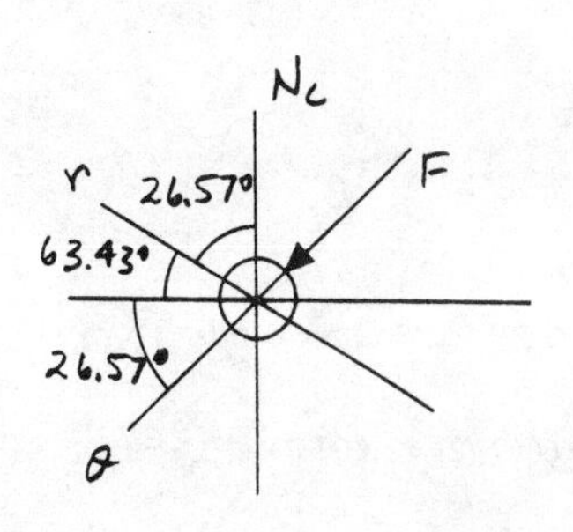

$a_r = \ddot{r} - r\dot{\theta}^2 = 0.5 - 1(2)^2 = -3.5$

$a_\theta = r\ddot{\theta} + 2\dot{r}\dot{\theta} = 1(1) + 2(1)(2) = 5$

$+\nwarrow \Sigma F_r = ma_r; \qquad -N_C\cos 26.57° = 0.5(-3.5)$

$N_C = 1.957 = 1.96$ N **Ans**

$+\swarrow \Sigma F_\theta = ma_\theta; \qquad F + 1.957\sin 26.57° = 0.5(5)$

$F = 1.62$ N **Ans**

13-90. Using a forked rod, a smooth cylinder C having a mass of 0.5 kg is forced to move along the *vertical slotted* path $r = (0.5\theta)$ m, where θ is in radians. If the angular position of the arm is $\theta = (0.5t^2)$ rad, where t is in seconds, determine the force of the forked rod on the cylinder and the normal force of the slot on the cylinder at the instant $t = 2$ s. The cylinder is in contact with only *one edge* of the rod and slot at any instant.

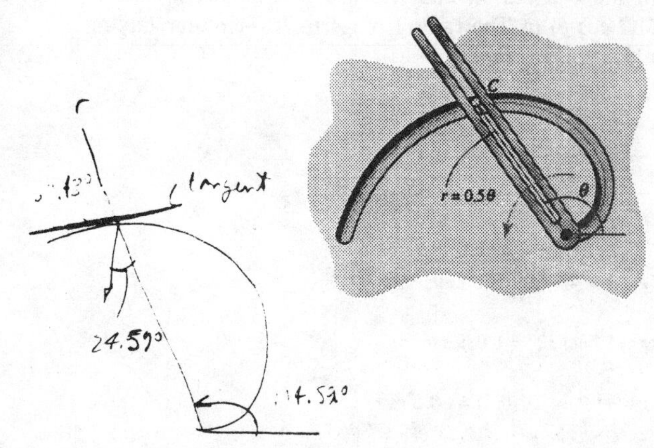

$r = 0.5\theta \qquad \dot{r} = 0.5\dot{\theta} \qquad \ddot{r} = 0.5\ddot{\theta}$

$\theta = 0.5t^2 \qquad \dot{\theta} = t \qquad \ddot{\theta} = 1$

At $t = 2$ s,

$\theta = 2 \text{ rad} = 114.59° \qquad \dot{\theta} = 2 \text{ rad/s} \qquad \ddot{\theta} = 1 \text{ rad/s}^2$

$r = 1 \text{ m} \qquad \dot{r} = 1 \text{ m/s} \qquad \ddot{r} = 0.5 \text{ m/s}^2$

$\tan\psi = \dfrac{r}{dr/d\theta} = \dfrac{0.5(2)}{0.5} \qquad \psi = 63.43°$

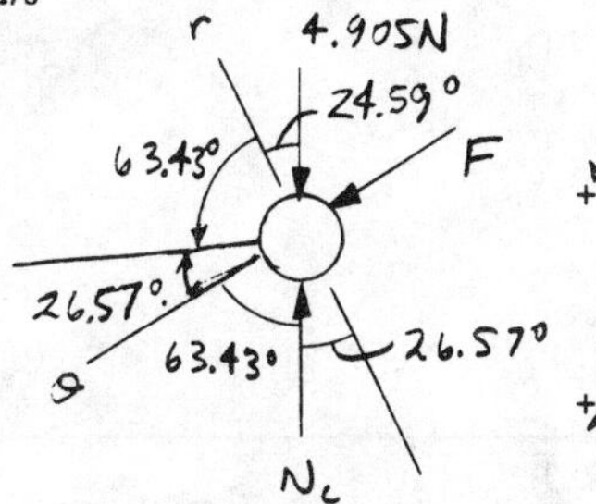

$a_r = \ddot{r} - r\dot{\theta}^2 = 0.5 - 1(2)^2 = -3.5$

$a_\theta = r\ddot{\theta} + 2\dot{r}\dot{\theta} = 1(1) + 2(1)(2) = 5$

$+\nwarrow \Sigma F_r = ma_r; \qquad N_C\cos 26.57° - 4.905\cos 24.59° = 0.5(-3.5)$

$N_C = 3.030 = 3.03$ N **Ans**

$+\swarrow \Sigma F_\theta = ma_\theta; \qquad F - 3.030\sin 26.57° + 4.905\sin 24.59° = 0.5(5)$

$F = 1.81$ N **Ans**

13-91. The particle has a mass of 0.5 kg and is confined to move along the smooth vertical slot due to the rotation of the arm OA. Determine the force of the rod on the particle and the normal force of the slot on the particle when $\theta = 30°$. The rod is rotating with a constant angular velocity of $\dot{\theta} = 2$ rad/s. Assume the particle contacts only one side of the slot at any instant.

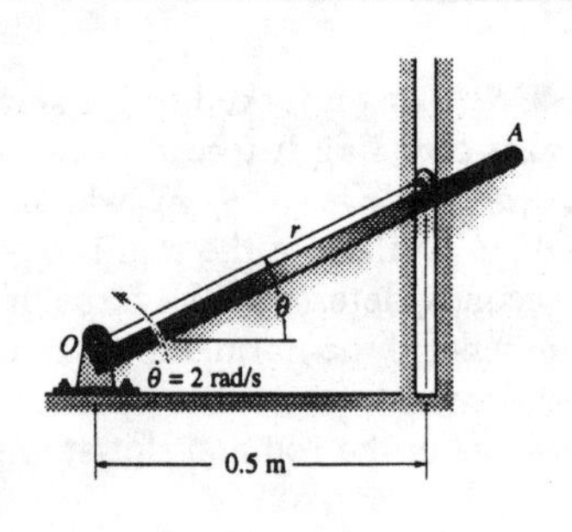

$r = \dfrac{0.5}{\cos\theta} = 0.5\sec\theta, \qquad \dot{r} = 0.5\sec\theta\tan\theta\dot{\theta}$

$\ddot{r} = 0.5\sec\theta\tan\theta\ddot{\theta} + 0.5\sec^3\theta\dot{\theta}^2 + 0.5\sec\theta\tan^2\theta\dot{\theta}^2$

At $\theta = 30°$,

$\dot{\theta} = 2$ rad/s

$\ddot{\theta} = 0$

$r = 0.5774$ m

$\dot{r} = 0.6667$ m/s

$\ddot{r} = 3.8490$ m/s^2

$a_r = \ddot{r} - r\dot{\theta}^2 = 3.8490 - 0.5774(2)^2 = 1.5396$ m/s^2

$a_\theta = r\ddot{\theta} + 2\dot{r}\dot{\theta} = 0 + 2(0.6667)(2) = 2.667$ m/s^2

$+\nearrow \Sigma F_r = ma_r;\qquad N_P\cos 30° - 0.5(9.81)\sin 30° = 0.5(1.5396)$

$N_P = 3.7208 = 3.72$ N **Ans**

$+\nwarrow \Sigma F_\theta = ma_\theta;\qquad F - 3.7208\sin 30° - 0.5(9.81)\cos 30° = 0.5(2.667)$

$F = 7.44$ N **Ans**

***13-92.** A smooth can C, having a mass of 3 kg, is lifted from a feed at A to a ramp at B by a rotating rod. If the rod maintains a constant angular velocity of $\dot{\theta} = 0.5$ rad/s, determine the force which the rod exerts on the can at the instant $\theta = 30°$. Neglect the effects of friction in the calculation and the size of the can so that $r = (1.2\cos\theta)$ m. The ramp from A to B is circular, having a radius of 600 mm.

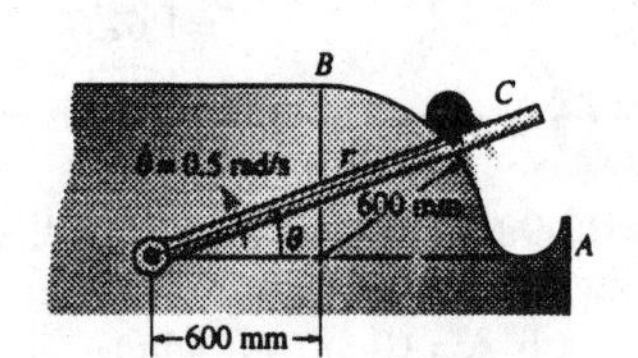

$r = 2(0.6\cos\theta) = 1.2\cos\theta$

$\dot{r} = -1.2\sin\theta\dot{\theta}$

$\ddot{r} = -1.2\cos\theta\dot{\theta}^2 - 1.2\sin\theta\ddot{\theta}$

At $\theta = 30°$, $\dot{\theta} = 0.5$ rad/s and $\ddot{\theta} = 0$

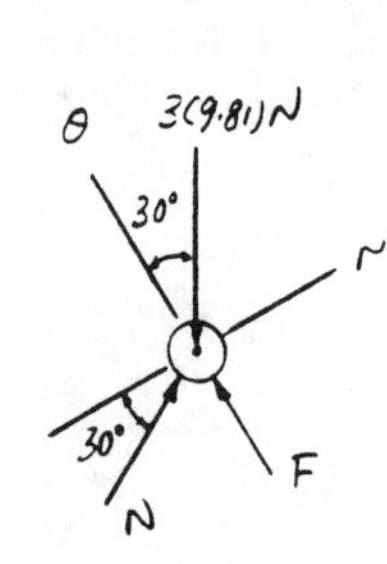

$r = 1.2\cos 30° = 1.0392$ m

$\dot{r} = -1.2\sin 30°(0.5) = -0.3$ m/s

$\ddot{r} = -1.2\cos 30°(0.5)^2 - 1.2\sin 30°(0) = -0.2598$ m/s^2

$a_r = \ddot{r} - r\dot{\theta}^2 = -0.2598 - 1.0392(0.5)^2 = -0.5196$ m/s^2

$a_\theta = r\ddot{\theta} + 2\dot{r}\dot{\theta} = 1.0392(0) + 2(-0.3)(0.5) = -0.3$ m/s^2

$+\nearrow \Sigma F_r = ma_r;\qquad N\cos 30° - 3(9.81)\sin 30° = 3(-0.5196) \qquad N = 15.19$ N

$+\nwarrow \Sigma F_\theta = ma_\theta;\qquad F + 15.19\sin 30° - 3(9.81)\cos 30° = 3(-0.3)$

$F = 17.0$ N **Ans**

13-93. The spool, which has a mass of 2 kg, slides along the smooth *horizontal* spiral rod, $r = (0.4\theta)$ m, where θ is in radians. If its angular rate of rotation is constant and equals $\dot{\theta} = 6$ rad/s, determine the horizontal tangential force P needed to cause the motion and the horizontal normal force component that the spool exerts on the rod at the instant $\theta = 45°$.

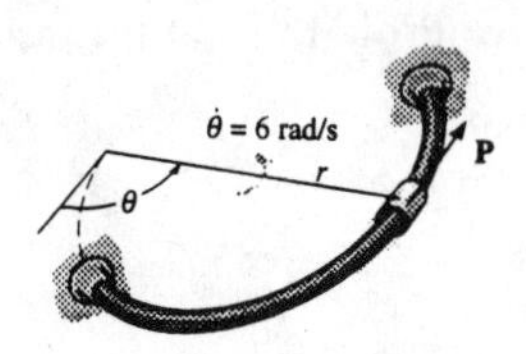

$r = 0.4\theta \qquad \dot{r} = 0.4\dot{\theta} \qquad \ddot{r} = 0.4\ddot{\theta}$

At $\theta = 45° = \dfrac{\pi}{4}$ rad, $\dot{\theta} = 6$ rad/s and $\ddot{\theta} = 0$

$r = 0.4\left(\dfrac{\pi}{4}\right) = (0.1)\ \pi$ m $\qquad \dot{r} = 0.4(6) = 2.4$ m/s $\qquad \ddot{r} = 0.4(0) = 0$

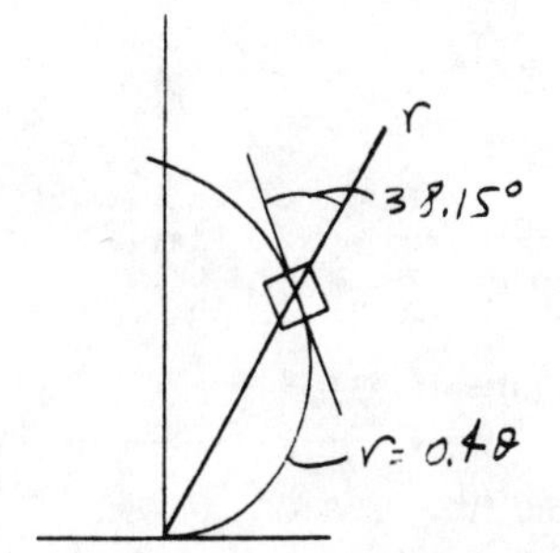

$a_r = \ddot{r} - r\dot{\theta}^2 = 0 - (0.1)\pi(6)^2 = -11.310$ m/s²

$a_\theta = r\ddot{\theta} + 2\dot{r}\dot{\theta} = (0.1)\pi(0) + 2(2.4)(6) = 28.8$ m/s²

$\tan\psi = \dfrac{r}{dr/d\theta} = \dfrac{0.4\theta}{0.4} = \theta = \dfrac{\pi}{4} \qquad \psi = 38.15°$

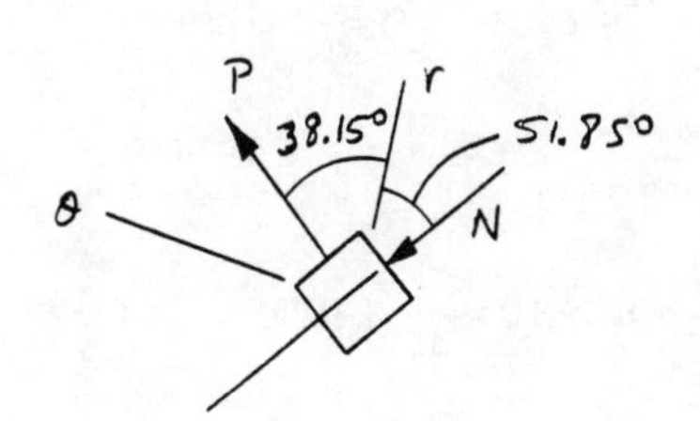

$+\nearrow \Sigma F_r = ma_r; \qquad P\cos 38.15° - N\cos 51.85° = 2(-11.310)$

$+\nwarrow \Sigma F_\theta = ma_\theta; \qquad P\sin 38.15° + N\sin 51.85° = 2(28.8)$

Solving

$N = 59.3$ N $\qquad P = 17.8$ N $\qquad$ **Ans**

13-94. The forked rod is used to move the smooth 2-lb particle around the horizontal path in the shape of a limaçon, $r = (2 + \cos\theta)$ ft. If at all times $\dot{\theta} = 0.5$ rad/s, determine the force which the rod exerts on the particle at the instant $\theta = 90°$. The fork and path contact the particle on only one side.

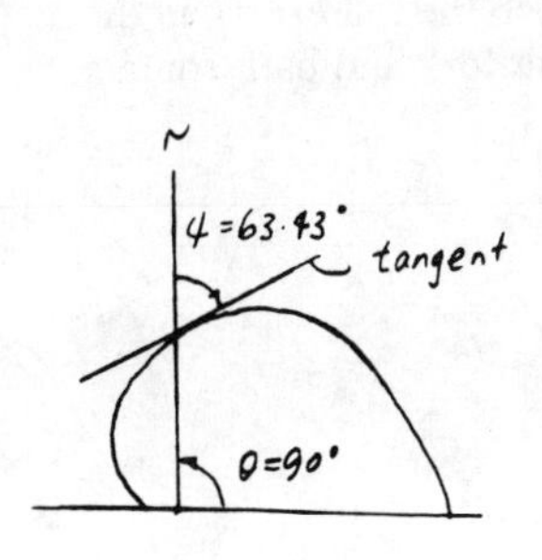

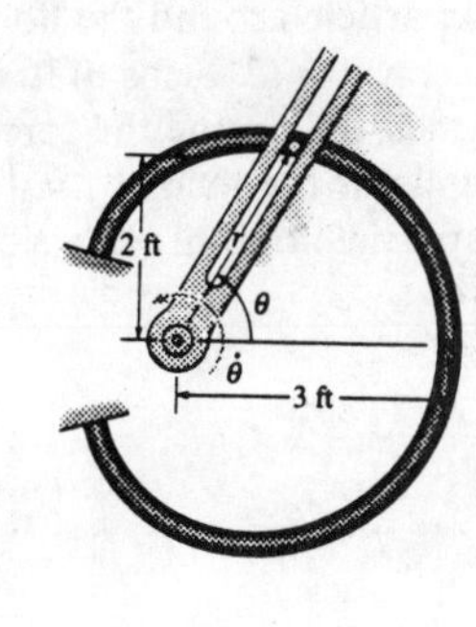

$r = 2 + \cos\theta$

$\dot{r} = -\sin\theta\dot{\theta}$

$\ddot{r} = -\cos\theta\dot{\theta}^2 - \sin\theta\ddot{\theta}$

At $\theta = 90°$, $\dot{\theta} = 0.5$ rad/s and $\ddot{\theta} = 0$

$r = 2 + \cos 90° = 2$ ft

$\dot{r} = -\sin 90°(0.5) = -0.5$ ft/s

$\ddot{r} = -\cos 90°(0.5)^2 - \sin 90°(0) = 0$

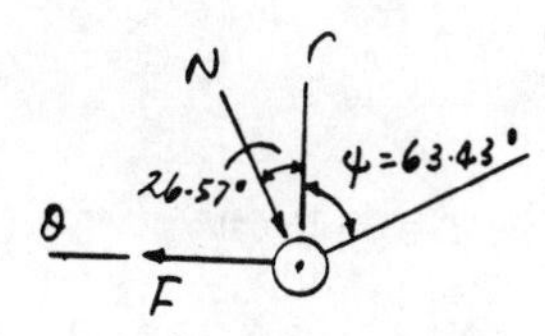

$a_r = \ddot{r} - r\dot{\theta}^2 = 0 - 2(0.5)^2 = -0.5$ ft/s²

$a_\theta = r\ddot{\theta} + 2\dot{r}\dot{\theta} = 2(0) + 2(-0.5)(0.5) = -0.5$ ft/s²

$\tan\psi = \dfrac{r}{dr/d\theta} = \left.\dfrac{2 + \cos\theta}{-\sin\theta}\right|_{\theta = 90°} = -2 \qquad \psi = -63.43°$

$+\uparrow \Sigma F_r = ma_r; \qquad -N\cos 26.57° = \dfrac{2}{32.2}(-0.5) \qquad N = 0.03472$ lb

$\overset{+}{\leftarrow} \Sigma F_\theta = ma_\theta; \qquad F - 0.03472\sin 26.57° = \dfrac{2}{32.2}(-0.5)$

$F = -0.0155$ lb $\qquad$ **Ans**

13-95. Solve Prob. 13-94 at the instant $\theta = 60°$.

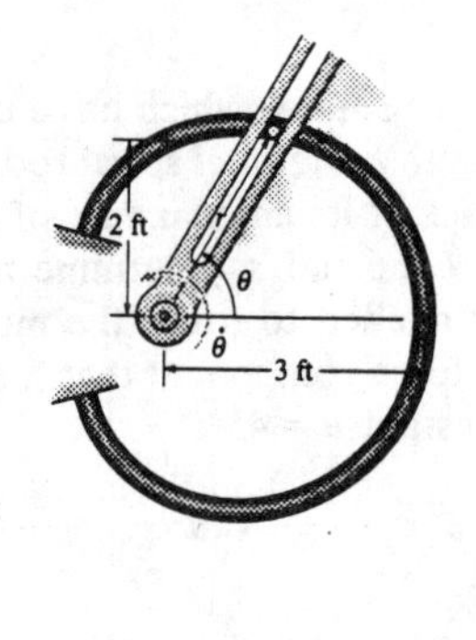

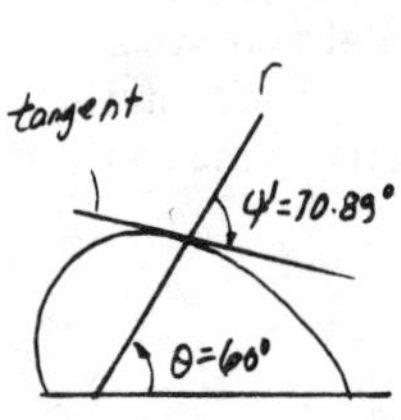
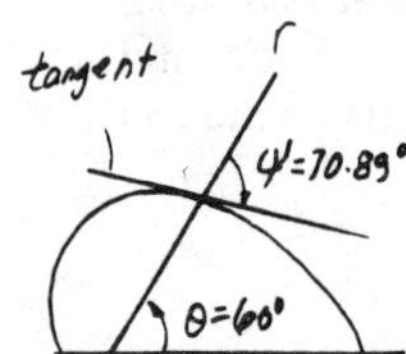

$r = 2+\cos\theta$

$\dot{r} = -\sin\theta\dot{\theta}$

$\ddot{r} = -\cos\theta\dot{\theta}^2 - \sin\theta\ddot{\theta}$

At $\theta = 60°$, $\dot{\theta} = 0.5$ rad/s and $\ddot{\theta} = 0$

$r = 2+\cos 60° = 2.5$ ft

$\dot{r} = -\sin 60°(0.5) = -0.4330$ ft/s

$\ddot{r} = -\cos 60°(0.5)^2 - \sin 60°(0) = -0.125$ ft/s²

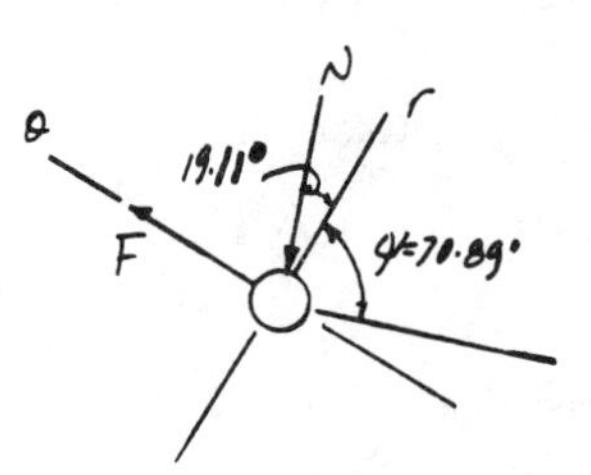

$a_r = \ddot{r} - r\dot{\theta}^2 = -0.125 - 2.5(0.5)^2 = -0.75$ ft/s²

$a_\theta = r\ddot{\theta} + 2\dot{r}\dot{\theta} = 2.5(0) + 2(-0.4330)(0.5) = -0.4330$ ft/s²

$\tan\psi = \dfrac{r}{dr/d\theta} = \dfrac{2+\cos\theta}{-\sin\theta}\bigg|_{\theta=60°} = -2.887 \qquad \psi = -70.89°$

$+\nearrow \Sigma F_r = ma_r; \qquad -N\cos 19.11° = \dfrac{2}{32.2}(-0.75) \qquad N = 0.04930$ lb

$+\nwarrow \Sigma F_\theta = ma_\theta; \qquad F - 0.04930\sin 19.11° = \dfrac{2}{32.2}(-0.4330)$

$F = -0.0108$ lb **Ans**

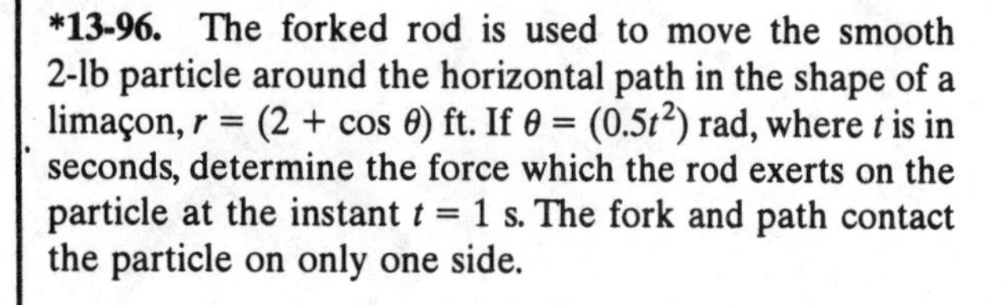

***13-96.** The forked rod is used to move the smooth 2-lb particle around the horizontal path in the shape of a limaçon, $r = (2 + \cos\theta)$ ft. If $\theta = (0.5t^2)$ rad, where t is in seconds, determine the force which the rod exerts on the particle at the instant $t = 1$ s. The fork and path contact the particle on only one side.

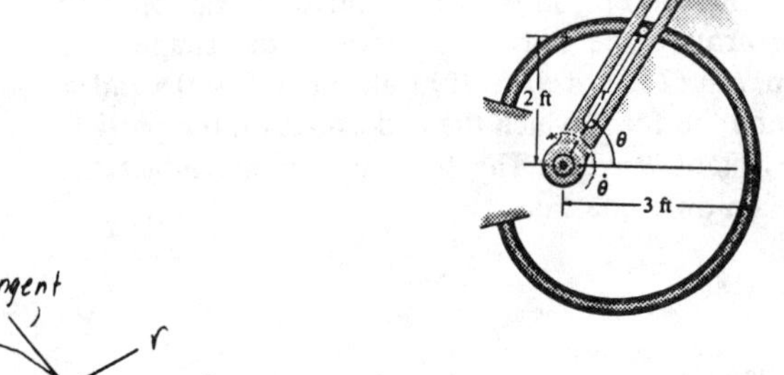

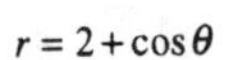

$r = 2+\cos\theta \qquad \theta = 0.5t^2$

$\dot{r} = -\sin\theta\dot{\theta} \qquad \dot{\theta} = t$

$\ddot{r} = -\cos\theta\dot{\theta}^2 - \sin\theta\ddot{\theta} \qquad \ddot{\theta} = 1$ rad/s²

At $t = 1$ s, $\theta = 0.5$ rad, $\dot{\theta} = 1$ rad/s and $\ddot{\theta} = 1$ rad/s²

$r = 2+\cos 0.5 = 2.8776$ ft

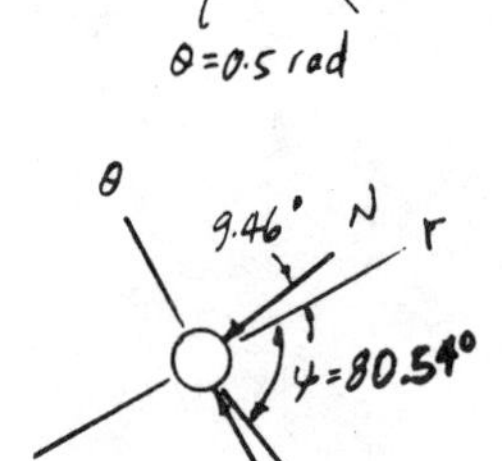

$\dot{r} = -\sin 0.5(1) = -0.4794$ ft/s

$\ddot{r} = -\cos 0.5(1)^2 - \sin 0.5(1) = -1.357$ ft/s²

$a_r = \ddot{r} - r\dot{\theta}^2 = -1.357 - 2.8776(1)^2 = -4.2346$ ft/s²

$a_\theta = r\ddot{\theta} + 2\dot{r}\dot{\theta} = 2.8776(1) + 2(-0.4794)(1) = 1.9187$ ft/s²

$\tan\psi = \dfrac{r}{dr/d\theta} = \dfrac{2+\cos\theta}{-\sin\theta}\bigg|_{\theta=0.5\text{ rad}} = -6.002 \qquad \psi = -80.54°$

$+\nearrow \Sigma F_r = ma_r; \qquad -N\cos 9.46° = \dfrac{2}{32.2}(-4.2346) \qquad N = 0.2666$ lb

$+\nwarrow \Sigma F_\theta = ma_\theta; \qquad F - 0.2666\sin 9.46° = \dfrac{2}{32.2}(1.9187)$

$F = 0.163$ lb **Ans**

13-97. The 2-lb collar slides along the smooth *horizontal* spiral rod, $r = (2\theta)$ ft, where θ is in radians. If its angular rate of rotation is constant and equals $\dot{\theta} = 4$ rad/s, determine the tangential force P needed to cause the motion and the normal force that the spool exerts on the rod at the instant $\theta = 90°$.

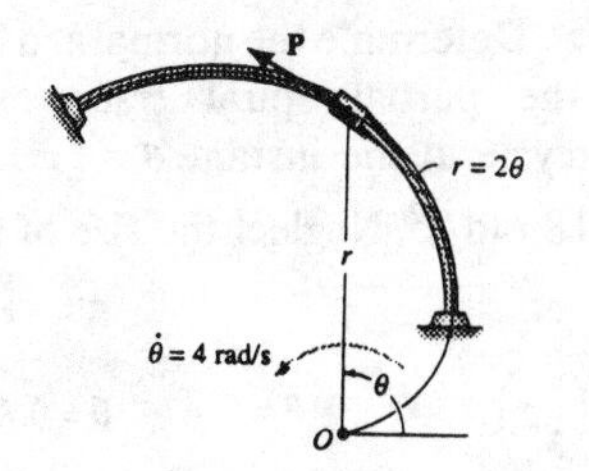

$r = 2\theta$

$\tan\psi = \dfrac{r}{dr/d\theta} = \dfrac{2\theta}{2} = \dfrac{\pi}{2} \qquad \psi = 57.52°$

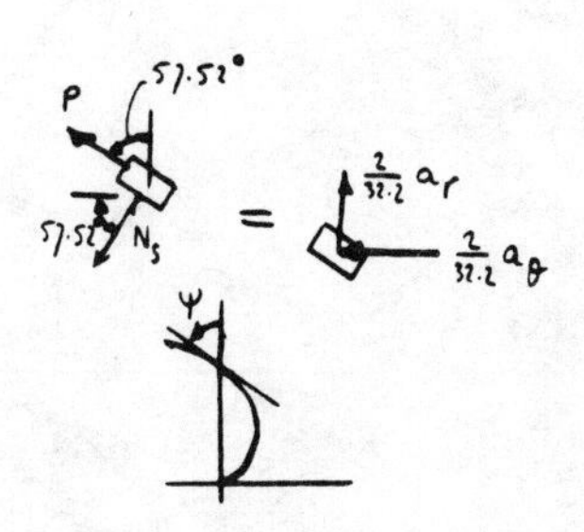

$\dot{\theta} = 4 \qquad \ddot{\theta} = 0$

$r = 2\theta = 2\left(\dfrac{\pi}{2}\right) = \pi$

$\dot{r} = 2\dot{\theta} = 2(4) = 8$

$\ddot{r} = 2\ddot{\theta} = 0$

$a_r = \ddot{r} - r\dot{\theta}^2 = 0 - \pi(4)^2 = -50.27$

$a_\theta = r\ddot{\theta} + 2\dot{r}\dot{\theta} = 0 + 2(8)(4) = 64$

$+\uparrow \Sigma F_r = ma_r; \qquad P\cos 57.52° - N_s \sin 57.52° = \left(\dfrac{2}{32.2}\right)(-50.27)$

$\overset{+}{\leftarrow} \Sigma F_\theta = ma_\theta; \qquad P\sin 57.52° + N_s \cos 57.52° = \left(\dfrac{2}{32.2}\right)(64)$

Solving :

$P = 1.68$ lb **Ans**

$N_s = 4.77$ lb **Ans**

13-98. Solve Prob. 13-97 if the spiral rod is *vertical.*

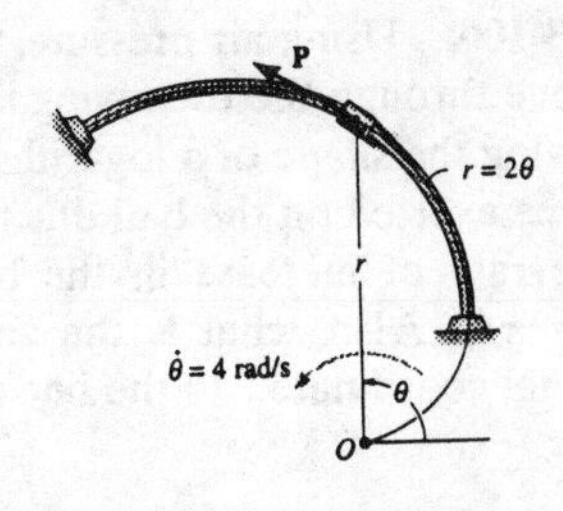

$r = 2\theta$

$\tan\psi = \dfrac{r}{dr/d\theta} = \dfrac{2\theta}{2} = \dfrac{\pi}{2} \qquad \psi = 57.52°$

$\dot{\theta} = 4 \qquad \ddot{\theta} = 0$

$r = 2\theta = 2\left(\dfrac{\pi}{2}\right) = \pi$

$\dot{r} = 2\dot{\theta} = 2(4) = 8$

$\ddot{r} = 2\ddot{\theta} = 0$

$a_r = \ddot{r} - r\dot{\theta}^2 = 0 - \pi(4)^2 = -50.27$

$a_\theta = r\ddot{\theta} + 2\dot{r}\dot{\theta} = 0 + 2(8)(4) = 64$

$+\uparrow \Sigma F_r = ma_r; \qquad P\cos 57.52° - N_s \sin 57.52° - 2 = \left(\dfrac{2}{32.2}\right)(-50.27)$

$\overset{+}{\leftarrow} \Sigma F_\theta = ma_\theta; \qquad P\sin 57.52° + N_s \cos 57.52° = \left(\dfrac{2}{32.2}\right)(64)$

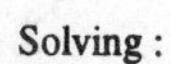

Solving :

$P = 2.75$ lb **Ans**

$N_s = 3.08$ lb **Ans**

13-99. Determine the normal and frictional driving forces that the partial spiral track exerts on the 200-kg motorcycle at the instant $\theta = \frac{5}{3}\pi$ rad, $\dot{\theta} = 0.4$ rad/s, and $\ddot{\theta} = 0.8$ rad/s^2. Neglect the size of the motorcycle.

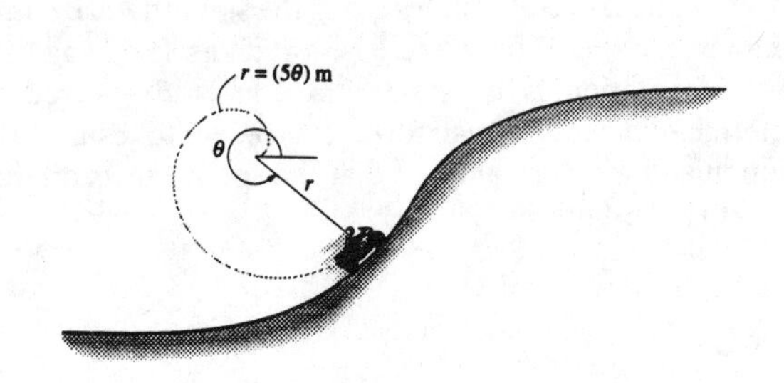

$$\theta = \left(\frac{5}{3}\pi\right) = 300° \qquad \dot{\theta} = 0.4 \qquad \ddot{\theta} = 0.8$$

$$r = 5\theta = 5\left(\frac{5}{3}\pi\right) = 26.18$$

$$\dot{r} = 5\dot{\theta} = 5(0.4) = 2$$

$$\ddot{r} = 5\ddot{\theta} = 5(0.8) = 4$$

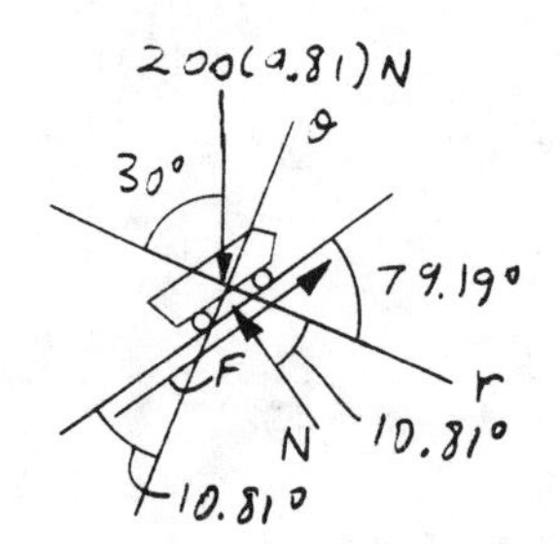

$$a_r = \ddot{r} - r\dot{\theta}^2 = 4 - 26.18(0.4)^2 = -0.1888$$

$$a_\theta = r\ddot{\theta} + 2\dot{r}\dot{\theta} = 26.18(0.8) + 2(2)(0.4) = 22.54$$

$$\tan\psi = \frac{r}{dr/d\theta} = \frac{5\left(\frac{5}{3}\pi\right)}{5} = 5.236 \qquad \psi = 79.19°$$

$+\searrow\Sigma F_r = ma_r;$ $\quad F\sin 10.81° - N\cos 10.81° + 200(9.81)\cos 30° = 200(-0.1888)$

$+\nearrow\Sigma F_\theta = ma_\theta;$ $\quad F\cos 10.81° - 200(9.81)\sin 30° + N\sin 10.81° = 200(22.54)$

$F = 5.07$ kN **Ans**

$N = 2.74$ kN **Ans**

***13-100.** Using air pressure, the 0.5-kg ball is forced to move through the tube lying in the horizontal plane and having the shape of a logarithmic spiral. If the tangential force exerted on the ball due to the air is 6 N, determine the rate of increase in the ball's speed at the instant $\theta = \pi/2$. Also, what is the angle ψ from the extended radial coordinate r to the line of action of the 6-N force?

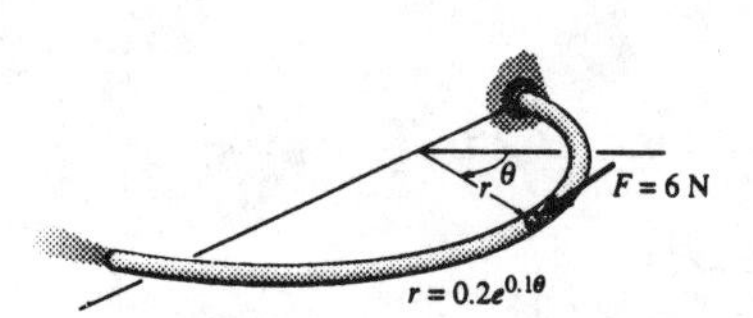

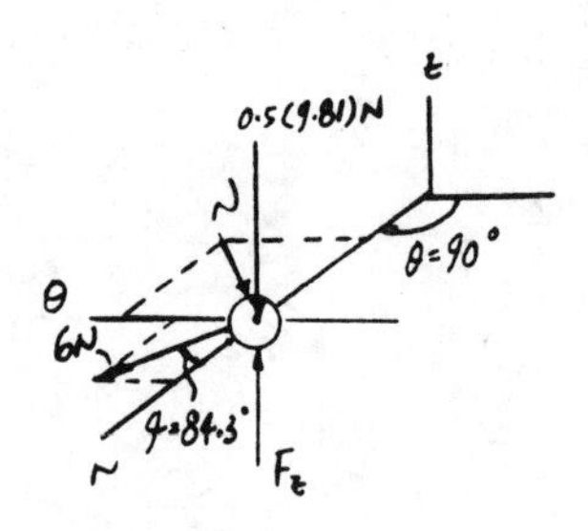

$$\tan\psi = \frac{r}{dr/d\theta} = \frac{0.2e^{0.1\theta}}{0.02e^{0.1\theta}} = 10 \qquad \psi = 84.3° \quad \textbf{Ans}$$

$\Sigma F_t = ma_t;$ $\quad 6 = 0.5a_t \quad a_t = 12$ m/s^2 **Ans**

13-101. The ball has a mass of 2 kg and a negligible size. It is originally traveling around the horizontal circular path of radius $r_0 = 0.5$ m such that the angular rate of rotation is $\dot{\theta}_0 = 1$ rad/s. If the attached cord ABC is drawn down through the hole at a constant speed of 0.2 m/s, determine the tension the cord exerts on the ball at the instant $r = 0.25$ m. Also, compute the angular velocity of the ball at this instant. Neglect the effects of friction between the ball and horizontal plane. *Hint:* First show that the equation of motion in the θ direction yields $a_\theta = r\ddot{\theta} + 2\dot{r}\dot{\theta} = (1/r)(d(r^2\dot{\theta})/dt) = 0$. When integrated, $r^2\dot{\theta} = c$, where the constant c is determined from the problem data.

$\Sigma F_\theta = m\,a_\theta;\qquad 0 = m\left[r\,\ddot{\theta} + 2\,\dot{r}\,\dot{\theta}\right] = m\left[\frac{1}{r}\frac{d}{dt}(r^2\,\dot{\theta})\right] = 0$

Thus,

$d(r^2\,\dot{\theta}) = 0$

$r^2\,\dot{\theta} = C$

$(0.5)^2(1) = C = (0.25)^2\,\dot{\theta}$

$\dot{\theta} = 4.00$ rad/s **Ans**

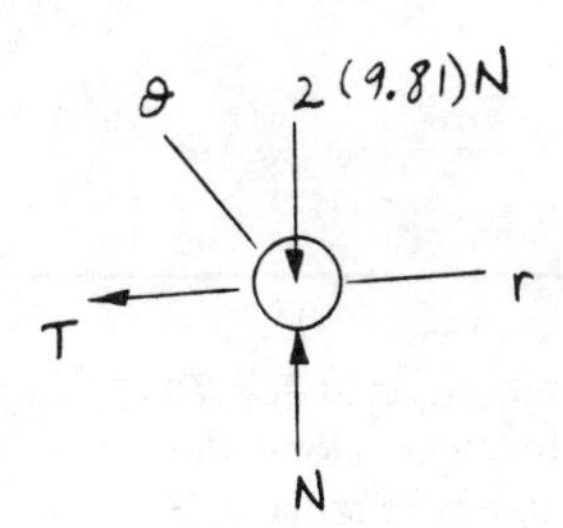

Since $\dot{r} = -0.2$ m/s, $\ddot{r} = 0$

$a_r = \ddot{r} - r\,(\dot{\theta})^2 = 0 - 0.25(4.00)^2 = -4\text{ m/s}^2$

$\Sigma F_r = m\,a_r;\qquad -T = 2(-4)$

$T = 8$ N **Ans**

13-102. The 0.5-lb ball is guided along the vertical circular path $r = 2r_c \cos\theta$ using the arm OA. If the arm has an angular velocity $\dot{\theta} = 0.4$ rad/s and an angular acceleration $\ddot{\theta} = 0.8$ rad/s^2 at the instant $\theta = 30°$, determine the force of the arm on the ball. Neglect friction and the size of the ball. Set $r_c = 0.4$ ft.

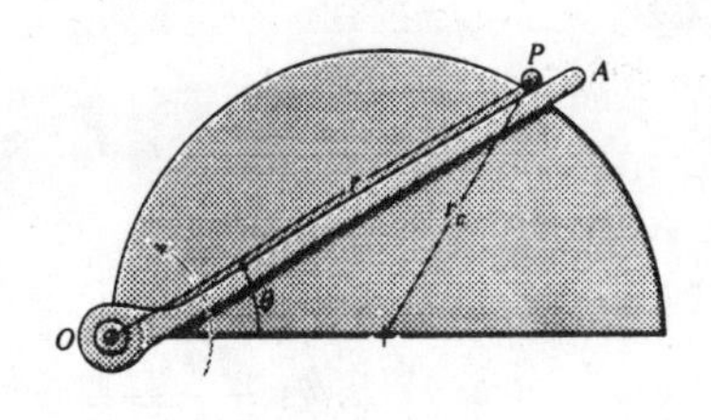

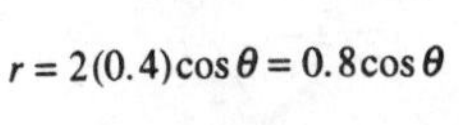

$r = 2(0.4)\cos\theta = 0.8\cos\theta$

$\dot{r} = -0.8\sin\theta\dot{\theta}$

$\ddot{r} = -0.8\cos\theta\dot{\theta}^2 - 0.8\sin\theta\ddot{\theta}$

At $\theta = 30°$, $\dot{\theta} = 0.4$ rad/s and $\ddot{\theta} = 0.8$ rad/s^2

$r = 0.8\cos 30° = 0.6928$ ft

$\dot{r} = -0.8\sin 30°(0.4) = -0.16$ ft/s

$\ddot{r} = -0.8\cos 30°(0.4)^2 - 0.8\sin 30°(0.8) = -0.4309$ ft/s^2

$a_r = \ddot{r} - r\dot{\theta}^2 = -0.4309 - 0.6928(0.4)^2 = -0.5417$ ft/s^2

$a_\theta = r\ddot{\theta} + 2\dot{r}\dot{\theta} = 0.6928(0.8) + 2(-0.16)(0.4) = 0.4263$ ft/s^2

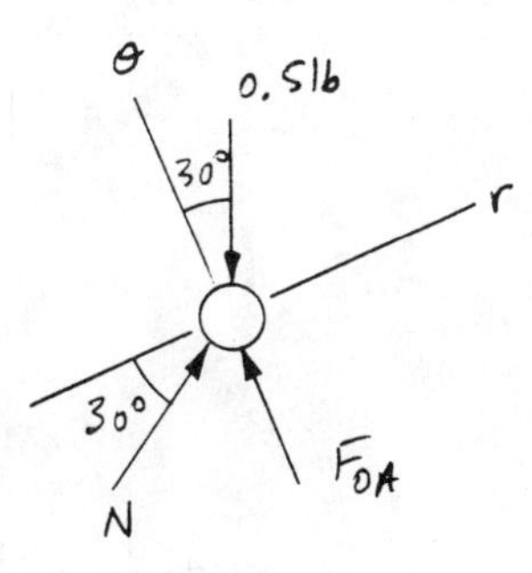

$\Sigma F_r = ma_r;\qquad N\cos 30° - 0.5\sin 30° = \frac{0.5}{32.2}(-0.5417)\qquad N = 0.2790$ lb

$\Sigma F_\theta = ma_\theta;\qquad F_{OA} + 0.2790\sin 30° - 0.5\cos 30° = \frac{0.5}{32.2}(0.4263)$

$F_{OA} = 0.300$ lb **Ans**

13-103. The ball of mass m is guided along the vertical circular path $r = 2r_c \cos\theta$ using the arm OA. If the arm has a constant angular velocity $\dot{\theta}_0$, determine the angle $\theta \leq 45°$ at which the ball starts to leave the surface of the semicylinder. Neglect friction and the size of the ball.

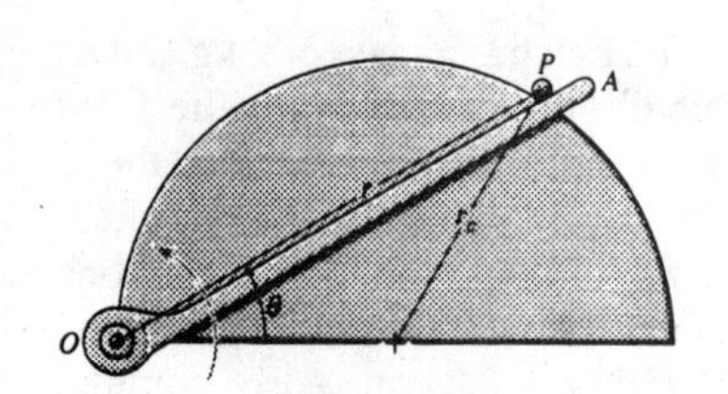

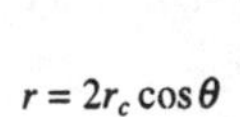

$$r = 2r_c \cos\theta$$

$$\dot{r} = -2r_c \sin\theta\dot{\theta}$$

$$\ddot{r} = -2r_c \cos\theta\dot{\theta}^2 - 2r_c \sin\theta\ddot{\theta}$$

Since $\dot{\theta}$ is constant, $\ddot{\theta} = 0$.

$$a_r = \ddot{r} - r\dot{\theta}^2 = -2r_c \cos\theta\dot{\theta}_0^2 - 2r_c \cos\theta\dot{\theta}_0^2 = -4r_c \cos\theta\dot{\theta}_0^2$$

$$\Sigma F_r = ma_r; \qquad -mg\sin\theta = m(-4r_c\cos\theta\dot{\theta}_0^2)$$

$$\tan\theta = \frac{4r_c\dot{\theta}_0^2}{g} \qquad \theta = \tan^{-1}\left(\frac{4r_c\dot{\theta}_0^2}{g}\right) \qquad \textbf{Ans}$$

***13-104.** Using a forked rod, a smooth cylinder P, having a mass of 0.4 kg, is forced to move along the *vertical slotted* path $r = (0.6\theta)$ m, where θ is in radians. If the cylinder has a constant speed of $v_C = 2$ m/s, determine the force of the rod and the normal force of the slot on the cylinder at the instant $\theta = \pi$ rad. Assume the cylinder is in contact with only *one edge* of the rod and slot at any instant. *Hint:* To obtain the time derivatives necessary to compute the cylinder's acceleration components a_r and a_θ, take the first and second time derivatives of $r = 0.6\theta$. Then, for further information, use Eq. 12–26 to determine $\dot{\theta}$. Also, take the time derivative of Eq. 12–26, noting that $\dot{v}_C = 0$, to determine $\ddot{\theta}$.

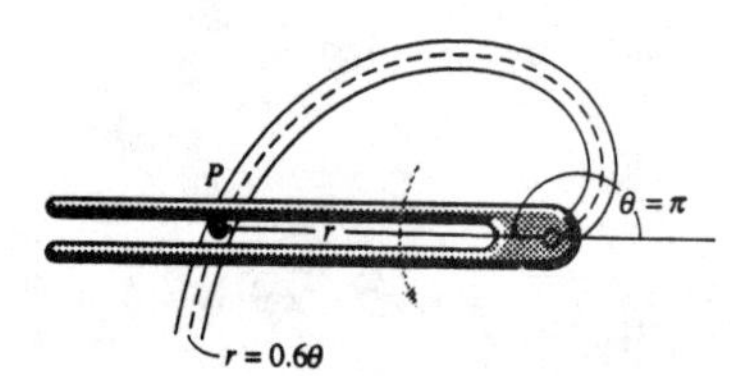

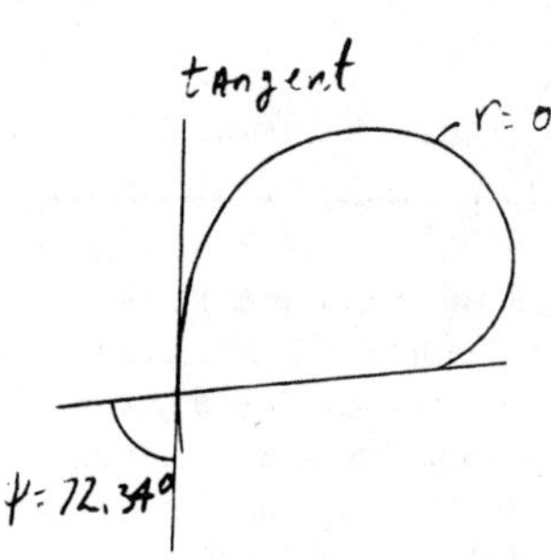

$$r = 0.6\theta \qquad \dot{r} = 0.6\dot{\theta} \qquad \ddot{r} = 0.6\ddot{\theta}$$

$$v_r = \dot{r} = 0.6\dot{\theta} \qquad v_\theta = r\dot{\theta} = 0.6\theta\dot{\theta}$$

$$v^2 = \dot{r}^2 + \left(r\dot{\theta}\right)^2$$

$$2^2 = \left(0.6\dot{\theta}\right)^2 + \left(0.6\theta\dot{\theta}\right)^2 \qquad \dot{\theta} = \frac{2}{0.6\sqrt{1+\theta^2}}$$

$$0 = 0.72\dot{\theta}\ddot{\theta} + 0.36\left(2\theta\dot{\theta}^3 + 2\theta^2\dot{\theta}\ddot{\theta}\right) \qquad \ddot{\theta} = -\frac{\theta\dot{\theta}^2}{1+\theta^2}$$

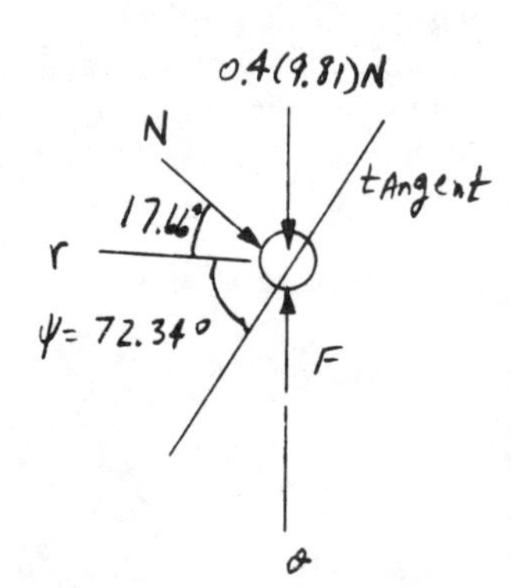

At $\theta = \pi$ rad, $\quad \dot{\theta} = \dfrac{2}{0.6\sqrt{1+\pi^2}} = 1.011$ rad/s

$$\ddot{\theta} = -\frac{(\pi)(1.011)^2}{1+\pi^2} = -0.2954 \text{ rad/s}^2$$

$$\tan\psi = \frac{r}{dr/d\theta} = \frac{0.6\theta}{0.6} = \theta = \pi \qquad \psi = 72.34°$$

$$r = 0.6(\pi) = 0.6\pi \text{ m} \qquad \dot{r} = 0.6(1.011) = 0.6066 \text{ m/s}$$

$$\ddot{r} = 0.6(-0.2954) = -0.1772 \text{ m/s}^2$$

$$a_r = \ddot{r} - r\dot{\theta}^2 = -0.1772 - 0.6\pi(1.011)^2 = -2.104 \text{ m/s}^2$$

$$a_\theta = r\ddot{\theta} + 2\dot{r}\dot{\theta} = 0.6\pi(-0.2954) + 2(0.6066)(1.011) = 0.6698 \text{ m/s}^2$$

$$\overset{+}{\leftarrow} \Sigma F_r = ma_r; \qquad -N\cos 17.66° = 0.4(-2.104) \qquad N = 0.883 \text{ N} \qquad \textbf{Ans}$$

$$+\downarrow \Sigma F_\theta = ma_\theta; \qquad -F + 0.4(9.81) + 0.883\sin 17.66° = 0.4(0.6698)$$

$$F = 3.92 \text{ N} \qquad \textbf{Ans}$$

13-105. A ride in an amusement park consists of a cart which is supported by small wheels. Initially the cart is traveling in a circular path of radius $r_0 = 16$ ft such that the angular rate of rotation is $\dot{\theta}_0 = 0.2$ rad/s. If the attached cable OC is drawn inward at a constant speed of $\dot{r} = -0.5$ ft/s, determine the tension it exerts on the cart at the instant $r = 4$ ft. The cart and its passengers have a total weight of 400 lb. Neglect the effects of friction. *Hint:* First show that the equation of motion in the θ direction yields $a_\theta = r\ddot{\theta} + 2\dot{r}\dot{\theta} = (1/r)d(r^2\dot{\theta})/dt = 0$. When integrated, $r^2\dot{\theta} = c$, where the constant c is determined from the problem data.

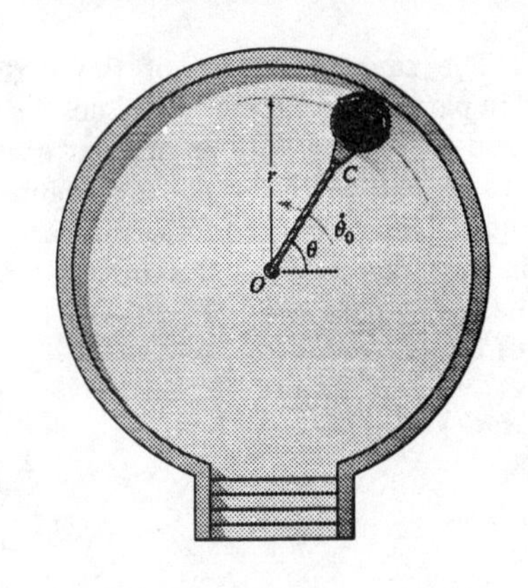

$+\nearrow \Sigma F_r = ma_r;\quad -T = \left(\dfrac{400}{32.2}\right)\left(\ddot{r} - r\dot{\theta}^2\right) \quad (1)$

$+\nwarrow \Sigma F_\theta = ma_\theta;\quad 0 = \left(\dfrac{400}{32.2}\right)\left(r\ddot{\theta} + 2\dot{r}\dot{\theta}\right) \quad (2)$

From Eq. (2), $\left(\dfrac{1}{r}\right)\dfrac{d}{dt}\left(r^2\dot{\theta}\right) = 0 \qquad r^2\dot{\theta} = c$

T

$\dfrac{400}{32.2}a_\theta$

$\dfrac{400}{32.2}a_r$

Since $\dot{\theta}_0 = 0.2$ rad/s when $r_0 = 16$ ft, $c = 51.2$.
Hence, when $r = 4$ ft,

$$\dot{\theta} = \left(\frac{51.2}{(4)^2}\right) = 3.2 \text{ rad/s}$$

Since $\dot{r} = -0.5$ ft/s, $\ddot{r} = 0$, Eq. (1) becomes

$$-T = \left(\frac{400}{32.2}\right)\left(0 - (4)(3.2)^2\right)$$

$T = 509$ lb **Ans**

13-106. The smooth surface of the vertical cam is defined in part by the curve $r = (0.2\cos\theta + 0.3)$ m. If the forked rod is rotating with a constant angular velocity of $\dot{\theta} = 4$ rad/s, determine the force the cam and the rod exert on the 2-kg roller when $\theta = 30°$. The attached spring has a stiffness $k = 30$ N/m and an unstretched length of 0.1 m.

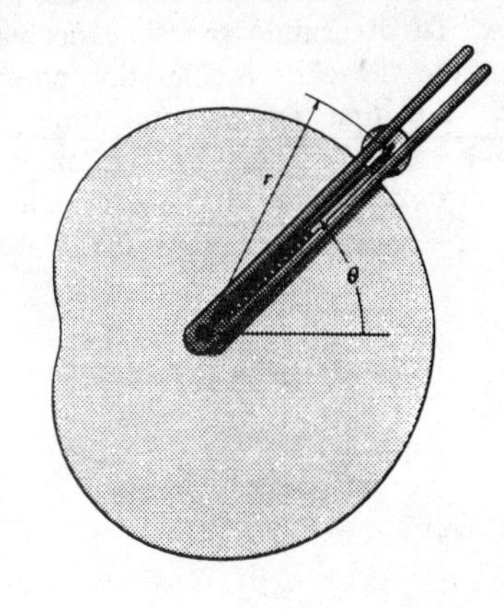

$r = 0.2\cos\theta + 0.3$

$\dot{r} = -0.2\sin\theta\dot{\theta}$

$\ddot{r} = -0.2\cos\theta\dot{\theta}^2 - 0.2\sin\theta\ddot{\theta}$

$\theta = 30° \qquad \dot{\theta} = 4 \qquad \ddot{\theta} = 0$

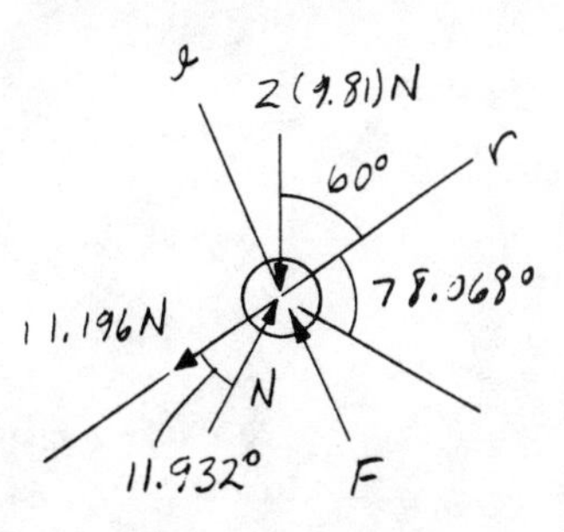

Thus,

$r = 0.47321 \qquad \dot{r} = -0.4 \qquad \ddot{r} = -2.77128$

$a_r = \ddot{r} - r\dot{\theta}^2 = -2.77128 - 0.47321(4)^2 = -10.3426$

$a_\theta = r\ddot{\theta} + 2\dot{r}\dot{\theta} = 0 + 2(-0.4)(4) = -3.20$

$F_s = kx = 30(0.47321 - 0.1) = 11.196$ N

$\tan\psi = \dfrac{r}{dr/d\theta} = \dfrac{0.47321}{-0.2\sin 30°} = -4.7321 \qquad \psi = -78.068°$

$+\nearrow \Sigma F_r = ma_r;\quad N\cos 11.932° - 11.196 - 2(9.81)\cos 60° = 2(-10.3426)$

$N = 0.3281$ N $= 0.328$ N **Ans**

$+\nwarrow \Sigma F_\theta = ma_\theta;\quad F + 0.3281\sin 11.932° - 2(9.81)\sin 60° = 2(-3.20)$

$F = 10.5$ N **Ans**

13-107. The smooth surface of the vertical cam is defined in part by the curve $r = (0.2\cos\theta + 0.3)$ m. The forked rod is rotating with an angular acceleration of $\ddot{\theta} = 2$ rad/s^2, and when $\theta = 45°$ the angular velocity is $\dot{\theta} = 6$ rad/s. Determine the force the cam and the rod exert on the 2-kg roller at this instant. The attached spring has a stiffness $k = 100$ N/m and an unstretched length of 0.1 m.

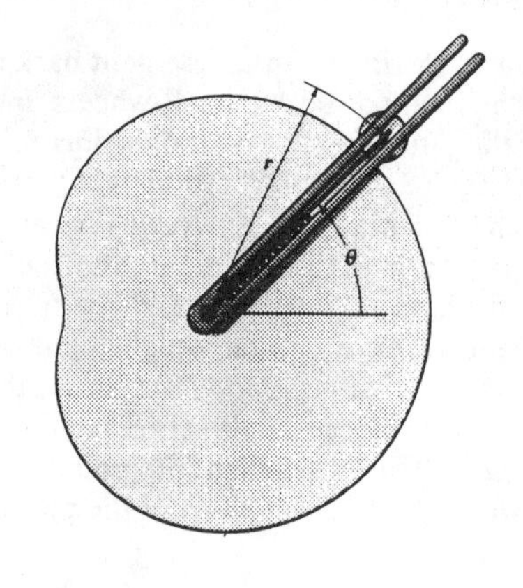

$r = 0.2\cos\theta + 0.3$

$\dot{r} = -0.2\sin\theta\dot{\theta}$

$\ddot{r} = -0.2\cos\theta\dot{\theta}^2 - 0.2\sin\theta\ddot{\theta}$

$\theta = 45°$ $\quad \dot{\theta} = 6$ $\quad \ddot{\theta} = 2$

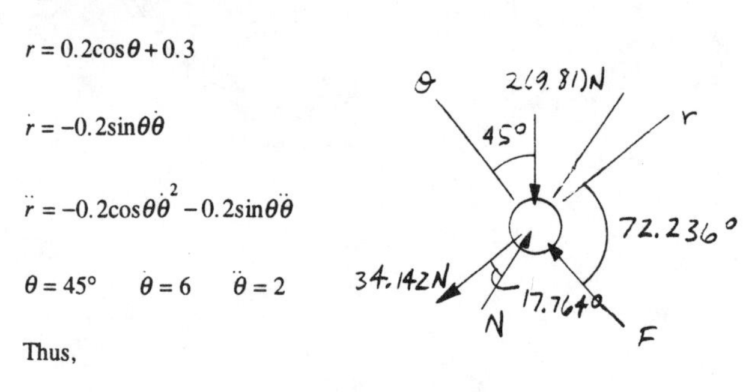

Thus,

$r = 0.44142$ $\quad \dot{r} = -0.84853$ $\quad \ddot{r} = -5.37401$

$a_r = \ddot{r} - r\dot{\theta}^2 = -5.37401 - 0.44142(6)^2 = -21.265$

$a_\theta = r\ddot{\theta} + 2\dot{r}\dot{\theta} = 0.44142(2) + 2(-0.84853)(6) = -9.2995$

$F_s = kx = 100(0.44142 - 0.1) = 34.142$ N

$\tan\psi = \dfrac{r}{dr/d\theta} = \dfrac{0.44142}{-0.2\sin 45°} = -3.1213$ $\quad \psi = -72.236°$

$+\nearrow \Sigma F_r = ma_r;$ $\quad N\cos 17.764° - 34.142 - 2(9.81)\sin 45° = 2(-21.265)$

$N = 5.7598$ N $= 5.76$ N **Ans**

$+\nwarrow \Sigma F_\theta = ma_\theta;$ $\quad 5.7598\sin 17.764° - 2(9.81)\cos 45° + F = 2(-9.2995)$

$F = -6.48$ N $= 6.48$ N **Ans**

***13-108.** The 4.20-Mg car C is traveling along a portion of a road defined by the cardioid $r = 400(1 + \cos\theta)$ m. If the car maintains a constant speed $v_C = 20$ m/s, determine the radial and transverse components of the frictional force which must be exerted by the road on all the wheels in order to maintain the motion when $\theta = 0°$. *Hint:* To determine the time derivatives necessary to compute the car's acceleration components a_r and a_θ, take the first and second time derivatives of $r = 400(1 + \cos\theta)$. Then, for further information, use Eq. 12–26 to determine $\dot{\theta}$. Also, take the time derivative of Eq. 12–26, noting that $\dot{v}_C = 0$, to determine $\ddot{\theta}$.

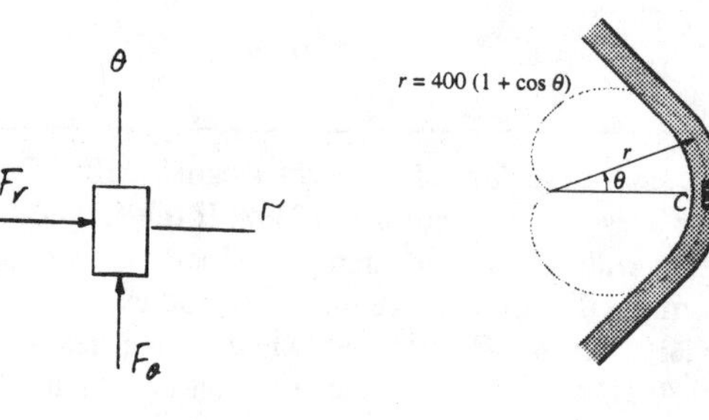

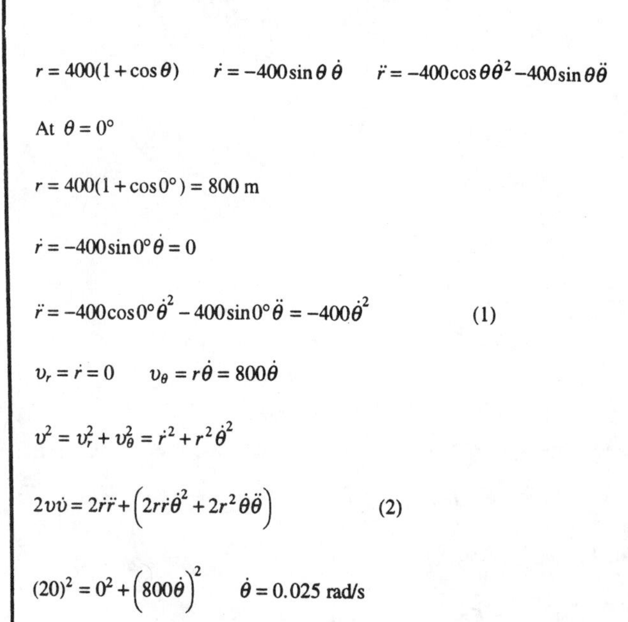

$r = 400(1 + \cos\theta)$ $\quad \dot{r} = -400\sin\theta\,\dot{\theta}$ $\quad \ddot{r} = -400\cos\theta\,\dot{\theta}^2 - 400\sin\theta\,\ddot{\theta}$

At $\theta = 0°$

$r = 400(1 + \cos 0°) = 800$ m

$\dot{r} = -400\sin 0°\dot{\theta} = 0$

$\ddot{r} = -400\cos 0°\dot{\theta}^2 - 400\sin 0°\ddot{\theta} = -400\dot{\theta}^2$ $\quad$ (1)

$v_r = \dot{r} = 0$ $\quad v_\theta = r\dot{\theta} = 800\dot{\theta}$

$v^2 = v_r^2 + v_\theta^2 = \dot{r}^2 + r^2\dot{\theta}^2$

$2v\dot{v} = 2\dot{r}\ddot{r} + \left(2r\dot{r}\dot{\theta}^2 + 2r^2\dot{\theta}\ddot{\theta}\right)$ $\quad$ (2)

$(20)^2 = 0^2 + \left(800\dot{\theta}\right)^2$ $\quad \dot{\theta} = 0.025$ rad/s

From Eq. (1)

$\ddot{r} = -400(0.025)^2 = -0.250$ m/s^2

From Eq. (2)

$2(20)(0) = 2(0)(-0.250) + \left[2(800)(0)(0.250)^2 + 2(800)^2(0.250)\ddot{\theta}\right]$

$\ddot{\theta} = 0$

$a_r = \ddot{r} - r\dot{\theta}^2 = -0.250 - 800(0.025)^2 = -0.75$ m/s^2

$a_\theta = r\ddot{\theta} + 2\dot{r}\dot{\theta} = 800(0) + 2(0)(0.025) = 0$

$\overset{+}{\leftarrow} \Sigma F_r = ma_r;$ $\quad F_r = 4.20(10^3)(-0.75)$ $\quad F_r = -3.51$ kN **Ans**

$+\uparrow \Sigma F_\theta = ma_\theta;$ $\quad F_\theta = 0$ **Ans**

13-109. The pilot of an airplane executes a vertical loop which in part follows the path of a "four-leaved rose," $r = (-600 \cos 2\theta)$ ft, where θ is in radians. If his speed at A is a constant $v_P = 80$ ft/s, determine the vertical reaction the seat of the plane exerts on the pilot when the plane is at A. He weighs 130 lb. See hint related to Prob. 13–108.

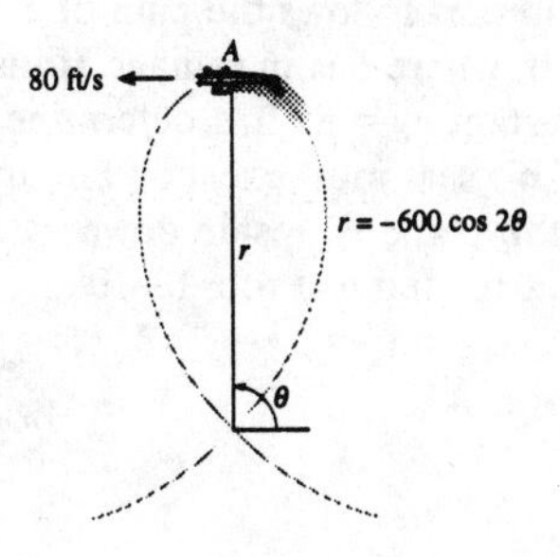

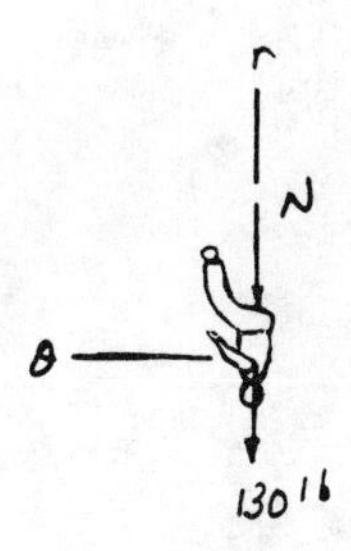

$r = -600\cos 2\theta \qquad \dot{r} = 1200\sin 2\theta\dot{\theta} \qquad \ddot{r} = 1200\left(2\cos 2\theta\dot{\theta}^2 + \sin 2\theta\ddot{\theta}\right)$

At $\theta = 90°$

$r = -600\cos 180° = 600$ ft $\qquad \dot{r} = 1200\sin 180°\dot{\theta} = 0$

$\ddot{r} = 1200\left(2\cos 180°\dot{\theta}^2 + \sin 180°\ddot{\theta}\right) = -2400\dot{\theta}^2$

$v_r = \dot{r} = 0 \qquad v_\theta = r\dot{\theta} = 600\dot{\theta}$

$v_p^2 = v_r^2 + v_\theta^2$

$80^2 = 0^2 + \left(600\dot{\theta}\right)^2 \qquad \dot{\theta} = 0.1333$ rad/s

$\ddot{r} = -2400(0.1333)^2 = -42.67$ ft/s^2

$a_r = \ddot{r} - r\dot{\theta}^2 = -42.67 - 600(0.1333)^2 = -53.33$ ft/s^2

$+\uparrow \Sigma F_r = ma_r; \qquad -N - 130 = \dfrac{130}{32.2}(-53.33) \qquad N = 85.3$ lb **Ans**

13-110. The pilot of an airplane executes a vertical loop which in part follows the path of a cardioid, $r = 600(1 + \cos\theta)$ ft, where θ is in radians. If his speed at A ($\theta = 0°$) is a constant $v_P = 80$ ft/s, determine the vertical force the belt of his seat must exert on him to hold him to his seat when the plane is upside down at A. He weighs 150 lb. See hint related to Prob. 13–108.

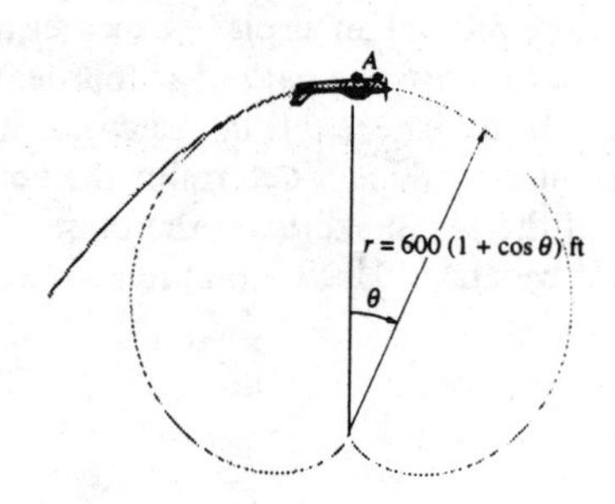

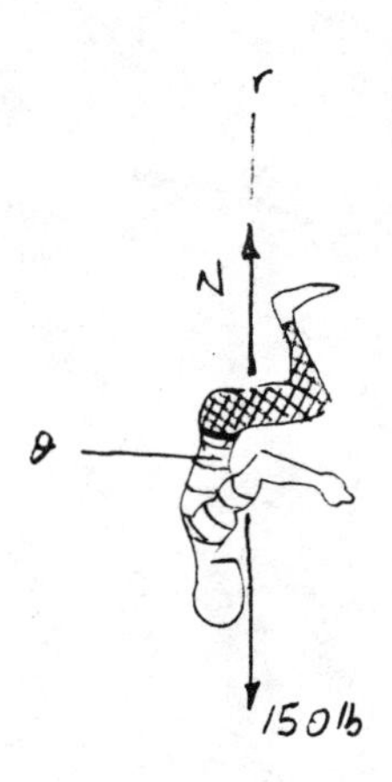

$$r = 600(1+\cos\theta)|_{\theta=0°} = 1200 \text{ ft}$$

$$\dot{r} = -600\sin\theta\dot{\theta}|_{\theta=0°} = 0$$

$$\ddot{r} = -600\sin\theta\ddot{\theta} - 600\cos\theta\dot{\theta}^2|_{\theta=0°} = -600\dot{\theta}^2$$

$$v_p^2 = \dot{r}^2 + \left(r\dot{\theta}\right)^2$$

$$(80)^2 = 0 + \left(1200\dot{\theta}\right)^2 \qquad \dot{\theta} = 0.06667$$

$$2v_p\dot{v}_p = 2\dot{r}\ddot{r} + 2\left(r\dot{\theta}\right)\left(\dot{r}\dot{\theta} + r\ddot{\theta}\right)$$

$$0 = 0 + 0 + 2r^2\dot{\theta}\ddot{\theta} \qquad \ddot{\theta} = 0$$

$$a_r = \ddot{r} - r\dot{\theta}^2 = -600(0.06667)^2 - 1200(0.06667)^2 = -8 \text{ ft/s}^2$$

$$a_\theta = r\ddot{\theta} + 2\dot{r}\dot{\theta} = 0 + 0 = 0$$

$$+\uparrow\Sigma F_r = ma_r; \qquad N - 150 = \left(\frac{150}{32.2}\right)(-8) \qquad N = 113 \text{ lb} \qquad \textbf{Ans}$$

13-111. The tube rotates in the horizontal plane at a constant rate of $\dot{\theta} = 4$ rad/s. If a 0.2-kg ball B starts at the origin O with an initial radial velocity of $\dot{r} = 1.5$ m/s and moves outward through the tube, determine the radial and transverse components of the ball's velocity at the instant it leaves the outer end at C, $r = 0.5$ m. *Hint:* Show that the equation of motion in the r direction is $\ddot{r} - 16r = 0$. The solution is of the form $r = Ae^{-4t} + Be^{4t}$. Evaluate the integration constants A and B, and determine the time t when $r = 0.5$ m. Proceed to obtain v_r and v_θ.

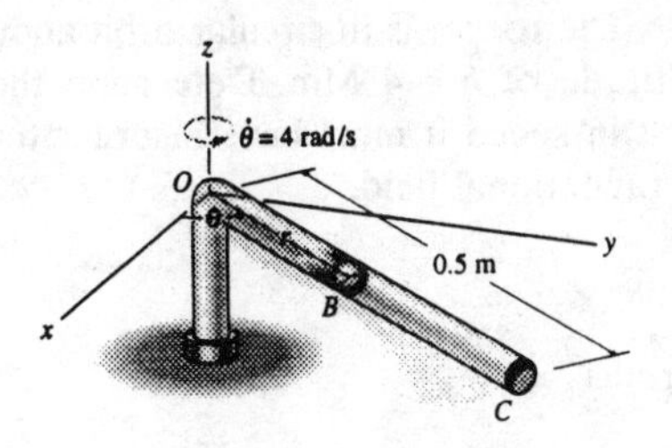

$\dot{\theta} = 4 \qquad \ddot{\theta} = 0$

$\Sigma F_r = ma_r; \qquad 0 = 0.2\left[\ddot{r} - r(4)^2\right]$

$$\ddot{r} - 16r = 0$$

Solving this second - order differential equation,

$r = Ae^{-4t} + Be^{4t} \qquad (1)$

$\dot{r} = -4Ae^{-4t} + 4Be^{4t} \qquad (2)$

At $t = 0$, $r = 0$, $\dot{r} = 1.5$

$0 = A + B \qquad \dfrac{1.5}{4} = -A + B$

$A = -0.1875 \qquad B = 0.1875$

From Eq. (1) at $r = 0.5$ m,

$0.5 = 0.1875\left(-e^{-4t} + e^{4t}\right)$

$\dfrac{2.667}{2} = \dfrac{(-e^{-4t} + e^{4t})}{2}$

$1.333 = \sinh(4t)$

$t = \dfrac{1}{4}\sinh^{-1}(1.333) \qquad t = 0.275$ s

Using Eq. (2),

$\dot{r} = 4(0.1875)\left(e^{-4t} + e^{4t}\right)$

$\dot{r} = 8(0.1875)\left(\dfrac{e^{-4t} + e^{4t}}{2}\right) = 8(0.1875)(\cosh(4t))$

At $t = 0.275$ s

$\dot{r} = 1.5\cosh[4(0.275)]$

$v_r = \dot{r} = 2.50$ m/s **Ans**

$v_\theta = r\dot{\theta} = 0.5(4) = 2$ m/s **Ans**

***13-112.** The rocket is in circular orbit about the earth at an altitude of $h = 4$ Mm. Determine the minimum increment in speed it must have in order to escape the earth's gravitational field.

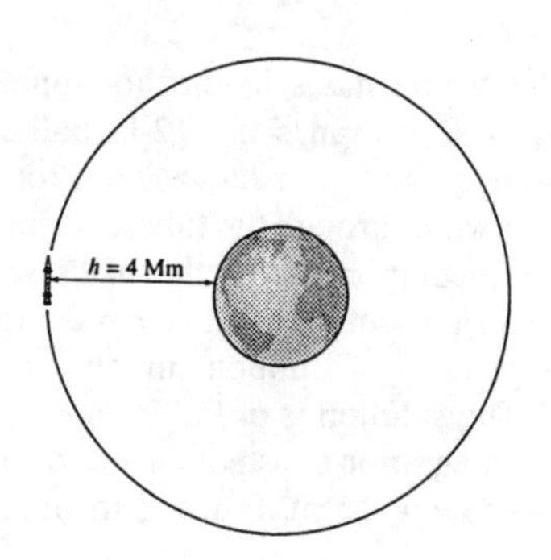

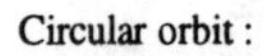

Circular orbit :

$$v_C = \sqrt{\frac{GM_e}{r_0}} = \sqrt{\frac{66.73(10^{-12})5.976(10^{24})}{4000(10^3)+6378(10^3)}} = 6198.8 \text{ m/s}$$

Parabolic orbit :

$$v_e = \sqrt{\frac{2GM_e}{r_0}} = \sqrt{\frac{2(66.73)(10^{-12})5.976(10^{24})}{4000(10^3)+6378(10^3)}} = 8766.4 \text{ m/s}$$

$$\Delta v = v_e - v_C = 8766.4 - 6198.8 = 2567.6 \text{ m/s}$$

$\Delta v = 2.57$ km/s **Ans**

13-113. Prove Kepler's third law of motion. *Hint:* Use Eqs. 13–19, 13–28, 13–29, and 13–31.

From Eq. 13 – 19,

$$\frac{1}{r} = C\cos\theta + \frac{GM_s}{h^2}$$

For $\theta = 0°$ and $\theta = 180°$

$$\frac{1}{r_p} = C + \frac{GM_s}{h^2}$$

$$\frac{1}{r_a} = -C + \frac{GM_s}{h^2}$$

Eliminating C,

From Eqs. 13 – 28 and 13 – 29,

$$\frac{2a}{b^2} = \frac{2GM_s}{h^2}$$

From Eq. 13 – 31,

$$T = \frac{\pi}{h}(2a)(b)$$

Thus,

$$b^2 = \frac{T^2h^2}{4\pi^2a^2}$$

$$\frac{4\pi^2a^3}{T^2h^2} = \frac{GM_s}{h^2}$$

$$T^2 = \left(\frac{4\pi^2}{GM_s}\right)a^3$$ **Q.E.D.**

13-114. The satellite is moving in an elliptical path with an eccentricity $e = 0.25$. Determine its speed when it is at its maximum distance A and minimum distance B from the earth.

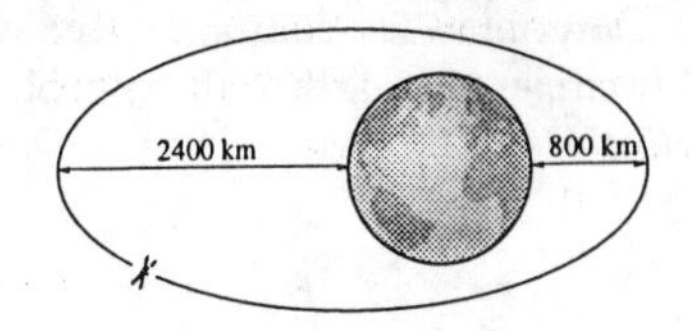

$$e = \frac{Ch^2}{GM_e} \quad \text{where } C = \frac{1}{r_0}\left(1 - \frac{GM_e}{r_0 v_0^2}\right) \text{ and } h = r_0 v_0$$

$$e = \frac{1}{GM_e r_0}\left(1 - \frac{GM_e}{r_0 v_0^2}\right)(r_0 v_0)^2$$

$$e = \left(\frac{r_0 v_0^2}{GM_e} - 1\right)$$

$$\frac{r_0 v_0^2}{GM_e} = e + 1 \qquad v_0 = \sqrt{\frac{GM_e(e+1)}{r_0}}$$

Where $r_0 = r_p = 2(10^6) + 6378(10^3) = 8.378(10^6)\text{m}$

$$v_B = v_0 = \sqrt{\frac{66.73(10^{-12})(5.976)(10^{24})(0.25+1)}{8.378(10^6)}} = 7713 \text{ m/s} = 7.71 \text{ km/s} \qquad \textbf{Ans}$$

$$r_a = \frac{r_0}{\frac{2GM_e}{r_0 v_0^2} - 1} = \frac{8.378(10^6)}{\frac{2(66.73)(10^{-12})(5.976)(10^{24})}{8.378(10^6)(7713)^2} - 1} = 13.96(10^6)$$

$$v_A = \frac{r_p}{r_a} v_B = \frac{8.378(10^6)}{13.96(10^6)}(7713) = 4628 \text{ m/s} = 4.63 \text{ km/s} \qquad \textbf{Ans}$$

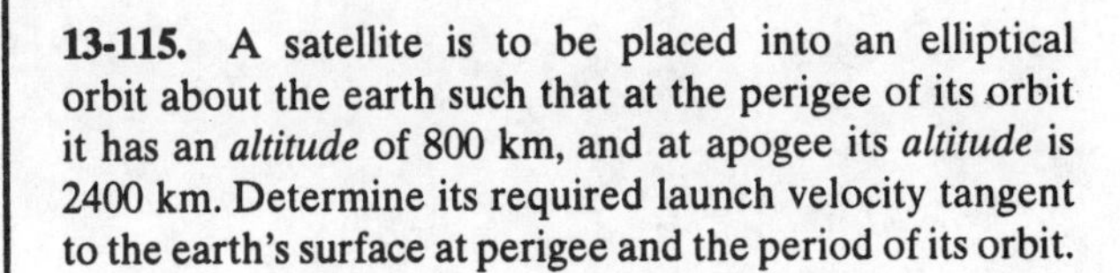

13-115. A satellite is to be placed into an elliptical orbit about the earth such that at the perigee of its orbit it has an *altitude* of 800 km, and at apogee its *altitude* is 2400 km. Determine its required launch velocity tangent to the earth's surface at perigee and the period of its orbit.

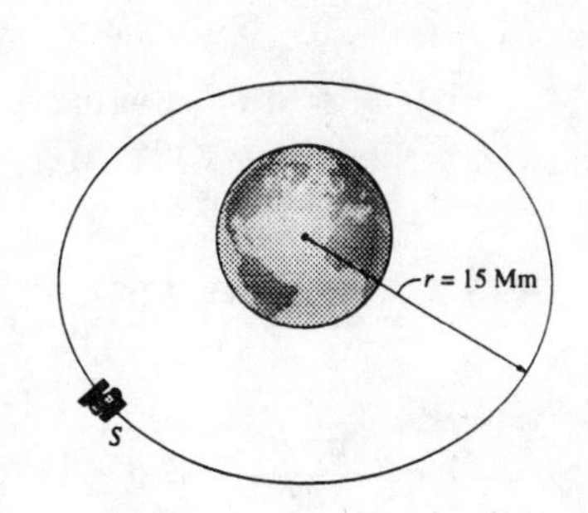

$$r_p = 800 + 6378 = 7178 \text{ km} \qquad r_a = 2400 + 6378 = 8778 \text{ km}$$

$$r_a = \frac{r_p}{\frac{2GM_e}{r_p v_0^2} - 1}$$

$$v_0 = \sqrt{\frac{2GM_e r_a}{r_p(r_a + r_p)}}$$

$$= \sqrt{\frac{2(66.73)(10^{-12})(5.976)(10^{24})(8778)(10^3)}{7178(10^3)[8778(10^3) + 7178(10^3)]}}$$

$$= 7818 \text{ m/s} = 7.82 \text{ km/s} \qquad \textbf{Ans}$$

$$h = r_p v_0 = 7178(10^3)(7818) = 56.12(10^9) \text{ m}^2/\text{s}$$

$$T = \frac{\pi}{h}(r_p + r_a)\sqrt{r_p r_a}$$

$$= \frac{\pi}{56.12(10^9)}[7178(10^3) + 8778(10^3)]\sqrt{[7178(10^3)][8778(10^3)]}$$

$$= 7090 \text{ s} = 1.97 \text{ h} \qquad \textbf{Ans}$$

***13-116.** Determine the constant speed of the satellite S so that it circles the earth with an orbit of radius r = 15 Mm. *Hint:* Use Eq. 13–1.

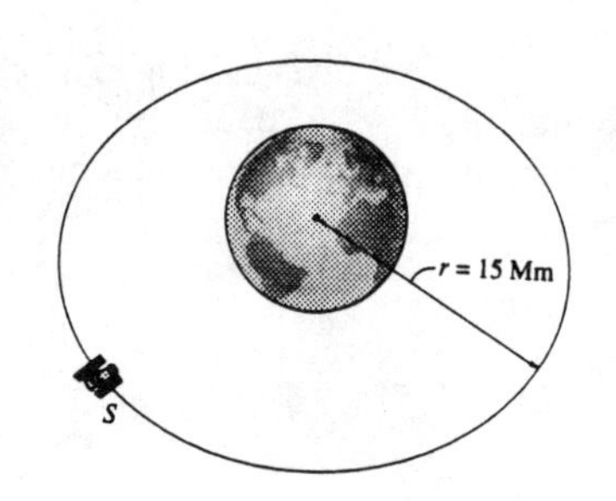

$$F = G\frac{m_s m_e}{r^2} \quad \text{Also} \quad F = m_s\left(\frac{v_s^2}{r}\right) \quad \text{Hence}$$

$$m_s\left(\frac{v_s^2}{r}\right) = G\frac{m_s m_e}{r^2}$$

$$v = \sqrt{G\frac{m_e}{r}} = \sqrt{66.73(10^{-12})\left(\frac{5.976(10^{24})}{15(10^6)}\right)} = 5156 \text{ m/s} = 5.16 \text{ km/s} \quad \textbf{Ans}$$

13-117. A satellite is launched with an initial velocity $v_0 = 2500$ mi/h parallel to the surface of the earth. Determine the required altitude (or range of altitudes) above the earth's surface for launching if the free-flight trajectory is to be (a) circular, (b) parabolic, (c) elliptical, and (d) hyperbolic. Take $G = 34.4(10^{-9})$ (lb·ft²)/slug², $M_e = 409(10^{21})$ slug, the earth's radius $r_e = 3960$ mi, and 1 mi = 5280 ft.

$$v_0 = 2500 \text{ mi/h} = 3.67(10^3) \text{ ft/s}$$

(a) $$e = \frac{C^2 h}{GM_e} = 0 \quad \text{or } C = 0$$

$$1 = \frac{GM_e}{r_0 v_0^2}$$

$$GM_e = 34.4(10^{-9})(409)(10^{21})$$

$$= 14.07(10^{15})$$

$$r_0 = \frac{GM_e}{v_0^2} = \frac{14.07(10^{15})}{[3.67(10^3)]^2} = 1.046(10^9) \text{ ft}$$

$$r = \frac{1.047(10^9)}{5280} - 3960 = 194(10^3) \text{ mi} \quad \textbf{Ans}$$

(b) $$e = \frac{C^2 h}{GM_e} = 1$$

$$\frac{1}{GM_e}(r_0^2 v_0^2)\left(\frac{1}{r_0}\right)\left(1 - \frac{GM_e}{r_0 v_0^2}\right) = 1$$

$$r_0 = \frac{2GM_e}{v_0^2} = \frac{2(14.07)(10^{15})}{[3.67(10^3)]^2} = 2.09(10^9) \text{ ft} = 396(10^3) \text{ mi}$$

$$r = 396(10^3) - 3960 = 392(10^3) \text{ mi} \quad \textbf{Ans}$$

(c) $$e < 1$$

$$194(10^3) \text{ mi} < r < 392(10^3) \text{ mi} \quad \textbf{Ans}$$

(d) $$e > 1$$

$$r > 392(10^3) \text{ mi} \quad \textbf{Ans}$$

13-118. The elliptical path of a satellite has an eccentricity $e = 0.130$. If the satellite has a speed of 15 Mm/h when it is at perigee, P, determine its speed when it arrives at apogee, A. Also, how far is it from the earth's surface when it is at A?

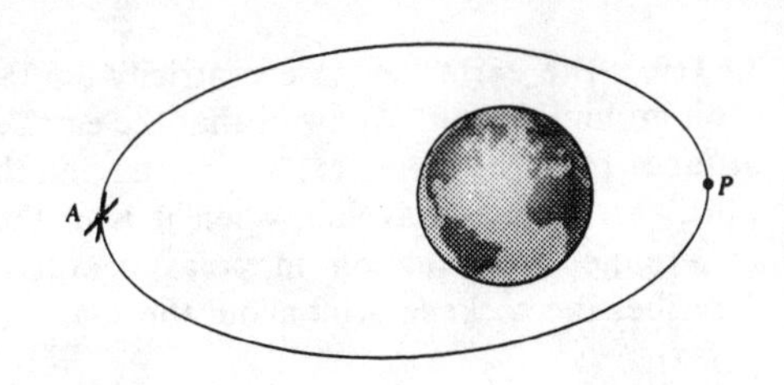

$$e = 0.130$$

$$v_p = v_0 = 15 \text{ Mm/h} = 4.167 \text{ km/s}$$

$$e = \frac{Ch^2}{GM_e} = \frac{1}{r_0}\left(1 - \frac{GM_e}{r_0 v_0^2}\right)\left(\frac{r_0^2 v_0^2}{GM_e}\right)$$

$$e = \left(\frac{r_0 v_0^2}{GM_e} - 1\right)$$

$$\frac{r_0 v_0^2}{GM_e} = e + 1$$

$$r_0 = \frac{(e+1)GM_e}{v_0^2}$$

$$= \frac{1.130(66.73)(10^{-12})(5.976)(10^{24})}{\left[4.167(10^3)\right]^2}$$

$$= 25.96 \text{ Mm}$$

$$\frac{GM_e}{r_0 v_0^2} = \frac{1}{e+1}$$

$$r_A = \frac{r_0}{\frac{2GM_e}{r_0 v_0^2} - 1} = \frac{r_0}{\left(\frac{2}{e+1}\right) - 1}$$

$$r_A = \frac{r_0(e+1)}{1-e}$$

$$= \frac{25.96(10^6)(1.130)}{0.870}$$

$$= 33.71\left(10^6\right)\text{m} = 33.7 \text{ Mm}$$

$$v_A = \frac{v_0 r_0}{r_A}$$

$$= \frac{15(25.96)(10^6)}{33.71(10^6)}$$

$$= 11.5 \text{ Mm/h} \qquad \textbf{Ans}$$

$$d = 33.71\left(10^6\right) - 6.378\left(10^6\right)$$

$$= 27.3 \text{ Mm} \qquad \textbf{Ans}$$

13-119. The earth has an eccentricity $e = 0.0821$ in its orbit around the sun. Knowing that the earth's minimum distance from the sun is $151.3(10^6)$ km, find the speed at which a rocket is traveling when it is at this distance. Determine the equation in polar coordinates which describes the rocket's orbit about the sun.

$$e = \frac{Ch^2}{GM_S} \quad \text{where } C = \frac{1}{r_0}\left(1 - \frac{GM_S}{r_0 v_0^2}\right) \text{ and } h = r_0 v_0$$

$$e = \frac{1}{GM_S r_0}\left(1 - \frac{GM_S}{r_0 v_0^2}\right)(r_0 v_0)^2 \qquad e = \left(\frac{r_0 v_0^2}{GM_S} - 1\right) \qquad \frac{r_0 v_0^2}{GM_S} = e + 1$$

$$v_0 = \sqrt{\frac{GM_S(e+1)}{r_0}}$$

$$= \sqrt{\frac{66.73(10^{-12})(1.99)(10^{30})(0.0821+1)}{151.3(10^9)}} = 30818 \text{ m/s} = 30.8 \text{ km/s} \qquad \textbf{Ans}$$

$$\frac{1}{r} = \frac{1}{r_0}\left(1 - \frac{GM_S}{r_0 v_0^2}\right)\cos\theta + \frac{GM_S}{r_0^2 v_0^2}$$

$$\frac{1}{r} = \frac{1}{151.3(10^9)}\left(1 - \frac{66.73(10^{-12})(1.99)(10^{30})}{151.3(10^9)(30818)^2}\right)\cos\theta + \frac{66.73(10^{-12})(1.99)(10^{30})}{\left[151.3(10^9)\right]^2(30818)^2}$$

$$\frac{1}{r} = 0.502(10^{-12})\cos\theta + 6.11(10^{-12}) \qquad \textbf{Ans}$$

***13-120.** The rocket is in a free-flight elliptical orbit about the earth such that the eccentricity of its orbit is e and its perigee is r_0. Determine the minimum increment of speed it should have in order to escape the earth's gravitational field when it is at this point along its orbit.

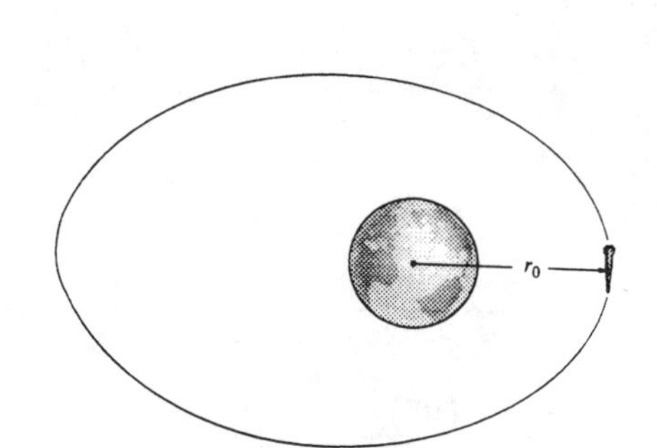

To escape the earth's gravitation field, the rocket has to make a parabolic trajectory.

Parabolic trajectory :

$$v_e = \sqrt{\frac{2GM_e}{r_0}}$$

Elliptical orbit :

$$e = \frac{Ch^2}{GM_e} \quad \text{where } C = \frac{1}{r_0}\left(1 - \frac{GM_e}{r_0 v_0^2}\right) \text{ and } h = r_0 v_0$$

$$e = \frac{1}{GM_e r_0}\left(1 - \frac{GM_e}{r_0 v_0^2}\right)(r_0 v_0)^2$$

$$e = \left(\frac{r_0 v_0^2}{GM_e} - 1\right)$$

$$\frac{r_0 v_0^2}{GM_e} = e + 1 \qquad v_0 = \sqrt{\frac{GM_e(e+1)}{r_0}}$$

$$\Delta v = \sqrt{\frac{2GM_e}{r_0}} - \sqrt{\frac{GM_e(e+1)}{r_0}} = \sqrt{\frac{GM_e}{r_0}}\left(\sqrt{2} - \sqrt{1+e}\right) \qquad \textbf{Ans}$$

The change in speed should occur at **perigee**. **Ans**

13-121. A communications satellite is to be placed into an equatorial circular orbit around the earth so that it always remains directly over a point on the earth's surface. If this requires the period to be 24 hours (approximately), determine the radius of the orbit and the satellite's velocity.

$$\frac{GM_eM_s}{r^2} = \frac{M_sv^2}{r}$$

$$\frac{GM_e}{r} = v^2$$

$$\frac{GM_e}{r} = \left[\frac{2\pi r}{24(3600)}\right]^2$$

$$\frac{66.73(10^{-12})(5.976)(10^{24})}{\left[\frac{2\pi}{24(3600)}\right]^2} = r^3$$

$r = 42.25(10^6)\text{ m} = 42.2\text{ Mm}$ **Ans**

$v = \dfrac{2\pi(42.25)(10^6)}{24(3600)} = 3.07\text{ km/s}$ **Ans**

13-122. The rocket is traveling in free flight along an elliptical trajectory $A'A$. The planet has no atmosphere, and its mass is 0.60 times that of the earth's. If the rocket has the apogee and perigee shown, determine the rocket's velocity when it is at point A. Take $G = 34.4(10^{-9})\ (\text{lb}\cdot\text{ft}^2)/\text{slug}^2$, $M_e = 409(10^{21})$ slug, 1 mi = 5280 ft.

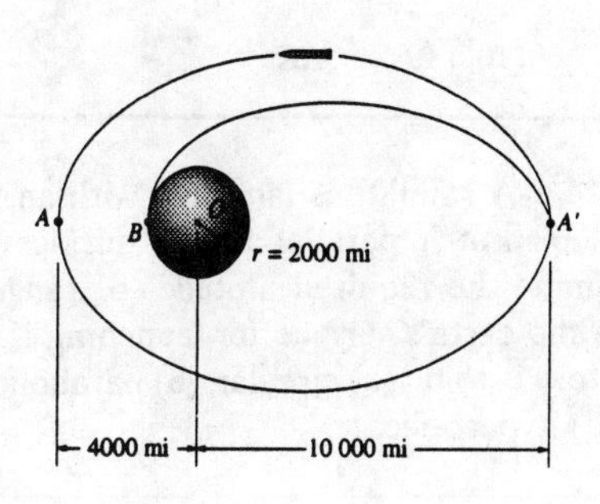

$r_0 = OA = (4000)(5280) = 21.12(10^6)\text{ ft}$ $OA' = (10\,000)(5280) = 52.80(10^6)\text{ ft}$

$M_P = (409(10^{21}))(0.6) = 245.4(10^{21})\text{ slug}$

$$OA' = \frac{OA}{\left(\frac{2GM_P}{OAv_0^2} - 1\right)}$$

$$v_0 = \sqrt{\frac{2GM_P}{OA\left(\frac{OA}{OA'} + 1\right)}} = \sqrt{\frac{2(34.4)(10^{-9})(245.4)(10^{21})}{21.12(10^6)\left(\frac{21.12}{52.80} + 1\right)}}$$

$v_0 = 23.9(10^3)\text{ ft/s}$ **Ans**

13-123. If the rocket in Prob. 13-122 is to land on the surface of the planet, determine the required free-flight speed it must have at A' so that the landing occurs at B. How long does it take for the rocket to land, in going from A' to B?

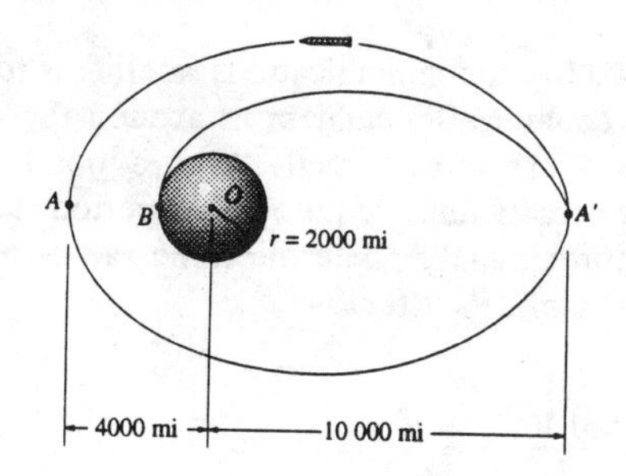

$M_P = 409(10^{21})(0.6) = 245.4(10^{21})$ slug

$OA' = (10\,000)(5280) = 52.80(10^6)$ ft $\quad OB = (2000)(5280) = 10.56(10^6)$ ft

$$OA' = \frac{OB}{\left(\frac{2GM_P}{OBv_0^2} - 1\right)}$$

$$v_0 = \sqrt{\frac{2GM_P}{OB\left(\frac{OB}{OA'}+1\right)}} = \sqrt{\frac{2(34.4(10^{-9}))245.4(10^{21})}{10.56(10^6)\left(\frac{10.56}{52.80}+1\right)}}$$

$v_0 = 36.50(10^3)$ ft/s (speed at B)

$$v_{A'} = \frac{OBv_0}{OA'}$$

$$v_{A'} = \frac{10.56(10^6)36.50(10^3)}{52.80(10)^6}$$

$v_{A'} = 7.30(10^3)$ ft/s **Ans**

$$T = \frac{\pi}{h}(OB + OA')\sqrt{(OB)(OA')}$$

$h = (OB)(v_0) = 10.56(10^6)36.50(10^3) = 385.4(10^9)$

Thus,

$$T = \frac{\pi(10.56 + 52.80)(10^6)}{385.5(10^9)}\left(\sqrt{(10.56)(52.80)}\right)(10^6)$$

$T = 12.20(10^3)$ s

$\frac{T}{2} = 6.10(10^3)$ s $= 1.69$ h **Ans**

***13-124.** A satellite is launched with an initial velocity $v_0 = 4000$ km/h parallel to the surface of the earth. Determine the required altitude (or range of altitudes) above the earth's surface for launching if the free-flight trajectory is to be (a) circular, (b) parabolic, (c) elliptical, and (d) hyperbolic.

$$v_0 = \frac{4000(10^3)}{3600} = 1111 \text{ m/s}$$

(a) For circular trajectory, $e = 0$

$$v_0 = \sqrt{\frac{GM_e}{r_0}} \qquad r_0 = \frac{GM_e}{v_0^2} = \frac{(66.73)(10^{-12})(5.976)(10^{24})}{(1111)^2} = 323(10^3) \text{ km}$$

$r = r_0 - 6378$ km $= 317(10^3)$ km $= 317$ Mm **Ans**

(b) For parabolic trajectory, $e = 1$

$$v_0 = \sqrt{\frac{2GM_e}{r_0}} \qquad r_0 = \frac{2GM_e}{v_0^2} = \frac{2(66.73)(10^{-12})(5.976)(10^{24})}{1111^2} = 646(10^3) \text{ km}$$

$r = r_0 - 6378$ km $= 640(10^3)$ km $= 640$ Mm **Ans**

(c) For elliptical trajectory, $e < 1$

317 Mm $< r <$ 640 Mm **Ans**

(c) For hyperbolic trajectory, $e > 1$

$r >$ 640 Mm **Ans**

13-125. The rocket is traveling in a free-flight elliptical orbit about the earth such that $e = 0.76$ as shown. Determine its speed when it is at point A. Also determine the sudden change in speed the rocket must experience at B in order to travel in free flight along the orbit indicated by the dashed path.

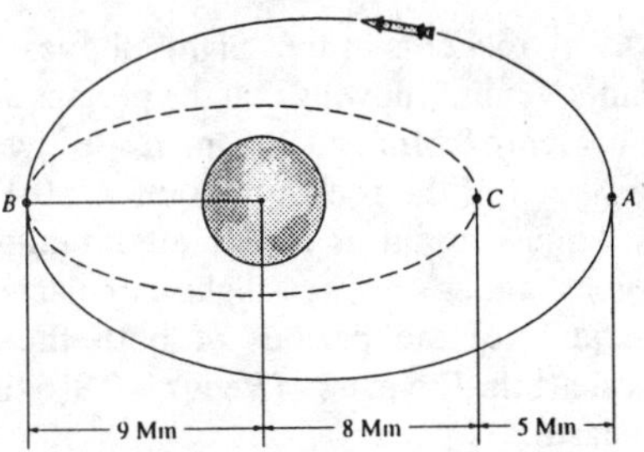

$$e = \frac{Ch^2}{GM_e} \quad \text{where } C = \frac{1}{r_0}\left(1 - \frac{GM_e}{r_0 v_0^2}\right) \text{ and } h = r_0 v_0$$

$$e = \frac{1}{GM_e r_0}\left(1 - \frac{GM_e}{r_0 v_0^2}\right)(r_0 v_0)^2$$

$$e = \left(\frac{r_0 v_0^2}{GM_e} - 1\right)$$

$$\frac{r_0 v_0^2}{GM_e} = e + 1 \quad \text{or} \quad \frac{GM_e}{r_0 v_0^2} = \frac{1}{e+1} \qquad (1)$$

$$r_a = \frac{r_0}{\frac{2GM_e}{r_0 v_0^2} - 1} \qquad (2)$$

Substituting Eq. (1) into (2) yields :

$$r_a = \frac{r_0}{2\left(\frac{1}{e+1}\right) - 1} = \frac{r_0(e+1)}{1-e} \qquad (3)$$

From Eq.(1) $\quad \frac{GM_e}{r_0 v_0^2} = \frac{1}{e+1} \qquad v_0 = \sqrt{\frac{GM_e(e+1)}{r_0}}$

$$v_B = v_0 = \sqrt{\frac{66.73(10^{-12})(5.976)(10^{24})(0.76+1)}{9(10^6)}} = 8831 \text{ m/s}$$

$$v_A = \frac{r_p}{r_a} v_B = \frac{9(10^6)}{13(10^6)}(8831) = 6113 \text{ m/s} = 6.11 \text{ km/s} \qquad \textbf{Ans}$$

From Eq. (3) $\quad r_a = \frac{r_0(e+1)}{1-e}$

$$9(10)^6 = \frac{8(10^6)(e+1)}{1-e} \qquad e = 0.05882$$

From Eq. (1) $\quad \frac{GM_e}{r_0 v_0^2} = \frac{1}{e+1} \qquad v_0 = \sqrt{\frac{GM_e(e+1)}{r_0}}$

$$v_C = v_0 = \sqrt{\frac{66.73(10^{-12})(5.976)(10^{24})(0.05882+1)}{8(10^6)}} = 7265 \text{ m/s}$$

$$v_B = \frac{r_p}{r_a} v_C = \frac{8(10^6)}{9(10^6)}(7265) = 6458 \text{ m/s}$$

$$\Delta v_B = 6458 - 8831 = -2374 \text{ m/s} = -2.37 \text{ km/s} \qquad \textbf{Ans}$$

13-126. A satellite is in an elliptical orbit around the earth such that $e = 0.156$. If its perigee is 5 Mm, determine its velocity at this point and also the distance OB when it is at point B, located 135° away as shown.

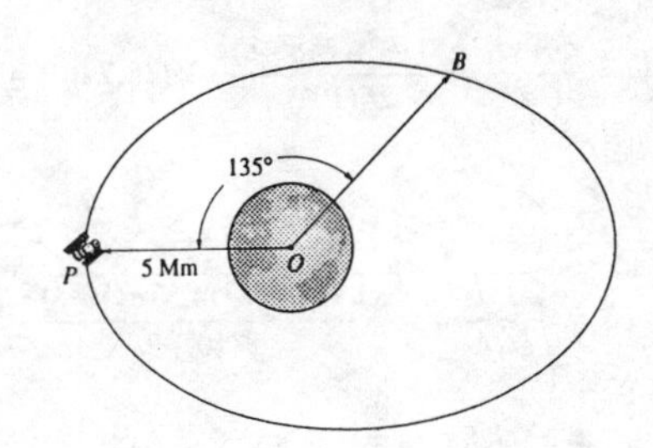

$$e = \frac{Ch^2}{GM_e} = \frac{1}{r_0}\left(1 - \frac{GM_e}{r_0 v_0^2}\right)\left(\frac{r_0^2 v_0^2}{GM_e}\right)$$

$$= \left(\frac{r_0 v_0^2}{GM_e} - 1\right)$$

$$\frac{r_0 v_0^2}{GM_e} = e + 1$$

$$\frac{5(10^6) v_0^2}{66.73(10^{-12})(5.976)(10^{24})} = 1.156$$

$v_0 = 9602 \text{ m/s} = 9.60 \text{ km/s}$ **Ans**

$$\frac{1}{r} = \frac{1}{r_0}\left(1 - \frac{GM_e}{r_0 v_0^2}\cos\theta\right) + \frac{GM_e}{r_0^2 v_0^2}$$

$$= \frac{1}{r_0}\left(1 - \frac{1}{e+1}\right)\cos\theta + \frac{1}{r_0}\left(\frac{1}{e+1}\right)$$

$$= \frac{1}{5(10^6)}\left(1 - \frac{1}{1.156}\right)\cos 135° + \frac{1}{5(10^6)}\left(\frac{1}{1.156}\right)$$

$r = 6.50 \text{ Mm}$ **Ans**

13-127. A rocket is in free-flight elliptical orbit around the planet Venus. Knowing that the periapsis and apoapsis of the orbit are 8 Mm and 26 Mm, respectively, determine (a) the speed of the rocket at point A', (b) the required speed it must attain at A just after braking so that it undergoes an 8-Mm free-flight circular orbit around Venus, and (c) the periods of both the circular and elliptical orbits. The mass of Venus is 0.816 times the mass of the earth.

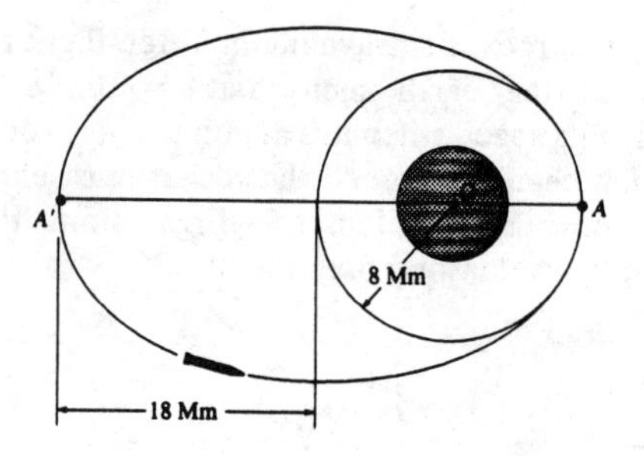

a)

$$M_v = 0.816\left(5.976\left(10^{24}\right)\right) = 4.876\left(10^{24}\right)$$

$$OA' = \frac{OA}{\left(\frac{2GM_v}{OA\, v_A^2} - 1\right)}$$

$$26(10)^6 = \frac{8(10^6)}{\left(\frac{2(66.73)(10^{-12})4.876(10^{24})}{8(10^6)v_A^2} - 1\right)}$$

$$\frac{81.34(10^6)}{v_A^2} = 1.307$$

$$v_A = 7886.8 \text{ m/s} = 7.89 \text{ km/s}$$

$$v_A' = \frac{OA\, v_A}{OA'} = \frac{8(10^6)(7886.8)}{26(10^6)} = 2426.7 \text{ m/s} = 2.43 \text{ km/s} \quad \textbf{Ans}$$

b)

$$v_A'' = \sqrt{\frac{GM_v}{OA}} = \sqrt{\frac{66.73(10^{-12})4.876(10^{24})}{8(10^6)}}$$

$$v_A'' = 6377.5 \text{ m/s} = 6.38 \text{ km/s} \quad \textbf{Ans}$$

c)

Circular orbit :

$$T_c = \frac{2\pi\, OA}{v_A''} = \frac{2\pi\, 8(10^6)}{6377.5} = 7881.75 \text{ s} = 2.19 \text{ h} \quad \textbf{Ans}$$

Elliptic orbit :

$$T_e = \frac{\pi}{OA\, v_A}(OA + OA')\sqrt{(OA)(OA')} = \frac{\pi}{8(10^6)(7886.8)}(8+26)\left(10^6\right)\left(\sqrt{(8)(26)}\right)\left(10^6\right)$$

$$T_e = 24\,415.7 \text{ s} = 6.78 \text{ h} \quad \textbf{Ans}$$

14-1. A woman having a mass of 70 kg stands in an elevator which has a downward acceleration of 4 m/s^2 starting from rest. Determine the work done by her weight and the work of the normal force which the floor exerts on her when the elevator descends 6 m. Explain why the work of these forces is different.

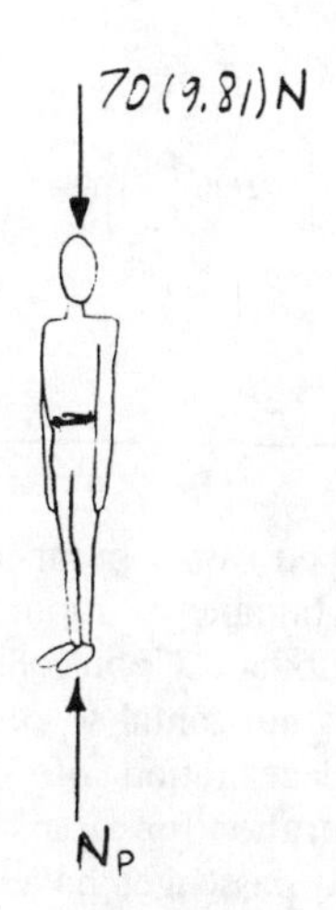

$+\downarrow \Sigma F_y = ma_y; \quad 70(9.81) - N_P = 70(4)$

$N_P = 406.7 \text{ N}$

$U_W = 6(686.7) = 4.12 \text{ kJ}$ **Ans**

$U_{N_P} = -6(406.7) = -2.44 \text{ kJ}$ **Ans**

The difference accounts for a change in kinetic energy. **Ans**

Note: $v^2 = v_0^2 + 2a_c(s - s_0)$

$v^2 = 0 + 2(4)(6-0)$

$v = 6.928 \text{ m/s}$

$\Delta T = \frac{1}{2}(70)(6.928)^2 = 1.68 \text{ kJ}$

Also, $T_1 + \Sigma U_{1-2} = T_2$

$\Delta T = \Sigma U_{1-2} = 4.12 - 2.44 = 1.68 \text{ kJ}$

14-2. A 1500-lb crate is pulled along the ground at a constant speed for a distance of 25 ft, using a cable that makes an angle of 15° with the horizontal. Determine the tension in the cable and the work done by this force. The coefficient of kinetic friction between the ground and the crate is $\mu_k = 0.55$.

$\xrightarrow{+} \Sigma F_x = 0; \quad T\cos 15° - 0.55N = 0$

$+\uparrow \Sigma F_y = 0; \quad N + T\sin 15° - 1500 = 0$

$N = 1307 \text{ lb}$

$T = 744.4 \text{ lb} = 744 \text{ lb}$ **Ans**

$U_T = (744.4\cos 15°)(25) = 18.0(10^3) \text{ ft} \cdot \text{lb}$ **Ans**

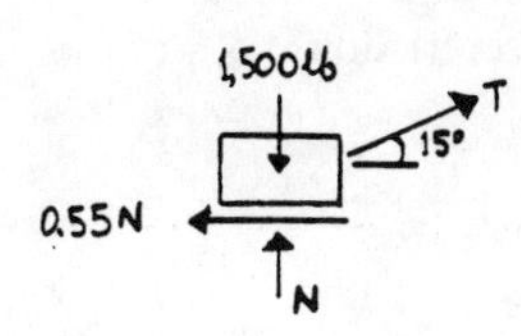

14-3. The 20-lb crate has a velocity of $v_A = 12$ ft/s when it is at A. Determine its velocity after it slides $s = 6$ ft down the plane. The coefficient of kinetic friction between the crate and the plane is $\mu_k = 0.2$.

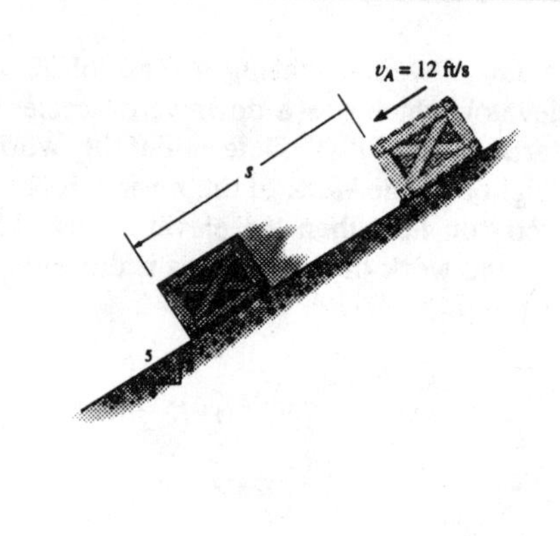

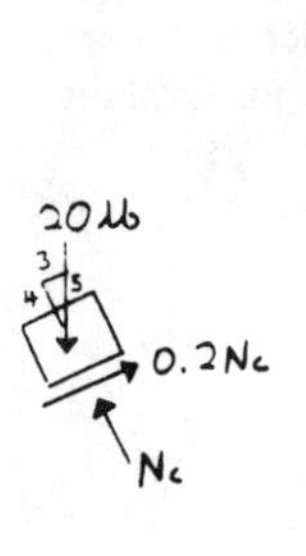

$$T_1 + \Sigma U_{1-2} = T_2$$

$$\frac{1}{2}\left(\frac{20}{32.2}\right)(12)^2 + 20\left(\frac{3}{5}\right)(6) - \left[0.2(20)\left(\frac{4}{5}\right)\right]6 = \frac{1}{2}\left(\frac{20}{32.2}\right)v_2^2$$

$v_2 = 17.7$ ft/s **Ans**

***14-4.** A car is equipped with a bumper B designed to absorb collisions. The bumper is mounted to the car using pieces of flexible tubing T. Upon collision with a rigid barrier at A, a constant horizontal force **F** is developed which causes a car deceleration of $3g = 29.43$ m/s^2 (the highest safe deceleration for a passenger without a seatbelt). If the car and passenger have a total mass of 1.5 Mg and the car is initially coasting with a speed of 1.5 m/s, determine the magnitude of **F** needed to stop the car and the deformation x of the bumper tubing.

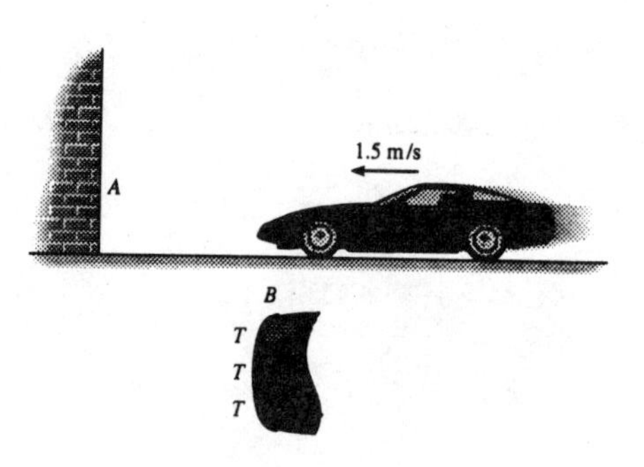

The average force needed to decelerate the car is

$$\overset{+}{\rightarrow} \Sigma F_x = ma_x; \quad F_{avg} = (1500)(29.43) = 44\,145 = 44.1 \text{ kN} \quad \textbf{Ans}$$

The deformation is

$$T_1 + \Sigma U_{1-2} = T_2$$

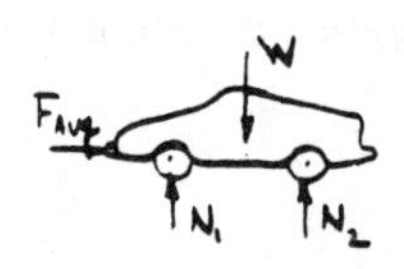

$$\frac{1}{2}(1500)(1.5)^2 - (44\,145)(x) = 0$$

$x = 0.0382 = 38.2$ mm **Ans**

14-5. The smooth plug has a weight of 20 lb and is pushed against a series of Belleville spring washers so that the compression in the spring is $s = 0.05$ ft. If the force of the spring on the plug is $F = (3s^{1/3})$ lb, where s is given in feet, determine the speed of the plug after it moves away from the spring. Neglect friction.

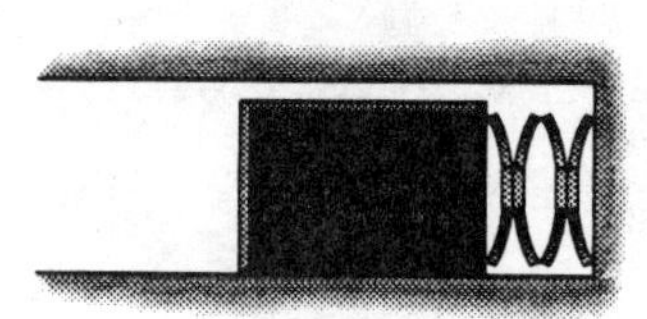

$$T_1 + \Sigma U_{1-2} = T_2$$

$$0 + \int_0^{0.05} 3s^{\frac{1}{3}}\, ds = \frac{1}{2}\left(\frac{20}{32.2}\right)v^2$$

20 lb

F = $3s^{\frac{1}{3}}$

N

$$3\left(\frac{3}{4}\right)(0.05)^{\frac{4}{3}} = \frac{1}{2}\left(\frac{20}{32.2}\right)v^2$$

$v = 0.365$ ft/s **Ans**

14-6. When a 7-kg projectile is fired from a cannon barrel that has a length of 2 m, the explosive force exerted on the projectile, while it is in the barrel, varies in the manner shown. Determine the approximate muzzle velocity of the projectile at the instant it leaves the barrel. Neglect the effects of friction inside the barrel and assume the barrel is horizontal.

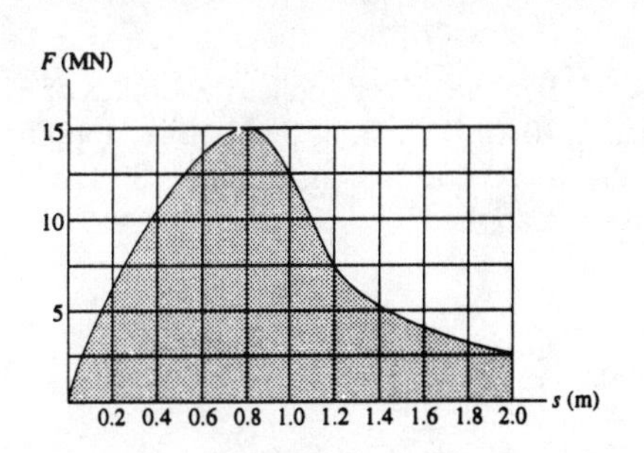

The work done is measured as the area under the force – displacement curve. This area is approximately 31.5 squares. Since each square has an area of $2.5(10^6)(0.2)$

F →

$$T_1 + \Sigma U_{1-2} = T_2$$

$$0 + \left[(31.5)(2.5)(10^6)(0.2)\right] = \frac{1}{2}(7)(v_2)^2$$

$v_2 = 2121 \text{ m/s} = 2.12 \text{ km/s}$ (approx.) **Ans**

14-7. The 100-kg crate is subjected to forces $F_1 = 800$ N and $F_2 = 1.5$ kN, as shown. If it is originally at rest, determine the distance it slides in order to attain a speed of $v = 6$ m/s. The coefficient of kinetic friction between the crate and the surface is $\mu_k = 0.2$.

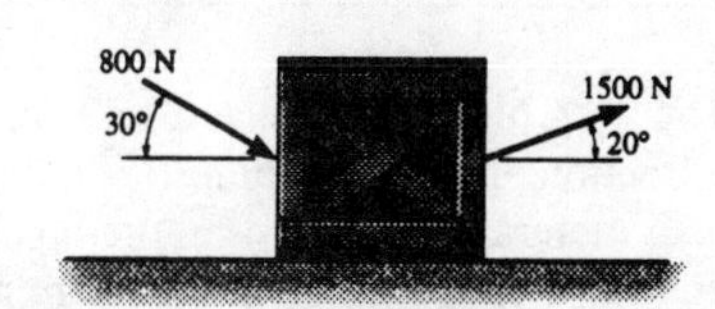

$+\uparrow \Sigma F_y = 0;\quad N_C - 800(\sin 30°) - 100(9.81) + 1500(\sin 20°) = 0$

$N_C = 867.97$ N

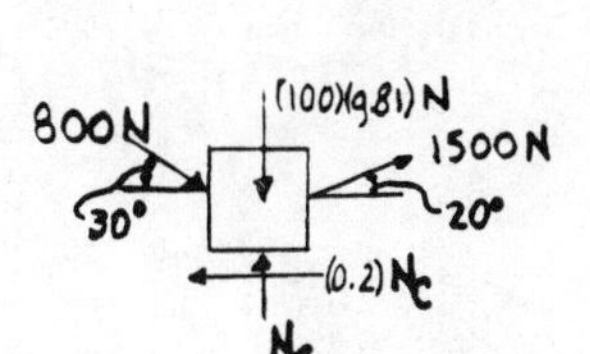

$$T_1 + \Sigma U_{1-2} = T_2$$

$$0 + (800\cos 30°)s - 0.2(867.97)s + (1500\cos 20°)s = \frac{1}{2}(100)(6)^2$$

$s(1928.8) = 1800$

$s = 0.933$ m **Ans**

***14-8.** Determine the required height h of the roller coaster so that when it is essentially at rest at the crest of the hill it will reach a speed of 100 km/h when it comes to the bottom. Also, what should be the minimum radius of curvature ρ for the track at B so that the passengers do not experience a normal force greater than $4\,mg = (39.24\,m)$ N? Neglect the size of the car and passenger.

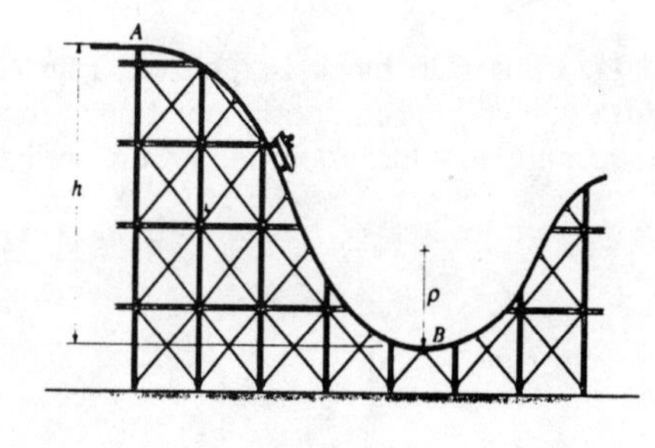

$$100 \text{ km/h} = \frac{100(10^3)}{3600} = 27.778 \text{ m/s}$$

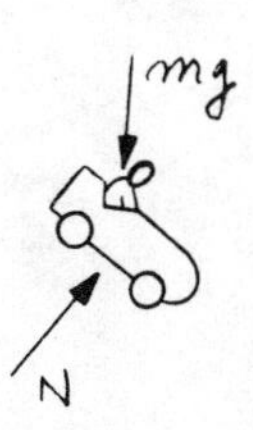

$$T_1 + \Sigma U_{1-2} = T_2$$

$$0 + m(9.81)h = \frac{1}{2}m(27.778)^2$$

$h = 39.3$ m **Ans**

$$+\uparrow \Sigma F_n = ma_n;\quad 39.24m - mg = m\left(\frac{(27.778)^2}{\rho}\right)$$

$\rho = 26.2$ m **Ans**

14-9. When the driver applies the brakes of a light truck traveling 40 km/h, it skids 3 m before stopping. How far will the truck skid if it is traveling 80 km/h when the brakes are applied?

$$40 \text{ km/h} = \frac{40(10^3)}{3600} = 11.11 \text{ m/s} \qquad 80 \text{ km/h} = 22.22 \text{ m/s}$$

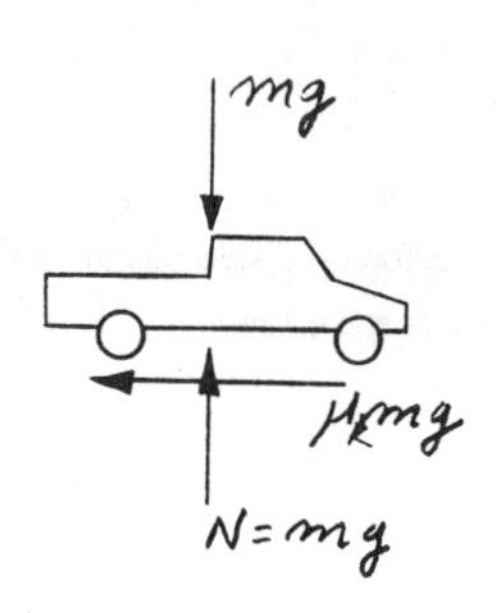

$$T_1 + \Sigma U_{1-2} = T_2$$

$$\frac{1}{2}m(11.11)^2 - \mu_k mg(3) = 0$$

$$\mu_k g = 20.576$$

$$T_1 + \Sigma U_{1-2} = T_2$$

$$\frac{1}{2}m(22.22)^2 - (20.576)m(d) = 0$$

$d = 12$ m **Ans**

14-10. The 2-kg block is subjected to a force having a constant direction and a magnitude $F = [300/(1 + s)]$ N, where s is in meters. When $s = 4$ m, the block is moving to the left with a speed of 8 m/s. Determine its speed when $s = 12$ m. The coefficient of kinetic friction between the block and the ground is $\mu_k = 0.25$.

$$+\uparrow \Sigma F_y = 0; \quad N_B = 2(9.81) + \left(\frac{300}{1+s}\right)\sin 30^\circ$$

$$T_1 + \Sigma U_{1-2} = T_2$$

$$\frac{1}{2}(2)(8)^2 - 0.25[2(9.81)(12-4)] - 0.25\int_4^{12} \frac{300 \sin 30^\circ}{1+s}\, ds + \int_4^{12} \left(\frac{300}{1+s}\right) \cos 30^\circ ds = \frac{1}{2}(2)v_2^2$$

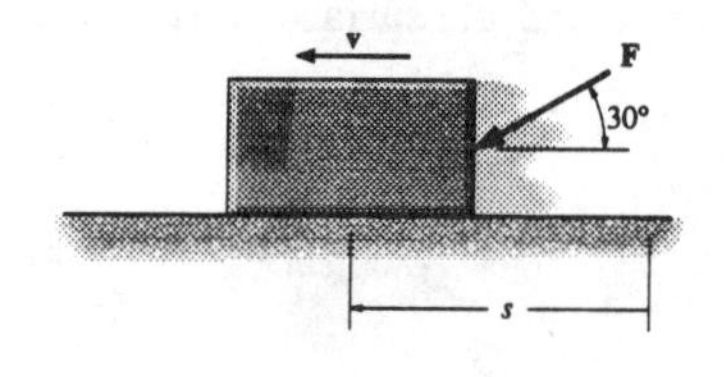

$$v_2^2 = 24.76 - 37.5\ln\left(\frac{1+12}{1+4}\right) + 259.81\ln\left(\frac{1+12}{1+4}\right)$$

$v_2 = 15.4$ m/s **Ans**

14-11. The 6-lb block is released from rest at A and slides down the smooth parabolic surface shown. Determine how far it compresses the spring.

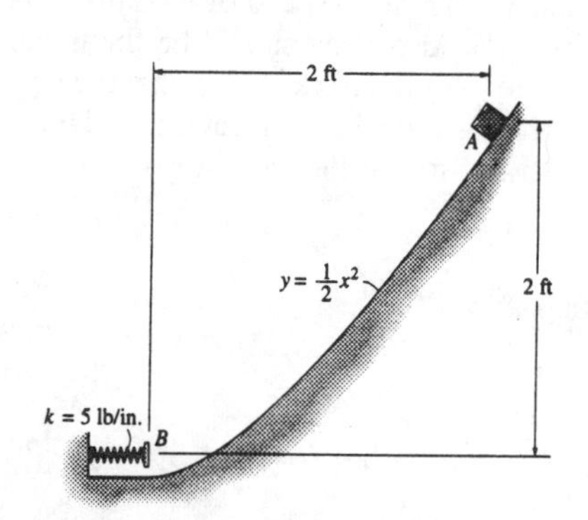

$$T_1 + \Sigma U_{1-2} = T_2$$

$$0 + 2(6) - \frac{1}{2}(5)(12)s^2 = 0$$

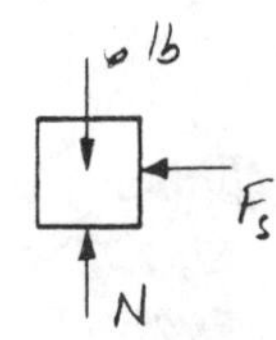

$s = 0.632$ ft $= 7.59$ in. **Ans**

***14-12.** As indicated by the derivation, the principle of work and energy is valid for observers in *any* inertial reference frame. Show that this is so, by considering the 10-kg block which rests on the smooth surface and is subjected to a horizontal force of 6 N. If observer A is in a *fixed* frame x, determine the final speed of the block if it has an initial speed of 5 m/s and travels 10 m, both directed to the right and measured from the fixed frame. Compare the result with that obtained by an observer B, attached to the x' axis and moving at a constant velocity of 2 m/s relative to A. *Hint:* The distance the block travels will first have to be computed for observer B before applying the principle of work and energy.

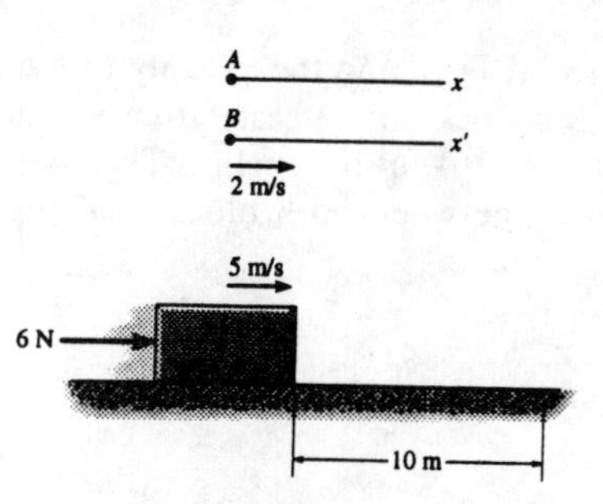

Observer A :

$$T_1 + \Sigma U_{1-2} = T_2$$

$$\frac{1}{2}(10)(5)^2 + 6(10) = \frac{1}{2}(10)v_2^2$$

$v_2 = 6.08$ m/s **Ans**

Observer B :

$$F = ma$$

$$6 = 10a \qquad a = 0.6 \text{ m/s}^2$$

$$(\overset{+}{\rightarrow}) \qquad s = s_0 + v_0 t + \frac{1}{2}a_c t^2$$

$$10 = 0 + 5t + \frac{1}{2}(0.6)t^2$$

$$t^2 + 16.67t - 33.33 = 0$$

$$t = 1.805 \text{ s}$$

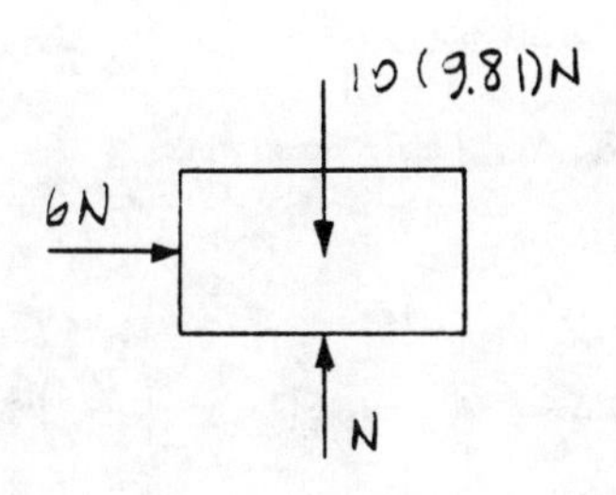

At $v = 2$ m/s, $s' = 2(1.805) = 3.609$ m

Block moves $10 - 3.609 = 6.391$ m

Thus

$$T_1 + \Sigma U_{1-2} = T_2$$

$$\frac{1}{2}(10)(3)^2 + 6(6.391) = \frac{1}{2}(10)v_2^2$$

$v_2 = 4.08$ m/s **Ans**

Note that this result is 2m/s less than that observed by A.

14-13. Determine the velocity of the 60-lb block A if the two blocks are released from rest and the 40-lb block B moves 2 ft up the incline. The coefficient of kinetic friction between both blocks and the inclined planes is $\mu_k = 0.10$.

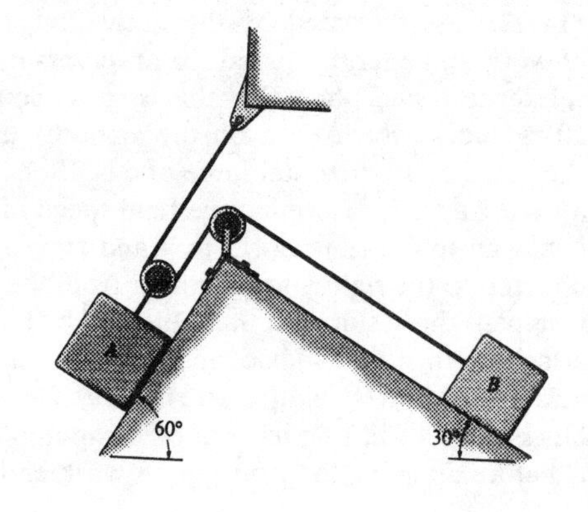

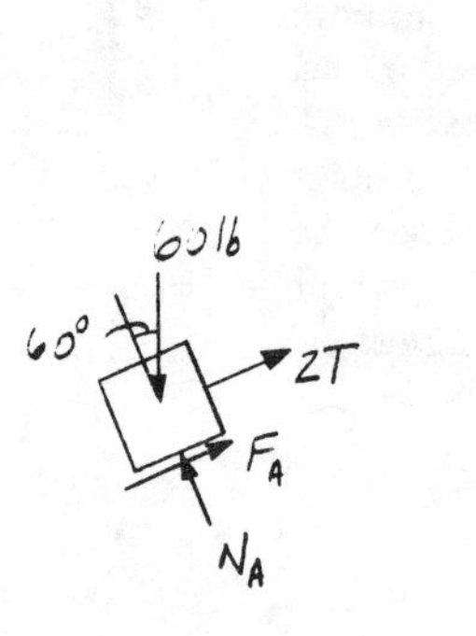

Block A :

$+\nwarrow \Sigma F_y = ma_y$; $\quad N_A - 60\cos 60° = 0$

$N_A = 30$ lb

$F_A = 0.1(30) = 3$ lb

Block B :

$+\nearrow \Sigma F_y = ma_y$; $\quad N_B - 40\cos 30° = 0$

$N_B = 34.64$ lb

$F_B = 0.1(34.64) = 3.464$ lb

Use the system of both blocks. N_A, N_B, T and R do no work.

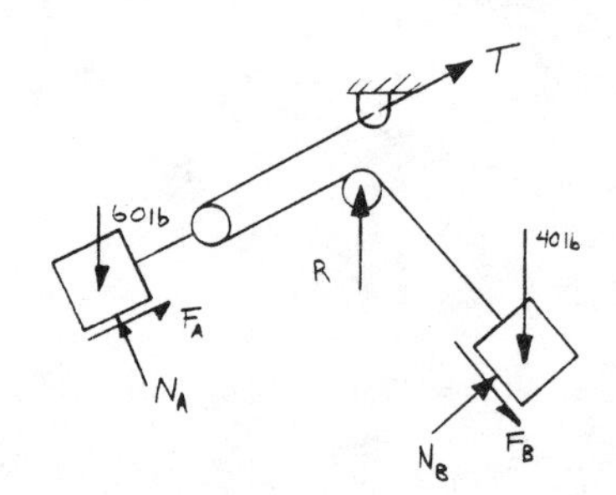

$$T_1 + \Sigma U_{1-2} = T_2$$

$$(0+0) + 60\sin 60°\Delta s_A - 40\sin 30°\Delta s_B - 3\Delta s_A - 3.464\Delta s_B = \frac{1}{2}\left(\frac{60}{32.2}\right)v_A^2 + \frac{1}{2}\left(\frac{40}{32.2}\right)v_B^2$$

$$2s_A + s_B = l$$

$$2\Delta s_A = -\Delta s_B$$

When $|\Delta s_B| = |2 \text{ ft}|$, $|\Delta s_A| = |1 \text{ ft}|$

Also,

$$2v_A = -v_B$$

Substituting and solving,

$v_A = 0.771$ ft/s **Ans**

$v_B = -1.54$ ft/s

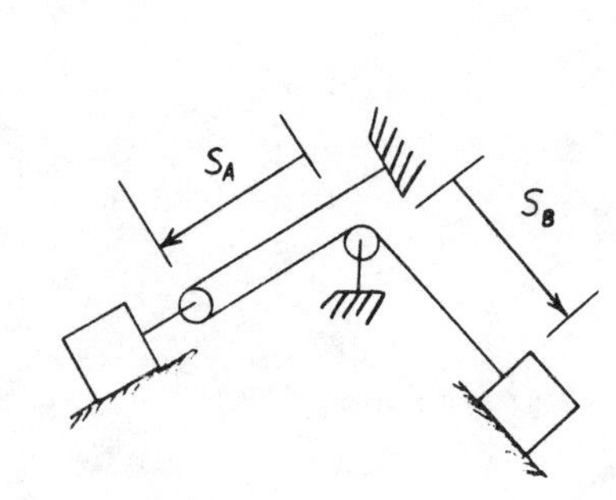

14-14. Determine the velocity of the 20-kg block A after it is released from rest and moves 2 m down the plane. Block B has a mass of 10 kg and the coefficient of kinetic friction between the plane and block A is $\mu_k = 0.2$. Also, what is the tension in the cord?

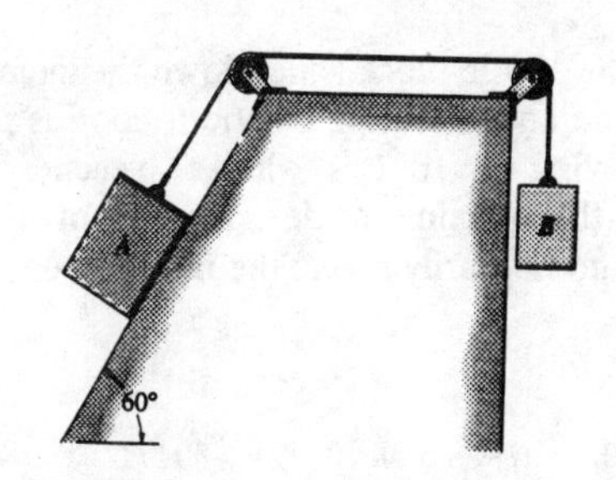

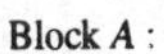

Block A :

$\nwarrow \Sigma F_y = 0; \quad N_A - 20(9.81)\cos 60° = 0$

$N_A = 98.1 \text{ N}$

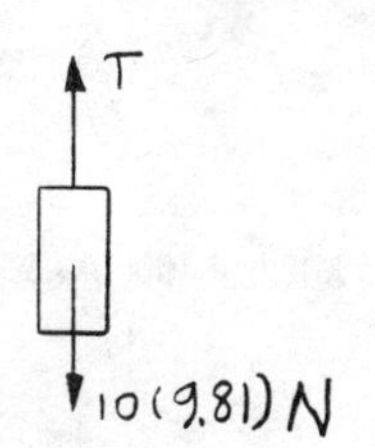

System :

$T_1 + \Sigma U_{1-2} = T_2$

$$(0+0) + 20(9.81)(\sin 60°)2 - 0.2(98.1)(2) - 10(9.81)(2) = \frac{1}{2}(20)v^2 + \frac{1}{2}(10)v^2$$

$v = 2.638 = 2.64 \text{ m/s}$ **Ans**

Block B :

$T_1 + \Sigma U_{1-2} = T_2$

$$0 + T(2) - 10(9.81)(2) = \frac{1}{2}(10)(2.638)^2$$

$T = 115 \text{ N}$ **Ans**

Also, block A :

$T_1 + \Sigma U_{1-2} = T_2$

$$0 + 20(9.81)(\sin 60°)(2) - T(2) - 0.2(98.1)(2) = \frac{1}{2}(20)(2.638)^2$$

$T = 115 \text{ N}$ **Ans**

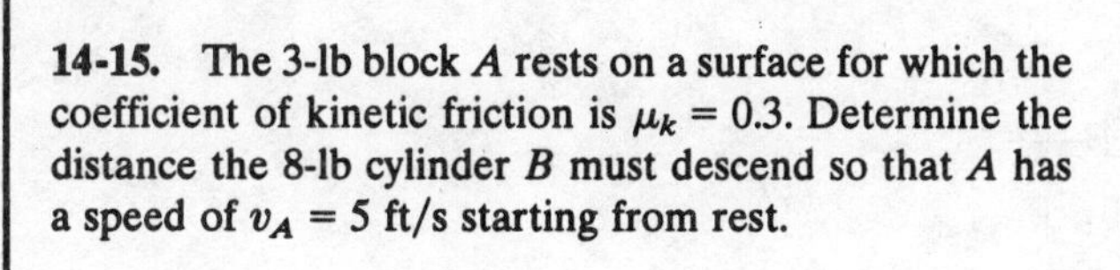

14-15. The 3-lb block A rests on a surface for which the coefficient of kinetic friction is $\mu_k = 0.3$. Determine the distance the 8-lb cylinder B must descend so that A has a speed of $v_A = 5$ ft/s starting from rest.

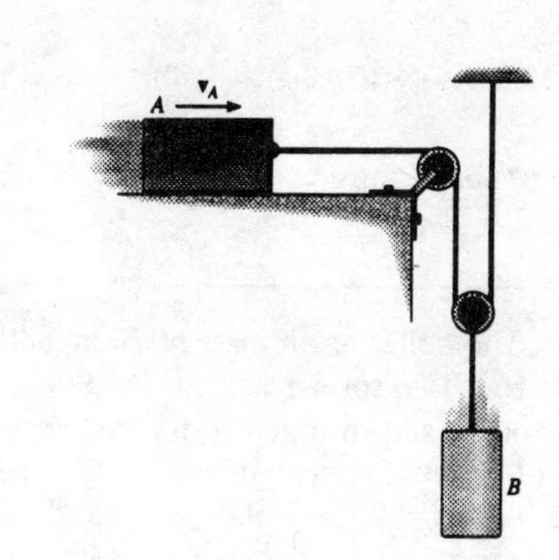

$+\uparrow \Sigma F_y = 0; \quad N_A - 3 = 0$

$N_A = 3 \text{ lb}$

$F_A = 0.3(3) = 0.9 \text{ lb}$

$s_A + 2s_B = l$

$\Delta s_A = -2\Delta s_B$

$v_A = -2v_B$

$T_1 + \Sigma U_{1-2} = T_2$

$$(0+0) + (8)(\Delta s_B) - 0.9(2\Delta s_B) = \frac{1}{2}\left(\frac{8}{32.2}\right)\left(\frac{5}{2}\right)^2 + \frac{1}{2}\left(\frac{3}{32.2}\right)(5)^2$$

$\Delta s_B = 0.313 \text{ ft}$ **Ans**

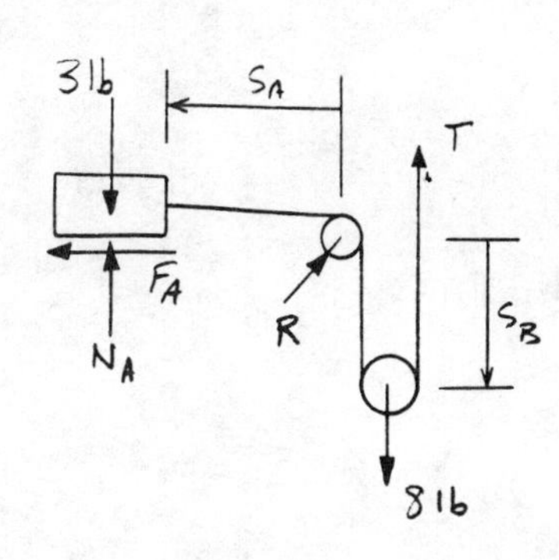

***14-16.** The 100-lb block slides down the inclined plane for which the coefficient of kinetic friction is $\mu_k = 0.25$. If it is moving at 10 ft/s when it reaches point A, determine the maximum deformation of the spring needed to momentarily arrest the motion.

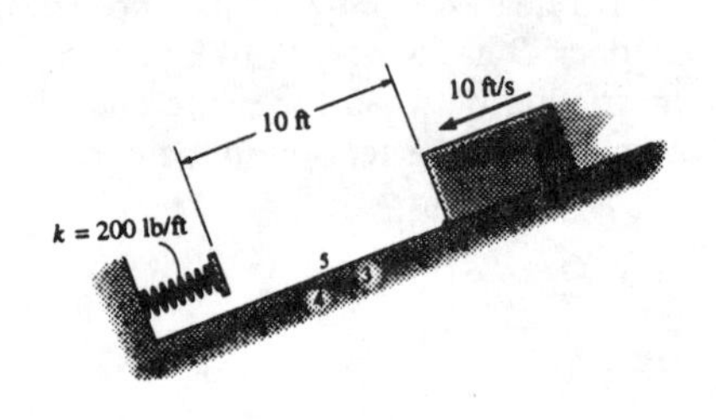

$$+\nwarrow \Sigma F_y = 0; \quad N - \frac{4}{5}(100) = 0 \qquad N = 80 \text{ lb}$$

$$T_1 + \Sigma U_{1-2} = T_2$$

$$\frac{1}{2}\left(\frac{100}{32.2}\right)(10)^2 - 0.25(80)(10+d) - \frac{1}{2}(200)(d)^2 + 100(10+d)\left(\frac{3}{5}\right) = 0$$

$$-100d^2 + 40d + 555.280 = 0$$

Use the positive root :

$d = 2.56$ ft **Ans**

14-17. The collar has a mass of 20 kg and rests on the smooth rod. Two springs are attached to it and the ends of the rod as shown. Each spring has an uncompressed length of 1 m. If the collar is displaced $s = 0.5$ m and released from rest, determine its velocity at the instant it returns to the point $s = 0$.

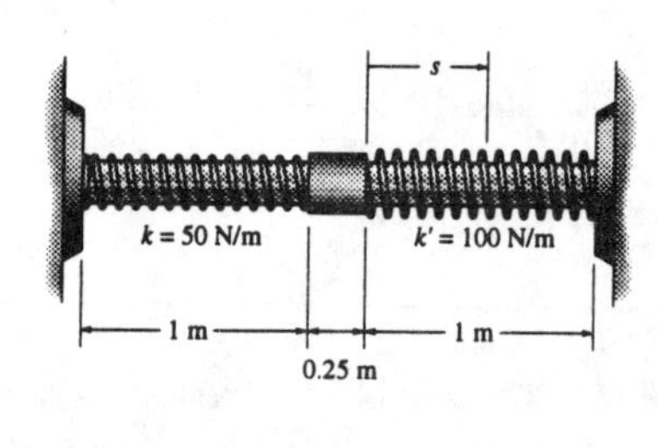

$$T_1 + \Sigma U_{1-2} = T_2$$

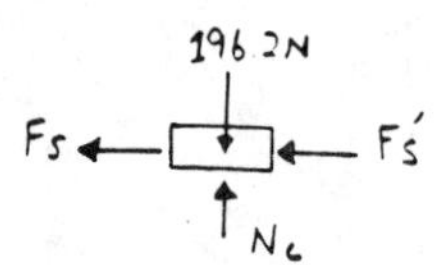

$$0 + \frac{1}{2}(50)(0.5)^2 + \frac{1}{2}(100)(0.5)^2 = \frac{1}{2}(20)v_C^2$$

$v_C = 1.37$ m/s **Ans**

14-18. The collar has a mass of 20 kg and rests on the smooth rod. Two springs are attached to it and the ends of the rod as shown. Each spring has an uncompressed length of 1.5 m. If the collar is held in the center position of the rod, $s = 0$, and then displaced $s = 0.5$ m and released from rest, determine its velocity at the instant it returns to the point $s = 0$.

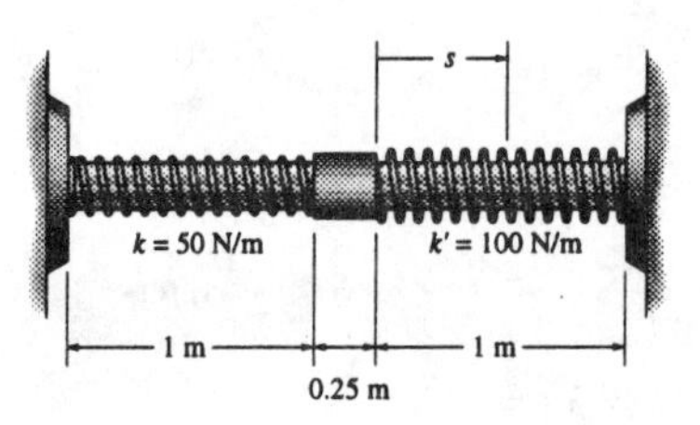

$$0 + \frac{1}{2}(50)\left[(0)^2 - (0.5)^2\right] + \frac{1}{2}(100)\left[(1)^2 - (0.5)^2\right] = \frac{1}{2}(20)v_C^2$$

$v_C = 1.77$ m/s **Ans**

14-19. The collar has a mass of 30 kg and is supported on the rod having a coefficient of kinetic friction $\mu_k = 0.4$. The attached spring has an unstretched length of 0.2 m and a stiffness $k = 50$ N/m. Determine the speed of the collar after the applied force $F = 200$ N causes it to be displaced $s = 1.5$ m from point A. When $s = 0$ the collar is held at rest.

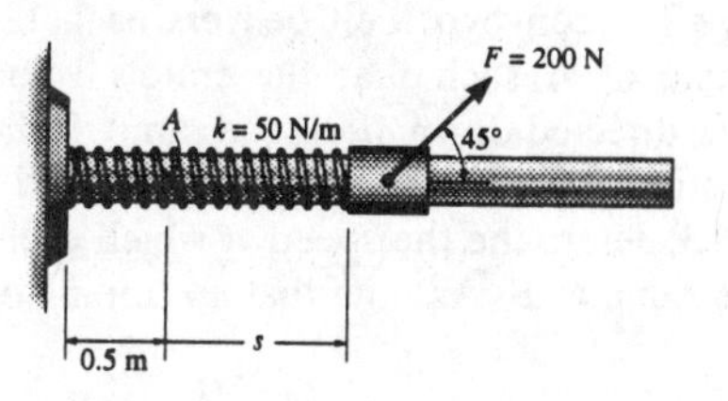

$+\uparrow\Sigma F_y = 0;\quad 200\sin 45° - 30(9.81) + N_C = 0$

$N_C = 152.88$ N

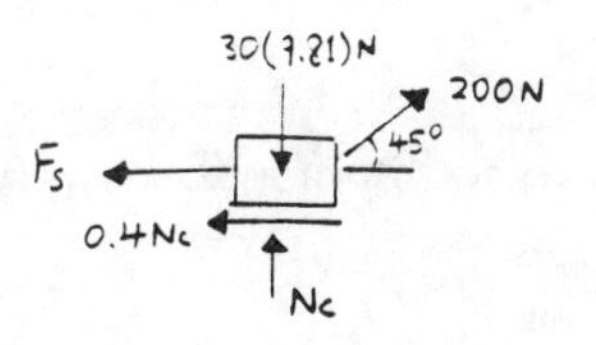

$T_1 + \Sigma U_{1-2} = T_2$

$$0 + 200\cos 45°(1.5) - 0.4(152.88)(1.5) - \left[\frac{1}{2}(50)(1.8)^2 - \frac{1}{2}(50)(0.3)^2\right] = \frac{1}{2}(30)v^2$$

$v = 1.67$ m/s **Ans**

***14-20.** The 5-lb block is released from rest at A and slides down the smooth circular surface AB. It then continues to slide along the horizontal rough surface until it strikes the spring. Determine how far it compresses the spring before stopping.

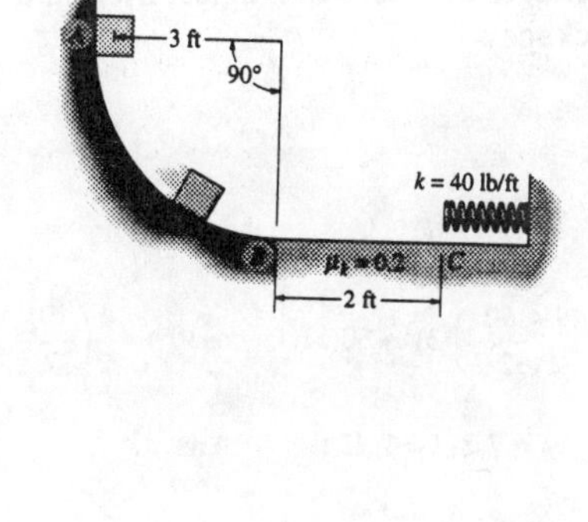

$+\uparrow\Sigma F_y = 0;\quad -5 + N_b = 0$

$N_b = 5$ lb

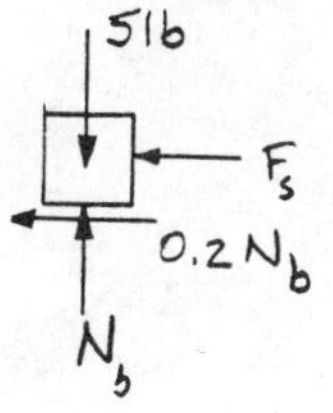

$T_1 + \Sigma U_{1-2} = T_2$

$$0 + 5(3) - 0.2(5)(2 + s) - \frac{1}{2}(40)(s)^2 = 0$$

$-20s^2 - s + 13 = 0$

Solving for the positive root :

$s = 0.782$ ft **Ans**

14-21. The 2-lb box slides on the smooth curved ramp. If the box has a velocity of 30 ft/s at A, determine the velocity of the box and normal force acting on the ramp when the box is located at B and C. Assume the radius of curvature of the path at C is still 5 ft.

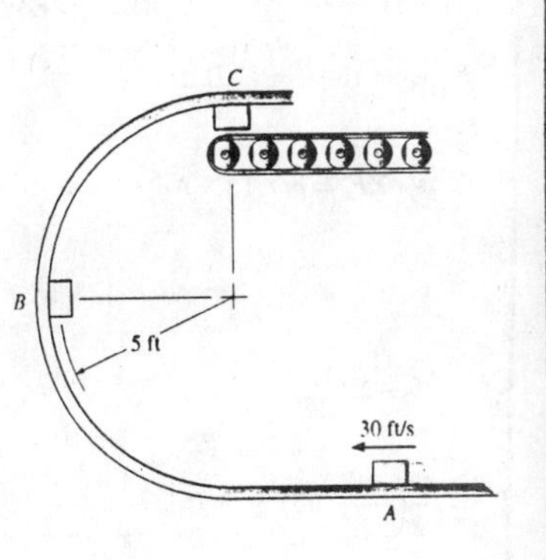

Point B :

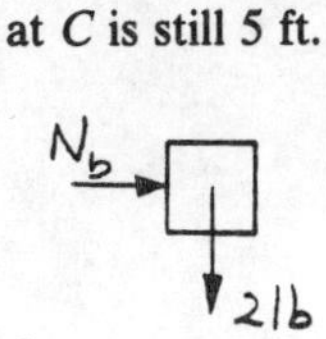

$T_1 + \Sigma U_{1-2} = T_2$

$$\frac{1}{2}\left(\frac{2}{32.2}\right)(30)^2 - 2(5) = \frac{1}{2}\left(\frac{2}{32.2}\right)(v_B)^2$$

$v_B = 24.0$ ft/s **Ans**

$\overset{+}{\rightarrow}\Sigma F_n = ma_n;\quad N_B = \left(\frac{2}{32.2}\right)\left(\frac{(24.0)^2}{5}\right)$

$N_B = 7.18$ lb **Ans**

Point C :

$T_1 + \Sigma U_{1-2} = T_2$

$$\frac{1}{2}\left(\frac{2}{32.2}\right)(30)^2 - 2(10) = \frac{1}{2}\left(\frac{2}{32.2}\right)(v_C)^2$$

$v_C = 16.0$ ft/s **Ans**

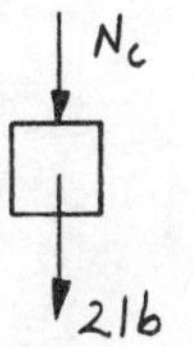

$+\downarrow\Sigma F_n = ma_n;\quad N_C + 2 = \left(\frac{2}{32.2}\right)\left(\frac{(16.0)^2}{5}\right)$

$N_C = 1.18$ lb **Ans**

14-22. The conveyor belt delivers each 12-kg crate to the ramp at A such that the crate's velocity is $v_A = 2.5$ m/s, directed down *along* the ramp. If the coefficient of kinetic friction between each crate and the ramp is $\mu_k = 0.3$, determine the speed at which each crate slides off the ramp at B. Assume that no tipping occurs.

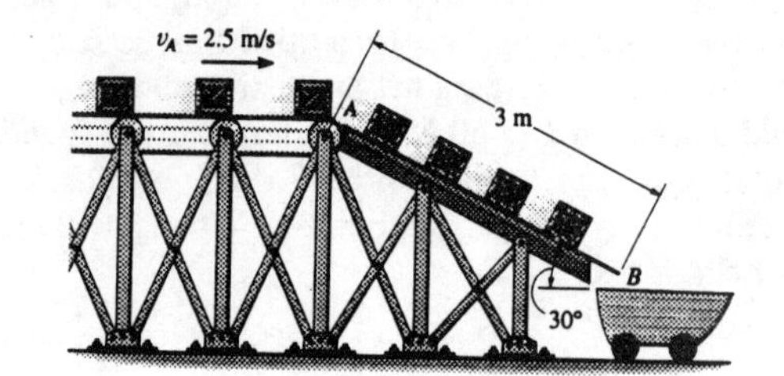

$+\nearrow \Sigma F_y = ma_y; \quad N_C - 12(9.81)\cos 30° = 0$

$N_C = 102.0$ N

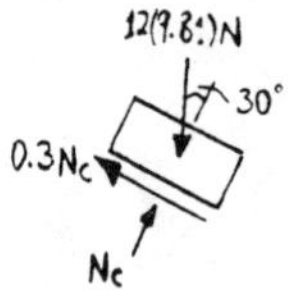

$T_A + \Sigma U_{A-B} = T_B$

$\frac{1}{2}(12)(2.5)^2 + 12(9.81)(3\sin 30°) - 0.3(102.0)(3) = \frac{1}{2}(12)v_B^2$

$v_B = 4.52$ m/s **Ans**

14-23. Packages having a weight of 50 lb are delivered to the chute at $v_A = 3$ ft/s using a conveyor belt. Determine their speeds when they reach points B, C, and D. Also calculate the normal force of the chute on the packages at B and C. Neglect friction and the size of the packages.

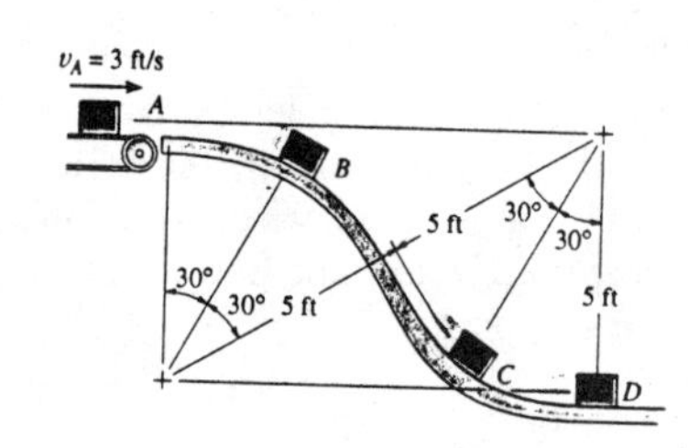

$T_A + \Sigma U_{A-B} = T_B$

$$\frac{1}{2}\left(\frac{50}{32.2}\right)(3)^2 + 50(5)(1 - \cos 30°) = \frac{1}{2}\left(\frac{50}{32.2}\right)v_B^2$$

$v_B = 7.221 = 7.22$ ft/s **Ans**

$+\swarrow \Sigma F_n = ma_n; \quad -N_B + 50\cos 30° = \left(\frac{50}{32.2}\right)\left[\frac{(7.221)^2}{5}\right]$

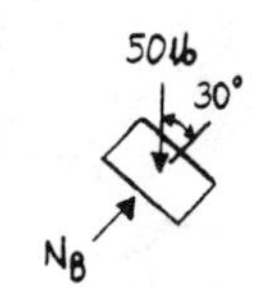

$N_B = 27.1$ lb **Ans**

$T_A + \Sigma U_{A-C} = T_C$

$$\frac{1}{2}\left(\frac{50}{32.2}\right)(3)^2 + 50(5\cos 30°) = \frac{1}{2}\left(\frac{50}{32.2}\right)v_C^2$$

$v_C = 16.97 = 17.0$ ft/s **Ans**

$+\nearrow \Sigma F_n = ma_n; \quad N_C - 50\cos 30° = \left(\frac{50}{32.2}\right)\left[\frac{(16.97)^2}{5}\right]$

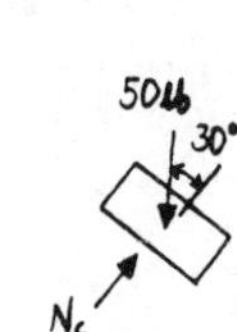

$N_C = 133$ lb **Ans**

$T_A + \Sigma U_{A-D} = T_D$

$$\frac{1}{2}\left(\frac{50}{32.2}\right)(3)^2 + 50(5) = \frac{1}{2}\left(\frac{50}{32.2}\right)v_D^2$$

$v_D = 18.2$ ft/s **Ans**

***14-24.** The 2-lb block slides down the smooth parabolic surface, such that when it is at A it has a speed of 10 ft/s. Determine the magnitude of the block's velocity and acceleration when it reaches point B, and the maximum height y_{max} reached by the block.

$y = 0.25x^2$

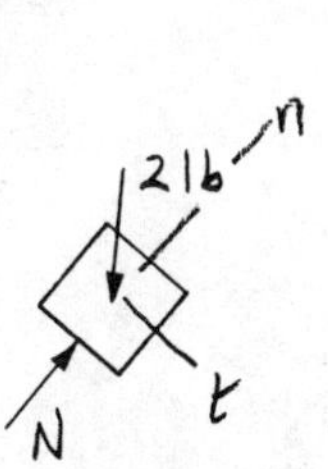

$y_A = 0.25(-4)^2 = 4$ ft

$y_B = 0.25(1)^2 = 0.25$ ft

$T_A + \Sigma U_{A-B} = T_B$

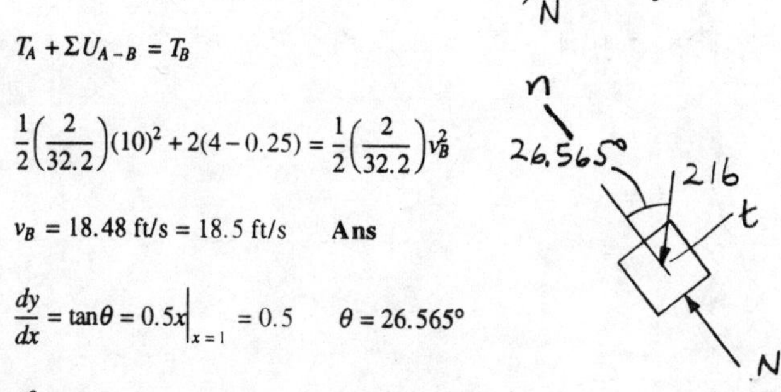

$$\frac{1}{2}\left(\frac{2}{32.2}\right)(10)^2 + 2(4 - 0.25) = \frac{1}{2}\left(\frac{2}{32.2}\right)v_B^2$$

$v_B = 18.48$ ft/s $= 18.5$ ft/s **Ans**

$$\frac{dy}{dx} = \tan\theta = 0.5x\Big|_{x=1} = 0.5 \qquad \theta = 26.565°$$

$$\frac{d^2y}{dx^2} = 0.5$$

$+\swarrow \Sigma F_t = ma_t;\qquad -2\sin 26.565° = \left(\frac{2}{32.2}\right)a_t$

$a_t = -14.4$ ft/s^2

$$\rho = \frac{\left[1 + \left(\frac{dy}{dx}\right)^2\right]^{\frac{3}{2}}}{\left|\frac{d^2y}{dx^2}\right|} = \frac{\left[1 + (0.5)^2\right]^{\frac{3}{2}}}{|0.5|} = 2.795 \text{ ft}$$

$$a_n = \frac{v_B^2}{\rho} = \frac{(18.48)^2}{2.795} = 122.2 \text{ ft/s}^2$$

$a_B = \sqrt{(-14.4)^2 + (122.2)^2} = 123$ ft/s^2 **Ans**

$T_A + \Sigma U_{A-C} = T_C$

$\frac{1}{2}\left(\frac{2}{32.2}\right)(10)^2 - 2(y_{max} - 4) = 0 \qquad y_{max} = 5.55$ ft **Ans**

14-25. When the 150-lb skier is at point A he has a speed of 5 ft/s. Determine his speed when he reaches point B on the smooth slope. For this distance the slope follows the cosine curve shown. Also, what is the normal force on his skis at B and his rate of increase in speed? Neglect friction and air resistance.

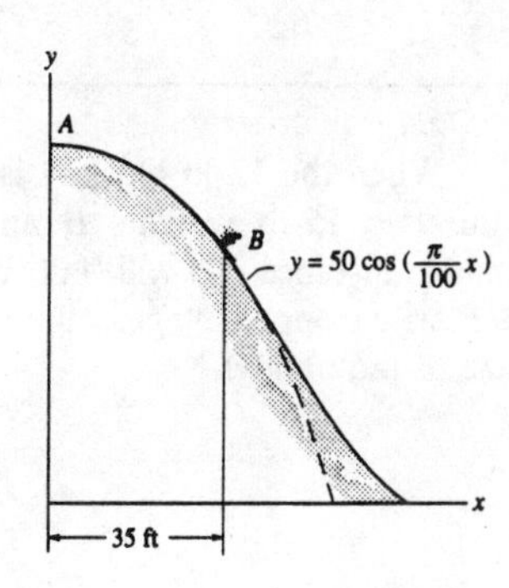

$$y = 50\cos\left(\frac{\pi}{100}\right)x\Big|_{x=35} = 22.70 \text{ ft}$$

$$\frac{dy}{dx} = \tan\theta = -50\left(\frac{\pi}{100}\right)\sin\left(\frac{\pi}{100}\right)x = -\left(\frac{\pi}{2}\right)\sin\left(\frac{\pi}{100}\right)x\Big|_{x=35} = -1.3996$$

$\theta = -54.45°$

$$\frac{d^2y}{dx^2} = -\left(\frac{\pi^2}{200}\right)\cos\left(\frac{\pi}{100}\right)x\Big|_{x=35} = -0.02240$$

$$\rho = \frac{\left[1 + \left(\frac{dy}{dx}\right)^2\right]^{\frac{3}{2}}}{\left|\frac{d^2y}{dx^2}\right|} = \frac{\left[1 + (-1.3996)^2\right]^{\frac{3}{2}}}{|-0.02240|} = 227.179$$

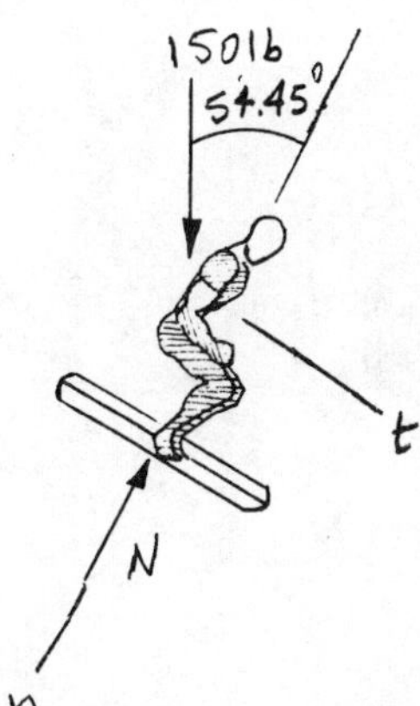

$T_A + \Sigma U_{A-B} = T_B$

$$\frac{1}{2}\left(\frac{150}{32.2}\right)(5)^2 + 150(50 - 22.70) = \frac{1}{2}\left(\frac{150}{32.2}\right)v_B^2$$

$v_B = 42.227$ ft/s $= 42.2$ ft/s **Ans**

$+\swarrow \Sigma F_n = ma_n;\qquad -N + 150\cos 54.45° = \left(\frac{150}{32.2}\right)\left(\frac{(42.227)^2}{227.179}\right)$

$N = 50.6$ lb **Ans**

$+\searrow \Sigma F_t = ma_t;\qquad 150\sin 54.45° = \left(\frac{150}{32.2}\right)a_t$

$a_t = 26.2$ ft/s^2 **Ans**

14-26. The skier starts from rest at A and travels down the ramp. If friction and air resistance can be neglected, determine his speed v_B when he reaches B. Also, find the distance s to where he strikes the ground at C, if he makes the jump traveling horizontally at B. Neglect the skier's size. He has a mass of 70 kg.

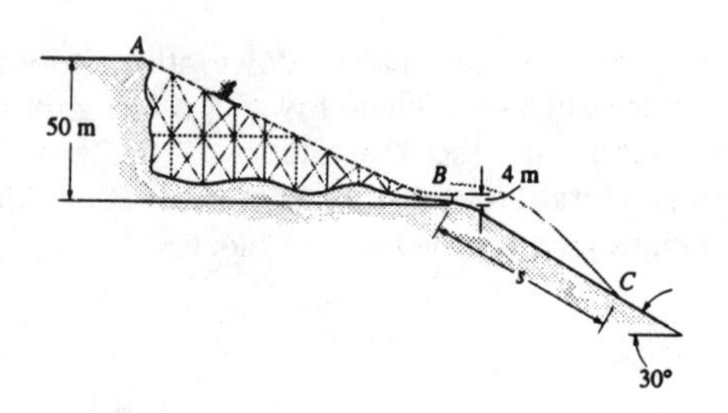

$T_A + \Sigma U_{A-B} = T_B$

$0 + 70(9.81)(46) = \frac{1}{2}(70)(v_B)^2$

$v_B = 30.04 \text{ m/s} = 30.0 \text{ m/s}$ **Ans**

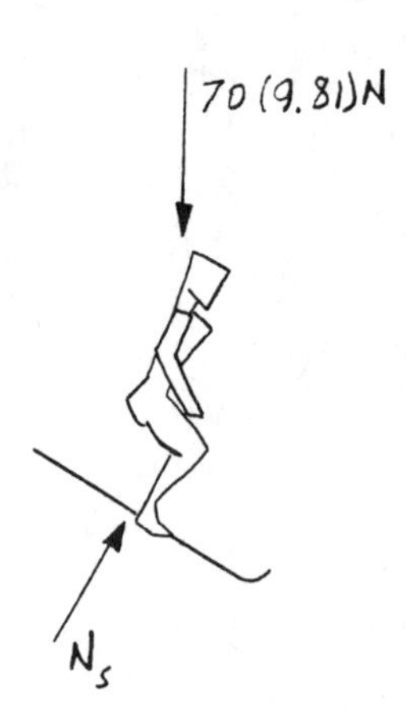

$(\overset{+}{\rightarrow})$ $s = s_0 + v_0 t$

$s \cos 30° = 0 + 30.04t$

$(+\downarrow)$ $s = s_0 + v_0 t + \frac{1}{2} a_c t^2$

$s \sin 30° + 4 = 0 + 0 + \frac{1}{2}(9.81)t^2$

Eliminating t,

$s^2 - 122.67s - 981.33 = 0$

Solving for the positive root

$s = 130 \text{ m}$ **Ans**

14-27. When the 12-lb block A is released from rest it lifts the two 15-lb weights B and C. Determine the maximum distance A will fall before its motion is momentarily stopped. Neglect the weight of the cord and the size of the pulleys.

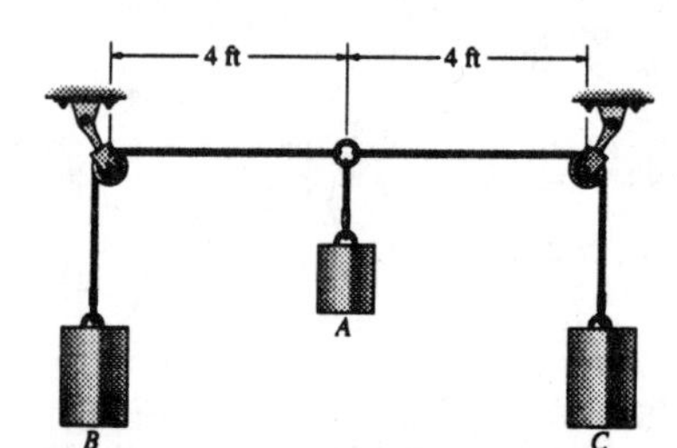

Consider the entire system :

$l = \sqrt{y^2 + 4^2}$

$T_1 + \Sigma U_{1-2} = T_2$

$(0+0+0) + 12y - 2(15)\left(\sqrt{y^2 + 4^2} - 4\right) = (0+0+0)$

$0.4y = \sqrt{y^2 + 16} - 4$

$(0.4y + 4)^2 = y^2 + 16$

$-0.84y^2 + 3.20y + 16 = 16$

$-0.84y + 3.20 = 0$

$y = 3.81 \text{ ft}$ **Ans**

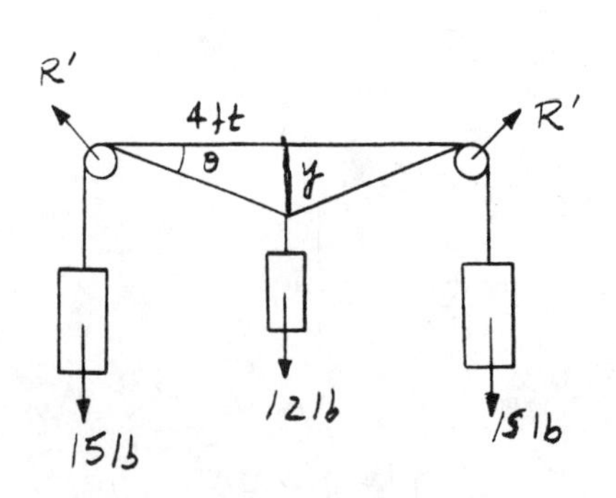

***14-28.** The 2-lb brick slides down a smooth roof, such that when it is at A it has a velocity of 5 ft/s. Determine the speed of the block just before it leaves the surface at B, the distance d from the wall to where it strikes the ground, and the speed at which it hits the ground.

$$T_A + \Sigma U_{A-B} = T_B$$

$$\frac{1}{2}\left(\frac{2}{32.2}\right)(5)^2 + 2(15) = \frac{1}{2}\left(\frac{2}{32.2}\right)v_B^2$$

$v_B = 31.48$ ft/s $= 31.5$ ft/s **Ans**

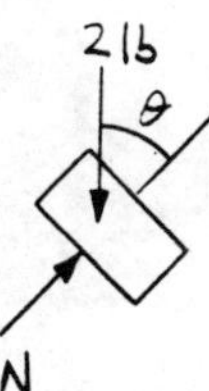

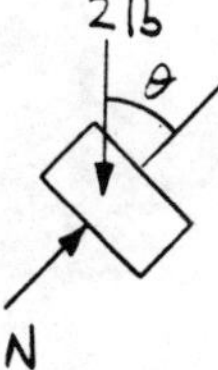

$(\overset{+}{\rightarrow})$ $s = s_0 + v_0 t$

$$d = 0 + 31.48\left(\frac{4}{5}\right)t$$

$(+\downarrow)$ $s = s_0 + v_0 t + \frac{1}{2}a_c t^2$

$$30 = 0 + 31.48\left(\frac{3}{5}\right)t + \frac{1}{2}(32.2)t^2$$

$$16.1t^2 + 18.888t - 30 = 0$$

Solving for the positive root,

$t = 0.89916$ s

$d = 31.48\left(\frac{4}{5}\right)(0.89916) = 22.6$ ft **Ans**

$$T_A + \Sigma U_{A-C} = T_C$$

$$\frac{1}{2}\left(\frac{2}{32.2}\right)(5)^2 + 2(45) = \frac{1}{2}\left(\frac{2}{32.2}\right)v_C^2$$

$v_C = 54.1$ ft/s **Ans**

14-29. The 10-lb block is pressed against the spring so as to compress it 2 ft when it is at A. If the plane is smooth, determine the distance d, measured from the wall, to where the block strikes the ground. Neglect the size of the block.

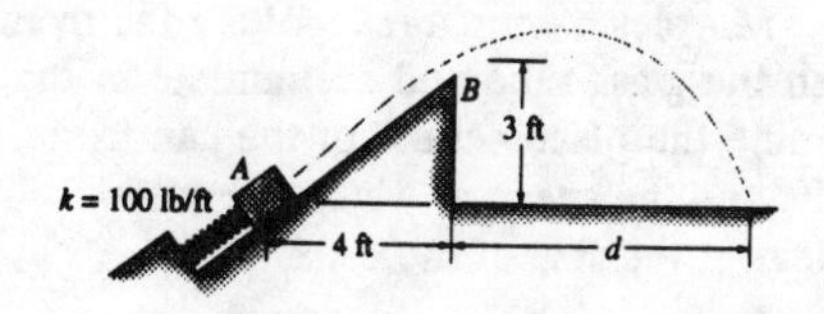

$$T_A + \Sigma U_{A-B} = T_B$$

$$0 + \frac{1}{2}(100)(2)^2 - (10)(3) = \frac{1}{2}\left(\frac{10}{32.2}\right)v_B^2$$

$v_B = 33.09$ ft/s

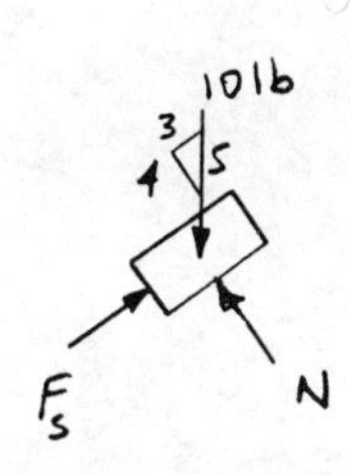

$(\overset{+}{\rightarrow})$ $s = s_0 + v_0 t$

$$d = 0 + 33.09\left(\frac{4}{5}\right)t$$

$(+\uparrow)$ $s = s_0 + v_0 t + \frac{1}{2}a_c t^2$

$$-3 = 0 + (+33.09)\left(\frac{3}{5}\right)t + \frac{1}{2}(-32.2)t^2$$

$$16.1t^2 - 19.853t - 3 = 0$$

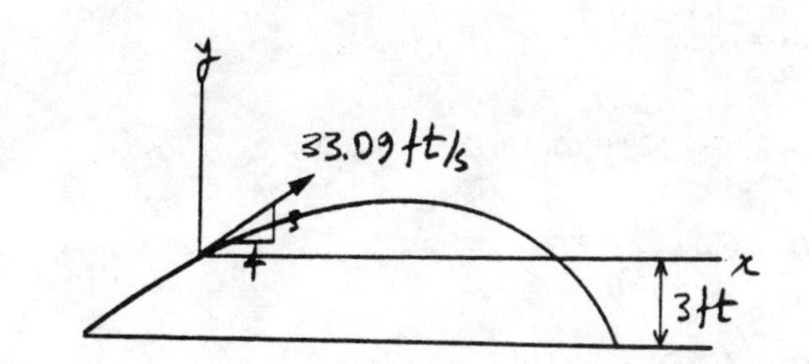

Solving for the positive root,

$t = 1.369$ s

$d = 33.09\left(\frac{4}{5}\right)(1.369) = 36.2$ ft **Ans**

14-30. The 120-lb man acts as a human cannonball by being "fired" from the spring-loaded cannon shown. If the greatest acceleration he can experience is $a = 10g = 322\ \text{ft/s}^2$, determine the required stiffness of the spring which is compressed 2 ft at the moment of firing. With what velocity will he exit the cannon barrel when the cannon is fired? When the spring is compressed $s = 2$ ft then $d = 8$ ft. Neglect friction and assume the man holds himself in a rigid position throughout the motion.

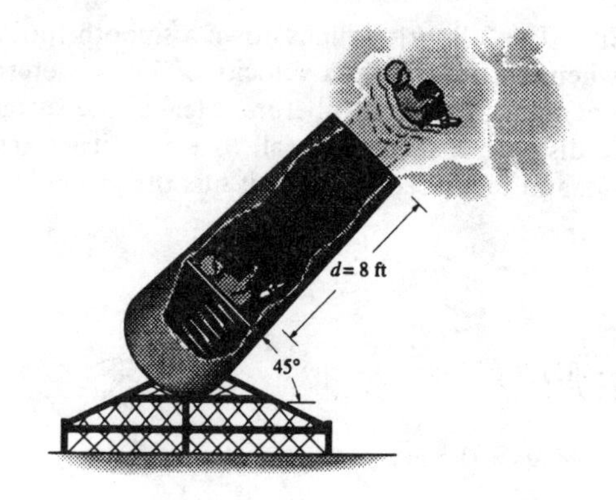

Initial acleeration is $10g = 322\ \text{ft/s}^2$

$+\nearrow \Sigma F_x = ma_x;\quad F_s - 120\sin 45° = \left(\frac{120}{32.2}\right)(322),\quad F_s = 1284.85\ \text{lb}$

For $s = 2$ ft: $\quad 1284.85 = k(2) \quad k = 642.4 = 642\ \text{lb/ft}$ **Ans**

$T_1 + \Sigma U_{1-2} = T_2$

$$0 + \left[\frac{1}{2}(642.4)(2)^2 - 120(8)\sin 45°\right] = \frac{1}{2}\left(\frac{120}{32.2}\right)v_2^2$$

$v_2 = 18.0\ \text{ft/s}$ **Ans**

14-31. Marbles having a mass of 5 g fall from rest at *A* through the glass tube and accumulate in the can at *C*. Determine the placement *R* of the can from the end of the tube and the speed at which the marbles fall into the can. Neglect the size of the can.

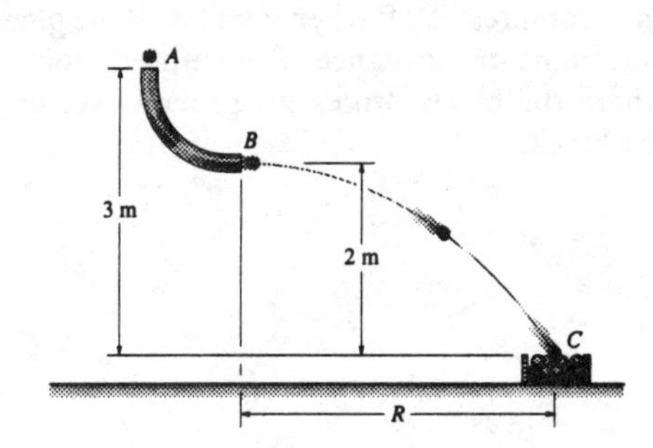

$T_A + \Sigma U_{A-B} = T_B$

$$0 + [0.005(9.81)(3-2)] = \frac{1}{2}(0.005)v_B^2$$

$v_B = 4.429\ \text{m/s}$

$(+\downarrow)\quad s = s_0 + v_0 t + \frac{1}{2}a_c t^2$

$$2 = 0 + 0 + \frac{1}{2}(9.81)t^2$$

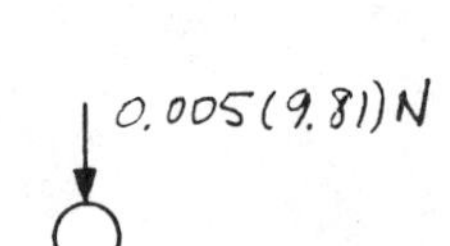

$t = 0.6386\ \text{s}$

$(\overset{+}{\rightarrow})\quad s = s_0 + v_0 t$

$R = 0 + 4.429(0.6386) = 2.83\ \text{m}$ **Ans**

$T_A + \Sigma U_{A-C} = T_C$

$$0 + [0.005(9.81)(3)] = \frac{1}{2}(0.005)v_C^2$$

$v_C = 7.67\ \text{m/s}$ **Ans**

***14-32.** The 100-kg stone is being dragged across the smooth surface by means of a truck T. If the towing cable passes over a small pulley at A, determine the amount of work that must be done by the truck in order to increase the cable angle θ from $\theta_1 = 30°$ to $\theta_2 = 45°$. The truck exerts a constant force $F = 500$ N on the cable at B. Neglect the mass of the pulley and cable.

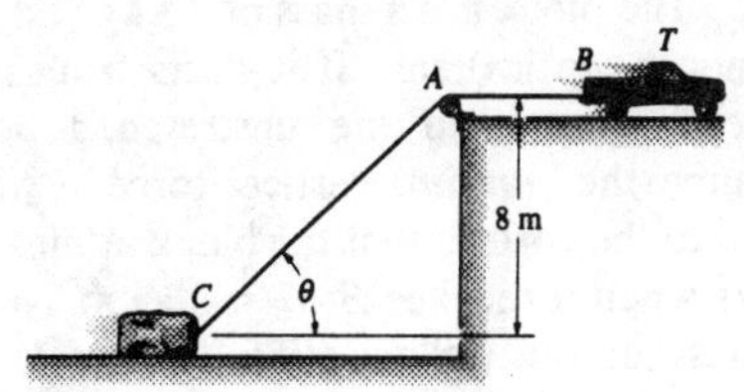

At $\theta_1 = 30°$: $\quad (AC)_1 = \dfrac{8}{\sin 30°}$

At $\theta_2 = 45°$: $\quad (AC)_2 = \dfrac{8}{\sin 45°}$

$$U_T = F\left[(AC)_1 - (AC)_2\right]$$

$$= 500\left(\frac{8}{\sin 30°} - \frac{8}{\sin 45°}\right)$$

$$= 2343.1 \text{ J} = 2.34 \text{ kJ} \quad \textbf{Ans}$$

14-33. A rocket of mass m is fired vertically from the surface of the earth, i.e., at $r = r_1$. Assuming no mass is lost as it travels upward, determine the work it must do against gravity to reach a distance r_2. The force of gravity is $F = GM_e m/r^2$ (Eq. 13–1), where M_e is the mass of the earth and r the distance between the rocket and the center of the earth.

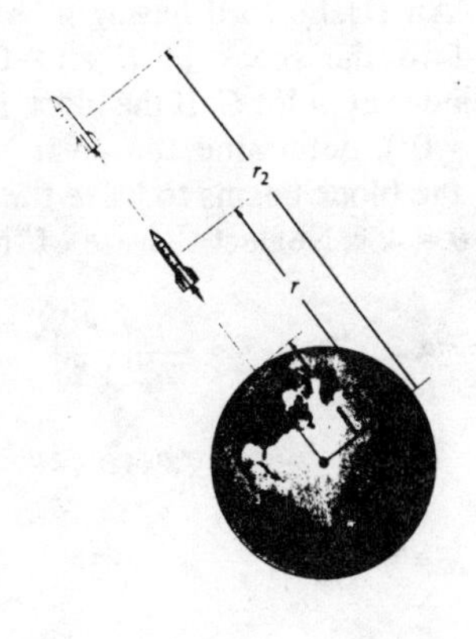

$$F = G\frac{M_e m}{r^2}$$

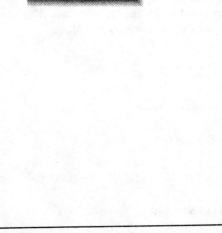

$$U_{1-2} = \int F\,dr = GM_e m \int_{r_1}^{r_2} \frac{dr}{r^2}$$

$$= GM_e m\left(\frac{1}{r_1} - \frac{1}{r_2}\right) \quad \textbf{Ans}$$

14-34. The spring has a stiffness $k = 50$ lb/ft and an *unstretched length* of 2 ft. As shown, it is confined by the plate and wall using cables so that its length is 1.5 ft. A 4-lb block is given a speed v_A when it is at A, and it slides down the incline having a coefficient of kinetic friction $\mu_k = 0.2$. If it strikes the plate and pushes it forward 0.25 ft before stopping, determine its speed at A. Neglect the mass of the plate and spring.

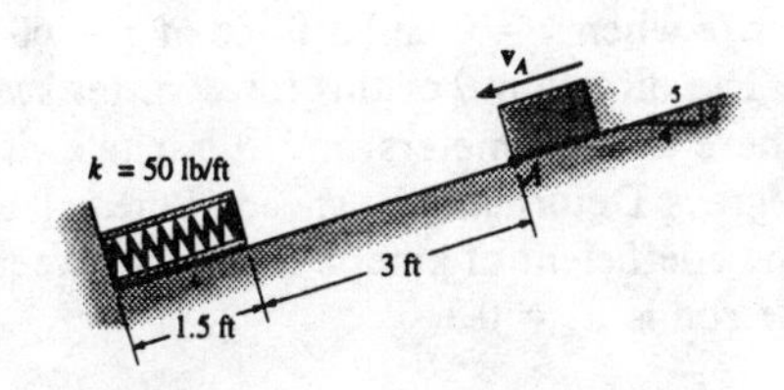

$$+\!\nwarrow \Sigma F_y = 0; \quad N_B - 4\left(\frac{4}{5}\right) = 0$$

$$N_B = 3.20 \text{ lb}$$

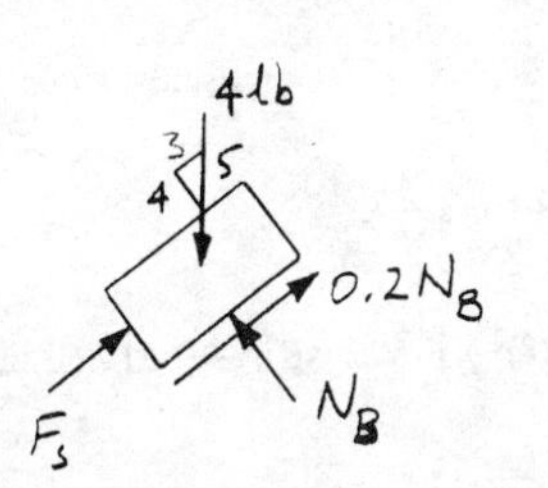

$$T_1 + \Sigma U_{1-2} = T_2$$

$$\frac{1}{2}\left(\frac{4}{32.2}\right)v_A^2 + (3+0.25)\left(\frac{3}{5}\right)(4) - 0.2(3.20)(3+0.25) - \left[\frac{1}{2}(50)(0.75)^2 - \frac{1}{2}(50)(0.5)^2\right] = 0$$

$$v_A = 5.80 \text{ ft/s} \quad \textbf{Ans}$$

14-35. The block has a mass of 0.8 kg and moves within the smooth vertical slot. If it starts from rest when the *attached* spring is in the unstretched position at A, determine the *constant* vertical force F which must be applied to the cord so that the block attains a speed $v_B = 2.5$ m/s when it reaches B; $s_B = 0.15$ m. Neglect the size and mass of the pulley. *Hint:* The work of **F** can be determined by finding the difference Δl in cord lengths AC and BC and using $U_F = F\,\Delta l$.

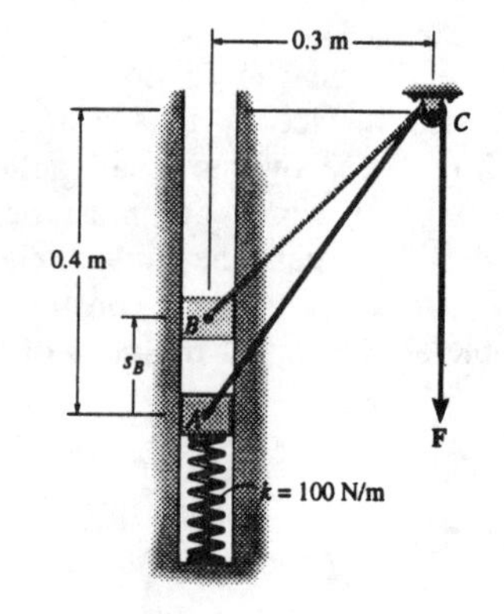

$$l_{AC} = \sqrt{(0.3)^2 + (0.4)^2} = 0.5 \text{ m}$$

$$l_{BC} = \sqrt{(0.4 - 0.15)^2 + (0.3)^2} = 0.3905 \text{ m}$$

$$T_A + \Sigma U_{A-B} = T_B$$

$$0 + F(0.5 - 0.3905) - \frac{1}{2}(100)(0.15)^2 - (0.8)(9.81)(0.15) = \frac{1}{2}(0.8)(2.5)^2$$

$F = 43.9$ N **Ans**

***14-36.** A 2-lb block rests on the smooth semicylindrical surface. An elastic cord having a stiffness $k = 2$ lb/ft is attached to the block at B and to the base of the semicylinder at point C. If the block is released from rest at $A(\theta = 0°)$, determine the unstretched length of the cord so the block begins to leave the semicylinder at the instant $\theta = 45°$. Neglect the size of the block.

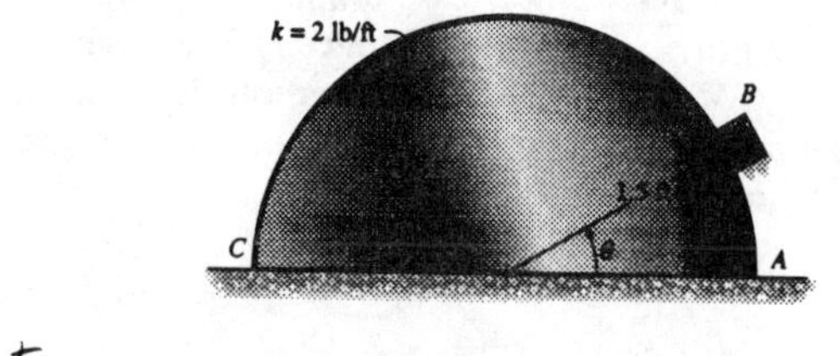

$$+\swarrow \Sigma F_n = ma_n; \quad 2\sin 45° = \frac{2}{32.2}\left(\frac{v^2}{1.5}\right)$$

$$v = 5.844 \text{ ft/s}$$

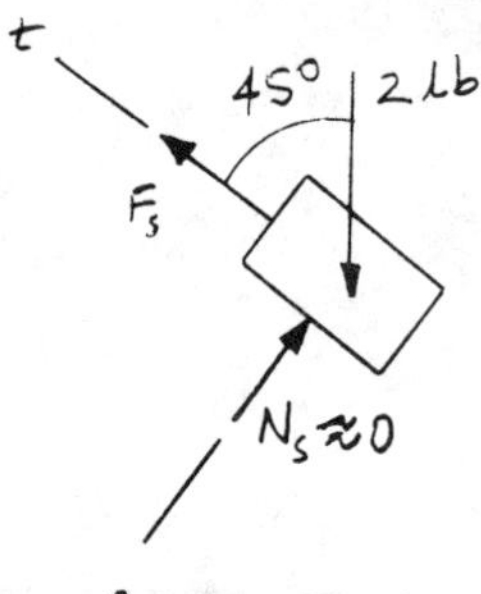

$$T_1 + \Sigma U_{1-2} = T_2$$

$$0 + \frac{1}{2}(2)\left[\pi(1.5) - l_0\right]^2 - \frac{1}{2}(2)\left[\frac{3\pi}{4}(1.5) - l_0\right]^2 - 2(1.5\sin 45°) = \frac{1}{2}\left(\frac{2}{32.2}\right)(5.844)^2$$

$l_0 = 2.77$ ft **Ans**

14-37. The collar has a mass of 5 kg and is moving at 8 m/s when $x = 0$ and a force of $F = 60$ N is applied to it. The direction θ of this force varies such that $\theta = 10x$, where x is in meters and θ is clockwise, measured in degrees. Determine the speed of the collar when $x = 3$ m. The coefficient of kinetic friction between the collar and the rod is $\mu_k = 0.3$.

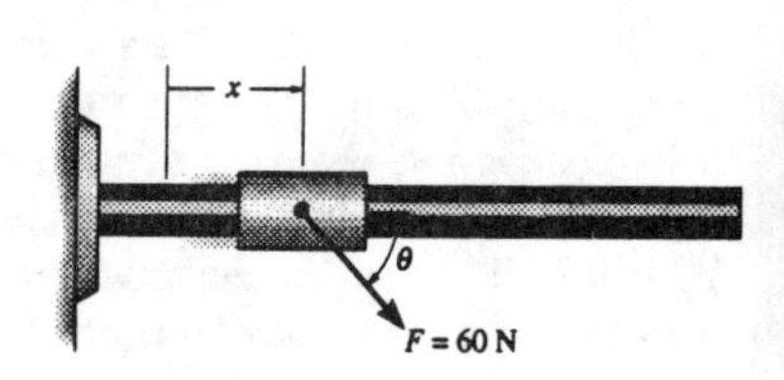

$$+\uparrow \Sigma F_y = 0; \quad N - 5(9.81) - 60\sin\theta = 0$$

$$N = 60\sin\theta + 49.05$$

$$T_1 + \Sigma U_{1-2} = T_2$$

$$\frac{1}{2}(5)(8)^2 + \int_0^3 60\cos\theta\,dx - 0.3\int_0^3 (60\sin\theta + 49.05)\,dx = \frac{1}{2}(5)v^2$$

$$160 + 60\int_0^3 \cos 10x\,dx - 18\int_0^3 \sin 10x\,dx - 14.715\int_0^3 dx = 2.5v^2$$

$$160 + 60\left[\frac{\sin 10x}{10}\right]_0^3 + 18\left[\frac{\cos 10x}{10}\right]_0^3 - 44.145 = 2.5v^2$$

$v = 6.89$ m/s **Ans**

14-38. The 0.5-kg ball of negligible size is fired up the vertical circular track using the spring plunger. The plunger keeps the spring compressed 0.08 m when $s = 0$. Determine how far s the plunger was pulled back and released if the ball begins to leave the track when $\theta = 135°$.

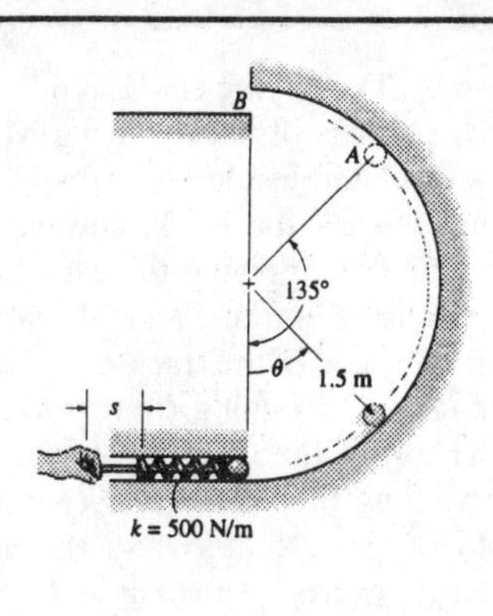

t
0.5(9.81)N
45°
$N_B \approx 0$
n

$$+\swarrow \Sigma F_n = ma_n; \quad 0.5(9.81)\cos 45° = 0.5\left(\frac{v_A^2}{1.5}\right)$$

$$v_A = 3.226 \text{ m/s}$$

$$T_1 + \Sigma U_{1-2} = T_2$$

$$0 + \left[\frac{1}{2}(500)(s')^2 - \frac{1}{2}(500)(0.08)^2 - 0.5(9.81)(1.5)(1+\cos 45°)\right] = \frac{1}{2}(0.5)(3.226)^2$$

$$s' = 0.2589$$

$$s = (0.2589 - 0.08) = 0.179 \text{ m} = 179 \text{ mm} \qquad \textbf{Ans}$$

14-39. The 0.5-kg ball of negligible size is fired up the vertical circular track using the spring plunger. The plunger keeps the spring compressed 0.08 m when $s = 0$. Determine how far s it must be pulled back and released so that the ball will just make it around the loop and land on the platform at B. What is the ball's speed when it reaches the platform?

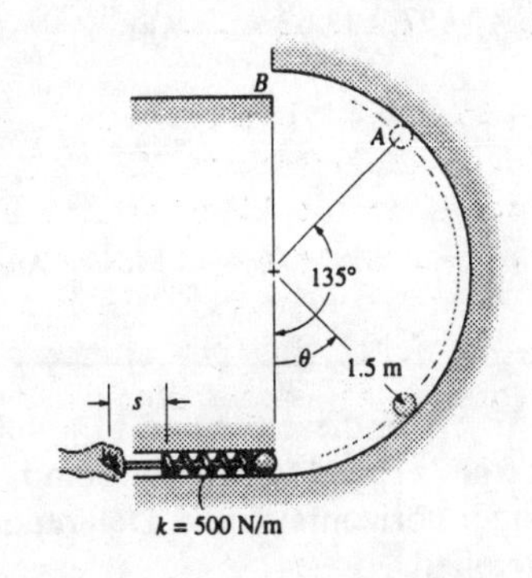

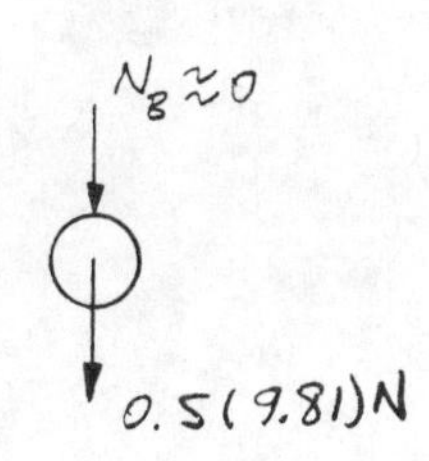

$$+\downarrow \Sigma F_n = ma_n; \quad 0.5(9.81) = 0.5\left(\frac{v_B^2}{1.5}\right)$$

$$v_B = 3.836 = 3.84 \text{ m/s} \qquad \textbf{Ans}$$

$$T_1 + \Sigma U_{1-2} = T_2$$

$$0 + \left[\frac{1}{2}(500)(s')^2 - \frac{1}{2}(500)(0.08)^2\right] - 0.5(9.81)(3) = \frac{1}{2}(0.5)(3.836)^2$$

$$s' = 0.2828 \text{ m}$$

$$s = 0.2828 - 0.08 = 0.2028 \text{ m} = 203 \text{ mm} \qquad \textbf{Ans}$$

***14-40.** The "flying car" is a ride at an amusement park which consists of a car having wheels that roll along a track mounted inside a rotating drum. By design the car cannot fall off the track, however motion of the car is developed by applying the car's brake, thereby gripping the car to the track and allowing it to move with a constant speed of the track, $v_t = 3$ m/s. If the rider applies the brake when going from B to A and then releases it at the top of the drum, A, so that the car coasts freely down along the track to B ($\theta = \pi$ rad), determine the speed of the car at B and the normal reaction which the drum exerts on the car at B. Neglect friction during the motion from A to B. The rider and car have a total mass of 250 kg and the center of mass of the car and rider moves along a circular path having a radius of 8 m.

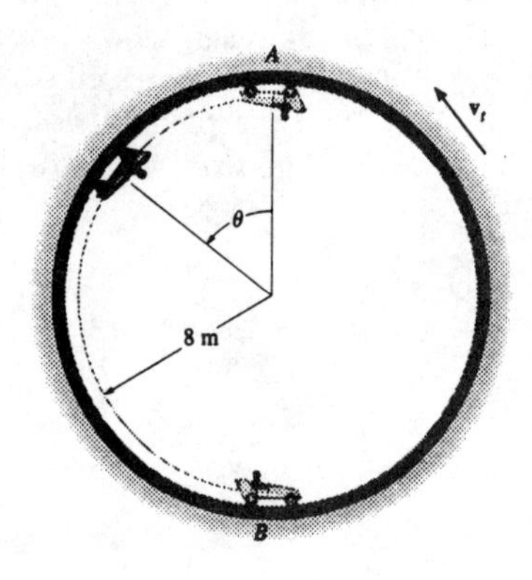

$$T_A + \Sigma U_{A-B} = T_B$$

N_T 250(9.81) N

$$\frac{1}{2}(250)(3)^2 + 250(9.81)(16) = \frac{1}{2}(250)(v_B)^2$$

$v_B = 17.97 = 18.0$ m/s **Ans**

250(9.81) N = $250\left(\frac{v_B^2}{8}\right)$ N_T

$$+\uparrow \Sigma F_n = ma_n; \quad N_B - 250(9.81) = 250\left(\frac{(17.97)^2}{8}\right)$$

$N_B = 12.5$ kN **Ans**

14-41. The diesel engine of a 400-Mg train increases the train's speed uniformly from rest to 10 m/s in 100 s along a horizontal track. Determine the average power developed.

$$T_1 + \Sigma U_{1-2} = T_2$$

$$0 + U_{1-2} = \frac{1}{2}(400)(10^3)(10)^2$$

$$U_{1-2} = 20(10^6)\ \text{J}$$

$P_{avg} = \frac{U_{1-2}}{t} = \frac{20(10^6)}{100} = 200$ kW **Ans**

Also,

$$v = v_0 + a_c t$$

$$10 = 0 + a_c(100)$$

$$a_c = 0.1\ \text{m/s}^2$$

$$\overset{+}{\rightarrow} \Sigma F_x = ma_x; \quad F = 400(10^3)(0.1) = 40(10^3)\ \text{N}$$

$P_{avg} = \mathbf{F} \cdot \mathbf{v}_{avg} = 40(10^3)\left(\frac{10}{2}\right) = 200$ kW **Ans**

14-42. A spring having a stiffness of 5 kN/m is compressed 400 mm. The stored energy in the spring is used to drive a machine which requires 90 W of power. Determine how long the spring can supply energy at the required rate.

$$U_{1-2} = \frac{1}{2}(5000)(0.4)^2 = 400 \text{ J}$$

$$P = \frac{U_{1-2}}{t}; \qquad 90 = \frac{400}{t}$$

$$t = 4.44 \text{ s} \qquad \textbf{Ans}$$

14-43. An electric streetcar has a weight of 15 000 lb and accelerates along a horizontal straight road from rest such that the power is always 100 hp. Determine how far it must travel to reach a speed of 40 ft/s.

$$F = ma = \frac{W}{g}\left(\frac{v\,dv}{ds}\right)$$

$$P = Fv = \left[\left(\frac{W}{g}\right)\left(\frac{v\,dv}{ds}\right)\right]v$$

$$\int_0^s P\,ds = \int_0^v \frac{W}{g}v^2\,dv$$

W

N_c

P = constant

$$Ps = \frac{W}{g}\left(\frac{1}{3}\right)v^3 \qquad s = \frac{W}{3gP}v^3$$

$$s = \frac{(15\,000)(40)^3}{3(32.2)(100)(550)} = 181 \text{ ft} \qquad \textbf{Ans}$$

***14-44.** The Milkin Aircraft Co. manufactures a turbo-jet engine that is placed in a plane having a weight of 13 000 lb. If the engine develops a constant thrust of 5200 lb, determine the power outputs of the plane when it is just ready to take off and when it is flying at 600 mi/h.

$$P = \mathbf{F} \cdot \mathbf{v}$$

Just before take off $v = 0$ and

$$P = 0 \qquad \textbf{Ans}$$

At 600 mi/h,

$$P = 5200(600)\left(\frac{88 \text{ ft/s}}{60 \text{ mi/h}}\right)\frac{1}{550} = 8.32(10^3) \text{ hp} \qquad \textbf{Ans}$$

14-45. A truck has a weight of 25 000 lb and an engine which transmits a power of 350 hp to *all* the wheels. Assuming that the wheels do not slip on the ground, determine the angle θ of the largest incline the truck can climb at a constant speed of $v = 50$ ft/s.

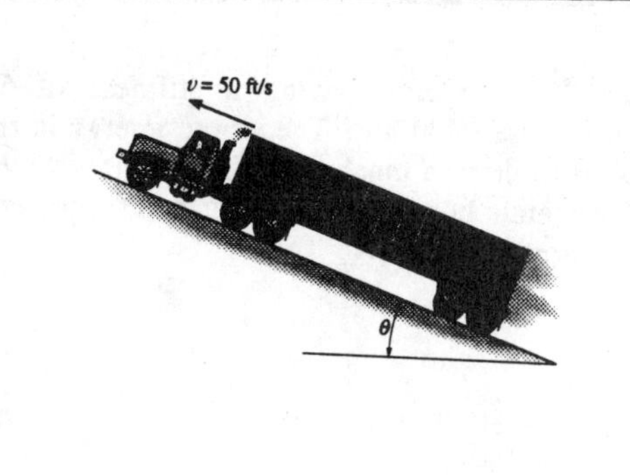

$+\nwarrow \Sigma F_x = ma_x;$ $\quad F - 25\,000\sin\theta = 0 \quad F = 25\,000\sin\theta$ lb

$P = \mathbf{F}\cdot\mathbf{v};$ $\quad 350(550) = 25\,000\sin\theta(50)$

$\theta = \sin^{-1}(0.1540) = 8.86°$ **Ans**

14-46. A loaded truck weighs $16(10^3)$ lb and accelerates uniformly on a level road from 15 ft/s to 30 ft/s during 4 s. If the frictional resistance to motion is 325 lb, determine the maximum power that must be delivered to the wheels.

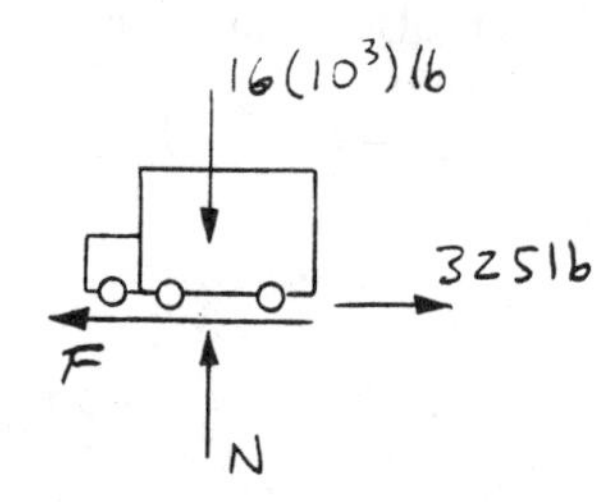

$$a = \frac{\Delta v}{\Delta t} = \frac{30-15}{4} = 3.75 \text{ ft/s}^2$$

$$\overset{+}{\leftarrow} \Sigma F_x = ma_x; \quad F - 325 = \left(\frac{16(10^3)}{32.2}\right)(3.75)$$

$$F = 2188.35 \text{ lb}$$

$$P_{max} = \mathbf{F}\cdot\mathbf{v}_{max} = \frac{2188.35(30)}{550} = 119 \text{ hp}$$ **Ans**

14-47. An automobile having a weight of 3500 lb travels up a 7° slope at a constant speed of $v = 40$ ft/s. If friction and wind resistance are neglected, determine the power developed by the engine if the automobile has a mechanical efficiency of $\epsilon = 0.65$.

$s = vt = 40(1) = 40$ ft

$U_{1-2} = (3500)(40\sin 7°) = 17.062(10^3)$ ft· lb

$$P_o = \frac{U_{1-2}}{t} = \frac{17.602(10^3)}{1} = 17.062(10^3) \text{ ft}\cdot\text{lb/s}$$

$$P_i = \frac{P_o}{e} = \frac{17.062(10^3)\text{ ft}\cdot\text{lb/s}}{0.65} = 26.249(10^3) \text{ ft}\cdot\text{lb/s} = 47.7 \text{ hp}$$ **Ans**

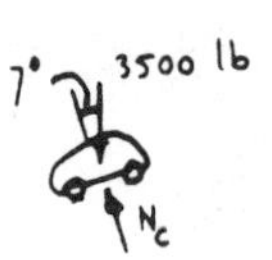

Also,

$F = 3500\sin 7° = 426.543$ lb

$P_o = \mathbf{F}\cdot\mathbf{v} = 426.543(40) = 17.062(10^3)$ ft· lb/s

$$P_i = \frac{P_o}{e} = \frac{17.062(10^3)\text{ ft}\cdot\text{lb/s}}{0.65} = 26.249(10^3) \text{ ft}\cdot\text{lb/s} = 47.7 \text{ hp}$$ **Ans**

***14-48.** The escalator steps move with a constant speed of 0.6 m/s. If the steps are 125 mm high and 250 mm in length, determine the power of a motor needed to lift an average mass of 150 kg per step. There are 32 steps.

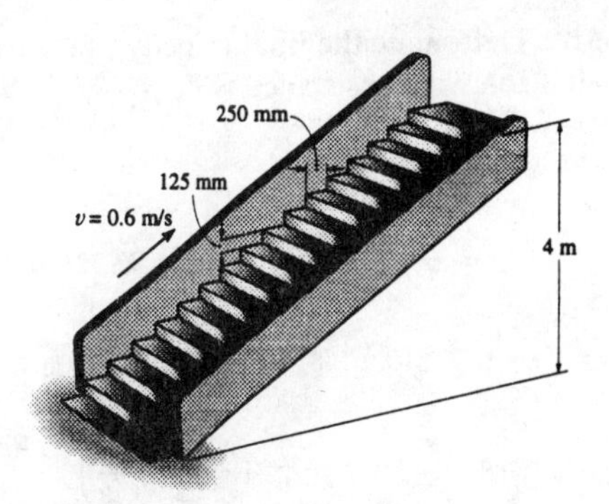

Step height : 0.125 m

The number of steps : $\frac{4}{0.125} = 32$

Total load : $32(150)(9.81) = 47\,088$ N

If load is placed at the center height, $h = \frac{4}{2} = 2$ m, then

$$U = 47\,088\left(\frac{4}{2}\right) = 94.18 \text{ kJ}$$

$$v_y = v\sin\theta = 0.6\left(\frac{4}{\sqrt{(32(0.25))^2 + 4^2}}\right) = 0.2683 \text{ m/s}$$

$$t = \frac{h}{v_y} = \frac{2}{0.2683} = 7.454 \text{ s}$$

$$P = \frac{U}{t} = \frac{94.18}{7.454} = 12.6 \text{ kW} \quad \textbf{Ans}$$

Also,

$$P = \mathbf{F}\cdot\mathbf{v} = 47\,088(0.2683) = 12.6 \text{ kW} \quad \textbf{Ans}$$

14-49. If the escalator in Prob. 14-48 is *not moving,* determine the constant speed at which a man having a mass of 80 kg must walk up the steps to generate 100 W of power—the same amount that is needed to power a standard light bulb.

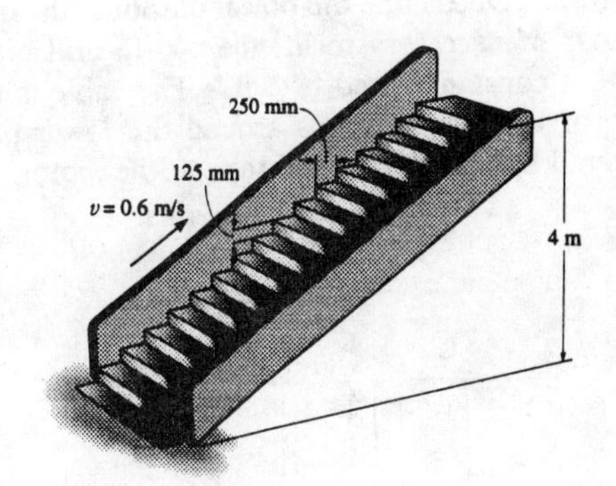

$$P = \frac{U_{1-2}}{t} = \frac{(80)(9.81)(4)}{t} = 100 \qquad t = 31.4 \text{ s}$$

$$v = \frac{s}{t} = \frac{\sqrt{(32(0.25))^2 + 4^2}}{31.4} = 0.285 \text{ m/s} \quad \textbf{Ans}$$

14-50. An electrically powered train car draws 30 kW of power. If the car weighs 40 000 lb and starts from rest, determine the maximum speed it attains in 30 s. The mechanical efficiency is $\epsilon = 0.8$.

$$P_i = 30\,000 \text{ W}\left(\frac{1 \text{ hp}}{746 \text{ W}}\right)\left(\frac{550 \text{ ft}\cdot\text{lb/s}}{1 \text{ hp}}\right) = 22.12(10^3) \text{ ft}\cdot\text{lb/s}$$

$$P_o = P_i e = 22.12(10^3)(0.8) = 17.694(10^3) \text{ ft}\cdot\text{lb/s}$$

$$\overset{+}{\rightarrow}\Sigma F_x = ma_x; \qquad F = \left(\frac{40\,000}{32.2}\right)\left(\frac{dv}{dt}\right)$$

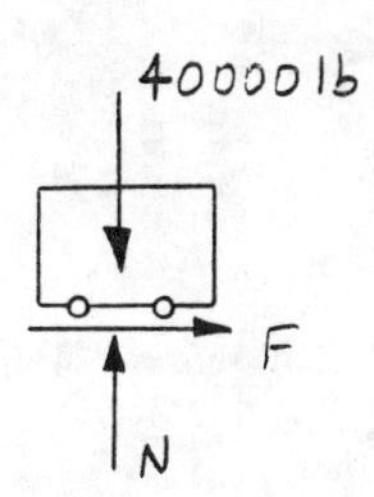

Since $P_o = Fv$, substituting,

$$\int_0^{30}\left(\frac{17\,694}{40\,000}\right)(32.2)\,dt = \int_0^v v\,dv$$

$$v = 29.2 \text{ ft/s} \quad \textbf{Ans}$$

14-51. Determine the final velocity of the train in Prob. 14-50 if the wind resistance is $F_w = (0.6v)$ lb, where v is in ft/s.

$$P_i = 30\,000 \text{ W}\left(\frac{1 \text{ hp}}{746 \text{ W}}\right)\left(\frac{550 \text{ ft}\cdot\text{lb/s}}{1 \text{ hp}}\right) = 22.12(10^3) \text{ ft}\cdot\text{lb/s}$$

$$P_o = P_i e = 22.12(10^3)(0.8) = 17.694(10^3) \text{ ft}\cdot\text{lb/s}$$

$$\overset{+}{\rightarrow}\Sigma F_x = ma_x; \qquad -0.6v + F = \left(\frac{40\,000}{32.2}\right)\left(\frac{dv}{dt}\right)$$

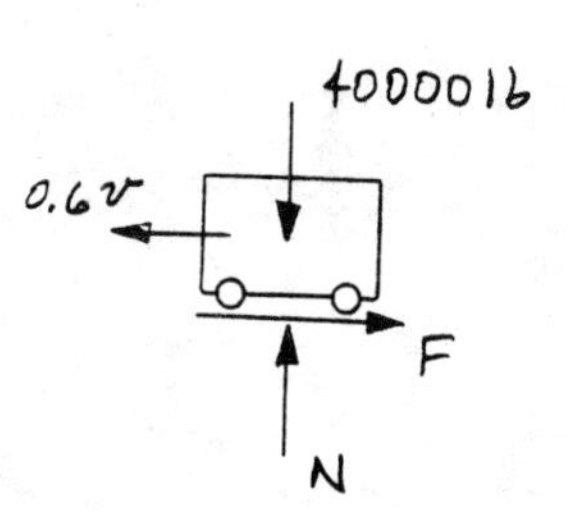

Since $P_o = Fv$, then substituting,

$$\int_0^v \frac{1242.2v\,dv}{17\,694 - 0.6v^2} = \int_0^{30} dt$$

$$-\left(\frac{1242.2}{2(0.6)}\right)\ln\left(17\,694 - 0.6v^2\right)\Big|_0^v = 30$$

$$-1035.2\ln\left(\frac{17\,694 - 0.6v^2}{17\,694}\right) = 30$$

$v = 29.0$ ft/s **Ans**

***14-52.** Determine the power output of the draw-works motor M necessary to lift the 600-lb drill pipe upward with a constant speed of 4 ft/s. The cable is tied to the top of the oil rig, wraps around the lower pulley, then around the top pulley and then to the motor.

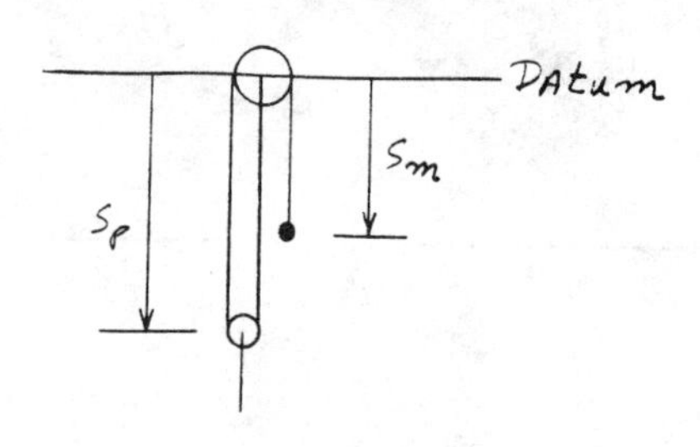

$$2s_p + s_M = l$$

$$2v_p = -v_M$$

$$2(-4) = -v_M$$

$$v_M = 8 \text{ ft/s}$$

$$P_o = Fv = \left(\frac{600}{2}\right)(8) = 2400 \text{ ft}\cdot\text{lb/s} = 4.36 \text{ hp}$$ **Ans**

14-53. The 50-lb crate is hoisted by the motor M. If the crate starts from rest and by constant acceleration attains a speed of 12 ft/s after rising $s = 10$ ft, determine the power that must be supplied to the motor at the instant $s = 10$ ft. The motor has an efficiency $\epsilon = 0.65$. Neglect the mass of the pulley and cable.

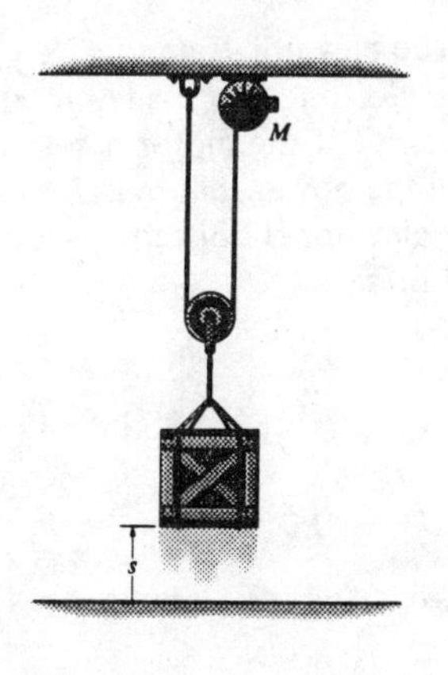

$+\uparrow \Sigma F_y = m\, a_y; \qquad 2T - 50 = \frac{50}{32.2}a$

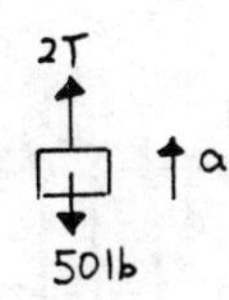

$(+\uparrow)\; v^2 = v_0^2 + 2\, a_c(s - s_0)$

$(12)^2 = 0 + 2(a)(10-0)$

$a = 7.20 \text{ ft/s}^2$

Thus, $T = 30.59$ lb

$s_C + (s_C - s_P) = l$

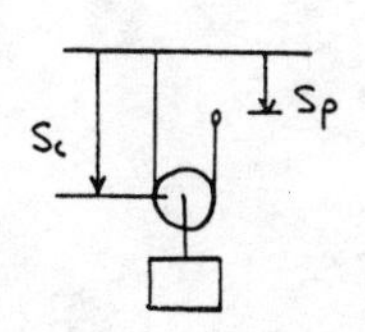

$2\, v_C = v_P$

$2(12) = v_P = 24$ ft/s

$P_o = 30.59(24) = 734.16$

$P_i = \frac{734.16}{0.65} = 1129.5 \text{ ft}\cdot\text{lb/s}$

$P_i = 2.05$ hp **Ans**

14-54. The 50-lb crate is given a speed of 10 ft/s in $t = 4$ s starting from rest. If the acceleration is constant, determine the power that must be supplied to the motor when $t = 2$ s. The motor has an efficiency $\epsilon = 0.65$. Neglect the mass of the pulley and cable.

$+\uparrow \Sigma F_y = m\, a_y; \qquad 2T - 50 = \frac{50}{32.2}a$

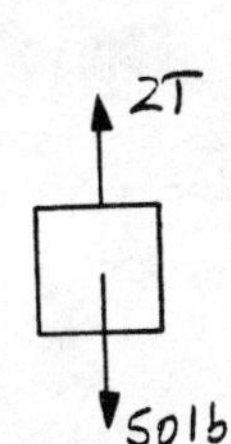

$(+\uparrow)\; v = v_0 + a_c t$

$10 = 0 + a\,(4)$

$a = 2.5 \text{ ft/s}^2$

$T = 26.94$ lb

In $t = 2$ s,

$(+\uparrow)\quad v = v_0 + a_c t$

$v = 0 + 2.5(2) = 5$ ft/s

$s_C + (s_C - s_P) = l$

$2\, v_C = v_P$

$2(5) = v_P = 10$ ft/s

$P_o = 26.94(10) = 269.4$

$P_i = \frac{269.4}{0.65} = 414.5 \text{ ft}\cdot\text{lb/s}$

$P_i = 0.754$ hp **Ans**

14-55. The elevator E and its freight have a total mass of 400 kg. Hoisting is provided by the motor M and the 60-kg block C. If the motor has an efficiency of $\epsilon = 0.6$, determine the power that must be supplied to the motor when the elevator is hoisted upward at a constant speed of $v_E = 4$ m/s.

Elevator :

Since $a = 0$,

$+\uparrow \Sigma F_y = 0; \quad 60(9.81) + 3T - 400(9.81) = 0$

$T = 1111.8$ N

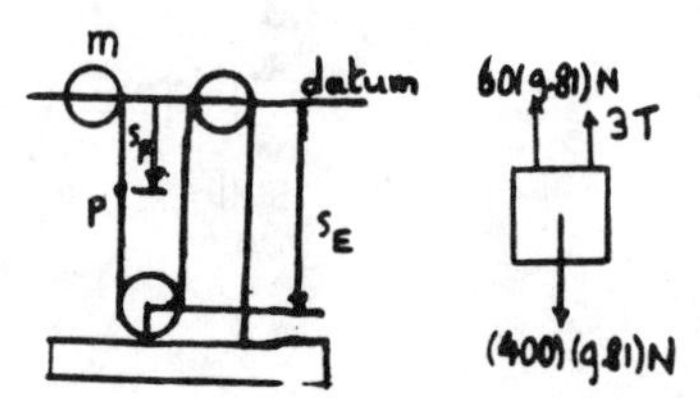

$2s_E + (s_E - s_P) = l$

$3v_E = v_P$

Since $v_E = -4$ m/s, $\quad v_P = -12$ m/s

$$P_i = \frac{\mathbf{F} \cdot \mathbf{v}_P}{e} = \frac{(1111.8)(12)}{0.6} = 22.2 \text{ kW} \quad \textbf{Ans}$$

***14-56.** The 50-kg crate is hoisted up the 30° incline by the pulley system and motor M. If the crate starts from rest and by constant acceleration attains a speed of 4 m/s after traveling 8 m along the plane, determine the power that must be supplied to the motor at this instant. Neglect friction along the plane. The motor has an efficiency of $\epsilon = 0.74$.

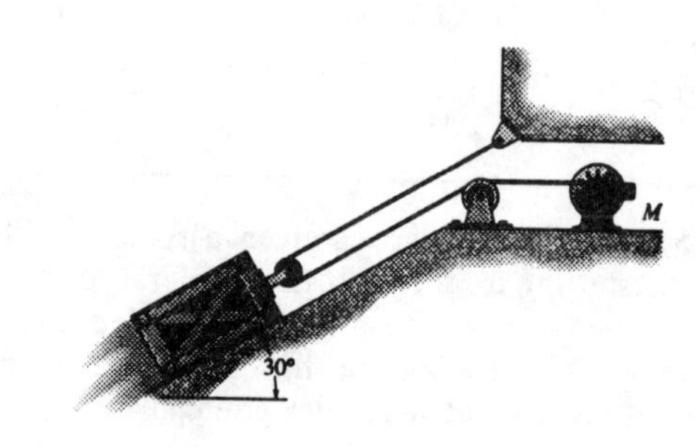

$v^2 = v_0^2 + 2a_c(s - s_0)$

$(4)^2 = 0 + 2a_c(8 - 0)$

$a_c = 1 \text{ m/s}^2$

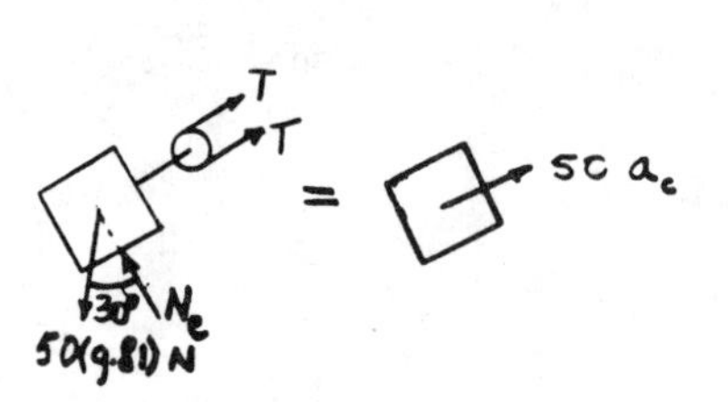

$+\nearrow \Sigma F_x = ma_x; \quad 2T - 50(9.81)\sin 30° = (50)(1) \quad T = 147.6$ N

$2s_C + s_P = l$

$2v_C = -v_P$

$(2)(-4) = -v_P$

$v_P = 8$ m/s

$P_o = \mathbf{T} \cdot \mathbf{v}_P = 147.6(8) = 1181$ W

$$P_i = \frac{P_o}{e} = \frac{1181}{0.74} = 1595.9 \text{ W} = 1.60 \text{ kW} \quad \textbf{Ans}$$

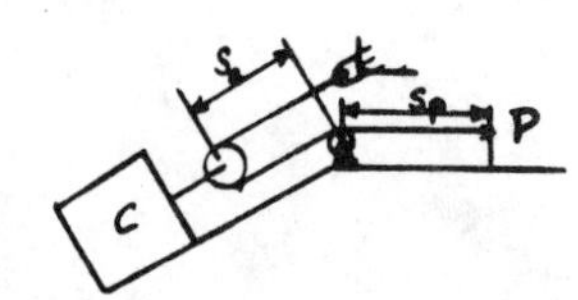

14-57. The 50-lb load is hoisted by the pulley system and motor M. If the crate starts from rest and by constant acceleration attains a speed of 15 ft/s after rising $s = 6$ ft, determine the power that must be supplied to the motor at this instant. The motor has an efficiency of $\epsilon = 0.76$. Neglect the mass of the pulleys and cable.

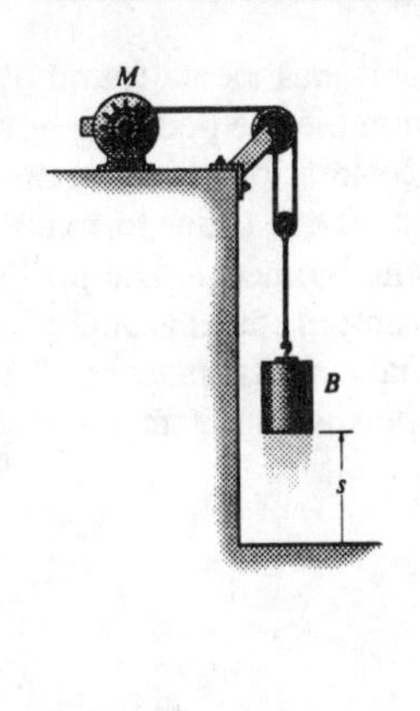

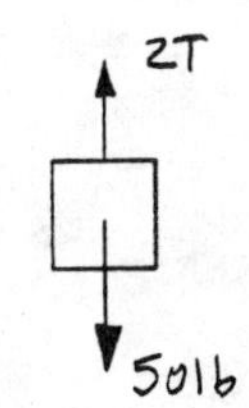

$$(+\uparrow) \quad v^2 = v_0^2 + 2\,a_c(s - s_0)$$

$$(15)^2 = 0 + 2\,a_c(6 - 0)$$

$$a_c = 18.75 \text{ ft/s}^2$$

$$+\uparrow \Sigma F_y = m\,a_y; \quad 2T - 50 = \frac{50}{32.2}(18.75)$$

$$T = 39.56 \text{ lb}$$

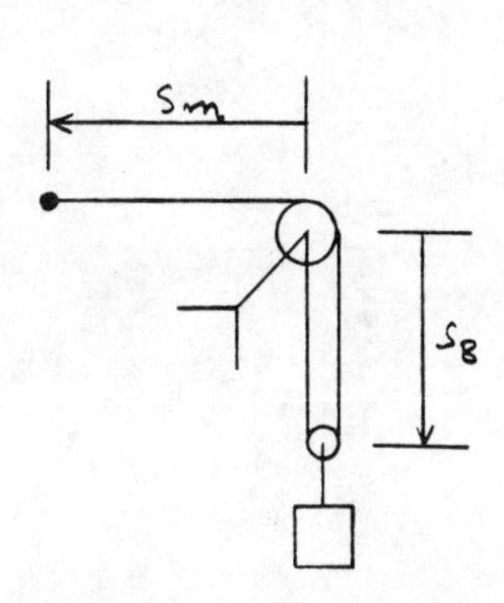

$$2\,s_B + s_M = l$$

$$2\,v_B = -\,v_M$$

$$2(-15) = -\,v_M$$

$$v_M = 30 \text{ ft/s}$$

$$P_o = 30(39.56) = 1186.7 \text{ ft}\cdot\text{lb/s} = 2.16 \text{ hp}$$

$$P_i = \frac{2.16}{0.76} = 2.84 \text{ hp} \qquad \textbf{Ans}$$

14-58. The 50-lb load is hoisted by the pulley system and motor M. If the motor exerts a constant force of 30 lb on the cable, determine the power that must be supplied to the motor if the load has been hoisted $s = 10$ ft starting from rest. The motor has an efficiency of $\epsilon = 0.76$.

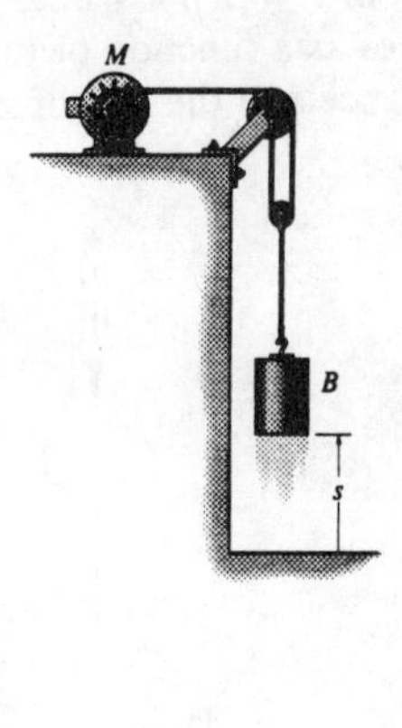

$$+\uparrow \Sigma F_y = m\,a_y; \quad 2(30) - 50 = \frac{50}{32.2}a_B$$

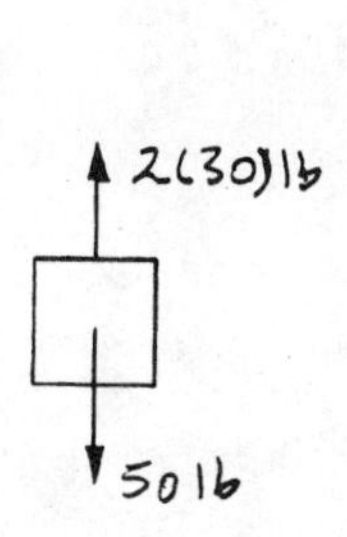

$$a_B = 6.44 \text{ ft/s}^2$$

$$(+\downarrow)\; v^2 = v_0^2 + 2a_c(s - s_0)$$

$$v_B^2 = 0 + 2(6.44)(10 - 0)$$

$$v_B = -11.349 \text{ ft/s}$$

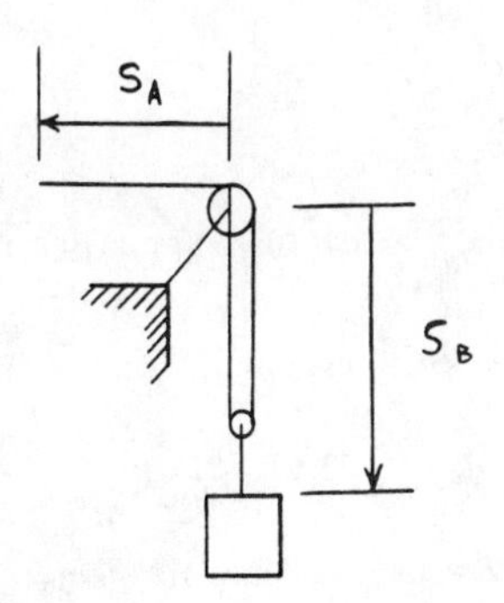

$$2s_B + s_M = l$$

$$2v_B = -\,v_M$$

$$v_M = -2(-11.349) = 22.698 \text{ ft/s}$$

$$P_o = \mathbf{F}\cdot\mathbf{v} = 30(22.698) = 680.94 \text{ ft}\cdot\text{lb/s}$$

$$P_i = \frac{680.94}{0.76} = 895.97 \text{ ft}\cdot\text{lb/s}$$

$$P_i = 1.63 \text{ hp} \qquad \textbf{Ans}$$

14-59. It has been found that the best strategy for maximum bicycle pedaling power is to initially provide a large acceleration, then increase the power gradually over the next 30 s in order to reach maximum cruising speed. Using the biomechanical power curve shown, determine the maximum speed attained by the rider and his bicycle, which have a total mass of 92 kg, as the rider ascends the 20° slope starting from rest.

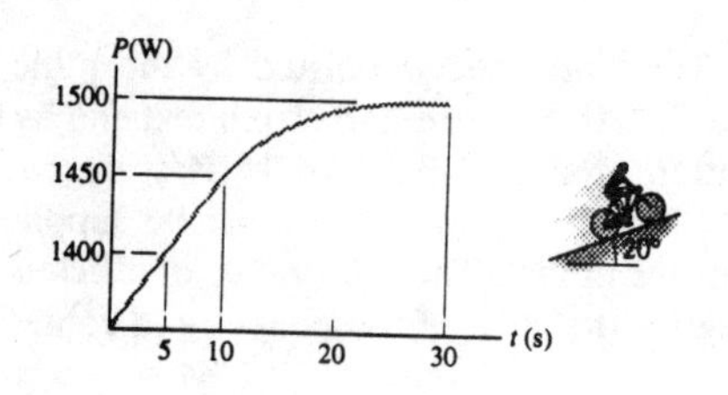

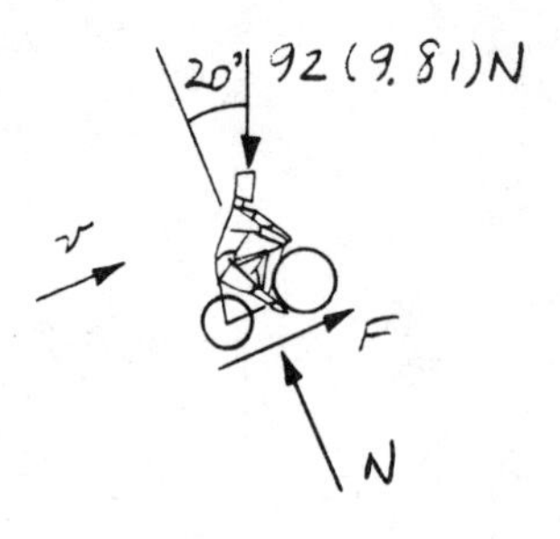

$F = 92(9.81)\sin 20° = 308.68 \text{ N}$

$P = \mathbf{F} \cdot \mathbf{v}; \qquad 1500 = 308.68v$

$v = 4.86 \text{ m/s}$ **Ans**

***14-60.** A rocket having a total mass of 8 Mg is fired vertically from rest. If the engines provide a constant thrust of $T = 300$ kN, determine the power output of the engines as a function of time. Neglect the effect of drag resistance and the loss of fuel mass and weight.

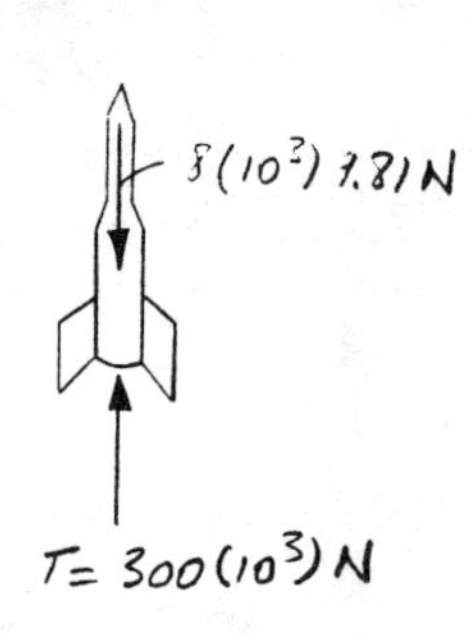

$+\uparrow \Sigma F_y = ma_y; \qquad 300(10^3) - 8(10^3)(9.81) = 8(10^3)a \qquad a = 27.69 \text{ m/s}^2$

$(+\uparrow) \qquad v = v_0 + a_c t$

$= 0 + 27.69t = 27.69t$

$P = \mathbf{T} \cdot \mathbf{v} = 300(10^3)(27.69t) = 8.31t \text{ MW}$ **Ans**

14-61. The block has a weight of 80 lb and rests on the floor for which $\mu_k = 0.4$. If the motor draws in the cable at a constant rate of 6 ft/s, determine the output of the motor at the instant $\theta = 30°$. Neglect the mass of the cable and pulleys.

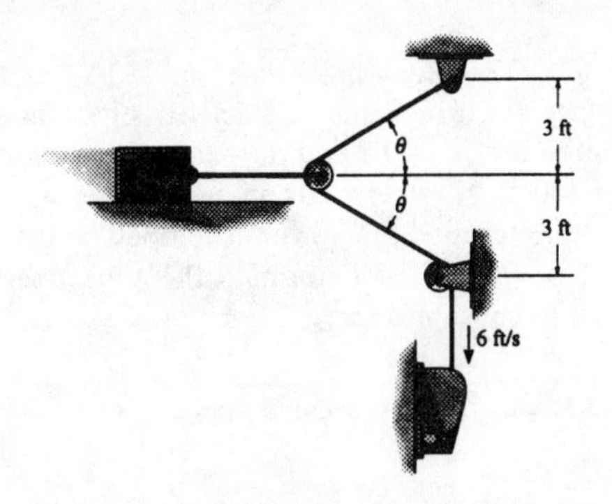

$$2\left(\sqrt{s_B^2 + 3^2}\right) + s_P = l \qquad (1)$$

Time derivative of Eq. (1) yields :

$$\frac{2s_B\dot{s}_B}{\sqrt{s_B^2+9}} + \dot{s}_P = 0 \qquad \text{Where } \dot{s}_B = v_B \text{ and } \dot{s}_P = v_P \qquad (2)$$

$$\frac{2s_B v_B}{\sqrt{s_B^2+9}} + v_P = 0 \qquad v_B = -\frac{\sqrt{s_B^2+9}}{2s_B}v_P \qquad (3)$$

Time derivative of Eq. (2) yields :

$$\frac{1}{(s_B^2+9)^{3/2}}\left[2(s_B^2+9)\dot{s}_B^2 - 2s_B^2\dot{s}_B^2 + 2s_B(s_B^2+9)\ddot{s}_B\right] + \ddot{s}_P = 0$$

where $\ddot{s}_P = a_P = 0$ and $\ddot{s}_B = a_B$

$$2(s_B^2+9)v_B^2 - 2s_B^2 v_B^2 + 2s_B(s_B^2+9)a_B = 0$$

$$a_B = \frac{s_B^2 v_B^2 - v_B^2(s_B^2+9)}{s_B(s_B^2+9)} \qquad (4)$$

At $\theta = 30°$, $\quad s_B = \frac{3}{\tan 30°} = 5.196$ ft

From Eq. (3) $\quad v_B = -\frac{\sqrt{5.196^2+9}}{2(5.196)}(6) = -3.464$ ft/s

From Eq. (4) $\quad a_B = \frac{5.196^2(-3.464)^2 - (-3.464^2)(5.196^2+9)}{5.196(5.196^2+9)} = -0.5773 \text{ ft/s}^2$

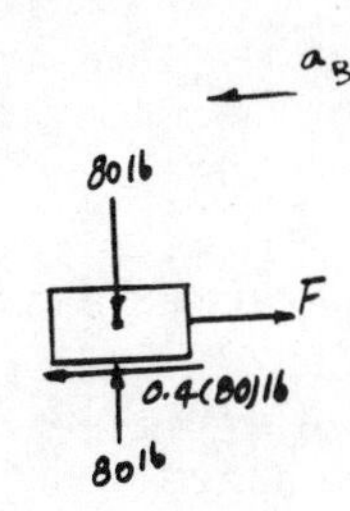

$\xrightarrow{+} \Sigma F_x = ma; \quad F - 0.4(80) = \frac{80}{32.2}(-0.5773) \quad F = 30.57$ lb

$P_o = \mathbf{F}\cdot\mathbf{v} = 30.57(3.464) = 105.9$ ft · lb/s $= 0.193$ hp **Ans**

Also,

$\xrightarrow{+} \Sigma F_x = 0; \quad -F + 2T\cos 30° = 0$

$$T = \frac{30.57}{2\cos 30°} = 17.65 \text{ lb}$$

$P_o = \mathbf{T}\cdot\mathbf{v}_P = 17.65(6) = 105.9$ ft · lb/s $= 0.193$ hp **Ans**

14-62. The block has a mass of 150 kg and rests on a surface for which the coefficients of static and kinetic friction are $\mu_s = 0.5$ and $\mu_k = 0.4$, respectively. If a force $F = (60t^2)$ N, where t is in seconds, is applied to the cable, determine the power developed by the force when $t = 5$ s. *Hint:* First determine the time needed for the force to cause motion.

$\xrightarrow{+} \Sigma F_x = 0; \quad 2F - 0.5(150)(9.81) = 0$

$F = 367.875 = 60t^2$

$t = 2.476$ s

$\xrightarrow{+} \Sigma F_x = ma_x; \quad 2(60t^2) - 0.4(150)(9.81) = 150a_p$

$a_p = 0.8t^2 - 3.924$

$dv = a\,dt$

$$\int_0^v dv = \int_{2.476}^{5} (0.8t^2 - 3.924)\,dt$$

$$v = \left(\frac{0.8}{3}\right)t^3 - 3.924t\Big|_{2.476}^{5} = 19.38 \text{ m/s}$$

$s_p + (s_p - s_F) = l$

$2v_p = v_F$

$v_F = 2(19.38) = 38.76$ m/s

$F = 60(5)^2 = 1500$ N

$P = \mathbf{F} \cdot \mathbf{v} = 1500(38.76) = 58.1$ kW **Ans**

14-63. The 50-lb block rests on the rough surface for which the coefficient of kinetic friction is $\mu_k = 0.2$. A force $F = (40 + s^2)$ lb, where s is in ft, acts on the block in the direction shown. If the spring is originally unstretched ($s = 0$) and the block is at rest, determine the power developed by the force the instant the block has moved $s = 1.5$ ft.

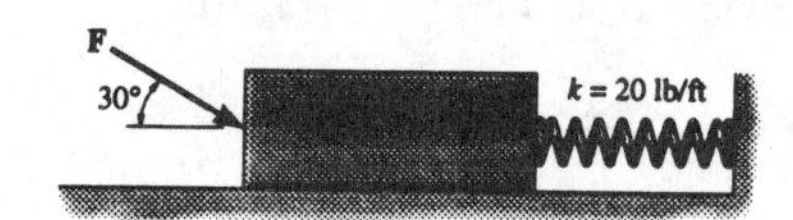

$+\uparrow \Sigma F_y = 0; \quad N_B - (40 + s^2)\sin 30° - 50 = 0$

$N_B = 70 + 0.5s^2$

$T_1 + \Sigma U_{1-2} = T_2$

$$0 + \int_0^{1.5} (40 + s^2)\cos 30°\,ds - \frac{1}{2}(20)(1.5)^2 - 0.2\int_0^{1.5} (70 + 0.5s^2)\,ds = \frac{1}{2}\left(\frac{50}{32.2}\right)v_2^2$$

$0 + 52.936 - 22.5 - 21.1125 = 0.7764v_2^2$

$v_2 = 3.465$ ft/s

When $s = 1.5$ ft,

$F = 40 + (1.5)^2 = 42.25$ lb

$P = \mathbf{F} \cdot \mathbf{v} = (42.25\cos 30°)(3.465)$

$P = 126.79$ ft · lb/s $= 0.231$ hp **Ans**

***14-64.** Solve Prob. 14-8 using the conservation of energy equation.

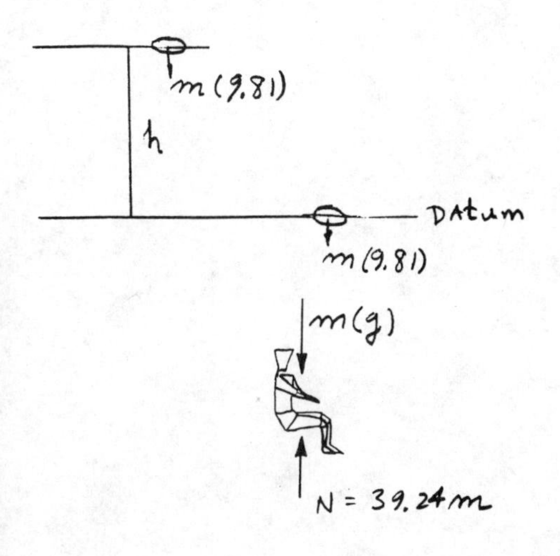

$$100 \text{ km/h} = \frac{100(10^3)}{3600} = 27.778 \text{ m/s}$$

$$T_1 + V_1 = T_2 + V_2$$

$$0 + m(9.81)h = \frac{1}{2}m(27.778)^2 + 0$$

$h = 39.3$ m **Ans**

$$+\uparrow \Sigma F_n = ma_n; \qquad 39.24\,m - mg = m\left(\frac{(27.778)^2}{\rho}\right)$$

$\rho = 26.2$ m **Ans**

14-65. Solve Prob. 14-11 using the conservation of energy equation.

Datum at B :

$$T_A + V_A = T_B + V_B$$

$$0 + 6(2) = 0 + \frac{1}{2}(5)(12)(x)^2$$

$x = 0.6325$ ft $= 7.59$ in. **Ans**

14-66. Solve Prob. 14-17 using the conservation of energy equation.

$$T_1 + V_1 = T_2 + V_2$$

$$0 + \frac{1}{2}(100)(0.5)^2 + \frac{1}{2}(50)(0.5)^2 = \frac{1}{2}(20)v^2 + 0$$

$v = 1.37$ m/s **Ans**

14-67. Solve Prob. 14-21 using the conservation of energy equation.

Datum at A :

$T_A + V_A = T_B + V_B$

$\frac{1}{2}\left(\frac{2}{32.2}\right)(30)^2 + 0 = \frac{1}{2}\left(\frac{2}{32.2}\right)v_B^2 + 2(5)$

$v_B = 24.042 = 24.0$ ft/s **Ans**

$\overset{+}{\rightarrow}\Sigma F_n = ma_n;$ $N_B = \left(\frac{2}{32.2}\right)\left(\frac{(24.042)^2}{5}\right)$

$N_B = 7.18$ lb **Ans**

$T_A + V_A = T_C + V_C$

$\frac{1}{2}\left(\frac{2}{32.2}\right)(30)^2 + 0 = \frac{1}{2}\left(\frac{2}{32.2}\right)v_C^2 + 2(10)$

$v_C = 16.0$ ft/s **Ans**

$+\downarrow\Sigma F_n = ma_n;$ $N_C + 2 = \left(\frac{2}{32.2}\right)\left(\frac{(16.0)^2}{5}\right)$

$N_C = 1.18$ lb **Ans**

***14-68.** Solve Prob. 14-36 using the conservation of energy equation.

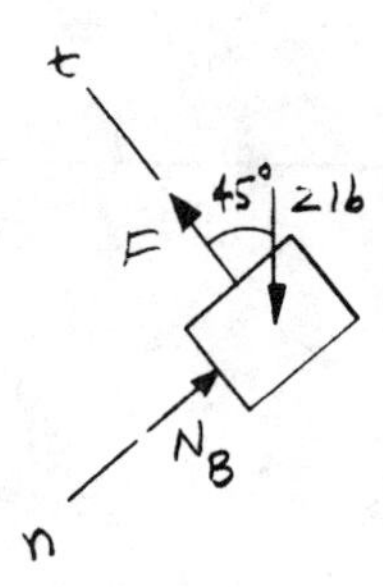

$+\swarrow\Sigma F_n = ma_n;$ $2\sin 45° = \left(\frac{2}{32.2}\right)\left(\frac{v_B^2}{1.5}\right)$

$v_B = 5.844$ ft/s

Datum at A :

$T_A + V_A = T_B + V_B$

$0 + \frac{1}{2}(2)(\pi(1.5) - l_0)^2 = \frac{1}{2}\left(\frac{2}{32.2}\right)(5.844)^2 + \frac{1}{2}(2)\left[\pi(1.5) - \frac{\pi}{4}(1.5) - l_0\right]^2 + 2(1.5\sin 45°)$

$l_0 = 2.77$ ft **Ans**

14-69. Solve Prob. 14-23 using the conservation of energy equation.

Datum at A :

$$T_A + V_A = T_B + V_B$$

$$\frac{1}{2}\left(\frac{50}{32.2}\right)(3)^2 + 0 = \frac{1}{2}\left(\frac{50}{32.2}\right)v_B^2 - 50(5)(1-\cos 30°)$$

$v_B = 7.221$ ft/s $= 7.22$ ft/s **Ans**

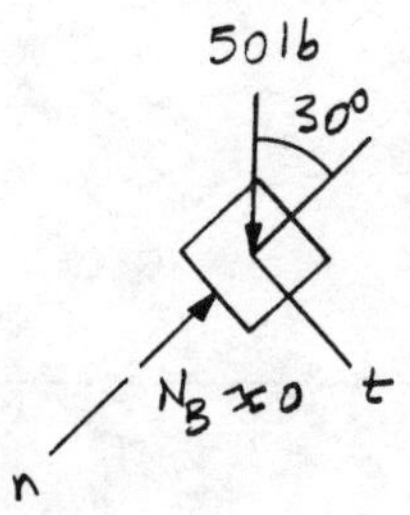

$$+\swarrow \Sigma F_n = ma_n; \qquad -N_B + 50\cos 30° = \left(\frac{50}{32.2}\right)\left(\frac{(7.221)^2}{5}\right)$$

$N_B = 27.1$ lb **Ans**

$$T_A + V_A = T_C + V_C$$

$$\frac{1}{2}\left(\frac{50}{32.2}\right)(3)^2 + 0 = \frac{1}{2}\left(\frac{50}{32.2}\right)v_C^2 - 50(5\cos 30°)$$

$v_C = 16.97$ ft/s $= 17.0$ ft/s **Ans**

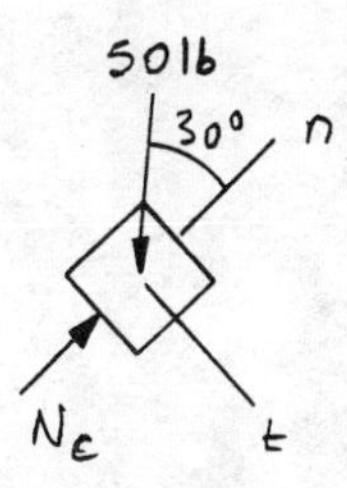

$$+\nearrow \Sigma F_n = ma_n; \qquad N_C - 50\cos 30° = \left(\frac{50}{32.2}\right)\left(\frac{(16.97)^2}{5}\right)$$

$N_C = 133$ lb **Ans**

$$T_A + V_A = T_D + V_D$$

$$\frac{1}{2}\left(\frac{50}{32.2}\right)(3)^2 + 0 = \frac{1}{2}\left(\frac{50}{32.2}\right)v_D^2 - 50(5)$$

$v_D = 18.2$ ft/s **Ans**

14-70. The bob of the pendulum has a mass of 0.2 kg and is released from rest when it is in the horizontal position shown. Determine its speed and the tension in the cord at the instant the bob passes through its lowest position.

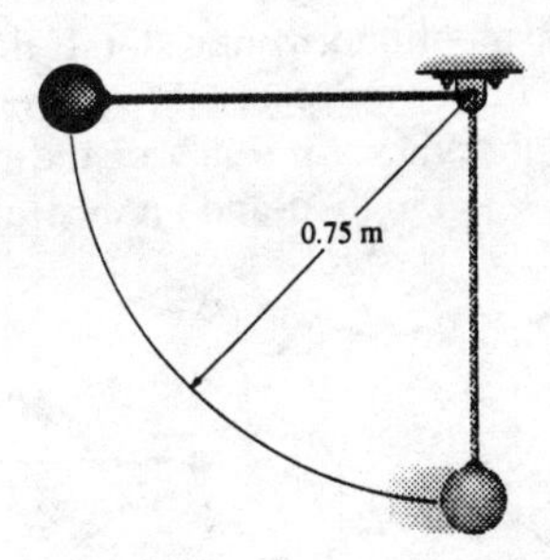

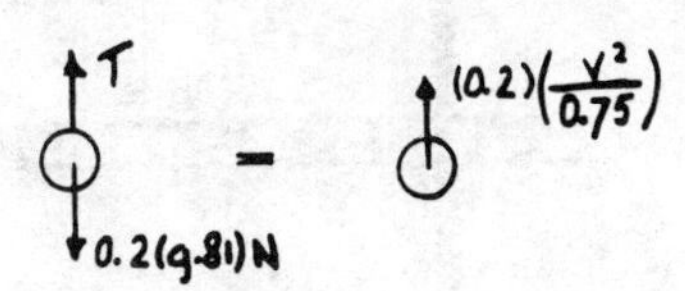

Datum at initial position.

$$T_1 + V_1 = T_2 + V_2$$

$$0 + 0 = \frac{1}{2}(0.2)(v_2)^2 - (0.2)(9.81)(0.75)$$

$v_2 = 3.836$ m/s $= 3.84$ m/s **Ans**

$$+\uparrow \Sigma F_n = ma_n; \qquad T - 0.2(9.81) = 0.2\left(\frac{(3.836)^2}{0.75}\right)$$

$T = 5.89$ N **Ans**

14-71. The block has a weight of 1.5 lb and slides along the smooth chute AB. It is released from rest at A, which has coordinates of A(5 ft, 0, 10 ft). Determine the speed at which it slides off at B, which has coordinates of B(0, 8 ft, 0).

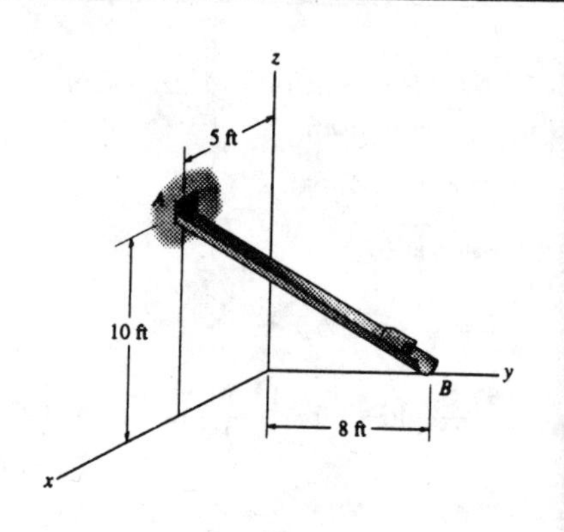

Datum at B :

$$T_A + V_A = T_B + V_B$$

$$0 + 1.5(10) = \frac{1}{2}\left(\frac{1.5}{32.2}\right)(v_B)^2 + 0$$

$v_B = 25.4$ ft/s **Ans**

***14-72.** The blocks A and B weigh 10 lb and 30 lb, respectively. They are connected together by a light cord and ride in the frictionless grooves. Determine the speed of both blocks after they are released from rest and block A moves 6 ft along the plane.

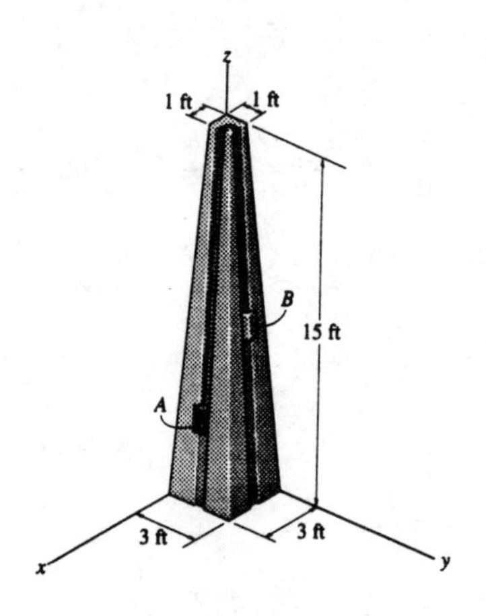

$$\frac{6}{z} = \frac{\sqrt{15^2 + 2^2}}{15} \qquad z = 5.947 \text{ ft}$$

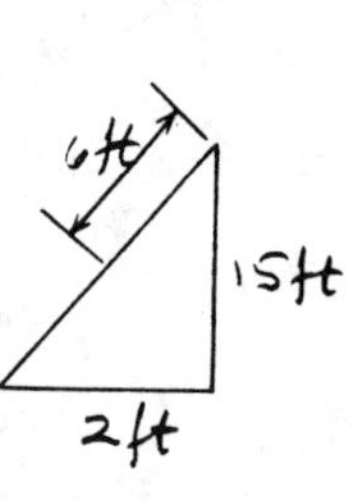

Put datum at each block in its initial position.

$$T_1 + V_1 = T_2 + V_2$$

$$0 + 0 = \frac{1}{2}\left(\frac{10}{32.2}\right)v^2 + \frac{1}{2}\left(\frac{30}{32.2}\right)v^2 + (10)(5.947) - 30(5.947)$$

Solving : $v = 13.8$ ft/s **Ans**

14-73. Block A has a weight of 1.5 lb and slides in the smooth horizontal slot. If the block is drawn back to s = 1.5 ft and released from rest, determine its speed at the instant s = 0. Each of the two springs has a stiffness of k = 150 lb/ft and an unstretched length of 0.5 ft.

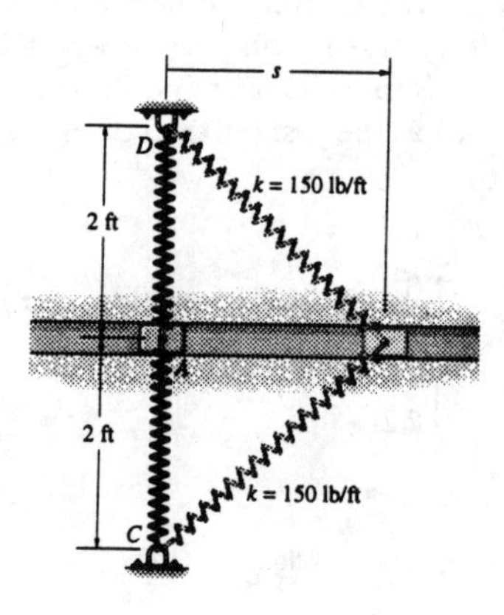

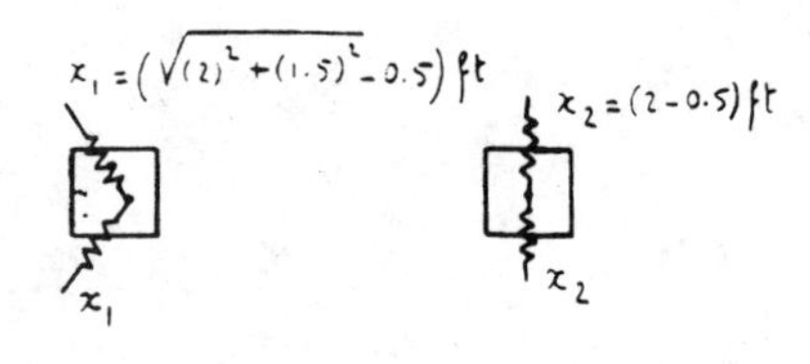

$$T_1 + V_1 = T_2 + V_2$$

$$0 + 2\left[\frac{1}{2}(150)\left(\sqrt{(2)^2 + (1.5)^2} - 0.5\right)^2\right] = \frac{1}{2}\left(\frac{1.5}{32.2}\right)(v_2)^2 + 2\left[\frac{1}{2}(150)(2 - 0.5)^2\right]$$

$v_2 = 106$ ft/s **Ans**

14-74. The 2-lb block A slides in the smooth horizontal slot. When $s = 0$ the block is given an initial velocity of 60 ft/s to the right. Determine the maximum horizontal displacement s of the block. Each of the two springs has a stiffness of $k = 150$ lb/ft and an unstretched length of 0.5 ft.

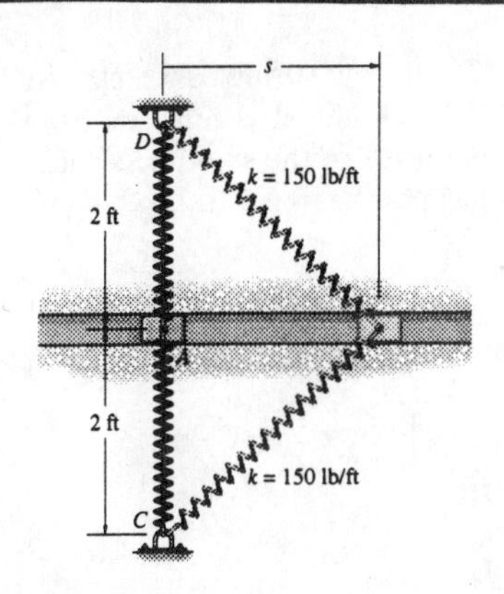

$$T_1 + V_1 = T_2 + V_2$$

$$\frac{1}{2}\left(\frac{2}{32.2}\right)(60)^2 + 2\left[\frac{1}{2}(150)(2-0.5)^2\right] = 0 + 2\left[\frac{1}{2}(150)\left(\sqrt{(2)^2+s^2}-0.5\right)^2\right]$$

Set $d = \sqrt{(2)^2 + s^2}$ then

$$d^2 - d - 2.745 = 0$$

Solving for the positive root,

$$d = 2.231$$

$$(2.231)^2 = (2)^2 + s^2$$

$s = 0.988$ ft **Ans**

14-75. The firing mechanism of a pinball machine consists of a plunger P having a mass of 0.25 kg and a spring of stiffness $k = 300$ N/m. When $s = 0$, the spring is compressed 50 mm. If the arm is pulled back such that $s = 100$ mm and released, determine the speed of the 0.3-kg pinball B *just before* the plunger strikes the stop, i.e., $s = 0$. Assume all surfaces of contact to be smooth. The ball moves in the horizontal plane. Neglect friction, the mass of the spring, and the rolling motion of the ball.

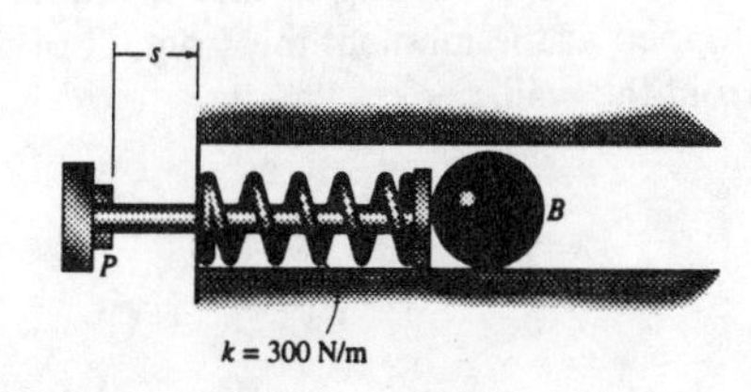

$$T_1 + V_1 = T_2 + V_2$$

$$0 + \frac{1}{2}(300)(0.1+0.05)^2 = \frac{1}{2}(0.25)(v_2)^2 + \frac{1}{2}(0.3)(v_2)^2 + \frac{1}{2}(300)(0.05)^2$$

$v_2 = 3.30$ m/s **Ans**

***14-76.** Determine the smallest amount the spring at B must be compressed against the 0.5-lb block so that when it is released from B it slides along the smooth surface and reaches point A.

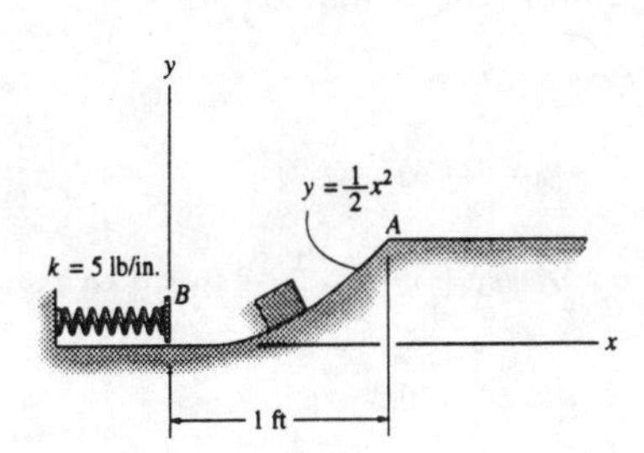

Datum at B :

$$y_A = \frac{1}{2}(1)^2 = 0.5 \text{ ft}$$

$$T_1 + V_1 = T_2 + V_2$$

$$0 + \frac{1}{2}[5(12)]x^2 = 0 + 0.5(0.5)$$

$x = 0.0913$ ft $= 1.10$ in. **Ans**

14-77. If the spring is compressed 3 in. against the 0.5-lb block and it is released from rest, determine the normal force of the smooth surface on the block when it reaches point $x = 0.5$ ft.

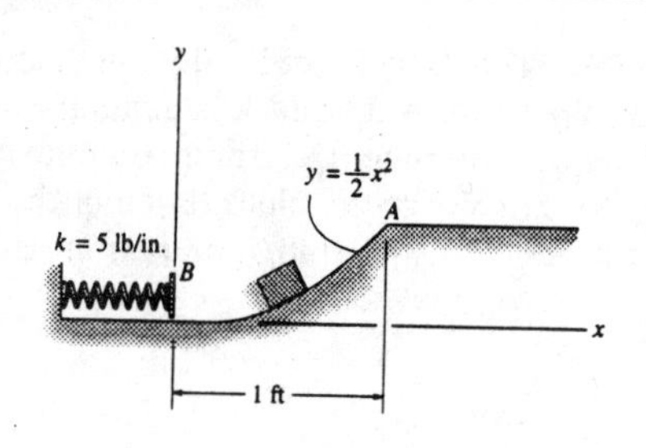

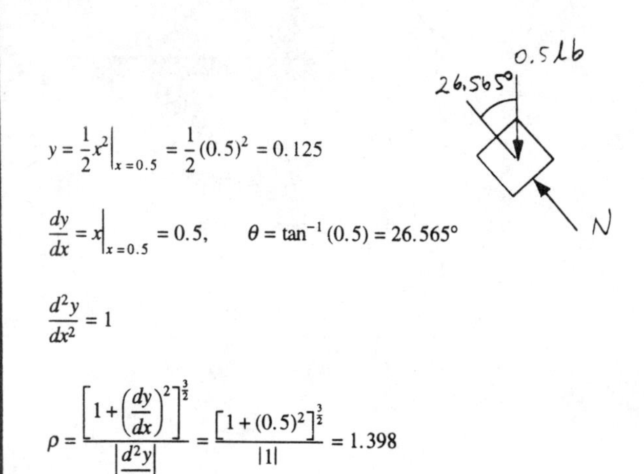

$$y = \frac{1}{2}x^2\Big|_{x=0.5} = \frac{1}{2}(0.5)^2 = 0.125$$

$$\frac{dy}{dx} = x\Big|_{x=0.5} = 0.5, \qquad \theta = \tan^{-1}(0.5) = 26.565°$$

$$\frac{d^2y}{dx^2} = 1$$

$$\rho = \frac{\left[1+\left(\frac{dy}{dx}\right)^2\right]^{\frac{3}{2}}}{\left|\frac{d^2y}{dx^2}\right|} = \frac{\left[1+(0.5)^2\right]^{\frac{3}{2}}}{|1|} = 1.398$$

Datum at B :

$$T_1 + V_1 = T_2 + V_2$$

$$0 + \frac{1}{2}[5(12)]\left(\frac{3}{12}\right)^2 = \frac{1}{2}\left(\frac{0.5}{32.2}\right)(v)^2 + 0.5(0.125)$$

$$v = 15.279 \text{ ft/s}$$

$$+\nwarrow \Sigma F_n = ma_n; \qquad N - 0.5\cos 26.565° = \frac{0.5}{32.2}\left(\frac{(15.279)^2}{1.398}\right)$$

$N = 3.04$ lb **Ans**

14-78. The steel ingot has a mass of 1800 kg. It travels along the conveyor at a speed $v = 0.5$ m/s when it collides with the "nested" spring assembly. If the stiffness of the outer spring is $k_A = 5$ kN/m, determine the required stiffness k_B of the inner spring so that the motion of the ingot is stopped at the moment the front, C, of the ingot is 0.3 m from the wall.

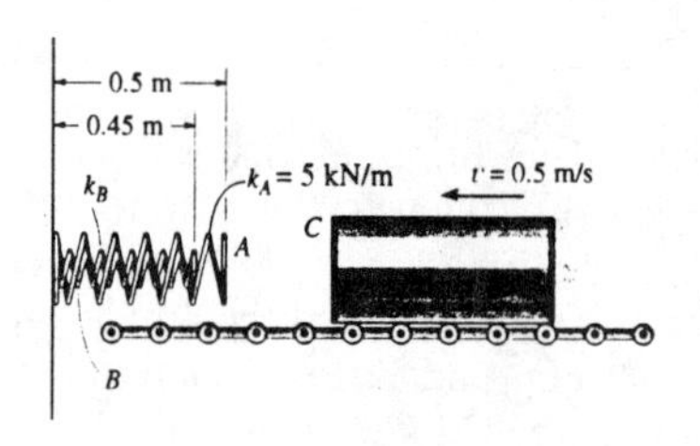

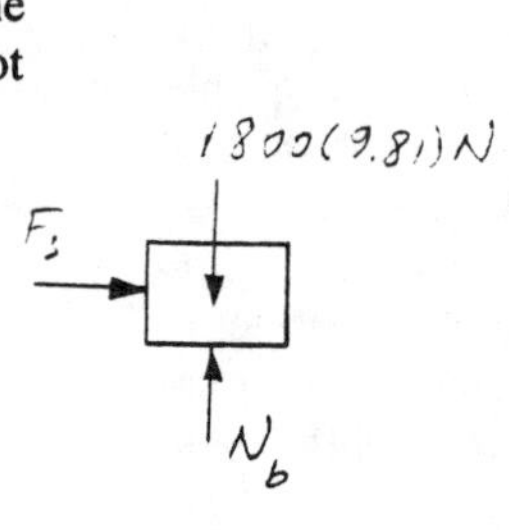

$$T_1 + V_1 = T_2 + V_2$$

$$\frac{1}{2}(1800)(0.5)^2 + 0 = 0 + \frac{1}{2}(5000)(0.5-0.3)^2 + \frac{1}{2}(k_B)(0.45-0.3)^2$$

$k_B = 11.1$ kN/m **Ans**

14-79. The 0.75-kg bob of a pendulum is fired from rest at position A by a spring which has a stiffness of $k = 6$ kN/m and is compressed 125 mm. Determine the speed of the bob and the tension in the cord when the bob is at positions B and C. Point B is located on the path where the radius of curvature is still 0.6 m, i.e., just before the cord becomes horizontal.

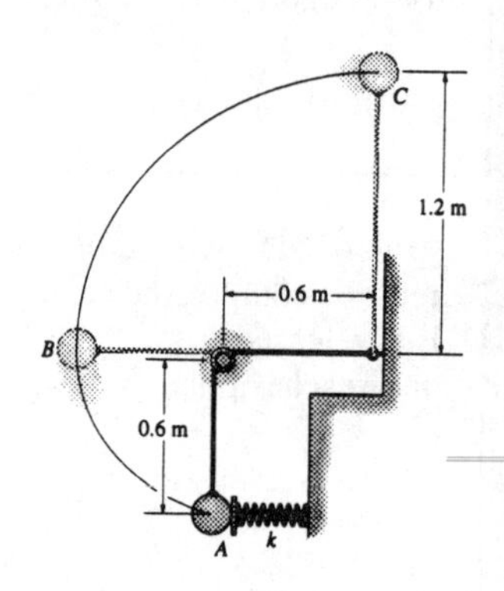

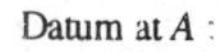

Datum at A :

$$T_A + V_A = T_B + V_B$$

$$0 + \frac{1}{2}(6000)(0.125)^2 = \frac{1}{2}(0.75)(v_B)^2 + 0.75(9.81)(0.6)$$

$v_B = 10.64 = 10.6$ m/s **Ans**

$$\overset{+}{\rightarrow}\Sigma F_n = ma_n; \qquad T_B = 0.75\left(\frac{(10.64)^2}{0.6}\right) = 142 \text{ N} \quad \textbf{Ans}$$

$$T_A + V_A = T_C + V_C$$

$$0 + \frac{1}{2}(6000)(0.125)^2 = \frac{1}{2}(0.75)(v_C)^2 + 0.75(9.81)(1.8)$$

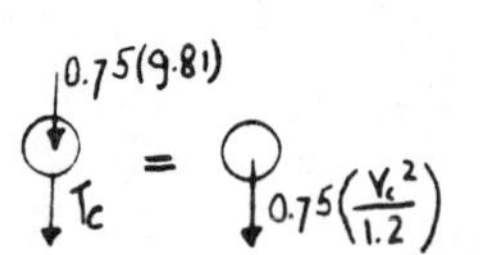

$v_C = 9.470 = 9.47$ m/s **Ans**

$$+\downarrow\Sigma F_n = ma_n; \qquad T_C + 0.75(9.81) = 0.75\left(\frac{(9.470)^2}{1.2}\right)$$

$T_C = 48.7$ N **Ans**

***14-80.** The 0.75-kg bob of a pendulum is fired from rest at position A. If the spring is compressed 50 mm and released, determine (a) its stiffness k so that the speed of the bob is zero when it reaches point B, where the radius of curvature is still 0.6 m, and (b) the stiffness k so that when the bob reaches point C the tension in the cord is zero.

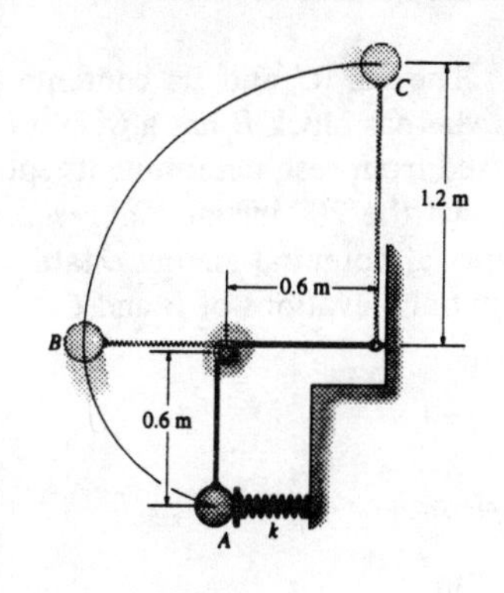

a)

Datum at A :

$$T_A + V_A = T_B + V_B$$

$$0 + \frac{1}{2}(k)(0.05)^2 = 0 + (0.75)(9.81)(0.6)$$

$k = 3532\ \text{N/m} = 3.53\ \text{kN/m}$ **Ans**

b)

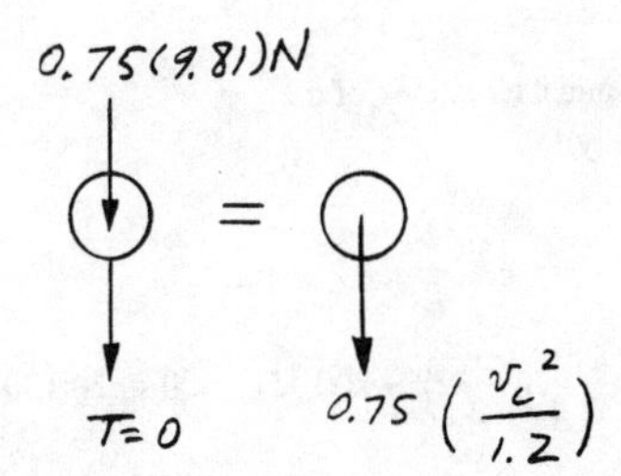

$$+\downarrow \Sigma F_n = ma_n; \quad 0.75(9.81) = 0.75\left(\frac{v_C^2}{1.2}\right)$$

$$v_C = 3.431\ \text{m/s}$$

$$T_A + V_A = T_C + V_C$$

$$0 + \frac{1}{2}(k)(0.05)^2 = \frac{1}{2}(0.75)(3.431)^2 + (0.75)(9.81)(1.8)$$

$k = 14\ 126 = 14.1\ \text{kN/m}$ **Ans**

14-81. Tarzan has a mass of 100 kg and from rest swings from the cliff by rigidly holding on to the tree vine, which is 10 m measured from the supporting limb A to his center of mass. Determine his speed just after the vine strikes the lower limb at B. Also, with what force must he hold on to the vine just before and just after the vine contacts the limb at B?

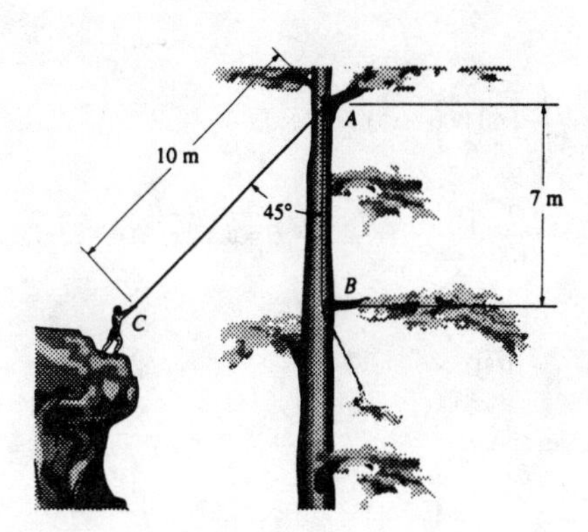

Datum at C :

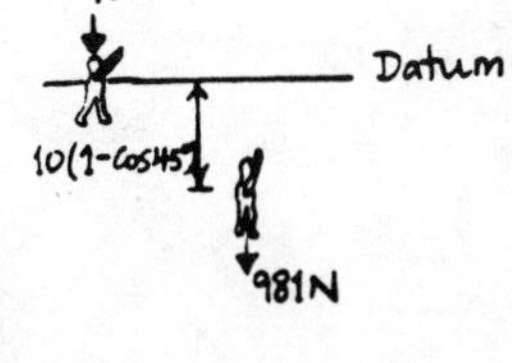

$$T_1 + V_1 = T_2 + V_2$$

$$0 + 0 = \frac{1}{2}(100)(v_C)^2 - 100(9.81)(10)(1 - \cos 45°)$$

$v_C = 7.581 = 7.58\ \text{m/s}$ **Ans**

Just before striking B, $\rho = 10\ \text{m}$

$$+\uparrow \Sigma F_n = ma_n; \quad T - 981 = 100\left(\frac{(7.581)^2}{10}\right)$$

$T = 1.56\ \text{kN}$ **Ans**

Just after striking B, $\rho = 3\ \text{m}$

$$+\uparrow \Sigma F_n = ma_n; \quad T - 981 = 100\left(\frac{(7.581)^2}{3}\right)$$

$T = 2.90\ \text{kN}$ **Ans**

14-82. The car C and its contents have a weight of 600 lb, whereas block B has a weight of 200 lb. If the car is released from rest, determine its speed when it travels 30 ft down the 20° incline. *Suggestion:* To measure the gravitational potential energy, establish separate datums at the initial elevations of B and C.

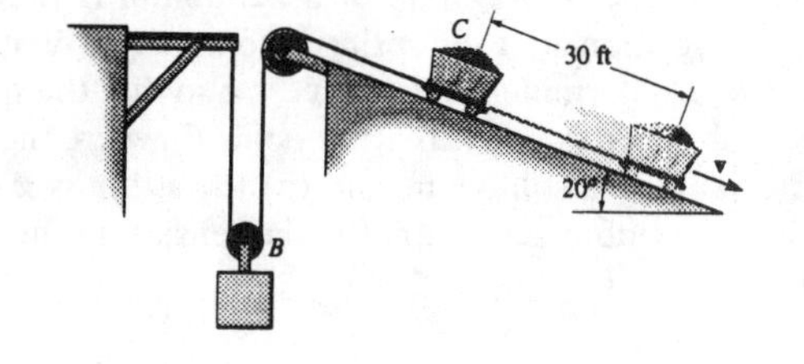

$$2s_B + s_C = l$$

$$2\Delta s_B = -\Delta s_C$$

$$\Delta s_B = -\frac{30}{2} = -15 \text{ ft}$$

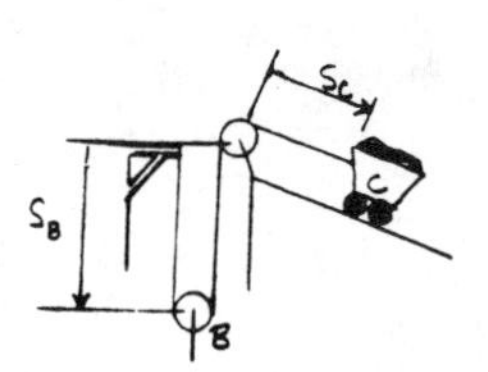

$$2v_B = -v_C$$

Establish two datums at the initial elevations of the car and the block, respectively.

$$T_1 + V_1 = T_2 + V_2$$

$$0 + 0 = \frac{1}{2}\left(\frac{600}{32.2}\right)(v_C)^2 + \frac{1}{2}\left(\frac{200}{32.2}\right)\left(\frac{-v_C}{2}\right)^2 + 200(15) - 600\sin 20°(30)$$

$v_C = 17.7$ ft/s **Ans**

14-83. The roller-coaster car has a speed of 15 ft/s when it is at the crest of a vertical parabolic track. Determine the car's velocity and the normal force it exerts on the track when it reaches point B. Neglect friction and the mass of the wheels. The total weight of the car and the passengers is 350 lb.

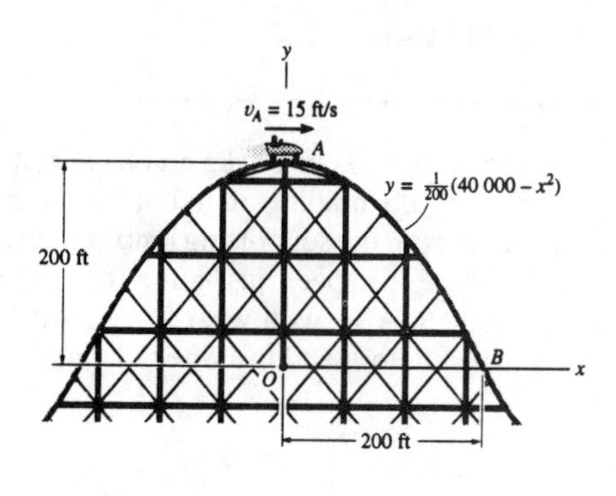

$$y = \frac{1}{200}(40\,000 - x^2)$$

$$\frac{dy}{dx} = -\frac{1}{100}x\Big|_{x=200} = -2, \qquad \theta = \tan^{-1}(-2) = -63.43°$$

$$\frac{d^2y}{dx^2} = -\frac{1}{100}$$

Datum at A:

$$T_A + V_A = T_B + V_B$$

$$\frac{1}{2}\left(\frac{350}{32.2}\right)(15)^2 + 0 = \frac{1}{2}\left(\frac{350}{32.2}\right)(v_B)^2 - 350(200)$$

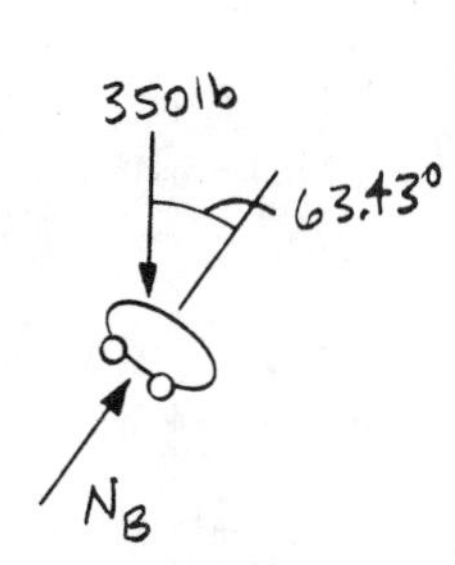

$v_B = 114.48 = 114$ ft/s **Ans**

$$\rho = \frac{\left[1 + \left(\frac{dy}{dx}\right)^2\right]^{\frac{3}{2}}}{\left|\frac{d^2y}{dx^2}\right|} = \frac{\left[1 + (-2)^2\right]^{\frac{3}{2}}}{\left|-\frac{1}{100}\right|} = 1118.0 \text{ ft}$$

$$+\swarrow \Sigma F_n = ma_n; \qquad 350\cos 63.43° - N_B = \left(\frac{350}{32.2}\right)\frac{(114.48)^2}{1118.0}$$

$N_B = 29.1$ lb **Ans**

***14-84.** Two equal-length springs having a stiffness $k_A = 300$ N/m and $k_B = 200$ N/m are "nested" together in order to form a shock absorber. If a 2-kg block is dropped from an at-rest position 0.6 m above the top of the springs, determine their deformation when the block momentarily stops.

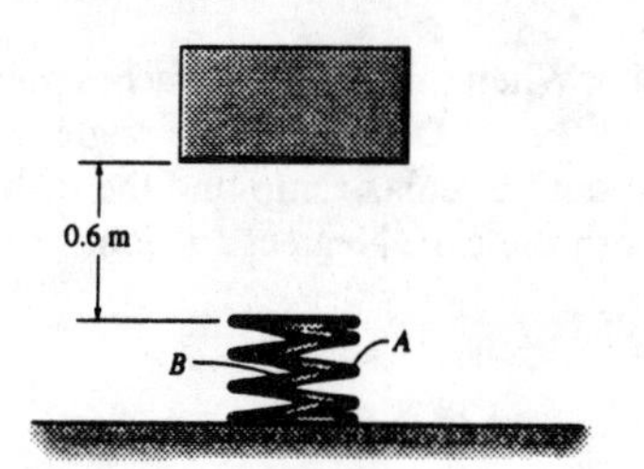

Datum at initial position :

$$T_1 + V_1 = T_2 + V_2$$

$$0 + 0 = 0 - 2(9.81)(0.6 + x) + \frac{1}{2}(300 + 200)(x)^2$$

$$250x^2 - 19.62x - 11.772 = 0$$

Solving for the positive root,

$x = 0.260$ m **Ans**

14-85. A block having a mass of 20 kg is attached to four springs. If each spring has a stiffness of $k = 2$ kN/m and an unstretched length of 150 mm, determine the *maximum* downward vertical displacement s_{max} of the block if it is released from rest when $s = 0$.

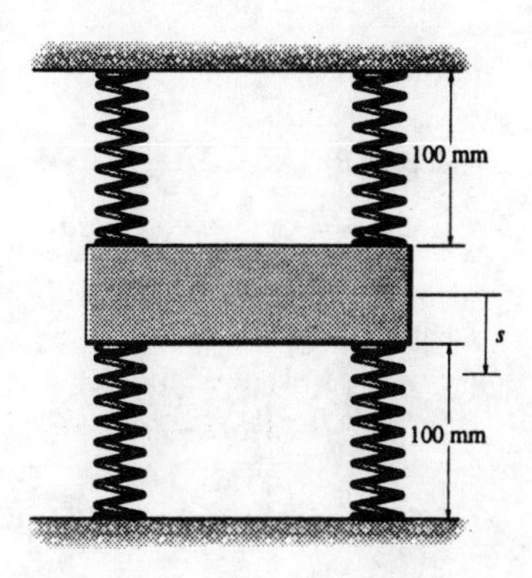

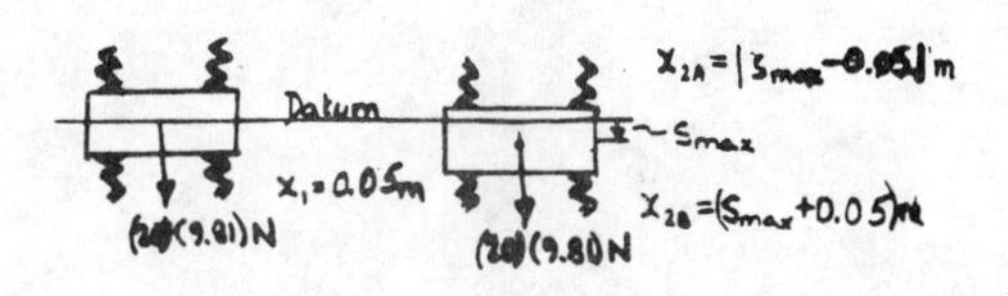

Place the datum at the initial elevation of the block.

$$T_1 + V_1 = T_2 + V_2$$

$$0 + 4\left[\frac{1}{2}(2000)(0.05)^2\right] = 0 + 2\left[\frac{1}{2}(2000)(s_{max} - 0.05)^2\right] + 2\left[\frac{1}{2}(2000)(s_{max} + 0.05)^2\right] - (20)(9.81)s_{max}$$

$$4000s_{max}^2 - 196.2s_{max} = 0$$

Solving,

$s_{max} = 0.0490$ m $= 49.0$ mm **Ans**

14-86. When the 6-kg box reaches point A it has a speed of $v_A = 2$ m/s. Determine the angle θ at which it leaves the smooth circular ramp and the distance s to where it falls into the cart. Neglect friction.

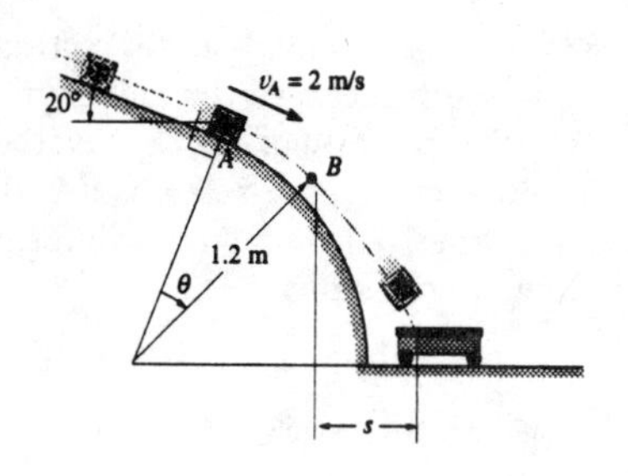

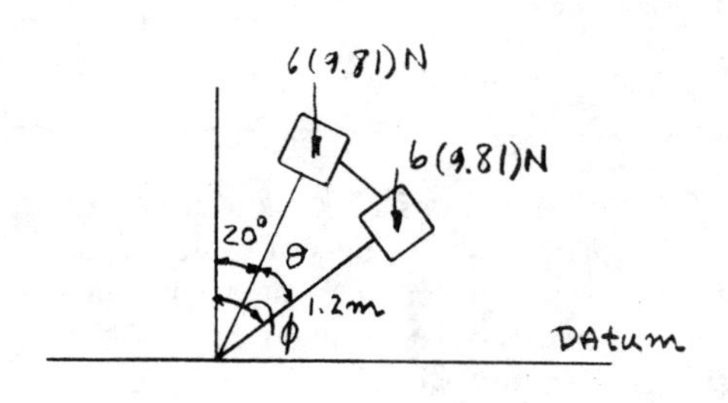

At point B :

$+\swarrow \Sigma F_n = ma_n;\quad 6(9.81)\cos\phi = 6\left(\dfrac{v_B^2}{1.2}\right) \quad (1)$

Datum at bottom of curve :

$T_A + V_A = T_B + V_B$

$\dfrac{1}{2}(6)(2)^2 + 6(9.81)(1.2\cos 20°) = \dfrac{1}{2}(6)(v_B)^2 + 6(9.81)(1.2\cos\phi)$

$13.062 = 0.5v_B^2 + 11.772\cos\phi \quad (2)$

Substitute Eq. (1) into Eq. (2), and solving for v_B,

$v_B = 2.951$ m/s

Thus, $\quad \phi = \cos^{-1}\left(\dfrac{(2.951)^2}{1.2(9.81)}\right) = 42.29°$

$\theta = \phi - 20° = 22.3°$ **Ans**

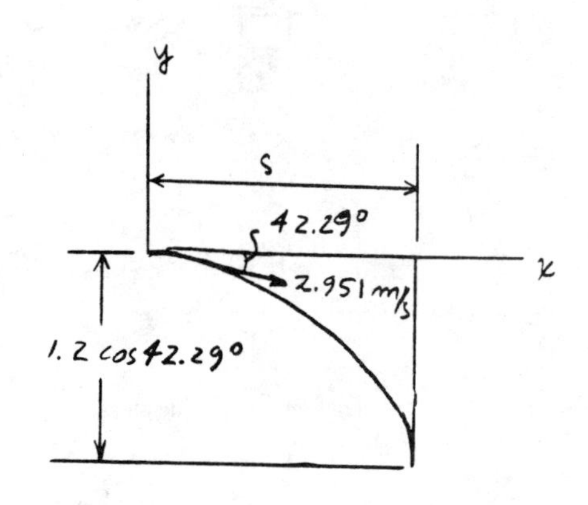

$(+\uparrow) \quad s = s_0 + v_0 t + \frac{1}{2}a_c t^2$

$-1.2\cos 42.29° = 0 - 2.951(\sin 42.29°)t + \dfrac{1}{2}(-9.81)t^2$

$4.905t^2 + 1.9857t - 0.8877 = 0$

Solving for the positive root : $\quad t = 0.2687$ s

$(\overset{+}{\rightarrow}) \quad s = s_0 + v_0 t$

$s = 0 + (2.951\cos 42.29°)(0.2687)$

$s = 0.587$ m **Ans**

14-87. The 2-lb box has a velocity of 5 ft/s when it begins to slide down the smooth inclined surface at A. Determine the point C (x, y) where it strikes the lower incline.

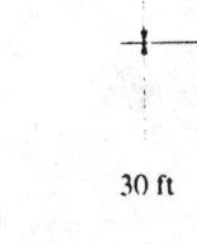
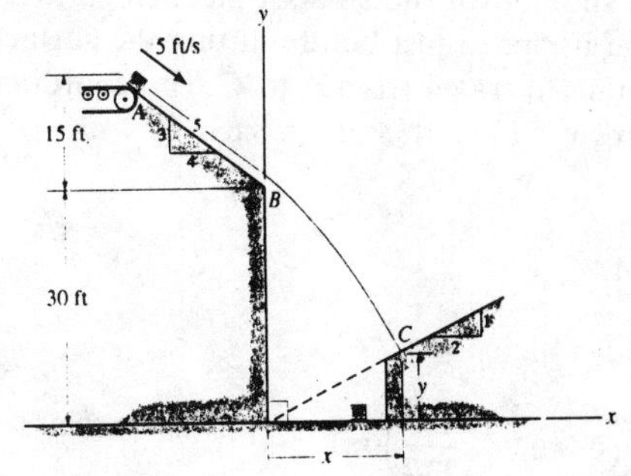

Datum at A :

$$T_A + V_A = T_B + V_B$$

$$\frac{1}{2}\left(\frac{2}{32.2}\right)(5)^2 + 0 = \frac{1}{2}\left(\frac{2}{32.2}\right)v_B^2 - 2(15)$$

$$v_B = 31.48 \text{ ft/s}$$

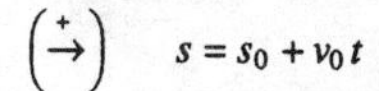

$(\overset{+}{\rightarrow})$ $\quad s = s_0 + v_0 t$

$$x = 0 + 31.48\left(\frac{4}{5}\right)t \qquad (1)$$

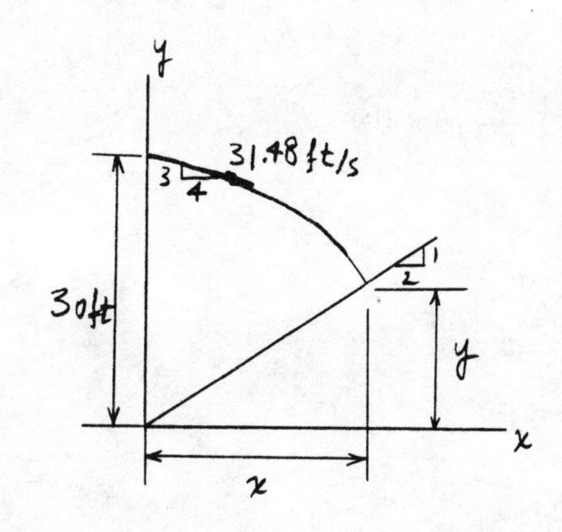

$(+\uparrow)$ $\quad s = s_0 + v_0 t + \frac{1}{2}a_c t^2$

$$y = 30 - 31.48\left(\frac{3}{5}\right)t + \frac{1}{2}(-32.2)t^2$$

Equation of inclined surface :

$$\frac{y}{x} = \frac{1}{2}; \qquad y = \frac{1}{2}x \qquad (2)$$

Thus,

$$30 - 18.888t - 16.1t^2 = 12.592t$$

$$-16.1t^2 - 31.480t + 30 = 0$$

Solving for the positive root,

$$t = 0.7014 \text{ s}$$

From Eqs. (1) and (2) :

$$x = 31.48\left(\frac{4}{5}\right)(0.7014) = 17.66 = 17.7 \text{ ft} \qquad \textbf{Ans}$$

$$y = \frac{1}{2}(17.664) = 8.832 = 8.83 \text{ ft} \qquad \textbf{Ans}$$

$$T_A + V_A = T_C + V_C$$

$$\frac{1}{2}\left(\frac{2}{32.2}\right)(5)^2 + 0 = \frac{1}{2}\left(\frac{2}{32.2}\right)(v_C)^2 - 2\left[15 + (30 - 8.83)\right]$$

$$v_C = 48.5 \text{ ft/s} \qquad \textbf{Ans}$$

***14-88.** The 2-lb box has a velocity of 5 ft/s when it begins to slide down the smooth inclined surface at A. Determine its speed just before hitting the surface at C and the time to travel from A to C. The coordinates of point C are $x = 17.66$ ft, and $y = 8.832$ ft.

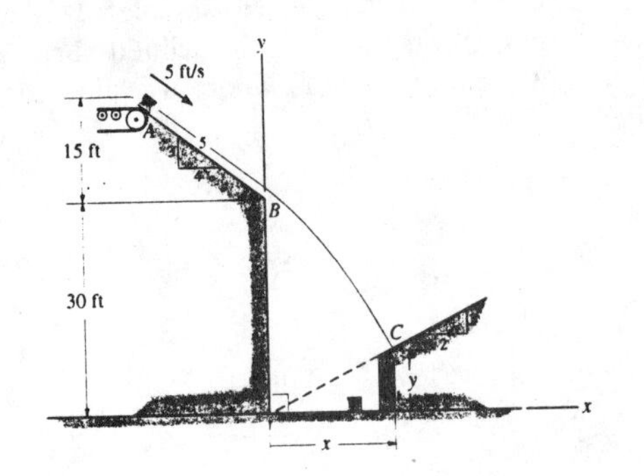

Datum at A :

$$T_A + V_A = T_C + V_C$$

$$\frac{1}{2}\left(\frac{2}{32.2}\right)(5)^2 + 0 = \frac{1}{2}\left(\frac{2}{32.2}\right)(v_C)^2 - 2[15 + (30 - 8.832)]$$

$v_C = 48.5$ ft/s **Ans**

$$+\searrow \Sigma F_{x'} = ma_{x'}; \quad 2\left(\frac{3}{5}\right) = \left(\frac{2}{32.2}\right)a_{x'}$$

$$a_{x'} = 19.32 \text{ ft/s}^2$$

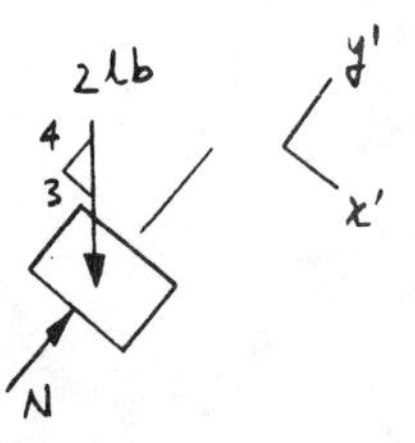

$$T_A + V_A = T_B + V_B$$

$$\frac{1}{2}\left(\frac{2}{32.2}\right)(5)^2 + 0 = \frac{1}{2}\left(\frac{2}{32.2}\right)v_B^2 - 2(15)$$

$$v_B = 31.48 \text{ ft/s}$$

$$(+\searrow) \quad v_B = v_A + a_c t$$

$$31.48 = 5 + 19.32 t_{AB}$$

$$t_{AB} = 1.371 \text{ s}$$

$$(\overset{+}{\rightarrow}) \quad s = s_0 + v_0 t$$

$$x = 0 + 31.48\left(\frac{4}{5}\right)t \qquad (1)$$

$$(+\uparrow) \quad s = s_0 + v_0 t + \frac{1}{2}a_c t^2$$

$$y = 30 - 31.48\left(\frac{3}{5}\right)t + \frac{1}{2}(-32.2)t^2$$

Equation of inclined surface :

$$\frac{y}{x} = \frac{1}{2}; \quad y = \frac{1}{2}x \qquad (2)$$

Thus

$$30 - 18.888t - 16.1t^2 = 12.592t$$

$$-16.1t^2 - 31.480t + 30 = 0$$

Solving for the positive root :

$$t = 0.7014 \text{ s}$$

Total time is

$t = 1.371 + 0.7014 = 2.07$ s **Ans**

14-89. The 2-kg ball of negligible size is fired from point A with an initial velocity of 10 m/s up the smooth inclined plane. Determine the distance from point C to where it hits the horizontal surface at D. Also, what is its velocity when it strikes the surface?

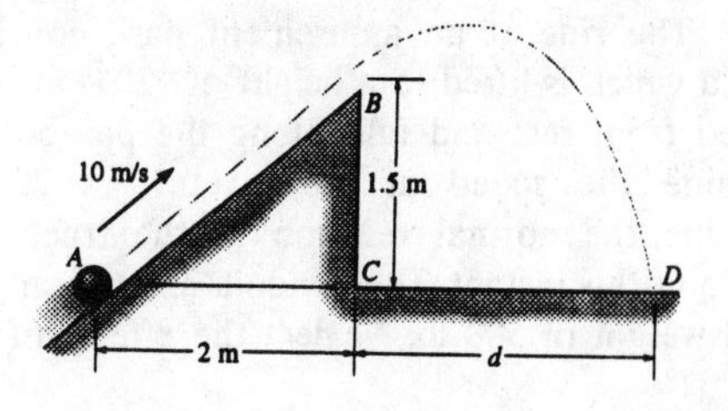

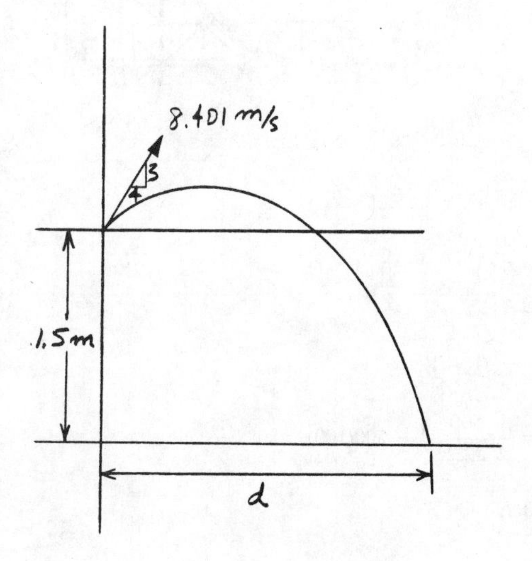

Datum at A :

$$T_A + V_A = T_B + V_B$$

$$\frac{1}{2}(2)(10)^2 + 0 = \frac{1}{2}(2)(v_B)^2 + 2(9.81)(1.5)$$

$$v_B = 8.401 \text{ m/s}$$

$(\overset{+}{\rightarrow})$ $\quad s = s_0 + v_0 t$

$$d = 0 + 8.401\left(\frac{4}{5}\right)t$$

$(+\uparrow)$ $\quad s = s_0 + v_0 t + \frac{1}{2}a_c t^2$

$$-1.5 = 0 + 8.401\left(\frac{3}{5}\right)t + \frac{1}{2}(-9.81)t^2$$

$$-4.905t^2 + 5.040t + 1.5 = 0$$

Solving for the positive root,

$$t = 1.269 \text{ s}$$

$d = 8.401\left(\frac{4}{5}\right)(1.269) = 8.53 \text{ m}$ **Ans**

Datum at A :

$$T_A + V_A = T_D + V_D$$

$$\frac{1}{2}(2)(10)^2 + 0 = \frac{1}{2}(2)(v_D)^2 + 0$$

$v_D = 10$ m/s **Ans**

14-90. The ride at an amusement park consists of a gondola which is lifted to a height of 120 ft at A. If it is released from rest and falls along the parabolic track, determine the speed at the instant $y = 20$ ft. Also determine the normal reaction of the tracks on the gondola at this instant. The gondola and passenger have a total weight of 500 lb. Neglect the effects of friction.

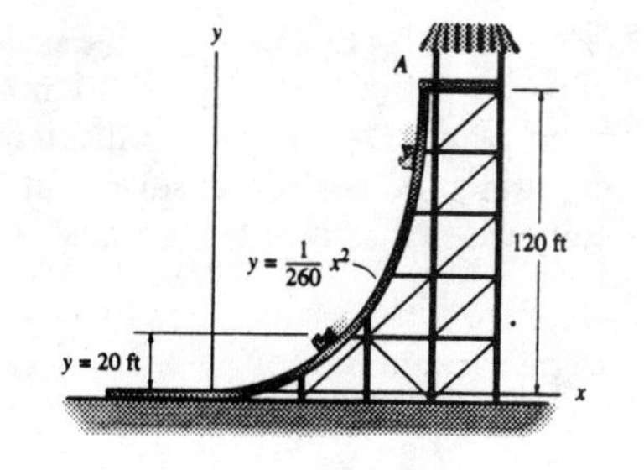

$$y = \frac{1}{260}x^2$$

$$\frac{dy}{dx} = \frac{1}{130}x$$

$$\frac{d^2y}{dx^2} = \frac{1}{130}$$

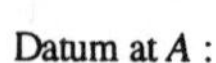

At $y = 120 - 100 = 20$ ft

$x = 72.11$ ft

$$\tan\theta = \frac{dy}{dx} = 0.555, \qquad \theta = 29.02°$$

$$\rho = \frac{[1+(0.555)^2]^{3/2}}{\frac{1}{130}} = 194.40 \text{ ft}$$

$$\nwarrow + \Sigma F_n = m\,a_n; \quad N_G - 500\cos 29.02° = \frac{500}{32.2}\left(\frac{v^2}{194.40}\right) \qquad (1)$$

Datum at A :

$$T_1 + V_1 = T_2 + V_2$$

$$0 + 0 = \frac{1}{2}\left(\frac{500}{32.2}\right)v^2 - 500(100)$$

$$v^2 = 6440$$

$v = 80.2$ ft/s **Ans**

Substituting into Eq. (1) yields

$N_G = 952$ lb **Ans**

14-91. The Raptor is an outside loop roller coaster in which riders are belted into seats resembling ski-lift chairs. Determine the minimum speed v_0 at which the cars should coast down from the top of the hill, so that passengers can just make the loop without leaving contact with their seats. Neglect friction, the size of the car and passenger, and assume each passenger and car has a mass m.

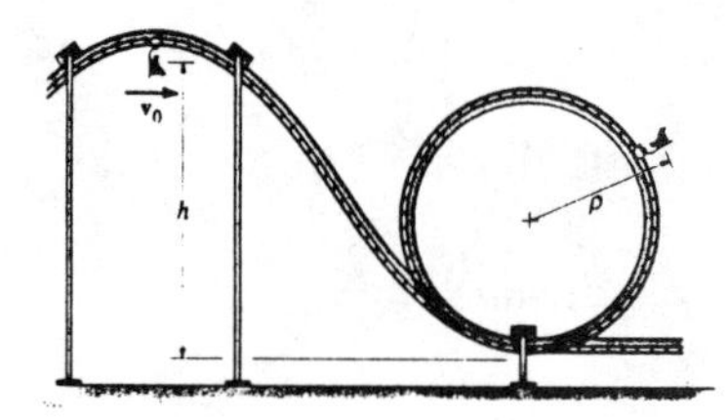

Datum at ground :

$$T_1 + V_1 = T_2 + V_2$$

$$\frac{1}{2}mv_0^2 + mgh = \frac{1}{2}mv_1^2 + mg2\rho$$

$$v_1 = \sqrt{v_0^2 + 2g(h - 2\rho)}$$

$$+\downarrow \Sigma F_n = ma_n; \qquad mg = m\left(\frac{v_1^2}{\rho}\right)$$

$$v_1 = \sqrt{g\rho}$$

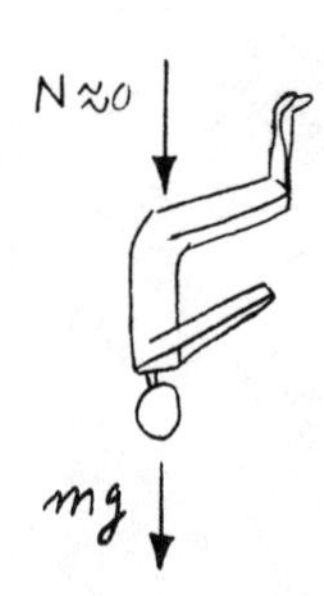

Thus,

$$g\rho = v_0^2 + 2gh - 4g\rho$$

$v_0 = \sqrt{g(5\rho - 2h)}$ **Ans**

***14-92.** The Raptor is an outside loop roller coaster in which riders are belted into seats resembling ski-lift chairs. If the cars travel at $v_0 = 4$ m/s when they are at the top of the hill, determine their speed when they are at the top of the loop and the reaction of the 70-kg passenger on his seat at this instant. The car has a mass of 50 kg. Take $h = 12$ m, $\rho = 5$ m. Neglect friction and the size of the car and passenger.

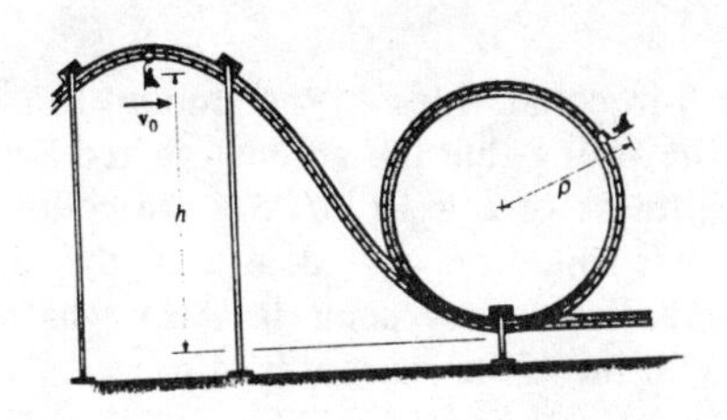

Datum at ground :

$$T_1 + V_1 = T_2 + V_2$$

$$\frac{1}{2}(120)(4)^2 + 120(9.81)(12) = \frac{1}{2}(120)(v_1)^2 + 120(9.81)(10)$$

$v_1 = 7.432$ m/s **Ans**

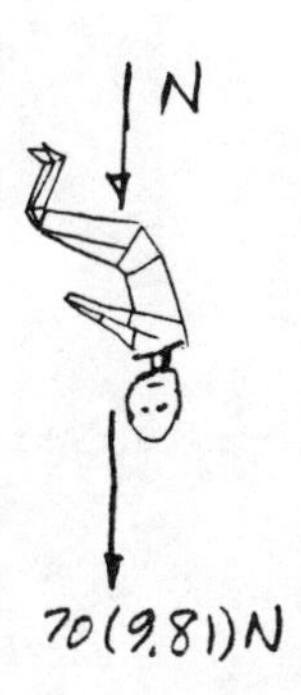

$$+\downarrow \Sigma F_n = ma_n; \quad 70(9.81) + N = 70\left(\frac{(7.432)^2}{5}\right)$$

$N = 86.7$ N **Ans**

14-93. The 2-lb collar has a speed of 5 ft/s at A. The attached spring has an unstretched length of 2 ft and a stiffness of $k = 10$ lb/ft. If the collar moves over the smooth rod, determine its speed when it reaches point B, the normal force of the rod on the collar, and the rate of decrease in its speed.

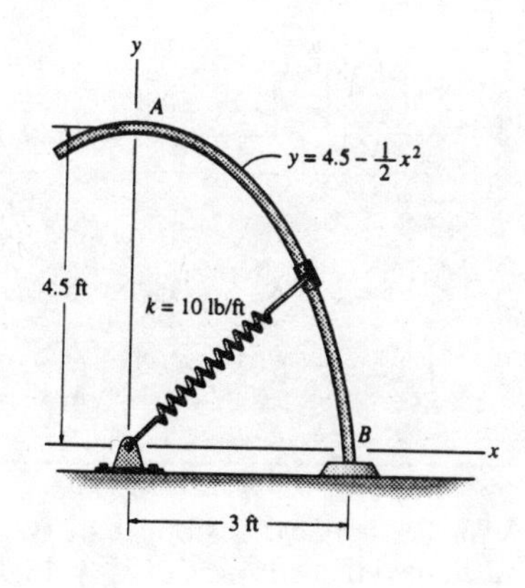

Datum at B :

$$T_A + V_A = T_B + V_B$$

$$\frac{1}{2}\left(\frac{2}{32.2}\right)(5)^2 + \frac{1}{2}(10)(4.5-2)^2 + 2(4.5) = \frac{1}{2}\left(\frac{2}{32.2}\right)(v_B)^2 + \frac{1}{2}(10)(3-2)^2 + 0$$

$v_B = 34.060$ ft/s $= 34.1$ ft/s **Ans**

$$y = 4.5 - \frac{1}{2}x^2$$

$$\frac{dy}{dx} = \tan\theta = -x\Big|_{x=3} = -3$$

$$\theta = -71.57^\circ \qquad \frac{d^2y}{dx^2} = -1$$

$$\rho = \frac{\left[1 + \left(\frac{dy}{dx}\right)^2\right]^{\frac{3}{2}}}{\left|\frac{d^2y}{dx^2}\right|} = \frac{\left[1 + (-3)^2\right]^{\frac{3}{2}}}{|-1|} = 31.623 \text{ ft}$$

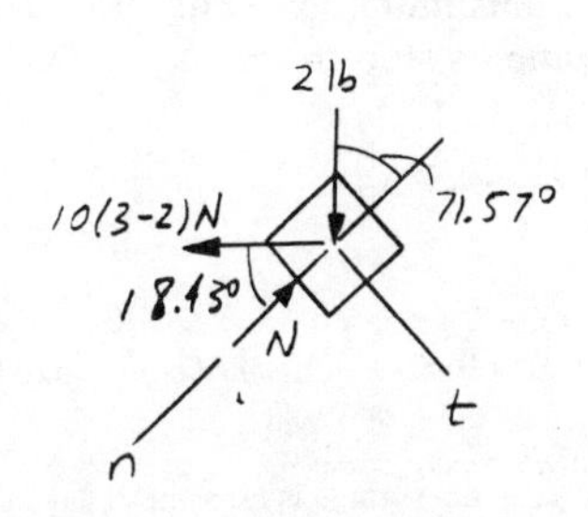

$$+\swarrow \Sigma F_n = ma_n; \quad -N + 10\cos 18.43^\circ + 2\cos 71.57^\circ = \left(\frac{2}{32.2}\right)\left(\frac{(34.060)^2}{31.623}\right)$$

$N = 7.84$ lb **Ans**

$$+\searrow \Sigma F_t = ma_t; \quad 2\sin 71.57^\circ - 10\sin 18.43^\circ = \left(\frac{2}{32.2}\right)a_t$$

$a_t = -20.4$ ft/s^2 **Ans**

14-94. The 5-lb collar rides on the smooth rod and is attached to the spring that has an unstretched length of 4 ft and a stiffness of $k = 10$ lb/ft. If the collar has a velocity of 5 ft/s when it is at A, determine the speed of the collar when it reaches point B. Also, what is the normal force on the collar when it is at point B?

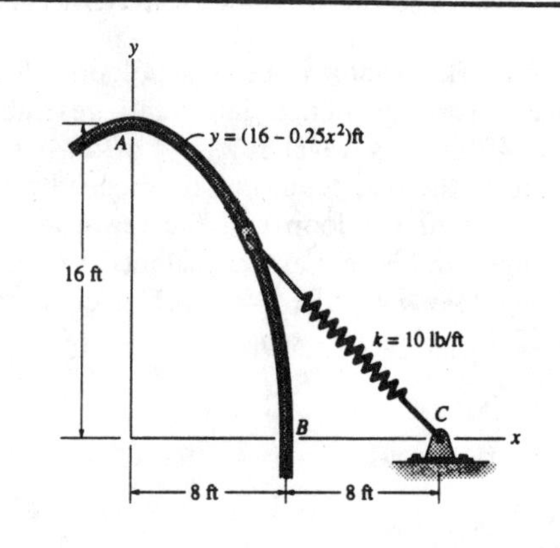

$CA = \sqrt{(16)^2 + (16)^2} = 22.627$ ft

Datum at B :

$T_A + V_A = T_B + V_B$

$$\frac{1}{2}\left(\frac{5}{32.2}\right)(5)^2 + 5(16) + \frac{1}{2}(10)(22.627 - 4)^2 = \frac{1}{2}\left(\frac{5}{32.2}\right)v_B^2 + 0 + \frac{1}{2}(10)(8-4)^2$$

$v_B = 149.56$ ft/s $= 150$ ft/s **Ans**

$y = 16 - 0.25x^2$

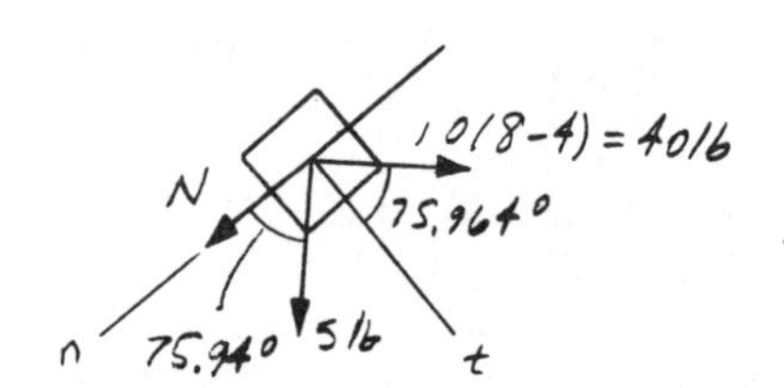

$$\frac{dy}{dx} = \tan\theta = -0.5x\bigg|_{x=8} = -4, \qquad \theta = -75.964°$$

$$\frac{d^2y}{dx^2} = -0.5$$

$$\rho = \frac{\left[1 + \left(\frac{dy}{dx}\right)^2\right]^{\frac{3}{2}}}{\left|\frac{d^2y}{dx^2}\right|} = \frac{\left[1 + (-4)^2\right]^{\frac{3}{2}}}{|-0.5|} = 140.19 \text{ ft}$$

$+\swarrow \Sigma F_n = ma_n; \qquad N + 5\cos 75.964° - 40\sin 75.964° = \left(\frac{5}{32.2}\right)\left(\frac{(149.56)^2}{140.19}\right)$

$N = 62.4$ lb **Ans**

14-95. A tank car is stopped by two spring bumpers A and B, having a stiffness of $k_A = 15(10^3)$ lb/ft and $k_B = 20(10^3)$ lb/ft, respectively. Bumper A is attached to the car, whereas bumper B is attached to the wall. If the car has a weight of $25(10^3)$ lb and is freely coasting at 3 ft/s, determine the maximum deflection of each spring at the instant the bumpers stop the car.

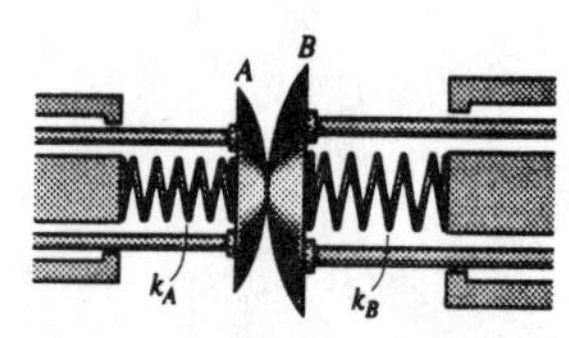

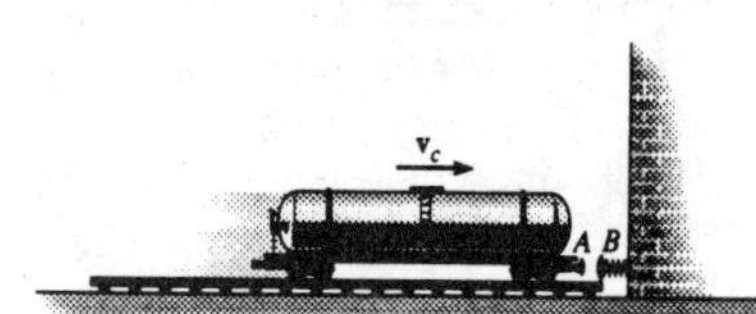

$T_1 + V_1 = T_2 + V_2$

$$\frac{1}{2}\left(\frac{25\,000}{32.2}\right)(3)^2 + 0 = 0 + \frac{1}{2}(15\,000)(x_A)^2 + \frac{1}{2}(20\,000)(x_B)^2$$

Since the force in the springs is the same at any instant,

$F = k_A x_A = k_B x_B \qquad 15\,000 x_A = 20\,000 x_B$

$x_A = 1.333 x_B$

Solving,

$x_A = 0.516$ ft **Ans**

$x_B = 0.387$ ft **Ans**

***14-96.** The block has a mass of 20 kg and is released from rest when $s = 0.5$ m. If the mass of the bumpers A and B can be neglected, determine the maximum deformation of each spring due to the collision.

Datum at initial position :

$$T_1 + V_1 = T_2 + V_2$$

$$0+0 = 0+\frac{1}{2}(500)s_A^2 + \frac{1}{2}(800)s_B^2 + 20(9.81)\left[-(s_A+s_B)-0.5\right] \qquad (1)$$

Also, $\quad F_s = 500s_A = 800s_B \qquad s_A = 1.6s_B \qquad (2)$

Solving Eqs. (1) and (2) yields :

$s_B = 0.638$ m **Ans**

$s_A = 1.02$ m **Ans**

14-97. If the mass of the earth is M_e, show that the gravitational potential energy of a body of mass m located a distance r from the center of the earth is $V_g = -GM_e m/r$. Recall that the gravitational force acting between the earth and the body is $F = G(M_e m/r^2)$, Eq. 13–1. For the calculation, locate the datum an "infinite" distance from the earth. Also, prove that **F** is a conservative force.

The work is computed by moving F from position r_1 to a farther position r_2.

$$V_g = -U = -\int F\,dr$$

$$= -G M_e m \int_{r_1}^{r_2} \frac{dr}{r^2}$$

$$= -G M_e m\left(\frac{1}{r_2} - \frac{1}{r_1}\right)$$

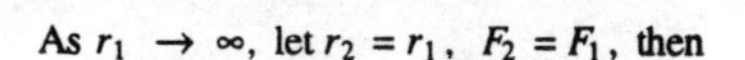

As $r_1 \to \infty$, let $r_2 = r_1$, $F_2 = F_1$, then

$$V_g \to \frac{-G M_e m}{r}$$

To be conservative, require

$$F = -\nabla V_g = -\frac{\partial}{\partial r}\left(-\frac{G M_e m}{r}\right)$$

$$= \frac{-G M_e m}{r^2} \qquad \textbf{Q.E.D.}$$

14-98. A 60-kg satellite is traveling in free flight along an elliptical orbit such that at A, where $r_A = 20$ Mm, it has a speed $v_A = 40$ Mm/h. What is the speed of the satellite when it reaches point B, where $r_B = 80$ Mm? *Hint:* See Prob. 14-97, where $M_e = 5.976(10^{24})$ kg and $G = 66.73(10^{-12})$ m³/(kg·s²).

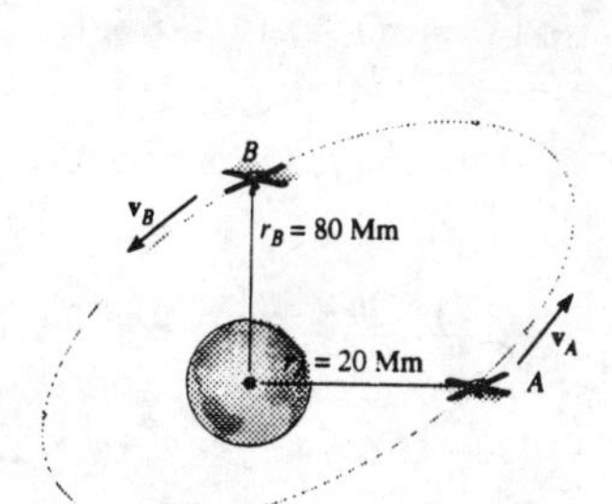

$v_A = 40$ Mm/h $= 11\,111.1$ m/s

Since $V = -\dfrac{GM_e m}{r}$

$$T_1 + V_1 = T_2 + V_2$$

$$\frac{1}{2}(60)(11\,111.1)^2 - \frac{66.73(10)^{-12}(5.976)(10)^{24}(60)}{20(10)^6} = \frac{1}{2}(60)v_B^2 - \frac{66.73(10)^{-12}(5.976)(10)^{24}(60)}{80(10)^6}$$

$v_B = 9672$ m/s $= 34.8$ Mm/h **Ans**

15-1. A 20-lb block slides down a 30° inclined plane with an initial velocity of 2 ft/s. Determine the velocity of the block in 3 s if the coefficient of kinetic friction between the block and the plane is $\mu_k = 0.25$.

$(+\nwarrow)\quad m(v_{y'})_1 + \Sigma\int_{t_1}^{t_2} F_{y'}\,dt = m(v_{y'})_2$

$0 + N(3) - 20\cos 30°(3) = 0 \qquad N = 17.32\text{ lb}$

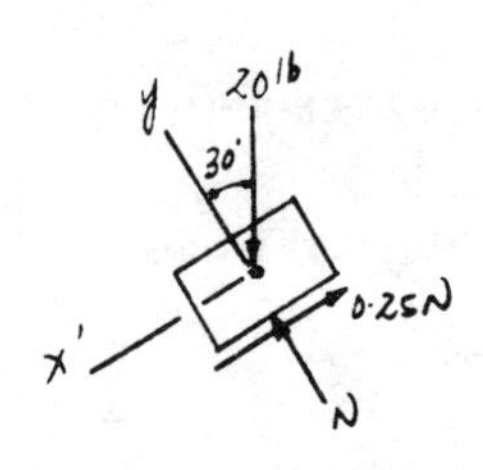

$(+\swarrow)\quad m(v_{x'})_1 + \Sigma\int_{t_1}^{t_2} F_{x'}\,dt = m(v_{x'})_2$

$$\frac{20}{32.2}(2) + 20\sin 30°(3) - 0.25(17.32)(3) = \frac{20}{32.2}v$$

$v = 29.4\text{ ft/s}$ **Ans**

15-2. The baseball has a horizontal speed of 35 m/s when it is struck by the bat B. If it then travels away at an angle of 60° from the horizontal and reaches a maximum height of 50 m, measured from the height of the bat, determine the magnitude of the net impulse of the bat on the ball. The ball has a mass of 400 g. Neglect the weight of the ball during the time the bat strikes the ball.

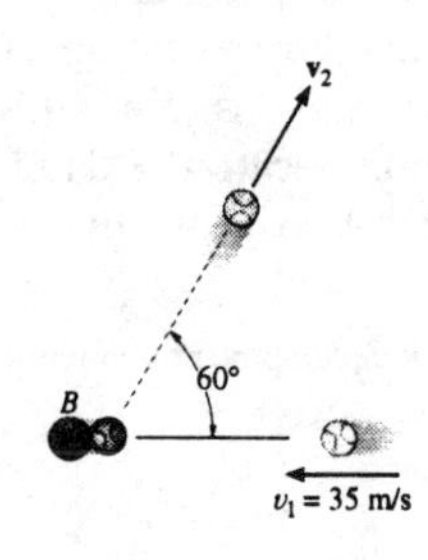

$(+\uparrow)\quad v^2 = v_0^2 + 2a_c(s - s_0)$

$0^2 = (v_2\sin 60°)^2 - 2(9.81)(50 - 0)$

$v_2 = 36.17\text{ m/s}$

$(\overset{+}{\rightarrow})\quad m(v_x)_1 + \Sigma\int F_x\,dt = m(v_x)_2$

$-0.4(35) + \int F_x\,dt = 0.4(36.17)\cos 60°$

$\int F_x\,dt = 21.23$

$(+\uparrow)\quad m(v_y)_1 + \Sigma\int F_y\,dt = m(v_y)_2$

$0 + \int F_y\,dt = 0.4(36.17)\sin 60°$

$\int F_y\,dt = 12.53$

$\int F\,dt = \sqrt{(21.23)^2 + (12.53)^2} = 24.7\text{ N}\cdot\text{s}$ **Ans**

15-3. A 5-lb block is given an initial velocity of 10 ft/s up a 45° smooth slope. Determine the time it will take to travel up the slope before it stops.

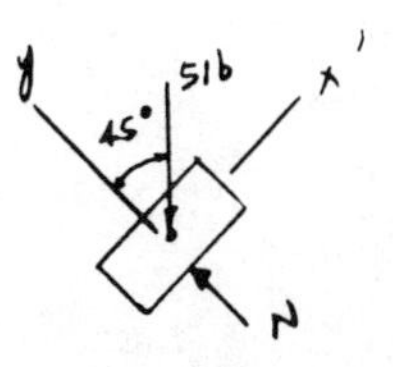

$(\nearrow +)$ $\quad m(v_{x'})_1 + \Sigma \int_{t_1}^{t_2} F_{x'}\, dt = m(v_{x'})_2$

$$\frac{5}{32.2}(10) + (-5\sin 45°)t = 0$$

$t = 0.439$ s **Ans**

***15-4.** The 12-Mg "jump jet" is capable of taking off vertically from the deck of a ship. If its jets exert a constant vertical force of 150 kN on the plane, determine its velocity and how high it goes in $t = 6$ s, starting from rest. Neglect the loss of fuel during the lift.

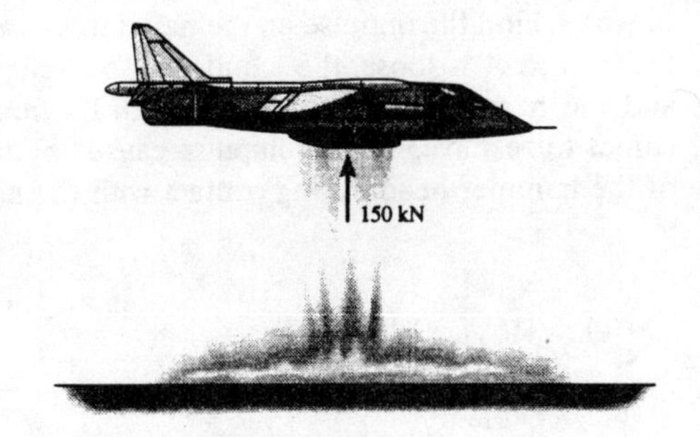

$(+\uparrow)$ $\quad m(v_y)_1 + \Sigma \int F_y\, dt = m(v_y)_2$

$$0 + 150(10^3)(6) - 12(10^3)(9.81)(6) = 12(10^3)v$$

$v = 16.14$ m/s $= 16.1$ m/s **Ans**

150(10³)

12(10³)9.81 N

$(+\uparrow)$ $\quad v = v_0 + a_c t$

$$16.14 = 0 + a(6)$$

$$a = 2.690 \text{ m/s}^2$$

$(+\uparrow)$ $\quad s = s_0 + v_0 t + \frac{1}{2}a_c t^2$

$$s = 0 + 0 + \frac{1}{2}(2.690)(6)^2$$

$s = 48.4$ m **Ans**

15-5. The choice of a seating material for moving vehicles depends upon its ability to resist shock and vibration. From the data shown in the graphs, determine the impulses created by a falling weight onto a sample of urethane foam and CONFOR foam.

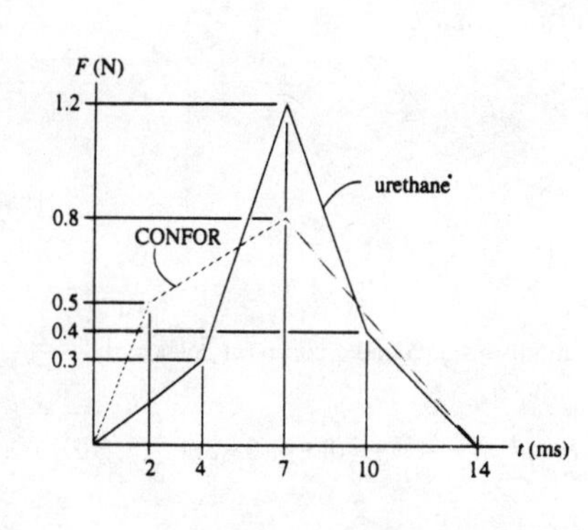

CONFOR foam :

$$I_C = \int F\, dt = \left[\frac{1}{2}(2)(0.5) + \frac{1}{2}(0.5 + 0.8)(7-2) + \frac{1}{2}(0.8)(14-7)\right](10^{-3})$$

$= 6.55$ N · ms **Ans**

Urethane foam :

$$I_U = \int F\, dt = \left[\frac{1}{2}(4)(0.3) + \frac{1}{2}(1.2 + 0.3)(7-4) + \frac{1}{2}(1.2 + 0.4)(10-7) + \frac{1}{2}(14-10)(0.4)\right](10^{-3})$$

$= 6.05$ N · ms **Ans**

15-6. The 28-Mg bulldozer is originally at rest. Determine its speed when $t = 4$ s if the horizontal traction **F** varies with time as shown in the graph.

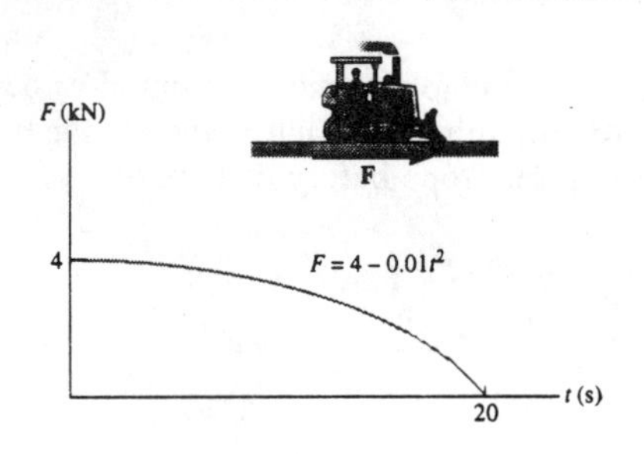

$(\overset{+}{\rightarrow})$ $\quad m(v_x)_1 + \Sigma \int_{t_1}^{t_2} F_x\, dt = m(v_x)_2$

$$0 + \int_0^4 \left(4 - 0.01t^2\right)\left(10^3\right) dt = 28\left(10^3\right) v$$

$v = 0.564$ m/s $\qquad$ **Ans**

15-7. A hammer head H having a weight of 0.25 lb is moving vertically downward at 40 ft/s when it strikes the head of a nail of negligible mass and drives it into a block of wood. Find the impulse on the nail if it is assumed that the grip at A is loose, the handle has a negligible mass, and the hammer stays in contact with the nail while it comes to rest. Neglect the impulse caused by the weight of the hammer head during contact with the nail.

$(+\downarrow)$ $\quad m(v_y)_1 + \Sigma \int F_y\, dt = m(v_y)_2$

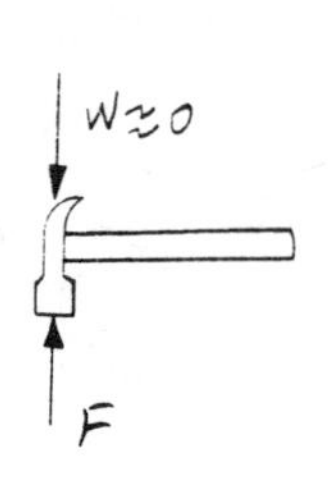

$$\left(\frac{0.25}{32.2}\right)(40) - \int F\, dt = 0$$

$\int F\, dt = 0.311$ lb · s $\qquad$ **Ans**

***15-8.** A solid-fueled rocket can be made using a fuel grain with either a hole (a), or starred cavity (b), in the cross section. From experiment the engine thrust-time curves (T vs. t) for the same amount of propellant using these geometrics are shown. Determine the total impulse in both cases.

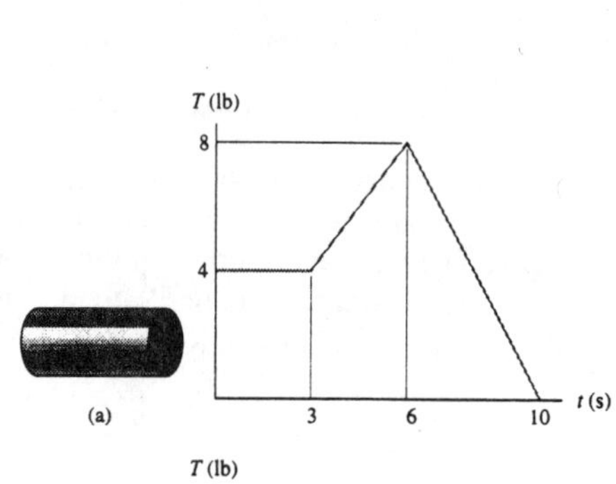

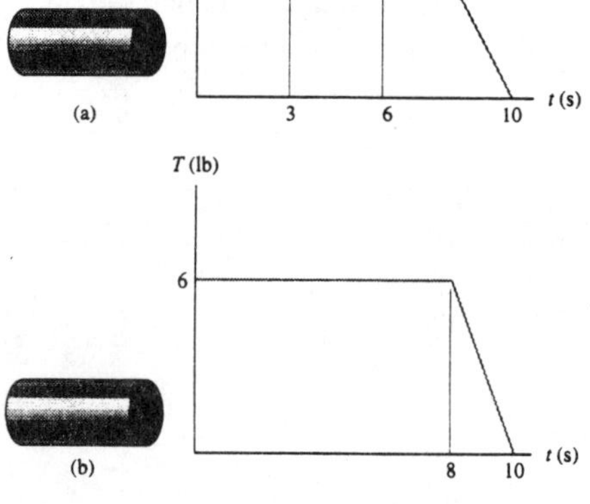

Impulse is area under curve for hole cavity.

$$I = \int F\, dt = 4(3) + \frac{1}{2}(8+4)(6-3) + \frac{1}{2}(8)(10-6)$$

$= 46$ lb · s $\qquad$ **Ans**

For starred cavity :

$$I = \int F\, dt = 6(8) + \frac{1}{2}(6)(10-8)$$

$= 54$ lb · s $\qquad$ **Ans**

15-9. The jet plane has a mass of 250 Mg and a horizontal velocity of 100 m/s when $t = 0$. If *both* engines provide a horizontal thrust which varies as shown in the graph, determine the plane's velocity in $t = 15$ s. Neglect air resistance and the loss of fuel during the motion.

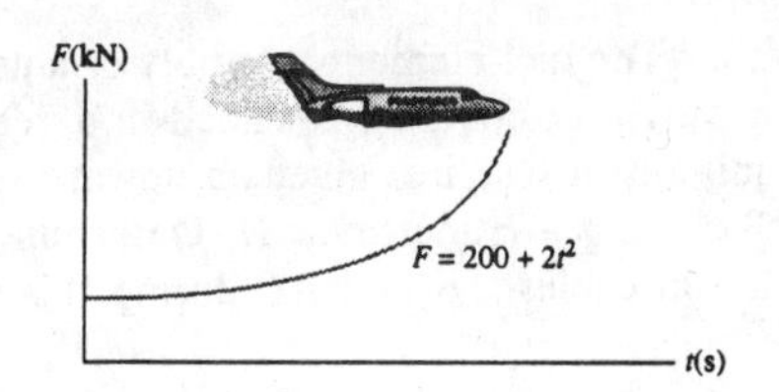

$(\overset{+}{\rightarrow})$ $\quad m(v_x)_1 + \Sigma \int_{t_1}^{t_2} F_x\, dt = m(v_x)_2$

$$250(10^3)(100) + \int_0^{15} 10^3(200 + 2t^2)\, dt = 250(10^3)\, v$$

$v = 121$ m/s **Ans**

15-10. The particle P is acted upon by its weight of 3 lb and forces $\mathbf{F}_1$ and $\mathbf{F}_2$, where t is in seconds. If the particle originally has a velocity of $\mathbf{v}_1 = \{3\mathbf{i} + 1\mathbf{j} + 6\mathbf{k}\}$ ft/s, determine its speed after 2 s.

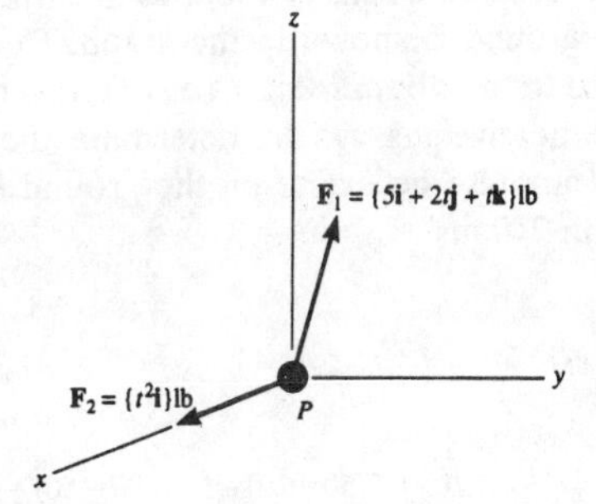

$m\mathbf{v}_1 + \Sigma \int \mathbf{F}\, dt = m\mathbf{v}_2$

$$\left(\frac{3}{32.2}\right)(3\mathbf{i} + 1\mathbf{j} + 6\mathbf{k}) + \int_0^2 t^2\mathbf{i}\, dt + \int_0^2 (5\mathbf{i} + 2t\mathbf{j} + t\mathbf{k})\, dt - \int_0^2 3\mathbf{k}\, dt = \left(\frac{3}{32.2}\right)(v_x\mathbf{i} + v_y\mathbf{j} + v_z\mathbf{k})$$

Expand and equate components :

$$\frac{9}{32.2} + \frac{1}{3}(2)^3 + 5(2) = \frac{3}{32.2}v_x$$

$v_x = 139.0$ ft/s

$$\frac{3}{32.2} + (2)^2 = \frac{3}{32.2}v_y$$

$v_y = 43.93$ ft/s

$$\frac{18}{32.2} + \frac{1}{2}(2)^2 - 6 = \frac{3}{32.2}v_z$$

$v_z = -36.93$ ft/s

$v = \sqrt{(139.0)^2 + (43.93)^2 + (-36.93)^2} = 150$ ft/s **Ans**

15-11. The 20-lb cabinet is subjected to the force $F = (3 + 2t)$ lb, where t is in seconds. If the cabinet is initially moving down the plane with a speed of 6 ft/s, determine how long it will take the force to bring the cabinet to rest. **F** always acts parallel to the plane.

$(+\nearrow)$ $\quad m(v_x)_1 + \Sigma \int F_x\, dt = m(v_x)_2$

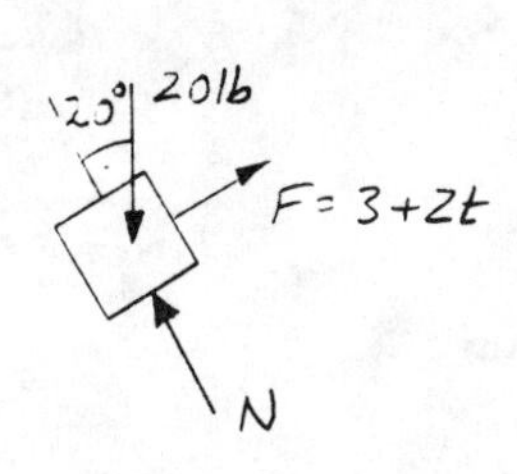

$$\left(\frac{20}{32.2}\right)(6) + 20(\sin 20°)t - \int_0^t (3 + 2t)\, dt = 0$$

$$3.727 + 3.840t - t^2 = 0$$

Solving for the positive root,

$t = 4.64$ s **Ans**

***15-12.** The fuel-element assembly of a nuclear reactor has a weight of 600 lb. Suspended in a vertical position and initially at rest, it is given an upward speed of 5 ft/s in 0.3 s using a crane hook H. Determine the average tension in cables AB and AC during this time interval.

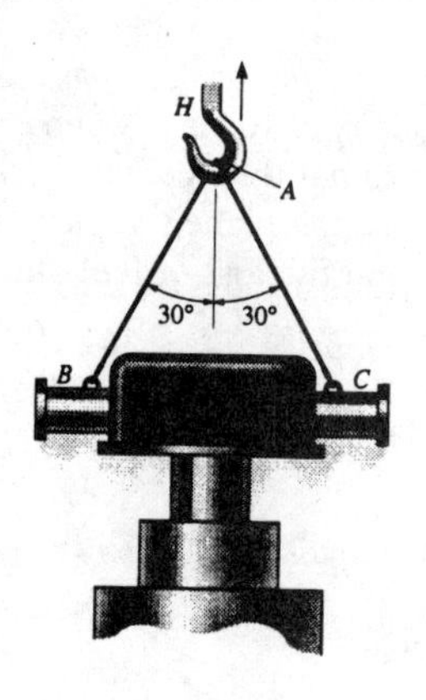

$(+\uparrow)\qquad m(v_y)_1 + \Sigma\int F_y\,dt = m(v_y)_2$

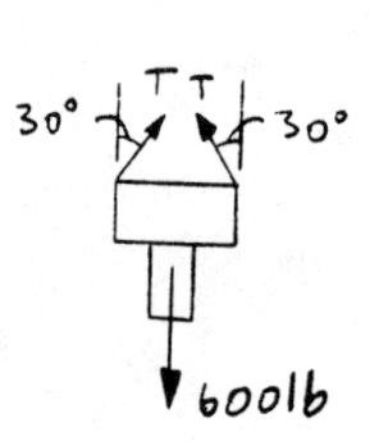

$$0 + 2(T\cos 30°)(0.3) - 600(0.3) = \left(\frac{600}{32.2}\right)(5)$$

$T = 526$ lb **Ans**

15-13. From experiments, the time variation of the vertical force on a runner's foot as he strikes and pushes off the ground is shown in the graph. These results are reported for a 1-lb *static* load, i.e., in terms of unit weight. If a runner weighs 175 lb, determine the approximate vertical impulse he exerts on the ground if the impulse occurs in 210 ms.

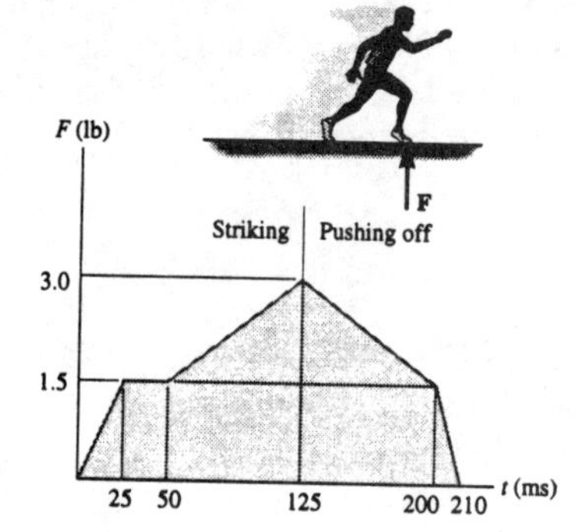

$$\int F\,dt = (175\text{ lb})(\text{area under curve})$$

$$\text{Area} = \left[\frac{1}{2}(25)(1.5) + 1.5(50-25) + 1.5(200-50) + \frac{1}{2}(210-200)(1.5) + \frac{1}{2}(3-1.5)(200-50)\right](10^{-3})$$

$$= 401.25(10^{-3})$$

$\int F\,dt = 175(401.25)(10^{-3}) = 70.2$ lb · s **Ans**

15-14. When the 0.4-lb ball is fired, it leaves the ground at an angle of 40° from the horizontal and strikes the ground at the same elevation a distance of 130 ft away. Determine the impulse given to the ball.

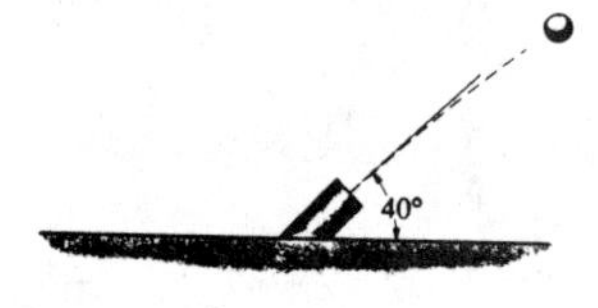

$(\overset{+}{\rightarrow})\qquad s = v_0 t$

$$130 = v_0 \cos 40°(t)$$

$(+\uparrow)\qquad v = v_0 - at$

$$-v_0 \sin 40° = v_0 \sin 40° - 32.2t$$

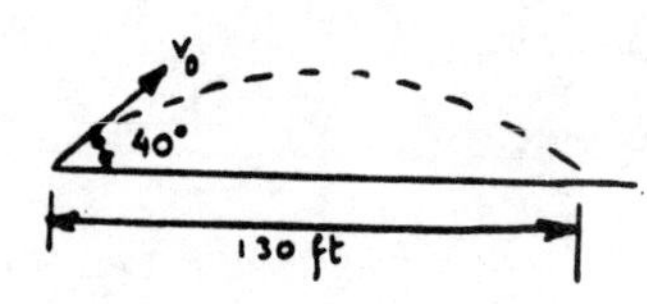

Solving,

$v_0 = 65.20$ ft/s, $t = 2.60$ s

$$m\mathbf{v}_1 + \Sigma\int \mathbf{F}\,dt = m\mathbf{v}_2$$

$0 + \Sigma\int F\,dt = \left(\frac{0.4}{32.2}\right)(65.20) = 0.810$ lb · s **Ans**

15-15. A 50-kg crate rests against a stop block s, which prevents the crate from moving down the plane. If the coefficients of static and kinetic friction between the plane and the crate are $\mu_s = 0.3$ and $\mu_k = 0.2$, respectively, determine the time needed for the force **F** to give the crate a speed of 2 m/s up the plane. The force always acts parallel to the plane and has a magnitude of $F = (300t)$ N, where t is in seconds. *Hint:* First determine the time needed to overcome static friction and start the crate moving.

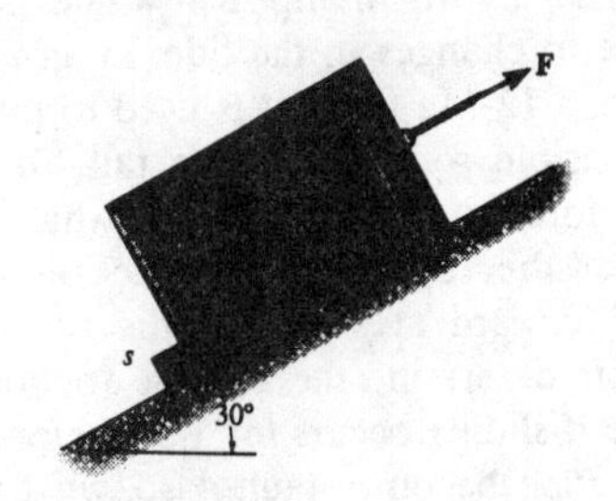

The time needed to overcome friction is determined from statics :

$+\nwarrow \Sigma F_y = 0; \quad -50(9.81)\cos 30° + N_C = 0$

$N_C = 424.79$ N

$+\nearrow \Sigma F_x = 0; \quad 300t - 0.3(424.79) - 50(9.81)\sin 30° = 0$

$t = 1.242$ s

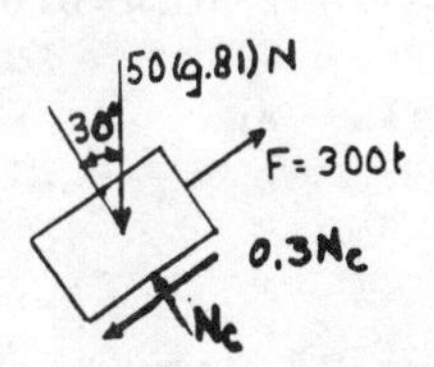

$(+\nearrow) \quad m(v_x)_1 + \Sigma \int F_x \, dt = m(v_x)_2$

$$0 + \int_{1.242}^{t} 300t \, dt - 0.2(424.79)(t - 1.242) - 50(9.81)\sin 30°(t - 1.242) = 50(2)$$

$$150\left[t^2 - (1.242)^2\right] - 84.959t + 105.541 - 245.25t + 304.67 = 100$$

$$150t^2 - 330.21t + 78.720 = 0$$

Solving for the positive root $t > 1.242$ s, yields

$t = 1.93$ s **Ans**

***15-16.** A train consists of a 50-Mg engine and three cars, each having a mass of 30 Mg. If it takes 80 s for the train to increase its speed uniformly to 40 km/h, starting from rest, determine the force T developed at the coupling between the engine E and the first car A. The wheels of the engine provide a resultant frictional tractive force **F** which gives the train forward motion, whereas the car wheels roll freely. Also, determine F acting on the engine wheels.

$(v_x)_2 = 40$ km/h $= 11.11$ m/s

Entire train :

$(\overset{+}{\rightarrow}) \quad m(v_x)_1 + \Sigma \int F_x \, dt = m(v_x)_2$

$$0 + F(80) = [50 + 3(30)](10^3)(11.11)$$

$F = 19.44$ kN **Ans**

Three cars :

$(\overset{+}{\rightarrow}) \quad m(v_x)_1 + \Sigma \int F_x \, dt = m(v_x)_2$

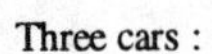

$0 + T(80) = 3(30)(10^3)(11.11) \qquad T = 12.5$ kN **Ans**

15-17. The 5.5-Mg humpback whale is stuck on the shore due to changes in the tide. In an effort to rescue the whale, a 12-Mg tugboat is used to pull it free using an inextensible rope tied to its tail. To overcome the frictional force of the sand on the whale, the tug backs up so that the rope becomes slack and then the tug proceeds forward at 3 m/s. If the tug then turns the engines off, determine the average frictional force **F** on the whale if sliding occurs for 1.5 s before the tug stops after the rope becomes taut. Also, what is the average force on the rope during the tow?

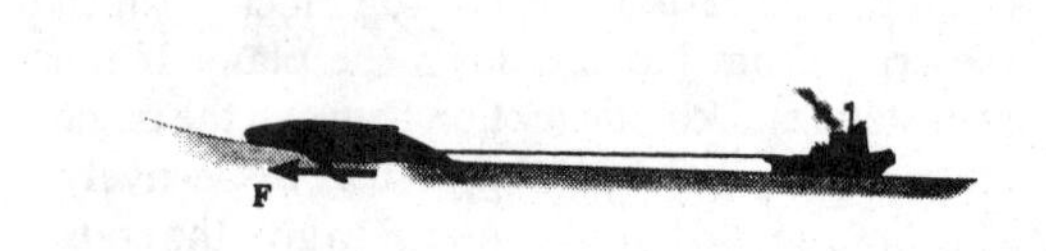

System :

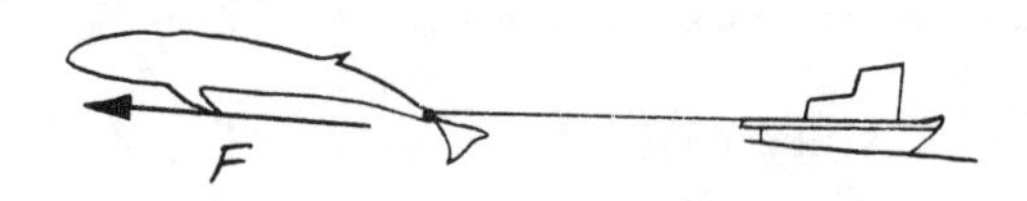

$(\overset{+}{\rightarrow})$ $\quad m_1 (v_x)_1 + \Sigma \int F_x \, dt = m_2 (v_x)_2$

$$0 + 12(10^3)(3) - F(1.5) = 0 + 0$$

$F = 24$ kN **Ans**

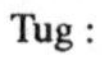

Tug :

$(\overset{+}{\rightarrow})$ $\quad m (v_x)_1 + \Sigma \int F_x \, dt = m (v_x)_2$

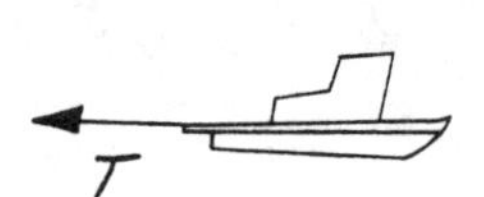

$$12(10^3)(3) - T(1.5) = 0$$

$T = 24$ kN **Ans**

15-18. The automobile has a weight of 2700 lb and is traveling forward at 4 ft/s when it crashes into the wall. If the impact occurs in 0.06 s, determine the average impulsive force acting on the car. Assume the brakes are *not applied.* If the coefficient of kinetic friction between the wheels and the pavement is $\mu_k = 0.3$, calculate the impulsive force on the wall if the brakes *were applied* during the crash. The brakes are applied to all four wheels so that all the wheels slip.

Without brakes :

$(\overset{+}{\leftarrow})$ $\quad m (v_x)_1 + \Sigma \int F_x \, dt = m (v_x)_2$

$$\left(\frac{2700}{32.2}\right)(4) - F_w(0.06) = 0$$

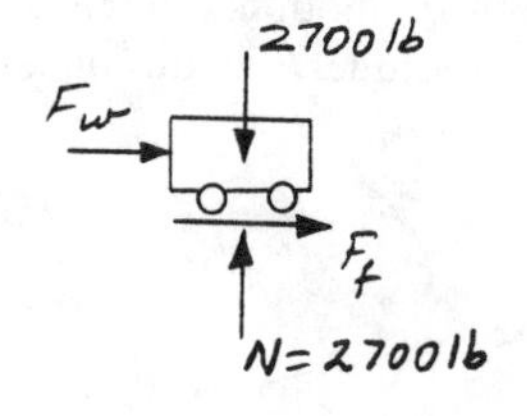

$F_w = 5590$ lb $= 5.59$ kip **Ans**

With brakes :

$F_f = 0.3(2700) = 810$ lb

$(\overset{+}{\leftarrow})$ $\quad m (v_x)_1 + \Sigma \int F_x \, dt = m (v_x)_2$

$$\left(\frac{2700}{32.2}\right)(4) - F_w(0.06) - 810(0.06) = 0$$

$F_w = 4780$ lb $= 4.78$ kip **Ans**

15-19. In case of emergency, the gas actuator is used to move a 75-kg block B by exploding a charge C near a pressurized cylinder of negligible mass. As a result of the explosion, the cylinder fractures and the released gas forces the front part of the cylinder, A, to move B forward, giving it a speed of 200 mm/s in 0.4 s. If the coefficient of kinetic friction between B and the floor is $\mu_k = 0.5$, determine the impulse that the actuator imparts to B.

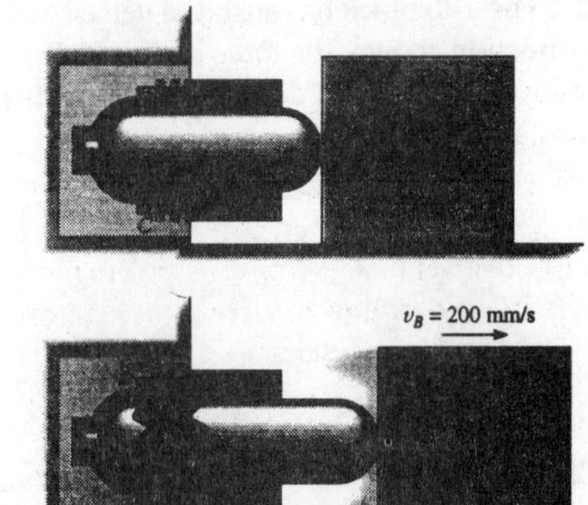

$$(\overset{+}{\rightarrow}) \quad m(v_x)_1 + \Sigma\int F_x\, dt = m(v_x)_2$$

$$0 + \int F\, dt - (0.5)(9.81)(75)(0.4) = 75(0.2)$$

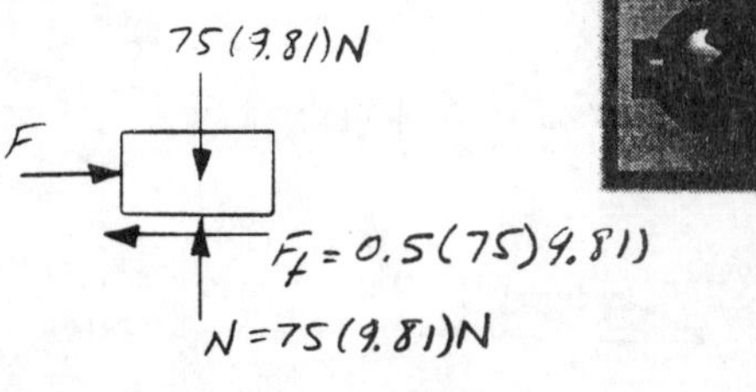

$$\int F\, dt = 162 \text{ N}\cdot\text{s} \quad \textbf{Ans}$$

***15-20.** The force acting on a projectile having a mass m as it passes horizontally through the barrel of the cannon is $F = C \sin(\pi t/t')$. Determine the projectile's velocity when $t = t'$. If the projectile reaches the end of the barrel at this instant, determine the length s.

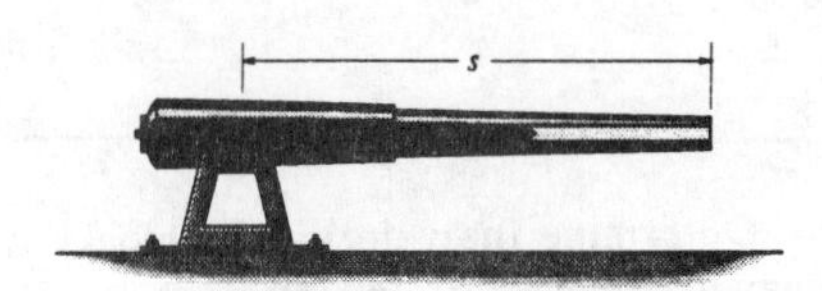

$$(\overset{+}{\rightarrow}) \quad m(v_x)_1 + \Sigma\int F_x\, dt = m(v_x)_2$$

F

$$0 + \int_0^t C\sin\left(\frac{\pi t}{t}\right)dt = mv$$

$$-C\left(\frac{t}{\pi}\right)\cos\left(\frac{\pi t}{t}\right)\Big|_0^t = mv$$

$$v = \frac{Ct}{\pi m}\left(1 - \cos\left(\frac{\pi t}{t'}\right)\right)$$

When $t = t'$,

$$v_2 = \frac{2Ct'}{\pi m} \quad \textbf{Ans}$$

$$ds = v\, dt$$

$$\int_0^s ds = \int_0^{t'}\left(\frac{Ct'}{\pi m}\right)\left(1 - \cos\left(\frac{\pi t}{t'}\right)\right) dt$$

$$s = \left(\frac{Ct'}{\pi m}\right)\left[t - \frac{t'}{\pi}\sin\left(\frac{\pi t}{t'}\right)\right]_0^{t'}$$

$$s = \frac{Ct'^2}{\pi m} \quad \textbf{Ans}$$

15-21. A 30-lb block is initially moving along a smooth horizontal surface with a speed of $v_1 = 6$ ft/s to the left. If it is acted upon by a force **F**, which varies in the manner shown, determine the velocity of the block in 15 s.

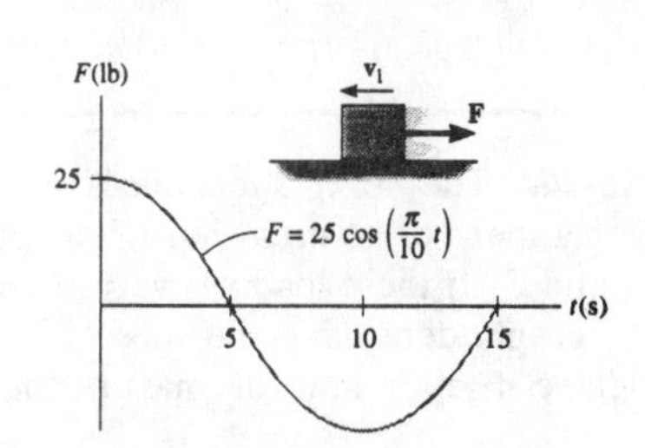

$$(\overset{+}{\rightarrow}) \quad m(v_x)_1 + \Sigma\int F_x\, dt = m(v_x)_2$$

$$-\left(\frac{30}{32.2}\right)(6) + \int_0^{15} 25\cos\left(\frac{\pi}{10}t\right) dt = \left(\frac{30}{32.2}\right)(v_x)_2$$

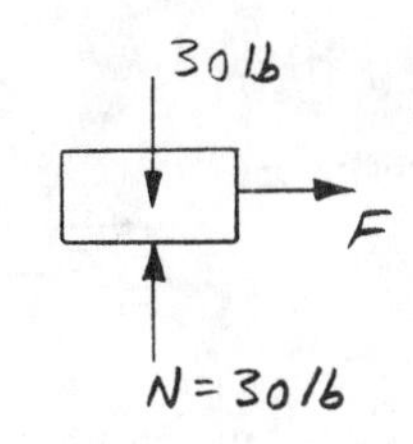

$$-5.59 + (25)\left[\sin\left(\frac{\pi}{10}t\right)\right]_0^{15}\left(\frac{10}{\pi}\right) = \left(\frac{30}{32.2}\right)(v_x)_2$$

$$-5.59 + (25)[-1]\left(\frac{10}{\pi}\right) = \left(\frac{30}{32.2}\right)(v_x)_2$$

$$(v_x)_2 = -91.4 = 91.4 \text{ ft/s} \leftarrow \quad \textbf{Ans}$$

15-22. The 2-lb block has an initial velocity of $v_1 = 10$ ft/s in the direction shown. If a force of $\mathbf{F} = \{0.5\mathbf{i} + 0.2\mathbf{j}\}$ lb acts on the block for $t = 5$ s, determine the final speed of the block. Neglect friction.

$$\mathbf{v}_1 = 10\left(-\frac{2}{\sqrt{13}}\mathbf{i} + \frac{3}{\sqrt{13}}\mathbf{j}\right) = \{-5.547\mathbf{i} + 8.321\mathbf{j}\} \text{ ft/s}$$

$$m(v_x)_1 + \Sigma\int F_x\,dt = m(v_x)_2$$

$$\left(\frac{2}{32.2}\right)(-5.547\mathbf{i} + 8.321\mathbf{j}) + (0.5\mathbf{i} + 0.2\mathbf{j})5 - 2(5)\mathbf{k} + 2(5)\mathbf{k} = \left(\frac{2}{32.2}\right)(v_x\mathbf{i} + v_y\mathbf{j} + v_z\mathbf{k})$$

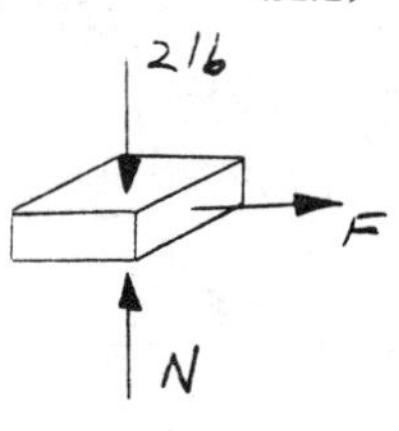

Expand and equate components :

$$-0.3445 + 2.5 = \frac{2}{32.2}v_x$$

$$v_x = 34.70 \text{ ft/s}$$

$$0.5168 + 1 = \frac{2}{32.2}v_y$$

$$v_y = 24.42 \text{ ft/s}$$

$$v_z = 0$$

Thus,

$$v = \sqrt{(34.70)^2 + (24.42)^2 + (0)^2} = 42.4 \text{ ft/s} \qquad \textbf{Ans}$$

15-23. Determine the velocities of blocks A and B 2 s after they are released from rest. Neglect the mass of the pulleys and cables.

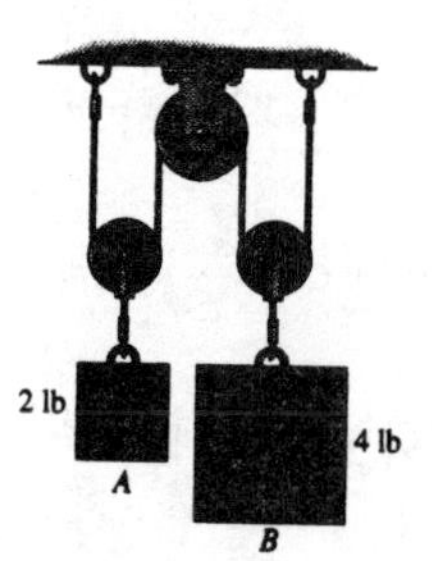

$$2s_A + 2s_B = l$$

$$v_A = -v_B$$

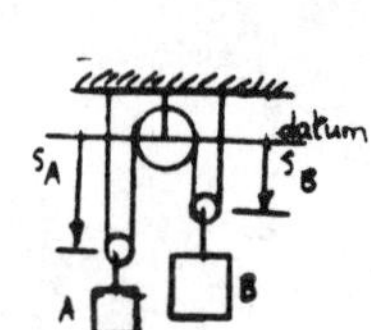

Block A :

$$(+\downarrow) \qquad m(v_y)_1 + \Sigma\int F_y\,dt = m(v_y)_2$$

$$0 - T(2) + 2(2) = \left(\frac{2}{32.2}\right)v_A$$

Block B :

$$(+\downarrow) \qquad m(v_y)_1 + \Sigma\int F_y\,dt = m(v_y)_2$$

$$0 + 4(2) - T(2) = \left(\frac{4}{32.2}\right)v_B$$

Solving,

$$T = 2.67 \text{ lb}$$

$$v_B = 21.5 \text{ ft/s} \downarrow \qquad \textbf{Ans}$$

$$v_A = -21.5 \text{ ft/s} = 21.5 \text{ ft/s} \uparrow \qquad \textbf{Ans}$$

***15-24.** The 40-kg slider block is moving to the right with a speed of 1.5 m/s when it is acted upon by the forces $\mathbf{F}_1$ and $\mathbf{F}_2$. If these loadings vary in the manner shown on the graph, determine the speed of the block at $t = 6$ s. Neglect friction and the mass of the pulleys and cords.

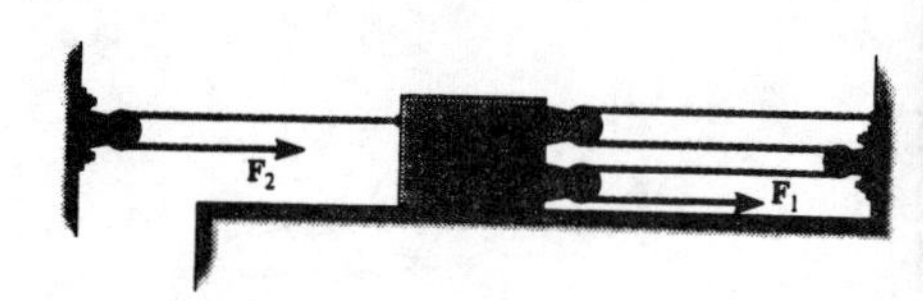

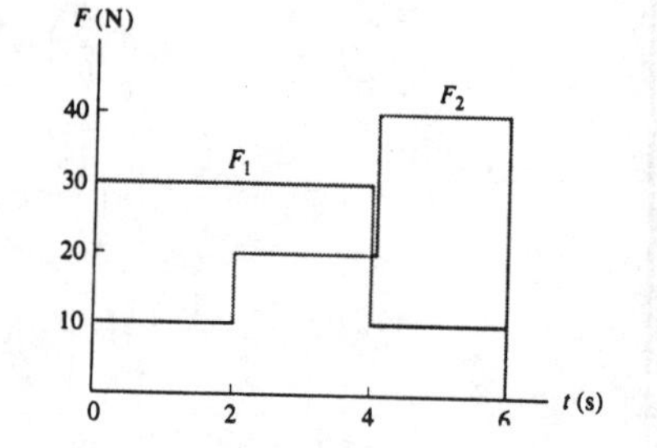

The impulses acting on the block are equal to the areas under the graph.

$$(\overset{+}{\rightarrow}) \qquad m(v_x)_1 + \Sigma\int F_x\,dt = m(v_x)_2$$

$$40(1.5) + 4[(30)4 + 10(6-4)] - [10(2) + 20(4-2) + 40(6-4)] = 40v_2$$

$$v_2 = 12.0 \text{ m/s}(\rightarrow) \qquad \textbf{Ans}$$

15-25. Block A has a mass of 3 kg and B has a mass of 5 kg. If the system is released from rest, determine the velocity of each block in $t = 4$ s. Neglect the mass of the pulleys.

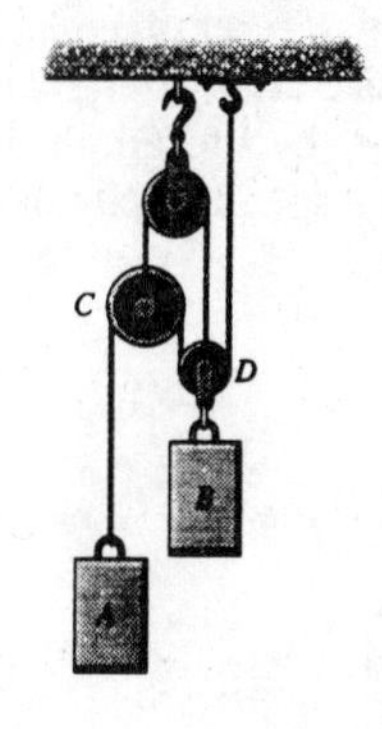

Block A :

$$(+\downarrow) \qquad m(v_y)_1 + \Sigma\int F_y\, dt = m(v_y)_2$$

$$0 + 3(9.81)(4) - T_A(4) = 3(v_A)_2 \qquad (1)$$

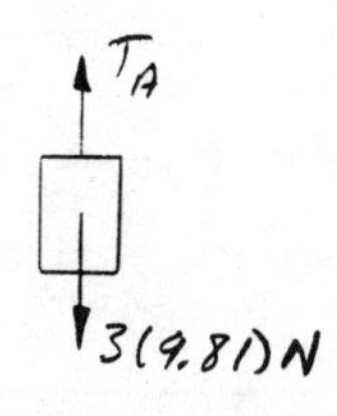

Block B :

$$(+\downarrow) \qquad m(v_y)_1 + \Sigma\int F_y\, dt = m(v_y)_2$$

$$0 + 5(9.81)(4) - T_B(4) = 5(v_B)_2 \qquad (2)$$

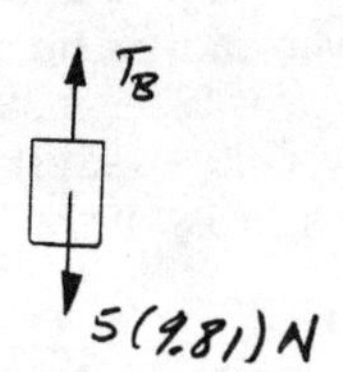

Pulley C :

$$+\uparrow \Sigma F_y = 0; \qquad T - 2T_A = 0$$

$$T = 2T_A$$

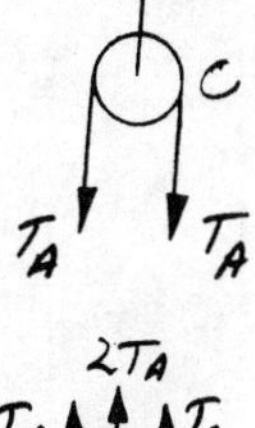

Pulley D :

$$+\uparrow \Sigma F_y = 0; \qquad 4T_A - T_B = 0$$

$$T_B = 4T_A \qquad (3)$$

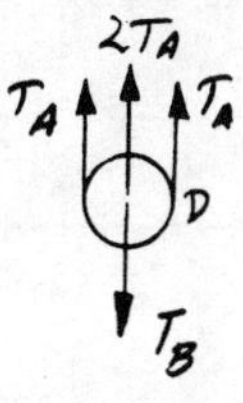

$$(s_A - s_C) + (s_B - s_C) + s_B = l$$

$$v_A = 2v_C - 2v_B$$

$$s_C + s_B = l'$$

$$v_C = -v_B$$

Thus,

$$v_A = -4v_B \qquad (4)$$

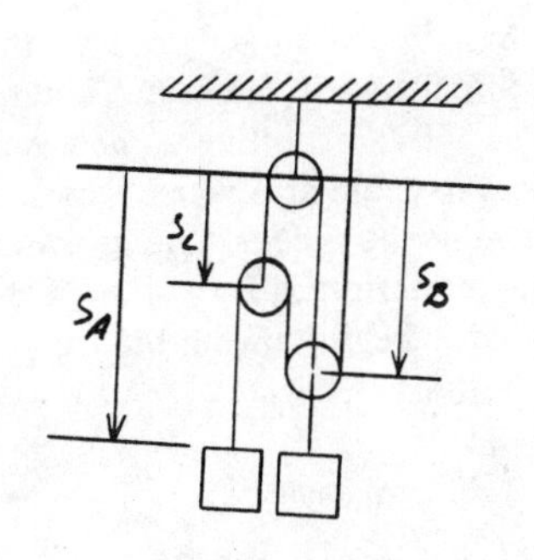

Solving Eqs. (1) – (4) :

$(v_A)_2 = 20.7$ m/s $\downarrow$ **Ans**

$(v_B)_2 = -5.18$ m/s $= 5.18$ m/s $\uparrow$ **Ans**

$T_A = 13.9$ N $\qquad T_B = 55.5$ N

15-26. A jet plane having a mass of 7 Mg takes off from an aircraft carrier such that the engine thrust varies as shown by the graph. If the carrier is traveling forward with a speed of 40 km/h, determine the plane's airspeed after 5 s.

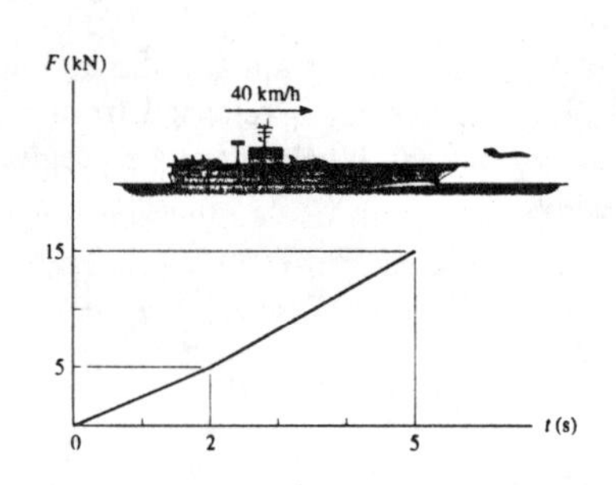

The impulse exerted on the plane is equal to the area under the graph.

$v_1 = 40 \text{ km/h} = 11.11 \text{ m/s}$

$(\overset{+}{\rightarrow}) \qquad m(v_x)_1 + \Sigma\int F_x\, dt = m(v_x)_2$

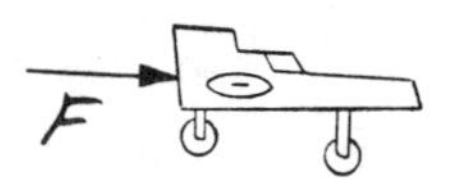

$$(7)(10^3)(11.11) + \frac{1}{2}(2)(5)(10^3) + \frac{1}{2}(15+5)(5-2)(10^3) = 7(10^3)v_2$$

$v_2 = 16.1 \text{ m/s}$ **Ans**

15-27. The 50-kg block is hoisted up the incline using the cable and motor arrangement shown. The coefficient of kinetic friction between the block and the surface is $\mu_k = 0.4$. If the block is initially moving up the plane at $v_0 = 2$ m/s, and at this instant ($t = 0$) the motor develops a tension in the cord of $T = (300 + 120\sqrt{t})$ N, where t is in seconds, determine the velocity of the block when $t = 2$ s.

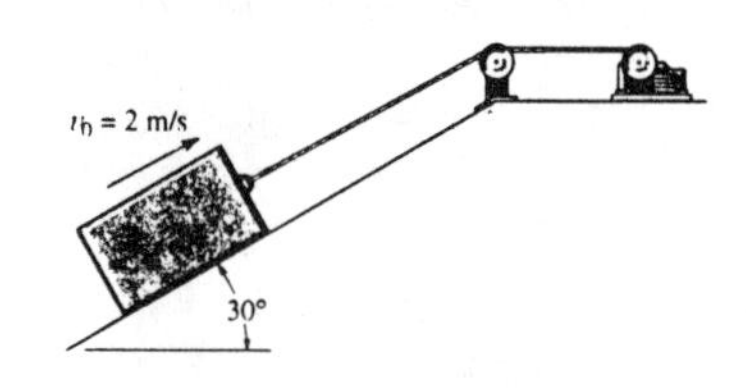

$+\nwarrow \Sigma F_y = 0; \qquad N_B - 50(9.81)\cos 30° = 0 \qquad N_B = 424.79 \text{ N}$

$(+\nearrow) \qquad m(v_x)_1 + \Sigma\int F_x\, dt = m(v_x)_2$

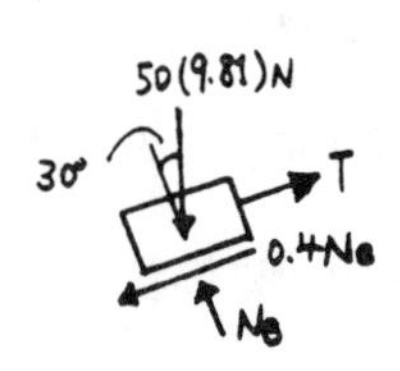

$$50(2) + \int_0^2 (300 + 120\sqrt{t})\,dt - 0.4(424.79)(2) - 50(9.81)\sin 30°(2) = 50v_2$$

$v_2 = 1.92 \text{ m/s}$ **Ans**

***15-28.** The log has a mass of 500 kg and rests on the ground for which the coefficients of static and kinetic friction are $\mu_s = 0.5$ and $\mu_k = 0.4$, respectively. The winch delivers a towing force **T** to its cable at *A* which varies as shown in the graph. Determine the speed of the log when $t = 5$ s. Originally the cable tension is zero. *Hint:* First determine the force needed to begin moving the log.

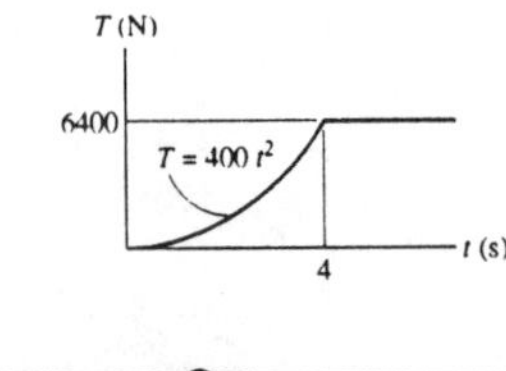

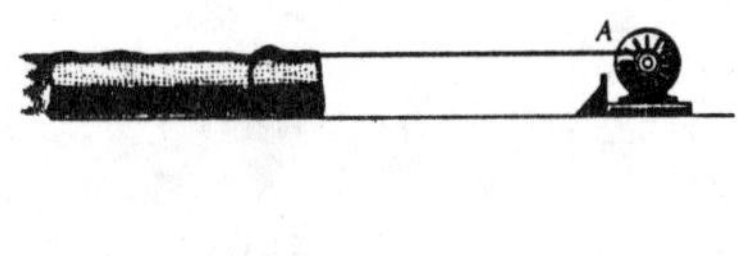

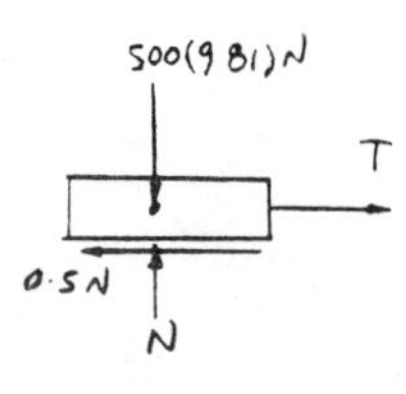

$+\uparrow \Sigma F_y = 0; \qquad N - 500(9.81) = 0 \qquad N = 4905 \text{ N}$

$\overset{+}{\rightarrow} \Sigma F_x = 0; \qquad T - 0.5(4905) = 0 \qquad T = 2452.5 \text{ N}$

Tension force needed to move the log is $T = 2452.5 = 400t^2$, therefore, the log moves after $t = 2.476 \text{ s} < 4 \text{ s}$.

$(\overset{+}{\rightarrow}) \qquad m(v_x)_1 + \Sigma\int_{t_1}^{t_2} F_x\, dt = m(v_x)_2$

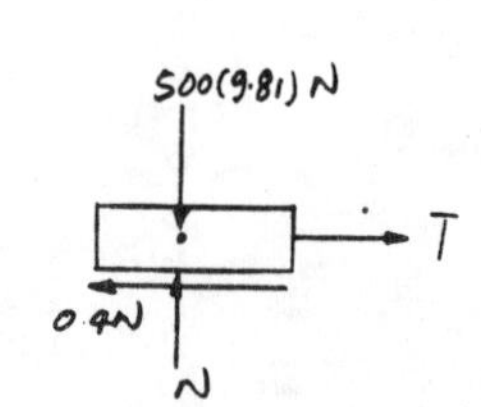

$$0 + \int_{2.476}^{4} 400t^2\,dt + 6400(5-4) - 0.4(4905)(5-2.476) = 500v$$

$v = 15.9 \text{ m/s}$ **Ans**

15-29. The log has a mass of 500 kg and rests on the ground for which the coefficients of static and kinetic friction are $\mu_s = 0.5$ and $\mu_k = 0.4$, respectively. The winch delivers a towing force **T** to its cable at A which varies as shown in the graph. How much time is required to give the log a speed of 15 m/s? Originally the cable tension is zero. *Hint:* First determine the force needed to begin moving the log.

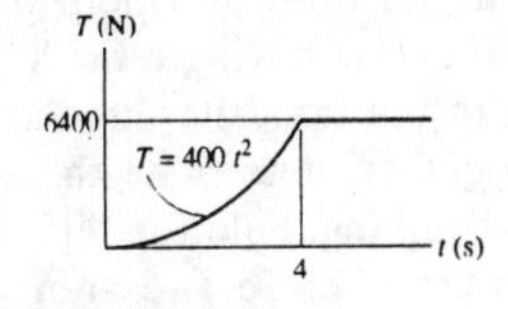

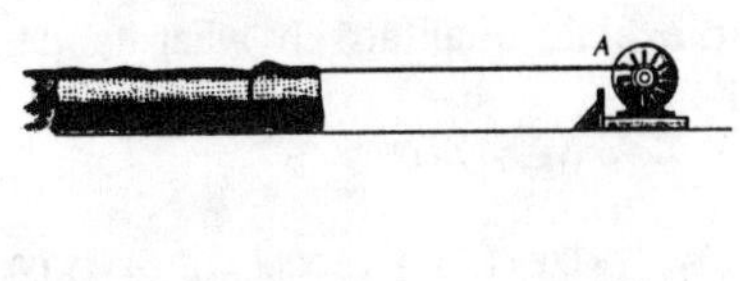

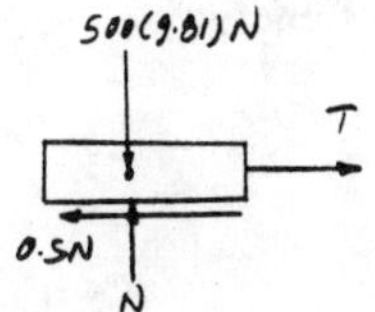

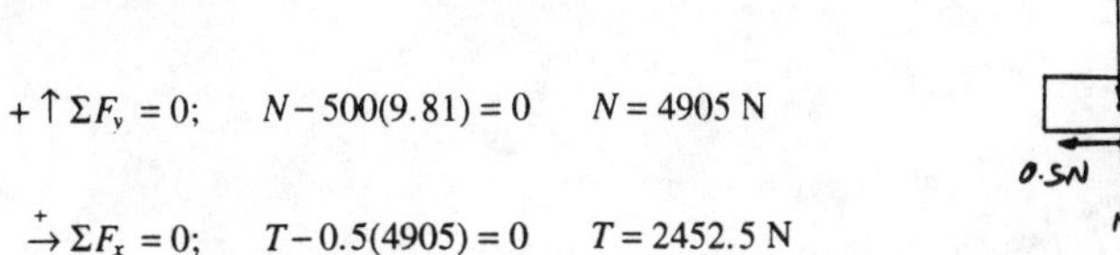

$+\uparrow \Sigma F_y = 0; \quad N - 500(9.81) = 0 \quad N = 4905 \text{ N}$

$\xrightarrow{+} \Sigma F_x = 0; \quad T - 0.5(4905) = 0 \quad T = 2452.5 \text{ N}$

Tension force needed to move the log is $T = 2452.5 = 400t^2$, therefore, the log moves after $t = 2.476 \text{ s} < 4 \text{ s}$.

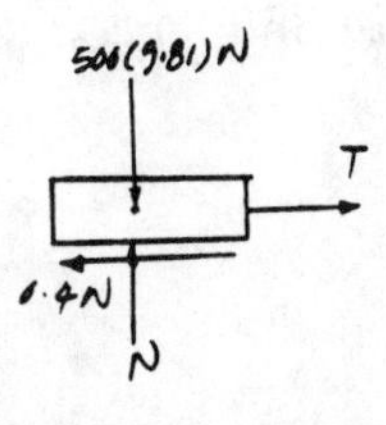

$(\xrightarrow{+}) \quad m(v_x)_1 + \Sigma \int_{t_1}^{t_2} F_x \, dt = m(v_x)_2$

$$0 + \int_{2.476}^{4} 400t^2 \, dt + 6400(t-4) - 0.4(4905)(t - 2.476) = 500(15)$$

$t = 4.90 \text{ s}$ **Ans**

15-30. The motor pulls on the cable at A with a force $F = (30 + t^2)$ lb, where t is in seconds. If the 17-lb crate is originally at rest at $t = 0$, determine its speed in $t = 4$ s. Neglect the mass of the cable and pulleys. *Hint:* First find the time needed to begin lifting the crate.

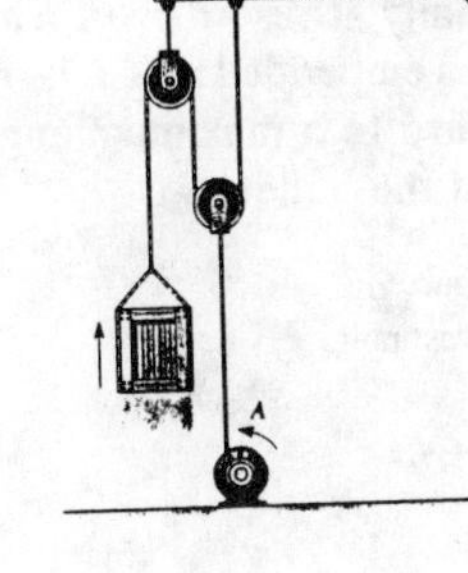

Time to begin lifting crate : $\frac{1}{2}(30 + t^2) = 17$

$t = 2 \text{ s}$

$(+\uparrow) \quad m(v_y)_1 + \Sigma \int F_y \, dt = m(v_y)_2$

$$0 + \frac{1}{2}\int_2^4 (30 + t^2) \, dt - 17(2) = \left(\frac{17}{32.2}\right) v_2$$

$$\frac{1}{2}\left[30t + \frac{1}{3}t^3\right]_2^4 - 34 = \left(\frac{17}{32.2}\right) v_2$$

F/2

17 lb

$v_2 = 10.1 \text{ ft/s}$ **Ans**

15-31. As a 4-lb sphere falls vertically from rest through a liquid, the drag force exerted on it is $F_D = (0.06v)$ lb, where v is the speed measured in ft/s. Using the differential form of the impulse and momentum principle ($\Sigma F \, dt = m \, dv$), determine the time required for the sphere to attain one fourth of its terminal velocity (the terminal velocity occurs when $t \rightarrow \infty$).

$+\downarrow \Sigma F \, dt = m \, dv$

$F_D = 0.06v$

4 lb

$$(4 - 0.06v) \, dt = \left(\frac{4}{32.2}\right) dv$$

$$\int_0^t dt = 0.1242 \int_0^v \frac{dv}{4 - 0.06v}$$

$$t = -\left(\frac{0.1242}{0.06}\right) \ln(4 - 0.06v) \Big|_0^v$$

$$t = -2.07 \ln\left(\frac{4 - 0.06v}{4}\right) \quad (1)$$

$$-0.483t = \ln\left(\frac{4 - 0.06v}{4}\right)$$

$$e^{-0.483t} = \left(\frac{4 - 0.06v}{4}\right)$$

$$v = 66.67\left(1 - e^{-0.483t}\right)$$

As $t \rightarrow \infty, \quad v \rightarrow v_t = 66.67 \text{ ft/s}$

$$\frac{1}{4} v_t = 16.67$$

Using Eq. (1),

$$t = -2.07 \ln\left(\frac{4 - 0.06(16.67)}{4}\right)$$

$= 0.596 \text{ s}$ **Ans**

***15-32.** A railroad car having a mass of 15 Mg is coasting at 1.5 m/s on a horizontal track. At the same time another car having a mass of 12 Mg is coasting at 0.75 m/s in the opposite direction. If the cars meet and couple together, determine the speed of both cars just after the coupling. Find the difference between the total kinetic energy before and after coupling has occurred, and explain qualitatively what happened to this energy.

$(\overset{+}{\rightarrow})\quad \Sigma m v_1 = \Sigma m v_2$

$$15\,000(1.5) - 12\,000(0.75) = 27\,000(v_2)$$

$v_2 = 0.5 \text{ m/s}$ **Ans**

$$T_1 = \frac{1}{2}(15\,000)(1.5)^2 + \frac{1}{2}(12\,000)(0.75)^2 = 20.25 \text{ kJ}$$

$$T_2 = \frac{1}{2}(27\,000)(0.5)^2 = 3.375 \text{ kJ}$$

$$\Delta T = T_2 - T_1$$

$$= 3.375 - 20.25 = -16.9 \text{ kJ}$$ **Ans**

This energy is dissipated as noise, shock, and heat during the coupling.

15-33. A ballistic pendulum consists of a 4-kg wooden block originally at rest, $\theta = 0°$. When a 2-g bullet strikes and becomes embedded in it, it is observed that the block swings upward to a maximum angle of $\theta = 6°$. Estimate the speed of the bullet.

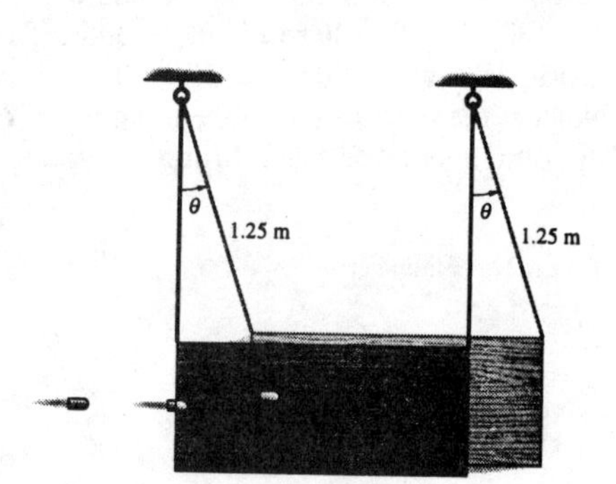

Just after impact :
Datum at lowest point.

$$T_2 + V_2 = T_3 + V_3$$

$$\frac{1}{2}(4+0.002)(v_B)_2^2 + 0 = 0 + (4+0.002)(9.81)(1.25)(1-\cos 6°)$$

$(v_B)_2 = 0.3665 \text{ m/s}$

For the system of bullet and block :

$(\overset{+}{\rightarrow})\quad \Sigma m v_1 = \Sigma m v_2$

$$0.002(v_B)_1 = (4+0.002)(0.3665)$$

$(v_B)_1 = 733 \text{ m/s}$ **Ans**

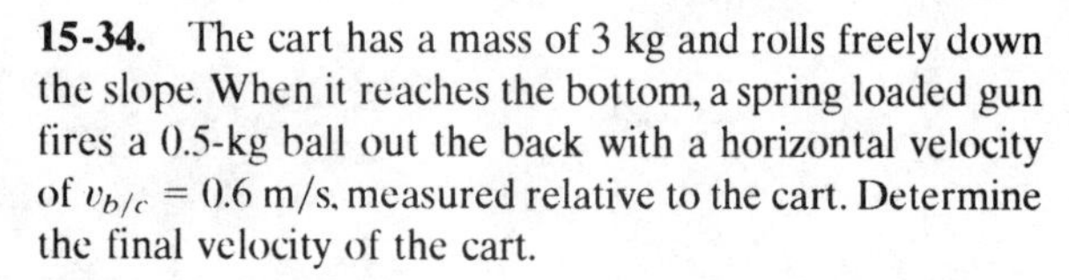
15-34. The cart has a mass of 3 kg and rolls freely down the slope. When it reaches the bottom, a spring loaded gun fires a 0.5-kg ball out the back with a horizontal velocity of $v_{b/c} = 0.6$ m/s, measured relative to the cart. Determine the final velocity of the cart.

Datum at B :

$$T_A + V_A = T_B + V_B$$

$$0 + (3+0.5)(9.81)(1.25) = \frac{1}{2}(3+0.5)(v_B)_2^2 + 0$$

$v_B = 4.952 \text{ m/s}$

$(\overset{+}{\leftarrow})\quad \Sigma m v_1 = \Sigma m v_2$

$$(3+0.5)(4.952) = (3)v_c - (0.5)v_b \qquad (1)$$

$(\overset{+}{\leftarrow})\quad v_b = v_c + v_{b/c}$

$$-v_b = v_c - 0.6 \qquad (2)$$

Solving Eqs. (1) and (2),

$v_c = 5.04 \text{ m/s} \leftarrow$ **Ans**

$v_b = -4.44 \text{ m/s} = 4.44 \text{ m/s} \leftarrow$

15-35. The two blocks A and B each have a mass of 5 kg and are suspended from parallel cords. A spring, having a stiffness of $k = 60$ N/m, is attached to B and is compressed 0.3 m against A and B as shown. Determine the maximum angles θ and ϕ of the cords when the blocks are released from rest and the spring becomes unstretched.

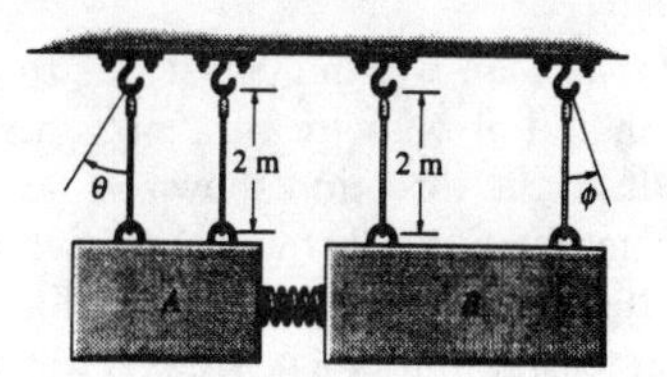

$(\overset{+}{\rightarrow})\quad \Sigma m v_1 = \Sigma m v_2$

$0 + 0 = -5v_A + 5v_B$

$v_A = v_B = v$

Just before the blocks begin to rise :

$T_1 + V_1 = T_2 + V_2$

$(0+0) + \frac{1}{2}(60)(0.3)^2 = \frac{1}{2}(5)(v)^2 + \frac{1}{2}(5)(v)^2 + 0$

$v = 0.7348$ m/s

For A or B :
Datum at lowest point.

$T_1 + V_1 = T_2 + V_2$

$\frac{1}{2}(5)(0.7348)^2 + 0 = 0 + 5(9.81)(2)(1 - \cos\theta)$

$\theta = \phi = 9.52°$ **Ans**

***15-36.** Block A has a mass of 4 kg and B has a mass of 6 kg. A spring, having a stiffness of $k = 40$ N/m, is attached to B and is compressed 0.3 m against A and B as shown. Determine the maximum angles θ and ϕ of the cords after the blocks are released from rest and the spring becomes unstretched.

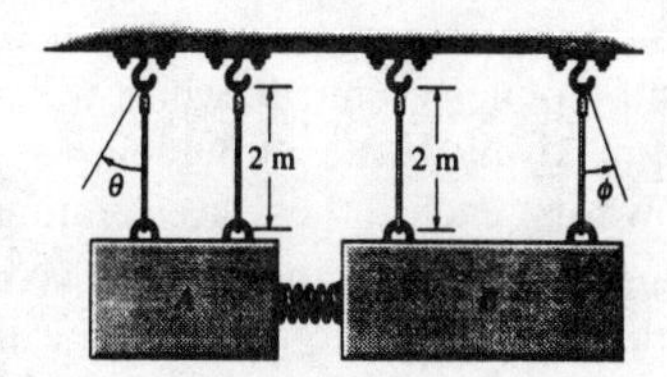

$(\overset{+}{\rightarrow})\quad \Sigma m_1 v_1 = \Sigma m_2 v_2$

$0 + 0 = 6v_B - 4v_A$

$v_A = 1.5v_B$

Just before the blocks begin to rise :

$T_1 + V_1 = T_2 + V_2$

$(0+0) + \frac{1}{2}(40)(0.3)^2 = \frac{1}{2}(4)(v_A)^2 + \frac{1}{2}(6)(v_B)^2 + 0$

$3.6 = 4v_A^2 + 6v_B^2$

$3.6 = 4(1.5v_B)^2 + 6v_B^2$

$v_B = 0.4899$ m/s $\quad v_A = 0.7348$ m/s

For A :
Datum at lowest point.

$T_1 + V_1 = T_2 + V_2$

$\frac{1}{2}(4)(0.7348)^2 + 0 = 0 + 4(9.81)(2)(1 - \cos\theta)$

$\theta = 9.52°$ **Ans**

For B :
Datum at lowest point.

$T_1 + V_1 = T_2 + V_2$

$\frac{1}{2}(6)(0.4899)^2 + 0 = 0 + 6(9.81)(2)(1 - \cos\phi)$

$\phi = 6.34°$ **Ans**

15-37. A man wearing ice skates throws an 8-kg block with an initial velocity of 2 m/s, measured relative to himself, in the direction shown. If he is originally at rest and completes the throw in 1.5 s while keeping his legs rigid, determine the horizontal velocity of the man just after releasing the block. What is the vertical reaction of both his skates on the ice during the throw? The man has a mass of 70 kg. Neglect friction and the motion of his arms.

$(\overset{+}{\rightarrow})$ $\quad 0 = -m_M\, v_M + m_B\,(v_B)_x \qquad (1)$

However, $\mathbf{v}_B = \mathbf{v}_M + \mathbf{v}_{B/M}$

$(\overset{+}{\rightarrow})$ $\quad (v_B)_x = -v_M + 2\cos 30° \qquad (2)$

$(+\uparrow)$ $\quad (v_B)_y = 0 + 2\sin 30° = 1 \text{ m/s}$

Substituting Eq. (2) into (1) yields :

$$0 = -m_M\, v_M + m_B\,(-v_M + 2\cos 30°)$$

$$v_M = \frac{2m_B\cos 30°}{m_B + m_M} = \frac{2(8)\cos 30°}{8+70} = 0.178 \text{ m/s} \qquad \textbf{Ans}$$

For the block

$(+\uparrow)$ $\quad m(v_y)_1 + \Sigma\int_{t_1}^{t_2} F_y\,dt = m(v_y)_2$

$$0 + F_y(1.5) - 8(9.81)(1.5) = 8(2\sin 30°) \qquad F_y = 83.81 \text{ N}$$

For the man

$(+\uparrow)$ $\quad m(v_y)_1 + \Sigma\int_{t_1}^{t_2} F_y\,dt = m(v_y)_2$

$$0 + N(1.5) - 70(9.81)(1.5) - 83.81(1.5) = 0$$

$$N = 771 \text{ N} \qquad \textbf{Ans}$$

15-38. The barge weighs 45 000 lb and supports two automobiles A and B, which weigh 4000 lb and 3000 lb, respectively. If the automobiles start from rest and drive towards each other, accelerating at $a_A = 4\ \text{ft/s}^2$ and $a_B = 8\ \text{ft/s}^2$ until they reach a constant speed of 6 ft/s relative to the barge, determine the speed of the barge just before the automobiles collide. How much time does this take? Originally the barge is at rest. Neglect water resistance.

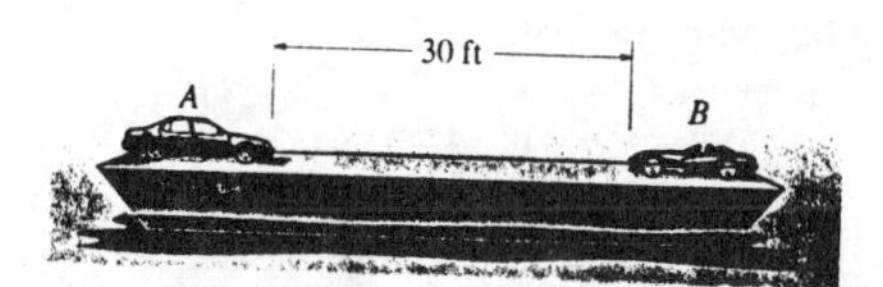

$(\overset{+}{\leftarrow})$ $\quad v_A = v_C + v_{A/C} = v_C - 6$

$(\overset{+}{\leftarrow})$ $\quad v_B = v_C + v_{B/C} = v_C + 6$

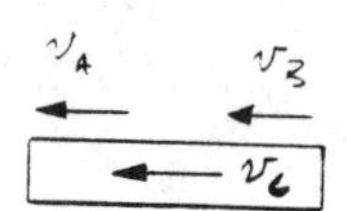

$(\overset{+}{\leftarrow})$ $\quad \Sigma m v_1 = \Sigma m v_2$

$$0 = m_A\,(v_C - 6) + m_B\,(v_C + 6) + m_C v_C$$

$$0 = \left(\frac{4000}{32.2}\right)(v_C - 6) + \left(\frac{3000}{32.2}\right)(v_C + 6) + \left(\frac{45\,000}{32.2}\right)v_C$$

$$v_C = 0.1154 \text{ ft/s} = 0.115 \text{ ft/s} \qquad \textbf{Ans}$$

For A :

$(\overset{+}{\rightarrow})$ $\quad v = v_0 + a_c t$

$$6 = 0 + 4t_A$$

$$t_A = 1.5 \text{ s}$$

$(\overset{+}{\rightarrow})$ $\quad s = s_0 + v_0 t + \frac{1}{2}a_c t^2$

$$s = 0 + 0 + \frac{1}{2}(4)(1.5)^2 = 4.5 \text{ ft}$$

For B :

$(\overset{+}{\leftarrow})$ $\quad v = v_0 + a_c t$

$$6 = 0 + 8t_B$$

$$t_B = 0.75 \text{ s}$$

$(\overset{+}{\leftarrow})$ $\quad s = s_0 + v_0 t + \frac{1}{2}a_c t^2$

$$s = 0 + 0 + \frac{1}{2}(8)(0.75)^2 = 2.25 \text{ ft}$$

For the remaining $(1.5 - 0.75)\text{s} = 0.75 \text{ s}$

$$s = vt = 6(0.75) = 4.5 \text{ ft}$$

Thus,

$$s = 30 - 4.5 - 4.5 - 2.25 = 18.75 \text{ ft}$$

$$t' = \frac{s/2}{v} = \frac{18.75/2}{6} = 1.5625$$

$$t = 1.5 + 1.5625 = 3.06 \text{ s} \qquad \textbf{Ans}$$

15-39. A 40-lb box slides from rest down the smooth ramp onto the surface of a 20-lb cart. Determine the speed of the box at the instant it stops sliding on the cart. If someone ties the cart to the ramp at B, determine the horizontal impulse the box will exert at C in order to stop its motion. Neglect friction and the size of the box.

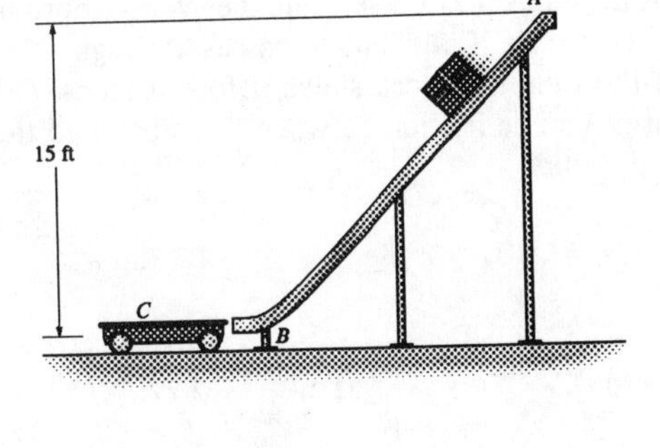

Datum at B :

$$T_A + V_A = T_B + V_B$$

$$0 + 40(15) = \frac{1}{2}\left(\frac{40}{32.2}\right)(v_B)^2 + 0$$

$$v_B = 31.08 \text{ ft/s}$$

Box and cart :

$$(\overset{+}{\leftarrow}) \qquad \Sigma m v_1 = \Sigma m v_2$$

$$0 + \left(\frac{40}{32.2}\right)(31.08) = \left(\frac{40+20}{32.2}\right)v_2$$

$$v_2 = 20.7 \text{ ft/s} \qquad \textbf{Ans}$$

Box :

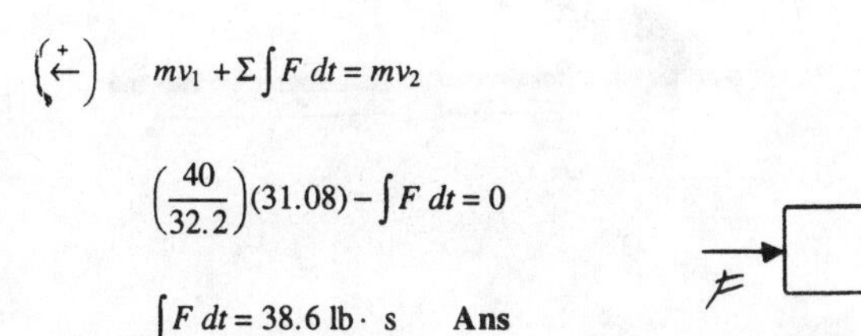

$$(\overset{+}{\leftarrow}) \qquad m v_1 + \Sigma \int F\,dt = m v_2$$

$$\left(\frac{40}{32.2}\right)(31.08) - \int F\,dt = 0$$

$$\int F\,dt = 38.6 \text{ lb} \cdot \text{s} \qquad \textbf{Ans}$$

***15-40.** A 0.03-lb bullet traveling at 1300 ft/s strikes the 10-lb wooden block and exits the other side at 50 ft/s as shown. Determine the speed of the block just after the bullet exits the block, and also determine how far the block slides before it stops. The coefficient of kinetic friction between the block and the surface is $\mu_k = 0.5$.

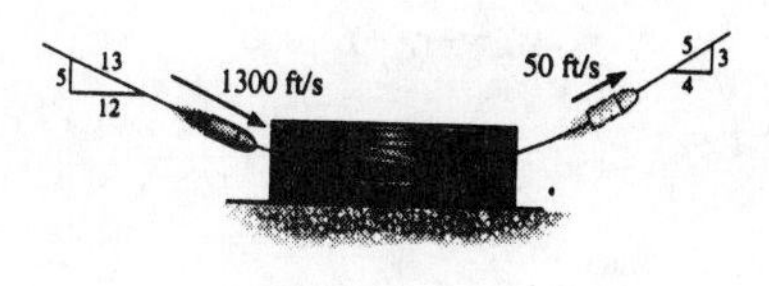

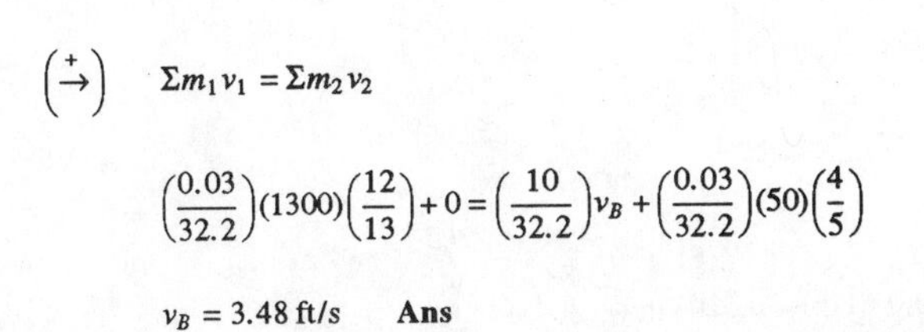

$$(\overset{+}{\rightarrow}) \qquad \Sigma m_1 v_1 = \Sigma m_2 v_2$$

$$\left(\frac{0.03}{32.2}\right)(1300)\left(\frac{12}{13}\right) + 0 = \left(\frac{10}{32.2}\right)v_B + \left(\frac{0.03}{32.2}\right)(50)\left(\frac{4}{5}\right)$$

$$v_B = 3.48 \text{ ft/s} \qquad \textbf{Ans}$$

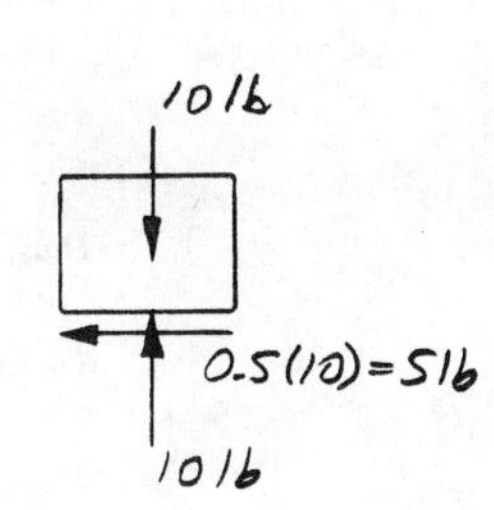

$$T_1 + \Sigma U_{1-2} = T_2$$

$$\frac{1}{2}\left(\frac{10}{32.2}\right)(3.48)^2 - 5(d) = 0$$

$$d = 0.376 \text{ ft} \qquad \textbf{Ans}$$

15-41. A 0.03-lb bullet traveling at 1300 ft/s strikes the 10-lb wooden block and exits the other side at 50 ft/s as shown. Determine the speed of the block just after the bullet exits the block. Also, determine the average normal force on the block if the bullet passes through it in 1 ms, and the time the block slides before it stops. The coefficient of kinetic friction between the block and the surface is $\mu_k = 0.5$.

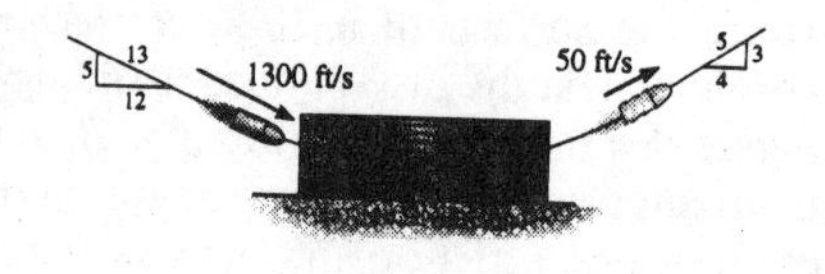

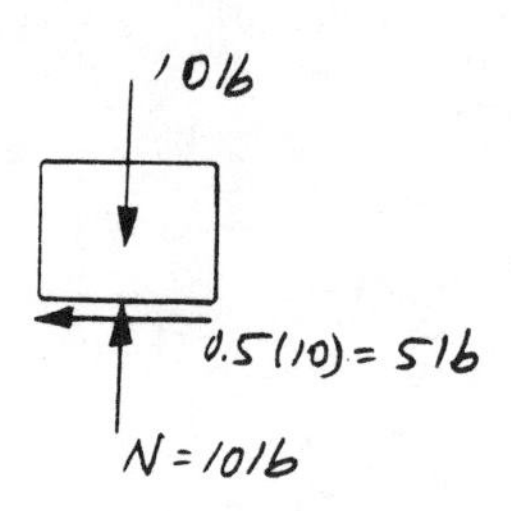

$(\overset{+}{\rightarrow})$ $\quad \Sigma m_1 v_1 = \Sigma m_2 v_2$

$$\left(\frac{0.03}{32.2}\right)(1300)\left(\frac{12}{13}\right) + 0 = \left(\frac{10}{32.2}\right)v_B + \left(\frac{0.03}{32.2}\right)(50)\left(\frac{4}{5}\right)$$

$v_B = 3.48$ ft/s $\quad$ **Ans**

$(+\uparrow)$ $\quad mv_1 + \Sigma\int F\,dt = mv_2$

$$-\left(\frac{0.03}{32.2}\right)(1300)\left(\frac{5}{13}\right) - 10(1)(10^{-3}) + N(1)(10^{-3}) = \left(\frac{0.03}{32.2}\right)(50)\left(\frac{3}{5}\right)$$

$N = 504$ lb $\quad$ **Ans**

$(\overset{+}{\rightarrow})$ $\quad mv_1 + \Sigma\int F\,dt = mv_2$

$$\left(\frac{10}{32.2}\right)(3.48) - 5(t) = 0$$

$t = 0.216$ s $\quad$ **Ans**

15-42. The 5-kg spring-loaded gun rests on the smooth surface. It fires a ball having a mass of 1 kg with a velocity of $v' = 6$ m/s relative to the gun in the direction shown. If the gun is originally at rest, determine the horizontal distance d the ball is from the gun at the instant the ball strikes the ground at D. Neglect the size of the gun.

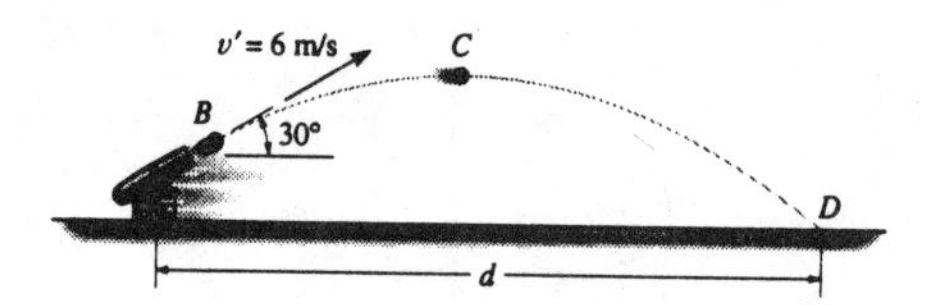

$(\overset{+}{\rightarrow})$ $\quad \Sigma m v_1 = \Sigma m v_2$

$$0 = 1(v_B)_x - 5v_G$$

$$(v_B)_x = 5v_G$$

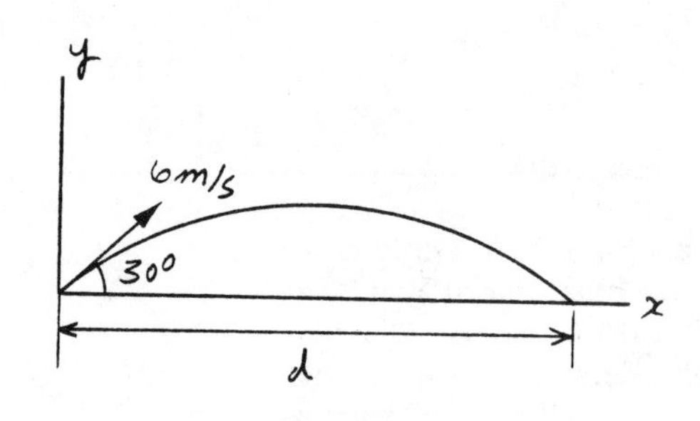

$(\overset{+}{\rightarrow})$ $\quad v_B = v_G + v_{B/G}$

$$5v_G = -v_G + 6\cos 30°$$

$$v_G = 0.8660 \text{ m/s} \leftarrow$$

So that,

$(v_B)_x = 4.330$ m/s $\rightarrow$

$(v_B)_y = 6\sin 30° = 3$ m/s $\uparrow$

Time of flight for ball :

$(+\uparrow)$ $\quad v = v_0 + a_c t$

$$-3 = 3 - 9.81t$$

$$t = 0.6116 \text{ s}$$

Distance ball travels :

$(\overset{+}{\rightarrow})$ $\quad s = v_0 t$

$$s = 4.330(0.6116) = 2.648 \text{ m} \rightarrow$$

Distance gun travels :

$(\overset{+}{\leftarrow})$ $\quad s = v_0 t$

$$s' = 0.8660(0.6116) = 0.5297 \text{ m} \leftarrow$$

Thus,

$d = 2.648 + 0.5297 = 3.18$ m $\quad$ **Ans**

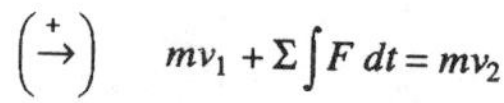

15-43. The 5-kg spring-loaded gun rests on the smooth surface. It fires a ball having a mass of 1 kg with a velocity of $v' = 6$ m/s relative to the gun in the direction shown. If the gun is originally at rest, determine the distance the ball is from the gun at the instant the ball reaches its highest elevation C. Neglect the size of the gun.

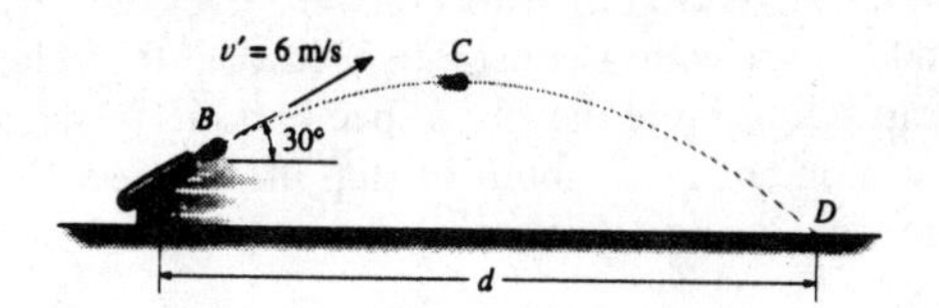

$(\overset{+}{\rightarrow})$ $\Sigma m v_1 = \Sigma m v_2$

$0 = 1(v_B)_x - 5v_G$

$(v_B)_x = 5v_G$

$(\overset{+}{\rightarrow})$ $v_B = v_G + v_{B/G}$

$5v_G = -v_G + 6\cos 30°$

$v_G = 0.8660 \text{ m/s} \leftarrow$

So that,

$(v_B)_x = 4.330 \text{ m/s} \rightarrow$

$(v_B)_y = 6\sin 30° = 3 \text{ m/s} \uparrow$

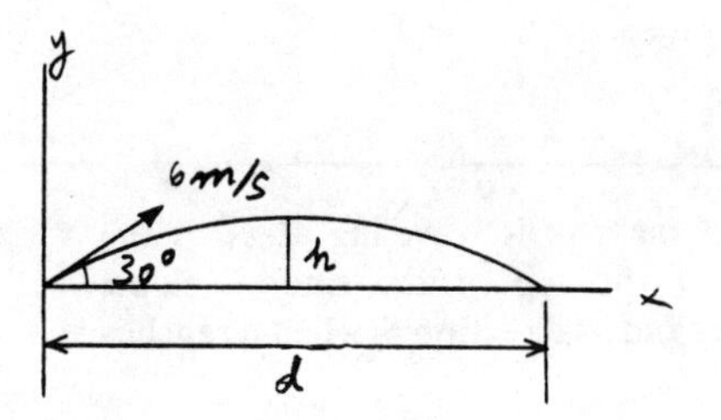

Time of flight for ball :

$(+\uparrow)$ $v = v_0 + a_c t;$

$0 = 3 - 9.81t$

$t = 0.3058 \text{ s}$

Height of ball :

$(+\uparrow)$ $v^2 = v_0^2 + 2a_c(s - s_0)$

$0 = (3)^2 - 2(9.81)(h - 0)$

$h = 0.4587 \text{ m}$

Distance ball travels :

$(\overset{+}{\rightarrow})$ $s = v_0 t$

$s = 4.330(0.3058) = 1.324 \text{ m} \rightarrow$

Distance gun travels :

$(\overset{+}{\leftarrow})$ $s = v_0 t$

$s' = 0.8660(0.3058) = 0.2648 \text{ m} \leftarrow$

Thus,

$d = 1.324 + 0.2648 = 1.589 \text{ m}$

Distance from cannon to ball :

$r = \sqrt{(0.4587)^2 + (1.589)^2} = 1.65 \text{ m}$ **Ans**

***15-44.** A 100-lb boy walks forward over the surface of the 60-lb cart with a constant speed of 3 ft/s relative to the cart. Determine the cart's speed and its displacement at the moment he is about to step off. Neglect the mass of the wheels and assume the cart and boy are originally at rest.

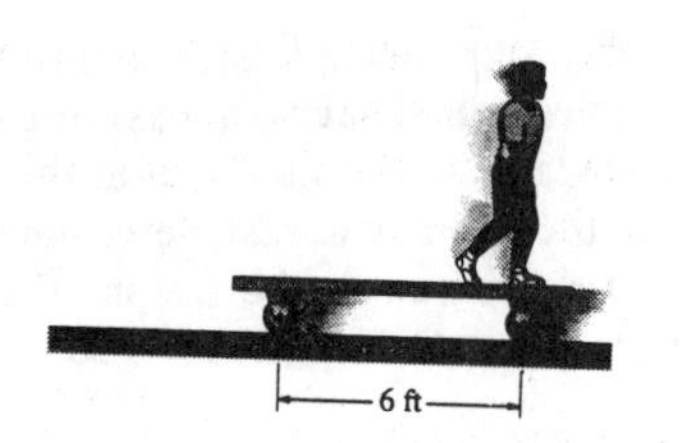

$(\xrightarrow{+})$ $\quad 0 = m_B v_B + m_C v_C$, however, $v_B = v_C + v_{B/C} = v_C + 3$

$$0 = \frac{100}{32.2}(v_C + 3) + \frac{60}{32.2}v_C$$

$v_C = -1.875 \text{ ft/s} = 1.88 \text{ ft/s} \leftarrow$ **Ans**

$$s_{B/C} = v_{B/C}t$$
$$6 = 3t$$
$$t = 2 \text{ s}$$

$s_C = v_C t = 1.875(2) = 3.75 \text{ ft}$ **Ans**

15-45. The block of mass m is traveling at v_1 in the direction θ_1 shown at the top of the smooth slope. Determine its speed v_2 and its direction θ_2 when it reaches the bottom.

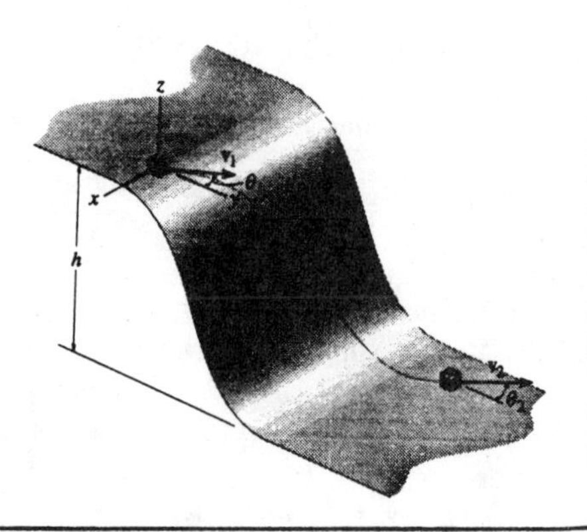

There are no impulses in the x direction :

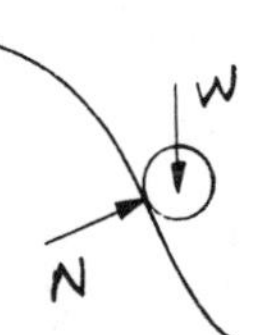

$$mv_1 \sin\theta_1 = mv_2 \sin\theta_2$$

$$T_1 + V_1 = T_2 + V_2$$

$$\frac{1}{2}mv_1^2 + mgh = \frac{1}{2}mv_2^2 + 0$$

$v_2 = \sqrt{v_1^2 + 2gh}$ **Ans**

$$\sin\theta_2 = \frac{v_1 \sin\theta_1}{\sqrt{v_1^2 + 2gh}}$$

$\theta_2 = \sin^{-1}\left(\frac{v_1 \sin\theta_1}{\sqrt{v_1^2 + 2gh}}\right)$ **Ans**

15-46. Two boxes A and B, each having a weight of 160 lb, sit on the 500-lb conveyor which is free to roll on the ground. If the belt starts from rest and begins to run with a speed of 3 ft/s, determine the final speed of the conveyor if (a) the boxes are not stacked and A falls off then B falls off, and (b) A is stacked on top of B and both fall off together.

a) Let v_b be the velocity of A and B.

$(\xrightarrow{+})$ $\quad \Sigma mv_1 = \Sigma mv_2$

$$0 = \left(\frac{320}{32.2}\right)(v_b) - \left(\frac{500}{32.2}\right)(v_c)$$

$(\xrightarrow{+})$ $\quad v_b = v_c + v_{b/c}$

$$v_b = -v_c + 3$$

Thus, $\quad v_b = 1.83 \text{ ft/s} \rightarrow \quad v_c = 1.17 \text{ ft/s} \leftarrow$

When a box falls off, it exerts no impulse on the conveyor, and so does not alter the momentum of the conveyor. Thus,

a) $\quad v_c = 1.17 \text{ ft/s} \leftarrow$ **Ans**

b) $\quad v_c = 1.17 \text{ ft/s} \leftarrow$ **Ans**

15-47. The winch on the back of the jeep A is turned on and pulls in the tow rope at $v_{rel} = 2$ m/s. If both the 1.25-Mg car B and the 2.5-Mg jeep A are free to roll, determine their velocities at the instant they meet. If the rope is 5 m long, how long will this take?

$(\overset{+}{\rightarrow})$ $\quad 0 + 0 = m_A\, v_A - m_B\, v_B \qquad (1)$

$$0 = 2.5(10^3)\, v_A - 1.25(10^3)\, v_B$$

However, $\quad \mathbf{v}_A = \mathbf{v}_B + \mathbf{v}_{A/B}$

$(\overset{+}{\rightarrow})$ $\quad v_A = -v_B + 2 \qquad (2)$

Substituting Eq. (2) into (1) yields :

$v_B = 1.33$ m/s **Ans** $\qquad v_A = 0.667$ m/s **Ans**

Kinematics :

$(\overset{+}{\rightarrow})$ $\quad s_{A/B} = v_{A/B}\, t$

$$5 = 2t$$

$t = 2.5$ s **Ans**

***15-48.** The block B has a weight of 75 lb and rests at the end of the 50-lb cart. If the cart is free to roll, and the rope is pulled in at 4 ft/s relative to the cart, determine how far d the cart has moved when the block has moved 8 ft on the cart. The coefficient of kinetic friction between the cart and the block is $\mu_k = 0.4$.

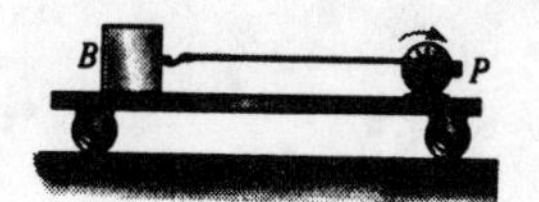

$(\overset{+}{\rightarrow})$ $\quad \Sigma m v_1 = \Sigma m v_2$

$$0 + 0 = \left(\frac{75}{32.2}\right) v_B - \left(\frac{50}{32.2}\right) v_C$$

$$v_C = 1.5 v_B$$

$$\mathbf{v}_B = \mathbf{v}_C + \mathbf{v}_{B/C}$$

$(\overset{+}{\rightarrow})$ $\quad v_B = -v_C + 4$

Thus, $\quad v_B = 1.6$ ft/s $\quad v_C = 2.4$ ft/s

Time to travel 8 ft :

$$t = \frac{s}{v_{B/C}} = \frac{8}{4} = 2 \text{ s}$$

$d = v_C t = 2.4(2) = 4.80$ ft **Ans**

15-49. The 20-lb cart B is supported on rollers of negligible size. If a 10-lb suitcase A is thrown horizontally on it at 10 ft/s, determine the length of time that A slides relative to B, and the final velocity of A and B. The coefficient of kinetic friction between A and B is $\mu_k = 0.4$.

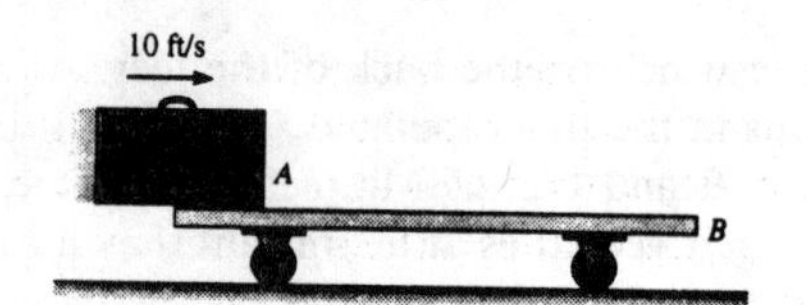

System :

$(\overset{+}{\rightarrow})$ $\qquad \Sigma m_1 v_1 = \Sigma m_2 v_2$

$$\left(\frac{10}{32.2}\right)(10) + 0 = \left(\frac{10+20}{32.2}\right)v$$

$v = 3.33$ ft/s **Ans**

10 lb

0.4(10) = 4 lb

10 lb

For A :

$$mv_1 + \Sigma\int F\,dt = mv_2$$

$$\left(\frac{10}{32.2}\right)(10) - 4t = \left(\frac{10}{32.2}\right)(3.33)$$

$t = 0.5176 = 0.518$ s **Ans**

15-50. The 20-lb cart B is supported on rollers of negligible size. If a 10-lb suitcase A is thrown horizontally on it at 10 ft/s, determine the time t and the distance B moves before A stops relative to B. The coefficient of kinetic friction between A and B is $\mu_k = 0.4$.

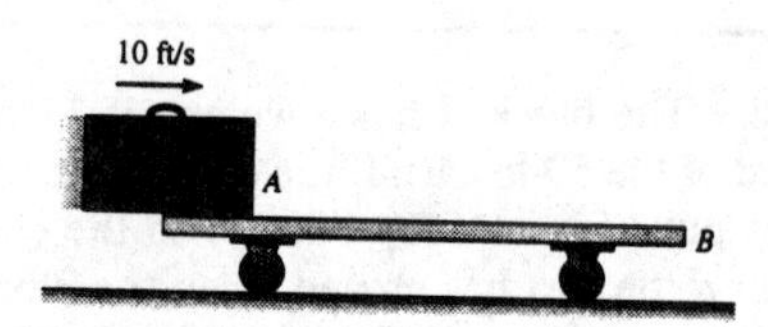

System :

$(\overset{+}{\rightarrow})$ $\qquad \Sigma m_1 v_1 = \Sigma m_2 v_2$

$$\left(\frac{10}{32.2}\right)(10) + 0 = \left(\frac{10+20}{32.2}\right)v$$

$v = 3.33$ ft/s

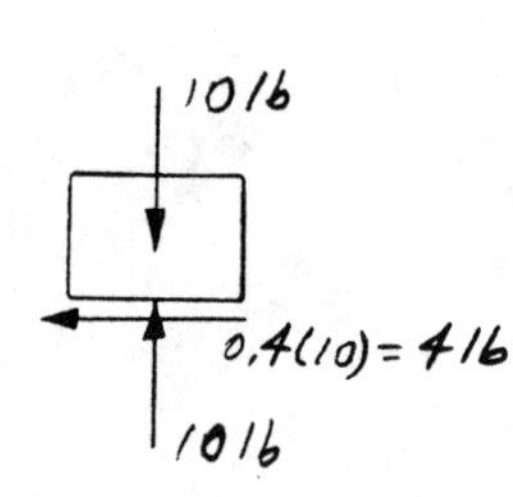

For A :

$$mv_1 + \Sigma\int F\,dt = mv_2$$

$$\left(\frac{10}{32.2}\right)(10) - 4t = \left(\frac{10}{32.2}\right)(3.33)$$

$t = 0.5176 = 0.518$ s **Ans**

For B :

$(\overset{+}{\rightarrow})$ $\qquad v = v_0 + a_c t$

$$3.33 = 0 + a_c\,(0.5176)$$

$$a_c = 6.440 \text{ ft/s}^2$$

$(\overset{+}{\rightarrow})$ $\qquad s = s_0 + v_0 t + \frac{1}{2}a_c t^2$

$s = 0 + 0 + \frac{1}{2}(6.440)(0.5176)^2 = 0.863$ ft **Ans**

15-51. A boy A having a weight of 80 lb and a girl B having a weight of 65 lb stand motionless at the ends of the toboggan, which has a weight of 20 lb. If they exchange positions, A going to B and then B going to A's original position, determine the final position of the toboggan just after the motion. Neglect friction.

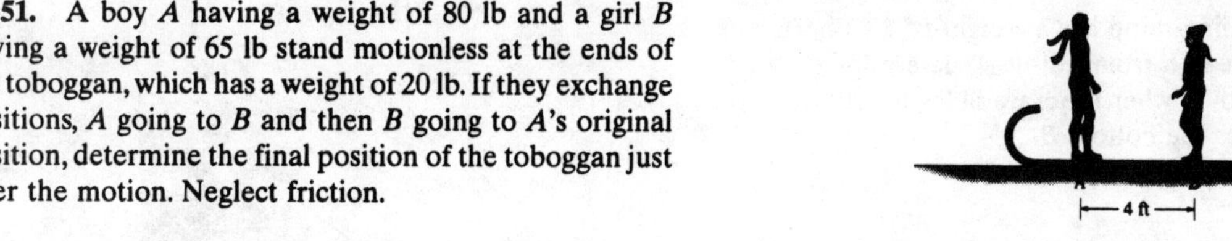

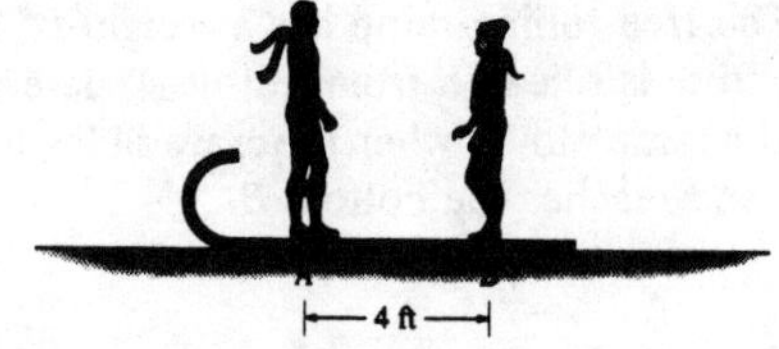

A goes to B :

$(\overset{+}{\rightarrow})$ $\Sigma m v_1 = \Sigma m v_2$

$$0 = m_A v_A - (m_t + m_B) v_B$$

$$0 = m_A s_A - (m_t + m_B) s_B$$

Assume B moves x to the left, then A moves $(4 - x)$ to the right.

$$0 = m_A (4 - x) - (m_t + m_B) x$$

$$x = \frac{4 m_A}{m_A + m_B + m_t}$$

$$= \frac{4(80)}{80 + 65 + 20} = 1.939 \text{ ft} \leftarrow$$

B goes to other end :

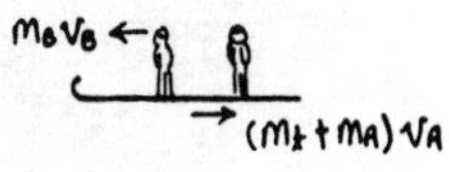

$(\overset{+}{\rightarrow})$ $\Sigma m v_1 = \Sigma m v_2$

$$0 = -m_B v_B + (m_t + m_A) v_A$$

$$0 = -m_B s_B + (m_t + m_A) s_A$$

Assume B moves x' to the right, then A moves $(4 - x')$ to the left.

$$0 = m_B (4 - x') - (m_t + m_A) x'$$

$$x = \frac{4 m_B}{m_A + m_B + m_t}$$

$$= \frac{4(65)}{80 + 65 + 20} = 1.576 \text{ ft} \rightarrow$$

Net movement of sled is

$x = 1.939 - 1.576 = 0.364$ ft $\leftarrow$ **Ans**

***15-52.** The free-rolling ramp has a weight of 120 lb. The crate whose weight is 80 lb slides from rest at A, 15 ft down the ramp to B. Determine the ramp's speed when the crate reaches B. Assume that the ramp is smooth, and neglect the mass of the wheels.

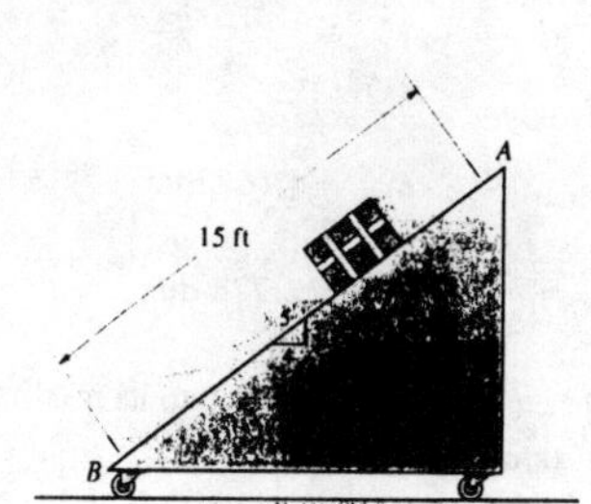

$$T_1 + V_1 = T_2 + V_2$$

$$0 + 80\left(\frac{3}{5}\right)(15) = \frac{1}{2}\left(\frac{80}{32.2}\right)v_B^2 + \frac{1}{2}\left(\frac{120}{32.2}\right)v_r^2 \qquad (1)$$

$(\overset{+}{\rightarrow})$ $\Sigma m v_1 = \Sigma m v_2$

$$0 + 0 = \frac{120}{32.2} v_r - \frac{80}{32.2}(v_B)_x$$

$$(v_B)_x = 1.5 v_r$$

$$\mathbf{v}_B = \mathbf{v}_r + \mathbf{v}_{B/r}$$

$$(\overset{+}{\rightarrow}) \; -(v_B)_x = v_r - \frac{4}{5} v_{B/r} \qquad (2)$$

$$(+\uparrow) \; (-v_B)_y = 0 - \frac{3}{5} v_{B/r} \qquad (3)$$

Eliminating $v_{B/r}$ from Eqs. (2) and (3) and substituting $(v_B)_x = 1.5 v_r$ results in

$$(v_B)_y = 1.875 v_r$$

$$v_B^2 = (v_B)_x^2 + (v_B)_y^2 = (1.5 v_r)^2 + (1.875 v_r)^2 = 5.7656 v_r^2 \qquad (4)$$

Substituting Eq. (4) into (1) yields :

$$80\left(\frac{3}{5}\right)(15) = \frac{1}{2}\left(\frac{80}{32.2}\right)(5.7656 v_r^2) + \frac{1}{2}\left(\frac{120}{32.2}\right)v_r^2$$

$v_r = 8.93$ ft/s **Ans**

15-53. The free-rolling ramp has a weight of 120 lb. If the 80-lb crate is released from rest at A, determine the distance the ramp moves when the crate slides 15 ft down the ramp and reaches the bottom B.

$(\overset{+}{\rightarrow})\quad \Sigma m v_1 = \Sigma m v_2$

$$0 = \frac{120}{32.2} v_r - \frac{80}{32.2}(v_B)_x$$

$$(v_B)_x = 1.5\, v_r$$

$$\mathbf{v}_B = \mathbf{v}_r + \mathbf{v}_{B/r}$$

$(\overset{+}{\rightarrow})\quad -(v_B)_x = v_r - (v_{B/r})_x$

$$-1.5 v_r = v_r - (v_{B/r})_x$$

$$2.5\, v_r = (v_{B/r})_x$$

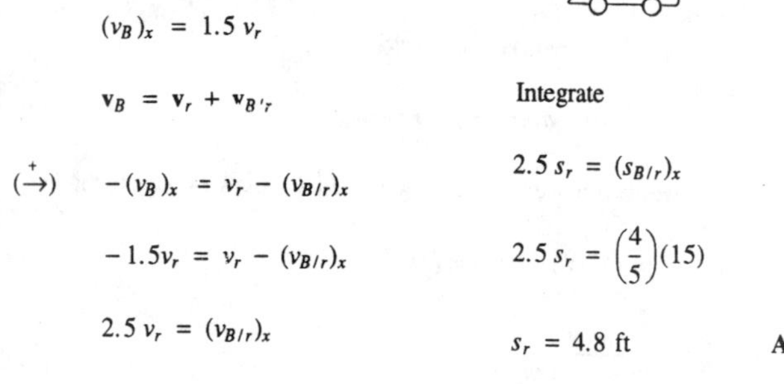

Integrate

$$2.5\, s_r = (s_{B/r})_x$$

$$2.5\, s_r = \left(\frac{4}{5}\right)(15)$$

$$s_r = 4.8 \text{ ft} \qquad \textbf{Ans}$$

15-54. A tugboat T having a mass of 19 Mg is tied to a barge B having a mass of 75 Mg. If the rope is "elastic" such that it has a stiffness $k = 600$ kN/m, determine the maximum stretch in the rope during the initial towing. Originally both the tugboat and barge are moving in the same direction with speeds $(v_T)_1 = 15$ km/h and $(v_B)_1 = 10$ km/h, respectively. Neglect the resistance of the water.

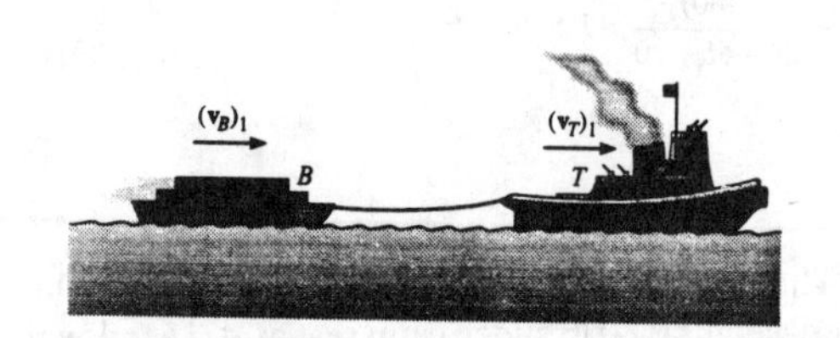

$(v_T)_1 = 15 \text{ km/h} = 4.167 \text{ m/s}$

$(v_B)_1 = 10 \text{ km/h} = 2.778 \text{ m/s}$

When the rope is stretched to its maximum, both the tug and barge have a common velocity. Hence,

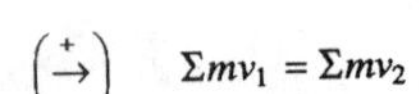

$(\overset{+}{\rightarrow})\quad \Sigma m v_1 = \Sigma m v_2$

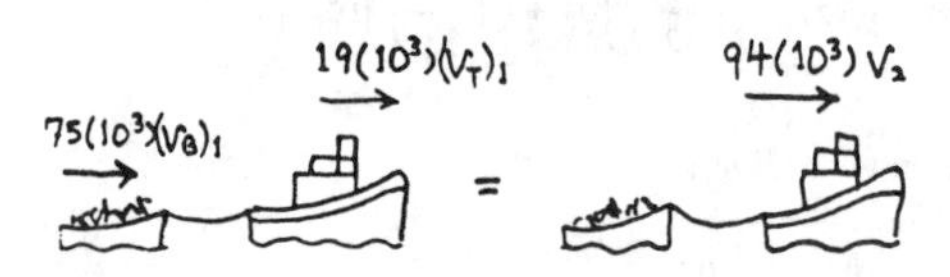

$$19\,000(4.167) + 75\,000(2.778) = (19\,000 + 75\,000)v_2$$

$$v_2 = 3.059 \text{ m/s}$$

$$T_1 + V_1 = T_2 + V_2$$

$$T_1 = \frac{1}{2}(19\,000)(4.167)^2 + \frac{1}{2}(75\,000)(2.778)^2 = 454.282 \text{ kJ}$$

$$T_2 = \frac{1}{2}(19\,000 + 75\,000)(3.059)^2 = 439.661 \text{ kJ}$$

Hence,

$$454.282\left(10^3\right) + 0 = 439.661\left(10^3\right) + \frac{1}{2}(600)\left(10^3\right)x^2$$

$$x = 0.221 \text{ m} \qquad \textbf{Ans}$$

15-55. An ivory ball having a mass of 200 g is released from rest at a height of 400 mm above a very large fixed metal surface. If the ball rebounds to a height of 325 mm above the surface, determine the coefficient of restitution between the ball and the surface.

Before impact

$$T_1 + V_1 = T_2 + V_2$$

$$0 + 0.2(9.81)(0.4) = \frac{1}{2}(0.2)v_1^2 + 0$$

$$v_1 = 2.801 \text{m/s}$$

After the impact

$$\frac{1}{2}(0.2)v_2^2 = 0 + 0.2(9.81)(0.325)$$

$$v_2 = 2.525 \text{ m/s}$$

Coefficient of restitution :

$$(+\downarrow) \qquad e = \frac{(v_A)_2 - (v_B)_2}{(v_B)_1 - (v_A)_1}$$

$$= \frac{0 - (-2.525)}{2.801 - 0}$$

$$= 0.901 \qquad \textbf{Ans}$$

***15-56.** Block B has a mass of 0.75 kg and is sliding forward on the *smooth* surface with a velocity $(v_B)_1 = 4$ m/s when it strikes the 2-kg block A, which is originally at rest. If the coefficient of restitution between the blocks is $e = 0.5$, determine the velocities of A and B just after collision.

$$(\overset{+}{\rightarrow}) \qquad m_B(v_B)_1 + m_A(v_A)_1 = m_B(v_B)_2 + m_A(v_A)_2$$

$$0.75(4) + 0 = 0.75(v_B)_2 + 2(v_A)_2 \qquad (1)$$

$$(\overset{+}{\rightarrow}) \qquad e = \frac{(v_A)_2 - (v_B)_2}{(v_B)_1 - (v_A)_1}$$

$$0.5 = \frac{(v_A)_2 - (v_B)_2}{4 - 0}$$

$$(v_A)_2 - (v_B)_2 = 2 \qquad (2)$$

Solving Eqs. (1) and (2) yields :

$(v_A)_2 = 1.64 \text{ m/s} \rightarrow$ **Ans** $\quad (v_B)_2 = -0.364 \text{ m/s} = 0.364 \text{ m/s} \leftarrow$ **Ans**

15-57. Disks A and B have a mass of 2 kg and 4 kg, respectively. If they have the velocities shown, and $e = 0.4$, determine their velocities just after direct central impact.

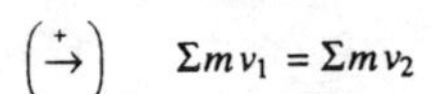

$(\xrightarrow{+})$ $\quad \Sigma m v_1 = \Sigma m v_2$

$$2(2) - 4(5) = 2(v_A)_2 + 4(v_B)_2$$

$(\xrightarrow{+})$ $\quad e = \dfrac{(v_B)_2 - (v_A)_2}{(v_A)_1 - (v_B)_1};$

$$0.4 = \frac{(v_B)_2 - (v_A)_2}{(2 - (-5))}$$

Solving :

$(v_B)_2 = -1.73 \text{ m/s} = 1.73 \text{ m/s} \leftarrow$ **Ans**

$(v_A)_2 = -4.53 \text{ m/s} = 4.53 \text{ m/s} \leftarrow$ **Ans**

15-58. The three balls each weigh 0.5 lb and have a coefficient of restitution of $e = 0.85$. If ball A is released from rest and strikes ball B and then ball B strikes ball C, determine the velocity of each ball after the second collision has occurred. Neglect the size of each ball.

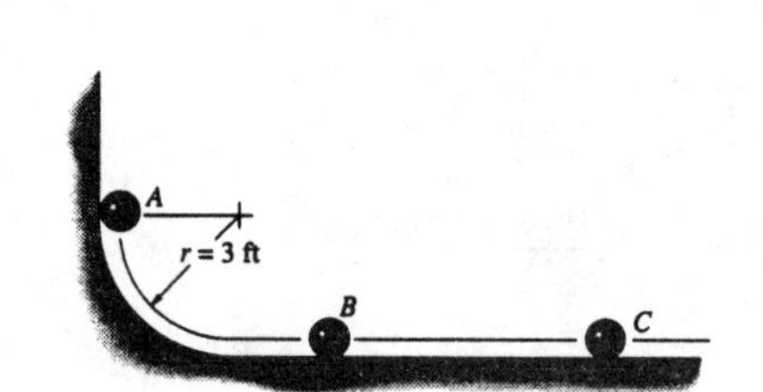

Ball A :
Datum at lowest point.

$$T_1 + V_1 = T_2 + V_2$$

$$0 + (0.5)(3) = \frac{1}{2}\left(\frac{0.5}{32.2}\right)(v_A)_1^2 + 0$$

$(v_A)_1 = 13.90$ ft/s

Balls A and B :

$(\xrightarrow{+})$ $\quad \Sigma m v_1 = \Sigma m v_2$

$$\left(\frac{0.5}{32.2}\right)(13.90) + 0 = \left(\frac{0.5}{32.2}\right)(v_A)_2 + \left(\frac{0.5}{32.2}\right)(v_B)_2$$

$(\xrightarrow{+})$ $\quad e = \dfrac{(v_B)_2 - (v_A)_2}{(v_A)_1 - (v_B)_1}$

$$0.85 = \frac{(v_B)_2 - (v_A)_2}{13.90 - 0}$$

Solving :

$(v_A)_2 = 1.04$ ft/s **Ans**

$(v_B)_2 = 12.86$ ft/s

Balls B and C :

$(\xrightarrow{+})$ $\quad \Sigma m v_2 = \Sigma m v_3$

$$\left(\frac{0.5}{32.2}\right)(12.86) + 0 = \left(\frac{0.5}{32.2}\right)(v_B)_3 + \left(\frac{0.5}{32.2}\right)(v_C)_3$$

$(\xrightarrow{+})$ $\quad e = \dfrac{(v_C)_3 - (v_B)_3}{(v_B)_2 - (v_C)_2}$

$$0.85 = \frac{(v_C)_3 - (v_B)_3}{12.86 - 0}$$

Solving :

$(v_B)_3 = 0.964$ ft/s **Ans**

$(v_C)_3 = 11.9$ ft/s **Ans**

15-59. If two disks A and B have the same mass and are subjected to direct central impact such that the collision is perfectly elastic ($e = 1$), prove that the kinetic energy before collision equals the kinetic energy after collision. The surface upon which they slide is smooth.

$(\overset{+}{\rightarrow})$ $\quad \Sigma m v_1 = \Sigma m v_2$

$$m_A (v_A)_1 + m_B (v_B)_1 = m_A (v_A)_2 + m_B (v_B)_2$$

$$m_A\left[(v_A)_1 - (v_A)_2\right] = m_B\left[(v_B)_2 - (v_B)_1\right] \qquad (1)$$

$(\overset{+}{\rightarrow})$ $\quad e = \dfrac{(v_B)_2 - (v_A)_2}{(v_A)_1 - (v_B)_1} = 1$

$$(v_B)_2 - (v_A)_2 = (v_A)_1 - (v_B)_1 \qquad (2)$$

Combining Eqs. (1) and (2) :

$$m_A\left[(v_A)_1 - (v_A)_2\right]\left[(v_A)_1 + (v_A)_2\right] = m_B\left[(v_B)_2 - (v_B)_1\right]\left[(v_B)_2 + (v_B)_1\right]$$

Expand and multiply by $\frac{1}{2}$:

$$\frac{1}{2}m_A (v_A)_1^2 + \frac{1}{2}m_B (v_B)_1^2 = \frac{1}{2}m_A (v_A)_2^2 + \frac{1}{2}m_B (v_B)_2^2 \qquad \textbf{Q.E.D.}$$

***15-60.** Each ball has a mass m and the coefficient of restitution between the balls is e. If they are moving towards one another with a velocity v, determine their speeds after collision. Also, determine their common velocity when they reach the state of maximum deformation. Neglect the size of each ball.

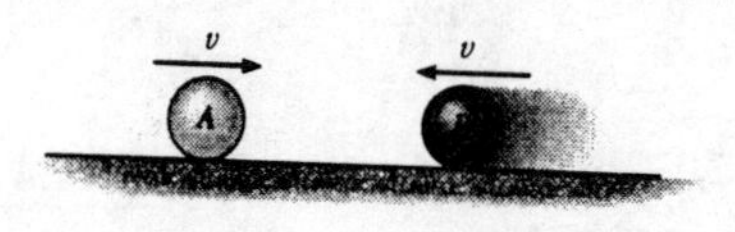

$(\overset{+}{\rightarrow})$ $\quad \Sigma m v_1 = \Sigma m v_2$

$$mv - mv = mv_A + mv_B$$

$$v_A = -v_B$$

$(\overset{+}{\rightarrow})$ $\quad e = \dfrac{(v_B)_2 - (v_A)_2}{(v_A)_1 - (v_B)_1} = \dfrac{v_B - v_A}{v - (-v)}$

$$2ve = 2v_B$$

$v_B = ve \rightarrow$ $\quad$ **Ans**

$v_A = -ve = ve \leftarrow$ $\quad$ **Ans**

At maximum deformation $v_A = v_B = v'$.

$(\overset{+}{\rightarrow})$ $\quad \Sigma m v_1 = \Sigma m v_2$

$$mv - mv = (2m)v'$$

$v' = 0$ $\quad$ **Ans**

15-61. The 1-lb ball A is thrown so that when it strikes the 10-lb block B it is traveling horizontally at 20 ft/s. If the coefficient of restitution between A and B is $e = 0.6$, and the coefficient of kinetic friction between the plane and the block is $\mu_k = 0.4$, determine the time before block B stops sliding.

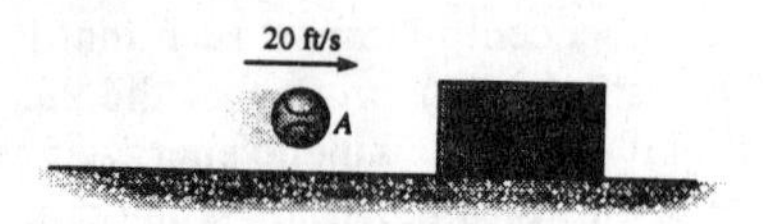

$(\overset{+}{\rightarrow})$ $\Sigma m_1 v_1 = \Sigma m_2 v_2$

$$\left(\frac{1}{32.2}\right)(20) + 0 = \left(\frac{1}{32.2}\right)(v_A)_2 + \left(\frac{10}{32.2}\right)(v_B)_2$$

$$(v_A)_2 + 10(v_B)_2 = 20$$

$(\overset{+}{\rightarrow})$ $$e = \frac{(v_B)_2 - (v_A)_2}{(v_A)_1 - (v_B)_1}$$

$$0.6 = \frac{(v_B)_2 - (v_A)_2}{20 - 0}$$

$$(v_B)_2 - (v_A)_2 = 12$$

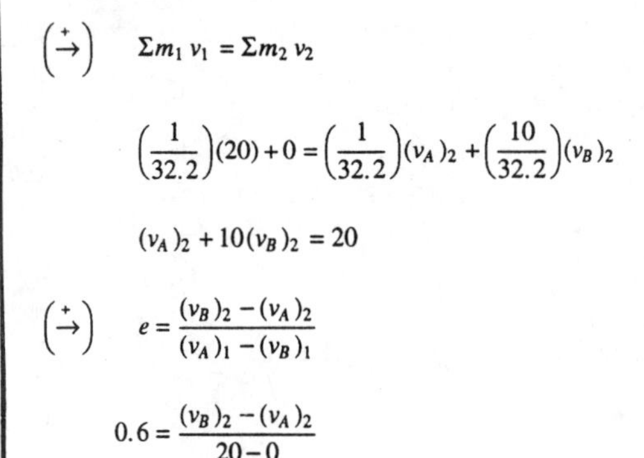

Thus,

$(v_B)_2 = 2.909$ ft/s $\rightarrow$

$(v_A)_2 = -9.091$ ft/s $= 9.091$ ft/s $\leftarrow$

Block B :

$(\overset{+}{\rightarrow})$ $m v_1 + \Sigma \int F\, dt = m v_2$

$$\left(\frac{10}{32.2}\right)(2.909) - 4t = 0$$

$t = 0.226$ s **Ans**

15-62. The 1-lb ball A is thrown so that when it strikes the 10-lb block B it is traveling horizontally at 20 ft/s. If the coefficient of restitution between A and B is $e = 0.6$, and the coefficient of kinetic friction between the plane and the block is $\mu_k = 0.4$, determine the distance block B slides on the plane before stopping.

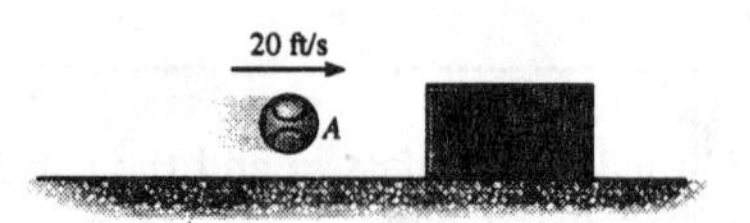

$(\overset{+}{\rightarrow})$ $\Sigma m_1 v_1 = \Sigma m_2 v_2$

$$\left(\frac{1}{32.2}\right)(20) + 0 = \left(\frac{1}{32.2}\right)(v_A)_2 + \left(\frac{10}{32.2}\right)(v_B)_2$$

$$(v_A)_2 + 10(v_B)_2 = 20$$

$(\overset{+}{\rightarrow})$ $$e = \frac{(v_B)_2 - (v_A)_2}{(v_A)_1 - (v_B)_1}$$

$$0.6 = \frac{(v_B)_2 - (v_A)_2}{20 - 0}$$

$$(v_B)_2 - (v_A)_2 = 12$$

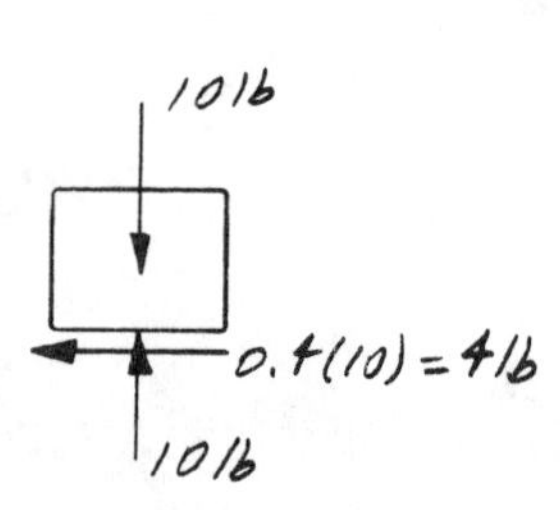

Thus,

$(v_B)_2 = 2.909$ ft/s $\rightarrow$

$(v_A)_2 = -9.091$ ft/s $= 9.091$ ft/s $\leftarrow$

Block B :

$$T_1 + \Sigma U_{1-2} = T_2$$

$$\frac{1}{2}\left(\frac{10}{32.2}\right)(2.909)^2 - 4d = 0$$

$d = 0.329$ ft **Ans**

15-63. The 1-lb ball A is thrown so that when it strikes the 10-lb block B it is traveling horizontally at 20 ft/s. Determine the average normal force exerted between A and B if the impact occurs in 0.02 s. The coefficient of restitution between A and B is $e = 0.6$.

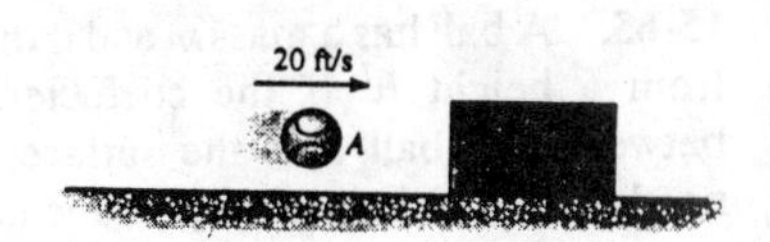

$(\overset{+}{\rightarrow})$ $\quad \Sigma m_1 v_1 = \Sigma m_2 v_2$

$$\left(\frac{1}{32.2}\right)(20) + 0 = \left(\frac{1}{32.2}\right)(v_A)_2 + \left(\frac{10}{32.2}\right)(v_B)_2$$

$$(v_A)_2 + 10(v_B)_2 = 20$$

$(\overset{+}{\rightarrow})$ $\quad e = \dfrac{(v_B)_2 - (v_A)_2}{(v_A)_1 - (v_B)_1}$

$$0.6 = \frac{(v_B)_2 - (v_A)_2}{20 - 0}$$

$$(v_B)_2 - (v_A)_2 = 12$$

Thus,

$$(v_B)_2 = 2.909 \text{ ft/s} \rightarrow$$

$$(v_A)_2 = -9.091 \text{ ft/s} = 9.091 \text{ ft/s} \leftarrow$$

Ball A :

$(\overset{+}{\rightarrow})$ $\quad m v_1 + \Sigma \int F\, dt = m v_2$

$$\left(\frac{1}{32.2}\right)(20) - F(0.02) = \left(\frac{1}{32.2}\right)(-9.091)$$

$F = 45.2$ lb **Ans**

***15-64.** The 0.2-lb ball bearing travels over the edge A with a velocity of $v_A = 3$ ft/s. Determine the speed at which it rebounds from the smooth inclined plane at B. Take $e = 0.8$.

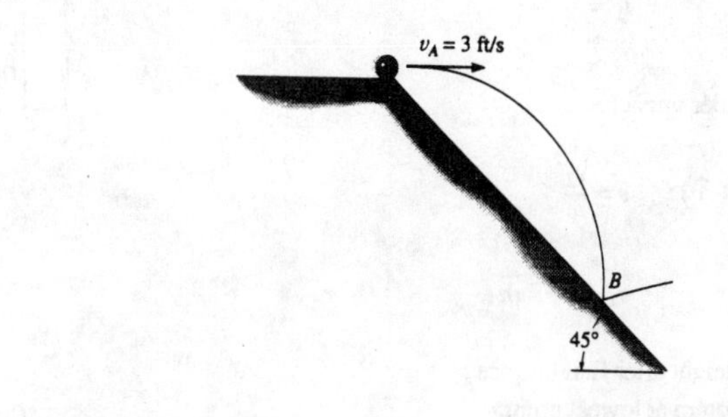

$(\overset{+}{\rightarrow})$ $\quad s = s_0 + v_0 t$

$$d = 0 + 3t$$

$(+\downarrow)$ $\quad s = s_0 + v_0 t + \frac{1}{2} a_c t^2$

$$d = 0 + 0 + \frac{1}{2}(32.2)t^2$$

$d = 0.559$ ft

$t = 0.1863$ s

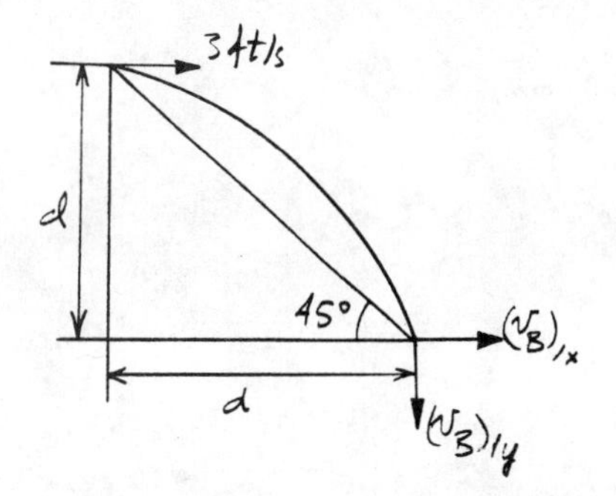

$(\overset{+}{\rightarrow})$ $\quad v = v_0$

$$(v_B)_{1x} = 3 \text{ ft/s} \rightarrow$$

$(+\downarrow)$ $\quad v = v_0 + a_c t$

$$(v_B)_{1y} = 0 + (32.2)(0.1863) = 6 \text{ ft/s} \downarrow$$

$(v_B)_1 = \sqrt{(3)^2 + (6)^2} = 6.708$ ft/s

$\theta = \tan^{-1}\left(\frac{6}{3}\right) = 63.43°$ $\qquad \phi = 63.43° - 45° = 18.43°$

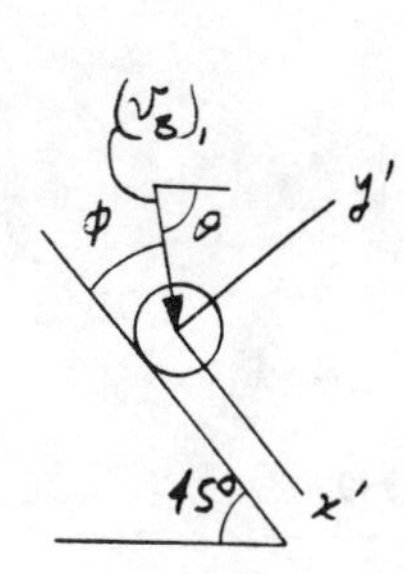

$(+\ )$ $\quad (v_B)_{2x'} = (v_B)_1 \cos\phi = 6.364$ ft/s

$(+\ )$ $\quad e = \dfrac{(v_B)_{2y'} - 0}{0 - (-6.708 \sin 18.43°)} = 0.8$

$$(v_B)_{2y'} = 1.697 \text{ ft/s}$$

$(v_B)_2 = \sqrt{(6.364)^2 + (1.697)^2} = 6.59$ ft/s **Ans**

15-65. A ball has a mass m and is dropped onto a surface from a height h. If the coefficient of restitution is e between the ball and the surface, determine the time needed for the ball to stop bouncing.

Just before impact :

$$T_1 + V_1 = T_2 + V_2$$

$$0 + mgh = \frac{1}{2}mv^2 + 0$$

$$v = \sqrt{2gh}$$

Time to fall :

$$(+\downarrow) \qquad v = v_0 + a_c t$$

$$v = v_0 + g t_1$$

$$\sqrt{2gh} = 0 + g t_1$$

$$t_1 = \sqrt{\frac{2h}{g}}$$

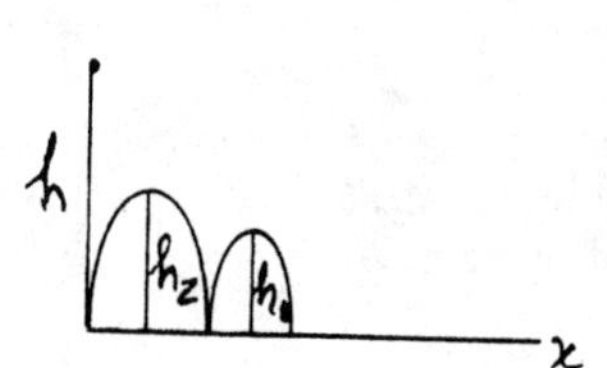

After impact :

$$(+\uparrow) \qquad e = \frac{v_2}{v}$$

$$v_2 = e\sqrt{2gh}$$

Height after first bounce :
Datum at lowest point.

$$T_2 + V_2 = T_3 + V_3$$

$$\frac{1}{2}m\left(e\sqrt{2gh}\right)^2 + 0 = 0 + mgh_2$$

$$h_2 = \frac{1}{2}e^2\left(\frac{2\,gh}{g}\right) = e^2 h$$

Time to rise to h_2 :

$$(+\uparrow) \qquad v = v_0 + a_c t$$

$$v_3 = v_2 - g t_2$$

$$0 = e\sqrt{2gh} - g t_2$$

$$t_2 = e\sqrt{\frac{2h}{g}}$$

Total time for first bounce :

$$t_{1b} = t_1 + t_2 = \sqrt{\frac{2h}{g}} + e\sqrt{\frac{2h}{g}} = \sqrt{\frac{2h}{g}}(1+e)$$

For the second bounce,

$$t_{2b} = \sqrt{\frac{2h_2}{g}}(1+e) = \sqrt{\frac{2h}{g}}(1+e)e$$

For the third bounce,

$$h_3 = e^2 h_2 = e^2\left(e^2 h\right) = e^4 h$$

$$t_{3b} = \sqrt{\frac{2h_3}{g}}(1+e) = \sqrt{\frac{2h}{g}}(1+e)e^2$$

Thus the total time for an infinite number of bounces :

$$t_{tot} = \sqrt{\frac{2h}{g}}(1+e)\left(1+e+e^2+e^3+\ldots\right)$$

$$t_{tot} = \sqrt{\frac{2h}{g}}\left(\frac{1+e}{1-e}\right) \qquad \textbf{Ans}$$

15-66. To test the manufactured properties of 2-lb steel balls, each ball is released from rest as shown and strikes the 45° smooth inclined surface. If the coefficient of restitution is to be $e = 0.8$, determine the distance s to where the ball must strike the horizontal plane at A. At what speed does the ball strike point A?

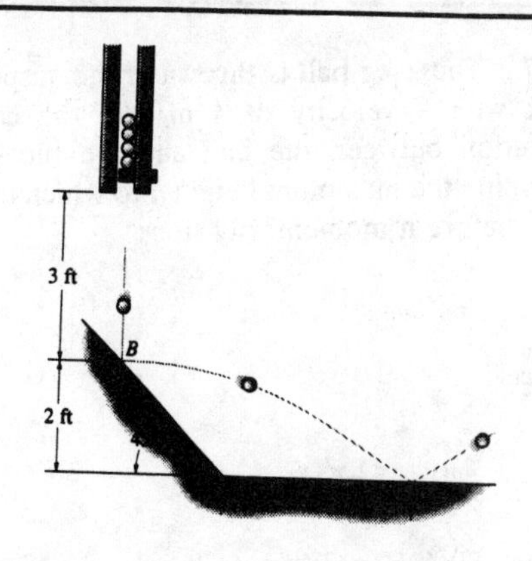

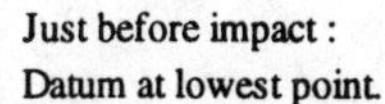

Just before impact :
Datum at lowest point.

$$T_1 + V_1 = T_2 + V_2$$

$$0 + (2)(3) = \frac{1}{2}\left(\frac{2}{32.2}\right)(v_B)_1^2 + 0$$

$$(v_B)_1 = 13.900 \text{ ft/s}$$

At B :

$$(+\searrow) \qquad \Sigma m(v_B)_{x_1} = \Sigma m(v_B)_{x_2}$$

$$\left(\frac{2}{32.2}\right)(13.900)\sin 45° = \left(\frac{2}{32.2}\right)(v_B)_2 \sin\theta$$

$$(v_B)_2 \sin\theta = 9.829 \text{ ft/s} \qquad (1)$$

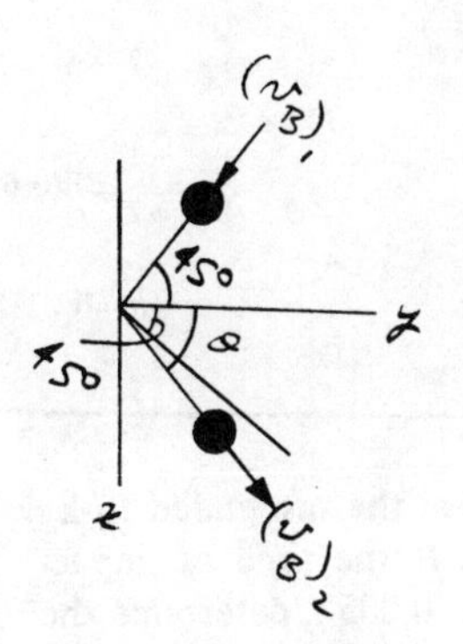

$$(+\nearrow) \qquad e = \frac{(v_B)_{y_2} - 0}{0 - (v_B)_{y_1}}$$

$$0.8 = \frac{(v_B)_2 \cos\theta - 0}{0 - (-13.900)\cos 45°}$$

$$(v_B)_2 \cos\theta = 7.863 \text{ ft/s} \qquad (2)$$

Solving Eqs. (1) and (2) :

$$(v_B)_2 = 12.587 \text{ ft/s} \qquad \theta = 51.34°$$

$$\phi = 51.34° - 45° = 6.34°$$

$$(+\downarrow) \qquad v^2 = v_0^2 + 2a_c(s - s_0)$$

$$(v_{A_y})^2 = [12.587 \sin 6.34°]^2 + 2(32.2)(2 - 0)$$

$$v_{A_y} = 11.434 \text{ ft/s}$$

$$(+\downarrow) \qquad v = v_0 + a_c t$$

$$11.434 = 12.587 \sin 6.34° + 32.2t$$

$$t = 0.3119 \text{ s}$$

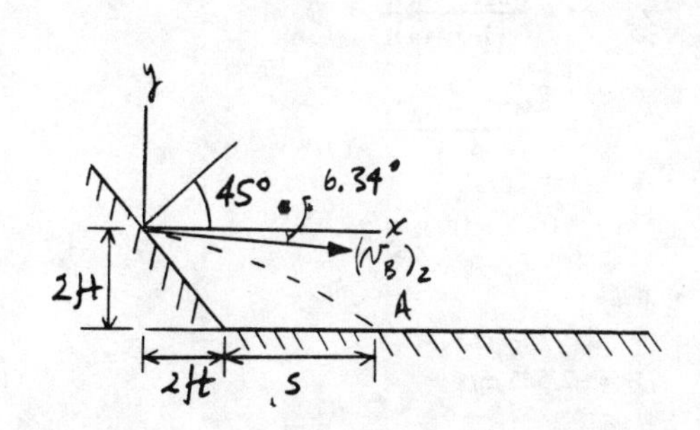

$$(\overset{+}{\rightarrow}) \qquad v_{Ax} = 12.587 \cos 6.34° = 12.510 \text{ ft/s}$$

$$s_t = v_B t$$

$$s + \frac{2}{\tan 45°} = (12.50)(0.3119)$$

$s = 1.90$ ft **Ans**

$v_A = \sqrt{(12.510)^2 + (11.434)^2} = 16.9$ ft/s **Ans**

15-67. The 2-kg ball is thrown at the suspended 20-kg block with a velocity of 4 m/s. If the coefficient of restitution between the ball and the block is $e = 0.8$, determine the maximum height h to which the block will swing before it momentarily stops.

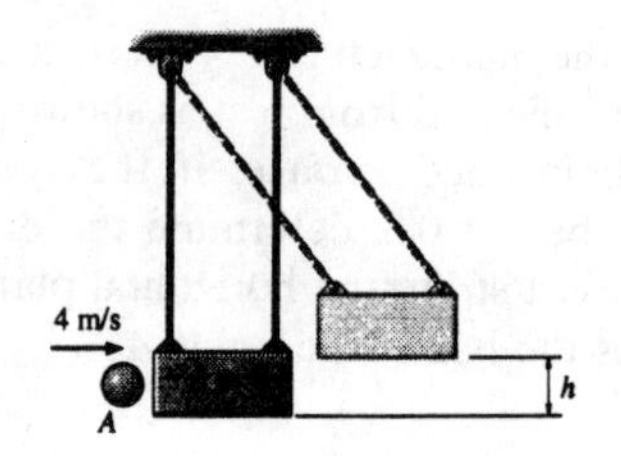

System :

$(\overset{+}{\rightarrow})$ $\Sigma m_1 v_1 = \Sigma m_2 v_2$

$$(2)(4) + 0 = (2)(v_A)_2 + (20)(v_B)_2$$

$$(v_A)_2 + 10(v_B)_2 = 4$$

$(\overset{+}{\rightarrow})$ $e = \dfrac{(v_B)_2 - (v_A)_2}{(v_A)_1 - (v_B)_1}$

$$0.8 = \frac{(v_B)_2 - (v_A)_2}{4 - 0}$$

$$(v_B)_2 - (v_A)_2 = 3.2$$

Solving :

$$(v_A)_2 = -2.545 \text{ m/s}$$

$$(v_B)_2 = 0.6545 \text{ m/s}$$

Block :
Datum at lowest point.

$$T_1 + V_1 = T_2 + V_2$$

$$\frac{1}{2}(20)(0.6545)^2 + 0 = 0 + 20(9.81)h$$

$h = 0.0218 \text{ m} = 21.8 \text{ mm}$ **Ans**

***15-68.** The 2-kg ball is thrown at the suspended 20-kg block with a velocity of 4 m/s. If the time of impact between the ball and the block is 0.005 s, determine the average normal force exerted on the block during this time. Take $e = 0.8$.

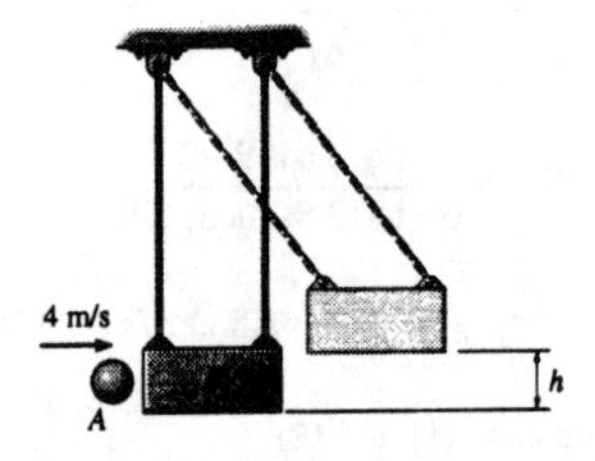

System :

$(\overset{+}{\rightarrow})$ $\Sigma m_1 v_1 = \Sigma m_2 v_2$

$$(2)(4) + 0 = (2)(v_A)_2 + (20)(v_B)_2$$

$$(v_A)_2 + 10(v_B)_2 = 4$$

$(\overset{+}{\rightarrow})$ $e = \dfrac{(v_B)_2 - (v_A)_2}{(v_A)_1 - (v_B)_1}$

$$0.8 = \frac{(v_B)_2 - (v_A)_2}{4 - 0}$$

$$(v_B)_2 - (v_A)_2 = 3.2$$

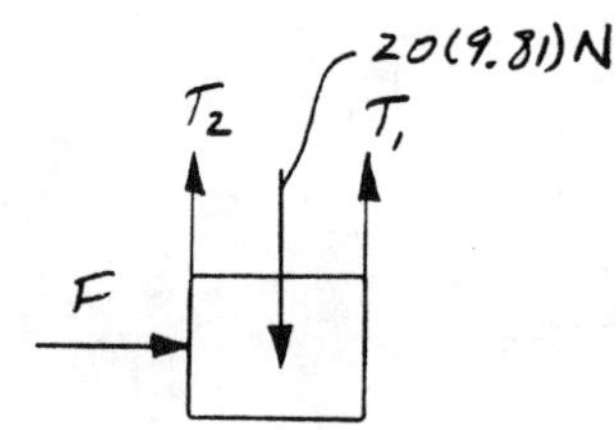

Solving :

$$(v_A)_2 = -2.545 \text{ m/s}$$

$$(v_B)_2 = 0.6545 \text{ m/s}$$

Block :

$(\overset{+}{\rightarrow})$ $m v_1 + \Sigma \int F\, dt = m v_2$

$$0 + F(0.005) = 20(0.6545)$$

$F = 2618 \text{ N} = 2.62 \text{ kN}$ **Ans**

15-69. The slider block B is confined to move within the smooth slot. It is connected to two springs, each of which has a stiffness of $k = 30$ N/m. They are originally stretched 0.5 m when $s = 0$ as shown. Determine the maximum distance, s_{max}, block B moves after it is hit by block A which is originally traveling at $(v_A)_1 = 8$ m/s. Take $e = 0.4$ and the mass of each block to be 1.5 kg.

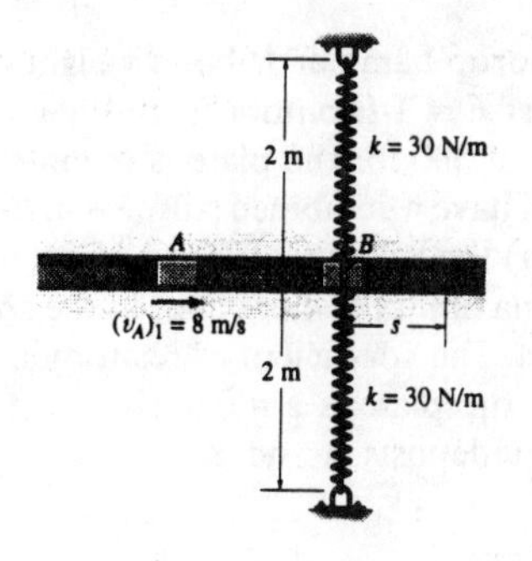

$(\overset{+}{\rightarrow})$ $\quad \Sigma m\, v_1 = \Sigma m\, v_2$

$$(1.5)(8) + 0 = (1.5)(v_A)_2 + (1.5)(v_B)_2$$

$(\overset{+}{\rightarrow})$ $\quad e = \dfrac{(v_B)_2 - (v_A)_2}{(v_A)_1 - (v_B)_1}$

$$0.4 = \frac{(v_B)_2 - (v_A)_2}{8 - 0}$$

Solving :

$(v_A)_2 = 2.40$ m/s

$(v_B)_2 = 5.60$ m/s

$T_1 + V_1 = T_2 + V_2$

$$\frac{1}{2}(1.5)(5.60)^2 + 2\left[\frac{1}{2}(30)(0.5)^2\right] = 0 + 2\left[\frac{1}{2}(30)\left(\sqrt{s_{max}^2 + 2^2} - 1.5\right)^2\right]$$

$s_{max} = 1.53$ m **Ans**

15-70. In Prob. 15-69 determine the average net force exerted between the two blocks A and B during impact if the impact occurs in 0.005 s.

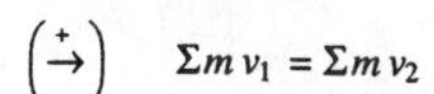

$(\overset{+}{\rightarrow})$ $\quad \Sigma m\, v_1 = \Sigma m\, v_2$

$$(1.5)(8) + 0 = (1.5)(v_A)_2 + (1.5)(v_B)_2$$

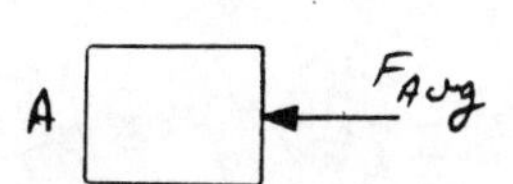

$(\overset{+}{\rightarrow})$ $\quad e = \dfrac{(v_B)_2 - (v_A)_2}{(v_A)_1 - (v_B)_1}$

$$0.4 = \frac{(v_B)_2 - (v_A)_2}{8 - 0}$$

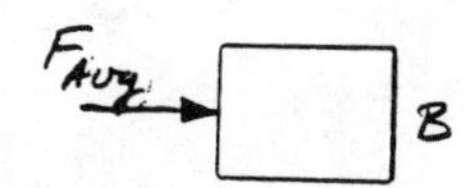

Solving :

$(v_A)_2 = 2.40$ m/s

$(v_B)_2 = 5.60$ m/s

Choosing block A :

$(\overset{+}{\rightarrow})$ $\quad m v_1 + \Sigma \int F\, dt = m v_2$

$$(1.5)(8) - F_{avg}(0.005) = 1.5(2.40)$$

$F_{avg} = 1.68$ kN **Ans**

Choosing block B :

$(\overset{+}{\rightarrow})$ $\quad m v_1 + \Sigma \int F\, dt = m v_2$

$$0 + F_{avg}(0.005) = 1.5(5.60)$$

$F_{avg} = 1.68$ kN **Ans**

15-71. The drop hammer H has a weight of 900 lb and falls from rest $h = 3$ ft onto a forged anvil plate P that has a weight of 500 lb. The plate is mounted on a set of springs which have a combined stiffness of $k_T = 500$ lb/ft. Determine (a) the velocity of P and H just after collision and (b) the maximum compression in the springs caused by the impact. The coefficient of restitution between the hammer and the plate is $e = 0.6$. Neglect friction along the vertical guideposts A and B.

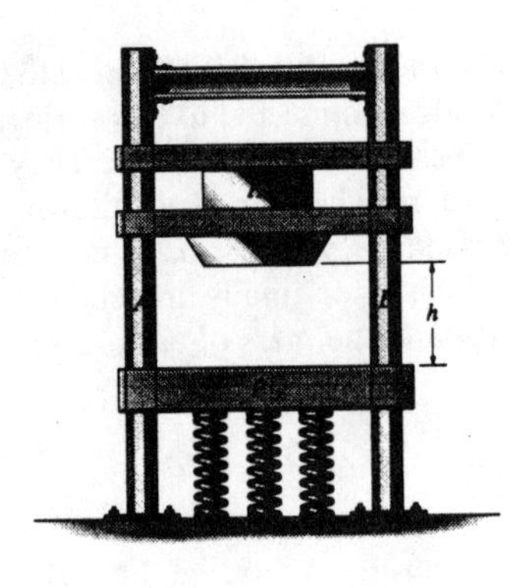

Just before impact :
Datum at lowest point.

$$T_0 + V_0 = T_1 + V_1$$

$$0 + 900(3) = \frac{1}{2}\left(\frac{900}{32.2}\right)(v_H)_1^2 + 0$$

$$(v_H)_1 = 13.90 \text{ ft/s}$$

Conservation of momentum will be applied since the force of the springs is nonimpulsive compared to the impact force.

$$(+\downarrow) \qquad m_H(v_H)_1 + m_P(v_P)_1 = m_H(v_H)_2 + m_P(v_P)_2$$

$$\left(\frac{900}{32.2}\right)(13.90) + 0 = \left(\frac{900}{32.2}\right)(v_H)_2 + \left(\frac{500}{32.2}\right)(v_P)_2$$

$$13.90 = (v_H)_2 + 0.556(v_P)_2 \qquad (1)$$

$$(+\downarrow) \qquad e = \frac{(v_P)_2 - (v_H)_2}{(v_H)_1 - (v_P)_1}$$

$$0.6 = \frac{(v_P)_2 - (v_H)_2}{13.90 - 0}$$

$$8.34 = (v_P)_2 - (v_H)_2 \qquad (2)$$

Solving Eqs. (1) and (2) :

$(v_P)_2 = 14.30 = 14.3$ ft/s **Ans**

$(v_H)_2 = 5.96$ ft/s **Ans**

The initial compression in the springs is

$F = kx; \qquad 500 = 500x_1 \qquad x_1 = 1$ ft

Datum at highest point :

$$T_1 + V_1 = T_2 + V_2$$

$$\frac{1}{2}\left(\frac{500}{32.2}\right)(14.30)^2 + \frac{1}{2}(500)(1)^2 = 0 - 500x_2 + \frac{1}{2}(500)(x_2 + 1)^2$$

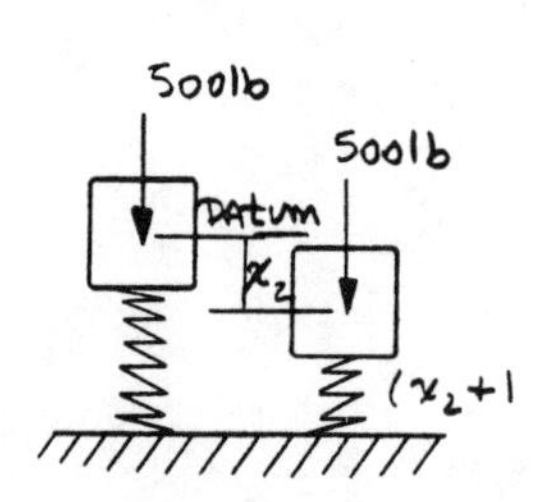

$$1837 = -500x_2 + 250x_2^2 + 500x_2 + 250$$

$$1587 = 250x_2^2$$

$$x_2 = 2.52 \text{ ft}$$

Total compression in springs is

$x = x_2 + 1 = 3.52$ ft **Ans**

***15-72.** The 8-lb ball is released from rest 10 ft from the surface of a flat plate P which weighs 6 lb. Determine the maximum compression in the spring if the impact is perfectly elastic.

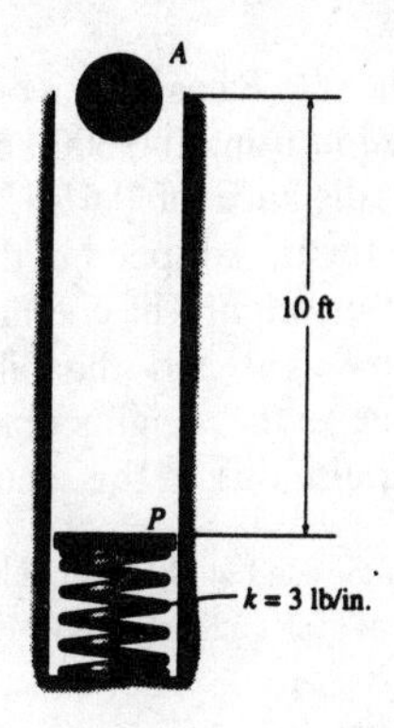

Datum at lowest point :

$$T_1 + V_1 = T_2 + V_2$$

$$0 + (8)(10) = \frac{1}{2}\left(\frac{8}{32.2}\right)(v_A)_1^2 + 0$$

$$(v_A)_1 = 25.377 \text{ ft/s}$$

$$(+\downarrow) \qquad \Sigma m v_1 = \Sigma m v_2$$

$$\frac{8}{32.2}(25.377) + 0 = \frac{8}{32.2}(v_A)_2 + \frac{6}{32.2}(v_P)_2$$

$$e = \frac{(v_P)_2 - (v_A)_2}{(v_A)_1 - (v_P)_1}$$

$$1 = \frac{(v_P)_2 - (v_A)_2}{25.377 - 0}$$

Solving,

$$(v_A)_2 = 3.625 \text{ ft/s}$$

$$(v_P)_2 = 29.00 \text{ ft/s}$$

Initially the plate is compressed

$$F_s = kx; \qquad 6 = (3)x, \qquad x = 2 \text{ in.} = \frac{1}{6} \text{ ft}$$

Datum at final plate position :

$$T_2 + V_2 = T_3 + V_3$$

$$\frac{1}{2}\left(\frac{6}{32.2}\right)(29.00)^2 + \frac{1}{2}(3)(12)\left(\frac{1}{6}\right)^2 + 6x = 0 + \frac{1}{2}(3)(12)\left(x + \frac{1}{6}\right)^2$$

$$18x^2 - 78.367 = 0$$

$$x = 2.087 \text{ ft}$$

Maximum spring compression is

$$x_{max} = 2.087 + \frac{1}{6} = 2.25 \text{ ft} \qquad \textbf{Ans}$$

15-73. The pile P has a mass of 800 kg and is being driven into *loose sand* using the 300-kg counterweight C which is dropped a distance of 0.5 m from the top of the pile. Determine the initial speed of the pile just after it is struck by the counterweight. The coefficient of restitution between the counterweight and the pile is $e = 0.1$. Neglect the impulses due to the weights of the pile and counterweight and the impulse due to the sand during the impact.

The force of the sand on the pile can be considered nonimpulsive, along with the weights of each colliding body. Hence,

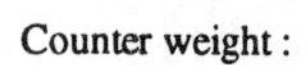

Counter weight :
Datum at lowest point.

$$T_1 + V_1 = T_2 + V_2$$

$$0 + 300(9.81)(0.5) = \frac{1}{2}(300)(v)^2 + 0$$

$$v = 3.1321 \text{ m/s}$$

System :

$$(+\downarrow) \qquad \Sigma m\, v_1 = \Sigma m\, v_2$$

$$300(3.1321) + 0 = 300(v_C)_2 + 800(v_P)_2$$

$$(v_C)_2 + 2.667(v_P)_2 = 3.1321$$

$$(+\downarrow) \qquad e = \frac{(v_P)_2 - (v_C)_2}{(v_C)_1 - (v_P)_1}$$

$$0.1 = \frac{(v_P)_2 - (v_C)_2}{3.1321 - 0}$$

$$(v_P)_2 - (v_C)_2 = 0.31321$$

Solving :

$(v_P)_2 = 0.940$ m/s **Ans**

$(v_C)_2 = 0.626$ m/s

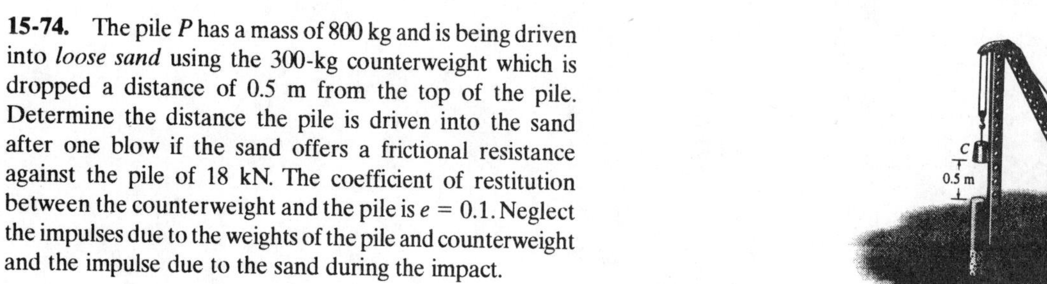

15-74. The pile P has a mass of 800 kg and is being driven into *loose sand* using the 300-kg counterweight which is dropped a distance of 0.5 m from the top of the pile. Determine the distance the pile is driven into the sand after one blow if the sand offers a frictional resistance against the pile of 18 kN. The coefficient of restitution between the counterweight and the pile is $e = 0.1$. Neglect the impulses due to the weights of the pile and counterweight and the impulse due to the sand during the impact.

The force of the sand on the pile can be considered nonimpulsive, along with the weights of each colliding body. Hence,

Counter weight :
Datum at lowest point.

$$T_1 + V_1 = T_2 + V_2$$

$$0 + 300(9.81)(0.5) = \frac{1}{2}(300)(v)^2 + 0$$

$$v = 3.1321 \text{ m/s}$$

System :

$$(+\downarrow) \qquad \Sigma m\, v_1 = \Sigma m\, v_2$$

$$300(3.1321) + 0 = 300(v_C)_2 + 800(v_P)_2$$

$$(v_C)_2 + 2.667(v_P)_2 = 3.1321$$

$$(+\downarrow) \qquad e = \frac{(v_P)_2 - (v_C)_2}{(v_C)_1 - (v_P)_1}$$

$$0.1 = \frac{(v_P)_2 - (v_C)_2}{3.1321 - 0}$$

$$(v_P)_2 - (v_C)_2 = 0.31321$$

Solving :

$(v_P)_2 = 0.9396$ m/s

$(v_C)_2 = 0.6264$ m/s

Pile :

$$T_2 + \Sigma U_{2-3} = T_3$$

$$\frac{1}{2}(800)(0.9396)^2 + 800(9.81)d - 18\,000d = 0$$

$d = 0.0348$ m $= 34.8$ mm **Ans**

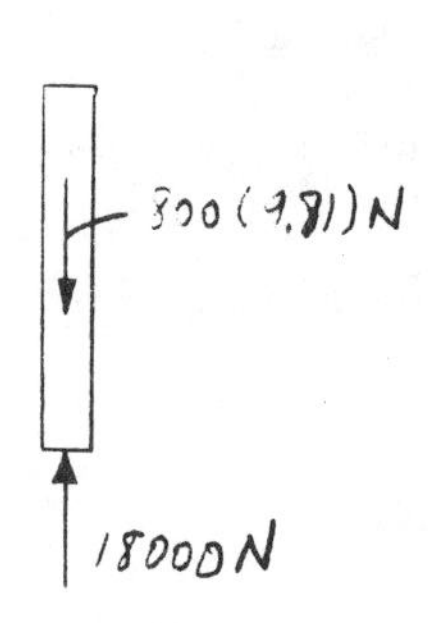

15-75. The 10-lb collar B is at rest, and when it is in the position shown the spring is unstretched. If another 1-lb collar A strikes it so that B slides 4 ft on the smooth rod before momentarily stopping, determine the velocity of A just after impact, and the average force exerted between A and B during the impact if the impact occurs in 0.002 s. The coefficient of restitution between A and B is $e = 0.5$.

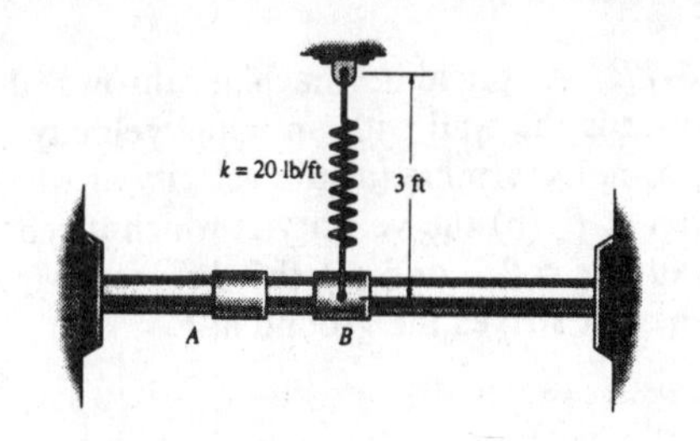

Collar B after impact :

$$T_2 + V_2 = T_3 + V_3$$

$$\frac{1}{2}\left(\frac{10}{32.2}\right)(v_B)_2^2 + 0 = 0 + \frac{1}{2}(20)(5-3)^2$$

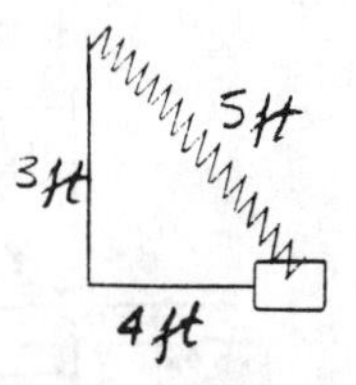

$$(v_B)_2 = 16.05 \text{ ft/s}$$

System :

$$(\overset{+}{\rightarrow}) \qquad \Sigma m_1 v_1 = \Sigma m_1 v_2$$

$$\frac{1}{32.2}(v_A)_1 + 0 = \frac{1}{32.2}(v_A)_2 + \frac{10}{32.2}(16.05)$$

$$(v_A)_1 - (v_A)_2 = 160.5$$

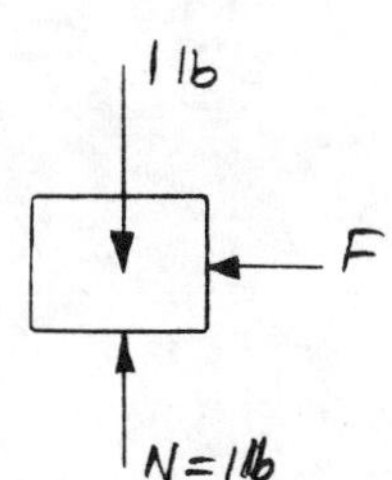

$$(\overset{+}{\rightarrow}) \qquad e = \frac{(v_B)_2 - (v_A)_2}{(v_A)_1 - (v_B)_1}$$

$$0.5 = \frac{16.05 - (v_A)_2}{(v_A)_1 - 0}$$

$$0.5(v_A)_1 + (v_A)_2 = 16.05$$

Solving :

$$(v_A)_1 = 117.7 \text{ ft/s} = 118 \text{ ft/s} \rightarrow$$

$$(v_A)_2 = -42.8 \text{ ft/s} = 42.8 \text{ ft/s} \leftarrow \qquad \textbf{Ans}$$

Collar A :

$$(\overset{+}{\rightarrow}) \qquad m v_1 + \Sigma\int F\,dt = m v_2$$

$$\left(\frac{1}{32.2}\right)(117.7) - F(0.002) = \left(\frac{1}{32.2}\right)(-42.8)$$

$$F = 2492.2 \text{ lb} = 2.49 \text{ kip} \qquad \textbf{Ans}$$

***15-76.** The ball is ejected from the tube with a horizontal velocity of $v_1 = 8$ ft/s as shown. If the coefficient of restitution between the ball and the ground is $e = 0.8$, determine (a) the velocity of the ball just after it rebounds from the ground and (b) the maximum height to which the ball rises after the first bounce.

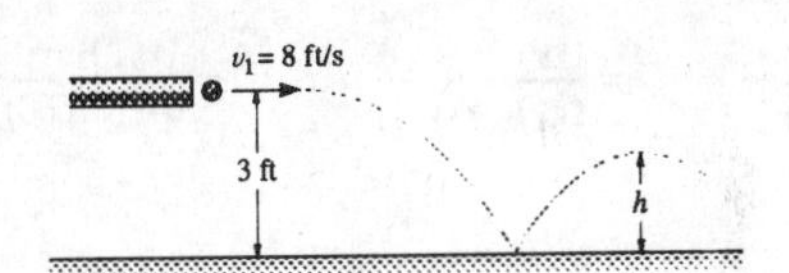

$$(+\uparrow) \qquad v^2 = v_0^2 + 2a_c(s - s_0)$$

$$(v_y)_1^2 = 0 + 2(-32.2)(-3-0)$$

$$(v_y)_1 = 13.90 \text{ ft/s} \downarrow$$

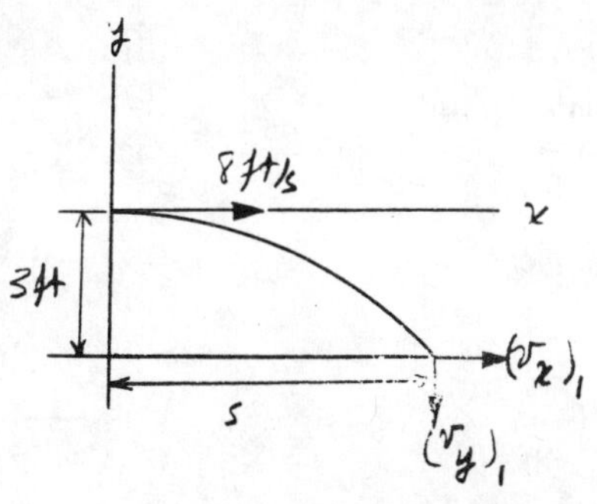

In y direction :

$$(+\uparrow) \qquad e = \frac{(v_B)_2 - (v_A)_2}{(v_A)_1 - (v_B)_1}; \qquad 0.8 = \frac{(v_y)_2 - 0}{0 - (-13.90)}$$

$$(v_y)_2 = 11.12 \text{ ft/s} \uparrow$$

$$(v_x)_2 = 8 \text{ ft/s} \rightarrow$$

$$v_2 = \sqrt{(8)^2 + (11.12)^2} = 13.7 \text{ ft/s} \qquad \textbf{Ans}$$

$$\theta = \tan^{-1}\left(\frac{11.12}{8}\right) = 54.3° \qquad \textbf{Ans}$$

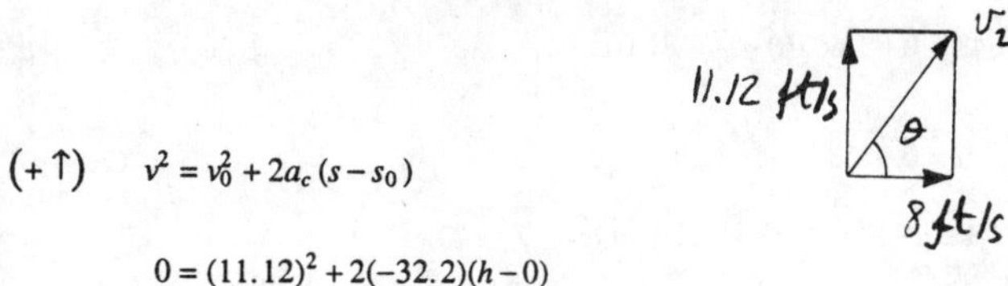

$$(+\uparrow) \qquad v^2 = v_0^2 + 2a_c(s - s_0)$$

$$0 = (11.12)^2 + 2(-32.2)(h - 0)$$

$$h = 1.92 \text{ ft} \qquad \textbf{Ans}$$

15-77. A pitching machine throws the 0.5-kg ball towards the wall with an initial velocity $v_A = 10$ m/s as shown. Determine (a) the velocity at which it strikes the wall at B, (b) the velocity at which it rebounds from the wall if $e = 0.5$, and (c) the distance s from the wall to where it strikes the ground at C.

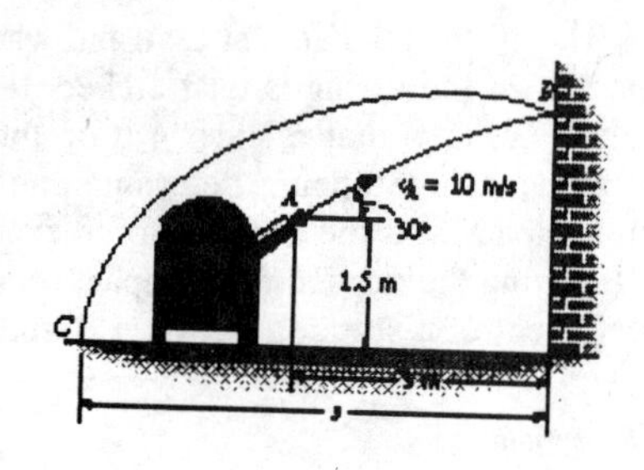

(a)

$(v_B)_{x1} = 10\cos 30° = 8.660$ m/s →

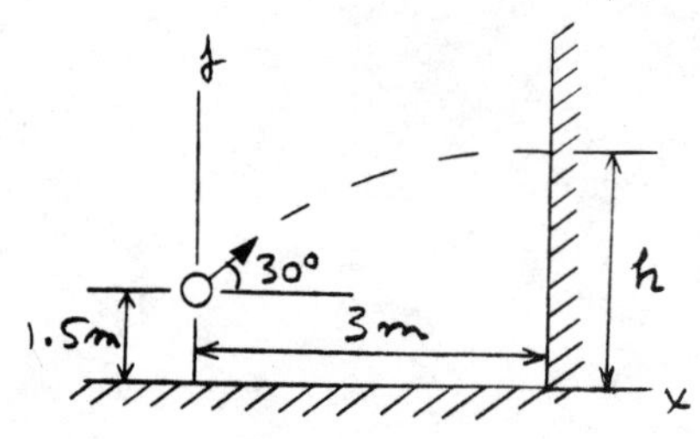

$(\overset{+}{\rightarrow})$ $\quad s = s_0 + v_0 t$

$$3 = 0 + 10\cos 30° t$$

$$t = 0.3464 \text{ s}$$

$(+\uparrow)$ $\quad v = v_0 + a_c t$

$$(v_B)_{y1} = 10\sin 30° - 9.81(0.3464) = 1.602 \text{ m/s} \uparrow$$

$(+\uparrow)$ $\quad s = s_0 + v_0 t + \frac{1}{2}a_c t^2$

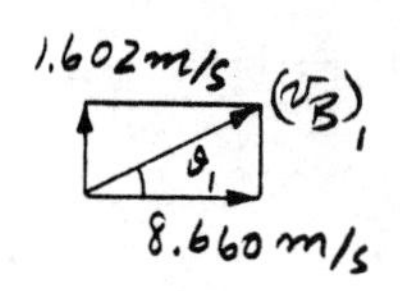

$$h = 1.5 + 10\sin 30°(0.3464) - \frac{1}{2}(9.81)(0.3464)^2$$

$$= 2.643 \text{ m}$$

$(v_B)_1 = \sqrt{(1.602)^2 + (8.660)^2} = 8.81$ m/s **Ans**

$\theta_1 = \tan^{-1}\left(\frac{1.602}{8.660}\right) = 10.5°$ ∠ **Ans**

(b)

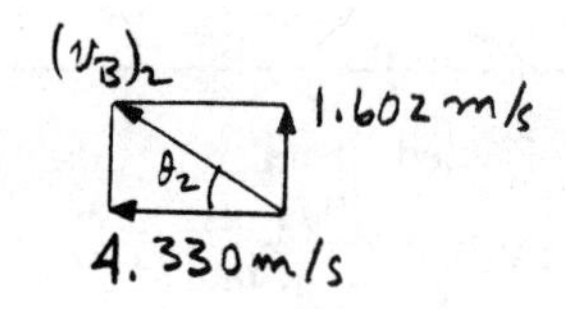

$(\overset{+}{\rightarrow})$ $\quad e = \frac{(v_B)_2 - (v_A)_2}{(v_A)_1 - (v_B)_1}; \qquad 0.5 = \frac{(v_{Bx})_2 - 0}{0 - (8.660)}$

$(v_{Bx})_2 = 4.330$ m/s ←

$(v_{By})_2 = (v_{By})_1 = 1.602$ m/s ↑

$(v_B)_2 = \sqrt{(4.330)^2 + (1.602)^2} = 4.62$ m/s **Ans**

$\theta_2 = \tan^{-1}\left(\frac{1.602}{4.330}\right) = 20.3°$ **Ans**

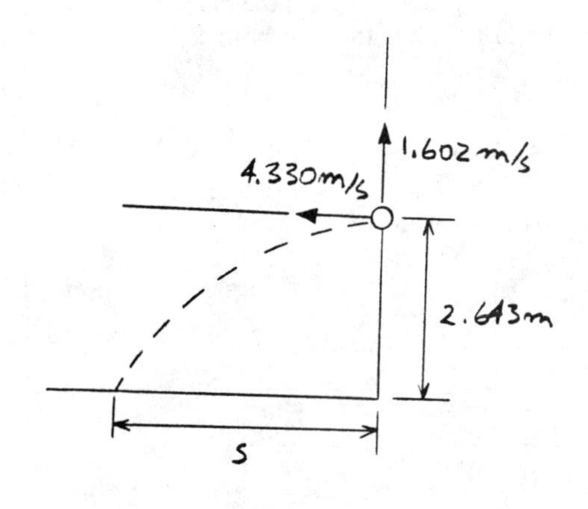

(c)

$(+\uparrow)$ $\quad s = s_0 + v_0 t + \frac{1}{2}a_c t^2$

$$-2.643 = 0 + 1.602(t) - \frac{1}{2}(9.81)(t)^2$$

$$t = 0.9153 \text{ s}$$

$(\overset{+}{\leftarrow})$ $\quad s = s_0 + v_0 t$

$s = 0 + 4.330(0.9153) = 3.96$ m **Ans**

15-78. The 5-lb box B is dropped from rest 5 ft from the top of the 10-lb plate P, which is supported by the spring having a stiffness of $k = 30$ lb/ft. If $e = 0.6$ between the box and plate, determine the maximum compression imparted to the spring. Neglect the mass of the spring.

Box :

Datum at P.

$T_1 + V_1 = T_2 + V_2$

$$0 + 5(5) = \frac{1}{2}\left(\frac{5}{32.2}\right)(v_B)_1^2 + 0$$

$(v_B)_1 = 17.94$ ft/s

System :

$(+\downarrow) \qquad \Sigma m v_1 = \Sigma m v_2$

$$\left(\frac{5}{32.2}\right)(17.94) + 0 = \left(\frac{5}{32.2}\right)(v_B)_2 + \left(\frac{10}{32.2}\right)(v_P)_2$$

$$(+\downarrow) \qquad e = \frac{(v_P)_2 - (v_B)_2}{(v_B)_1 - (v_P)_1}; \qquad 0.6 = \frac{(v_P)_2 - (v_B)_2}{17.94 - 0}$$

Solving :

$(v_P)_2 = 9.570$ ft/s $\downarrow$

$(v_B)_2 = 1.196$ ft/s $\uparrow$

Initial compression of spring :

$$F_s = kx; \qquad x_1 = \frac{F_s}{k} = \frac{10}{30} = 0.333 \text{ ft}$$

Datum at initial position of plate :

$T_1 + V_1 = T_2 + V_2$

$$\frac{1}{2}\left(\frac{10}{32.2}\right)(9.570)^2 + \frac{1}{2}(30)(0.333)^2 = 0 + \frac{1}{2}(30)(x' + 0.333)^2 - 10x'$$

$x' = 0.974$ ft

Thus,

$x_{max} = 0.974 + 0.333 = 1.31$ ft **Ans**

15-79. The 200-g billiard ball is moving with a speed of 2.5 m/s when it strikes the side of the pool table at A. If the coefficient of restitution between the ball and the side of the table is $e = 0.6$, determine the speed of the ball just after striking the table twice, i.e., at A, then at B. Neglect the size of the ball.

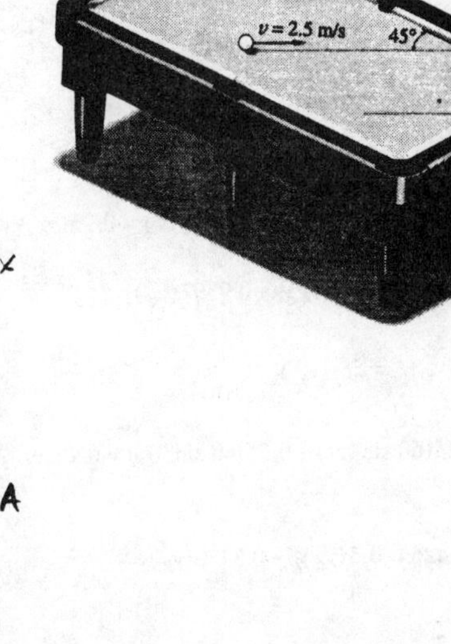

At A :

$(v_A)_{y1} = 2.5(\sin 45°) = 1.7678$ m/s $\rightarrow$

$$e = \frac{(v_{A_y})_2}{(v_{A_y})_1}; \qquad 0.6 = \frac{(v_{A_y})_2}{1.7678}$$

$(v_{A_y})_2 = 1.061$ m/s $\leftarrow$

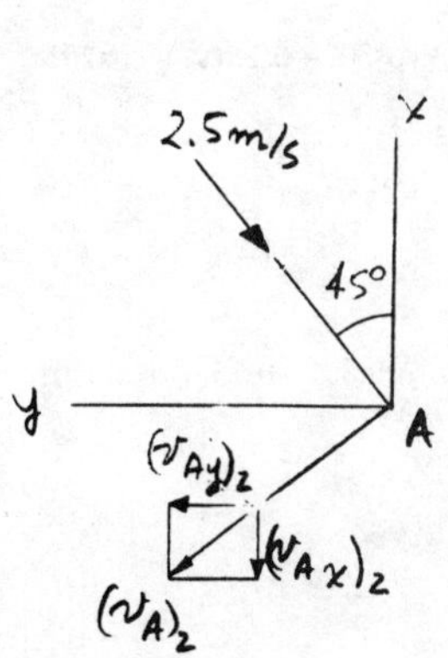

$(v_{A_x})_2 = (v_{A_x})_1 = 2.5 \cos 45° = 1.7678$ m/s $\downarrow$

At B :

$$e = \frac{(v_{B_x})_3}{(v_{B_x})_2}; \qquad 0.6 = \frac{(v_{B_x})_3}{1.7678}$$

$(v_{B_x})_3 = 1.061$ m/s

$(v_{B_y})_3 = (v_{A_y})_2 = 1.061$ m/s

Hence,

$(v_B)_3 = \sqrt{(1.061)^2 + (1.061)^2} = 1.50$ m/s **Ans**

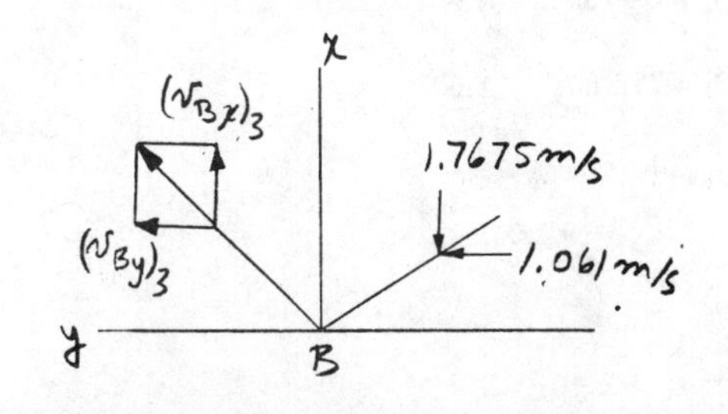

***15-80.** Two smooth billiard balls A and B each have a mass of 200 g. If A strikes B with a velocity $(v_A)_1 =$ 1.5 m/s as shown, determine their final velocities just after collision. Ball B is originally at rest and the coefficient of restitution is $e = 0.85$. Neglect the size of each ball.

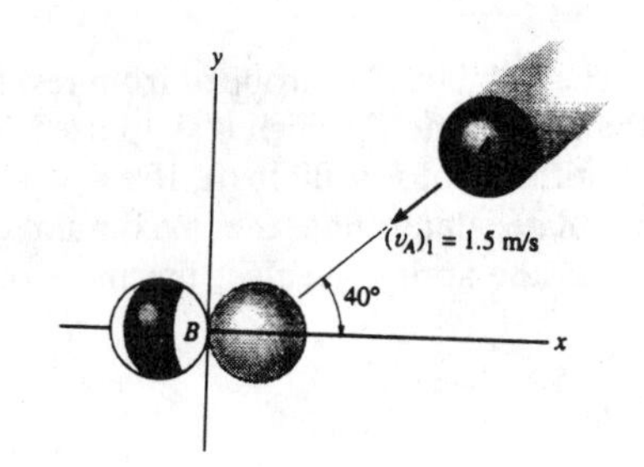

$(v_{A_x})_1 = -1.5\cos 40° = -1.1491$ m/s

$(v_{A_y})_1 = -1.5\sin 40° = -0.9642$ m/s

$(\overset{+}{\rightarrow})\quad m_A(v_{A_x})_1 + m_B(v_{B_x})_1 = m_A(v_{A_x})_2 + m_B(v_{B_x})_2$

$-0.2(1.1491) + 0 = 0.2(v_{A_x})_2 + 0.2(v_{B_x})_2$

$(\overset{+}{\rightarrow})\quad e = \dfrac{(v_{A_x})_2 - (v_{B_x})_2}{(v_{B_x})_1 - (v_{A_x})_1};\qquad 0.85 = \dfrac{(v_{A_x})_2 - (v_{B_x})_2}{1.1491}$

Solving ,

$(v_{A_x})_2 = -0.08618$ m/s

$(v_{B_x})_2 = -1.0629$ m/s

For A :

$(+\downarrow)\quad m_A(v_{A_y})_1 = m_A(v_{A_y})_2$

$(v_{A_y})_2 = 0.9642$ m/s

For B :

$(+\uparrow)\quad m_B(v_{B_y})_1 = m_B(v_{B_y})_2$

Hence, $\quad (v_{B_y})_2 = 0$

$(v_B)_2 = (v_{B_x})_2 = 1.06$ m/s $\leftarrow$ **Ans**

$(v_A)_2 = \sqrt{(-0.08618)^2 + (0.9642)^2} = 0.968$ m/s **Ans**

$(\theta_A)_2 = \tan^{-1}\left(\dfrac{0.08618}{0.9642}\right) = 5.11°$ **Ans**

15-81. The two hockey pucks A and B each have a mass of 250 g. If they collide at O and are deflected along the colored paths, determine their speeds just after impact. Assume that the icy surface over which they slide is smooth. *Hint:* Since the y' axis is *not* along the line of impact, apply the conservation of momentum along the x' and y' axes.

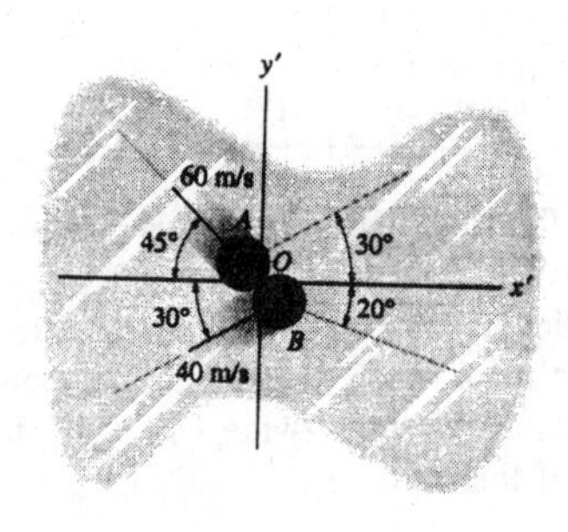

$(\overset{+}{\rightarrow})\quad \Sigma m(v_{x'})_1 = \Sigma m(v_{x'})_2$

$0.25(60\cos 45°) + 0.25(40\cos 30°) = 0.25(v_A)_2\cos 30° + 0.25(v_B)_2\cos 20°$

$77.067 = 0.8660(v_A)_2 + 0.9397(v_B)_2$

$(+\uparrow)\quad \Sigma m(v_{y'})_1 = \Sigma m(v_{y'})_2$

$-0.25(60\sin 45°) + 0.25(40\sin 30°) = 0.25(v_A)_2\sin 30° - 0.25(v_B)_2\sin 20°$

$-22.426 = 0.5(v_A)_2 - 0.3420(v_B)_2$

Solving ,

$(v_A)_2 = 6.90$ m/s **Ans**

$(v_B)_2 = 75.7$ m/s **Ans**

15-82. The two disks A and B have a mass of 3 kg and 5 kg, respectively. If they collide with the initial velocities shown, determine their velocities just after impact. The coefficient of restitution is $e = 0.65$.

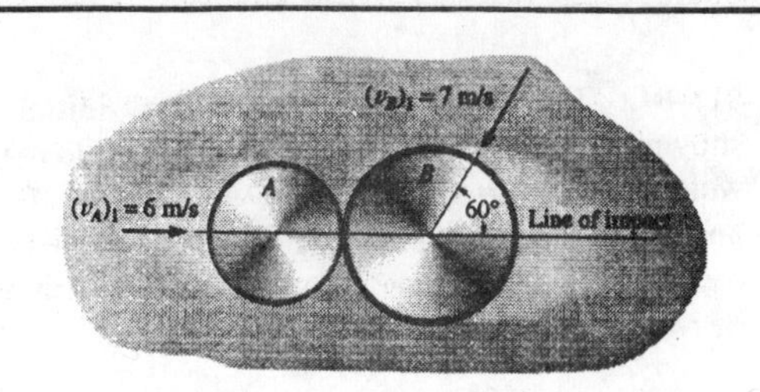

$(v_{A_x})_1 = 6$ m/s $\quad (v_{A_y})_1 = 0$

$(v_{B_x})_1 = -7\cos 60° = -3.5$ m/s $\quad (v_{B_y})_1 = -7\sin 60° = -6.062$ m/s

$(\overset{+}{\rightarrow}) \quad m_A(v_{A_x})_1 + m_B(v_{B_x})_1 = m_A(v_{A_x})_2 + m_B(v_{B_x})_2$

$3(6) - 5(3.5) = 3(v_A)_{x2} + 5(v_B)_{x2}$

$(\overset{+}{\rightarrow}) \quad e = \dfrac{(v_{B_x})_2 - (v_{A_x})_2}{(v_{A_x})_1 - (v_{B_x})_1}; \quad 0.65 = \dfrac{(v_{B_x})_2 - (v_{A_x})_2}{6-(-3.5)}$

$(v_{B_x})_2 - (v_{A_x})_2 = 6.175$

Solving ,

$(v_{A_x})_2 = -3.80$ m/s $\quad (v_{B_x})_2 = 2.378$ m/s

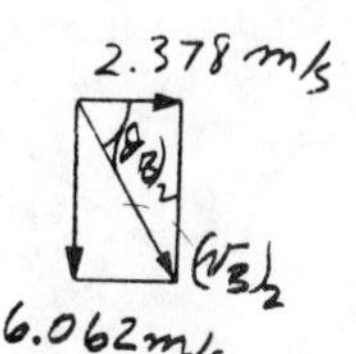

$(+\uparrow) \quad m_A(v_{A_y})_1 = m_A(v_{A_y})_2$

$(v_{A_y})_2 = 0$

$(+\uparrow) \quad m_B(v_{B_y})_1 = m_B(v_{B_y})_2$

$(v_{B_y})_2 = -6.062$ m/s

$(v_A)_2 = \sqrt{(3.80)^2 + (0)^2} = 3.80$ m/s $\leftarrow$ **Ans**

$(v_B)_2 = \sqrt{(2.378)^2 + (-6.062)^2} = 6.51$ m/s **Ans**

$(\theta_B)_2 = \tan^{-1}\left(\dfrac{6.062}{2.378}\right) = 68.6°$ **Ans**

15-83. Two cars A and B, each having a weight of 4000 lb, collide on the icy pavement of an intersection. The direction of motion of each car after collision is measured from snow tracks as shown. If the driver in car A states that he was going 44 ft/s (30 mi/h) just before collision and that after collision he applied the brakes so that his car skidded 10 ft before stopping, determine the approximate speed of car B just before the collision. Assume that the coefficient of kinetic friction between the car wheels and the pavement is $\mu_k = 0.15$. *Note:* The line of impact has not been defined; however, this information is not needed for the solution.

Velocity of A just after collision :

$T_1 + \Sigma U_{1-2} = T_2$

$\dfrac{1}{2}\left(\dfrac{4000}{32.2}\right)(v_A)_2^2 - (0.15)(4000)(10) = 0$

$(v_A)_2 = 9.829$ ft/s

$(v_A)_1 = 44$ ft/s

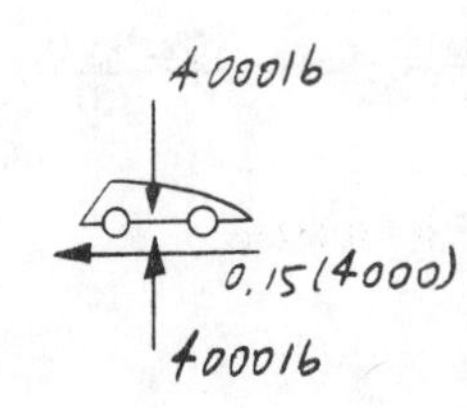

$(\overset{+}{\rightarrow}) \quad \Sigma m_1(v_x)_1 = \Sigma m_2(v_x)_2$

$\left(\dfrac{4000}{32.2}\right)(44) + 0 = \left(\dfrac{4000}{32.2}\right)(9.829)\sin 40° + \left(\dfrac{4000}{32.2}\right)(v_B)_2\cos 30°$

$(v_B)_2 = 43.51$ ft/s

$(+\uparrow) \quad \Sigma m_1(v_y)_1 = \Sigma m_2(v_y)_2$

$0 - \left(\dfrac{4000}{32.2}\right)(v_B)_1 = -\left(\dfrac{4000}{32.2}\right)(9.829)\cos 40° - \left(\dfrac{4000}{32.2}\right)(43.51)\sin 30°$

$(v_B)_1 = 29.3$ ft/s **Ans**

***15-84.** Two coins A and B have the initial velocities shown just before they collide at point O. If they have weights of $W_A = 13.2(10^{-3})$ lb and $W_B = 6.60(10^{-3})$ lb and the surface upon which they slide is smooth, determine their speeds just after impact. The coefficient of restitution is $e = 0.65$.

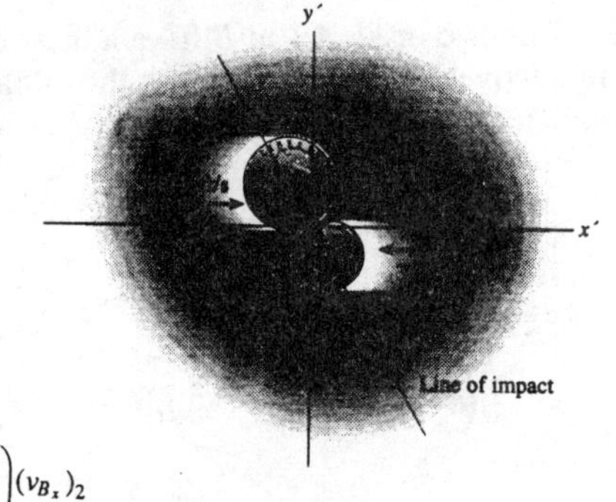

$(+\searrow)\quad m_A\,(v_{A_x})_1 + m_B\,(v_{B_x})_1 = m_A\,(v_{A_x})_2 + m_B\,(v_{B_x})_2$

$$\left(\frac{13.2(10^{-3})}{32.2}\right)2\sin 30^\circ - \left(\frac{6.6(10^{-3})}{32.2}\right)3\sin 30^\circ = \left(\frac{13.2(10^{-3})}{32.2}\right)(v_{A_x})_2 + \left(\frac{6.6(10^{-3})}{32.2}\right)(v_{B_x})_2$$

$(+\searrow)\quad e = \dfrac{(v_{B_x})_2 - (v_{A_x})_2}{(v_{A_x})_1 - (v_{B_x})_1};\qquad 0.65 = \dfrac{(v_{B_x})_2 - (v_{A_x})_2}{2\sin 30^\circ - (-3\sin 30^\circ)}$

Solving :

$(v_{A_x})_2 = -0.3750$ ft/s

$(v_{B_x})_2 = 1.250$ ft/s

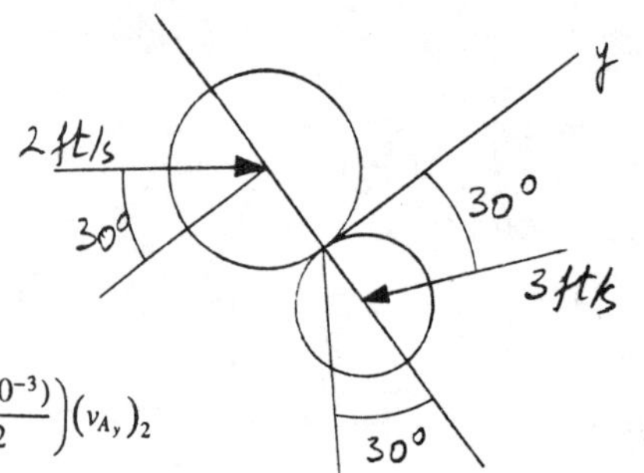

$(+\nearrow)\quad m_A\left(v_{A_y}\right)_2 = m_A\left(v_{A_y}\right)_2$

$$\left(\frac{13.2(10^{-3})}{32.2}\right)2\cos 30^\circ = \left(\frac{13.2(10^{-3})}{32.2}\right)\left(v_{A_y}\right)_2$$

$\left(v_{A_y}\right)_2 = 1.732$ ft/s

$(+\nearrow)\quad m_B\left(v_{B_y}\right)_1 = m_B\left(v_{B_y}\right)_2$

$$\left(\frac{6.6(10^{-3})}{32.2}\right)3\cos 30^\circ = \left(\frac{6.6(10^{-3})}{32.2}\right)\left(v_{B_y}\right)_2$$

$\left(v_{B_y}\right)_2 = 2.598$ ft/s

Thus,

$(v_B)_2 = \sqrt{(1.250)^2 + (2.598)^2} = 2.88$ ft/s **Ans**

$(v_A)_2 = \sqrt{(-0.3750)^2 + (1.732)^2} = 1.77$ ft/s **Ans**

15-85. The 2-kg ball is thrown so that it is traveling horizontally at 10 m/s when it strikes the 6-kg block as it is traveling down the inclined plane at 1 m/s. If the coefficient of restitution between the ball and the block is $e = 0.6$, determine the speeds of the ball and the block just after the impact. Also, what distance does B slide up the plane before it stops? The coefficient of kinetic friction between the block and the plane is $\mu_k = 0.4$.

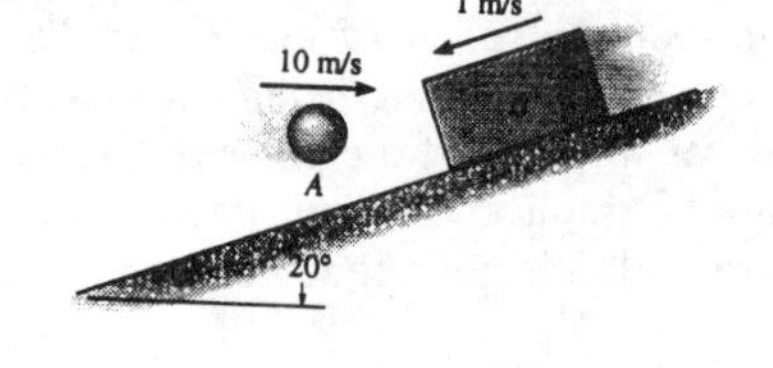

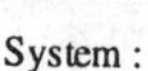

System :

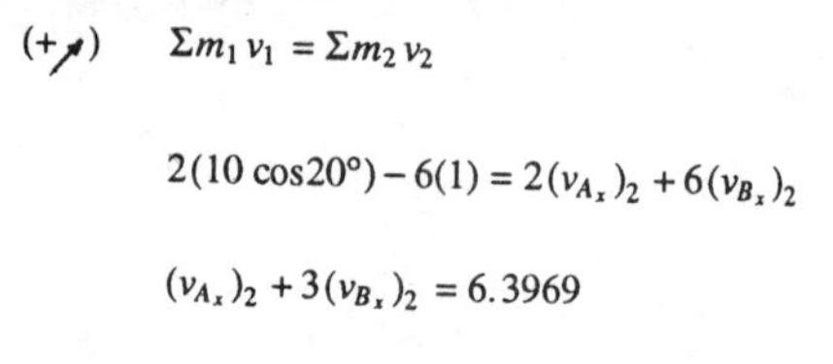

$(+\nearrow)\quad \Sigma m_1 v_1 = \Sigma m_2 v_2$

$2(10\cos 20^\circ) - 6(1) = 2(v_{A_x})_2 + 6(v_{B_x})_2$

$(v_{A_x})_2 + 3(v_{B_x})_2 = 6.3969$

$(+\nearrow)\quad e = \dfrac{(v_{B_x})_2 - (v_{A_x})_2}{(v_{A_x})_1 - (v_{B_x})_1};\qquad 0.6 = \dfrac{(v_{B_x})_2 - (v_{A_x})_2}{10\cos 20^\circ - (-1)}$

$(v_{B_x})_2 - (v_{A_x})_2 = 6.23816$

Solving :

$(v_{A_x})_2 = -3.0794$ m/s

$(v_{B_x})_2 = 3.1588$ m/s

Ball A :

$(+\nwarrow)\quad m_A\left(v_{A_y}\right)_1 = m_A\left(v_{A_y}\right)_2$

$m_A(-10\sin 20^\circ) = m_A\left(v_{A_y}\right)_2$

$\left(v_{A_y}\right)_2 = -3.4202$ m/s

Thus,

$(v_A)_2 = \sqrt{(-3.0794)^2 + (-3.4202)^2} = 4.60$ m/s **Ans**

$(v_B)_2 = 3.1588 = 3.16$ m/s **Ans**

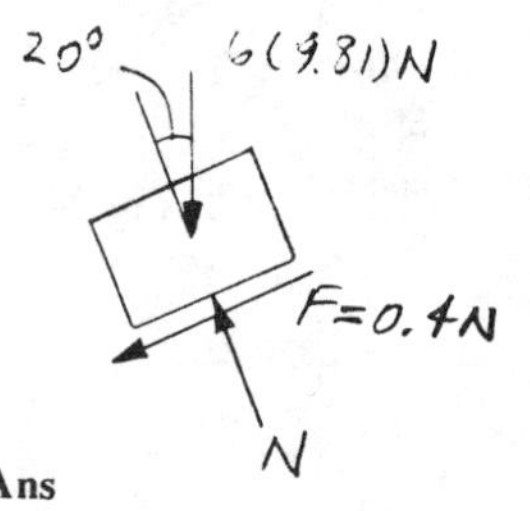

$+\nwarrow\ \Sigma F_y = 0;\qquad -6(9.81)\cos 20^\circ + N = 0\qquad N = 55.31$ N

$T_1 + \Sigma U_{1-2} = T_2$

$\frac{1}{2}(6)(3.1588)^2 - 6(9.81)\sin 20^\circ d - 0.4(55.31)d = 0$

$d = 0.708$ m **Ans**

15-86. The 2-kg ball is thrown so that it is traveling horizontally at 10 m/s when it strikes the 6-kg block as it is traveling down the smooth inclined plane at 1 m/s. If the coefficient of restitution between the ball and the block is $e = 0.6$, and the impact occurs in 0.006 s, determine the average impulsive force between the ball and block.

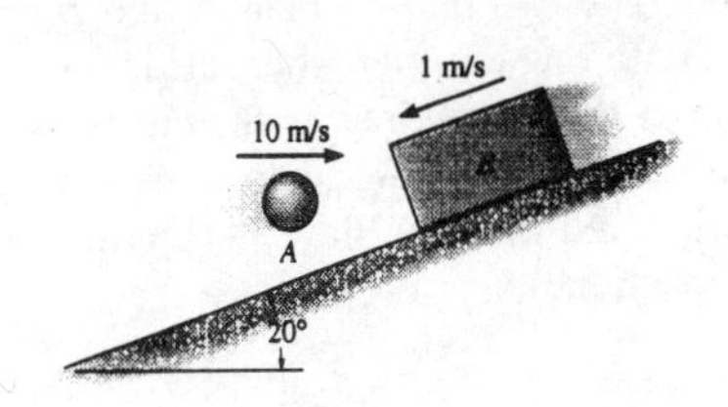

System :

$(+\nearrow)\quad \Sigma m_1 v_1 = \Sigma m_2 v_2$

$$2(10\cos 20°) - 6(1) = 2(v_{A_x})_2 + 6(v_{B_x})_2$$

$$(v_{A_x})_2 + 3(v_{B_x})_2 = 6.3969$$

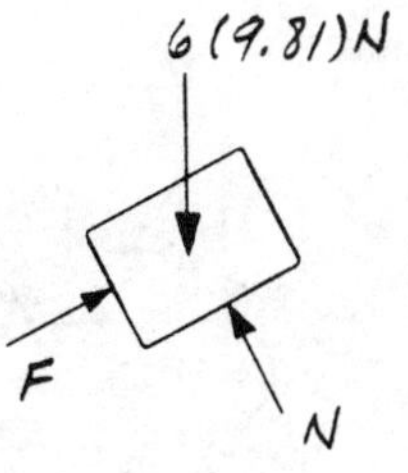

$(+\nearrow)\quad e = \dfrac{(v_{B_x})_2 - (v_{A_x})_2}{(v_{A_x})_1 - (v_{B_x})_1};\qquad 0.6 = \dfrac{(v_{B_x})_2 - (v_{A_x})_2}{10\cos 20° - (-1)}$

$$(v_{B_x})_2 - (v_{A_x})_2 = 6.23816$$

Solving :

$(v_{A_x})_2 = -3.0794$ m/s

$(v_{B_x})_2 = 3.1588$ m/s

Block B :
Neglect impulse of weight.

$(+\nearrow)\quad m v_1 + \Sigma \int F\, dt = m v_2$

$$-6(1) + F(0.006) = 6(3.1588)$$

$F = 4.16$ kN **Ans**

15-87. Two disks A and B each have a weight of 2 lb and the initial velocities shown just before they collide. If the coefficient of restitution is $e = 0.5$, determine their speeds just after impact.

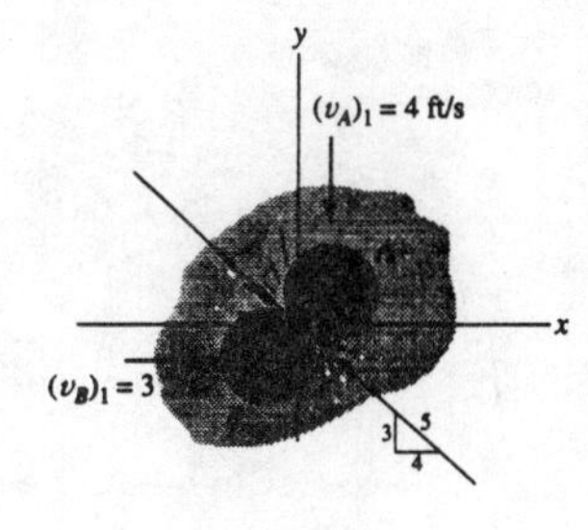

System :

$(+\nearrow)\quad \Sigma m v_{1x} = \Sigma m v_{2x}$

$$\frac{2}{32.2}(3)\left(\frac{3}{5}\right) - \frac{2}{32.2}(4)\left(\frac{4}{5}\right) = \frac{2}{32.2}(v_B)_{2x} + \frac{2}{32.2}(v_A)_{2x}$$

$(+\nearrow)\quad e = \dfrac{(v_{A_x})_2 - (v_{B_x})_2}{(v_{B_x})_1 - (v_{A_x})_1};\qquad 0.5 = \dfrac{(v_{A_x})_2 - (v_{B_x})_2}{3\left(\frac{3}{5}\right) - \left[-4\left(\frac{4}{5}\right)\right]}$

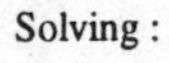

Solving,

$(v_{A_x})_2 = 0.550$ ft/s

$(v_{B_x})_2 = -1.95$ ft/s $= 1.95$ ft/s $\leftarrow$

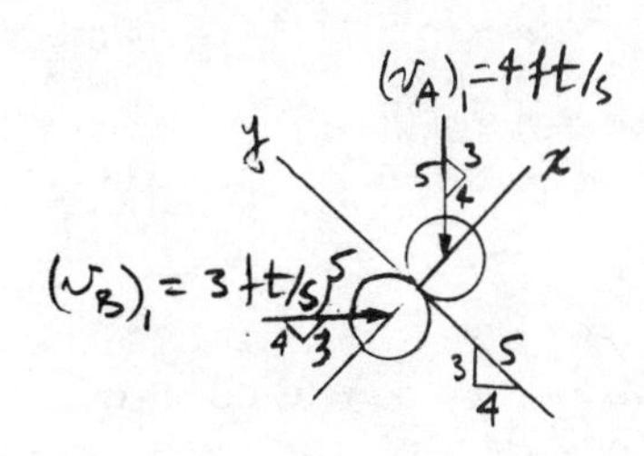

Ball A :

$(+\nwarrow)\quad m_A (v_{A_y})_1 = m_A (v_{A_y})_2$

$$-\frac{2}{32.2}(4)\left(\frac{3}{5}\right) = \frac{2}{32.2}(v_{A_y})_2$$

$$(v_{A_y})_2 = -2.40 \text{ ft/s}$$

Ball B :

$(+\nwarrow)\quad m_B (v_{B_y})_1 = m_B (v_{B_y})_2$

$$-\frac{2}{32.2}(3)\left(\frac{4}{5}\right) = \frac{2}{32.2}(v_{B_y})_2$$

$$(v_{B_y})_2 = -2.40 \text{ ft/s}$$

Thus,

$(v_A)_2 = \sqrt{(0.550)^2 + (2.40)^2} = 2.46$ ft/s **Ans**

$(v_B)_2 = \sqrt{(1.95)^2 + (2.40)^2} = 3.09$ ft/s **Ans**

***15-88.** The two billiard balls A and B are originally in contact with one another when a third ball C strikes each of them at the same time as shown. If ball C remains at rest after the collision, determine the coefficient of restitution. All the balls have the same mass. Neglect the size of each ball.

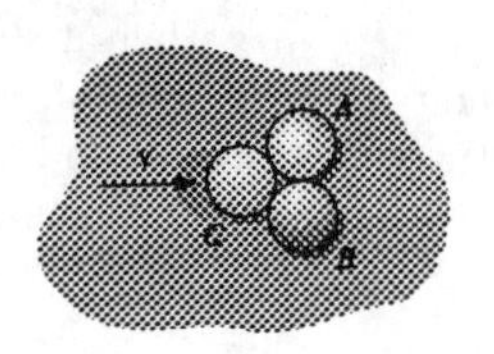

Conservation of " x" momentum :

$$(\overset{+}{\rightarrow}) \qquad mv = 2mv' \cos 30°$$

$$v = 2v' \cos 30° \qquad (1)$$

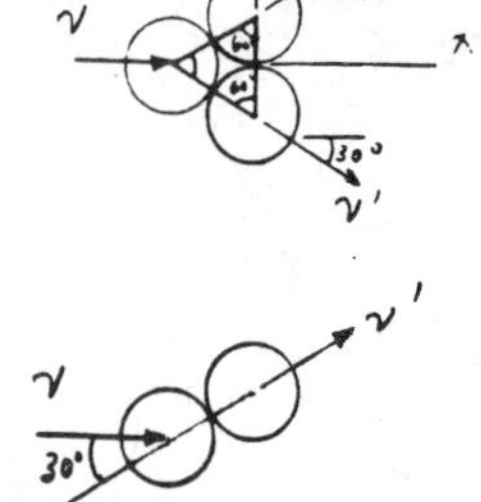

Coefficient of restitution :

$$(+\nearrow) \qquad e = \frac{v'}{v \cos 30°} \qquad (2)$$

Substituting Eq. (1) into Eq. (2) yields :

$$e = \frac{v'}{2v' \cos^2 30°} = \frac{2}{3} \qquad \textbf{Ans}$$

15-89. Ball A strikes ball B with an initial velocity of $(\mathbf{v}_A)_1$ as shown. If both balls have the same mass and the collision is perfectly elastic, determine the angle θ after collision. Ball B is originally at rest. Neglect the size of each ball.

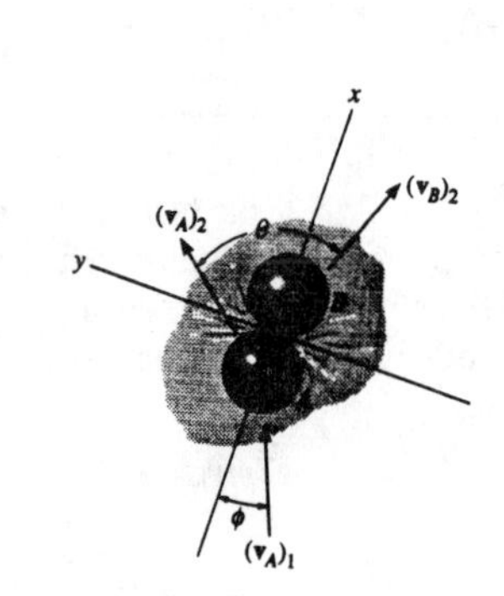

Velocity before impact :

$$(v_{Ax})_1 = (v_A)_1 \cos\phi \qquad (v_{Ay})_1 = (v_A)_1 \sin\phi$$
$$(v_{Bx})_1 = 0 \qquad (v_{By})_1 = 0$$

Velocity after impact :

$$(v_{Ax})_2 = (v_A)_2 \cos\theta_1 \qquad (v_{Ay})_2 = (v_A)_2 \sin\theta_1$$
$$(v_{Bx})_2 = (v_B)_2 \cos\theta_2 \qquad (v_{By})_2 = -(v_B)_2 \sin\theta_2$$

Conservation of " y " momentum :

$$m_B\,(v_{By})_1 = m_B\,(v_{By})_2$$
$$0 = m\left[-(v_B)_2 \sin\theta_2\right] \qquad \theta_2 = 0°$$

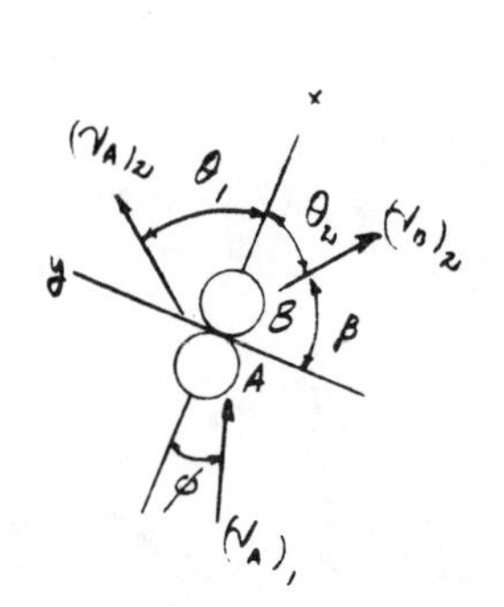

Conservation of " x " momentum :

$$m_A\,(v_{Ax})_1 + m_B\,(v_{Bx})_1 = m_A\,(v_{Ax})_2 + m_B\,(v_{Bx})_2$$
$$m(v_A)_1 \cos\phi + 0 = m(v_A)_2 \cos\theta_1 + m(v_B)_2 \cos 0°$$
$$(v_A)_1 \cos\phi = (v_A)_2 \cos\theta_1 + (v_B)_2 \qquad (1)$$

Coefficient of Restitution (x direction) :

$$e = \frac{(v_{Bx})_2 - (v_{Ax})_2}{(v_{Ax})_1 - (v_{Bx})_1}; \qquad 1 = \frac{(v_B)_2 \cos 0° - (v_A)_2 \cos\theta_1}{(v_A)_1 \cos\phi - 0}$$
$$(v_A)_1 \cos\phi = -(v_A)_2 \cos\theta_1 + (v_B)_2 \qquad (2)$$

Subtracting Eq. (1) from Eq. (2) yields :

$$2(v_A)_2 \cos\theta_1 = 0 \qquad \text{Since } 2(v_A)_2 \neq 0$$
$$\cos\theta_1 = 0 \qquad \theta_1 = 90°$$
$$\theta = \theta_1 + \theta_2 = 90° + 0° = 90° \qquad \textbf{Ans}$$

15-90. Determine the angular momentum $\mathbf{H}_O$ of each of the particles about point O.

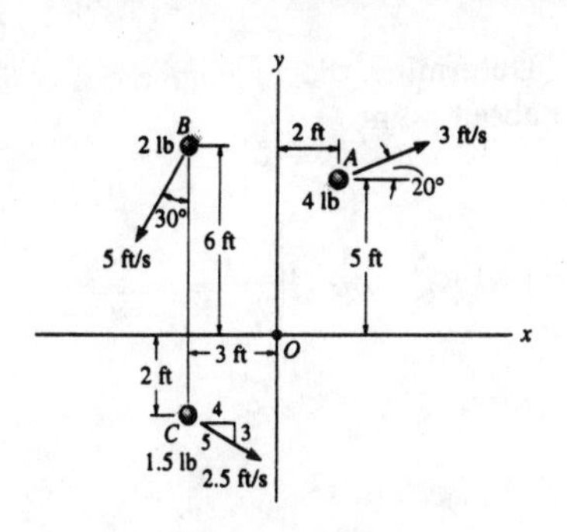

$$(H_A)_O = -5\left(\frac{4}{32.2}\right)(3\cos 20°) + 2\left(\frac{4}{32.2}\right)(3\sin 20°) = -1.50 \text{ slug}\cdot\text{ft}^2/\text{s} \quad \textbf{Ans}$$

$$(H_B)_O = 6\left(\frac{2}{32.2}\right)(5\sin 30°) + 3\left(\frac{2}{32.2}\right)(5\cos 30°) = 1.74 \text{ slug}\cdot\text{ft}^2/\text{s} \quad \textbf{Ans}$$

$$(H_C)_O = 2\left(\frac{1.5}{32.2}\right)\left(\frac{4}{5}\right)(2.5) + 3\left(\frac{1.5}{32.2}\right)\left(\frac{3}{5}\right)(2.5) = 0.396 \text{ slug}\cdot\text{ft}^2/\text{s} \quad \textbf{Ans}$$

15-91. Determine the angular momentum $\mathbf{H}_O$ of the particle about point O.

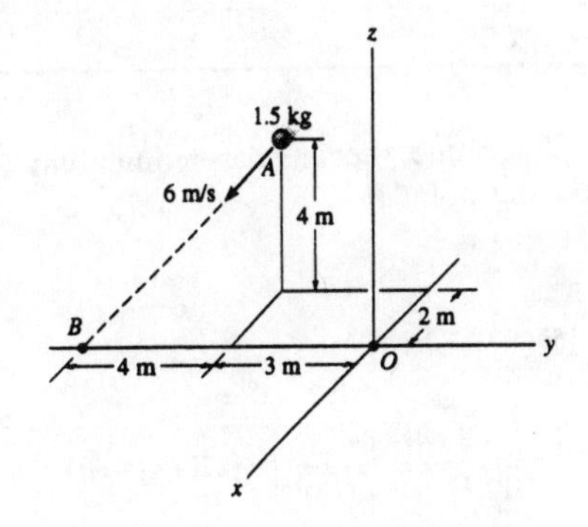

$$\mathbf{r}_{OB} = \{-7\mathbf{j}\} \text{ m} \qquad \mathbf{v}_A = 6\left(\frac{2\mathbf{i} - 4\mathbf{j} - 4\mathbf{k}}{\sqrt{2^2 + (-4)^2 + (-4)^2}}\right) = \{2\mathbf{i} - 4\mathbf{j} - 4\mathbf{k}\} \text{ m/s}$$

$$\mathbf{H}_O = \mathbf{r}_{OB} \times m\mathbf{v}_A$$

$$\mathbf{H}_O = \begin{vmatrix} \mathbf{i} & \mathbf{j} & \mathbf{k} \\ 0 & -7 & 0 \\ 1.5(2) & 1.5(-4) & 1.5(-4) \end{vmatrix} = \{42\mathbf{i} + 21\mathbf{k}\} \text{ kg}\cdot\text{m}^2/\text{s} \quad \textbf{Ans}$$

***15-92.** Determine the angular momentum $\mathbf{H}_O$ of each of the particles about point O.

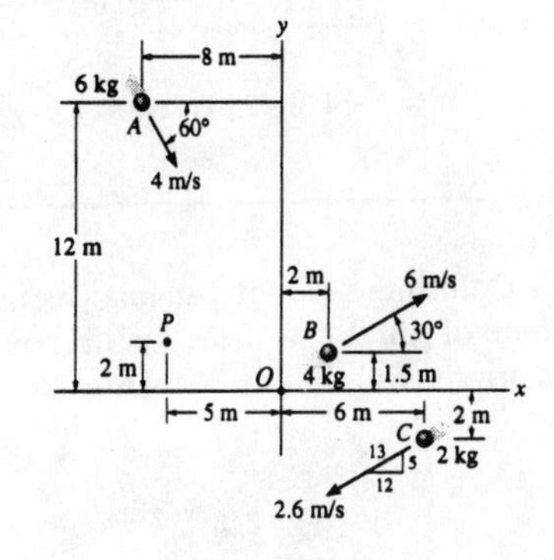

$$(H_A)_O = 8(6)(4\sin 60°) - 12(6)(4\cos 60°) = 22.3 \text{ kg}\cdot\text{m}^2/\text{s} \quad \textbf{Ans}$$

$$(H_B)_O = -1.5(4)(6\cos 30°) + 2(4)(6\sin 30°) = -7.18 \text{ kg}\cdot\text{m}^2/\text{s} \quad \textbf{Ans}$$

$$(H_C)_O = -2(2)\left(\frac{12}{13}\right)(2.6) - 6(2)\left(\frac{5}{13}\right)(2.6) = -21.6 \text{ kg}\cdot\text{m}^2/\text{s} \quad \textbf{Ans}$$

15-93. Determine the angular momentum $\mathbf{H}_P$ of each of the particles about point P.

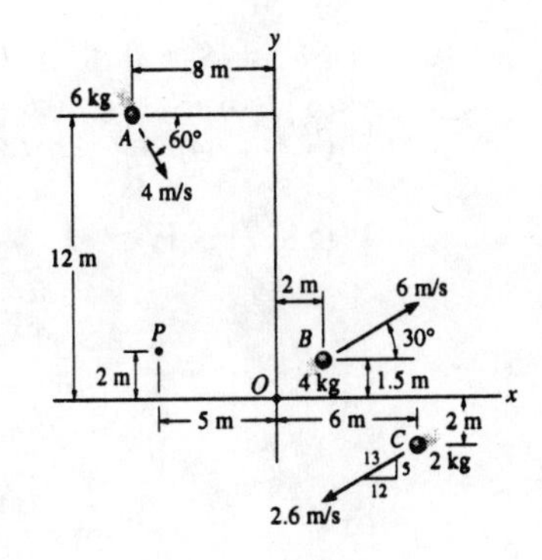

$$(H_A)_P = -6(4\cos 60°)(10) + 6(4\sin 60°)(3) = -57.6 \text{ kg}\cdot\text{m}^2/\text{s} \quad \textbf{Ans}$$

$$(H_B)_P = 4(6\cos 30°)(0.5) + 4(6\sin 30°)(7) = 94.4 \text{ kg}\cdot\text{m}^2/\text{s} \quad \textbf{Ans}$$

$$(H_C)_P = -2(2.6)\left(\frac{5}{13}\right)(11) - 2(2.6)\left(\frac{12}{13}\right)(4) = -41.2 \text{ kg}\cdot\text{m}^2/\text{s} \quad \textbf{Ans}$$

15-94. Determine the angular momentum $\mathbf{H}_O$ of the particle about point O.

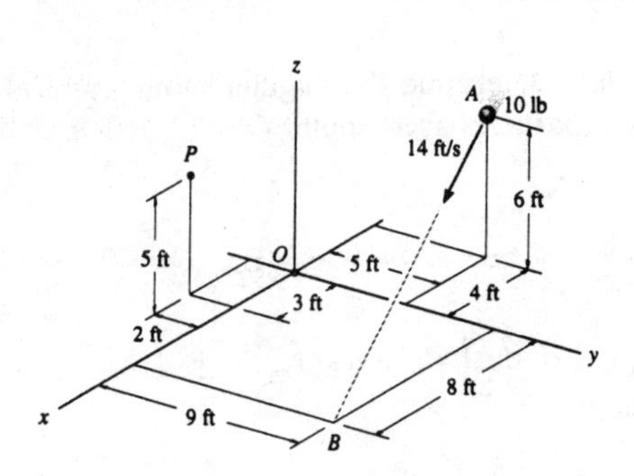

$$\mathbf{r}_{OB} = \{8\mathbf{i} + 9\mathbf{j}\}\text{ ft} \qquad \mathbf{v}_A = 14\left(\frac{12\mathbf{i}+4\mathbf{j}-6\mathbf{k}}{\sqrt{12^2+4^2+(-6)^2}}\right) = \{12\mathbf{i}+4\mathbf{j}-6\mathbf{k}\}\text{ft/s}$$

$$\mathbf{H}_O = \mathbf{r}_{OB} \times m\mathbf{v}_A$$

$$\mathbf{H}_O = \begin{vmatrix} \mathbf{i} & \mathbf{j} & \mathbf{k} \\ 8 & 9 & 0 \\ \left(\frac{10}{32.2}\right)(12) & \left(\frac{10}{32.2}\right)(4) & \left(\frac{10}{32.2}\right)(-6) \end{vmatrix}$$

$$= \{-16.8\mathbf{i} + 14.9\mathbf{j} - 23.6\mathbf{k}\}\text{ slug}\cdot\text{ft}^2/\text{s} \qquad \textbf{Ans}$$

15-95. Determine the angular momentum $\mathbf{H}_P$ of the particle about point P.

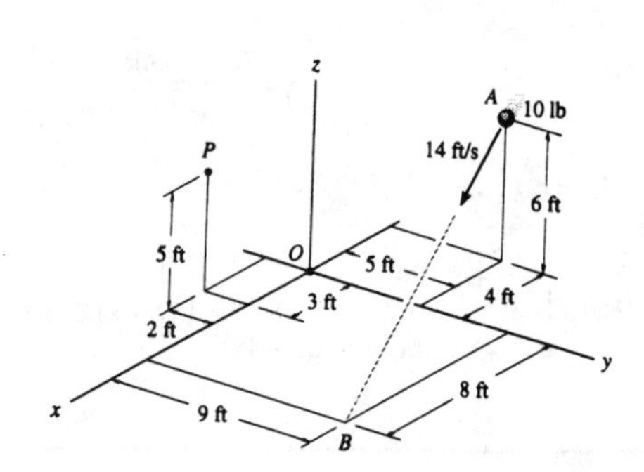

$$\mathbf{r}_{PB} = \{5\mathbf{i} + 11\mathbf{j} - 5\mathbf{k}\}\text{ ft}$$

$$\mathbf{v}_A = 14\left(\frac{12\mathbf{i}+4\mathbf{j}-6\mathbf{k}}{\sqrt{(12)^2+(4)^2+(-6)^2}}\right) = \{12\mathbf{i}+4\mathbf{j}-6\mathbf{k}\}\text{ ft/s}$$

$$\mathbf{H}_P = \mathbf{r}_{PB} \times m\mathbf{v}_A = \begin{vmatrix} \mathbf{i} & \mathbf{j} & \mathbf{k} \\ 5 & 11 & -5 \\ \left(\frac{10}{32.2}\right)(12) & \left(\frac{10}{32.2}\right)(4) & \left(\frac{10}{32.2}\right)(-6) \end{vmatrix}$$

$$= \{-14.3\mathbf{i} - 9.32\mathbf{j} - 34.8\mathbf{k}\}\text{ slug}\cdot\text{ft}^2/\text{s} \qquad \textbf{Ans}$$

***15-96.** Determine the total angular momentum $\mathbf{H}_O$ for the system of three particles about point O. All the particles are moving in the x–y plane.

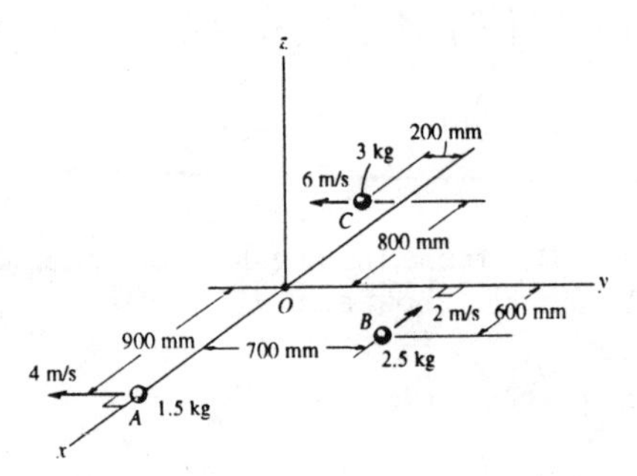

$$\mathbf{H}_O = \Sigma\mathbf{r}\times m\mathbf{v}$$

$$= \begin{vmatrix} \mathbf{i} & \mathbf{j} & \mathbf{k} \\ 0.9 & 0 & 0 \\ 0 & -1.5(4) & 0 \end{vmatrix} + \begin{vmatrix} \mathbf{i} & \mathbf{j} & \mathbf{k} \\ 0.6 & 0.7 & 0 \\ -2.5(2) & 0 & 0 \end{vmatrix} + \begin{vmatrix} \mathbf{i} & \mathbf{j} & \mathbf{k} \\ -0.8 & -0.2 & 0 \\ 0 & 3(-6) & 0 \end{vmatrix}$$

$$= \{12.5\mathbf{k}\}\text{ kg}\cdot\text{m}^2/\text{s} \qquad \textbf{Ans}$$

15-97. Determine the angular momentum $\mathbf{H}_O$ of each of the two particles about point O. Use a scalar solution.

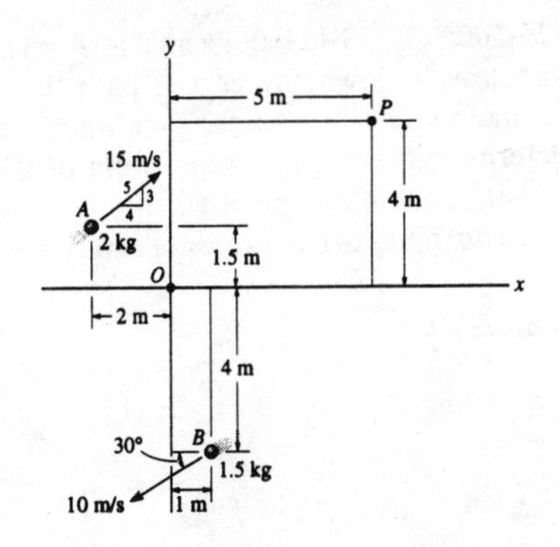

$$\circlearrowleft + (H_A)_O = -2(15)(\frac{4}{5})(1.5) - 2(15)(\frac{3}{5})(2)$$

$$= -72.0 \text{ kg}\cdot\text{m}^2/\text{s} = 72.0 \text{ kg}\cdot\text{m}^2/\text{s} \circlearrowright \quad \textbf{Ans}$$

$$\circlearrowleft + (H_B)_O = -1.5(10)(\cos 30°)(4) - 1.5(10)(\sin 30°)(1)$$

$$= -59.5 \text{ kg}\cdot\text{m}^2/\text{s} = 59.5 \text{ kg}\cdot\text{m}^2/\text{s} \circlearrowright \quad \textbf{Ans}$$

15-98. Determine the angular momentum $\mathbf{H}_P$ of each of the two particles about point P. Use a scalar solution.

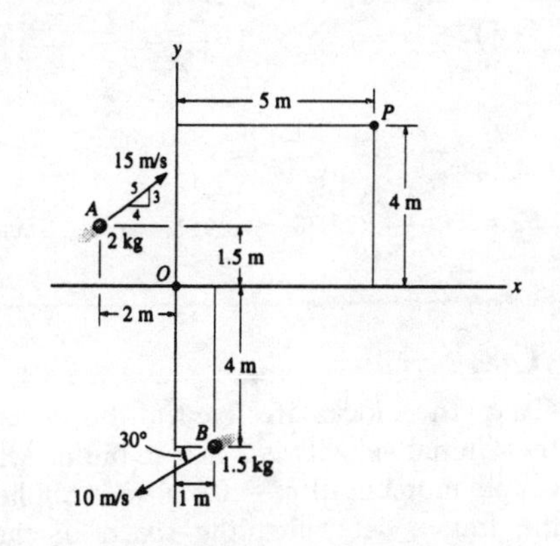

$$\circlearrowleft + (H_A)_P = 2(15)\left(\frac{4}{5}\right)(2.5) - 2(15)\left(\frac{3}{5}\right)(7)$$

$$= -66.0 \text{ kg}\cdot\text{m}^2/\text{s} = 66.0 \text{ kg}\cdot\text{m}^2/\text{s} \circlearrowright \quad \textbf{Ans}$$

$$\circlearrowleft + (H_B)_P = -1.5(10)(\cos 30°)(8) + 1.5(10)(\sin 30°)(4)$$

$$= -73.9 \text{ kg}\cdot\text{m}^2/\text{s} = 73.9 \text{ kg}\cdot\text{m}^2/\text{s} \circlearrowright \quad \textbf{Ans}$$

15-99. The projectile having a mass of 3 kg is fired from a cannon with a muzzle velocity of $v_0 = 500$ m/s. Determine the projectile's angular momentum about point O at the instant it is at the maximum height of its trajectory.

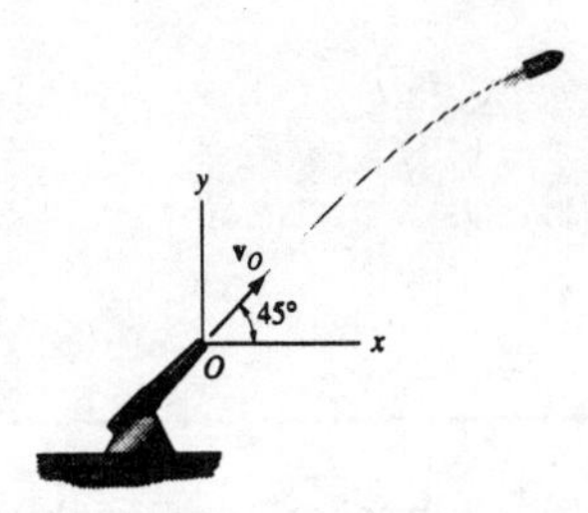

At the maximum height, the projectile travels with a horizontal speed of $v = v_x = 500\cos 45° = 353.6$ m/s.

$$(+\uparrow) \quad v_y^2 = (v_0)_y^2 + 2a_c[s_y - (s_0)_y]$$

$$0 = (500\sin 45°)^2 + 2(-9.81)[(s_y)_{max} - 0]$$

$$(s_y)_{max} = 6371 \text{ m}$$

$$H_O = (d)(mv) = 6371(3)(353.6) = 6.76(10^6) \text{ kg}\cdot\text{m}^2/\text{s} \quad \textbf{Ans}$$

***15-100.** The 3-lb ball located at A is released from rest and travels down the curved path. If the ball exerts a normal force of 5 lb on the path when it reaches point B, determine the angular momentum of the ball about the center of curvature, point O. *Hint:* Neglect the size of the ball. The radius of curvature at point B must be determined.

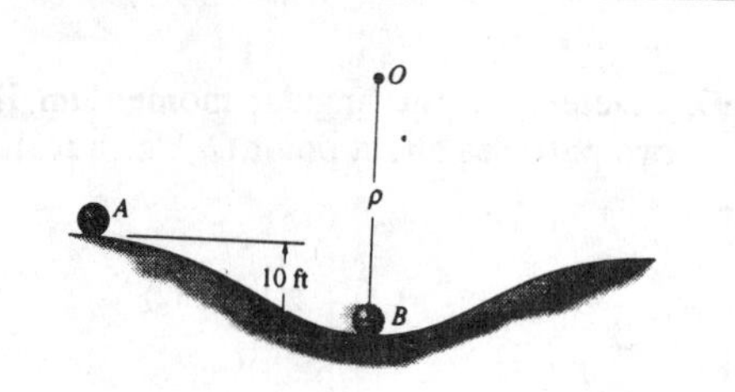

Datum at B :

$$T_1 + V_1 = T_2 + V_2$$

$$0 + 3(10) = \frac{1}{2}\left(\frac{3}{32.2}\right)(v_B)^2 + 0$$

$$v_B = 25.38 \text{ ft/s}$$

$$(+\uparrow)\Sigma F_n = ma_n; \qquad 5 - 3 = \left(\frac{3}{32.2}\right)\left(\frac{(25.38)^2}{\rho}\right)$$

$$\rho = 30 \text{ ft}$$

$$H_B = 30\left(\frac{3}{32.2}\right)(25.38) = 70.9 \text{ slug} \cdot \text{ft}^2/\text{s} \qquad \textbf{Ans}$$

15-101. The two blocks A and B each have a mass of 400 g. The blocks are fixed to the horizontal rods, and their initial velocity is 2 m/s in the direction shown. If a couple moment of $M = 0.6$ N·m is applied about CD of the frame, determine the speed of the blocks when $t = 3$ s. The mass of the frame is negligible, and it is free to rotate about CD. Neglect the size of the blocks.

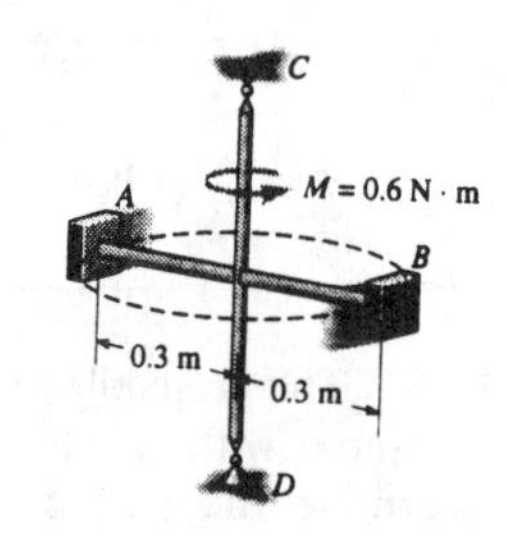

$$(H_O)_1 + \Sigma\int_{t_1}^{t_2} M_O dt = (H_O)_2$$

$$2[0.3(0.4)(2)] + 0.6(3) = 2[0.3(0.4)v]$$

$$v = 9.50 \text{ m/s} \qquad \textbf{Ans}$$

15-102. The four 5-lb spheres are rigidly attached to the crossbar frame having a negligible weight. If a couple moment $M = (0.5t + 0.8)$ lb·ft, where t is in seconds, is applied as shown, determine the speed of each of the spheres in 4 seconds starting from rest. Neglect the size of the spheres.

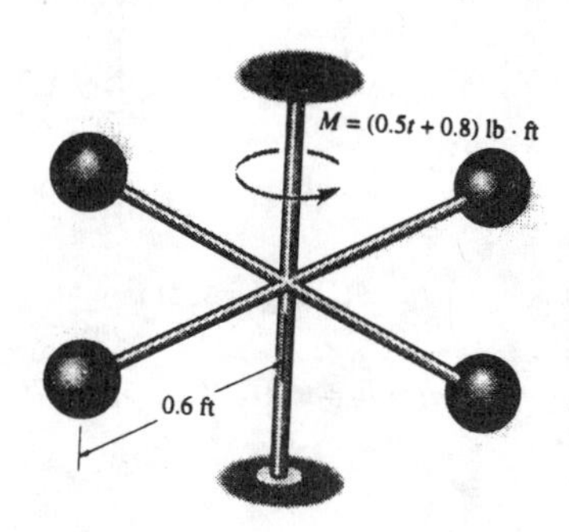

$$(H_z)_1 + \Sigma\int M_z \, dt = (H_z)_2$$

$$0 + \int_0^4 (0.5t + 0.8) \, dt = 4\left[\left(\frac{5}{32.2}\right)(0.6v_2)\right]$$

$$7.2 = 0.37267 \, v_2$$

$$v_2 = 19.3 \text{ ft/s} \qquad \textbf{Ans}$$

15-103. An earth satellite of mass 700 kg is launched into a free-flight trajectory about the earth with an initial speed of $v_A = 10$ km/s when the distance from the center of the earth is $r_A = 15$ Mm. If the launch angle at this position is $\phi_A = 70°$, determine the speed v_B of the satellite and its closest distance r_B from the center of the earth. The earth has a mass $M_e = 5.976(10^{24})$ kg. *Hint:* Under these conditions, the satellite is subjected only to the earth's gravitational force, $F = GM_e m_s/r^2$, Eq. 13–1. For part of the solution, use the conservation of energy (see Prob. 14-97).

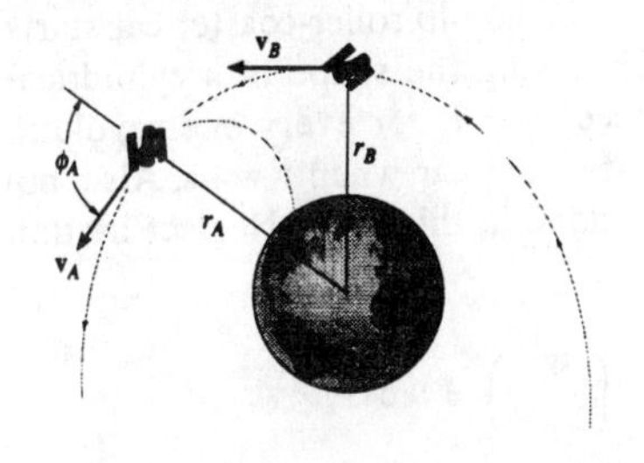

$$(H_O)_1 = (H_O)_2$$

$$m_s(v_A \sin\phi_A)r_A = m_s(v_B)r_B$$

$$700[10(10^3)\sin 70°](15)(10^6) = 700(v_B)(r_B) \quad (1)$$

$$T_A + V_A = T_B + V_B$$

$$\frac{1}{2}m_s(v_A)^2 - \frac{GM_e m_s}{r_A} = \frac{1}{2}m_s(v_B)^2 - \frac{GM_e m_s}{r_B}$$

$$\frac{1}{2}(700)[10(10^3)]^2 - \frac{66.73(10^{-12})(5.976)(10^{24})(700)}{[15(10^6)]} = \frac{1}{2}(700)(v_B)^2 - \frac{66.73(10^{-12})(5.976)(10^{24})(700)}{r_B} \quad (2)$$

Solving,

$v_B = 10.2$ km/s **Ans**

$r_B = 13.8$ Mm **Ans**

***15-104.** A box having a weight of 8 lb is moving around in a circle of radius $r_A = 2$ ft with a speed of $(v_A)_1 = 5$ ft/s while connected to the end of a rope. If the rope is pulled inward with a constant speed of $v_r = 4$ ft/s, determine the speed of the box at the instant $r_B = 1$ ft. How much work is done after pulling in the rope from A to B? Neglect friction and the size of the box.

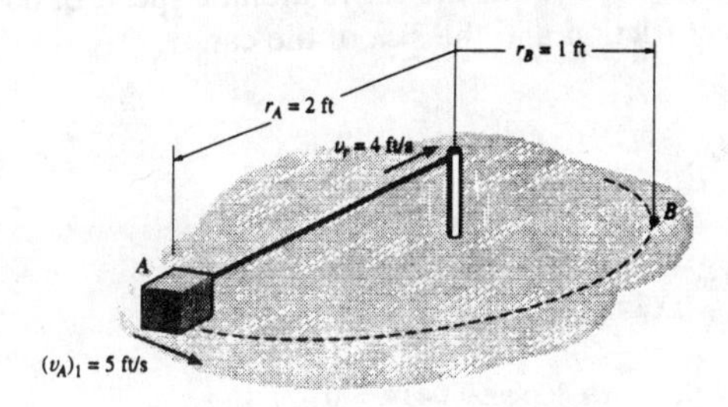

$$(H_z)_A = (H_z)_B; \quad \left(\frac{8}{32.2}\right)(2)(5) = \left(\frac{8}{32.2}\right)(1)(v_B)_{\text{tangent}}$$

$$(v_B)_{\text{tangent}} = 10 \text{ ft/s}$$

$$v_B = \sqrt{(10)^2 + (4)^2} = 10.77 = 10.8 \text{ ft/s} \quad \textbf{Ans}$$

$$\sum U_{AB} = T_B - T_A \quad U_{AB} = \frac{1}{2}\left(\frac{8}{32.2}\right)(10.77)^2 - \frac{1}{2}\left(\frac{8}{32.2}\right)(5)^2$$

$$U_{AB} = 11.3 \text{ ft} \cdot \text{lb} \quad \textbf{Ans}$$

15-105. The 800-lb roller-coaster car starts from rest on the track having the shape of a cylindrical helix. If the helix descends 8 ft for every one revolution, determine the speed of the car when $t = 4$ s. Also, how far has the car descended in this time? Neglect friction and the size of the car.

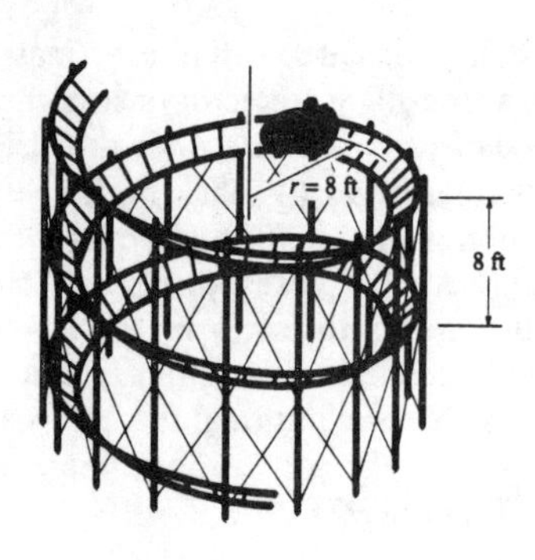

$$\theta = \tan^{-1}\left(\frac{8}{2\pi(8)}\right) = 9.043°$$

$$\Sigma F_y = 0; \quad N - 800\cos 9.043° = 0$$

$$N = 790.1 \text{ lb}$$

$$H_1 + \int M\,dt = H_2$$

$$0 + \int_0^4 8(790.1\sin 9.043°)dt = \frac{800}{32.2}(8)v_t$$

$$v_t = 20.0 \text{ ft/s}$$

$$v = \frac{20.0}{\cos 9.043°} = 20.2 \text{ ft/s} \qquad \textbf{Ans}$$

$$T_1 + \Sigma U_{1-2} = T_2$$

$$0 + 800h = \frac{1}{2}\left(\frac{800}{32.2}\right)(20.2)^2$$

$$h = 6.36 \text{ ft} \qquad \textbf{Ans}$$

15-106. The 800-lb roller-coaster car starts from rest on the track having the shape of a cylindrical helix. If the helix descends 8 ft for every one revolution, determine the time required for the car to attain a speed of 60 ft/s. Neglect friction and the size of the car.

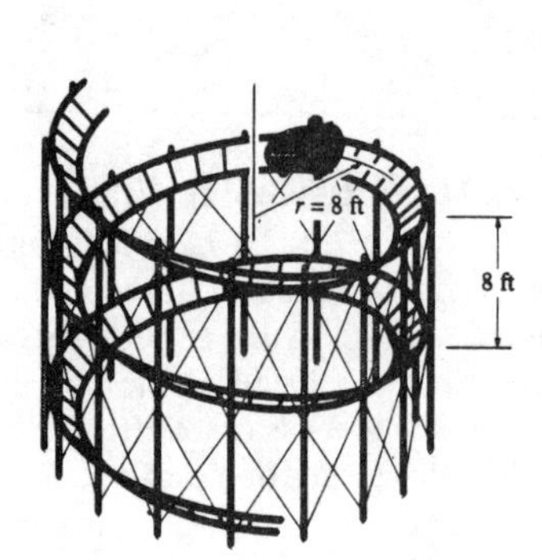

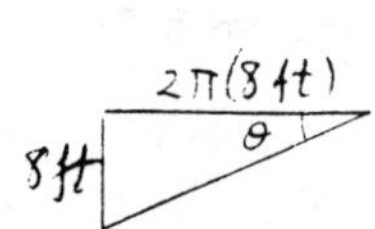

$$\theta = \tan^{-1}\left(\frac{8}{2\pi(8)}\right) = 9.043°$$

$$\Sigma F_y = 0; \quad N - 800\cos 9.043° = 0$$

$$N = 790.1 \text{ lb}$$

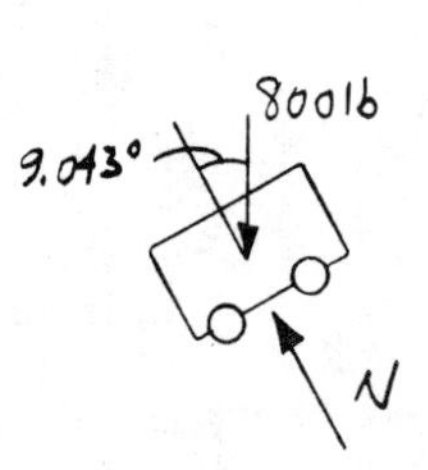

$$v = \frac{v_t}{\cos 9.043°}$$

$$60 = \frac{v_t}{\cos 9.043°}$$

$$v_t = 59.254 \text{ ft/s}$$

$$H_1 + \int M\,dt = H_2$$

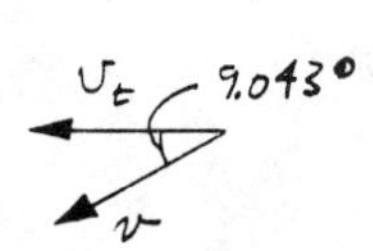

$$0 + \int_0^t 8(790.1\sin 9.043°)dt = \frac{800}{32.2}(8)(59.254)$$

$$t = 11.9 \text{ s} \qquad \textbf{Ans}$$

■15-107. The 5-lb ball B is rotating in a circle as shown when the distance AB is 3 ft. If 1.5 ft of cord is pulled through the hole at A, determine the speed of the ball when it moves in a circular path at C.

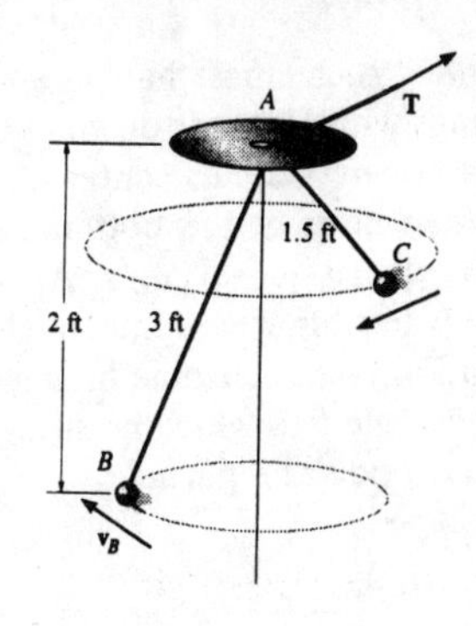

$$\phi = \sin^{-1}\frac{2}{3} = 41.81^\circ$$

$$+\uparrow \Sigma F_z = 0; \quad T\sin 41.81^\circ - 5 = 0$$

$$T = 7.50 \text{ lb}$$

$$\xrightarrow{+} \Sigma F_n = ma_n; \quad 7.50\cos 41.81^\circ = \left(\frac{5}{32.2}\right)\left(\frac{v_B^2}{\sqrt{3^2 - 2^2}}\right)$$

$$v_B = 8.972 \text{ ft/s}$$

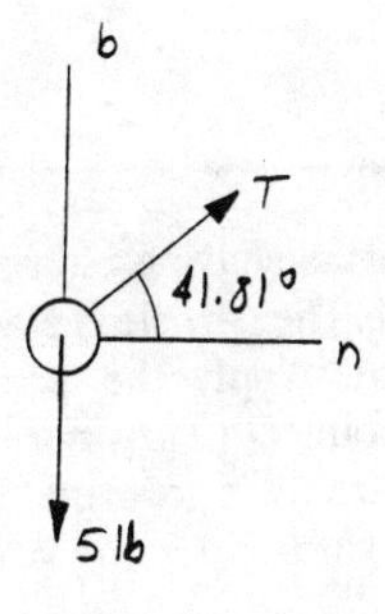

$$(H_B)_z = (H_C)_z$$

$$\left(\frac{5}{32.2}\right)(8.972)\left(\sqrt{3^2 - 2^2}\right) = \left(\frac{5}{32.2}\right)(r)(v_C); \quad rv_C = 20.062 \quad (1)$$

$$\cos\theta = \frac{r}{1.5} \qquad \sin\theta = \frac{\sqrt{(1.5)^2 - r^2}}{1.5}$$

$$+\uparrow \Sigma F_b = 0; \quad T'\left(\frac{\sqrt{(1.5)^2 - r^2}}{1.5}\right) - 5 = 0$$

$$T' = \frac{7.5}{\sqrt{(1.5)^2 - r^2}}$$

$$\xrightarrow{+} \Sigma F_n = ma_n; \quad T'\left(\frac{r}{1.5}\right) = \left(\frac{5}{32.2}\right)\left(\frac{v_C^2}{r}\right)$$

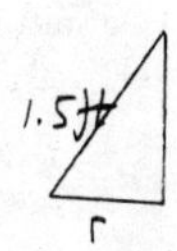

$$T' = \frac{7.50 v_C^2}{32.2\, r^2}$$

Using Eq. (1),

$$T' = \frac{93.75}{r^4} = \frac{7.5}{\sqrt{(1.5)^2 - r^2}}$$

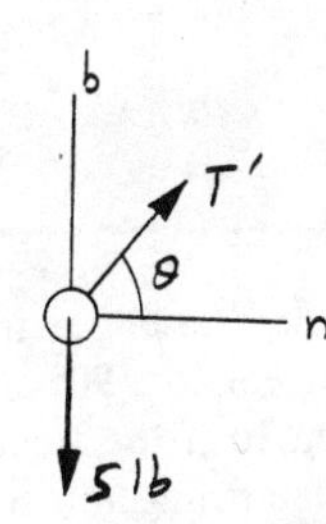

$$r^4 = 12.50\sqrt{2.25 - r^2}$$

$$r^8 = 156.25\left(2.25 - r^2\right)$$

$$r^8 + 156.25r^2 - 351.56 = 0$$

Solving for the positive real root :

$$r = 1.456 \text{ ft}$$

Thus,

$$v_C = \frac{20.062}{1.456} = 13.8 \text{ ft/s} \quad \textbf{Ans}$$

***15-108.** A gymnast having a mass of 80 kg holds the two rings with his arms down in the position shown as he swings downward. His center of mass is located at point G_1. When he is at the bottom position of his swing, his velocity is $(v_G)_1 = 5$ m/s. At this lower position he *suddenly* lets his arms come up, shifting his center of mass to position G_2. Determine his new velocity in the upswing and the angle θ to which he swings before coming to rest. Treat his body as a particle.

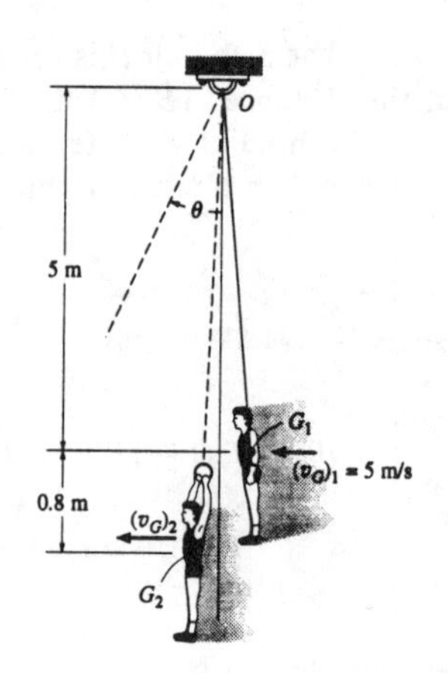

$(\mathbf{H}_O)_1 = (\mathbf{H}_O)_2$

$5(80)(5) = 5.8(80)\upsilon_2 \qquad \upsilon_2 = 4.310\text{ m/s} = 4.31\text{m/s}$ **Ans**

$T_1 + V_1 = T_2 + V_2$

$\frac{1}{2}(80)(4.310)^2 + 0 = 0 + 80(9.81)[5.8(1-\cos\theta)]$

$\theta = 33.2°$ **Ans**

15-109. A small particle having a mass m is placed inside the semicircular tube. The particle is placed at the position shown and released. Apply the principle of angular momentum about point O ($\Sigma M_O = \dot{H}_O$), and show that the motion of the particle is governed by the differential equation $\ddot{\theta} + (g/R)\sin\theta = 0$.

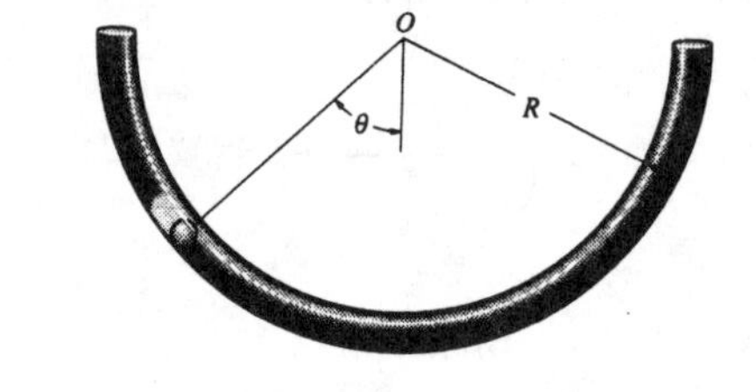

$\curvearrowleft + \Sigma M_O = \frac{dH_O}{dt}; \qquad -Rmg\sin\theta = \frac{d}{dt}(mvR)$

$g\sin\theta = -\frac{dv}{dt} = -\frac{d^2s}{dt^2}$

O
R θ
N θ W = mg

But, $s = R\theta$

Thus, $g\sin\theta = -R\ddot{\theta}$

or, $\ddot{\theta} + \left(\frac{g}{R}\right)\sin\theta = 0$ **Q.E.D.**

15-110. A toboggan and rider, having a total mass of 150 kg, enter horizontally tangent to a 90° circular curve with a velocity of $v_A = 70$ km/h. If the track is flat and banked at an angle of 60°, determine the speed v_B and the angle θ of "descent," measured from the horizontal in a vertical x–z plane, at which the toboggan exits at B. Neglect friction in the calculation.

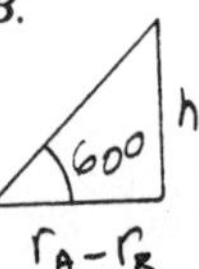

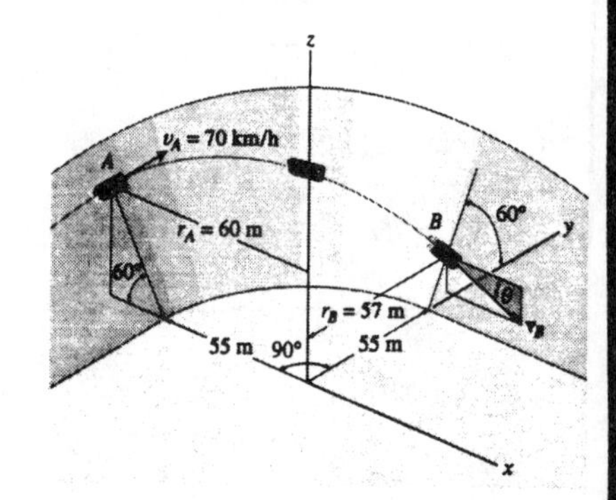

$v_A = 70\text{ km/h} = 19.44\text{ m/s}$

$(H_A)_z = (H_B)_z$

$150(19.44)(60) = 150(v_B)\cos\theta(57) \quad (1)$

Datum at B :

$T_A + V_A = T_B + V_B$

$\frac{1}{2}(150)(19.44)^2 + 150(9.81)h = \frac{1}{2}(150)(v_B)^2 + 0 \quad (2)$

Since $h = (r_A - r_B)\tan 60° = (60-57)\tan 60° = 5.196$

Solving Eq. (1) and Eq (2) :

$v_B = 21.9$ m/s **Ans**

$\theta = 20.9°$ **Ans**

15-111. The 150-lb fireman is holding a hose which has a nozzle diameter of 1 in. and hose diameter of 2 in. If the velocity of the water at discharge is 60 ft/s, determine the resultant normal and frictional force acting on the man's feet at the ground. Neglect the weight of the hose and the water within it. $\gamma_w = 62.4\ \text{lb/ft}^3$.

Originally, the water flow is horizontal. The fireman alters the direction of flow to 40° from the horizontal.

$$\frac{dm}{dt} = \rho v_B A_B = \frac{62.4}{32.2}(60)\left(\frac{\pi\left(\frac{1}{2}\right)^2}{(12)^2}\right) = 0.6342 \text{ slug/s}$$

Also, the velocity of the water through the hose is

$$\rho v_A A_A = \rho v_B A_B$$

$$\rho v_A\left(\frac{\pi(1)^2}{(12)^2}\right) = \rho(60)\left(\frac{\pi\left(\frac{1}{2}\right)^2}{(12)^2}\right)$$

$$v_A = 15 \text{ ft/s}$$

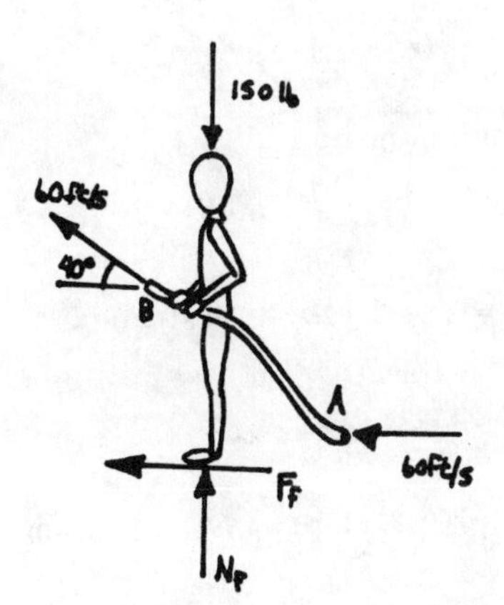

$$\overset{+}{\leftarrow} \Sigma F_x = \frac{dm}{dt}((v_B)_x - (v_A)_x)$$

$$F_f = 0.6342[60\cos 40° - 15]$$

$F_f = 19.6$ lb **Ans**

$$+\uparrow \Sigma F_y = \frac{dm}{dt}((v_B)_y - (v_A)_y)$$

$$N_f - 150 = 0.6342[60\sin 40° - 0]$$

$N_f = 174$ lb **Ans**

***15-112.** The nozzle discharges water at a constant rate of 2 ft^3/s. The cross-sectional area of the nozzle at A is 4 in^2, and at B the cross-sectional area is 12 in^2. If the static gauge pressure due to the water at B is 2 lb/in^2, determine the magnitude of force which must be applied by the coupling at B to hold the nozzle in place. Neglect the weight of the nozzle and the water within it. $\gamma_w = 62.4\ \text{lb/ft}^3$.

$$\frac{dm}{dt} = \rho Q = \left(\frac{62.4}{32.2}\right)(2) = 3.876 \text{ slug/s}$$

$$(v_{Bx}) = \frac{Q}{A_B} = \frac{2}{12/144} = 24 \text{ ft/s} \qquad (v_{By}) = 0$$

$$(v_{Ay}) = \frac{Q}{A_A} = \frac{2}{4/144} = 72 \text{ ft/s} \qquad (v_{Ax}) = 0$$

$$F_B = p_B A_B = 2(12) = 24 \text{ lb}$$

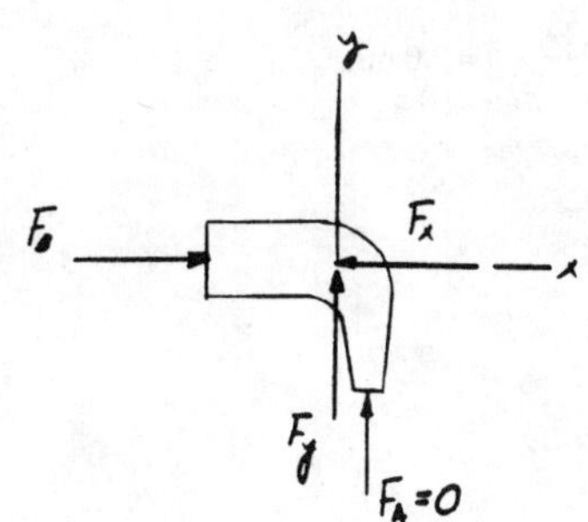

Equations of steady flow :

$$\overset{+}{\rightarrow} \Sigma F_x = \frac{dm}{dt}(v_{Ax} - v_{Bx}); \quad 24 - F_x = 3.876(0 - 24) \quad F_x = 117.01 \text{ lb}$$

$$+\uparrow \Sigma F_y = \frac{dm}{dt}(v_{Ay} - v_{By}); \quad F_y = 3.876(72 - 0) = 279.06 \text{ lb}$$

$F = \sqrt{F_x^2 + F_y^2} = \sqrt{117.01^2 + 279.06^2} = 303$ lb **Ans**

15-113. When operating, the air-jet fan discharges air with a speed of $v_B = 20$ m/s into a slipstream having a diameter of 0.5 m. If air has a density of 1.22 kg/m³, determine the horizontal and vertical components of reaction at C and the vertical reaction at each of the two wheels, D, when the fan is in operation. The fan and motor have a mass of 20 kg and a center of mass at G. Neglect the weight of the frame. Due to symmetry, both of the wheels support an equal load. Assume the air entering the fan at A is essentially at rest.

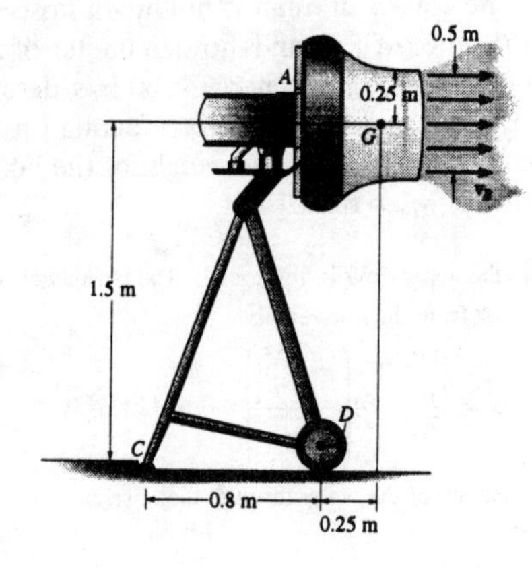

$$\frac{dm}{dt} = \rho v A = 1.22(20)(\pi)(0.25)^2 = 4.791 \text{ kg/s}$$

$$\xrightarrow{+} \Sigma F_x = \frac{dm}{dt}(v_{B_x} - v_{A_x})$$

$$C_x = 4.791(20 - 0)$$

$C_x = 95.8$ N **Ans**

$$+\uparrow \Sigma F_y = 0; \quad C_y + 2D_y - 20(9.81) = 0$$

$$\circlearrowleft + \Sigma M_C = \frac{dm}{dt}(d_{CG} v_B - d_{CG} v_A)$$

$$2D_y(0.8) - 20(9.81)(1.05) = 4.791(-1.5(20) - 0)$$

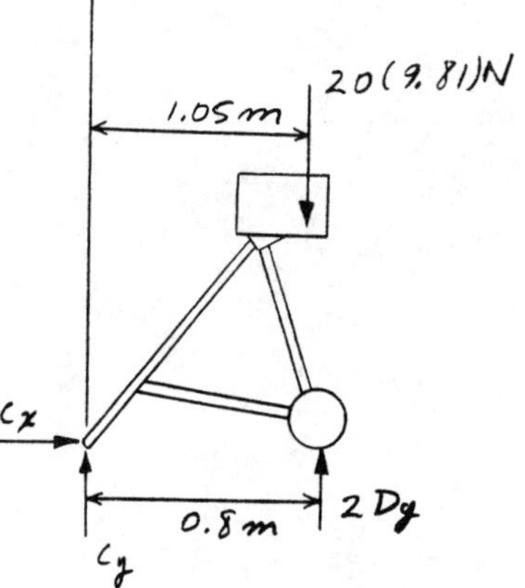

Solving :

$D_y = 38.9$ N **Ans**

$C_y = 118$ N **Ans**

15-114. A snowblower having a scoop S with a cross-sectional area of $A_S = 0.12$ m² is pushed into snow with a speed of $v_S = 0.5$ m/s. The machine discharges the snow through a tube T that has a cross-sectional area of $A_T = 0.03$ m² and is directed 60° from the horizontal. If the density of snow is $\rho_s = 104$ kg/m³, determine the horizontal force P required to push the blower forward, and the resultant frictional force F of the wheels on the ground, necessary to prevent the blower from moving sideways. The wheels roll freely.

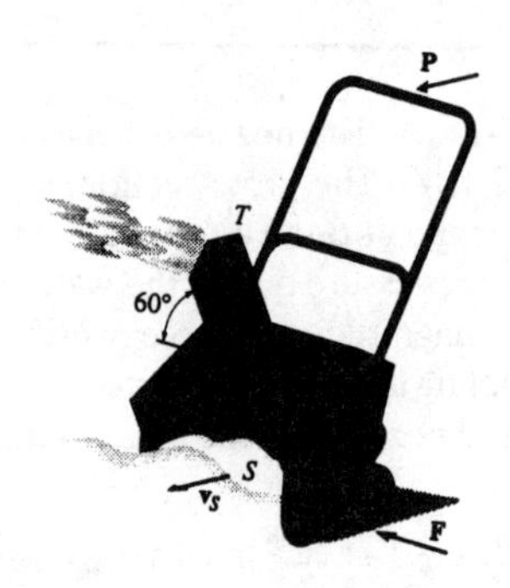

$$\frac{dm}{dt} = \rho v_S A_S = (104)(0.5)(0.12) = 6.24 \text{ kg/s}$$

$$v_T = \frac{dm}{dt}\left(\frac{1}{\rho A_T}\right) = \left(\frac{6.24}{104(0.03)}\right) = 2.0 \text{ m/s}$$

$$\Sigma F_x = \frac{dm}{dt}(v_{T_x} - v_{S_x})$$

$$-F = 6.24(-2\cos 60° - 0)$$

$F = 6.24$ N **Ans**

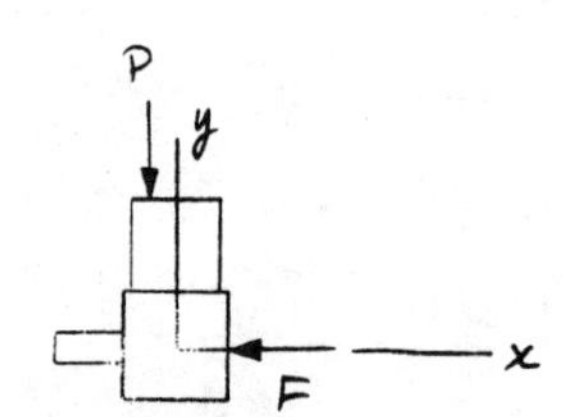

$$\Sigma F_y = \frac{dm}{dt}(v_{T_y} - v_{S_y})$$

$$-P = 6.24(0 - 0.5)$$

$P = 3.12$ N **Ans**

15-115. The nozzle has a diameter of 40 mm. If it discharges water uniformly with a downward velocity of 20 m/s against the fixed blade, determine the vertical force exerted by the water on the blade. $\rho_w = 1\ \text{Mg/m}^3$.

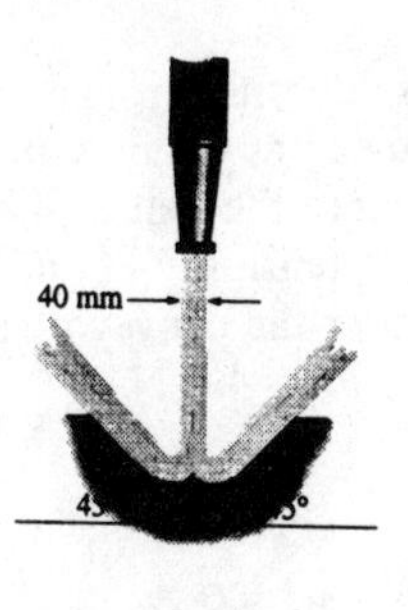

$$\frac{dm}{dt} = \rho v A = (1000)(20)(\pi)(0.02)^2 = 25.13\ \text{kg/s}$$

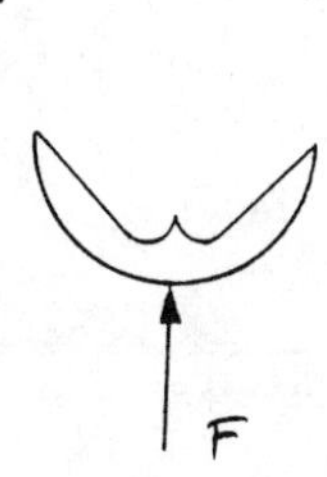

$$+\uparrow \Sigma F_y = \frac{dm}{dt}\left(v_{B_y} - v_{A_y}\right)$$

$$F = (25.13)(20\sin 45° - (-20))$$

$F = 858$ N **Ans**

***15-116.** A jet of water has a cross-sectional area of 5 in². If it strikes the fixed blade with a constant speed of 60 ft/s, determine the magnitude of force which the water exerts on the blade. $\gamma_w = 62.4\ \text{lb/ft}^3$.

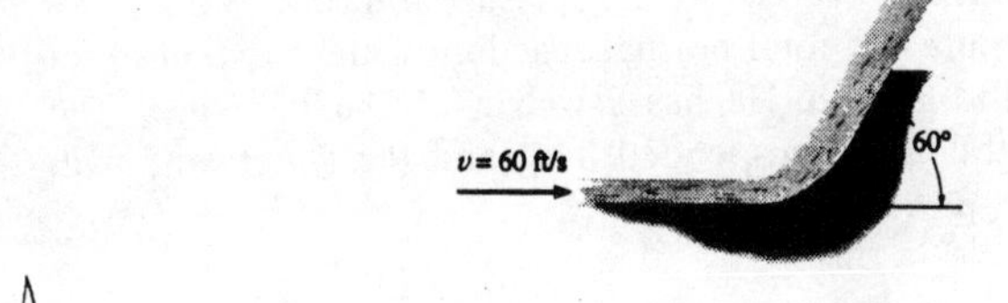

$$\frac{dm}{dt} = \rho v A = \left(\frac{62.4}{32.2}\right)(60)\left(\frac{5}{144}\right) = 4.037\ \text{slug/s}$$

$$\overset{+}{\rightarrow}\Sigma F_x = \frac{dm}{dt}(v_{B_x} - v_{A_x})$$

$$F_x = 4.037(60\cos 60° - 60) = -121.1\ \text{lb}$$

$$+\uparrow \Sigma F_y = \frac{dm}{dt}\left(v_{B_y} - v_{A_y}\right)$$

$$F_y = 4.037(60\sin 60° - 0) = 209.8\ \text{lb}$$

$F = \sqrt{(-121.1)^2 + (209.8)^2} = 242$ lb **Ans**

15-117. The static pressure of water at C is 40 lb/in². If water flows out of the pipe at A and B with velocities $v_A = 12$ ft/s and $v_B = 25$ ft/s, determine the horizontal and vertical components of force exerted on the elbow at C necessary to hold the pipe assembly in equilibrium. Neglect the weight of water within the pipe and the weight of the pipe. The pipe has a diameter of 0.75 in. at C, and at A and B the diameter is 0.5 in. $\gamma_w = 62.4\ \text{lb/ft}^3$.

$$\frac{dm_A}{dt} = \frac{62.4}{32.2}(12)(\pi)\left(\frac{0.25}{12}\right)^2 = 0.03171\ \text{slug/s}$$

$$\frac{dm_B}{dt} = \frac{62.4}{32.2}(25)(\pi)\left(\frac{0.25}{12}\right)^2 = 0.06606\ \text{slug/s}$$

$$\frac{dm_C}{dt} = 0.03171 + 0.06606 = 0.09777\ \text{slug/s}$$

$$v_C A_C = v_A A_A + v_B A_B$$

$$v_C(\pi)\left(\frac{0.375}{12}\right)^2 = 12(\pi)\left(\frac{0.25}{12}\right)^2 + 25(\pi)\left(\frac{0.25}{12}\right)^2$$

$$v_C = 16.44\ \text{ft/s}$$

$$\overset{+}{\rightarrow}\Sigma F_x = \frac{dm_B}{dt}v_{B_x} + \frac{dm_A}{dt}v_{A_x} - \frac{dm_C}{dt}v_{C_x}$$

$$40(\pi)(0.375)^2 - F_x = 0 - 0.03171(12)\left(\frac{3}{5}\right) - 0.09777(16.44)$$

$F_x = 19.5$ lb **Ans**

$$+\uparrow \Sigma F_y = \frac{dm_B}{dt}v_{B_y} + \frac{dm_A}{dt}v_{A_y} - \frac{dm_C}{dt}v_{C_y}$$

$$F_y = 0.06606(25) + 0.03171\left(\frac{4}{5}\right)(12) - 0$$

$F_y = 1.9559 = 1.96$ lb **Ans**

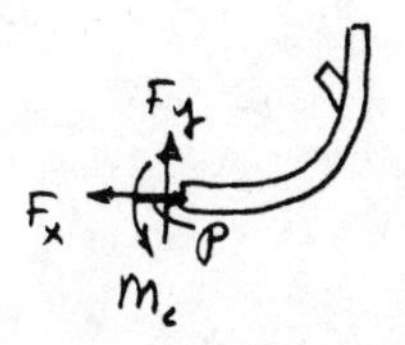

15-118. Sand is deposited from a chute onto a conveyor belt which is moving at 0.5 m/s. If the sand is assumed to fall vertically onto the belt at A at the rate of 4 kg/s, determine the belt tension F_B to the right of A. The belt is free to move over the conveyor rollers and its tension to the left of A is $F_C = 400$ N.

$$\xrightarrow{+} \Sigma F_x = \frac{dm}{dt}(v_{Bx} - v_{Ax})$$

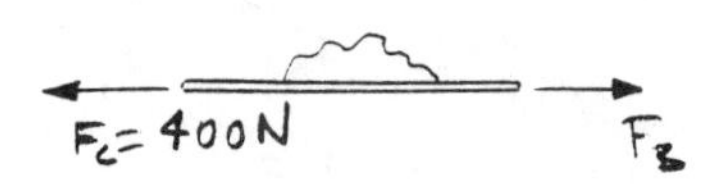

$$F_B - 400 = 4(0.5 - 0)$$

$$F_B = 2 + 400 = 402 \text{ N} \quad \textbf{Ans}$$

15-119. The fireman sprays a 2-in.-diameter jet of water from a hose at the building. If the water is discharged at 0.80 ft^3/s, determine the elevation y to where the centerline of the water splashes on the wall. Also, compute the total normal reaction of the fireman's feet on the ground. He has a weight of 180 lb. Neglect the weight of the suspended hose and the water within it. $\gamma_w = 62.4$ lb/ft^3.

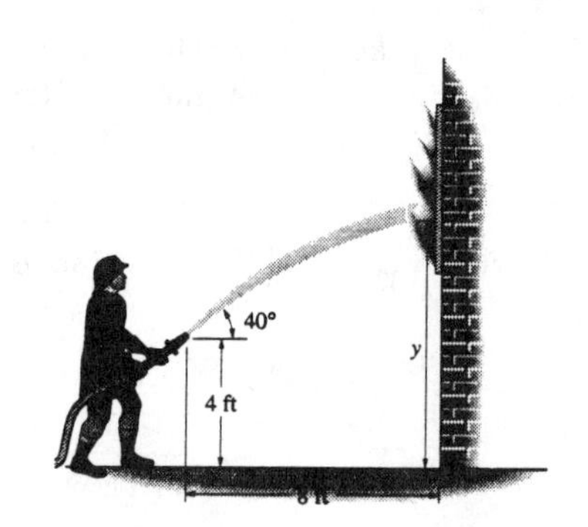

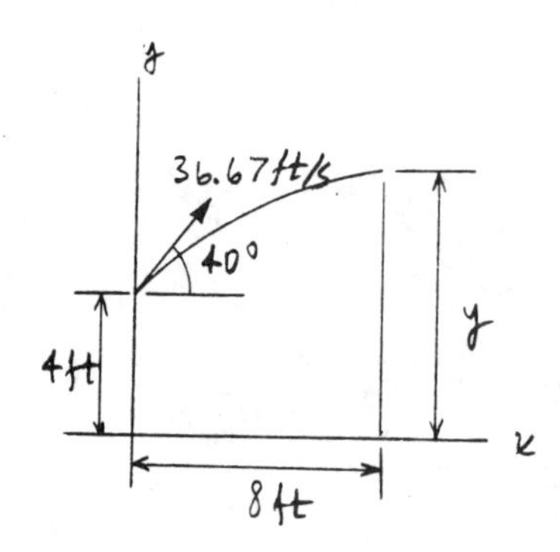

$$v = \frac{Q}{A} = \frac{0.80}{\pi(\frac{1}{12})^2} = 36.67 \text{ ft/s}$$

$(\xrightarrow{+})$ $\quad s = s_0 + v_0 t$

$$8 = 0 + 36.67 \cos 40°(t)$$

$$t = 0.2848 \text{ s}$$

$(+\uparrow)$ $\quad s = s_0 + v_0 t + \frac{1}{2} a_c t^2$

$$y = 4 + 36.67 \sin 40°(0.2848) - \frac{1}{2}(32.2)(0.2848)^2$$

$$y = 9.41 \text{ ft} \quad \textbf{Ans}$$

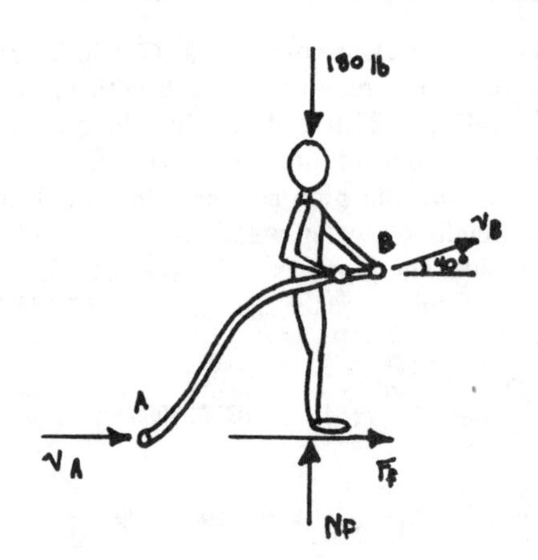

$$\frac{dm}{dt} = \rho v A = \frac{62.4}{32.2}(36.67)(\pi)\left(\frac{1}{12}\right)^2 = 1.550 \text{ slug/s}$$

Note that the velocity of the water at A is horizontal.

$$+\uparrow \Sigma F_y = \frac{dm}{dt}(v_{B_y} - v_{A_y})$$

$$N_f - 180 = 1.550[36.67 \sin 40° - 0]$$

$$N_f = 217 \text{ lb} \quad \textbf{Ans}$$

***15-120.** The fountain shoots water in the direction shown. If the water is discharged at 30° from the horizontal, and the cross-sectional area of the water stream is approximately 2 in^2, determine the force it exerts on the concrete wall at B. $\gamma_w = 62.4$ lb/ft^3.

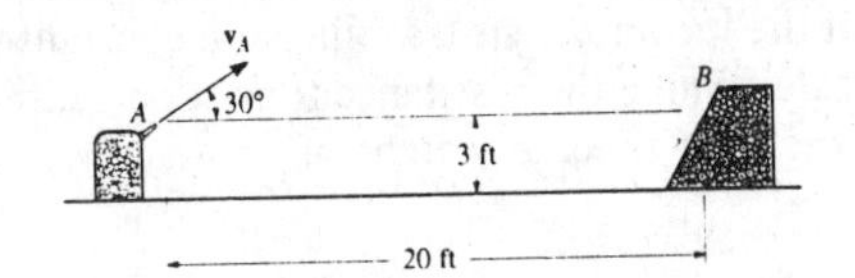

$(\overset{+}{\rightarrow})$ $\quad s = s_0 + v_0 t$

$$20 = 0 + v_A \cos 30° t$$

$(+\uparrow)$ $\quad v = v_0 + a_c t$

$$-(v_A \sin 30°) = (v_A \sin 30°) - 32.2t$$

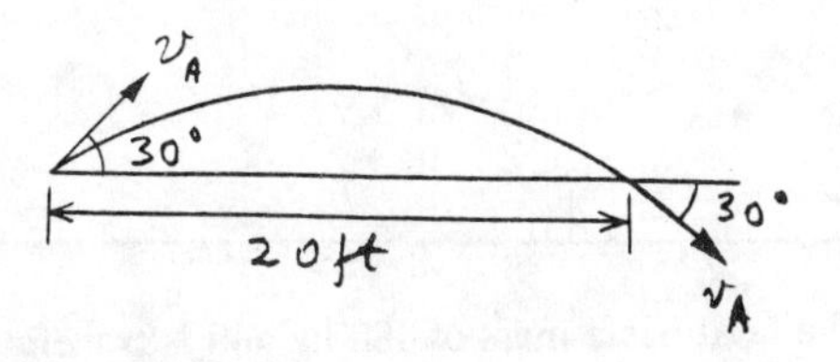

Solving,

$t = 0.8469$ s

$v_A = v_B = 27.27$ ft/s

At B :

$$\frac{dm}{dt} = \rho v A = \left(\frac{62.4}{32.2}\right)(27.27)\left(\frac{2}{144}\right) = 0.7340 \text{ slug/s}$$

$$+\searrow \Sigma F = \frac{dm}{dt}(v_s - v_B)$$

$$-F = 0.7340(0 - 27.27)$$

$F = 20.0$ lb **Ans**

15-121. The vane on a turbine is moving at 80 ft/s when a jet of water having a velocity of $v_A = 150$ ft/s strikes it. If the cross-sectional area of the jet is 1.5 in^2, and it is diverted as shown, determine the power developed by the water on the blade. $\gamma_w = 62.4$ lb/ft^3.

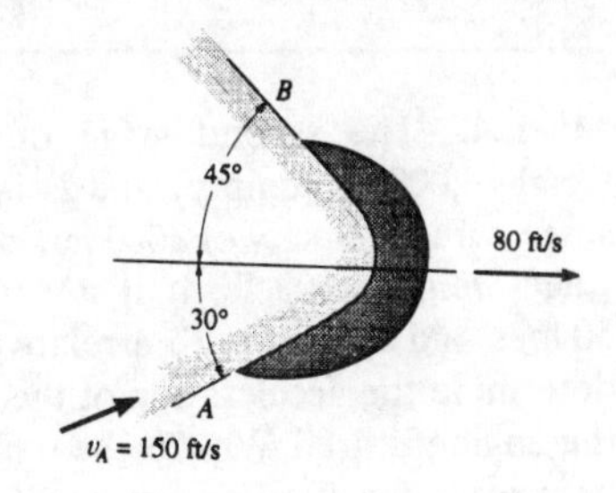

Relative flow of water into the blade :

$$\mathbf{v}_A = 150\cos 30°\mathbf{i} + 150\sin 30°\mathbf{j} = \{129.90\mathbf{i} + 75\mathbf{j}\} \text{ ft/s}$$

$$\mathbf{v}_{w/bl} = \mathbf{v}_w - \mathbf{v}_{bl} = 150\cos 30°\mathbf{i} + 150\sin 30°\mathbf{j} - 80\mathbf{i} = \{49.90\mathbf{i} + 75\mathbf{j}\} \text{ ft/s}$$

$$v_{w/bl} = \sqrt{(49.90)^2 + (75)^2} = 90.08 \text{ ft/s}$$

$$\mathbf{v}_B = \mathbf{v}_{bl} + \mathbf{v}_{w/bl} = 80\mathbf{i} - 90.08\cos 45°\mathbf{i} + 90.08\sin 45°\mathbf{j} = \{16.30\mathbf{i} + 63.70\mathbf{j}\} \text{ ft/s}$$

$$\frac{dm}{dt} = \rho v A = \left(\frac{62.4}{32.2}\right)(90.08)\left(\frac{1.5}{144}\right) = 1.818 \text{ slug/s}$$

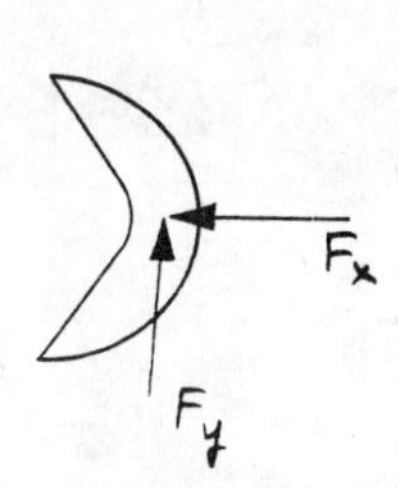

$$\overset{+}{\rightarrow} \Sigma F_x = \frac{dm}{dt}((v_B)_x - (v_A)_x)$$

$$-F_x = 1.818(16.30 - 129.90)$$

$$F_x = 206.59 \text{ lb}$$

$P = \mathbf{F} \cdot \mathbf{v} = \dfrac{206.59(80)}{550} = 30.0$ hp **Ans**

15-122. A plow located on the front of a locomotive scoops up snow at the rate of 10 ft^3/s and stores it in the train. If the locomotive is traveling at a constant speed of 12 ft/s, determine the resistance to motion caused by the shoveling. The specific weight of snow is $\gamma_s = 6$ lb/ft^3.

$$\Sigma F_s = m\frac{dv}{dt} + v_{D/i}\frac{dm_i}{dt}$$

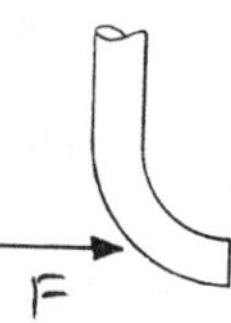

$$F = 0 + (12 - 0)\left(\frac{10(6)}{32.2}\right)$$

$F = 22.4$ lb **Ans**

15-123. The boat has a mass of 180 kg and is traveling forward on a river with a constant velocity of 70 km/h, measured *relative* to the river. The river is flowing in the opposite direction at 5 km/h. If a tube is placed in the water, as shown, and it collects 40 kg of water in the boat in 80 s, determine the horizontal thrust T on the tube that is required to overcome the resistance to the water collection. $\rho_w = 1$ Mg/m^3.

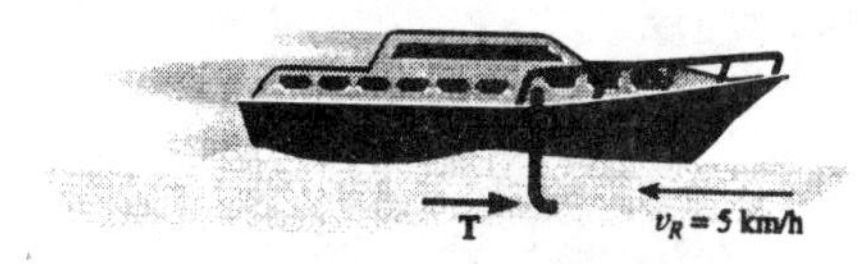

$$\frac{dm}{dt} = \frac{40}{80} = 0.5 \text{ kg/s}$$

$$v_{D/i} = (70)\left(\frac{1000}{3600}\right) = 19.444 \text{ m/s}$$

$$\Sigma F_s = m\frac{dv}{dt} + v_{D/i}\frac{dm_i}{dt}$$

$T = 0 + 19.444(0.5) = 9.72$ N **Ans**

***15-124.** The second stage of the two-stage rocket weighs 2000 lb (empty) and is launched from the first stage with a velocity of 3000 mi/h. The fuel in the second stage weighs 1000 lb. If it is consumed at the rate of 50 lb/s and ejected with a relative velocity of 8000 ft/s, determine the acceleration of the second stage just after the engine is fired. What is the rocket's acceleration after before all the fuel is consumed? Neglect the effect of gravitation.

Initially,

$$\Sigma F_s = m\frac{dv}{dt} - v_{D/e}\left(\frac{dm_e}{dt}\right)$$

$$0 = \frac{3000}{32.2}a - 8000\left(\frac{50}{32.2}\right)$$

$a = 133$ ft/s^2 **Ans**

Finally,

$$0 = \frac{2000}{32.2}a - 8000\left(\frac{50}{32.2}\right)$$

$a = 200$ ft/s^2 **Ans**

15-125. The missile weighs 40 000 lb. The constant thrust provided by the turbojet engine is $T = 15\,000$ lb. Additional thrust is provided by *two* rocket boosters B. The propellant in each booster is burned at a constant rate of 150 lb/s, with a relative exhaust velocity of 3000 ft/s. If the mass of the propellant lost by the turbojet engine can be neglected, determine the velocity of the missile after the 4-s burn time of the boosters. The initial velocity of the missile is 300 mi/h.

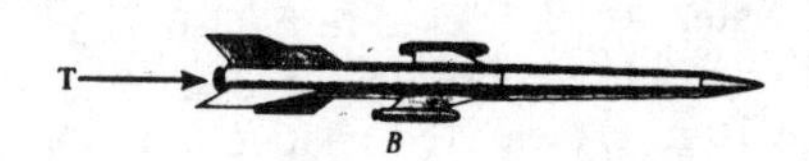

$$\overset{+}{\rightarrow} \Sigma F_s = m\frac{dv}{dt} - v_{D/e}\frac{dm_e}{dt}$$

At a time t, $m = m_0 - ct$, where $c = \dfrac{dm_e}{dt}$.

$$T = (m_0 - ct)\frac{dv}{dt} - v_{D/e}c$$

$$\int_{v_0}^{v} dv = \int_0^t \left(\frac{T + cv_{D/e}}{m_0 - ct}\right)dt$$

$$v = \left(\frac{T + cv_{D/e}}{c}\right)\ln\left(\frac{m_0}{m_0 - ct}\right) + v_0 \qquad (1)$$

Here, $m_0 = \dfrac{40\,000}{32.2} = 1242.24$ slug, $c = 2\left(\dfrac{150}{32.2}\right) = 9.3168$ slug/s, $v_{D/e} = 3000$ ft/s,

$t = 4$ s, $v_0 = \dfrac{300(5280)}{3600} = 440$ ft/s.

Substitute the numerical values into Eq. (1) :

$$v_{max} = \left(\frac{15\,000 + 9.3168(3000)}{9.3168}\right)\ln\left(\frac{1242.24}{1242.24 - 9.3168(4)}\right) + 440$$

$$v_{max} = 580 \text{ ft/s}$$ **Ans**

15-126. The rocket car has a mass of 2 Mg (empty) and carries 120 kg of fuel. If the fuel is consumed at a constant rate of 6 kg/s and ejected from the car with a relative velocity of 800 m/s, determine the maximum speed attained by the car starting from rest. The drag resistance due to the atmosphere is $F_D = (6.8v^2)$ N, where v is the speed in m/s.

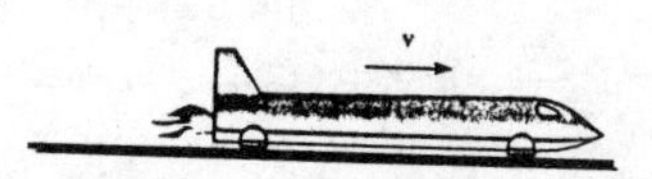

$$\Sigma F_s = m\frac{dv}{dt} - v_{D/e}\left(\frac{dm_e}{dt}\right)$$

At time t, the mass of the car is $m_0 - ct$, where $c = \dfrac{dm_e}{dt} = 6$ kg/s

Set $F = kv^2$, then

$$-kv^2 = (m_0 - ct)\frac{dv}{dt} - v_{D/e}c$$

$$\int_0^v \frac{dv}{(cv_{D/e} - kv^2)} = \int_0^t \frac{dt}{(m_0 - ct)}$$

$$\left(\frac{1}{2\sqrt{cv_{D/e}k}}\right)\ln\left[\frac{\sqrt{\frac{cv_{D/e}}{k}} + v}{\sqrt{\frac{cv_{D/e}}{k}} - v}\right]_0^v = -\frac{1}{c}\ln(m_0 - ct)\Big|_0^t$$

$$\left(\frac{1}{2\sqrt{cv_{D/e}k}}\right)\ln\left(\frac{\sqrt{\frac{cv_{D/e}}{k}} + v}{\sqrt{\frac{cv_{D/e}}{k}} - v}\right) = -\frac{1}{c}\ln\left(\frac{m_0 - ct}{m_0}\right)$$

Maximum speed occurs at the instant the fuel runs out.

$$t = \frac{120}{6} = 20 \text{ s}$$

Thus,

$$\left(\frac{1}{2\sqrt{(6)(800)(6.8)}}\right)\ln\left(\frac{\sqrt{\frac{(6)(800)}{6.8}} + v}{\sqrt{\frac{(6)(800)}{6.8}} - v}\right) = -\frac{1}{6}\ln\left(\frac{2120 - 6(20)}{2120}\right)$$

Solving,

$$v = 25.0 \text{ m/s}$$ **Ans**

15-127. The 10-Mg helicopter carries a bucket containing 500 kg of water, which is used to fight fires. If it hovers over the land in a fixed position and then releases 50 kg/s of water at 10 m/s, measured relative to the helicopter, determine the initial upward acceleration the helicopter experiences as the water is being released.

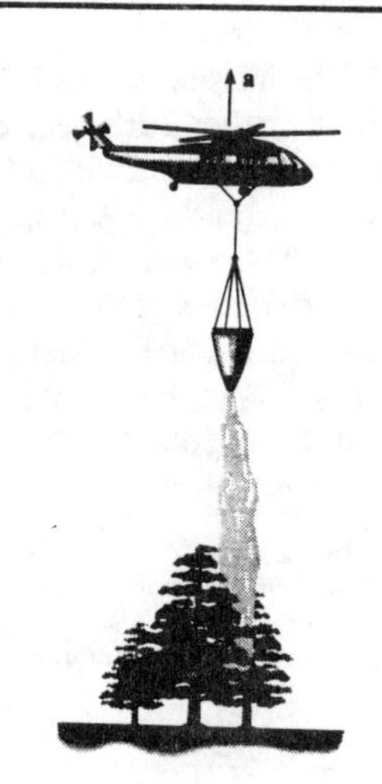

$$+\uparrow \Sigma F_s = m\frac{dv}{dt} - v_{D/e}\frac{dm_e}{dt}$$

Initially, the bucket is full of water, hence $m = 10(10^3) + 0.5(10^3) = 10.5(10^3)$ kg

$$0 = 10.5(10^3)\,a - (10)(50)$$

$a = 0.0476\ \text{m/s}^2$ **Ans**

***15-128.** A rocket burns 2000 lb of fuel in 2 min. If the velocity of the gas is 6500 ft/s with respect to the rocket, determine the thrust of the rocket engine.

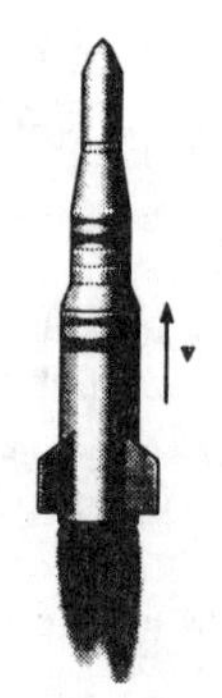

$$\frac{dm_e}{dt} = \frac{2000/32.2}{2(60)} = 0.5176\ \text{slug/s} \qquad v_{D/e} = 6500\ \text{ft/s}$$

$T = v_{D/e}\dfrac{dm_e}{dt} = 6500(0.5176) = 3364\ \text{lb} = 3.36\ \text{kip}$ **Ans**

15-129. A rocket has an empty weight of 500 lb and carries 300 lb of fuel. If the fuel is burned at the rate of 15 lb/s and ejected with a relative velocity of 4400 ft/s, determine the maximum speed attained by the rocket starting from rest. Neglect the effect of gravitation on the rocket.

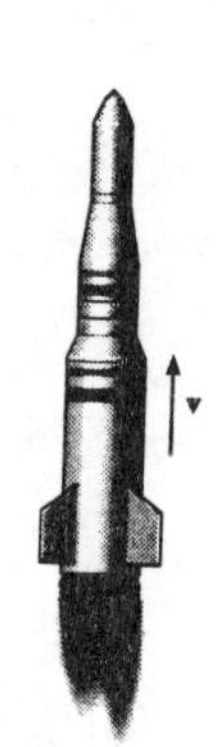

$$+\uparrow \Sigma F_s = m\frac{dv}{dt} - v_{D/e}\frac{dm_e}{dt}$$

At a time t, $m = m_0 - ct$, where $c = \dfrac{dm_e}{dt}$. In space the weight of the rocket is zero.

$$0 = (m_0 - ct)\frac{dv}{dt} - v_{D/e}c$$

$$\int_0^v dv = \int_0^t \left(\frac{cv_{D/e}}{m_0 - ct}\right)dt$$

$$v = v_{D/e}\ln\left(\frac{m_0}{m_0 - ct}\right) \qquad (1)$$

The maximum speed occurs when all the fuel is consumed, that is, when $t = \frac{300}{15} = 20$ s.
Here, $m_0 = \frac{500+300}{32.2} = 24.8447$ slug, $c = \frac{15}{32.2} = 0.4658$ slug/s, $v_{D/e} = 4400$ ft/s.
Substitute the numerical values into Eq.(1) :

$$v_{max} = 4400\ln\left(\frac{24.8447}{24.8447 - 0.4658(20)}\right)$$

$v_{max} = 2068\ \text{ft/s} = 2.07(10^3)\ \text{ft/s}$ **Ans**

15-130. The 12-Mg jet airplane has a constant speed of 950 km/h when it is flying along a horizontal straight line. Air enters the intake scoops S at the rate of 50 m^3/s. If the engine burns fuel at the rate of 0.4 kg/s and the gas (air and fuel) is exhausted relative to the plane with a speed of 450 m/s, determine the resultant drag force exerted on the plane by air resistance. Assume that air has a constant density of 1.22 kg/m^3. *Hint:* Since mass both enters and exits the plane, Eqs. 15–29 and 15–30 must be combined to yield

$$\Sigma F_s = m\frac{dv}{dt} - v_{D/e}\frac{dm_e}{dt} + v_{D/i}\frac{dm_i}{dt}$$

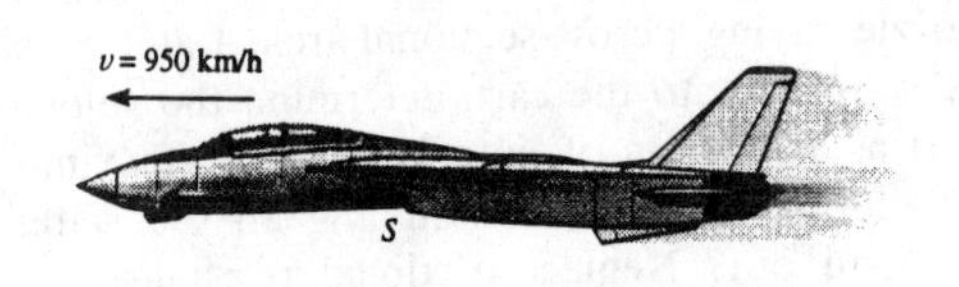

$$\Sigma F_s = m\frac{dv}{dt} - \frac{dm_e}{dt}(v_{D/E}) + \frac{dm_i}{dt}(v_{D/i}) \quad (1)$$

$v = 950 \text{ km/h} = 0.2639 \text{ km/s}, \qquad \frac{dv}{dt} = 0$

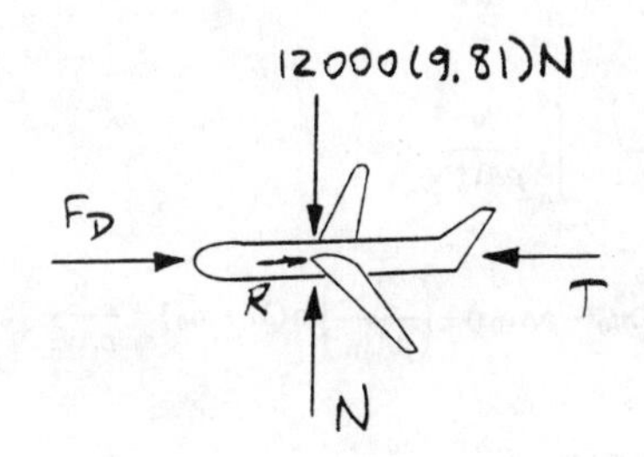

$v_{D/E} = 0.45 \text{ km/s}$

$v_{D/i} = 0.2639 \text{ km/s}$

$\frac{dm_i}{dt} = 50(1.22) = 61.0 \text{ kg/s}$

$\frac{dm_e}{dt} = 0.4 + 61.0 = 61.4 \text{ kg/s}$

Forces T and R are incorporated into Eq. (1) as the last two terms in the equation.

$(\overset{+}{\leftarrow}) \; -F_D = 0 - (0.45)(61.4) + (0.2639)(61)$

$F_D = 11.5 \text{ kN}$ **Ans**

15-131. The chain has a mass per unit length of m. Determine the magnitude of force, $F = F(t)$, which must be applied to the end of the chain to raise it with a constant speed v. Initially, the entire chain is at rest on the ground.

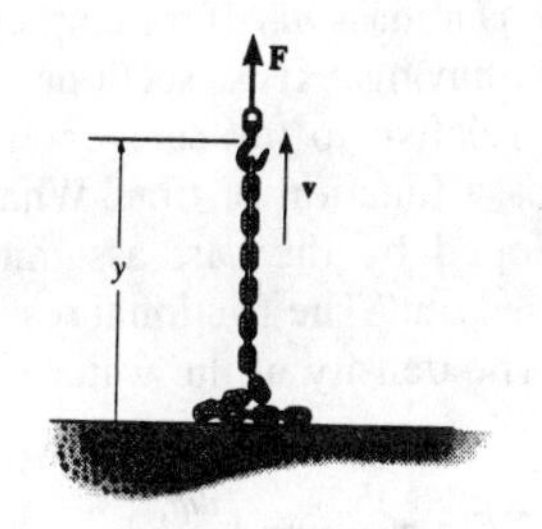

$\frac{dv}{dt} = 0$

$y = vt$

$m_i = my = mvt$

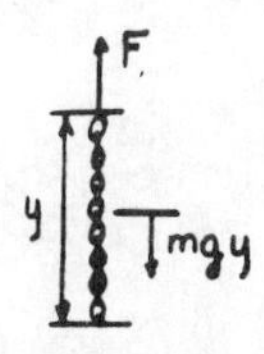

$\frac{dm_i}{dt} = mv$

$+\uparrow \Sigma F_s = m\frac{dv}{dt} + v_{D/i}\left(\frac{dm_i}{dt}\right)$

$F - mgvt = 0 + v(mv)$

$F = mv(gt + v)$ **Ans**

***15-132.** The cart has a mass M and is filled with water that has a mass m_0. If a pump ejects the water through a nozzle having a cross-sectional area A at a constant rate of v_0 relative to the cart, determine the velocity of the cart as a function of time. What is the maximum speed developed by the cart assuming all the water can be pumped out? Neglect frictional resistance to forward motion. The density of the water is ρ.

$$\overset{+}{\leftarrow} \Sigma F_s = m\frac{dv}{dt} - v_{D/e}\left(\frac{dm_e}{dt}\right)$$

$$\frac{dm_e}{dt} = \rho A v_0$$

$$0 = (M + m_0 - \rho A v_0 t)\frac{dv}{dt} - v_0\,(\rho A v_0)$$

$$\int_0^t \frac{dt}{M + m_0 - \rho A v_0 t} = \int_0^v \frac{dv}{\rho A v_0^2}$$

$$-\left(\frac{1}{\rho A v_0}\right)\ln(M + m_0 - \rho A v_0 t) + \left(\frac{1}{\rho A v_0}\right)\ln(M + m_0) = \frac{v}{\rho A v_0^2}$$

$$\left(\frac{1}{\rho A v_0}\right)\ln\left(\frac{M + m_0}{M + m_0 - \rho A v_0 t}\right) = \frac{v}{\rho A v_0^2}$$

$$v = v_0 \ln\left(\frac{M + m_0}{M + m_0 - \rho A v_0 t}\right) \qquad \textbf{Ans}$$

v_{max} occurs when $t = \dfrac{m_0}{\rho A v_0}$, i.e.,

$$v_{max} = v_0 \ln\left(\frac{M + m_0}{M}\right) \qquad \textbf{Ans}$$

15-133. The cart has a mass M and is filled with water that has a mass m_0. If a pump ejects the water through a nozzle having a cross-sectional area A at a constant rate of v_0 relative to the cart, determine the velocity of the cart as a function of time. What is the maximum speed developed by the cart assuming all the water can be pumped out? The frictional resistance to forward motion is F. The density of the water is ρ.

$$\overset{+}{\leftarrow} \Sigma F_s = m\frac{dv}{dt} - v_{D/e}\left(\frac{dm_e}{dt}\right)$$

$$\frac{dm_e}{dt} = \rho A v_0$$

$$-F = (M + m_0 - \rho A v_0 t)\frac{dv}{dt} - v_0\,(\rho A v_0)$$

$$\int_0^t \frac{dt}{M + m_0 - \rho A v_0 t} = \int_0^v \frac{dv}{\rho A v_0^2 - F}$$

$$-\left(\frac{1}{\rho A v_0}\right)\ln(M + m_0 - \rho A v_0 t) + \left(\frac{1}{\rho A v_0}\right)\ln(M + m_0) = \frac{v}{\rho A v_0^2 - F}$$

$$\left(\frac{1}{\rho A v_0}\right)\ln\left(\frac{M + m_0}{M + m_0 - \rho A v_0 t}\right) = \frac{v}{\rho A v_0^2 - F}$$

$$v = \left(\frac{\rho A v_0^2 - F}{\rho A v_0}\right)\ln\left(\frac{M + m_0}{M + m_0 - \rho A v_0 t}\right) \qquad \textbf{Ans}$$

v_{max} occurs when $t = \dfrac{m_0}{\rho A v_0}$, or,

$$v_{max} = \left(\frac{\rho A v_0^2 - F}{\rho A v_0}\right)\ln\left(\frac{M + m_0}{M}\right) \qquad \textbf{Ans}$$

15-134. The chain has a total length $L < d$ and a mass per unit length of m'. If a portion h of the chain is suspended over the table and released, determine the velocity of its end A as a function of its position y. Neglect friction.

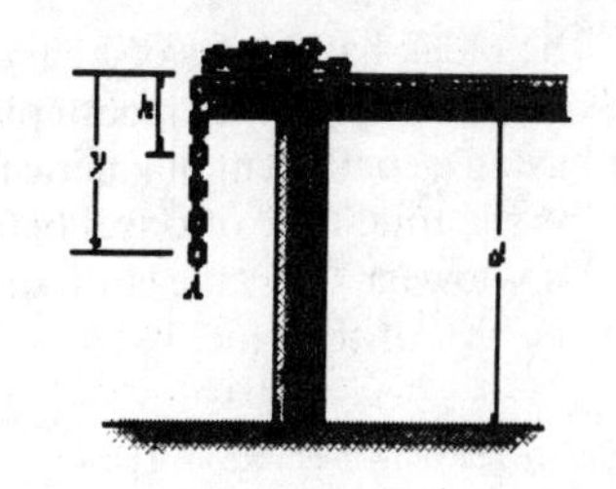

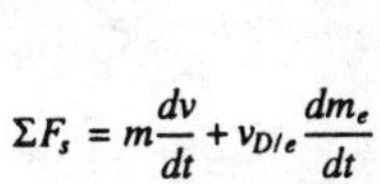

$$\Sigma F_s = m\frac{dv}{dt} + v_{D/e}\frac{dm_e}{dt}$$

$$m' gy = m' y\frac{dv}{dt} + v(m' v)$$

$$m' gy = m'\left(y\frac{dv}{dt} + v^2\right)$$

Since $dt = \frac{dy}{v}$, we have

$$gy = vy\frac{dy}{dt} + v^2$$

Multiply by $2y$ and integrate :

$$\int 2gy^2\, dy = \int\left(2vy^2\,\frac{dv}{dy} + 2yv^2\right) dy$$

$$\frac{2}{3}g^3 + C = v^2y^2$$

When $v = 0$, $y = h$, so that $C = -\frac{2}{3}gh^3$

Thus, $\quad v^2 = \frac{2}{3}g\left(\frac{y^3 - h^3}{y^2}\right)$

$$v = \sqrt{\frac{2}{3}g\left(\frac{y^3 - h^3}{y^2}\right)} \qquad \textbf{Ans}$$

R1-1. The block has a mass of 5 kg and is released from rest at A. It slides down the smooth plane onto the rough surface having a coefficient of kinetic friction of $\mu_k = 0.2$. Determine the total time of travel before the block stops sliding. Also, how far s does the block slide before stopping? Neglect the size of the block.

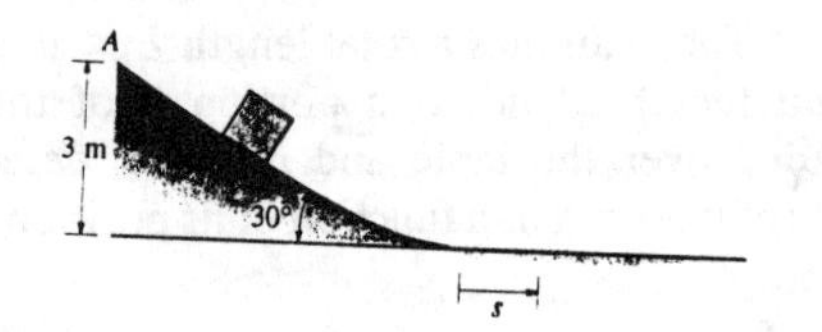

Just before reaching the horizontal plane

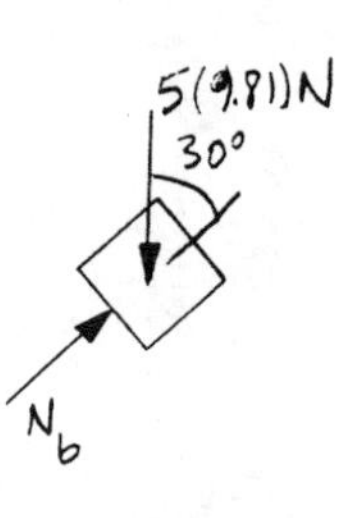

$$T_1 + V_1 = T_2 + V_2$$

$$0 + (5)(9.81)(3) = \frac{1}{2}(5)v_2^2$$

$$v_2 = 7.672 \text{ m/s}$$

Time to travel down the incline

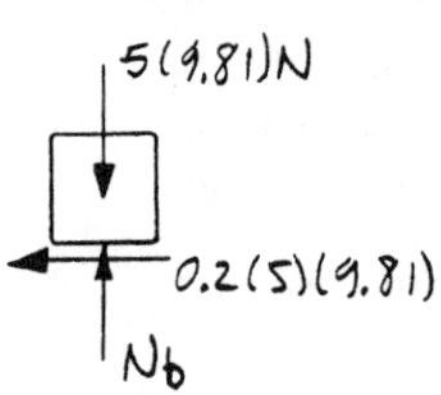

$$(+\searrow) \quad m\mathbf{v}_1 + \Sigma\int \mathbf{F}\,dt = m\,\mathbf{v}_2$$

$$0 + 5(9.81)\sin 30°(t_1) = 5(7.672)$$

$$t_1 = 1.564 \text{ s}$$

Time to travel along horizontal plane

$$(\overset{+}{\rightarrow}) \quad m\mathbf{v}_2 + \Sigma\int \mathbf{F}\,dt = m\mathbf{v}_3$$

$$5(7.672) - 0.2(5)(9.81)t_2 = 0$$

$$t_2 = 3.910 \text{ s}$$

Thus,

$$t = t_1 + t_2 = 1.564 + 3.910 = 5.47 \text{ s} \qquad \textbf{Ans}$$

$$T_1 + \Sigma U_{1-2} = T_2$$

$$\frac{1}{2}(5)(7.672)^2 - 0.2(5)(9.81)s = 0$$

$$s = 15.0 \text{ m} \qquad \textbf{Ans}$$

R1-2. Cartons having a mass of 5 kg are required to move along the assembly line at a constant speed of 8 m/s. Determine the smallest radius of curvature, ρ, for the conveyor so the cartons do not slip. The coefficients of static and kinetic friction between a carton and the conveyor are $\mu_s = 0.7$ and $\mu_k = 0.5$, respectively.

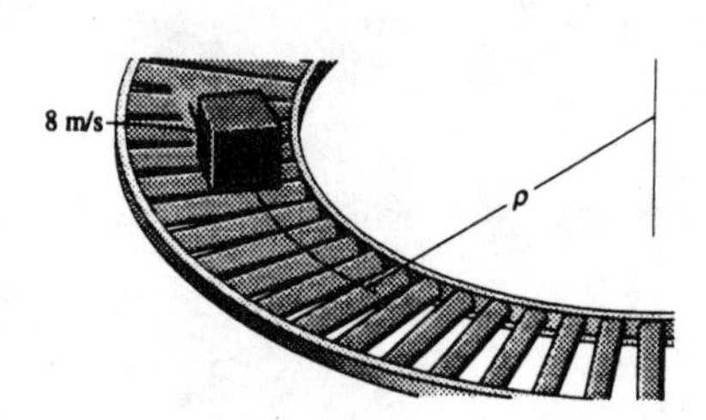

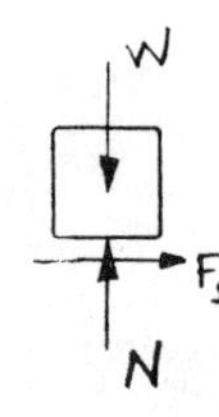

$$+\uparrow \Sigma F_b = m\,a_b; \quad N - W = 0$$

$$N = W$$

$$F_s = 0.7W$$

$$\overset{+}{\rightarrow} \Sigma F_n = m\,a_n; \quad 0.7W = \frac{W}{9.81}\left(\frac{8^2}{\rho}\right)$$

$$\rho = 9.32 \text{ m} \qquad \textbf{Ans}$$

R1-3. A small metal particle passes downward through a fluid medium while being subjected to the attraction of a magnetic field such that its position is observed to be $s = (15t^3 - 3t)$ mm, where t is in seconds. Determine (a) the particle's displacement from $t = 2$ s to $t = 4$ s, and (b) the velocity and acceleration of the particle when $t = 5$ s.

a) $s = 15t^3 - 3t$

At $t = 2$ s, $s_1 = 114$ mm

At $t = 4$ s, $s_3 = 948$ mm

$\Delta s = 948 - 114 = 834$ mm **Ans**

b) $v = \dfrac{ds}{dt} = 45t^2 - 3\Big|_{t=5} = 1122 \text{ mm/s} = 1.12 \text{ m/s}$ **Ans**

$a = \dfrac{dv}{dt} = 90t\Big|_{t=5} = 450 \text{ mm/s}^2 = 0.450 \text{ m/s}^2$ **Ans**

***R1-4.** The device shown is designed to produce the experience of weightlessness in a passenger when he reaches point A, $\theta = 90°$, along the path. If the passenger has a mass of 75 kg, determine the minimum speed he should have when he reaches A so that he does not exert a normal reaction on the seat. The chair is pin connected to the frame BC so that he is always seated in an upright position. During the motion his speed remains constant.

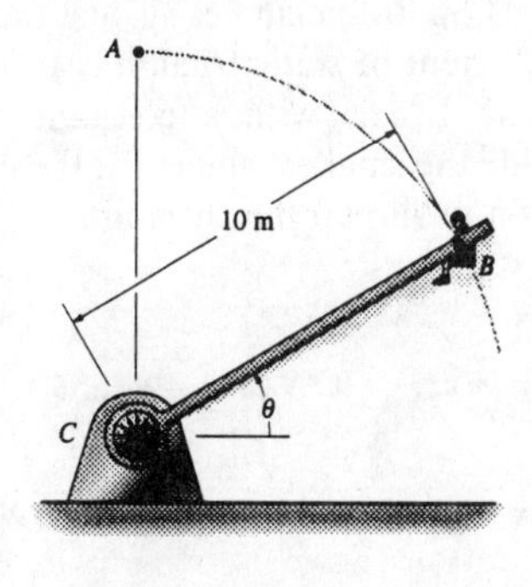

$+\downarrow \Sigma F_n = m\,a_n; \quad mg = m\left(\dfrac{v^2}{10}\right)$

$a_n\downarrow$ mg N=0

$9.81 = \dfrac{v^2}{10}$

$v = 9.90$ m/s **Ans**

R1-5. The passenger has a mass of 75 kg and always sits in an upright position on the chair. At the instant $\theta = 30°$, he has a speed of 5 m/s and an increase of speed of 2 m/s². Determine the horizontal and vertical forces that the chair exerts on him in order to produce this motion.

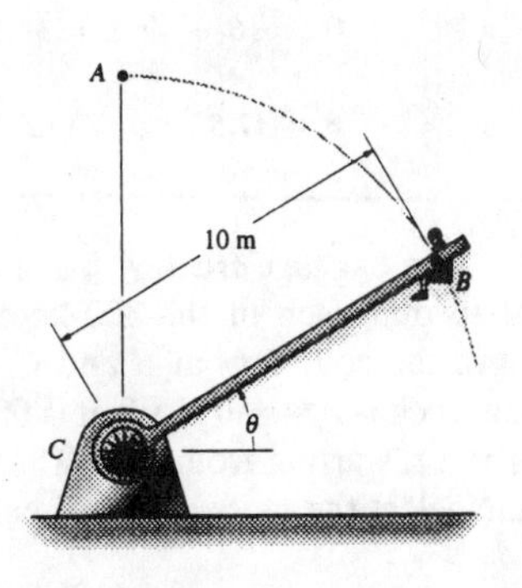

$\overset{+}{\leftarrow} \Sigma F_x = m\,a_x; \quad F_x = 75\left(\dfrac{(5)^2}{10}\right)\cos 30° + 75(2)\sin 30°$

$F_x = 237 \text{ N} \leftarrow$ **Ans**

$+\uparrow \Sigma F_y = m\,a_y; \quad F_y - 75(9.81) = -75\left(\dfrac{(5)^2}{10}\right)\sin 30° + 75(2)\cos 30°$

$F_y = 772 \text{ N} \uparrow$ **Ans**

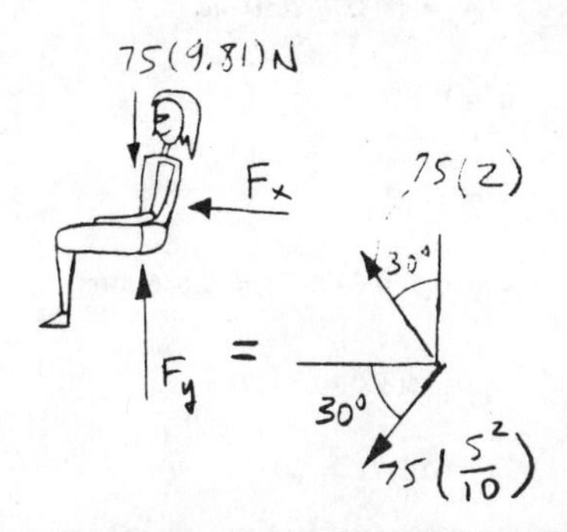

R1-6. The 150-lb man lies against the cushion for which the coefficient of static friction is $\mu_s = 0.5$. Determine the resultant normal and frictional forces the cushion exerts on him if, due to rotation about the z axis, he has a constant speed $v = 20$ ft/s. Neglect the size of the man. Take $\theta = 60°$.

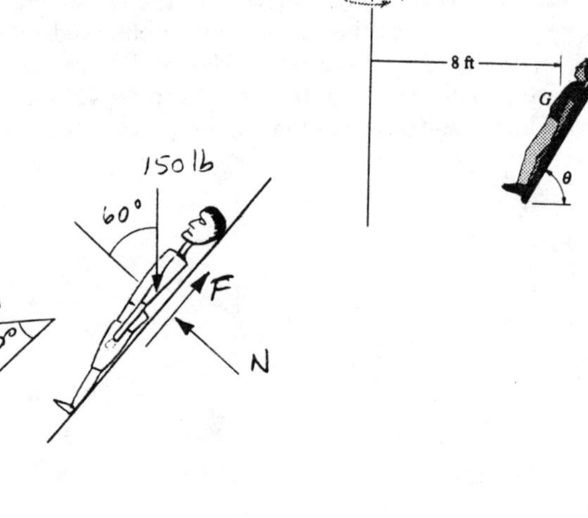

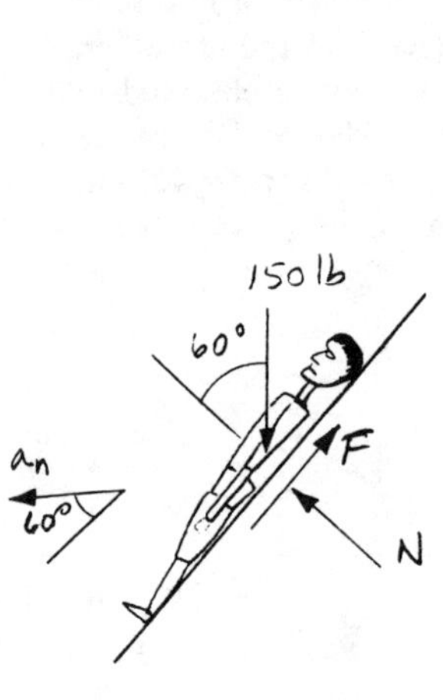

$$+\nwarrow \Sigma F_y = m(a_n)_y; \quad N - 150\cos 60° = \frac{150}{32.2}\left(\frac{20^2}{8}\right)\sin 60°$$

$$N = 277 \text{ lb} \qquad \textbf{Ans}$$

$$+\swarrow \Sigma F_x = m(a_n)_x; \quad -F + 150\sin 60° = \frac{150}{32.2}\left(\frac{20^2}{8}\right)\cos 60°$$

$$F = 13.4 \text{ lb} \qquad \textbf{Ans}$$

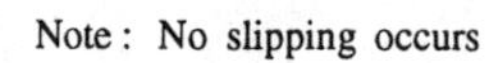

Note : No slipping occurs

Since $\mu_s N = 138.4 \text{ lb} > 13.4 \text{ lb}$

R1-7. The 150-lb man lies against the cushion for which the coefficient of static friction is $\mu_s = 0.5$. If he rotates about the z axis with a constant speed $v = 30$ ft/s, determine the smallest angle θ of the cushion at which he will begin to slip up the cushion.

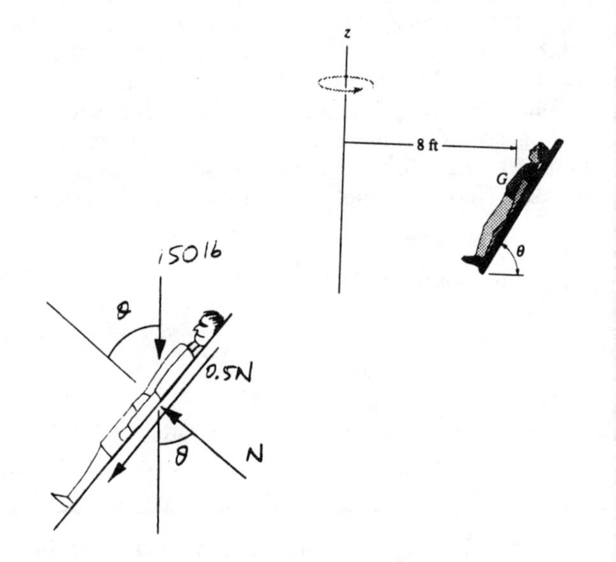

$$\overset{+}{\leftarrow} \Sigma F_n = m\,a_n; \quad 0.5N\cos\theta + N\sin\theta = \frac{150}{32.2}\left(\frac{(30)^2}{8}\right)$$

$$+\uparrow \Sigma F_b = 0; \quad -150 + N\cos\theta - 0.5\,N\sin\theta = 0$$

$$N = \frac{150}{\cos\theta - 0.5\sin\theta}$$

$$\frac{(0.5\cos\theta + \sin\theta)150}{(\cos\theta - 0.5\sin\theta)} = \frac{150}{32.2}\left(\frac{(30)^2}{8}\right)$$

$$0.5\cos\theta + \sin\theta = 3.493\,79\cos\theta - 1.746\,89\sin\theta$$

$$\theta = 47.5° \qquad \textbf{Ans}$$

***R1-8.** The baggage truck A has a mass of 800 kg and is used to pull each of the 300-kg cars. Determine the tension in the couplings at B and C if the tractive force **F** on the truck is $F = 480$ N. What is the speed of the truck when $t = 2$ s, starting from rest? The car wheels are free to roll. Neglect the mass of the wheels.

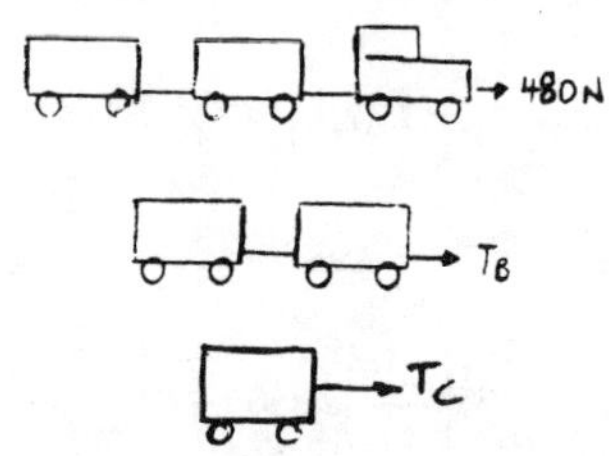

$$\overset{+}{\rightarrow} \Sigma F_x = m\,a_x; \quad 480 = [800 + 2(300)]a$$

$$a = 0.3429 \text{ m/s}^2$$

$$(\overset{+}{\rightarrow}) \qquad v = v_0 + a_c t$$

$$v = 0 + 0.3429(2) = 0.686 \text{ m/s} \qquad \textbf{Ans}$$

$$\overset{+}{\rightarrow} \Sigma F_x = m\,a_x; \quad T_B = 2(300)(0.3429)$$

$$T_B = 205.71 = 206 \text{ N} \qquad \textbf{Ans}$$

$$\overset{+}{\rightarrow} \Sigma F_x = m\,a_x; \quad T_C = (300)(0.3429)$$

$$T_C = 102.86 = 103 \text{ N} \qquad \textbf{Ans}$$

R1-9. The baggage truck A has a mass of 800 kg and is used to pull each of the 300-kg cars. If the tractive force **F** on the truck is $F = 480$ N, determine the initial acceleration of the truck. What is the acceleration of the truck if the coupling at C suddenly fails? The car wheels are free to roll. Neglect the mass of the wheels.

$\overset{+}{\rightarrow} \Sigma F_x = m\,a_x;\qquad 480 = [800 + 2(300)]a$

480N

$a = 0.3429 = 0.343 \text{ m/s}^2$ **Ans**

$\overset{+}{\rightarrow} \Sigma F_x = m\,a_x;\qquad 480 = (800+300)a$

480N

$a = 0.436 \text{ m/s}^2$ **Ans**

R1-10. A car is traveling at 80 ft/s when the brakes are suddenly applied, causing a constant deceleration of 10 ft/s^2. Determine the time required to stop the car and the distance traveled before stopping.

$(\overset{+}{\rightarrow})\qquad v = v_0 + a_c t$

$0 = 80 + (-10)t$

$t = 8 \text{ s}$ **Ans**

$(\overset{+}{\rightarrow})\qquad v^2 = v_0^2 + 2a_c(s - s_0)$

$0 = (80)^2 + 2(-10)(s - 0)$

$s = 320 \text{ ft}$ **Ans**

R1-11. Determine the speed of block B if the end of the cable at C is pulled downward with a speed of 10 ft/s. What is the relative velocity of the block with respect to C?

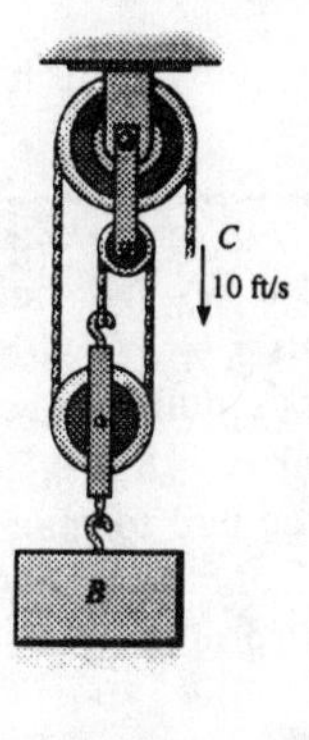

$3s_B + s_C = l$

$3v_B = -v_C$

$3v_B = -(10)$

$v_B = -3.33 \text{ ft/s} = 3.33 \text{ ft/s} \uparrow$ **Ans**

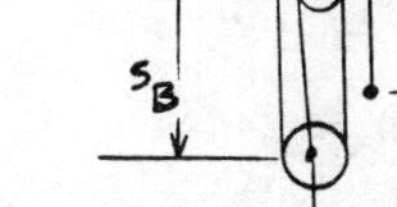
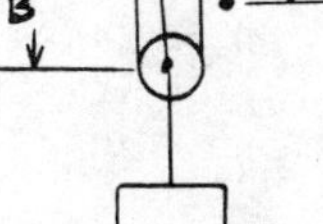

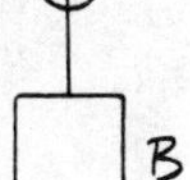

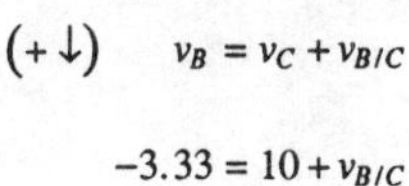

$(+\downarrow)\qquad v_B = v_C + v_{B/C}$

$-3.33 = 10 + v_{B/C}$

$v_{B/C} = -13.3 \text{ ft/s} = 13.3 \text{ ft/s} \uparrow$ **Ans**

***R1-12.** During operation the breaker hammer develops on the concrete surface a force which is indicated in the graph. To achieve this the 2-lb spike S is fired from rest into the surface at 200 ft/s. Determine the speed of the spike just after rebounding.

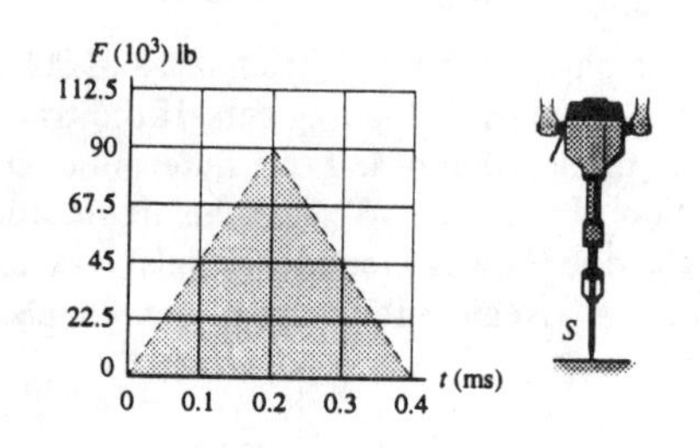

$(+\downarrow) \qquad mv_1 + \int F\,dt = mv_2$

$$\frac{2}{32.2}(200) + 2(0.0004) - Area = \frac{-2}{32.2}(v)$$

$$Area = \frac{1}{2}(90)(10^3)(0.4)(10^{-3}) = 18 \text{ lb}\cdot\text{s}$$

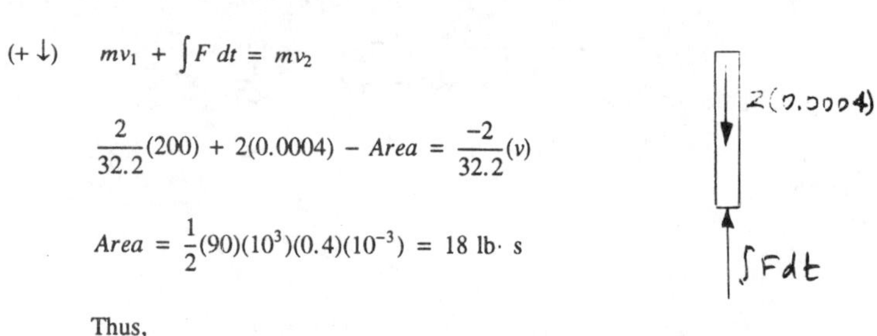

Thus,

$v = 89.8$ ft/s **Ans**

R1-13. A 4-lb ball B is traveling around in a circle of radius $r_1 = 3$ ft with a speed $(v_B)_1 = 6$ ft/s. If the attached cord is pulled down through the hole with a constant speed $v_r = 2$ ft/s, determine the ball's speed at the instant $r_2 = 2$ ft. How much work has to be done to pull down the cord? Neglect friction and the size of the ball.

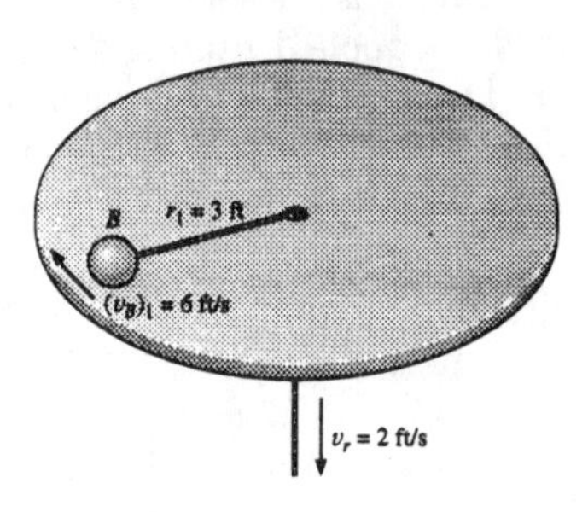

$H_1 = H_2$

$$\frac{4}{32.2}(6)(3) = \frac{4}{32.2}v_\theta(2)$$

$v_\theta = 9$ ft/s

$v_2 = \sqrt{9^2 + 2^2} = 9.22$ ft/s **Ans**

$T_1 + \Sigma U_{1-2} = T_2$

$$\frac{1}{2}\left(\frac{4}{32.2}\right)(6)^2 + \Sigma U_{1-2} = \frac{1}{2}\left(\frac{4}{32.2}\right)(9.22)^2$$

$\Sigma U_{1-2} = 3.04$ ft·lb **Ans**

R1-14. A 4-lb ball B is traveling around in a circle of radius $r_1 = 3$ ft with a speed $(v_B)_1 = 6$ ft/s. If the attached cord is pulled down through the hole with a constant speed $v_r = 2$ ft/s, determine how much time is required for the ball to reach a speed of 12 ft/s. How far r_2 is the ball from the hole when this occurs? Neglect friction and the size of the ball.

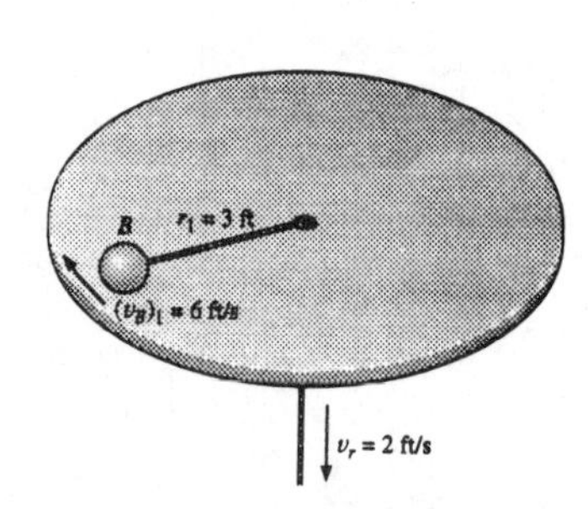

$v = \sqrt{(v_\theta)^2 + (2)^2}$

$12 = \sqrt{(v_\theta)^2 + (2)^2}$

$v_\theta = 11.832$ ft/s

$H_1 = H_2$

$$\frac{4}{32.2}(6)(3) = \frac{4}{32.2}(11.832)(r_2)$$

$r_2 = 1.5213 = 1.52$ ft **Ans**

$\Delta r = v_r t$

$(3 - 1.5213) = 2t$

$t = 0.739$ s **Ans**

R1-15. The block has a mass of 50 kg and rests on the surface of the cart having a mass of 75 kg. If the spring which is attached to the cart and not the block is compressed 0.2 m and the system is released from rest, determine the speed of the block after the spring becomes undeformed. Neglect the mass of the cart's wheels and the spring in the calculation. Also neglect friction. Take $k = 300$ N/m.

$$T_1 + V_1 = T_2 + V_2$$

$$[0+0] + \frac{1}{2}(300)(0.2)^2 = \frac{1}{2}(50)v_b^2 + \frac{1}{2}(75)v_c^2$$

$$12 = 50\,v_b^2 + 75\,v_c^2$$

$$(\overset{+}{\rightarrow}) \quad \Sigma m v_1 = \Sigma m v_2$$

$$0 + 0 = 50\,v_b - 75\,v_c$$

$$v_b = 1.5 v_c$$

$$v_c = 0.253 \text{ m/s} \leftarrow$$

$$v_b = 0.379 \text{ m/s} \rightarrow \qquad \textbf{Ans}$$

***R1-16.** The block has a mass of 50 kg and rests on the surface of the cart having a mass of 75 kg. If the spring which is attached to the cart and not the block is compressed 0.2 m and the system is released from rest, determine the speed of the block with respect to the cart after the spring becomes undeformed. Neglect the mass of the cart's wheels and the spring in the calculation. Also neglect friction. Take $k = 300$ N/m.

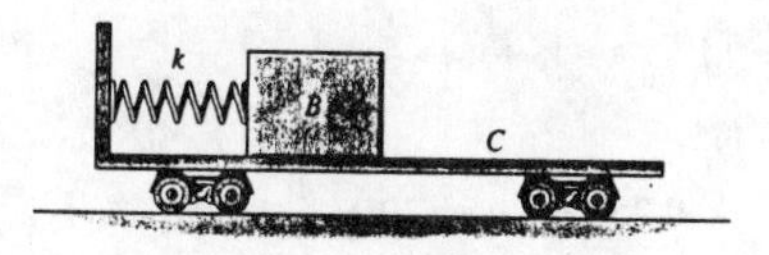

$$T_1 + V_1 = T_2 + V_2$$

$$[0+0] + \frac{1}{2}(300)(0.2)^2 = \frac{1}{2}(50)v_b^2 + \frac{1}{2}(75)v_c^2$$

$$12 = 50\,v_b^2 + 75\,v_c^2$$

$$(\overset{+}{\rightarrow}) \quad \Sigma m v_1 = \Sigma m v_2$$

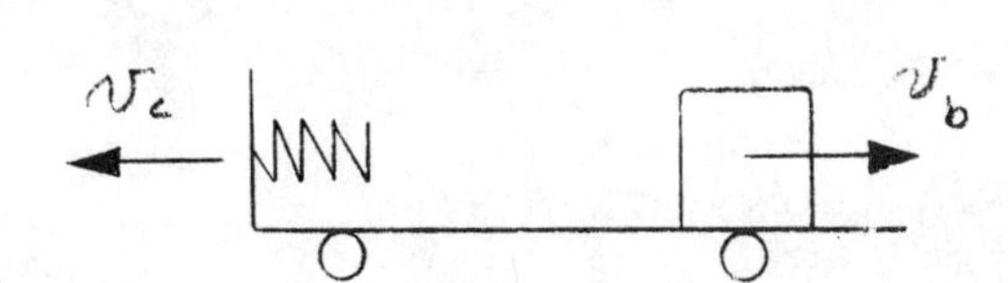

$$0 + 0 = 50\,v_b - 75\,v_c$$

$$v_b = 1.5 v_c$$

$$v_c = 0.253 \text{ m/s} \leftarrow$$

$$v_b = 0.379 \text{ m/s} \rightarrow$$

$$\mathbf{v}_b = \mathbf{v}_c + \mathbf{v}_{b/c}$$

$$(\overset{+}{\rightarrow}) \quad 0.379 = -0.253 + v_{b/c}$$

$$v_{b/c} = 0.632 \text{ m/s} \rightarrow \qquad \textbf{Ans}$$

R1-17. A ball is launched from point A at an angle of 30°. Determine the maximum and minimum speed v_A it can have so that it lands in the container.

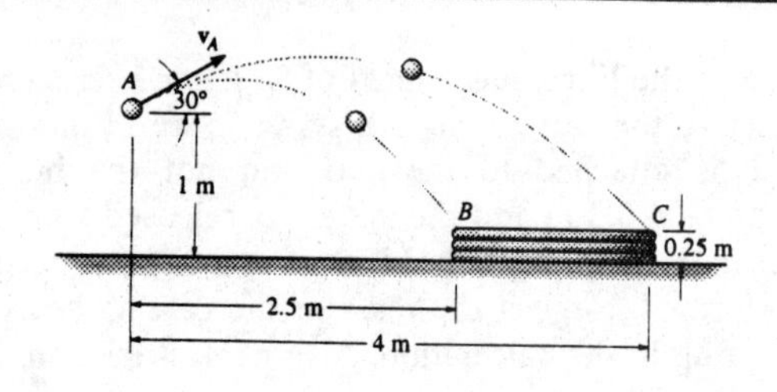

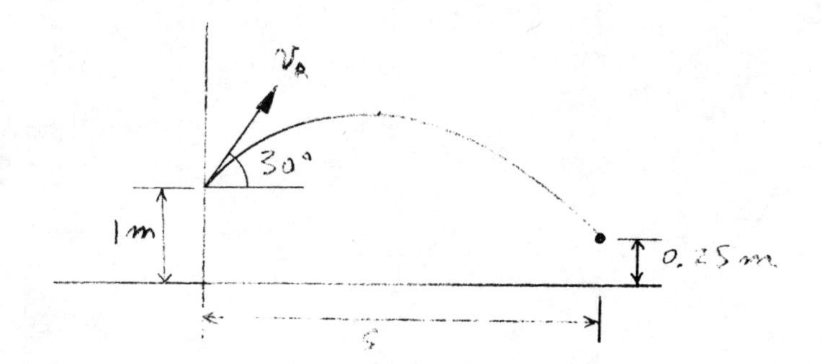

Min. speed :

$(\overset{+}{\rightarrow})$ $\quad s = s_0 + v_0 t$

$$2.5 = 0 + v_A \cos 30° t$$

$(+\uparrow)$ $\quad s = s_0 + v_0 t + \frac{1}{2} a_c t^2$

$$0.25 = 1 + v_A \sin 30° t - \frac{1}{2}(9.81)t^2$$

Solving

$t = 0.669$ s

$v_A = (v_A)_{min} = 4.32$ m/s **Ans**

Max. speed :

$(\overset{+}{\rightarrow})$ $\quad s = s_0 + v_0 t$

$$4 = 0 + v_A \cos 30° t$$

$(+\uparrow)$ $\quad s = s_0 + v_0 t + \frac{1}{2} a_c t^2$

$$0.25 = 1 + v_A \sin 30° t - \frac{1}{2}(9.81)t^2$$

Solving :

$t = 0.790$ s

$v_A = (v_A)_{max} = 5.85$ m/s **Ans**

R1-18. At the instant shown, cars A and B are traveling at speeds of 55 mi/h and 40 mi/h, respectively. If B is increasing its speed by 1200 mi/h^2, while A maintains a constant speed, determine the velocity and acceleration of B with respect to A. Car B moves along a curve having a radius of curvature of 0.5 mi.

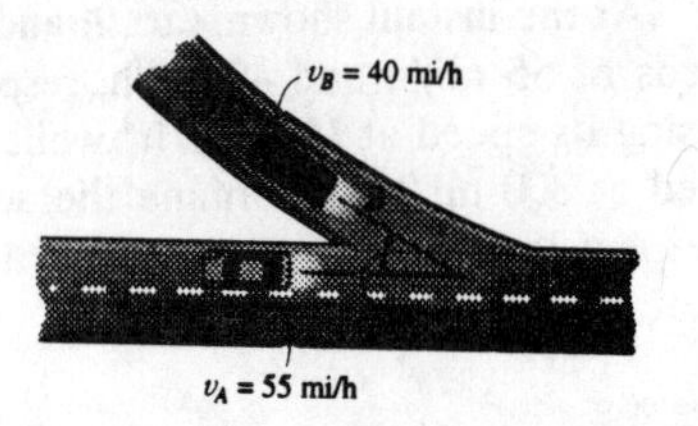

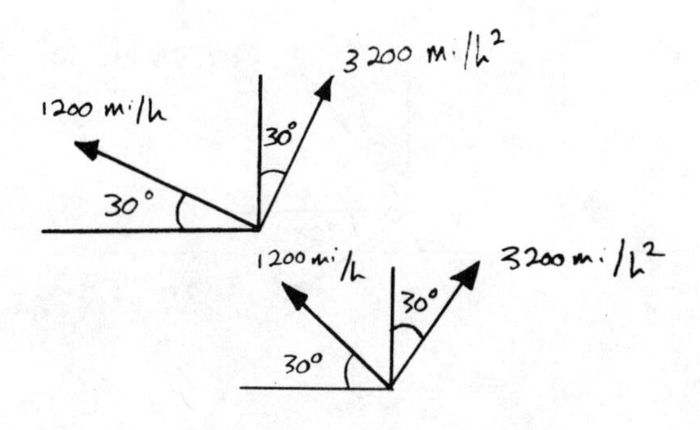

$\mathbf{v}_B = -40\cos 30°\mathbf{i} + 40\sin 30°\mathbf{j} = \{-34.64\mathbf{i} + 20\mathbf{j}\}$ mi/h

$\mathbf{v}_A = \{-55\mathbf{i}\}$ mi/h

$\mathbf{v}_{B/A} = \mathbf{v}_B - \mathbf{v}_A$

$= (-34.64\mathbf{i} + 20\mathbf{j}) - (-55\mathbf{i}) = \{20.36\mathbf{i} + 20\mathbf{j}\}$ mi/h

$v_{B/A} = \sqrt{20.36^2 + 20^2} = 28.5$ mi/h **Ans**

$\theta = \tan^{-1}\left(\dfrac{20}{20.36}\right) = 44.5°$ **Ans**

$(a_B)_n = \dfrac{v_B^2}{\rho} = \dfrac{40^2}{0.5} = 3200$ mi/h^2 $\qquad (a_B)_t = 1200$ mi/h^2

$\mathbf{a}_B = (3200\sin 30° - 1200\cos 30°)\mathbf{i} + (3200\cos 30° + 1200\sin 30°)\mathbf{j}$

$= \{560.77\mathbf{i} + 3371.28\mathbf{j}\}$ mi/h^2

$\mathbf{a}_A = 0$

$\mathbf{a}_B = \mathbf{a}_A + \mathbf{a}_{B/A}$

$560.77\mathbf{i} + 3371.28\mathbf{j} = \mathbf{0} + \mathbf{a}_{B/A}$

$\mathbf{a}_{B/A} = \{560.77\mathbf{i} + 3371.28\mathbf{j}\}$ mi/h^2

$a_{B/A} = \sqrt{(560.77)^2 + (3371.28)^2} = 3418$ mi/h^2 $= 3.42(10^3)$ mi/h^2 **Ans**

$\theta = \tan^{-1}\left(\dfrac{3371.28}{560.77}\right) = 80.6°$ **Ans**

R1-19. At the instant shown, cars A and B are traveling at speeds of 55 mi/h and 40 mi/h, respectively. If B is decreasing its speed at 1500 mi/h^2 while A is increasing its speed at 800 mi/h^2, determine the acceleration of B with respect to A. Car B moves along a curve having a radius of curvature of 0.75 mi.

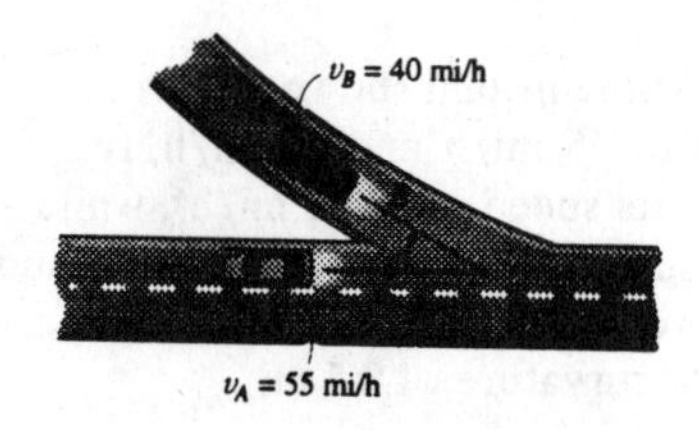

$$(a_B)_n = \frac{v_B^2}{\rho} = \frac{(40)^2}{0.75} = 2133.33 \text{ mi/h}^2$$

$$\mathbf{a}_B = \mathbf{a}_A + \mathbf{a}_{B/A}$$

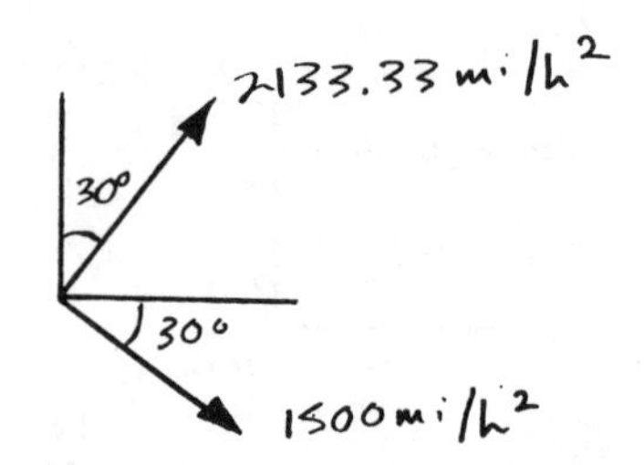

$$2133.33\sin30°\mathbf{i} + 2133.33\cos30°\mathbf{j} + 1500\cos30°\mathbf{i} - 1500\sin30°\mathbf{j}$$

$$= -800\mathbf{i} + (a_{B/A})_x\,\mathbf{i} + (a_{B/A})_y\,\mathbf{j}$$

$(\overset{+}{\rightarrow})$ $\quad 2133.33\sin30° + 1500\cos30° = -800 + (a_{B/A})_x$

$$(a_{B/A})_x = 3165.705 \rightarrow$$

$(+\uparrow)$ $\quad 2133.33\cos30° - 1500\sin30° = (a_{B/A})_y$

$$(a_{B/A})_y = 1097.521 \uparrow$$

$$(a_{B/A}) = \sqrt{(1097.521)^2 + (3165.705)^2}$$

$$a_{B/A} = 3351 \text{ mi/h}^2 = 3.35(10^3)\text{mi/h}^2 \qquad \textbf{Ans}$$

$$\theta = \tan^{-1}\left(\frac{1097.521}{3165.705}\right) = 19.1° \nearrow \qquad \textbf{Ans}$$

***R1-20.** Four inelastic cables C are attached to a plate P and hold the 1-ft-long spring 0.25 ft in compression when *no weight* is on the plate. There is also an undeformed spring nested within this compressed spring. If the block, having a weight of 10 lb, is moving downward at $v = 4$ ft/s, when it is 2 ft above the plate, determine the maximum compression in each spring after it strikes the plate. Neglect the mass of the plate and springs and any energy lost in the collision.

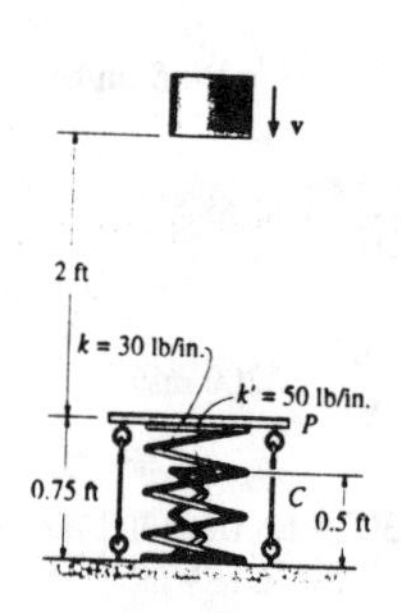

$$k = 30(12) = 360 \text{ lb/ft}$$

$$k' = 50(12) = 600 \text{ lb/ft}$$

Assume both springs compress;

$$T_1 + V_1 = T_2 + V_2$$

$$\frac{1}{2}\left(\frac{10}{32.2}\right)(4)^2 + 0 + \frac{1}{2}(360)(0.25)^2 = 0 + \frac{1}{2}(360)(s+0.25)^2 + \frac{1}{2}(600)(s-0.25)^2 - 10(s+2)$$

$$13.73 = 180(s+0.25)^2 + 300(s-0.25)^2 - 10s - 20 \qquad (1)$$

$$33.73 = 180(s+0.25)^2 + 300(s-0.25)^2 - 10s$$

$$480s^2 - 70s - 3.73 = 0$$

Choose the positive root;

$$s = 0.1874 \text{ ft} < 0.25 \text{ ft} \qquad \text{NG!}$$

The nested spring does not deform.

Thus Eq. (1) becomes

$$13.73 = 180(s+0.25)^2 - 10s - 20$$

$$180\,s^2 + 80\,s - 22.48 = 0$$

$$s = 0.195 \text{ ft} \qquad \textbf{Ans}$$

R1-21. Four inelastic cables C are attached to a plate P and hold the 1-ft-long spring 0.25 ft in compression when *no weight* is on the plate. There is also a 0.5-ft long undeformed spring nested within this compressed spring. Determine the speed v of the 10-lb block when it is 2 ft above the plate, so that after it strikes the plate, it compresses the nested spring, having a stiffness of 50 lb/in., an amount of 0.20 ft. Neglect the mass of the plate and springs and any energy lost in the collision.

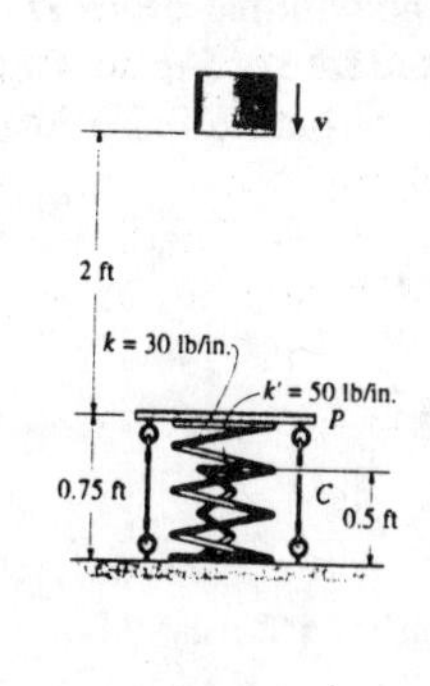

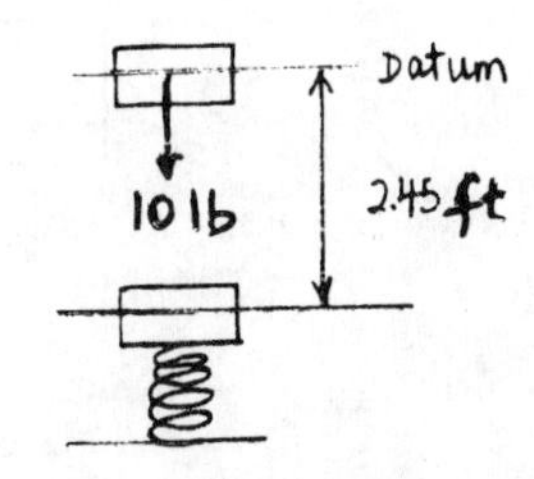

$k = 30(12) = 360$ lb/ft

$k' = 50(12) = 600$ lb/ft

$T_1 + V_1 = T_2 + V_2$

$$\frac{1}{2}\left(\frac{10}{32.2}\right)v^2 + \frac{1}{2}(360)(0.25)^2 = \frac{1}{2}(360)(0.25 + 0.25 + 0.20)^2 + \frac{1}{2}(600)(0.20)^2 - 10(2+0.25+0.20)$$

$v = 20.4$ ft/s **Ans**

R1-22. The 2-kg spool S fits loosely on the rotating inclined rod for which the coefficient of static friction is $\mu_s = 0.2$. If the spool is located 0.25 m from A, determine the minimum constant speed the spool can have so that it does not slip down the rod.

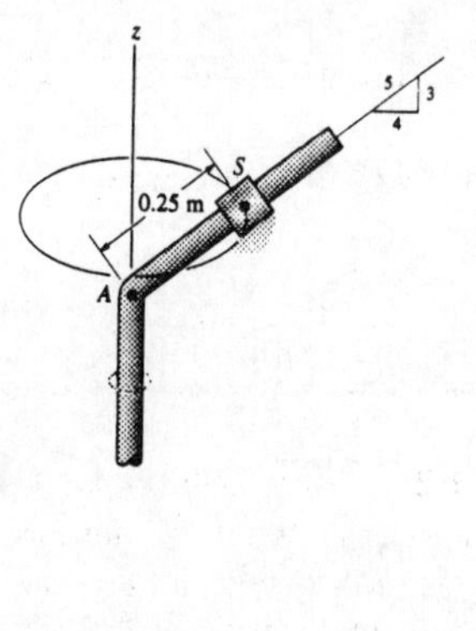

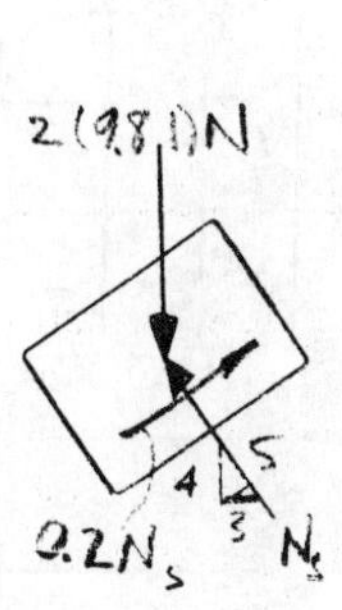

$$\rho = 0.25\left(\frac{4}{5}\right) = 0.2 \text{ m}$$

$$\overset{+}{\leftarrow}\Sigma F_n = m\,a_n; \quad N_s\left(\frac{3}{5}\right) - 0.2N_s\left(\frac{4}{5}\right) = 2\left(\frac{v^2}{0.2}\right)$$

$$+\uparrow\Sigma F_b = m\,a_b; \quad N_s\left(\frac{4}{5}\right) + 0.2N_s\left(\frac{3}{5}\right) - 2(9.81) = 0$$

$N_s = 21.3$ N

$v = 0.969$ m/s **Ans**

R1-23. The 2-kg spool S fits loosely on the inclined rod for which the coefficient of static friction is $\mu_s = 0.2$. If the spool is located 0.25 m from A, determine the maximum constant speed the spool can have so that it does not slip up the rod.

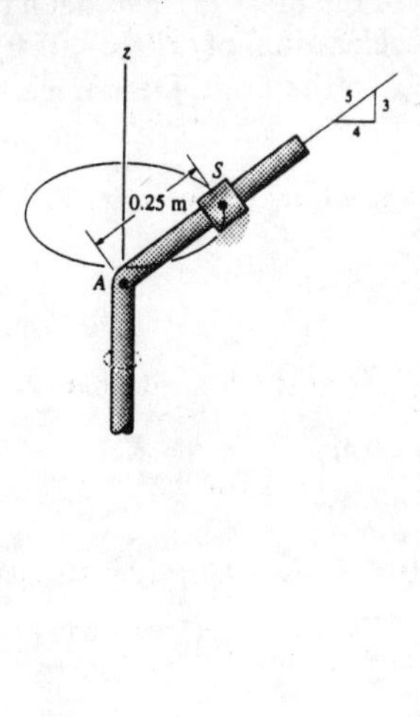

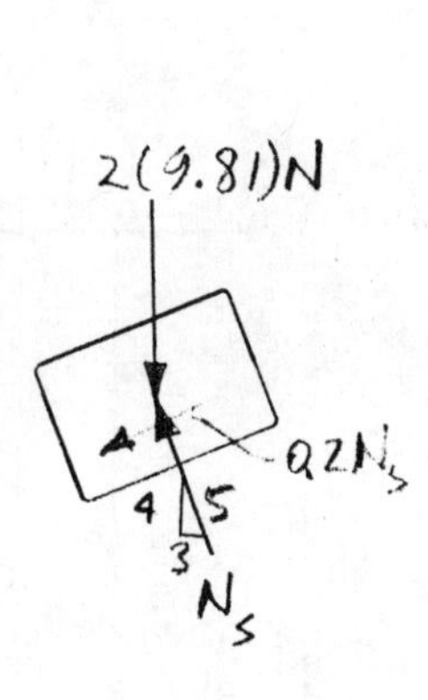

$$\rho = 0.25\left(\frac{4}{5}\right) = 0.2 \text{ m}$$

$$\overset{+}{\leftarrow}\Sigma F_n = m\,a_n; \quad N_s\left(\frac{3}{5}\right) + 0.2N_s\left(\frac{4}{5}\right) = 2\left(\frac{v^2}{0.2}\right)$$

$$+\uparrow\Sigma F_b = m\,a_b; \quad N_s\left(\frac{4}{5}\right) - 0.2N_s\left(\frac{3}{5}\right) - 2(9.81) = 0$$

$N_s = 28.85$ N

$v = 1.48$ m/s **Ans**

***R1-24.** The winding drum D is drawing in the cable at an accelerated rate of $5\ \text{m/s}^2$. Determine the cable tension if the suspended crate has a mass of 800 kg.

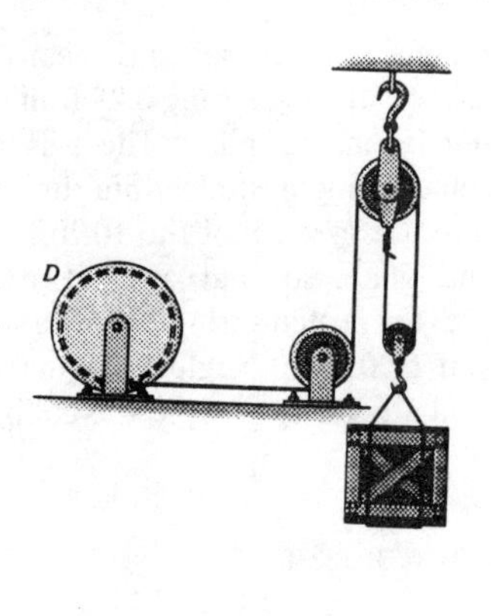

$$s_A + 2s_B = l$$

$$a_A = -2a_B$$

$$5 = -2a_B$$

$$a_B = -2.5\ \text{m/s}^2 = 2.5\ \text{m/s}^2 \uparrow$$

$$+\uparrow \Sigma F_y = ma_y; \qquad 2T - 800(9.81) = 800(2.5)$$

$$T = 4924\ \text{N} = 4.92\ \text{kN} \qquad \textbf{Ans}$$

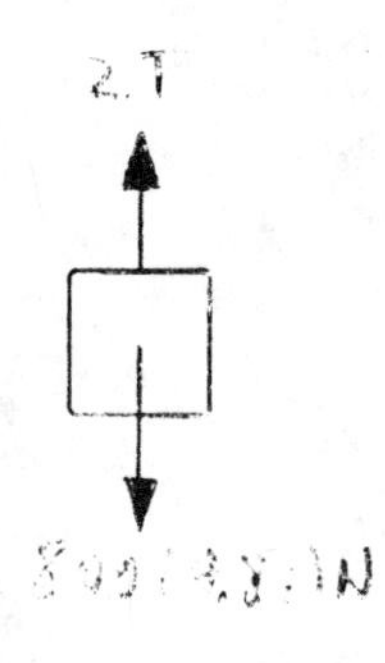

R1-25. If the hoist H is moving upward at 6 ft/s, determine the speed at which the motor M must draw in the supporting cable.

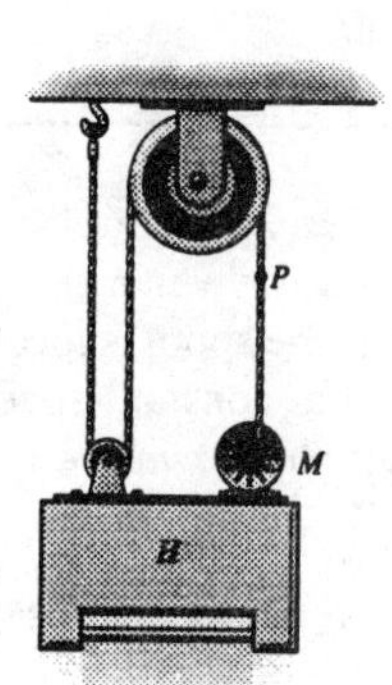

$$2s_H + s_P = l$$

$$2v_H = -v_P$$

$$v_P = -2(-6) = 12\ \text{ft/s}$$

$$(+\downarrow) \qquad v_P = v_H + v_{P/H}$$

$$12 = -6 + v_{P/H}$$

$$v_{P/H} = 18\ \text{ft/s} \downarrow \qquad \textbf{Ans}$$

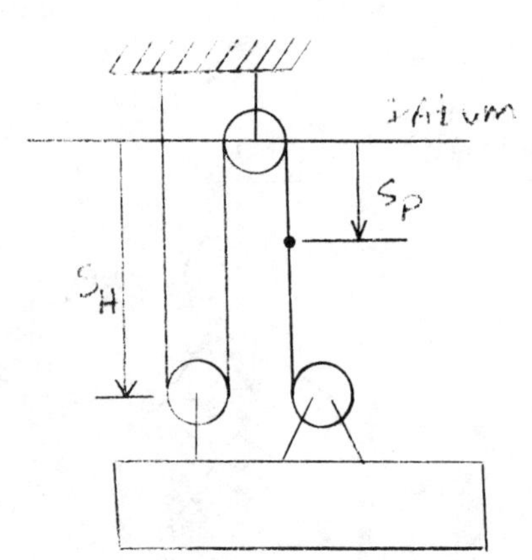

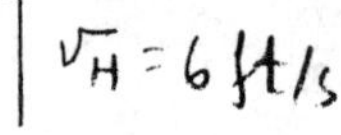

R1-26. Starting from rest, the motor M can draw in the cable of the hoist H such that a point P on the cable has an acceleration of $a_{P/H} = (0.4t)\ \text{ft/s}^2$, measured with respect to the hoist. Determine the velocity of the hoist when $t = 3$ s.

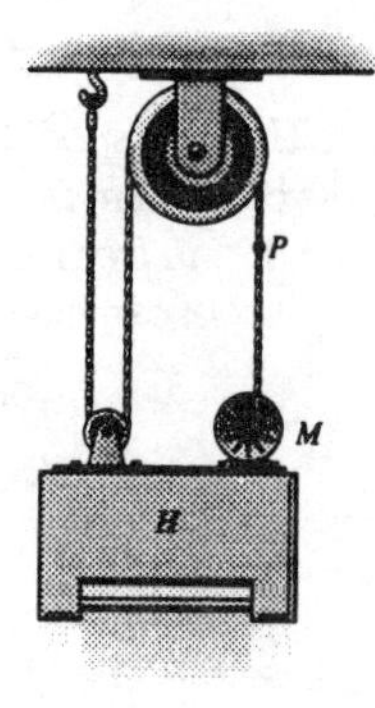

$$2s_H + s_P = l$$

$$2v_H = -v_P$$

$$2a_H = -a_P$$

$$a_{P/H} = 0.4t$$

$$dv = a\,dt$$

$$v_{P/H} = \int_0^3 0.4t\,dt = 0.2(3)^2 = 1.8\ \text{ft/s} \downarrow$$

$$(+\downarrow) \qquad v_P = v_H + v_{P/H}$$

$$-2v_H = v_H + 1.8$$

$$v_H = -0.6\ \text{ft/s} = 0.6\ \text{ft/s} \uparrow \qquad \textbf{Ans}$$

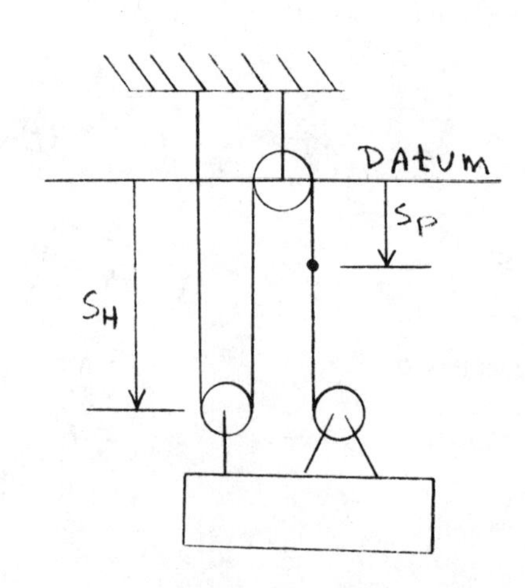

R1-27. The winch delivers a horizontal towing force F to its cable at A which varies as shown in the graph. Determine the speed of the 70-kg block B when $t = 18$ s. Originally the block is moving upward at $v_1 = 3$ m/s.

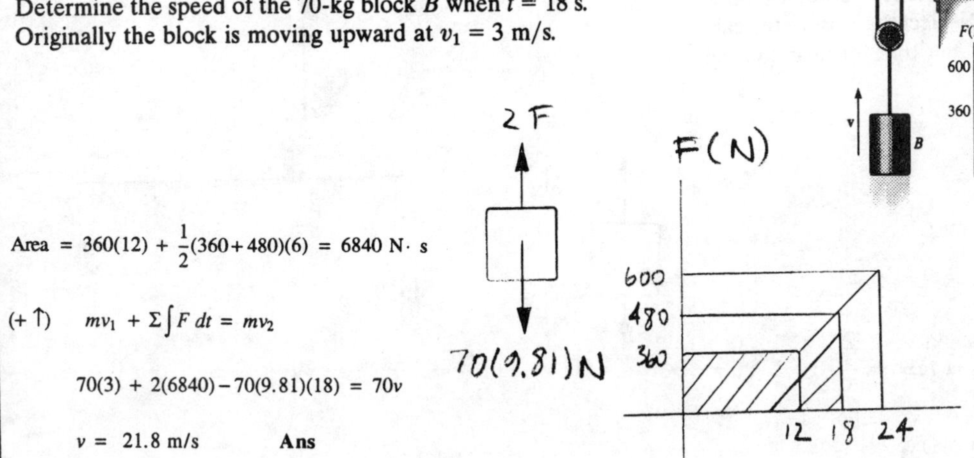

$$\text{Area} = 360(12) + \frac{1}{2}(360+480)(6) = 6840 \text{ N}\cdot\text{s}$$

$$(+\uparrow) \qquad mv_1 + \Sigma\int F\,dt = mv_2$$

$$70(3) + 2(6840) - 70(9.81)(18) = 70v$$

$$v = 21.8 \text{ m/s} \qquad \textbf{Ans}$$

***R1-28.** The winch delivers a horizontal towing force F to its cable at A which varies as shown in the graph. Determine the speed of the 40-kg block B when $t = 24$ s. Originally the block is resting on the ground.

$$\text{Area} = 360(12) + \frac{1}{2}(360+600)(12) = 10\,080 \text{ N}\cdot\text{s}$$

$$(+\uparrow) \qquad mv_1 + \Sigma\int F\,dt = mv_2$$

$$0 + 2(10\,080) - 40(9.81)(24) = 40v$$

$$v = 269 \text{ m/s} \qquad \textbf{Ans}$$

R1-29. The motor pulls on the cable at A with a force $F = (30 + t^2)$ lb, where t is in seconds. If the 34-lb crate is originally at rest on the ground when $t = 0$, determine its speed when $t = 4$ s. Neglect the mass of the cable and pulleys. *Hint:* First find the time needed to begin lifting the crate.

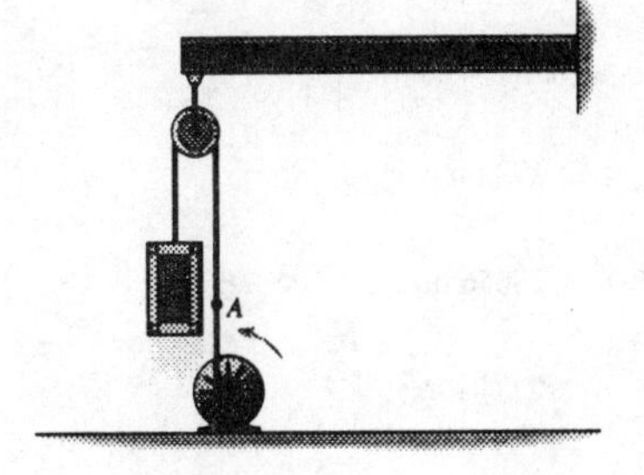

$$30 + t^2 = 34$$

$t = 2$ s for crate to start moving

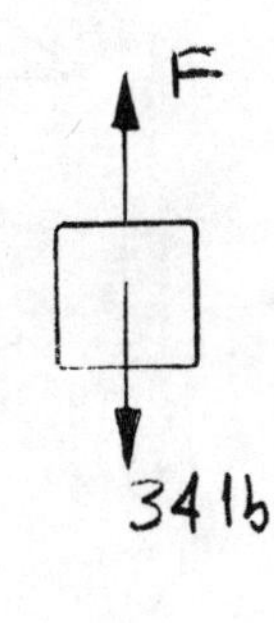

$$(+\uparrow) \qquad mv_1 + \Sigma\int F\,dt = mv_2$$

$$0 + \int_2^4 (30+t^2)\,dt - 34(4-2) = \frac{34}{32.2}v_2$$

$$\left[30t + \frac{1}{3}t^3\right]_2^4 - 68 = \frac{34}{32.2}v_2$$

$$v_2 = 10.1 \text{ ft/s} \qquad \textbf{Ans}$$

R1-30. The motor pulls on the cable at A with a force $F = (e^{2t})$ lb, where t is in seconds. If the 34-lb crate is originally at rest on the ground when $t = 0$, determine the crate's velocity when $t = 2$ s. Neglect the mass of the cable and pulleys. *Hint:* First find the time needed to begin lifting the crate.

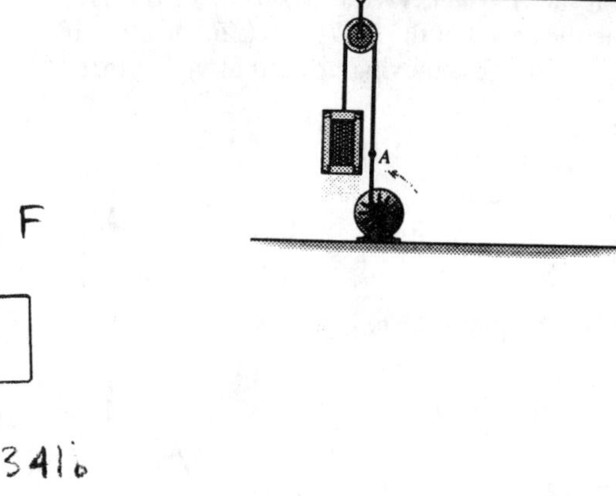

$F = e^{2t} = 34$

$t = 1.7632$ s for crate to start moving

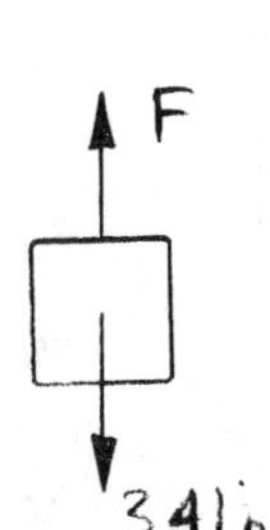

$(+\uparrow) \quad mv_1 + \Sigma\int F\,dt = mv_2$

$$0 + \int_{1.7632}^{2} e^{2t}dt - 34(2-1.7632) = \frac{34}{32.2}v_2$$

$$\frac{1}{2}e^{2t}\Big|_{1.7632}^{2} - 8.0519 = 1.0559\,v_2$$

$v_2 = 2.13$ ft/s **Ans**

R1-31. The collar has a mass of 2 kg and travels along the smooth horizontal rod defined by the equiangular spiral $r = (e^{\theta})$ m, where θ is in radians. Determine the tangential force F and the normal force N acting on the collar when $\theta = 45°$, if the force F maintains a constant angular motion $\dot{\theta} = 2$ rad/s.

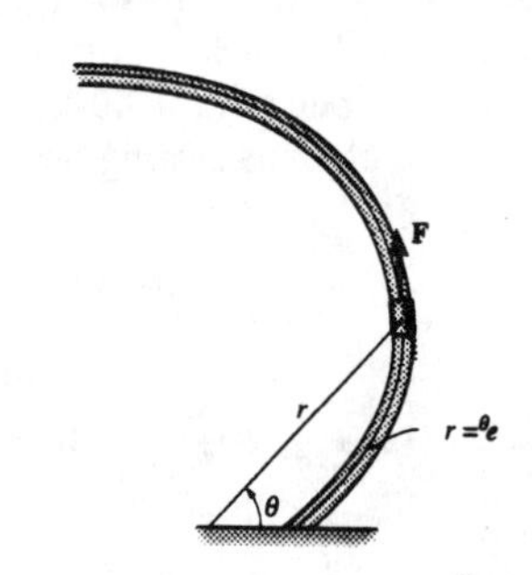

$r = e^{\theta}$

$\dot{r} = e^{\theta}\dot{\theta}$

$\ddot{r} = e^{\theta}(\dot{\theta})^2 + e^{\theta}\ddot{\theta}$

At $\theta = 45°$

$\dot{\theta} = 2$ rad/s

$\ddot{\theta} = 0$

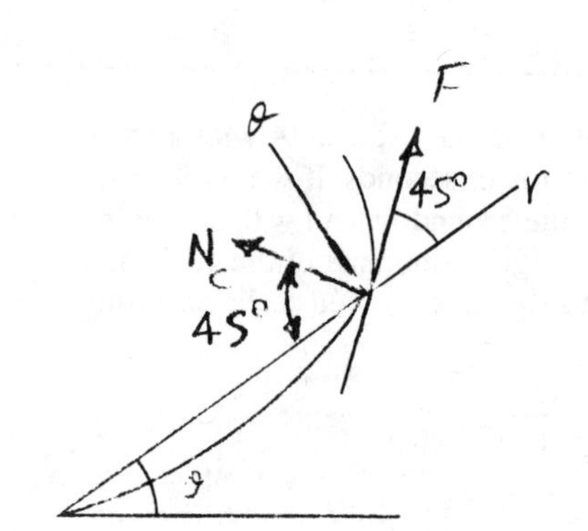

$r = 2.1933$ m

$\dot{r} = 4.38656$ m/s

$\ddot{r} = 8.7731$ m/s^2

$a_r = \ddot{r} - r(\dot{\theta})^2 = 8.7731 - 2.1933(2)^2 = 0$

$a_{\theta} = r\ddot{\theta} + 2\dot{r}\dot{\theta} = 0 + 2(4.38656)(2) = 17.5462$ m/s^2

$$\tan\psi = \frac{r}{\left(\frac{dr}{d\theta}\right)} = e^{\theta}/e^{\theta} = 1$$

$\psi = \theta = 45°$

$\Sigma F_r = m\,a_r; \quad -N_C\cos 45° + F\cos 45° = 2(0) \qquad N_C = 24.8$ N **Ans**

$\Sigma F_{\theta} = m\,a_{\theta}; \quad F\sin 45° + N_C\sin 45° = 2(17.5462) \qquad F = 24.8$ N **Ans**

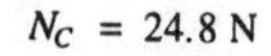

***R1-32.** The collar has a mass of 2 kg and travels along the smooth horizontal rod defined by the equiangular spiral $r = (e^{\theta})$ m, where θ is in radians. Determine the tangential force F and the normal force N acting on the collar when $\theta = 90°$, if the force F maintains a constant angular motion $\dot{\theta} = 2$ rad/s.

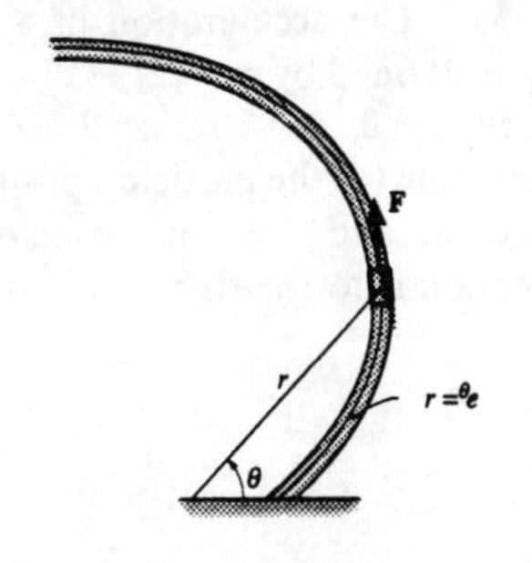

$r = e^{\theta}$

$\dot{r} = e^{\theta}\dot{\theta}$

$\ddot{r} = e^{\theta}(\dot{\theta})^2 + e^{\theta}\ddot{\theta}$

At $\theta = 90°$

$\dot{\theta} = 2$ rad/s

$\ddot{\theta} = 0$

$r = 4.8105$ m

$\dot{r} = 9.6210$ m/s

$\ddot{r} = 19.242$ m/s^2

$a_r = \ddot{r} - r(\dot{\theta})^2 = 19.242 - 4.8105(2)^2 = 0$

$a_\theta = r\ddot{\theta} + 2\dot{r}\dot{\theta} = 0 + 2(9.6210)(2) = 38.4838$ m/s^2

$\tan\psi = \dfrac{r}{\left(\frac{dr}{d\theta}\right)} = e^{\theta}/e^{\theta} = 1$

$\psi = \theta = 45°$

$+\uparrow \Sigma F_r = m a_r;$ $\quad -N_C \cos 45° + F\cos 45° = 2(0)$

$\overset{+}{\leftarrow} \Sigma F_\theta = m a_\theta;$ $\quad F\sin 45° + N_C \sin 45° = 2(38.4838)$

$N_C = 54.4$ N **Ans**

$F = 54.4$ N **Ans**

R1-33. The acceleration of a particle along a straight line is defined by $a = (2t - 9)$ m/s², where t is in seconds. When $t = 0$, $s = 1$ m and $v = 10$ m/s. When $t = 9$ s, determine (a) the particle's position, (b) the total distance traveled, and (c) the velocity. Assume the positive direction is to the right.

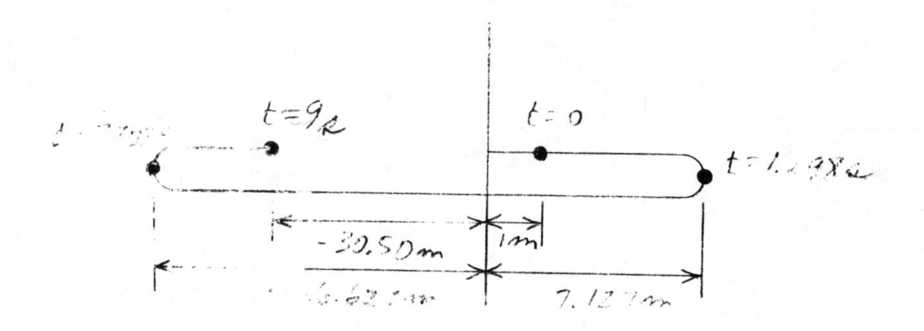

$a = (2t - 9)$

$dv = a\,dt$

$\int_{10}^{v} dv = \int_0^t (2t-9)\,dt$

$v - 10 = t^2 - 9t$

$v = t^2 - 9t + 10$

$ds = v\,dt$

$\int_1^s ds = \int_0^t (t^2 - 9t + 10)\,dt$

$s - 1 = \frac{1}{3}t^3 - 4.5t^2 + 10t$

$s = \frac{1}{3}t^3 - 4.5t^2 + 10t + 1$

Note $v = 0$ at $t^2 - 9t + 10 = 0$

$t = 1.298$ s and $t = 7.702$ s

At $t = 1.298$ s, $s = 7.127$ m

At $t = 7.702$ s, $s = -36.627$ m

At $t = 9$ s, $s = -30.50$ m

a) $s = -30.5$ m **Ans**

b) $s_{tot} = (7.127 - 1) + 7.127 + 36.627 + (36.627 - 30.50) = 56.0$ m **Ans**

c) $v|_{t=9} = (9)^2 - 9(9) + 10 = 10$ m/s **Ans**

R1-34. The 400-kg mine car is hoisted up the incline using the cable and motor M. For a short time, the force in the cable is $F = (3200t^2)$ N, where t is in seconds. If the car has an initial velocity $v_1 = 2$ m/s when $t = 0$, determine its velocity when $t = 2$ s.

$+\nearrow \Sigma F_{x'} = ma_{x'};\quad 3200t^2 - 400(9.81)\left(\frac{8}{17}\right) = 400a \qquad a = 8t^2 - 4.616$

$dv = a\,dt$

$\int_2^v dv = \int_0^2 \left(8t^2 - 4.616\right) dt$

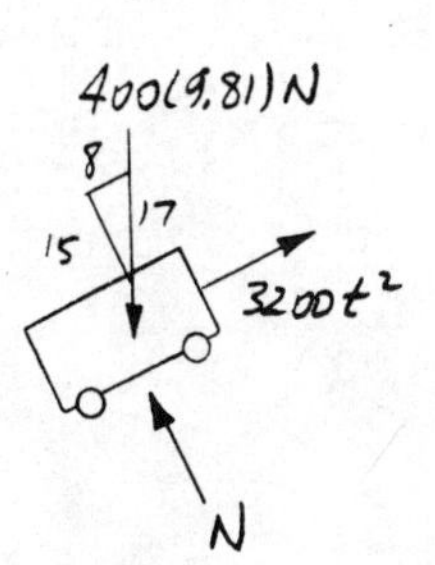

$v = 14.1$ m/s **Ans**

Also,

$m\mathbf{v}_1 + \Sigma \int \mathbf{F}\,dt = m\mathbf{v}_2$

$+\nearrow \quad 400(2) + \int_0^2 3200\,t^2\,dt - 400(9.81)(2-0)\left(\frac{8}{17}\right) = 400v_2$

$800 + 8533.33 - 3693.18 = 40v_2$

$v_2 = 14.1$ m/s

R1-35. The 400-kg mine car is hoisted up the incline using the cable and motor M. For a short time, the force in the cable is $F = (3200t^2)$ N, where t is in seconds. If the car has an initial velocity $v_1 = 2$ m/s at $s = 0$ and $t = 0$, determine the distance it moves up the plane when $t = 2$ s.

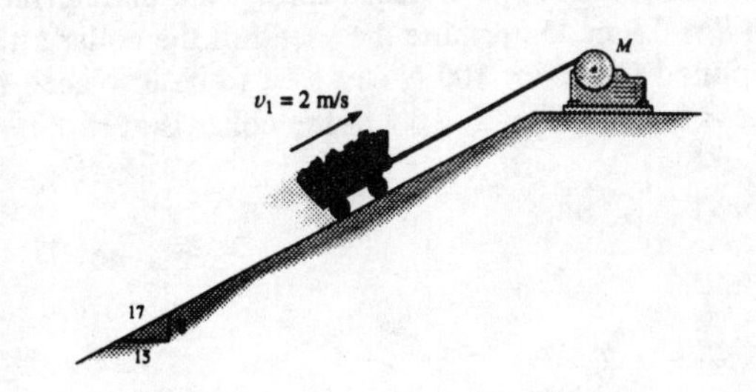

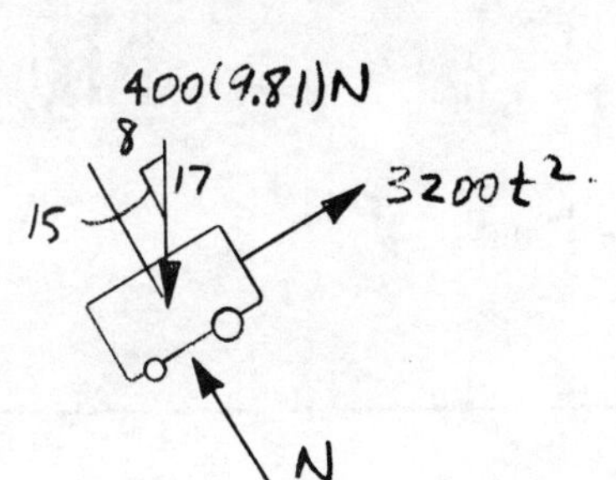

$\Sigma F_{x'} = ma_{x'};\quad 3200t^2 - 400(9.81)\left(\frac{8}{17}\right) = 400a \qquad a = 8t^2 - 4.616$

$dv = a\,dt$

$\int_2^v dv = \int_0^t \left(8t^2 - 4.616\right) dt$

$v = \frac{ds}{dt} = 2.667t^3 - 4.616t + 2$

$\int_0^s ds = \int_0^2 \left(2.667t^3 - 4.616t + 2\right) dt$

$s = 5.43$ m **Ans**

***R1-36.** The rocket sled has a mass of 4 Mg and travels from rest along the smooth horizontal track such that it maintains a constant power output of 450 kW. Neglect the loss of fuel mass and air resistance, and determine how far it must travel to reach a speed of $v = 60$ m/s.

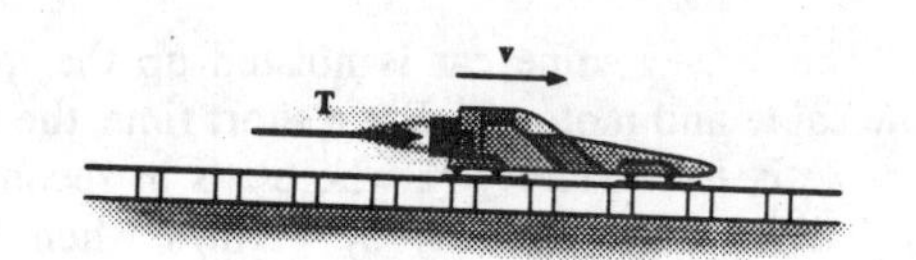

$$\xrightarrow{+} \Sigma F_x = m\,a_x; \qquad F = m\,a = m\left(\frac{v\,dv}{ds}\right)$$

$$P = F\,v = m\left(\frac{v^2\,dv}{ds}\right)$$

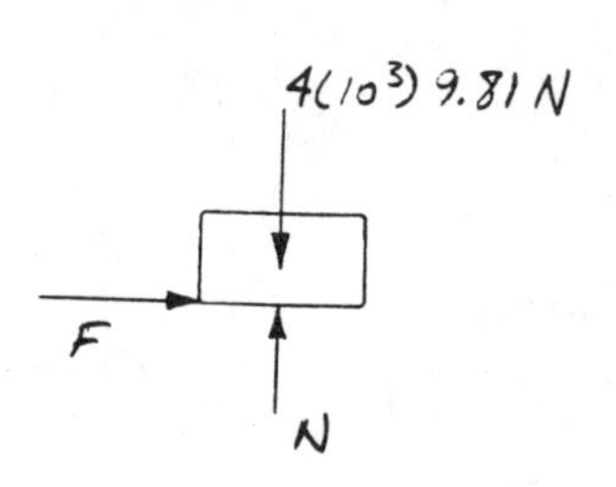

$$\int P\,ds = m\int v^2\,dv$$

$$P\int_0^s ds = m\int_0^v v^2\,dv$$

$$Ps = \frac{m\,v^3}{3}$$

$$s = \frac{m\,v^3}{3\,P}$$

$$s = \frac{4(10^3)(60)^3}{3(450)(10^3)} = 640 \text{ m} \qquad \textbf{Ans}$$

R1-37. The collar has a mass of 20 kg and is supported on the smooth rod. The attached springs are undeformed when $d = 0.5$ m. Determine the speed of the collar after the applied force $F = 100$ N causes it to be displaced so that $d = 0.3$ m. When $d = 0.5$ m the collar is at rest.

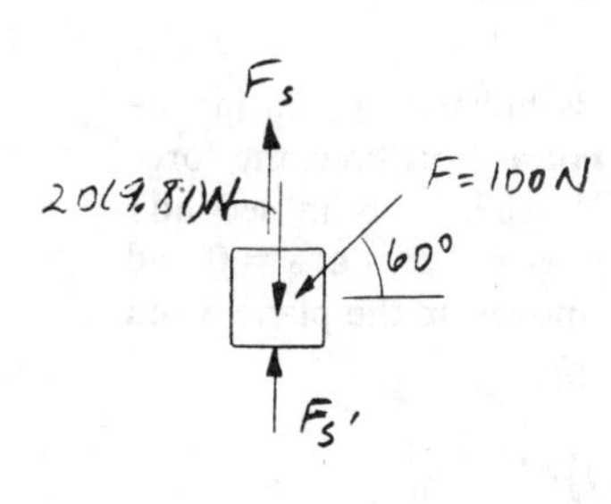

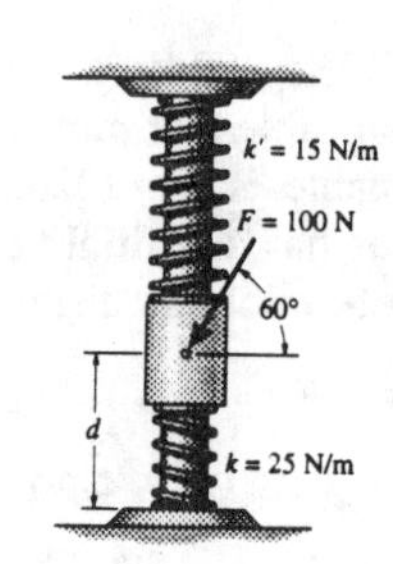

$$T_1 + \Sigma U_{1-2} = T_2$$

$$0 + 100\,\sin 60°(0.5-0.3) + 20(9.81)(0.5-0.3) - \frac{1}{2}(15)(0.5-0.3)^2 - \frac{1}{2}(25)(0.5-0.3)^2 = \frac{1}{2}(20)v_C^2$$

$v_C = 2.36$ m/s **Ans**

R1-38. The collar has a mass of 20 kg and is supported on the smooth rod. The attached springs are both compressed 0.4 m when $d = 0.5$ m. Determine the speed of the collar after the applied force $F = 100$ N causes it to be displaced so that $d = 0.3$ m. When $d = 0.5$ m the collar is at rest.

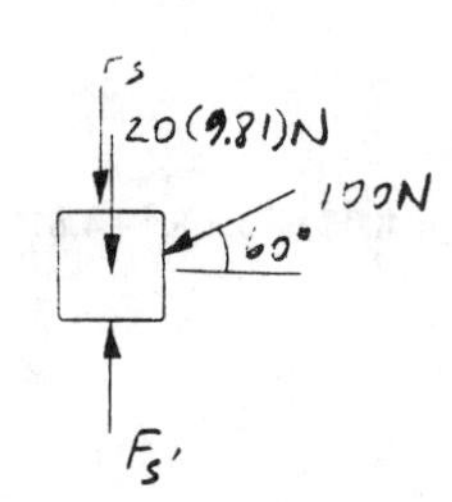

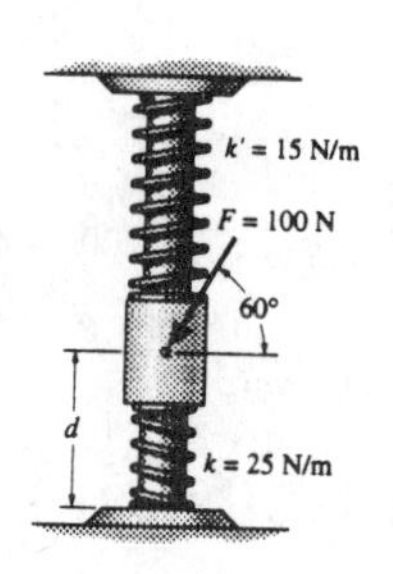

$$T_1 + \Sigma U_{1-2} = T_2$$

$$0 + 100\,\sin 60°(0.5-0.3) + 196.2(0.5-0.3) - [\frac{1}{2}(25)[0.4 + 0.2]^2 - \frac{1}{2}(25)(0.4)^2]$$

$$- [\frac{1}{2}(15)[0.4 - 0.2]^2 - \frac{1}{2}(15)(0.4)^2] = \frac{1}{2}(20)v_C^2$$

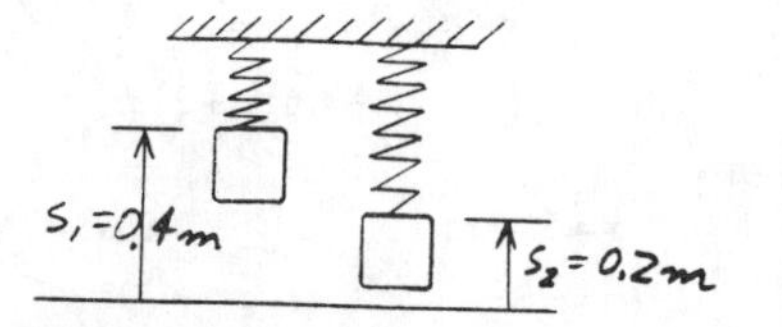

$v_C = 2.34$ m/s **Ans**

R1-39. The assembly consists of two blocks A and B which have a mass of 20 kg and 30 kg, respectively. Determine the speed of each block when B descends 1.5 m. The blocks are released from rest. Neglect the mass of the pulleys and cords.

$$3\,s_A + s_B = l$$

$$3\,\Delta s_A = -\,\Delta s_B$$

$$3\,v_A = -\,v_B$$

$$T_1 + V_1 = T_2 + V_2$$

$$(0+0) + (0+0) = \frac{1}{2}(20)(v_A)^2 + \frac{1}{2}(30)(-3v_A)^2 + 20(9.81)\left(\frac{1.5}{3}\right) - 30(9.81)(1.5)$$

$v_A = 1.54$ m/s **Ans**

$v_B = 4.62$ m/s **Ans**

***R1-40.** The assembly consists of two blocks A and B, which have a mass of 20 kg and 30 kg, respectively. Determine the distance B must descend in order for A to achieve a speed of 3 m/s starting from rest.

$$3\,s_A + s_B = l$$

$$3\,\Delta s_A = -\,\Delta s_B$$

$$3\,v_A = -\,v_B$$

$$v_B = -9 \text{ m/s}$$

$$T_1 + V_1 = T_2 + V_2$$

$$(0+0) + (0+0) = \frac{1}{2}(20)(3)^2 + \frac{1}{2}(30)(-9)^2 + 20(9.81)\left(\frac{s_B}{3}\right) - 30(9.81)(s_B)$$

$s_B = 5.70$ m **Ans**

R1-41. Block A, having a mass m, is released from rest, falls a distance h and strikes the plate B having a mass $2m$. If the coefficient of restitution between A and B is e, determine the velocity of the plate just after collision. The spring has a stiffness k.

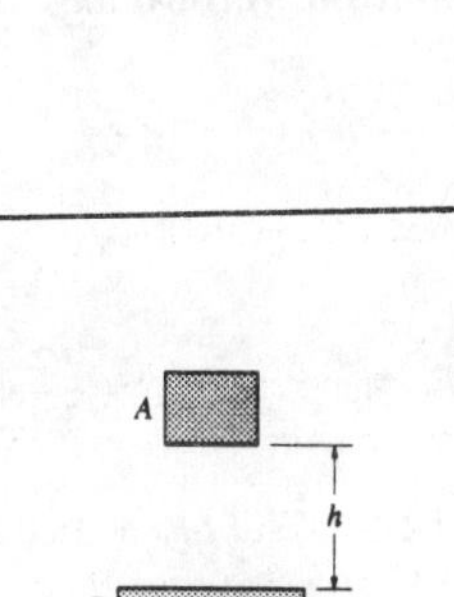

Just before impact, the velocity of A is

$$T_1 + V_1 = T_2 + V_2$$

$$0 + 0 = \frac{1}{2}mv_A^2 - mgh$$

$$v_A = \sqrt{2gh}$$

$$(+\downarrow) \qquad e = \frac{(v_B)_2 - (v_A)_2}{\sqrt{2gh}}$$

$$e\sqrt{2gh} = (v_B)_2 - (v_A)_2 \qquad (1)$$

$$(+\downarrow) \qquad \Sigma mv_1 = \Sigma mv_2$$

$$m(v_A) + 0 = m(v_A)_2 + 2m(v_B)_2 \qquad (2)$$

Solving Eqs. (1) and (2) for $(v_B)_2$ yields

$$(v_B)_2 = \frac{1}{3}\sqrt{2gh}(1+e)$$ **Ans**

R1-42. Block A, having a mass of 2 kg, is released from rest, falls a distance $h = 0.5$ m, and strikes the plate B having a mass of 3 kg. If the coefficient of restitution between A and B is $e = 0.6$, determine the velocity of the block just after collision. The spring has a stiffness $k = 30$ N/m.

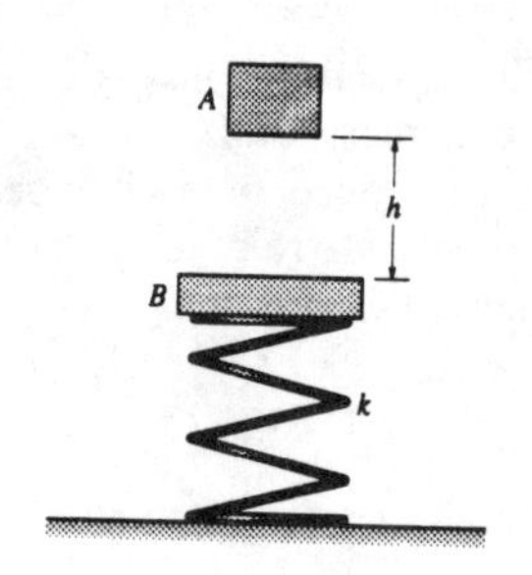

Just before impact, the velocity of A is

$$T_1 + V_1 = T_2 + V_2$$

$$0 + 0 = \frac{1}{2}(2)(v_A)_2^2 - 2(9.81)(0.5)$$

$$(v_A)_2 = \sqrt{2(9.81)(0.5)} = 3.132 \text{ m/s}$$

$$(+\downarrow) \quad e = \frac{(v_B)_3 - (v_A)_3}{3.132 - 0}$$

$$0.6(3.132) = (v_B)_3 - (v_A)_3$$

$$1.879 = (v_B)_3 - (v_A)_3 \qquad (1)$$

$$(+\downarrow) \quad \Sigma m v_2 = \Sigma m v_3$$

$$2(3.132) + 0 = 2(v_A)_3 + 3(v_B)_3 \qquad (2)$$

Solving Eqs. (1) and (2) for $(v_A)_3$ yields

$$(v_B)_3 = 2.00 \text{ m/s}$$

$$(v_A)_3 = 0.125 \text{ m/s} \qquad \textbf{Ans}$$

R1-43. The cylindrical plug has a weight of 2 lb and it is free to move within the confines of the smooth pipe. The spring has a stiffness $k = 14$ lb/ft and when no motion occurs the distance $d = 0.5$ ft. Determine the force of the spring on the plug when the plug is at rest with respect to the pipe. The plug is traveling with a constant speed of 15 ft/s, which is caused by the rotation of the pipe about the vertical axis. Neglect the size of the plug.

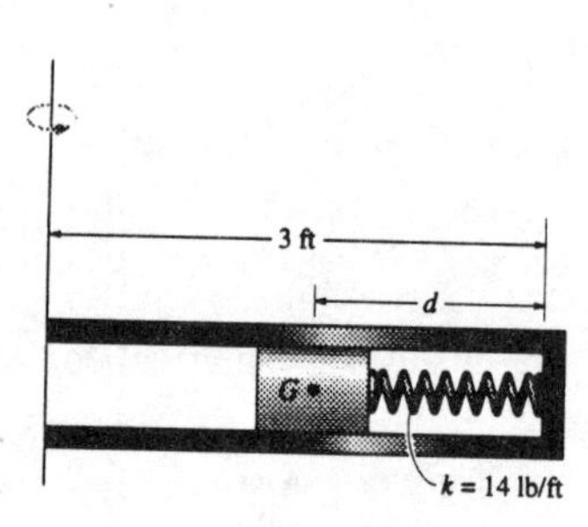

$$\overset{+}{\leftarrow} \Sigma F_n = m\, a_n; \qquad F_s = \frac{2}{32.2}\left[\frac{(15)^2}{3-d}\right]$$

$$F_s = ks; \qquad F_s = 14(0.5 - d)$$

Thus,

$$14(0.5 - d) = \frac{2}{32.2}\left[\frac{(15)^2}{3-d}\right]$$

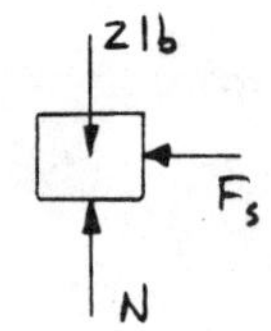

$$(0.5 - d)(3 - d) = 0.9982$$

$$1.5 - 3.5d + d^2 = 0.9982$$

$$d^2 - 3.5d + 0.5018 = 0$$

Choosing the root < 0.5 ft

$$d = 0.1498 \text{ ft}$$

$$F_s = 14(0.5 - 0.1498) = 4.90 \text{ lb} \qquad \textbf{Ans}$$

***R1-44.** A 20-g bullet is fired horizontally into the 300-g block which rests on the smooth surface. After the bullet becomes embedded into the block, the block moves to the right 0.3 m before momentarily coming to rest. Determine the speed $(v_B)_1$ of the bullet. The spring has a stiffness $k = 200$ N/m and is originally unstretched.

After collision

$$T_1 + \Sigma U_{1-2} = T_2$$

$$\frac{1}{2}(0.320)(v_2)^2 - \frac{1}{2}(200)(0.3)^2 = 0$$

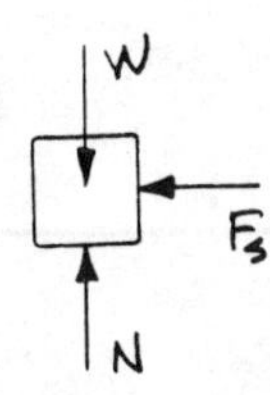

$$v_2 = 7.50 \text{ m/s}$$

Impact

$$\Sigma mv_1 = \Sigma mv_2$$

$$0.02(v_B)_1 + 0 = 0.320(7.50)$$

$$(v_B)_1 = 120 \text{ m/s} \qquad \textbf{Ans}$$

R1-45. The 20-g bullet is fired horizontally at $(v_B)_1 = 1200$ m/s into the 300-g block which rests on the smooth surface. Determine the distance the block moves to the right before momentarily coming to rest. The spring has a stiffness $k = 200$ N/m and is originally unstretched.

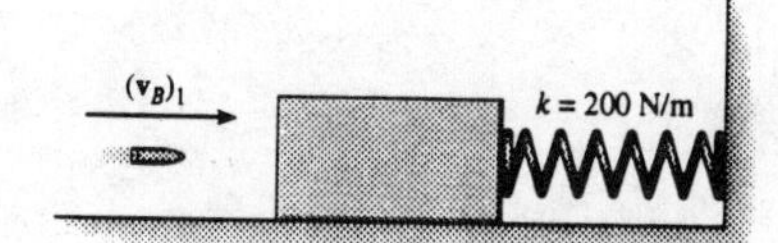

Impact

$$\Sigma mv_1 = \Sigma mv_2$$

$$0.02(1200) + 0 = 0.320(v_2)$$

$$v_2 = 75 \text{ m/s}$$

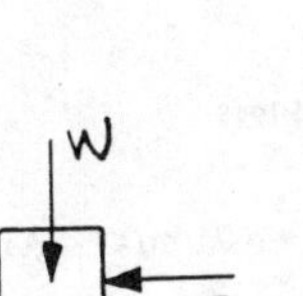

After collision;

$$T_1 + \Sigma U_{1-2} = T_2$$

$$\frac{1}{2}(0.320)(75)^2 - \frac{1}{2}(200)(x^2) = 0$$

$$x = 3 \text{ m} \qquad \textbf{Ans}$$

R1-46. The collar has a weight of 8 lb. If it is pushed down so as to compress the spring 2 ft and then released from rest ($h = 0$), determine its speed when it is displaced $h = 4.5$ ft. The spring is not attached to the collar. Neglect friction.

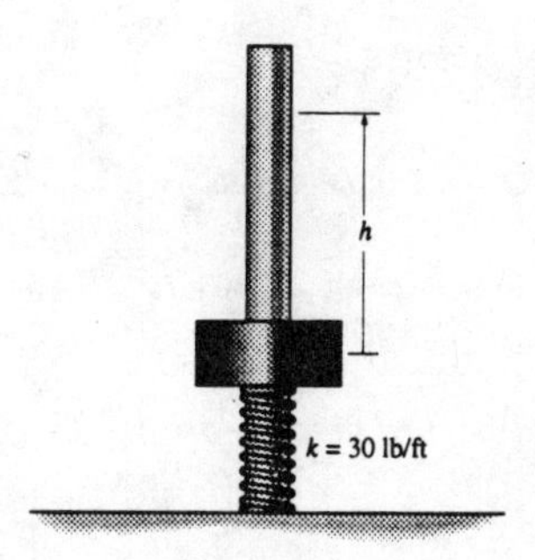

$$T_1 + V_1 = T_2 + V_2$$

$$0 + \frac{1}{2}(30)(2)^2 = \frac{1}{2}\left(\frac{8}{32.2}\right)v_2^2 + 8(4.5)$$

$$v_2 = 13.9 \text{ ft/s} \qquad \textbf{Ans}$$

R1-47. The collar has a weight of 8 lb. If it is released from rest at a height of $h = 2$ ft from the top of the uncompressed spring, determine the speed of the collar after it falls and compresses the spring 0.3 ft.

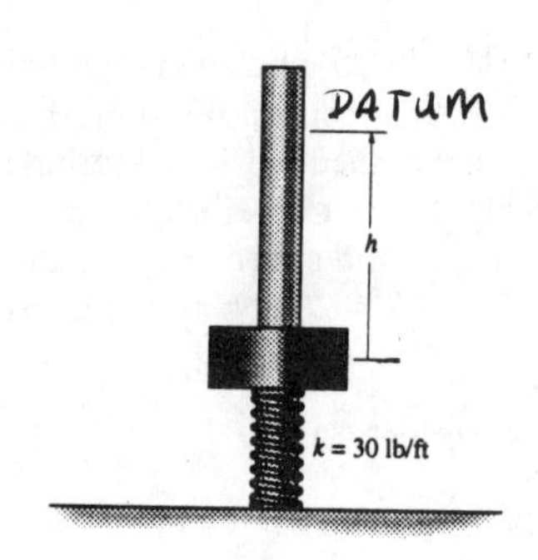

$$T_1 + V_1 = T_2 + V_2$$

$$0 + 0 = \frac{1}{2}\left(\frac{8}{32.2}\right)v_2^2 - 8(2.3) + \frac{1}{2}(30)(0.3)^2$$

$v_2 = 11.7$ ft/s **Ans**

***R1-48.** The position of particles A and B are $\mathbf{r}_A = \{3t\mathbf{i} + 9t(2-t)\mathbf{j}\}$ m and $\mathbf{r}_B = \{3(t^2 - 2t + 2)\mathbf{i} + 3(t-2)\mathbf{j}\}$ m, respectively, where t is in seconds. Determine the point where the particles collide and their speeds just before the collision. How long does it take before the collision occurs?

When collision occurs, $\mathbf{r}_A = \mathbf{r}_B$.

$$3t = 3(t^2 - 2t + 2)$$

$$t^2 - 3t + 2 = 0$$

$t = 1$ s, $t = 2$ s

Also,

$$9t(2-t) = 3(t-2)$$

$$3t^2 - 5t - 2 = 0$$

The positive root is $t = 2$ s
Thus,

$t = 2$ s **Ans**

$x = 3(2) = 6$ m $\quad y = 9(2)(2-2) = 0,$

Hence, (6 m, 0) **Ans**

$$\mathbf{v}_A = \frac{d\mathbf{r}_A}{dt} = 3\mathbf{i} + (18 - 18t)\mathbf{j}$$

$$\mathbf{v}_A|_{t=2} = \{3\mathbf{i} - 18\mathbf{j}\} \text{ m/s}$$

$v_A = \sqrt{(3)^2 + (-18)^2} = 18.2$ m/s **Ans**

$$\mathbf{v}_B = \frac{d\mathbf{r}_B}{dt} = 3(2t-2)\mathbf{i} + 3\mathbf{j}$$

$$\mathbf{v}_B|_{t=2} = \{6\mathbf{i} + 3\mathbf{j}\} \text{ m/s}$$

$v_B = \sqrt{(6)^2 + (3)^2} = 6.71$ m/s **Ans**

R1-49. Determine the speed of the automobile if it has the acceleration shown and is traveling on a road which has a radius of curvature of $\rho = 50$ m. Also, what is the automobile's rate of increase in speed?

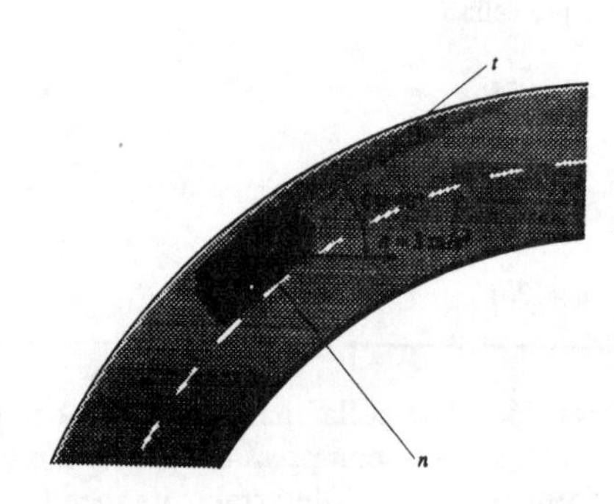

$$a_n = \frac{v^2}{\rho}$$

$$3\sin 40^\circ = \frac{v^2}{50}$$

$v = 9.82$ m/s **Ans**

$a_t = 3\cos 40^\circ = 2.30$ m/s^2 **Ans**

R1-50. If block A of the pulley system is moving downward at 6 ft/s while block C is moving down at 18 ft/s, determine the relative velocity of block B with respect to C.

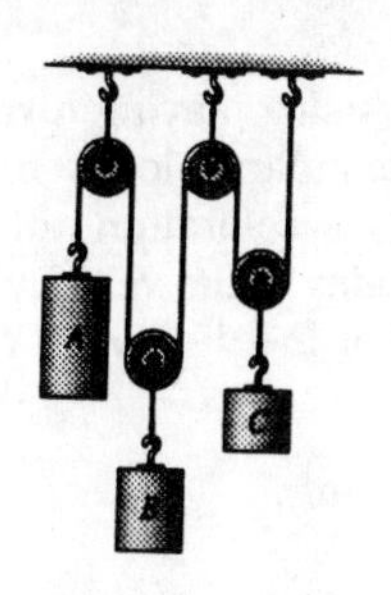

$(+\downarrow)\qquad s_A + 2s_B + 2s_C = l$

$v_A + 2v_B + 2v_C = 0$

$6 + 2v_B + 2(18) = 0$

$v_B = -21 \text{ ft/s} = 21 \text{ ft/s} \uparrow$

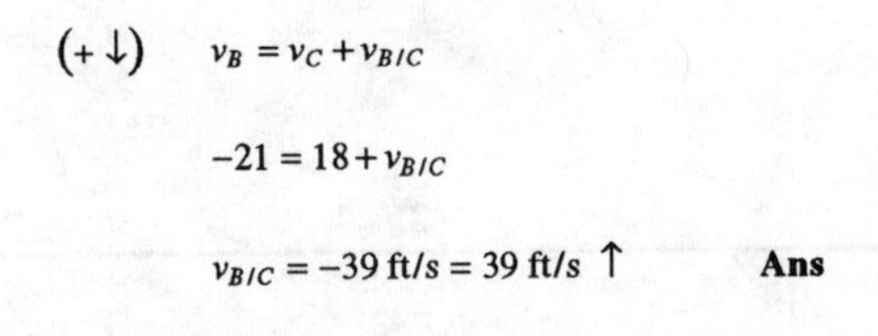

$(+\downarrow)\qquad v_B = v_C + v_{B/C}$

$-21 = 18 + v_{B/C}$

$v_{B/C} = -39 \text{ ft/s} = 39 \text{ ft/s} \uparrow$ **Ans**

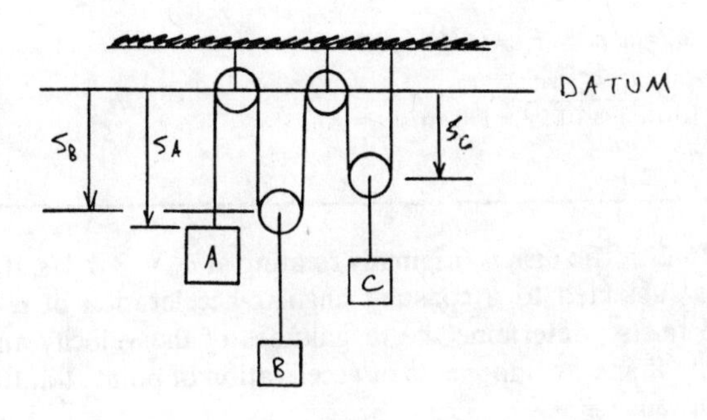

16-1. A disk having a radius of 0.5 ft rotates with an initial angular velocity of 2 rad/s and has a constant angular acceleration of 1 rad/s^2. Determine the magnitudes of the velocity and acceleration of a point on the rim of the disk when $t = 2$ s.

$\omega = \omega_0 + \alpha_c t;$

$\omega = 2 + 1(2) = 4 \text{ rad/s}$

$v = r\omega; \quad v = 0.5(4) = 2 \text{ ft/s}$ **Ans**

$a_t = r\alpha; \quad a_t = 0.5(1) = 0.5 \text{ ft/s}^2$

$a_n = \omega^2 r; \quad a_n = (4)^2(0.5) = 8 \text{ ft/s}^2$

$a = \sqrt{8^2 + (0.5)^2} = 8.02 \text{ ft/s}^2$ **Ans**

16-2. The disk is originally rotating at $\omega_0 = 8$ rad/s. If it is subjected to a constant angular acceleration of $\alpha = 6$ rad/s^2, determine the magnitudes of the velocity and the n and t components of acceleration of point A at the instant $t = 0.5$ s.

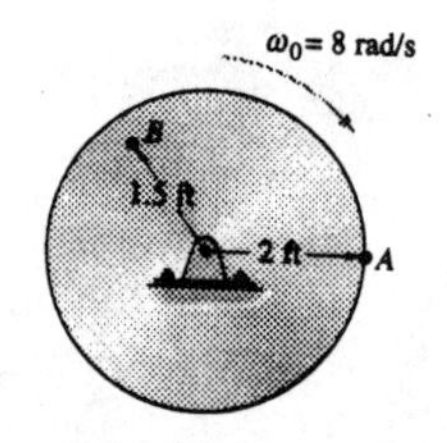

$\omega = \omega_0 + \alpha_c t$

$\omega = 8 + 6(0.5) = 11 \text{ rad/s}$

$v = r\omega; \quad v_A = 2(11) = 22 \text{ ft/s}$ **Ans**

$a_t = r\alpha; \quad (a_A)_t = 2(6) = 12.0 \text{ ft/s}^2$ **Ans**

$a_n = \omega^2 r; \quad (a_A)_n = (11)^2(2) = 242 \text{ ft/s}^2$ **Ans**

16-3. The disk is originally rotating at $\omega_0 = 8$ rad/s. If it is subjected to a constant angular acceleration of $\alpha = 6$ rad/s^2, determine the magnitudes of the velocity and the n and t components of acceleration of point B just after the wheel undergoes 2 revolutions.

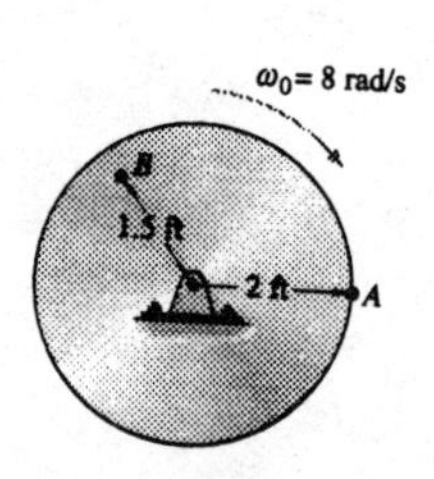

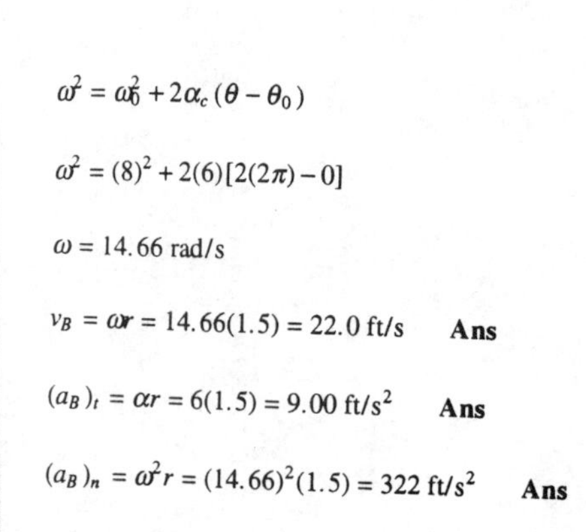

$\omega^2 = \omega_0^2 + 2\alpha_c(\theta - \theta_0)$

$\omega^2 = (8)^2 + 2(6)[2(2\pi) - 0]$

$\omega = 14.66 \text{ rad/s}$

$v_B = \omega r = 14.66(1.5) = 22.0 \text{ ft/s}$ **Ans**

$(a_B)_t = \alpha r = 6(1.5) = 9.00 \text{ ft/s}^2$ **Ans**

$(a_B)_n = \omega^2 r = (14.66)^2(1.5) = 322 \text{ ft/s}^2$ **Ans**

***16-4.** Just after the fan is turned on, the motor gives the blade an angular acceleration $\alpha = (20e^{-0.6t})$ rad/s^2, where t is in seconds. Determine the speed of the tip P of one of the blades when $t = 3$ s. How many revolutions has the blade turned in 3 s? When $t = 0$ the blade is at rest.

$$d\omega = \alpha\, dt$$

$$\int_0^{\omega} d\omega = \int_0^t 20e^{-0.6t}\, dt$$

$$\omega = -\frac{20}{0.6}e^{-0.6t}\Big|_0^t = 33.3\left(1 - e^{-0.6t}\right)$$

$$\omega = 27.82 \text{ rad/s}$$

$$v_P = \omega r = 27.82(1.75) = 48.7 \text{ ft/s} \qquad \textbf{Ans}$$

$$d\theta = \omega\, dt$$

$$\int_0^{\theta} d\theta = \int_0^t 33.3\left(1 - e^{-0.6t}\right) dt$$

$$\theta = 33.3\left(t + \left(\frac{1}{0.6}\right)e^{-0.6t}\right)\Big|_0^3 = 33.3\left[3 + \left(\frac{1}{0.6}\right)\left(e^{-0.6(3)} - 1\right)\right]$$

$$\theta = 53.63 \text{ rad} = 8.54 \text{ rev} \qquad \textbf{Ans}$$

16-5. Due to an increase in power, the motor M rotates the shaft A with an angular acceleration of $\alpha = (0.06\theta^2)$ rad/s^2, where θ is in radians. If the shaft is initially turning at $\omega_0 = 50$ rad/s, determine the angular velocity of gear B after the shaft undergoes an angular displacement $\Delta\theta = 10$ rev.

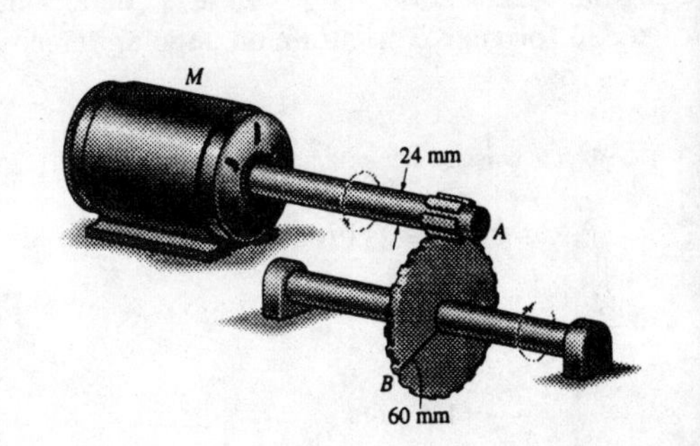

$$\omega\, d\omega = \alpha\, d\theta$$

$$\int_{50}^{\omega} \omega\, d\omega = \int_0^{2\pi(10)} 0.06\theta^2\, d\theta$$

$$\frac{1}{2}\omega^2\Big|_{50}^{\omega} = 0.02\theta^3\Big|_0^{2\pi(10)}$$

$$0.5\omega^2 - 1250 = 4961$$

$$\omega = 111.45 \text{ rad/s}$$

$$\omega_A r_A = \omega_B r_B$$

$$(111.45)(12) = \omega_B(60)$$

$$\omega_B = 22.3 \text{ rad/s} \qquad \textbf{Ans}$$

16-6. The figure shows the internal gearing of a "spinner" used for drilling wells. With constant angular acceleration, the motor M rotates the shaft S to 100 rev/min in $t = 2$ s starting from rest. Determine the angular acceleration of the drill-pipe connection D and the number of revolutions it makes during the 2-s start up.

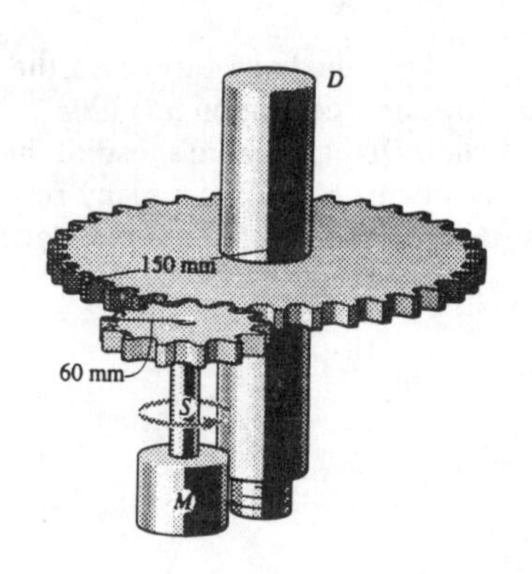

$\omega_M = 100 \text{ rev/min}(1\text{min}/60\text{ s})(2\pi \text{ rad}/1 \text{ rev}) = 10.472 \text{ rad/s}$

$\omega = \omega_0 + \alpha_c t$

$10.472 = 0 + \alpha_c(2)$

$\alpha_M = 5.236 \text{ rad/s}^2$

$\alpha_M r_M = \alpha_D r_D$

$(5.236)(60) = \alpha_D(150)$

$\alpha_D = 2.094 = 2.09 \text{ rad/s}^2$ **Ans**

$\theta = \theta_0 + \omega_0 t + \frac{1}{2}\alpha_c t^2$

$\theta = 0 + 0 + \frac{1}{2}(2.094)(2)^2$

$\theta = 4.19 \text{ rad} = 0.667 \text{ rev}$ **Ans**

16-7. If gear A starts from rest and has a constant angular acceleration of $\alpha_A = 2 \text{ rad/s}^2$, determine the time needed for gear B to attain an angular velocity of $\omega_B = 50$ rad/s.

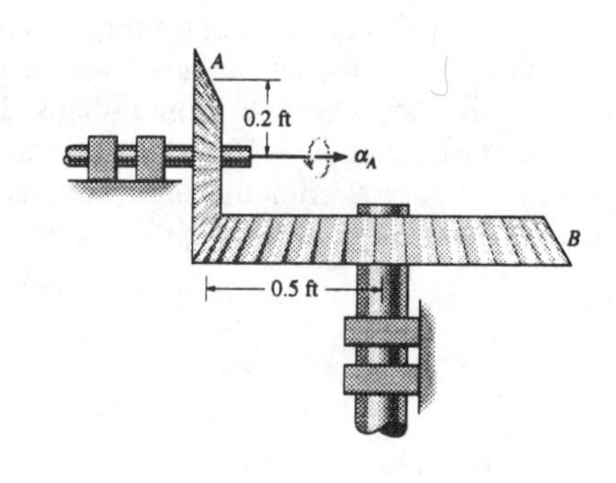

The point in contact with both gears has a speed of

$v_P = \omega_B r_B = 50(0.5) = 25 \text{ ft/s}$

Thus,

$\omega_A = \frac{v_P}{r_A} = \frac{25}{0.2} = 125 \text{ rad/s}$

So that

$\omega = \omega_0 + \alpha_c t$

$125 = 0 + 2t$

$t = 62.5 \text{ s}$ **Ans**

***16-8.** If the armature A of the electric motor in the drill has a constant angular acceleration of $\alpha_A = 20 \text{ rad/s}^2$, determine its angular velocity and angular displacement when $t = 3$ s. The motor starts from rest.

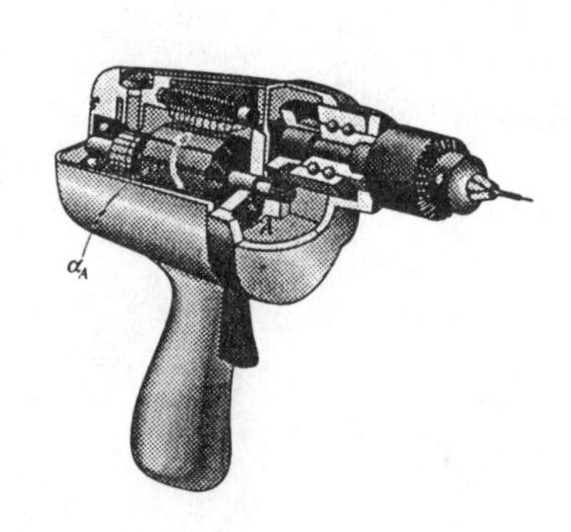

$\omega = \omega_0 + \alpha_c t$

$= 0 + 20(3) = 60 \text{ rad/s}$ **Ans**

$\theta = \theta_0 + \omega_0 t + \frac{1}{2}\alpha_c t^2$

$= 0 + 0 + \frac{1}{2}(20)(3)^2$

$= 90 \text{ rad}$ **Ans**

16-9. The mechanism for a car window winder is shown in the figure. Here the handle turns the small cog C, which rotates the spur gear S, thereby rotating the fixed-connected lever AB which raises track D in which the window rests. The window is free to slide on the track. If the handle is wound at 0.5 rad/s, determine the speed of points A and E and the speed v_w of the window at the instant $\theta = 30°$.

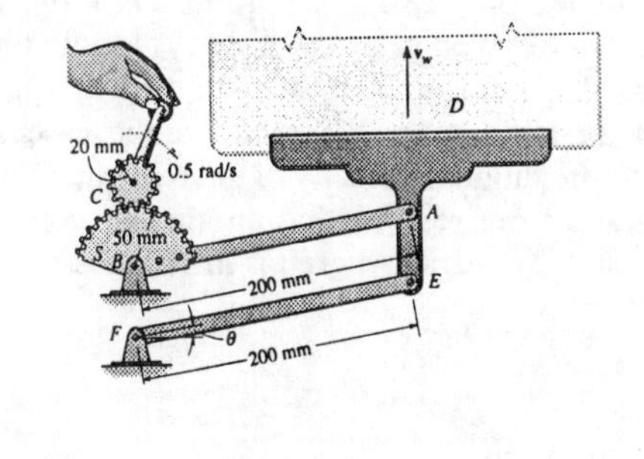

$v_C = \omega_C r_C = 0.5(0.02) = 0.01\text{ m/s}$

$$\omega_S = \frac{v_C}{r_S} = \frac{0.01}{0.05} = 0.2\text{ rad/s}$$

$v_A = v_E = \omega_S r_A = 0.2(0.2) = 0.04\text{ m/s} = 40\text{ mm/s}$ **Ans**

Points A and E move along circular paths. The vertical component closes the window.

$v_w = 40\cos 30° = 34.6\text{ mm/s}$ **Ans**

16-10. The blade on the horizontal-axis windmill is turning with an angular velocity of $\omega_0 = 2$ rad/s. Determine the distance point P on the tip of the blade has traveled if the blade attains an angular velocity of $\omega = 5$ rad/s in 3 s. The angular acceleration is constant. Also, what is the magnitude of the acceleration of this point when $t = 3$ s?

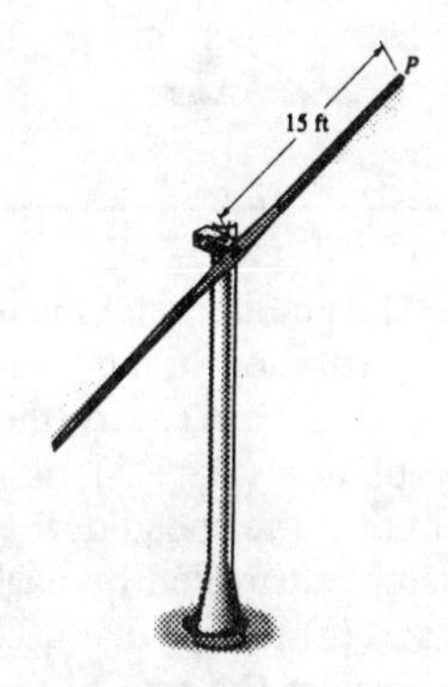

$\omega = \omega_0 + \alpha_c t$

$5 = 2 + \alpha_c(3) \qquad \alpha_c = 1\text{ rad/s}^2$

$$\theta = \theta_0 + \omega_0 t + \frac{1}{2}\alpha_c t^2$$

$$\theta = 0 + 2(3) + \frac{1}{2}(1)(3)^2 = 10.5\text{ rad}$$

$s_P = \theta r_P = 10.5(15) = 157.5\text{ ft} = 158\text{ ft}$ **Ans**

$a_t = \alpha r \qquad a_n = \omega^2 r$

$(a_P)_t = 1(15) = 15\text{ ft/s}^2 \qquad (a_P)_n = (5)^2(15) = 375\text{ ft/s}^2$

$a_P = \sqrt{(a_P)_t^2 + (a_P)_n^2} = \sqrt{(15)^2 + (375)^2} = 375\text{ ft/s}^2$ **Ans**

16-11. The blade on the horizontal-axis windmill is turning with an angular velocity of $\omega_0 = 2$ rad/s. If it is given an angular acceleration of $\alpha_c = 0.6\text{ rad/s}^2$, determine the angular velocity and the magnitude of acceleration of point P on the tip of the blade when $t = 3$ s.

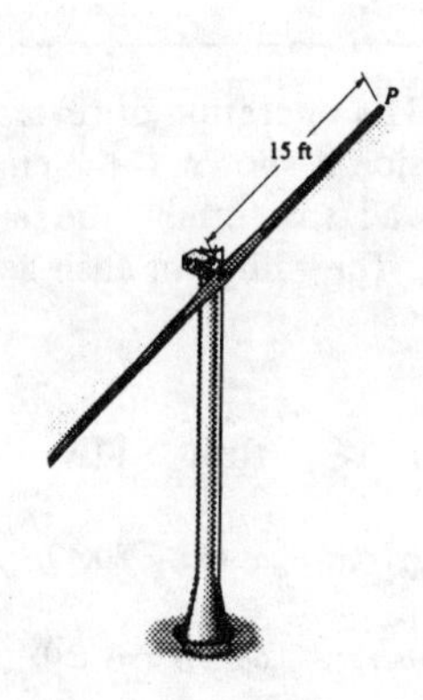

$\omega = \omega_0 + \alpha_c t$

$\omega = 2 + 0.6(3) = 3.80\text{ rad/s}$ **Ans**

$(a_P)_t = \alpha r = 0.6(15) = 9\text{ ft/s}^2$

$(a_P)_n = \omega^2 r = (3.80)^2(15) = 216.60\text{ ft/s}^2$

$a_P = \sqrt{(a_P)_t^2 + (a_P)_n^2} = \sqrt{(9)^2 + (216.60)^2} = 217\text{ ft/s}^2$ **Ans**

***16-12.** When only two gears are in mesh, the driving gear A and the driven gear B will always turn in opposite directions. In order to get them to turn in the *same direction* an idler gear C is used. In the case shown, determine the angular velocity of gear B when $t = 5$ s, if gear A starts from rest and has an angular acceleration of $\alpha_A = (3t + 2)$ rad/s^2, where t is in seconds.

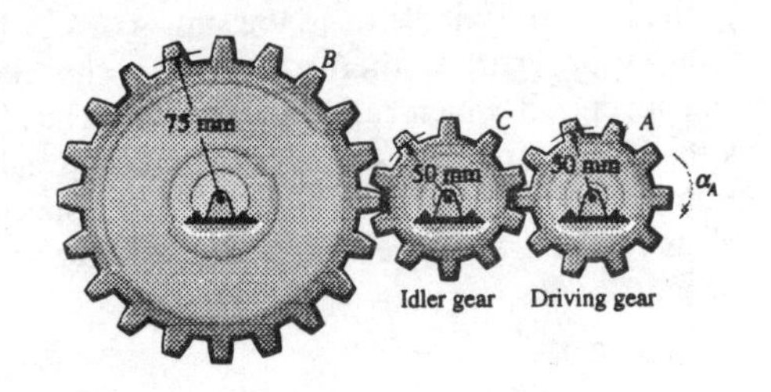

$d\omega = \alpha\, dt$

$\int_0^{\omega_A} d\omega_A = \int_0^t (3t+2)\, dt$

$\omega_A = 1.5t^2 + 2t|_{t=5} = 47.5$ rad/s

$(47.5)(50) = \omega_C(50)$

$\omega_C = 47.5$ rad/s

$\omega_B(75) = 47.5(50)$

$\omega_B = 31.7$ rad/s **Ans**

16-13. The anemometer measures the speed of the wind due to the rotation of the three cups. If during a 3-s time period a wind gust causes the cups to have an angular velocity of $\omega = (2t^2 + 3)$ rad/s, where t is in seconds, determine (a) the speed of the cups when $t = 2$ s, (b) the total distance traveled by each cup during the 3-s time period, and (c) the angular acceleration of the cups when $t = 2$ s. Neglect the size of the cups for the calculation.

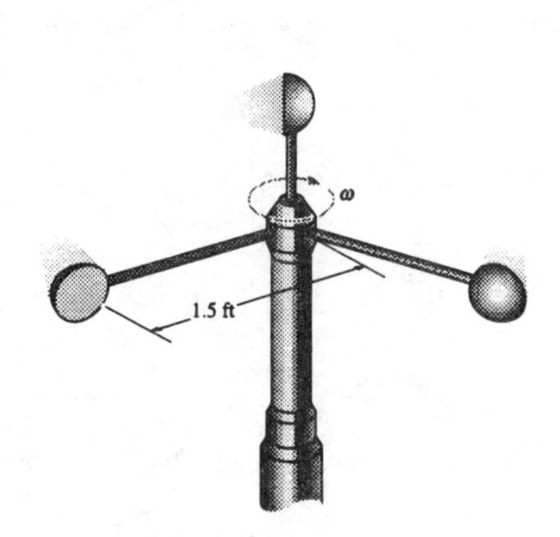

$\omega = 2t^2 + 3|_{t=2\text{ s}} = 11$ rad/s

$v = \omega r = 11(1.5) = 16.5$ ft/s **Ans**

$d\theta = \omega\, dt$

$\int_0^{\theta} d\theta = \int_0^3 (2t^2 + 3)\, dt \qquad \theta = 27$ rad

$s = \theta r = 27(1.5) = 40.5$ ft **Ans**

$\alpha = \frac{d\omega}{dt} = 4t\Big|_{t=2\text{ s}} = 8$ rad/s^2 **Ans**

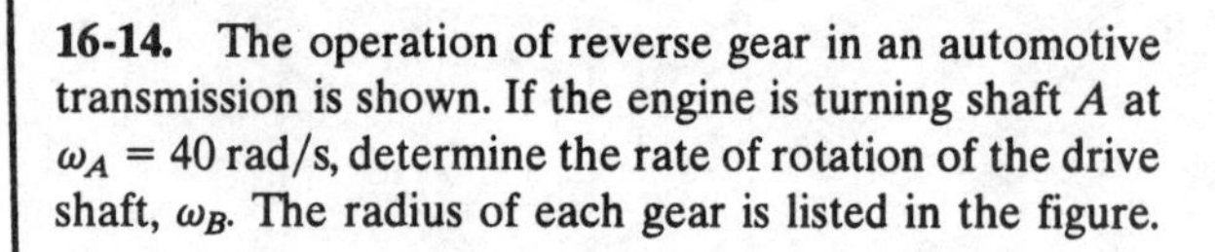

16-14. The operation of reverse gear in an automotive transmission is shown. If the engine is turning shaft A at $\omega_A = 40$ rad/s, determine the rate of rotation of the drive shaft, ω_B. The radius of each gear is listed in the figure.

$r_G\omega_A = r_C\omega_C; \qquad 80(40) = 40\omega_C \qquad \omega_C = \omega_D = 80$ rad/s

$\omega_E r_E = \omega_D r_D; \qquad \omega_E(50) = 80(40) \qquad \omega_E = \omega_F = 64$ rad/s

$\omega_F r_F = \omega_B r_B; \qquad 64(70) = \omega_B(50) \qquad \omega_B = 89.6$ rad/s

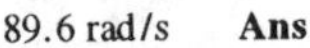

$\omega_B = 89.6$ rad/s **Ans**

G
$\omega_A = 40$ rad/s
B
ω_B
D
A
C
F
E
$r_G = 80$ mm
$r_C = r_D = 40$ mm
$r_E = r_B = 50$ mm
$r_F = 70$ mm

16-15. The turntable T is driven by the frictional idler wheel A, which simultaneously bears against the inner rim of the turntable and the motor-shaft spindle B. Determine the required diameter d of the spindle if the motor turns it at 25 rad/s and it is required that the turntable rotate at 2 rad/s.

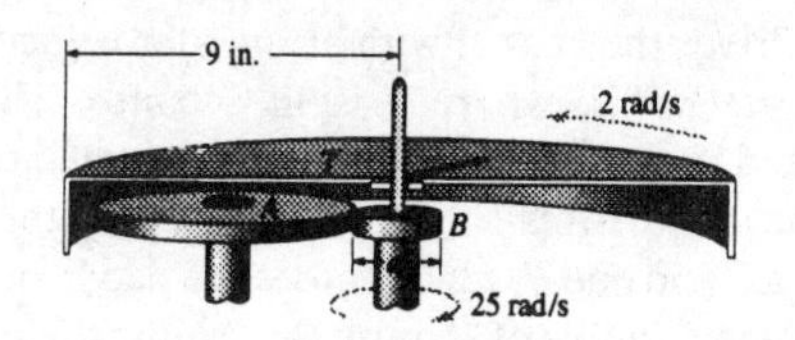

$$\omega_B r_B = \omega_A r_A; \qquad 25(0.5d) = \omega_A \left(\frac{9-0.5d}{2}\right)$$

$$\omega_A = \frac{25d}{9-0.5d}$$

$$\omega_A r_A = \omega_T r_T; \qquad \left(\frac{25d}{9-0.5d}\right)\left(\frac{9-0.5d}{2}\right) = 2(9)$$

$$d = 1.44 \text{ in.} \qquad \textbf{Ans}$$

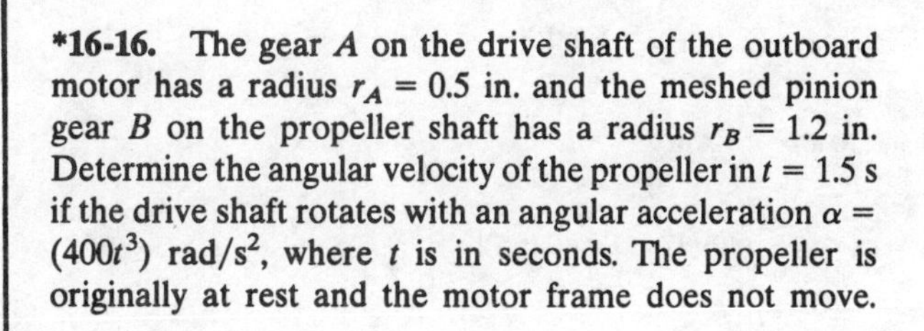

***16-16.** The gear A on the drive shaft of the outboard motor has a radius $r_A = 0.5$ in. and the meshed pinion gear B on the propeller shaft has a radius $r_B = 1.2$ in. Determine the angular velocity of the propeller in $t = 1.5$ s if the drive shaft rotates with an angular acceleration $\alpha = (400t^3)$ rad/s^2, where t is in seconds. The propeller is originally at rest and the motor frame does not move.

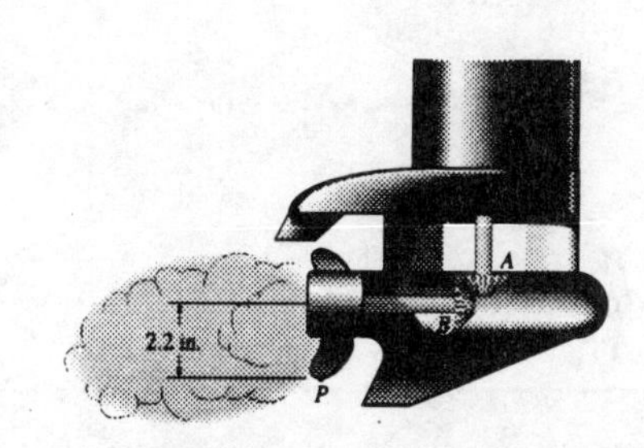

$$\alpha_A r_A = \alpha_B r_B$$

$$(400t^3)(0.5) = \alpha_P(1.2)$$

$$\alpha_P = 166.7t^3$$

$$d\omega = \alpha\, dt$$

$$\int_0^{\omega} d\omega = \int_0^t 166.7t^3\, dt$$

$$\omega = 41.67t^4\big|_{t=1.5} = 210.9 = 211 \text{ rad/s} \qquad \textbf{Ans}$$

16-17. For the outboard motor in Prob. 16-16, determine the magnitudes of the velocity and acceleration of a point P located on the tip of the propeller at the instant $t = 0.75$ s.

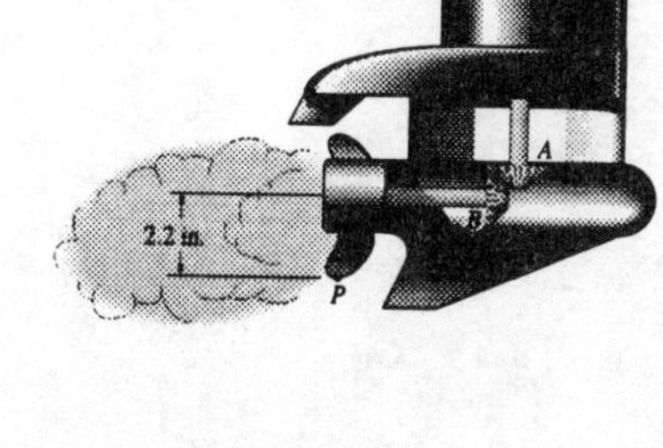

$$\alpha_A r_A = \alpha_P r_B$$

$$(400t^3)(0.5) = \alpha_P(1.2)$$

$$\alpha_P = 166.7t^3$$

$$d\omega = \alpha\, dt$$

$$\int_0^{\omega} d\omega = \int_0^t 166.7t^3\, dt$$

$$\omega = 41.67t^4$$

$$v_P = \omega r = \left[41.67(0.75)^4\right](2.2)$$

$$v_P = 29.0 \text{ in./s} \qquad \textbf{Ans}$$

$$(a_P)_t = \alpha r = \left[166.7(0.75)^3\right](2.2)$$

$$(a_P)_t = 154.7 \text{ in./s}^2$$

$$(a_P)_n = \omega^2 r = \left[41.67(0.75)^4\right]^2(2.2)$$

$$(a_P)_n = 382.4 \text{ in./s}^2$$

$$a_P = \sqrt{(154.7)^2 + (382.4)^2} = 412 \text{ in./s}^2 \qquad \textbf{Ans}$$

16-18. For a short time a motor of the random-orbit sander drives the gear A with an angular velocity of $\omega_A = 40(t^3 + 6t)$ rad/s, where t is in seconds. This gear is connected to gear B, which is fixed connected to the shaft CD. The end of this shaft is connected to the eccentric spindle EF and pad P, which causes the pad to orbit around shaft CD at a radius of 15 mm. Determine the magnitudes of the velocity and the tangential and normal components of acceleration of the spindle EF when $t = 2$ s after starting from rest.

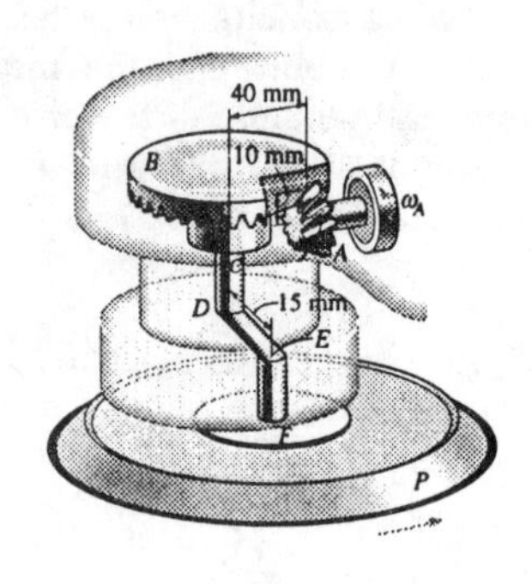

$$\omega_A r_A = \omega_B r_B$$

$$\omega_A (10) = \omega_B (40)$$

$$\omega_B = \frac{1}{4}\omega_A$$

$$v_E = \omega_B r_E = \frac{1}{4}\omega_A (0.015) = \frac{1}{4}(40)\left(t^3 + 6t\right)(0.015)\Big|_{t=2}$$

$v_E = 3$ m/s **Ans**

$$\alpha_A = \frac{d\omega_A}{dt} = \frac{d}{dt}\left[40\left(t^3 + 6t\right)\right] = 120t^2 + 240$$

$$\alpha_A r_A = \alpha_B r_B$$

$$\alpha_A (10) = \alpha_B (40)$$

$$\alpha_B = \frac{1}{4}\alpha_A$$

$$(a_E)_t = \alpha_B r_E = \frac{1}{4}\left(120t^2 + 240\right)(0.015)\Big|_{t=2}$$

$(a_E)_t = 2.70$ m/s^2 **Ans**

$$(a_E)_n = \omega_B^2 r_E = \left[\frac{1}{4}(40)\left(t^3 + 6t\right)\right]^2 (0.015)\Big|_{t=2}$$

$(a_E)_n = 600$ m/s^2 **Ans**

16-19. For a short time the motor of the random-orbit sander drives the gear A with an angular velocity of $\omega_A = (5\theta^2)$ rad/s, where θ is in radians. This gear is connected to gear B, which is fixed connected to the shaft CD. The end of this shaft is connected to the eccentric spindle EF and pad P, which causes the pad to orbit around shaft CD at a radius of 15 mm. Determine the magnitudes of the velocity and the tangential and normal components of acceleration of the spindle EF when $\theta = 0.5$ revolutions starting from rest.

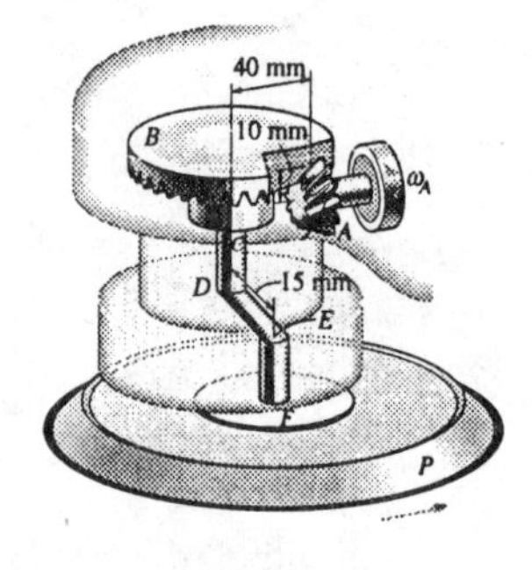

$$\omega_A (10) = \omega_B (40)$$

$$\omega_B = \frac{1}{4}\omega_A$$

0.5 rev = π rad

$$v_E = \omega_B r_E = \frac{1}{4}\omega_A (0.015) = \frac{1}{4}\left(5\theta^2\right)(0.015)\Big|_{\theta=\pi}$$

$v_E = 0.185$ m/s **Ans**

$$\alpha_A\, d\theta = \omega_A\, d\omega_A$$

$$\alpha_A\, d\theta = \left(5\theta^2\right)(10\theta\, d\theta)$$

$$\alpha_A = 50\theta^3$$

$$\alpha_A r_A = \alpha_B r_B$$

$$\alpha_A (10) = \alpha_B (40)$$

$$\alpha_B = \frac{1}{4}\alpha_A$$

$$(a_E)_t = \alpha_B r_E = \frac{1}{4}\left(50\theta^3\right)(0.015)\Big|_{\theta=\pi}$$

$(a_E)_t = 5.81$ m/s^2 **Ans**

$$(a_E)_n = \omega_B^2 r_E = \left[\frac{1}{4}\left(5\theta^2\right)\right]^2 (0.015)\Big|_{\theta=\pi}$$

$(a_E)_n = 2.28$ m/s^2 **Ans**

***16-20.** The aerobic machine manufactured by Precor, Inc. transfers pedaling power to a flywheel F, which develops resistance using an AC-powered electromagnet. If the operator initially drives the pedals at 20 rev/min, and then begins an angular acceleration of 30 rev/min^2, determine the angular velocity of the flywheel when t = 3 s. Note that the pedal arm is fixed-connected to the chain wheel A, which in turn drives the sheave B using the fixed-connected clutch gear D. The poly-V belt wrapped around the sheave, then drives the pulley E and fixed-connected flywheel.

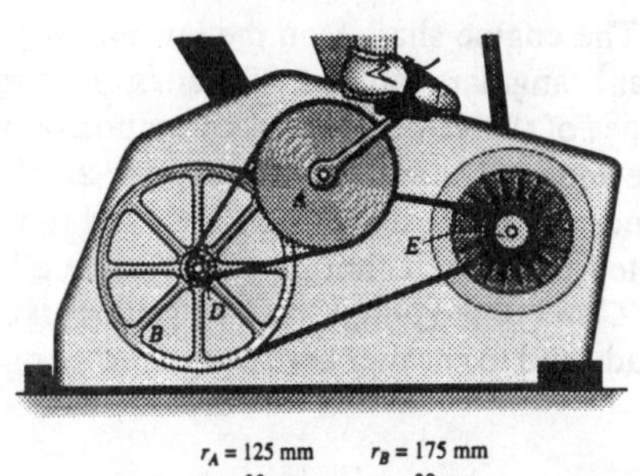

r_A = 125 mm r_B = 175 mm
r_D = 20 mm r_E = 30 mm

$$\omega = \omega_0 + \alpha_c t$$

$$\omega_A = 20 + 30\left(\frac{3}{60}\right) = 21.5 \text{ rev/min}$$

$$\omega_A r_A = \omega_D r_D$$

$$21.5(125) = \omega_D(20)$$

$$\omega_D = \omega_B = 134.375$$

$$\omega_B r_B = \omega_E r_E$$

$$134.375(175) = \omega_E(30)$$

$$\omega_E = 783.9 \text{ rev/min}$$

$$\omega_F = 784 \text{ rev/min}$$ **Ans**

16-21. The aerobic machine manufactured by Precor, Inc. transfers pedaling power to a flywheel F, which develops resistance using an AC-powered electromagnet. If the operator initially drives the pedals at 12 rev/min, and then begins an angular acceleration of 8 rev/min^2, determine the angular velocity of the flywheel after 2 revolutions of the pedal arm. Note that the pedal arm is fixed connected to the chain wheel A, which in turn drives the sheave B using the fixed-connected clutch gear D. The poly-V belt wrapped around the sheave, then drives the pulley E and fixed-connected flywheel.

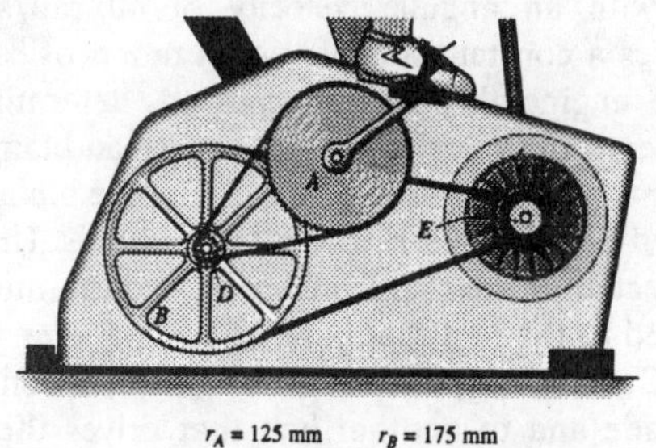

r_A = 125 mm r_B = 175 mm
r_D = 20 mm r_E = 30 mm

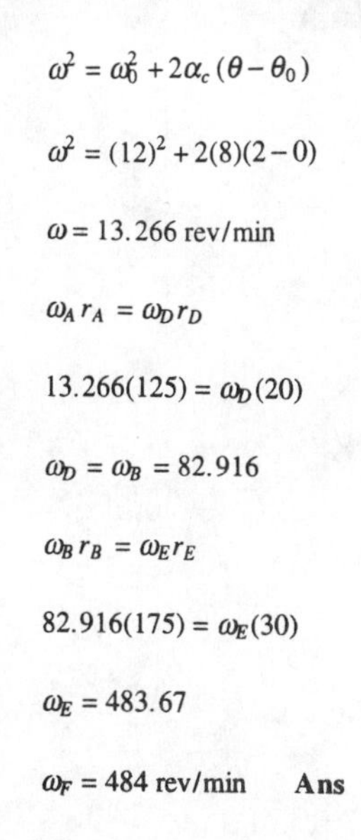

$$\omega^2 = \omega_0^2 + 2\alpha_c(\theta - \theta_0)$$

$$\omega^2 = (12)^2 + 2(8)(2-0)$$

$$\omega = 13.266 \text{ rev/min}$$

$$\omega_A r_A = \omega_D r_D$$

$$13.266(125) = \omega_D(20)$$

$$\omega_D = \omega_B = 82.916$$

$$\omega_B r_B = \omega_E r_E$$

$$82.916(175) = \omega_E(30)$$

$$\omega_E = 483.67$$

$$\omega_F = 484 \text{ rev/min}$$ **Ans**

16-22. The engine shaft S on the lawnmower rotates at a constant angular rate of 40 rad/s. Determine the magnitudes of the velocity and acceleration of point P on the blade and the distance P travels in 3 seconds. The shaft S is connected to the driver pulley A, and the motion is transmitted to the belt that passes over the idler pulleys at B and C and to the pulley at D. This pulley is connected to the blade and to another belt that drives the other blade.

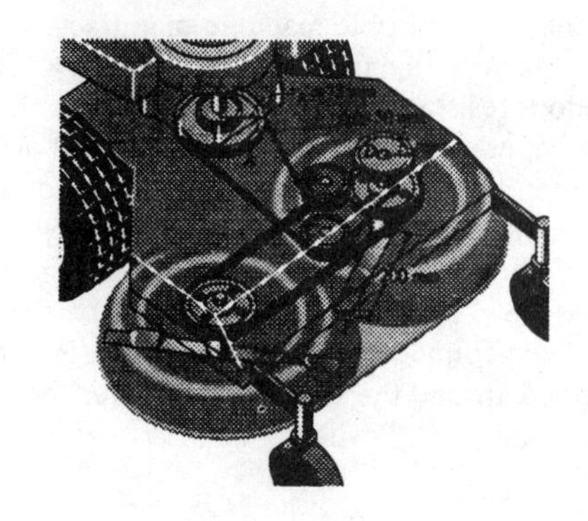

$\omega_A = 40 \text{ rad/s} \qquad \alpha_A = 0$

$\theta = \theta_0 + \omega t$

$\theta_A = 0 + 40(3) = 120 \text{ rad}$

$\theta_A r_A = \theta_D r_D$

$120(75) = \theta_D(50)$

$\theta_D = 180 \text{ rad}$

$s_P = r_P \theta_D = 0.2(180) = 36 \text{ m}$ **Ans**

$\omega_A r_A = \omega_D r_D$

$40(75) = \omega_D(50)$

$\omega_D = 60 \text{ rad/s}$

$v_P = r_P \omega_D = 0.2(60) = 12 \text{ m/s}$ **Ans**

Also,

$v_P = \frac{s_P}{t} = \frac{36 \text{ m}}{3 \text{ s}} = 12 \text{ m/s}$ **Ans**

$\alpha_D = 0$

Thus,

$(a_P)_t = \alpha_D r_P = 0$

$(a_P)_n = \omega_D^2 r_P = (60)^2(0.2) = 720 \text{ m/s}^2$

$a_P = 720 \text{ m/s}^2$ **Ans**

16-23. The engine shaft S on the lawnmower is initially rotating with an angular velocity of 40 rad/s. If it experiences a constant angular acceleration of 3 rad/s^2 while the engine throttle is increased, determine the magnitudes of the velocity and the normal and tangential components of acceleration of point P on the blade when $t = 2$ s, and the distance P travels in 2 seconds. The shaft S is connected to the driver pulley A, and the motion is transmitted to the belt that passes over the idler pulleys at B and C and to the pulley at D. This pulley is connected to the blade and to another belt that drives the other blade.

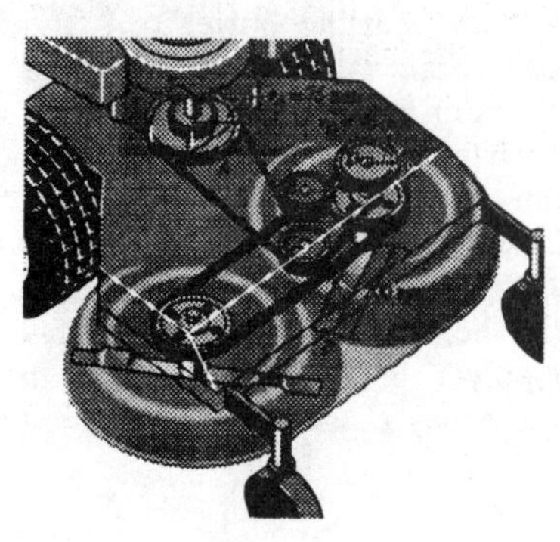

$\omega_A = 40 \text{ rad/s} \qquad \alpha_A = 3 \text{ rad/s}^2$

$\theta = \theta_0 + \omega_0 t + \frac{1}{2}\alpha_c t^2$

$\theta_A = 0 + 40(2) + \frac{1}{2}(3)(2)^2 = 86 \text{ rad}$

$\omega = \omega_0 + \alpha_c t$

$\omega_A = 40 + 3(2) = 46 \text{ rad/s}$

$\theta_A r_A = \theta_D r_D$

$86(75) = \theta_D(50)$

$\theta_D = 129 \text{ rad}$

$s_P = r_P \theta_D = 0.2(129) = 25.8 \text{ m}$ **Ans**

$\omega_A r_A = \omega_D r_D$

$46(75) = \omega_D(50)$

$\omega_D = 69 \text{ rad/s}$

$v_P = r_P \omega_D = 0.2(69) = 13.8 \text{ m/s}$ **Ans**

$\alpha_A r_A = \alpha_D r_D$

$3(75) = \alpha_D(50)$

$\alpha_D = 4.5 \text{ rad/s}^2$

Thus,

$(a_P)_t = \alpha_D r_P = 4.5(0.2) = 0.9 \text{ m/s}^2$ **Ans**

$(a_P)_n = \omega_D^2 r_P = (69)^2(0.2) = 952 \text{ m/s}^2$ **Ans**

***16-24.** The disk starts from rest and is given an angular acceleration $\alpha = (10\theta^{1/3})$ rad/s², where θ is in radians. Determine the angular velocity of the disk and the angular displacement when $t = 4$ s.

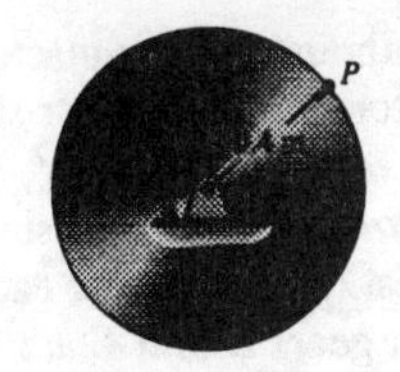

$\alpha = 10\theta^{\frac{1}{3}}$

$\omega\, d\omega = \alpha\, d\theta$

$\int_0^{\omega} \omega\, d\omega = \int_0^{\theta} 10\theta^{\frac{1}{3}}\, d\theta$

$\frac{1}{2}\omega^2 = 10\left(\frac{3}{4}\theta^{\frac{4}{3}}\right) = 7.5\theta^{\frac{4}{3}}$

$\omega = \frac{d\theta}{dt} = \sqrt{15}\theta^{\frac{2}{3}}$

$\int_0^{\theta} \theta^{-\frac{2}{3}}\, d\theta = \int_0^{t} \sqrt{15}\, dt$

$3\theta^{\frac{1}{3}} = \sqrt{15}\, t$

$\theta = 2.152t^3|_{t=4} = 138$ rad **Ans**

$\omega = \frac{d\theta}{dt} = 6.455t^2\Big|_{t=4} = 103$ rad/s **Ans**

16-25. The disk starts from rest and is given an angular acceleration $\alpha = (10\theta^{1/3})$ rad/s², where θ is in radians. Determine the magnitudes of the normal and tangential components of acceleration of a point P on the rim of the disk when $t = 4$ s.

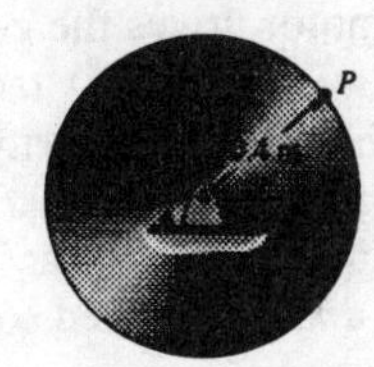

$\alpha = 10\theta^{\frac{1}{3}}$

$\omega\, d\omega = \alpha\, d\theta$

$\int_0^{\omega} \omega\, d\omega = \int_0^{\theta} 10\theta^{\frac{1}{3}}\, d\theta$

$\frac{1}{2}\omega^2 = 10\left(\frac{3}{4}\theta^{\frac{4}{3}}\right) = 7.5\theta^{\frac{4}{3}}$

$\omega = \frac{d\theta}{dt} = \sqrt{15}\theta^{\frac{2}{3}}$

$\int_0^{\theta} \theta^{-\frac{2}{3}}\, d\theta = \int_0^{t} \sqrt{15}\, dt$

$3\theta^{\frac{1}{3}} = \sqrt{15}\, t$

$\theta = 2.152t^3|_{t=4} = 137.71$

$\omega = \frac{d\theta}{dt} = 6.455t^2\Big|_{t=4} = 103.28$

$(a_P)_n = \omega^2 r = (103.28)^2(0.4) = 4267$ m/s² **Ans**

$(a_P)_t = \alpha r = \left(10(137.71)^{\frac{1}{3}}\right)(0.4) = 20.7$ m/s² **Ans**

16-26. Morse Industrial manufactures the speed reducer shown. If a motor drives the gear shaft S with an angular acceleration of $\alpha = (0.4e^t)$ rad/s^2, where t is in seconds, determine the angular velocity of shaft E when $t = 2$ s after starting from rest. The radius of each gear is listed in the figure. Note that gears B and C are fixed connected to the same shaft.

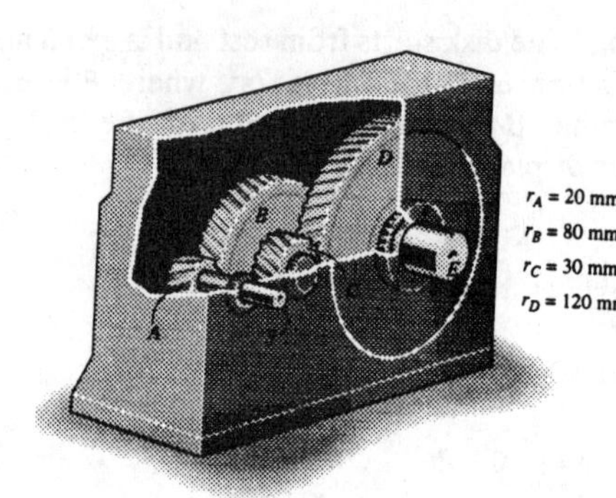

$$\alpha = \frac{d\omega}{dt} = 0.4e^t$$

$$\int_1^{\omega} d\omega = \int_0^t 0.4e^t\, dt$$

$$\omega_S = 0.4e^t|_0^2 = 0.4\left(e^2 - 1\right) = 2.556 \text{ rad/s}$$

$$\omega_S r_A = \omega_B r_B$$

$$2.556(20) = \omega_B(80)$$

$$\omega_B = 0.6389 \text{ rad/s}$$

$$\omega_B r_C = \omega_D r_D$$

$$0.6389(30) = \omega_D(120)$$

$$\omega_D = 0.160 \text{ rad/s}$$

$$\omega_E = 0.160 \text{ rad/s} \qquad \textbf{Ans}$$

16-27. Morse Industrial manufactures the speed reducer shown. If a motor drives the gear shaft S with an angular acceleration of $\alpha = (4\omega^{-3})$ rad/s^2, where ω is in rad/s, determine the angular velocity of shaft E when $t = 2$ s after starting from an angular velocity of 1 rad/s when $t = 0$. The radius of each gear is listed in the figure. Note that gears B and C are fixed connected to the same shaft.

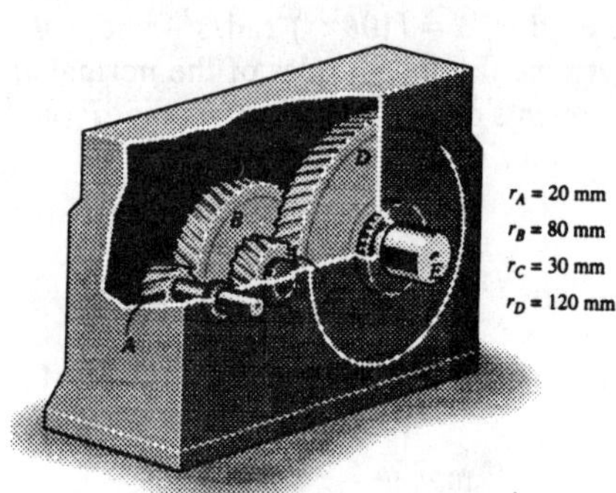

$$\alpha = \frac{d\omega}{dt} = 4\omega^{-3}$$

$$\int_1^{\omega} \omega^3\, d\omega = \int_0^t 4\, dt$$

$$\frac{1}{4}\omega^4\Big|_1^{\omega} = 4t\Big|_0^t$$

$$\frac{1}{4}\left(\omega^4 - 1\right) = 4t$$

$$\omega_S = (16t + 1)^{\frac{1}{4}}\Big|_{t=2} = 2.397 \text{ rad/s}$$

$$\omega_S r_A = \omega_B r_B$$

$$2.397(20) = \omega_B(80)$$

$$\omega_B = 0.5992 \text{ rad/s}$$

$$\omega_B r_C = \omega_D r_D$$

$$0.5992(30) = \omega_D(120)$$

$$\omega_D = 0.150 \text{ rad/s}$$

$$\omega_E = 0.150 \text{ rad/s} \qquad \textbf{Ans}$$

***16-28.** The sphere starts from rest at $\theta = 0°$ and rotates with an angular acceleration of $\alpha = (4\theta)$ rad/s², where θ is in radians. Determine the magnitudes of the velocity and acceleration of point P on the sphere at the instant $\theta = 6$ rad.

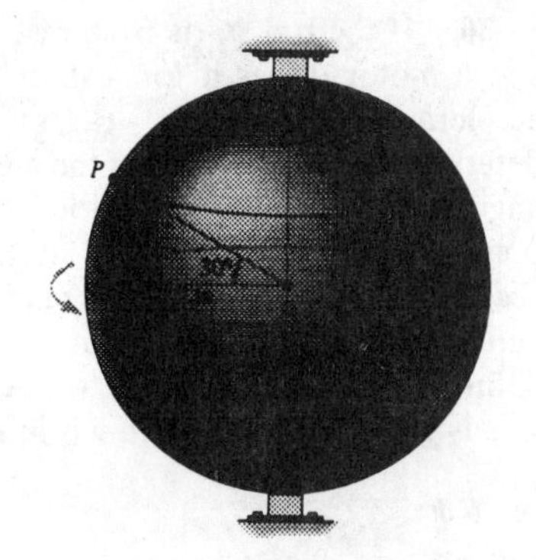

$\omega\, d\omega = \alpha\, d\theta$

$\int_0^{\omega} \omega\, d\omega = \int_0^{\theta} 4\theta\, d\theta$

$\omega = 2\theta$

At $\theta = 6$ rad,

$\alpha = 4(6) = 24 \text{ rad/s}^2, \qquad \omega = 2(6) = 12 \text{ rad/s}$

$v = \omega r' = 12(8\cos 30°) = 83.14 \text{ in./s}$

$v = 6.93$ ft/s **Ans**

$a_r = \dfrac{v^2}{r'} = \dfrac{(83.14)^2}{(8\cos 30°)} = 997.66 \text{ in./s}^2$

$a_t = \alpha r' = 24(8\cos 30°) = 166.28 \text{ in./s}^2$

$a = \sqrt{(997.66)^2 + (166.28)^2} = 1011.42 \text{ in./s}^2$

$a = 84.3 \text{ ft/s}^2$ **Ans**

16-29. At the instant shown, gear A is rotating with a constant angular velocity of $\omega_A = 6$ rad/s. Determine the largest angular velocity of gear B and the maximum speed of point C.

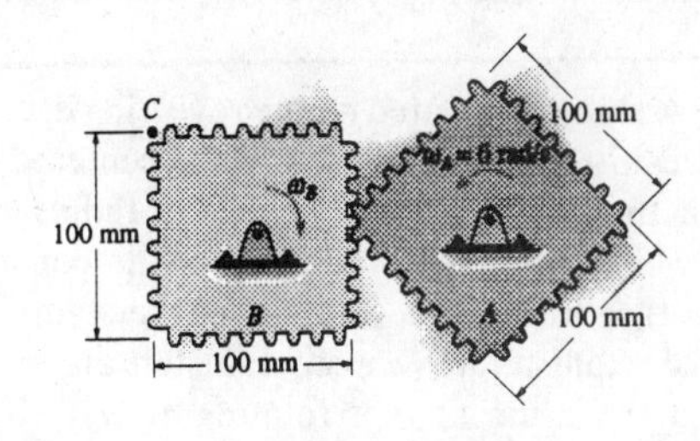

$(r_B)_{max} = (r_A)_{max} = 50\sqrt{2}$ mm

$(r_B)_{min} = (r_A)_{min} = 50$ mm

When r_A is max., r_B is min.

$\omega_B (r_B) = \omega_A r_A$

$(\omega_B)_{max} = 6\left(\dfrac{r_A}{r_B}\right) = 6\left(\dfrac{50\sqrt{2}}{50}\right)$

$(\omega_B)_{max} = 8.49$ rad/s **Ans**

$v_C = (\omega_B)_{max}\, r_C = 8.49\left(0.05\sqrt{2}\right)$

$v_C = 0.6$ m/s **Ans**

16-30. If the rod starts from rest in the position shown and a motor drives it for a short time with an angular acceleration of $\alpha = (1.5e^t)$ rad/s^2, where t is in seconds, determine the magnitude of the angular velocity and the angular displacement of the rod when $t = 3$ s. Locate the point on the rod which has the greatest velocity and acceleration, and compute the magnitudes of the velocity and acceleration of this point when $t = 3$ s. The rod is defined by $z = 0.25 \sin(\pi y)$, where the argument for the sine is given in radians and y is in meters.

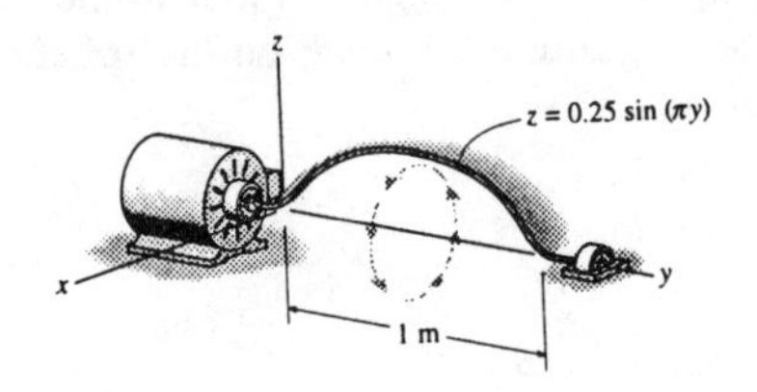

$d\omega = \alpha\, dt$

$\int_0^{\omega} d\omega = \int_0^t 1.5e^t\, dt$

$\omega = 1.5e^t|_0^t = 1.5[e^t - 1]$

$d\theta = \omega\, dt$

$\int_0^{\theta} d\theta = 1.5\int_0^t [e^t - 1]\, dt$

$\theta = 1.5[e^t - t]_0^t = 1.5[e^t - t - 1]$

When $t = 3$ s,

$\omega = 1.5[e^3 - 1] = 28.63 = 28.6$ rad/s **Ans**

$\theta = 1.5[e^3 - 3 - 1] = 24.1$ rad **Ans**

The point having the greatest velocity and acceleration is located furthest from the axis of rotation. This is at $y = 0.5$ m, where $z = 0.25\sin(\pi\, 0.5) = 0.25$ m.

Hence,

$v_P = \omega(z) = 28.63(0.25) = 7.16$ m/s **Ans**

$(a_t)_P = \alpha(z) = (1.5e^3)(0.25) = 7.532$ m/s^2

$(a_n)_P = \omega^2(z) = (28.63)^2(0.25) = 204.89$ m/s^2

$a_P = \sqrt{(a_t)_P^2 + (a_n)_P^2} = \sqrt{(7.532)^2 + (204.89)^2}$

$a_P = 205$ m/s^2 **Ans**

16-31. A stamp S, located on the revolving drum, is used to label canisters. If the canisters are centered 200 mm apart on the conveyor, determine the radius r_A of the driving wheel A and the radius r_B of the conveyor belt drum so that for each revolution of the stamp it marks the top of a canister. How many canisters are marked per minute if the drum at B is rotating at $\omega_B = 0.2$ rad/s? Note that the driving belt is twisted as it passes between the wheels.

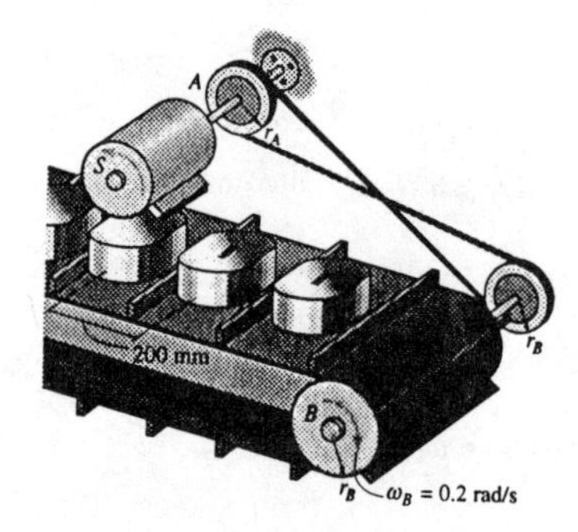

For the wheel at A :

$l = 2\pi(r_A)$

$r_A = \dfrac{200}{2\pi} = 31.8$ mm **Ans**

For the drum at B :

$l = 2\pi(r_B)$

$r_B = \dfrac{200}{2\pi} = 31.8$ mm **Ans**

In $t = 60$ s,

$\theta = \theta_0 + \omega_0 t$

$\theta = 0 + 0.2(60) = 12$ rad

$l = \theta r_B = 12(31.8) = 382.0$ mm

Hence,

$n = \dfrac{382.0}{200} = 1.91$ canisters marked per minute **Ans**

***16-32.** A tape having a thickness s wraps around the wheel which is turning at a constant rate ω. Assuming the unwrapped portion of tape remains horizontal, determine the acceleration of point P of the unwrapped tape when the radius of the wrapped tape is r. *Hint:* Since $v_P = \omega r$, take the time derivative and note that $dr/dt = \omega(s/2\pi)$.

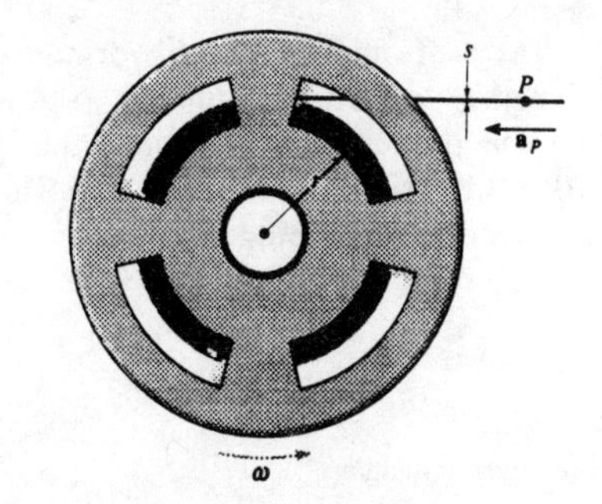

$v_P = \omega r$

$$a = \frac{dv_P}{dt} = \frac{d\omega}{dt}r + \omega\frac{dr}{dt}$$

Since $\frac{d\omega}{dt} = 0$,

$$a = \omega\left(\frac{dr}{dt}\right)$$

In one revolution r is increased by s, so that

$$\frac{2\pi}{\theta} = \frac{s}{\Delta r}$$

Hence,

$$\Delta r = \frac{s}{2\pi}\theta$$

$$\frac{dr}{dt} = \frac{s}{2\pi}\omega$$

$$a = \frac{s}{2\pi}\omega^2 \qquad \textbf{Ans}$$

16-33. At the instant shown, $\theta = 60°$, and rod AB is subjected to a deceleration of 16 m/s² when the velocity is 10 m/s. Determine the angular velocity and angular acceleration of link CD at this instant.

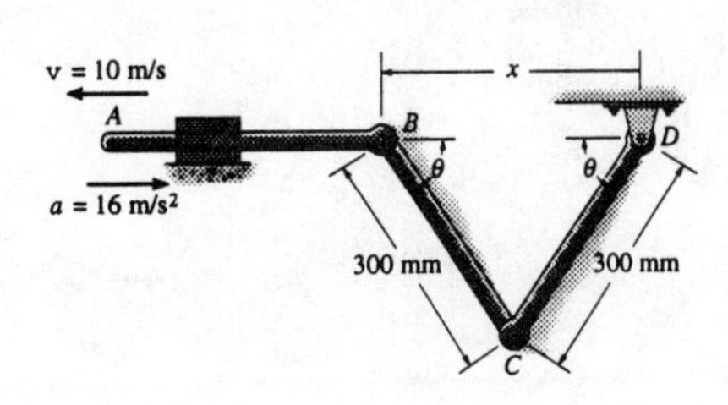

$$x = 2(0.3)\cos\theta$$

$$\dot{x} = -0.6\sin\theta(\dot{\theta}) \qquad (1)$$

$$\ddot{x} = -0.6\cos\theta(\dot{\theta})^2 - 0.6\sin\theta(\ddot{\theta}) \qquad (2)$$

Using Eqs. (1) and (2) at $\theta = 60°$, $\dot{x} = 10$ m/s, $\ddot{x} = -16$ m/s²,

$$10 = -0.6\sin 60°(\omega)$$

$$\omega = -19.245 = -19.2 \text{ rad/s} \qquad \textbf{Ans}$$

$$-16 = -0.6\cos 60°(-19.245)^2 - 0.6\sin 60°(\alpha)$$

$$\alpha = -183 \text{ rad/s}^2 \qquad \textbf{Ans}$$

16-34. The scaffold S is raised hydraulically by moving the roller at A toward the pin at B. If A is approaching B with a speed of 1.5 ft/s, determine the speed at which the platform is rising as a function of θ. The 4-ft links are pin-connected at their midpoint.

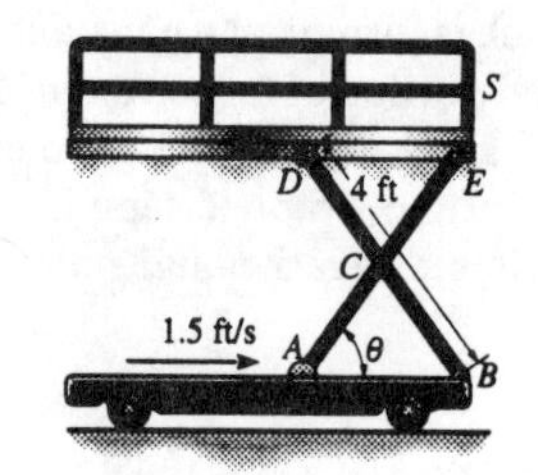

Position coordinate equation :

$$x = 4\cos\theta \qquad y = 4\sin\theta$$

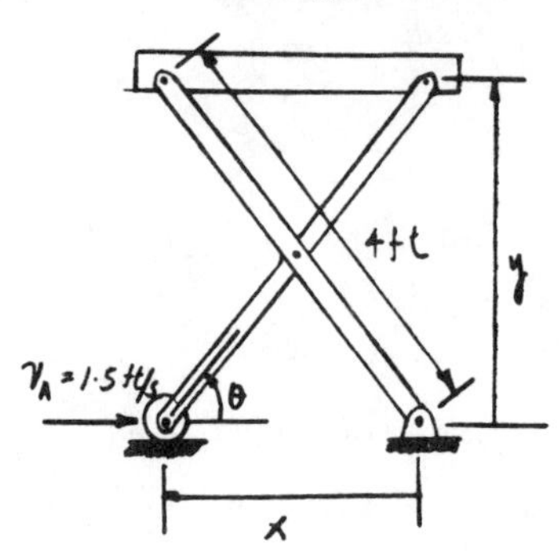

Time derivatives :

$$\dot{x} = -4\sin\theta\dot{\theta} \qquad \text{However,} \quad \dot{x} = -v_A = -1.5 \text{ ft/s}$$

$$-1.5 = -4\sin\theta\dot{\theta} \qquad \dot{\theta} = \frac{0.375}{\sin\theta}$$

$$\dot{y} = v_y = 4\cos\theta\dot{\theta} = 4\cos\theta\left(\frac{0.375}{\sin\theta}\right) = 1.5\cot\theta \qquad \textbf{Ans}$$

16-35. The 2-m-long bar is confined to move in the horizontal and vertical slots A and B. If the velocity of the slider block at A is 8 m/s, determine the bar's angular velocity and the velocity of block B at the instant $\theta = 60°$.

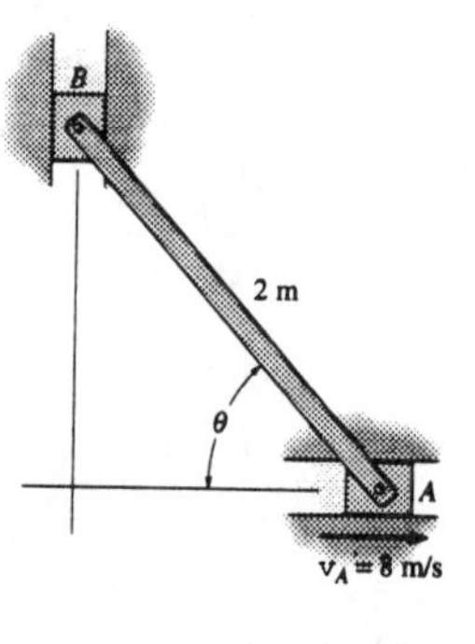

$x = 2\cos\theta$

$v_A = -2\sin\theta\omega$

$8 = -2\sin 60°\omega$

$\omega = -4.619 = -4.62 \text{ rad/s} = 4.62 \text{ rad/s}$ ↺ **Ans**

$y = 2\sin\theta$

$v_B = 2\cos\theta\omega$

$v_B = 2(\cos 60°)(-4.619)$

$v_B = -4.62 \text{ m/s} = 4.62 \text{ m/s}$ ↓ **Ans**

***16-36.** Determine the angular velocity of rod AB when $\theta = 30°$. The shaft and the center of the roller C move forward at a constant rate $v = 5$ m/s.

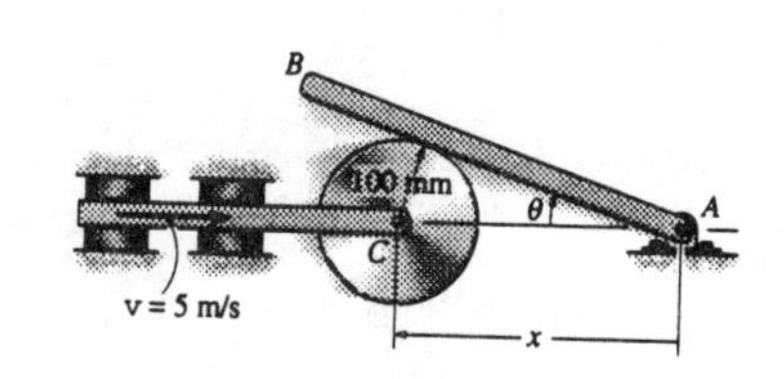

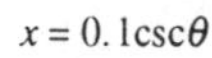

$x = 0.1\csc\theta$

$v = 0.1(-\csc\theta\cot\theta)\omega$

$-5 = 0.1(-\csc 30°\cot 30°)\omega$

$\omega = 14.4 \text{ rad/s}$ **Ans**

16-37. The inclined plate moves to the left with a constant velocity **v**. Determine the angular velocity and angular acceleration of the slender rod of length l. The rod pivots about the step at C as it slides on the plate.

$$\frac{x}{\sin(\phi-\theta)} = \frac{l}{\sin(180^\circ-\phi)} = \frac{l}{\sin\phi}$$

$$x\sin\phi = l\sin(\phi-\theta)$$

$$\dot{x}\sin\phi = -l\cos(\phi-\theta)\,\dot{\theta}$$

Thus

$$\omega = \frac{-v(\sin\phi)}{l\cos(\phi-\theta)} \qquad \textbf{Ans}$$

$$\ddot{x}\sin\phi = -l\cos(\phi-\theta)\ddot{\theta} - l\sin(\phi-\theta)(\dot{\theta})^2$$

$$0 = -\cos(\phi-\theta)\alpha - \sin(\phi-\theta)\,\omega^2$$

$$\alpha = \frac{-\sin(\phi-\theta)}{\cos(\phi-\theta)}\left(\frac{v^2\sin^2\phi}{l^2\cos^2(\phi-\theta)}\right)$$

$$\alpha = \frac{-v^2\sin^2\phi\sin(\phi-\theta)}{l^2\cos^3(\phi-\theta)} \qquad \textbf{Ans}$$

16-38. The crankshaft AB is rotating at a constant angular velocity of $\omega = 150$ rad/s. Determine the velocity of the piston P at the instant $\theta = 30°$.

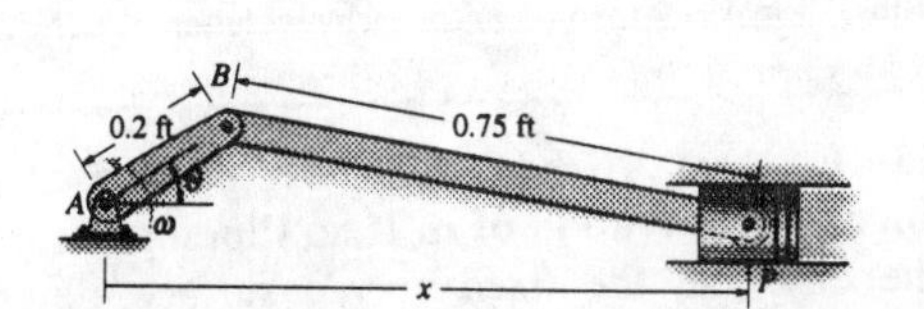

$$x = 0.2\cos\theta + \sqrt{(0.75)^2 - (0.2\sin\theta)^2}$$

$$\dot{x} = -0.2\sin\theta\dot{\theta} + \frac{1}{2}\left[(0.75)^2 - (0.2\sin\theta)^2\right]^{-\frac{1}{2}}(-2)(0.2\sin\theta)(0.2\cos\theta)\dot{\theta}$$

$$v_P = -0.2\omega\sin\theta - \left(\frac{1}{2}\right)\frac{(0.2)^2\omega\sin 2\theta}{\sqrt{(0.75)^2 - (0.2\sin\theta)^2}}$$

At $\theta = 30°$, $\omega = 150$ rad/s

$$v_P = -0.2(150)\sin 30^\circ - \left(\frac{1}{2}\right)\frac{(0.2)^2(150)\sin 60^\circ}{\sqrt{(0.75)^2 - (0.2\sin 30^\circ)^2}}$$

$v_P = -18.5$ ft/s $= 18.5$ ft/s $\leftarrow$ **Ans**

16-39. Determine the velocity of the rod R for any angle θ of cam C if the cam rotates with a constant angular velocity ω. The pin connection at O does not cause an interference with the motion of plate A on C.

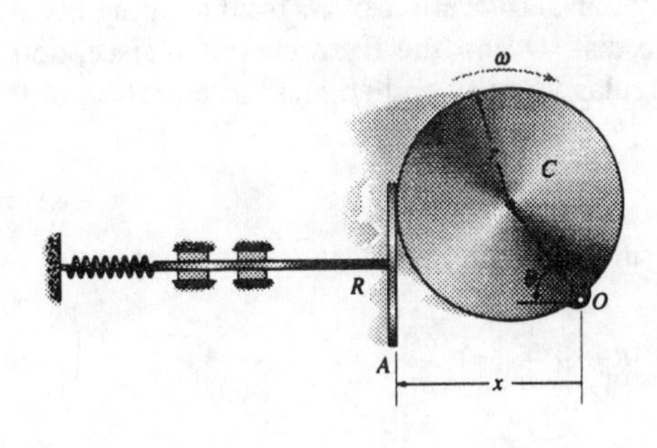

$$x = r + r\cos\theta$$

$$\dot{x} = -r\sin\theta\dot{\theta}$$

$v = -r\omega\sin\theta$ **Ans**

***16-40.** Disk A rolls without slipping over the surface of the *fixed* cylinder B. Determine the angular velocity of A if its center C has a speed $v_C = 5$ m/s. How many revolutions will A have made about its center just after link DC completes one revolution?

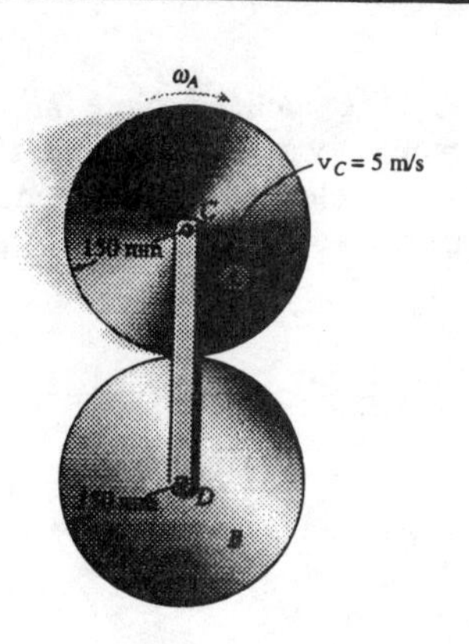

As shown by the construction, as A rolls through the arc $s = \theta_A r$, the center of the disk moves through the same distance $s' = s$. Hence,

$s = \theta_A r$

$\dot{s} = \dot{\theta}_A r$

$5 = \omega_A (0.15)$

$\omega_A = 33.3$ rad/s **Ans**

Link :

$s' = 2r\theta_{CD} = s = \theta_A r$

$2\theta_{CD} = \theta_A$

Thus, A makes 1 revolution for each revolution of CD. **Ans**

16-41. Arm AB has an angular velocity of $\boldsymbol{\omega}$ and an angular acceleration of $\boldsymbol{\alpha}$. If no slipping occurs between the disk and the fixed curved surface, determine the angular velocity and angular acceleration of the disk.

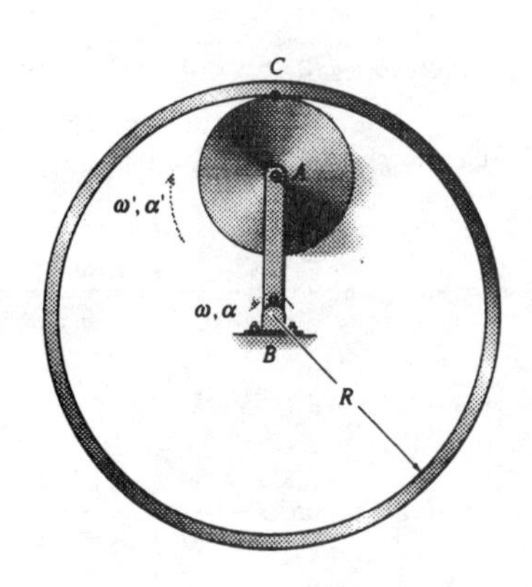

$ds = (R - r)\, d\theta = -r\, d\phi$

$(R-r)\left(\dfrac{d\theta}{dt}\right) = -r\left(\dfrac{d\phi}{dt}\right)$

$\omega' = -\dfrac{(R-r)\omega}{r}$ **Ans**

$\alpha' = -\dfrac{(R-r)\alpha}{r}$ **Ans**

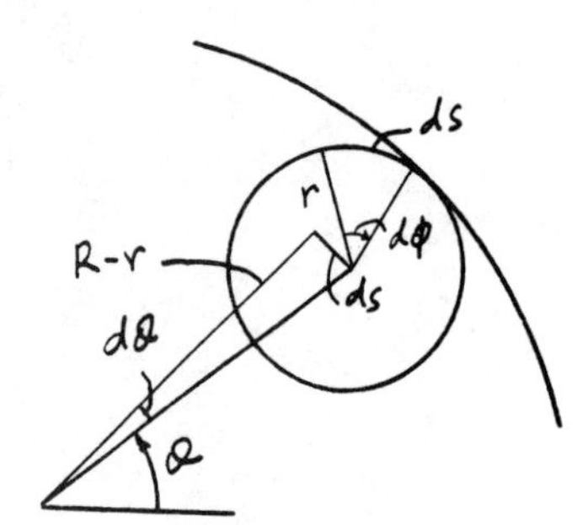

16-42. Arm AB has an angular velocity of $\boldsymbol{\omega}$ and an angular acceleration of $\boldsymbol{\alpha}$. If no slipping occurs between the disk D and the fixed curved surface, determine the angular velocity and angular acceleration of the disk.

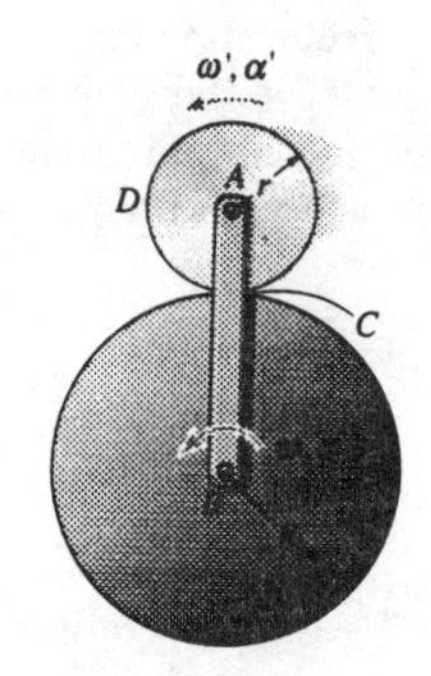

$ds = (R + r)\, d\theta = r\, d\phi$

$(R+r)\left(\dfrac{d\theta}{dt}\right) = r\left(\dfrac{d\phi}{dt}\right)$

$\omega' = \dfrac{(R+r)\omega}{r}$ **Ans**

$\alpha' = \dfrac{(R+r)\alpha}{r}$ **Ans**

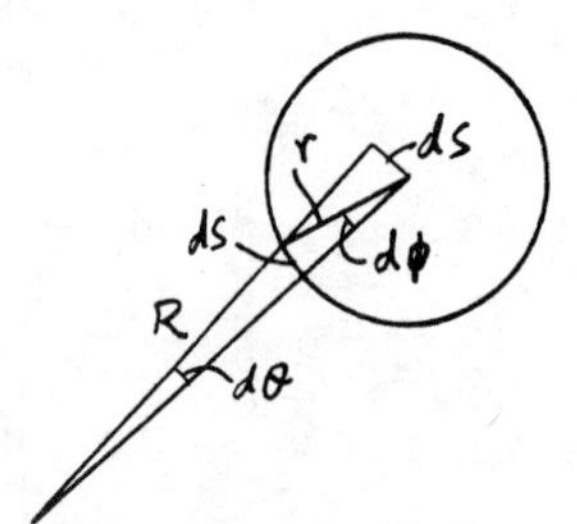

■16-43. The end A of the bar is moving downward along the slotted guide with a constant velocity $\mathbf{v}_A$. Determine the angular velocity $\boldsymbol{\omega}$ and angular acceleration $\boldsymbol{\alpha}$ of the bar as a function of its position y.

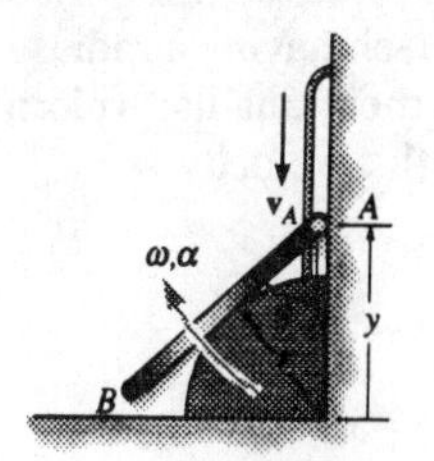

Position coordinate equation :

$$\sin\theta = \frac{r}{y}$$

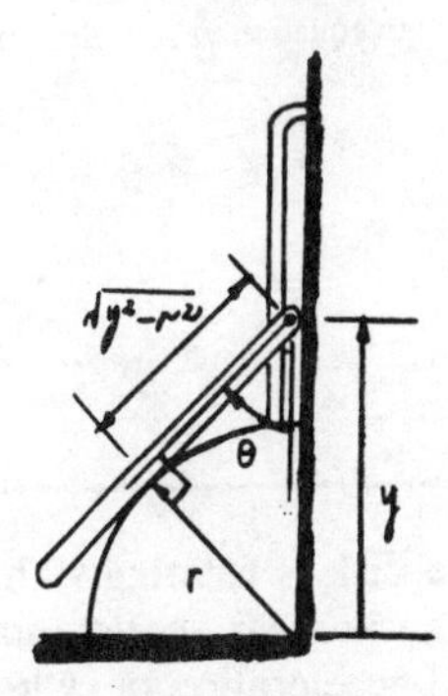

Time derivatives :

$$\cos\theta\dot{\theta} = -\frac{r}{y^2}\dot{y} \qquad \text{however, } \cos\theta = \frac{\sqrt{y^2-r^2}}{y} \text{ and } \dot{y} = -v_A,\ \dot{\theta} = \omega$$

$$\left(\frac{\sqrt{y^2-r^2}}{y}\right)\omega = \frac{r}{y^2}v_A \qquad \omega = \frac{rv_A}{y\sqrt{y^2-r^2}} \qquad \textbf{Ans}$$

$$\alpha = \dot{\omega} = rv_A\left[-y^{-2}\dot{y}(y^2-r^2)^{-\frac{1}{2}} + (y^{-1})\left(-\tfrac{1}{2}\right)(y^2-r^2)^{-\frac{3}{2}}(2y\dot{y})\right]$$

$$\alpha = \frac{rv_A^2\,(2y^2-r^2)}{y^2\,(y^2-r^2)^{\frac{3}{2}}} \qquad \textbf{Ans}$$

***16-44.** The pins at A and B are confined to move in the vertical and horizontal tracks. If the slotted arm is causing A to move downward at $\mathbf{v}_A$, determine the velocity of B at the instant shown.

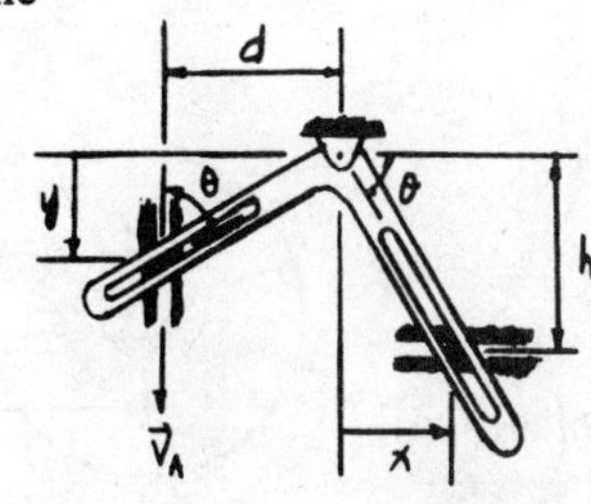

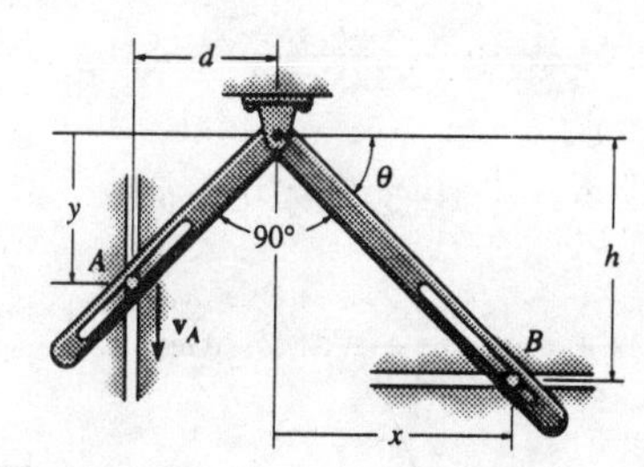

Position coordinate equation :

$$\tan\theta = \frac{h}{x} = \frac{d}{y}$$

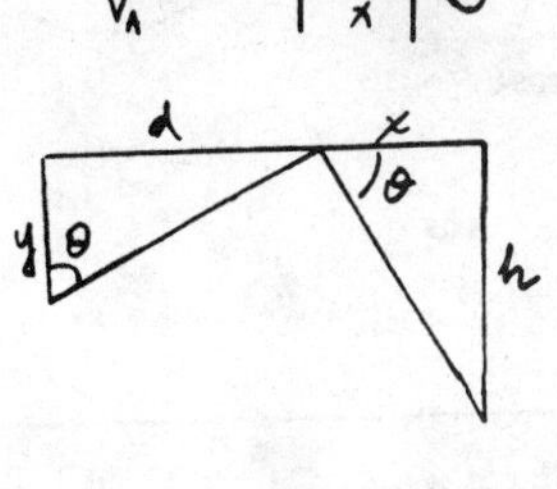

$$x = \left(\frac{h}{d}\right)y$$

Time derivatives :

$$\dot{x} = \left(\frac{h}{d}\right)\dot{y}$$

$$v_B = \left(\frac{h}{d}\right)v_A \qquad \textbf{Ans}$$

16-45. The bar remains in contact with the floor and with point A. If point B moves to the right with a constant velocity $\mathbf{v}_B$, determine the angular velocity and angular acceleration of the bar as a function of x.

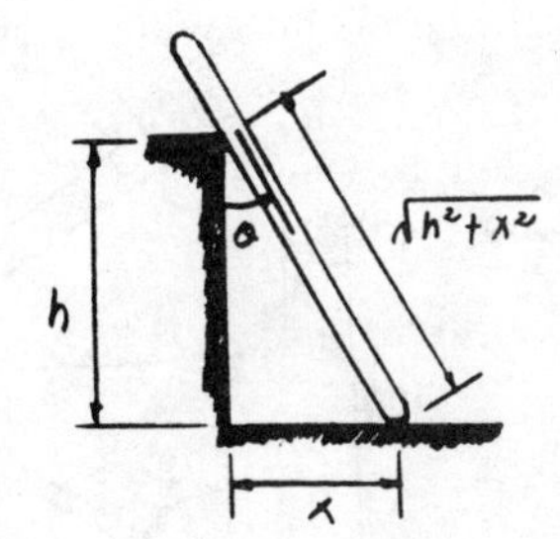

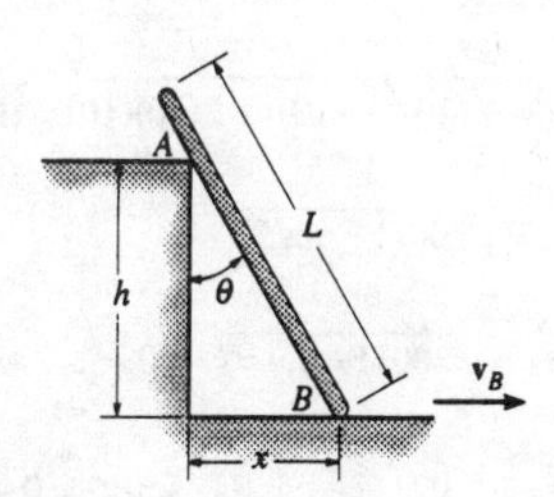

Position coordinate equation :

$$\tan\theta = \frac{x}{h}$$

Time derivatives :

$$\sec^2\theta\dot{\theta} = \frac{1}{h}\dot{x} \qquad \text{However, } \sec\theta = \frac{\sqrt{h^2+x^2}}{h} \text{ and } \dot{x} = v_B,\ \dot{\theta} = \omega$$

$$\left(\frac{\sqrt{h^2+x^2}}{h}\right)^2\omega = \frac{1}{h}v_B \qquad \omega = \frac{h}{h^2+x^2}v_B \qquad \textbf{Ans}$$

$$\alpha = \dot{\omega} = v_B h\left[-(h^2+x^2)^{-2}(2x\dot{x})\right] \qquad \alpha = \frac{-2xh}{(h^2+x^2)^2}v_B^2 \qquad \textbf{Ans}$$

16-46. The crate is transported on a platform which rests on rollers, each having a radius r. If the rollers do not slip, determine their angular velocity if the platform moves forward with a velocity $\mathbf{v}$.

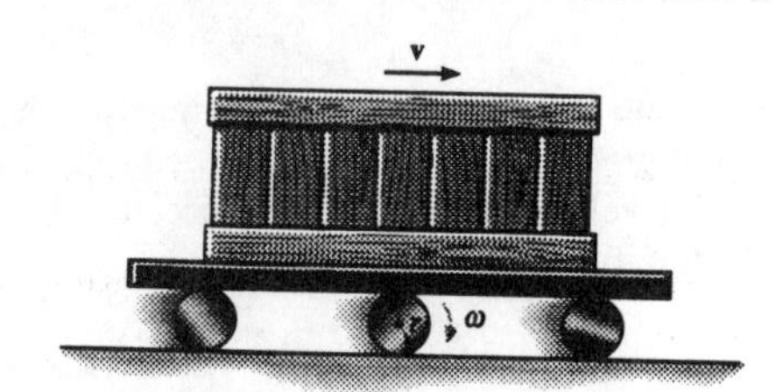

Position coordinate equation : From Example 16 – 3, $s_G = r\theta$. Using similar triangles

$$s_A = 2s_G = 2r\theta$$

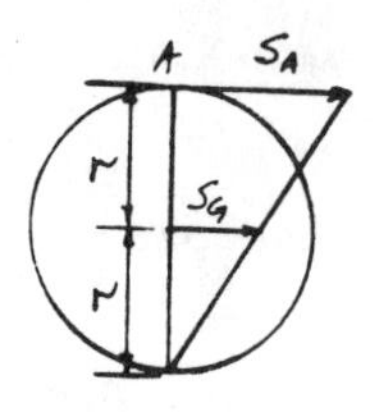

Time derivatives :

$$\dot{s}_A = v = 2r\dot{\theta} \qquad \text{Where } \dot{\theta} = \omega$$

$$\omega = \frac{v}{2r} \qquad \textbf{Ans}$$

16-47. The disk is rotating with an angular velocity of $\boldsymbol{\omega}$ and has an angular acceleration of $\boldsymbol{\alpha}$. Determine the velocity and acceleration of cylinder B. Neglect the size of the pulley at C.

$$s = \sqrt{3^2 + 5^2 - 2(3)(5)\cos\theta}$$

$$v_B = \dot{s} = \frac{1}{2}(34 - 30\cos\theta)^{-\frac{1}{2}}(30\sin\theta)\dot{\theta}$$

$$v_B = \frac{15\omega\sin\theta}{(34 - 30\cos\theta)^{\frac{1}{2}}} \qquad \textbf{Ans}$$

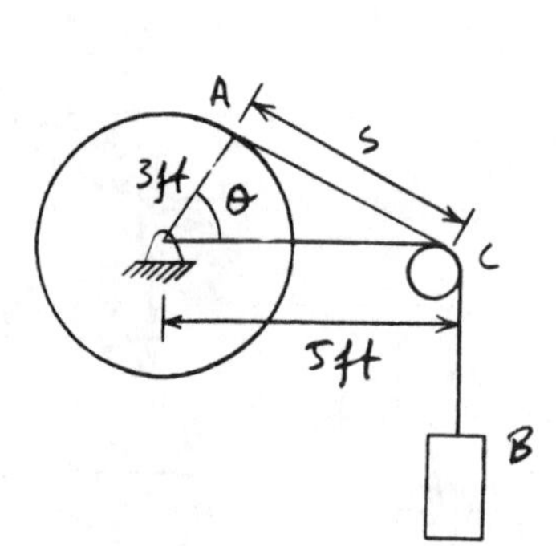

$$a_B = \ddot{s} = \frac{15\omega\cos\theta\dot{\theta} + 15\dot{\omega}\sin\theta}{\sqrt{34 - 30\cos\theta}} + \frac{\left(-\frac{1}{2}\right)(15\omega\sin\theta)\left(30\sin\theta\dot{\theta}\right)}{(34 - 30\cos\theta)^{\frac{3}{2}}}$$

$$= \frac{15(\omega^2\cos\theta + \alpha\sin\theta)}{(34 - 30\cos\theta)^{\frac{1}{2}}} - \frac{225\omega^2\sin^2\theta}{(34 - 30\cos\theta)^{\frac{3}{2}}} \qquad \textbf{Ans}$$

***16-48.** When the bar is at the angle θ, the rod is rotating clockwise at $\boldsymbol{\omega}$ and has an angular acceleration of $\boldsymbol{\alpha}$. Determine the velocity and acceleration of the weight A at this instant. The cord is 20 ft long.

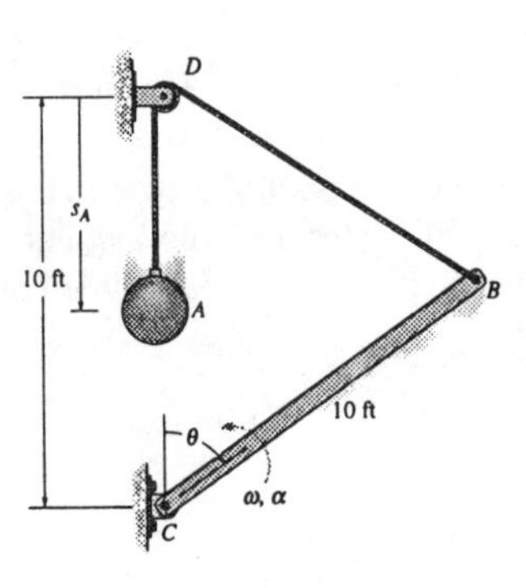

$$s = \sqrt{(10)^2 + (10)^2 - 2(10)(10)\cos\theta}$$

$$s = \sqrt{200(1 - \cos\theta)}$$

$$s_A + \sqrt{200(1 - \cos\theta)} = 20$$

$$\dot{s}_A + 7.071(1 - \cos\theta)^{-\frac{1}{2}}\sin\theta\dot{\theta} = 0$$

$$v_A = \dot{s}_A = \frac{-7.071\omega\sin\theta}{(1 - \cos\theta)^{\frac{1}{2}}} \qquad \textbf{Ans}$$

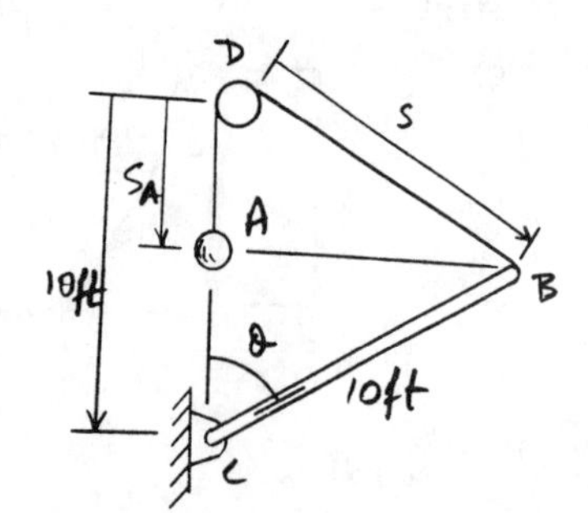

$$\ddot{s}_A = -7.071\left(\dot{\omega}\sin\theta(1 - \cos\theta)^{-\frac{1}{2}} + \omega\cos\theta\dot{\theta}(1 - \cos\theta)^{-\frac{1}{2}} + \omega\sin\theta\left(-\frac{1}{2}\right)(1 - \cos\theta)^{-\frac{3}{2}}(\sin\theta)\dot{\theta}\right)$$

$$a_A = \ddot{s}_A = 7.07\left[\frac{(\omega\sin\theta)^2}{2(1 - \cos\theta)^{\frac{3}{2}}} - \frac{(\alpha\sin\theta + \omega^2\cos\theta)}{(1 - \cos\theta)^{\frac{1}{2}}}\right] \qquad \textbf{Ans}$$

■16-49. The crank AB has a constant angular velocity ω. Determine the velocity and acceleration of the slider at C as a function of θ. *Suggestion:* Use the x coordinate to express the motion of C and the ϕ coordinate for CB. $x = 0$ when $\phi = 0°$.

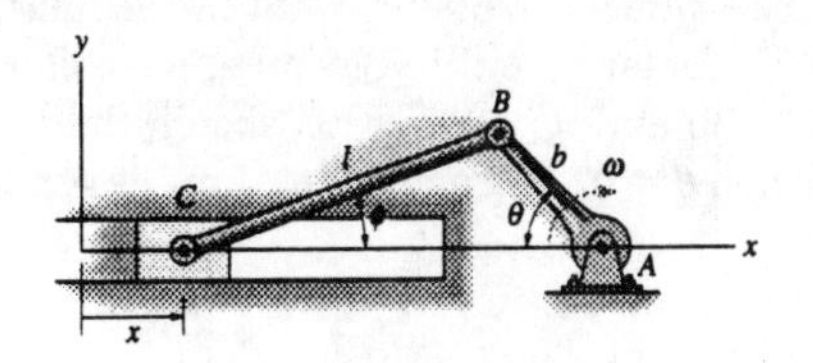

$$x = l + b - (L\cos\phi + b\cos\theta)$$

$$l\sin\phi = b\sin\theta \quad \text{or} \quad \sin\phi = \frac{b}{l}\sin\theta$$

$$v_C = \dot{x} = l\sin\phi\dot{\phi} + b\sin\theta\dot{\theta} \qquad (1)$$

$$\cos\phi\dot{\phi} = \frac{b}{l}\cos\theta\dot{\theta} \qquad (2)$$

Since $\cos\phi = \sqrt{1-\sin^2\phi} = \sqrt{1-\left(\frac{b}{l}\right)^2\sin^2\theta}$

then,

$$\dot{\phi} = \frac{\left(\frac{b}{l}\right)\cos\theta\omega}{\sqrt{1-\left(\frac{b}{l}\right)^2\sin^2\theta}} \qquad (3)$$

$$v_C = b\omega\left[\frac{\left(\frac{b}{l}\right)\sin\theta\cos\theta}{\sqrt{1-\left(\frac{b}{l}\right)^2\sin^2\theta}}\right] + b\omega\sin\theta \qquad \textbf{Ans}$$

From Eq. (1) and (2):

$$a_C = \dot{v}_C = l\ddot{\phi}\sin\phi + l\dot{\phi}\cos\phi\dot{\phi} + b\cos\theta\left(\dot{\theta}\right)^2 \qquad (4)$$

$$-\sin\phi\dot{\phi}^2 + \cos\phi\ddot{\phi} = -\left(\frac{b}{l}\right)\sin\theta\dot{\theta}^2$$

$$\ddot{\phi} = \frac{\dot{\phi}^2\sin\phi - \frac{b}{l}\omega^2\sin\theta}{\cos\phi} \qquad (5)$$

Substituting Eqs. (1), (2), (3) and (5) into Eq. (4) and simplifying yields

$$a_C = b\omega^2\left[\frac{\left(\frac{b}{l}\right)\left(\cos 2\theta + \left(\frac{b}{l}\right)^2\sin^4\theta\right)}{\left(1-\left(\frac{b}{l}\right)^2\sin^2\theta\right)^{\frac{3}{2}}} + \cos\theta\right] \qquad \textbf{Ans}$$

16-50. If h and θ are known, and the speed of A and B is $v_A = v_B = v$, determine the angular velocity ω of the body and the direction ϕ of $\mathbf{v}_B$.

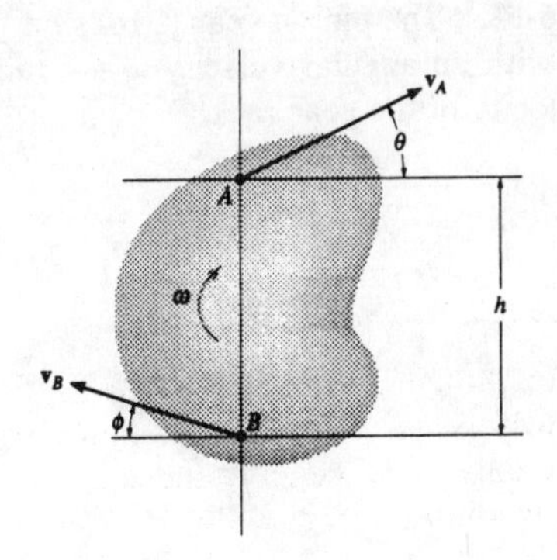

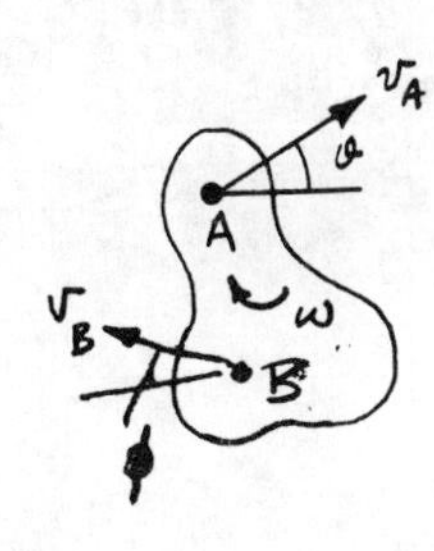

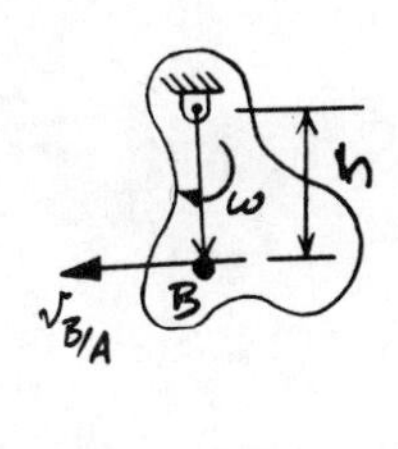

$$\mathbf{v}_B = \mathbf{v}_A + \omega \times \mathbf{r}_{B/A}$$

$$-v\cos\phi\mathbf{i} + v\sin\phi\mathbf{j} = v\cos\theta\mathbf{i} + v\sin\theta\mathbf{j} + (-\omega\mathbf{k})\times(-h\mathbf{j})$$

$(\overset{+}{\rightarrow})$ $$-v\cos\phi = v\cos\theta - \omega h \qquad (1)$$

$(+\uparrow)$ $$v\sin\phi = v\sin\theta \qquad (2)$$

From Eq. (2), $\phi = \theta$ **Ans**

From Eq. (1), $\omega = \dfrac{2v}{h}\cos\theta$ **Ans**

16-51. The wheel is rotating with an angular velocity $\omega = 8$ rad/s. Determine the velocity of the collar A at the instant $\theta = 30°$ and $\phi = 60°$. Also, sketch the location of bar AB when $\theta = 0°, 30°$, and $60°$ to show its general plane motion.

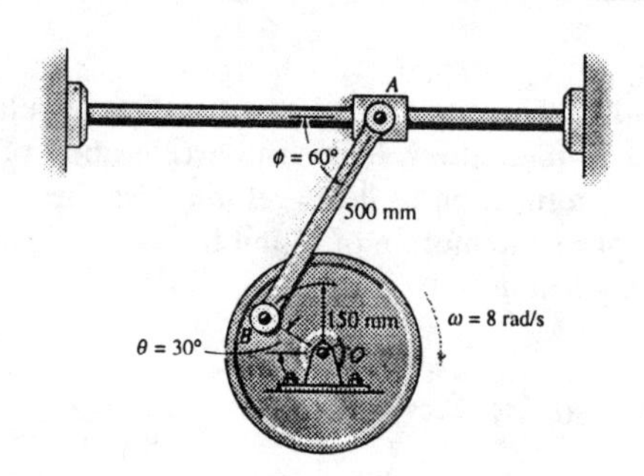

$$\mathbf{v}_A = \mathbf{v}_B + \mathbf{v}_{A/B}$$

$$\underset{\rightarrow}{v_A} = \underset{\angle 60°}{1.2} + \underset{\nwarrow 30°}{0.5\omega_{AB}}$$

$(\overset{+}{\rightarrow})\quad v_A = 1.2\cos 60° + 0.5\omega_{AB}\cos 30°$

$(+\uparrow)\quad 0 = 1.2\sin 60° - 0.5\omega_{AB}\sin 30°$

$\omega_{AB} = 4.16$ rad/s

$v_A = 2.40$ m/s $\rightarrow$ **Ans**

Also, $\mathbf{v}_B = \omega \times \mathbf{r}_B$

$\mathbf{v}_A = \mathbf{v}_B + \omega_{AB} \times \mathbf{r}_{A/B}$

$v_A\mathbf{i} = (-8\mathbf{k})\times(-0.15\cos 30°\mathbf{i} + 0.15\sin 30°\mathbf{j}) + (-\omega_{AB}\mathbf{k})\times(0.5\cos 60°\mathbf{i} + 0.5\sin 60°\mathbf{j})$

$v_A = 0.60 + 0.433\omega_{AB}$

$0 = 1.039 - 0.25\omega_{AB}$

$\omega_{AB} = 4.16$ rad/s

$v_A = 2.40$ m/s $\rightarrow$ **Ans**

***16-52.** The pinion gear A rolls on the fixed gear rack B with an angular velocity $\omega = 4$ rad/s. Determine the velocity of the gear rack C.

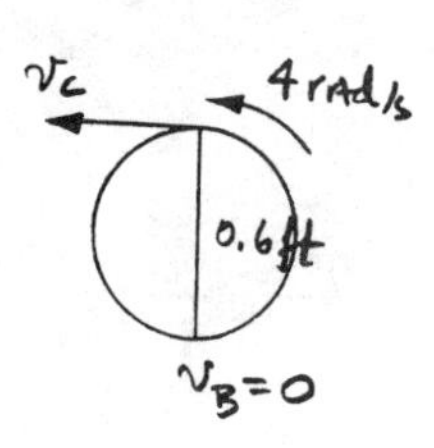

$\mathbf{v}_C = \mathbf{v}_B + \mathbf{v}_{C/B}$

$(\overset{+}{\leftarrow})\quad v_C = 0 + 4(0.6)$

$v_C = 2.40$ ft/s **Ans**

Also :

$\mathbf{v}_C = \mathbf{v}_B + \omega \times \mathbf{r}_{C/B}$

$-v_C\mathbf{i} = 0 + (4\mathbf{k})\times(0.6\mathbf{j})$

$v_C = 2.40$ ft/s **Ans**

16-53. The pinion gear rolls on the gear racks. If B is moving to the right at 8 ft/s and C is moving to the left at 4 ft/s, determine the angular velocity of the pinion gear and the velocity of its center A.

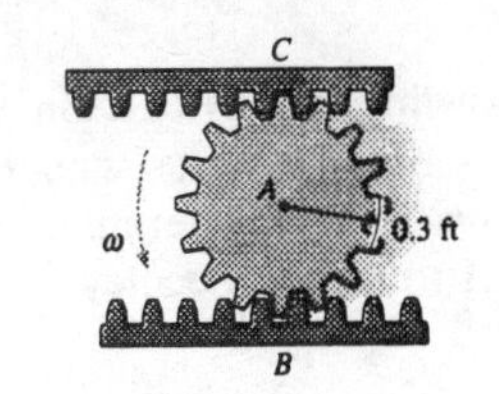

$$\mathbf{v}_C = \mathbf{v}_B + \mathbf{v}_{C/B}$$

$(\overset{+}{\rightarrow})$ $\quad -4 = 8 - 0.6(\omega)$

$\omega = 20$ rad/s **Ans**

$$\mathbf{v}_A = \mathbf{v}_B + \mathbf{v}_{A/B}$$

$(\overset{+}{\rightarrow})$ $\quad v_A = 8 - 20(0.3)$

$v_A = 2$ ft/s $\rightarrow$ **Ans**

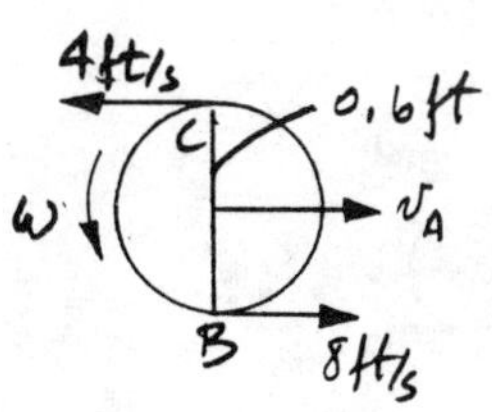

Also :

$$\mathbf{v}_C = \mathbf{v}_B + \omega \times \mathbf{r}_{C/B}$$

$$-4\mathbf{i} = 8\mathbf{i} + (\omega\mathbf{k}) \times (0.6\mathbf{j})$$

$$-4 = 8 - 0.6\omega$$

$\omega = 20$ rad/s **Ans**

$$\mathbf{v}_A = \mathbf{v}_B + \omega \times \mathbf{r}_{A/B}$$

$$v_A\mathbf{i} = 8\mathbf{i} + 20\mathbf{k} \times (0.3\mathbf{j})$$

$v_A = 2$ ft/s $\rightarrow$ **Ans**

16-54. The gear rests in a fixed horizontal rack. A cord is wrapped around the inner core of the gear so that it remains horizontally tangent to the inner core at A. If the cord is pulled to the right with a constant velocity of 2 ft/s, determine the velocity of the center of the gear, C.

$$\mathbf{v}_A = \mathbf{v}_D + \mathbf{v}_{A/D}$$

$$\left[\underset{\rightarrow}{2}\right] = 0 + \left[\underset{\rightarrow}{\omega(1.5)}\right]$$

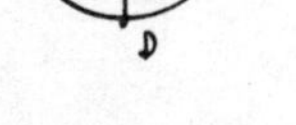

$(\overset{+}{\rightarrow})$ $\quad 2 = 1.5\omega \qquad \omega = 1.33$ rad/s

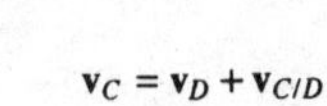

$$\mathbf{v}_C = \mathbf{v}_D + \mathbf{v}_{C/D}$$

$$\left[\underset{\rightarrow}{v_C}\right] = 0 + \left[\underset{\rightarrow}{1.33(1)}\right]$$

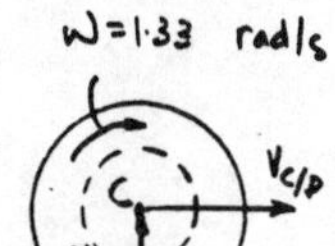

$(\overset{+}{\rightarrow})$ $\quad v_C = 1.33$ ft/s $\rightarrow$ **Ans**

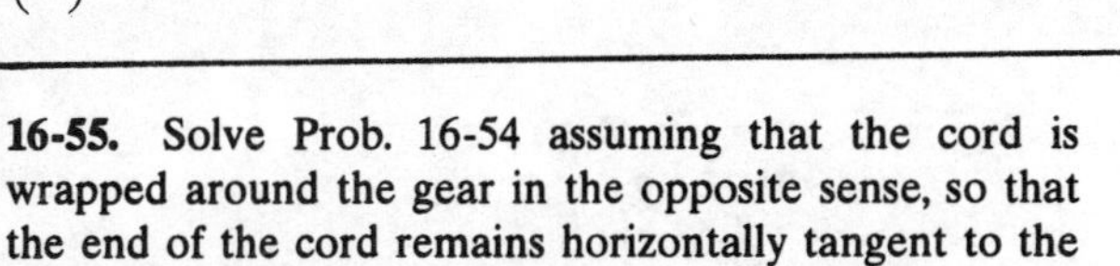

16-55. Solve Prob. 16-54 assuming that the cord is wrapped around the gear in the opposite sense, so that the end of the cord remains horizontally tangent to the inner core at B and is pulled to the right at 2 ft/s.

$$\mathbf{v}_B = \mathbf{v}_D + \mathbf{v}_{B/D}$$

$$\left[\underset{\rightarrow}{2}\right] = 0 + \left[\underset{\rightarrow}{\omega(0.5)}\right]$$

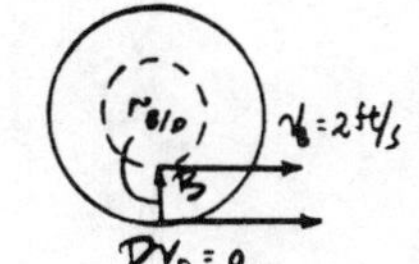

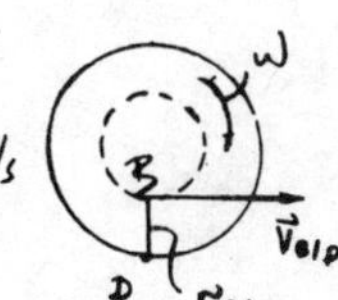

$(\overset{+}{\rightarrow})$ $\quad 2 = 0.5\omega \qquad \omega = 4$ rad/s

$$\mathbf{v}_C = \mathbf{v}_D + \mathbf{v}_{C/D}$$

$$\left[\underset{\rightarrow}{v_C}\right] = 0 + \left[\underset{\rightarrow}{4(1)}\right]$$

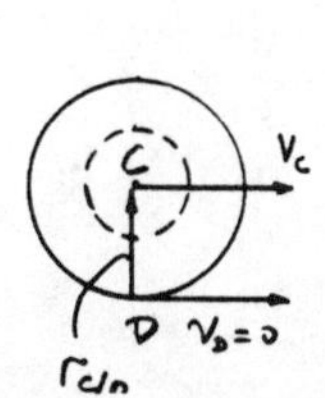

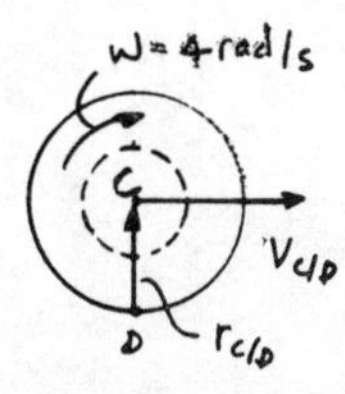

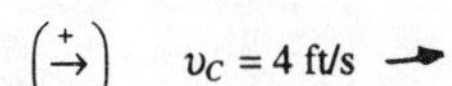

$(\overset{+}{\rightarrow})$ $\quad v_C = 4$ ft/s $\rightarrow$ **Ans**

***16-56.** A bowling ball is cast on the "alley" with a backspin of $\omega = 10$ rad/s while its center O has a forward velocity of $v_O = 8$ m/s. Determine the velocity of the contact point A touching the alley.

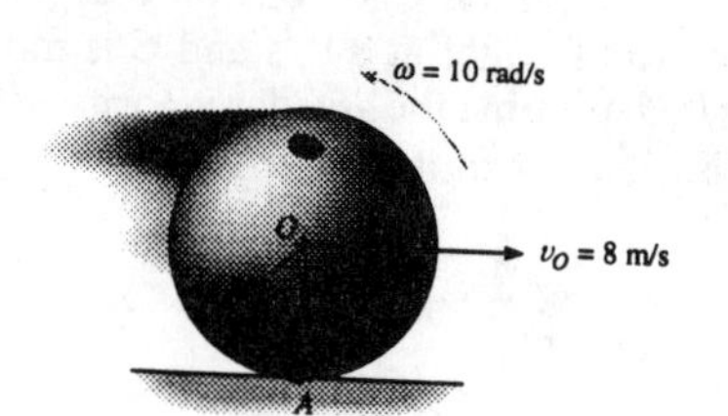

$$\mathbf{v}_A = \mathbf{v}_O + \mathbf{v}_{A/O}$$

$(\overset{+}{\rightarrow})$ $\quad v_A = 8 + 10(0.12)$

$v_A = 9.20$ m/s $\rightarrow$ **Ans**

10 rad/s
O → V_o = 8 m/s
A → V_A

Also,

$$\mathbf{v}_A = \mathbf{v}_O + \omega \times \mathbf{r}_{A/O}$$

$$v_A\mathbf{i} = 8\mathbf{i} + (10\mathbf{k}) \times (-0.12\mathbf{j})$$

$(\overset{+}{\rightarrow})$ $\quad v_A = 9.20$ m/s $\rightarrow$ **Ans**

16-57. Rod AB is rotating with an angular velocity $\omega_{AB} = 5$ rad/s. Determine the velocity of the collar C at the instant $\theta = 60°$ and $\phi = 45°$. Also, sketch the location of bar BC when $\theta = 30°$, 60° and 45° to show its general plane motion.

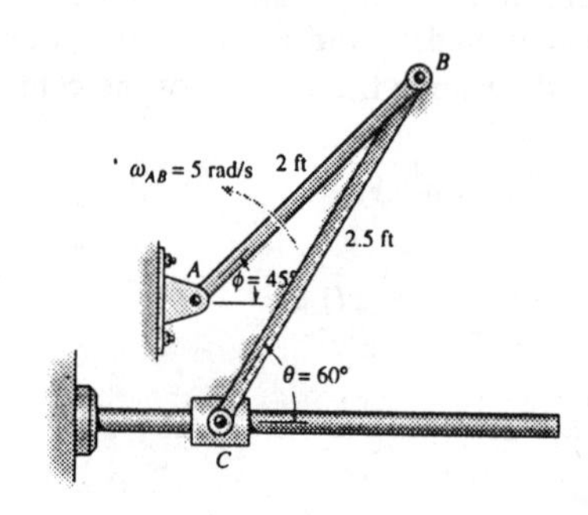

$$\mathbf{v}_C = \mathbf{v}_B + \mathbf{v}_{C/B}$$

$$\begin{bmatrix} v_C \\ \rightarrow \end{bmatrix} = \begin{bmatrix} 10 \\ \nwarrow 45° \end{bmatrix} + \begin{bmatrix} \omega_{CB}(2.5) \\ \nwarrow 30° \end{bmatrix}$$

$(\overset{+}{\rightarrow})$ $\quad v_C = -10\cos 45° + \omega_{CB}(2.5)(\cos 30°)$

$(+\uparrow)$ $\quad 0 = 10\sin 45° - \omega_{CB}(2.5)(\sin 30°)$

$v_C = 5.18$ ft/s $\rightarrow$ **Ans**

$\omega_{CB} = 5.66$ rad/s

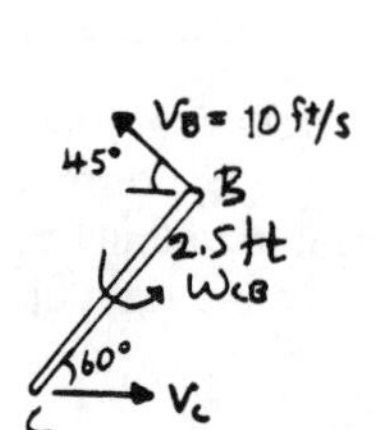

Also,

$$\mathbf{v}_B = \omega_{AB} \times \mathbf{r}_{B/A}$$

$$\mathbf{v}_C = \mathbf{v}_B + \omega_{CB} \times \mathbf{r}_{C/B}$$

$$v_C\mathbf{i} = (5\mathbf{k}) \times (2\cos 45°\mathbf{i} + 2\sin 45°\mathbf{j}) + (\omega_{CB}\mathbf{k}) \times (-2.5\cos 60°\mathbf{i} - 2.5\sin 60°\mathbf{j})$$

$(\overset{+}{\rightarrow})$ $\quad v_C = -7.07 + 2.17\omega_{CB}$

$(+\uparrow)$ $\quad 0 = 7.07 - 1.25\omega_{CB}$

$\omega_{CB} = 5.66$ rad/s

$v_C = 5.18$ ft/s $\rightarrow$ **Ans**

16-58. The angular velocity of link AB is $\omega_{AB} = 4$ rad/s. Determine the velocity of the collar at C and the angular velocity of link CB at the instant $\theta = 60°$ and $\phi = 45°$. Link CB is horizontal at this instant. Also, sketch the location of link CB when $\theta = 30°$, $60°$, and $90°$ to show its general plane motion.

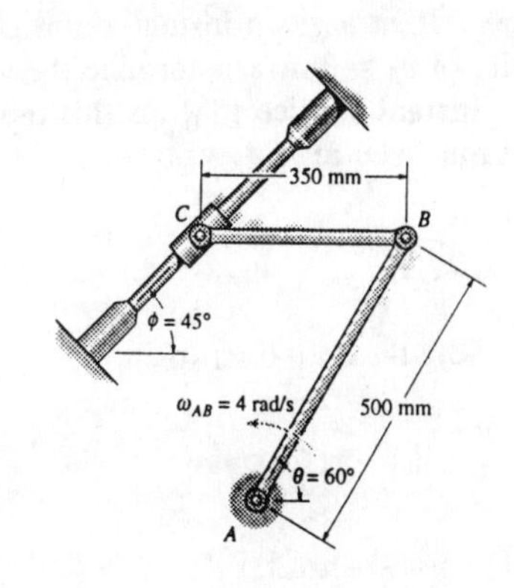

For link AB : Link AB rotates about the fixed point A. Hence

$$v_B = \omega_{AB} r_{AB}$$

$$= 4(0.5) = 2 \text{ m/s}$$

For link CB

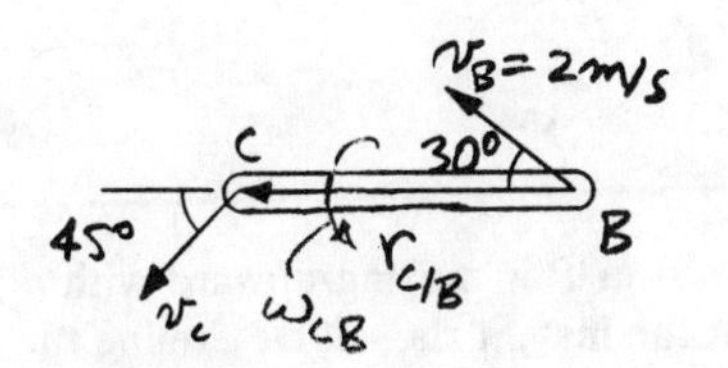

$\mathbf{v}_B = \{-2\cos 30°\mathbf{i} + 2\sin 30°\mathbf{j}\}$ m/s $\qquad \mathbf{v}_C = -v_C\cos 45°\mathbf{i} - v_C\sin 45°\mathbf{j}$

$\omega = \omega_{CB}\mathbf{k} \qquad \mathbf{r}_{C/B} = \{-0.35\mathbf{i}\}$ m

$$\mathbf{v}_C = \mathbf{v}_B + \omega \times \mathbf{r}_{C/B}$$

$$-v_C\cos 45°\mathbf{i} - v_C\sin 45°\mathbf{j} = (-2\cos 30°\mathbf{i} + 2\sin 30°\mathbf{j}) + (\omega_{CB}\mathbf{k}) \times (-0.35\mathbf{i})$$

$$-v_C\cos 45°\mathbf{i} - v_C\sin 45°\mathbf{j} = -2\cos 30°\mathbf{i} + (2\sin 30° - 0.35\omega_{CB})\mathbf{j}$$

Equating the **i** and **j** components yields :

$-v_C\cos 45° = -2\cos 30° \qquad v_C = 2.45$ m/s ↙ **Ans**

$-2.45\sin 45° = 2\sin 30° - 0.35\omega_{CB} \qquad \omega_{CB} = 7.81$ rad/s ↻ **Ans**

16-59. The link AB has a clockwise angular velocity of 2 rad/s. Determine the velocity of block C at the instant $\theta = 45°$. Also, sketch the location of link BC when $\theta = 60°$, $45°$, and $30°$ to show its general plane motion.

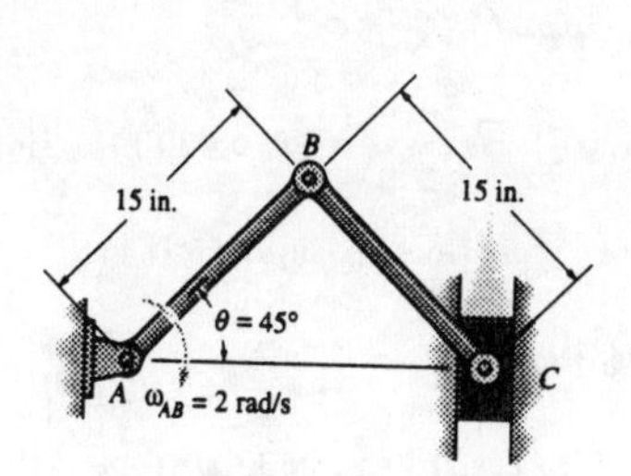

For link AB : Link AB rotates about the fixed point A. Hence

$$v_B = \omega_{AB} r_{AB}$$

$$= 2\left(\frac{15}{12}\right) = 2.5 \text{ ft/s}$$

For link BC

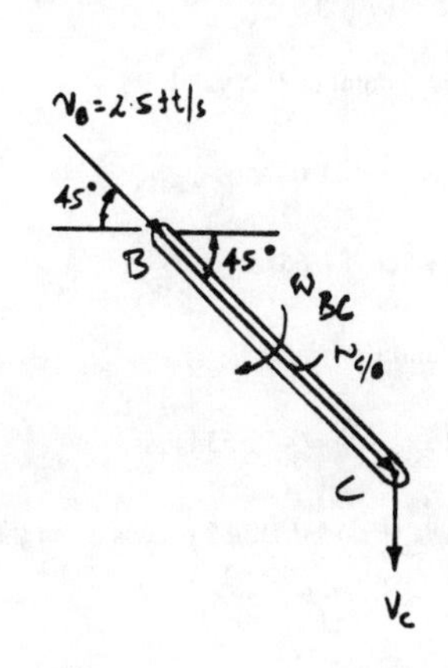

$\mathbf{v}_B = \{2.5\cos 45°\mathbf{i} - 2.5\sin 45°\mathbf{j}\}$ ft/s $\qquad \mathbf{v}_C = -v_C\mathbf{j} \qquad \omega = -\omega_{BC}\mathbf{k}$

$\mathbf{r}_{C/B} = \{1.25\cos 45°\mathbf{i} - 1.25\sin 45°\mathbf{j}\}$ ft

$$\mathbf{v}_C = \mathbf{v}_B + \omega \times \mathbf{r}_{C/B}$$

$$-v_C\mathbf{j} = (2.5\cos 45°\mathbf{i} - 2.5\sin 45°\mathbf{j}) + (-\omega_{BC}\mathbf{k}) \times (1.25\cos 45°\mathbf{i} - 1.25\sin 45°\mathbf{j})$$

$$-v_C\mathbf{j} = (2.5\cos 45° - 1.25\sin 45°\omega_{BC})\mathbf{i} - (2.5\sin 45° + 1.25\cos 45°\omega_{BC})\mathbf{j}$$

Equating the **i** and **j** components yields :

$0 = 2.5\cos 45° - 1.25\sin 45°\omega_{BC} \qquad \omega_{BC} = 2$ rad/s

$-v_C = -[2.5\sin 45° + 1.25\cos 45°(2)] \qquad v_C = 3.54$ ft/s ↓ **Ans**

***16-60.** If, at a given instant, point B has a downward velocity of $v_B = 3$ m/s, determine the velocity of point A at this instant. Notice that for this motion to occur, the wheel must slip at A.

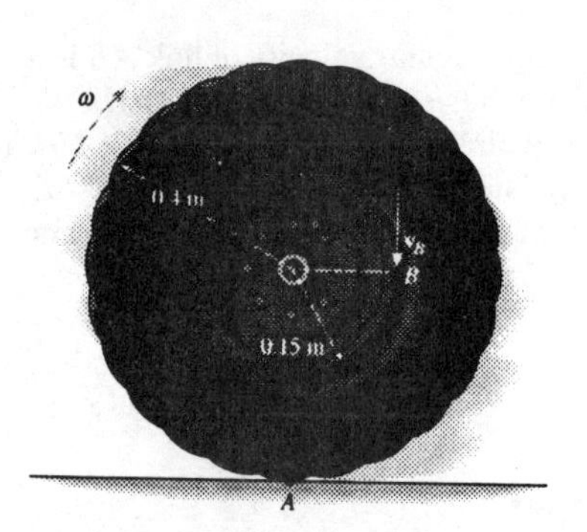

$\mathbf{v}_A = \mathbf{v}_B + \boldsymbol{\omega} \times \mathbf{r}_{A/B}$

$-v_A\mathbf{i} = -3\mathbf{j} + (-\omega\mathbf{k}) \times (-0.15\mathbf{i} - 0.4\mathbf{j})$

$(\overset{+}{\rightarrow}) \qquad -v_A = -\omega(0.4)$

$(+\uparrow) \qquad 0 = -3 + \omega(0.15)$

$\omega = 20$ rad/s

$v_A = 8$ m/s $\leftarrow$ **Ans**

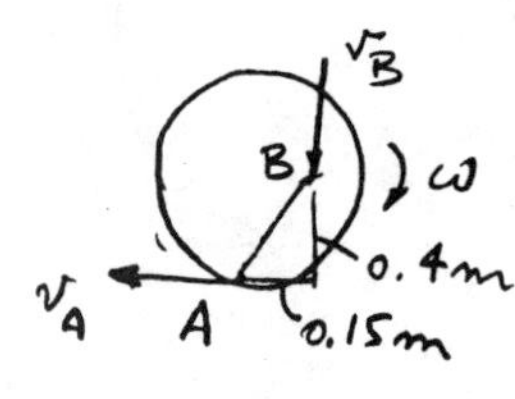

16-61. The piston P is moving upward with a velocity of 300 in./s at the instant shown. Determine the angular velocity of the crankshaft AB at this instant.

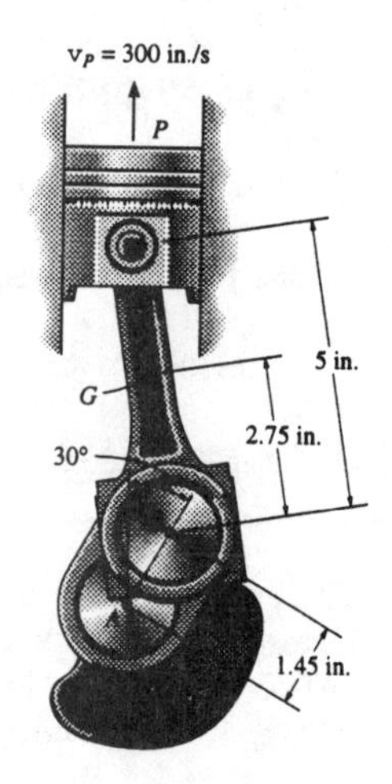

From the geometry :

$\cos\theta = \dfrac{1.45\sin 30^\circ}{5} \qquad \theta = 81.66^\circ$

For link BP

$\mathbf{v}_P = \{300\mathbf{j}\}$ in/s $\qquad \mathbf{v}_B = -v_B\cos 30^\circ\mathbf{i} + v_B\sin 30^\circ\mathbf{j} \qquad \boldsymbol{\omega} = -\omega_{BP}\mathbf{k}$

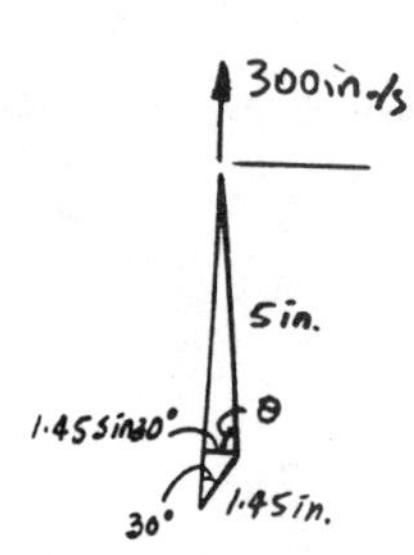

$\mathbf{r}_{P/B} = \{-5\cos 81.66^\circ\mathbf{i} + 5\sin 81.66^\circ\mathbf{j}\}$ in.

$\mathbf{v}_P = \mathbf{v}_B + \boldsymbol{\omega} \times \mathbf{r}_{P/B}$

$300\mathbf{j} = (-v_B\cos 30^\circ\mathbf{i} + v_B\sin 30^\circ\mathbf{j}) + (-\omega_{BP}\mathbf{k}) \times (-5\cos 81.66^\circ\mathbf{i} + 5\sin 81.66^\circ\mathbf{j})$

$300\mathbf{j} = (-v_B\cos 30^\circ + 5\sin 81.66^\circ\omega_{BP})\mathbf{i} + (v_B\sin 30^\circ + 5\cos 81.66^\circ\omega_{BP})\mathbf{j}$

Equating the $\mathbf{i}$ and $\mathbf{j}$ components yields :

$0 = -v_B\cos 30^\circ + 5\sin 81.66^\circ\omega_{BP} \qquad (1)$

$300 = v_B\sin 30^\circ + 5\cos 81.66^\circ\omega_{BP} \qquad (2)$

Solving Eqs. (1) and (2) yields :

$\omega_{BP} = 83.77$ rad/s $\qquad v_B = 478.53$ in./s

For crankshaft AB : Crankshaft AB rotates about the fixed point A. Hence

$v_B = \omega_{AB} r_{AB}$

$478.53 = \omega_{AB}(1.45) \qquad \omega_{AB} = 330$ rad/s ↻ **Ans**

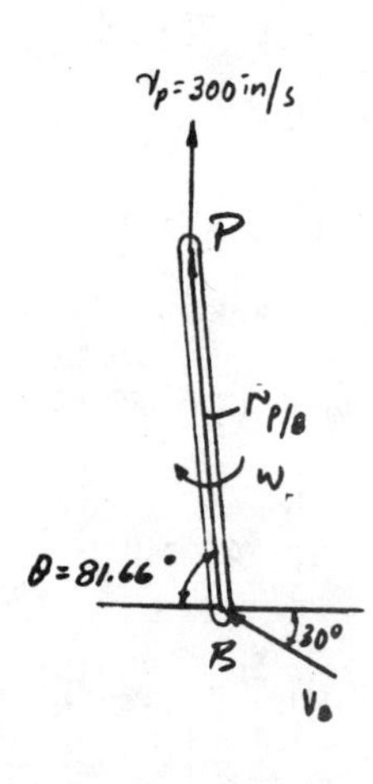

16-62. Determine the velocity of the center of gravity G of the connecting rod at the instant shown. The piston is moving upward with a velocity of 300 in./s.

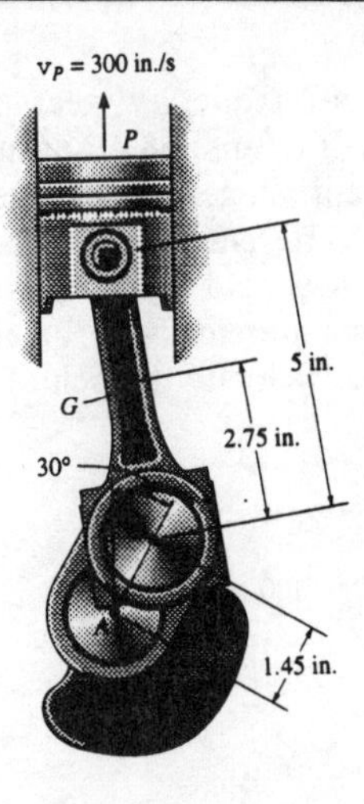

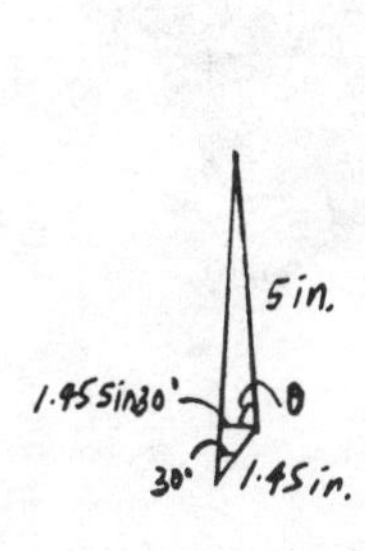

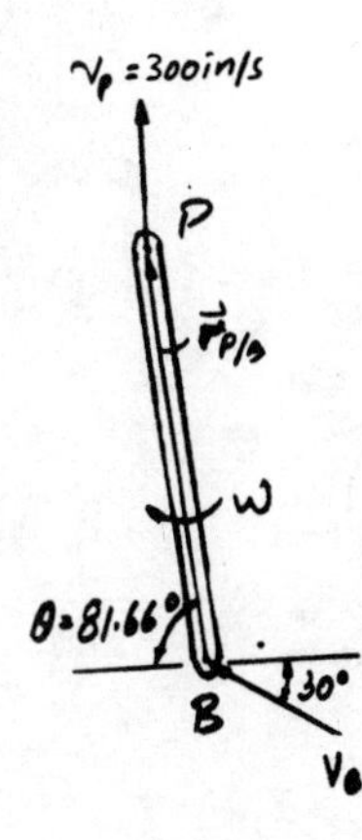

From the geometry :

$$\cos\theta = \frac{1.45\sin 30^\circ}{5} \qquad \theta = 81.66^\circ$$

For link BP

$$\mathbf{v}_P = \{300\mathbf{j}\}\ \text{in/s} \qquad \mathbf{v}_B = -v_B\cos 30^\circ\mathbf{i} + v_B\sin 30^\circ\mathbf{j} \qquad \omega = -\omega_{BP}\mathbf{k}$$

$$\mathbf{r}_{P/B} = \{-5\cos 81.66^\circ\mathbf{i} + 5\sin 81.66^\circ\mathbf{j}\}\ \text{in.}$$

$$\mathbf{v}_P = \mathbf{v}_B + \omega \times \mathbf{r}_{P/B}$$

$$300\mathbf{j} = (-v_B\cos 30^\circ\mathbf{i} + v_B\sin 30^\circ\mathbf{j}) + (-\omega_{BP}\mathbf{k}) \times (-5\cos 81.66^\circ\mathbf{i} + 5\sin 81.66^\circ\mathbf{j})$$

$$300\mathbf{j} = (-v_B\cos 30^\circ + 5\sin 81.66^\circ\,\omega_{BP})\mathbf{i} + (v_B\sin 30^\circ + 5\cos 81.66^\circ\,\omega_{BP})\mathbf{j}$$

Equating the **i** and **j** components yields :

$$0 = -v_B\cos 30^\circ + 5\sin 81.66^\circ\,\omega_{BP} \qquad (1)$$

$$300 = v_B\sin 30^\circ + 5\cos 81.66^\circ\,\omega_{BP} \qquad (2)$$

Solving Eqs. (1) and (2) yields :

$$\omega_{BP} = 83.77\ \text{rad/s} \qquad v_B = 478.53\ \text{in./s}$$

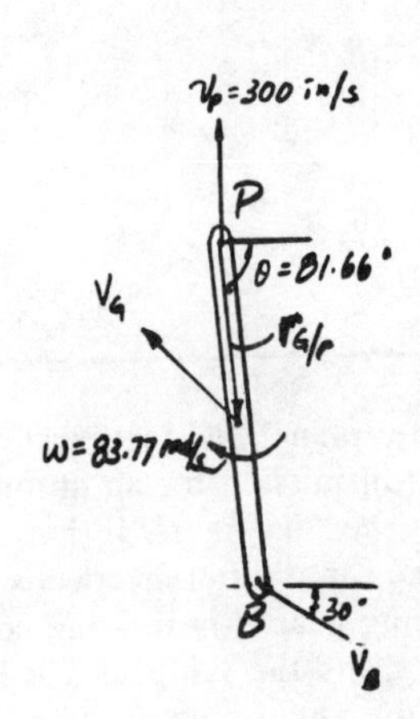

$$\mathbf{v}_P = \{300\mathbf{j}\}\ \text{in/s} \qquad \omega = \{-83.77\mathbf{k}\}\ \text{rad/s}$$

$$\mathbf{r}_{G/P} = \{2.25\cos 81.66^\circ\mathbf{i} - 2.25\sin 81.66^\circ\mathbf{j}\}\ \text{in.}$$

$$\mathbf{v}_G = \mathbf{v}_P + \omega \times \mathbf{r}_{G/P}$$

$$= 300\mathbf{j} + (-83.77\mathbf{k}) \times (2.25\cos 81.66^\circ\mathbf{i} - 2.25\sin 81.66^\circ\mathbf{j})$$

$$= \{-186.49\mathbf{i} + 272.67\mathbf{j}\}\ \text{in./s}$$

$$v_G = \sqrt{(-186.49)^2 + 272.67^2} = 330\ \text{in./s} \qquad \textbf{Ans}$$

$$\theta = \tan^{-1}\left(\frac{272.67}{186.49}\right) = 55.6^\circ \qquad \textbf{Ans}$$

16-63. The planetary gear system is used in an automatic transmission for an automobile. By locking or releasing certain gears, it has the advantage of operating the car at different speeds. Consider the case where the ring gear R is held fixed, $\omega_R = 0$, and the sun gear S is rotating counterclockwise at $\omega_S = 5$ rad/s. Determine the angular velocity of each of the planet gears P and shaft A.

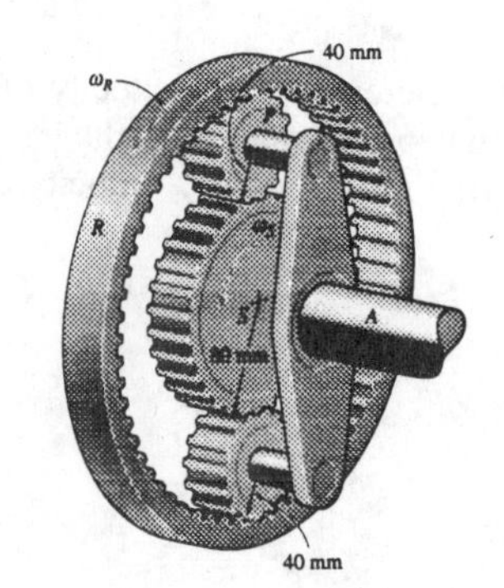

$v_A = 5(80) = 400$ mm/s $\leftarrow$

$v_B = 0$

$\mathbf{v}_B = \mathbf{v}_A + \omega \times \mathbf{r}_{B/A}$

$0 = -400\mathbf{i} + (\omega_p \mathbf{k}) \times (80\mathbf{j})$

$0 = -400\mathbf{i} - 80\omega_p$

$\omega_P = -5$ rad/s $= 5$ rad/s **Ans**

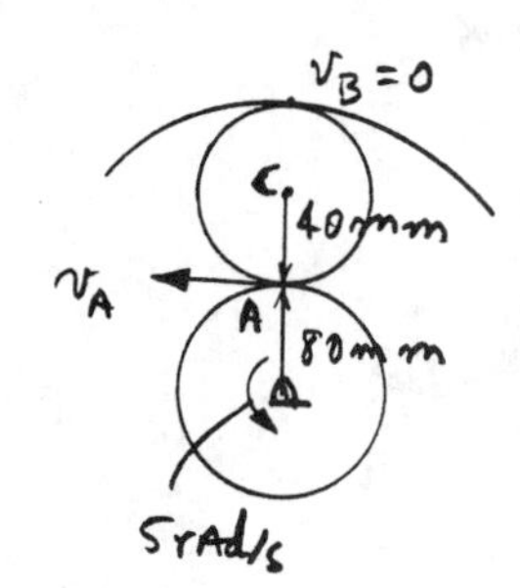

$\mathbf{v}_C = \mathbf{v}_B + \omega \times \mathbf{r}_{C/B}$

$\mathbf{v}_C = 0 + (-5\mathbf{k}) \times (-40\mathbf{j}) = -200\mathbf{i}$

$\omega_A = \dfrac{200}{120} = 1.67$ rad/s **Ans**

***16-64.** The planetary gear system is used in an automatic transmission for an automobile. By locking or releasing certain gears, it has the advantage of operating the car at different speeds. Consider the case where the ring gear R is rotating counterclockwise at $\omega_R = 3$ rad/s, and the sun gear S is held fixed, $\omega_S = 0$. Determine the angular velocity of each of the planet gears P and shaft A.

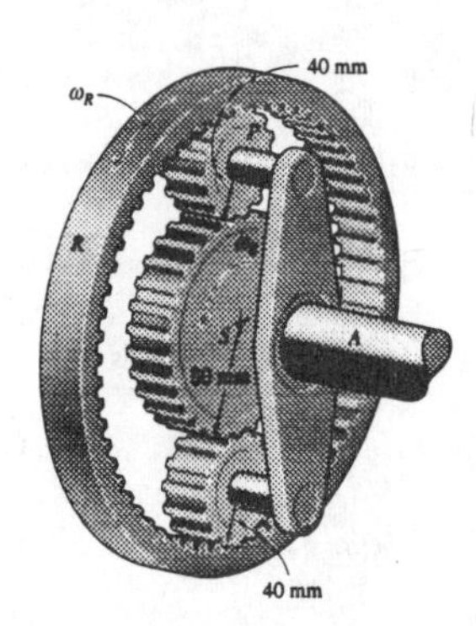

$v_A = 3(160) = 480$ mm/s

$\mathbf{v}_B = \mathbf{v}_A + \omega \times \mathbf{r}_{B/A}$

$0 = -480\mathbf{i} + (\omega_P \mathbf{k}) \times (-80\mathbf{j})$

$0 = -480\mathbf{i} + 80\omega_P$

$\omega_p = 6$ rad/s ↺ **Ans**

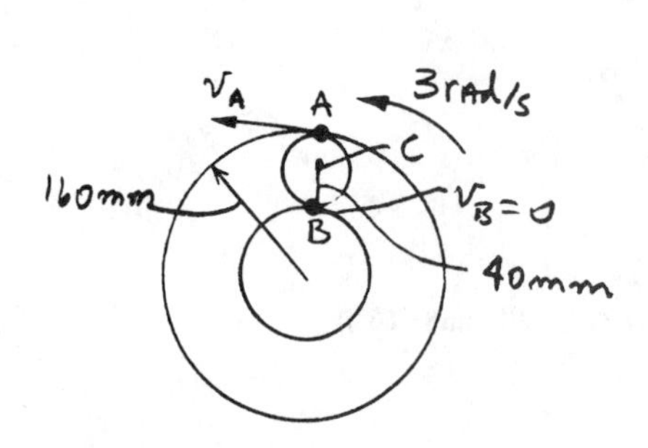

$\mathbf{v}_C = \mathbf{v}_B + \omega \times \mathbf{r}_{C/B}$

$\mathbf{v}_C = 0 + (6\mathbf{k}) \times (40\mathbf{j}) = 240\mathbf{i}$

$\omega_A = \dfrac{240}{120} = 2$ rad/s ↺ **Ans**

16-65. If bar AB has an angular velocity $\omega_{AB} = 6$ rad/s, determine the velocity of the slider block C at the instant $\theta = 45°$ and $\phi = 30°$. Also, sketch the location of bar BC when $\theta = 30°, 45°$, and $60°$ to show its general plane motion.

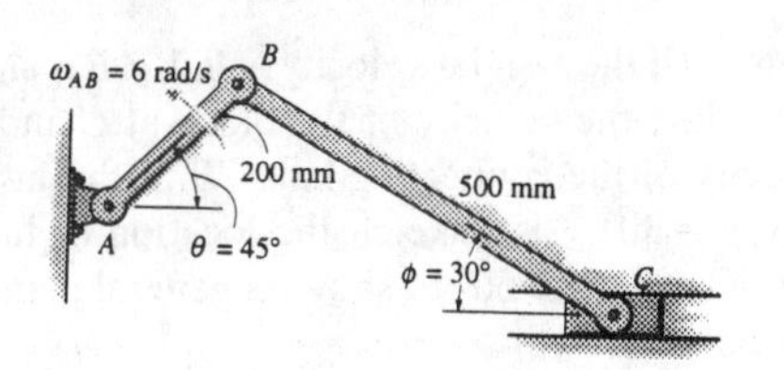

$\mathbf{v}_C = \mathbf{v}_B + \mathbf{v}_{C/B}$

$$\begin{bmatrix} v_C \\ \leftarrow \end{bmatrix} = \begin{bmatrix} 1.2 \\ \nwarrow 45° \end{bmatrix} + \begin{bmatrix} 0.5\omega \\ \swarrow 30° \end{bmatrix}$$

$(\overset{+}{\rightarrow})$ $\quad -v_C = -1.2\cos 45° - 0.5\omega\sin 30°$

$(+\uparrow)$ $\quad 0 = 1.2\sin 45° - 0.5\omega\cos 30°$

$\omega = 1.96$ rad/s

$v_C = 1.34$ m/s $\leftarrow$ **Ans**

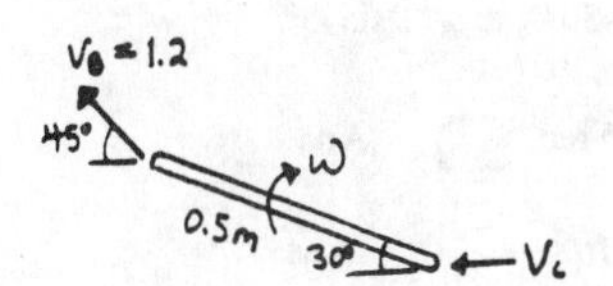

Also,

$\mathbf{v}_B = \omega_{AB} \times \mathbf{r}_{B/A}$

$\mathbf{v}_C = \mathbf{v}_B + \omega \times \mathbf{r}_{C/B}$

$v_C\mathbf{i} = (6\mathbf{k}) \times (0.2\cos 45°\mathbf{i} + 0.2\sin 45°\mathbf{j}) + (\omega\mathbf{k}) \times (0.5\cos 30°\mathbf{i} - 0.5\sin 30\mathbf{j})$

$(\overset{+}{\rightarrow})$ $\quad v_C = -0.8485 + \omega(0.25)$

$(+\uparrow)$ $\quad 0 = 0.8485 + 0.433\omega$

$\omega = -1.96$ rad/s

$v_C = 1.34$ m/s $\leftarrow$ **Ans**

16-66. The bicycle has a velocity $v = 4$ ft/s, and at the same instant the rear wheel has a clockwise angular velocity $\omega = 3$ rad/s, which causes it to slip at its contact point A. Determine the velocity of point A.

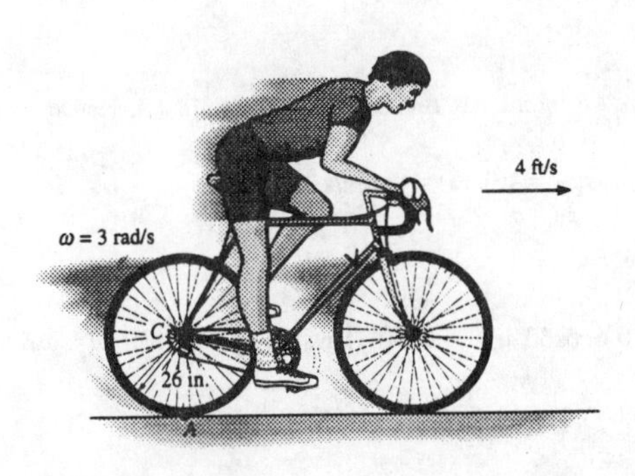

$\mathbf{v}_A = \mathbf{v}_C + \mathbf{v}_{A/C}$

$$\begin{bmatrix} v_A \\ \leftarrow \end{bmatrix} = \begin{bmatrix} 4 \\ \rightarrow \end{bmatrix} + \begin{bmatrix} \left(\frac{26}{12}\right)(3) \\ \leftarrow \end{bmatrix}$$

$v_A = 2.5$ ft/s **Ans**

Also,

$\mathbf{v}_A = \mathbf{v}_C + \omega \times \mathbf{r}_{A/C}$

$\mathbf{v}_A = 4\mathbf{i} + (-3\mathbf{k}) \times \left(-\frac{26}{12}\mathbf{j}\right)$

$\mathbf{v}_A = 4\mathbf{i} - 6.5\mathbf{i} = -2.5\mathbf{i}$

$v_A = 2.5$ ft/s $\leftarrow$ **Ans**

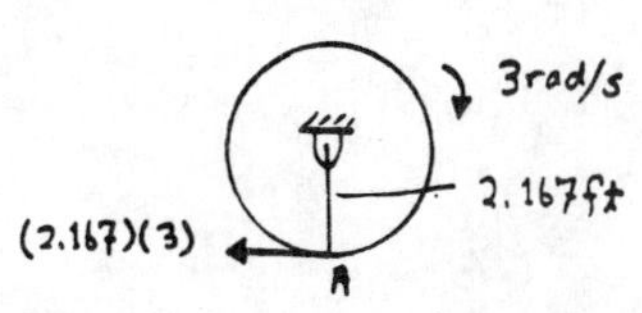

16-67. If the angular velocity of link AB is $\omega_{AB} = 3\text{ rad/s}$, determine the velocity of the block at C and the angular velocity of the connecting link CB at the instant $\theta = 45°$ and $\phi = 30°$. Also, sketch the location of link BC when $\theta = 30°, 45°$, and $60°$ to show its general plane motion.

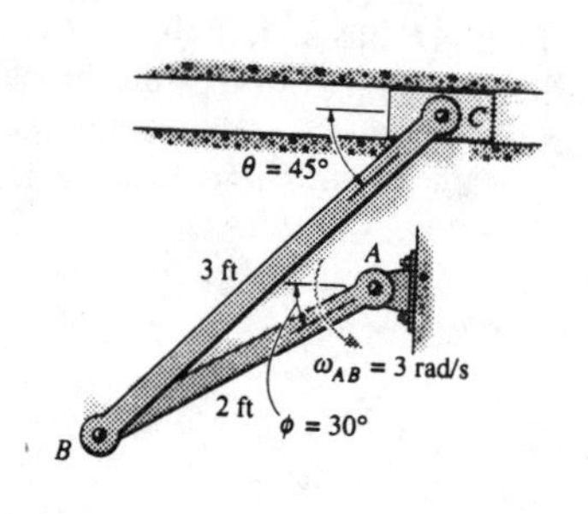

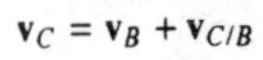

$\mathbf{v}_C = \mathbf{v}_B + \mathbf{v}_{C/B}$

$$\begin{bmatrix} v_C \\ \leftarrow \end{bmatrix} = \begin{bmatrix} 6 \\ \searrow 30° \end{bmatrix} + \begin{bmatrix} \omega_{CB}(3) \\ \nwarrow 45° \end{bmatrix}$$

$(\overset{+}{\rightarrow})\quad -v_C = 6\sin 30° - \omega_{CB}(3)\cos 45°$

$(+\uparrow)\quad 0 = -6\cos 30° + \omega_{CB}(3)\sin 45°$

$\omega_{CB} = 2.45\text{ rad/s}$ ↺ **Ans**

$v_C = 2.20\text{ ft/s} \leftarrow$ **Ans**

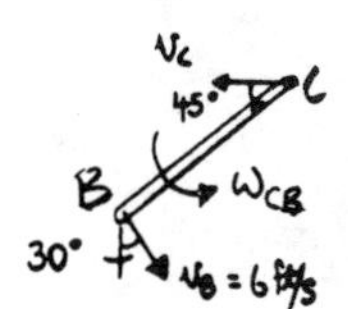

Also,

$\mathbf{v}_C = \mathbf{v}_B + \boldsymbol{\omega} \times \mathbf{r}_{C/B}$

$-v_C\mathbf{i} = (6\sin 30°\mathbf{i} - 6\cos 30°\mathbf{j}) + (\omega_{CB}\mathbf{k}) \times (3\cos 45°\mathbf{i} + 3\sin 45°\mathbf{j})$

$(\overset{+}{\rightarrow})\quad -v_C = 3 - 2.12\omega_{CB}$

$(+\uparrow)\quad 0 = -5.196 + 2.12\omega_{CB}$

$\omega_{CB} = 2.45\text{ rad/s}$ ↺ **Ans**

$v_C = 2.20\text{ ft/s} \leftarrow$ **Ans**

***16-68.** If bar AB has an angular velocity $\omega_{AB} = 4\text{ rad/s}$, determine the velocity of the slider block C at the instant shown.

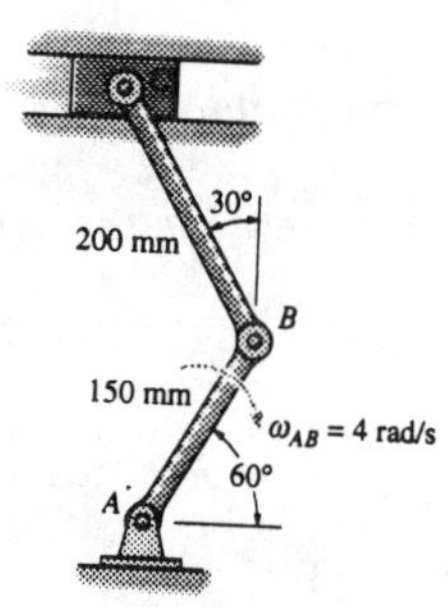

For link AB: Link AB rotates about a fixed point A. Hence

$v_B = \omega_{AB} r_{AB} = 4(0.15) = 0.6\text{ m/s}$

For link BC

$\mathbf{v}_B = \{0.6\cos 30°\mathbf{i} - 0.6\sin 30°\mathbf{j}\}\text{ m/s} \qquad \mathbf{v}_C = v_C\mathbf{i} \qquad \boldsymbol{\omega} = \omega_{BC}\mathbf{k}$

$\mathbf{r}_{C/B} = \{-0.2\sin 30°\mathbf{i} + 0.2\cos 30°\mathbf{j}\}\text{ m}$

$\mathbf{v}_C = \mathbf{v}_B + \boldsymbol{\omega} \times \mathbf{r}_{C/B}$

$v_C\mathbf{i} = (0.6\cos 30°\mathbf{i} - 0.6\sin 30°\mathbf{j}) + (\omega_{BC}\mathbf{k}) \times (-0.2\sin 30°\mathbf{i} + 0.2\cos 30°\mathbf{j})$

$v_C\mathbf{i} = (0.5196 - 0.1732\omega_{BC})\mathbf{i} - (0.3 + 0.1\omega_{BC})\mathbf{j}$

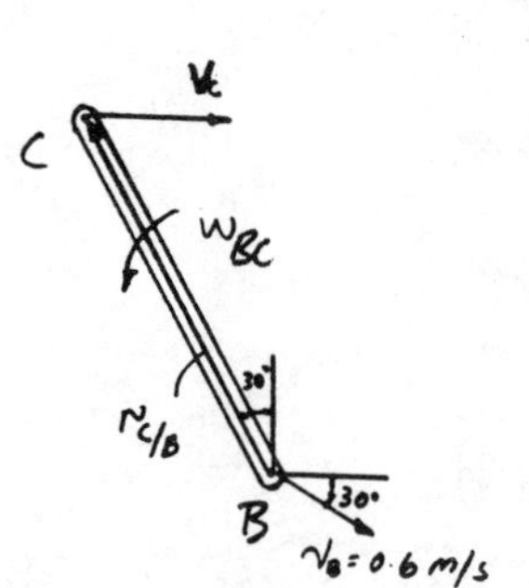

Equating the **i** and **j** components yields :

$0 = 0.3 + 0.1\omega_{BC} \qquad \omega_{BC} = -3\text{ rad/s}$

$v_C = 0.5196 - 0.1732(-3) = 1.04\text{ m/s} \rightarrow$ **Ans**

16-69. At the instant shown, the truck is traveling to the right at 3 m/s, while the pipe is rolling counterclockwise at $\omega = 8$ rad/s without slipping at B. Determine the velocity of the pipe's center G.

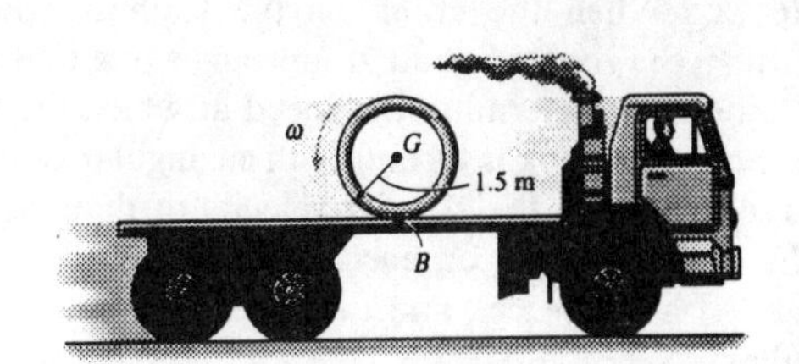

$$\mathbf{v}_G = \mathbf{v}_B + \mathbf{v}_{G/B}$$

$$\left[\underset{\rightarrow}{v_G}\right] = \left[\underset{\rightarrow}{3}\right] + \left[\underset{\leftarrow}{1.5(8)}\right]$$

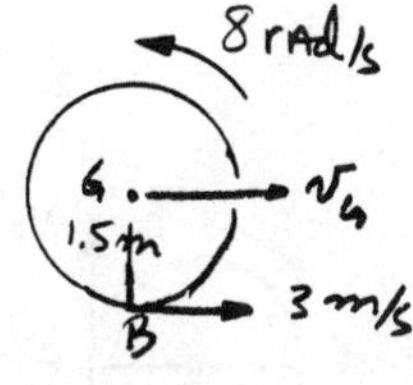

$v_G = 9$ m/s $\leftarrow$ **Ans**

Also :

$$\mathbf{v}_G = \mathbf{v}_B + \omega \times \mathbf{r}_{G/B}$$

$$v_G\mathbf{i} = 3\mathbf{i} + (8\mathbf{k}) \times (1.5\mathbf{j})$$

$$v_G = 3 - 12$$

$v_G = -9$ m/s $= 9$ m/s $\leftarrow$ **Ans**

16-70. At the instant shown, the truck is traveling to the right at 8 m/s. If the pipe does not slip at B, determine its angular velocity if its mass center G appears to an observer on the ground to remain stationary.

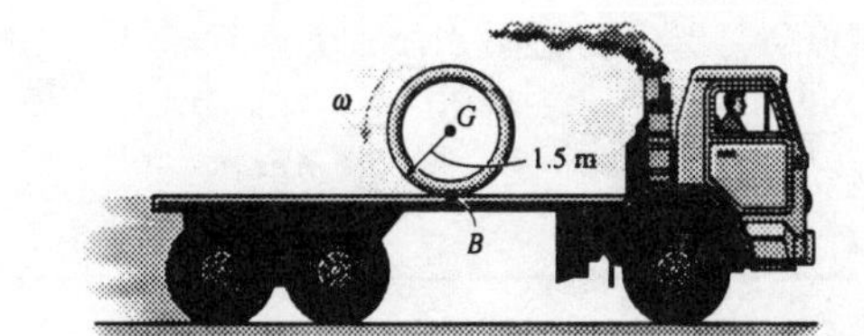

$$\mathbf{v}_G = \mathbf{v}_B + \mathbf{v}_{G/B}$$

$$0 = \left[\underset{\rightarrow}{8}\right] + \left[\underset{\leftarrow}{1.5\omega}\right]$$

$\omega = \dfrac{8}{1.5} = 5.33$ rad/s ↺ **Ans**

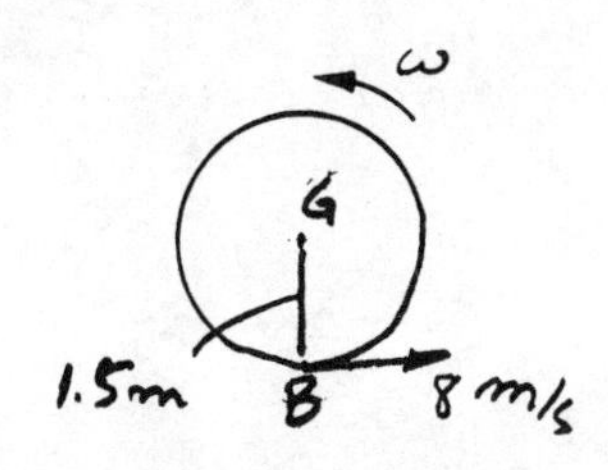

Also :

$$\mathbf{v}_G = \mathbf{v}_B + \omega \times \mathbf{r}_{G/B}$$

$$0\mathbf{i} = 8\mathbf{i} + (\omega\mathbf{k}) \times (1.5\mathbf{j})$$

$$0 = 8 - 1.5\omega$$

$\omega = \dfrac{8}{1.5} = 5.33$ rad/s ↺ **Ans**

16-71. The pinion gear A rolls on the fixed gear rack B with an angular velocity $\omega = 4$ rad/s. Determine the velocity of the gear rack C.

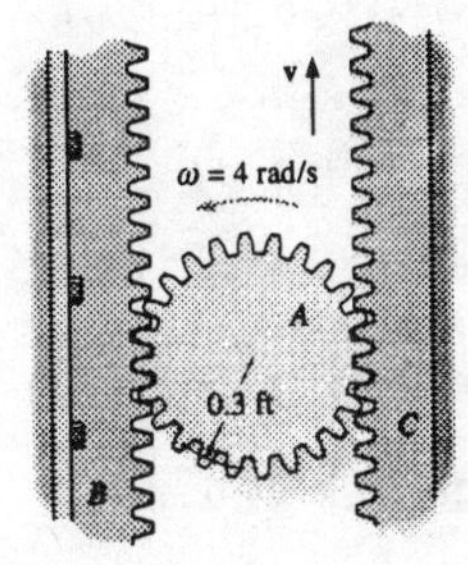

$$\mathbf{v}_C = \mathbf{v}_B + \mathbf{v}_{C/B}$$

$$\left[\underset{\uparrow}{v_C}\right] = 0 + \left[\underset{\uparrow}{4(0.6)}\right]$$

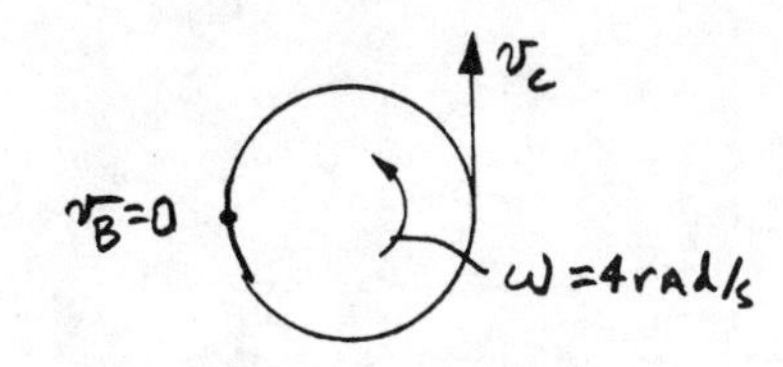

$v_C = 2.40$ ft/s ↑ **Ans**

Also,

$$\mathbf{v}_C = \mathbf{v}_B + \omega \times \mathbf{r}_{C/B}$$

$$v_C\mathbf{j} = 0 + (4\mathbf{k}) \times (0.6\mathbf{i})$$

$v_C = 2.40$ ft/s ↑ **Ans**

***16-72.** When the crank on the Chinese windlass is turning, the rope on shaft A unwinds while that on shaft B winds up. Determine the speed at which the block D lowers if the crank is turning with an angular velocity ω = 4 rad/s. What is the angular velocity of the pulley at C? The rope segments on each side of the pulley are both parallel and vertical, and the rope does not slip on the pulley.

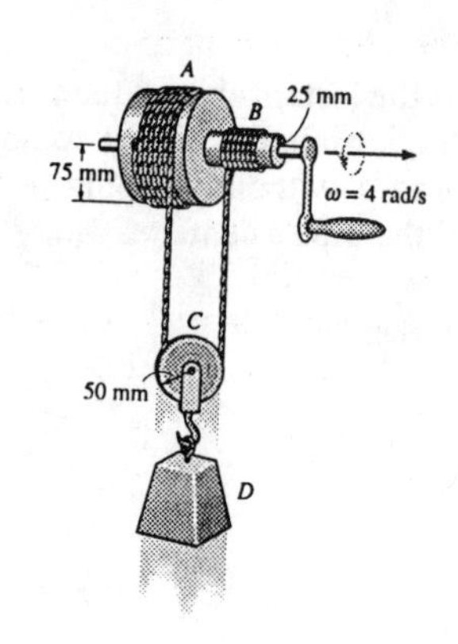

$v_P = \omega r_A = 4(75) = 300 \text{ mm/s} \downarrow$

$v_{P'} = \omega r_B = 4(25) = 100 \text{ mm/s} \uparrow$

$\mathbf{v}_{P'} = \mathbf{v}_P + \mathbf{v}_{P'/P}$

$100\mathbf{j} = -300\mathbf{j} + \omega(100)\mathbf{j}$

$(+\uparrow) \qquad 100 = -300 + \omega(100)$

$\omega = 4 \text{ rad/s}$ ↺ **Ans**

$\mathbf{v}_D = \mathbf{v}_P + \mathbf{v}_{D/P}$

$-v_D\mathbf{j} = -300\mathbf{j} + 4(50)\mathbf{j}$

$(+\uparrow) \qquad -v_D = -300 + 200$

$v_D = 100 \text{ mm/s} \downarrow$ **Ans**

16-73. The cylinder B rolls on the *fixed cylinder A* without slipping. If the connected bar CD is rotating with an angular velocity of $\omega_{CD} = 5$ rad/s, determine the angular velocity of cylinder B.

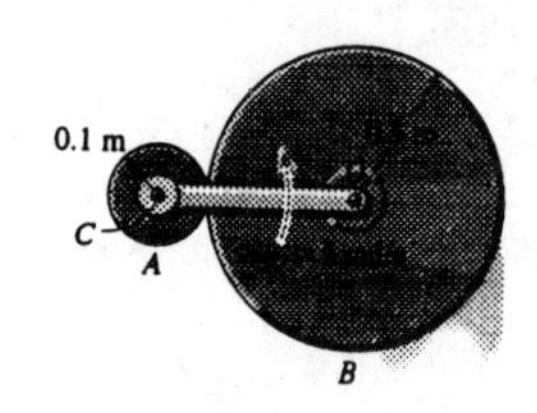

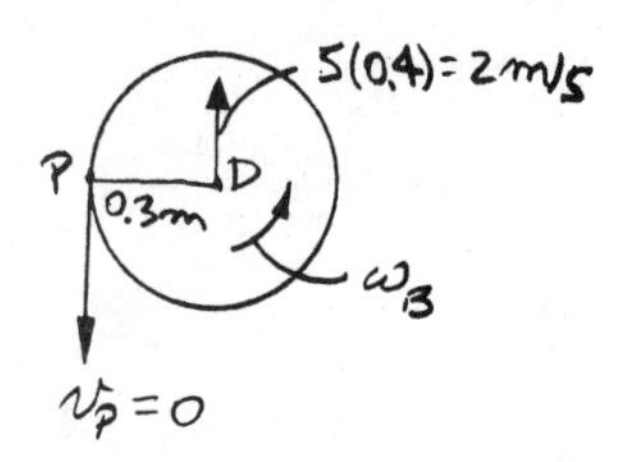

The contact point P between the cylinders has zero velocity

$\mathbf{v}_P = \mathbf{v}_D + \mathbf{v}_{P/D}$

$$0 = \left[\underset{\uparrow}{2}\right] + \left[\underset{\downarrow}{\omega_B (0.3)}\right]$$

$\omega_B = 6.67 \text{ rad/s}$ ↻ **Ans**

Also,

$\mathbf{v}_P = \mathbf{v}_D + \boldsymbol{\omega}_B \times \mathbf{r}_{P/D}$

$0 = 2\mathbf{j} + (\omega_B\mathbf{k}) \times (-0.3\mathbf{i})$

$(+\uparrow) \qquad 0 = 2 - 0.3\omega_B$

$\omega_B = 6.67 \text{ rad/s}$ ↻ **Ans**

16-74. The slider mechanism is used to increase the stroke of travel of one slider with respect to that of another. As shown, when the slider A is moving forward, the attached pinion F rolls on the *fixed* rack D, forcing slider C to move forward. This in turn causes the attached pinion G to roll on the *fixed* rack E, thereby moving slider B. If A has a velocity of $v_A = 4$ ft/s at the instant shown, determine the velocity of B. $r = 0.2$ ft.

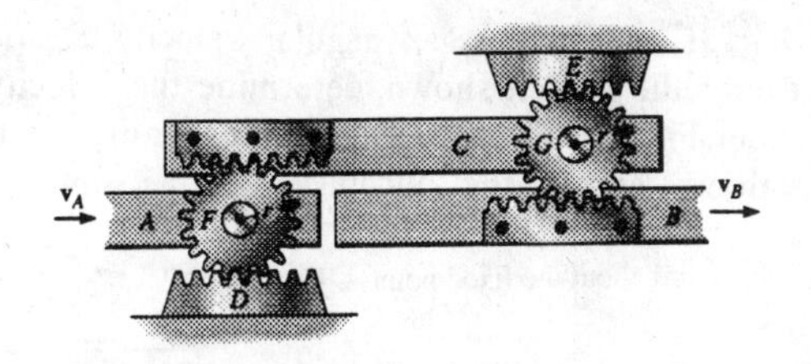

$\mathbf{v}_A = \mathbf{v}_D + \mathbf{v}_{A/D}$

$4\mathbf{i} = \mathbf{0} + \omega_F(0.2)\mathbf{i}$

$\omega_F = 20$ rad/s

$\mathbf{v}_C = \mathbf{v}_D + \mathbf{v}_{C/D}$

$v_C\mathbf{i} = \mathbf{0} + 20(0.4)\mathbf{i}$

$v_C = 8$ ft/s

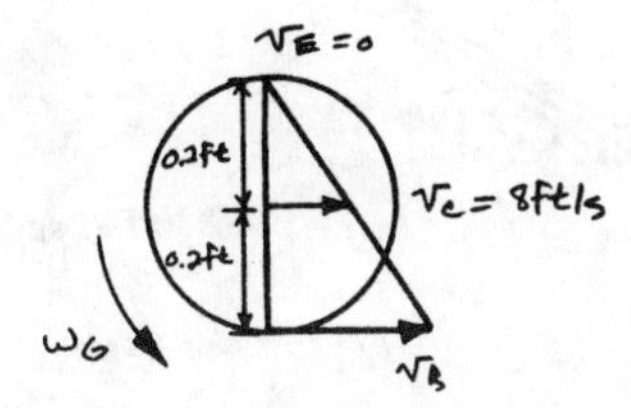

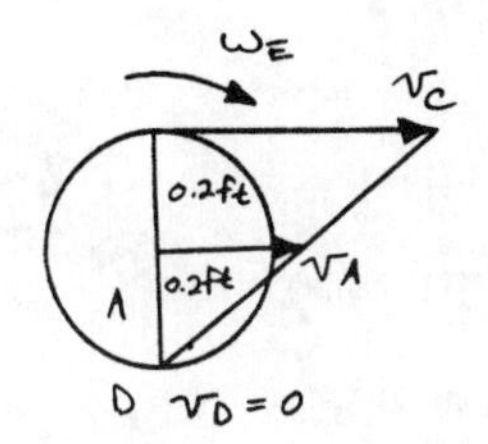

$\mathbf{v}_C = \mathbf{v}_E + \mathbf{v}_{C/E}$

$8\mathbf{i} = \mathbf{0} + \omega_G(0.2)\mathbf{i}$

$\omega_G = 40$ rad/s

$\mathbf{v}_B = \mathbf{v}_E + \mathbf{v}_{B/E}$

$v_B\mathbf{i} = \mathbf{0} + 40(0.4)\mathbf{i}$

$v_B = 16$ ft/s $\rightarrow$ **Ans**

16-75. The epicyclic gear train consists of the sun gear A which is in mesh with the planet gear B. This gear has an inner hub C which is fixed to B and in mesh with the fixed ring gear R. If the connecting link DE attached to B and C is rotating at $\omega_{DE} = 18$ rad/s, determine the angular velocities of the planet and sun gears.

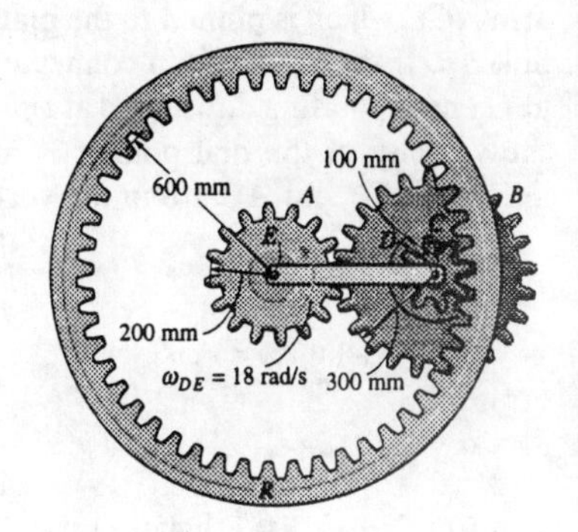

$v_D = r_{DE}\omega_{DE} = (0.5)(18) = 9$ m/s $\uparrow$

The velocity of the contact point P with the ring is zero.

$\mathbf{v}_D = \mathbf{v}_P + \omega \times \mathbf{r}_{D/P}$

$9\mathbf{j} = \mathbf{0} + (-\omega_B\mathbf{k}) \times (-0.1\mathbf{i})$

$\omega_B = 90$ rad/s $\circlearrowright$ **Ans**

Let P' be the contact point between A and B.

$\mathbf{v}_{P'} = \mathbf{v}_P + \omega \times \mathbf{r}_{P'/P}$

$v_{P'}\mathbf{j} = \mathbf{0} + (-90\mathbf{k}) \times (-0.4\mathbf{i})$

$v_{P'} = 36$ m/s $\uparrow$

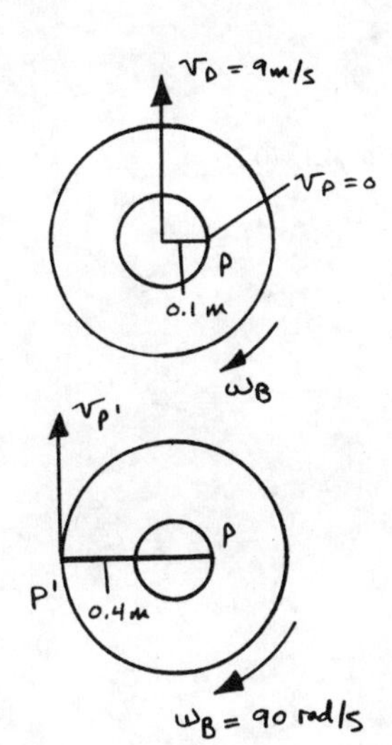

$\omega_A = \frac{v_{P'}}{r_A} = \frac{36}{0.2} = 180$ rad/s $\circlearrowleft$ **Ans**

***16-76.** If link AB has an angular velocity of $\omega_{AB} = 4$ rad/s at the instant shown, determine the velocity of the slider block E at this instant. Also, identify the type of motion of each of the four links.

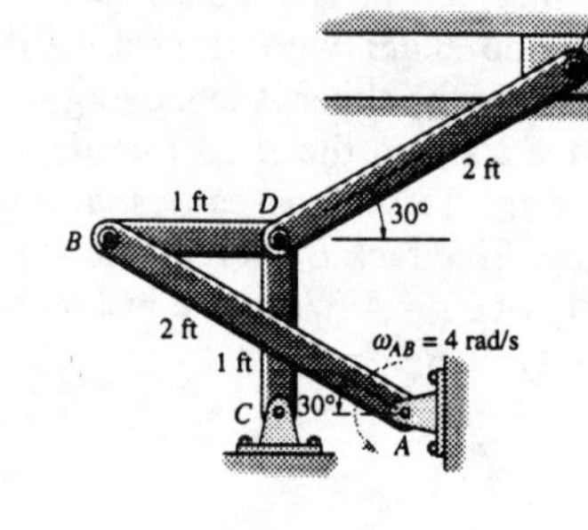

Link AB rotates about the fixed point A. Hence

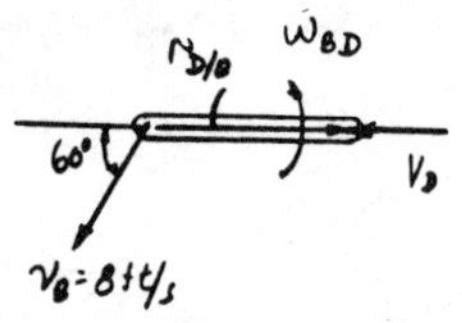

$v_B = \omega_{AB} r_{AB} = 4(2) = 8$ ft/s

For link BD

$\mathbf{v}_B = \{-8\cos 60°\mathbf{i} - 8\sin 60°\mathbf{j}\}$ ft/s $\quad \mathbf{v}_D = -v_D\mathbf{i} \quad \omega_{BD} = \omega_{BD}\mathbf{k}$

$\mathbf{r}_{D/B} = \{1\mathbf{i}\}$ ft

$\mathbf{v}_D = \mathbf{v}_B + \omega_{BD} \times \mathbf{r}_{D/B}$

$-v_D\mathbf{i} = (-8\cos 60°\mathbf{i} - 8\sin 60°\mathbf{j}) + (\omega_{BD}\mathbf{k}) \times (1\mathbf{i})$

$-v_D\mathbf{i} = -8\cos 60°\mathbf{i} + (\omega_{BD} - 8\sin 60°)\mathbf{j}$

$(\overset{+}{\rightarrow}) \quad -v_D = -8\cos 60° \qquad v_D = 4$ ft/s

$(+\uparrow) \quad 0 = \omega_{BD} - 8\sin 60° \qquad \omega_{BD} = 6.928$ rad/s

For Link DE

$\mathbf{v}_D = \{-4\mathbf{i}\}$ ft/s $\quad \omega_{DE} = \omega_{DE}\mathbf{k} \quad \mathbf{v}_E = -v_E\mathbf{i}$

$\mathbf{r}_{E/D} = \{2\cos 30°\mathbf{i} + 2\sin 30°\mathbf{j}\}$ ft

$\mathbf{v}_E = \mathbf{v}_D + \omega_{DE} \times \mathbf{r}_{E/D}$

$-v_E\mathbf{i} = -4\mathbf{i} + (\omega_{DE}\mathbf{k}) \times (2\cos 30°\mathbf{i} + 2\sin 30°\mathbf{j})$

$-v_E\mathbf{i} = (-4 - 2\sin 30°\,\omega_{DE})\mathbf{i} + 2\cos 30°\,\omega_{DE}\mathbf{j}$

$(\overset{+}{\rightarrow}) \quad 0 = 2\cos 30°\,\omega_{DE} \qquad \omega_{DE} = 0$

$(+\uparrow) \quad -v_E = -4 - 2\sin 30°(0) \qquad v_E = 4$ ft/s $\leftarrow$ **Ans**

16-77. The gauge is used to indicate the safe load acting at the end of the boom, B, when it is in any angular position. It consists of a fixed dial plate D and an indicator arm ACE which is pinned to the plate at C and to a short link EF. If the boom is pin-connected to the trunk frame at G and is rotating downward at $\omega_B = 4$ rad/s, determine the velocity of the dial pointer A at the instant shown, i.e., when EF and AC are in the vertical position.

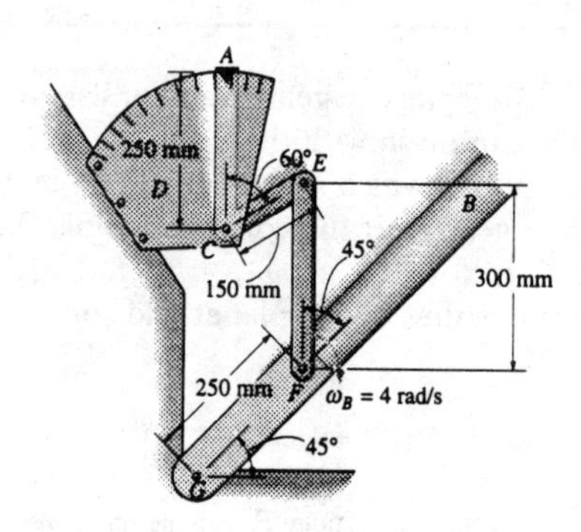

$v_F = \omega_B r_{GF} = (4)(0.25) = 1$ m/s

$\mathbf{v}_E = \mathbf{v}_F + \omega_{EF} \times \mathbf{r}_{E/F}$

$v_E\cos 60°\mathbf{i} - v_E\sin 60°\mathbf{j} = 1\cos 45°\mathbf{i} - 1\sin 45°\mathbf{j} + (\omega_{EF}\mathbf{k}) \times (0.3\mathbf{i})$

$(\overset{+}{\rightarrow}) \quad v_E\cos 60° = 1\cos 45° - \omega_{EF}(0.3)$

$(+\uparrow) \quad -v_E\sin 60° = -1\sin 45° + 0$

Solving,

$v_E = 0.8165$ m/s, $\quad \omega_{EF} = 0.996$ rad/s

$$\omega_{ACE} = \frac{v_E}{r_{EC}} = \frac{0.8164}{0.150} = 5.44 \text{ rad/s}$$

$v_A = \omega_{ACE} r_{AC} = (5.44)(0.250) = 1.36$ m/s $\rightarrow$ **Ans**

16-78. Solve Prob. 16-51 using the method of instantaneous center of zero velocity.

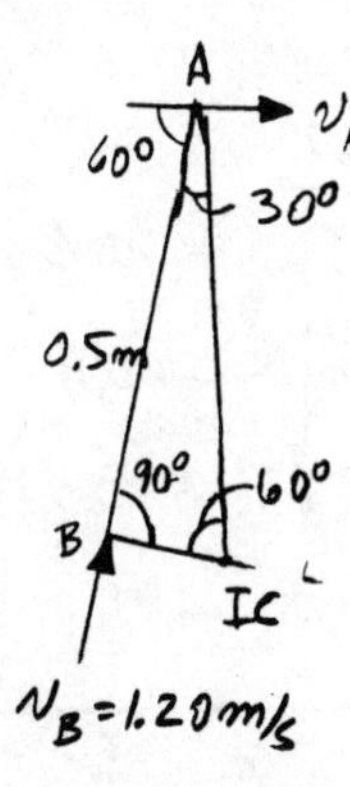

$v_B = 8(0.150) = 1.20$ m/s

$r_{B/IC} = 0.5 \tan 30° = 0.28868$ m

$\omega_{AB} = \dfrac{1.20}{0.28868} = 4.157$ rad/s

$r_{A/IC} = \dfrac{0.5}{\sin 60°} = 0.5774$ m

$v_A = 0.5774(4.157) = 2.40$ m/s $\rightarrow$ **Ans**

16-79. Solve Prob. 16-54 using the method of instantaneous center of zero velocity.

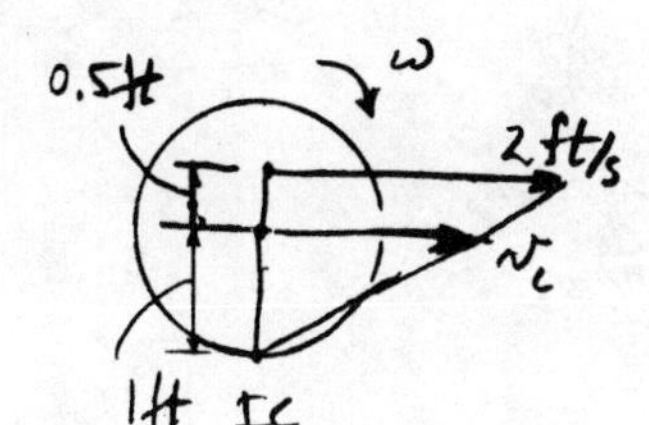

$\omega = \dfrac{2}{1.5} = 1.33$ rad/s

$v_C = 1(1.33) = 1.33$ ft/s $\rightarrow$ **Ans**

***16-80.** Solve Prob. 16-56 using the method of instantaneous center of zero velocity.

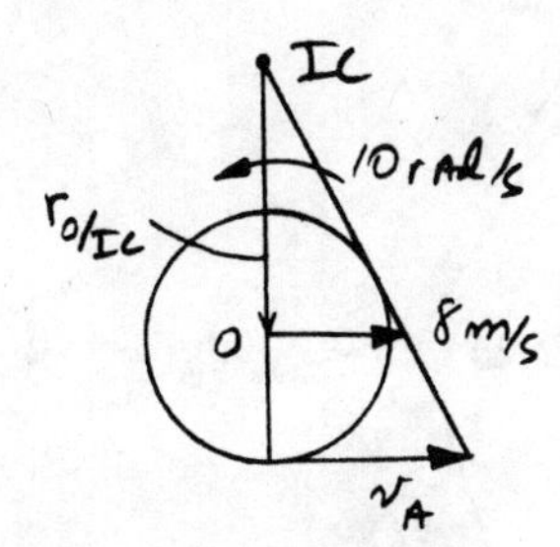

$r_{O/IC} = \dfrac{8}{10} = 0.8$ m

$v_A = 10(0.8 + 0.120) = 9.20$ m/s **Ans**

16-81. Solve Prob. 16-58 using the method of instantaneous center of zero velocity.

$v_B = 4(0.5) = 2$ m/s

$\dfrac{\sin 75°}{0.350} = \dfrac{\sin 45°}{r_{B/IC}}$

$r_{B/IC} = 0.2562$ m

$\dfrac{\sin 75°}{0.350} = \dfrac{\sin 60°}{r_{C/IC}}$

$r_{C/IC} = 0.31380$ m

$\omega_{CB} = \dfrac{2}{0.2562} = 7.806$ rad/s $= 7.81$ rad/s ↷ **Ans**

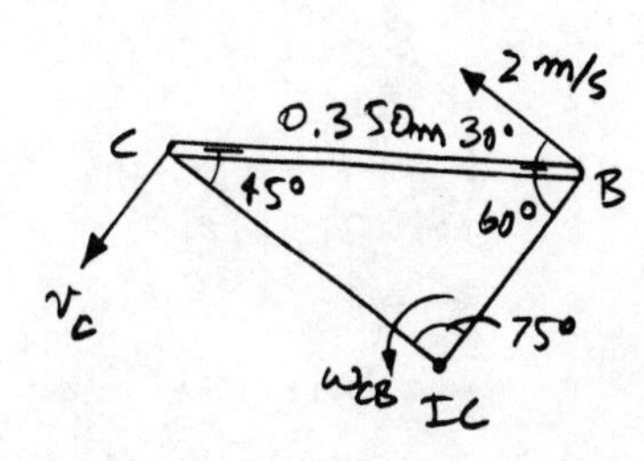

$v_C = 7.806(0.31380) = 2.45$ m/s ↙ **Ans**

16-82. Solve Prob. 16-60 using the method of instantaneous center of zero velocity.

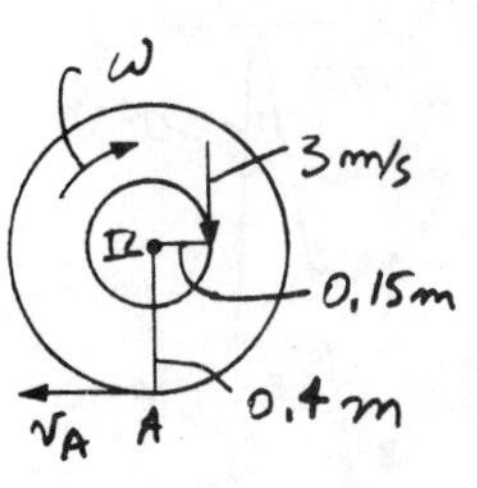

$$\omega = \frac{3 \text{ m/s}}{0.15 \text{ m}} = 20 \text{ rad/s}$$

$v_A = (0.4)(20) = 8 \text{ m/s} \leftarrow$ **Ans**

16-83. Solve Prob. 16-63 using the method of instantaneous center of zero velocity.

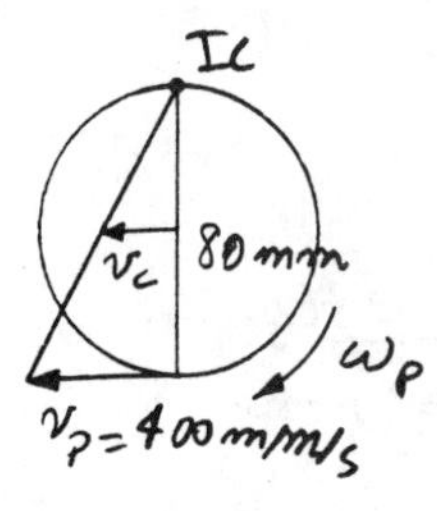

$v_P = (80)(5) = 400 \text{ mm/s}$

$$\omega_P = \frac{400}{80} = 5 \text{ rad/s}$$ **Ans**

$v_C = (5)(40) = 200 \text{ mm/s}$

$$\omega_A = \frac{200}{(80+40)} = 1.67 \text{ rad/s}$$ **Ans**

***16-84.** Solve Prob. 16-68 using the method of instantaneous center of zero velocity.

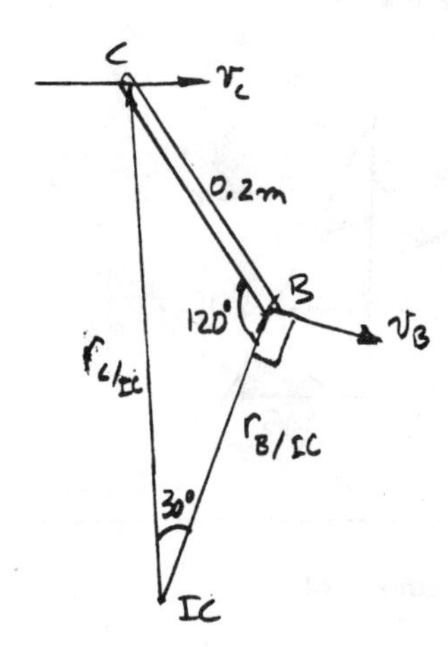

$v_B = 4(0.150) = 0.6 \text{ m/s}$

$$\frac{r_{C/IC}}{\sin 120°} = \frac{0.2}{\sin 30°}$$

$r_{C/IC} = 0.34641 \text{ m}$

$$\frac{r_{B/IC}}{\sin 30°} = \frac{0.2}{\sin 30°}$$

$r_{B/IC} = 0.2 \text{ m}$

$$\omega = \frac{0.6}{0.2} = 3 \text{ rad/s}$$

$v_C = 0.34641(3) = 1.04 \text{ m/s} \rightarrow$ **Ans**

16-85. Solve Prob. 16-66 using the method of instantaneous center of zero velocity.

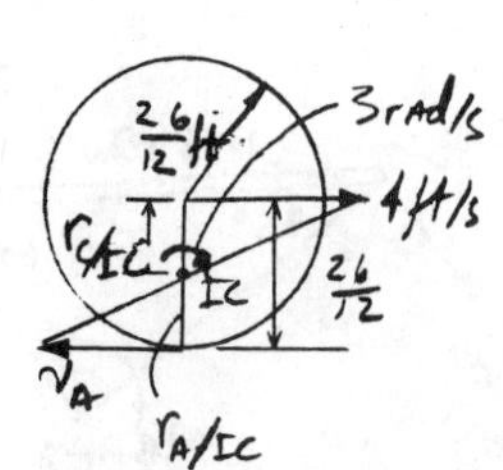

$$r_{C/IC} = \frac{4}{3} = 1.33 \text{ ft}$$

$$r_{A/IC} = \frac{26}{12} - 1.33 \text{ ft} = 0.833 \text{ ft}$$

$v_A = 3(0.833) = 2.5 \text{ ft/s} \leftarrow$ **Ans**

16-86. The instantaneous center of zero velocity for the body is located at point IC (0.5 m, 2 m). If the body has an angular velocity of 4 rad/s, as shown, determine the velocity of B with respect to A.

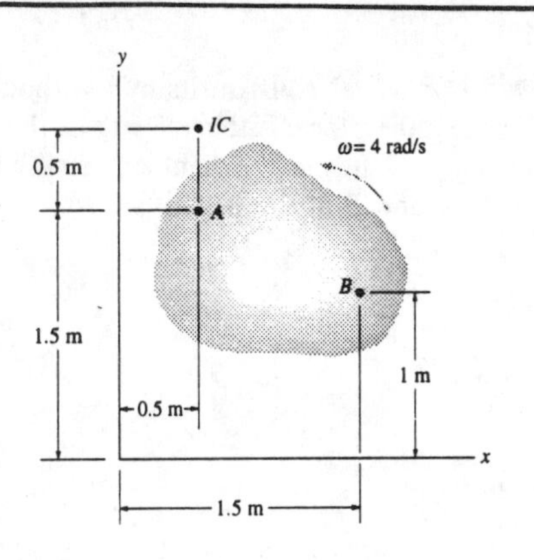

$v_A = \omega r_{A/IC} = 4(0.5) = 2 \text{ m/s} \rightarrow$

$v_B = \omega r_{B/IC} = 4\left(\sqrt{(1)^2 + (1)^2}\right) = 5.66 \text{ m/s}$

$5.66 \cos 45°\mathbf{i} + 5.66 \sin 45°\mathbf{j} = 2\mathbf{i} + v_{B/A} \cos\theta\mathbf{i} + v_{B/A} \sin\theta\mathbf{j}$

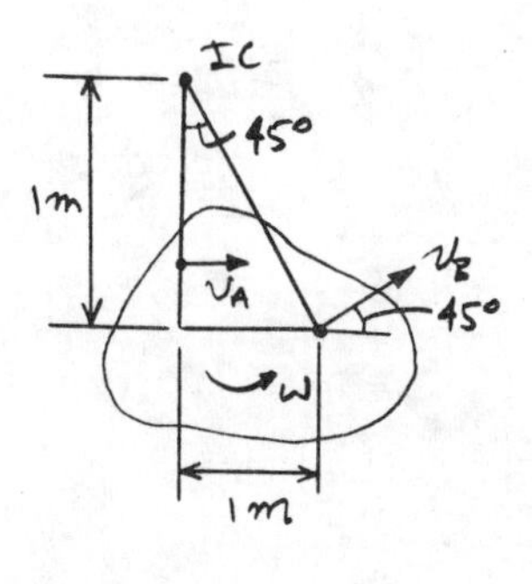

$(\overset{+}{\rightarrow})$ $(5.66)\cos 45° = 2 + v_{B/A} \cos\theta$

$(+\uparrow)$ $(5.66)\sin 45° = 0 + v_{B/A} \sin\theta$

Solving,

$\theta = 63.4°$ ∠ **Ans**

$v_{B/A} = 4.47 \text{ m/s}$ **Ans**

16-87. The slider block C is moving 4 ft/s up the incline. Determine the angular velocities of links AB and BC and the velocity of point B at the instant shown.

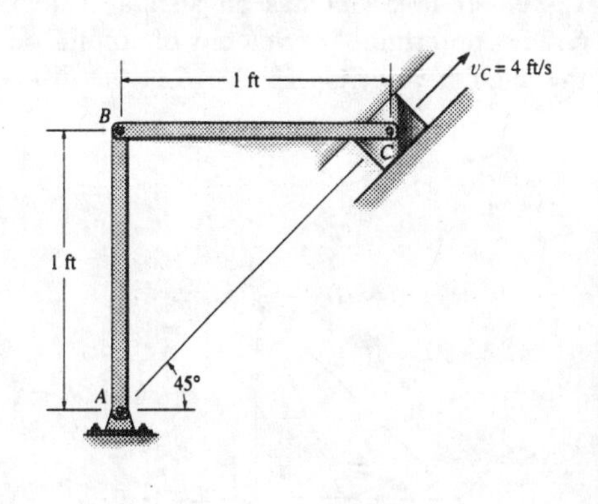

$v_C = \omega_{BC} (r_{C/IC})$

$4 = \omega_{BC} \sqrt{2}$

$\omega_{BC} = 2.83 \text{ rad/s}$ **Ans**

$v_B = 1(2.83) = 2.83 \text{ ft/s}$ **Ans**

Thus,

$v_B = \omega_{AB} r_{AB}$

$2.83 = \omega_{AB} (1)$

$\omega_{AB} = 2.83 \text{ rad/s}$ **Ans**

***16-88.** As the cord unravels from the wheel's inner hub, the wheel is rotating at $\omega = 2$ rad/s at the instant shown. Determine the velocities of points A and B.

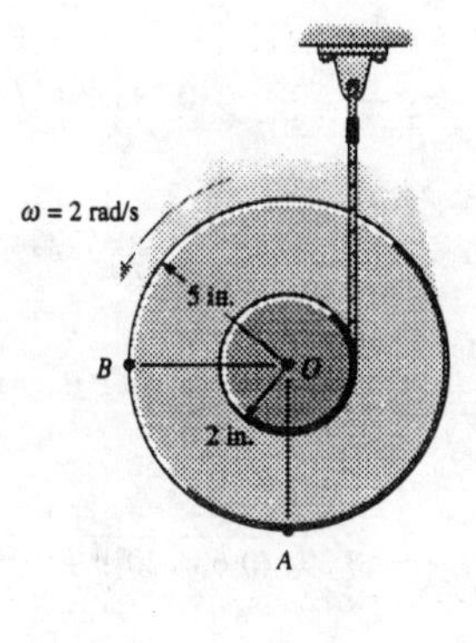

$r_{B/IC} = 5 + 2 = 7 \text{ in.}$ $\quad r_{A/IC} = \sqrt{2^2 + 5^2} = \sqrt{29} \text{ in.}$

$v_B = \omega r_{B/IC} = 2(7) = 14 \text{ in./s} \downarrow$ **Ans**

$v_A = \omega r_{A/IC} = 2\left(\sqrt{29}\right) = 10.8 \text{ in./s}$ **Ans**

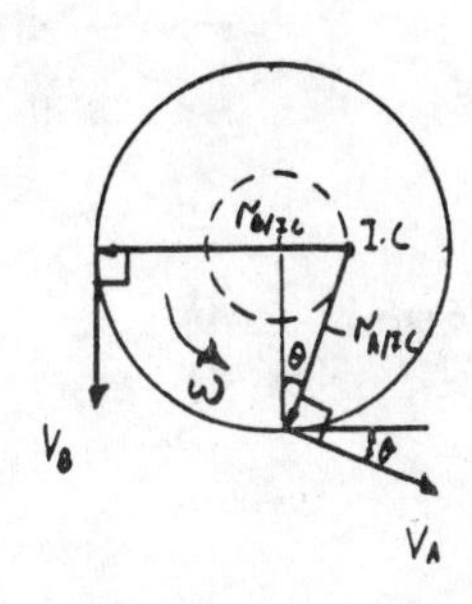

$\theta = \tan^{-1}\left(\frac{2}{5}\right) = 21.8°$ **Ans**

16-89. The wheel rolls on its hub without slipping on the horizontal surface. If the velocity of the center of the wheel is $v_C = 2$ ft/s to the right, determine the velocities of points A and B at the instant shown.

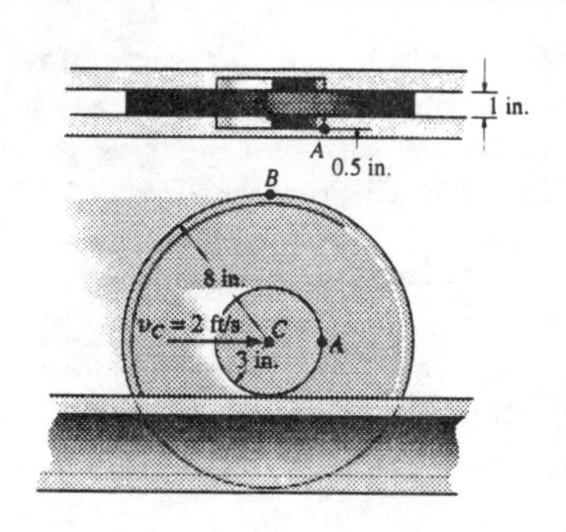

$v_C = \omega r_{C/IC}$

$2 = \omega\left(\frac{3}{12}\right)$

$\omega = 8$ rad/s

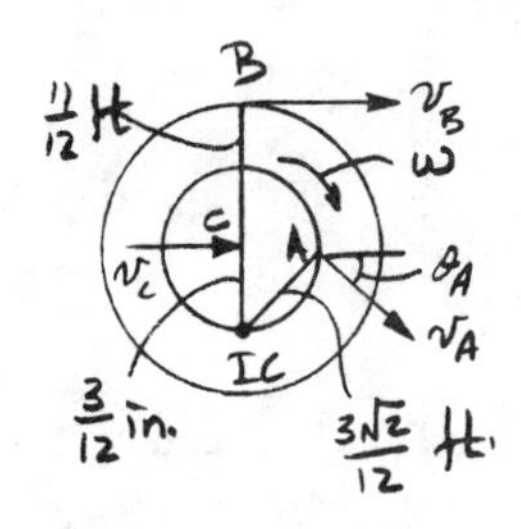

$v_B = \omega r_{B/IC} = 8\left(\frac{11}{12}\right) = 7.33$ ft/s $\rightarrow$ **Ans**

$v_A = \omega r_{A/IC} = 8\left(\frac{3\sqrt{2}}{12}\right) = 2.83$ ft/s **Ans**

$\theta_A = \tan^{-1}\left(\frac{3}{3}\right) = 45°$ **Ans**

16-90. If link CD has an angular velocity of $\omega_{CD} = 6$ rad/s, determine the velocity of point E on link BC and the angular velocity of link AB at the instant shown.

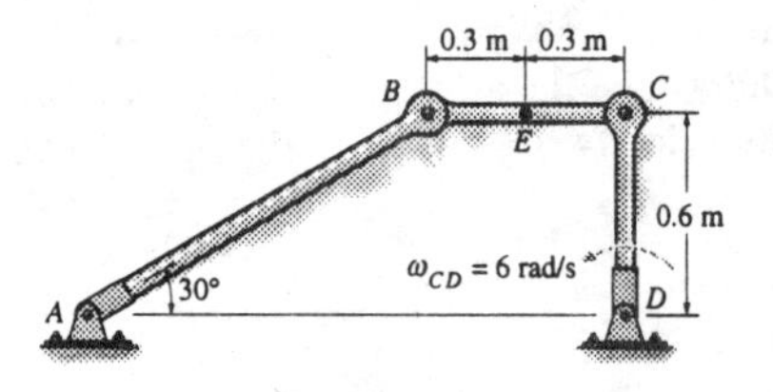

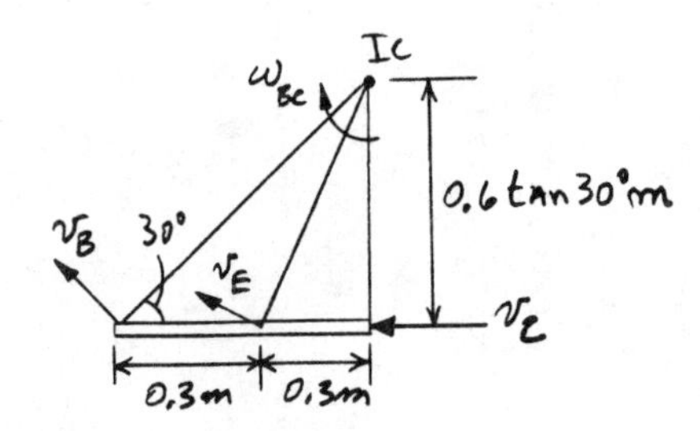

$v_C = \omega_{CD}(r_{CD}) = (6)(0.6) = 3.60$ m/s

$\omega_{BC} = \frac{v_C}{r_{C/IC}} = \frac{3.60}{0.6\tan 30°} = 10.39$ rad/s

$v_B = \omega_{BC} r_{B/IC} = (10.39)\left(\frac{0.6}{\cos 30°}\right) = 7.20$ m/s

$\omega_{AB} = \frac{v_B}{r_{AB}} = \frac{7.20}{\left(\frac{0.6}{\sin 30°}\right)} = 6$ rad/s **Ans**

$v_E = \omega_{BC} r_{E/IC} = 10.39\sqrt{(0.6\tan 30°)^2 + (0.3)^2} = 4.76$ m/s **Ans**

$\theta = \tan^{-1}\left(\frac{0.3}{0.6\tan 30°}\right) = 40.9°$ **Ans**

16-91. The disk of radius r is confined to roll without slipping at A and B. If the plates have the velocities shown, determine the angular velocity of the disk.

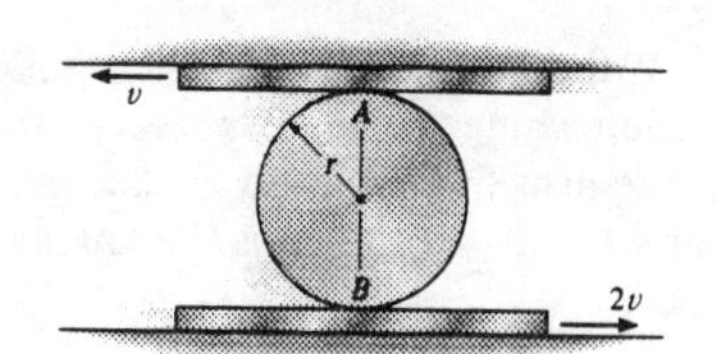

$$\frac{v}{2r-x} = \frac{2v}{x}$$

$$x = 4r - 2x$$

$$3x = 4r$$

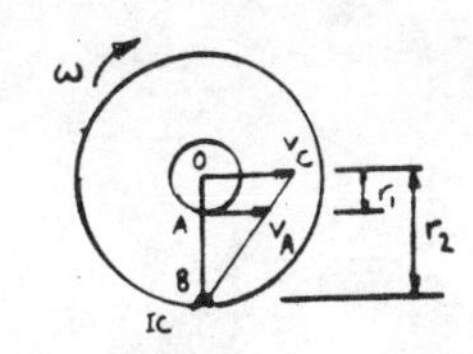

$$x = \frac{4}{3}r = 1.33r$$

$$\omega = \frac{2v}{1.33r} = 1.5\frac{v}{r}$$

***16-92.** Show that if the rim of the wheel and its hub maintain contact with the three tracks as the wheel rolls, it is necessary that slipping occurs at the hub A if no slipping occurs at B. Under these conditions, what is the speed at A if the wheel has an angular velocity $\boldsymbol{\omega}$?

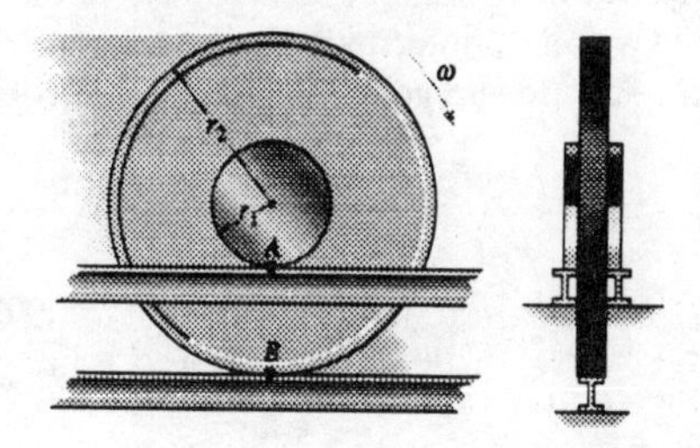

IC is at B.

$v_A = \omega(r_2 - r_1) \rightarrow$ **Ans**

16-93. In an automobile transmission the planet pinions A and B rotate on shafts that are mounted on the planet-pinion carrier CD. As shown, CD is attached to a shaft at E which is aligned with the center of the *fixed* sun gear S. This shaft is not attached to the sun gear. If CD is rotating at $\omega_{CD} = 8$ rad/s, determine the angular velocity of the ring gear R.

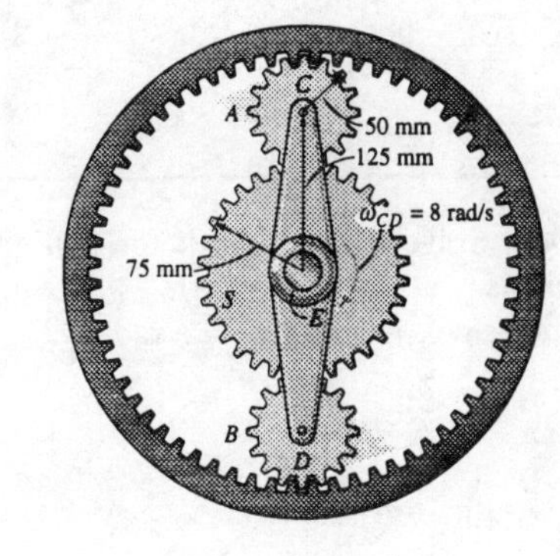

Pinion A :

$$\omega_A = \frac{1}{0.05} = 20 \text{ rad/s}$$

$$v_R = (20)(0.1) = 2 \text{ m/s}$$

$$\omega_R = \frac{2}{(0.125 + 0.05)} = 11.4 \text{ rad/s} \quad \textbf{Ans}$$

16-94. If the hub gear H and ring gear R have angular velocities $\omega_H = 5$ rad/s and $\omega_R = 20$ rad/s, respectively, determine the angular velocity ω_S of the spur gear S and the angular velocity of its attached arm OA.

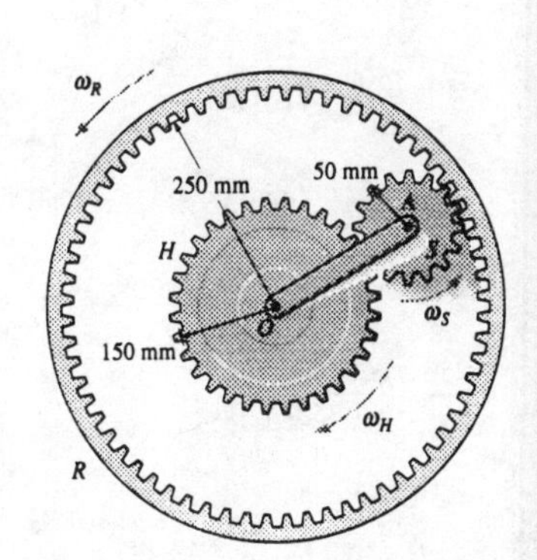

$$\frac{5}{0.1-x} = \frac{0.75}{x}$$

$$x = 0.01304 \text{ m}$$

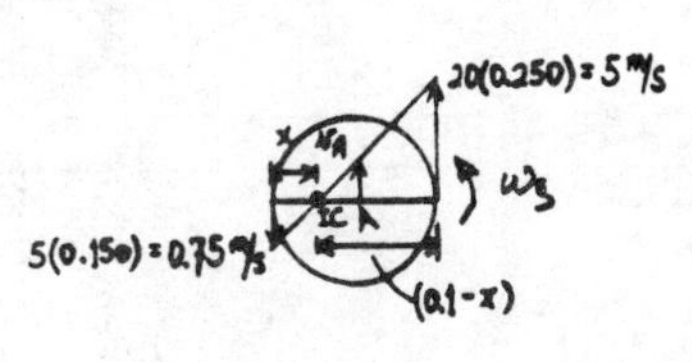

$$\omega_S = \frac{0.75}{0.01304} = 57.5 \text{ rad/s} \quad \textbf{Ans}$$

$$v_A = 57.5(0.05 - 0.01304) = 2.125 \text{ m/s}$$

$$\omega_{OA} = \frac{2.125}{0.2} = 10.6 \text{ rad/s} \quad \textbf{Ans}$$

16-95. If the hub gear H has an angular velocity $\omega_H = 5$ rad/s, determine the angular velocity of the ring gear R so that the arm OA attached to the spur gear S remains stationary ($\omega_{OA} = 0$). What is the angular velocity of the spur gear?

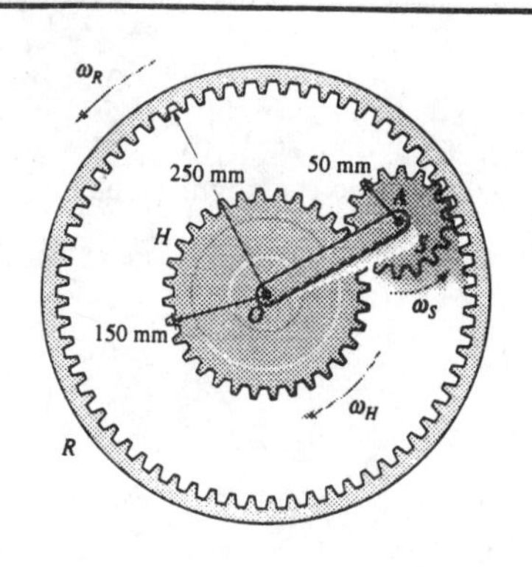

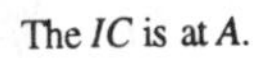
The IC is at A.

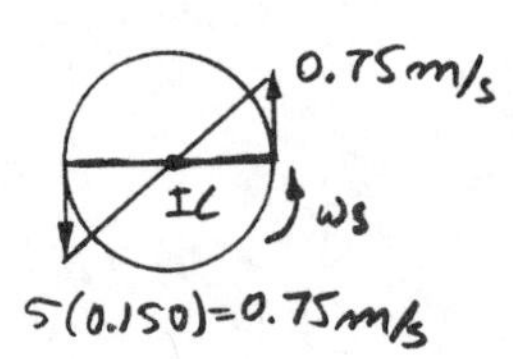

$$\omega_S = \frac{0.75}{0.05} = 15.0 \text{ rad/s} \qquad \textbf{Ans}$$

$$\omega_R = \frac{0.75}{0.250} = 3.00 \text{ rad/s} \qquad \textbf{Ans}$$

***16-96.** The planet gear A is pin connected to the end of the link BC. If the link rotates about the fixed point B at 4 rad/s, determine the angular velocity of the ring gear R. The sun gear D is fixed from rotating.

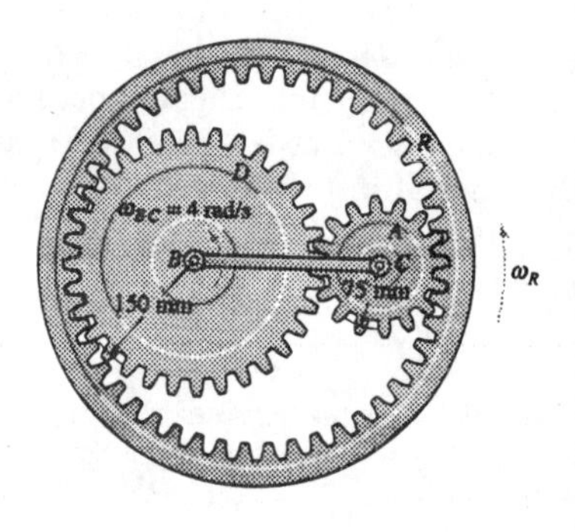

Gear A :

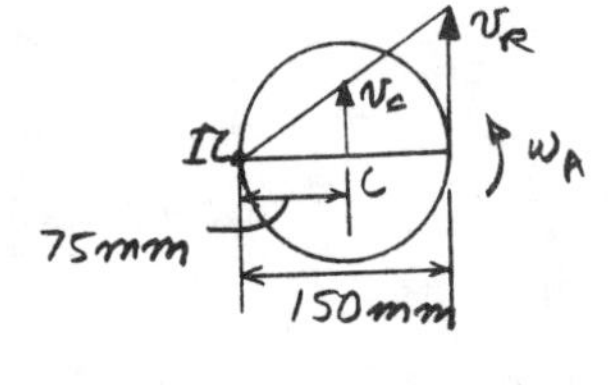

$$v_C = 4(225) = 900 \text{ mm/s}$$

$$\omega_A = \frac{900}{75} = \frac{v_R}{150}$$

$$v_R = 1800 \text{ mm/s}$$

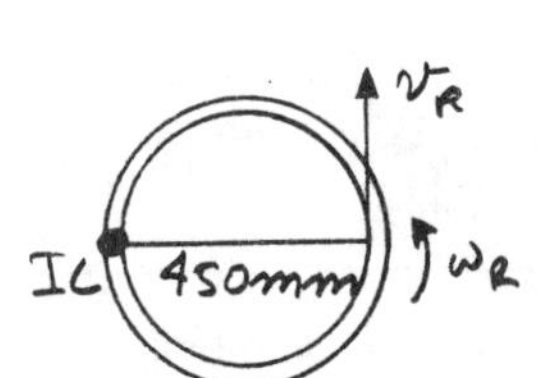

Ring gear :

$$\omega_R = \frac{1800}{450} = 4 \text{ rad/s} \qquad \textbf{Ans}$$

16-97. Solve Prob. 16-96 if the sun gear D is rotating clockwise at $\omega_D = 5$ rad/s while link BC rotates counterclockwise at $\omega_{BC} = 4$ rad/s.

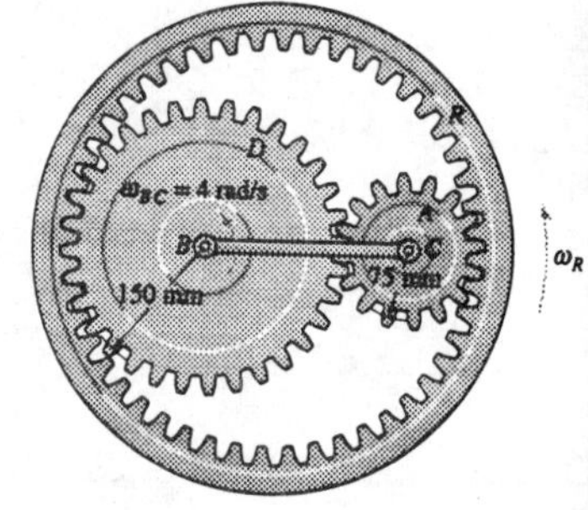

Gear A :

$$v_P = 5(150) = 750 \text{ mm/s}$$

$$v_C = 4(225) = 900 \text{ mm/s}$$

$$\frac{x}{750} = \frac{75 - x}{900}$$

$$x = 34.09 \text{ mm}$$

$$\omega = \frac{750}{34.09} = 22.0 \text{ rad/s}$$

$$v_R = [75 + (75 - 34.09)](22) = 2550 \text{ mm/s}$$

Ring gear :

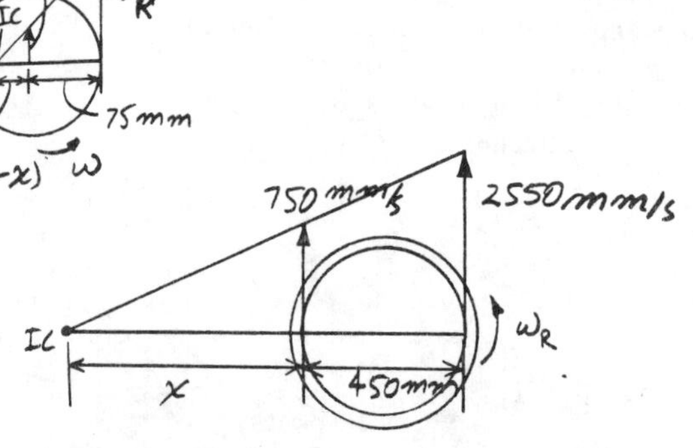

$$\frac{750}{x} = \frac{2550}{x + 450}$$

$$x = 187.5 \text{ mm}$$

$$\omega_R = \frac{750}{187.5} = 4 \text{ rad/s} \qquad \textbf{Ans}$$

16-98. The mechanism used in a marine engine consists of a single crank AB and two connecting rods BC and BD. Determine the velocity of the piston at C the instant the crank is in the position shown and has an angular velocity of 5 rad/s.

$v_B = 0.2(5) = 1 \text{ m/s} \rightarrow$

Member BC:

$$\frac{r_{C/IC}}{\sin 60^\circ} = \frac{0.4}{\sin 45^\circ}$$

$r_{C/IC} = 0.4899 \text{ m}$

$$\frac{r_{B/IC}}{\sin 75^\circ} = \frac{0.4}{\sin 45^\circ}$$

$r_{B/IC} = 0.5464 \text{ m}$

$$\omega_{BC} = \frac{1}{0.5464} = 1.830 \text{ rad/s}$$

$v_C = 0.4899(1.830) = 0.897 \text{ m/s} \nearrow$ **Ans**

16-99. The mechanism used in a marine engine consists of a single crank AB and two connecting rods BC and BD. Determine the velocity of the piston at D the instant the crank is in the position shown and has an angular velocity of 5 rad/s.

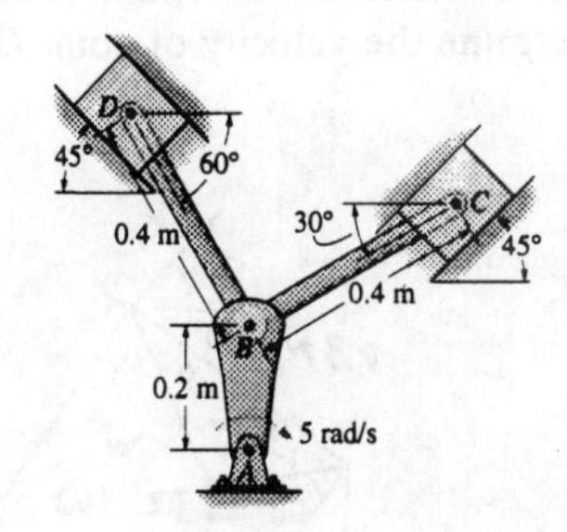

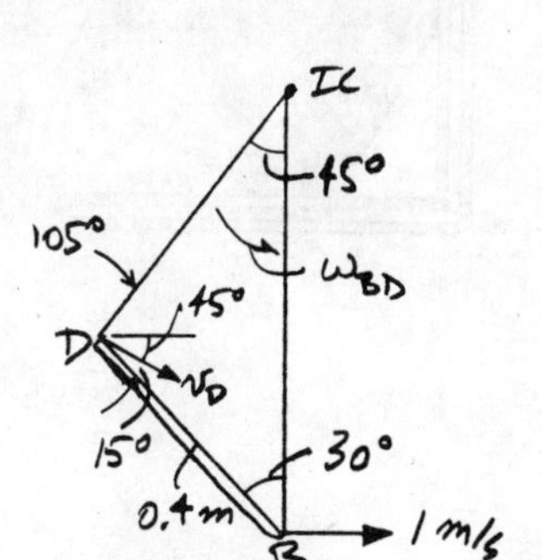

$v_B = 0.2(5) = 1 \text{ m/s} \rightarrow$

Member BD:

$$\frac{r_{B/IC}}{\sin 105^\circ} = \frac{0.4}{\sin 45^\circ}$$

$r_{B/IC} = 0.54641 \text{ m}$

$$\frac{r_{D/IC}}{\sin 30^\circ} = \frac{0.4}{\sin 45^\circ}$$

$r_{D/IC} = 0.28284 \text{ m}$

$$\omega_{BD} = \frac{1}{0.54641} = 1.830 \text{ rad/s}$$

$v_D = 1.830(0.28284) = 0.518 \text{ m/s} \searrow$ **Ans**

***16-100.** The square plate is confined within the slots at A and B. When $\theta = 30°$, point A is moving at $v_A = 8$ m/s. Determine the velocity of point C at this instant.

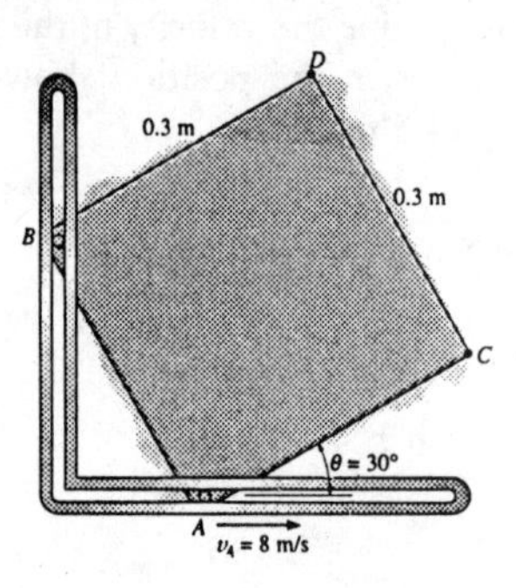

$r_{A/IC} = 0.3\cos 30° = 0.2598$ m

$$\omega = \frac{8}{0.2598} = 30.792 \text{ rad/s}$$

$r_{C/IC} = \sqrt{(0.2598)^2 + (0.3)^2 - 2(0.2598)(0.3)\cos 60°} = 0.2821$ m

$v_C = (0.2821)(30.792) = 8.69$ m/s **Ans**

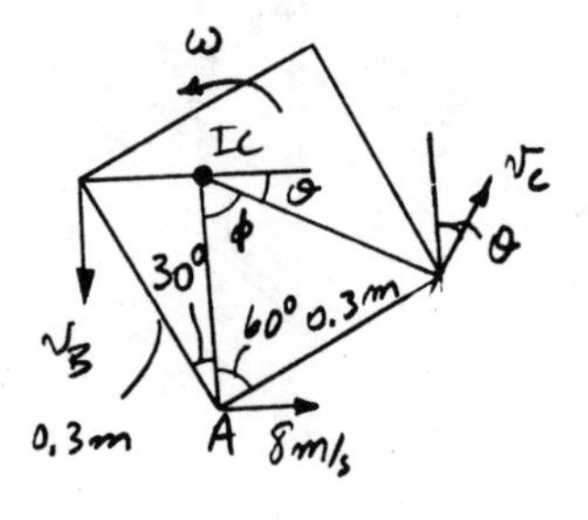

$$\frac{\sin\phi}{0.3} = \frac{\sin 60°}{0.2821}$$

$\phi = 67.09°$

$\theta = 90° - 67.09° = 22.9°$ **Ans**

16-101. The square plate is confined within the slots at A and B. When $\theta = 30°$, point A is moving at $v_A = 8$ m/s. Determine the velocity of point D at this instant.

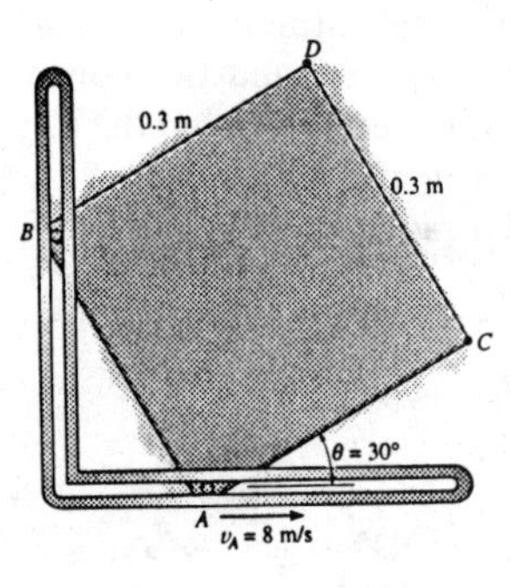

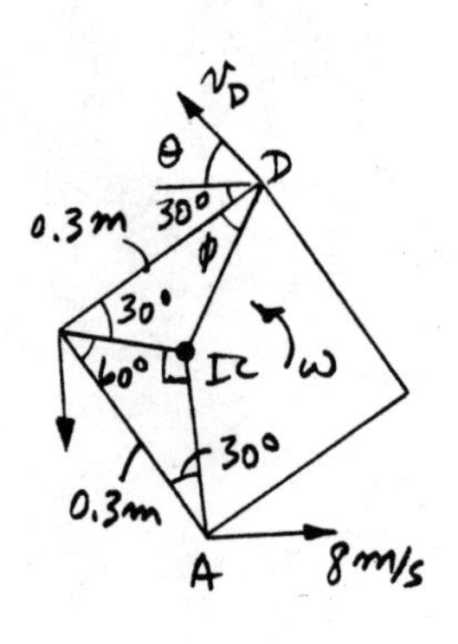

$r_{A/IC} = 0.3\cos 30° = 0.2598$ m

$$\omega = \frac{8}{0.2598} = 30.792 \text{ rad/s}$$

$r_{B/IC} = 0.3\sin 30° = 0.15$ m

$r_{D/IC} = \sqrt{(0.3)^2 + (0.15)^2 - 2(0.3)(0.15)\cos 30°} = 0.1859$ m

$v_D = (30.792)(0.1859) = 5.72$ m/s **Ans**

$$\frac{\sin\phi}{0.15} = \frac{\sin 30°}{0.1859}$$

$\phi = 23.794°$

$\theta = 90° - 30° - 23.794° = 36.2°$ **Ans**

16-102. If the slider block A is moving to the right at $v_A = 8$ ft/s, determine the velocities of blocks B and C at the instant shown.

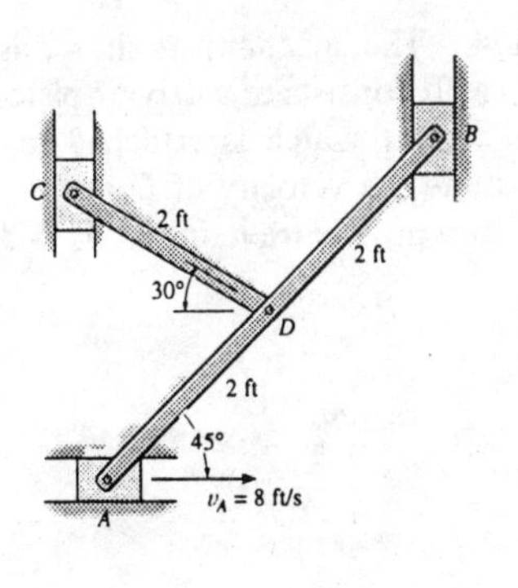

Bar AB :

$r_{A/IC} = 4\cos 45° = 2.828$

$\omega = \dfrac{8}{2.828} = 2.83$ rad/s

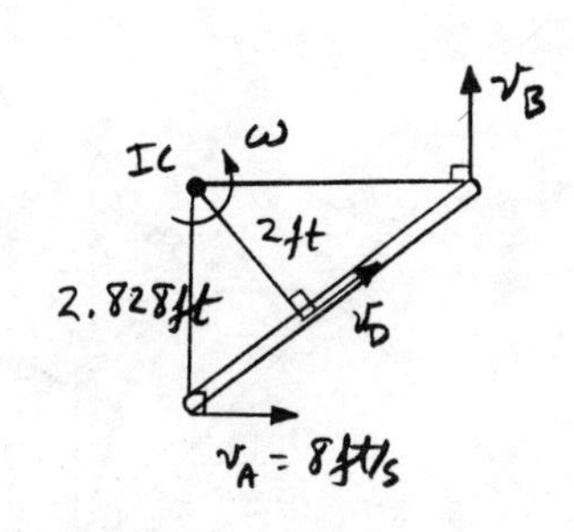

$v_B = 2.83(2.828) = 8$ ft/s ↑ **Ans**

$v_D = 2(2.83) = 5.657$ ft/s

Bar CD :

$\dfrac{r_{D/IC}}{\sin 30°} = \dfrac{2}{\sin 135°}$

$r_{D/IC} = 1.414$ ft

$\dfrac{r_{C/IC}}{\sin 15°} = \dfrac{2}{\sin 135°}$

$r_{C/IC} = 0.7321$ ft

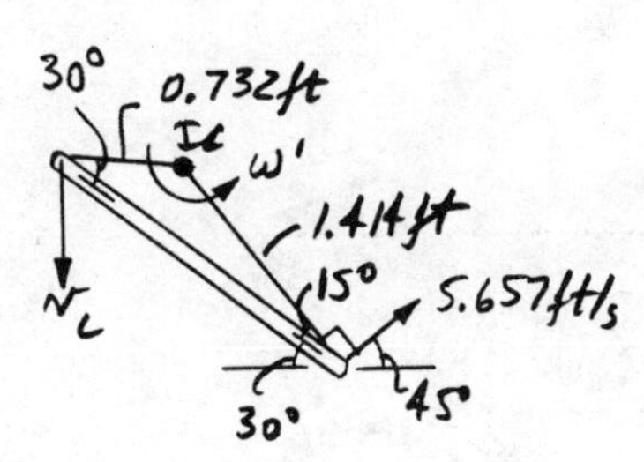

$\omega' = \dfrac{5.657}{1.414} = 4.00$ rad/s

$v_C = 0.7321(4.00) = 2.93$ ft/s ↓ **Ans**

16-103. The crankshaft AB rotates at $\omega_{AB} = 50$ rad/s about the fixed axis through point A, and the disk at C is held fixed in its support at E. Determine the angular velocity of rod CD at the instant shown.

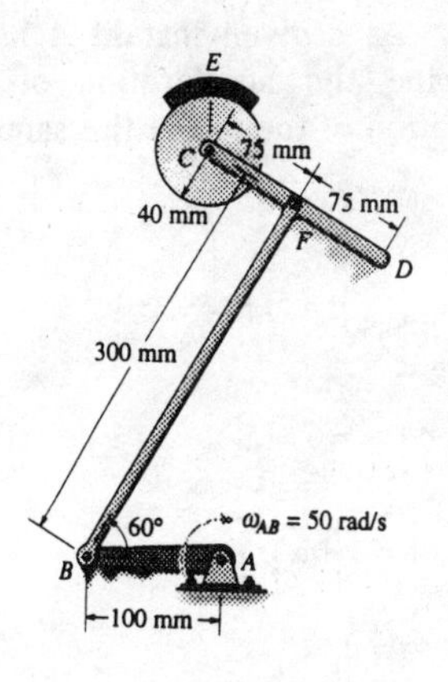

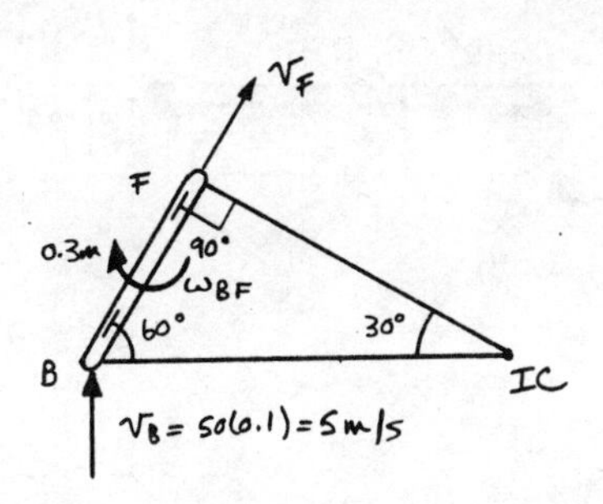

$r_{B/IC} = \dfrac{0.3}{\sin 30°} = 0.6$ m

$r_{F/IC} = \dfrac{0.3}{\tan 30°} = 0.5196$ m

$\omega_{BF} = \dfrac{5}{0.6} = 8.333$ rad/s

$v_F = 8.333(0.5196) = 4.330$ m/s

Thus,

$\omega_{CD} = \dfrac{4.330}{0.075} = 57.7$ rad/s ↻ **Ans**

***16-104.** The mechanism shown is used in a riveting machine. It consists of a driving piston A, three members, and a riveter which is attached to the slider block D. Determine the velocity of D at the instant shown, when the piston at A is traveling at $v_A = 30$ m/s.

Link AC :

$$\omega_{AC} = \frac{v_A}{r_{A/IC}} = \frac{30}{\left(\frac{0.3}{\sin 15^\circ}\right)} = 25.88 \text{ rad/s}$$

$$v_C = \omega_{AC} r_{C/IC} = (25.88)\left(\frac{0.3}{\tan 15^\circ}\right) = 28.98 \text{ m/s}$$

Link DC :

$$\frac{r_{C/IC}}{\sin 45^\circ} = \frac{0.15}{\sin 120^\circ}$$

$$r_{C/IC} = 0.1225 \text{ m}$$

$$\frac{r_{D/IC}}{\sin 15^\circ} = \frac{0.15}{\sin 120^\circ}$$

$$r_{D/IC} = 0.04483 \text{ m}$$

$$\omega_{DC} = \frac{v_C}{r_{C/IC}} = \frac{28.98}{0.1225} = 236.60 \text{ rad/s}$$

$$v_D = \omega_{DC} r_{D/IC} = 236.60(0.04483)$$

$v_D = 10.6 \text{ m/s} \downarrow$ **Ans**

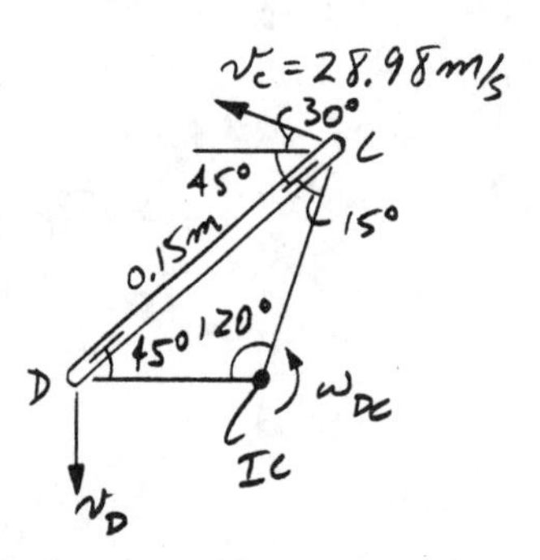

16-105. At a given instant A has the motion shown. Determine the acceleration of B and the angular acceleration of the bar at this same instant.

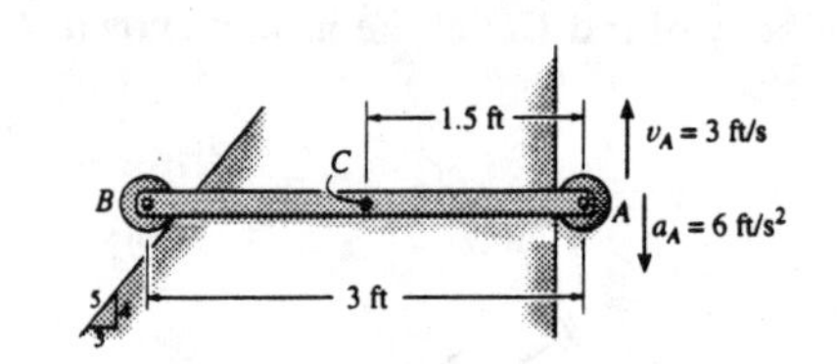

$$\mathbf{v}_B = \mathbf{v}_A + \boldsymbol{\omega} \times \mathbf{r}_{B/A}$$

$$v_B\left(\frac{3}{5}\right)\mathbf{i} + v_B\left(\frac{4}{5}\right)\mathbf{j} = 3\mathbf{j} + \omega\mathbf{k} \times (-3\mathbf{i})$$

$(\overset{+}{\rightarrow})$ $\quad 0.6v_B = 0$

$(+\uparrow)$ $\quad 0 = 3 - \omega(3)$

$$v_B = 0$$

$$\omega = 1 \text{ rad/s}$$

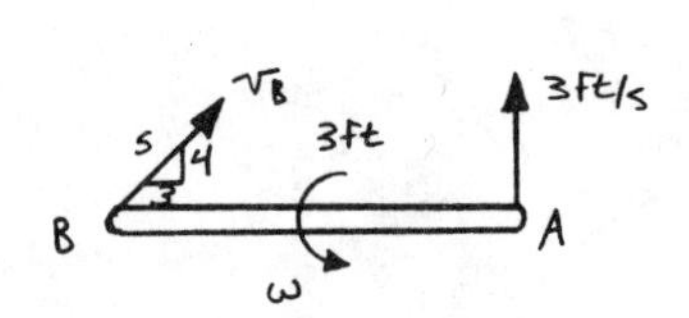

$$\mathbf{a}_B = \mathbf{a}_A + \boldsymbol{\alpha} \times \mathbf{r}_{B/A} - \omega^2 \mathbf{r}_{B/A}$$

$$a_B\left(\frac{3}{5}\right)\mathbf{i} + a_B\left(\frac{4}{5}\right)\mathbf{j} = -6\mathbf{j} + \alpha\mathbf{k} \times (-3\mathbf{i}) - (1)^2(-3\mathbf{i})$$

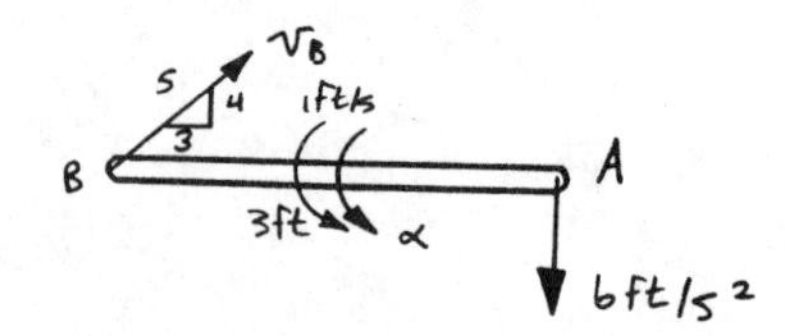

$(\overset{+}{\rightarrow})$ $\quad 0.6a_B = 3$

$(+\uparrow)$ $\quad 0.8(5) = -6 - 3\alpha$

$a_B = 5 \text{ ft/s}^2 \nearrow$ **Ans**

$\alpha = -3.33 \text{ rad/s}^2 = 3.33 \text{ rad/s}^2$ **Ans**

16-106. At a given instant A has the motion shown. Determine the acceleration of point C at this same instant.

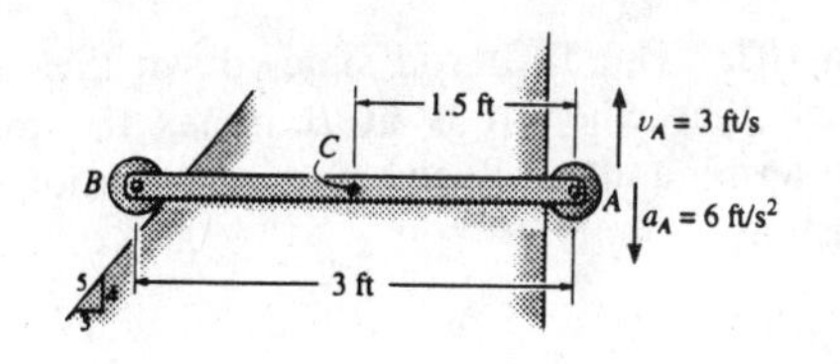

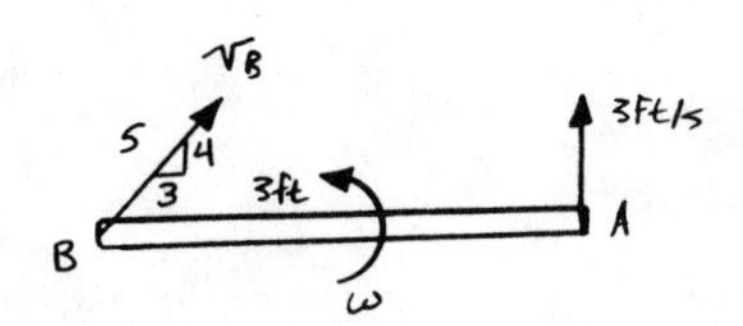

$\mathbf{v}_B = \mathbf{v}_A + \omega \times \mathbf{r}_{B/A}$

$v_B\left(\frac{3}{5}\right)\mathbf{i} + v_B\left(\frac{4}{5}\right)\mathbf{j} = 3\mathbf{j} + \omega\mathbf{k} \times (-3\mathbf{i})$

$(\overset{+}{\rightarrow})$ $\quad 0.6v_B = 0$

$(+\uparrow)$ $\quad 0 = 3 - \omega(3)$

$v_B = 0$

$\omega = 1$ rad/s

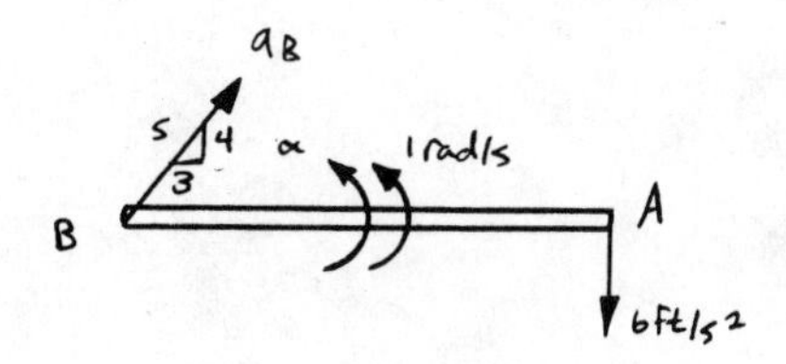

$\mathbf{a}_B = \mathbf{a}_A + \alpha \times \mathbf{r}_{B/A} - \omega^2 \mathbf{r}_{B/A}$

$a_B\left(\frac{3}{5}\right)\mathbf{i} + a_B\left(\frac{4}{5}\right)\mathbf{j} = -6\mathbf{j} + \alpha\mathbf{k} \times (-3\mathbf{i}) - (1)^2(-3\mathbf{i})$

$(\overset{+}{\rightarrow})$ $\quad 0.6a_B = 3$

$(+\uparrow)$ $\quad 0.8(5) = -6 - 3\alpha$

$a_B = 5$ ft/s²

$\alpha = -3.33$ rad/s² $= 3.33$ rad/s²

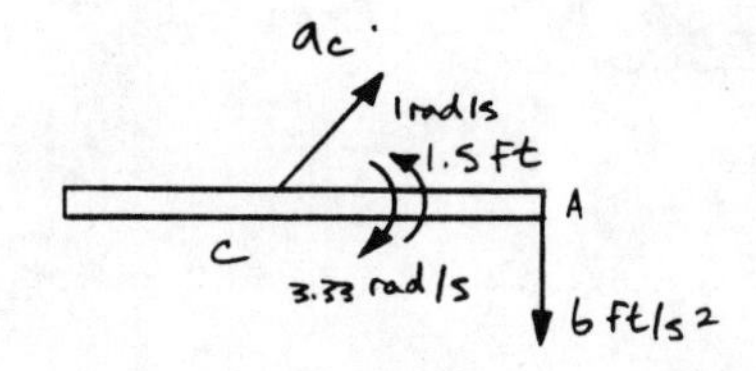

$\mathbf{a}_C = \mathbf{a}_A + \alpha \times \mathbf{r}_{C/A} - \omega^2 \mathbf{r}_{C/A}$

$\mathbf{a}_C = -6\mathbf{j} + (-3.33)\mathbf{k} \times (-1.5\mathbf{i}) - (1)^2(-1.5\mathbf{i})$

$\mathbf{a}_C = \{1.5\mathbf{i} - 1\mathbf{j}\}$ ft/s²

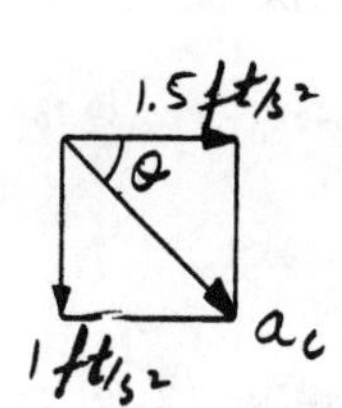

$a_C = \sqrt{(1.5)^2 + (-1)^2} = 1.80$ ft/s² **Ans**

$\theta = \tan^{-1}\left(\frac{1}{1.5}\right) = 33.7°$ **Ans**

16-107. The 10-ft rod slides down the inclined plane, such that when it is at B it has the motion shown. Determine the velocity and acceleration of A at this instant.

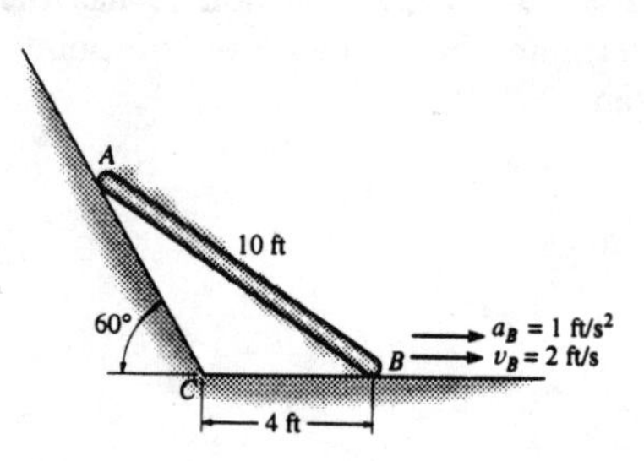

$(10)^2 = (4)^2 + (AC)^2 - 2(AC)(4)\cos 120°$

$(AC)^2 + 4(AC) - 84 = 0$

Solving for the positive root :

$AC = 7.381$ ft

$\frac{\sin\theta}{7.381} = \frac{\sin 120°}{10}$ $\theta = 39.732°$

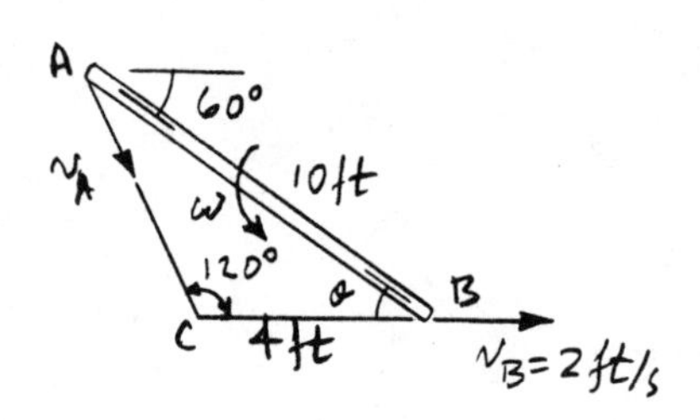

$\mathbf{v}_A = \mathbf{v}_B + \omega \times \mathbf{r}_{A/B}$

$v_A \cos 60°\mathbf{i} - v_A \sin 60°\mathbf{j} = 2\mathbf{i} + \omega\mathbf{k} \times (-10\cos 39.732°\mathbf{i} + 10\sin 39.732°\mathbf{j})$

$(\overset{+}{\rightarrow})$ $0.5v_A = 2 - 6.39199\omega$

$(+\uparrow)$ $-0.86603v_A = -7.6904\omega$

Solving :

$\omega = 0.1846$ rad/s

$v_A = 1.64$ ft/s **Ans**

$\mathbf{a}_A = \mathbf{a}_B + \alpha \times \mathbf{r}_{A/B} - \omega^2 \mathbf{r}_{A/B}$

$a_A \cos 60°\mathbf{i} - a_A \sin 60°\mathbf{j} = 1\mathbf{i} + (\alpha\mathbf{k}) \times (-10\cos 39.732°\mathbf{i} + 10\sin 39.732°\mathbf{j})$

$- (0.1846)^2(-10\cos 39.732°\mathbf{i} + 10\sin 39.732°\mathbf{j})$

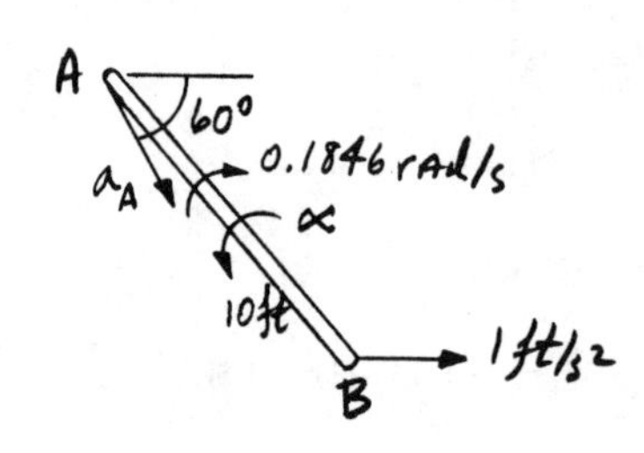

$(\overset{+}{\rightarrow})$ $0.5a_A = 1 - 6.3920\alpha + 0.2621$

$(+\uparrow)$ $-0.86603a_A = -7.69042\alpha - 0.21791$

Solving :

$a_A = 1.18$ ft/s² **Ans**

$\alpha = 0.105$ rad/s²

***16-108.** At a given instant, the slider block A has the velocity and deceleration shown. Determine the acceleration of block B and the angular acceleration of the link at this instant.

$$\omega_{AB} = \frac{v_B}{r_{A/IC}} = \frac{1.5}{0.3\cos 45^\circ} = 7.07 \text{ rad/s}$$

$$\mathbf{a}_B = \mathbf{a}_A + \alpha \times \mathbf{r}_{B/A} - \omega^2 \mathbf{r}_{B/A}$$

$$-a_B\mathbf{j} = 16\mathbf{i} + (\alpha\mathbf{k}) \times (0.3\cos 45^\circ\mathbf{i} + 0.3\sin 45^\circ\mathbf{j}) - (7.07)^2(0.3\cos 45^\circ\mathbf{i} + 0.3\sin 45^\circ\mathbf{j})$$

$(\overset{+}{\rightarrow})$ $\quad 0 = 16 - \alpha(0.3)\sin 45^\circ - (7.07)^2(0.3)\cos 45^\circ$

$(+\downarrow)$ $\quad a_B = 0 - \alpha(0.3)\cos 45^\circ + (7.07)^2(0.3)\sin 45^\circ$

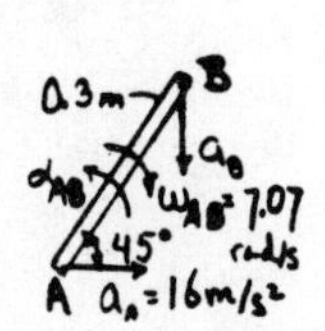

Solving :

$\alpha_{AB} = 25.4 \text{ rad/s}^2$ ↻ **Ans**

$a_B = 5.21 \text{ m/s}^2$ ↓ **Ans**

16-109. Determine the angular acceleration of link AB at the instant $\theta = 90^\circ$ if the collar C has a velocity of $v_C = 4$ ft/s and deceleration of $a_C = 3$ ft/s² as shown.

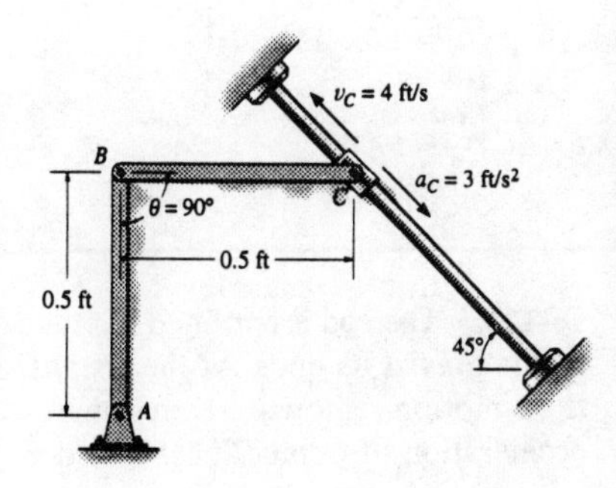

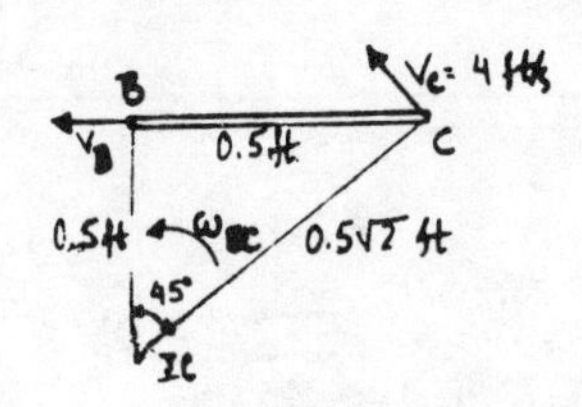

$$\omega_{BC} = \frac{v_C}{r_{C/IC}} = \frac{4}{0.5\sqrt{2}} = 5.657 \text{ rad/s}$$

$$v_B = \omega_{BC} r_{B/IC} = 5.657(0.5) = 2.828 \text{ ft/s}$$

$$\omega_{AB} = \frac{v_B}{r_{B/A}} = \frac{2.828}{0.5} = 5.657 \text{ rad/s}$$

$$(\mathbf{a}_B)_t + (\mathbf{a}_B)_n = \mathbf{a}_C + \alpha \times \mathbf{r}_{B/C} - \omega^2 \mathbf{r}_{B/C}$$

$$\alpha_{AB}(0.5)\mathbf{i} - (5.657)^2(0.5)\mathbf{j} = 3\cos 45^\circ\mathbf{i} - 3\sin 45^\circ\mathbf{j} + (-\alpha_{BC}\mathbf{k}) \times (-0.5\mathbf{i}) - (5.657)^2(-0.5\mathbf{i})$$

$(\overset{+}{\rightarrow})$ $\quad \alpha_{AB}(0.5) = 3\cos 45^\circ + (5.657)^2(0.5)$

$(+\uparrow)$ $\quad -(5.657)^2(0.5) = -3\sin 45^\circ - \alpha_{BC}(0.5)$

Solving,

$\alpha_{AB} = 36.2 \text{ rad/s}^2$ ↻ **Ans**

$\alpha_{BC} = 27.8 \text{ rad/s}^2$ ↻

16-110. At a given instant the slider block B is moving to the right with the motion shown. Determine the angular acceleration of link AB and the acceleration of point A at this instant.

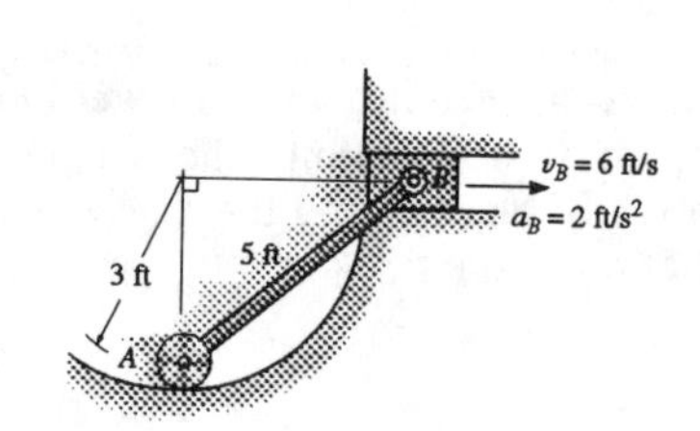

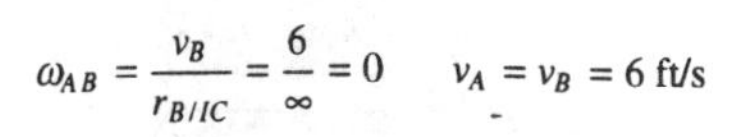

$$\omega_{AB} = \frac{v_B}{r_{B/IC}} = \frac{6}{\infty} = 0 \qquad v_A = v_B = 6 \text{ ft/s}$$

$$\omega_{AC} = \frac{v_A}{r_{AC}} = \frac{6}{3} = 2 \text{ rad/s}$$

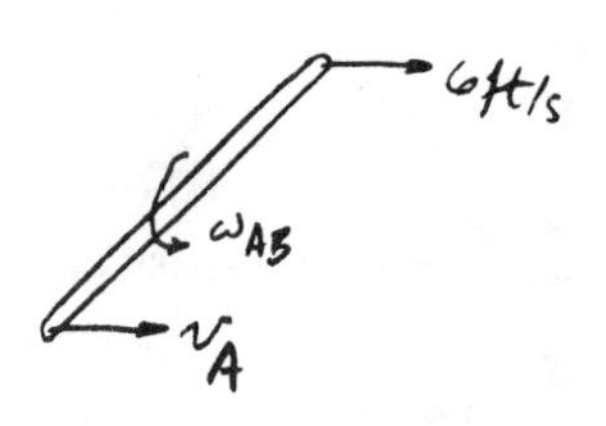

$$\mathbf{a}_B = \{2\mathbf{i}\} \text{ ft/s}^2 \qquad \mathbf{a}_A = (a_A)_x\,\mathbf{i} + (2)^2(3)\mathbf{j} = (a_A)_x\,\mathbf{i} + 12\mathbf{j}$$

$$\alpha_{AB} = -\alpha_{AB}\mathbf{k} \qquad \mathbf{r}_{B/A} = \{4\mathbf{i} + 3\mathbf{j}\} \text{ ft}$$

$$\mathbf{a}_B = \mathbf{a}_A + \alpha_{AB} \times \mathbf{r}_{B/A} - \omega^2 \mathbf{r}_{B/A}$$

$$2\mathbf{i} = \left[(a_A)_x\,\mathbf{i} + 12\mathbf{j}\right] + (-\alpha_{AB}\mathbf{k}) \times (4\mathbf{i} + 3\mathbf{j}) - \mathbf{0}$$

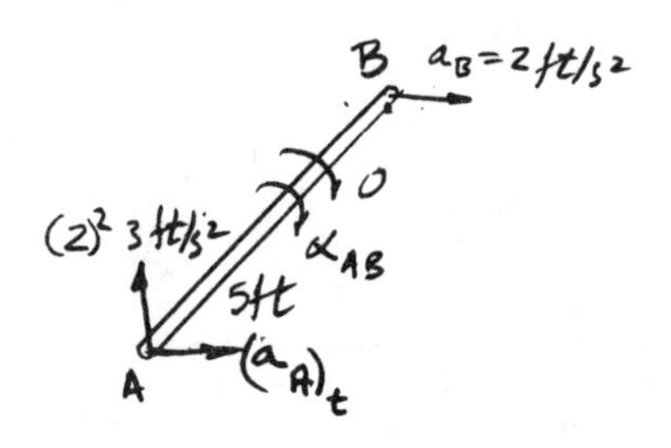

$(\overset{+}{\rightarrow})$ $\quad 2 = (a_A)_x + 3(3) \qquad (a_A)_x = -7 \text{ ft/s}^2$

$(+\uparrow)$ $\quad 0 = 12 - 4\alpha_{AB} \qquad \alpha_{AB} = 3 \text{ rad/s}^2$ ↺ **Ans**

$$\mathbf{a}_A = \{-7\mathbf{i} + 12\mathbf{j}\} \text{ ft/s}^2$$

$a_A = \sqrt{(-7)^2 + 12^2} = 13.9 \text{ ft/s}^2$ **Ans**

$\theta = \tan^{-1}\frac{12}{7} = 59.7°$ ↘ **Ans**

16-111. The rod is confined to move along the path due to the pins at its ends. At the instant shown, point A has the motion shown. Determine the velocity and acceleration of point B at this instant.

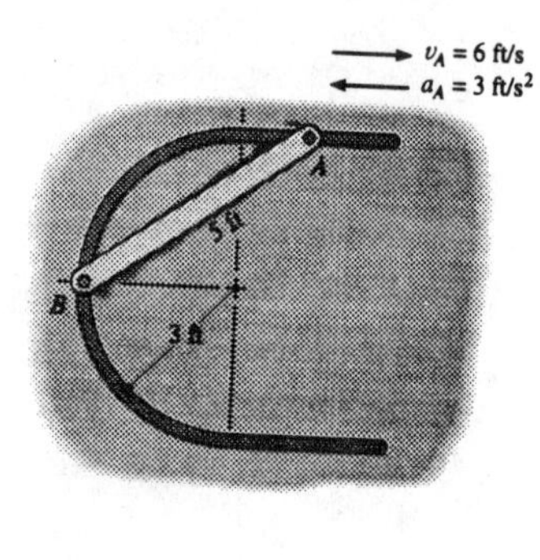

$$\mathbf{v}_B = \mathbf{v}_A + \omega \times \mathbf{r}_{B/A}$$

$$v_B\mathbf{j} = 6\mathbf{i} + (-\omega\mathbf{k}) \times (-4\mathbf{i} - 3\mathbf{j})$$

$$0 = 6 - 3\omega; \qquad \omega = 2 \text{ rad/s}$$

$v_B = 4\omega = 4(2) = 8 \text{ ft/s} \uparrow$ **Ans**

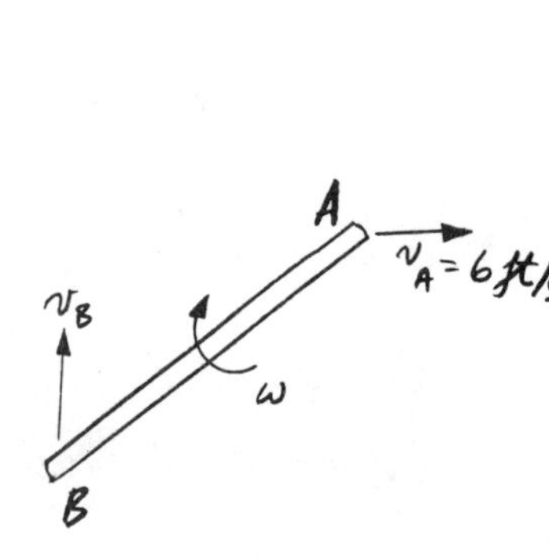

$$\mathbf{a}_B = \mathbf{a}_A + \alpha \times \mathbf{r}_{B/A} - \omega^2 \mathbf{r}_{B/A}$$

$$21.33\mathbf{i} + (a_B)_t\,\mathbf{j} = -3\mathbf{i} + \alpha\mathbf{k} \times (-4\mathbf{i} - 3\mathbf{j}) - (-2)^2(-4\mathbf{i} - 3\mathbf{j})$$

$(\overset{+}{\rightarrow})$ $\quad 21.33 = -3 + 3\alpha + 16; \qquad \alpha = 2.778 \text{ rad/s}^2$

$(+\uparrow)$ $\quad (a_B)_t = -(2.778)(4) + 12 = 0.8889 \text{ ft/s}^2$

$a_B = \sqrt{(21.33)^2 + (0.8889)^2} = 21.4 \text{ ft/s}^2$ **Ans**

$\theta = \tan^{-1}\left(\frac{0.8889}{21.33}\right) = 2.39°$ ∠ **Ans**

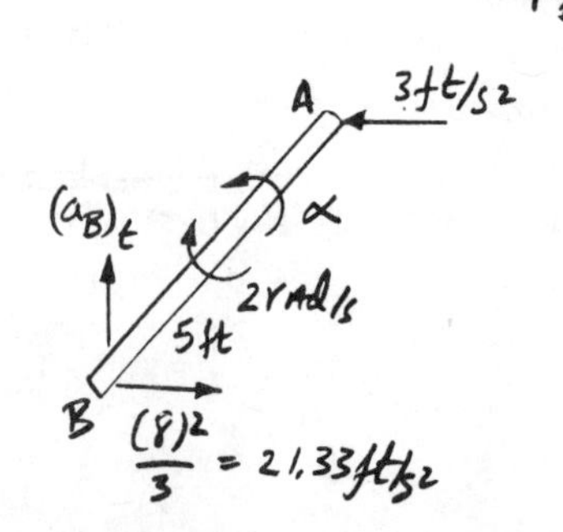

***16-112.** The closure is manufactured by the LCN Company and is used to control the restricted motion of a heavy door. If the door to which is it connected has an angular acceleration of 3 rad/s^2, determine the angular acceleration of links BC and CD. Originally the door is not rotating but is hinged at A.

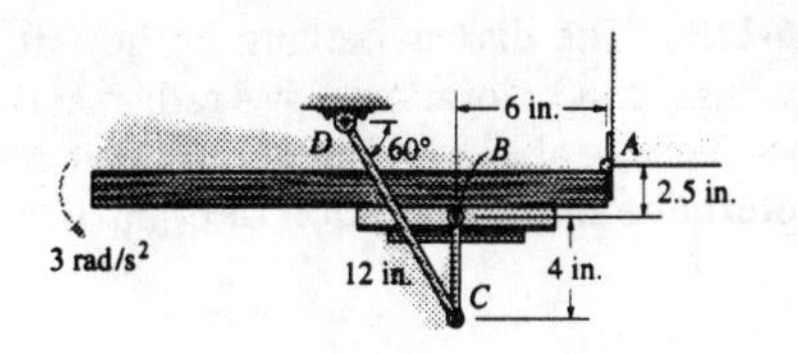

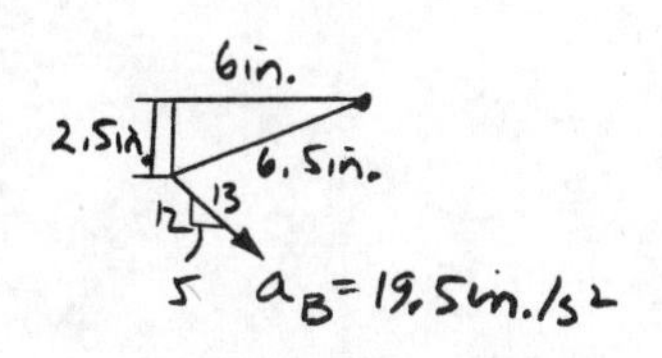

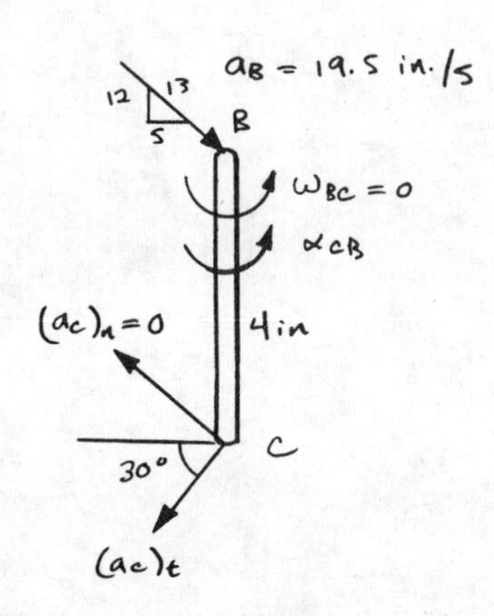

$\mathbf{a}_B = (\mathbf{a}_B)_t + (\mathbf{a}_B)_n$

$(a_B)_t = \alpha r_{AB} = 3(6.5) = 19.5 \text{ in./s}^2$

$(a_B)_n = \omega^2 r_{AB} = 0^2(6.5) = 0$

$(a_C)_n = \omega^2 r_{DC} = 0(12) = 0$

$\mathbf{a}_C = \mathbf{a}_B + \boldsymbol{\alpha} \times \mathbf{r}_{C/B} - \omega^2 \mathbf{r}_{C/B}$

$-(a_C)_t \cos 30°\mathbf{i} - (a_C)_t \sin 30°\mathbf{j} = 19.5\left(\frac{5}{13}\right)\mathbf{i} - 19.5\left(\frac{12}{13}\right)\mathbf{j} + (\alpha_{CB}\mathbf{k}) \times (-4\mathbf{j}) - (0)^2(-4\mathbf{j})$

$(\xrightarrow{+})$ $\quad -0.8660(a_C)_t = 7.5 + 4\alpha_{CB}$

$(+\uparrow)$ $\quad -0.5(a_C)_t = -18$

Solving :

$(a_C)_t = 36 \text{ in./s}^2$

$\alpha_{CB} = -9.67 \text{ rad/s}^2 = 9.67 \text{ rad/s}^2$ ↻ **Ans**

$\alpha_{DC} = \frac{36}{12} = 3 \text{ rad/s}^2$ ↻ **Ans**

16-113. The disk is moving to the left such that it has an angular acceleration $\alpha = 8\ \text{rad/s}^2$ and angular velocity $\omega = 3\ \text{rad/s}$ at the instant shown. If it does not slip at A, determine the acceleration of point B.

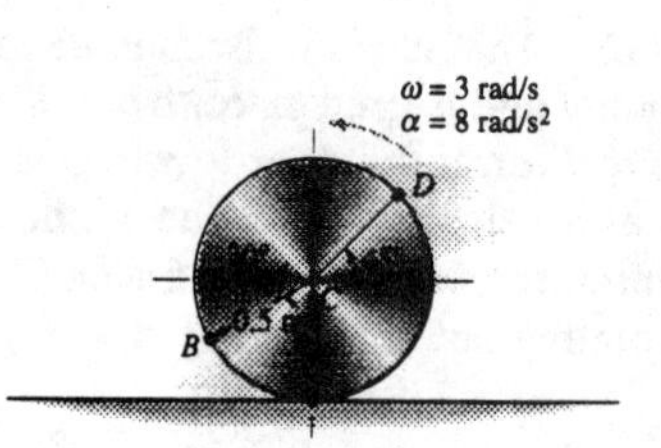

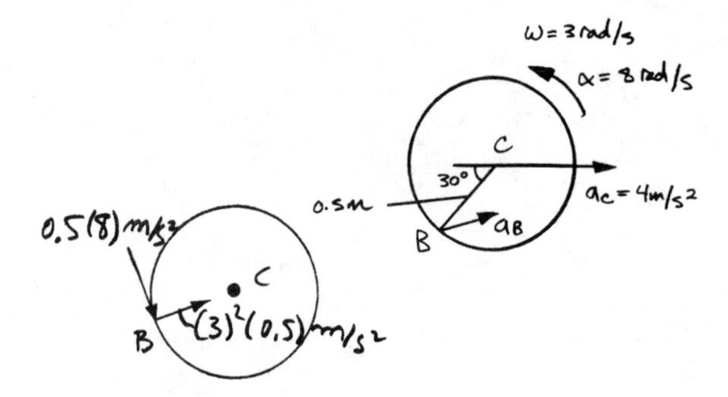

$a_C = 0.5(8) = 4\ \text{m/s}^2$

$\mathbf{a}_B = \mathbf{a}_C + \mathbf{a}_{B/C}$

$$\mathbf{a}_B = \left[\underset{\leftarrow}{4}\right] + \left[\begin{matrix}(3)^2(0.5)\\ \measuredangle 30^\circ\end{matrix}\right] + \left[\begin{matrix}(0.5)(8)\\ \nwarrow 30^\circ\end{matrix}\right]$$

$(\overset{+}{\rightarrow})$ $\quad (a_B)_x = -4 + 4.5\cos 30^\circ + 4\sin 30^\circ = 1.897\ \text{m/s}^2$

$(+\uparrow)$ $\quad (a_B)_y = 0 + 4.5\sin 30^\circ - 4\cos 30^\circ = -1.214\ \text{m/s}^2$

$$a_B = \sqrt{(1.897)^2 + (-1.214)^2} = 2.25\ \text{m/s}^2 \qquad \textbf{Ans}$$

$$\theta = \tan^{-1}\left(\frac{1.214}{1.897}\right) = 32.6^\circ \searrow \qquad \textbf{Ans}$$

Also,

$\mathbf{a}_B = \mathbf{a}_C + \alpha \times \mathbf{r}_{B/C} - \omega^2 \mathbf{r}_{B/C}$

$(a_B)_x\mathbf{i} + (a_B)_y\mathbf{j} = -4\mathbf{i} + (8\mathbf{k}) \times (-0.5\cos 30^\circ\mathbf{i} - 0.5\sin 30^\circ\mathbf{j}) - (3)^2(-0.5\cos 30^\circ\mathbf{i} - 0.5\sin 30^\circ\mathbf{j})$

$(\overset{+}{\rightarrow})$ $\quad (a_B)_x = -4 + 8(0.5\sin 30^\circ) + (3)^2(0.5\cos 30^\circ) = 1.897\ \text{m/s}^2$

$(+\uparrow)$ $\quad (a_B)_y = 0 - 8(0.5\cos 30^\circ) + (3)^2(0.5\sin 30^\circ) = -1.214\ \text{m/s}^2$

1.897 m/s²

1.214 m/s²

$$\theta = \tan^{-1}\left(\frac{1.214}{1.897}\right) = 32.6^\circ \searrow \qquad \textbf{Ans}$$

$$a_B = \sqrt{(1.897)^2 + (-1.214)^2} = 2.25\ \text{m/s}^2 \qquad \textbf{Ans}$$

16-114. The disk is moving to the left such that it has an angular acceleration $\alpha = 8\ \text{rad/s}^2$ and angular velocity $\omega = 3$ rad/s at the instant shown. If it does not slip at A, determine the acceleration of point D.

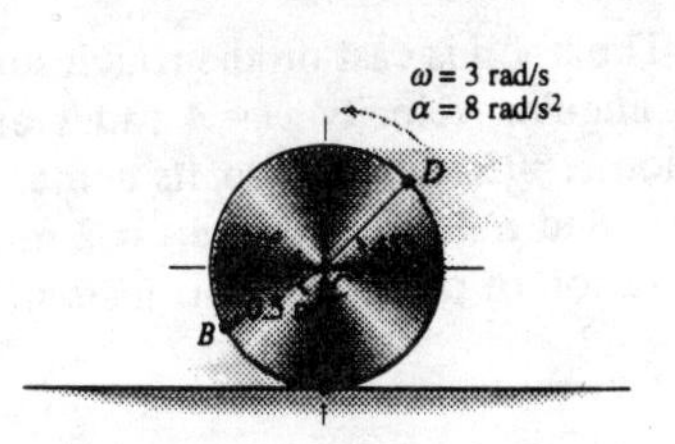

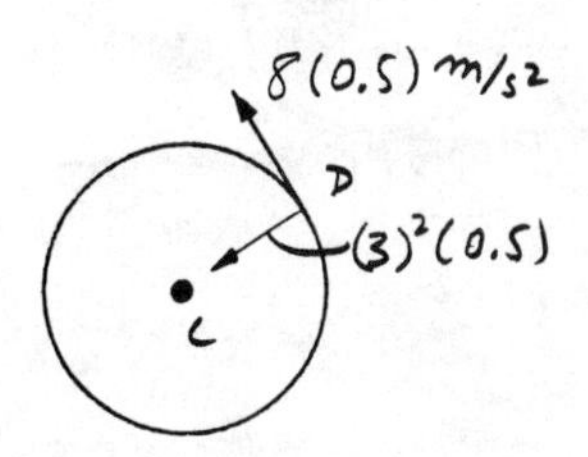

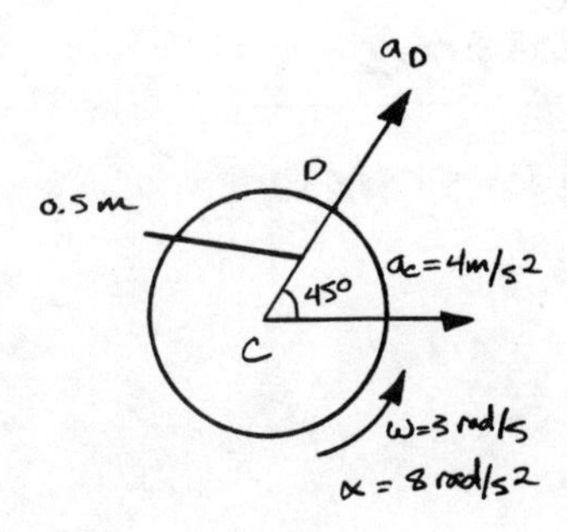

$a_C = 0.5(8) = 4\ \text{m/s}^2$

$\mathbf{a}_D = \mathbf{a}_C + \mathbf{a}_{D/C}$

$$\mathbf{a}_D = \left[\underset{\leftarrow}{4}\right] + \left[\underset{\swarrow 45^\circ}{(3)^2(0.5)}\right] + \left[\underset{45^\circ \searrow}{8(0.5)}\right]$$

$(\overset{+}{\rightarrow})\qquad (a_D)_x = -4 - 4.5\sin 45^\circ - 4\cos 45^\circ = -10.01\ \text{m/s}^2$

$(+\uparrow)\qquad (a_D)_y = 0 - 4.5\cos 45^\circ + 4\sin 45^\circ = -0.3536\ \text{m/s}^2$

$$\theta = \tan^{-1}\left(\frac{0.3536}{10.01}\right) = 2.02^\circ \qquad \textbf{Ans}$$

$$a_D = \sqrt{(-10.01)^2 + (-0.3536)^2} = 10.0\ \text{m/s}^2 \ \swarrow \qquad \textbf{Ans}$$

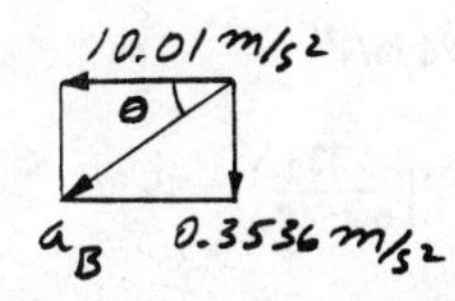

Also,

$\mathbf{a}_D = \mathbf{a}_C + \boldsymbol{\alpha} \times \mathbf{r}_{D/C} - \omega^2 \mathbf{r}_{D/C}$

$(a_D)_x\,\mathbf{i} + (a_D)_y\,\mathbf{j} = -4\mathbf{i} + (8\mathbf{k}) \times (0.5\cos 45^\circ\mathbf{i} + 0.5\sin 45^\circ\mathbf{j}) - (3)^2(0.5\cos 45^\circ\mathbf{i} + 0.5\sin 45^\circ\mathbf{j})$

$(\overset{+}{\rightarrow})\qquad (a_D)_x = -4 - 8(0.5\sin 45^\circ) - (3)^2(0.5\cos 45^\circ) = -10.01\ \text{m/s}^2$

$(+\uparrow)\qquad (a_D)_y = +8(0.5\cos 45^\circ) - (3)^2(0.5\sin 45^\circ) = -0.3536\ \text{m/s}^2$

$$\theta = \tan^{-1}\left(\frac{0.3536}{10.01}\right) = 2.02^\circ \ \swarrow \qquad \textbf{Ans}$$

$$a_D = \sqrt{(-10.01)^2 + (-0.3536)^2} = 10.0\ \text{m/s}^2 \qquad \textbf{Ans}$$

16-115. The hoop is cast on the rough surface such that it has an angular velocity $\omega = 4$ rad/s and an angular deceleration $\alpha = 5$ rad/s^2. Also, its center has a velocity $v_O = 5$ m/s and a deceleration $a_O = 2$ m/s^2. Determine the acceleration of point A at this instant.

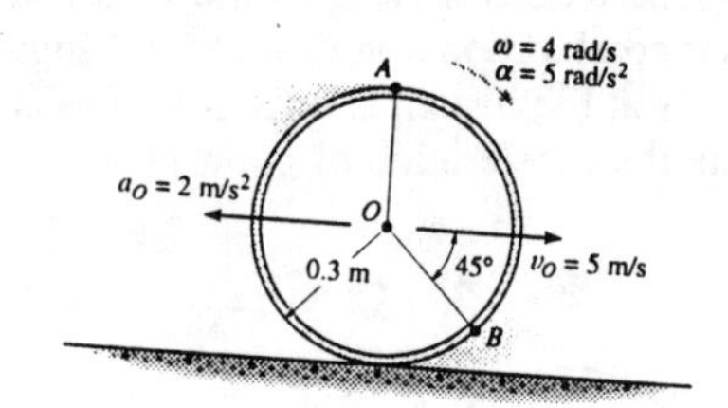

$\mathbf{a}_A = \mathbf{a}_O + \mathbf{a}_{A/O}$

$\mathbf{a}_A = \left[\underset{\leftarrow}{2}\right] + \left[\underset{\downarrow}{(4)^2(0.3)}\right] + \left[\underset{\rightarrow}{5(0.3)}\right]$

$\mathbf{a}_A = \left[\underset{\leftarrow}{0.5}\right] + \left[\underset{\downarrow}{4.8}\right]$

$a_A = 4.83$ m/s^2 **Ans**

$\theta = \tan^{-1}\left(\frac{4.8}{0.5}\right) = 84.1°$ **Ans**

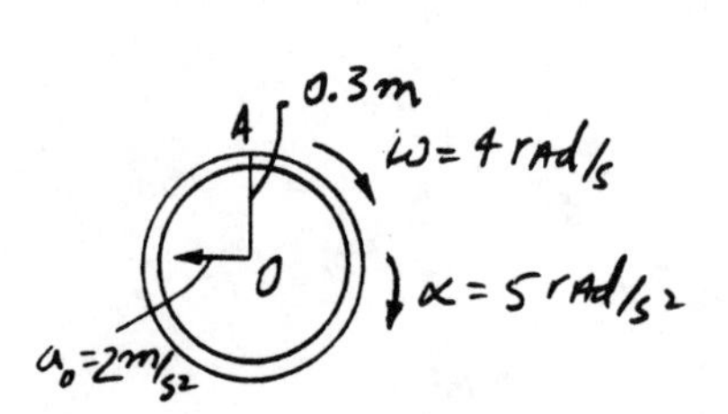

Also :

$\mathbf{a}_A = \mathbf{a}_O - \omega^2 \mathbf{r}_{A/O} + \alpha \times \mathbf{r}_{A/O}$

$\mathbf{a}_A = -2\mathbf{i} - (4)^2(0.3\mathbf{j}) + (-5\mathbf{k}) \times (0.3\mathbf{j})$

$\mathbf{a}_A = \{-0.5\mathbf{i} - 4.8\mathbf{j}\}$ m/s^2

$a_A = 4.83$ m/s^2 **Ans**

$\theta = \tan^{-1}\left(\frac{4.8}{0.5}\right) = 84.1°$ **Ans**

***16-116.** The hoop is cast on the rough surface such that it has an angular velocity $\omega = 4$ rad/s and an angular deceleration $\alpha = 5$ rad/s^2. Also, its center has a velocity of $v_O = 5$ m/s and a deceleration $a_O = 2$ m/s^2. Determine the acceleration of point B at this instant.

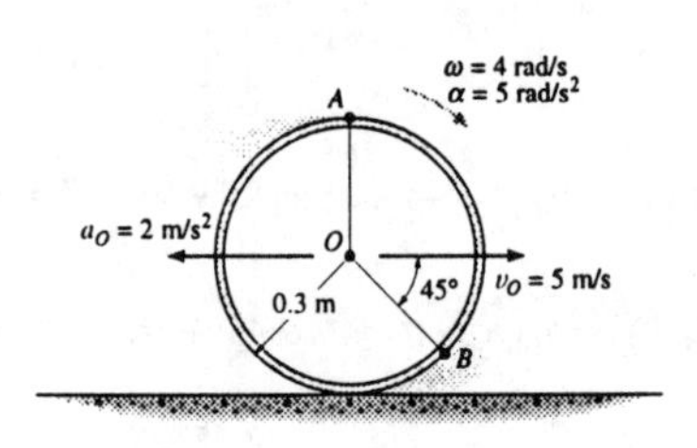

$\mathbf{a}_B = \mathbf{a}_O + \mathbf{a}_{B/O}$

$\mathbf{a}_B = \left[\underset{\leftarrow}{2}\right] + \left[5(0.3)\right] + \left[(4)^2(0.3)\right]$

$\mathbf{a}_B = \left[\underset{\leftarrow}{6.4548}\right] + \left[\underset{\uparrow}{2.333}\right]$

$a_B = 6.86$ m/s^2 **Ans**

$\theta = \tan^{-1}\left(\frac{2.333}{6.4548}\right) = 19.9°$

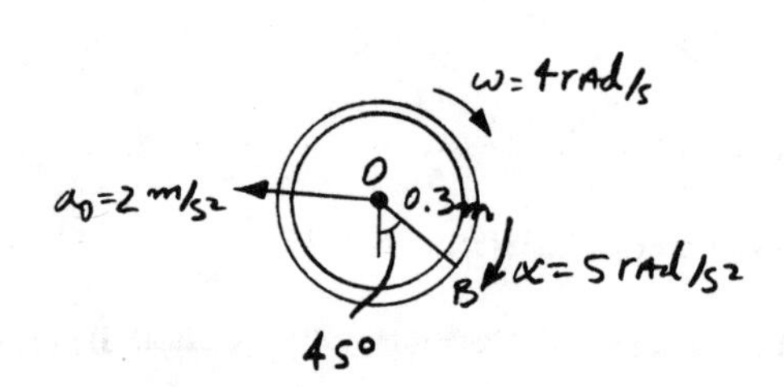

Also :

$\mathbf{a}_B = \mathbf{a}_O + \alpha \times \mathbf{r}_{B/O} - \omega^2 \mathbf{r}_{B/O}$

$\mathbf{a}_B = -2\mathbf{i} + (-5\mathbf{k}) \times (0.3\cos 45°\mathbf{i} - 0.3\sin 45°\mathbf{j}) - (4)^2(0.3\cos 45°\mathbf{i} - 0.3\sin 45°\mathbf{j})$

$\mathbf{a}_B = \{-6.4548\mathbf{i} + 2.333\mathbf{j}\}$ m/s^2

$a_B = 6.86$ m/s^2 **Ans**

$\theta = \tan^{-1}\left(\frac{2.333}{6.4548}\right) = 19.9°$

16-117. Rod AB has the angular motion shown. Determine the acceleration of block C at this instant.

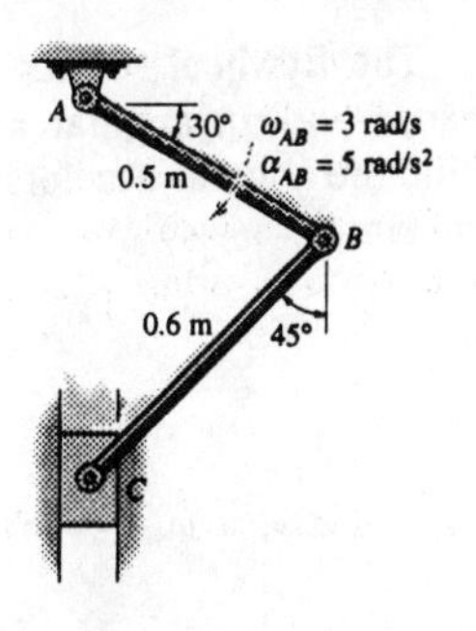

$\mathbf{v}_C = \mathbf{v}_B + \mathbf{v}_{C/B}$

$$\begin{bmatrix} v_C \\ \downarrow \end{bmatrix} = \begin{bmatrix} 1.5 \\ \measuredangle 30^\circ \end{bmatrix} + \begin{bmatrix} 0.6\omega_{CB} \\ 45^\circ \end{bmatrix}$$

$(\xrightarrow{+})$ $\quad 0 = -1.5\sin 30^\circ + 0.6\omega_{CB}\sin 45^\circ$

$\omega_{CB} = 1.768 \text{ rad/s}$

$\mathbf{a}_C = \mathbf{a}_B + \mathbf{a}_{C/B}$

$$\begin{bmatrix} a_C \\ \downarrow \end{bmatrix} = \begin{bmatrix} 5\,(0.5) \\ 60^\circ \end{bmatrix} + \begin{bmatrix} (3)^2(0.5) \\ 30^\circ \end{bmatrix} + \begin{bmatrix} (1.768)^2(0.6) \\ 45^\circ \end{bmatrix} + \begin{bmatrix} \alpha_{CB}\,(0.6) \\ 45^\circ \end{bmatrix}$$

$(\xrightarrow{+})$ $\quad 0 = -2.5\cos 60^\circ - 4.5\cos 30^\circ + 1.875\cos 45^\circ + \alpha_{CB}(0.6)\cos 45^\circ$

$(+\uparrow)$ $\quad -a_C = -2.5\sin 60^\circ + 4.5\sin 30^\circ + 1.875\sin 45^\circ - \alpha_{CB}(0.6)\sin 45^\circ$

$a_C = 2.41 \text{ m/s}^2 \downarrow$ **Ans**

$\alpha_{CB} = 9.01 \text{ rad/s}^2$

Also,

$\mathbf{v}_C = \mathbf{v}_B + \boldsymbol{\omega} \times \mathbf{r}_{C/B}$

$-v_C\mathbf{j} = (-1.5\sin 30^\circ\mathbf{i} - 1.5\cos 30^\circ\mathbf{j}) + (\omega_{CB}\mathbf{k}) \times (-0.6\sin 45^\circ\mathbf{i} - 0.6\cos 45^\circ\mathbf{j})$

$(\xrightarrow{+})$ $\quad 0 = -1.5\sin 30^\circ + \omega_{CB}(0.6\cos 45^\circ)$

$\omega_{CB} = 1.768 \text{ rad/s}$

$\mathbf{a}_C = \mathbf{a}_B + \boldsymbol{\alpha} \times \mathbf{r}_{C/B} - \omega^2 \mathbf{r}_{C/B}$

$-a_C\mathbf{j} = (-4.5\cos 30^\circ\mathbf{i} + 4.5\sin 30^\circ\mathbf{j}) + (-2.5\cos 60^\circ\mathbf{i} - 2.5\sin 60^\circ\mathbf{j})$

$\quad + (\alpha_{CB}\mathbf{k}) \times (-0.6\sin 45^\circ\mathbf{i} - 0.6\cos 45^\circ\mathbf{j}) - (1.768)^2(-0.6\sin 45^\circ\mathbf{i} - 0.6\cos 45^\circ\mathbf{j})$

$(\xrightarrow{+})$ $\quad 0 = -4.5\cos 30^\circ - 2.5\cos 60^\circ + \alpha_{CB}(0.6\cos 45^\circ) + (1.768)^2(0.6\sin 45^\circ)$

$(+\uparrow)$ $\quad -a_C = 4.5\sin 30^\circ - 2.5\sin 60^\circ - \alpha_{CB}(0.6\sin 45^\circ) + (1.768)^2(0.6\cos 45^\circ)$

$a_C = 2.41 \text{ m/s}^2 \downarrow$ **Ans**

$\alpha_{CB} = 9.01 \text{ rad/s}^2$

16-118. The flywheel rotates with an angular velocity $\omega = 2$ rad/s and an angular acceleration $\alpha = 6$ rad/s^2. Determine the angular accelerations of links AB and BC at this instant.

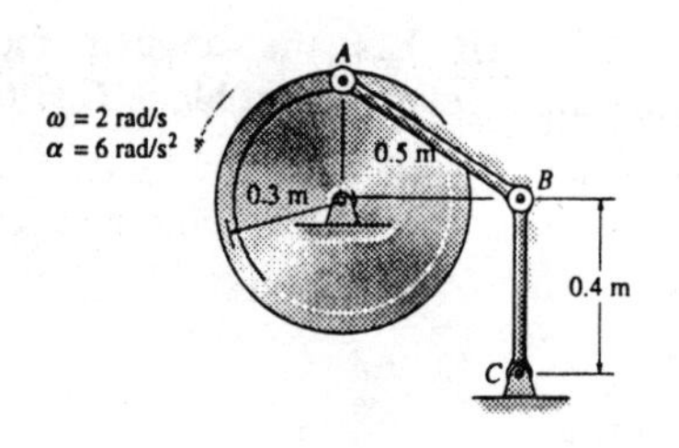

For AB the IC is at ∞, so $\omega_{AB} = 0$ and $v_A = 2(0.3) = 0.6$ m/s $= v_B$. Thus,

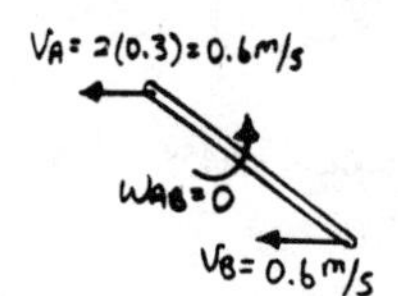

$$(a_B)_n = \frac{(0.6)^2}{0.4} = 0.9 \text{ m/s}^2$$

$$(a_A)_n = (2)^2(0.3) = 1.2 \text{ m/s}^2$$

$$(a_A)_t = 6(0.3) = 1.8 \text{ m/s}^2$$

$$\mathbf{a}_B = \mathbf{a}_A + \mathbf{a}_{B/A}$$

$$\left[\underset{\leftarrow}{(a_B)_t}\right] + \left[\underset{\downarrow}{0.9}\right] = \left[\underset{\leftarrow}{1.8}\right] + \left[\underset{\downarrow}{1.2}\right] + \left[\alpha_{AB}(0.5)\right]$$

$(\overset{+}{\rightarrow})$ $\quad -(a_B)_t = -1.8 + \frac{3}{5}(\alpha_{AB})(0.5)$

$(+\uparrow)$ $\quad -0.9 = -1.2 + \frac{4}{5}(\alpha_{AB})(0.5)$

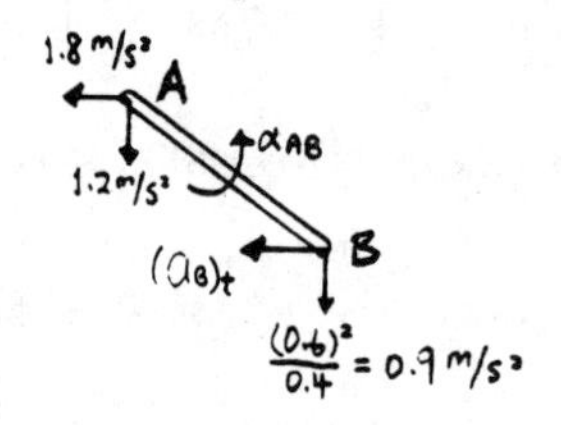

$\alpha_{AB} = 0.75$ rad/s^2 ↻ **Ans**

$(a_B)_t = 1.575$ m/s^2

$\alpha_{BC} = \frac{1.575}{0.4} = 3.94$ rad/s^2 ↻ **Ans**

Also,

$$\mathbf{a}_B = \mathbf{a}_A + \boldsymbol{\alpha}_{AB} \times \mathbf{r}_{B/A} - \omega^2 \mathbf{r}_{B/A}$$

$$-(a_B)_t\mathbf{i} - \frac{(0.6)^2}{0.4}\mathbf{j} = -6(0.3)\mathbf{i} - (2)^2(0.3)\mathbf{j} + (\alpha_{AB}\mathbf{k}) \times (0.4\mathbf{i} - 0.3\mathbf{j}) - \mathbf{0}$$

$(\overset{+}{\rightarrow})$ $\quad -(a_B)_t = -1.8 + 0.3\alpha_{AB}$

$(+\uparrow)$ $\quad -0.9 = -1.2 + 0.4\alpha_{AB}$

$\alpha_{AB} = 0.75$ rad/s^2 ↻ **Ans**

$(a_B)_t = 1.575$ m/s^2

$\alpha_{BC} = \frac{1.575}{0.4} = 3.94$ rad/s^2 ↻ **Ans**

16-119. The ends of the bar AB are confined to move along the paths shown. At a given instant, A has a velocity of 8 ft/s and an acceleration of 3 ft/s^2. Determine the angular velocity and angular acceleration of AB at this instant.

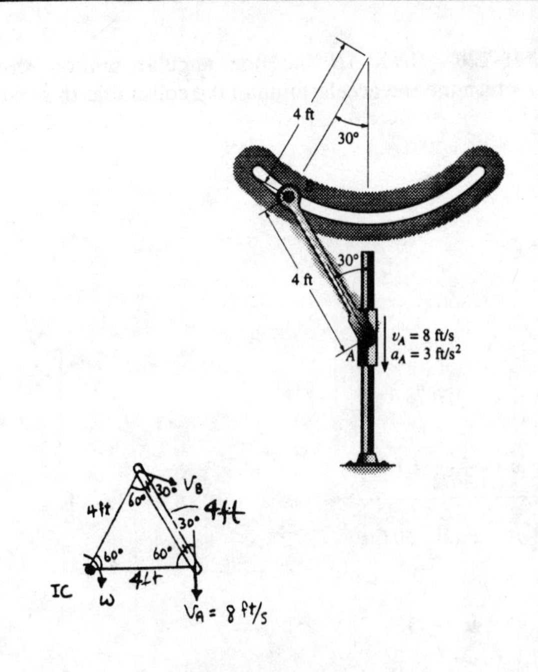

$$\omega = \frac{8}{4} = 2 \text{ rad/s} \quad \text{Ans}$$

$$v_B = 4(2) = 8 \text{ ft/s}$$

$$(a_B)_n = \frac{(8)^2}{4} = 16 \text{ ft/s}^2$$

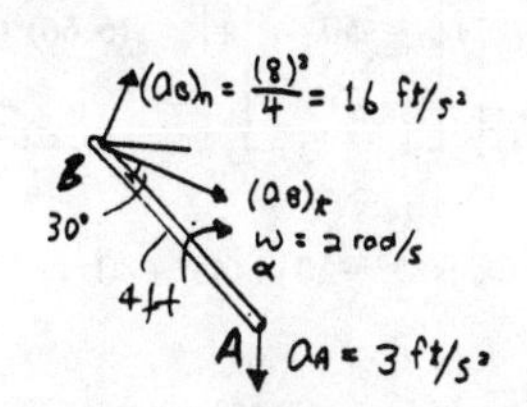

$$\mathbf{a}_B = \mathbf{a}_A + \mathbf{a}_{B/A}$$

$$\begin{bmatrix} 16 \\ \nearrow 30^\circ \end{bmatrix} + \begin{bmatrix} (a_B)_t \\ \searrow 30^\circ \end{bmatrix} = \begin{bmatrix} -3 \\ \downarrow \end{bmatrix} + \begin{bmatrix} \alpha(4) \\ \nearrow 60^\circ \end{bmatrix} + \begin{bmatrix} (2)^2(4) \\ \searrow 60^\circ \end{bmatrix}$$

$(\overset{+}{\rightarrow})$ $\quad 16\sin 30^\circ + (a_B)_t \cos 30^\circ = 0 + \alpha(4)\sin 60^\circ + 16\cos 60^\circ$

$(+\uparrow)$ $\quad 16\cos 30^\circ - (a_B)_t \sin 30^\circ = -3 + \alpha(4)\cos 60^\circ - 16\sin 60^\circ$

$\alpha = 7.68 \text{ rad/s}^2$ $\quad$ **Ans**

$(a_B)_t = 30.7 \text{ ft/s}^2$

Also :

$$\mathbf{a}_B = \mathbf{a}_A + \alpha_{AB} \times \mathbf{r}_{B/A} - \omega^2 \mathbf{r}_{B/A}$$

$$(a_B)_t \cos 30^\circ\mathbf{i} - (a_B)_t \sin 30^\circ\mathbf{j} + \left(\frac{(8)^2}{4}\right)\sin 30^\circ\mathbf{i} + \left(\frac{(8)^2}{4}\right)\cos 30^\circ\mathbf{j} = -3\mathbf{j}$$
$$+ (\alpha\mathbf{k}) \times (-4\sin 30^\circ\mathbf{i} + 4\cos 30^\circ\mathbf{j}) - (2)^2(-4\sin 30^\circ\mathbf{i} + 4\cos 30^\circ\mathbf{j})$$

$(\overset{+}{\rightarrow})$ $\quad (a_B)_t \cos 30^\circ + 8 = -3.464\alpha + 8$

$(+\uparrow)$ $\quad -(a_B)_t \sin 30^\circ + 13.8564 = -3 - 2\alpha - 13.8564$

$\alpha = 7.68 \text{ rad/s}^2$ $\quad$ **Ans**

$(a_B)_t = 30.7 \text{ ft/s}^2$

***16-120.** Rod AB has the angular motion shown. Determine the acceleration of the collar C at this instant.

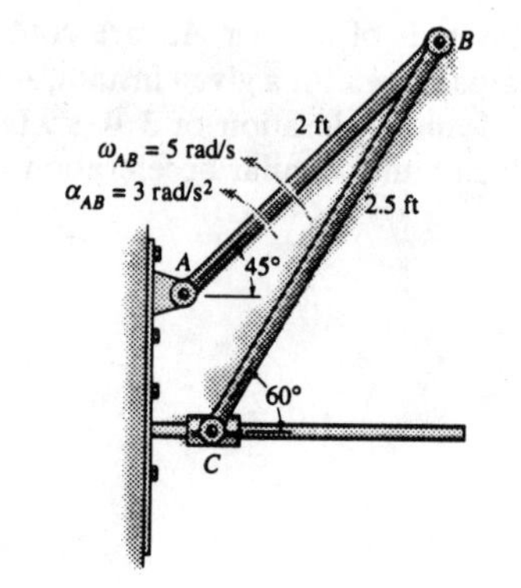

$$\frac{r_{B/IC}}{\sin 30^\circ} = \frac{2.5}{\sin 135^\circ}$$

$r_{B/IC} = 1.7678$ ft

$$\omega = \frac{10}{1.7678} = 5.66 \text{ rad/s}$$

$a_B = 25(2) = 50$ ft/s²

$\mathbf{a}_C = \mathbf{a}_B + \mathbf{a}_{C/B}$

$$\left[\underset{\rightarrow}{a_C}\right] = \left[\underset{45^\circ}{6}\right] + \left[\underset{45^\circ}{50}\right] + \left[\underset{60^\circ}{(5.66)^2(2.5)}\right] + \left[\underset{30^\circ}{\alpha(2.5)}\right]$$

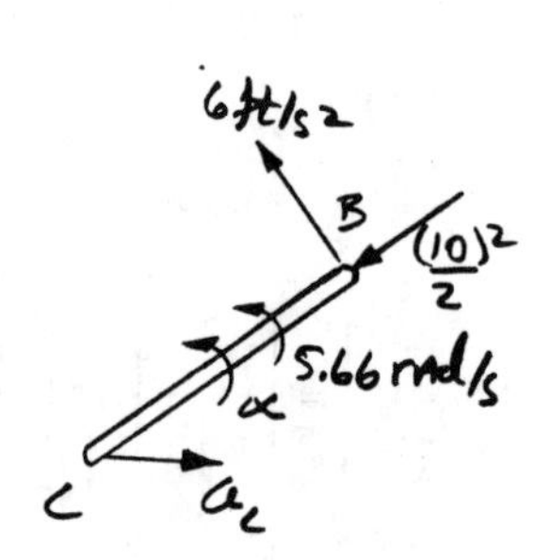

$(\overset{+}{\rightarrow})$ $a_C = -6\cos 45^\circ - 50\cos 45^\circ + 80\cos 60^\circ + \alpha(2.5)\cos 30^\circ$

$(+\uparrow)$ $0 = 6\sin 45^\circ - 50\sin 45^\circ + 80\sin 60^\circ - \alpha(2.5)\sin 30^\circ$

$\alpha = 30.5$ rad/s²

$a_C = 66.5$ ft/s² $\rightarrow$ **Ans**

Also,

$v_B = 5(2) = 10$ ft/s

$\mathbf{v}_C = \mathbf{v}_B + \mathbf{v}_{C/B}$

$-v_C\mathbf{i} = -10\cos 45^\circ\mathbf{i} + 10\sin 45^\circ\mathbf{j} + \omega\mathbf{k} \times (-2.5\sin 30^\circ\mathbf{i} - 2.5\cos 30^\circ\mathbf{j})$

$(+\uparrow)$ $0 = 10\sin 45^\circ - 2.5\omega\sin 30^\circ$

$\omega = 5.66$ rad/s

$\mathbf{a}_C = \mathbf{a}_B + \alpha \times \mathbf{r}_{C/B} - \omega^2\mathbf{r}_{C/B}$ -

$$a_C\mathbf{i} = -\frac{(10)^2}{2}\cos 45^\circ\mathbf{i} - \frac{(10)^2}{2}\sin 45^\circ\mathbf{j} - 6\cos 45^\circ\mathbf{i} + 6\sin 45^\circ\mathbf{j}$$

$$+(\alpha\mathbf{k}) \times (-2.5\cos 60^\circ\mathbf{i} - 2.5\sin 60^\circ\mathbf{j}) - (5.66)^2(-2.5\cos 60^\circ\mathbf{i} - 2.5\sin 60^\circ\mathbf{j})$$

$(\overset{+}{\rightarrow})$ $a_C = -35.355 - 4.243 + 2.165\alpha + 40$

$(+\uparrow)$ $0 = -35.355 + 4.243 - 1.25\alpha + 69.282$

$\alpha = 30.5$ rad/s²

$a_C = 66.5$ ft/s² $\rightarrow$ **Ans**

16-121. At the given instant member AB has the angular motions shown. Determine the velocity and acceleration of the slider block C at this instant.

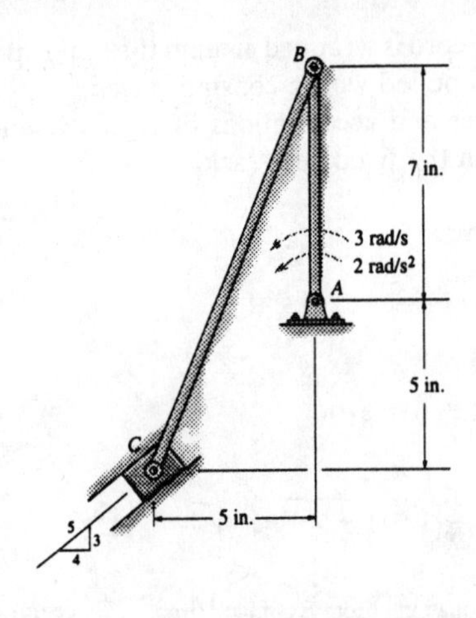

$v_B = 3(7) = 21$ in./s $\leftarrow$

$\mathbf{v}_C = \mathbf{v}_B + \omega \times \mathbf{r}_{C/B}$

$-v_C\left(\frac{4}{5}\right)\mathbf{i} - v_C\left(\frac{3}{5}\right)\mathbf{j} = -21\mathbf{i} + \omega\mathbf{k} \times (-5\mathbf{i} - 12\mathbf{j})$

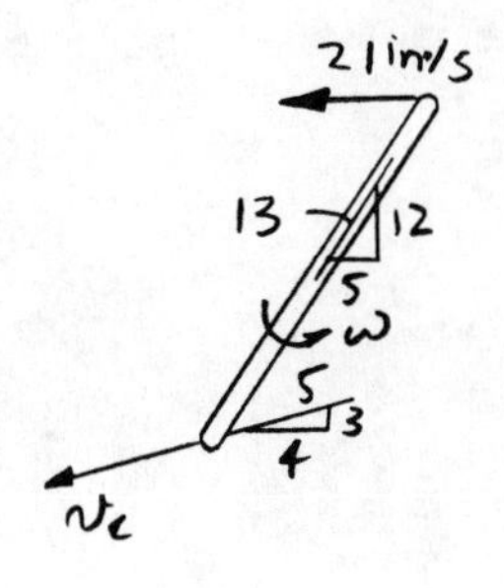

$(\overset{+}{\rightarrow}) \qquad -0.8v_C = -21 + 12\omega$

$(+\uparrow) \qquad -0.6v_C = -5\omega$

Solving :

$\omega = 1.125$ rad/s

$v_C = 9.375$ in./s $= 9.38$ in./s $\swarrow$ (5-3-4) **Ans**

$(a_B)_n = (3)^2(7) = 63$ in./s^2 $\downarrow$

$(a_B)_t = (2)(7) = 14$ in./s^2 $\leftarrow$

$\mathbf{a}_C = \mathbf{a}_B + \alpha \times \mathbf{r}_{C/B} - \omega^2 \mathbf{r}_{C/B}$

$-a_C\left(\frac{4}{5}\right)\mathbf{i} - a_C\left(\frac{3}{5}\right)\mathbf{j} = -14\mathbf{i} - 63\mathbf{j} + (\alpha\mathbf{k}) \times (-5\mathbf{i} - 12\mathbf{j}) - (1.125)^2(-5\mathbf{i} - 12\mathbf{j})$

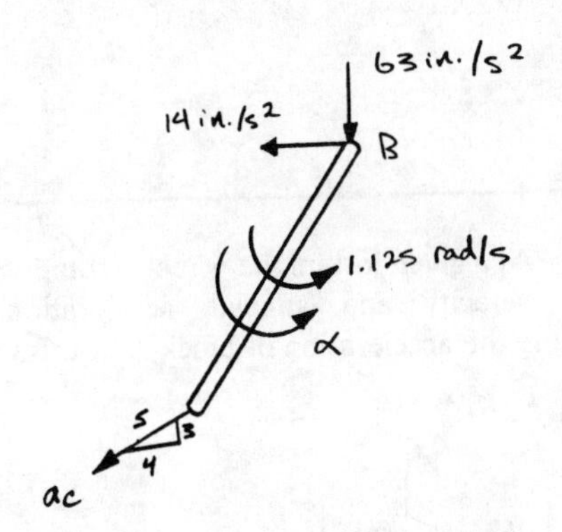

$(\overset{+}{\rightarrow}) \qquad -0.8a_C = -14 + 12\alpha + 6.328$

$(+\uparrow) \qquad -0.6a_C = -63 - 5\alpha + 15.1875$

$a_C = 54.7$ in./s^2 $\swarrow$ (5-3-4) **Ans**

$\alpha = -3.00$ rad/s^2

16-122. A cord is wrapped around the inner spool of the gear. If it is pulled with a constant velocity **v**, determine the velocities and accelerations of points A and B. The gear rolls on the fixed gear rack.

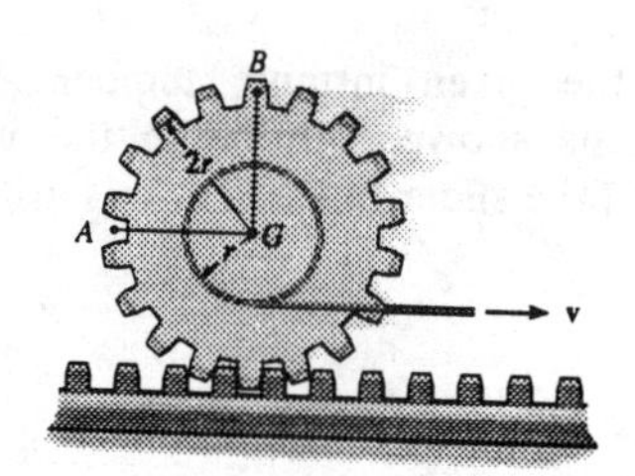

Velocity analysis :

$$\omega = \frac{v}{r}$$

$$v_B = \omega r_{B/IC} = \frac{v}{r}(4r) = 4v \rightarrow \qquad \textbf{Ans}$$

$$v_A = \omega r_{A/IC} = \frac{v}{r}\left(\sqrt{(2r)^2 + (2r)^2}\right) = 2\sqrt{2}v \quad 45^\circ \qquad \textbf{Ans}$$

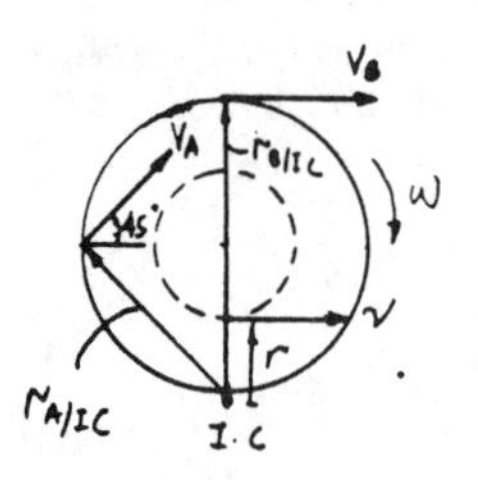

Acceleration equation : From Example 16 – 3 , Since $a_G = 0$, $\alpha = 0$

$\mathbf{r}_{B/G} = 2r\,\mathbf{j} \qquad \mathbf{r}_{A/G} = -2r\,\mathbf{i}$

$$\mathbf{a}_B = \mathbf{a}_G + \alpha \times \mathbf{r}_{B/G} - \omega^2 \mathbf{r}_{B/G}$$

$$= \mathbf{0} + \mathbf{0} - \left(\frac{v}{r}\right)^2 (2r\mathbf{j}) = -\frac{2v^2}{r}\mathbf{j}$$

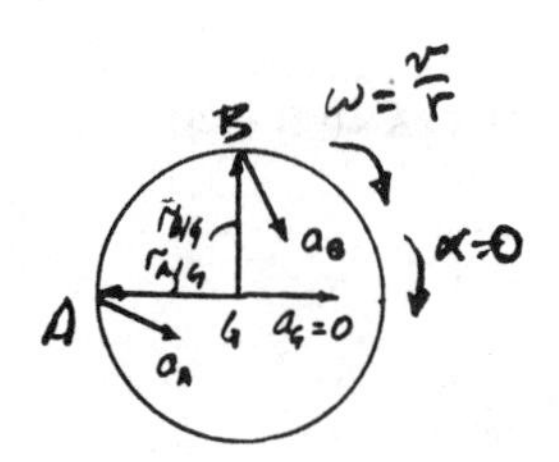

$$a_B = \frac{2v^2}{r} \downarrow \qquad \textbf{Ans}$$

$$\mathbf{a}_A = \mathbf{a}_G + \alpha \times \mathbf{r}_{A/G} - \omega^2 \mathbf{r}_{A/G}$$

$$= \mathbf{0} + \mathbf{0} - \left(\frac{v}{r}\right)^2 (-2r\mathbf{i}) = \frac{2v^2}{r}\mathbf{i}$$

$$a_A = \frac{2v^2}{r} \rightarrow \qquad \textbf{Ans}$$

16-123. At a given instant the wheel is rotating with the angular velocity and angular acceleration shown. Determine the acceleration of block B at this instant.

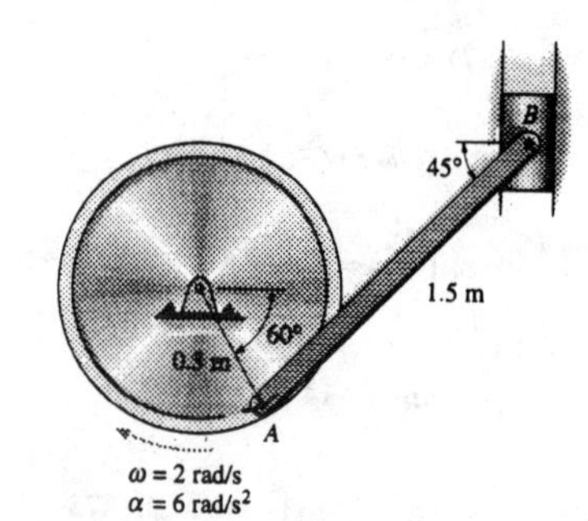

$$\mathbf{v}_B = \mathbf{v}_A + \mathbf{v}_{B/A}$$

$$\begin{bmatrix} v_B \\ \downarrow \end{bmatrix} = \begin{bmatrix} 0.6 \\ \swarrow 30^\circ \end{bmatrix} + \begin{bmatrix} \omega(1.5) \\ \searrow 45^\circ \end{bmatrix}$$

1.5 m $\quad v_B \quad \omega \quad 2(0.3) = 0.6$ m/s

$(\overset{+}{\rightarrow}) \qquad 0 = -0.6\cos 30^\circ + \omega(1.5)\cos 45^\circ$

$\omega = 0.4899$ rad/s

$(a_A)_n = (2)^2(0.3) = 1.2 \text{ m/s}^2$

$(a_A)_t = 6(0.3) = 1.8 \text{ m/s}^2$

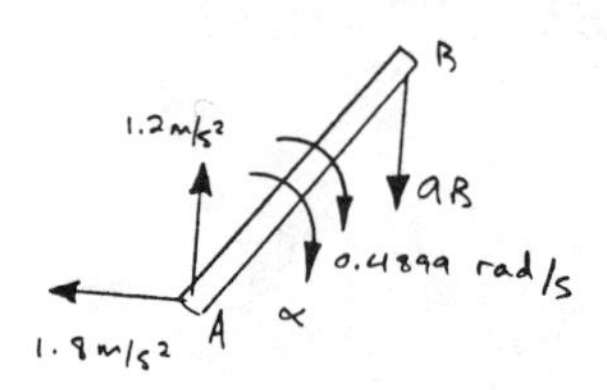

$$\mathbf{a}_B = \mathbf{a}_A + \mathbf{a}_{B/A}$$

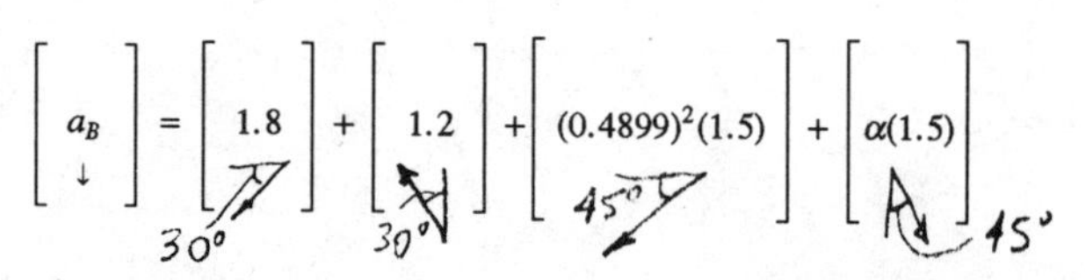

a_B ↓ = 1.8 (30°) + 1.2 (30°) + (0.4899)²(1.5) (45°) + α(1.5) (45°)

$\alpha = 2.28 \text{ rad/s}^2$

$(\overset{+}{\rightarrow}) \qquad 0 = -1.8\cos 30^\circ - 1.2\sin 30^\circ - 0.4899(1.5)\sin 45^\circ + \alpha(1.5)\cos 45^\circ$

$a_B = 2.53 \text{ m/s}^2 \qquad$ **Ans**

$(+\downarrow) \qquad a_B = 1.8\sin 30^\circ - 1.2\cos 30^\circ + (0.4899)^2(1.5)\sin 45^\circ + \alpha(1.5)\sin 45^\circ$

Also:

$$\mathbf{a}_B = \mathbf{a}_A + \alpha \times \mathbf{r}_{B/A} - \omega^2 \mathbf{r}_{B/A}$$

$$-a_B\mathbf{j} = (-1.2\sin 30°\mathbf{i} + 1.2\cos 30°\mathbf{j}) + (-1.8\cos 30°\mathbf{i} - 1.8\sin 30°\mathbf{j})$$

$$+(-\alpha\mathbf{k}) \times (1.5\cos 45°\mathbf{i} + 1.5\sin 45°\mathbf{j})$$

$$-(0.4899)(1.5\cos 45°\mathbf{i} + 1.5\sin 45°\mathbf{j})$$

$$0 = -1.2\sin 30° - 1.8\cos 30° + \alpha(1.5\sin 45°) - (0.4899)^2(1.5\cos 45°)$$

$$-a_B = 1.2\cos 30° - 1.8\sin 30° - \alpha(1.5\cos 45°) - (0.4899)^2(1.5\sin 45°)$$

$$\alpha = 2.28 \text{ rad/s}$$

$$a_B = 2.53 \text{ m/s}^2 \qquad \textbf{Ans}$$

***16-124.** As the cord unravels from the cylinder, the cylinder has an angular acceleration of $\alpha = 4$ rad/s^2 and an angular velocity of $\omega = 2$ rad/s at the instant shown. Determine the accelerations of points A and B at this instant.

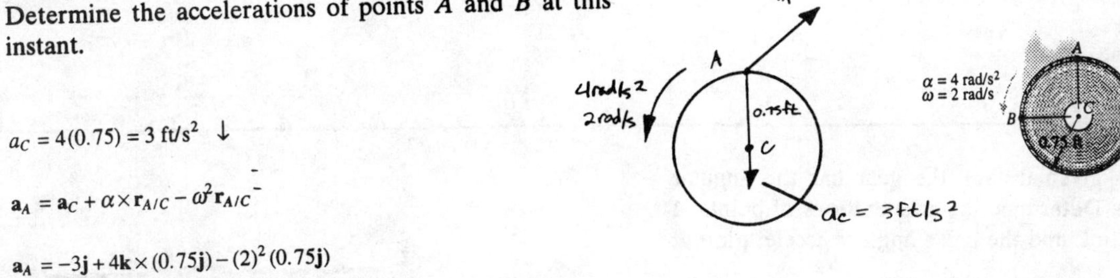

$$a_C = 4(0.75) = 3 \text{ ft/s}^2 \downarrow$$

$$\mathbf{a}_A = \mathbf{a}_C + \alpha \times \mathbf{r}_{A/C} - \omega^2 \mathbf{r}_{A/C}$$

$$\mathbf{a}_A = -3\mathbf{j} + 4\mathbf{k} \times (0.75\mathbf{j}) - (2)^2(0.75\mathbf{j})$$

$$\mathbf{a}_A = \{-3\mathbf{i} - 6\mathbf{j}\} \text{ ft/s}^2$$

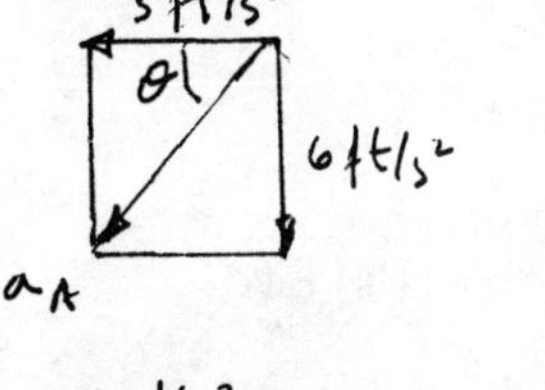

$$a_A = \sqrt{(-3)^2 + (-6)^2} = 6.71 \text{ ft/s}^2 \qquad \textbf{Ans}$$

$$\theta = \tan^{-1}\left(\frac{6}{3}\right) = 63.4° \qquad \textbf{Ans}$$

$$\mathbf{a}_B = \mathbf{a}_C + \alpha \times \mathbf{r}_{B/C} - \omega^2 \mathbf{r}_{B/C}$$

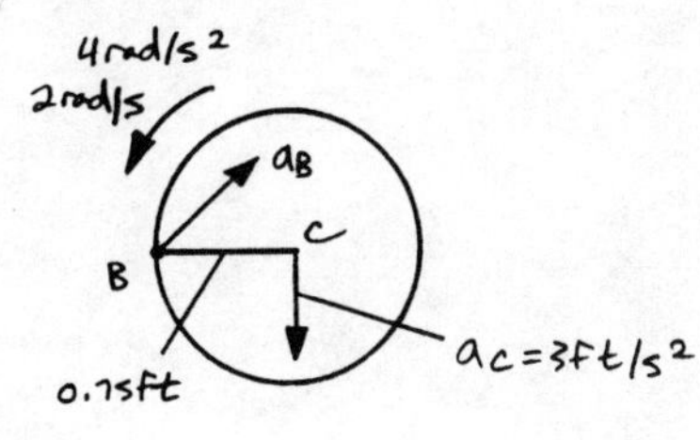

$$\mathbf{a}_B = -3\mathbf{j} + 4\mathbf{k} \times (-0.75\mathbf{i}) - (2)^2(-0.75\mathbf{i})$$

$$\mathbf{a}_B = \{3\mathbf{i} - 6\mathbf{j}\} \text{ ft/s}^2$$

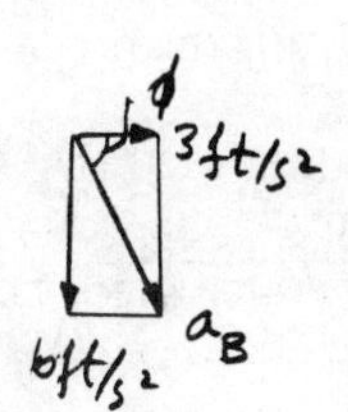

$$a_B = \sqrt{(3)^2 + (-6)^2} = 6.71 \text{ ft/s}^2 \qquad \textbf{Ans}$$

$$\phi = \tan^{-1}\left(\frac{6}{3}\right) = 63.4° \qquad \textbf{Ans}$$

16-125. The disk rolls without slipping such that it has an angular acceleration of $\alpha = 4$ rad/s^2 and angular velocity of $\omega = 2$ rad/s at the instant shown. Determine the accelerations of points A and B on the link and the link's angular acceleration at this instant. Assume point A lies on the periphery of the disk, 150 mm from C.

The IC is at ∞, so $\omega = 0$.

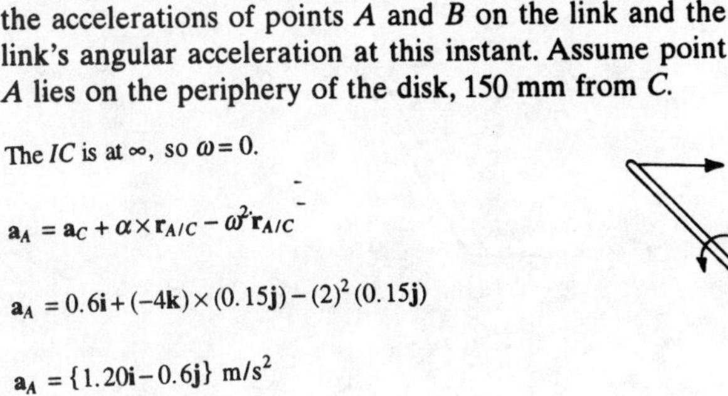

$$\mathbf{a}_A = \mathbf{a}_C + \alpha \times \mathbf{r}_{A/C} - \omega^2 \mathbf{r}_{A/C}$$

$$\mathbf{a}_A = 0.6\mathbf{i} + (-4\mathbf{k}) \times (0.15\mathbf{j}) - (2)^2(0.15\mathbf{j})$$

$$\mathbf{a}_A = \{1.20\mathbf{i} - 0.6\mathbf{j}\} \text{ m/s}^2$$

$$a_A = \sqrt{(1.20)^2 + (-0.6)^2} = 1.34 \text{ m/s}^2 \qquad \textbf{Ans}$$

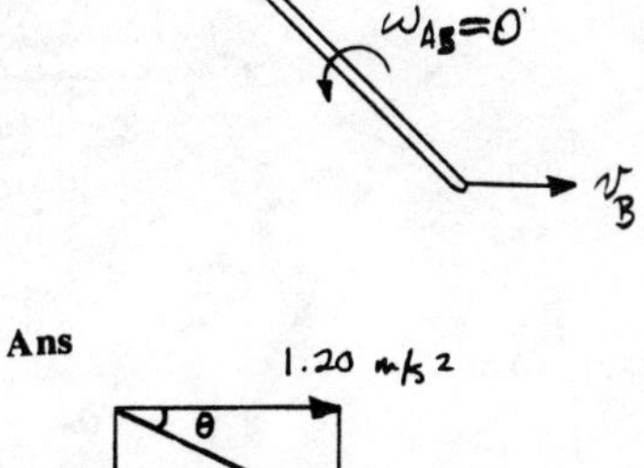

$$\theta = \tan^{-1}\left(\frac{0.6}{1.20}\right) = 26.6° \qquad \textbf{Ans}$$

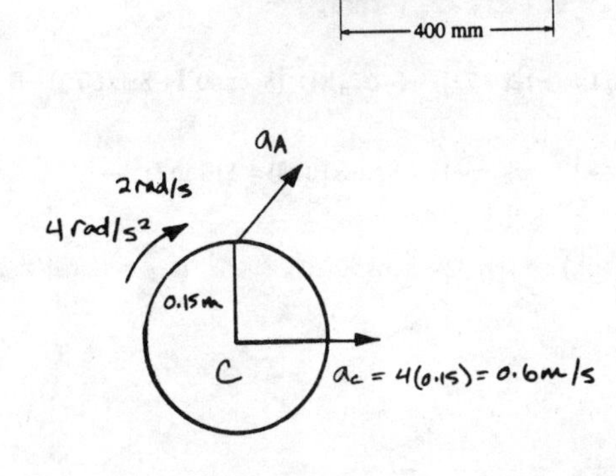

$$\mathbf{a}_B = \mathbf{a}_A + \alpha \times \mathbf{r}_{B/A} - \omega^2 \mathbf{r}_{B/A}$$

$$a_B \mathbf{i} = 1.20\mathbf{i} - 0.6\mathbf{j} + \alpha_{AB}\mathbf{k} \times (0.4\mathbf{i} - 0.3\mathbf{j}) - \mathbf{0}$$

$(\overset{+}{\rightarrow})$ $\quad a_B = 1.20 + 0.3\alpha_{AB}$

$(+\uparrow)$ $\quad 0 = -0.6 + 0.4\alpha_{AB}$

$\alpha_{AB} = 1.5 \text{ rad/s}^2$ **Ans**

$a_B = 1.65 \text{ m/s}^2 \rightarrow$ **Ans**

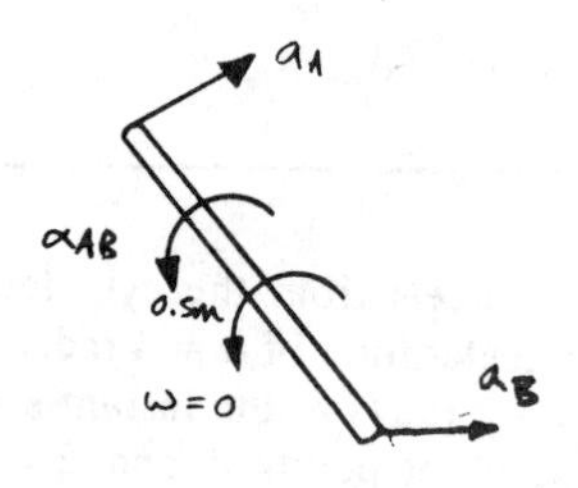

16-126. At a given instant, the gear has the angular motion shown. Determine the accelerations of points *A* and *B* on the link and the link's angular acceleration at this instant.

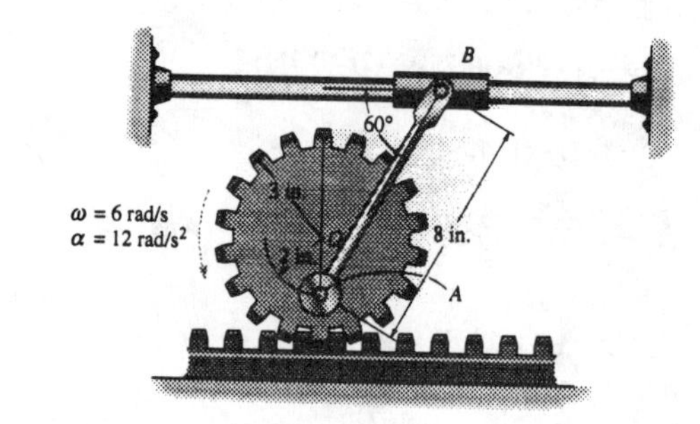

For the gear

$v_A = \omega r_{A/IC} = 6(1) = 6 \text{ in./s}$

$\mathbf{a}_O = -12(3)\mathbf{i} = \{-36\mathbf{i}\} \text{ in./s}^2 \quad \mathbf{r}_{A/O} = \{-2\mathbf{j}\} \text{ in.} \quad \alpha = \{12\mathbf{k}\} \text{ rad/s}^2$

$$\mathbf{a}_A = \mathbf{a}_O + \alpha \times \mathbf{r}_{A/O} - \omega^2 \mathbf{r}_{A/O}$$

$$= -36\mathbf{i} + (12\mathbf{k}) \times (-2\mathbf{j}) - (6)^2(-2\mathbf{j})$$

$$= \{-12\mathbf{i} + 72\mathbf{j}\} \text{ in./s}^2$$

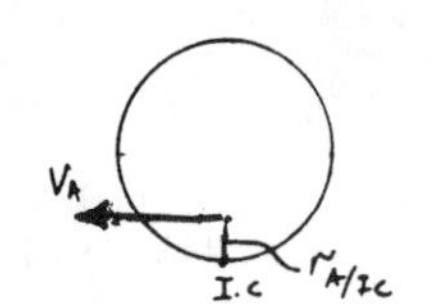

$a_A = \sqrt{(-12)^2 + 72^2} = 73.0 \text{ in./s}^2$ **Ans**

$\theta = \tan^{-1}\left(\frac{72}{12}\right) = 80.5°$ **Ans**

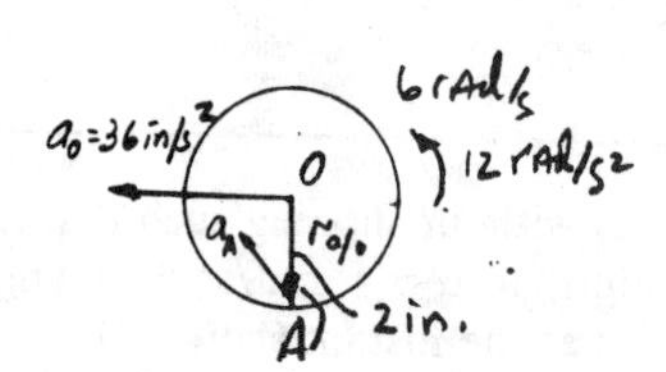

For link *AB*

The *IC* is at ∞, so $\omega_{AB} = 0$, i.e.,

$$\omega_{AB} = \frac{v_A}{r_{A/IC}} = \frac{6}{\infty} = 0$$

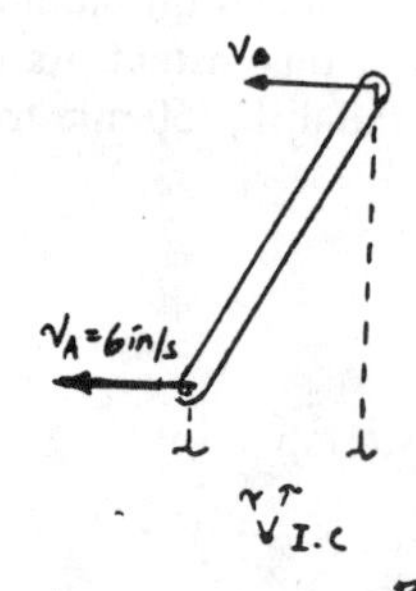

$\mathbf{a}_B = a_B\mathbf{i} \quad \alpha_{AB} = -\alpha_{AB}\mathbf{k} \quad \mathbf{r}_{B/A} = \{8\cos 60°\mathbf{i} + 8\sin 60°\mathbf{j}\} \text{ in.}$

$$\mathbf{a}_B = \mathbf{a}_A + \alpha_{AB} \times \mathbf{r}_{B/A} - \omega^2 \mathbf{r}_{B/A}$$

$$a_B\mathbf{i} = (-12\mathbf{i} + 72\mathbf{j}) + (-\alpha_{AB}\mathbf{k}) \times (8\cos 60°\mathbf{i} + 8\sin 60°\mathbf{j}) - \mathbf{0}$$

$(\overset{+}{\rightarrow})$ $\quad a_B = -12 + 8\sin 60°(18) = 113 \text{ in./s}^2 \rightarrow$ **Ans**

$(+\uparrow)$ $\quad 0 = 72 - 8\cos 60°\alpha_{AB} \quad \alpha_{AB} = 18 \text{ rad/s}^2$ **Ans**

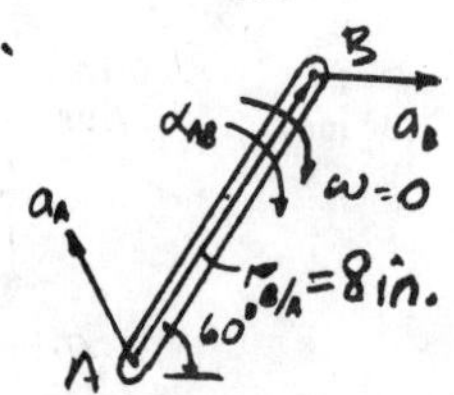

16-127. Determine the angular acceleration of link AB if link CD has the angular velocity and angular deceleration shown.

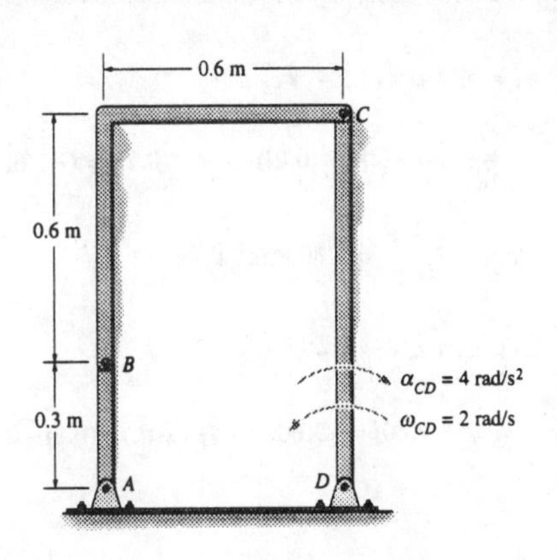

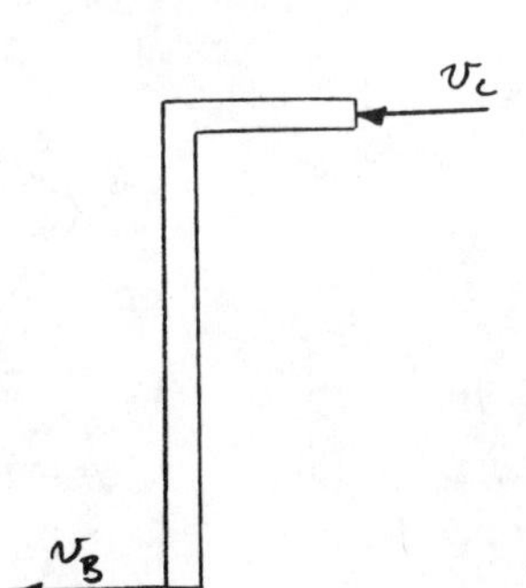

IC is at ∞, thus

$\omega_{BC} = 0$

$v_B = v_C = (0.9)(2) = 1.8\ \text{m/s}$

$(a_C)_n = (2)^2(0.9) = 3.6\ \text{m/s}^2 \downarrow$

$(a_C)_t = 4(0.9) = 3.6\ \text{m/s}^2 \rightarrow$

$(a_B)_n = \dfrac{(1.8)^2}{0.3} = 10.8\ \text{m/s}^2 \downarrow$

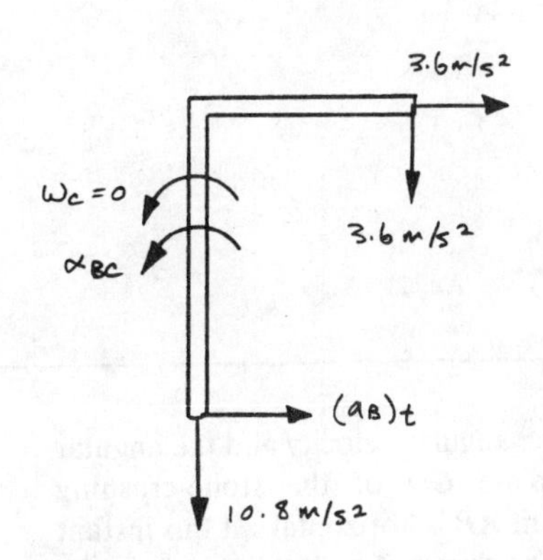

$\mathbf{a}_B = \mathbf{a}_C + \alpha_{BC} \times \mathbf{r}_{B/C} - \omega_{BC}^2 \mathbf{r}_{B/C}$

$(a_B)_t\mathbf{i} - 10.8\mathbf{j} = 3.6\mathbf{i} - 3.6\mathbf{j} + (\alpha_{BC}\mathbf{k}) \times (-0.6\mathbf{i} - 0.6\mathbf{j}) - \mathbf{0}$

$(\overset{+}{\rightarrow})\qquad (a_B)_t = 3.6 + 0.6\alpha_{BC}$

$(+\uparrow)\qquad -10.8 = -3.6 - 0.6\alpha_{BC}$

$\alpha_{BC} = 12\ \text{rad/s}^2$

$(a_B)_t = 10.8\ \text{m/s}^2$

$\alpha_{AB} = \dfrac{10.8}{0.3} = 36\ \text{rad/s}^2$ ↻ **Ans**

***16-128.** The gear at A is subjected to the angular motion shown. Determine the angular velocity and angular acceleration of link CD at this instant.

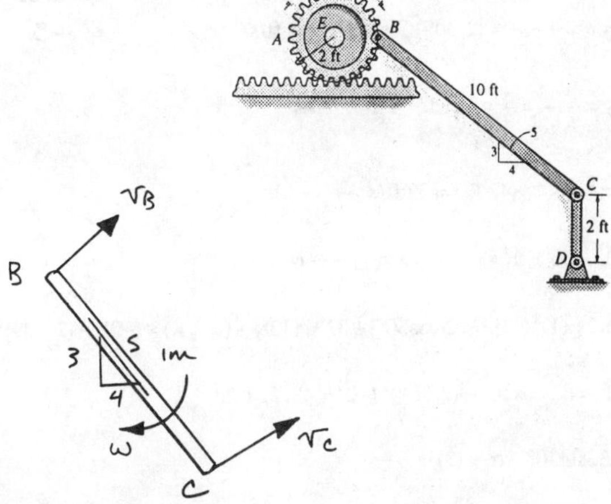

$v_E = 0.2(4) = 0.8\ \text{m/s} \rightarrow$

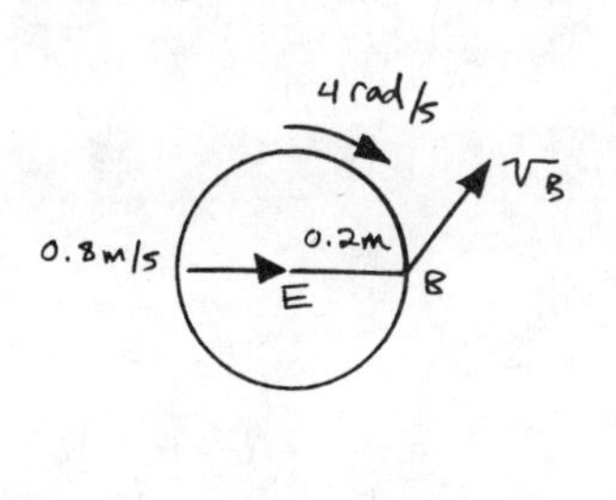

$\mathbf{v}_B = \mathbf{v}_E + \omega \times \mathbf{r}_{B/E}$

$= 0.8\mathbf{i} + (-4\mathbf{k}) \times (0.2\mathbf{i})$

$= \{0.8\mathbf{i} - 0.8\mathbf{j}\}\ \text{m/s}$

$\mathbf{v}_C = \mathbf{v}_B + \omega \times \mathbf{r}_{C/B}$

$v_C\mathbf{i} = 0.8\mathbf{i} - 0.8\mathbf{j} + (\omega\mathbf{k}) \times (0.8\mathbf{i} - 0.6\mathbf{j})$

$(\overset{+}{\rightarrow})\qquad v_C = 0.8 + 0.6\omega$

$(+\uparrow)\qquad 0 = -0.8 + 0.8\omega$

$\omega = 1\ \text{rad/s}, \qquad v_C = 1.4\ \text{m/s}$

$\omega_{CD} = \dfrac{1.4}{0.2} = 7\ \text{rad/s}$ ↻ **Ans**

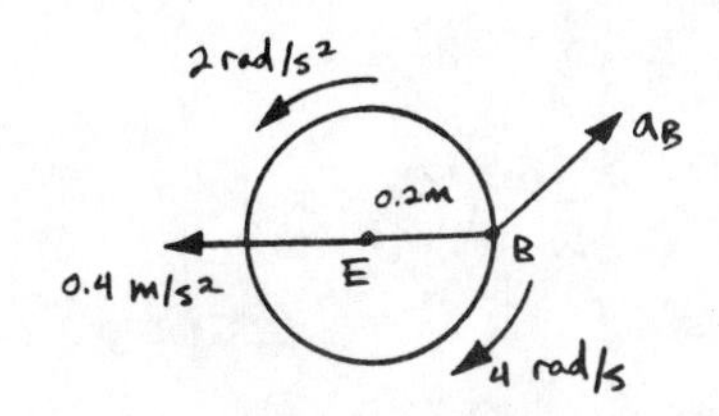

$a_E = 2(0.2) = 0.4\ \text{m/s}^2 \leftarrow$

$\mathbf{a}_B = \mathbf{a}_E + \alpha \times \mathbf{r}_{B/E} - \omega^2 \mathbf{r}_{B/E}$

$= -0.4\mathbf{i} + (2\mathbf{k}) \times (0.2\mathbf{i}) - (-4)^2(0.2\mathbf{i}) = \{-3.6\mathbf{i} + 0.4\mathbf{j}\}\ \text{ft/s}^2$

$(a_C)_n = \dfrac{(1.4)^2}{0.2} = 9.80\ \text{m/s}^2 \downarrow$

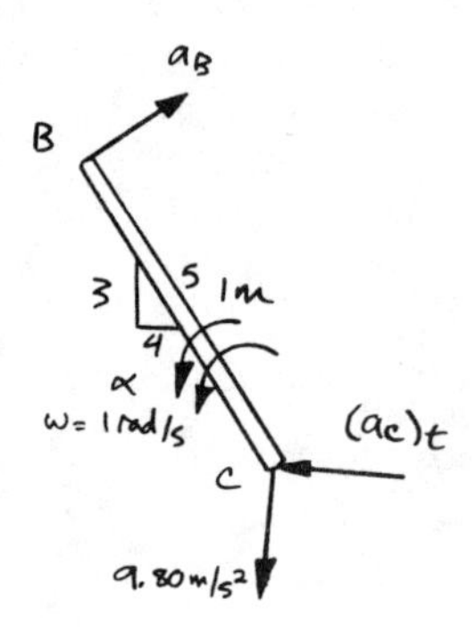

$\mathbf{a}_C = \mathbf{a}_B + \alpha \times \mathbf{r}_{C/B} - \omega^2 \mathbf{r}_{C/B}$

$-(a_C)_t \mathbf{i} - 9.80\mathbf{j} = -3.60\mathbf{i} + 0.4\mathbf{j} + (\alpha\mathbf{k}) \times (0.8\mathbf{i} - 0.6\mathbf{j}) - (1)^2(0.8\mathbf{i} - 0.6\mathbf{j})$

$(\overset{+}{\rightarrow}) \qquad -(a_C)_t = -3.60 + 0.6\alpha - 0.8$

$(+\uparrow) \qquad -9.80 = 0.4 + 0.8\alpha + 0.6$

$\alpha = -13.5\ \text{rad/s}^2 = 13.5\ \text{rad/s}^2$ ↻

$(a_C)_t = 12.5\ \text{m/s}^2 \leftarrow$

$\alpha_{CD} = \dfrac{12.5}{0.2} = 62.5\ \text{rad/s}^2$ ↺ **Ans**

16-129. Determine the angular velocity and the angular acceleration of the plate CD of the stone-crushing mechanism at the instant AB is horizontal. At this instant $\theta = 30°$ and $\phi = 90°$. The driving link AB is turning with a constant angular velocity of $\omega_{AB} = 4$ rad/s.

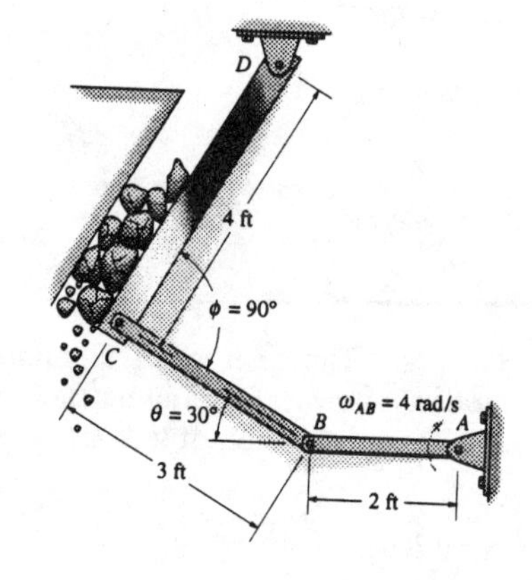

$v_B = \omega_{AB} r_{BA} = (4)(2) = 8\ \text{ft/s} \uparrow$

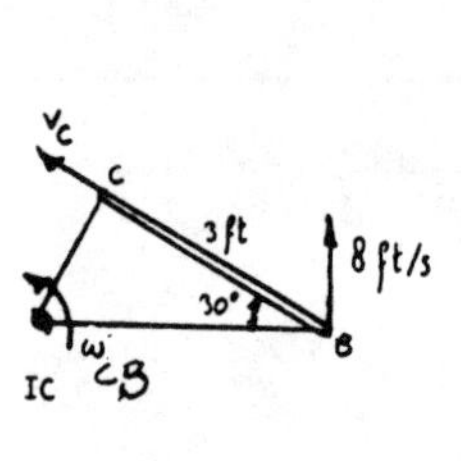

$\omega_{CB} = \dfrac{v_B}{r_{B/IC}} = \dfrac{8}{3/\cos 30°} = 2.309\ \text{rad/s}$

$v_C = \omega_{BC} r_{C/IC} = (2.309)(3\tan 30°) = 4\ \text{ft/s}$

$\omega_{CD} = \dfrac{v_C}{r_{CD}} = \dfrac{4}{4} = 1\ \text{rad/s}$ ↻ **Ans**

$a_B = (a_B)_n = (4)^2(2) = 32\ \text{ft/s}^2 \rightarrow$

$(\mathbf{a}_C)_t + (\mathbf{a}_C)_n = \mathbf{a}_B + \alpha_{CB} \times \mathbf{r}_{C/B} - \omega^2 \mathbf{r}_{C/B}$

$(a_C)_t \mathbf{i} + (1)^2(4)\mathbf{j} = 32\cos 30°\mathbf{i} + 32\sin 30°\mathbf{j} + (\alpha_{CB}\mathbf{k}) \times (-3\mathbf{i}) - (2.309)^2(-3\mathbf{i})$

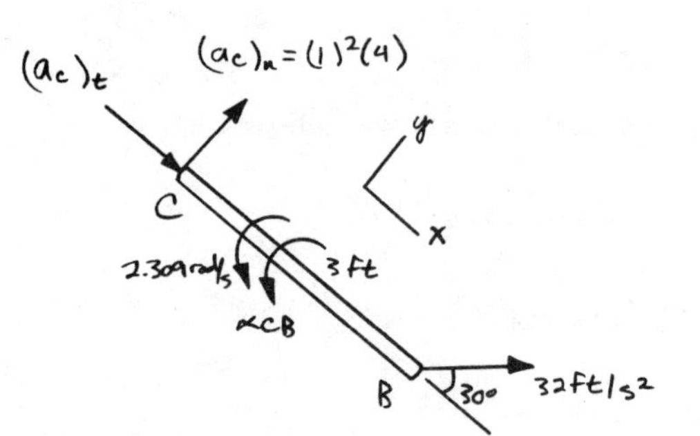

$(a_C)_t = 32\cos 30° - (2.309)^2(-3) = 43.71\ \text{ft/s}^2$

$4 = 32\sin 30° - \alpha_{CB}(3)$

$\alpha_{CB} = 4\ \text{rad/s}^2$ ↺

$\alpha_{DC} = \dfrac{43.71}{4} = 10.9\ \text{rad/s}^2$ ↺ **Ans**

16-130. The mechanism produces intermittent motion of link AB. If the sprocket S is turning with an angular acceleration $\alpha_S = 2$ rad/s^2 and has an angular velocity $\omega_S = 6$ rad/s at the instant shown, determine the angular velocity and angular acceleration of link AB at this instant. The sprocket S is mounted on a shaft which is *separate* from a collinear shaft attached to AB at A. The pin at C is attached to one of the chain links such that it moves vertically downward.

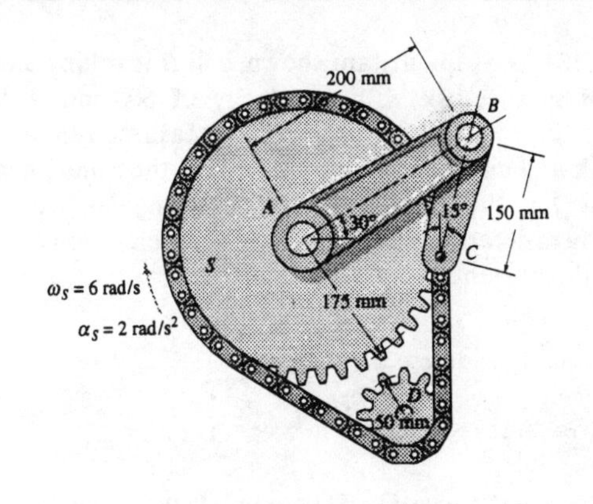

$$\omega_{BC} = \frac{1.05}{0.2121} = 4.950 \text{ rad/s}$$

$$v_B = (4.95)(0.2898) = 1.434 \text{ m/s}$$

$$\omega_{AB} = \frac{1.435}{0.2} = 7.1722 \text{ rad/s} = 7.17 \text{ rad/s} \circlearrowright \qquad \textbf{Ans}$$

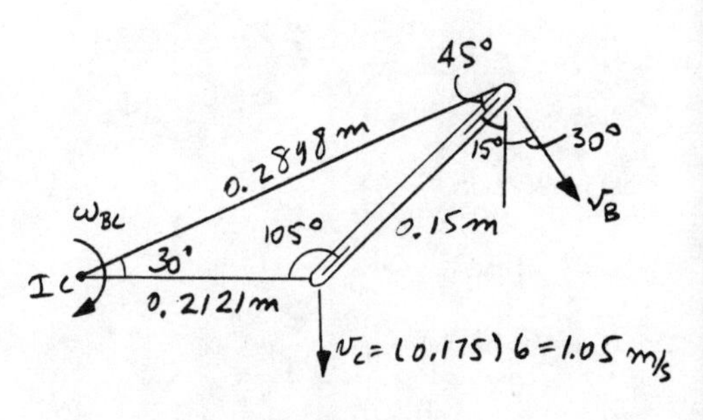

$$a_C = \alpha_S r_S = 2(0.175) = 0.350 \text{ m/s}^2$$

$$(\mathbf{a}_B)_n + (\mathbf{a}_B)_t = \mathbf{a}_C + (\mathbf{a}_{B/C})_n + (\mathbf{a}_{B/C})_t$$

$$\left[\begin{matrix}(7.172)^2(0.2) \\ 30° \searrow\end{matrix}\right] + \left[\begin{matrix}(a_B)_t \\ \nwarrow 30°\end{matrix}\right] = \left[\begin{matrix}0.350 \\ \downarrow\end{matrix}\right] + \left[\begin{matrix}(4.949)^2(0.15) \\ 15° \swarrow\end{matrix}\right] + \left[\begin{matrix}\alpha_{BC}(0.15) \\ 15° \searrow\end{matrix}\right]$$

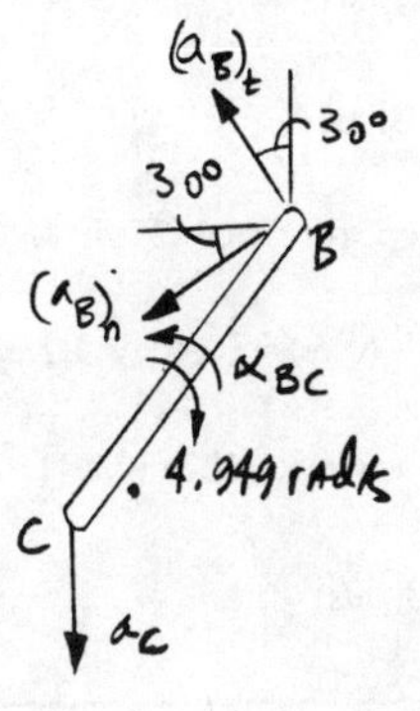

$(\overset{+}{\rightarrow})$ $\quad -(10.29)\cos 30° - (a_B)_t \sin 30° = 0 - (4.949)^2(0.15)\sin 15° - \alpha_{BC}(0.15)\cos 15°$

$(+\uparrow)$ $\quad -(10.29)\sin 30° + (a_B)_t \cos 30° = -0.350 - (4.949)^2(0.15)\cos 15° + \alpha_{BC}(0.15)\sin 15°$

$\alpha_{BC} = 70.8$ rad/s^2, $\quad (a_B)_t = 4.61$ m/s^2

Hence,

$$\alpha_{AB} = \frac{(a_B)_t}{r_{B/A}} = \frac{4.61}{0.2} = 23.1 \text{ rad/s}^2 \circlearrowright \qquad \textbf{Ans}$$

Also,

$v_C = \omega_S r_S = 6(0.175) = 1.05$ m/s $\downarrow$

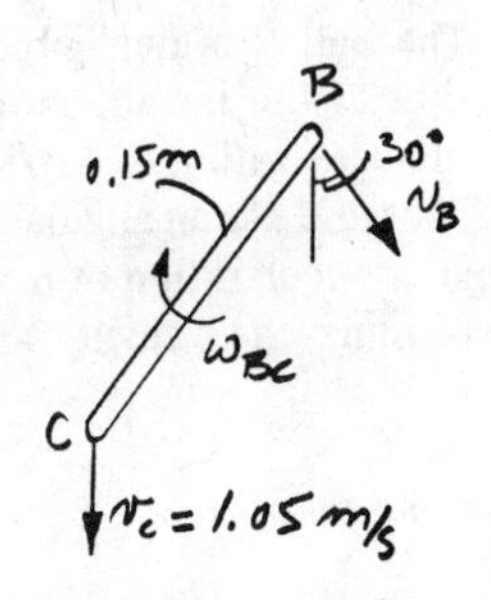

$$\mathbf{v}_B = \mathbf{v}_C + \omega_{BC} \times \mathbf{r}_{B/C}$$

$$v_B \sin 30°\mathbf{i} - v_B \cos 30°\mathbf{j} = -1.05\mathbf{j} + (-\omega_{BC}\mathbf{k}) \times (0.15 \sin 15°\mathbf{i} + 0.15 \cos 15°\mathbf{j})$$

$(\overset{+}{\rightarrow})$ $\quad v_B \sin 30° = 0 + \omega_{BC}(0.15)\cos 15°$

$(+\uparrow)$ $\quad -v_B \cos 30° = -1.05 - \omega_{BC}(0.15)\sin 15°$

$v_B = 1.434$ m/s, $\quad \omega_{BC} = 4.950$ rad/s

$$\omega_{AB} = \frac{v_B}{r_{B/A}} = \frac{1.434}{0.2} = 7.172 = 7.17 \text{ rad/s} \circlearrowright \qquad \textbf{Ans}$$

$$\mathbf{a}_B = \mathbf{a}_C + \alpha_{BC} \times \mathbf{r}_{B/C} - \omega^2 \mathbf{r}_{B/C}$$

$$(\alpha_{AB}\mathbf{k}) \times (0.2\cos 30°\mathbf{i} + 0.2\sin 30°\mathbf{j}) - (7.172)^2(0.2\cos 30°\mathbf{i} + 0.2\sin 30°\mathbf{j})$$

$$= -(2)(0.175)\mathbf{j} + (\alpha_{BC}\mathbf{k}) \times (0.15\sin 15°\mathbf{i} + 0.15\cos 15°\mathbf{j}) - (4.950)^2(0.15\sin 15°\mathbf{i} + 0.15\cos 15°\mathbf{j})$$

$(\overset{+}{\rightarrow})$ $\quad -\alpha_{AB}(0.1) - 8.9108 = -0.1449\alpha_{BC} - 0.9512$

$(+\uparrow)$ $\quad \alpha_{AB}(0.1732) - 5.143 = -0.350 + 0.0388\alpha_{BC} - 3.550$

$$\alpha_{AB} = 23.1 \text{ rad/s}^2 \circlearrowright \qquad \textbf{Ans}$$

$$\alpha_{BC} = 70.8 \text{ rad/s}^2$$

16-131. At the instant shown, ball B is rolling along the slot in the disk with a velocity of 600 mm/s and an acceleration of 150 mm/s^2, both measured relative to the disk and directed away from O. If at the same instant the disk has the angular velocity and angular acceleration shown, determine the velocity and acceleration of the ball at this instant.

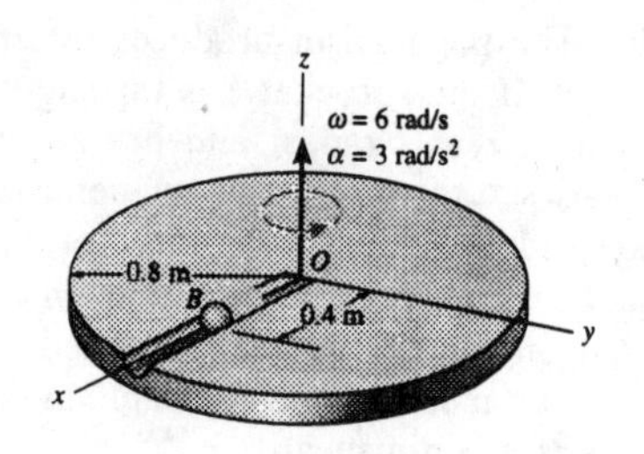

Kinematic Equations :

$$\mathbf{v}_B = \mathbf{v}_O + \Omega \times \mathbf{r}_{B/O} + (\mathbf{v}_{B/O})_{rel} \qquad (1)$$

$$\mathbf{a}_B = \mathbf{a}_O + \dot{\Omega} \times \mathbf{r}_{B/O} + \Omega \times (\Omega \times \mathbf{r}_{B/O}) + 2\Omega \times (\mathbf{v}_{B/O})_{xyz} + (\mathbf{a}_{B/O})_{xyz} \qquad (2)$$

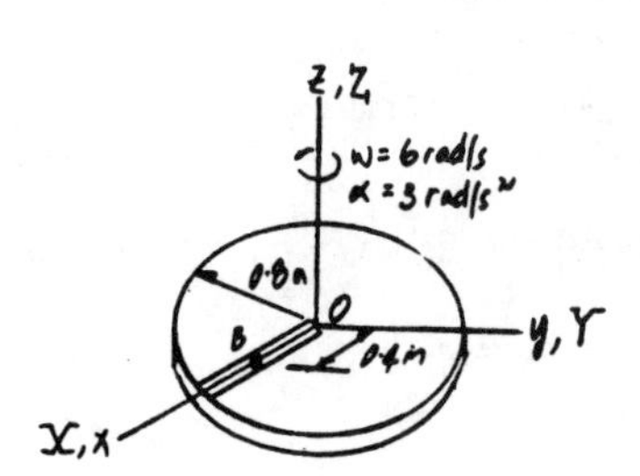

$\mathbf{v}_O = \mathbf{0}$

$\mathbf{a}_O = \mathbf{0}$

$\Omega = \{6\mathbf{k}\}$ rad/s

$\dot{\Omega} = \{3\mathbf{k}\}$ rad/s^2

$\mathbf{r}_{B/O} = \{0.4\mathbf{i}\}$ m

$(\mathbf{v}_{B/O})_{xyz} = \{0.6\mathbf{i}\}$ m/s

$(\mathbf{a}_{B/O})_{xyz} = \{0.15\mathbf{i}\}$ m/s^2

Substitute the data into Eqs. (1) and (2) yields :

$$\mathbf{v}_B = \mathbf{0} + (6\mathbf{k}) \times (0.4\mathbf{i}) + (0.6\mathbf{i}) = \{0.6\mathbf{i} + 2.4\mathbf{j}\}\ \text{m/s} \qquad \textbf{Ans}$$

$$\mathbf{a}_B = \mathbf{0} + (3\mathbf{k}) \times (0.4\mathbf{i}) + (6\mathbf{k}) \times \left[(6\mathbf{k}) \times (0.4\mathbf{i})\right] + 2(6\mathbf{k}) \times (0.6\mathbf{i}) + (0.15\mathbf{i})$$

$$= \{-14.2\mathbf{i} + 8.40\mathbf{j}\}\ \text{m/s}^2 \qquad \textbf{Ans}$$

***16-132.** The ball B of negligible size rolls through the tube such that at the instant shown it has a velocity of 5 ft/s and an acceleration of 3 ft/s^2, measured relative to the tube. If the tube has an angular velocity of $\omega = 3$ rad/s and an angular acceleration of $\alpha = 5$ rad/s^2 at this same instant, determine the velocity and acceleration of the ball.

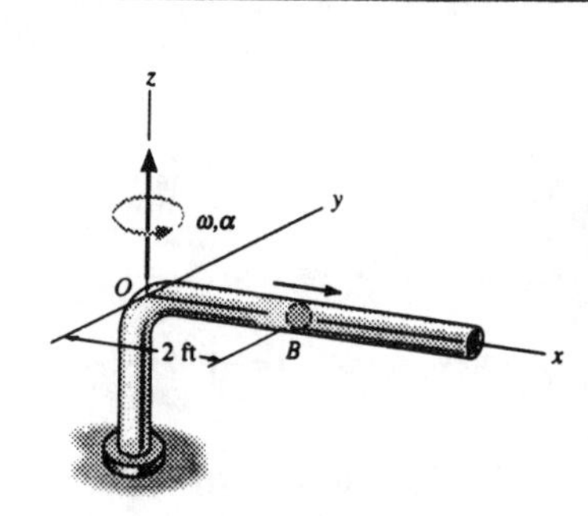

Kinematic Equations :

$$\mathbf{v}_B = \mathbf{v}_O + \Omega \times \mathbf{r}_{B/O} + (\mathbf{v}_{B/O})_{rel} \qquad (1)$$

$$\mathbf{a}_B = \mathbf{a}_O + \dot{\Omega} \times \mathbf{r}_{B/O} + \Omega \times (\Omega \times \mathbf{r}_{B/O}) + 2\Omega \times (\mathbf{v}_{B/O})_{xyz} + (\mathbf{a}_{B/O})_{xyz} \qquad (2)$$

$\mathbf{v}_O = \mathbf{0}$

$\mathbf{a}_O = \mathbf{0}$

$\Omega = \{3\mathbf{k}\}$ rad/s

$\dot{\Omega} = \{5\mathbf{k}\}$ rad/s^2

$\mathbf{r}_{B/O} = \{2\mathbf{i}\}$ ft

$(\mathbf{v}_{B/O})_{xyz} = \{5\mathbf{i}\}$ ft/s

$(\mathbf{a}_{B/O})_{xyz} = \{3\mathbf{i}\}$ ft/s^2

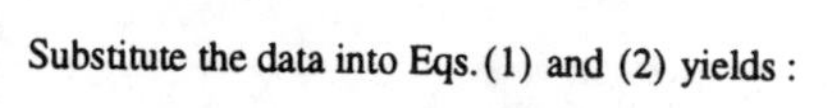

Substitute the data into Eqs. (1) and (2) yields :

$$\mathbf{v}_B = \mathbf{0} + (3\mathbf{k}) \times (2\mathbf{i}) + (5\mathbf{i}) = \{5\mathbf{i} + 6\mathbf{j}\}\ \text{ft/s} \qquad \textbf{Ans}$$

$$\mathbf{a}_B = \mathbf{0} + (5\mathbf{k}) \times (2\mathbf{i}) + (3\mathbf{k}) \times \left[(3\mathbf{k}) \times (2\mathbf{i})\right] + 2(3\mathbf{k}) \times (5\mathbf{i}) + (3\mathbf{i})$$

$$= \{-15\mathbf{i} + 40\mathbf{j}\}\ \text{ft/s}^2 \qquad \textbf{Ans}$$

16-133. The man stands on the platform at O and runs out toward the edge such that when he is at A, $y = 5$ ft, his mass center has a velocity of 2 ft/s and an acceleration of 3 ft/s^2, both measured with respect to the platform and directed along the y axis. If the platform has the angular motions shown, determine the velocity and acceleration of his mass center at this instant.

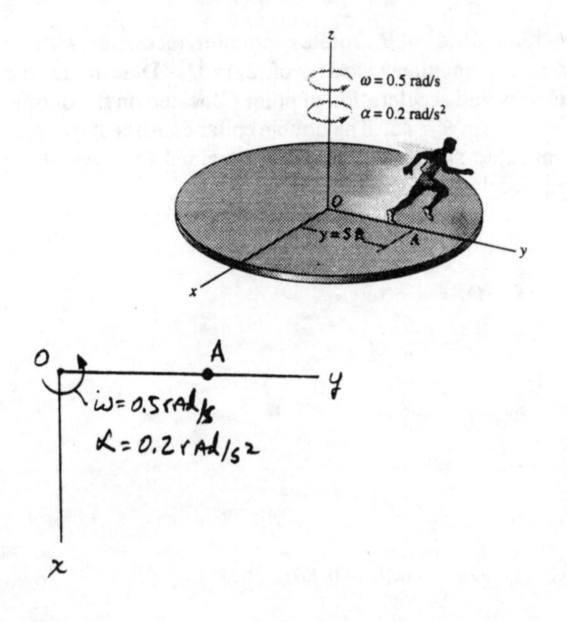

$$\mathbf{v}_A = \mathbf{v}_O + \Omega \times \mathbf{r}_{A/O} + (\mathbf{v}_{A/O})_{xyz}$$

$$\mathbf{v}_A = \mathbf{0} + (0.5\mathbf{k}) \times (5\mathbf{j}) + 2\mathbf{j}$$

$$\mathbf{v}_A = \{-2.50\mathbf{i} + 2.00\mathbf{j}\} \text{ ft/s} \qquad \textbf{Ans}$$

$$\mathbf{a}_A = \mathbf{a}_O + \dot{\Omega} \times \mathbf{r}_{A/O} + \Omega \times (\Omega \times \mathbf{r}_{A/O}) + 2\Omega \times (\mathbf{v}_{A/O})_{xyz} + (\mathbf{a}_{A/O})_{xyz}$$

$$\mathbf{a}_A = \mathbf{0} + (0.2\mathbf{k}) \times (5\mathbf{j}) + (0.5\mathbf{k}) \times (0.5\mathbf{k} \times 5\mathbf{j}) + 2(0.5\mathbf{k}) \times (2\mathbf{j}) + 3\mathbf{j}$$

$$\mathbf{a}_A = -1\mathbf{i} - 1.25\mathbf{j} - 2\mathbf{i} + 3\mathbf{j}$$

$$\mathbf{a}_A = \{-3.00\mathbf{i} + 1.75\mathbf{j}\} \text{ ft/s}^2 \qquad \textbf{Ans}$$

16-134. The dumpster pivots about C and is operated by the hydraulic cylinder AB. If the cylinder is extending at a constant rate of 0.5 ft/s, determine the angular velocity $\boldsymbol{\omega}$ of the container at the instant it becomes horizontal as shown.

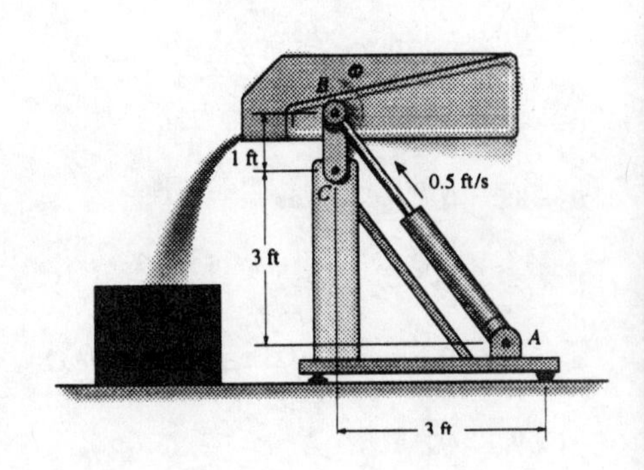

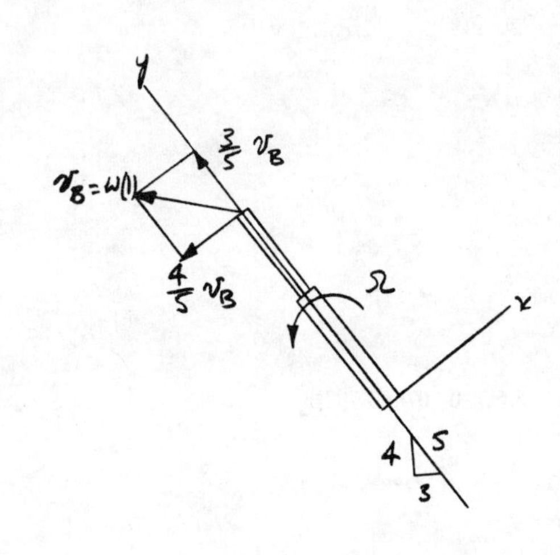

$$\mathbf{r}_{B/A} = 5\,\mathbf{j}$$

$$\mathbf{v}_{B/A} = 0.5\mathbf{j}$$

$$\mathbf{v}_B = -\frac{4}{5}\omega(1)\mathbf{i} + \frac{3}{5}\omega(1)\mathbf{j}$$

$$\mathbf{v}_B = \mathbf{v}_A + \Omega \times \mathbf{r}_{B/A} + (\mathbf{v}_{B/A})_{xyz}$$

$$-\frac{4}{5}\omega(1)\mathbf{i} + \frac{3}{5}\omega(1)\mathbf{j} = \mathbf{0} + (\Omega\mathbf{k}) \times (5\,\mathbf{j}) + 0.5\mathbf{j}$$

$$-\frac{4}{5}\omega(1)\mathbf{i} + \frac{3}{5}\omega(1)\mathbf{j} = -\Omega(5)\mathbf{i} + 0.5\mathbf{j}$$

Thus,

$$\omega = 0.833 \text{ rad/s} \qquad \textbf{Ans}$$

$$\Omega = 0.133 \text{ rad/s}$$

16-135. Rod AB rotates counterclockwise with a constant angular velocity of 2 rad/s. Determine the velocity and acceleration of point C located on the double collar when $\theta = 45°$. The double collar consists of two pin-connected sliders which are constrained to move along the circular shaft and the rod AB.

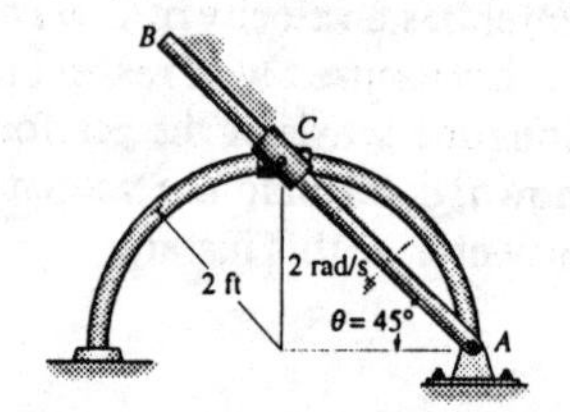

$$\mathbf{v}_C = \mathbf{v}_A + \Omega \times \mathbf{r}_{C/A} + (\mathbf{v}_{C/A})_{xyz}$$

Set x, y axes fixed in AB,

$\Omega = 2\mathbf{k}$

$\mathbf{v}_C = -v_C\mathbf{i}$

$\mathbf{v}_A = \mathbf{0}$

$(\mathbf{v}_{C/A})_{xyz} = v_{C/A}(-0.707\mathbf{i} + 0.707\mathbf{j})$

$\mathbf{r}_{C/A} = -2\mathbf{i} + 2\mathbf{j}$

$-v_C\mathbf{i} = \mathbf{0} + 2\mathbf{k} \times (-2\mathbf{i} + 2\mathbf{j}) - 0.707v_{C/A}\mathbf{i} + 0.707v_{C/A}\mathbf{j}$

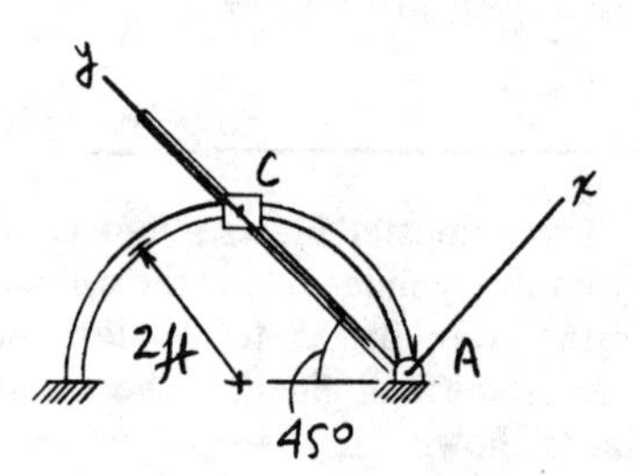

Thus,

$-v_C = -4 - 0.707v_{C/A}$

$0 = -4 + 0.707v_{C/A}$

$v_C = 8 \text{ ft/s} \leftarrow$ **Ans**

$v_{C/A} = 5.66 \text{ ft/s}$

$$\mathbf{a}_C = \mathbf{a}_A + \dot{\Omega} \times \mathbf{r}_{C/A} + \Omega \times (\Omega \times \mathbf{r}_{C/A}) + 2\Omega \times (\mathbf{v}_{C/A})_{xyz} + (\mathbf{a}_{C/A})_{xyz}$$

$\mathbf{a}_A = \mathbf{0}, \quad \alpha = \mathbf{0}$

$\mathbf{a}_{C/A} = a_{C/A}(-0.707\mathbf{i} + 0.707\mathbf{j})$

$$\mathbf{a}_C = -(a_C)_t\mathbf{i} - \frac{v_C^2}{\rho}\mathbf{j} = -(a_C)_t\mathbf{i} - 32\mathbf{j}$$

$$-(a_C)_t\mathbf{i} - 32\mathbf{j} = \mathbf{0} + \mathbf{0} + (2\mathbf{k}) \times (2\mathbf{k} \times (-2\mathbf{i} + 2\mathbf{j})) + 2[(2\mathbf{k}) \times 5.66(-0.707\mathbf{i} + 0.707\mathbf{j})]$$
$$- 0.707a_{C/A}\mathbf{i} + 0.707a_{C/A}\mathbf{j}$$

Thus,

$-(a_C)_t = -8 - 0.707a_{C/A}$

$0 = 8 + 0.707a_{C/A}$

$(a_C)_t = 0, \quad a_{C/A} = -11.32 \text{ ft/s}^2$

so that

$a_C = 32 \text{ ft/s}^2 \downarrow$ **Ans**

***16-136.** The collar E is attached to, and pivots about, rod AB while it slides on rod CD. If rod AB has an angular velocity of 6 rad/s and an angular acceleration of 1 rad/s^2, both acting clockwise, determine the angular velocity and the angular acceleration of rod CD at the instant shown.

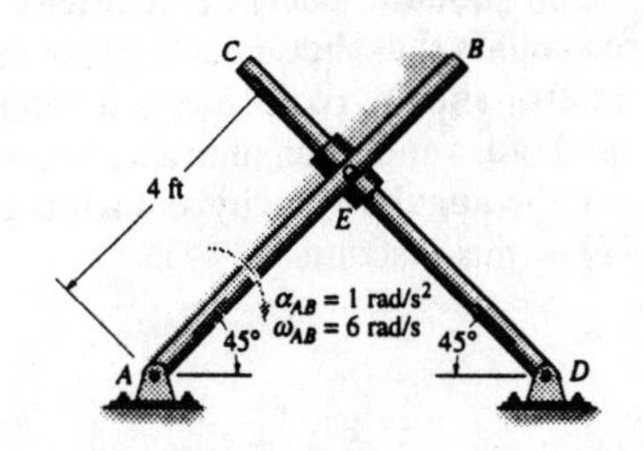

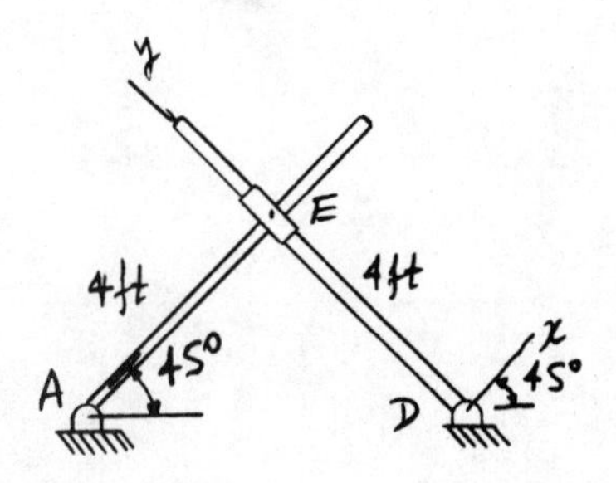

Fix axes to ED.

$$\Omega = \omega_{CD}\mathbf{k}$$

$$\dot{\Omega} = \alpha_{CD}\mathbf{k}$$

$$\mathbf{r}_{E/D} = 4\mathbf{j}$$

$$\mathbf{v}_{E/D} = v_{E/D}\mathbf{j}$$

$$\mathbf{a}_{E/D} = a_{E/D}\mathbf{j}$$

$$\mathbf{v}_E = -6(4)\mathbf{j} = -24\mathbf{j}$$

$$\mathbf{a}_E = -(6)^2(4)\mathbf{i} - 1(4)\mathbf{j} = -144\mathbf{i} - 4\mathbf{j}$$

$$\mathbf{v}_E = \mathbf{v}_D + \Omega \times \mathbf{r}_{E/D} + (\mathbf{v}_{E/D})_{xyz}$$

$$-24\mathbf{j} = \mathbf{0} + \omega_{CD}\mathbf{k} \times 4\mathbf{j} + v_{E/D}\mathbf{j}$$

$$-24\mathbf{j} = -4\omega_{CD}\mathbf{i} + v_{E/D}\mathbf{j}$$

Thus,

$$\omega_{CD} = 0 \qquad \textbf{Ans}$$

$$v_{E/D} = -24 \text{ ft/s}$$

$$\mathbf{a}_E = \mathbf{a}_D + \dot{\Omega} \times \mathbf{r}_{E/D} + \Omega \times (\Omega \times \mathbf{r}_{E/D}) + 2\Omega \times (\mathbf{v}_{E/D})_{xyz} + (\mathbf{a}_{E/D})_{xyz}$$

$$-144\mathbf{i} - 4\mathbf{j} = \mathbf{0} + \alpha_{CD}\mathbf{k} \times 4\mathbf{j} + \mathbf{0} + \mathbf{0} + a_{E/D}\mathbf{j}$$

$$-144\mathbf{i} - 4\mathbf{j} = -\alpha_{CD}(4)\mathbf{i} + a_{E/D}\mathbf{j}$$

$$\alpha_{CD} = \frac{144}{4} = 36 \text{ rad/s}^2 \qquad \textbf{Ans}$$

$$a_{E/D} = -4 \text{ ft/s}^2$$

16-137. The double collar E consists of two pin-connected collars that slide freely over rods AB and CD. At the instant shown, rod AB has a clockwise angular velocity of 5 rad/s and an angular acceleration of 3 rad/s². Determine the angular velocity and angular acceleration of rod CD at this instant.

$\Omega = \omega_{CD}\mathbf{k}$

$\dot{\Omega} = \alpha_{CD}\mathbf{k}$

$\mathbf{r}_{E/C} = 2\mathbf{j}$

$\mathbf{v}_{E/C} = -v_{E/C}\mathbf{j}$

$\mathbf{a}_{E/C} = -a_{E/C}\mathbf{j}$

$\mathbf{v}_E = -5(3)\mathbf{j} = -15\mathbf{j}$

$\mathbf{a}_E = -(5)^2(3)\mathbf{i} - 3(3)\mathbf{j} = -75\mathbf{i} - 9\mathbf{j}$

$\mathbf{v}_E = \mathbf{v}_C + \Omega \times \mathbf{r}_{E/C} + (\mathbf{v}_{E/C})_{xyz}$

$-15\mathbf{j} = \mathbf{0} + \omega_{CD}\mathbf{k} \times 2\mathbf{j} + (-v_{E/C}\mathbf{j})$

$-15\mathbf{j} = -2\omega_{CD}\mathbf{i} - v_{E/C}\mathbf{j}$

Thus,

$\omega_{CD} = 0$ **Ans**

$v_{E/C} = 15$ ft/s

$\mathbf{a}_E = \mathbf{a}_C + \dot{\Omega} \times \mathbf{r}_{E/C} + \Omega \times (\Omega \times \mathbf{r}_{E/C}) + 2\Omega \times (\mathbf{v}_{E/C})_{xyz} + (\mathbf{a}_{E/C})_{xyz}$

$-75\mathbf{i} - 9\mathbf{j} = \mathbf{0} + \alpha_{CD}\mathbf{k} \times 2\mathbf{j} + \mathbf{0} + \mathbf{0} + (-a_{E/C}\mathbf{j})$

$-75\mathbf{i} - 9\mathbf{j} = -2\alpha_{CD}\mathbf{i} - a_{E/D}\mathbf{j}$

$\alpha_{CD} = \frac{75}{2} = 37.5 \text{ rad/s}^2$ ↻ **Ans**

$a_{E/D} = 9 \text{ ft/s}^2$

16-138. At the instant $\theta = 45°$, link DC has an angular velocity of $\omega_{DC} = 4$ rad/s and an angular acceleration of $\alpha_{DC} = 2$ rad/s^2. Determine the angular velocity and angular acceleration of rod AB at this instant. The collar at C is pin connected to DC and slides over AB.

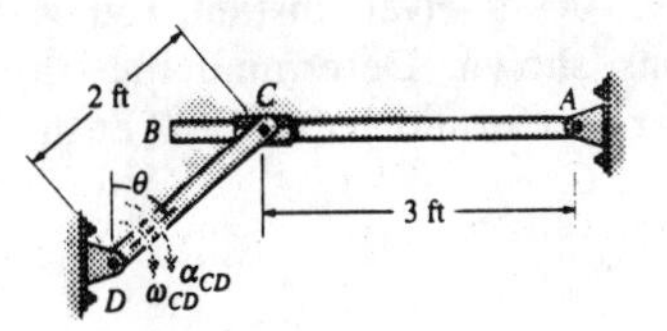

$\mathbf{v}_A = \mathbf{0}$

$\mathbf{a}_A = \mathbf{0}$

$\Omega = \omega_{AB}\,\mathbf{k}$

$\dot{\Omega} = \alpha_{AB}\,\mathbf{k}$

$\mathbf{r}_{C/A} = \{-3\mathbf{i}\}$ ft

$(\mathbf{v}_{C/A})_{xyz} = (v_{C/A})_{rel}\,\mathbf{i}$

$(\mathbf{a}_{C/A})_{xyz} = (a_{C/A})_{rel}\,\mathbf{i}$

$\mathbf{v}_C = \omega_{CD} \times \mathbf{r}_{C/D} = (-4\mathbf{k}) \times (2\sin 45°\mathbf{i} + 2\cos 45°\mathbf{j}) = \{5.6569\mathbf{i} - 5.6569\mathbf{j}\}$ ft/s

$\mathbf{a}_C = \alpha_{CD} \times \mathbf{r}_{C/D} - \omega_{CD}^2\mathbf{r}_{C/D}$

$= (-2\mathbf{k}) \times (2\sin 45°\mathbf{i} + 2\cos 45°\mathbf{j}) - (4)^2(2\sin 45°\mathbf{i} + 2\cos 45°\mathbf{j})$

$= \{-19.7990\mathbf{i} - 25.4558\mathbf{j}\}$ ft/s^2

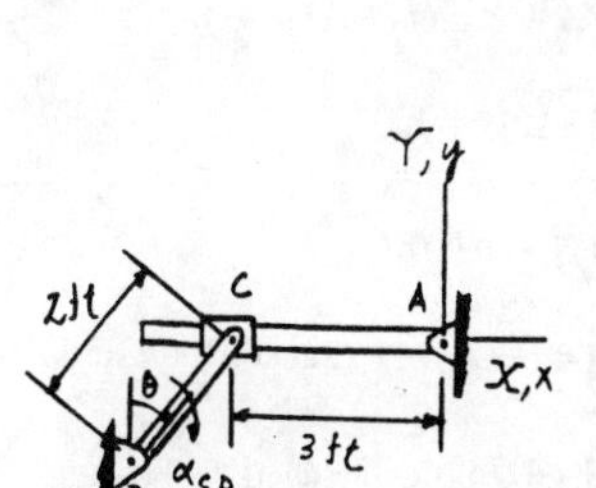

$\mathbf{v}_C = \mathbf{v}_A + \Omega \times \mathbf{r}_{C/A} + (\mathbf{v}_{C/A})_{xyz}$

$5.6569\mathbf{i} - 5.6569\mathbf{j} = \mathbf{0} + (\omega_{AB}\mathbf{k}) \times (-3\mathbf{i}) + (v_{C/A})_{xyz}\,\mathbf{i}$

$5.6569\mathbf{i} - 5.6569\mathbf{j} = (v_{C/A})_{xyz}\,\mathbf{i} - 3\omega_{AB}\,\mathbf{j}$

Solving

$(v_{C/A})_{xyz} = 5.6569$ ft/s

$\omega_{AB} = 1.89$ rad/s ↻ **Ans**

$\mathbf{a}_C = \mathbf{a}_A + \dot{\Omega} \times \mathbf{r}_{C/A} + \Omega \times (\Omega \times \mathbf{r}_{C/A}) + 2\Omega \times (\mathbf{v}_{C/A})_{xyz} + (\mathbf{a}_{C/A})_{xyz}$

$-19.7990\mathbf{i} - 25.4558\mathbf{j} = \mathbf{0} + (\alpha_{AB}\mathbf{k}) \times (-3\mathbf{i}) + (1.89\mathbf{k}) \times [(1.89\mathbf{k}) \times (-3\mathbf{i})]$

$+2(1.89\mathbf{k}) \times (5.6569\mathbf{i}) + (a_{C/A})_{xyz}\,\mathbf{i}$

$-19.7990\mathbf{i} - 25.4558\mathbf{j} = [10.6667 + (a_{C/A})_{xyz}]\mathbf{i} + (21.334 - 3\alpha_{AB})\mathbf{j}$

Solving

$(a_{C/A})_{xyz} = -30.47$ ft/s^2

$\alpha_{AB} = 15.6$ rad/s^2 ↻ **Ans**

16-139. At a given instant, rod AB has the angular motions shown. Determine the angular velocity and angular acceleration of rod CD at this instant. There is a collar at C.

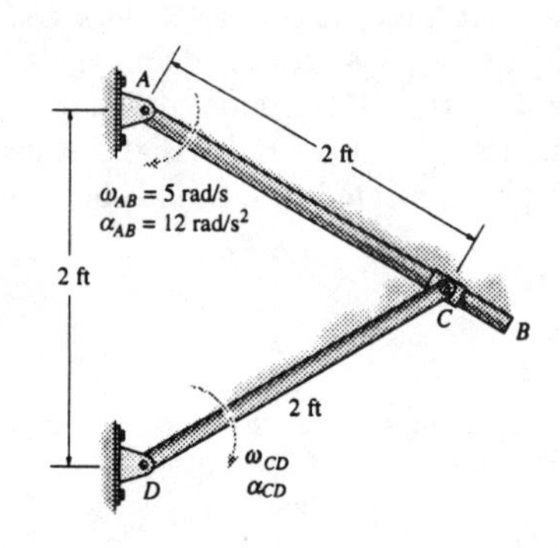

$\mathbf{v}_A = \mathbf{0}$

$\mathbf{a}_A = \mathbf{0}$

$\mathbf{\Omega} = \{-5\mathbf{k}\}$ rad/s

$\dot{\mathbf{\Omega}} = \{-12\mathbf{k}\}$ rad/s^2

$\mathbf{r}_{C/A} = \{2\mathbf{i}\}$ ft

$(\mathbf{v}_{C/A})_{xyz} = (v_{C/A})_{xyz}\mathbf{i}$

$(\mathbf{a}_{C/A})_{xyz} = (a_{C/A})_{xyz}\mathbf{i}$

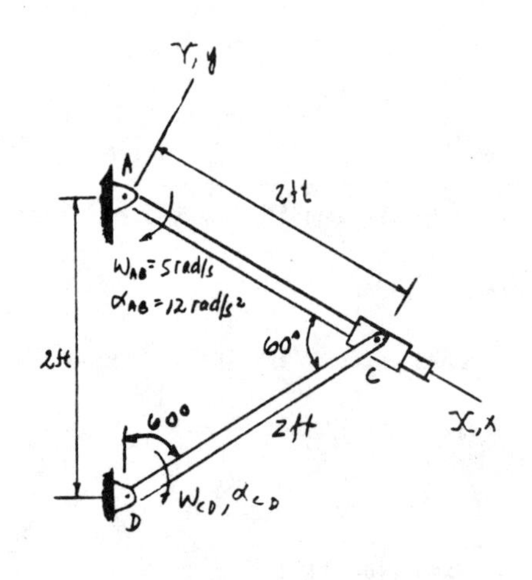

$\mathbf{v}_C = \mathbf{v}_A + \mathbf{\Omega}\times\mathbf{r}_{C/A} + (\mathbf{v}_{C/A})_{xyz}$

$\mathbf{v}_C = \mathbf{0} + (-5\mathbf{k})\times(2\mathbf{i}) + (v_{C/A})_{xyz}\mathbf{i}$

$= (v_{C/A})_{xyz}\mathbf{i} - 10\mathbf{j}$

$\mathbf{v}_C = \omega_{CD}\times\mathbf{r}_{CD}$

$(v_{C/A})_{xyz}\mathbf{i} - 10\mathbf{j} = (-\omega_{CD}\mathbf{k})\times(2\cos 60°\mathbf{i} + 2\sin 60°\mathbf{j})$

$(v_{C/A})_{xyz}\mathbf{i} - 10\mathbf{j} = 1.732\omega_{CD}\mathbf{i} - \omega_{CD}\mathbf{j}$

Solving :

$\omega_{CD} = 10$ rad/s ↻ **Ans**

$(v_{C/A})_{xyz} = 1.732(10) = 17.32$ ft/s

$\mathbf{a}_C = \mathbf{a}_A + \dot{\mathbf{\Omega}}\times\mathbf{r}_{C/A} + \mathbf{\Omega}\times(\mathbf{\Omega}\times\mathbf{r}_{C/A}) + 2\mathbf{\Omega}\times(\mathbf{v}_{C/A})_{xyz} + (\mathbf{a}_{C/A})_{xyz}$

$\mathbf{a}_C = \mathbf{0} + (-12\mathbf{k})\times(2\mathbf{i}) + (-5\mathbf{k})\times[(-5\mathbf{k})\times(2\mathbf{i})] + 2(-5\mathbf{k})\times[(v_{C/A})_{xyz}\mathbf{i}] + (a_{C/A})_{xyz}\mathbf{i}$

$= [(a_{C/A})_{xyz} - 50]\mathbf{i} - [10(v_{C/A})_{xyz} + 24]\mathbf{j}$

$\mathbf{a}_C = \alpha_{CD}\times\mathbf{r}_{C/D} - \omega_{CD}^2\mathbf{r}_{C/D}$

$[(a_{C/A})_{xyz} - 50]\mathbf{i} - [10(17.32) + 24]\mathbf{j} = (-\alpha_{CD}\mathbf{k})\times(2\cos 60°\mathbf{i} + 2\sin 60°\mathbf{j})$

$- (10)^2(2\cos 60°\mathbf{i} + 2\sin 60°\mathbf{j})$

$[(a_{C/A})_{xyz} - 50]\mathbf{i} - [10(17.32) + 24]\mathbf{j} = (1.732\alpha_{CD} - 100)\mathbf{i} - (\alpha_{CD} + 173.2)\mathbf{j}$

Solving :

$-[10(17.32) + 24] = -(\alpha_{CD} + 173.2)$ $\alpha_{CD} = 24$ rad/s^2 ↻ **Ans**

$(a_{C/A})_{xyz} - 50 = 1.732(24) - 100$ $(a_{C/A})_{xyz} = -8.43$ ft/s^2

***16-140.** The disk rolls without slipping and at a given instant has the angular motion shown. Determine the angular velocity and angular acceleration of the slotted link BC at this instant. The peg at A is fixed to the disk.

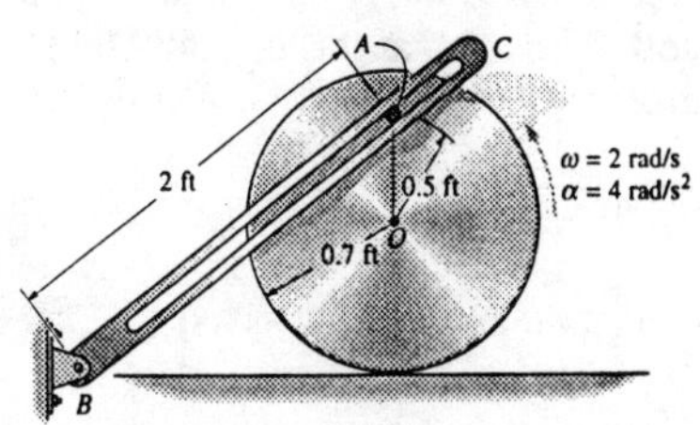

$\mathbf{v}_A = -(1.2)(2)\mathbf{i} = -2.4\mathbf{i}$ ft/s

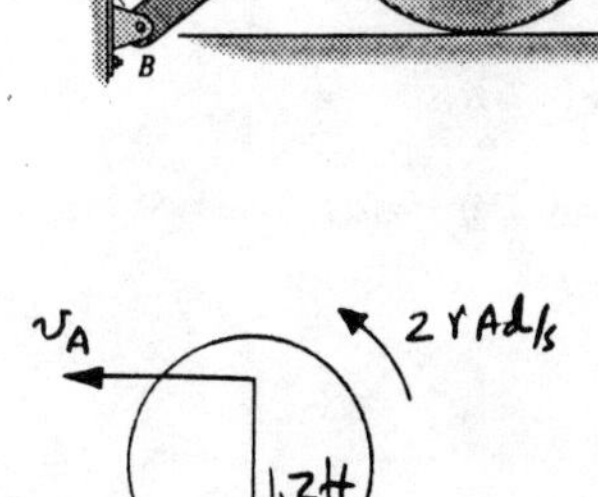

$\mathbf{a}_A = \mathbf{a}_O + \alpha \times \mathbf{r}_{A/O} - \omega^2 \mathbf{r}_{A/O}$

$\mathbf{a}_A = -4(0.7)\mathbf{i} + (4\mathbf{k}) \times (0.5\mathbf{j}) - (2)^2(0.5\mathbf{j})$

$\mathbf{a}_A = -4.8\mathbf{i} - 2\mathbf{j}$

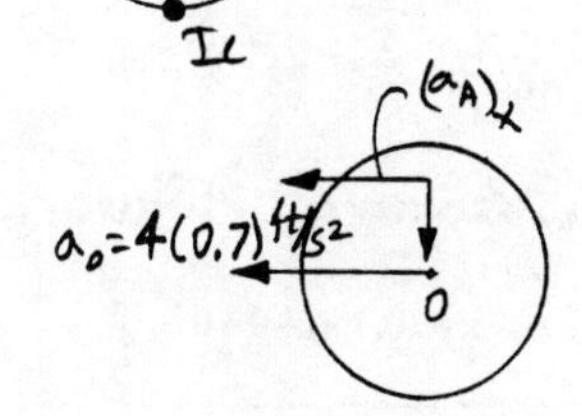

$\mathbf{v}_A = \mathbf{v}_B + \Omega \times \mathbf{r}_{A/B} + (\mathbf{v}_{A/B})_{xyz}$

$-2.4\mathbf{i} = \mathbf{0} + (\omega_{BC}\mathbf{k}) \times (1.6\mathbf{i} + 1.2\mathbf{j}) + v_{A/B}\left(\frac{4}{5}\right)\mathbf{i} + v_{A/B}\left(\frac{3}{5}\right)\mathbf{j}$

$-2.4\mathbf{i} = 1.6\omega_{BC}\mathbf{j} - 1.2\omega_{BC}\mathbf{i} + 0.8v_{A/B}\mathbf{i} + 0.6v_{A/B}\mathbf{j}$

$-2.4 = -1.2\omega_{BC} + 0.8v_{A/B}$

$0 = 1.6\omega_{BC} + 0.6v_{A/B}$

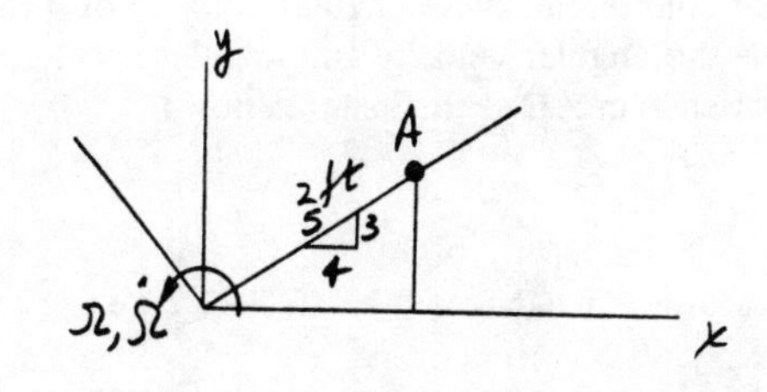

Solving,

$\omega_{BC} = 0.720$ rad/s ↻ **Ans**

$v_{A/B} = -1.92$ ft/s

$\mathbf{a}_A = \mathbf{a}_B + \dot{\Omega} \times \mathbf{r}_{A/B} + \Omega \times (\Omega \times \mathbf{r}_{A/B}) + 2\Omega \times (\mathbf{v}_{A/B})_{xyz} + (\mathbf{a}_{A/B})_{xyz}$

$-4.8\mathbf{i} - 2\mathbf{j} = \mathbf{0} + (\alpha_{BC}\mathbf{k}) \times (1.6\mathbf{i} + 1.2\mathbf{j}) + (0.72\mathbf{k}) \times (0.72\mathbf{k} \times (1.6\mathbf{i} + 1.2\mathbf{j}))$

$+ 2(0.72\mathbf{k}) \times \left[-(0.8)(1.92)\mathbf{i} - 0.6(1.92)\mathbf{j}\right] + 0.8a_{B/A}\mathbf{i} + 0.6a_{B/A}\mathbf{j}$

$-4.8\mathbf{i} - 2\mathbf{j} = 1.6\alpha_{BC}\mathbf{j} - 1.2\alpha_{BC}\mathbf{i} - 0.8294\mathbf{i} - 0.6221\mathbf{j} - 2.2118\mathbf{j} + 1.6589\mathbf{i} + 0.8a_{B/A}\mathbf{i} + 0.6a_{B/A}\mathbf{j}$

$-4.8 = -1.2\alpha_{BC} - 0.8294 + 1.6589 + 0.8a_{B/A}$

$-2 = 1.6\alpha_{BC} - 0.6221 - 2.2118 + 0.6a_{B/A}$

Solving,

$\alpha_{BC} = 2.02$ rad/s² ↻ **Ans**

$a_{B/A} = -4.00$ ft/s²

16-141. The disk rotates with the angular motion shown. Determine the angular velocity and angular acceleration of the slotted link AC at this instant. The peg at B is fixed to the disk.

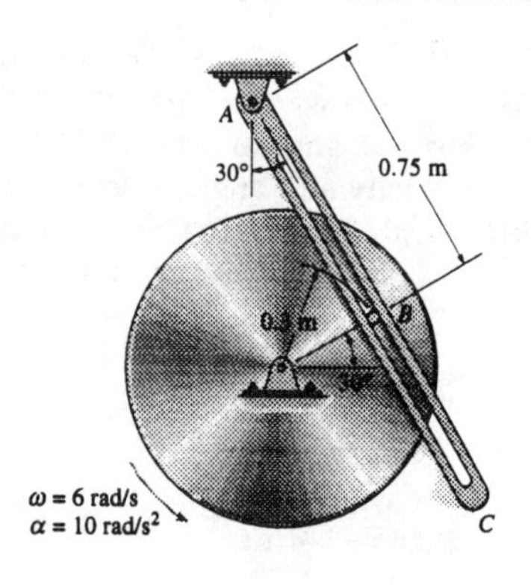

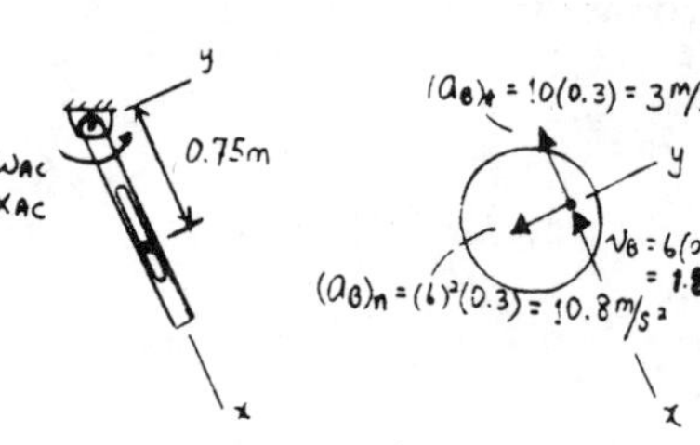

$\mathbf{v}_B = -6(0.3)\mathbf{i} = -1.8\mathbf{i}$

$\mathbf{a}_B = -10(0.3)\mathbf{i} - (6)^2(0.3)\mathbf{j} = -3\mathbf{i} - 10.8\mathbf{j}$

$\mathbf{v}_B = \mathbf{v}_A + \mathbf{\Omega} \times \mathbf{r}_{B/A} + (\mathbf{v}_{B/A})_{xyz}$

$-1.8\mathbf{i} = \mathbf{0} + (\omega_{AC}\mathbf{k}) \times (0.75\mathbf{i}) - (v_{B/A})_{xyz}\,\mathbf{i}$

$-1.8\mathbf{i} = -(v_{B/A})_{xyz}$

$(v_{B/A})_{xyz} = 1.8 \text{ m/s}$

$0 = \omega_{AC}(0.75)$

$\omega_{AC} = 0$ **Ans**

$\mathbf{a}_B = \mathbf{a}_A + \dot{\mathbf{\Omega}} \times \mathbf{r}_{B/A} + \mathbf{\Omega} \times (\mathbf{\Omega} \times \mathbf{r}_{B/A}) + 2\mathbf{\Omega} \times (\mathbf{v}_{B/A})_{xyz} + (\mathbf{a}_{B/A})_{xyz}$

$-3\mathbf{i} - 10.8\mathbf{j} = \mathbf{0} + \alpha_{AC}\mathbf{k} \times (0.75\mathbf{i}) + \mathbf{0} + \mathbf{0} - a_{A/B}\,\mathbf{i}$

$-3 = -a_{A/B}$

$a_{A/B} = 3 \text{ m/s}^2$

$-10.8 = \alpha_{AC}(0.75)$

$\alpha_{AC} = 14.4 \text{ rad/s}^2$ ↻ **Ans**

16-142. If the slider block C is fixed to the disk that has a constant counterclockwise angular velocity of 4 rad/s, determine the angular velocity and angular acceleration of the slotted arm AB at the instant shown.

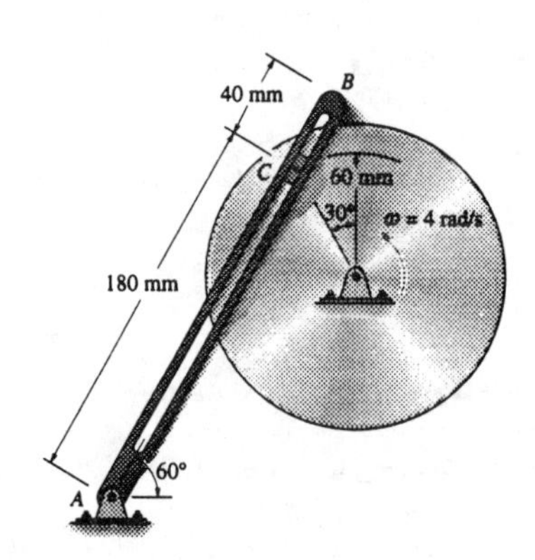

$\mathbf{v}_C = -(4)(60)\sin 30°\mathbf{i} - 4(60)\cos 30°\mathbf{j} = -120\mathbf{i} - 207.85\mathbf{j}$

$\mathbf{a}_C = (4)^2(60)\sin 60°\mathbf{i} - (4)^2(60)\cos 60°\mathbf{j} = 831.38\mathbf{i} - 480\mathbf{j}$

Thus,

$\mathbf{v}_C = \mathbf{v}_A + \mathbf{\Omega} \times \mathbf{r}_{C/A} + (\mathbf{v}_{C/A})_{xyz}$

$-120\mathbf{i} - 207.85\mathbf{j} = \mathbf{0} + (\omega_{AB}\mathbf{k}) \times (180\mathbf{j}) - v_{C/A}\,\mathbf{j}$

$-120 = -180\omega_{AB}$

$\omega_{AB} = 0.667 \text{ rad/s}$ ↺ **Ans**

$-207.85 = -v_{C/A}$

$v_{C/A} = 207.85 \text{ mm/s}$

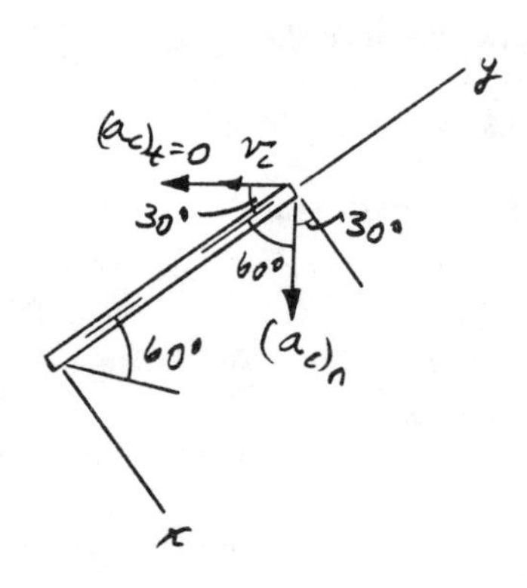

$\mathbf{a}_C = \mathbf{a}_A + \dot{\mathbf{\Omega}} \times \mathbf{r}_{C/A} + \mathbf{\Omega} \times (\mathbf{\Omega} \times \mathbf{r}_{C/A}) + 2\mathbf{\Omega} \times (\mathbf{v}_{C/A})_{xyz} + (\mathbf{a}_{C/A})_{xyz}$

$831.38\mathbf{i} - 480\mathbf{j} = \mathbf{0} + (\alpha_{AB}\mathbf{k}) \times (180\mathbf{j}) + (0.667\mathbf{k}) \times [(0.667\mathbf{k}) \times (180\mathbf{j})]$

$+ 2(0.667\mathbf{k}) \times (-207.85\mathbf{j}) - a_{C/A}\,\mathbf{j}$

$831.38\mathbf{i} - 480\mathbf{j} = -180\alpha_{AB}\mathbf{i} - 80\mathbf{j} + 277.13\mathbf{i} - a_{C/A}\,\mathbf{j}$

$831.38 = -180\alpha_{AB} + 277.13$

$\alpha_{AB} = -3.08$

Thus,

$\alpha_{AB} = 3.08 \text{ rad/s}^2$ ↻ **Ans**

$-480 = -80 - a_{C/A}$

$a_{C/A} = 400 \text{ mm/s}^2$

16-143. The two-link mechanism serves to amplify angular motion. Link AB has a pin at B which is confined to move within the slot of link CD. If at the instant shown, AB (input) has an angular velocity of $\omega_{AB} = 2.5$ rad/s and an angular acceleration of $\alpha_{AB} = 3$ rad/s^2, determine the angular velocity and angular acceleration of CD (output) at this instant.

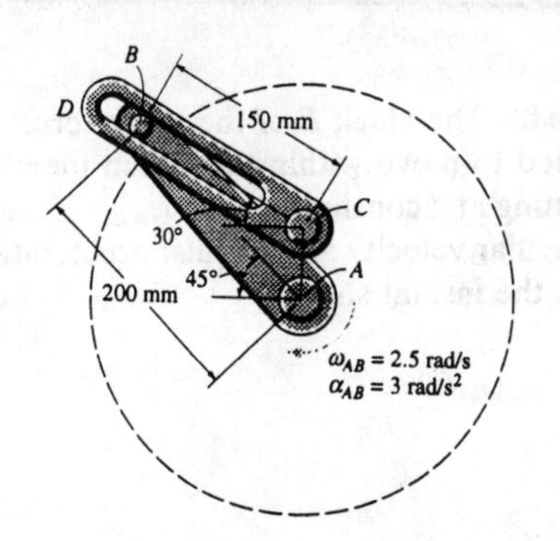

$\mathbf{v}_C = \mathbf{0}$

$\mathbf{a}_C = \mathbf{0}$

$\Omega = -\omega_{DC}\mathbf{k}$

$\dot{\Omega} = -\alpha_{DC}\mathbf{k}$

$\mathbf{r}_{B/C} = \{-0.15\mathbf{i}\}$ m

$(\mathbf{v}_{B/C})_{xyz} = (v_{B/C})_{xyz}\,\mathbf{i}$

$(\mathbf{a}_{B/C})_{xyz} = (a_{B/C})_{xyz}\,\mathbf{i}$

$\mathbf{v}_B = \omega_{AB} \times \mathbf{r}_{B/A} = (-2.5\mathbf{k}) \times (-0.2\cos 15°\mathbf{i} + 0.2\sin 15°\mathbf{j})$

$= \{0.1294\mathbf{i} + 0.4830\mathbf{j}\}$ m/s

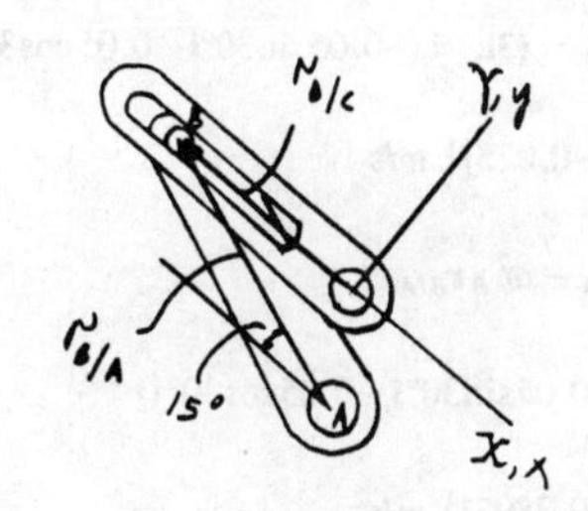

$\mathbf{a}_B = \alpha_{AB} \times \mathbf{r}_{B/A} - \omega_{AB}^2 \mathbf{r}_{B/A}$

$= (-3\mathbf{k}) \times (-0.2\cos 15°\mathbf{i} + 0.2\sin 15°\mathbf{j}) - (2.5)^2(-0.2\cos 15°\mathbf{i} + 0.2\sin 15°\mathbf{j})$

$= \{1.3627\mathbf{i} + 0.2560\mathbf{j}\}$ m/s^2

$\mathbf{v}_B = \mathbf{v}_C + \Omega \times \mathbf{r}_{B/C} + (\mathbf{v}_{B/C})_{xyz}$

$0.1294\mathbf{i} + 0.4830\mathbf{j} = \mathbf{0} + (-\omega_{DC}\mathbf{k}) \times (-0.15\mathbf{i}) + (v_{B/C})_{xyz}\,\mathbf{i}$

$0.1294\mathbf{i} + 0.4830\mathbf{j} = (v_{B/C})_{xyz}\,\mathbf{i} + 0.15\omega_{DC}\mathbf{j}$

Solving :

$(v_{B/C})_{xyz} = 0.1294$ m/s

$\omega_{DC} = 3.22$ rad/s ↻ **Ans**

$\mathbf{a}_B = \mathbf{a}_C + \dot{\Omega} \times \mathbf{r}_{B/C} + \Omega \times (\Omega \times \mathbf{r}_{B/C}) + 2\Omega \times (\mathbf{v}_{B/C})_{xyz} + (\mathbf{a}_{B/C})_{xyz}$

$1.3627\mathbf{i} + 0.2560\mathbf{j} = \mathbf{0} + (-\alpha_{DC}\mathbf{k}) \times (-0.15\mathbf{i}) + (-3.22\mathbf{k}) \times [(-3.22\mathbf{k}) \times (-0.15\mathbf{i})]$

$+ 2(-3.22\mathbf{k}) \times (0.1294\mathbf{i}) + (a_{B/C})_{xyz}\,\mathbf{i}$

$1.3627\mathbf{i} + 0.2560\mathbf{j} = [1.5550 + (a_{B/C})_{xyz}]\mathbf{i} + (0.15\alpha_{DC} - 0.8333)\mathbf{j}$

Solving :

$(a_{C/A})_{xyz} = -0.1923$ m/s^2

$\alpha_{DC} = 7.26$ rad/s^2 ↻ **Ans**

***16-144.** The block B of the quick-return mechanism is confined to move within the slot in member CD. If AB is rotating at a constant rate of $\omega_{AB} = 3$ rad/s, determine the angular velocity and angular acceleration of member CD at the instant shown.

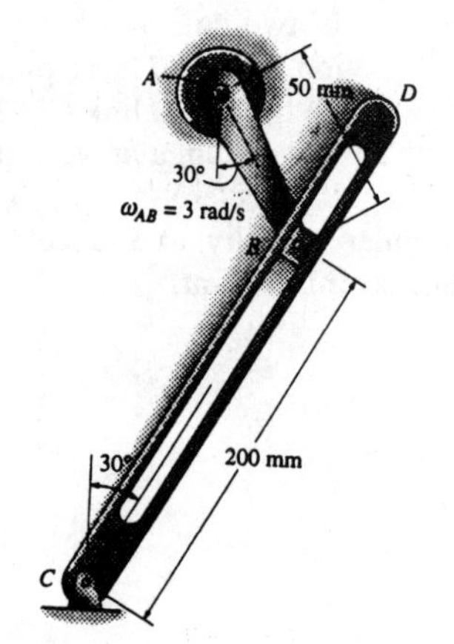

$\mathbf{v}_C = \mathbf{0}$

$\mathbf{a}_C = \mathbf{0}$

$\Omega = -\omega_{CD}\mathbf{k}$

$\dot{\Omega} = \alpha_{CD}\mathbf{k}$

$\mathbf{r}_{B/C} = \{0.2\mathbf{i}\}\text{ m}$

$(\mathbf{v}_{B/C})_{xyz} = (v_{B/C})_{xyz}\,\mathbf{i}$

$(\mathbf{a}_{B/C})_{xyz} = (a_{B/C})_{xyz}\,\mathbf{i}$

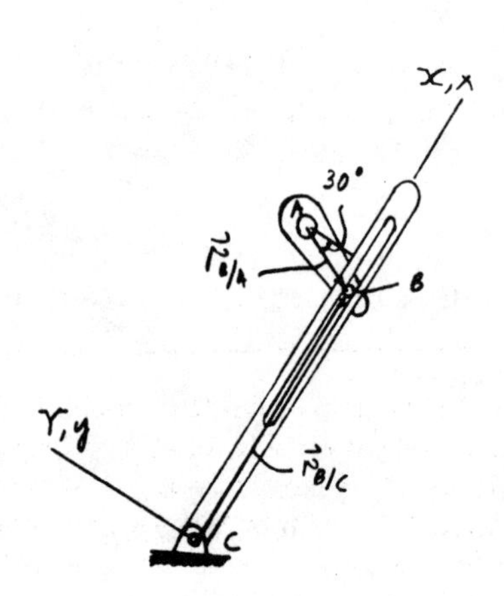

$\mathbf{v}_B = \omega_{AB} \times \mathbf{r}_{B/A} = (3\mathbf{k}) \times (-0.05\sin 30°\mathbf{i} - 0.05\cos 30°\mathbf{j})$

$= \{0.1299\mathbf{i} - 0.075\mathbf{j}\}$ m/s

$\mathbf{a}_B = \alpha_{AB} \times \mathbf{r}_{B/A} - \omega_{AB}^2\mathbf{r}_{B/A}$

$= \mathbf{0} - (3)^2(-0.05\sin 30°\mathbf{i} - 0.05\cos 30°\mathbf{j})$

$= \{0.225\mathbf{i} + 0.3897\mathbf{j}\}$ m/s^2

$\mathbf{v}_B = \mathbf{v}_C + \Omega \times \mathbf{r}_{B/C} + (\mathbf{v}_{B/C})_{xyz}$

$0.1299\mathbf{i} - 0.075\mathbf{j} = \mathbf{0} + (-\omega_{CD}\mathbf{k}) \times (0.2\mathbf{i}) + (v_{B/C})_{xyz}\,\mathbf{i}$

$0.1299\mathbf{i} - 0.075\mathbf{j} = (v_{B/C})_{rel}\,\mathbf{i} - 0.2\omega_{CD}\mathbf{j}$

Solving :

$(v_{B/C})_{xyz} = 0.1299$ m/s

$\omega_{CD} = 0.375$ rad/s ↻ **Ans**

$\mathbf{a}_B = \mathbf{a}_C + \dot{\Omega} \times \mathbf{r}_{B/C} + \Omega \times (\Omega \times \mathbf{r}_{B/C}) + 2\Omega \times (\mathbf{v}_{B/C})_{xyz} + (\mathbf{a}_{B/C})_{xyz}$

$0.225\mathbf{i} + 0.3897\mathbf{j} = \mathbf{0} + (\alpha_{CD}\mathbf{k}) \times (0.2\mathbf{i}) + (-0.375\mathbf{k}) \times [(-0.375\mathbf{k}) \times (0.2\mathbf{i})]$

$+ 2(-0.375\mathbf{k}) \times (0.1299\mathbf{i}) + (a_{B/C})_{xyz}\,\mathbf{i}$

$0.225\mathbf{i} + 0.3897\mathbf{j} = [(a_{B/C})_{xyz} - 0.028125]\mathbf{i} + (0.2\alpha_{CD} - 0.097428)\mathbf{j}$

Equating the **i** and **j** components and solving,

$(a_{B/C})_{xyz} = 0.2531$ m/s^2

$\alpha_{CD} = 2.44$ rad/s^2 ↻ **Ans**

16-145. The quick-return mechanism consists of a crank AB, slider block B, and slotted link CD. If the crank has the angular motions shown, determine the angular motions of the slotted link at this instant.

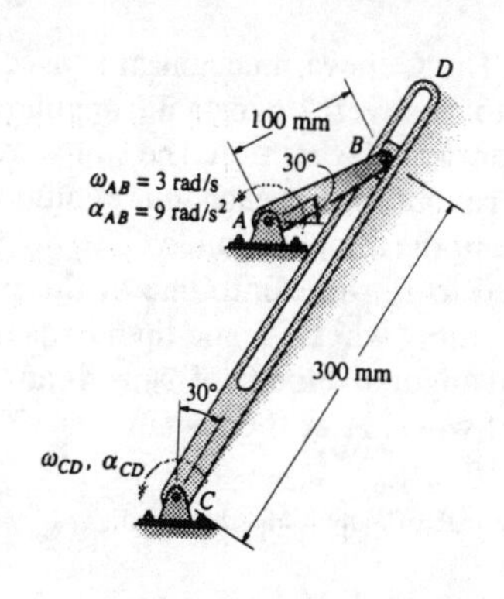

$v_B = 3(0.1) = 0.3$ m/s

$(a_B)_t = 9(0.1) = 0.9$ m/s^2

$(a_B)_n = (3)^2(0.1) = 0.9$ m/s^2

$\mathbf{v}_B = \mathbf{v}_C + \boldsymbol{\Omega}\times\mathbf{r}_{B/C} + (\mathbf{v}_{B/C})_{xyz}$

$0.3\cos60°\mathbf{i} + 0.3\sin60°\mathbf{j} = \mathbf{0} + (\omega_{CD}\mathbf{k})\times(0.3\mathbf{i}) + v_{B/C}\mathbf{i}$

$v_{B/C} = 0.15$ m/s

$\omega_{CD} = 0.866$ rad/s ↺ **Ans**

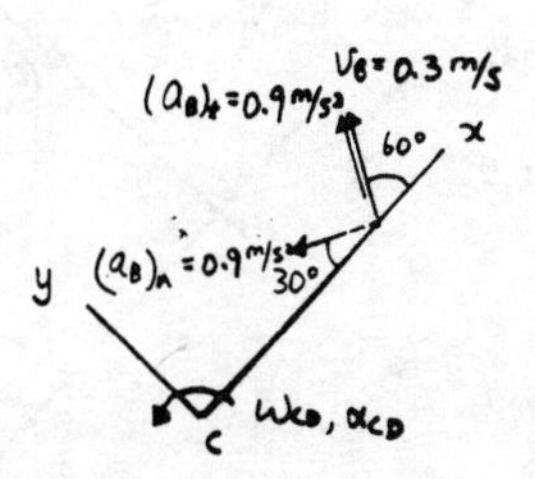

$\mathbf{a}_B = \mathbf{a}_C + \dot{\boldsymbol{\Omega}}\times\mathbf{r}_{B/C} + \boldsymbol{\Omega}\times(\boldsymbol{\Omega}\times\mathbf{r}_{B/C}) + 2\boldsymbol{\Omega}\times(\mathbf{v}_{B/C})_{xyz} + (\mathbf{a}_{B/C})_{xyz}$

$0.9\cos60°\mathbf{i} - 0.9\cos30°\mathbf{i} + 0.9\sin60°\mathbf{j} + 0.9\sin30°\mathbf{j} = \mathbf{0} + (\alpha_{CD}\mathbf{k})\times(0.3\mathbf{i})$

$+ (0.866\mathbf{k})\times(0.866\mathbf{k}\times0.3\mathbf{i}) + 2(0.866\mathbf{k}\times0.15\mathbf{i}) + a_{B/C}\mathbf{i}$

$-0.3294\mathbf{i} + 1.2294\mathbf{j} = 0.3\alpha_{CD}\mathbf{j} - 0.225\mathbf{i} + 0.2598\mathbf{j} + a_{B/C}\mathbf{i}$

$a_{B/C} = -0.104$ m/s^2

$\alpha_{CD} = 3.23$ rad/s^2 ↺ **Ans**

16-146. Particles B and A move along the parabolic and circular paths, respectively. If B has a velocity of 7 m/s in the direction shown and its speed is increasing at 4 m/s^2, while A has a velocity of 8 m/s in the direction shown and its speed is decreasing at 6 m/s^2, determine the relative velocity and relative acceleration of B with respect to A.

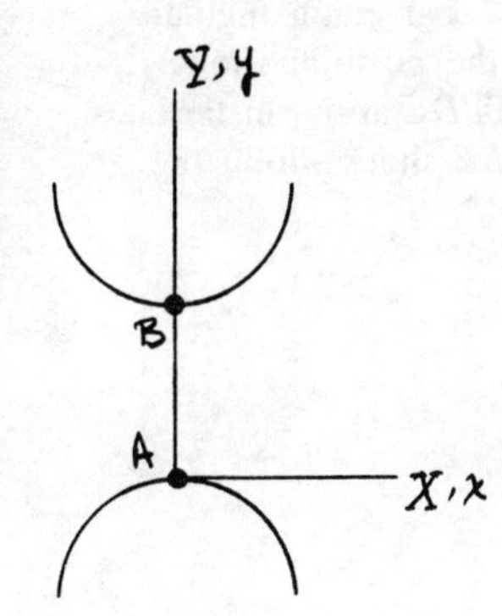

$\Omega = \frac{8}{1} = 8$ rad/s, $\quad \boldsymbol{\Omega} = \{8\mathbf{k}\}$ rad/s

$\mathbf{v}_B = \mathbf{v}_A + \boldsymbol{\Omega}\times\mathbf{r}_{B/A} + (\mathbf{v}_{B/A})_{xyz}$

$7\mathbf{i} = -8\mathbf{i} + (8\mathbf{k})\times(2\mathbf{j}) + (\mathbf{v}_{B/A})_{xyz}$

$7\mathbf{i} = -8\mathbf{i} - 16\mathbf{i} + (\mathbf{v}_{B/A})_{xyz}$

$(\mathbf{v}_{B/A})_{xyz} = \{31.0\mathbf{i}\}$ m/s **Ans**

$\dot{\Omega} = \frac{6}{1} = 6$ rad/s^2, $\quad \dot{\boldsymbol{\Omega}} = \{-6\mathbf{k}\}$ rad/s^2

$(a_A)_n = \frac{(v_A)^2}{1} = \frac{(8)^2}{1} = 64$ m/s^2 ↓

$y = x^2$

$\frac{dy}{dx} = 2x\Big|_{x=0} = 0$

$\frac{d^2y}{dx^2} = 2$

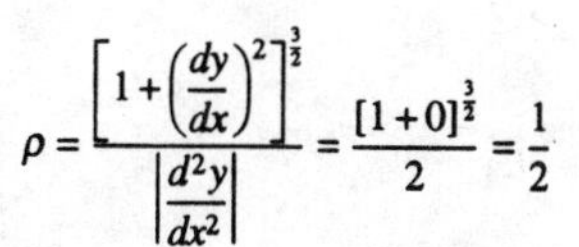

$$\rho = \frac{\left[1+\left(\frac{dy}{dx}\right)^2\right]^{\frac{3}{2}}}{\left|\frac{d^2y}{dx^2}\right|} = \frac{[1+0]^{\frac{3}{2}}}{2} = \frac{1}{2}$$

$(a_B)_n = \frac{(7)^2}{\frac{1}{2}} = 98$ m/s^2 ↑

$\mathbf{a}_B = \mathbf{a}_A + \dot{\boldsymbol{\Omega}}\times\mathbf{r}_{B/A} + \boldsymbol{\Omega}\times(\boldsymbol{\Omega}\times\mathbf{r}_{B/A}) + 2\boldsymbol{\Omega}\times(\mathbf{v}_{B/A})_{xyz} + (\mathbf{a}_{B/A})_{xyz}$

$4\mathbf{i} + 98\mathbf{j} = 6\mathbf{i} - 64\mathbf{j} + (-6\mathbf{k})\times(2\mathbf{j}) + (8\mathbf{k})\times(8\mathbf{k}\times2\mathbf{j}) + 2(8\mathbf{k})\times(31\mathbf{i}) + (\mathbf{a}_{B/A})_{xyz}$

$4\mathbf{i} + 98\mathbf{j} = 6\mathbf{i} - 64\mathbf{j} + 12\mathbf{i} - 128\mathbf{j} + 496\mathbf{j} + (\mathbf{a}_{B/A})_{xyz}$

$(\mathbf{a}_{B/A})_{xyz} = \{-14.0\mathbf{i} - 206\mathbf{j}\}$ m/s^2 **Ans**

16-147. The Geneva mechanism is used in a packaging system to convert constant angular motion into intermittent angular motion. The star wheel A makes one sixth of a revolution for each full revolution of the driving wheel B and the attached guide C. To do this, pin P, which is attached to B, slides into one of the radial slots of A, thereby turning wheel A, and then exits the slot. If B has a constant angular velocity of $\omega_B = 4$ rad/s, determine $\boldsymbol{\omega}_A$ and $\boldsymbol{\alpha}_A$ of wheel A at the instant $\theta = 30°$ as shown.

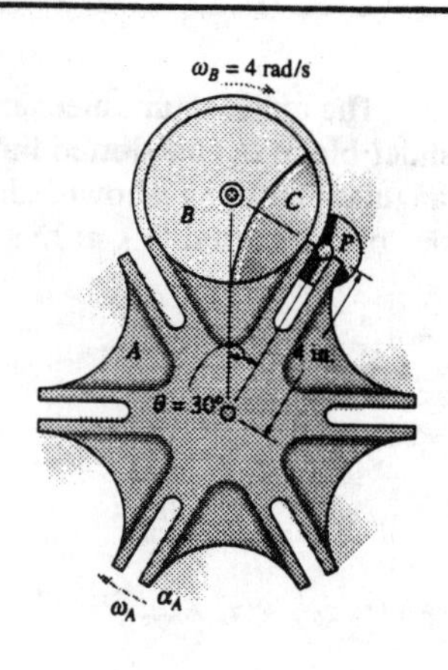

The circular path of motion of P has a radius of

$$r_P = 4\tan 30° = 2.309 \text{ in.}$$

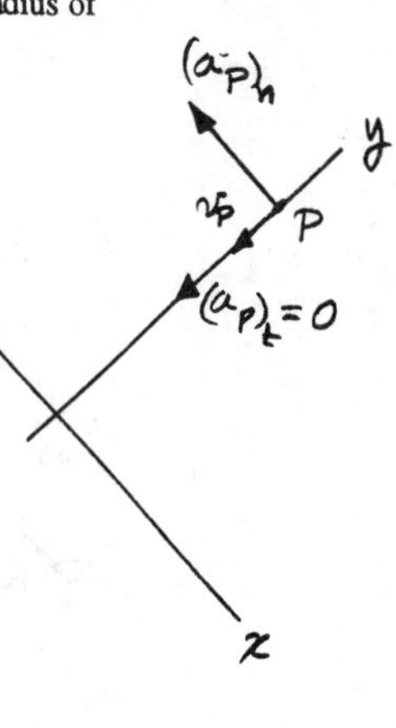

Thus,

$$\mathbf{v}_P = -4(2.309)\mathbf{j} = -9.238\mathbf{j}$$

$$\mathbf{a}_P = -(4)^2(2.309)\mathbf{i} = -36.95\mathbf{i}$$

Thus,

$$\mathbf{v}_P = \mathbf{v}_A + \boldsymbol{\Omega}\times\mathbf{r}_{P/A} + (\mathbf{v}_{P/A})_{xyz}$$

$$-9.238\mathbf{j} = \mathbf{0} + (\omega_A\mathbf{k})\times(4\mathbf{j}) - v_{P/A}\mathbf{j}$$

Solving,

$\omega_A = 0$ **Ans**

$v_{P/A} = 9.238$ in./s

$$\mathbf{a}_P = \mathbf{a}_A + \dot{\boldsymbol{\Omega}}\times\mathbf{r}_{P/A} + \boldsymbol{\Omega}\times(\boldsymbol{\Omega}\times\mathbf{r}_{P/A}) + 2\boldsymbol{\Omega}\times(\mathbf{v}_{P/A})_{xyz} + (\mathbf{a}_{P/A})_{xyz}$$

$$-36.95\mathbf{i} = \mathbf{0} + (\alpha_A\mathbf{k})\times(4\mathbf{j}) + \mathbf{0} + \mathbf{0} - a_{P/A}\mathbf{j}$$

Solving,

$$-36.95 = -4\alpha_A$$

$\alpha_A = 9.24 \text{ rad/s}^2$ ↻ **Ans**

$a_{P/A} = 0$

***16-148.** The combination hinge and support for a lid uses the mechanism shown. If the lid is given an angular velocity of 4 rad/s when it is in the position shown, determine the angular velocity of link DC at this instant. Note that the end E of member FBE slides within the grooved bracket on the lid.

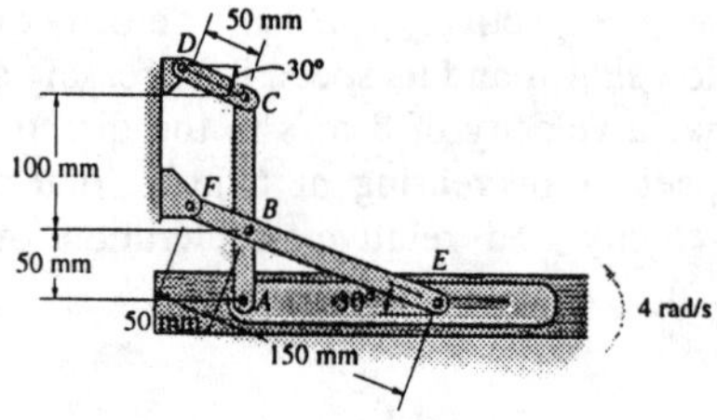

Member FBE:

$$\omega_{FBE} = \frac{v_B}{50} = \frac{v_E}{150}$$

$$v_E = 3v_B$$

Member ABC:

v_B and v_C are parallel. IC is at ∞. $\omega_{ABC} = 0$ and $v_B = v_C = v_A$

$$v_E = 3v_A$$

Lid:

$$\mathbf{v}_E = \mathbf{v}_A + \boldsymbol{\Omega}\times\mathbf{r}_{E/A} + (\mathbf{v}_{E/A})_{xyz}$$

$$3v_A(\sin 30°\mathbf{i} + \cos 30°\mathbf{j}) = v_A(\sin 30°\mathbf{i} + \cos 30°\mathbf{j}) + 4\mathbf{k}\times(0.15\cos 30°\mathbf{i}) + v_{E/A}\mathbf{i}$$

$$1.5v_A\mathbf{i} + 2.598v_A\mathbf{j} = 0.5v_A\mathbf{i} + 0.86603v_A\mathbf{j} + 0.5196\mathbf{j} + v_{E/A}\mathbf{i}$$

$$v_A = v_{E/A}$$

$$1.732v_A = 0.5196$$

$$v_A = v_{E/A} = 0.3 \text{ m/s}$$

$\omega_{DC} = \dfrac{0.3}{0.05} = 6$ rad/s ↻ **Ans**

17-1. Determine the moment of inertia I_y for the slender rod. The rod's density ρ and cross-sectional area A are constant. Express the result in terms of the rod's total mass m.

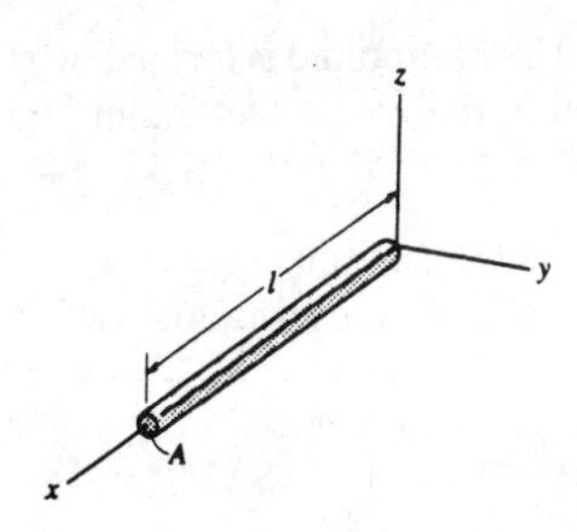

$$I_y = \int_M x^2\, dm$$

$$= \int_0^l x^2\, (\rho A\, dx)$$

$$= \frac{1}{3}\rho A l^3$$

$$m = \rho A l$$

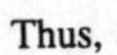

Thus,

$$I_y = \frac{1}{3} m l^2 \qquad \textbf{Ans}$$

17-2. Determine the moment of inertia of the thin ring about the z axis. The ring has a mass m.

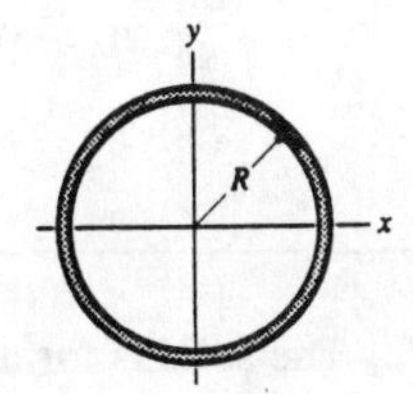

$$I_z = \int_0^{2\pi} \rho A\, (R\, d\theta)\, R^2 = 2\pi\rho A R^3$$

$$m = \int_0^{2\pi} \rho A\, R\, d\theta = 2\pi\rho A R$$

Thus,

$$I_z = m R^2 \qquad \textbf{Ans}$$

17-3. The right circular cone is formed by revolving the shaded area around the x axis. Determine the moment of inertia I_x and express the result in terms of the total mass m of the cone. The cone has a constant density ρ.

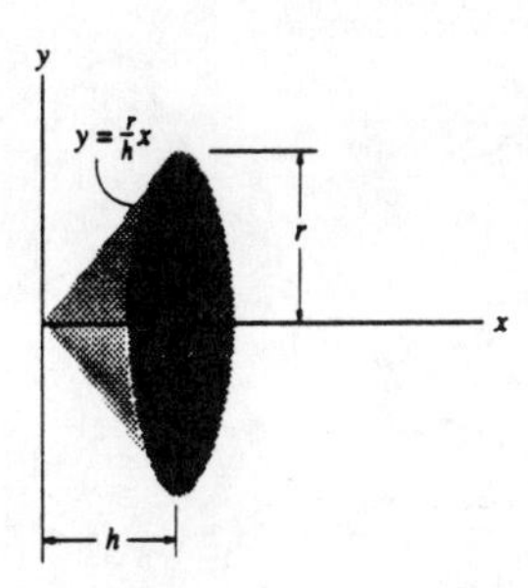

$$dm = \rho\, dV = \rho(\pi y^2\, dx)$$

$$m = \int_0^h \rho(\pi)\left(\frac{r^2}{h^2}\right)x^2\, dx = \rho\pi\left(\frac{r^2}{h^2}\right)\left(\frac{1}{3}\right)h^3 = \frac{1}{3}\rho\pi r^2 h$$

$$dI_x = \frac{1}{2}y^2\, dm$$

$$= \frac{1}{2}y^2\,(\rho\pi y^2\, dx)$$

$$= \frac{1}{2}\rho(\pi)\left(\frac{r^4}{h^4}\right)x^4\, dx$$

$$I_x = \int_0^h \frac{1}{2}\rho(\pi)\left(\frac{r^4}{h^4}\right)x^4\, dx = \frac{1}{10}\rho\pi r^4 h$$

Thus

$$I_x = \frac{3}{10} m r^2 \qquad \textbf{Ans}$$

***17-4.** The paraboloid is formed by revolving the shaded area around the x axis. Determine the radius of gyration k_x. The density of the material is $\rho = 5\ \text{Mg/m}^3$.

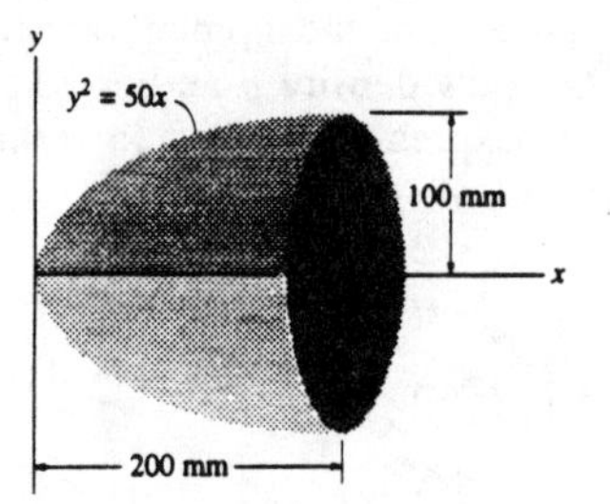

$$dm = \rho\,\pi\,y^2\,dx = \rho\,\pi\,(50x)\,dx$$

$$I_x = \int \frac{1}{2}y^2\,dm = \frac{1}{2}\int_0^{200} 50\,x\,\{\pi\,\rho\,(50x)\}\,dx$$

$$= \rho\,\pi\left(\frac{50^2}{2}\right)\left[\frac{1}{3}x^3\right]_0^{200}$$

$$= \rho\,\pi\left(\frac{50^2}{6}\right)(200)^3$$

$$m = \int dm = \int_0^{200} \pi\,\rho\,(50x)\,dx$$

$$= \rho\,\pi\,(50)\left[\frac{1}{2}x^2\right]_0^{200}$$

$$= \rho\,\pi\,(\frac{50}{2})(200)^2$$

$$k_x = \sqrt{\frac{I_x}{m}} = \sqrt{\frac{50}{3}(200)} = 57.7\ \text{mm} \qquad \textbf{Ans}$$

17-5. The solid is formed by revolving the shaded area around the y axis. Determine the radius of gyration k_y. The specific weight of the material is $\gamma = 380\ \text{lb/ft}^3$.

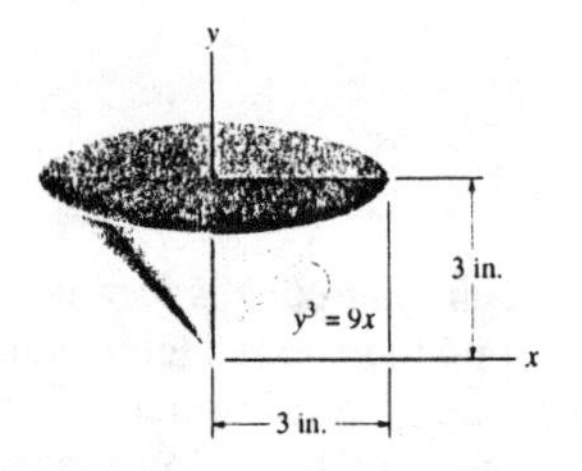

The moment of inertia of the solid : The mass of the disk element $dm = \rho\pi x^2\,dy = \frac{1}{81}\rho\pi y^6\,dy$.

$$dI_y = \frac{1}{2}dm x^2$$

$$= \frac{1}{2}(\rho\pi x^2\,dy)\,x^2$$

$$= \frac{1}{2}\rho\pi x^4\,dy = \frac{1}{2(9^4)}\rho\pi y^{12}\,dy$$

$$I_y = \int dI_y = \frac{1}{2(9^4)}\rho\pi\int_0^3 y^{12}\,dy$$

$$= 29.632\rho$$

The mass of the solid :

$$m = \int_m dm = \frac{1}{81}\rho\pi\int_0^3 y^6\,dy = 12.117\rho$$

$$k_y = \sqrt{\frac{I_y}{m}} = \sqrt{\frac{29.632\rho}{12.117\rho}} = 1.56\ \text{in.} \qquad \textbf{Ans}$$

17-6. The sphere is formed by revolving the shaded area around the x axis. Determine the moment of inertia I_x and express the result in terms of the total mass m of the sphere. The material has a constant density ρ.

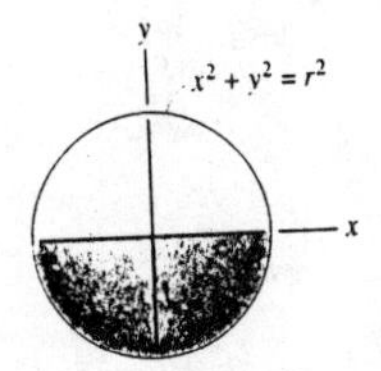

$$dI_x = \frac{y^2\, dm}{2}$$

$$dm = \rho\, dV = \rho\,(\pi y^2\, dx) = \rho\,\pi\,(r^2 - x^2)\, dx$$

$$dI_x = \frac{1}{2}\rho\,\pi\,(r^2 - x^2)^2\, dx$$

$$I_x = \int_{-r}^{r} \frac{1}{2}\rho\,\pi\,(r^2 - x^2)^2\, dx$$

$$= \frac{8}{15}\pi\,\rho\, r^5$$

$$m = \int_{-r}^{r} \rho\,\pi\,(r^2 - x^2)\, dx$$

$$= \frac{4}{3}\rho\,\pi\, r^3$$

Thus,

$$I_x = \frac{2}{5} m\, r^2 \qquad \textbf{Ans}$$

17-7. The paraboloid is formed by revolving the shaded area around the x axis. Determine the moment of inertia with respect to the x axis and express the result in terms of the mass m of the paraboloid. The material has a constant density ρ.

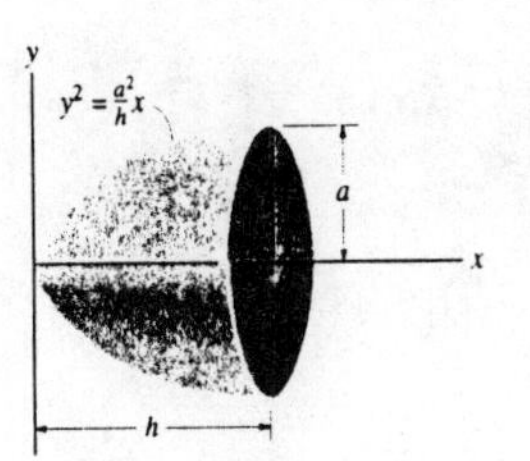

$$dm = \rho\, dV = \rho\,(\pi y^2\, dx)$$

$$dI_x = \frac{1}{2}\, dm\, y^2 = \frac{1}{2}\rho\,\pi\, y^4\, dx$$

$$I_x = \int_0^h \frac{1}{2}\rho\,\pi\left(\frac{a^4}{h^2}\right)x^2\, dx$$

$$= \frac{1}{6}\pi\,\rho a^4\, h$$

$$m = \int_0^h \frac{1}{2}\rho\,\pi\left(\frac{a^2}{h}\right)x\, dx$$

$$= \frac{1}{2}\rho\,\pi\, a^2\, h$$

$$I_x = \frac{1}{3} m a^2 \qquad \textbf{Ans}$$

***17-8.** The hemisphere is formed by rotating the shaded area about the y axis. Determine the moment of inertia I_y and express the result in terms of the total mass m of the hemisphere. The material has a constant density ρ.

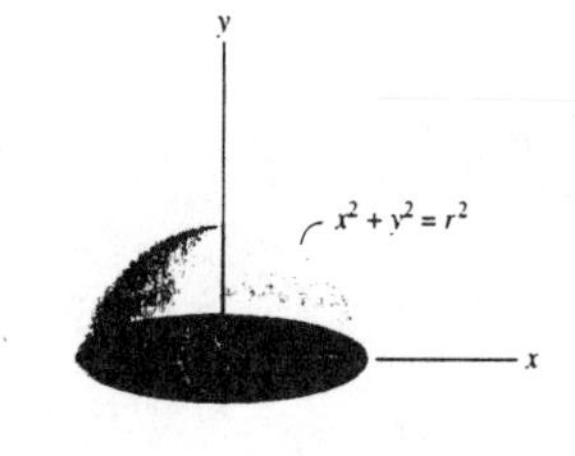

$$m = \int_V \rho\, dV = \rho \int_0^r \pi x^2\, dy = \rho\pi \int_0^r \left(r^2 - y^2\right) dy$$

$$= \rho\pi\left[r^2 y - \frac{1}{3}y^3\right]_0^r = \frac{2}{3}\rho\pi r^3$$

$$I_y = \int_m \frac{1}{2}(dm)\, x^2 = \frac{\rho}{2}\int_0^r \pi x^4\, dy = \frac{\rho\pi}{2}\int_0^r \left(r^2 - y^2\right)^2 dy$$

$$= \frac{\rho\pi}{2}\left[r^4 y - \frac{2}{3}r^2 y^3 + \frac{y^5}{5}\right]_0^r = \frac{4\rho\pi}{15}r^4$$

Thus,

$$I_y = \frac{2}{5}m\, r^2 \qquad \textbf{Ans}$$

17-9. The concrete shape is formed by rotating the shaded area about the y axis. Determine the moment of inertia I_y. The specific weight of concrete is $\gamma = 150\ \text{lb/ft}^3$.

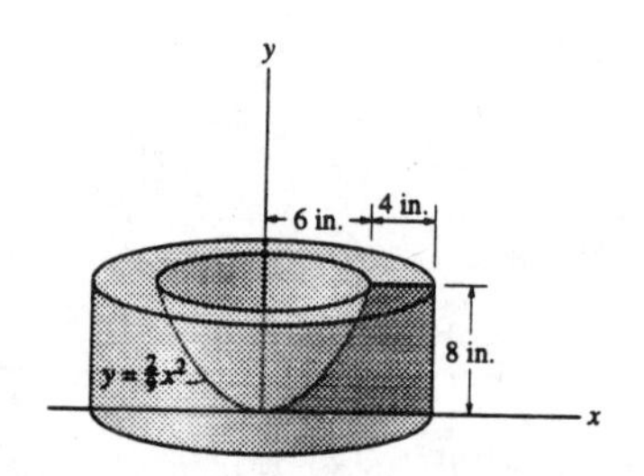

$$d I_y = \frac{1}{2}(dm)(10)^2 - \frac{1}{2}(dm)x^2$$

$$= \frac{1}{2}\{\pi\rho\,(10)^2\, dy\}(10)^2 - \frac{1}{2}\,\pi\rho\, x^2\, dy\, x^2$$

$$I_y = \frac{1}{2}\,\pi\rho\left[\int_0^8 (10)^4\, dy - \int_0^8 \left(\frac{9}{2}\right)^2 y^2\, dy\right]$$

$$= \frac{\frac{1}{2}\,\pi\,(150)}{32.2(12)^3}\left[(10)^4(8) - \left(\frac{9}{2}\right)^2\left(\frac{1}{3}\right)(8)^3\right]$$

$$= 324.1\ \text{slug}\cdot\text{in}^2$$

$$I_y = 2.25\ \text{slug}\cdot\text{ft}^2 \qquad \textbf{Ans}$$

17-10. The frustum is formed by rotating the shaded area around the x axis. Determine the moment of inertia I_x and express the result in terms of the total mass m of the frustum. The frustum has a constant density.

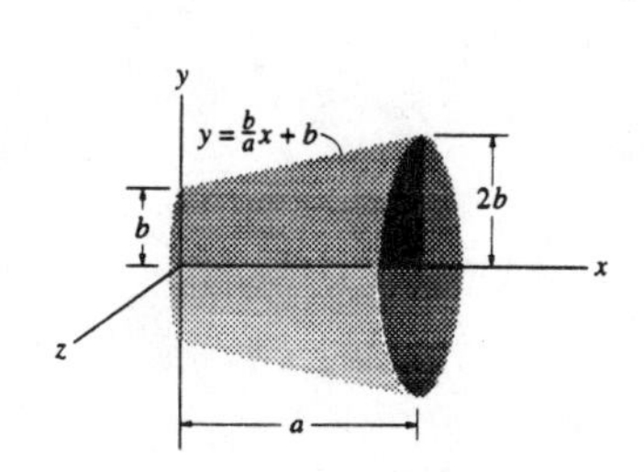

$$dm = \rho dV = \rho\pi y^2 dx = \rho\pi\left(\tfrac{b^2}{a^2}x^2 + \tfrac{2b^2}{a}x + b^2\right)dx$$

$$dI_x = \tfrac{1}{2}dmy^2 = \tfrac{1}{2}\rho\pi y^4 dx$$

$$dI_x = \tfrac{1}{2}\rho\pi\left(\tfrac{b^4}{a^4}x^4 + \tfrac{4b^4}{a^3}x^3 + \tfrac{6b^4}{a^2}x^2 + \tfrac{4b^4}{a}x + b^4\right)dx$$

$$I_x = \int dI_x = \tfrac{1}{2}\rho\pi\int_0^a\left(\tfrac{b^4}{a^4}x^4 + \tfrac{4b^4}{a^3}x^3 + \tfrac{6b^4}{a^2}x^2 + \tfrac{4b^4}{a}x + b^4\right)dx$$

$$= \tfrac{31}{10}\rho\pi ab^4$$

$$m = \int_m dm = \rho\pi\int_0^a\left(\tfrac{b^2}{a^2}x^2 + \tfrac{2b^2}{a}x + b^2\right)dx = \tfrac{7}{3}\rho\pi ab^2 \qquad I_x = \tfrac{93}{70}mb^2 \qquad \textbf{Ans}$$

17-11. Determine the moment of inertia of the homogeneous pyramid of mass m with respect to the z axis. The density of the material is ρ. *Suggestion:* Use a rectangular plate element having a volume of $dV = (2x)(2y)dz$.

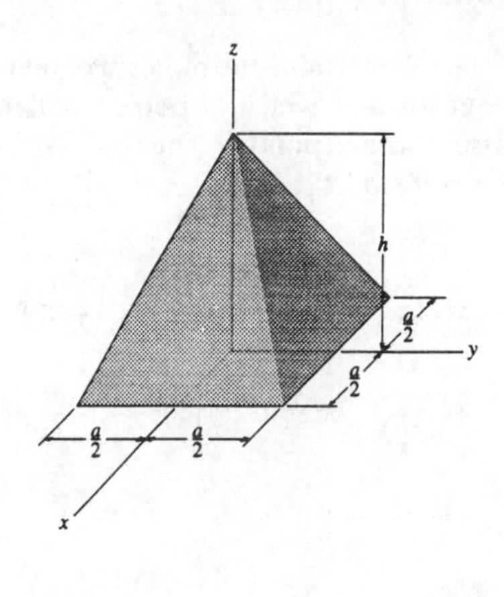

$$dI_z = \frac{dm}{12}\left[(2y)^2 + (2y)^2\right] = \frac{2}{3}y^2\,dm$$

$$dm = \rho y^2\,dz$$

$$dI_z = \frac{2}{3}\rho y^4\,dz = \frac{2}{3}\rho(h-z)^4\left(\frac{a^4}{16h^4}\right)dz$$

$$I_z = \frac{\rho}{24}\left(\frac{a^4}{h^4}\right)\int_0^h\left(h^4 - 4h^3z + 6h^2z^2 - 4hz^3 + z^4\right)dz = \frac{\rho}{24}\left(\frac{a^4}{h^4}\right)\left[h^5 - 2h^5 + 2h^5 - h^5 + \frac{1}{5}h^5\right]$$

$$= \frac{\rho a^4 h}{120}$$

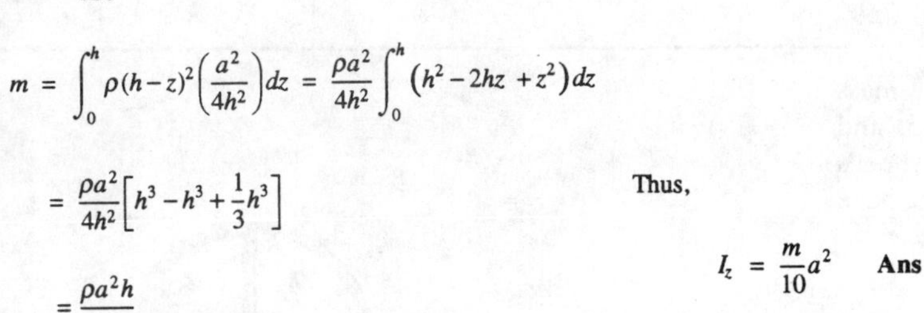

$$m = \int_0^h \rho(h-z)^2\left(\frac{a^2}{4h^2}\right)dz = \frac{\rho a^2}{4h^2}\int_0^h\left(h^2 - 2hz + z^2\right)dz$$

$$= \frac{\rho a^2}{4h^2}\left[h^3 - h^3 + \frac{1}{3}h^3\right]$$

$$= \frac{\rho a^2 h}{12}$$

Thus,

$$I_z = \frac{m}{10}a^2 \qquad \textbf{Ans}$$

***17-12.** Determine the moment of inertia of the thin plate about an axis perpendicular to the page and passing through the pin at O. The plate has a hole in its center. Its thickness is 50 mm and the material has a density of $\rho = 50\ \text{kg/m}^3$.

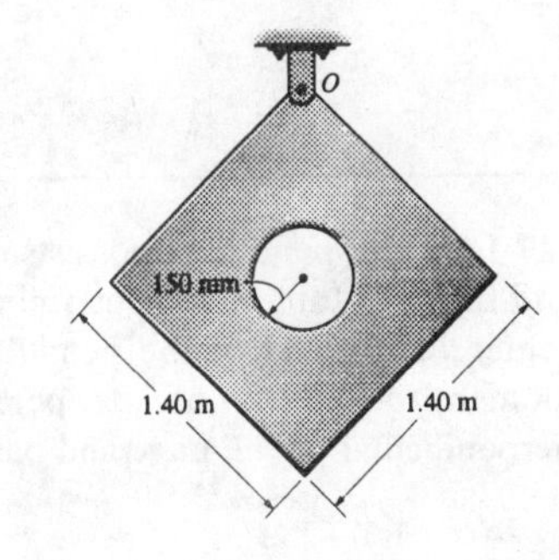

$$I_G = \frac{1}{12}\left[50(1.4)(1.4)(0.05)\right]\left[(1.4)^2 + (1.4)^2\right] - \frac{1}{2}\left[50(\pi)(0.15)^2(0.05)\right](0.15)^2$$

$$= 1.5987\ \text{kg}\cdot\text{m}^2$$

$$I_O = I_G + md^2$$

$$m = 50(1.4)(1.4)(0.05) - 50(\pi)(0.15)^2(0.05) = 4.7233\ \text{kg}$$

$$I_O = 1.5987 + 4.7233(1.4\sin 45^\circ)^2 = 6.23\ \text{kg}\cdot\text{m}^2 \qquad \textbf{Ans}$$

17-13. Determine the moment of inertia of the assembly about an axis which is perpendicular to the page and passes through the center of mass G. The material has a specific weight of $\gamma = 90\ \text{lb/ft}^3$.

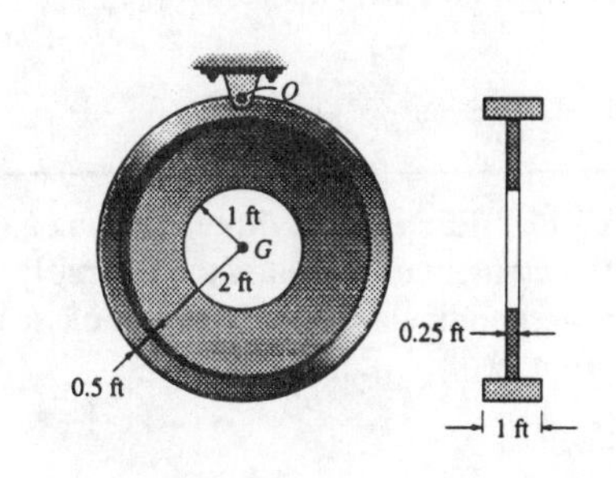

$$I_G = \frac{1}{2}\left[\left(\frac{90}{32.2}\right)\pi(2.5)^2(1)\right](2.5)^2 - \frac{1}{2}\left[\left(\frac{90}{32.2}\right)\pi(2)^2(1)\right](2)^2$$

$$+\frac{1}{2}\left[\left(\frac{90}{32.2}\right)\pi(2)^2(0.25)\right](2)^2 - \frac{1}{2}\left[\left(\frac{90}{32.2}\right)\pi(1)^2(0.25)\right](1)^2$$

$$= 118\ \text{slug}\cdot\text{ft}^2 \qquad \textbf{Ans}$$

17-14. Determine the moment of inertia of the assembly about an axis which is perpendicular to the page and passes through point O. The material has a specific weight of $\gamma = 90\ \text{lb/ft}^3$.

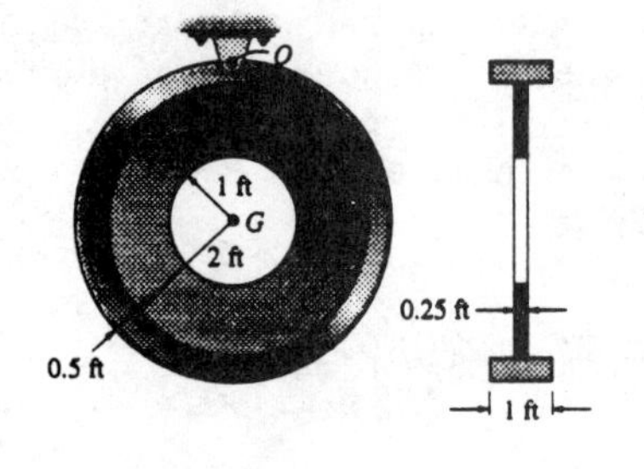

$$I_G = \frac{1}{2}\left[\left(\frac{90}{32.2}\right)\pi(2.5)^2(1)\right](2.5)^2 - \frac{1}{2}\left[\left(\frac{90}{32.2}\right)\pi(2)^2(1)\right](2)^2$$

$$+\frac{1}{2}\left[\left(\frac{90}{32.2}\right)\pi(2)^2(0.25)\right](2)^2 - \frac{1}{2}\left[\left(\frac{90}{32.2}\right)\pi(1)^2(0.25)\right](1)^2$$

$$= 117.72\ \text{slug}\cdot\text{ft}^2$$

$$I_O = I_G + md^2$$

$$m = \left(\frac{90}{32.2}\right)\pi(2^2 - 1^2)(0.25) + \left(\frac{90}{32.2}\right)\pi(2.5^2 - 2^2)(1) = 26.343\ \text{slug}$$

$$I_O = 117.72 + 26.343(2.5)^2 = 282\ \text{slug}\cdot\text{ft}^2$$ **Ans**

17-15. The wheel consists of a thin ring having a mass of 10 kg and four spokes made from slender rods and each having a mass of 2 kg. Determine the wheel's moment of inertia about an axis perpendicular to the page and passing through point A.

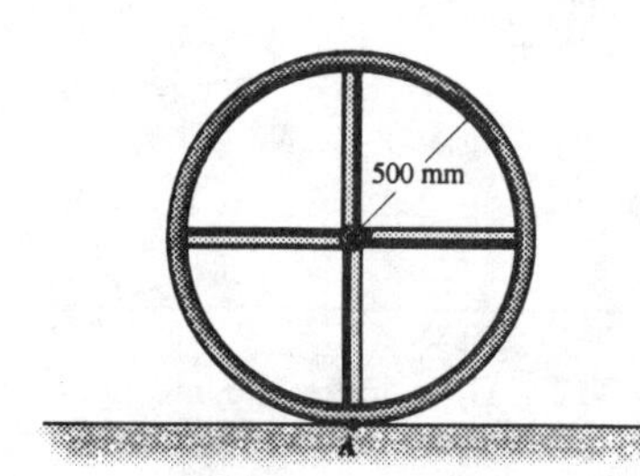

$$I_A = I_O + md^2$$

$$= \left[2\left[\frac{1}{12}(4)(1)^2\right] + 10(0.5)^2\right] + 18(0.5)^2$$

$$= 7.67\ \text{kg}\cdot\text{m}^2$$ **Ans**

***17-16.** The pendulum consists of the 3-kg slender rod and the 5-kg thin plate. Determine the location $\bar{y}$ of the center of mass G of the pendulum; then calculate the moment of inertia of the pendulum about an axis perpendicular to the page and passing through G.

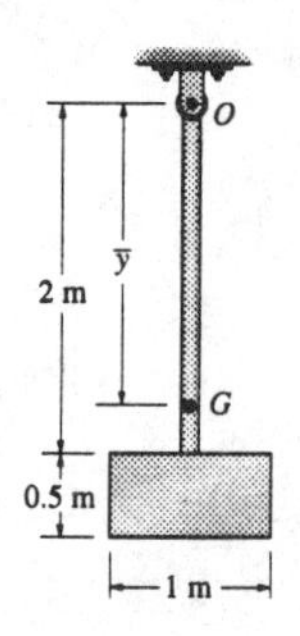

$$\bar{y} = \frac{\Sigma \tilde{y} m}{\Sigma m} = \frac{1(3) + 2.25(5)}{3+5} = 1.781\ \text{m} = 1.78\ \text{m}$$ **Ans**

$$I_G = \Sigma \bar{I}_{G'} + md^2$$

$$= \frac{1}{12}(3)(2)^2 + 3(1.781 - 1)^2 + \frac{1}{12}(5)(0.5^2 + 1^2) + 5(2.25 - 1.781)^2$$

$$= 4.45\ \text{kg}\cdot\text{m}^2$$ **Ans**

17-17. Each of the three rods has a mass m. Determine the moment of inertia of the assembly about an axis which is perpendicular to the page and passes through the center point O.

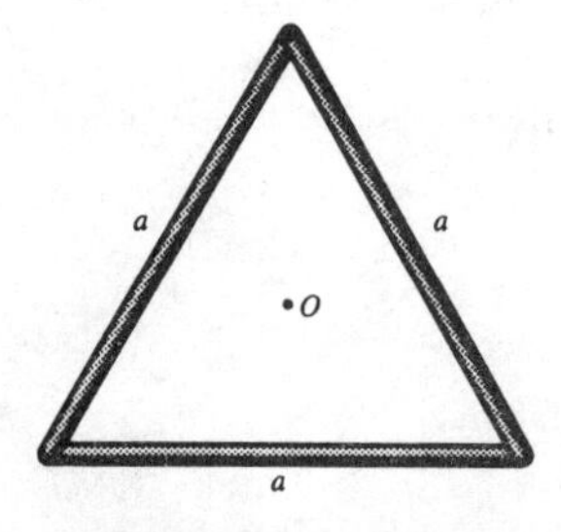

$$I_O = 3\left[\frac{1}{12}ma^2 + m\left(\frac{a\sin 60^\circ}{3}\right)^2\right] = \frac{1}{2}ma^2$$ **Ans**

17-18. The slender rods have a weight of 3 lb/ft. Determine the moment of inertia of the assembly about an axis perpendicular to the page and passing through point A.

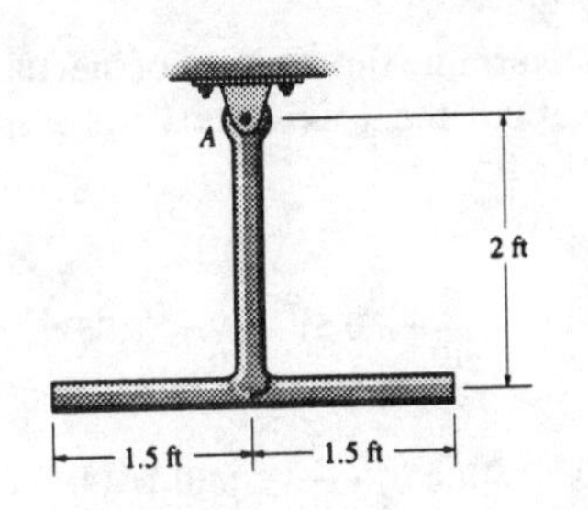

$$I_A = \frac{1}{3}\left[\frac{3(2)}{32.2}\right](2)^2 + \frac{1}{12}\left[\frac{3(3)}{32.2}\right](3)^2 + \left[\frac{3(3)}{32.2}\right](2)^2 = 1.58 \text{ slug} \cdot \text{ft}^2 \quad \textbf{Ans}$$

17-19. The pendulum consists of a plate having a weight of 12 lb and a slender rod having a weight of 4 lb. Determine the radius of gyration of the pendulum about an axis perpendicular to the page and passing through point O.

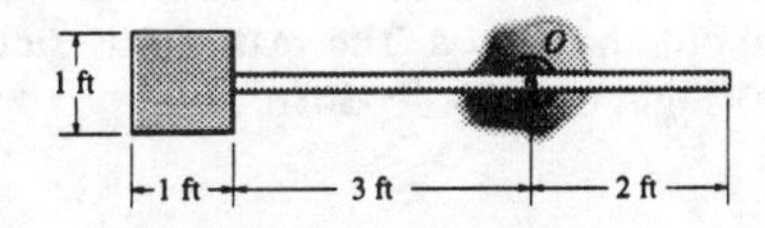

$$I_O = \Sigma I_G + md^2$$

$$= \frac{1}{12}\left(\frac{4}{32.2}\right)(5)^2 + \left(\frac{4}{32.2}\right)(0.5)^2 + \frac{1}{12}\left(\frac{12}{32.2}\right)(1^2 + 1^2) + \left(\frac{12}{32.2}\right)(3.5)^2$$

$$= 4.917 \text{ slug} \cdot \text{ft}^2$$

$$m = \left(\frac{4}{32.2}\right) + \left(\frac{12}{32.2}\right) = 0.4969 \text{ slug}$$

$$k_O = \sqrt{\frac{I_O}{m}} = \sqrt{\frac{4.917}{0.4969}} = 3.15 \text{ ft} \quad \textbf{Ans}$$

***17-20.** Determine the moment of inertia of the overhung crank about the x axis. The material is steel for which the density is $\rho = 7.85 \text{ Mg/m}^3$.

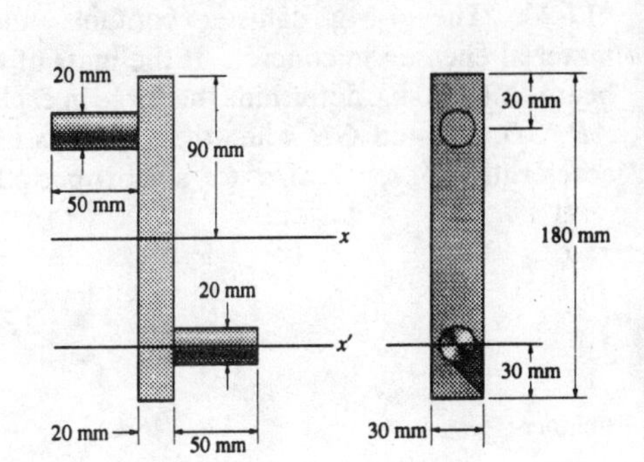

$$m_c = 7.85(10^3)\left((0.05)\pi(0.01)^2\right) = 0.1233 \text{ kg}$$

$$m_p = 7.85(10^3)((0.03)(0.180)(0.02)) = 0.8478 \text{ kg}$$

$$I_x = 2\left[\frac{1}{2}(0.1233)(0.02)^2 + (0.1233)(0.06)^2\right] + \left[\frac{1}{12}(0.8478)\left((0.03)^2 + (0.180)^2\right)\right]$$

$$= 0.00329 \text{ kg} \cdot \text{m}^2 = 3.29 \text{ g} \cdot \text{m}^2 \quad \textbf{Ans}$$

17-21. Determine the moment of inertia of the overhung crank about the x' axis. The material is steel for which the density is $\rho = 7.85 \text{ Mg/m}^3$.

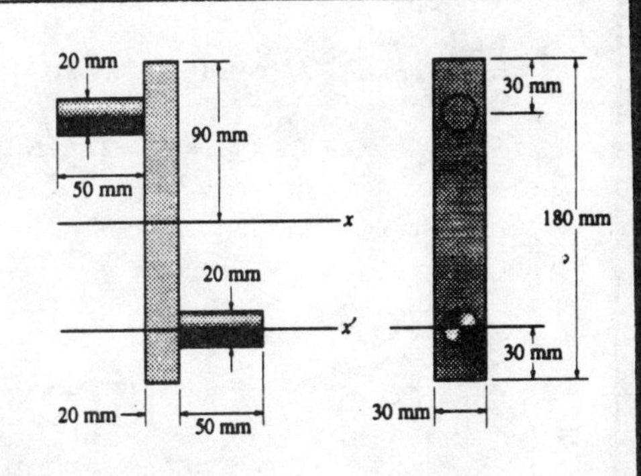

$$m_c = 7.85(10^3)\left((0.05)\pi(0.01)^2\right) = 0.1233 \text{ kg}$$

$$m_p = 7.85(10^3)((0.03)(0.180)(0.02)) = 0.8478 \text{ kg}$$

$$I_{x'} = \left[\frac{1}{2}(0.1233)(0.02)^2\right] + \left[\frac{1}{2}(0.1233)(0.02)^2 + (0.1233)(0.120)^2\right]$$

$$+ \left[\frac{1}{12}(0.8478)\left((0.03)^2 + (0.180)^2\right) + (0.8478)(0.06)^2\right] = 0.00723 \text{ kg} \cdot \text{m}^2 = 7.23 \text{ g} \cdot \text{m}^2 \quad \textbf{Ans}$$

17-22. Determine the moment of inertia of the solid steel assembly about the x axis. Steel has a specific weight of $\gamma_{st} = 490\ \text{lb/ft}^3$.

$$I_x = \frac{1}{2}m_1(0.5)^2 + \frac{3}{10}m_2(0.5)^2 - \frac{3}{10}m_3(0.25)^2$$

$$= \left[\frac{1}{2}\pi(0.5)^2(3)(0.5)^2 + \frac{3}{10}\left(\frac{1}{3}\right)\pi(0.5)^2(4)(0.5)^2 - \frac{3}{10}\left(\frac{1}{3}\right)\pi(0.25)^2(2)(0.25)^2\right]\left(\frac{490}{32.2}\right)$$

$$= 5.64\ \text{slug}\cdot\text{ft}^2 \qquad \textbf{Ans}$$

17-23. Determine the moment of inertia of the center crank about the x axis. The material is steel having a specific weight of $\gamma_{st} = 490\ \text{lb/ft}^3$.

$$m_s = \frac{490}{32.2}\left(\frac{\pi(0.25)^2(1)}{(12)^3}\right) = 0.0017291\ \text{slug}$$

$$m_p = \frac{490}{32.2}\left(\frac{(6)(1)(0.5)}{(12)^3}\right) = 0.02642\ \text{slug}$$

$$I_x = 2\left[\frac{1}{12}(0.02642)\left((1)^2+(6)^2\right) + (0.02642)(2)^2\right] + 2\left[\frac{1}{2}(0.0017291)(0.25)^2\right]$$

$$+ \left[\frac{1}{2}(0.0017291)(0.25)^2 + (0.0017291)(4)^2\right]$$

$$= 0.402\ \text{slug}\cdot\text{ft}^2 \qquad \textbf{Ans}$$

***17-24.** The 4-Mg canister contains nuclear waste material encased in concrete. If the mass of the spreader beam BD is 50 kg, determine the force in each of the links AB, CD, EF, and GH when the system is lifted with an acceleration of $a = 2\ \text{m/s}^2$ for a short period of time.

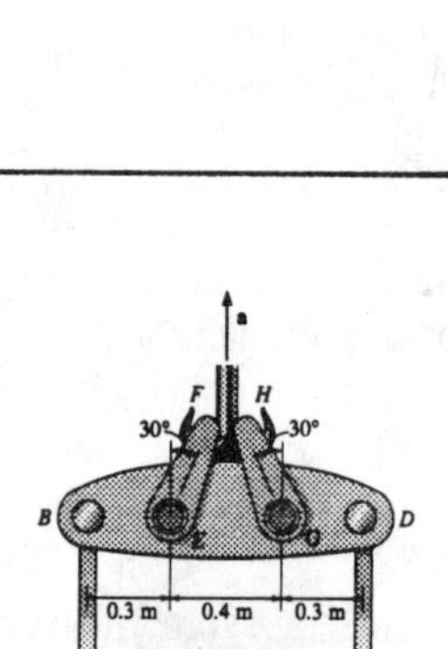

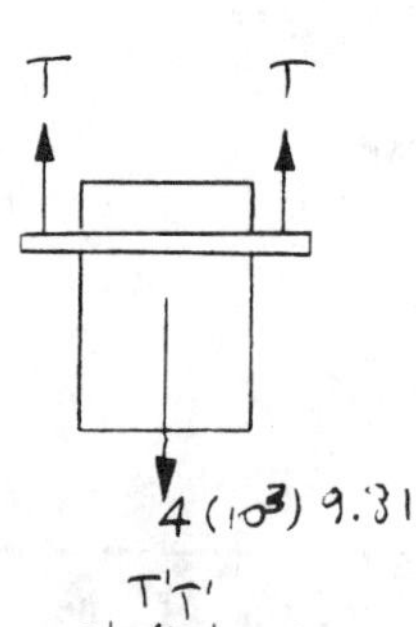

Canister :

$$+\uparrow \Sigma F_y = m(a_G)_y; \qquad 2T - 4(10^3)(9.81) = 4(10^3)(2)$$

$$T_{AB} = T_{CD} = T = 23.6\ \text{kN} \qquad \textbf{Ans}$$

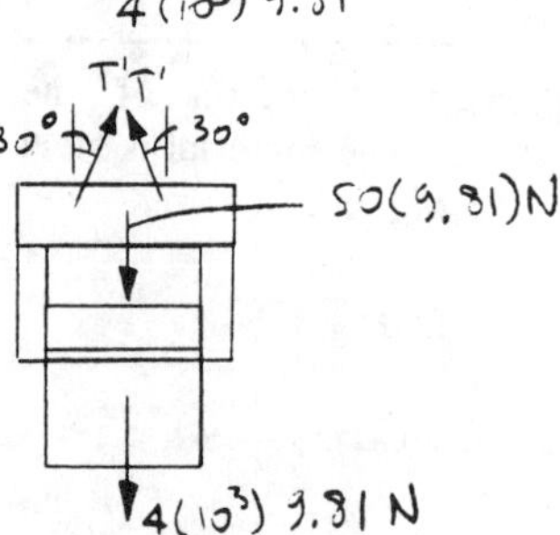

System :

$$+\uparrow \Sigma F_y = m(a_G)_y; \qquad 2T'\cos 30° - 4050(9.81) = 4050(2)$$

$$T_{EF} = T_{GH} = T' = 27.6\ \text{kN} \qquad \textbf{Ans}$$

17-25. The 4-Mg canister contains nuclear waste material encased in concrete. If the mass of the spreader beam BD is 50 kg, determine the largest vertical acceleration **a** of the system so that each of the links AB and CD are not subjected to a force greater than 30 kN and links EF and GH are not subjected to a force greater than 34 kN.

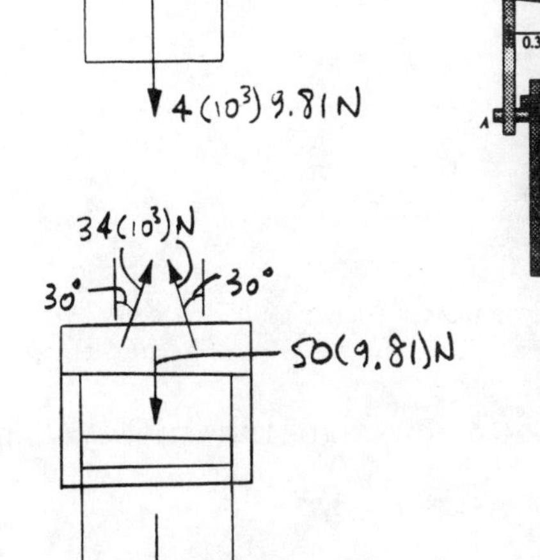

Canister :

$$+\uparrow \Sigma F_y = m(a_G)_y; \quad 2(30)(10^3) - 4(10^3)(9.81) = 4(10^3)a$$

$$a = 5.19 \text{ m/s}^2$$

System :

$$+\uparrow \Sigma F_y = m(a_G)_y; \quad 2\left[34(10^3)\cos 30^\circ\right] - 4050(9.81) = 4050a$$

$$a = 4.73 \text{ m/s}^2$$

Thus,

$a_{max} = 4.73 \text{ m/s}^2$ **Ans**

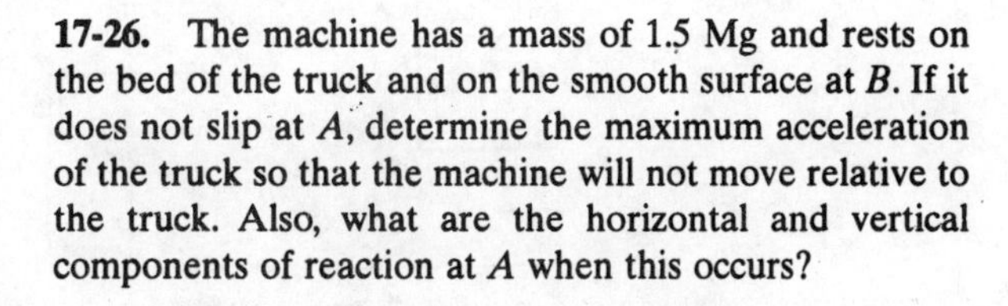

17-26. The machine has a mass of 1.5 Mg and rests on the bed of the truck and on the smooth surface at B. If it does not slip at A, determine the maximum acceleration of the truck so that the machine will not move relative to the truck. Also, what are the horizontal and vertical components of reaction at A when this occurs?

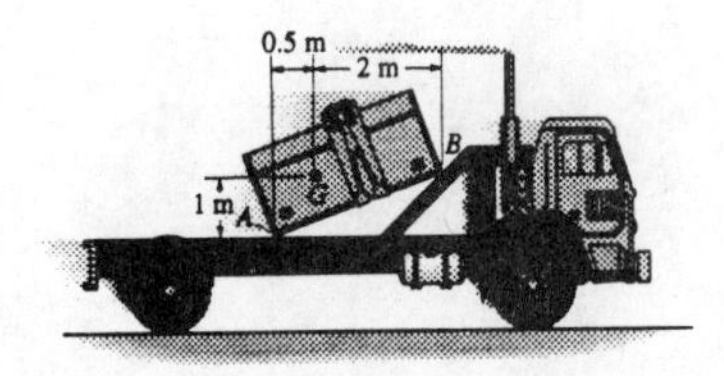

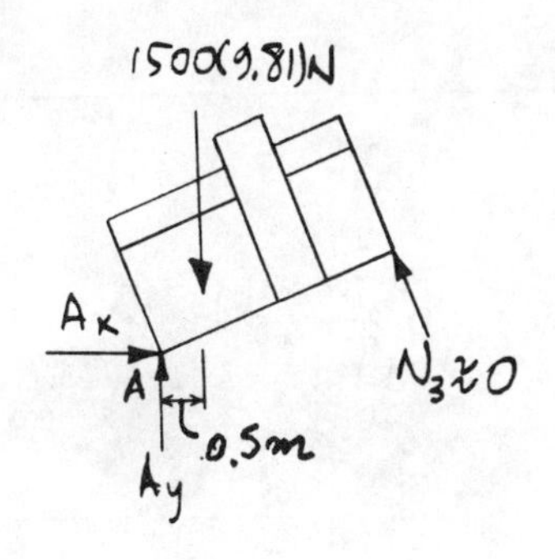

$$\circlearrowleft + \Sigma M_A = \Sigma (M_k)_A; \quad 1500(9.81)(0.5) = 1500(1)a_G$$

$a_G = 4.90 \text{ m/s}^2$ **Ans**

$\xrightarrow{+} \Sigma F_x = m(a_G)_x; \quad A_x = 1500(4.90) = 7.36 \text{ kN}$ **Ans**

$$+\uparrow \Sigma F_y = m(a_G)_y; \quad A_y - 1500(9.81) = 0$$

$A_y = 14.7 \text{ kN}$ **Ans**

17-27. The fork lift has a boom with a mass of 800 kg and a mass center at G. If the vertical acceleration of the boom is 4 m/s^2, determine the horizontal and vertical reactions at the pin A and on the short link BC when the 1.25-Mg load is lifted.

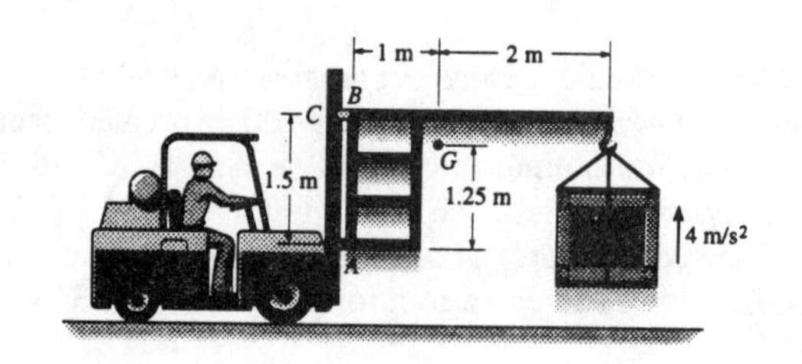

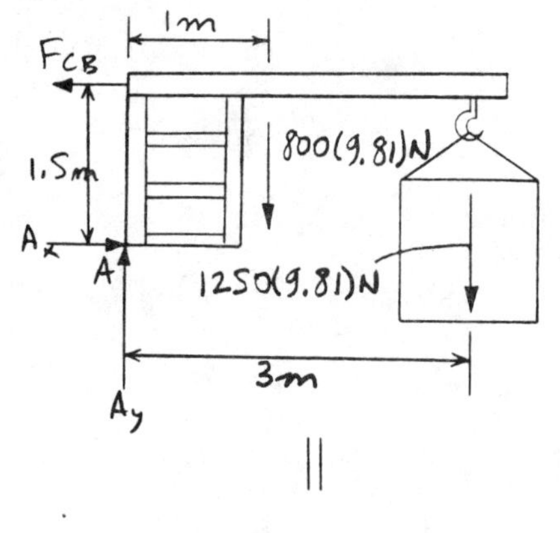

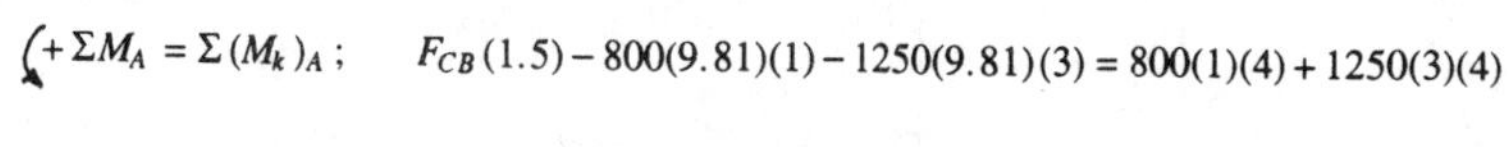

$$\circlearrowleft + \Sigma M_A = \Sigma (M_k)_A\,; \qquad F_{CB}(1.5) - 800(9.81)(1) - 1250(9.81)(3) = 800(1)(4) + 1250(3)(4)$$

$$F_{CB} = 41\,890 \text{ N} = 41.9 \text{ kN} \qquad \textbf{Ans}$$

$$\xrightarrow{+} \Sigma F_x = m(a_G)_x\,; \qquad A_x - 41.9 = 0$$

$$A_x = 41.9 \text{ kN} \qquad \textbf{Ans}$$

$$+\uparrow \Sigma F_y = m(a_G)_y\,; \qquad A_y - 800(9.81) - 1250(9.81) = (800 + 1250)(4)$$

$$A_y = 28.3 \text{ kN} \qquad \textbf{Ans}$$

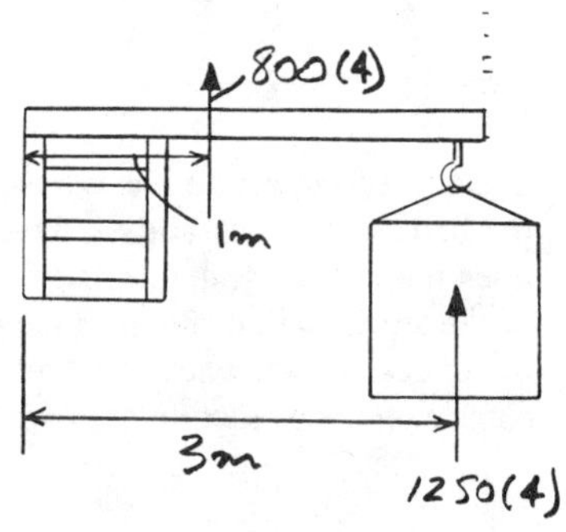

***17-28.** The pipe has a mass of 460 kg and is held in place on the truck bed using the two boards A and B. Determine the greatest acceleration of the truck so that the pipe begins to lose contact at A and the bed of the truck and starts to pivot about B. Assume board B will not slip on the bed of the truck, and the pipe is smooth. Also, what force does board B exert on the pipe during the acceleration?

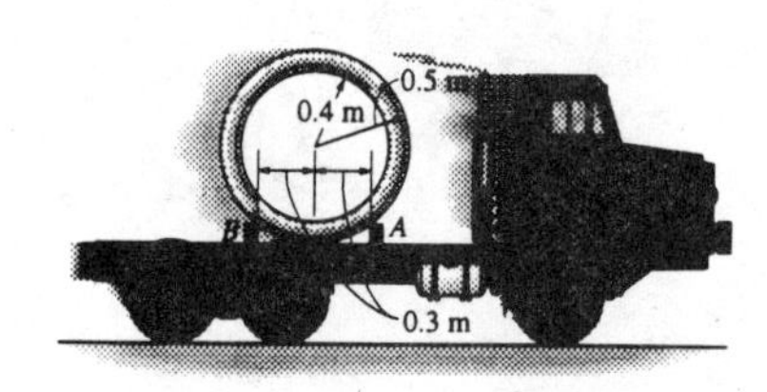

$$\circlearrowleft + \Sigma M_B = \Sigma (M_k)_B\,; \qquad 460(9.81)(0.30) = 460(a_G)(0.4)$$

$$a_G = 7.3575 = 7.36 \text{ m/s}^2 \qquad \textbf{Ans}$$

$$\xrightarrow{+} \Sigma F_x = m(a_G)_x\,; \qquad (N_B)_x = 460(7.3575) = 3384.45 \text{ N}$$

$$+\uparrow \Sigma F_y = m(a_G)_y\,; \qquad (N_B)_y - 460(9.81) = 4512.6 \text{ N}$$

$$N_B = \sqrt{(3384.45)^2 + (4512.6)^2} = 5.64 \text{ kN} \qquad \textbf{Ans}$$

460(9.81)N   0.5m $(N_B)_x$ B   0.1m   0.1m $N_C \simeq 0$ $(N_B)_y$ 0.3m $N_A \simeq 0$

17-29. The lift truck has a mass of 70 kg and mass center at G. If it lifts the 120-kg spool with an acceleration of 3 m/s^2, determine the reactions of each of the four wheels on the ground. The loading is symmetric. Neglect the mass of the movable arm CD.

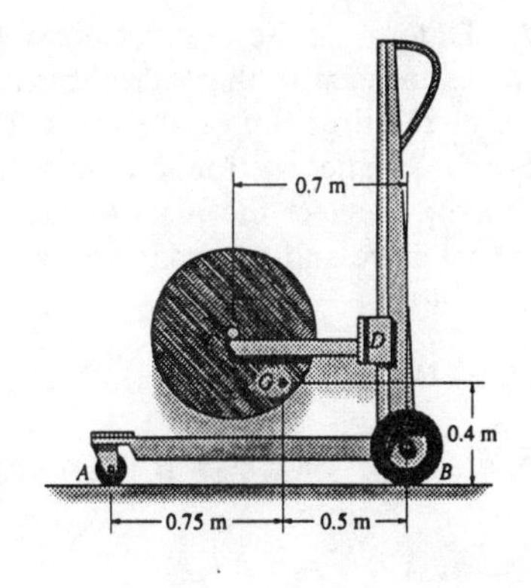

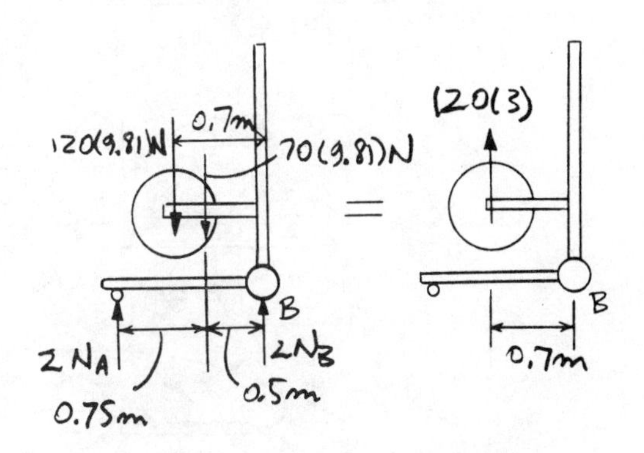

$$\circlearrowleft + \Sigma M_B = \Sigma (M_k)_B; \quad 70(9.81)(0.5) + 120(9.81)(0.7) - 2N_A(1.25) = -120(3)(0.7)$$

$$N_A = 567.76 \text{ N} = 568 \text{ N} \qquad \textbf{Ans}$$

$$+\uparrow \Sigma F_y = m(a_G)_y; \quad 2(567.76) + 2N_B - 120(9.81) - 70(9.81) = 120(3)$$

$$N_B = 544 \text{ N} \qquad \textbf{Ans}$$

17-30. The lift truck has a mass of 70 kg and mass center at G. Determine the largest upward acceleration of the 120-kg spool so that no reaction of the wheels on the ground exceeds 600 N.

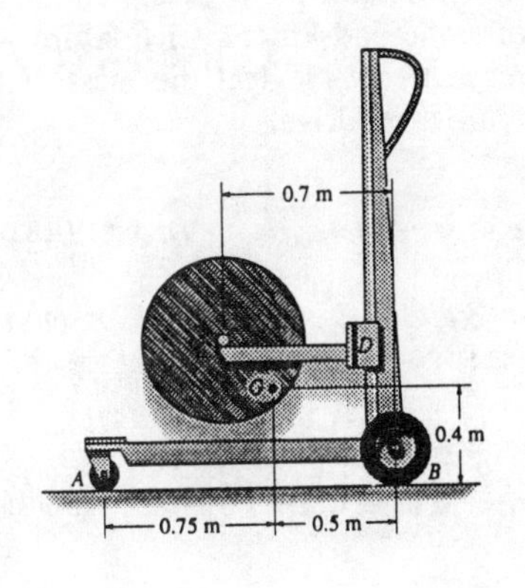

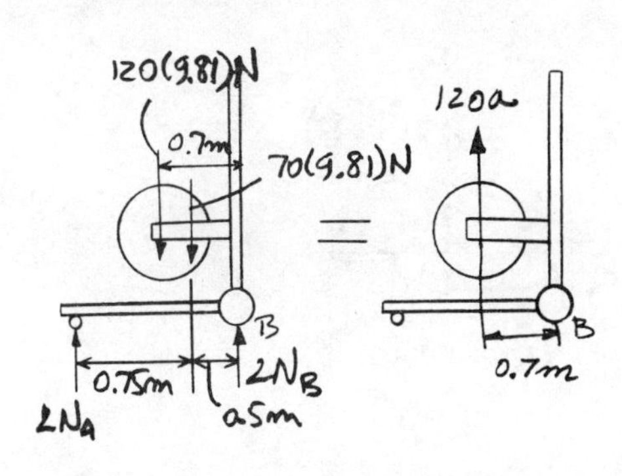

Assume $N_A = 600$ N,

$$\circlearrowleft + \Sigma M_B = \Sigma (M_k)_B; \quad 70(9.81)(0.5) + 120(9.81)(0.7) - 2(600)(1.25) = -120a(0.7)$$

$$a = 3.960 \text{ m/s}^2$$

$$+\uparrow \Sigma F_y = m(a_G)_y; \quad 2(600) + 2N_B - 120(9.81) - 70(9.81) = 120(3.960)$$

$$N_B = 570 \text{ N} < 600\text{N} \qquad \text{OK}$$

Thus $\quad a = 3.96 \text{ m/s}^2 \qquad$ **Ans**

17-31. Determine the greatest possible acceleration of the 975-kg race car so that its front tires do not leave the ground or the tires slip on the track. The coefficients of static and kinetic friction are $\mu_s = 0.8$ and $\mu_k = 0.6$, respectively. Neglect the mass of the tires. The car has rear-wheel drive and the front tires are free to roll.

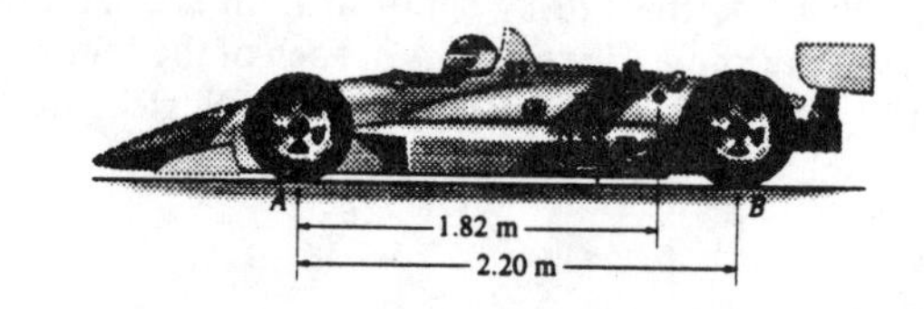

$\circlearrowleft + \Sigma M_A = \Sigma (M_k)_A ; \quad -975(9.81)(1.82) + N_B(2.20) = 975(a_G)(0.55) \quad (1)$

$+\uparrow \Sigma F_y = m(a_G)_y ; \quad N_A + N_B - 975(9.81) = 0 \quad (2)$

$\overset{+}{\leftarrow} \Sigma F_x = m(a_G)_x ; \quad F_B = 975 a_G \quad (3)$

Assume the rear wheels are on the verge of slipping.

$F_B = 0.8 N_B$

Solving,

$a_G = 8.12 \text{ m/s}^2$

$N_B = 9890.8 \text{ N}$

$N_A = -326.1 \text{ N} < 0$

Front wheels lift off ground

Assume $N_A = 0$. Solving Eqs. (1) – (3),

$N_B = 9565 \text{ N}$

$a_G = 6.78 \text{ m/s}^2$ **Ans**

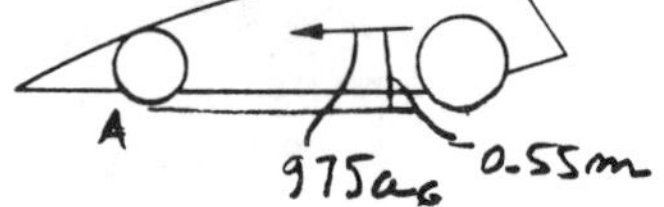

***17-32.** Determine the greatest possible acceleration of the 975-kg race car so that its front wheels do not leave the ground or the tires slip on the track. The coefficients of static and kinetic friction are $\mu_s = 0.8$ and $\mu_k = 0.6$, respectively. Neglect the mass of the tires. The car has four-wheel drive.

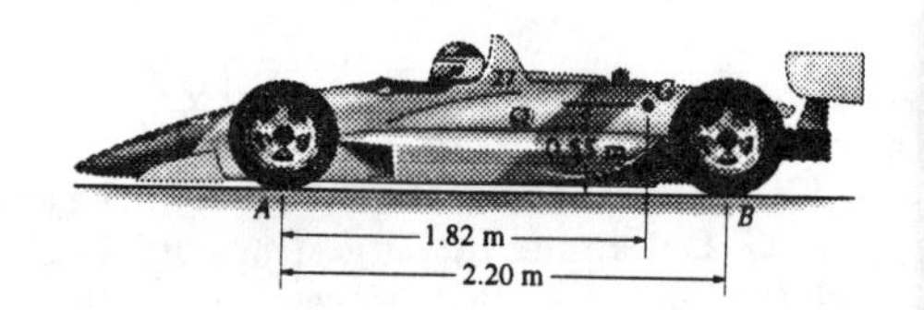

$\circlearrowleft + \Sigma M_A = \Sigma (M_k)_A ; \quad -975(9.81)(1.82) + N_B(2.20) = 975(a_G)(0.55)$

$+\uparrow \Sigma F_y = m(a_G)_y ; \quad N_A + N_B - 975(9.81) = 0$

$\overset{+}{\leftarrow} \Sigma F_x = m(a_G)_x ; \quad F_A + F_B = 975 a_G$

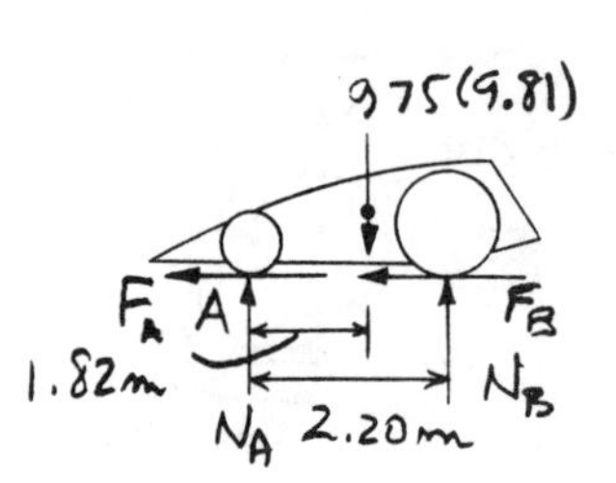

Assume all the wheels are on the verge of slipping.

$F_A = 0.8 N_A$

$F_B = 0.8 N_B$

Solving,

$a_G = 7.848 \text{ m/s}^2$

$N_B = 9825.6 \text{ N}$

$N_A = -261 \text{ N} < 0$

Front wheels lift off ground.

Assume $N_A = 0$ $(F_A = 0)$. Solving,

$N_B = 9565 \text{ N}$

$a_G = 6.78 \text{ m/s}^2$ **Ans**

17-33. The 1.6-Mg car shown has been "raked" by increasing the height of its center of mass to $h = 0.2$ m. This was done by raising the springs on the rear axle. If the coefficient of kinetic friction between the rear wheels and the ground is $\mu_k = 0.3$, show that the car can accelerate slightly faster than its counterpart for which $h = 0$. Neglect the mass of the wheels and driver and assume the front wheels at B are free to roll while the rear wheels slip.

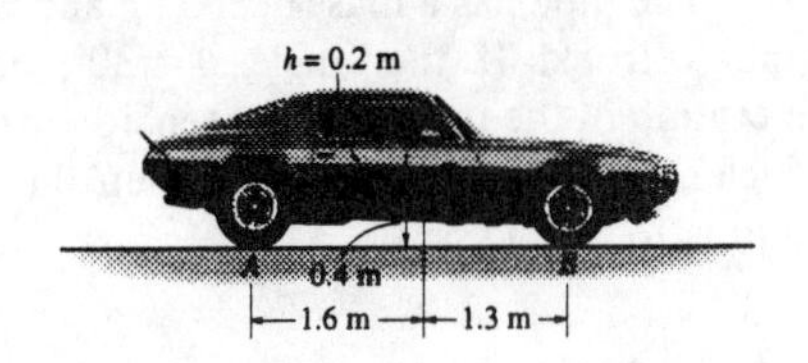

$\overset{+}{\rightarrow} \Sigma F_x = m(a_G)_x; \qquad 0.3N_A = 1600a_G$

$+\uparrow \Sigma F_y = m(a_G)_y; \qquad N_A + N_B - 1600(9.81) = 0$

$\circlearrowleft + \Sigma M_A = \Sigma (M_k)_A; \qquad -1600(9.81)(1.6) + N_B(2.9) = -1600a_G(h + 0.4)$

Set $h = 0.2$ m

$a_G = 1.41\ \text{m/s}^2$ **Ans**

$N_A = 7.50$ kN

$N_B = 8.19$ kN

Set $h = 0$

$a_G = 1.38\ \text{m/s}^2$ **Ans**

$N_A = 7.34$ kN

$N_B = 8.36$ kN

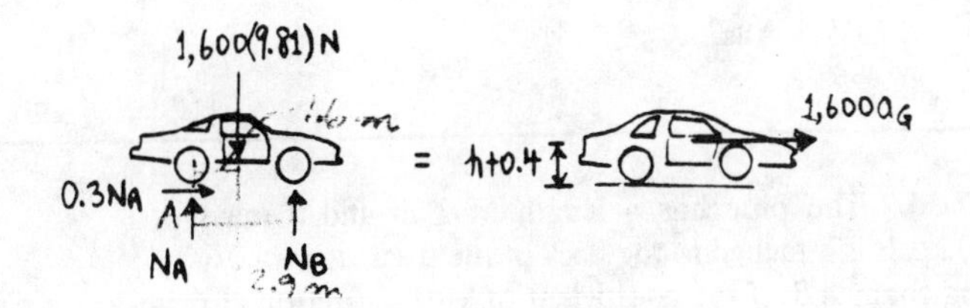

17-34. The pipe has a mass of 800 kg and is being towed behind the truck. If the acceleration of the truck is $a_t = 0.5\ \text{m/s}^2$, determine the angle θ and the tension in the cable. The coefficient of kinetic friction between the pipe and the ground is $\mu_k = 0.1$.

$\overset{+}{\rightarrow} \Sigma F_x = ma_x; \qquad -0.1N_C + T\cos 45° = 800(0.5)$

$+\uparrow \Sigma F_y = ma_y; \qquad N_C - 800(9.81) + T\sin 45° = 0$

$\circlearrowleft + \Sigma M_G = 0; \qquad -0.1N_C(0.4) + T\sin\phi(0.4) = 0$

$N_C = 6770.9$ N

$T = 1523.24\ \text{N} = 1.52\ \text{kN}$ **Ans**

$\sin\phi = \dfrac{0.1(6770.9)}{1523.24} \qquad \phi = 26.39°$

$\theta = 45° - \phi = 18.6°$ **Ans**

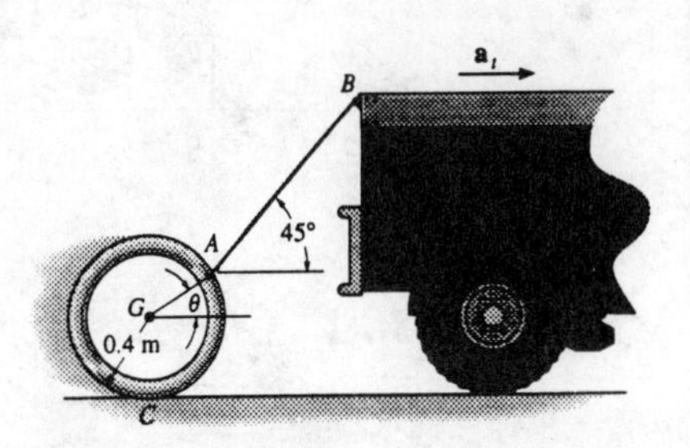

17-35. The pipe has a mass of 800 kg and is being towed behind a truck. If the angle $\theta = 30°$, determine the acceleration of the truck and the tension in the cable. The coefficient of kinetic friction between the pipe and the ground is $\mu_k = 0.1$.

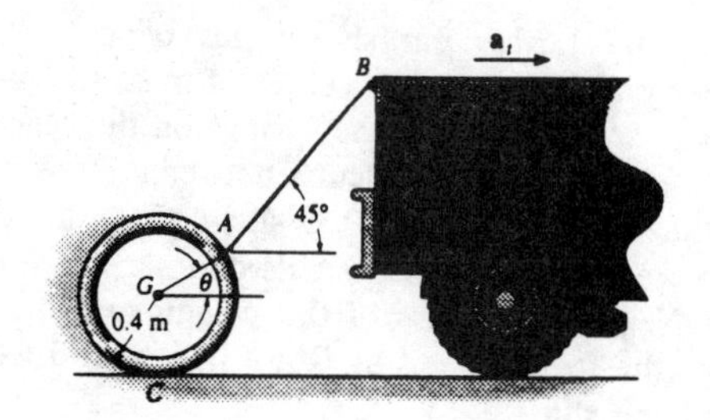

$\xrightarrow{+} \Sigma F_x = ma_x;\quad T\cos45° - 0.1N_C = 800a$

$+\uparrow \Sigma F_y = ma_y;\quad N_C - 800(9.81) + T\sin45° = 0$

$\circlearrowleft + \Sigma M_G = 0;\quad T\sin15°(0.4) - 0.1N_C(0.4) = 0$

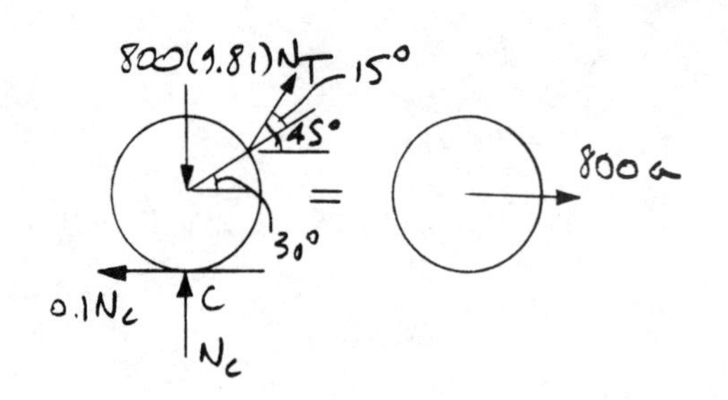

$N_C = 6164$ N

$T = 2382$ N $= 2.38$ kN **Ans**

$a = 1.33\ \text{m/s}^2$ **Ans**

***17-36.** The pipe has a length of 3 m and a mass of 500 kg. It is attached to the back of the truck using a 0.6-m-long chain AB. If the coefficient of kinetic friction at C is $\mu_k = 0.4$, determine the acceleration of the truck if the angle $\theta = 10°$ with the road as shown.

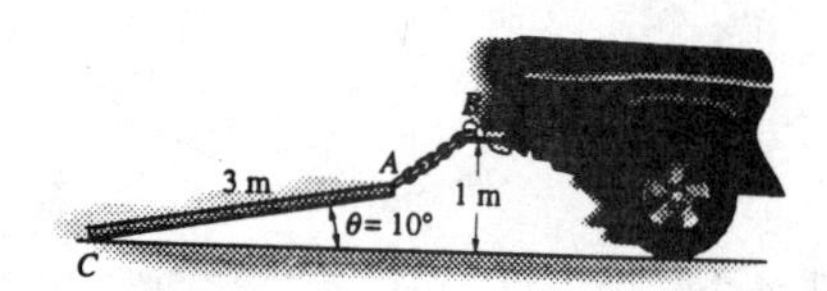

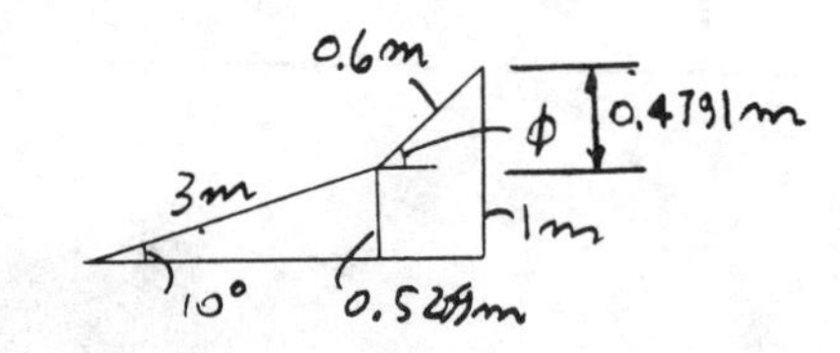

500(9.81)N

T

52.98°

10°

C

0.4N_c

N_c

=

500a_G

C 0.2605m

$\phi = \sin^{-1}\left(\frac{0.4791}{0.6}\right) = 52.98°$

$\xrightarrow{+} \Sigma F_x = m(a_G)_x;\quad T\cos52.98° - 0.4N_C = 500a_G$

$+\uparrow \Sigma F_y = m(a_G)_y;\quad N_C - 500(9.81) + T\sin52.98° = 0$

$\circlearrowleft + \Sigma M_C = \Sigma(M_k)_C;\quad -500(9.81)(1.5\cos10°) + T\sin(52.98° - 10°)(3) = -500a_G(0.2605)$

$T = 3.39$ kN

$N_C = 2.19$ kN

$a_G = 2.33\ \text{m/s}^2$ **Ans**

17-37. Block A weighs 50 lb and the platform weighs 10 lb. If $P = 100$ lb, determine the normal force exerted by block A on B. Neglect the weight of the pulleys and bars of the triangular frame.

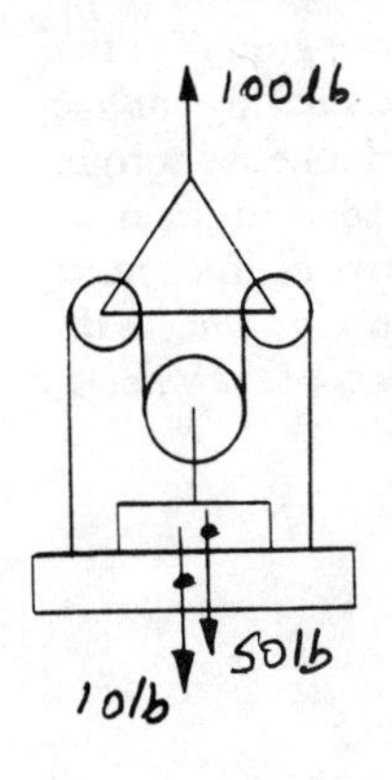

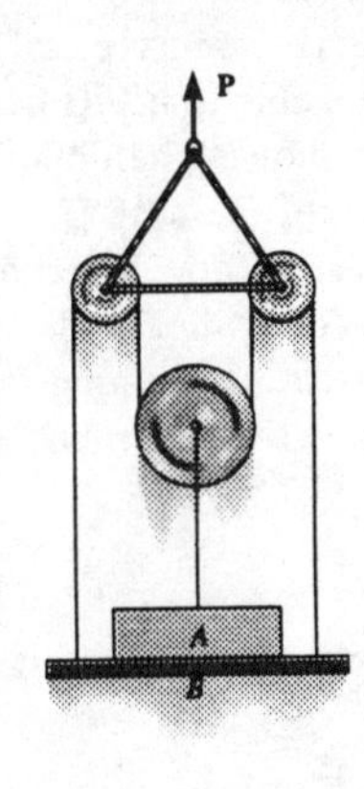

Assembly :

$$\uparrow +\Sigma F_y = m(a_G)_y; \qquad 100-60 = \frac{60}{32.2}a_G$$

$$a_G = 21.47 \text{ ft/s}^2$$

Block A :

$$+\uparrow \Sigma F_y = m(a_G)_y; \qquad 2T+R-50 = \frac{50}{32.2}(21.47)$$

Block B :

$$+\uparrow \Sigma F_y = m(a_G)_y; \qquad 2T-R-10 = \frac{10}{32.2}(21.47)$$

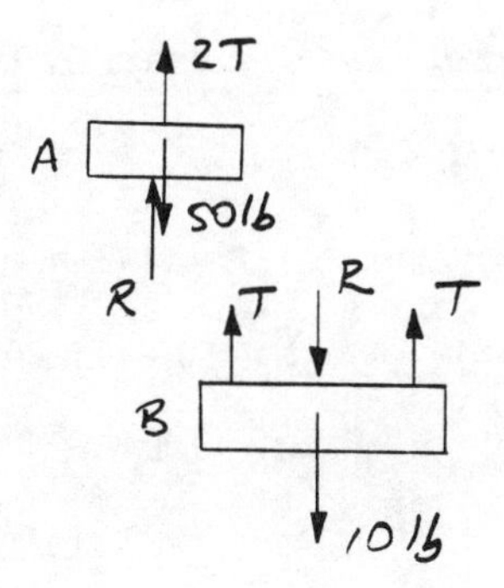

Solving,

$R = 33.3$ lb **Ans**

$T = 25$ lb

17-38. The sports car has a mass of 1.5 Mg and a center of mass at G. Determine the shortest time it takes for it to reach a speed of 80 km/h, starting from rest, if the engine only drives the rear wheels, whereas the front wheels are free rolling. The coefficient of static friction between the wheels and the road is $\mu_s = 0.2$. Neglect the mass of the wheels for the calculation. If driving power could be supplied to all four wheels, what would be the shortest time for the car to reach a speed of 80 km/h?

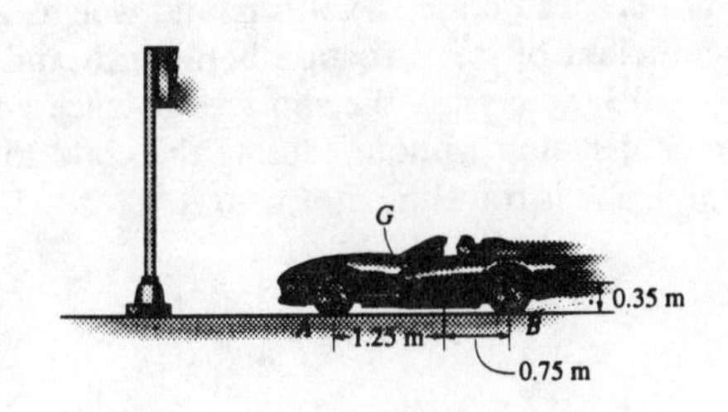

$$\overset{+}{\leftarrow} \Sigma F_x = m(a_G)_x; \qquad 0.2N_A + 0.2N_B = 1500a_G \qquad (1)$$

$$+\uparrow \Sigma F_y = m(a_G)_y; \qquad N_A + N_B - 1500(9.81) = 0 \qquad (2)$$

$$\circlearrowleft +\Sigma M_G = 0; \qquad -N_A(1.25) + N_B(0.75) - (0.2N_A + 0.2N_B)(0.35) = 0 \qquad (3)$$

For rear – wheel drive :

Set the friction force $0.2N_A = 0$ in Eqs. (1) and (3)

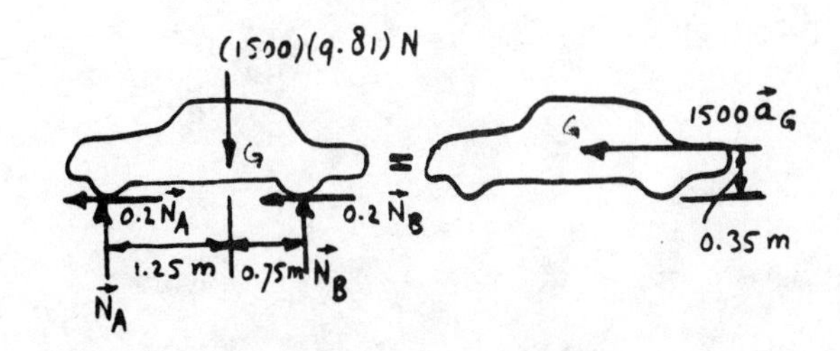

Solving yields :

$N_A = 5.18$ kN > 0 (OK); $N_B = 9.53$ kN; $a_G = 1.271$ m/s^2

Since $v = 80$ km/h $= 22.22$ m/s, then

$$(\overset{+}{\rightarrow}) \qquad v = v_0 + a_G t$$

$$22.22 = 0 + 1.271t$$

$t = 17.5$ s **Ans**

For 4 – wheel drive :

$N_A = 5.00$ kN > 0 (OK); $N_B = 9.71$ kN; $a_G = 1.962$ m/s^2

Since $v_2 = 80$ km/h $= 22.22$ m/s, then

$v_2 = v_1 + a_G t;$ $\qquad 22.22 = 0 + 1.962t$

$t = 11.3$ s **Ans**

17-39. The "muscle car" is designed to do a "wheeley," i.e., to be able to lift its front wheels off the ground in the manner shown when it accelerates. If the 1.35-Mg car has a center of mass at G, determine the minimum torque that must be developed at both rear wheels in order to do this. Also, what is the smallest necessary coefficient of static friction assuming the thick-walled rear wheels do not slip on the pavement? Neglect the mass of the wheels.

$\overset{+}{\rightarrow} \Sigma F_x = m(a_G)_x; \qquad F_A = 1350 a_G$

$+\uparrow \Sigma F_y = m(a_G)_y; \qquad N_A - 1350(9.81) = 0$

$\circlearrowleft + \Sigma M_A = \Sigma (M_k)_A; \qquad 1350(9.81)(1.10) = 1350 a_G (0.67)$

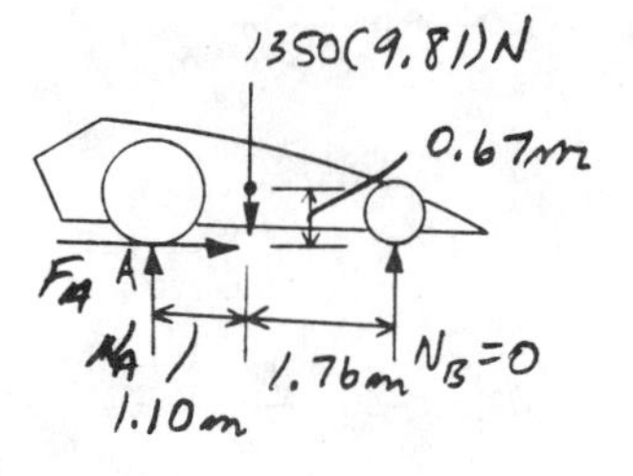

Solving,

$a_G = 16.11 \text{ m/s}^2; \qquad F_A = 21\,743 \text{ N}; \qquad N_A = 13\,244 \text{ N}$

$\circlearrowleft + \Sigma M_C = 0; \qquad 21\,743(0.31) - M = 0$

$M = 6.74 \text{ kN} \cdot \text{m}$ **Ans**

$\mu_{min} = \dfrac{F_A}{N_A} = \dfrac{21\,743}{13\,244} = 1.64$ **Ans**

***17-40.** The crate is uniform and weighs 200 lb. If the coefficient of static friction between it and the truck is $\mu_s = 0.3$, determine the shortest distance s in which the truck can stop without causing the crate to tip or slide. The truck is traveling at $v = 20$ ft/s.

Assume crate is on the verge of sliding.

$\overset{+}{\rightarrow} \Sigma F_x = m(a_G)_x; \qquad -0.3N_c = -\dfrac{200}{32.2} a_G$

$+\uparrow \Sigma F_y = m(a_G)_y; \qquad -200 + N_c = 0$

$\circlearrowleft + \Sigma M_G = 0; \qquad N_c x - 0.3N_c(2) = 0$

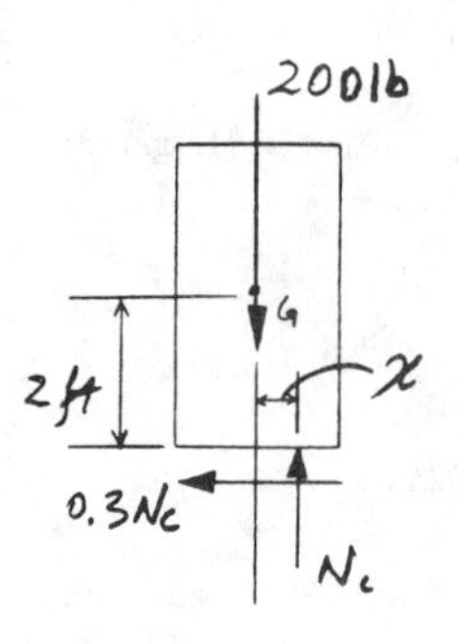

Solving :

$N_c = 200 \text{ lb}$

$a = 9.66 \text{ ft/s}^2$

$x = 0.6 \text{ ft} < 1 \text{ ft}$ The assumption is valid.

$v^2 = v_0^2 + 2a(s - s_0)$

$0 = (20)^2 + 2(-9.66)(s - 0)$

$s = 20.7 \text{ ft}$ **Ans**

17-41. The crate of mass m is supported on a cart of negligible mass. Determine the maximum force P that can be applied a distance d from the cart bottom without causing the crate to tip on the cart.

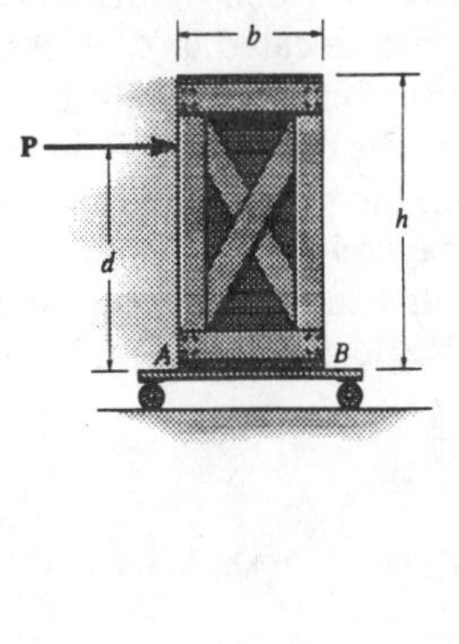

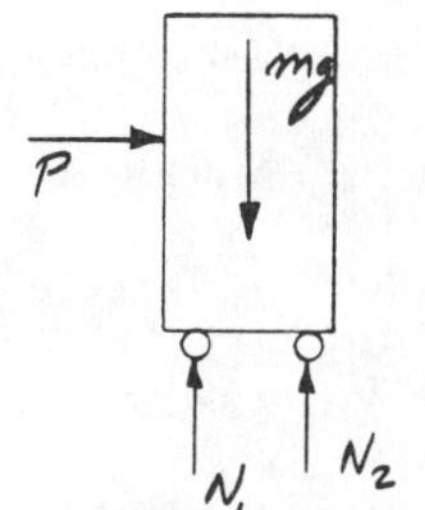

Crate :

Require N_c to act at corner B for tipping.

$$\curvearrowleft + \Sigma M_B = \Sigma (M_k)_B; \quad P(d) - mg\left(\frac{b}{2}\right) = m(a_G)\left(\frac{h}{2}\right) \quad (1)$$

System :

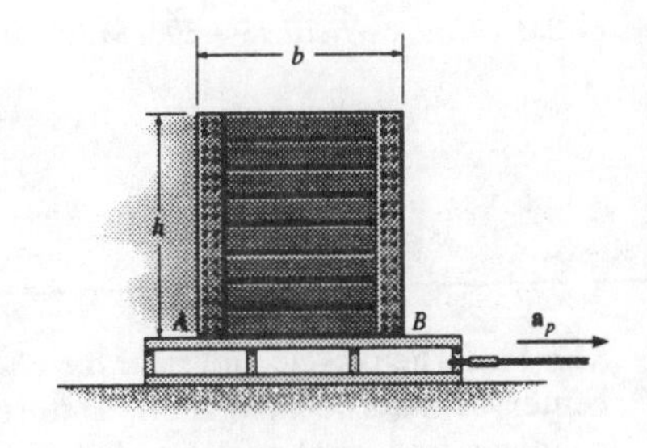

$$\xrightarrow{+} \Sigma F_x = m(a_G)_x; \quad P = ma_G$$

From Eq. (1) :

$$Pd - mg\left(\frac{b}{2}\right) = P\left(\frac{h}{2}\right)$$

$$P_{max} = \frac{mgb}{2\left(d - \frac{h}{2}\right)} \quad \textbf{Ans}$$

17-42. The uniform crate has a mass m and rests on a rough pallet for which the coefficient of static friction between the crate and pallet is μ_s. If the pallet is given an acceleration of a_p, show that the crate will tip and slip at the same time provided $\mu_s = b/h$.

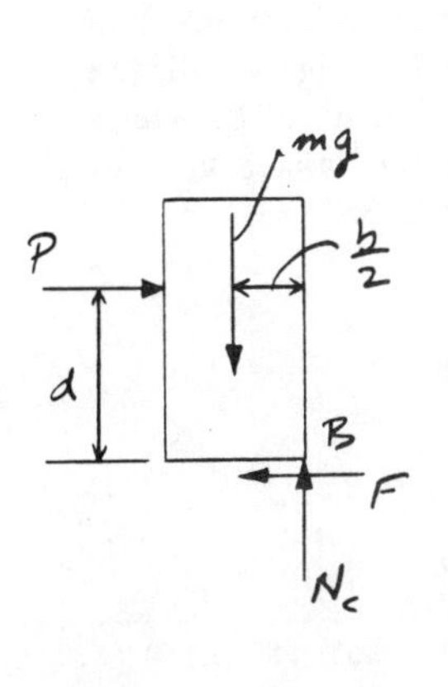

$$\xrightarrow{+} \Sigma F_x = m(a_G)_x; \quad \mu_s N_c = ma_p$$

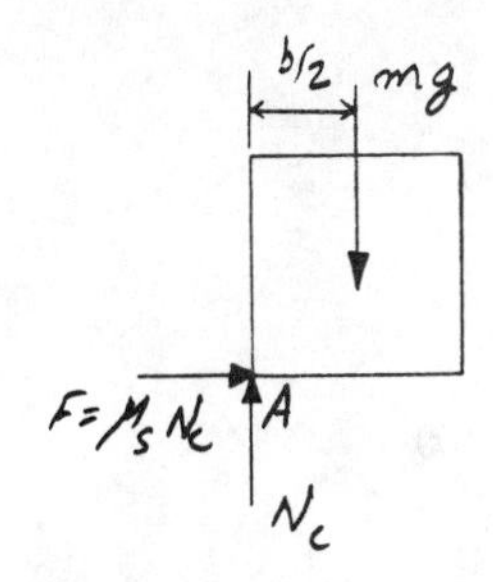

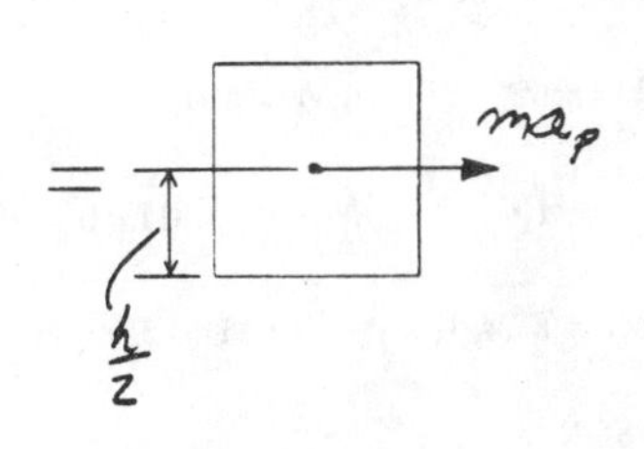

$$\curvearrowleft + \Sigma M_A = \Sigma (M_k)_A; \quad mg\left(\frac{b}{2}\right) = ma_p\left(\frac{h}{2}\right)$$

$$a_p = g\left(\frac{b}{h}\right)$$

$$+\uparrow \Sigma F_y = m(a_G)_y; \quad N_c - mg = 0$$

$$N_c = mg$$

$$\mu_s(mg) = m\left(g\frac{b}{h}\right)$$

$$\mu_s = \frac{b}{h} \quad \textbf{Q.E.D.}$$

17-43. The bicycle and rider have a mass of 80 kg with center of mass located at G. If the coefficient of kinetic friction at the rear tire is $\mu_B = 0.8$, determine the normal reactions at the tires A and B, and the deceleration of the rider, when the rear wheel locks for braking. What is the normal reaction at the rear wheel when the bicycle is traveling at constant velocity and the brakes are not applied? Neglect the mass of the wheels.

Deceleration :

$\xrightarrow{+} \Sigma F_x = m(a_G)_x; \quad 0.8N_B = 80a_G$

$+\uparrow \Sigma F_y = m(a_G)_y; \quad N_A + N_B - 80(9.81) = 0$

$\curvearrowleft + \Sigma M_A = \Sigma (M_k)_A; \quad -N_B(0.95) + 80(9.81)(0.55) = 80a_G(1.2)$

$a_G = 2.26 \text{ m/s}^2$ **Ans**

$N_B = 226 \text{ N}$ **Ans**

$N_A = 559 \text{ N}$ **Ans**

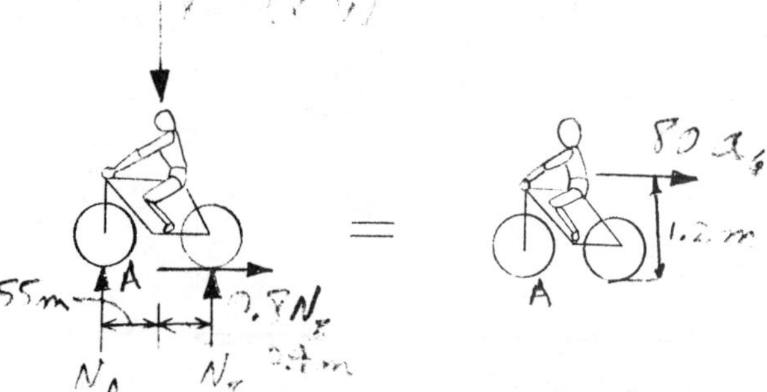

Equilibrium,

$+\uparrow \Sigma F_y = 0; \quad N_A + N_B - 80(9.81) = 0$

$\curvearrowleft + \Sigma M_A = 0; \quad -N_B(0.95) + 80(9.81)(0.55) = 0$

$N_A = 330 \text{ N}$

$N_B = 454 \text{ N}$ **Ans**

***17-44.** The bicycle and rider have a mass of 80 kg with center of mass located at G. Determine the minimum coefficient of kinetic friction between the road and the wheels so that the rear wheel B starts to lift off the ground when the rider applies the brakes to the front wheel. Neglect the mass of the wheels.

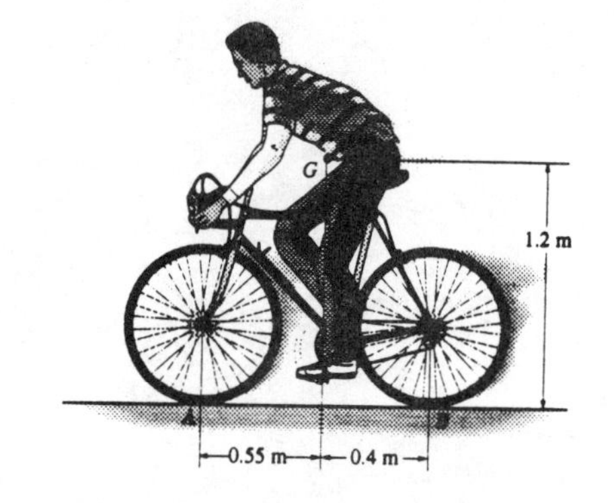

$\xrightarrow{+} \Sigma F_x = m(a_G)_x; \quad \mu_k N_A = 80a_G$

$+\uparrow \Sigma F_y = m(a_G)_y; \quad N_A - 80(9.81) = 0$

$\curvearrowleft + \Sigma M_A = \Sigma (M_k)_A; \quad 80(9.81)(0.55) = 80a_G(1.2)$

$N_A = 785 \text{ N}$

$a_G = 4.50 \text{ m/s}^2$

$\mu_k = 0.458$ **Ans**

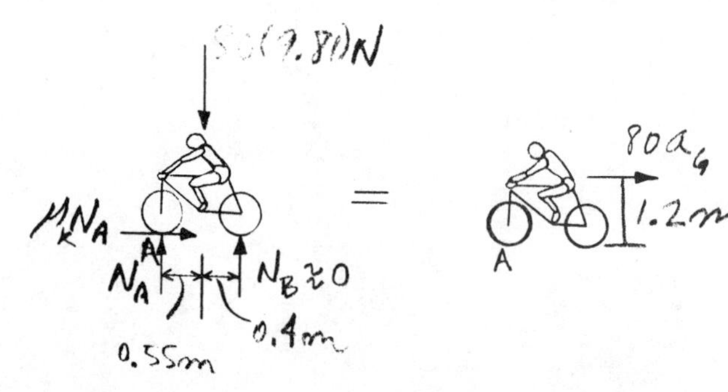

Also :

$\curvearrowleft + \Sigma M_G = 0; \quad N_A(0.55) - \mu_k N_A(1.2) = 0$

$\mu_k = 0.458$ **Ans**

17-45. The dresser has a weight of 80 lb and is pushed along the floor. If the coefficient of static friction at A and B is $\mu_s = 0.3$ and the coefficient of kinetic friction is $\mu_k = 0.2$, determine the smallest horizontal force P needed to cause motion. If this force is increased slightly, determine the acceleration of the dresser. Also, what are the normal reactions at A and B when it begins to move?

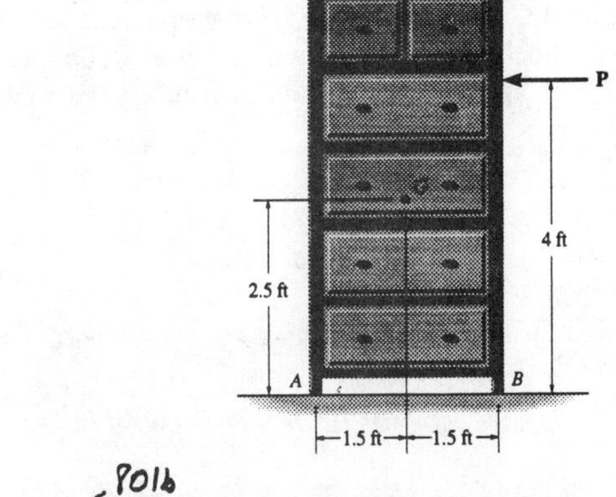

For slipping :

$\overset{+}{\rightarrow}\Sigma F_x = 0;\quad -P + 0.3(N_A + N_B) = 0$

$+\uparrow \Sigma F_y = 0;\quad N_A + N_B - 80 = 0$

$P = 24$ lb **Ans**

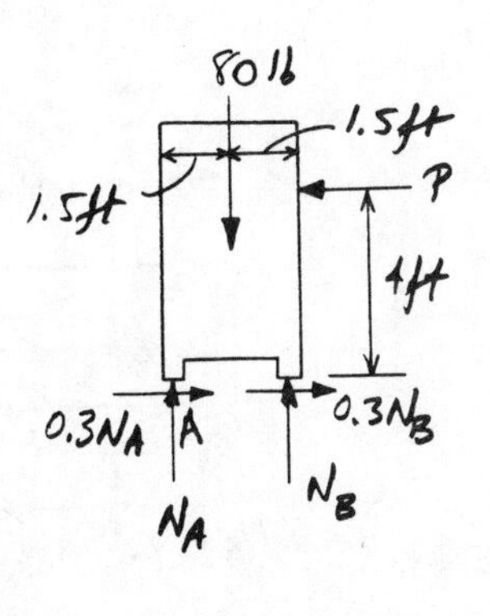

For tipping $N_B = 0$, $N_A = 80$ lb.

$\circlearrowleft + \Sigma M_A = 0;\quad P(4) - 80(1.5) = 0$

$P = 30 \text{ lb} > 24 \text{ lb}$

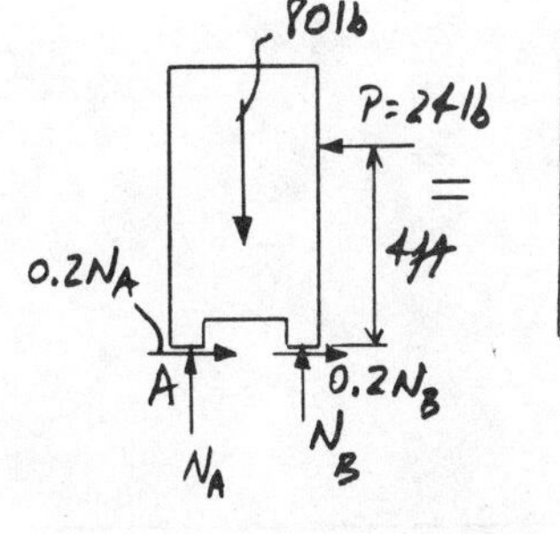

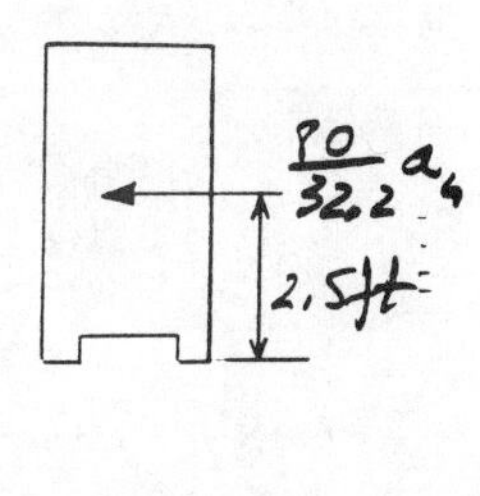

Dresser slips.

$\overset{+}{\leftarrow}\Sigma F_x = m(a_G)_x;\quad 24 - 0.2N_A - 0.2N_B = \left(\frac{80}{32.2}\right)a_G$

$+\uparrow \Sigma F_y = 0;\quad N_A + N_B - 80 = 0$

$\circlearrowleft + \Sigma M_A = \Sigma (M_k)_A;\quad 24(4) + N_B(3) - 80(1.5) = \left(\frac{80}{32.2}\right)a_G(2.5)$

$a_G = 3.22 \text{ ft/s}^2$ **Ans**

$N_B = 14.7$ lb **Ans**

$N_A = 65.3$ lb **Ans**

17-46. The dresser has a weight of 80 lb and is pushed along the floor. If the coefficient of static friction at A and B is $\mu_s = 0.3$ and the coefficient of kinetic friction is $\mu_k = 0.2$, determine the maximum horizontal force P that can be applied without causing the dresser to tip over.

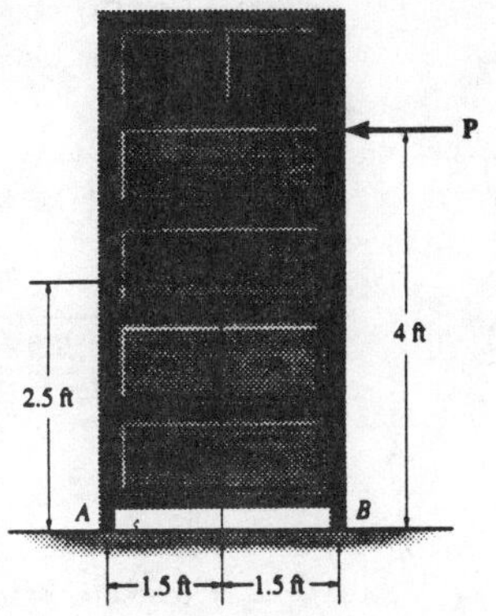

When force P is applied, dresser will slide before tipping. See Prob. 17 – 45.

$+\uparrow \Sigma F_y = 0;\quad N_A - 80 = 0$

$\overset{+}{\leftarrow}\Sigma F_x = m(a_G)_x;\quad P - 0.2N_A = \left(\frac{80}{32.2}\right)a_G$

$\circlearrowleft + \Sigma M_A = \Sigma (M_k)_A;\quad P(4) - 80(1.5) = \left(\frac{80}{32.2}\right)a_G(2.5)$

$P = 53.3$ lb **Ans**

$a_G = 15.0 \text{ ft/s}^2$

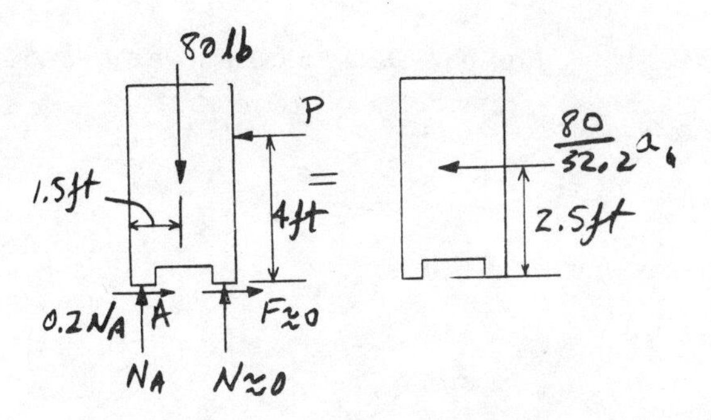

17-47. The handcart has a mass of 200 kg and center of mass at G. Determine the normal reactions at each of the two wheels at A and the two wheels at B if a force of $P = 50$ N is applied to the handle. Neglect the mass of the wheels.

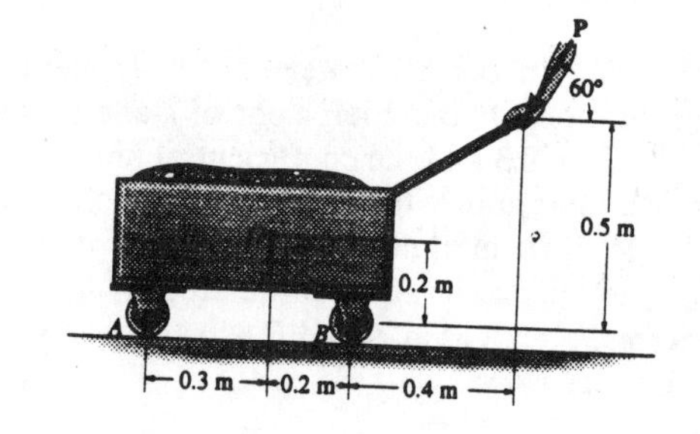

$\overset{+}{\leftarrow} \Sigma F_x = m(a_G)_x ;\quad 50 \cos 60° = 200 a_G$

$+\uparrow \Sigma F_y = m(a_G)_y ;\quad N_A + N_B - 200(9.81) - 50 \sin 60° = 0$

$\circlearrowleft + \Sigma M_G = 0;\quad -N_A(0.3) + N_B(0.2) + 50 \cos 60°(0.3) - 50 \sin 60°(0.6) = 0$

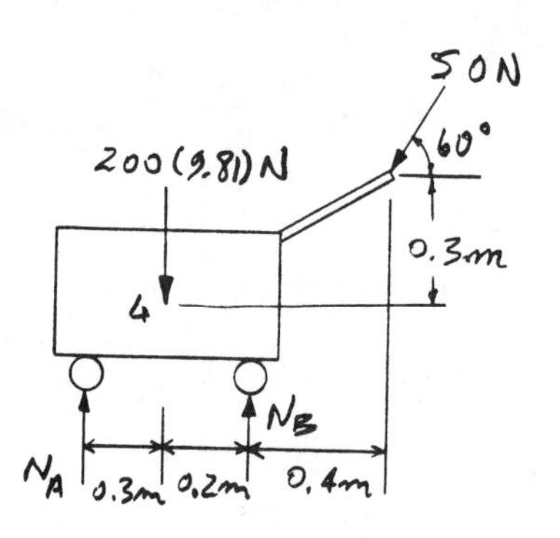

$a_G = 0.125 \text{ m/s}^2;\quad N_A = 765.2 \text{ N};\quad N_B = 1240 \text{ N}$

At each wheel,

$N_A' = \dfrac{N_A}{2} = 383 \text{ N}$ **Ans**

$N_B' = \dfrac{N_B}{2} = 620 \text{ N}$ **Ans**

***17-48.** The handcart has a mass of 200 kg and center of mass at G. Determine the magnitude of the largest force **P** that can be applied to the handle so that the wheels at A or B continue to maintain contact with the ground. Neglect the mass of the wheels.

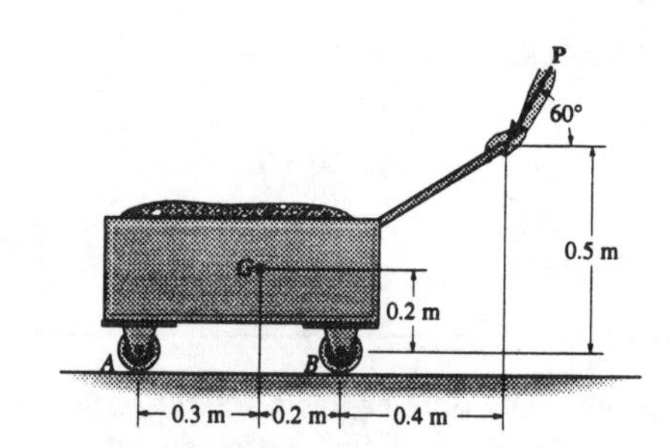

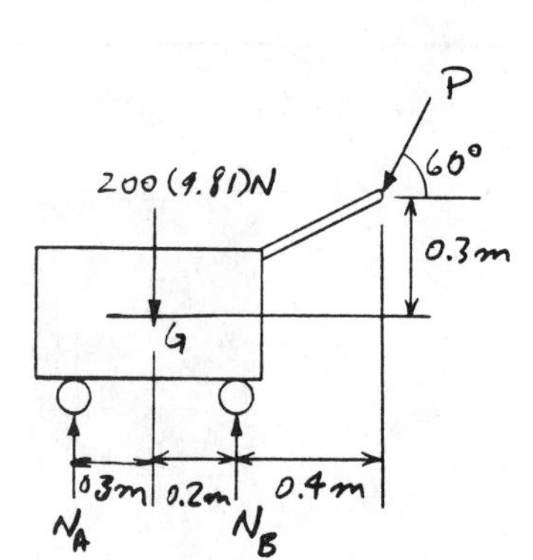

$\overset{+}{\leftarrow} \Sigma F_x = m(a_G)_x ;\quad P \cos 60° = 200 a_G$

$+\uparrow \Sigma F_y = m(a_G)_y ;\quad N_A + N_B - 200(9.81) - P \sin 60° = 0$

$\circlearrowleft + \Sigma M_G = 0;\quad -N_A(0.3) + N_B(0.2) + P \cos 60°(0.3) - P \sin 60°(0.6) = 0$

For P_{max}, require

$N_A = 0$

$P = 1998 \text{ N} = 2.00 \text{ kN}$ **Ans**

$N_B = 3692 \text{ N}$

$a_G = 4.99 \text{ m/s}^2$

17-49. The 50-kg uniform crate rests on the platform for which the coefficient of static friction is $\mu_s = 0.5$. If the supporting links have an angular velocity $\omega = 1$ rad/s, determine the greatest angular acceleration α they can have so that the crate does not slip or tip at the instant $\theta = 30°$ as shown.

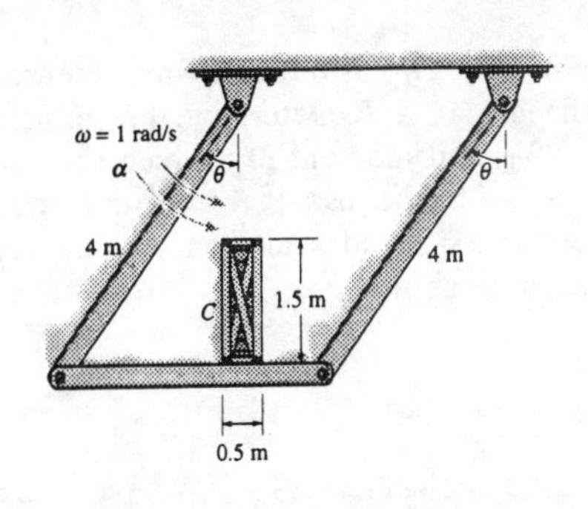

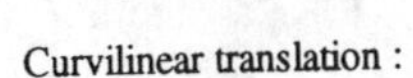

Curvilinear translation :

$(a_G)_n = (1)^2(4) = 4 \text{ m/s}^2$

$(a_G)_t = \alpha(4) \text{ m/s}^2$

$\xrightarrow{+} \Sigma F_x = m(a_G)_x;\quad F_C = 50(4)\sin 30° + 50(\alpha)(4)\cos 30°$

$+\uparrow \Sigma F_y = m(a_G)_y;\quad N_C - 50(9.81) = 50(4)\cos 30° - 50(\alpha)(4)\sin 30°$

$\circlearrowleft + \Sigma M_G = \Sigma(M_k)_G;\quad N_C(x) - F_C(0.75) = 0$

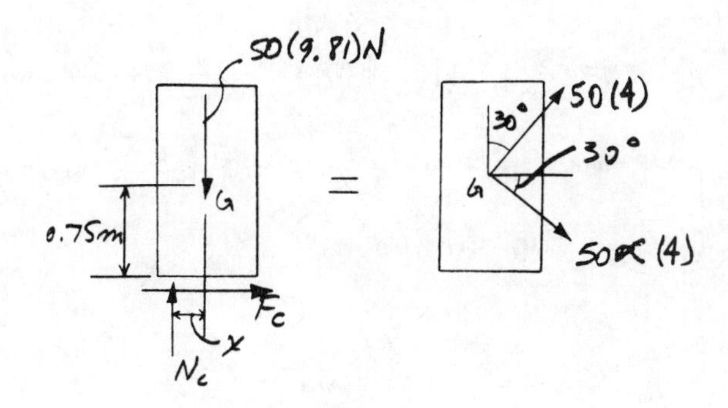

Assume crate is about to slip. $\quad F_C = 0.5N_C$

Thus,

$x = 0.375 \text{ m} > 0.25 \text{ m}$

Crate must tip. Set $x = 0.25$ m.

$N_C = 605$ N; $\quad F_C = 202$ N

$\alpha = 0.587 \text{ rad/s}^2$ **Ans**

Note : $(F_C)_{max} = 0.5(605) = 303 \text{ N} > 202 \text{ N}$ O.K.

17-50. The 50-kg uniform crate rests on the platform for which the coefficient of static friction is $\mu_s = 0.5$. If at the instant $\theta = 30°$ the supporting links have an angular velocity $\omega = 1$ rad/s and angular acceleration $\alpha = 0.5 \text{ rad/s}^2$, determine the friction force on the crate.

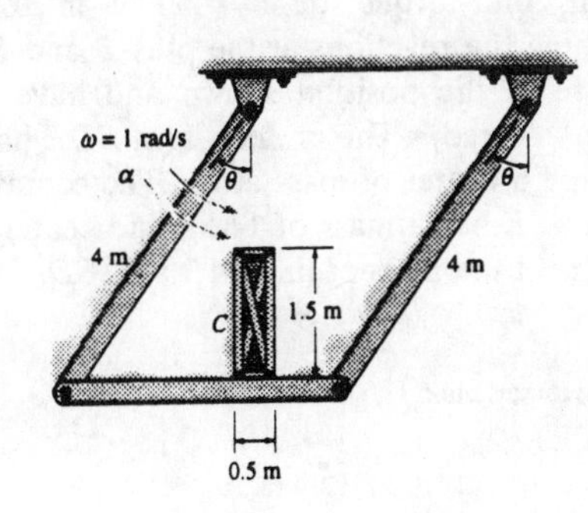

Curvilinear translation :

$(a_G)_n = (1)^2(4) = 4 \text{ m/s}^2$

$(a_G)_t = 0.5(4) \text{ m/s}^2 = 2 \text{ m/s}^2$

$\xrightarrow{+} \Sigma F_x = m(a_G)_x;\quad F_C = 50(4)\sin 30° + 50(2)\cos 30°$

$+\uparrow \Sigma F_y = m(a_G)_y;\quad N_C - 50(9.81) = 50(4)\cos 30° - 50(2)\sin 30°$

Solving,

$F_C = 186.6$ N

$N_C = 613.7$ N

$(F_C)_{max} = 0.5(613.7) = 306.9 \text{ N} > 186.6 \text{ N}$ OK

$\circlearrowleft + \Sigma M_G = \Sigma(M_k)_G;\quad N_C(x) - F_C(0.75) = 0$

$613.7(x) - 186.6(0.75) = 0$

$x = 0.228 \text{ m} < 0.25 \text{ m}$ OK

Thus $\quad F_C = 187$ N **Ans**

17-51. The two 3-lb rods EF and HI are fixed (welded) to the link AC at E. Determine the normal force N_E, shear force V_E, and moment M_E, which the bar AC exerts on FE at E if at the instant $\theta = 30°$ link AB has an angular velocity $\omega = 5$ rad/s and an angular acceleration $\alpha = 8$ rad/s^2 as shown.

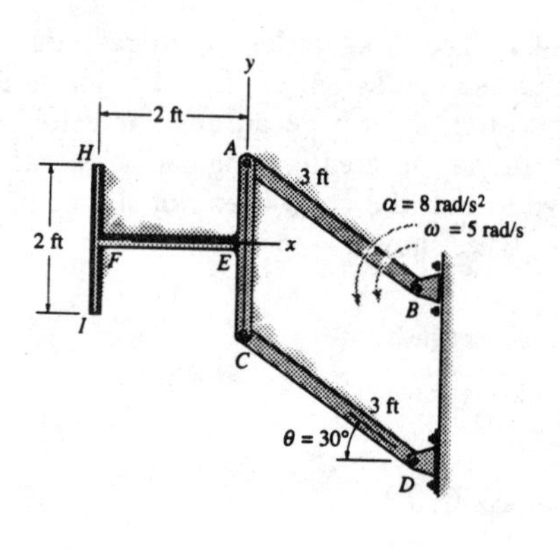

Curvilinear translation :

$(a_G)_n = (5)^2(3) = 75 \text{ ft/s}^2$ ∠30°

$(a_G)_t = 8(3) = 24 \text{ ft/s}^2$ ∠30°

$$\bar{x} = \frac{\Sigma \tilde{x} m}{\Sigma m} = \frac{1(3) + 2(3)}{6} = 1.5 \text{ ft}$$

$\xrightarrow{+} \Sigma F_x = m(a_G)_x; \quad N_E = \left(\frac{6}{32.2}\right)(75)\cos 30° - \left(\frac{6}{32.2}\right)(24)\sin 30°$

$+\downarrow \Sigma F_y = m(a_G)_y; \quad V_E + 6 = \left(\frac{6}{32.2}\right)(24)\cos 30° + \left(\frac{6}{32.2}\right)(75)\sin 30°$

$\circlearrowleft + \Sigma M_G = 0; \quad M_E - V_E(1.5) = 0$

$N_E = 9.87$ lb **Ans**

$V_E = 4.86$ lb **Ans**

$M_E = 7.29$ lb · ft **Ans**

***17-52.** The arm BDE of the industrial robot manufactured by Cincinnati Milacron is activated by applying the torque of $M = 50$ N·m to link CD. Determine the reactions at the pins B and D when the links are in the position shown and have an angular velocity of 2 rad/s. The uniform arm BDE has a mass of 10 kg and a center of mass at G_1. The container held in its grip at E has a mass of 12 kg and center of mass at G_2. Neglect the mass of links AB and CD.

Curvilinear translation :

$(a_D)_n = (a_G)_n = (2)^2(0.6) = 2.4 \text{ m/s}^2$

Member DC :

$\circlearrowleft + \Sigma M_C = 0; \quad -D_x(0.6) + 50 = 0$

$D_x = 83.33 \text{ N} = 83.3 \text{ N}$ **Ans**

Member BDE :

$\circlearrowleft + \Sigma M_D = \Sigma (M_k)_D; \quad -F_{BA}(0.220) + 10(9.81)(0.365) + 12(9.81)(1.10)$

$= 10(2.4)(0.365) + 12(2.4)(1.10)$

$F_{BA} = 567.54 \text{ N} = 568 \text{ N}$ **Ans**

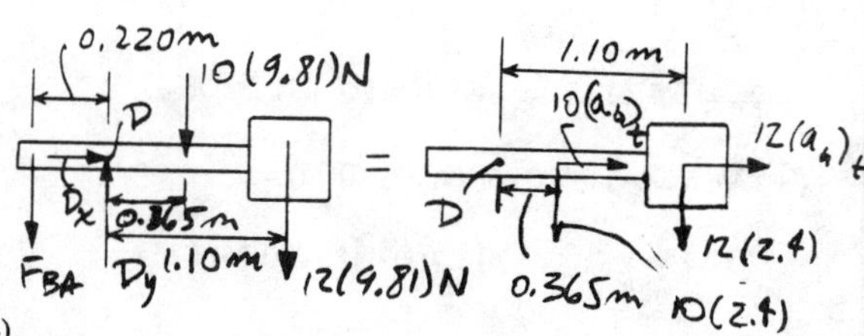

$+\uparrow \Sigma F_y = m(a_G)_y; \quad -567.54 + D_y - 10(9.81) - 12(9.81) = -10(2.4) - 12(2.4)$

$D_y = 731$ N **Ans**

17-53. The 10-lb rod is pin connected to its support at A and has an angular velocity $\omega = 4$ rad/s when it is in the horizontal position shown. Determine its angular acceleration and the horizontal and vertical components of reaction which the pin exerts on the rod at this instant.

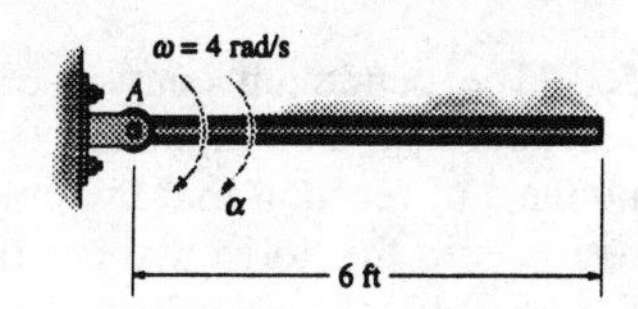

Equations of motion :

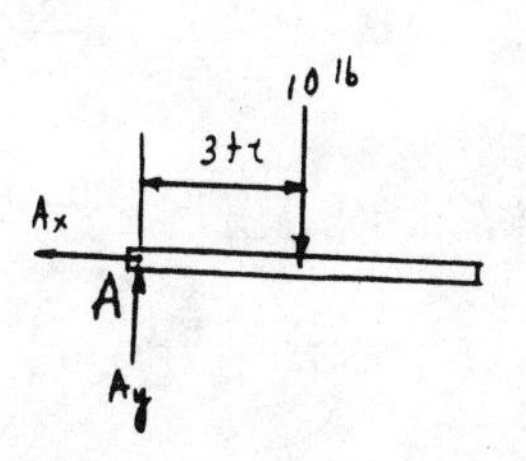

$\overset{+}{\leftarrow} \Sigma F_n = m\omega^2 r_G;\quad A_x = \left(\frac{10}{32.2}\right)(4)^2(3)\qquad A_x = 14.9$ lb **Ans**

$\curvearrowleft + \ \Sigma M_A = I_A\,\alpha;\quad 10(3) = \frac{1}{3}\left(\frac{10}{32.2}\right)(6)^2\alpha\qquad \alpha = 8.05\ \text{rad/s}^2$ **Ans**

$+\downarrow \Sigma F_t = m\alpha r_G;\quad 10 - A_y = \left(\frac{10}{32.2}\right)(8.05)(3)\qquad A_y = 2.50$ lb **Ans**

17-54. The pendulum consists of a 20-lb sphere and a 5-lb slender rod. Determine the reaction at the pin O just after the pendulum is released from the position shown.

$$I_O = \left[\frac{2}{5}\left(\frac{20}{32.2}\right)(0.25)^2 + \left(\frac{20}{32.2}\right)(2.25)^2\right] + \left[\frac{1}{3}\left(\frac{5}{32.2}\right)(2)^2\right] = 3.367\ \text{slug}\cdot\text{ft}^2$$

$\overset{+}{\rightarrow} \Sigma F_x = m(a_G)_x;\quad O_x = 0$

$+\downarrow \Sigma F_y = m(a_G)_y;\quad 20 + 5 - O_y = \left(\frac{20}{32.2}\right)a_b + \left(\frac{5}{32.2}\right)a_R$

$+\Sigma M_O = I_O\alpha;\quad (20)(2.25) + (5)(1) = 3.367\alpha$

$a_b = (2.25)\alpha$

$a_R = (1)\alpha$

Solving,

$O_x = 0$ **Ans**

$O_y = 1.94$ lb **Ans**

$\alpha = 14.9\ \text{rad/s}^2$ **Ans**

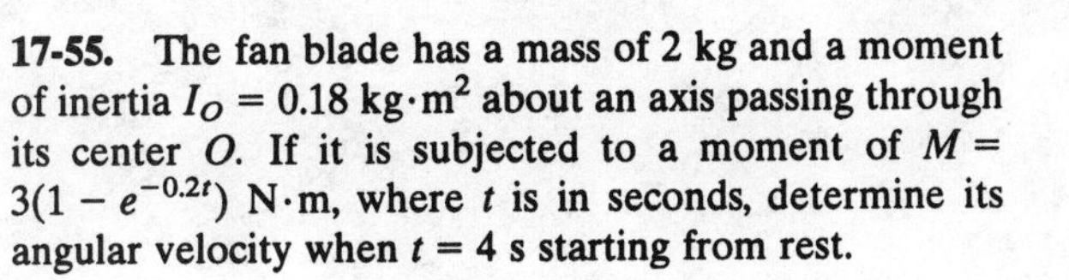

17-55. The fan blade has a mass of 2 kg and a moment of inertia $I_O = 0.18\ \text{kg}\cdot\text{m}^2$ about an axis passing through its center O. If it is subjected to a moment of $M = 3(1 - e^{-0.2t})$ N·m, where t is in seconds, determine its angular velocity when $t = 4$ s starting from rest.

$\curvearrowleft + \Sigma M_O = I_O\alpha;\quad 3\left(1 - e^{-0.2t}\right) = 0.18\alpha$

$$\alpha = 16.67\left(1 - e^{-0.2t}\right)$$

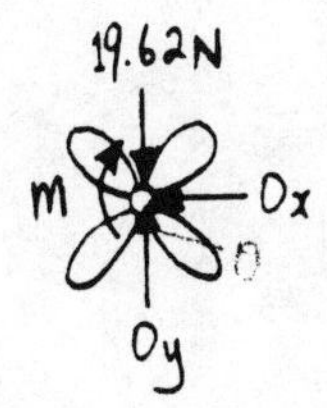

$d\omega = \alpha\, dt$

$$\int_0^{\omega} d\omega = \int_0^4 16.67\left(1 - e^{-0.2t}\right) dt$$

$$\omega = 16.67\left[t + \frac{1}{0.2}e^{-0.2t}\right]_0^4$$

$\omega = 20.8$ rad/s **Ans**

***17-56.** The pendulum consists of a 15-lb disk and a 10-lb slender rod. Determine the horizontal and vertical components of reaction that the pin O exerts on the rod just as it passes the horizontal position, at which time its angular velocity is $\omega = 8$ rad/s.

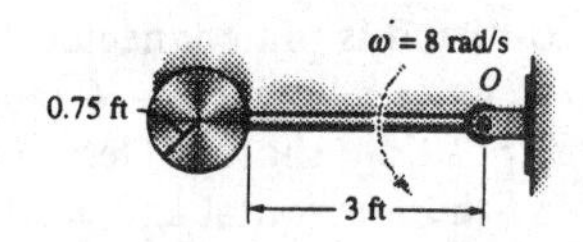

$+\uparrow \Sigma F_y = m(a_G)_y; \quad O_y - 15 - 10 = -\left(\frac{15}{32.2}\right)[3.75\alpha] - \left(\frac{10}{32.2}\right)[1.5\alpha]$

$\xrightarrow{+} \Sigma F_x = m(a_G)_x; \quad O_x = \left(\frac{15}{32.2}\right)(3.75)(8)^2 + \left(\frac{10}{32.2}\right)(1.5)(8)^2$

$\circlearrowleft + \Sigma M_O = \Sigma(M_k)_O; \quad 15(3.75) + 10(1.5) = \left[\frac{1}{2}\left(\frac{15}{32.2}\right)(0.75)^2\right]\alpha + \left(\frac{15}{32.2}\right)(3.75)\alpha(3.75)$

$+\left[\frac{1}{12}\left(\frac{10}{32.2}\right)(3)^2\right]\alpha + \left(\frac{10}{32.2}\right)(1.5)\alpha(1.5)$

$\alpha = 9.36 \text{ rad/s}^2$

$O_x = 142 \text{ lb} \rightarrow$ **Ans**

$O_y = 4.29 \text{ lb} \uparrow$ **Ans**

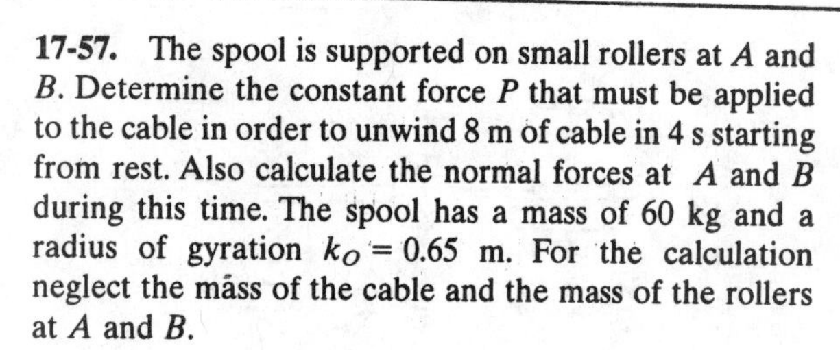

17-57. The spool is supported on small rollers at A and B. Determine the constant force P that must be applied to the cable in order to unwind 8 m of cable in 4 s starting from rest. Also calculate the normal forces at A and B during this time. The spool has a mass of 60 kg and a radius of gyration $k_O = 0.65$ m. For the calculation neglect the mass of the cable and the mass of the rollers at A and B.

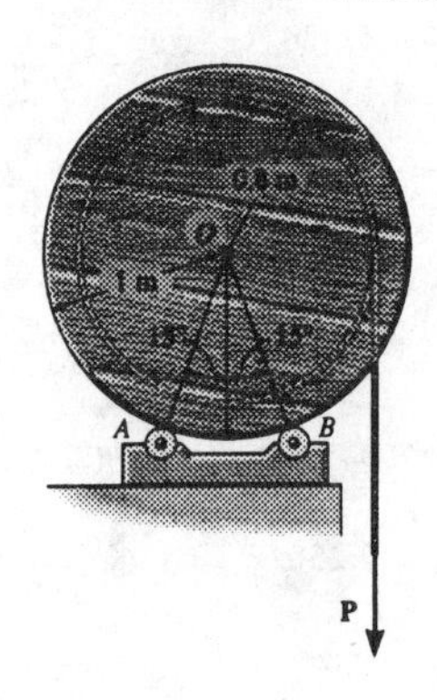

$(\downarrow +) \quad s = s_0 + v_0 t + \frac{1}{2} a_c t^2$

$8 = 0 + 0 + \frac{1}{2} a_c (4)^2$

$a_c = 1 \text{ m/s}^2$

$\alpha = \frac{1}{0.8} = 1.25 \text{ rad/s}^2$

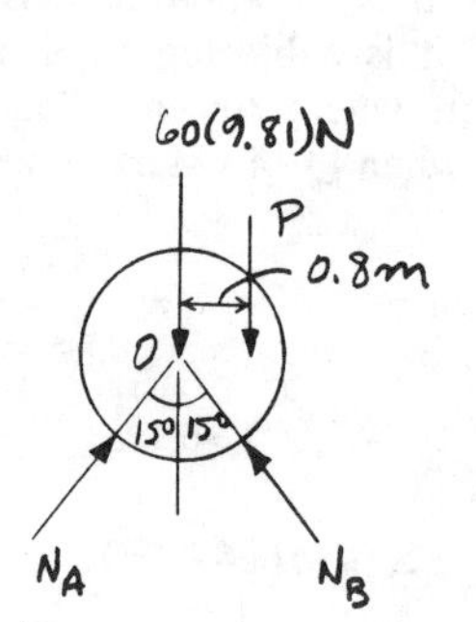

$\circlearrowleft + \Sigma M_O = I_O \alpha; \quad P(0.8) = 60(0.65)^2(1.25)$

$P = 39.6 \text{ N}$ **Ans**

$\xrightarrow{+} \Sigma F_x = m a_x; \quad N_A \sin 15° - N_B \sin 15° = 0$

$+\uparrow \Sigma F_y = m a_y; \quad N_A \cos 15° + N_B \cos 15° - 39.6 - 588.6 = 0$

$N_A = N_B = 325 \text{ N}$ **Ans**

17-58. A cord is wrapped around the inner core of a spool. If the cord is pulled with a constant tension of 30 lb and the spool is originally at rest, determine the spool's angular velocity when $s = 8$ ft of cord has unwound. Neglect the weight of the 8-ft portion of cord. The spool and the entire cord have a total weight of 400 lb, and the radius of gyration about the axle A is $k_A = 1.30$ ft.

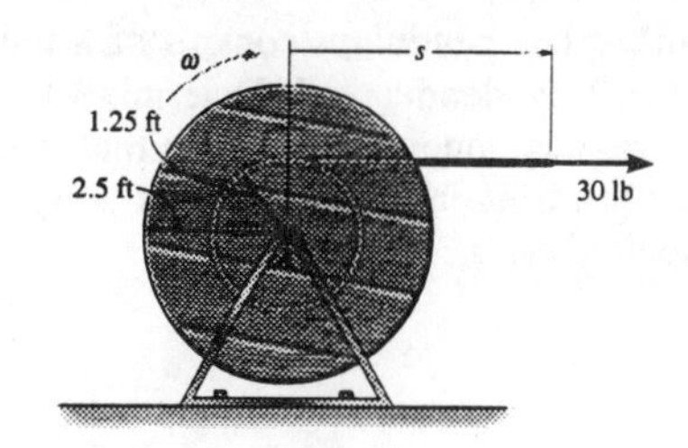

$I_A = mk_A^2 = \left(\frac{400}{32.2}\right)(1.30)^2 = 20.99 \text{ slug} \cdot \text{ft}^2$

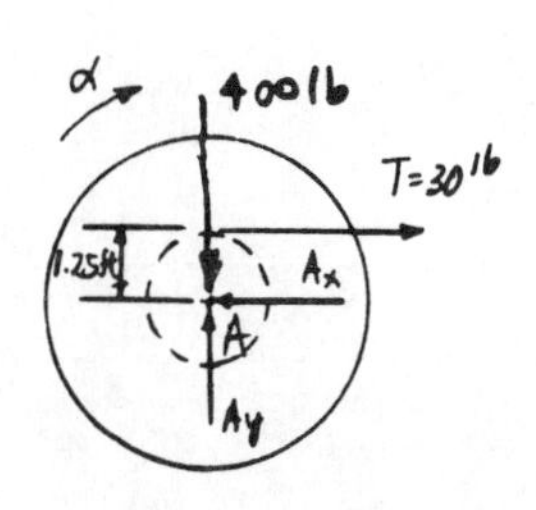

$\circlearrowleft + \Sigma M_A = I_A \alpha;$ $\quad 30(1.25) = 20.99(\alpha)$ $\quad \alpha = 1.786 \text{ rad/s}^2$

The angular displacement is $\theta = \frac{s}{r} = \frac{8}{1.25} = 6.4$ rad.

$\omega^2 = \omega_0^2 + 2\alpha(\theta - \theta_0)$

$\omega^2 = 0 + 2(1.786)(6.4 - 0)$

$\omega = 4.78$ rad/s $\quad$ **Ans**

17-59. A motor supplies a constant torque $M = 2$ N·m to a 50-mm-diameter shaft O connected to the center of the 30-kg flywheel. The resultant bearing friction **F**, which the pin exerts on the shaft, acts tangent to the supporting shaft and has a magnitude of 50 N. Determine how long the torque must be applied to the shaft to increase the flywheel's rotational speed from 4 rad/s to 15 rad/s. The flywheel has a radius of gyration $k_O = 0.15$ m about its center.

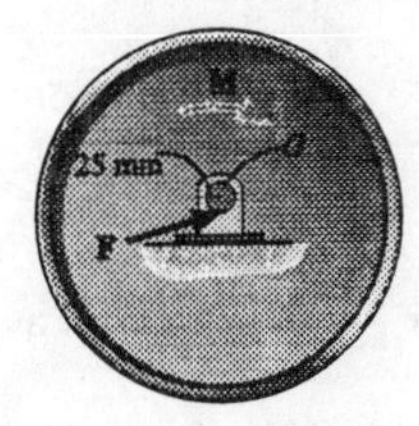

$\circlearrowleft + \Sigma M_O = I_O \alpha;$ $\quad 2 - 50(0.025) = 30(0.15)^2 \alpha$

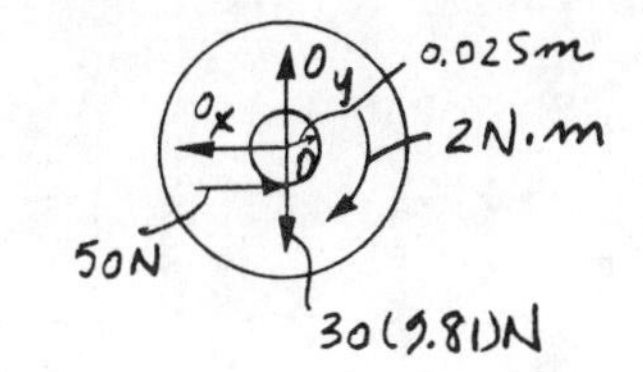

$\alpha = 1.11 \text{ rad/s}^2$

$\circlearrowleft + \omega = \omega_0 + \alpha_c t$

$15 = 4 + (1.11)t$

$t = 9.90$ s $\quad$ **Ans**

***17-60.** If the motor in Prob. 17-59 is disengaged from the shaft once the flywheel is rotating at 15 rad/s, so that $M = 0$, determine how long it will take before the resultant bearing frictional force $F = 50$ N stops the flywheel from rotating.

$\circlearrowleft + \Sigma M_O = I_O \alpha;$ $\quad 50(0.025) = 30(0.15)^2 \alpha$

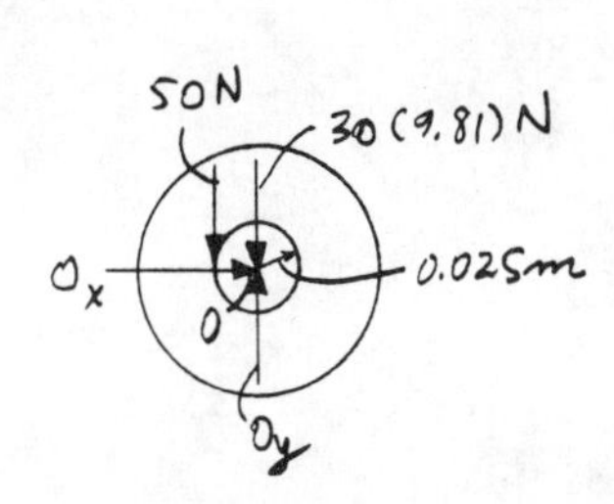

$\alpha = 1.852 \text{ rad/s}^2$

$\circlearrowleft + \omega = \omega_0 + \alpha_c t$

$0 = -15 + (1.852)t$

$t = 8.10$ s $\quad$ **Ans**

17-61. The pendulum consists of a uniform 5-kg plate and a 2-kg slender rod. Determine the horizontal and vertical components of reaction that the pin O exerts on the rod at the instant $\theta = 30°$, at which time its angular velocity is $\omega = 3$ rad/s.

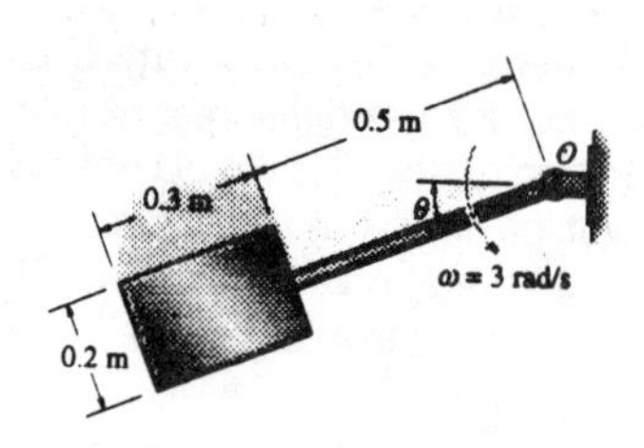

$$I_O = \frac{1}{12}(5)\left[(0.3)^2 + (0.2)^2\right] + 5(0.65)^2 + \frac{1}{3}(2)(0.5)^2 = 2.333 \text{ kg}\cdot\text{ m}^2$$

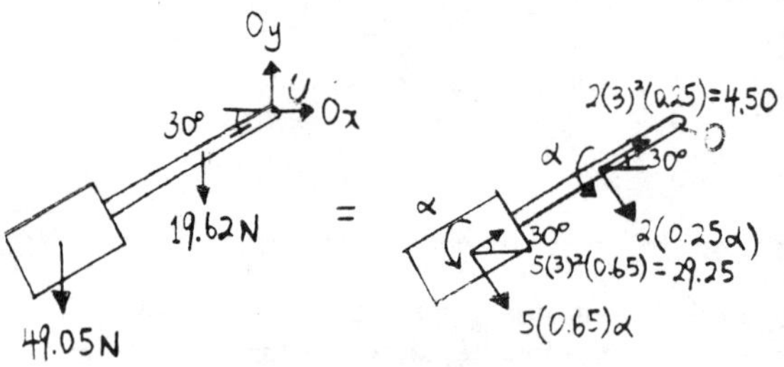

$\circlearrowleft + \Sigma M_O = I_O\alpha;\quad 19.62(0.25\cos 30°) + 49.05(0.65\cos 30°) = 2.333\alpha$

$$\alpha = 13.65 \text{ rad/s}^2$$

$\overset{+}{\rightarrow} \Sigma F_x = m(a_G)_x;\quad O_x = 4.50\cos 30° + 2(0.25)(13.65)\sin 30° + 29.25\cos 30° + 5(0.65)(13.65)\sin 30°$

$O_x = 54.8$ N **Ans**

$+\uparrow \Sigma F_y = m(a_G)_y;\quad O_y - 49.05 - 19.62 = (29.25 + 4.50)\sin 30° - [5(0.65)(13.65) + 2(0.25)(13.65)]\cos 30°$

$O_y = 41.2$ N **Ans**

Also, the problem can be solved as follows :

$$\bar{r} = \frac{\Sigma \tilde{r} m}{\Sigma m} = \frac{(0.25)(2) + 0.65(5)}{7} = 0.5357 \text{ m}$$

$+\Sigma M_O = I_O\alpha;\quad 7(9.81)(0.5357\cos 30°) = 2.333\alpha$

$$\alpha = 13.65 \text{ rad/s}^2$$

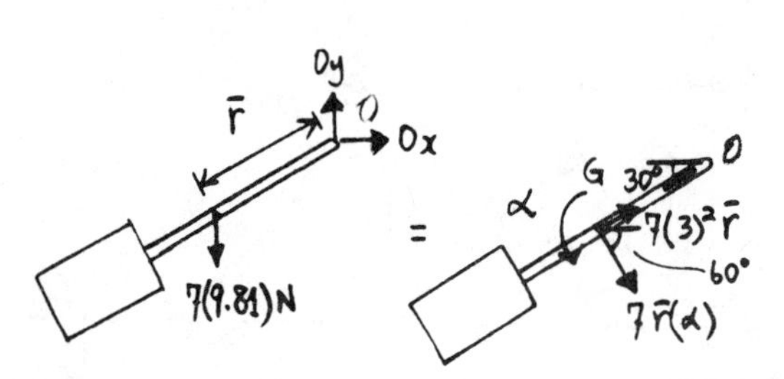

$\overset{+}{\rightarrow} \Sigma F_x = m(a_G)_x;\quad O_x = 7(3)^2(0.5357\cos 30°) + 7(0.5357)(13.65)\cos 60°$

$O_x = 54.8$ N **Ans**

$+\uparrow \Sigma F_y = m(a_G)_y;\quad O_y - 7(9.81) = 7(3)^2(0.5357)(\sin 30°) - 7(0.5357)(13.65)\sin 60°$

$O_y = 41.2$ N **Ans**

17-62. The cylinder has a radius r and mass m and rests in the trough for which the coefficient of kinetic friction at A and B is μ_k. If a horizontal force **P** is applied to the cylinder, determine the cylinder's angular acceleration when it begins to spin.

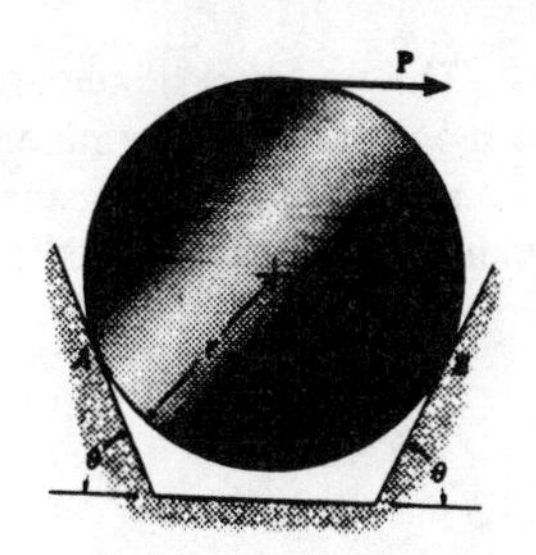

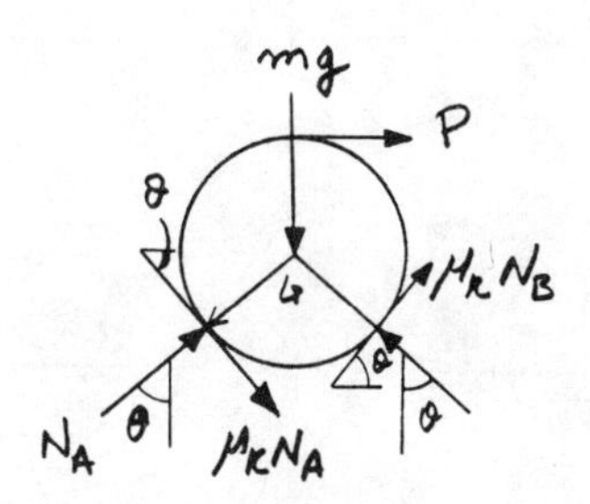

$\overset{+}{\rightarrow} \Sigma F_x = m(a_G)_x$; $\quad P + N_A \sin\theta + \mu_k N_A \cos\theta + \mu_k N_B \cos\theta - N_B \sin\theta = 0$

$+\uparrow \Sigma F_y = m(a_G)_y$; $\quad N_A \cos\theta - \mu_k N_A \sin\theta + N_B \cos\theta + \mu_k N_B \sin\theta - mg = 0$

$\circlearrowleft + \Sigma M_G = I_G\alpha$; $\quad \mu_k N_A (r) + \mu_k N_B (r) - P(r) = -\left(\frac{1}{2}mr^2\right)\alpha$

$$P - (N_B - N_A)\sin\theta + \mu_k \cos\theta (N_A + N_B) = 0 \qquad (2)$$

$$(N_A + N_B)\cos\theta + \mu_k \sin\theta (N_B - N_A) = mg \qquad (1)$$

$$\mu_k (N_A + N_B) = P - \frac{1}{2}mr\alpha \qquad (3)$$

From Eqs. (1) and (2),

$$(N_A + N_B)\cos\theta + \mu_k\left[P + \mu_k \cos\theta (N_A + N_B)\right] = mg$$

or,

$$(N_A + N_B)\left(\cos\theta + \mu_k^2 2\cos\theta\right) = mg - \mu_k P$$

$$(N_A + N_B) = \frac{mg - \mu_k P}{\cos\theta(1 + \mu_k{}^2)}$$

From Eq. (3),

$$\mu_k\left(\frac{mg - \mu_k P}{\cos\theta(1 + \mu_k{}^2)}\right) = P - \frac{1}{2}mr\alpha$$

$$\alpha = -\frac{2\mu_k}{mr}\left(\frac{mg - \mu_k P}{\cos\theta(1 + \mu_k{}^2)}\right) + \frac{2P}{mr} \qquad \textbf{Ans}$$

17-63. The uniform slender rod has a mass of 5 kg. If the cord at A is cut, determine the reaction at the pin O, (a) when the rod is still in the horizontal position, and (b) when the rod swings to the vertical position.

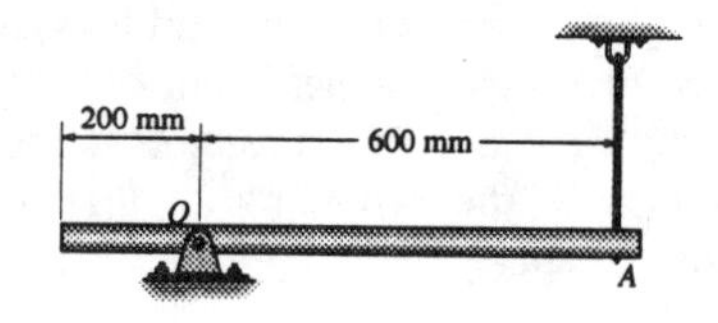

(a) $\omega = 0, \quad (a_G)_n = 0, \quad (a_G)_t = 0.2\alpha$

$\circlearrowleft + \Sigma M_O = I_O\alpha; \quad (0.2)(5)(9.81) = \left[\frac{1}{12}(5)(0.8)^2 + 5(0.2)^2\right]\alpha$

$$\alpha = 21.02 \text{ rad/s}^2$$

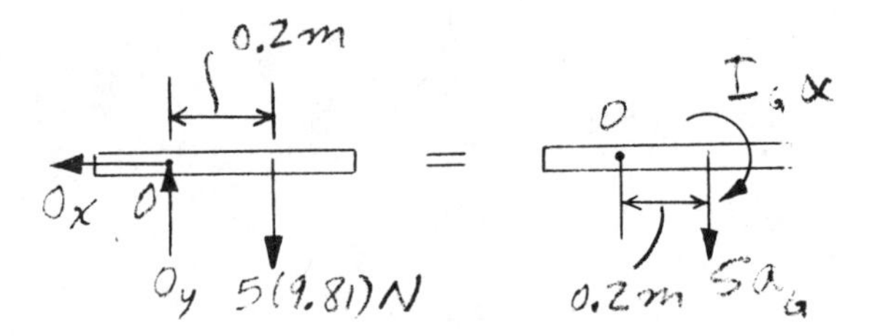

$\overset{+}{\rightarrow} \Sigma F_n = m(a_G)_n; \quad O_x = 0$

$+\uparrow \Sigma F_t = m(a_G)_t; \quad O_y - 5(9.81) = -5(0.2\alpha)$

Thus, $\quad O_y = 28.0 \text{ N}$

And

$$F_O = \sqrt{(0)^2 + (28.0)^2} = 28.0 \text{ N} \qquad \textbf{Ans}$$

(b)

$\circlearrowleft + \Sigma M_O = I_O\alpha; \quad 5(9.81)(0.2)\cos\theta = \left[\frac{1}{12}(5)(0.8)^2 + 5(0.2)^2\right]\alpha$

$$\alpha = 21.02\cos\theta \qquad (1)$$

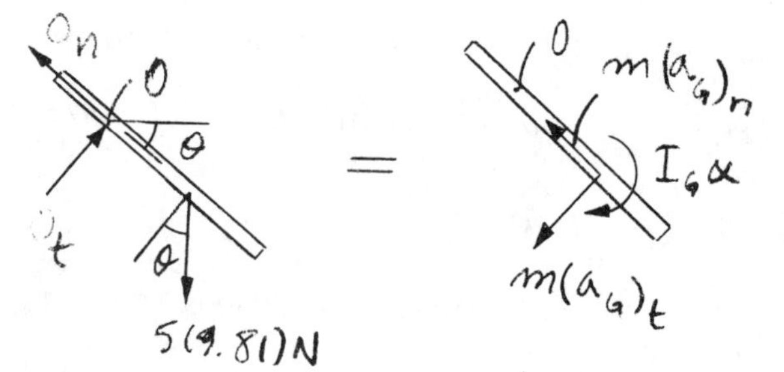

$\nwarrow + \Sigma F_n = m(a_G)_n; \quad O_n - 5(9.81)\sin\theta = 5(\omega^2)(0.2) \qquad (2)$

$\swarrow + \Sigma F_t = m(a_G)_t; \quad -O_t + 5(9.81)\cos\theta = 5(\alpha)(0.2) \qquad (3)$

$\omega\, d\omega = \alpha\, d\theta$

$\int_0^{\omega} \omega\, d\omega = \int_0^{90^\circ} 21.02\cos\theta\, d\theta$

$\frac{1}{2}\omega^2 = 21.02\sin\theta\Big|_0^{90^\circ}$

$\omega = 6.484 \text{ rad/s}$

Substituting into Eq. (2) and solving Eqs. (1) – (3) with $\theta = 90^\circ$ yields

$O_n = 91.09 \text{ N} \uparrow$

$O_t = 0$

$\alpha = 0$

Thus,

$$F_O = \sqrt{(0)^2 + (91.1)^2} = 91.1 \text{ N} \qquad \textbf{Ans}$$

***17-64.** The bar has a mass m and length l. If it is released from rest from the position $\theta = 30°$, determine its angular acceleration and the horizontal and vertical components of reaction at the pin O.

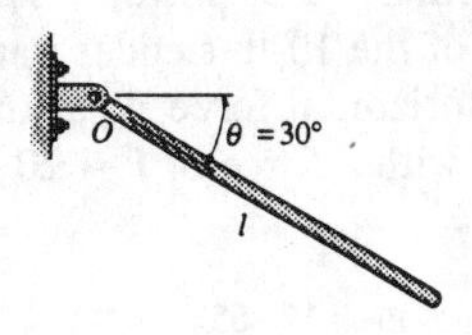

$$\circlearrowleft + \Sigma M_O = I_O\alpha; \quad (mg)\left(\frac{l}{2}\right)\cos 30° = \frac{1}{3}ml^2\,\alpha$$

$$\alpha = \frac{1.299g}{l} = \frac{1.30g}{l} \qquad \textbf{Ans}$$

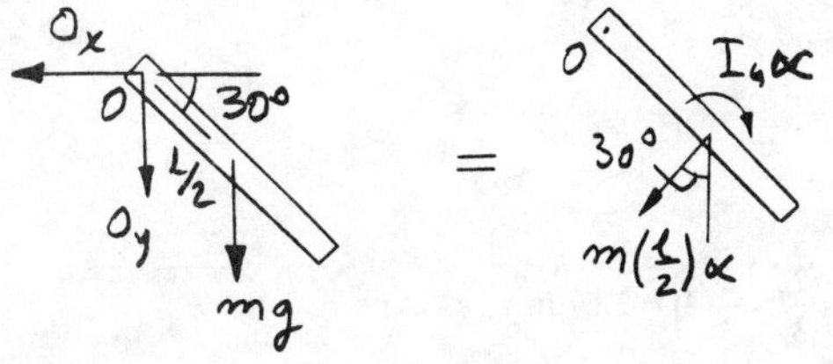

$$\overset{+}{\leftarrow} \Sigma F_x = m(a_G)_x; \quad O_x = m\left(\frac{l}{2}\right)\left(\frac{1.299g}{l}\right)\sin 30°$$

$$O_x = 0.325mg \qquad \textbf{Ans}$$

$$+\uparrow \Sigma F_y = m(a_G)_y; \quad O_y - mg = -m\left(\frac{l}{2}\right)\left(\frac{1.299g}{l}\right)\cos 30°$$

$$O_y = 0.438mg \qquad \textbf{Ans}$$

17-65. The kinetic diagram representing the general rotational motion of a rigid body about a fixed axis at O is shown in the figure. Show that $I_G\boldsymbol{\alpha}$ may be eliminated by moving the vectors $m(\mathbf{a}_G)_t$ and $m(\mathbf{a}_G)_n$ to point P, located a distance $r_{GP} = k_G^2/r_{OG}$ from the center of mass G of the body. Here k_G represents the radius of gyration of the body about G. The point P is called the *center of percussion* of the body.

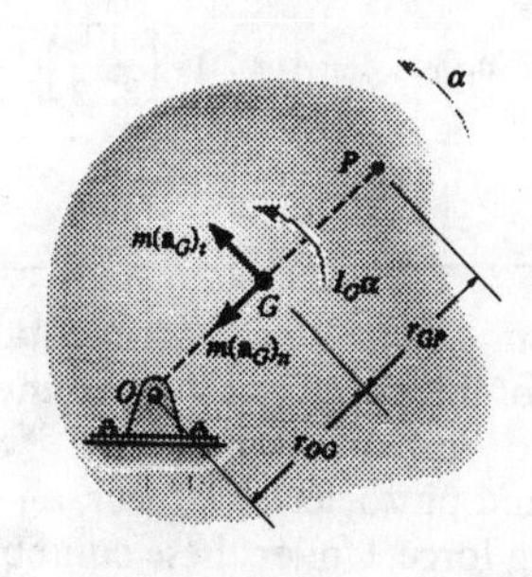

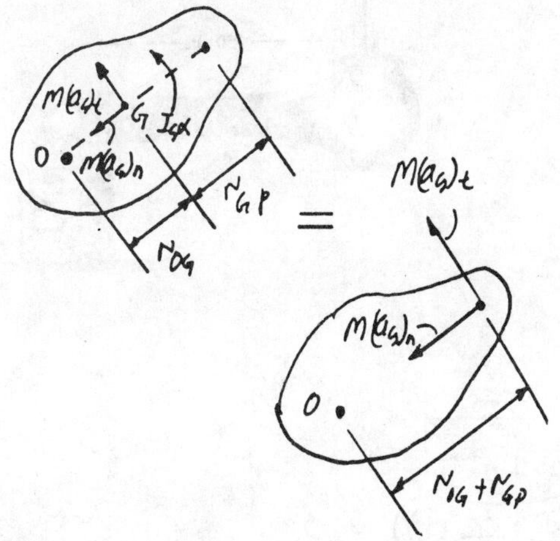

$$m(a_G)_t\, r_{OG} + I_G\alpha = m(a_G)_t\, r_{OG} + \left(mk_G^2\right)\alpha$$

However, $\quad k_G^2 = r_{OG}r_{GP}$ and $\alpha = \dfrac{(a_G)_t}{r_{OG}}$

$$m(a_G)_t\, r_{OG} + I_G\alpha = m(a_G)_t\, r_{OG} + (m r_{OG} r_{GP})\left[\frac{(a_G)_t}{r_{OG}}\right]$$

$$= m(a_G)_t\,(r_{OG} + r_{GP}) \qquad \textbf{Q.E.D.}$$

17-66. Determine the position r_P of the center of percussion P of the 10-lb slender bar. (See Prob. 17-65.) What is the horizontal force A_x at the pin when the bar is struck at P with a force of $F = 20$ lb?

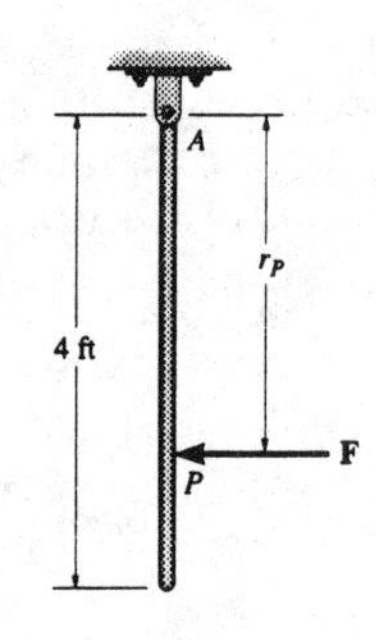

Using the result of Prob 17 - 65

$$r_{GP} = \frac{k_G^2}{r_{AG}} = \frac{\left[\sqrt{\frac{1}{12}\left(\frac{ml^2}{m}\right)}\right]^2}{\frac{l}{2}} = \frac{1}{6}l$$

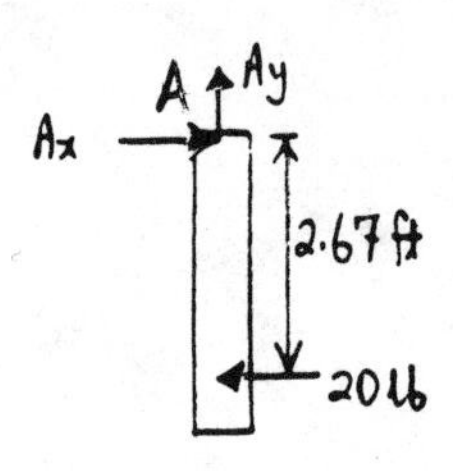

Thus,

$$r_P = \frac{1}{6}l + \frac{1}{2}l = \frac{2}{3}l = \frac{2}{3}(4) = 2.67 \text{ ft} \qquad \textbf{Ans}$$

$$\circlearrowleft + \Sigma M_A = I_A\,\alpha; \qquad 20(2.667) = \left[\frac{1}{3}\left(\frac{10}{32.2}\right)(4)^2\right]\alpha$$

$$\alpha = 32.2 \text{ rad/s}^2$$

$$(a_G)_t = 2(32.2) = 64.4 \text{ ft/s}^2$$

$$\overset{+}{\leftarrow} \Sigma F_x = m(a_G)_t; \qquad -A_x + 20 = \left(\frac{10}{32.2}\right)(64.4)$$

$$A_x = 0 \qquad \textbf{Ans}$$

17-67. In order to experimentally determine the moment of inertia I_G of a 4-kg connecting rod, the rod is suspended horizontally at A by a cord and at B by a bearing and piezoelectric sensor, an instrument used for measuring force. Under these equilibrium conditions, the force at B is measured as 14.6 N. If, at the instant the cord is released, the reaction at B is measured as 9.3 N, determine the value of I_G. The support at B does not move when the measurement is taken. For the calculation, the horizontal location of G must be determined.

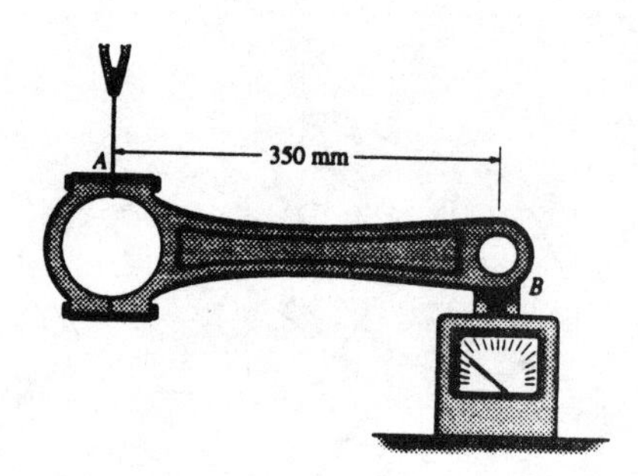

The location of G is :

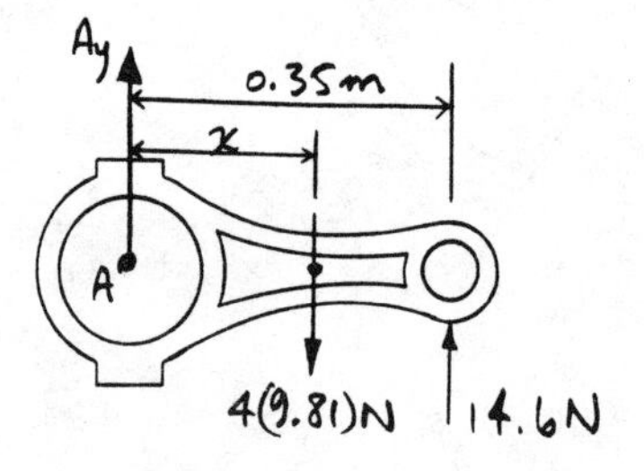

$$\circlearrowleft + \Sigma M_A = 0; \qquad 14.6(0.35) - 4(9.81)(x) = 0$$

$$x = 0.1302 \text{ m}$$

$$+\downarrow \Sigma F_y = m(a_G)_y; \qquad 4(9.81) - 9.3 = 4a_G$$

$$a_G = 7.485 \text{ m/s}^2$$

Since $a_G = (0.350 - 0.1302)\alpha$, $\alpha = 34.06 \text{ rad/s}^2$

$$\circlearrowleft + \Sigma M_G = I_G\alpha; \qquad 9.3(0.350 - 0.1302) = I_G(34.06)$$

$$I_G = 0.0600 \text{ kg}\cdot\text{m}^2 \qquad \textbf{Ans}$$

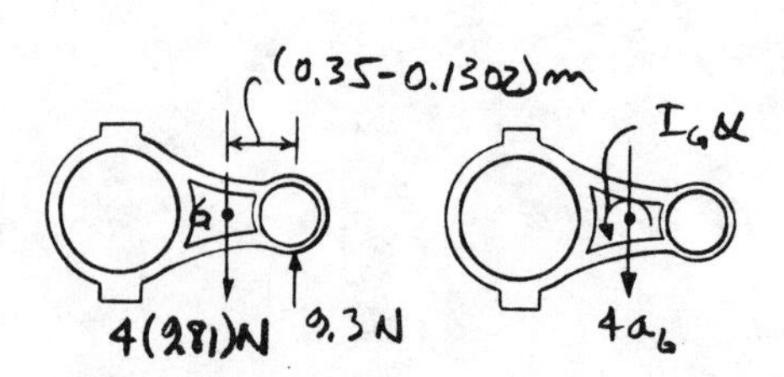

***17-68.** The 4-kg slender rod is supported horizontally by a spring at A and a cord at B. Determine the angular acceleration of the rod and the acceleration of the rod's mass center at the instant the cord at B is cut. *Hint:* The stiffness of the spring is not needed for the calculation.

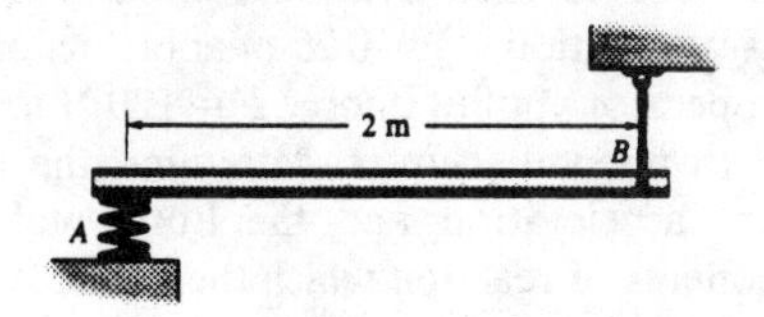

Since the deflection of the spring is unchanged at the instant the cord is cut, the reaction at A is

$$F_A = \frac{4}{2}(9.81) = 19.62 \text{ N}$$

$\overset{+}{\leftarrow} \Sigma F_x = m(a_G)_x; \qquad 0 = 4(a_G)_x$

$+\downarrow \Sigma F_y = m(a_G)_y; \qquad 4(9.81) - 19.62 = 4(a_G)_y$

$\circlearrowleft + \Sigma M_G = I_G\alpha; \qquad (19.62)(1) = \left[\frac{1}{12}(4)(2)^2\right]\alpha$

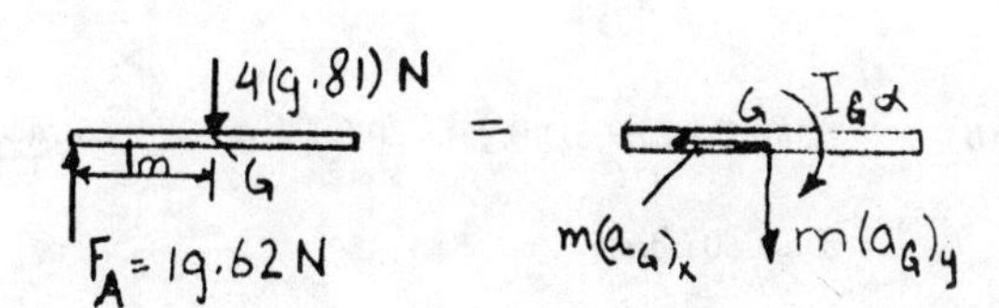

Solving :

$(a_G)_x = 0$

$(a_G)_y = 4.905 \text{ m/s}^2$

$\alpha = 14.7 \text{ rad/s}^2$ **Ans**

Thus,

$(a_G) = 4.90 \text{ m/s}^2$ **Ans**

17-69. The 10-lb disk D is subjected to a counterclockwise moment of $M = (10t)$ lb·ft, where t is in seconds. Determine the angular velocity of the disk 2 s after the moment is applied. Due to the spring the plate P exerts a constant force of 100 lb on the disk. The coefficients of static and kinetic friction between the disk and the plate are $\mu_s = 0.3$ and $\mu_k = 0.2$, respectively. *Hint:* First find the time needed to start the disk rotating.

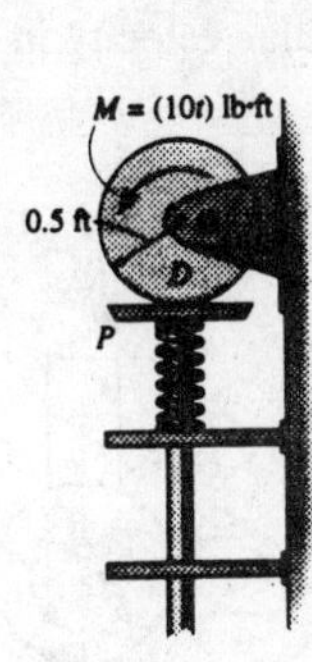

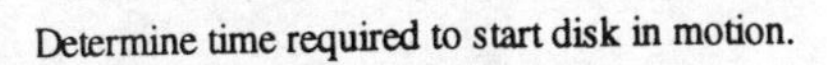

Determine time required to start disk in motion.

$F = 0.3(100) = 30 \text{ lb}$

$\circlearrowleft + \Sigma M_O = 0; \qquad 10t - 30(0.5) = 0$

$$t = 1.5 \text{ s}$$

Thus,

$F = 0.2(100) = 20 \text{ lb}$

$\circlearrowleft + \Sigma M_O = I_O\alpha; \qquad 10t - 20(0.5) = \left[\left(\frac{1}{2}\right)\left(\frac{10}{32.2}\right)(0.5)^2\right]\alpha$

$$\alpha = 257.6(t-1)$$

Since $\alpha = \dfrac{d\omega}{dt}$,

$\int_0^{\omega} d\omega = \int_{1.5}^{2} 257.6(t-1)\, dt$

$\omega = 257.6\left(\dfrac{t^2}{2} - t\right)\Big|_{1.5}^{2}$

$\omega = 96.6 \text{ rad/s}$ **Ans**

17-70. The furnace cover has a mass of 20 kg and a radius of gyration $k_G = 0.25$ m about its mass center G. If an operator applies a force $F = 120$ N to the handle in order to open the cover, determine the cover's initial angular acceleration and the horizontal and vertical components of reaction which the pin at A exerts on the cover at the instant the cover begins to open. Neglect the mass of the handle BAC in the calculation.

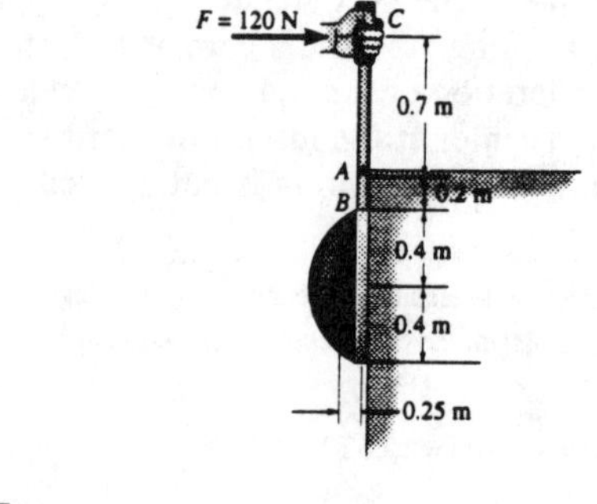

$\overset{+}{\leftarrow} \Sigma F_x = m(a_G)_x\,; \qquad A_x - 120 = 20(0.65)\alpha\cos 22.62°$

$+\uparrow \Sigma F_y = m(a_G)_y\,; \qquad A_y - 20(9.81) = 20(0.65)\alpha\sin 22.62°$

$\circlearrowleft + \Sigma M_A = \Sigma (M_k)_A\,; \qquad 120(0.7) - 20(9.81)(0.25) = 20(0.25)^2\alpha + 20(0.65)\alpha(0.65)$

$\alpha = 3.60 \text{ rad/s}^2$ **Ans**

$A_x = 163$ N **Ans**

$A_y = 214$ N **Ans**

17-71. The variable-reluctance motor is often used for appliances, pumps, and blowers. By applying a current through the stator S, an electromagnetic field is created that "pulls in" the nearest rotor poles. The result of this is to create a torque of 4 N·m about the bearing A. If the rotor is made from iron and has a 3-kg cylindrical core of 50-mm diameter and eight extended slender rods, each having a mass of 1 kg and 100-mm length, determine its angular velocity in 5 seconds starting from rest.

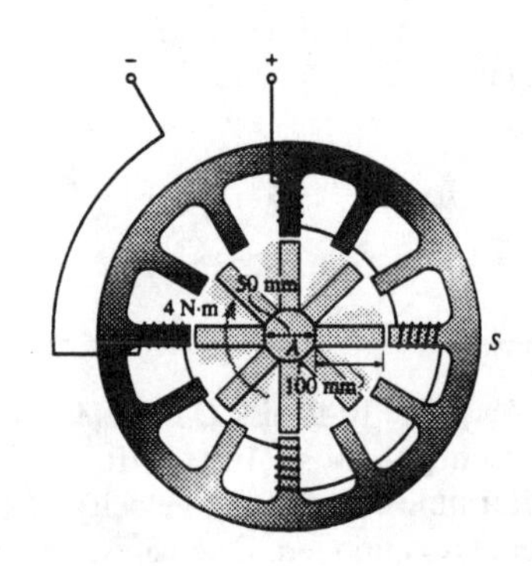

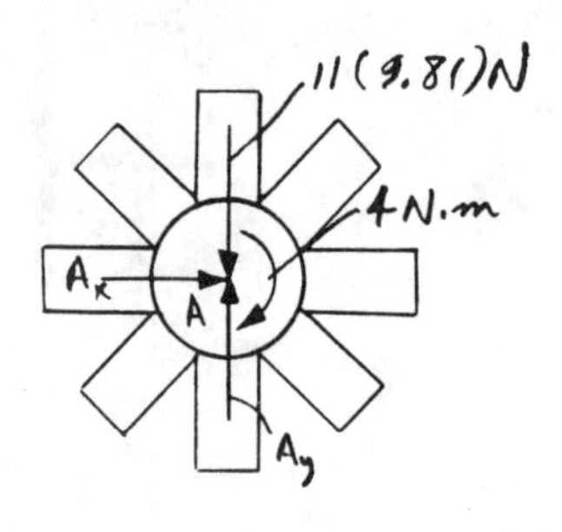

$$I_A = \frac{1}{2}(3)(0.025)^2 + 8\left[\frac{1}{12}(1)(0.1)^2 + (1)(0.075)^2\right] = 0.052604 \text{ kg}\cdot\text{m}^2$$

$\circlearrowleft + \Sigma M_A = I_A\,\alpha; \qquad 4 = 0.052604\alpha$

$\alpha = 76.04 \text{ rad/s}^2$

$\circlearrowleft + \omega = \omega_0 + \alpha_c t$

$\omega = 0 + 76.04(5) = 380$ rad/s **Ans**

***17-72.** The variable-reluctance motor is often used for appliances, pumps, and blowers. By applying a current through the stator S, an electromagnetic field is created that "pulls in" the nearest rotor poles. The result of this is to create a torque of 4 N·m about the bearing A. If the rotor is made from iron and has a 3-kg cylindrical core and eight extended slender rods, each having a mass of 1 kg, determine its angular velocity at the instant the rotor has undergone 15 revolutions, starting from rest.

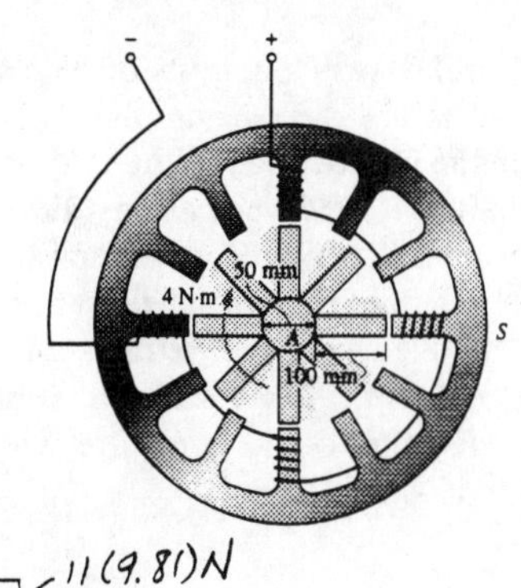

$$I_A = \frac{1}{2}(3)(0.025)^2 + 8\left[\frac{1}{12}(1)(0.1)^2 + (1)(0.075)^2\right] = 0.052604 \text{ kg}\cdot\text{m}^2$$

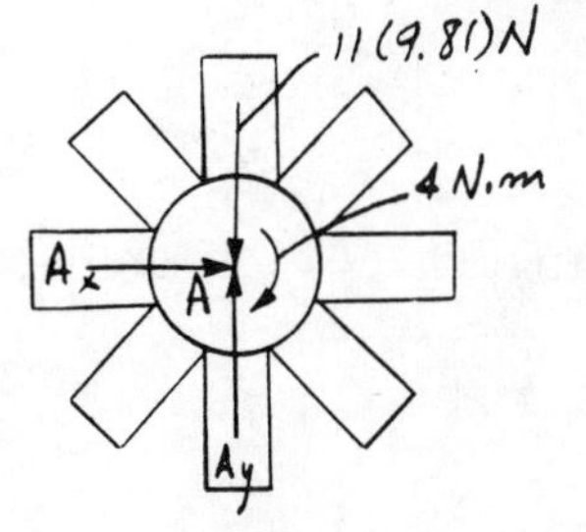

$\curvearrowleft + \Sigma M_A = I_A\,\alpha;\quad 4 = 0.052604\alpha$

$\alpha = 76.04 \text{ rad/s}^2$

$\omega^2 = \omega_0^2 + 2\alpha_C(\theta - \theta_0)$

$\omega^2 = 0 + 2(76.04)(15(2\pi) - 0)$

$\omega = 120$ rad/s **Ans**

17-73. The disk has a mass of 20 kg and is originally spinning at the end of the strut with an angular velocity of $\omega = 60$ rad/s. If it is then placed against the wall, for which the coefficient of kinetic friction is $\mu_k = 0.3$, determine the time required for the motion to stop. What is the force in strut BC during this time?

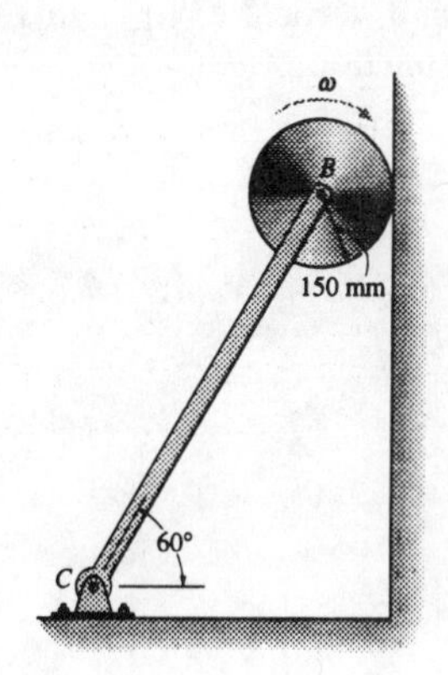

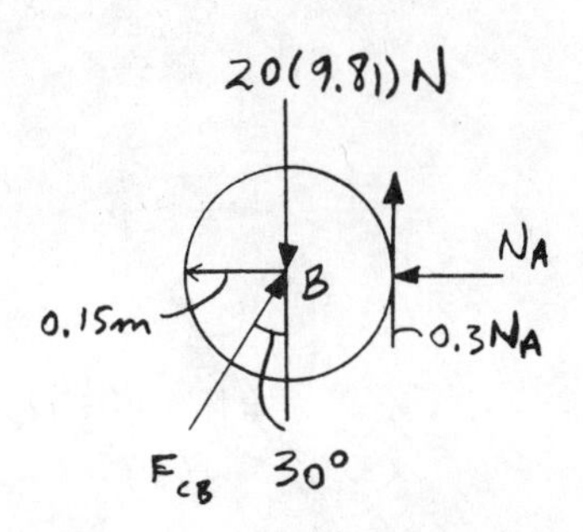

$\overset{+}{\rightarrow} \Sigma F_x = m(a_G)_x;\quad F_{CB}\sin 30° - N_A = 0$

$+\uparrow \Sigma F_y = m(a_G)_y;\quad F_{CB}\cos 30° - 20(9.81) + 0.3N_A = 0$

$\curvearrowleft + \Sigma M_B = I_B\,\alpha;\quad 0.3N_A(0.15) = \left[\frac{1}{2}(20)(0.15)^2\right]\alpha$

$N_A = 96.6$ N

$F_{CB} = 193$ N **Ans**

$\alpha = 19.3 \text{ rad/s}^2$

$\curvearrowleft + \omega = \omega_0 + \alpha_c t$

$0 = 60 + (-19.3)t$

$t = 3.11$ s **Ans**

17-74. The relay switch consists of an electromagnet E and a 20-g armature AB (slender bar) which is pinned at A and lies in the vertical plane. When the current is turned off, the armature is held open against the smooth stop at B by the spring CD, which exerts an upward vertical force $F_s = 0.85$ N on the armature at C. When the current is turned on, the electromagnet attracts the armature at E with a vertical force $F = 0.8$ N. Determine the initial angular acceleration of the armature when the contact BF begins to close.

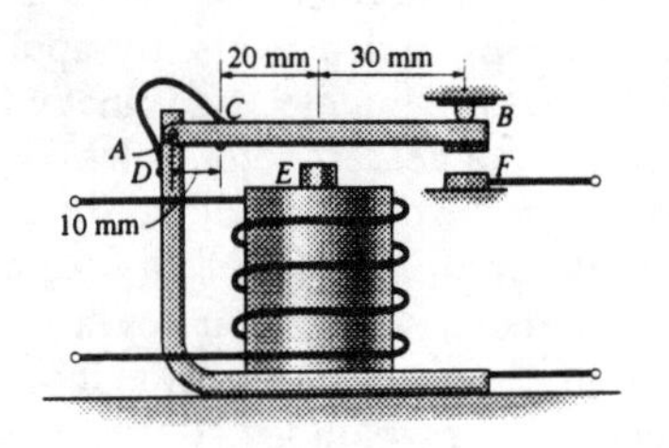

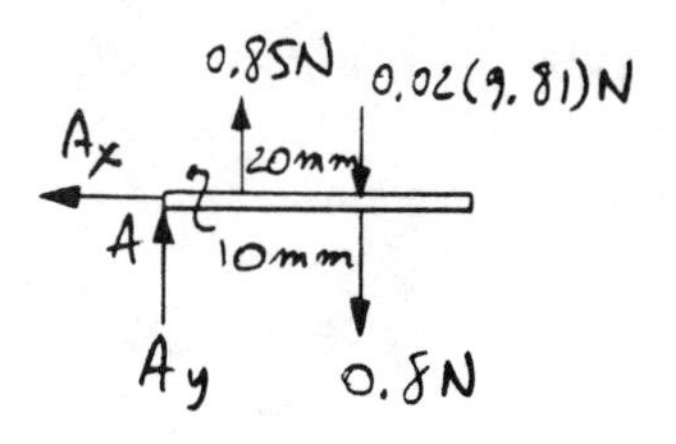

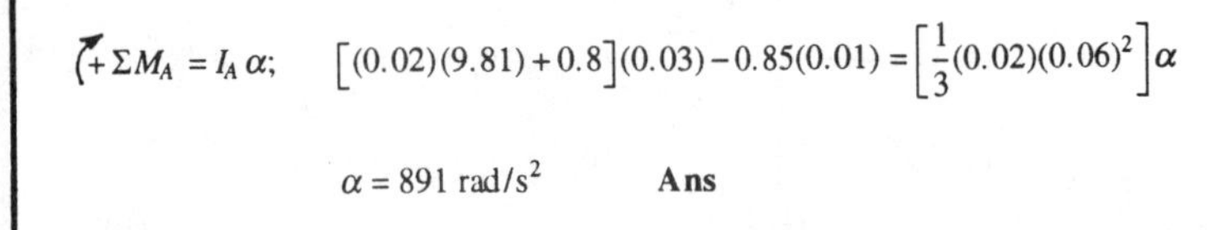

$$\curvearrowleft + \Sigma M_A = I_A\,\alpha; \quad \left[(0.02)(9.81)+0.8\right](0.03)-0.85(0.01)=\left[\frac{1}{3}(0.02)(0.06)^2\right]\alpha$$

$\alpha = 891\ \text{rad/s}^2$ **Ans**

17-75. The rod has a length L and mass m. If it is released from rest when $\theta \approx 0°$, determine its angular velocity as a function of θ. Also, express the horizontal and vertical components of reaction at the pin O as a function of θ.

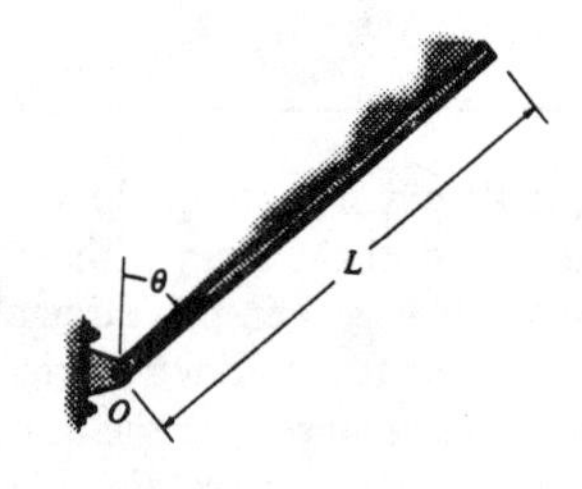

$$\curvearrowleft + \Sigma M_O = \Sigma (M_k)_O; \quad mg\left(\frac{L}{2}\right)\sin\theta = m\left(\frac{L}{2}\right)(\alpha)\left(\frac{L}{2}\right)+\left(\frac{1}{12}mL^2\right)\alpha$$

$$mg\left(\frac{L}{2}\right)\sin\theta = \left(\frac{1}{3}mL^2\right)\alpha$$

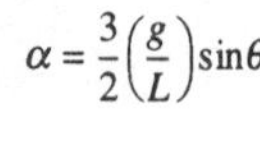

$$\alpha = \frac{3}{2}\left(\frac{g}{L}\right)\sin\theta$$

$$\alpha\, d\theta = \omega\, d\omega$$

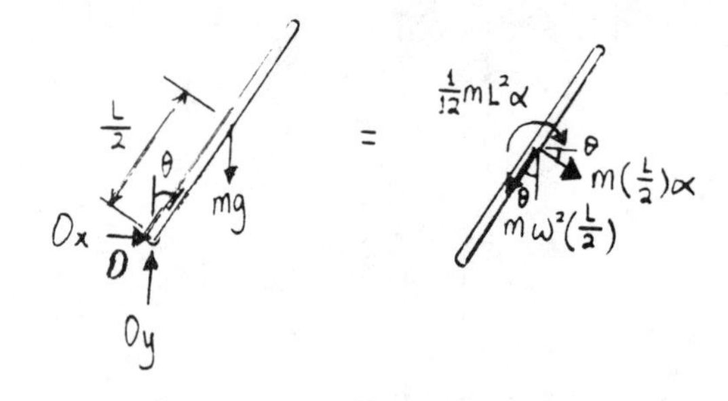

$$\int_0^\theta \frac{3}{2}\left(\frac{g}{L}\right)\sin\theta\, d\theta = \int_0^\omega \omega\, d\omega$$

$$-\left(\frac{3}{2}\right)\left(\frac{g}{L}\right)\cos\theta\Big|_0^\theta = \frac{1}{2}\omega^2$$

$$\omega = \sqrt{\frac{3g}{L}(1-\cos\theta)}$$ **Ans**

$$+\uparrow \Sigma F_y = m(a_G)_y; \quad O_y - mg = -m\left(\frac{L}{2}\right)\left(\frac{3g}{L}\right)(1-\cos\theta)\cos\theta - m\left(\frac{L}{2}\right)\left(\frac{3}{2}\right)\left(\frac{g}{L}\right)\sin\theta(\sin\theta)$$

$$O_y = mg\left[1-\frac{3}{2}(1-\cos\theta)\cos\theta-\frac{3}{4}\sin^2\theta\right]$$

$$O_y = mg\left(1-1.5\cos\theta+1.5\cos^2\theta-0.75\sin^2\theta\right)$$ **Ans**

$$\overset{+}{\rightarrow}\Sigma F_x = m(a_G)_x; \quad O_x = m\left(\frac{L}{2}\right)\left(\frac{3}{2}\right)\left(\frac{g}{L}\sin\theta\right)\cos\theta - m\left(\frac{L}{2}\right)\left(\frac{3g}{L}\right)(1-\cos\theta)\sin\theta$$

$$O_x = mg(0.75\sin\theta\cos\theta - 1.5\sin\theta + 1.5\sin\theta\cos\theta)$$

$$O_x = mg\sin\theta(2.25\cos\theta - 1.5)$$ **Ans**

***17-76.** The lightweight turbine consists of a rotor which is powered from a torque appled at its center. At the instant the rotor is horizontal it has an angular velocity of 15 rad/s and an angular acceleration of 8 rad/s². Determine the internal normal force, shear force, and moment at a section through A. Assume the rotor is a 50-m-long slender rod, having a mass of 3 kg/m.

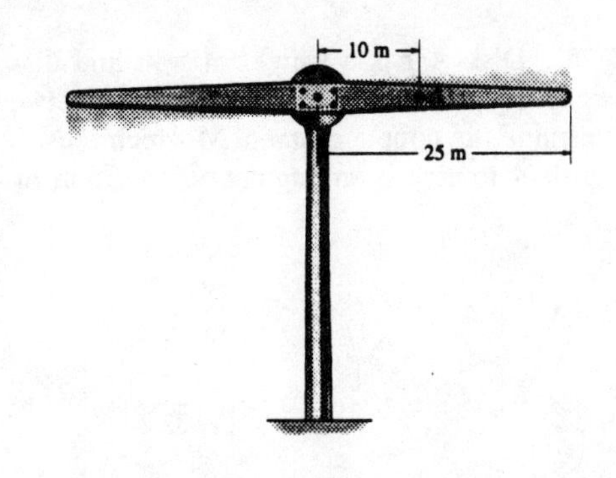

$\overset{+}{\leftarrow}\Sigma F_n = m(a_G)_n;\qquad N_A = 45(15)^2(17.5) = 177\text{ kN}$ **Ans**

$+\downarrow\Sigma F_t = m(a_G)_t;\qquad V_A + 45(9.81) = 45(8)(17.5)$

$V_A = 5.86\text{ kN}$ **Ans**

$\circlearrowleft+\Sigma M_A = \Sigma(M_k)_A;\qquad M_A + 45(9.81)(7.5) = \left[\frac{1}{12}(45)(15)^2\right](8) + [45(8)(17.5)](7.5)$ **Ans**

$M_A = 50.7\text{ kN}\cdot\text{m}$ **Ans**

17-77. The bar has a weight per length of w. If it is rotating in the vertical plane at a constant rate ω about point O, determine the internal normal force, shear force, and moment as a function of x and θ.

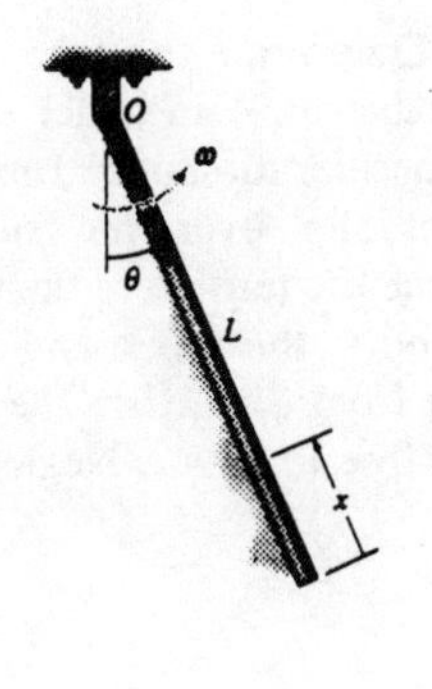

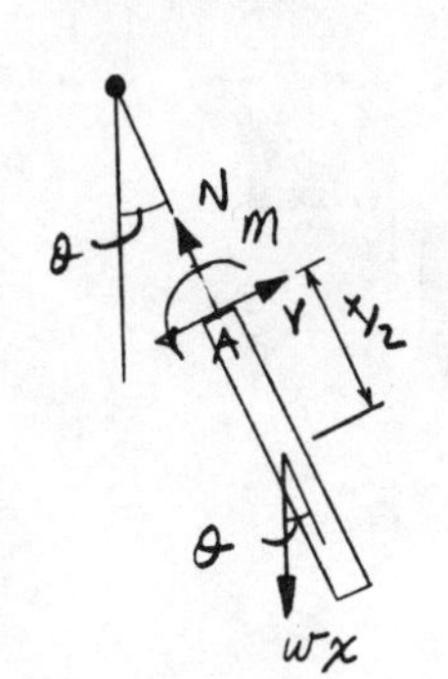

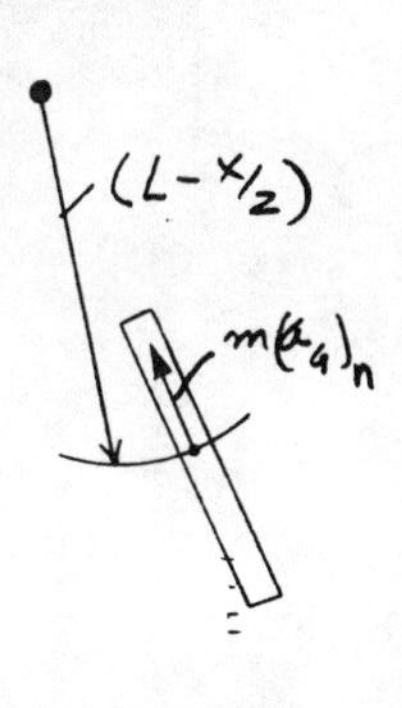

$\nwarrow\Sigma F_n = m(a_G)_n;\qquad N - wx\cos\theta = \frac{wx}{g}(\omega)^2\left(L - \frac{x}{2}\right)$

$N = wx\left[\cos\theta + \frac{\omega^2}{2g}(2L - x)\right]$ **Ans**

$\nearrow\Sigma F_t = m(a_G)_t;\qquad wx\sin\theta - V = 0$

$V = wx\sin\theta$ **Ans**

$\circlearrowleft+\Sigma M_A = I_A\,\alpha;\qquad M - (wx\sin\theta)\left(\frac{x}{2}\right) = 0$

$M = \frac{wx^2}{2}\sin\theta$ **Ans**

17-78. Disk A has a weight of 5 lb and disk B has a weight of 10 lb. If no slipping occurs between them, determine the couple moment **M** which must be applied to disk A to give it an angular acceleration of 4 rad/s^2.

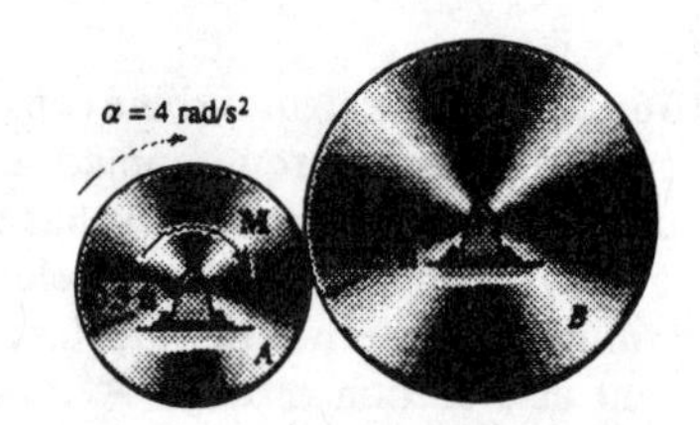

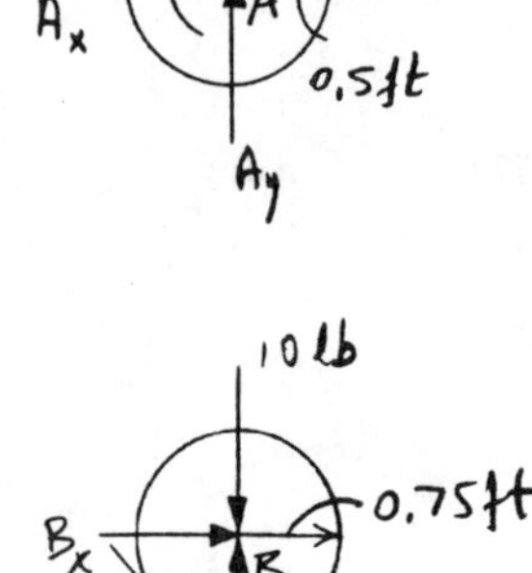

Disk A :

$$\curvearrowleft + \Sigma M_A = I_A\,\alpha_A; \qquad M - F_D(0.5) = \left[\frac{1}{2}\left(\frac{5}{32.2}\right)(0.5)^2\right](4)$$

Disk B :

$$+\Sigma M_B = I_B\,\alpha_B; \qquad F_D(0.75) = \left[\frac{1}{2}\left(\frac{10}{32.2}\right)(0.75)^2\right]\alpha_B$$

$$r_A\,\alpha_A = r_B\,\alpha_B$$

$$0.5(4) = 0.75\alpha_B$$

Solving :

$\alpha_B = 2.67$ rad/s^2; $\quad F_D = 0.311$ lb

$M = 0.233$ lb · ft **Ans**

17-79. Determine the force $\mathbf{T}_A$ which must be applied to the cable at A in order to give the 10-kg block an upward acceleration of 200 mm/s^2. Assume that the cable does not slip over the surface of the 20-kg disk. Determine the tension in the vertical segment of the cord that supports the block and explain why this tension is different from that at A. The disk is pinned at its center C and is free to rotate. Neglect the mass of the cable.

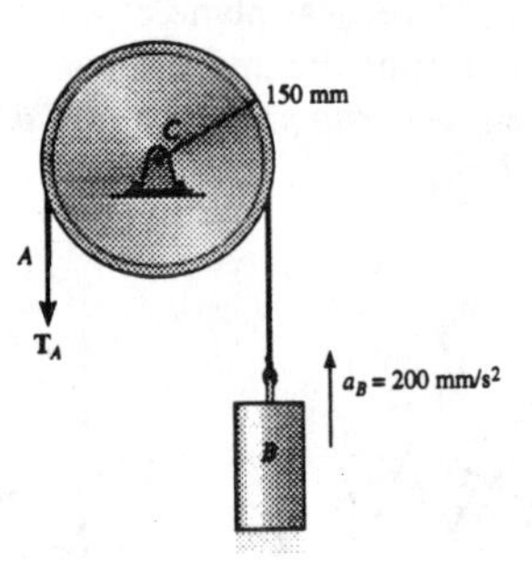

Block :

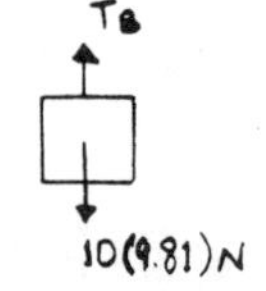

$$+\uparrow \Sigma F_y = m(a_G)_y; \qquad T_B - 10(9.81) = 10(0.2)$$

$$T_B = 100.1 = 100 \text{ N} \qquad \textbf{Ans}$$

Disk :

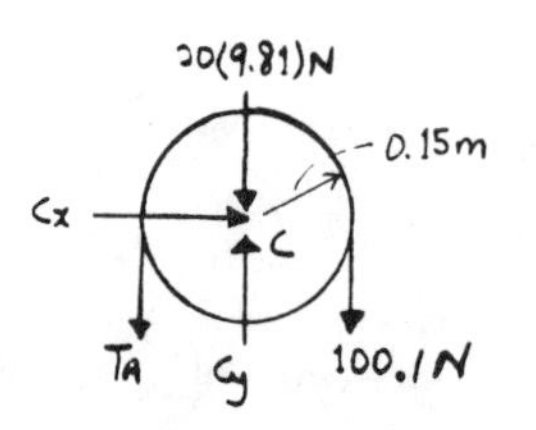

$$\curvearrowleft + \Sigma M_C = I_C\,\alpha; \qquad T_A(0.15) - 100.1(0.15) = \left[\frac{1}{2}(20)(0.15)^2\right]\alpha$$

Since $\alpha = \dfrac{a}{r} = \dfrac{0.2}{0.15} = 1.33$ rad/s^2, then

$T_A = 102$ N **Ans**

The difference in tension is applied as a frictional force along the periphery of the disk in order to overcome its inertia and give it an angular acceleration α.

***17-80.** The cord is wrapped around the inner core of the spool. If a 5-lb block B is suspended from the cord and released from rest, determine the spool's angular velocity when $t = 3$ s. Neglect the mass of the cord. The spool has a weight of 180 lb and the radius of gyration about the axle A is $k_A = 1.25$ ft. Solve the problem in two ways, first by considering the "system" consisting of the block and spool, and then by considering the block and spool separately.

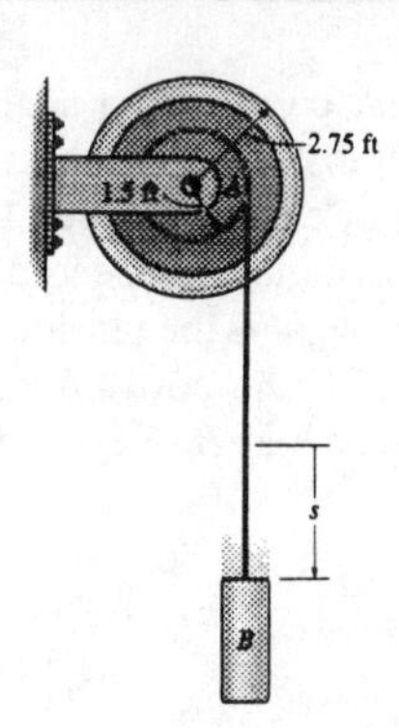

System :

$$\circlearrowleft + \Sigma M_A = \Sigma (M_k)_A ; \quad 5(1.5) = \left(\frac{180}{32.2}\right)(1.25)^2 \alpha + \left(\frac{5}{32.2}\right)(1.5\alpha)(1.5)$$

$$\alpha = 0.8256 \text{ rad/s}^2$$

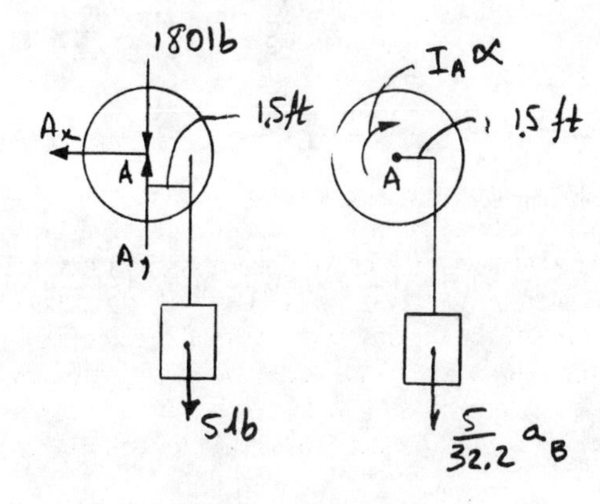

$(\circlearrowleft +) \quad \omega = \omega_0 + \alpha_c t$

$\omega = 0 + (0.8256)(3)$

$\omega = 2.48$ rad/s **Ans**

Also,

Spool :

$$\circlearrowleft + \Sigma M_A = I_A \alpha; \quad T(1.5) = \left(\frac{180}{32.2}\right)(1.25)^2 \alpha$$

Weight :

$$+\downarrow \Sigma F_y = m(a_G)_y ; \quad 5 - T = \left(\frac{5}{32.2}\right)(1.5\alpha)$$

$$\alpha = 0.8256 \text{ rad/s}^2$$

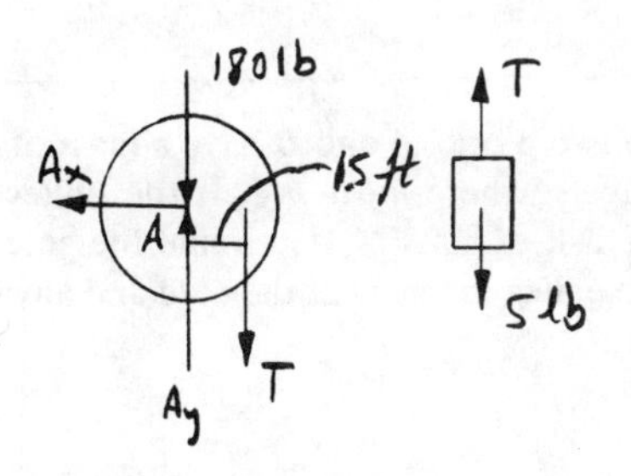

$(\circlearrowleft +) \quad \omega = \omega_0 + \alpha_c t$

$\omega = 0 + (0.8256)(3)$

$\omega = 2.48$ rad/s **Ans**

17-81. A cord having negligible mass is wrapped over the 15-lb disk at A and passes over the 5-lb disk at B. If a 3-lb block C is attached to its end and released from rest, determine the speed of the block after it descends 3 ft. Also, what is the tension in the horizontal and vertical segments of the cord? Assume no slipping of the cord over the disk at B. Neglect friction at the pins D and E.

Disk A :

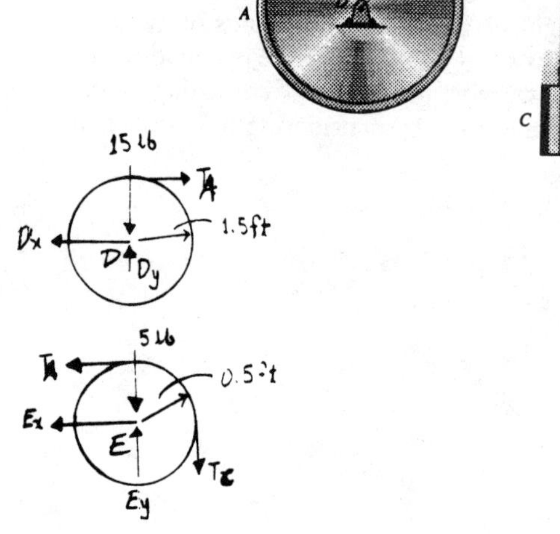

$$(\circlearrowleft + \Sigma M_D = I_D \alpha_A; \quad T_A(1.5) = \left[\frac{1}{2}\left(\frac{15}{32.2}\right)(1.5)^2\right]\alpha_A \quad (1)$$

Disk B :

$$(\circlearrowleft + \Sigma M_E = I_E \alpha_B; \quad T_C(0.5) - T_A(0.5) = \left[\frac{1}{2}\left(\frac{5}{32.2}\right)(0.5)^2\right]\alpha_B$$

Block C :

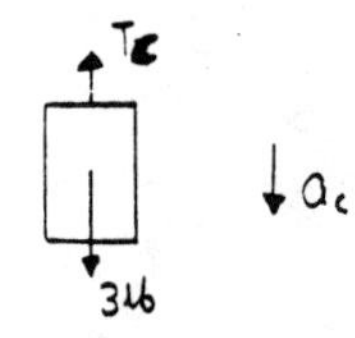

$$+\downarrow \Sigma F_y = m(a_G)_y; \quad 3 - T_C = \left(\frac{3}{32.2}\right)a_C \quad (3)$$

$$\alpha_B = \frac{a_C}{0.5} \quad (4)$$

$$\alpha_A = \frac{a_C}{1.5} \quad (5)$$

Substituting Eqs. 1, 3, 4 and 5 into Eq. 2,

$$\left(3 - \frac{3}{32.2}a_C\right)(0.5) - \frac{1}{2}\left(\frac{15}{32.2}\right)(1.5)\left(\frac{a_C}{1.5}\right)(0.5) = \left[\frac{1}{2}\left(\frac{5}{32.2}\right)(0.5)^2\right]\left(\frac{a_C}{0.5}\right)$$

$a_C = 7.43$ ft/s^2

Thus

$T_A = 1.73$ lb **Ans**

$T_C = 2.31$ lb **Ans**

$$(+\downarrow) \quad v^2 = v_0^2 + 2a_c(s - s_0)$$

$$v_C^2 = 0 + 2(7.43)(3 - 0)$$

$v_C = 6.68$ ft/s **Ans**

17-82. The two blocks A and B have a mass of m_A and m_B, respectively, where $m_B > m_A$. If the pulley can be treated as a disk of mass M, determine the acceleration of block A. Neglect the mass of the cord and any slipping on the pulley.

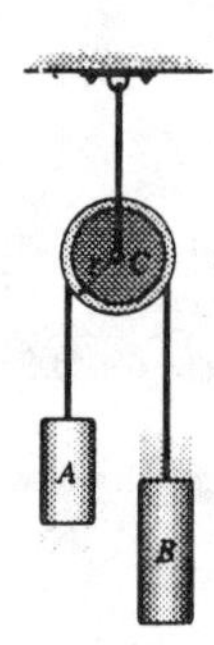

$a = \alpha r$

$$(\circlearrowleft + \Sigma M_C = \Sigma(M_k)_C; \quad m_B g(r) - m_A g(r) = \left(\frac{1}{2}Mr^2\right)\alpha + m_B r^2 \alpha + m_A r^2 \alpha$$

$$\alpha = \frac{g(m_B - m_A)}{r\left(\frac{1}{2}M + m_B + m_A\right)}$$

$$a = \frac{g(m_B - m_A)}{\left(\frac{1}{2}M + m_B + m_A\right)} \quad \textbf{Ans}$$

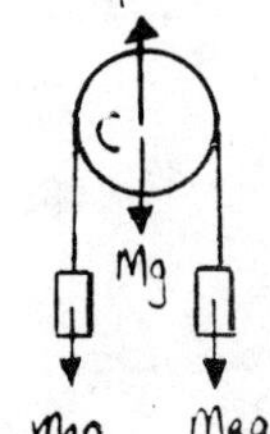

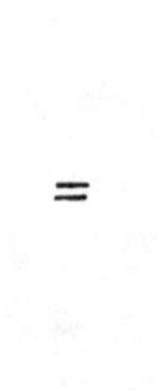

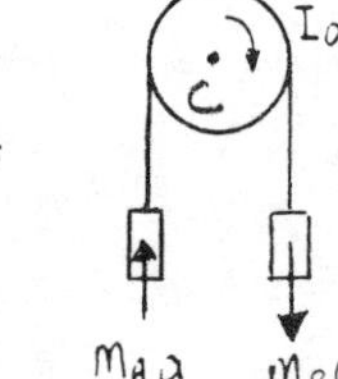

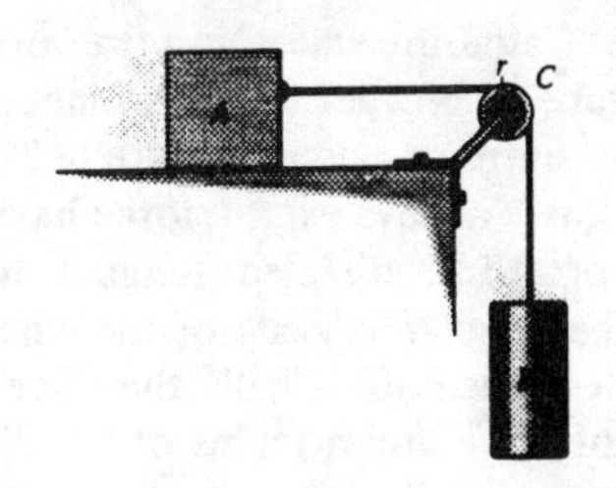

17-83. Block A has a mass m and rests on a surface having a coefficient of kinetic friction μ_k. The cord attached to A passes over a pulley at C and is attached to a block B having a mass $2m$. If B is released, determine the acceleration of A. Assume that the cord does not slip over the pulley. The pulley can be approximated as a thin disk of radius r and mass $\frac{1}{4}m$. Neglect the mass of the cord.

Block A :

$$\overset{+}{\rightarrow} \Sigma F_x = ma_x; \qquad T_1 - \mu_k mg = ma \qquad (1)$$

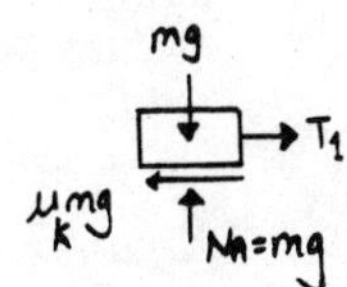

Block B :

$$+\downarrow \Sigma F_y = ma_y; \qquad 2mg - T_2 = 2ma \qquad (2)$$

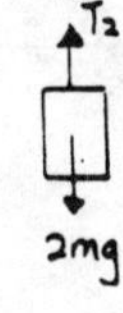

Pulley C :

$$\circlearrowleft + \Sigma M_C = I_G\alpha; \qquad T_2 r - T_1 r = \left[\frac{1}{2}\left(\frac{1}{4}m\right)r^2\right]\left(\frac{a}{r}\right)$$

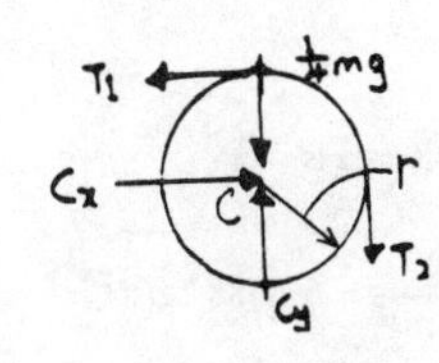

$$T_2 - T_1 = \frac{1}{8}ma \qquad (3)$$

Substituting Eqs. (1) and (2) into (3),

$$2mg - 2ma - (ma + \mu_k mg) = \frac{1}{8}ma$$

$$2mg - \mu_k mg = \frac{1}{8}ma + 3ma$$

$$(2 - \mu_k)g = \frac{25}{8}a$$

$$a = \frac{8}{25}(2 - \mu_k)g \qquad \textbf{Ans}$$

***17-84.** The slender rod of mass m is released from rest when $\theta = 45°$. At the same instant ball B having the same mass m is released. Will B or the end A of the rod have the greatest speed when they pass the horizontal ($\theta = 0°$)? What is the difference in their speeds?

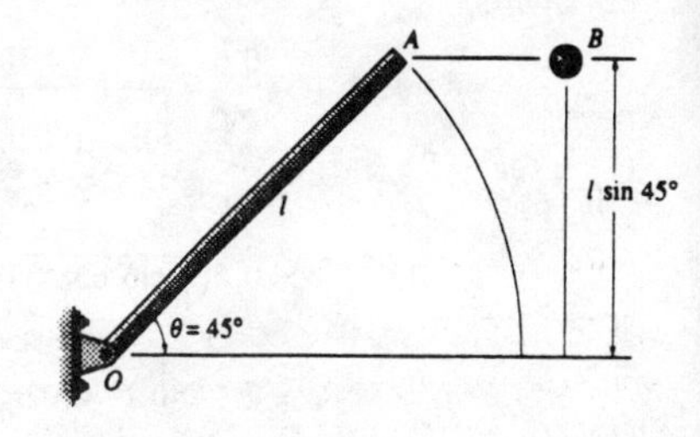

Rod :

$$\circlearrowleft + \Sigma M_O = \Sigma (M_k)_O; \qquad mg\left(\frac{l}{2}\right)\cos\theta = -\frac{1}{12}(m)(l)^2\alpha - m\left(\frac{l}{2}\right)\alpha\left(\frac{l}{2}\right)$$

$$mg\left(\frac{l}{2}\right)\cos\theta = -\frac{1}{3}(m)(l)^2\alpha$$

$$-\frac{3g}{2l}\cos\theta = \alpha$$

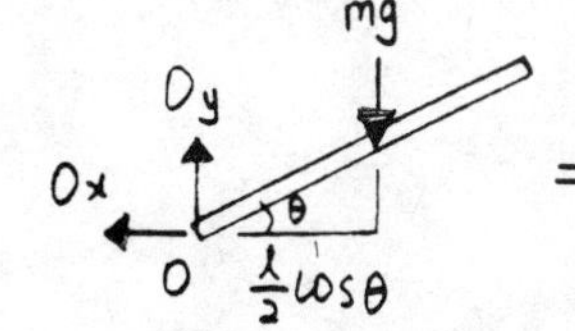

=

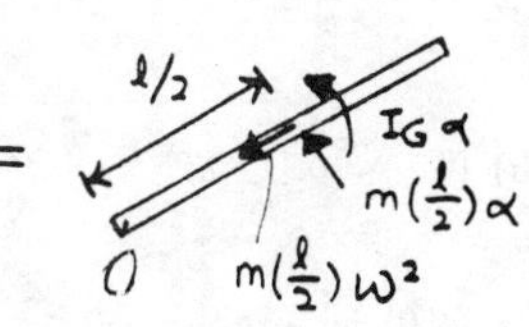

$$\omega\, d\omega = \alpha\, d\theta$$

$$\int_0^{\omega} \omega\, d\omega = -\int_{\frac{\pi}{4}}^{0} \frac{3g}{2l}\cos\theta\, d\theta$$

$$\frac{1}{2}\omega^2 = -\left(\frac{3g}{2l}\right)\sin\theta\Big|_{\frac{\pi}{4}}^{0}$$

$$\omega = \sqrt{\frac{3g}{l}\left(\frac{1}{\sqrt{2}}\right)}$$

$$v_A = \sqrt{\frac{3g}{l\sqrt{2}}}\,(l) = \frac{\sqrt{3gl}}{1.1892}$$

$$v_A = 1.46\sqrt{gl}$$

Ball :

$$(+\downarrow) \qquad v^2 = v_0^2 + 2a_c(s - s_0)$$

$$v_B^2 = 0 + 2g(l\sin 45° - 0)$$

$$v_B = 1.19\sqrt{gl}$$

Point A has the greatest speed.

$$\Delta v = 0.267\sqrt{gl} \qquad \textbf{Ans}$$

17-85. The "Catherine wheel" is a firework that consists of a coiled tube of powder which is pinned at its center. If the powder burns at a constant rate of 20 g/s such that the exhaust gases always exert a force having a constant magnitude of 0.3 N, directed tangent to the wheel, determine the angular velocity of the wheel when 75% of the mass is burned off. Initially, the wheel is at rest and has a mass of 100 g and a radius of $r = 75$ mm. For the calculation, consider the wheel to always be a thin disk.

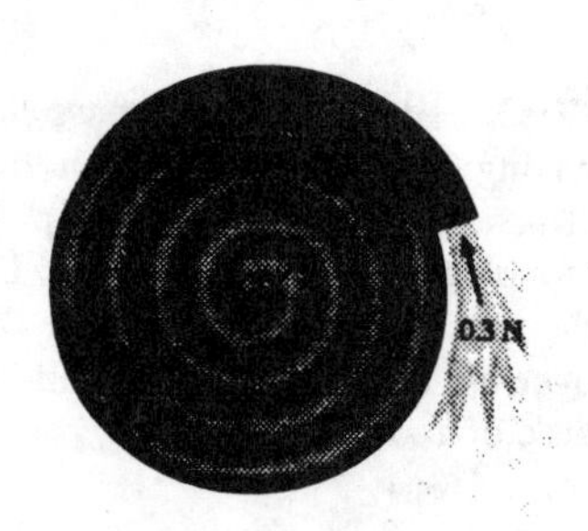

Mass of wheel when 75% of the powder is burned = 0.025 kg

Time to burn off 75% $= \dfrac{0.075 \text{ kg}}{0.02 \text{ kg/s}} = 3.75$ s

$m(t) = 0.1 - 0.02t$

Mass of disk per unit area is

$$\rho_0 = \frac{m}{A} = \frac{0.1 \text{ kg}}{\pi(0.075 \text{ m})^2} = 5.6588 \text{ kg/m}^2$$

At any time t,

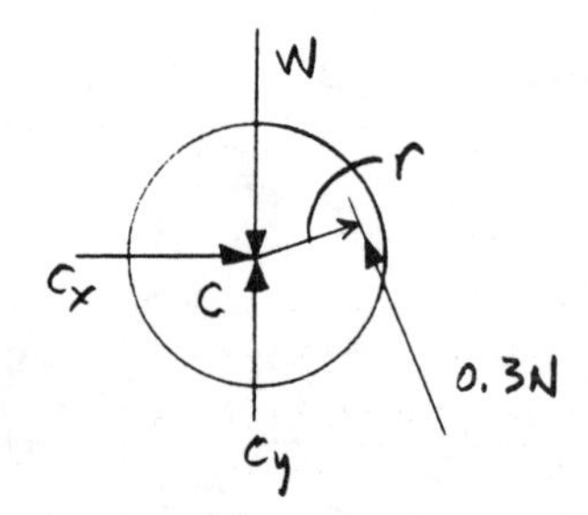

$$5.6588 = \frac{0.1 - 0.02t}{\pi r^2}$$

$$r(t) = \sqrt{\frac{0.1 - 0.02t}{\pi(5.6588)}}$$

$$+\Sigma M_C = I_C\alpha; \quad 0.3r = \frac{1}{2}mr^2\alpha$$

$$\alpha = \frac{0.6}{mr} = \frac{0.6}{(0.1-0.02t)\sqrt{\dfrac{0.1-0.02t}{\pi(5.6588)}}}$$

$$\alpha = 0.6\left(\sqrt{\pi(5.6588)}\right)[0.1 - 0.02t]^{-\frac{3}{2}}$$

$$\alpha = 2.530[0.1 - 0.02t]^{-\frac{3}{2}}$$

$d\omega = \alpha\, dt$

$$\int_0^\omega d\omega = 2.530\int_0^t [0.1 - 0.02t]^{-\frac{3}{2}}\, dt$$

$$\omega = 253\left[(0.1 - 0.02t)^{-\frac{1}{2}} - 3.162\right]$$

For $t = 3.75$ s,

$\omega = 800$ rad/s **Ans**

17-86. The drum has a weight of 50 lb and a radius of gyration $k_A = 0.4$ ft. A 35-ft-long chain having a weight of 2 lb/ft is wrapped around the outer surface of the drum so that a chain length of $s = 3$ ft is suspended as shown. If the drum is originally at rest, determine its angular velocity after the end B has descended $s = 13$ ft. Neglect the thickness of the chain.

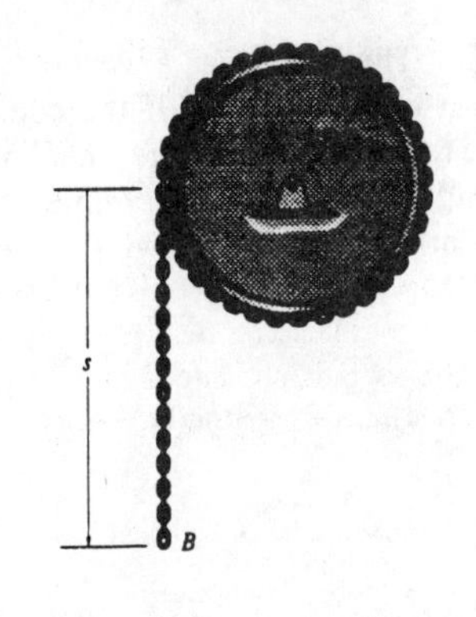

$$\curvearrowleft + \Sigma M_A = \Sigma (M_k)_A; \quad 2s(0.6) = \left(\frac{2s}{32.2}\right)[(\alpha)(0.6)](0.6) + \left[\left(\frac{50}{32.2}\right)(0.4)^2 + \frac{2(35-s)}{32.2}(0.6)^2\right]\alpha$$

$$1.2s = 0.02236s\alpha + (0.24845 + 0.7826 - 0.02236s)\alpha$$

$$1.164s = \alpha$$

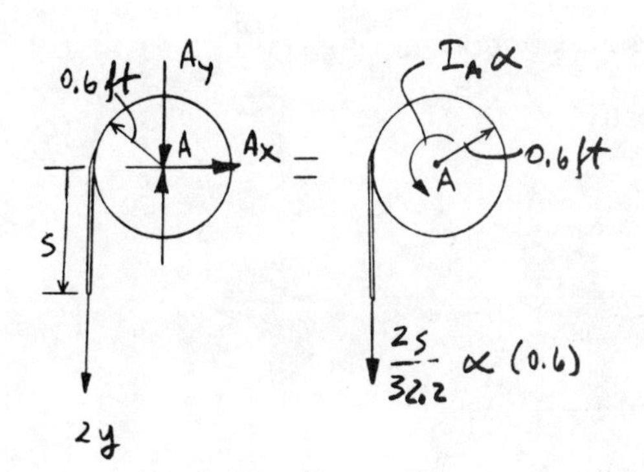

$$\alpha\, d\theta = \alpha\left(\frac{ds}{0.6}\right) = \omega\, d\omega$$

$$1.164s\left(\frac{ds}{0.6}\right) = \omega\, d\omega$$

$$1.9398\int_3^{13} s\, ds = \int_0^{\omega} \omega\, d\omega$$

$$1.9398\left[\frac{(13)^2}{2} - \frac{(3)^2}{2}\right] = \frac{1}{2}\omega^2$$

$\omega = 17.6$ rad/s **Ans**

17-87. If the disk in Fig. 17-21*a* *rolls without slipping,* show that when moments are summed about the instantaneous center of zero velocity, *IC*, it is possible to use the moment equation $\Sigma M_{IC} = I_{IC}\alpha$, where I_{IC} represents the moment of inertia of the disk calculated about the instantaneous axis of zero velocity.

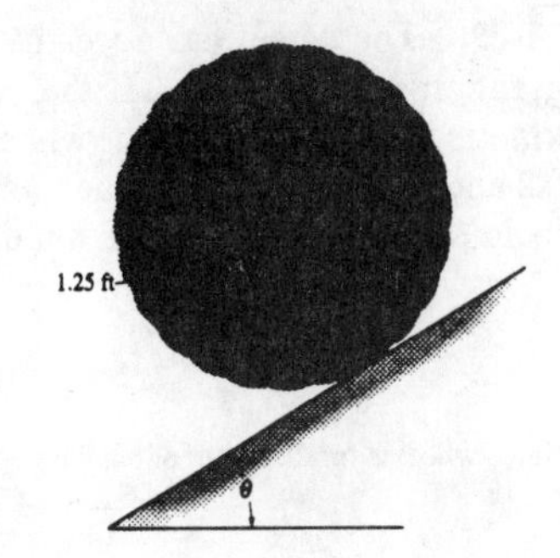

$$\curvearrowleft + \Sigma M_{IC} = \Sigma (M_k)_{IC}; \quad \Sigma M_{IC} = I_G\alpha + (ma_G)r$$

Since there is no slipping, $a_G = \alpha r$

Thus, $\Sigma M_{IC} = \left(I_G + mr^2\right)\alpha$

By the parallel – axis thoerem, the term in parenthesis represents I_{IC}. Thus,

$\Sigma M_{IC} = I_{IC}\alpha$ **Q.E.D.**

***17-88.** The wheel has a weight of 30 lb and a radius of gyration of $k_G = 0.6$ ft. If the coefficients of static and kinetic friction between the wheel and the plane are $\mu_s = 0.2$ and $\mu_k = 0.15$, determine the wheel's angular acceleration as it rolls down the incline. Set $\theta = 12°$.

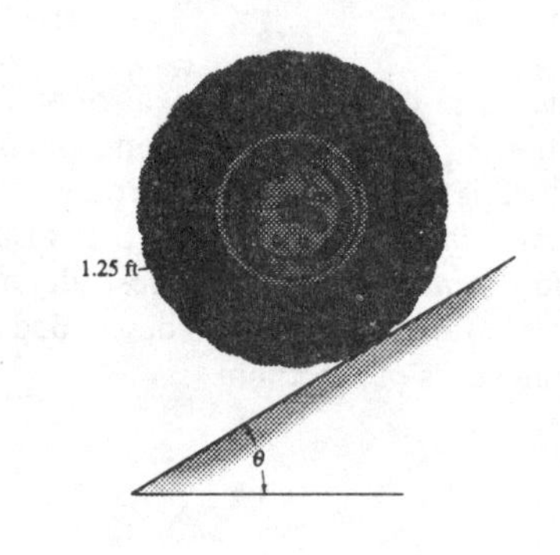

$+\swarrow \Sigma F_x = m(a_G)_x$; $\quad 30\sin 12° - F = \left(\frac{30}{32.2}\right)a_G$

$+\nwarrow \Sigma F_y = m(a_G)_y$; $\quad N - 30\cos 12° = 0$

$\circlearrowleft + \Sigma M_G = I_G\alpha$; $\quad F(1.25) = \left[\left(\frac{30}{32.2}\right)(0.6)^2\right]\alpha$

Assume the wheel does not slip.

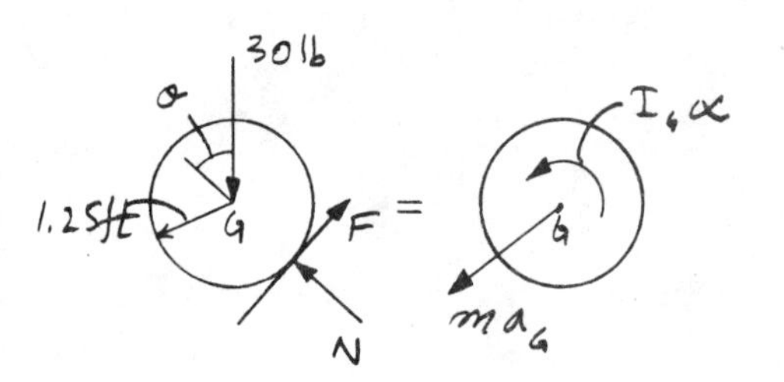

$a_G = (1.25)\alpha$

Solving,

$F = 1.17$ lb

$N = 29.34$ lb

$a_G = 5.44$ ft/s^2

$\alpha = 4.35$ rad/s^2 **Ans**

$F_{max} = 0.2(29.34) = 5.87 \text{ lb} > 1.17 \text{ lb}$ OK

17-89. The wheel has a weight of 30 lb and a radius of gyration of $k_G = 0.6$ ft. If the coefficients of static and kinetic friction between the wheel and the plane are $\mu_s = 0.2$ and $\mu_k = 0.15$, determine the maximum angle θ of the inclined plane so that the wheel rolls without slipping.

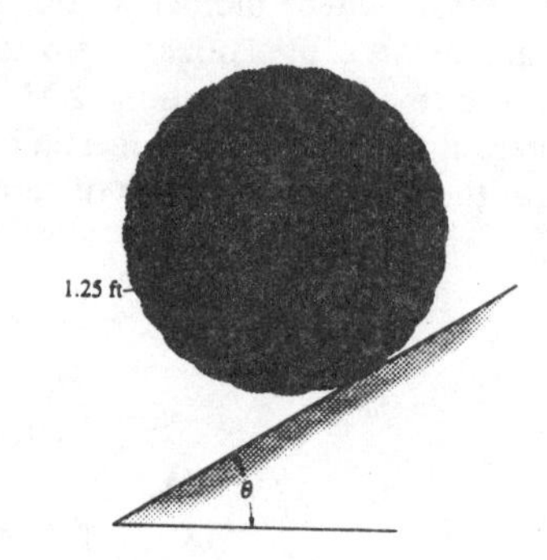

Since wheel is on the verge of slipping.

$+\swarrow \Sigma F_x = m(a_G)_x$; $\quad 30\sin\theta - 0.2N = \left(\frac{30}{32.2}\right)(1.25\alpha) \qquad (1)$

$+\nwarrow \Sigma F_y = m(a_G)_y$; $\quad N - 30\cos\theta = 0 \qquad (2)$

$\circlearrowleft + \Sigma M_G = I_G\alpha$; $\quad 0.2N(1.25) = \left[\left(\frac{30}{32.2}\right)(0.6)^2\right]\alpha \qquad (3)$

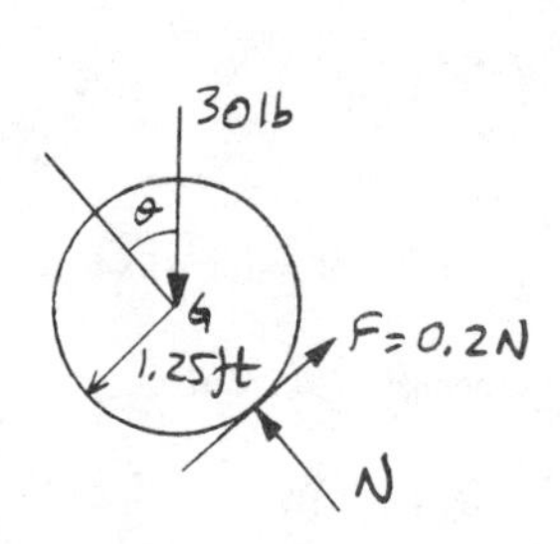

Subsituting Eqs. (2) and (3) into Eq. (1),

$30\sin\theta - 6\cos\theta = 26.042\cos\theta$

$30\sin\theta = 32.042\cos\theta$

$\tan\theta = 1.068$

$\theta = 46.9°$ **Ans**

■17-90. A rocket CD, having a mass of 20 Mg with center of mass at G, is located in deep space so that the effect of gravitation (weight) can be neglected. The smaller rockets A and B each have a mass of 4 Mg and center of mass at G_A and G_B, respectively. If these rockets travel in a straight line and exert a constant thrust of $T = 7$ kN perpendicular to CD, determine the angular acceleration of CD and the accelerations of rockets A and B. Assume that CD is initially at rest and that its radius of gyration about an axis passing through G and perpendicular to the plane of motion is $k_G = 4.60$ m.

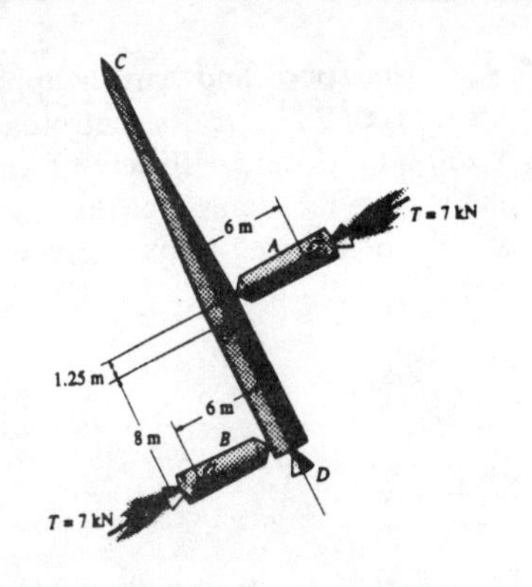

$+\uparrow \Sigma F_y = m(a_G)_y;\qquad 0 = 20\,000(a_G)_y$

$(a_G)_y = 0$

$\overset{+}{\rightarrow} \Sigma F_x = m(a_G)_x;\qquad 7000 - 7000 = 4000a_B - 4000a_A - 20\,000(a_G)_x$

$5(a_G)_x = a_B - a_A \qquad (1)$

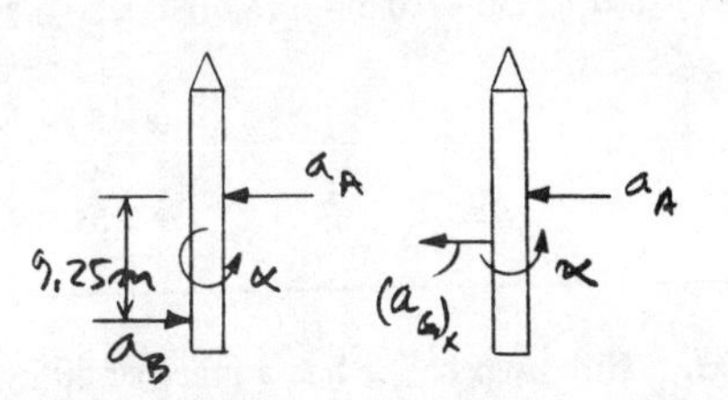

$\circlearrowleft + \Sigma M_G = \Sigma (M_k)_G;\qquad 7000(1.25) + 7000(8) = 20\,000(4.60)^2\alpha + 4000a_A(1.25) + 4000a_B(8)$

$64.75 = 423.2\alpha + 5a_A + 32a_B \qquad (2)$

$\mathbf{a}_B = \mathbf{a}_A + \boldsymbol{\alpha} \times \mathbf{r}_{B/A} - \omega^2 \mathbf{r}_{B/A}$

$(\overset{+}{\rightarrow})\qquad a_B = -a_A + \alpha(9.25) - 0 \qquad (3)$

$\mathbf{a}_G = \mathbf{a}_A + \boldsymbol{\alpha} \times \mathbf{r}_{G/A} - \omega^2 \mathbf{r}_{G/A}$

$(\overset{+}{\rightarrow})\qquad -(a_G)_x = -a_A + \alpha(1.25) - 0 \qquad (4)$

Solving Eqs. (1) – (4) yields

$\alpha = 0.0982 \text{ rad/s}^2$ **Ans**

$(a_G)_x = 0.0947 \text{ m/s}^2$

$a_A = 0.217 \text{ m/s}^2$ **Ans**

$a_B = 0.691 \text{ m/s}^2$ **Ans**

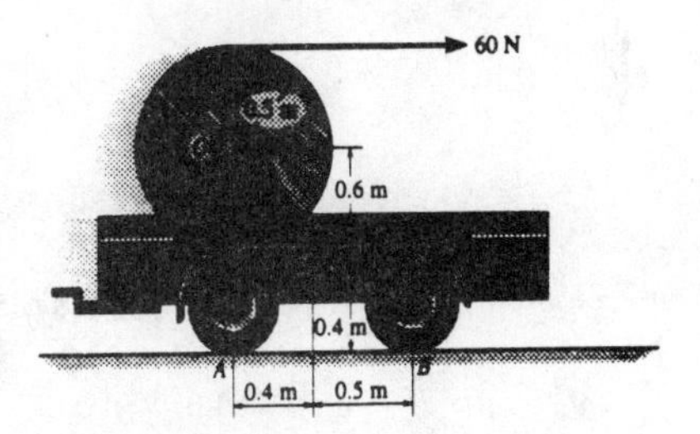

17-91. The trailer has a mass of 580 kg and a mass center at G, whereas the spool has a mass of 200 kg, mass center at O, and a radius of gyration about an axis passing through O of $k_O = 0.45$ m. If a force of 60 N is applied to the cable, determine the angular acceleration of the spool and the acceleration of the trailer. The wheels have negligible mass and are free to roll.

System :

$\overset{+}{\rightarrow} \Sigma F_x = m(a_G)_x;\qquad 60 = 200a + 580a$

$a = 0.0769 \text{ m/s}^2$ **Ans**

Spool :

$\circlearrowleft + \Sigma M_O = I_O\alpha;\qquad 60(0.5) = 200(0.45)^2\alpha$

$\alpha = 0.741 \text{ rad/s}^2$ **Ans**

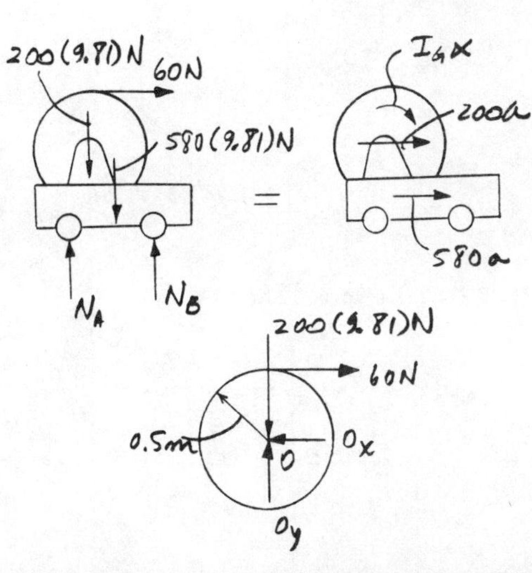

***17-92.** The spool and wire wrapped around its core have a mass of 20 kg and a centroidal radius of gyration $k_G = 250$ mm. If the coefficient of kinetic friction at the ground is $\mu_k = 0.1$, determine the angular acceleration of the spool when the 30-N·m couple is applied.

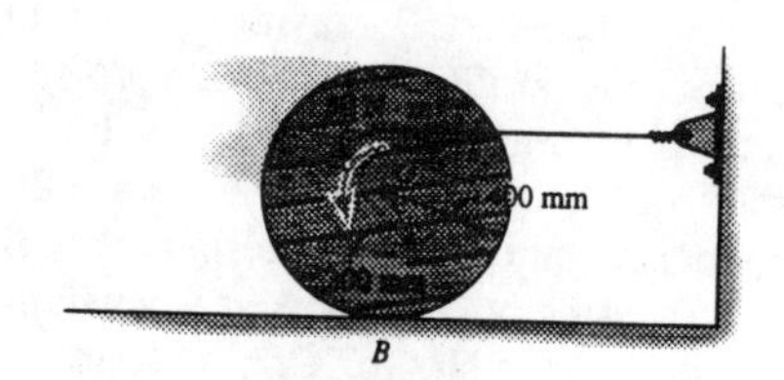

$$a_G = \alpha(0.2)$$

$$+\uparrow \Sigma F_y = m(a_G)_y; \quad N_B - 20(9.81) = 0, \quad N_B = 196.2 \text{ N}$$

$$\circlearrowleft + \Sigma M_P = \Sigma (M_k)_P; \quad 30 - 0.1(196.2)(0.6) = 20(0.2)\left[\alpha(0.2)\right] + \left[20(0.25)^2\right]\alpha$$

$$\alpha = 8.89 \text{ rad/s}^2 \quad \textbf{Ans}$$

Also,

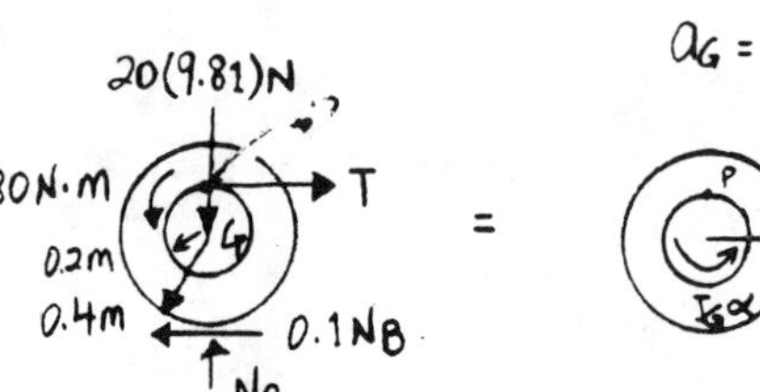

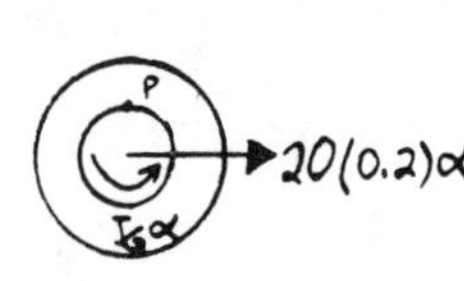

$$\xrightarrow{+} \Sigma F_x = m(a_G)_x; \quad T - 0.1N_B = 20(0.2)\alpha$$

$$+\uparrow \Sigma F_y = m(a_G)_y; \quad N_B - 20(9.81) = 0, \quad N_B = 196.2 \text{ N}$$

$$\circlearrowleft + \Sigma M_G = I_G\alpha; \quad 30 - T(0.2) - 0.1N_B(0.4) = 20(0.25)^2\alpha$$

$$\alpha = 8.89 \text{ rad/s}^2 \quad \textbf{Ans}$$

$$T = 55.2 \text{ N}$$

17-93. The lawn roller has a mass of 80 kg and a radius of gyration $k_G = 0.175$ m. If it is pushed forward with a force of 200 N when the handle is at 45°, determine its angular acceleration. The coefficients of static and kinetic friction between the ground and the roller are $\mu_s = 0.12$ and $\mu_k = 0.1$, respectively.

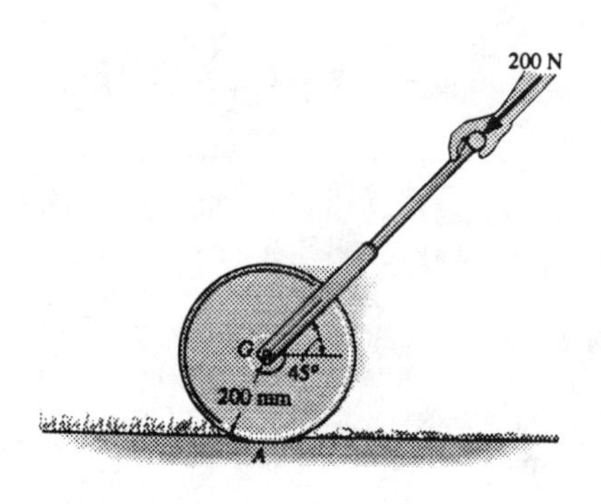

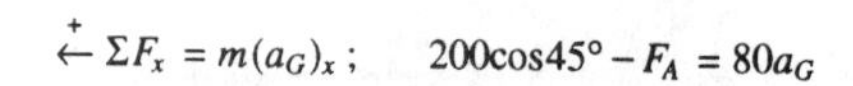

$$\xleftarrow{+} \Sigma F_x = m(a_G)_x; \quad 200\cos 45° - F_A = 80a_G$$

$$+\uparrow \Sigma F_y = m(a_G)_y; \quad N_A - 80(9.81) - 200\sin 45° = 0$$

$$\circlearrowleft + \Sigma M_G = I_G\alpha; \quad F_A(0.2) = 80(0.175)^2\alpha$$

Assume no slipping : $a_G = 0.2\alpha$

$F_A = 61.32$ N

$N_A = 926.2$ N

$\alpha = 5.01 \text{ rad/s}^2$ **Ans**

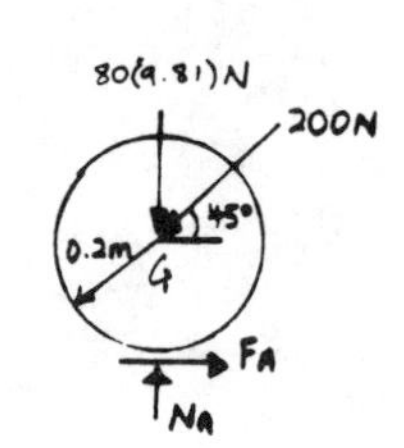

$(F_A)_{max} = \mu_s N_A = 0.12(926.2) = 111.1 \text{ N} > 61.32 \text{ N}$ OK

17-94. Solve Prob. 17-93 if $\mu_s = 0.6$ and $\mu_k = 0.45$.

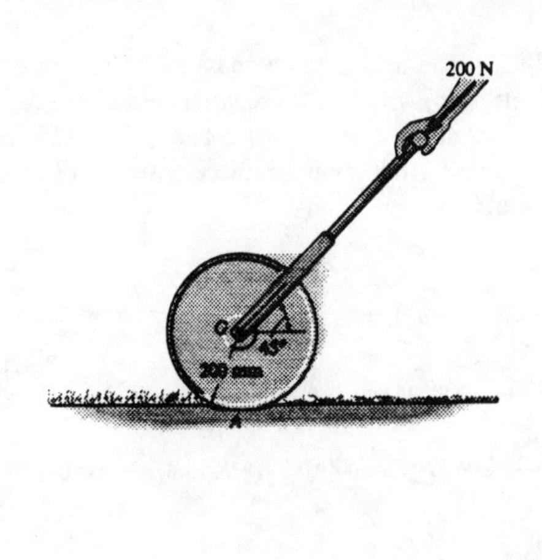

$\overset{+}{\leftarrow} \Sigma F_x = m(a_G)_x;$ $\quad 200\cos45° - F_A = 80a_G$

$+\uparrow \Sigma F_y = m(a_G)_y;$ $\quad N_A - 80(9.81) - 200\sin45° = 0$

$\circlearrowleft + \Sigma M_G = I_G\alpha;$ $\quad F_A(0.2) = 80(0.175)^2\alpha$

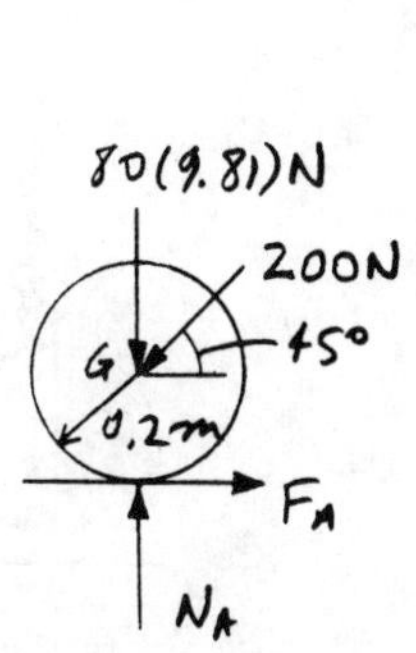

Assume no slipping : $\quad a_G = 0.2\alpha$

$F_A = 61.32$ N

$N_A = 926.2$ N

$\alpha = 5.01$ rad/s^2 **Ans**

$(F_A)_{max} = \mu_s N_A = 0.6(926.2 \text{ N}) = 555.7 \text{ N} > 61.32 \text{ N}$ OK

17-95. The spool has a mass of 100 kg and a radius of gyration of $k_G = 0.3$ m. If the coefficients of static and kinetic friction at A are $\mu_s = 0.2$ and $\mu_k = 0.15$, respectively, determine the angular acceleration of the spool if $P = 50$ N.

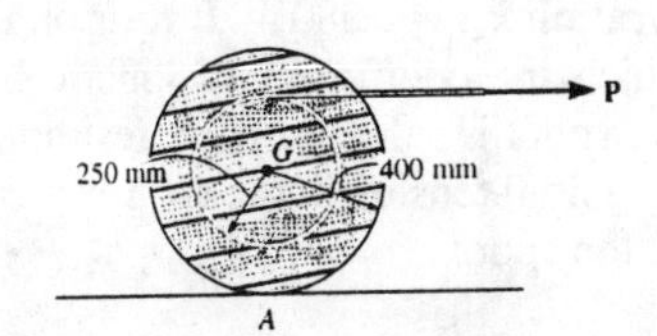

$\overset{+}{\rightarrow} \Sigma F_x = m(a_G)_x;$ $\quad 50 + F_A = 100a_G$

$+\uparrow \Sigma F_y = m(a_G)_y;$ $\quad N_A - 100(9.81) = 0$

$\circlearrowleft + \Sigma M_G = I_G\alpha;$ $\quad 50(0.25) - F_A(0.4) = \left[100(0.3)^2\right]\alpha$

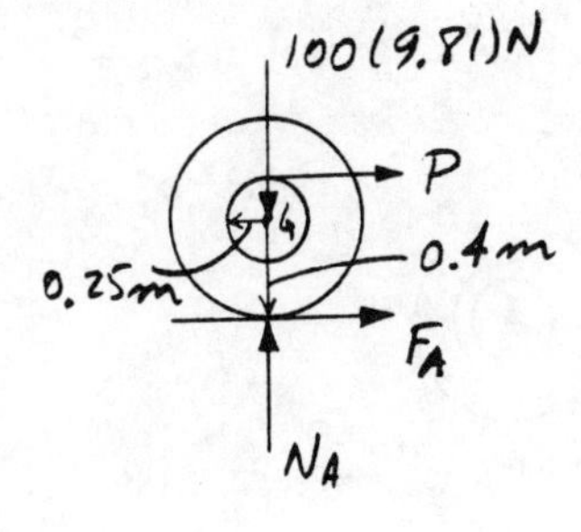

Assume no slipping : $\quad a_G = 0.4\alpha$

$\alpha = 1.30$ rad/s^2 **Ans**

$a_G = 0.520$ m/s^2 $\quad N_A = 981$ N $\quad F_A = 2.00$ N

Since $(F_A)_{max} = 0.2(981) = 196.2 \text{ N} > 2.00 \text{ N}$ OK

***17-96.** Solve Prob. 17-95 if the cord and force $P =$ 50 N are directed vertically upwards.

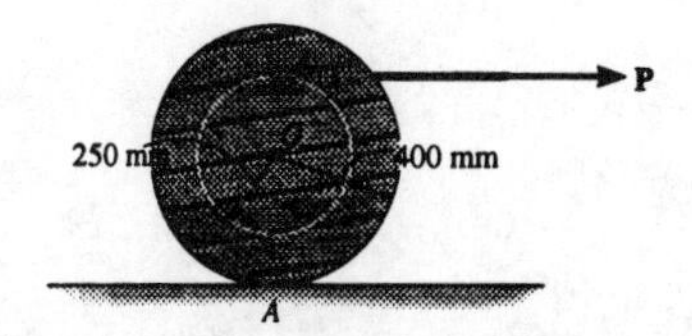

$\overset{+}{\rightarrow} \Sigma F_x = m(a_G)_x;$ $\quad F_A = 100a_G$

$+\uparrow \Sigma F_y = m(a_G)_y;$ $\quad N_A + 50 - 100(9.81) = 0$

$+\Sigma M_G = I_G\alpha;$ $\quad 50(0.25) - F_A(0.4) = \left[100(0.3)^2\right]\alpha$

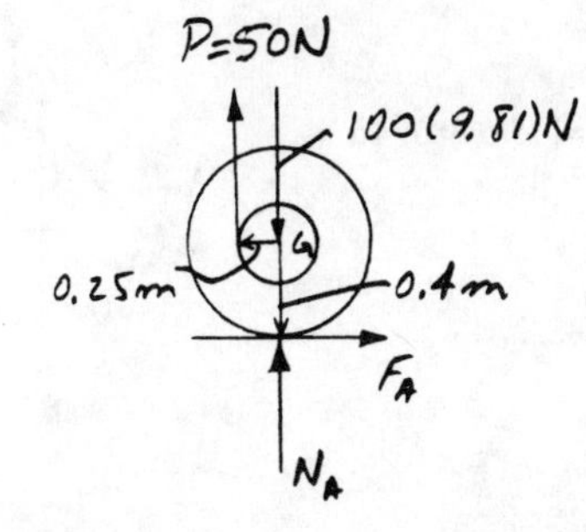

Assume no slipping : $\quad a_G = 0.4\alpha$

$\alpha = 0.500$ rad/s^2 **Ans**

$a_G = 0.2$ m/s^2 $\quad N_A = 931$ N $\quad F_A = 20$ N

Since $(F_A)_{max} = 0.2(931) = 186.2 \text{ N} > 20 \text{ N}$ OK

17-97. The spool has a mass of 100 kg and a radius of gyration $k_G = 0.3$ m. If the coefficients of static and kinetic friction at A are $\mu_s = 0.2$ and $\mu_k = 0.15$, respectively, determine the angular acceleration of the spool if $P = 600$ N.

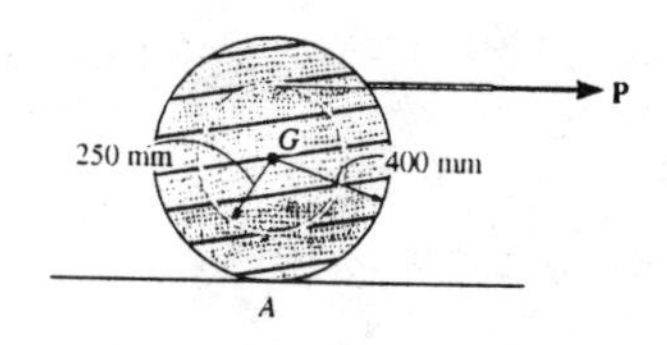

$\overset{+}{\rightarrow} \Sigma F_x = m(a_G)_x ; \quad 600 + F_A = 100 a_G$

$+\uparrow \Sigma F_y = m(a_G)_y ; \quad N_A - 100(9.81) = 0$

$\circlearrowleft + \Sigma M_G = I_G \alpha ; \quad 600(0.25) - F_A(0.4) = \left[100(0.3)^2\right]\alpha$

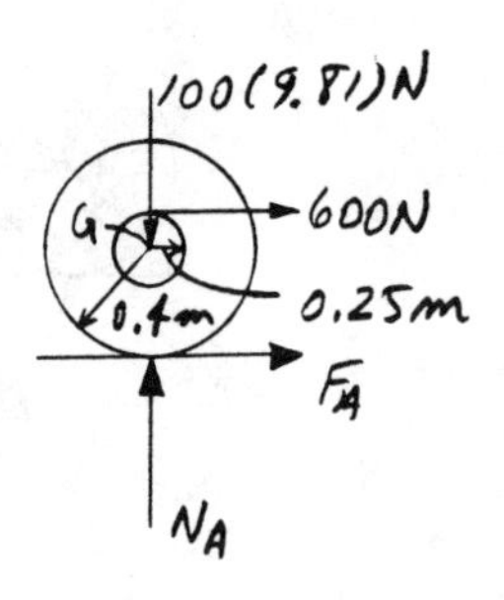

Assume no slipping : $\quad a_G = 0.4\alpha$

$\alpha = 15.6 \text{ rad/s}^2$ **Ans**

$a_G = 6.24 \text{ m/s}^2 \quad N_A = 981 \text{ N} \quad F_A = 24.0 \text{ N}$

Since $(F_A)_{max} = 0.2(981) = 196.2 \text{ N} > 24.0 \text{ N}$ OK

17-98. The spool has a mass of 75 kg and a radius of gyration $k_G = 0.380$ m. It rests on the inclined surface for which the coefficient of kinetic friction is $\mu_k = 0.15$. If the spool is released from rest and slips at A, determine the initial tension in the cord and the angular acceleration of the spool.

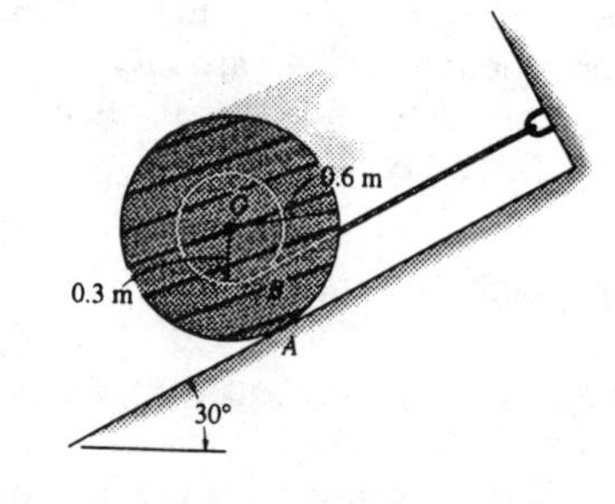

$F_A = 0.15 N_A$

$a_G = 0.3\alpha$

$\swarrow + \Sigma F_{x'} = m(a_G)_{x'} ; \quad 75(9.81)\sin 30° - T + 0.15 N_A = 75(0.3\alpha)$

$\nwarrow + \Sigma F_{y'} = m(a_G)_{y'} ; \quad N_A - 75(9.81)\cos 30° = 0$

$\circlearrowleft + \Sigma M_G = I_G \alpha ; \quad T(0.3) - (0.15 N_A)(0.6) = \left[75(0.380)^2\right]\alpha$

Solving,

$N_A = 637$ N

$\alpha = 4.65 \text{ rad/s}^2$ **Ans**

$T = 359$ N **Ans**

17-99. The wheel has a mass of 80 kg and a radius of gyration $k_G = 0.25$ m. If it is subjected to a couple moment of $M = 50$ N·m, determine its angular acceleration. The coefficients of static and kinetic friction between the ground and the wheel are $\mu_s = 0.2$ and $\mu_k = 0.15$, respectively.

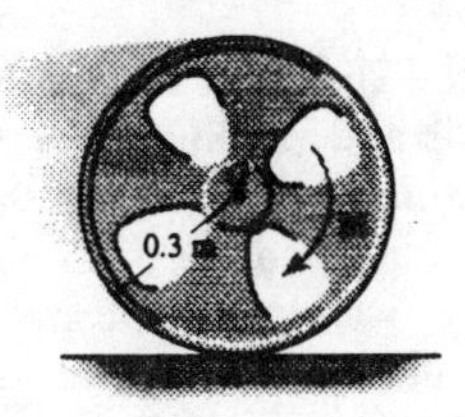

Assume no slipping,

$a_G = 0.3\alpha$

$\overset{+}{\rightarrow} \Sigma F_x = m(a_G)_x;\qquad F_A = 80a_G$

$+\uparrow \Sigma F_y = m(a_G)_y;\qquad N_A - 80(9.81) = 0$

$\circlearrowleft + \Sigma M_G = I_G\alpha;\qquad 50 - F_A(0.3) = \left[80(0.25)^2\right]\alpha$

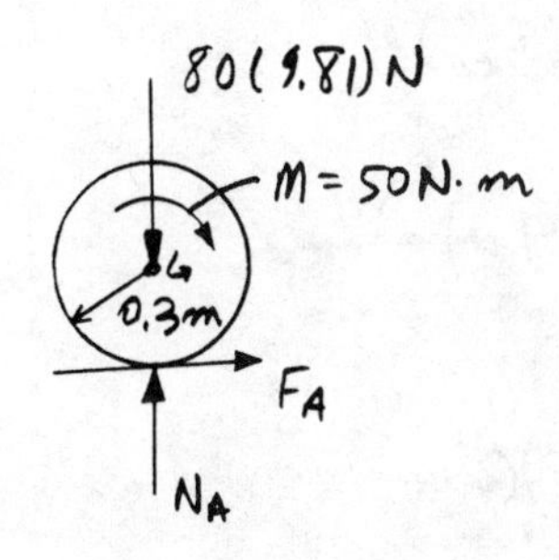

Solving,

$N_A = 785$ N

$\alpha = 4.10 \text{ rad/s}^2$ **Ans**

$F_A = 98.4$ N

$(F_A)_{max} = 0.2(784.8) = 157 \text{ N} > 98.4 \text{ N}$ O.K.

***17-100.** The truck carries the spool which has a weight of 500 lb and a radius of gyration of $k_G = 2$ ft. Determine the angular acceleration of the spool if it is not tied down on the truck and the truck begins to accelerate at 3 ft/s^2. Assume the spool does not slip on the bed of the truck.

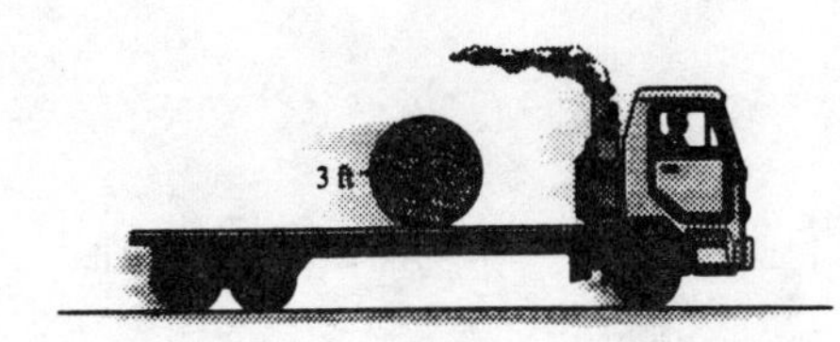

$\overset{+}{\rightarrow} \Sigma F_x = m(a_G)_x;\qquad F = \left(\frac{500}{32.2}\right)a_G \qquad (1)$

$\circlearrowleft + \Sigma M_G = I_G\alpha;\qquad F(3) = \left(\frac{500}{32.2}\right)(2)^2\alpha \qquad (2)$

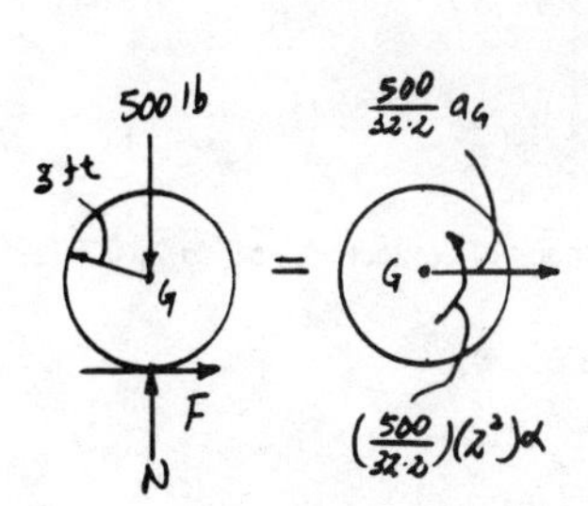

$$\mathbf{a}_A = \mathbf{a}_G + (\mathbf{a}_{A/G})_t + (\mathbf{a}_{A/G})_n$$

$$\left[\underset{\rightarrow}{(a_A)_t}\right] + \left[\underset{\uparrow}{(a_A)_n}\right] = \left[\underset{\rightarrow}{a_G}\right] + \left[\underset{\rightarrow}{3\alpha}\right] + \left[\underset{\uparrow}{(a_{A/G})_n}\right]$$

$(\overset{+}{\rightarrow})\qquad 3 = a_G + 3\alpha \qquad (3)$

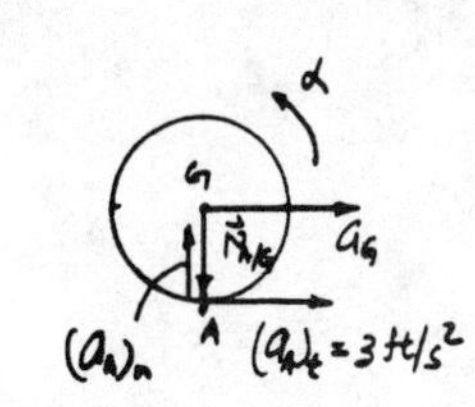

Solving Eqs. (1), (2) and (3) yields:

$F = 14.33$ lb $\qquad a_G = 0.923 \text{ ft/s}^2$

$\alpha = 0.692 \text{ rad/s}^2$ **Ans**

17-101. The truck carries the spool which has a weight of 200 lb and a radius of gyration of $k_G = 2$ ft. Determine the angular acceleration of the spool if it is not tied down on the truck and the truck begins to accelerate at 5 ft/s^2. The coefficients of static and kinetic friction between the spool and the truck bed are $\mu_s = 0.15$ and $\mu_k = 0.1$, respectively.

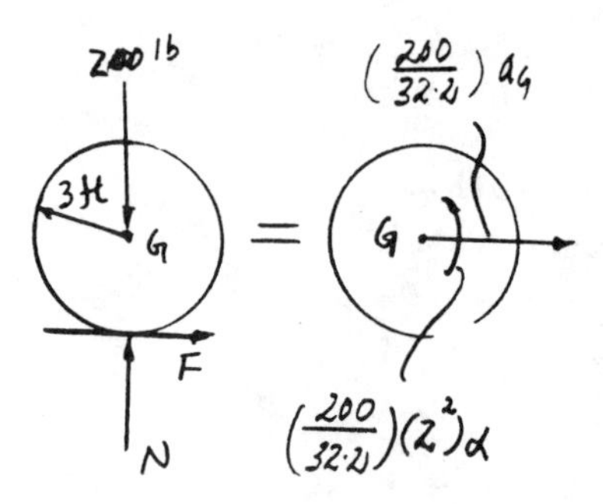

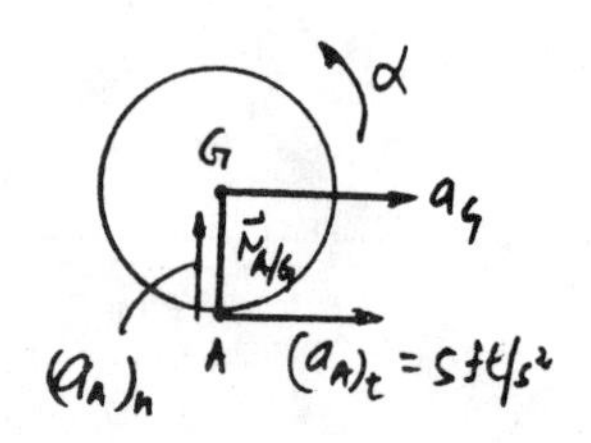

$+\uparrow \Sigma F_y = m(a_G)_y;\qquad N - 200 = 0 \qquad N = 200 \text{ lb}$

$\overset{+}{\rightarrow} \Sigma F_x = m(a_G)_x;\qquad F = \left(\dfrac{200}{32.2}\right) a_G \qquad (1)$

$\circlearrowleft + \Sigma M_G = I_G \alpha;\qquad F(3) = \left(\dfrac{200}{32.2}\right)(2)^2 \alpha \qquad (2)$

Assume no slipping occurs at the point of contact. Hence $(a_A)_t = 5$ ft/s^2.

$$\mathbf{a}_A = \mathbf{a}_G + (\mathbf{a}_{A/G})_t + (\mathbf{a}_{A/G})_n$$

$$\left[\underset{\rightarrow}{(a_A)_t}\right] + \left[\underset{\uparrow}{(a_A)_n}\right] = \left[\underset{\rightarrow}{a_G}\right] + \left[\underset{\rightarrow}{3\alpha}\right] + \left[\underset{\uparrow}{(a_{A/G})_n}\right]$$

$\left(\overset{+}{\rightarrow}\right) \qquad 5 = a_G + 3\alpha \qquad (3)$

Solving Eqs. (1), (2) and (3) yields :

$F = 9.556$ lb $\qquad a_G = 1.538$ ft/s^2

$\alpha = 1.15$ rad/s^2 $\qquad$ **Ans**

Since $F = 9.556 \text{ lb} < \mu_s N = 0.15(200) = 30$ lb, the assumption is OK

17-102. The uniform beam has a weight W. If it is originally at rest while being supported at A and B by cables, determine the tension in cable A if cable B suddenly fails. Assume the beam is a slender rod.

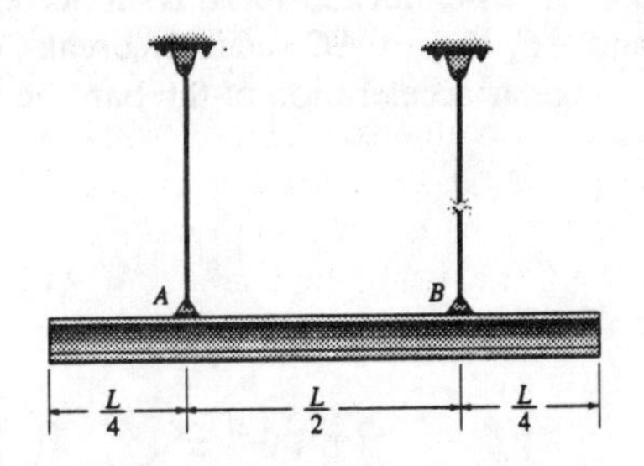

$+\uparrow \Sigma F_y = m(a_G)_y ; \quad T_A - W = -\frac{W}{g} a_G$

$\circlearrowleft + \Sigma M_A = I_A \alpha; \quad W\left(\frac{L}{4}\right) = \left[\frac{1}{12}\left(\frac{W}{g}\right)L^2\right]\alpha + \frac{W}{g}\left(\frac{L}{4}\right)\alpha\left(\frac{L}{4}\right)$

$$1 = \frac{1}{g}\left(\frac{L}{4} + \frac{L}{3}\right)\alpha$$

Since $a_G = \alpha\left(\frac{L}{4}\right)$,

$$\alpha = \frac{12}{7}\left(\frac{g}{L}\right)$$

$$T_A = W - \frac{W}{g}(\alpha)\left(\frac{L}{4}\right) = W - \frac{W}{g}\left(\frac{12}{7}\right)\left(\frac{g}{L}\right)\left(\frac{L}{4}\right)$$

$$T_A = \frac{4}{7}W \qquad \textbf{Ans}$$

Also,

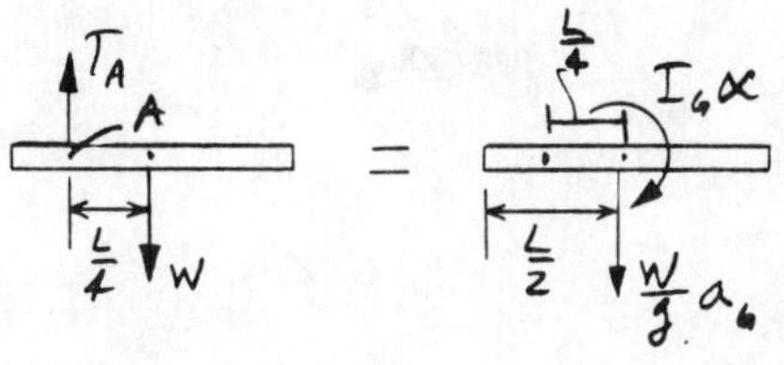

$+\uparrow \Sigma F_y = m(a_G)_y ; \quad T_A - W = -\frac{W}{g} a_G$

$\circlearrowleft + \Sigma M_G = I_G \alpha; \quad T_A\left(\frac{L}{4}\right) = \left[\frac{1}{12}\left(\frac{W}{g}\right)L^2\right]\alpha$

Since $a_G = \frac{L}{4}\alpha$

$$T_A = \frac{1}{3}\left(\frac{W}{g}\right)L\alpha$$

$$\frac{1}{3}\left(\frac{W}{g}\right)L\alpha - W = -\frac{W}{g}\left(\frac{L}{4}\right)\alpha$$

$$\alpha = \frac{12}{7}\left(\frac{g}{L}\right)$$

$$T_A = \frac{1}{3}\left(\frac{W}{g}\right)L\left(\frac{12}{7}\right)\left(\frac{g}{L}\right)$$

$$T_A = \frac{4}{7}W \qquad \textbf{Ans}$$

17-103. The slender 150-lb bar is supported by two cords AB and AC. If cord AC suddenly breaks, determine the initial angular acceleration of the bar and the tension in cord AB.

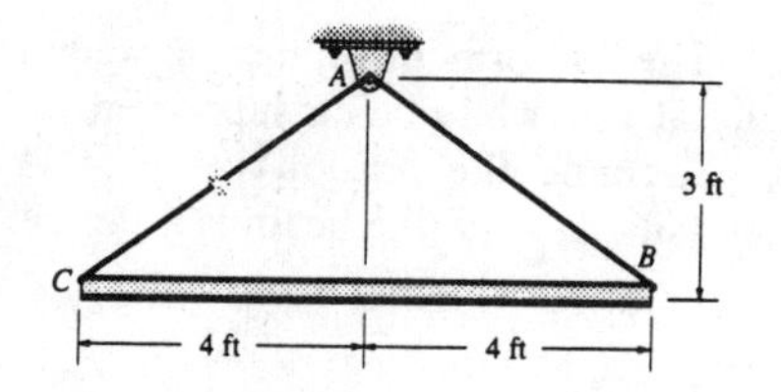

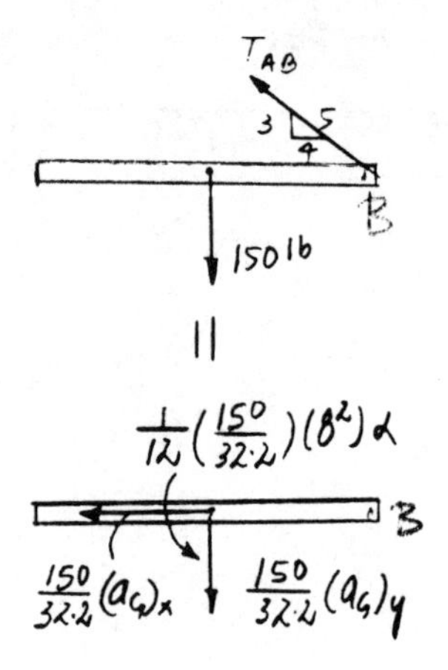

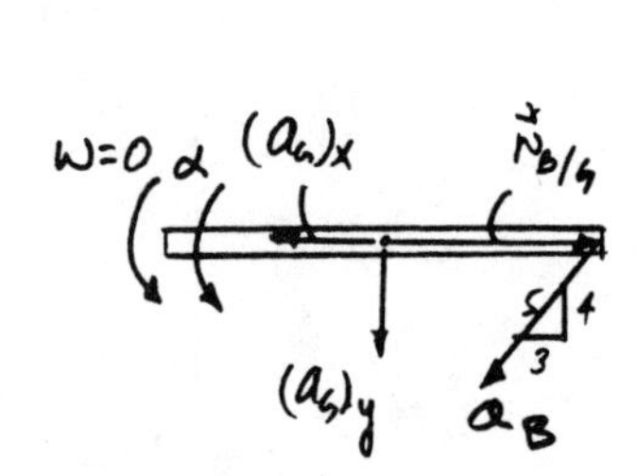

Equations of motion :

$\overset{+}{\leftarrow} \Sigma F_x = m(a_G)_x\,;\qquad \frac{4}{5}T_{AB} = \left(\frac{150}{32.2}\right)(a_G)_x \qquad (1)$

$+\uparrow \Sigma F_y = m(a_G)_y\,;\qquad \frac{3}{5}T_{AB} - 150 = -\left(\frac{150}{32.2}\right)(a_G)_y \qquad (2)$

$\circlearrowleft + \Sigma M_B = \Sigma (M_k)_B\,;\qquad 150(4) = \frac{1}{12}\left(\frac{150}{32.2}\right)(8)^2\alpha + \left(\frac{150}{32.2}\right)(a_G)_y\,(4) \qquad (3)$

Kinematics :

$\mathbf{a}_B = \mathbf{a}_G + (\mathbf{a}_{B/G})_t + (\mathbf{a}_{B/G})_n$

$$\left[\begin{matrix} a_B \\ \swarrow \end{matrix}\right] = \left[\begin{matrix}(a_G)_x \\ \leftarrow\end{matrix}\right] + \left[\begin{matrix}(a_G)_y \\ \downarrow\end{matrix}\right] + \left[\begin{matrix}4\alpha \\ \uparrow\end{matrix}\right] + [0]$$

$(\overset{+}{\leftarrow}) \qquad \frac{3}{5}a_B = (a_G)_x \qquad (4)$

$(+\downarrow) \qquad \frac{4}{5}a_B = (a_G)_y - 4\alpha \qquad (5)$

Solving Eqs. (1) – (5) yields :

$\alpha = 4.18\ \text{rad/s}^2 \qquad T_{AB} = 43.3\ \text{lb}$ **Ans**

$(a_G)_y = 26.63\ \text{ft/s}^2 \qquad (a_G)_x = 7.43\ \text{ft/s}^2 \qquad a_B = 12.38\ \text{ft/s}^2$

***17-104.** A long strip of paper is wrapped into two rolls, each having a mass of 8 kg. Roll A is pin supported about its center whereas roll B is not centrally supported. If B is brought into contact with A and released from rest, determine the initial tension in the paper between the rolls and the angular acceleration of each roll. For the calculation, assume the rolls to be approximated by cylinders.

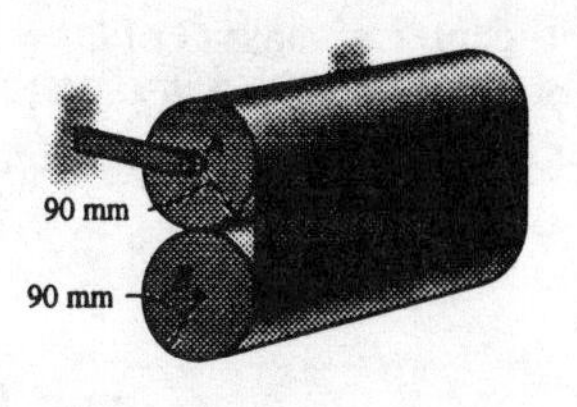

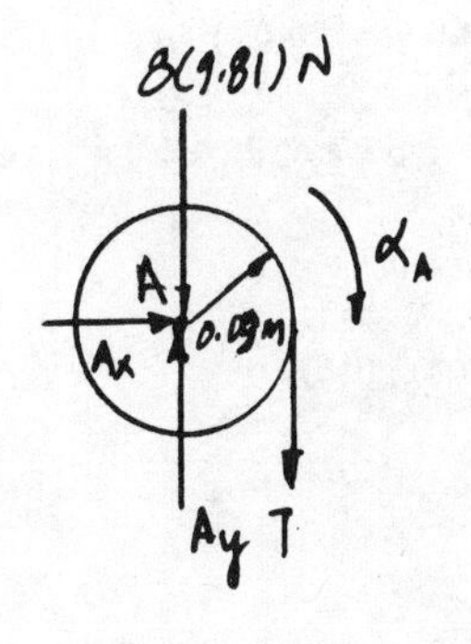

For roll A.

$\curvearrowleft + \Sigma M_A = I_A\,\alpha;\qquad T(0.09) = \frac{1}{2}(8)(0.09)^2\alpha_A\qquad (1)$

For roll B

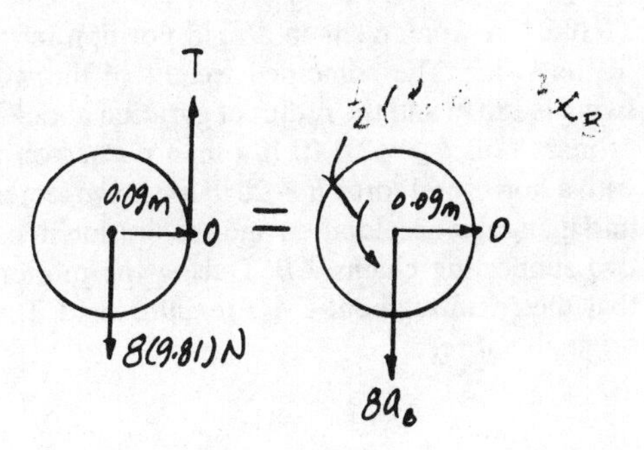

$\curvearrowleft + \Sigma M_O = \Sigma (M_k)_O;\qquad 8(9.81)(0.09) = \frac{1}{2}(8)(0.09)^2\alpha_B + 8a_B\,(0.09)\qquad (2)$

$+\uparrow \Sigma F_y = m(a_G)_y;\qquad T - 8(9.81) = -8a_B\qquad (3)$

Kinematics :

$$\mathbf{a}_B = \mathbf{a}_O + (\mathbf{a}_{B/O})_t + (\mathbf{a}_{B/O})_n$$

$$\left[\underset{\downarrow}{a_B}\right] = \left[\underset{\downarrow}{a_O}\right] + \left[\underset{\downarrow}{\alpha_B\,(0.09)}\right] + [0]$$

$(+\downarrow)\qquad a_B = a_O + 0.09\alpha_B\qquad (4)$

also, $\qquad (+\downarrow)\qquad a_O = \alpha_A\,(0.09)\qquad (5)$

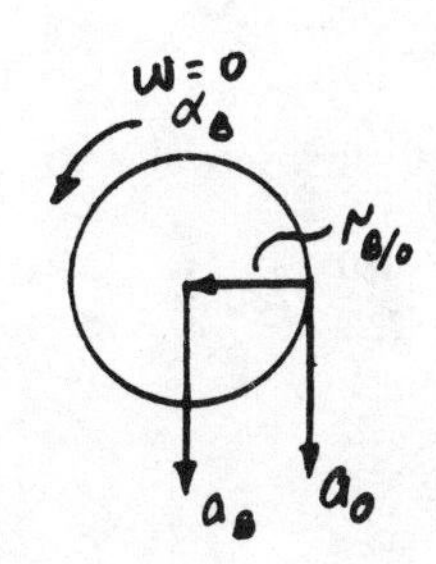

Solving Eqs. (1) – (5) yields :

$\alpha_A = 43.6\ \text{rad/s}^2$ **Ans**

$\alpha_B = 43.6\ \text{rad/s}^2$ **Ans**

$T = 15.7\ \text{N}$ **Ans**

$a_B = 7.85\ \text{m/s}^2\qquad a_O = 3.92\ \text{m/s}^2$

17-105. The 20-kg canister has a radius of gyration about its center of mass G of $k_G = 0.4$ m. If it is subjected to a horizontal force of $F = 30$ N, determine the initial angular acceleration of the canister and the tension in the supporting cable AB.

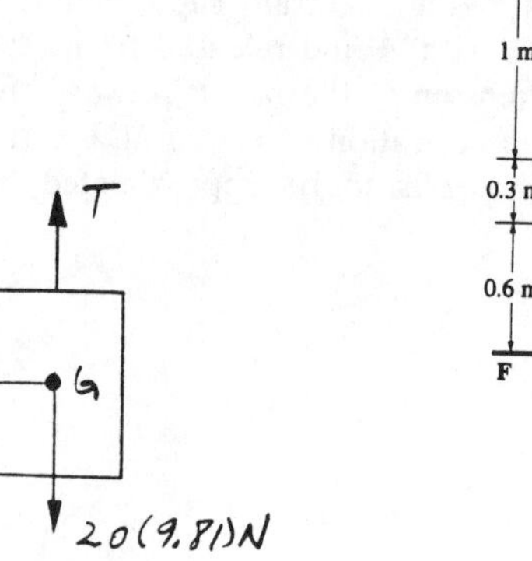

$\xrightarrow{+} \Sigma F_x = m(a_G)_x; \qquad 30 = 20(a_G)_x$

$(a_G)_x = 1.5 \text{ m/s}^2$

$+\uparrow \Sigma F_y = m(a_G)_y; \qquad T - 20(9.81) = 20(a_G)_y$

$\circlearrowleft + \Sigma M_G = I_G\alpha; \qquad 30(0.6) = \left[20(0.4)^2\right]\alpha$

$\alpha = 5.62 \text{ rad/s}^2$ **Ans**

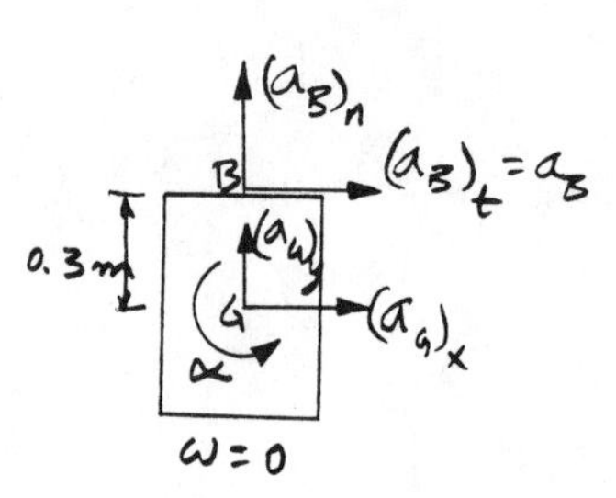

$\mathbf{a}_B = \mathbf{a}_G + \alpha \times \mathbf{r}_{B/G} - \omega^2 \mathbf{r}_{B/G}$

$a_B \mathbf{i} = 1.5\mathbf{i} + (a_G)_y \mathbf{j} + (\alpha \mathbf{k}) \times (0.3\mathbf{j}) - \mathbf{0}$

$(+\uparrow) \qquad (a_G)_y = 0$

Thus,

$T = 196$ N **Ans**

17-106. A woman sits in a rigid position in the middle of the swing. The combined weight of the woman and swing is 180 lb and the radius of gyration about the center of mass G is $k_G = 2.5$ ft. If a man pushes on the swing with a horizontal force $F = 20$ lb as shown, determine the initial angular acceleration and the tension in each of the two supporting chains AB. During the motion, assume that the chain segment CAD remains rigid. The swing is originally at rest.

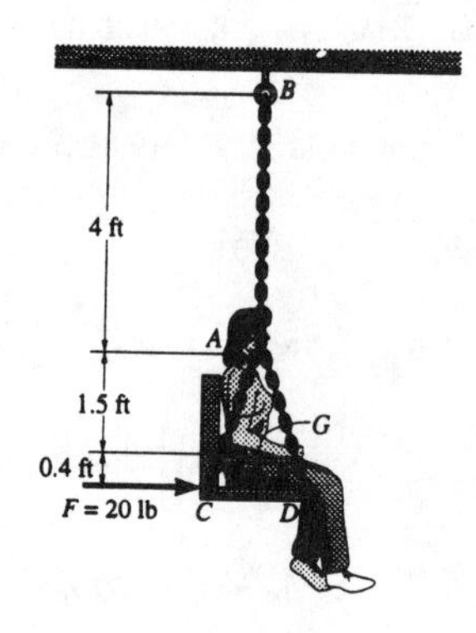

$\xrightarrow{+} \Sigma F_t = m(a_G)_t; \qquad 20 = \left(\frac{180}{32.2}\right)(a_G)_t$

$+\uparrow \Sigma F_n = m(a_G)_n; \qquad 2T - 180 = 0 \qquad (\omega = 0)$

$\circlearrowleft + \Sigma M_G = I_G\alpha; \qquad 20(0.4) = \left[\left(\frac{180}{32.2}\right)(2.5)^2\right]\alpha$

Solving,

$T = 90$ lb **Ans**

$\alpha = 0.229 \text{ rad/s}^2$ **Ans**

$(a_G)_t = 3.58 \text{ ft/s}^2$

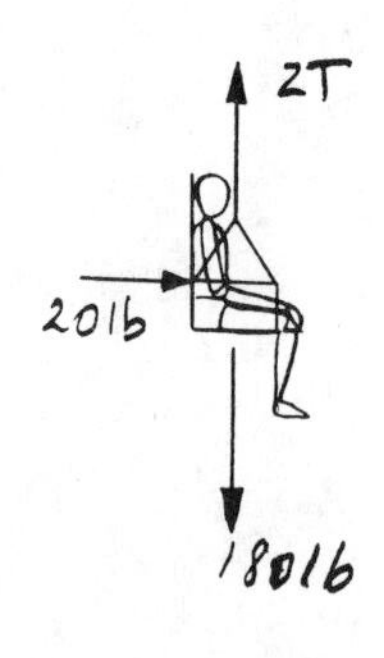

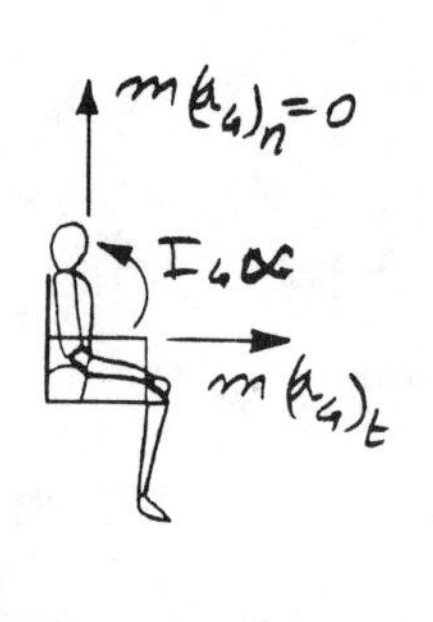

17-107. The 16-lb bowling ball is cast horizontally onto a lane such that initially $\omega = 0$ and its mass center has a velocity $v = 8$ ft/s. If the coefficient of kinetic friction between the lane and the ball is $\mu_k = 0.12$, determine the distance the ball travels before it rolls without slipping. For the calculation, neglect the finger holes in the ball and assume the ball has a uniform density.

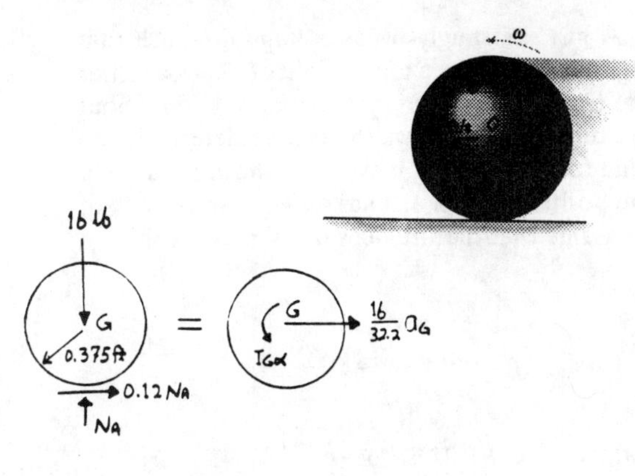

$\xrightarrow{+} \Sigma F_x = m(a_G)_x\,; \quad 0.12N_A = \frac{16}{32.2}a_G$

$+\uparrow \Sigma F_y = m(a_G)_y\,; \quad N_A - 16 = 0$

$\circlearrowleft + \Sigma M_G = I_G\alpha; \quad 0.12N_A\,(0.375) = \left[\frac{2}{5}\left(\frac{16}{32.2}\right)(0.375)^2\right]\alpha$

Solving,

$N_A = 16$ lb; $\quad a_G = 3.864$ ft/s^2; $\quad \alpha = 25.76$ rad/s^2

When the ball rolls without slipping $v = \omega(0.375)$,

$(\circlearrowleft +) \quad \omega = \omega_0 + \alpha_c t$

$\frac{v}{0.375} = 0 + 25.76t$

$v = 9.660t$

$(\overset{+}{\leftarrow}) \quad v = v_0 + a_c t$

$9.660t = 8 - 3.864t$

$t = 0.592$ s

$(\overset{+}{\leftarrow}) \quad s = s_0 + v_0 t + \frac{1}{2}a_c t^2$

$s = 0 + 8(0.592) - \frac{1}{2}(3.864)(0.592)^2$

$s = 4.06$ ft **Ans**

***17-108.** By pressing down with the finger at B, a thin ring having a mass m is given an initial velocity $\mathbf{v}_0$ and a backspin $\boldsymbol{\omega}_0$ when the finger is released. If the coefficient of kinetic friction between the table and the ring is μ_k, determine the distance the ring travels forward before backspinning stops.

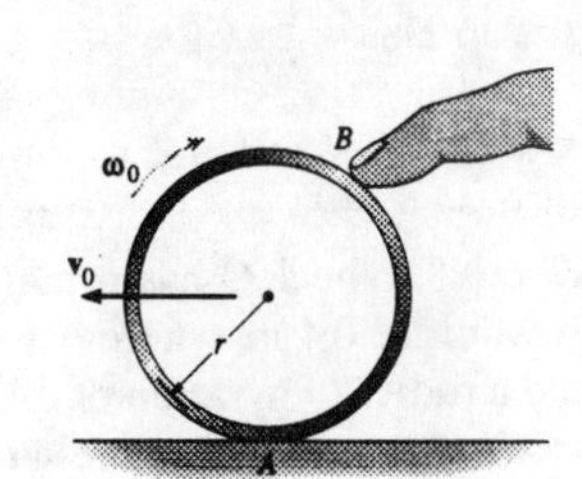

$+\uparrow \Sigma F_y = 0; \quad N_A - mg = 0$

$N_A = mg$

$\xrightarrow{+} \Sigma F_x = m(a_G)_x\,; \quad \mu_k(mg) = m(a_G)$

$a_G = \mu_k g$

$\circlearrowleft + \Sigma M_G = I_G\alpha; \quad \mu_k(mg)r = mr^2\alpha$

$\alpha = \frac{\mu_k g}{r}$

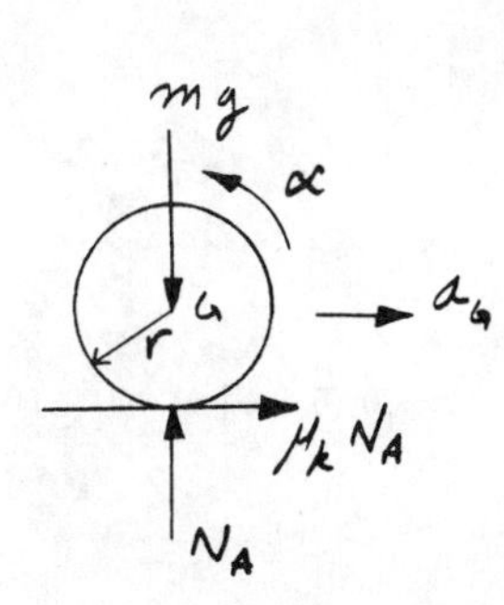

$(\circlearrowleft +) \quad \omega = \omega_0 + \alpha_c t$

$0 = \omega_0 - \left(\frac{\mu_k g}{r}\right)t$

$t = \frac{\omega_0 r}{\mu_k g}$

$(\overset{+}{\leftarrow}) \quad s = s_0 + v_0 t + \frac{1}{2}a_c t^2$

$s = 0 + v_0\left(\frac{\omega_0 r}{\mu_k g}\right) - \left(\frac{1}{2}\right)(\mu_k g)\left(\frac{\omega_0^2 r^2}{\mu_k^2 g^2}\right)$

$s = \left(\frac{\omega_0 r}{\mu_k g}\right)\left(v_0 - \frac{1}{2}\omega_0 r\right)$ **Ans**

17-109. A girl sits snugly inside a large tire such that both the girl and tire have a total weight of 185 lb, a center of mass at G, and a radius of gyration $k_G = 1.65$ ft about G. If the tire rolls freely down the incline, determine the normal and frictional forces it exerts on the ground when it is in the position shown and has an angular velocity of 6 rad/s. Assume that the tire does not slip as it rolls.

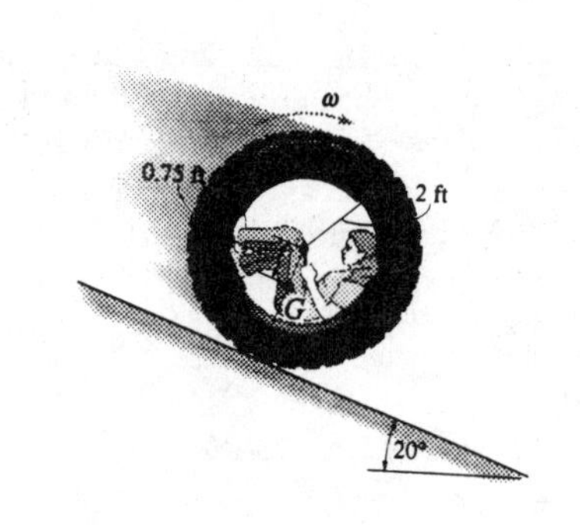

$+\nearrow \Sigma F_y = m(a_G)_y; \quad N_T - 185\cos 20° = \left(\frac{185}{32.2}\right)(a_G)_y$

$+\searrow \Sigma F_x = m(a_G)_x; \quad -F_T + 185\sin 20° = \left(\frac{185}{32.2}\right)(a_G)_x$

$\curvearrowleft + \Sigma M_G = I_G\alpha; \quad F_T(1.25) = \left(\frac{185}{32.2}\right)(1.65)^2\alpha$

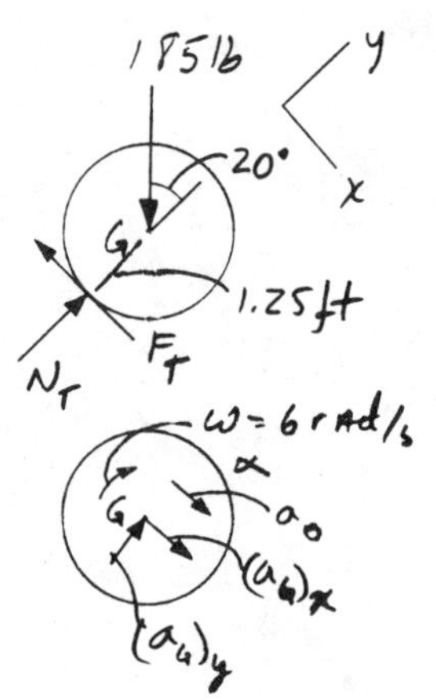

$\mathbf{a}_G = \mathbf{a}_O + \boldsymbol{\alpha} \times \mathbf{r}_{G/O} - \omega^2 \mathbf{r}_{G/O}$

$(a_G)_x\mathbf{i} + (a_G)_y\mathbf{j} = 2\alpha\mathbf{i} + (-\alpha\mathbf{k}) \times (-0.75\mathbf{j}) - (6)^2(-0.75\mathbf{j})$

$(+\searrow) \quad (a_G)_x = 1.25\alpha$

$(+\nearrow) \quad (a_G)_y = 27 \text{ ft/s}^2$

Thus,

$\alpha = 3.21 \text{ rad/s}^2$

$F_T = 40.2$ lb **Ans**

$N_T = 329$ lb **Ans**

17-110. Wheel C has a mass of 60 kg and a radius of gyration of 0.4 m, whereas wheel D has a mass of 40 kg and a radius of gyration of 0.35 m. Determine the angular acceleration of each wheel at the instant shown. Neglect the mass of the link and assume that the assembly does not slip on the plane.

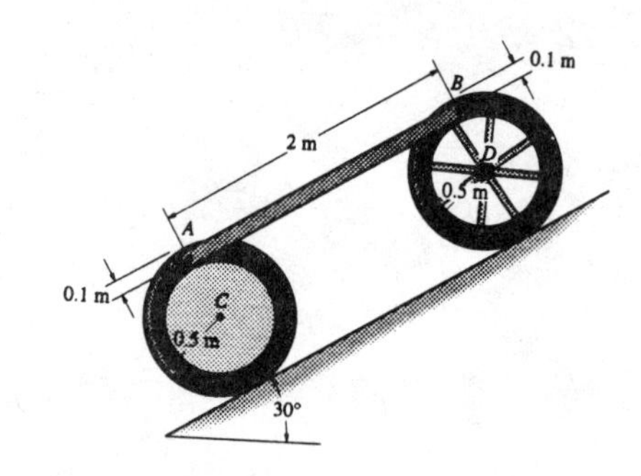

Each wheel has the same α.

Wheel C:

$\curvearrowleft + \Sigma M_{IC} = I_{IC}\alpha; \quad -F_{AB}(0.9) + 60(9.81)\sin 30°(0.5) = \left[60(0.4)^2 + 60(0.5)^2\right]\alpha$

Wheel D:

$\curvearrowleft + \Sigma M_{IC} = I_{IC}\alpha; \quad F_{AB}(0.9) + 40(9.81)\sin 30°(0.5) = \left[40(0.35)^2 + 40(0.5)^2\right]\alpha$

Solving,

$F_{AB} = -6.21$ N

$\alpha = 6.21 \text{ rad/s}^2$ **Ans**

***17-111.** The assembly consists of an 8-kg disk and a 10-kg bar which is pin connected to the disk. If the system is released from rest, determine the angular acceleration of the disk. The coefficients of static and kinetic friction between the disk and the inclined plane are $\mu_s = 0.6$ and $\mu_k = 0.4$, respectively. Neglect friction at B.

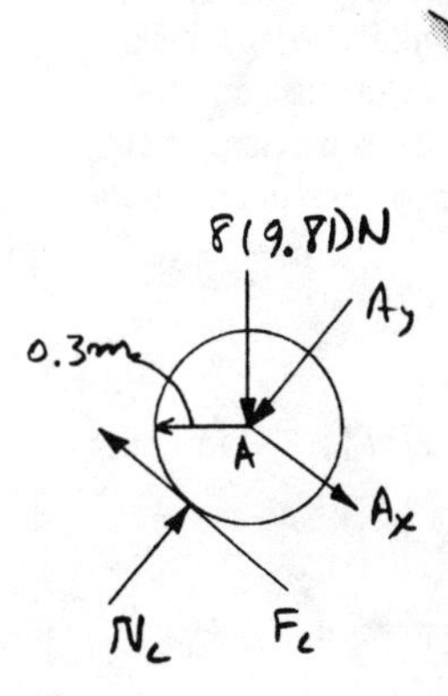

Disk :

$+\searrow \Sigma F_x = m(a_G)_x\,;\quad A_x - F_C + 8(9.81)\sin 30° = 8a_G \quad (1)$

$+\nearrow \Sigma F_y = m(a_G)_y\,;\quad N_C - A_y - 8(9.81)\cos 30° = 0 \quad (2)$

$+\Sigma M_A = I_A\alpha;\quad F_C(0.3) = \left[\frac{1}{2}(8)(0.3)^2\right]\alpha \quad (3)$

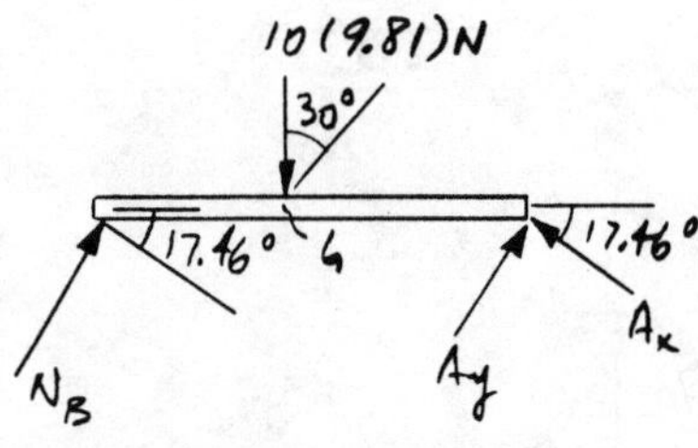

Bar :

$+\searrow \Sigma F_x = m(a_G)_x\,;\quad 10(9.81)\sin 30° - A_x = 10a_G \quad (4)$

$+\nearrow \Sigma F_y = m(a_G)_y\,;\quad N_B + A_y - 10(9.81)\cos 30° = 0 \quad (5)$

$+\Sigma M_G = I_G\alpha;\quad -N_B(0.5\cos 17.46°) + A_x(0.5\sin 17.46°) + A_y(0.5\cos 17.46°) = 0 \quad (6)$

Assume no slipping of the disk,

$a_G = 0.3\alpha \quad (7)$

Solving, Eqs. (1) – (7),

$A_x = 8.92$ N, $\quad A_y = 41.1$ N, $\quad N_B = 43.9$ N

$a_G = 4.01\ \text{m/s}^2$

$\alpha = 13.4\ \text{rad/s}^2$ **Ans**

$N_C = 109$ N

$F_C = 16.1$ N

$(F_C)_{max} = 0.6(109) = 65.4\ \text{N} > 16.1\ \text{N}$ **OK**

***17-112.** Solve Prob. 17-111 if the bar is removed. The coefficients of static and kinetic friction between the disk and inclined plane are $\mu_s = 0.15$ and $\mu_k = 0.1$, respectively.

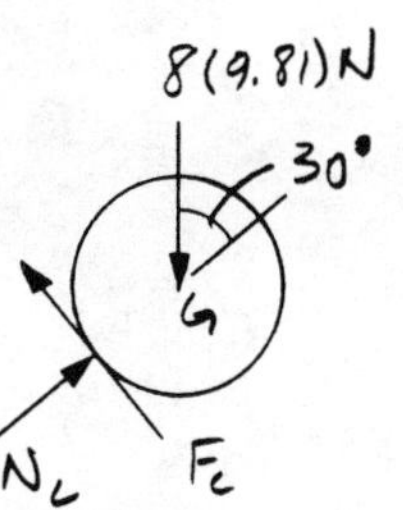

$+\searrow \Sigma F_x = m(a_G)_x\,;\quad 8(9.81)\sin 30° - F_C = 8a_G \quad (1)$

$+\nearrow \Sigma F_y = m(a_G)_y\,;\quad -8(9.81)\cos 30° + N_C = 0 \quad (2)$

$+\Sigma M_G = I_G\alpha;\quad F_C(0.3) = \left[\frac{1}{2}(8)(0.3)^2\right]\alpha \quad (3)$

Assume no slipping : $\quad a_G = 0.3\alpha$

Solving Eqs. (1) – (3) :

$N_C = 67.97$ N

$a_G = 3.27\ \text{m/s}^2$

$\alpha = 10.9\ \text{rad/s}^2$

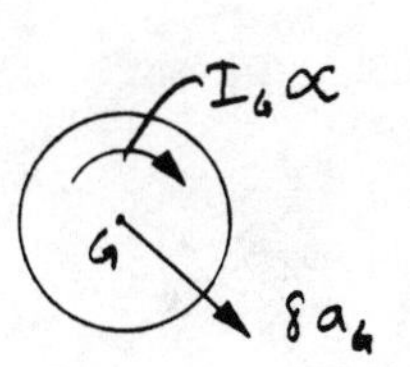

$F_C = 13.08$ N

$(F_C)_{max} = 0.15(67.97) = 10.2\ \text{N} < 13.08\ \text{N}$ NG

Slipping occurs

$F_C = 0.1N_C$

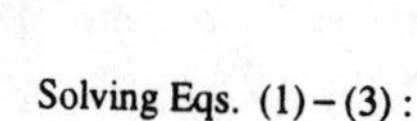

Solving Eqs. (1) – (3) :

$N_C = 67.97$ N

$\alpha = 5.66\ \text{rad/s}^2$ **Ans**

$a_G = 4.06\ \text{m/s}^2$

17-113. The 20-kg disk A is attached to the 10-kg block B using the cable and pulley system shown. If the disk rolls without slipping, determine its angular acceleration and the acceleration of the block when they are released. Also, what is the tension in the cable? Neglect the mass of the pulleys.

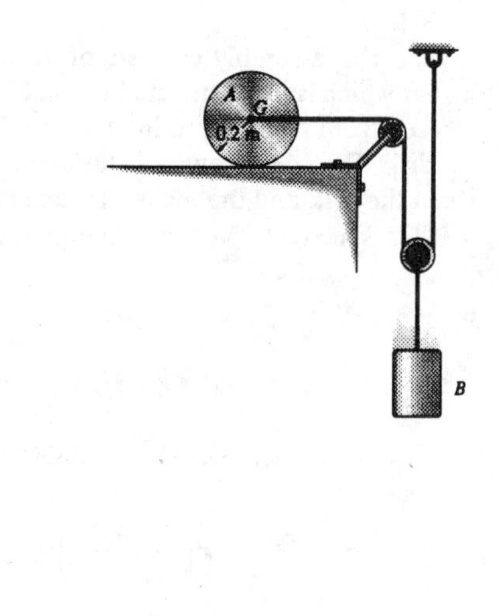

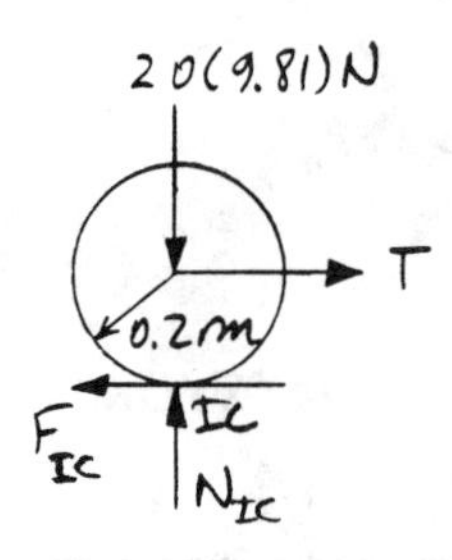

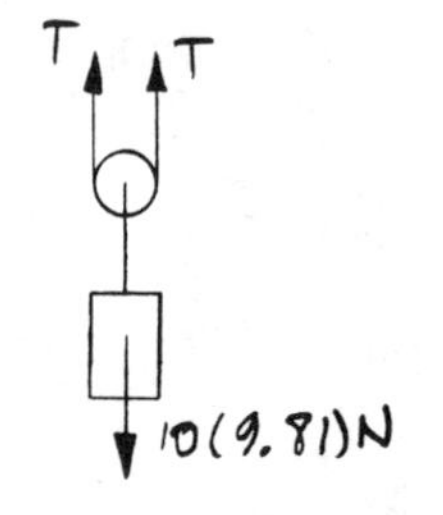

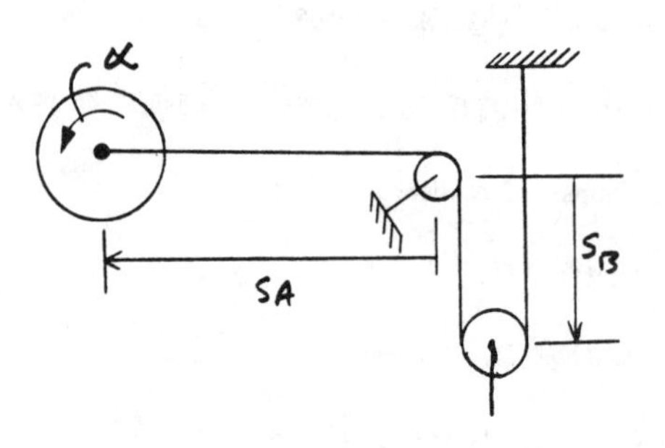

Disk :

$\circlearrowleft + \Sigma M_{IC} = \Sigma(M_k)_{IC}$; $T(0.2) = -\left[\frac{1}{2}(20)(0.2)^2 + 20(0.2)^2\right]\alpha$ (1)

Block :

$+\downarrow \Sigma F_y = m(a_G)_y$; $10(9.81) - 2T = 10a_B$ (2)

Kinematics :

$2s_B + s_A = l$

$2a_B = -a_A$

Also,

$a_A = 0.2\alpha$

Thus,

$a_B = -0.1\alpha$ (3)

Note the directions for α and a_B are the same for all equations.
Solving Eqs. (1) – (3) :

$a_B = 0.755 \text{ m/s}^2 = 0.755 \text{ m/s}^2 \downarrow$ **Ans**

$\alpha = -7.55 \text{ rad/s}^2 = 7.55 \text{ rad/s}^2$ ↻ **Ans**

$T = 45.3 \text{ N}$ **Ans**

17-114. Determine the minimum coefficient of static friction between the disk and the surface in Prob. 17-113 so that the disk will roll without slipping. Neglect the mass of the pulleys.

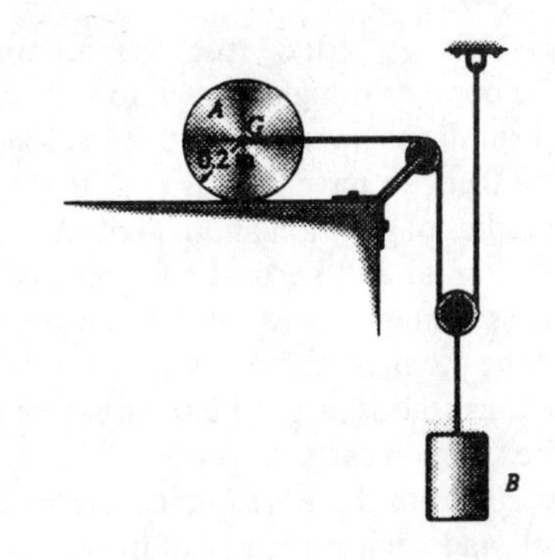

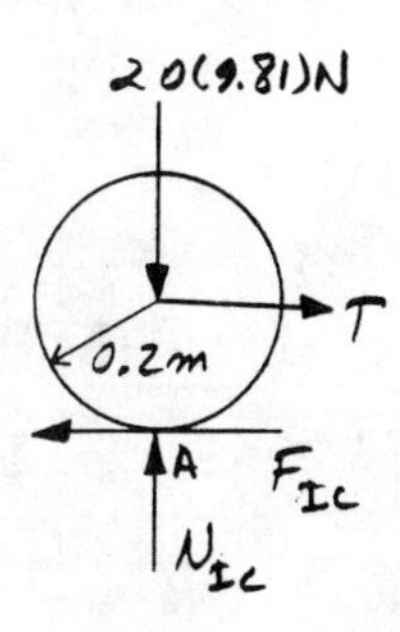

Disk :

$$\circlearrowleft + \Sigma M_{IC} = \Sigma(M_k)_{IC}; \qquad T(0.2) = -\left[\frac{1}{2}(20)(0.2)^2 + 20(0.2)^2\right]\alpha \qquad (1)$$

$$\overset{+}{\leftarrow} \Sigma F_x = m(a_G)_x; \qquad -T + F_A = 20a_A$$

$$+\uparrow \Sigma F_y = m(a_G)_y; \qquad N_A - 20(9.81) = 0$$

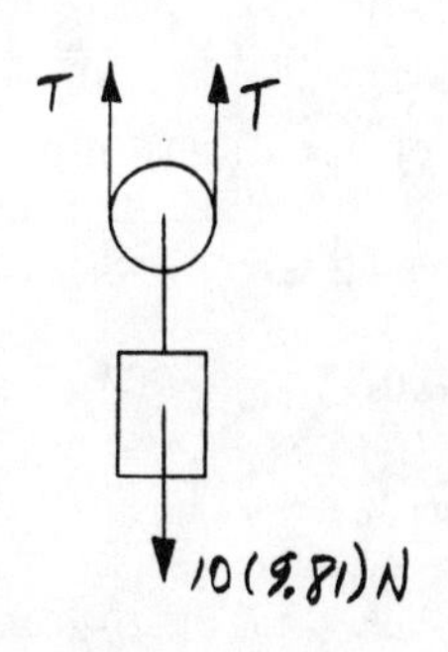

Block :

$$+\downarrow \Sigma F_y = m(a_G)_y; \qquad 10(9.81) - 2T = 10a_B \qquad (2)$$

Kinematics :

$$2s_B + s_A = l$$

$$2a_B = -a_A$$

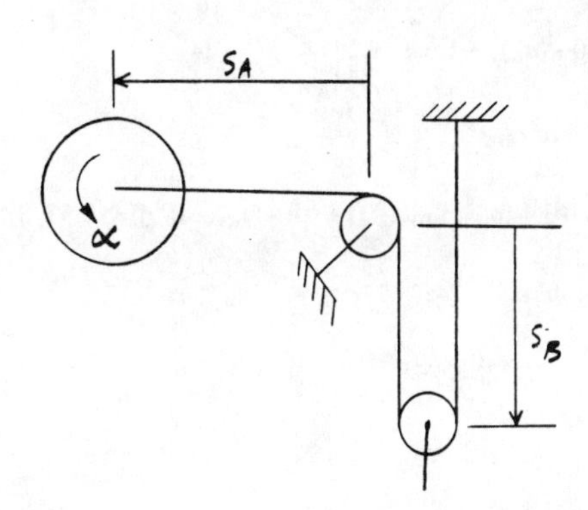

Also,

$$a_A = 0.2\alpha$$

Thus,

$$a_B = -0.1\alpha \qquad (3)$$

Note the directions for α and a_B are the same for all equations.
Solving Eqs. (1) – (3) :

$$a_B = 0.755 \text{ m/s}^2$$

$$\alpha = -7.55 \text{ rad/s}^2$$

$$T = 45.3 \text{ N}$$

Also,

$$a_A = 0.2(-7.55) = -1.509 \text{ m/s}^2, \qquad N_A = 196.2 \text{ N}, \qquad F_A = 15.09 \text{ N}$$

$$\mu_{min} = \frac{15.09}{196.2} = 0.0769 \qquad \textbf{Ans}$$

17-115. A "lifted" truck can become a road hazard since the bumper is high enough to ride up a standard car in the event the car is rear-ended. As a model of this case consider the truck to have a mass of 2.70 Mg, a mass center G, and a radius of gyration about G of $k_G = 1.45$ m. Determine the horizontal and vertical components of acceleration of the mass center G, and the angular acceleration of the truck, at the moment its front wheels at C have just left the ground and its smooth front bumper begins to ride up the back of the stopped car so that point B has a velocity of $v_B = 8$ m/s at 20° from the horizontal. Assume the wheels are free to roll, and neglect the size of the wheels and the deformation of the material.

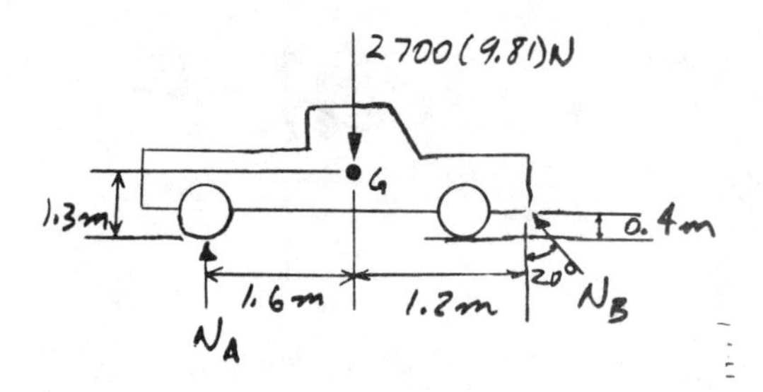

$\xrightarrow{+} \Sigma F_x = m(a_G)_x; \qquad -N_B \sin 20° = 2700(a_G)_x \qquad (1)$

$+\uparrow \Sigma F_y = m(a_G)_y; \qquad N_A + N_B \cos 20° - 2700(9.81) = 2700(a_G)_y \qquad (2)$

$\circlearrowleft + \Sigma M_G = I_G \alpha; \qquad N_B \cos 20°(1.2) - N_B \sin 20°(0.9) - N_A(1.6) = 2700(1.45)^2 \alpha \qquad (3)$

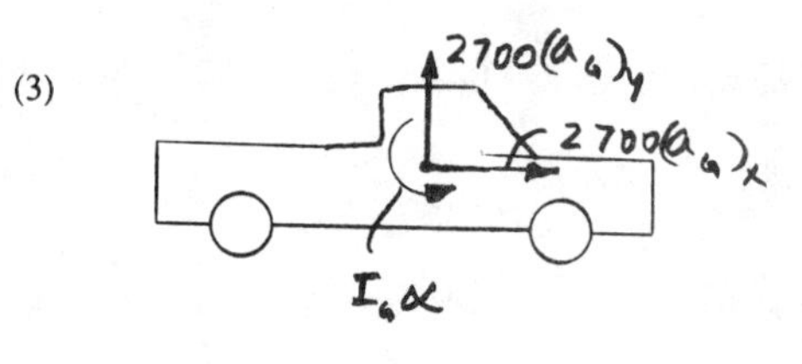

$\mathbf{v}_A = \mathbf{v}_B + \mathbf{v}_{A/B}$

$v_A \mathbf{i} = 8\cos 20°\mathbf{i} + 8\sin 20°\mathbf{j} + (\omega\mathbf{k}) \times (-2.8\mathbf{i} - 0.4\mathbf{j})$

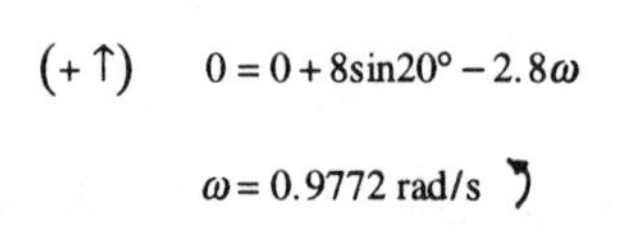

$(+\uparrow) \qquad 0 = 0 + 8\sin 20° - 2.8\omega$

$\omega = 0.9772$ rad/s ↻

$\mathbf{a}_A = \mathbf{a}_G + \alpha \times \mathbf{r}_{A/G} - \omega^2 \mathbf{r}_{A/G}$

$a_A \mathbf{i} = (a_G)_x \mathbf{i} + (a_G)_y \mathbf{j} + \alpha\mathbf{k} \times (-1.6\mathbf{i} - 1.3\mathbf{j}) - (0.9772)^2(-1.6\mathbf{i} - 1.3\mathbf{j})$

$(+\uparrow) \qquad 0 = 0 + (a_G)_y - 1.6\alpha + 1.2414 \qquad (4)$

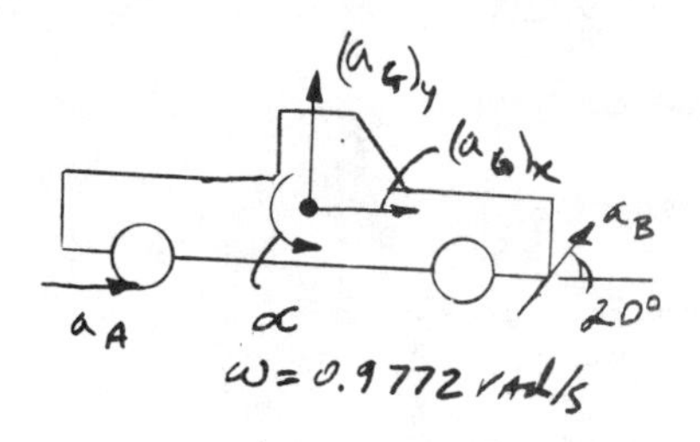

$\mathbf{a}_B = \mathbf{a}_G + \alpha \times \mathbf{r}_{B/G} - \omega^2 \mathbf{r}_{B/G}$

$a_B \cos 20°\mathbf{i} + a_B \sin 20°\mathbf{j} = (a_G)_x \mathbf{i} + (a_G)_y \mathbf{j} + \alpha\mathbf{k} \times (1.2\mathbf{i} - 0.9\mathbf{j}) - (0.9772)^2(1.2\mathbf{i} - 0.9\mathbf{j})$

$a_B \cos 20° = (a_G)_x + 0.9\alpha - 1.1459$

$a_B \sin 20° = (a_G)_y + 1.2\alpha + 0.85943$

or,

$1.0642(a_G)_x - 2.924(a_G)_y - 2.5508\alpha = 0 \qquad (5)$

Solving Eqs (1)–(5),

$N_A = 8.38$ kN $\qquad N_B = 14.4$ kN

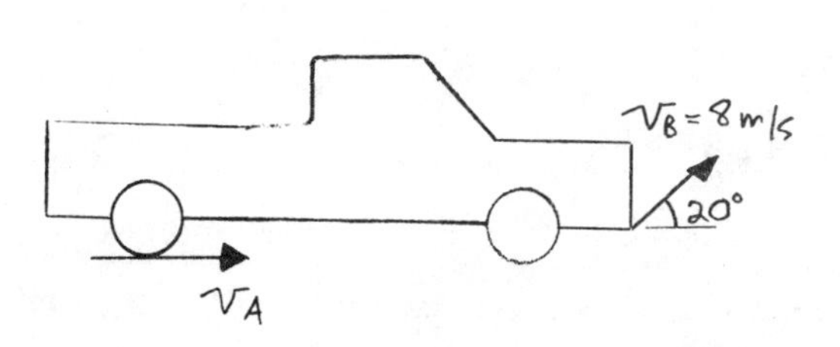

$(a_G)_x = -1.82$ m/s^2 = 1.82 m/s^2 ← **Ans**

$(a_G)_y = -1.69$ m/s^2 = 1.69 m/s^2 ↓ **Ans**

$\alpha = -0.283$ rad/s^2 = 0.283 rad/s^2 ↺ **Ans**

***17-116.** The solid ball of radius r and mass m rolls without slipping down the 60° trough. Determine its angular acceleration.

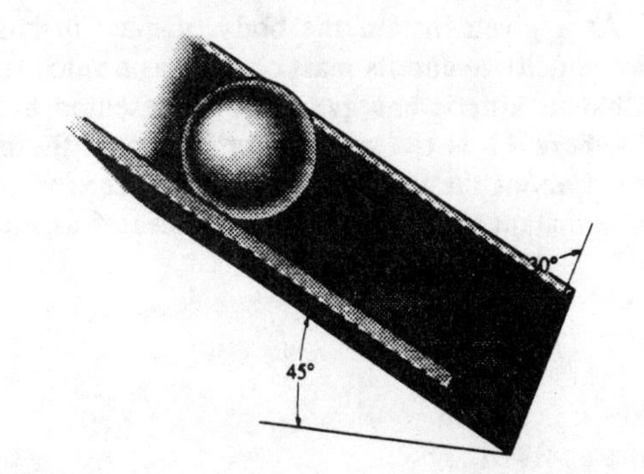

$$d = r\sin 30° = \frac{r}{2}$$

$$\Sigma M_{a-a} = \Sigma (M_k)_{a-a}; \quad mg\sin 45°\left(\frac{r}{2}\right) = \left[\frac{2}{5}mr^2 + m\left(\frac{r}{2}\right)^2\right]\alpha$$

$$\alpha = \frac{10g}{13\sqrt{2}\,r} \qquad \textbf{Ans}$$

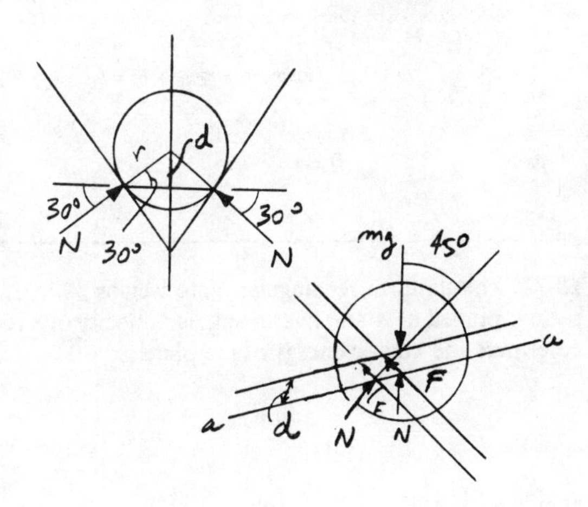

18-1. At a given instant the body of mass m has an angular velocity ω and its mass center has a velocity $\mathbf{v}_G$. Show that its kinetic energy can be represented as $T = \frac{1}{2}I_{IC}\omega^2$, where I_{IC} is the moment of inertia of the body computed about the instantaneous axis of zero velocity, located a distance $r_{G/IC}$ from the mass center as shown.

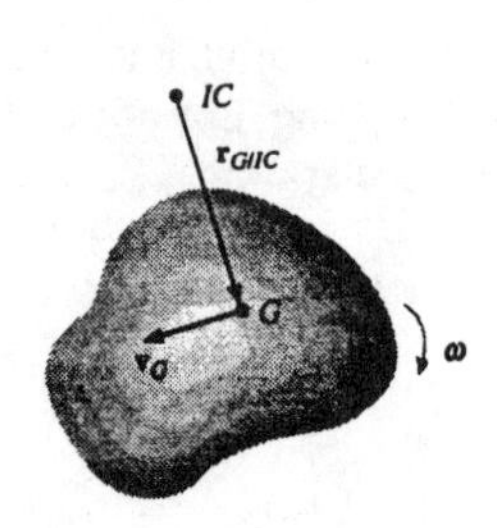

$$T = \frac{1}{2}mv_G^2 + \frac{1}{2}I_G\omega^2 \qquad \text{where } v_G = \omega r_{G/IC}$$

$$= \frac{1}{2}m(\omega r_{G/IC})^2 + \frac{1}{2}I_G\omega^2$$

$$= \frac{1}{2}\left(mr_{G/IC}^2 + I_G\right)\omega^2 \qquad \text{However } mr_{G/IC}^2 + I_G = I_{IC}$$

$$= \frac{1}{2}I_{IC}\omega^2 \qquad \textbf{Q.E.D.}$$

18-2. The uniform rectangular plate weighs 30 lb. If the plate is pinned at A and has an angular velocity of 3 rad/s, determine the kinetic energy of the plate.

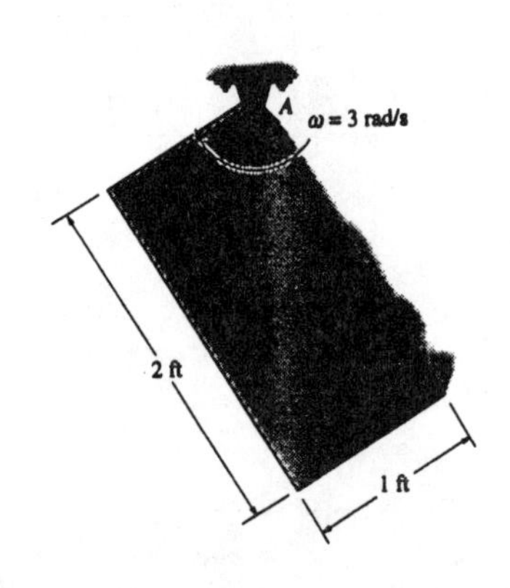

$$T = \frac{1}{2}mv_G^2 + \frac{1}{2}I_G\omega^2$$

$$T = \frac{1}{2}\left(\frac{30}{32.2}\right)\left((3)\sqrt{1^2 + (0.5)^2}\right)^2 + \frac{1}{2}\left[\frac{1}{12}\left(\frac{30}{32.2}\right)(1^2 + 2^2)\right](3)^2 = 6.99 \text{ ft}\cdot\text{lb} \qquad \textbf{Ans}$$

18-3. Determine the kinetic energy of the system of three links. Links AB and CD each weigh 10 lb, and link BC weighs 20 lb.

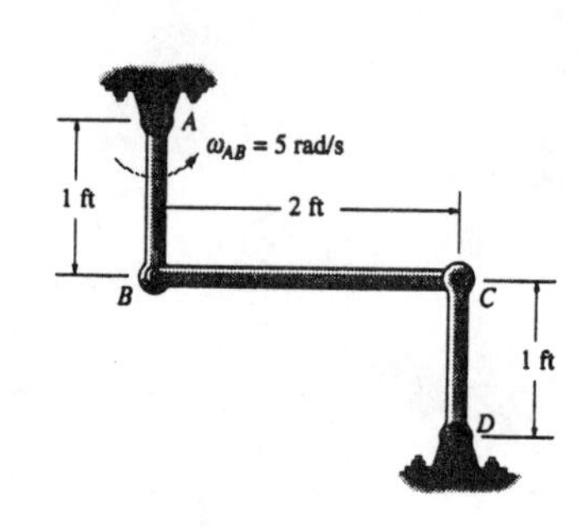

Link BC is subjected to general plane motion. Using the IC

$$r_{B/IC} = r_{C/IC} = r_{G/IC} = \infty$$

$$\omega_{BC} = \frac{v_B}{r_{B/IC}} = \frac{5(1)}{\infty} = 0$$

$$v_C = v_G = v_B = 5(1) = 5 \text{ ft/s}$$

$$\omega_{CD} = \frac{v_C}{r_{CD}} = \frac{5}{1} = 5 \text{rad/s}$$

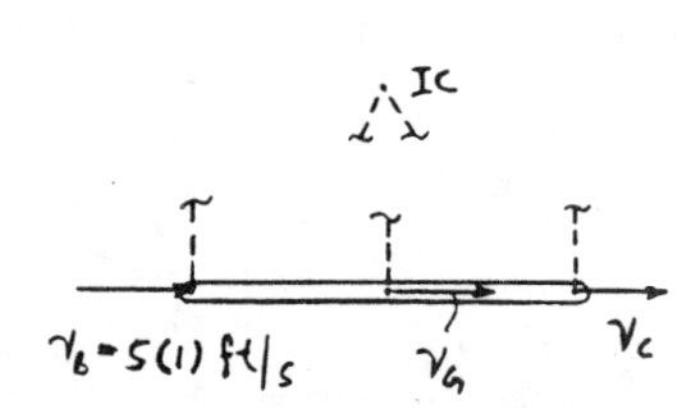

Kinetic energy :

$$T_{AB} = \frac{1}{2}I_A\omega_{AB}^2 = \frac{1}{2}\left[\frac{1}{3}\left(\frac{10}{32.2}\right)(1)^2\right](5)^2 = 1.2940 \text{ ft}\cdot\text{lb}$$

$$T_{BC} = \frac{1}{2}mv_G^2 + \frac{1}{2}I_G\omega_{BC}^2 = \frac{1}{2}\left[\frac{20}{32.2}\right](5)^2 + 0 = 7.7640 \text{ ft}\cdot\text{lb}$$

$$T_{CD} = \frac{1}{2}I_D\omega_{CD}^2 = \frac{1}{2}\left[\frac{1}{3}\left(\frac{10}{32.2}\right)(1)^2\right](5)^2 = 1.2940 \text{ ft}\cdot\text{lb}$$

$$T_T = 1.2940 + 7.7640 + 1.2940 = 10.4 \text{ ft}\cdot\text{lb} \qquad \textbf{Ans}$$

***18-4.** The double pulley consists of two parts that are attached to one another. It has a weight of 50 lb and a centroidal radius of gyration of $k_O = 0.6$ ft and is turning with an angular velocity of 20 rad/s clockwise. Determine the kinetic energy of the system. Assume that neither cable slips on the pulley.

$$T = \frac{1}{2}I_O\omega_O^2 + \frac{1}{2}m_A v_A^2 + \frac{1}{2}m_B v_B^2$$

$$T = \frac{1}{2}\left(\frac{50}{32.2}(0.6)^2\right)(20)^2 + \frac{1}{2}\left(\frac{20}{32.2}\right)\left[(20)(1)\right]^2 + \frac{1}{2}\left(\frac{30}{32.2}\right)\left[(20)(0.5)\right]^2 = 283 \text{ ft}\cdot\text{lb} \qquad \textbf{Ans}$$

18-5. The mechanism consists of two rods, AB and BC, which weigh 10 lb and 20 lb, respectively, and a 4-lb block at C. Determine the kinetic energy of the system at the instant shown, when the block is moving at 3 ft/s.

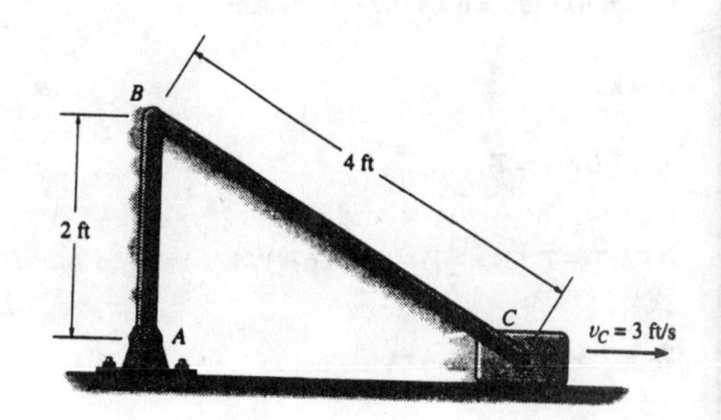

Link BC is subjected to general plane motion. Using the IC

$$r_{B/IC} = r_{C/IC} = r_{G/IC} = \infty$$

$$\omega_{BC} = \frac{v_C}{r_{C/IC}} = \frac{3}{\infty} = 0$$

$$v_B = v_G = v_C = 3 \text{ ft/s}$$

$$\omega_{AB} = \frac{v_B}{r_{AB}} = \frac{3}{2} = 1.5 \text{ rad/s}$$

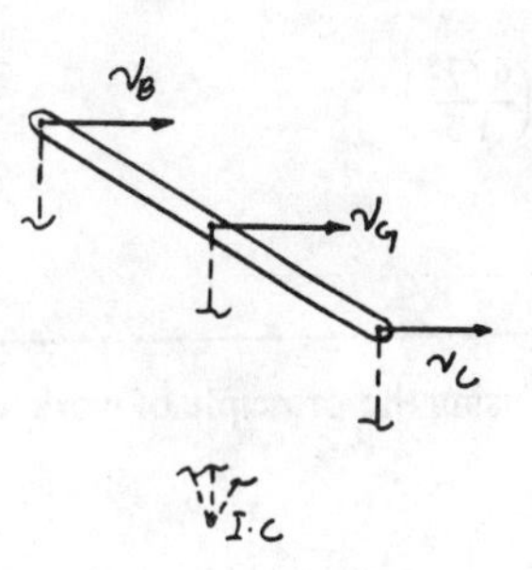

Kinetic energy : For the links

$$T_{AB} = \frac{1}{2}I_A\omega_{AB}^2 = \frac{1}{2}\left[\frac{1}{3}\left(\frac{10}{32.2}\right)(2)^2\right](1.5)^2 = 0.4658 \text{ ft}\cdot\text{lb}$$

$$T_{BC} = \frac{1}{2}mv_G^2 + \frac{1}{2}I_G\omega_{BC}^2 = \frac{1}{2}\left[\frac{20}{32.2}\right](3)^2 + 0 = 2.7950 \text{ ft}\cdot\text{lb}$$

For the block

$$T_C = \frac{1}{2}m_C v_C^2 = \frac{1}{2}\left(\frac{4}{32.2}\right)(3)^2 = 0.5590 \text{ ft}\cdot\text{lb}$$

$$T_T = 0.4658 + 2.7950 + 0.5590 = 3.82 \text{ ft}\cdot\text{lb} \qquad \textbf{Ans}$$

18-6. Solve Prob. 17-58 using the principle of work and energy.

$$T_1 + \Sigma U_{1-2} = T_2$$

$$0 + 30(8) = \frac{1}{2}\left[\left(\frac{400}{32.2}\right)(1.30)^2\right]\omega^2$$

$\omega = 4.78$ rad/s **Ans**

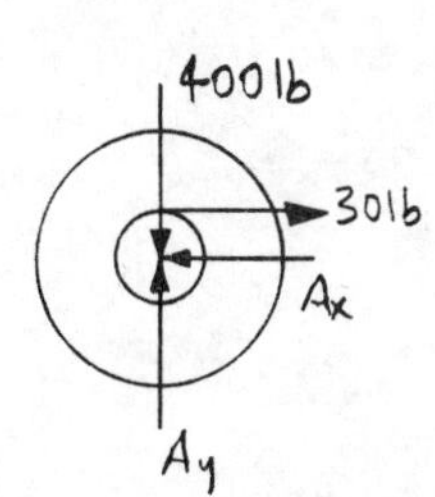

18-7. Solve Prob. 17-81 using the principle of work and energy.

System :

$T_1 + \Sigma U_{1-2} = T_2$

$$[0+0+0]+3(3) = \frac{1}{2}\left[\frac{1}{2}\left(\frac{15}{32.2}\right)(1.5)^2\right]\omega_A^2 + \frac{1}{2}\left[\frac{1}{2}\left(\frac{5}{32.2}\right)(0.5)^2\right]\omega_B^2 + \frac{1}{2}\left(\frac{3}{32.2}\right)v_C^2$$

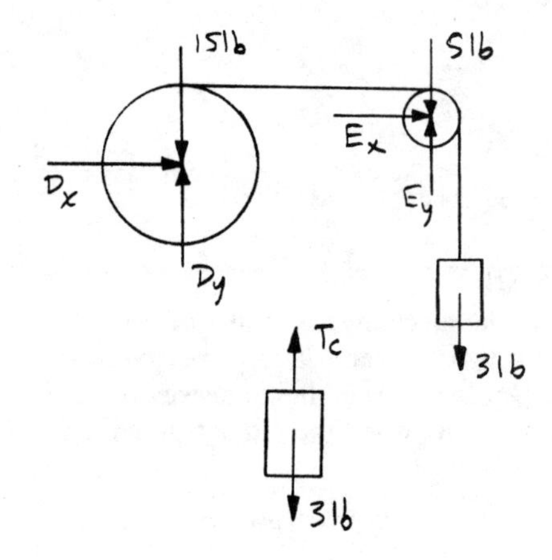

Kinematics :

$v_C = \omega_A(1.5) = \omega_B(0.5)$

Solving :

$v_C = 6.677$ ft/s $= 6.68$ ft/s **Ans**

Block :

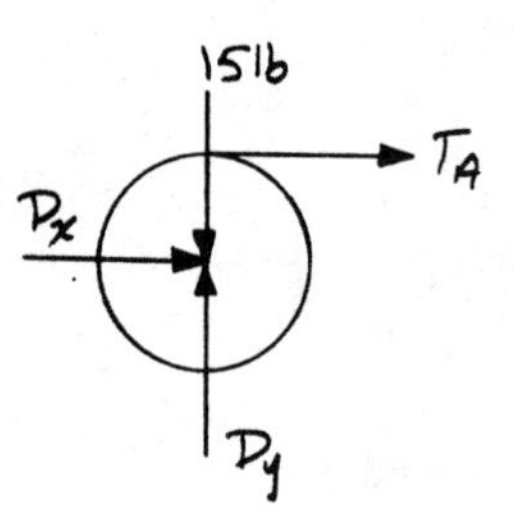

$T_1 + \Sigma U_{1-2} = T_2$

$$0+3(3) - T_C(3) = \frac{1}{2}\left(\frac{3}{32.2}\right)(6.677)^2$$

$T_C = 2.31$ lb **Ans**

Disk A :

$T_1 + \Sigma U_{1-2} = T_2$

$$0 + T_A(3) = \frac{1}{2}\left[\frac{1}{2}\left(\frac{15}{32.2}\right)(1.5)^2\right]\left(\frac{6.677}{1.5}\right)^2$$

$T_A = 1.73$ lb **Ans**

***18-8.** Solve Prob. 17-72 using the principle of work and energy.

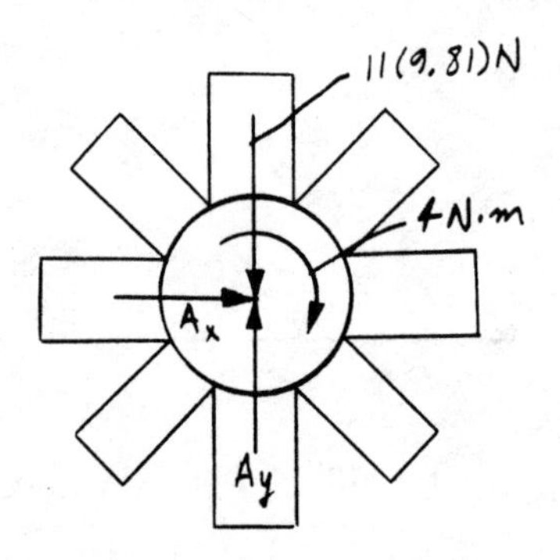

$$I = \frac{1}{2}(3)(0.025)^2 + 8\left[\frac{1}{12}(1)(0.1)^2 + 1(0.075)^2\right] = 0.052604 \text{ kg}\cdot\text{m}^2$$

$T_1 + \Sigma U_{1-2} = T_2$

$$0 + 4(2\pi)(15) = \frac{1}{2}(0.052604)\omega^2$$

$\omega = 120$ rad/s **Ans**

18-9. A force of $P = 20$ N is applied to the cable, which causes the 175-kg reel to turn since it is resting on the two rollers A and B of the dispenser. Determine the angular velocity of the reel after it has made two revolutions starting from rest. Neglect the mass of the rollers and the mass of the cable. The radius of gyration of the reel about its center axis is $k_G = 0.42$ m.

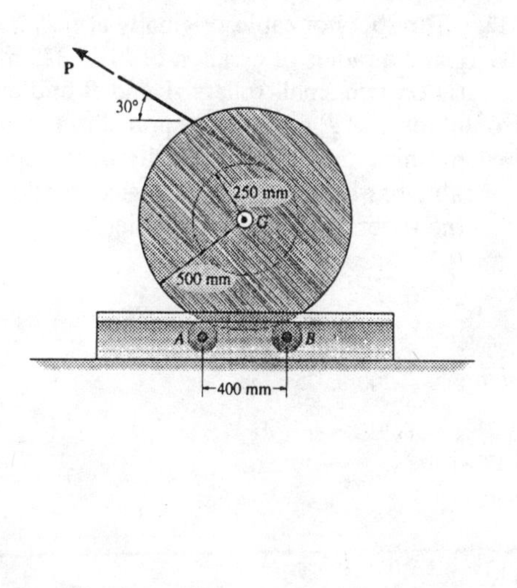

$T_1 + \Sigma U_{1-2} = T_2$

$0 + 20(2)(2\pi)(0.250) = \frac{1}{2}\left[175(0.42)^2\right]\omega^2$

$\omega = 2.02$ rad/s **Ans**

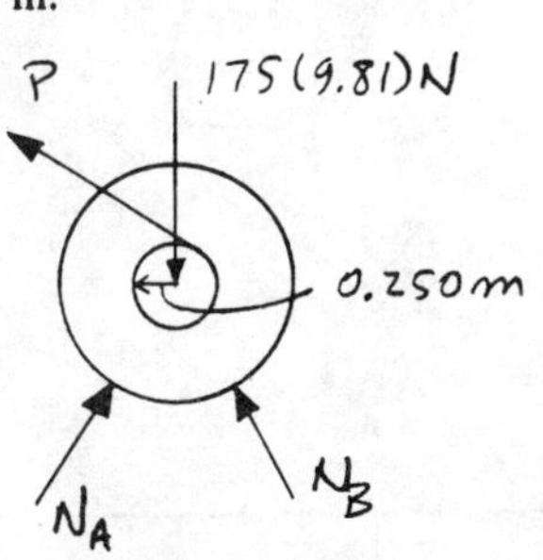

18-10. A force of $P = 20$ N is applied to the cable, which causes the 175-kg reel to turn without slipping on the two rollers A and B of the dispenser. Determine the angular velocity of the reel after it has made two revolutions starting from rest. Neglect the mass of the cable. Each roller can be considered as an 18-kg cylinder, having a radius of 0.1 m. The radius of gyration of the reel about its center axis is $k_G = 0.42$ m.

System :

$T_1 + \Sigma U_{1-2} = T_2$

$[0+0+0] + 20(2)(2\pi)(0.250) = \frac{1}{2}\left[175(0.42)^2\right]\omega^2 + 2\left[\frac{1}{2}\left(\frac{1}{2}(18)(0.1)^2\right]\omega_r^2$

$v = \omega_r(0.1) = \omega(0.5)$

$\omega_r = 5\omega$

Solving :

$\omega = 1.78$ rad/s **Ans**

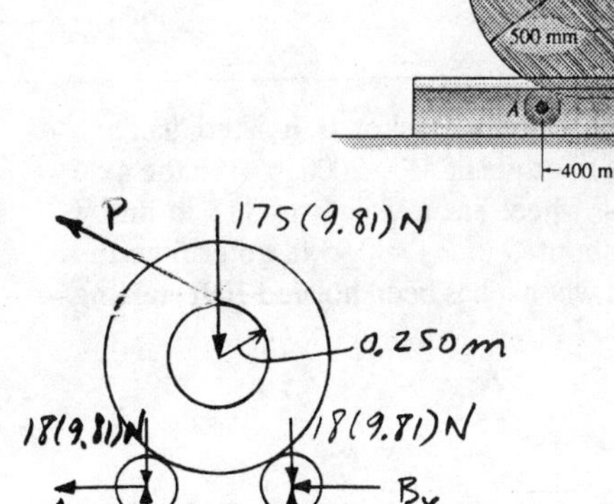

18-11. The rotary screen S is used to wash limestone. When empty it has a mass of 800 kg and a radius of gyration of $k_G = 1.75$ m. Rotation is achieved by applying a torque of $M = 280$ N·m about the drive wheel A. If no slipping occurs at A and the supporting wheel at B is free to roll, determine the angular velocity of the screen after it has rotated 5 revolutions. Neglect the mass of A and B.

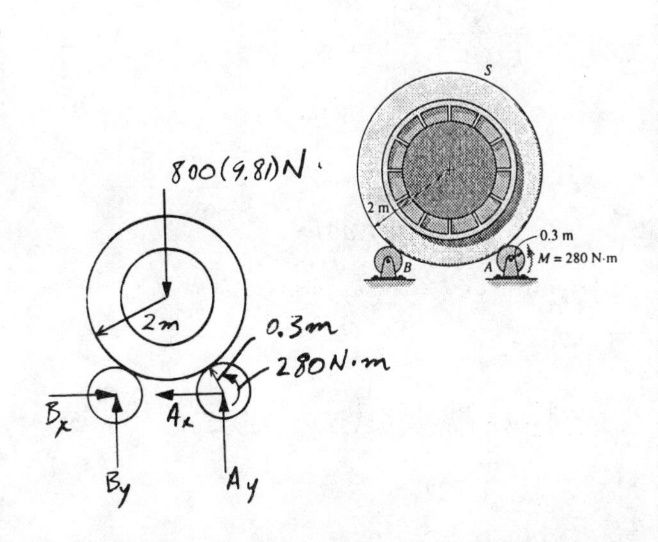

$T_1 + \Sigma U_{1-2} = T_2$

$0 + 280(\theta_A) = \frac{1}{2}\left[800(1.75)^2\right]\omega^2$

$\theta_S(2) = \theta_A(0.3)$

$5(2\pi)(2) = \theta_A(0.3)$ Thus

$\theta_A = 209.4$ rad $\omega = 6.92$ rad/s **Ans**

***18-12.** The spool of cable, originally at rest, has a mass of 200 kg and a radius of gyration of $k_G = 325$ mm. If the spool rests on two small rollers A and B and a constant horizontal force of $P = 400$ N is applied to the end of the cable, determine the angular velocity of the spool when 8 m of cable has been unwound. Neglect friction and the mass of the rollers and unwound cable.

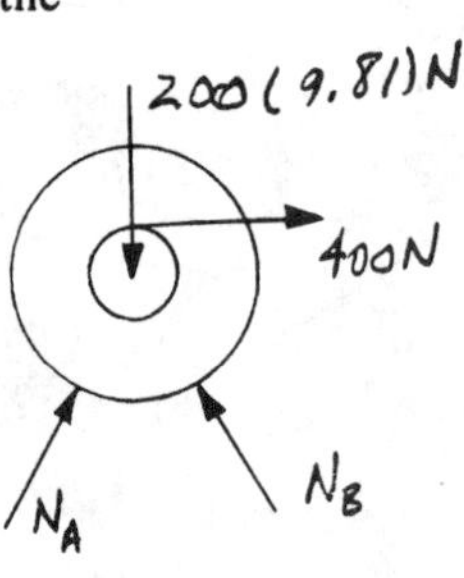

$$T_1 + \Sigma U_{1-2} = T_2$$

$$0 + (400)(8) = \frac{1}{2}\left[200(0.325)^2\right]\omega_2^2$$

$\omega_2 = 17.4$ rad/s **Ans**

18-13. The pendulum of the Charpy impact machine has a mass of 50 kg and a radius of gyration of $k_A = 1.75$ m. If it is released from rest when $\theta = 0°$, determine its angular velocity just before it strikes the specimen S, $\theta = 90°$.

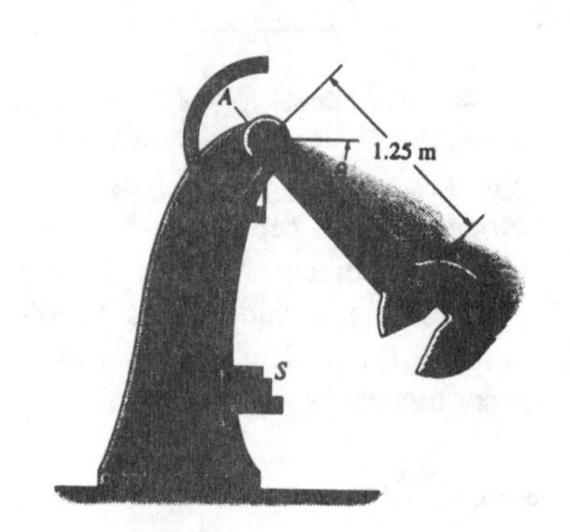

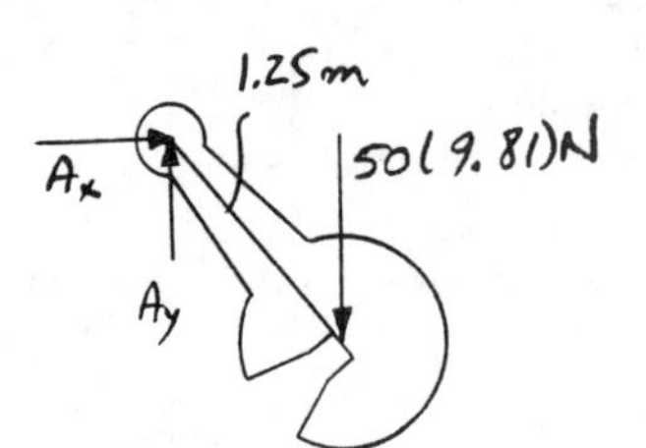

$$T_1 + \Sigma U_{1-2} = T_2$$

$$0 + (50)(9.81)(1.25) = \frac{1}{2}\left[(50)(1.75)^2\right]\omega_2^2$$

$\omega_2 = 2.83$ rad/s **Ans**

18-14. The 1500-lb cement bucket is hoisted using a motor that supplies a torque of $M = 2000$ lb·ft to the axle of the wheel. If the wheel has a weight of 115 lb and a radius of gyration about O of $k_O = 0.95$ ft, determine the speed of the bucket when it has been hoisted 10 ft starting from rest.

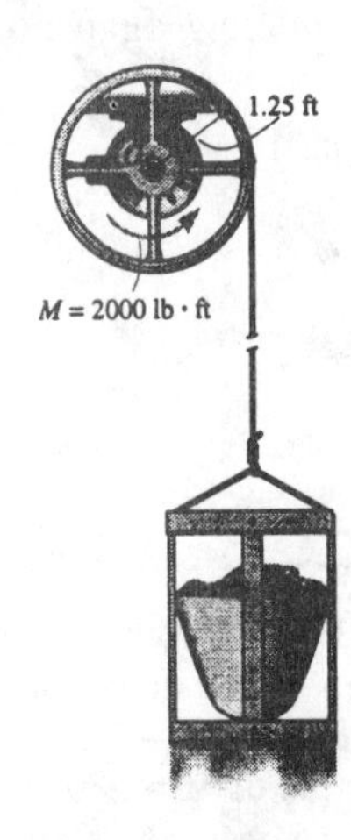

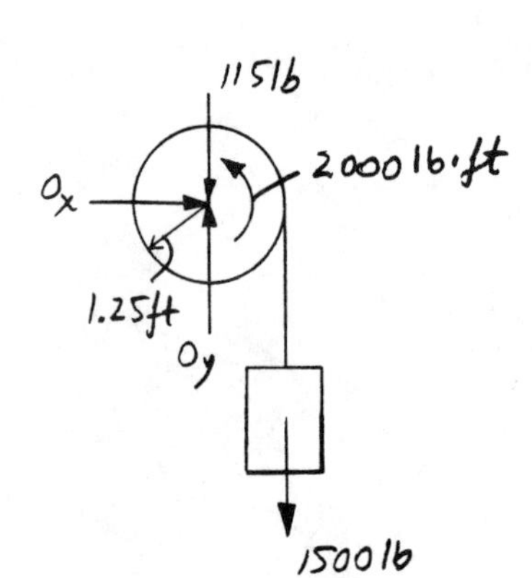

$$T_1 + \Sigma U_{1-2} = T_2$$

$$0 + 2000\left(\frac{10}{1.25}\right) - 1500(10) = \frac{1}{2}\left(\frac{1500}{32.2}\right)v^2 + \frac{1}{2}\left[\left(\frac{115}{32.2}\right)(0.95)^2\right]\left(\frac{v}{1.25}\right)^2$$

$v = 6.41$ ft/s **Ans**

18-15. The 10-kg pulley has a radius of gyration about O of $k_O = 0.21$ m. If a motor M supplies a force to the cable of $P = 800(3 - 2e^{-x})$ N, where x is the amount of cable wound up in meters, determine the speed of the 50-kg crate when it has been hoisted 2 m starting from rest. Neglect the mass of the cable and assume the cable does not slip on the pulley.

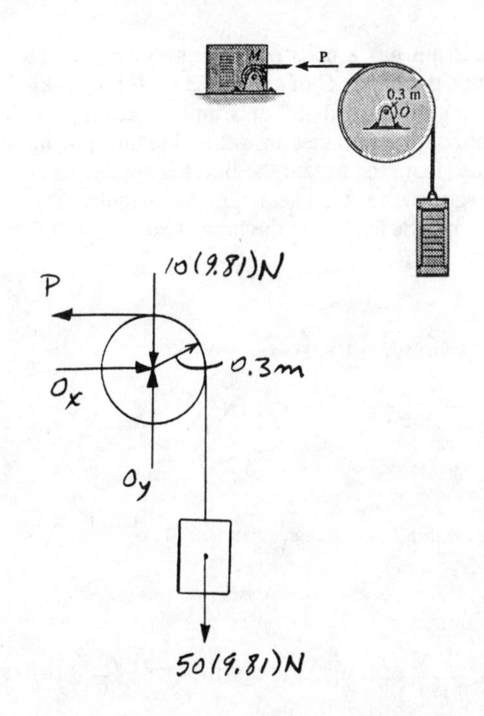

$$T_1 + \Sigma U_{1-2} = T_2$$

$$0 + \int_0^2 800(3 - 2e^{-x})\,dx - 50(9.81)(2) = \frac{1}{2}[(10)(0.21)^2]\left(\frac{v}{0.3}\right)^2 + \frac{1}{2}(50)(v)^2$$

$$800(3x + 2e^{-x})|_0^2 - 981 = 27.45v^2$$

$$2435.54 = 27.45v^2$$

$v = 9.42$ m/s **Ans**

***18-16.** The drum has a mass of 50 kg and a radius of gyration about the pin at O of $k_O = 0.23$ m. Starting from rest, the suspended 15-kg block B is allowed to fall 3 m without applying the brake ACD. Determine the speed of the block at this instant. If the coefficient of kinetic friction at the brake pad C is $\mu_k = 0.5$, determine the force P that must be applied at the brake handle which will then stop the block after it descends *another* 3 m. Neglect the thickness of the handle.

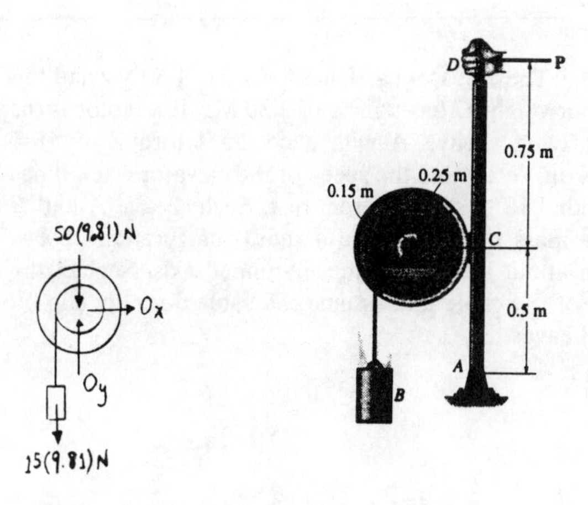

Before braking :

$$T_1 + \Sigma U_{1-2} = T_2$$

$$0 + 15(9.81)(3) = \frac{1}{2}(15)v_B^2 + \frac{1}{2}[50(0.23)^2]\left(\frac{v_B}{0.15}\right)^2$$

$v_B = 2.58$ m/s **Ans**

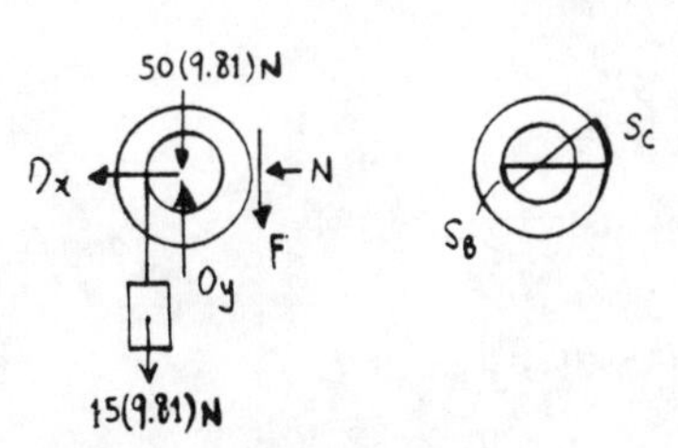

$$\frac{s_B}{0.15} = \frac{s_C}{0.25}$$

Set $s_B = 3$ m, then $s_C = 5$ m

$$T_1 + \Sigma U_{1-2} = T_2$$

$$0 - F(5) + 15(9.81)(6) = 0$$

$$F = 176.6 \text{ N}$$

$$N = \frac{176.6}{0.5} = 353.2 \text{ N}$$

Brake arm :

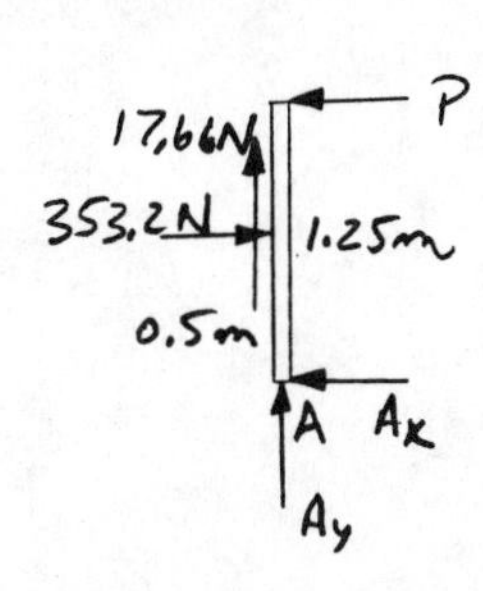

$\circlearrowleft + \Sigma M_A = 0; \quad -353.2(0.5) + P(1.25) = 0$

$P = 141$ N **Ans**

18-17. The drum has a mass of 50 kg and a radius of gyration about the pin at O of $k_O = 0.23$ m. If the 15-kg block is moving downward at 3 m/s, and a force of $P =$ 100 N is applied to the brake arm, determine how far the block descends from the instant the brake is applied until it stops. Neglect the thickness of the handle. The coefficient of kinetic friction at the brake pad is $\mu_k = 0.5$.

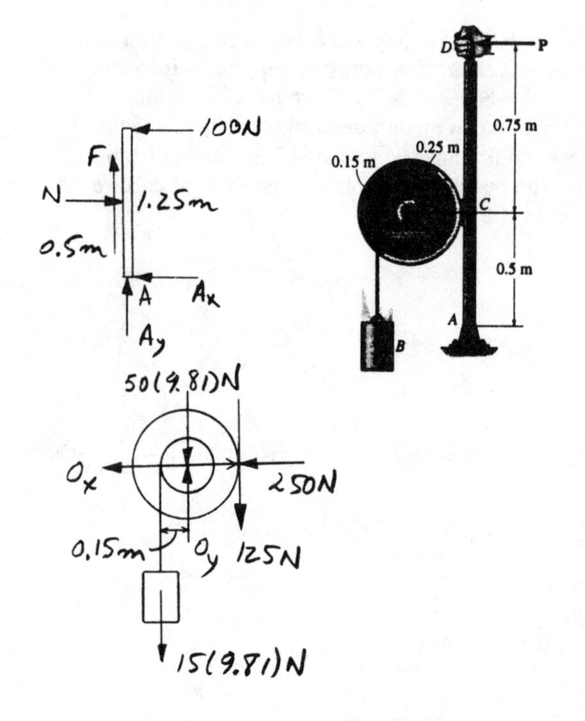

Brake arm :

$$+\Sigma M_A = 0; \qquad -N(0.5) + 100(1.25) = 0$$

$$N = 250 \text{ N}$$

$$F = 0.5(250) = 125 \text{ N}$$

If block descends s, then F acts through a distance $s' = s\left(\frac{0.25}{0.15}\right)$

$$T_1 + \Sigma U_{1-2} = T_2$$

$$\frac{1}{2}\left[(50)(0.23)^2\right]\left(\frac{3}{0.15}\right)^2 + \frac{1}{2}(15)(3)^2 + 15(9.81)(s) - 125(s)\left(\frac{0.25}{0.15}\right) = 0$$

$s = 9.75$ m **Ans**

18-18. The elevator car E has a mass of 1.80 Mg and the counterweight C has a mass of 2.30 Mg. If a motor turns the driving sheave A with a constant torque of $M =$ 100 N·m, determine the speed of the elevator when it has ascended 10 m starting from rest. Each sheave A and B has a mass of 150 kg and a radius of gyration of $k =$ 0.2 m about its mass center or pinned axis. Neglect the mass of the cable and assume the cable does not slip on the sheaves.

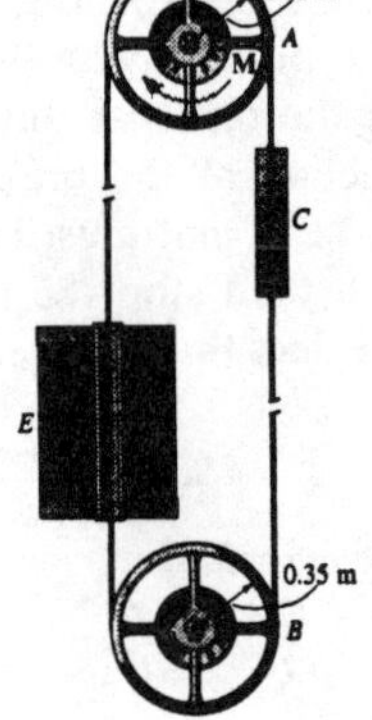

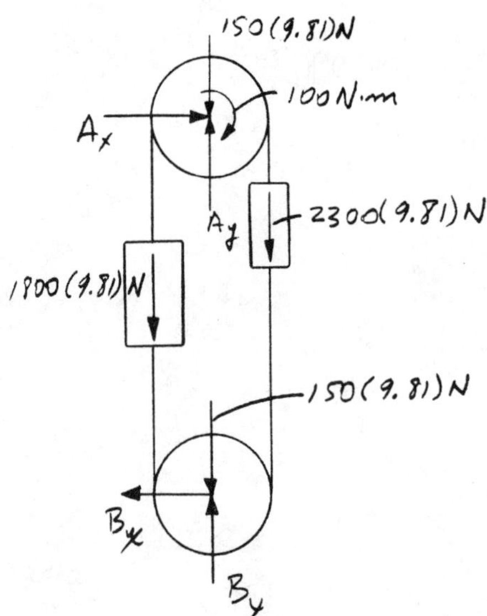

$$\theta = \frac{10}{0.35} = 28.57 \text{ rad}$$

$$T_1 + \Sigma U_{1-2} = T_2$$

$$0 + 2300(9.81)(10) - 1800(9.81)(10) + 100(28.57)$$

$$= \frac{1}{2}(1800)(v)^2 + \frac{1}{2}(2300)(v)^2 + (2)\frac{1}{2}\left[150(0.2)^2\right]\left(\frac{v}{0.35}\right)^2$$

$$51\,907.1 = 2099v^2$$

$v = 4.97$ m/s **Ans**

18-19. The elevator car E has a mass of 1.80 Mg and the counterweight C has a mass of 2.30 Mg. If a motor turns the driving sheave A with a torque of $M = (0.06\theta^2 + 7.5)$ N·m, where θ is in radians, determine the speed of the elevator when it has ascended 12 m starting from rest. Each sheave A and B has a mass of 150 kg and a radius of gyration of $k = 0.2$ m about its mass center or pinned axis. Neglect the mass of the cable and assume the cable does not slip on the sheaves.

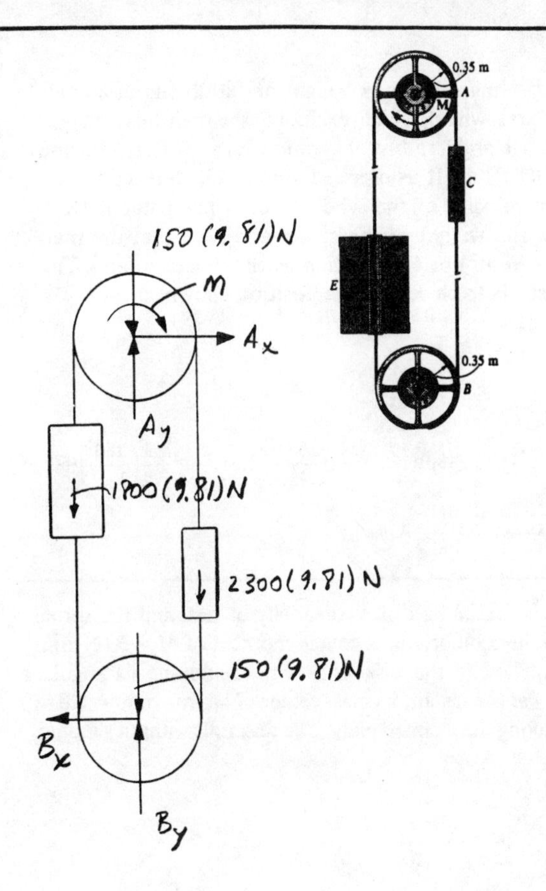

$$\theta = \frac{12}{0.35} = 34.29 \text{ rad}$$

$$T_1 + \Sigma U_{1-2} = T_2$$

$$0 + 2300(9.81)(12) - 1800(9.81)(12) + \int_0^{34.29} (0.06\theta^2 + 7.5)\,d\theta$$

$$= \frac{1}{2}(1800)(v)^2 + \frac{1}{2}(2300)(v)^2 + (2)\frac{1}{2}\left[150(0.2)^2\right]\left(\frac{v}{0.35}\right)^2$$

$$58860 + \left(0.02\theta^3 + 7.5\theta\right)\Big|_0^{34.29} = 2098.98v^2$$

$v = 5.34$ m/s **Ans**

***18-20.** The wheel has a mass of 100 kg and a radius of gyration $k_O = 0.2$ m. A motor supplies a torque $M = (40\theta + 900)$ N·m, where θ is in radians, about the drive shaft at O. Determine the speed of the loading car, which has a mass of 300 kg, after it travels $s = 4$ m. Initially the car is at rest when $s = 0$ and $\theta = 0°$. Neglect the mass of the attached cable and the mass of the car's wheels.

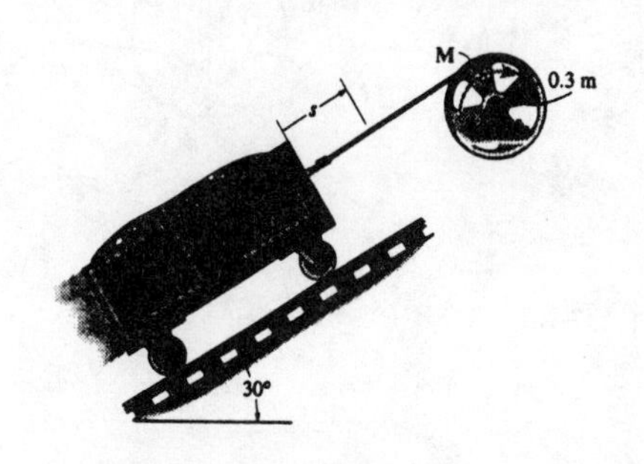

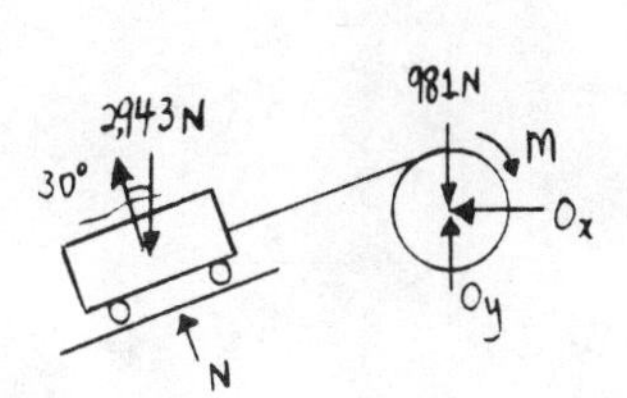

$$s = 0.3\theta = 4$$

$$\theta = 13.33 \text{ rad}$$

$$T_1 + \Sigma U_{1-2} = T_2$$

$$[0+0] + \int_0^{13.33} (40\theta + 900)\,d\theta - 300(9.81)\sin 30°(4) = \frac{1}{2}(300)v_C^2 + \frac{1}{2}\left[100(0.20)^2\right]\left(\frac{v_C}{0.3}\right)^2$$

$v_C = 7.49$ m/s **Ans**

18-21. A man having a weight of 180 lb sits in a chair of the Ferris wheel, which, excluding the man, has a weight of 15 000 lb and a radius of gyration $k_O = 37$ ft. If a torque $M = 80(10^3)$ lb·ft is applied about O, determine the angular velocity of the wheel after it has rotated 180°. Neglect the weight of the chairs and note that the man remains in an upright position as the wheel rotates. The wheel starts from rest in the position shown.

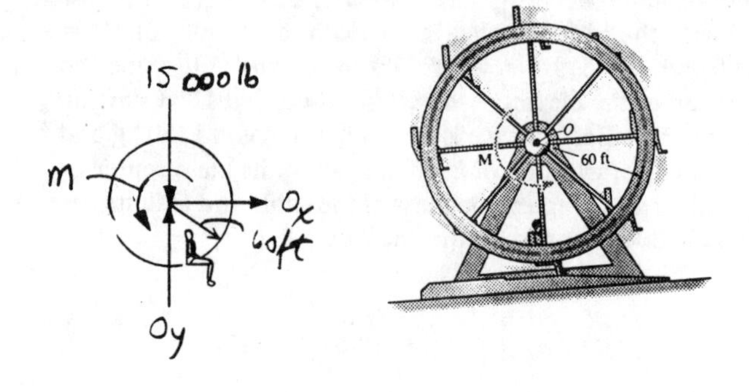

$$T_1 + \Sigma U_{1-2} = T_2$$

$$0 + 80(10^3)(\pi) - (180)(120) = \frac{1}{2}\left[\left(\frac{15\,000}{32.2}\right)(37)^2\right]\omega^2 + \frac{1}{2}\left(\frac{180}{32.2}\right)(60\omega)^2$$

$\omega = 0.836$ rad/s **Ans**

18-22. The 20-kg disk is originally at rest, and the spring holds it in equilibrium. A couple moment of $M = 30$ N·m is then applied to the disk as shown. Determine its angular velocity at the instant its mass center G has moved $s = 0.8$ m down along the inclined plane. The disk rolls without slipping.

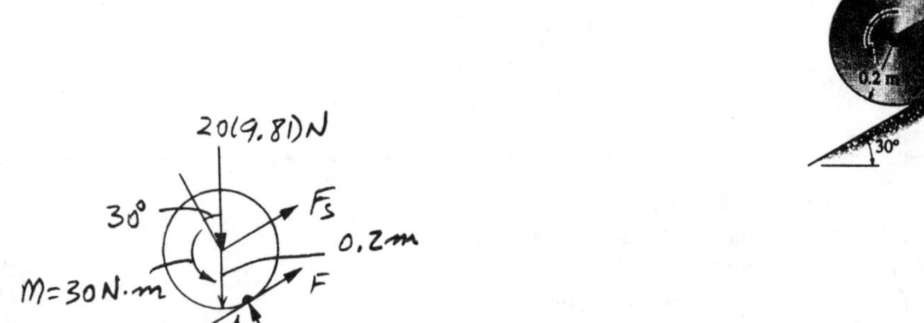

Initial tension in spring :

$$+\Sigma M_A = 0; \quad -F_s(0.2) + 20(9.81)\sin 30°(0.2) = 0$$

$$F_s = 98.1 \text{ N}$$

$$s_1 = \frac{98.1}{150} = 0.654 \text{ m}$$

When $s = 0.8$ m the disk rotates $\theta = \frac{0.8}{0.2} = 4$ rad

$$T_1 + \Sigma U_{1-2} = T_2$$

$$0 + 20(9.81)(0.8\sin 30°) + 30(4) - \left[\frac{1}{2}(150)(0.8 + 0.654)^2 - \frac{1}{2}(150)(0.654)^2\right]$$

$$= \frac{1}{2}\left[\frac{1}{2}(20)(0.2)^2\right]\omega^2 + \frac{1}{2}(20)(0.2\omega)^2$$

$$72 = 0.6\omega^2$$

$\omega = 11.0$ rad/s **Ans**

18-23. The 20-kg disk is originally at rest, and the spring holds it in equilibrium. A couple moment of $M = 30\ \text{N}\cdot\text{m}$ is then applied to the disk as shown. Determine how far the center of mass of the disk travels down along the incline, measured from the equilibrium position, before it stops. The disk rolls without slipping.

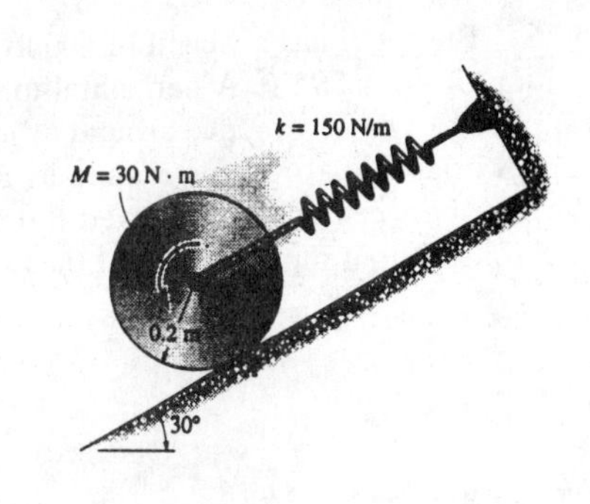

$+\Sigma M_A = 0; \qquad -F_s(0.2) + 20(9.81)\sin 30°(0.2) = 0$

$F_s = 98.1\ \text{N}$

$s_1 = \dfrac{98.1}{150} = 0.654\ \text{m}$

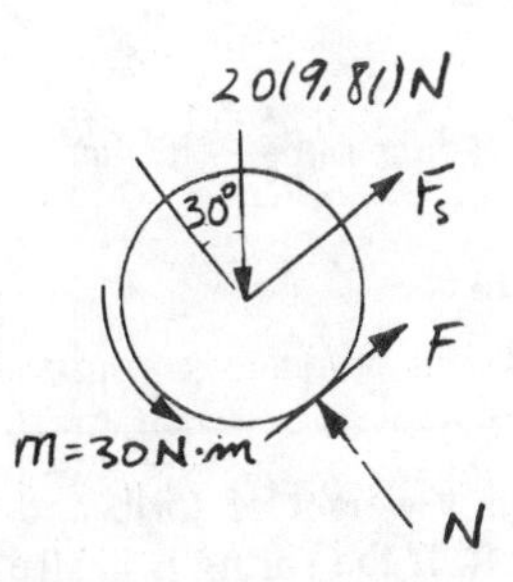

When G moves s, the disk rotates $\theta = \dfrac{s}{0.2}$

$T_1 + \Sigma U_{1-2} = T_2$

$$0 + 20(9.81)(s)\sin 30° + 30\left(\frac{s}{0.2}\right) - \left[\frac{1}{2}(150)(s+0.654)^2 - \frac{1}{2}(150)(0.654)^2\right] = 0$$

$248.1s = 75\left(s^2 + 1.308s + 0.4277\right) - 32.08$

$248.1 = 75s + 98.1$

$s = 2.00\ \text{m}$ **Ans**

***18-24.** The spool has a mass of 60 kg and a radius of gyration $k_G = 0.3$ m. If it is released from rest, determine how far its center descends down the smooth plane before it attains an angular velocity of $\omega = 6$ rad/s. Neglect friction and the mass of the cord which is wound around the central core.

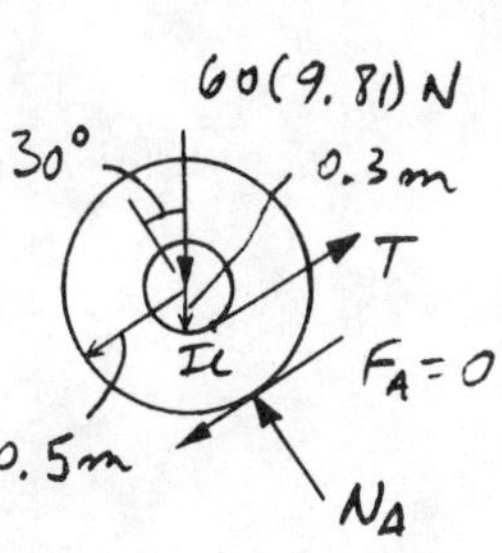

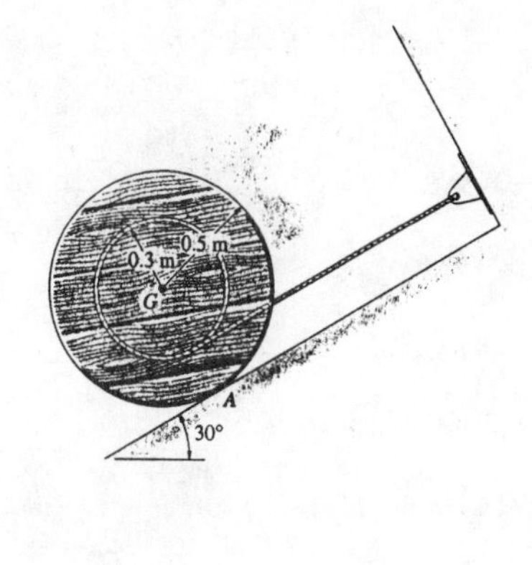

$T_1 + \Sigma U_{1-2} = T_2$

$$0 + 60(9.81)\sin 30°(s) = \frac{1}{2}\left[60(0.3)^2\right](6)^2 + \frac{1}{2}(60)\left[0.3(6)\right]^2$$

$s = 0.661\ \text{m}$ **Ans**

18-25. Solve Prob. 18-24 if the coefficient of kinetic friction between the spool and plane at A is $\mu_k = 0.2$.

$\dfrac{s_G}{0.3} = \dfrac{s_A}{(0.5-0.3)}$

$s_A = 0.6667 s_G$

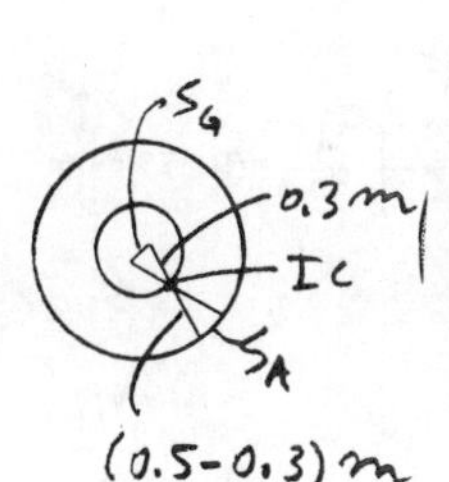

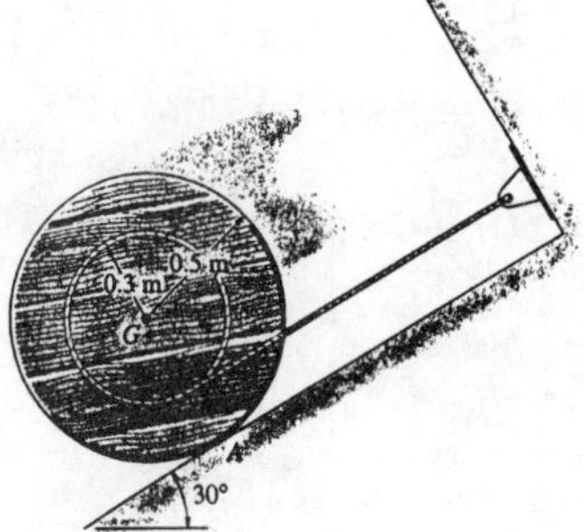

$+\nwarrow\Sigma F_y = 0; \qquad N_A - 60(9.81)\cos 30° = 0$

$N_A = 509.7\ \text{N}$

$T_1 + \Sigma U_{1-2} = T_2$

$$0 + 60(9.81)\sin 30°(s_G) - 0.2(509.7)(0.6667 s_G) = \frac{1}{2}\left[60(0.3)^2\right](6)^2 + \frac{1}{2}(60)\left[(0.3)(6)\right]^2$$

$s_G = 0.859\ \text{m}$ **Ans**

18-26. The spool has a weight of 500 lb and a radius of gyration of $k_G = 1.75$ ft. A horizontal force of $P = 15$ lb is applied to a cable wrapped around its inner core. If the spool is originally at rest, determine its angular velocity after the mass center G has moved 6 ft to the left. The spool rolls without slipping. Neglect the mass of the cable.

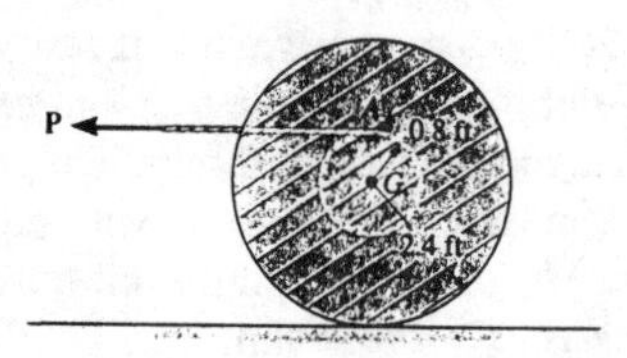

$$\frac{s_G}{2.4} = \frac{s_A}{3.2}$$

For $s_G = 6$ ft, then $s_A = 8$ ft

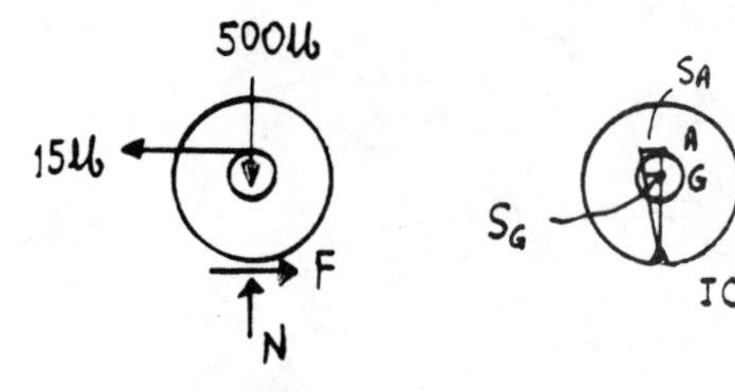

$$T_1 + \Sigma U_{1-2} = T_2$$

$$0 + 15(8) = \frac{1}{2}\left[\left(\frac{500}{32.2}\right)(1.75)^2\right]\omega^2 + \frac{1}{2}\left(\frac{500}{32.2}\right)(2.4\omega)^2$$

$\omega = 1.32$ rad/s **Ans**

18-27. The gear has a weight of 15 lb and a radius of gyration $k_G = 0.375$ ft. If the spring is unstretched when the torque $M = 6$ lb·ft is applied, determine the gear's angular velocity after its mass center G has moved to the left $s = 2$ ft.

$$\Delta s_G = 0.5\Delta\theta$$

$$\Delta s_{sp} = 0.9\Delta\theta$$

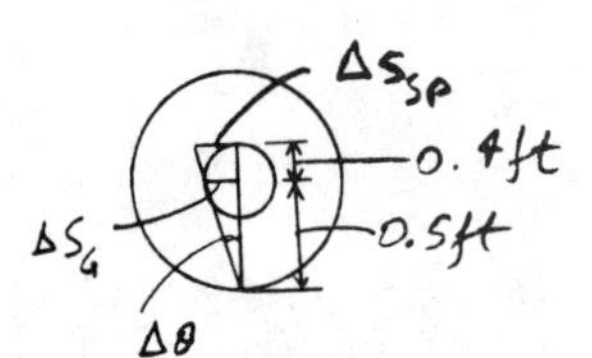

$$\Delta s_{sp} = 1.8\Delta s_G$$

For $\Delta s_G = 2$ ft, then

$$\Delta s_{sp} = 3.6 \text{ ft}$$

Also,

$$\theta = \frac{2}{0.5} = 4 \text{ rad}$$

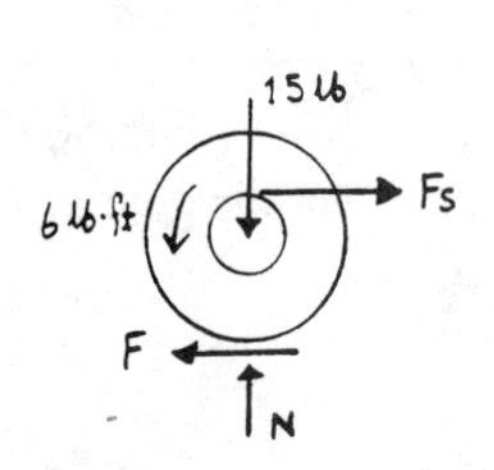

$$T_1 + \Sigma U_{1-2} = T_2$$

$$0 + 6(4) - \frac{1}{2}(3)(3.6)^2 = \frac{1}{2}\left(\frac{15}{32.2}\right)(0.5\omega)^2 + \frac{1}{2}\left[\left(\frac{15}{32.2}\right)(0.375)^2\right]\omega^2$$

$\omega = 7.08$ rad/s **Ans**

***18-28.** A man having a weight of 150 lb crouches down on the end of a diving board as shown. In this position the radius of gyration about his center of gravity is $k_G = 1.2$ ft. While holding this position at $\theta = 0°$, he rotates about his toes at A until he loses contact with the board when $\theta = 90°$. If he remains rigid, determine approximately how many revolutions he then makes before striking the water after falling 30 ft.

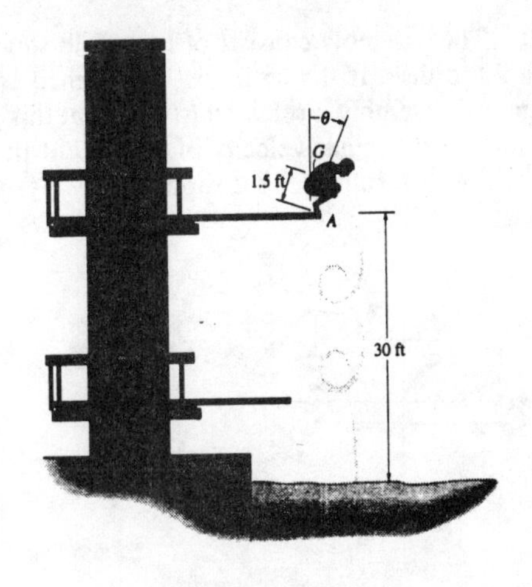

$$T_1 + \Sigma U_{1-2} = T_2$$

$$0 + 150(1.5) = \frac{1}{2}\left(\frac{150}{32.2}\right)(1.5\omega)^2 + \frac{1}{2}\left[\left(\frac{150}{32.2}\right)(1.2)^2\right]\omega^2$$

$\omega = 5.117$ rad/s

$v_G = (1.5)(5.117) = 7.675$ ft/s

During the fall no forces act on the man to cause an angular acceleration, so $\alpha = 0$.

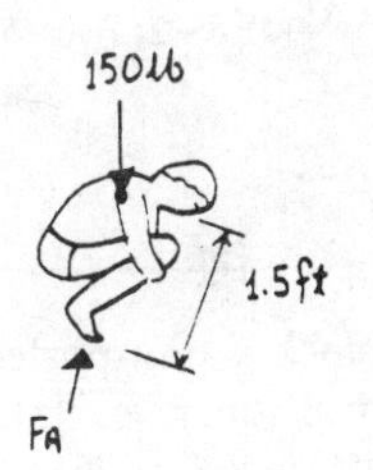

$(+\downarrow)$ $s = s_0 + v_0 t + \frac{1}{2} a_c t^2$

$$30 = 0 + 7.675t + \frac{1}{2}(32.2)t^2$$

Choosing the positive root,

$t = 1.147$ s

$(\circlearrowleft +)$ $\theta = \theta_0 + \omega_0 t + \frac{1}{2}\alpha_c t^2$

$$\theta = 0 + 5.117(1.147) + 0$$

$\theta = 5.870$ rad $= 0.934$ rev. **Ans**

18-29 The two 2-kg gears A and B are attached to the ends of a 3-kg slender bar. The gears roll within the fixed ring gear C, which lies in the horizontal plane. If a 10-N·m torque is applied to the center of the bar as shown, determine the number of revolutions the bar must rotate starting from rest in order for it to have an angular velocity of $\omega_{AB} = 20$ rad/s. For the calculation, assume the gears can be approximated by thin disks. What is the result if the gears lie in the vertical plane?

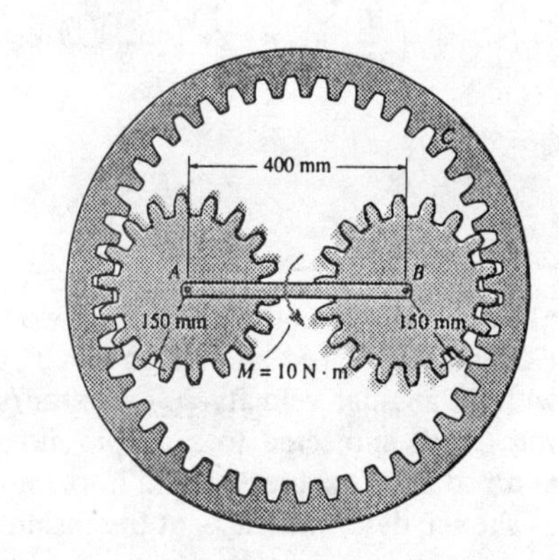

For each gear,

$v_G = 20(0.2) = 4$ m/s

$$\omega_G = \frac{4}{0.15} = 26.67 \text{ rad/s}$$

$$T_1 + \Sigma U_{1-2} = T_2$$

$$0 + 10(\theta) = \frac{1}{2}\left[\frac{1}{12}(3)(0.4)^2\right](20)^2 + 2\left[\frac{1}{2}\left(\frac{1}{2}(2)(0.15)^2\right)(26.67)^2 + \frac{1}{2}(2)(4)^2\right]$$

$\theta = 5.60$ rad $= 0.891$ rev. **Ans**

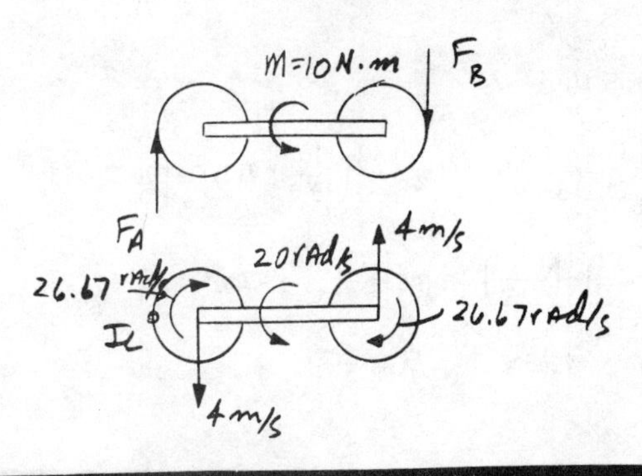

Same result **Ans**

since one gear rising by h causes the other gear to lower by h so that the sum of the positive and negative gravitational work of the gears cancels.

18-30. The assembly consists of two 15-lb slender rods and a 20-lb disk. If the spring is unstretched when $\theta = 45°$ and the assembly is released from rest at this position, determine the angular velocity of rod AB at the instant $\theta = 0°$. The disk rolls without slipping.

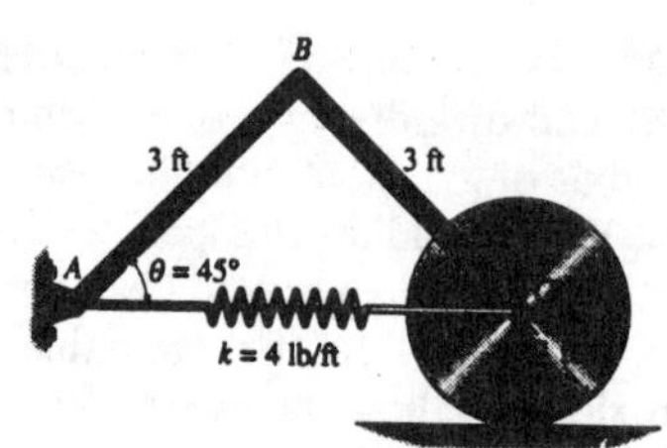

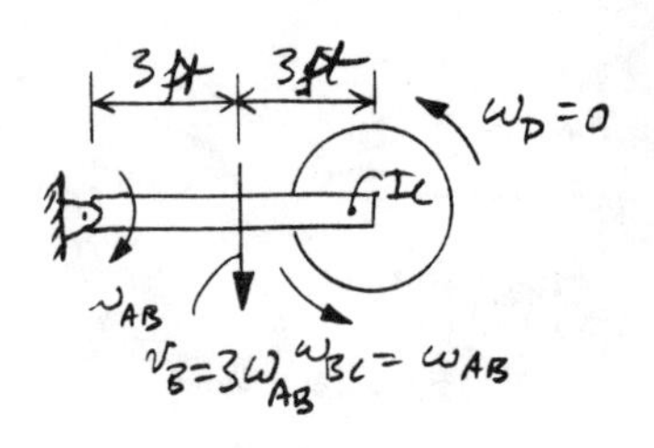

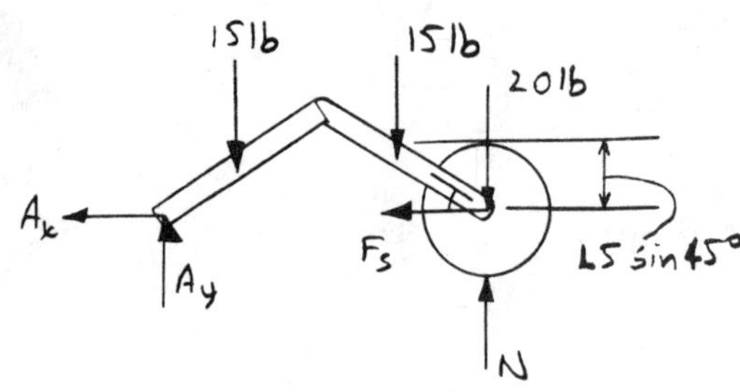

$$T_1 + \Sigma U_{1-2} = T_2$$

$$[0+0]+2(15)(1.5)\sin 45° - \frac{1}{2}(4)\left[6-2(3)\cos 45°\right]^2 = 2\left[\frac{1}{2}\left(\frac{1}{3}\left(\frac{15}{32.2}\right)(3)^2\right)\omega_{AB}^2\right]$$

$\omega_{AB} = 4.28$ rad/s **Ans**

18-31. The 100-lb block is transported a short distance by using two cylindrical rollers, each having a weight of 35 lb. If a horizontal force $P = 25$ lb is applied to the block, determine the block's speed after it has been displaced 2 ft to the left. Originally the block is at rest. No slipping occurs.

$$T_1 + \Sigma U_{1-2} = T_2$$

$$0+25(2) = \frac{1}{2}\left(\frac{100}{32.2}\right)(v_B)^2 + 2\left[\frac{1}{2}\left(\frac{35}{32.2}\right)\left(\frac{v_B}{2}\right)^2 + \frac{1}{2}\left(\frac{1}{2}\left(\frac{35}{32.2}\right)(1.5)^2\right)\left(\frac{v_B}{3}\right)^2\right]$$

$v_B = 5.05$ ft/s **Ans**

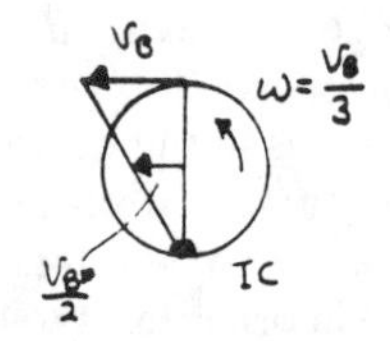

***18-32.** The linkage consists of two 8-lb rods AB and CD and a 10-lb rod AD. When $\theta = 0°$, rod AB is rotating with an angular velocity $\omega_{AB} = 2$ rad/s. If at this instant rod CD is subjected to a couple moment $M = 15$ lb·ft and rod AD is subjected to a horizontal force $P = 20$ lb as shown, determine ω_{AB} at the instant $\theta = 90°$.

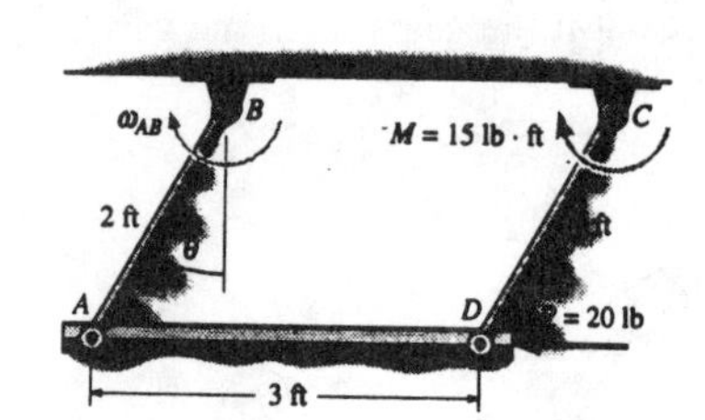

$$T_1 + \Sigma U_{1-2} = T_2$$

$$2\left[\frac{1}{2}\left(\frac{1}{3}\left(\frac{8}{32.2}\right)(2)^2\right)(2)^2\right] + \frac{1}{2}\left(\frac{10}{32.2}\right)[2(2)]^2 + \left[20(2)+15\left(\frac{\pi}{2}\right)-2(8)(1)-10(2)\right]$$

$$= 2\left[\frac{1}{2}\left(\frac{1}{3}\left(\frac{8}{32.2}\right)(2)^2\right)\omega^2\right] + \frac{1}{2}\left(\frac{10}{32.2}\right)(2\omega)^2$$

$\omega = 5.74$ rad/s **Ans**

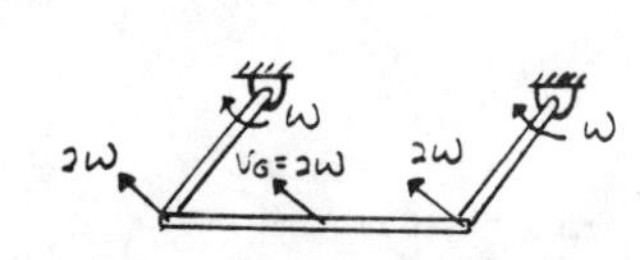

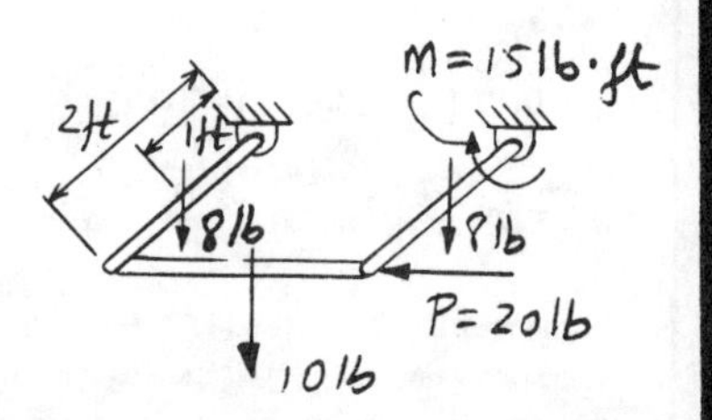

18-33. The 10-lb sphere starts from rest at $\theta = 0°$ and rolls without slipping down the cylindrical surface which has a radius of 10 ft. Determine the speed of the sphere's center of mass at the instant $\theta = 45°$.

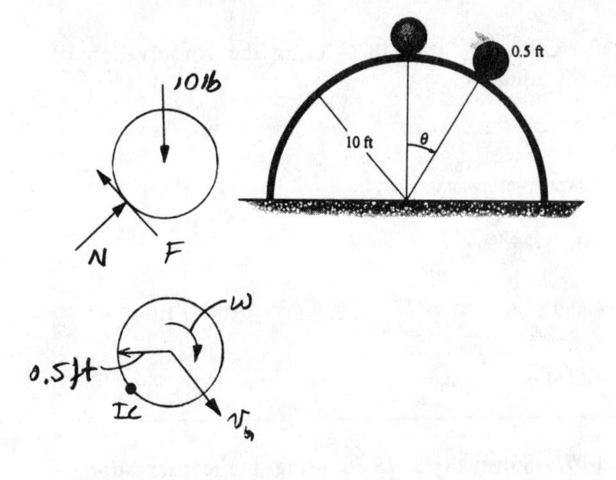

Kinematics :

$v_G = 0.5\omega_G$

$T_1 + \Sigma U_{1-2} = T_2$

$$0 + 10(10.5)(1 - \cos 45°) = \frac{1}{2}\left(\frac{10}{32.2}\right)v_G^2 + \frac{1}{2}\left[\frac{2}{5}\left(\frac{10}{32.2}\right)(0.5)^2\right]\left(\frac{v_G}{0.5}\right)^2$$

$v_G = 11.9$ ft/s **Ans**

18-34. The slender beam having a weight of 150 lb is supported by two cables. If the cable at end *B* is cut so that the beam is released from rest when $\theta = 30°$, determine the speed at which end *A* strikes the wall. Neglect friction at *B*.

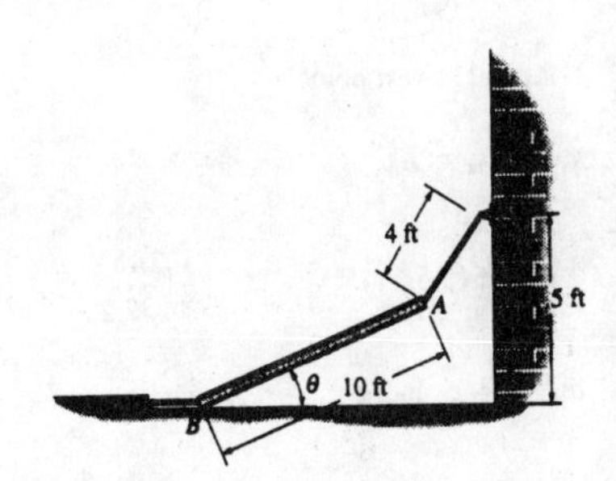

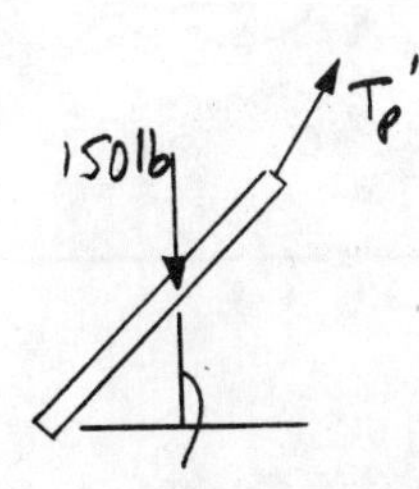

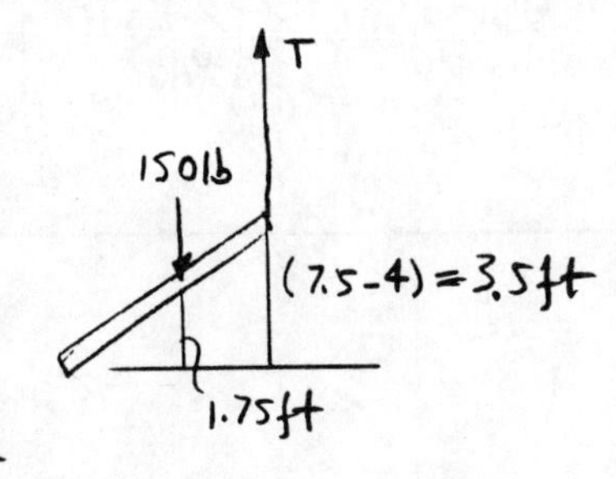

In the final position, the rod is in translation since the *IC* is at infinity.

$T_1 + \Sigma U_{1-2} = T_2$

$$0 + 150(2.5 - 1.75) = \frac{1}{2}\left(\frac{150}{32.2}\right)v_G^2$$

$v_G = v_A = 6.95$ ft/s **Ans**

18-35. Solve Prob. 18-13 using the conservation of energy equation.

Datum at lowest point.

$T_1 + V_1 = T_2 + V_2$

$$0 + 50(9.81)(1.25) = \frac{1}{2}\left[50(1.75)^2\right]\omega^2 + 0$$

$\omega = 2.83$ rad/s **Ans**

***18-36.** Solve Prob. 18-24 using the conservation of energy equation.

Datum at lowest point through G.

$$T_1 + V_1 = T_2 + V_2$$

$$0 + 60(9.81)(s \sin 30^\circ) = \frac{1}{2}\left[60(0.3)^2\right](6)^2 + \frac{1}{2}(60)\left[(0.3)(6)\right]^2 + 0$$

$s = 0.661$ m **Ans**

18-37. Solve Prob. 18-30 using the conservation of energy equation.

Datum at lowest point.

$$T_1 + V_1 = T_2 + V_2$$

$$0 + 2\left[15(1.5\sin 45^\circ)\right] = 2\left[\frac{1}{2}\left(\frac{1}{3}\left(\frac{15}{32.2}\right)(3)^2\right)\omega_{AB}^2\right] + \frac{1}{2}(4)\left[6 - 2(3\cos 45^\circ)\right]^2 + 0$$

$\omega_{AB} = 4.28$ rad/s **Ans**

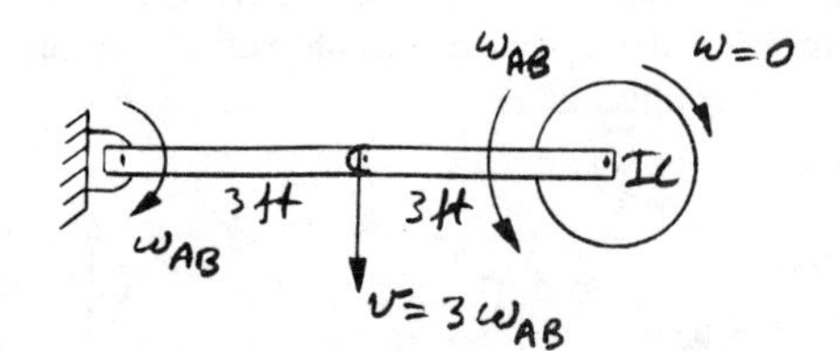

18-38. Solve Prob. 18-33 using the conservation of energy equation.

Datum at lowest point.

$$T_1 + V_1 = T_2 + V_2$$

$$0 + 10(10.5)(1 - \cos 45^\circ) = \frac{1}{2}\left[\frac{2}{5}\left(\frac{10}{32.2}\right)(0.5)^2\right]\left(\frac{v_G}{0.5}\right)^2 + \frac{1}{2}\left(\frac{10}{32.2}\right)v_G^2 + 0$$

$v_G = 11.9$ ft/s **Ans**

18-39. Solve Prob. 18-34 using the conservation of energy equation.

$$T_1 + V_1 = T_2 + V_2$$

$$0 + 150(2.5) = \frac{1}{2}\left(\frac{150}{32.2}\right)v_G^2 + 150(1.75)$$

$v_G = 6.95$ ft/s

Since the rod is in translation at the final instant, then

$v_A = 6.95$ ft/s **Ans**

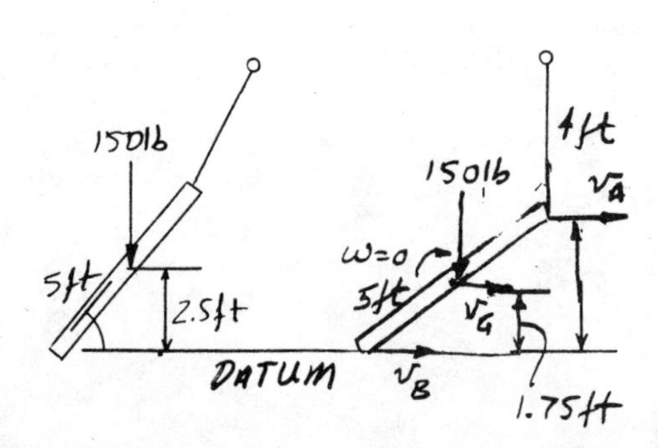

***18-40.** Solve Prob. 18-28 using the conservation of energy equation.

Datum at A.

$$T_1 + V_1 = T_2 + V_2$$

$$0 + 150(1.5) = \frac{1}{2}\left[\left(\frac{150}{32.2}\right)(1.2)^2\right]\omega^2 + \frac{1}{2}\left(\frac{150}{32.2}\right)(1.5\omega)^2 + 0$$

$\omega = 5.117$ rad/s

Time to fall :

$$s = s_0 + v_0 t + \tfrac{1}{2} a_c t^2$$

$$30 = 0 + 1.5(5.117)t + \frac{1}{2}(32.2)t^2$$

Choosing the positive root : $t = 1.147$ s

$$\theta = \theta_0 + \omega_0 t + \tfrac{1}{2}\alpha_c t^2$$

$\theta = 0 + 5.117(1.147) + 0 = 5.870 \text{ rad} = 0.934 \text{ rev.}$ **Ans**

18-41. The 15-lb disk is rotating about pin A in the vertical plane with an angular velocity $\omega = 2$ rad/s when $\theta = 0°$. Determine its angular velocity at the instant $\theta = 90°$. Also, compute the horizontal and vertical components of reaction at A at this instant.

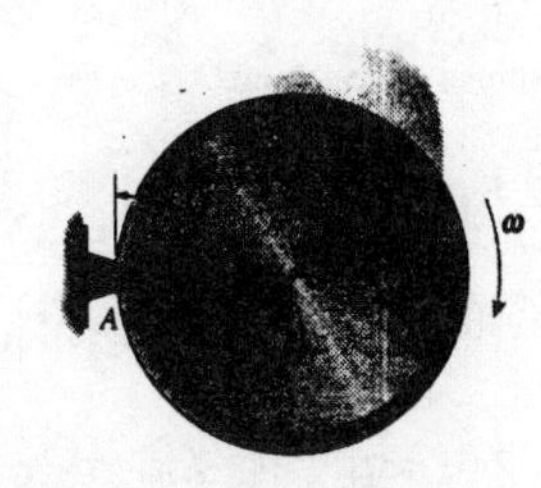

Datum through G at initial point.

$$T_1 + V_1 = T_2 + V_2$$

$$\frac{1}{2}\left[\frac{3}{2}\left(\frac{15}{32.2}\right)(0.5)^2\right](2)^2 + 0 = \frac{1}{2}\left[\frac{3}{2}\left(\frac{15}{32.2}\right)(0.5)^2\right]\omega^2 - 15(0.5)$$

$\omega = 9.48$ rad/s **Ans**

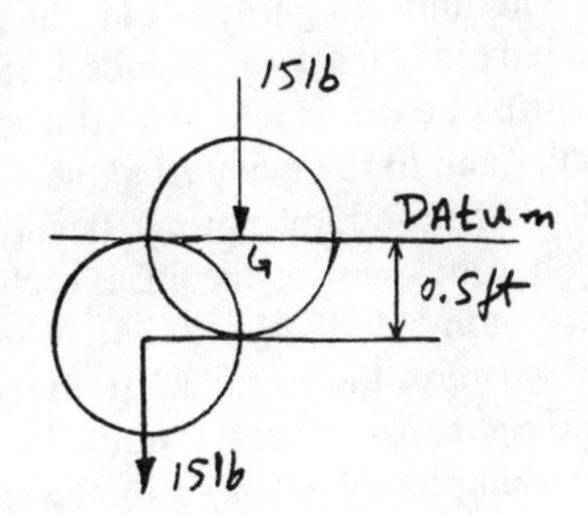

$\overset{+}{\leftarrow} \Sigma F_n = m(a_G)_n;$ $\quad A_x = \left(\frac{15}{32.2}\right)(9.48)^2(0.5)$

$A_x = 20.9$ lb **Ans**

$\circlearrowleft + \Sigma M_A = I_A \alpha;$ $\quad 15(0.5) = \frac{3}{2}\left(\frac{15}{32.2}\right)(0.5)^2 \alpha$

$\alpha = 42.93$ rad/s^2

$+\downarrow \Sigma F_t = m(a_G)_t;$ $\quad 15 - A_y = \left(\frac{15}{32.2}\right)(42.93)(0.5)$

$A_y = 5.00$ lb **Ans**

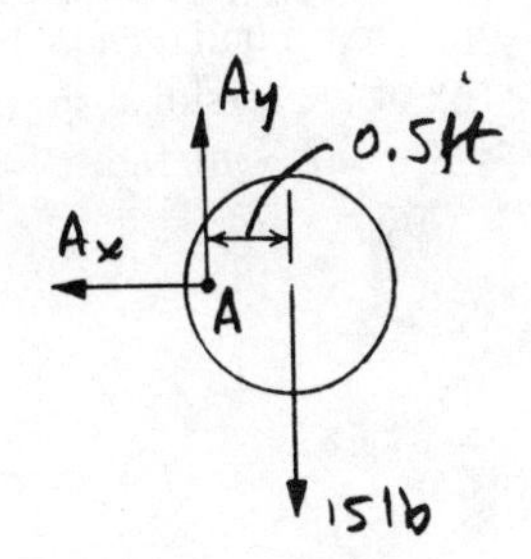

18-42. The link is rotating about point O in the vertical plane with an angular velocity $\omega = 5$ rad/s when $\theta = 0°$. If it has a mass of 20 kg, a mass center at G, and a radius of gyration $k_G = 800$ mm, determine its angular velocity at the instant $\theta = 90°$.

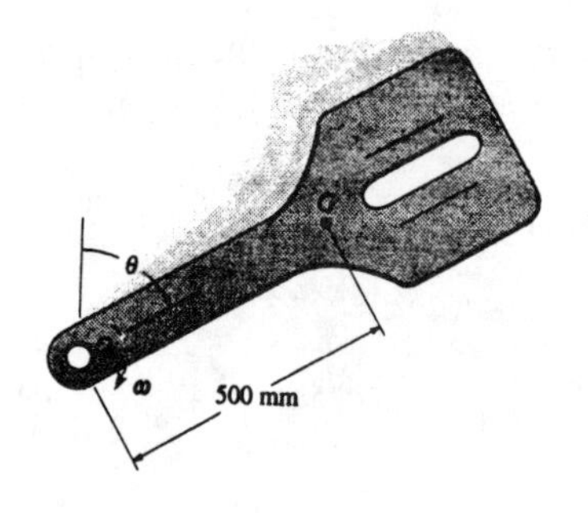

$v_G = 0.5\omega$

Datum through O at lowest point.

$T_1 + V_1 = T_2 + V_2$

$$\frac{1}{2}(20)\left[0.5(5)\right]^2 + \frac{1}{2}\left[(20)(0.8)^2\right](5)^2 + 0.5(20)(9.81) = \frac{1}{2}(20)(0.5\omega)^2 + \frac{1}{2}\left[(20)(0.8)^2\right]\omega^2 + 0$$

$\omega = 6.00$ rad/s **Ans**

18-43. The overhead door BC is pushed slightly from its open position and then rotates downward about the pin at A. Determine its angular velocity just before its end B strikes the floor. Assume the door is a thin plate having a mass of 180 kg and length of 6 m. Neglect the mass of the supporting frame AB and AC.

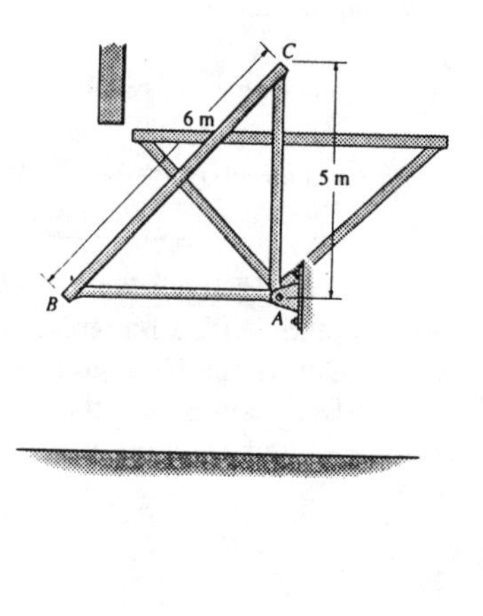

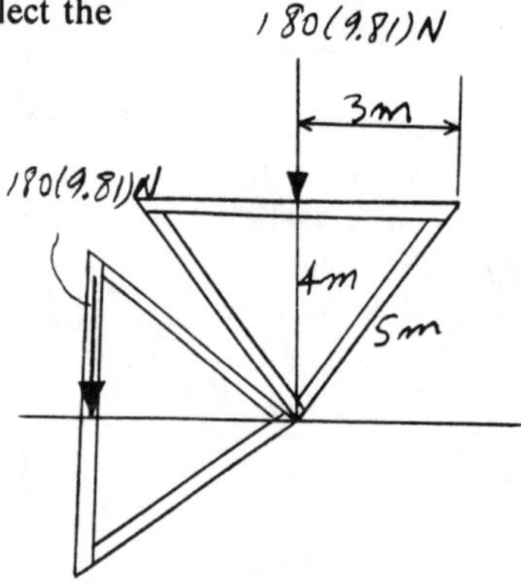

Datum at lowest point.

$T_1 + V_1 = T_2 + V_2$

$$0 + 180(9.81)4 = \frac{1}{2}\left[\frac{1}{12}(180)(6)^2\right]\omega^2 + \frac{1}{2}(180)(4\omega)^2 + 0$$

$\omega = 2.03$ rad/s **Ans**

***18-44.** The uniform garage door has a mass of 150 kg and is guided along tracks at its ends. Lifting is done using the two springs, each of which is attached to the anchor bracket at A and to the counterbalance shaft at B and C. As the door is raised, the springs begin to unwind from the shaft, thereby assisting the lift. If each spring provides a torsional moment of $M = (0.7\theta)$ N·m, where θ is in radians, determine the angle θ_0 at which both the left-wound and right-wound spring should be attached so that the door is completely balanced by the springs, i.e., when the door is in the vertical position and is given a slight force upwards, the springs will lift it along the side tracks to the horizontal plane with no final angular velocity. *Note:* The elastic potential energy of the torsional spring is $V_e = \frac{1}{2}k\theta^2$, where $M = k\theta$ and $k = 0.7$ N·m/rad.

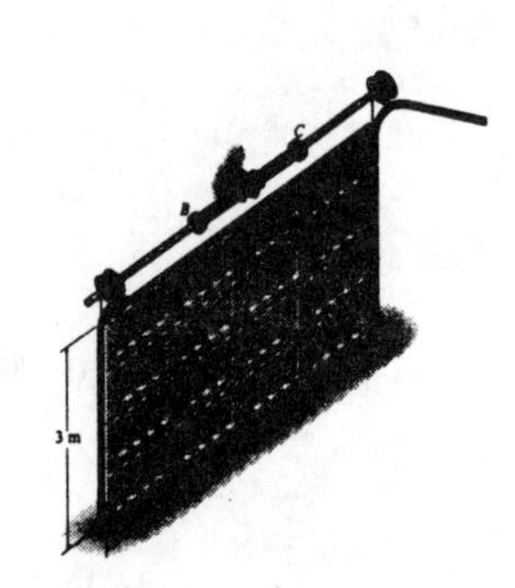

Datum at initial position.

$T_1 + V_1 = T_2 + V_2$

$$0 + 2\left[\frac{1}{2}(0.7)\theta_0^2\right] + 0 = 0 + 150(9.81)(1.5)$$

$\theta_0 = 56.15$ rad $= 8.94$ rev. **Ans**

18-45. The 80-lb cylinder is attached to the 10-lb slender rod which is pinned at point A. At the instant $\theta = 30^\circ$ the rod has an angular velocity of $\omega_0 = 1$ rad/s as shown. Determine the angle θ to which the rod swings before it momentarily stops.

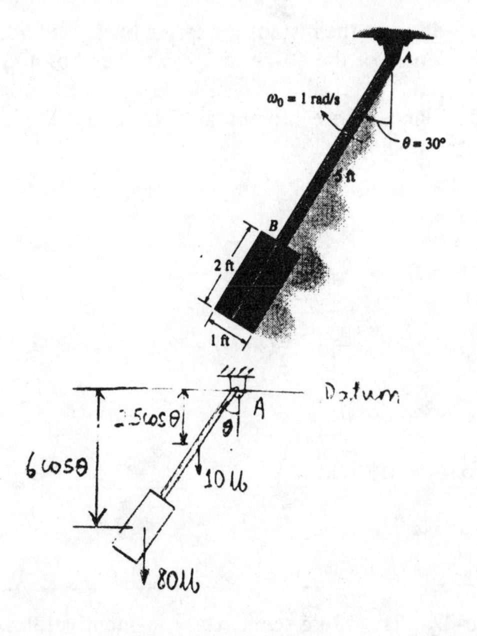

$$I_A = \frac{1}{12}\left(\frac{80}{32.2}\right)\left[3(0.5)^2 + (2)^2\right] + \left(\frac{80}{32.2}\right)(6)^2 + \frac{1}{3}\left(\frac{10}{32.2}\right)(5)^2 = 93.01 \text{ slug}\cdot\text{ft}^2$$

$$T_1 + V_1 = T_2 + V_2$$

$$\frac{1}{2}(93.01)(1)^2 - (10)(2.5\cos 30^\circ) - 80(6\cos 30^\circ) = 0 - (10)(2.5\cos\theta) - 80(6\cos\theta)$$

$\theta = 39.3^\circ$ **Ans**

18-46. An automobile tire has a mass of 7 kg and radius of gyration $k_G = 0.3$ m. If it is released from rest at A on the incline, determine its angular velocity when it reaches the horizontal plane. The tire rolls without slipping.

$$v_G = 0.4\omega$$

Datum at lowest point.

$$T_1 + V_1 = T_2 + V_2$$

$$0 + 7(9.81)(5) = \frac{1}{2}(7)(0.4\omega)^2 + \frac{1}{2}\left[7(0.3)^2\right]\omega^2 + 0$$

$\omega = 19.8$ rad/s **Ans**

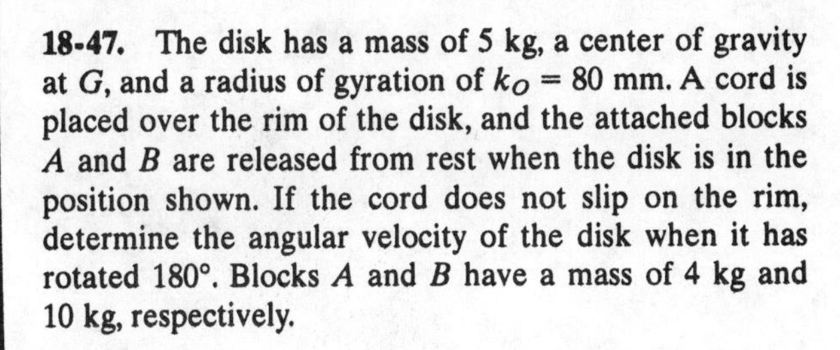

18-47. The disk has a mass of 5 kg, a center of gravity at G, and a radius of gyration of $k_O = 80$ mm. A cord is placed over the rim of the disk, and the attached blocks A and B are released from rest when the disk is in the position shown. If the cord does not slip on the rim, determine the angular velocity of the disk when it has rotated 180°. Blocks A and B have a mass of 4 kg and 10 kg, respectively.

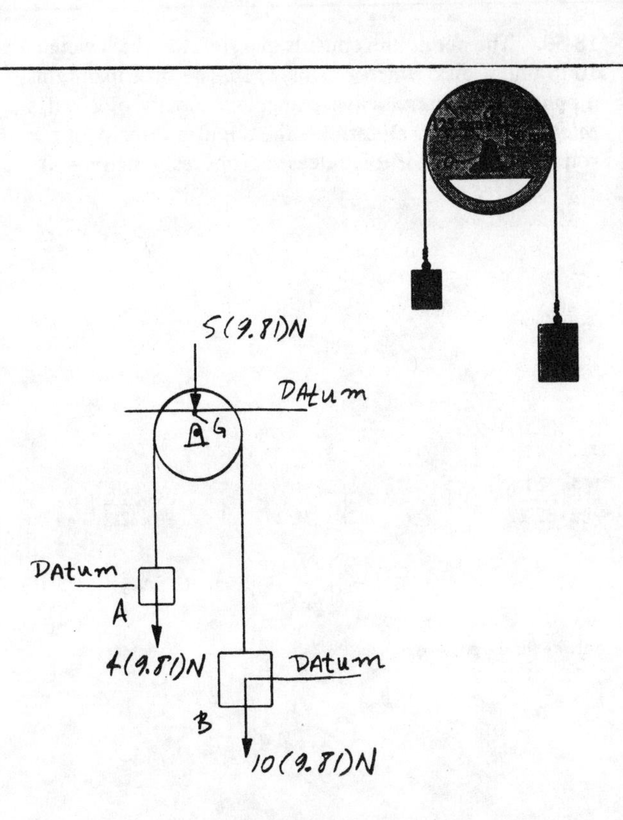

Select three datums which are located at the center of mass of blocks A and B and at G, respectively, when the system is in the initial position. Then, the blocks move $h = \pi(0.125)$ m

$$T_1 + V_1 = T_2 + V_2$$

$$0 + 0 = \frac{1}{2}(4)(v_A)^2 + \frac{1}{2}(10)(v_B)^2 + \frac{1}{2}\left[5(0.08)^2\right]\omega^2 + 4(9.81)\left[\pi(0.125)\right]$$

$$- 10(9.81)\left[\pi(0.125)\right] - 5(9.81)(0.10)$$

Since $v_A = v_B = 0.125\omega$, then

$\omega = 14.9$ rad/s **Ans**

***18-48.** At the instant the spring becomes undeformed, the center of the 40-kg disk has a speed of 4 m/s. From this point determine the distance d the disk moves down the plane before momentarily stopping. The disk rolls without slipping.

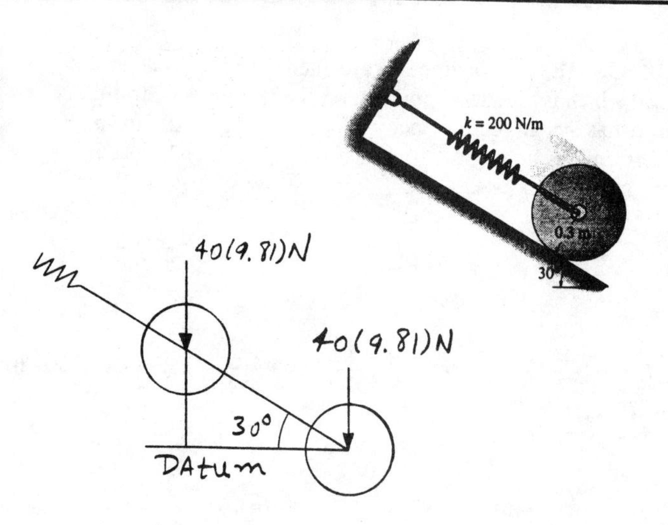

Datum at lowest point.

$$T_1 + V_1 = T_2 + V_2$$

$$\frac{1}{2}\left[\frac{1}{2}(40)(0.3)^2\right]\left(\frac{4}{0.3}\right)^2 + \frac{1}{2}(40)(4)^2 + 40(9.81)d\sin 30° = 0 + \frac{1}{2}(200)d^2$$

$$100d^2 - 196.2d - 480 = 0$$

Solving for the positive root

$d = 3.38$ m **Ans**

18-49. The 15-kg semicircular segment is released from rest in the position shown. Determine the velocity of point A when it has rotated counterclockwise 90°. Assume that the segment rolls without slipping on the surface. The moment of inertia about its mass center is $I_G = 0.25\ \text{kg}\cdot\text{m}^2$.

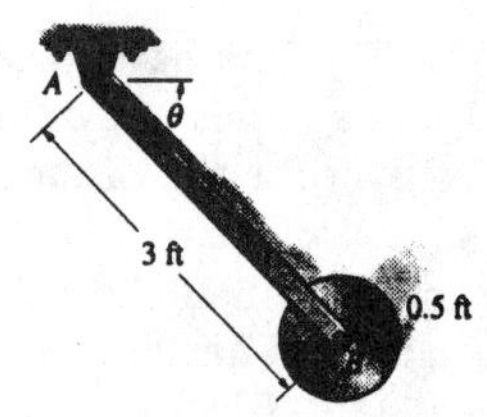

$$T_1 + V_1 = T_2 + V_2$$

$$0 + 15(9.81)(0.2) = \frac{1}{2}(0.25)\omega^2 + \frac{1}{2}(15)\left[(0.2 - 0.15)\omega\right]^2 + 15(9.81)(0.2 - 0.15)$$

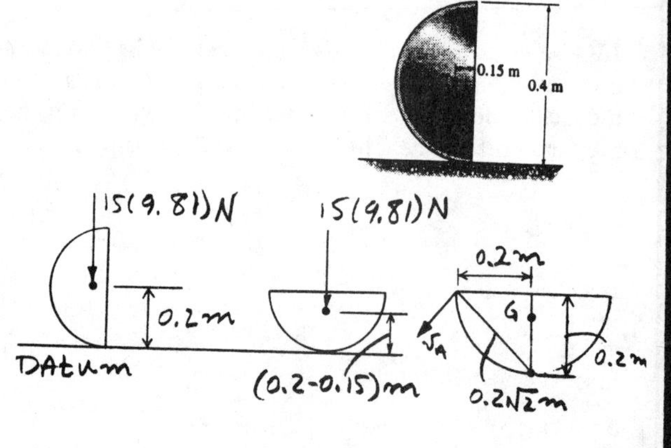

$\omega = 12.39$ rad/s

$v_A = 12.39\left(0.2\sqrt{2}\right) = 3.50$ m/s $\measuredangle 45°$ **Ans**

18-50. The pendulum consists of a rod AB which weighs 10 lb and a disk which weighs 20 lb. The disk maintains a *constant* counterclockwise angular velocity of 2 rad/s *relative to the rod.* Determine the angular velocity of the rod when $\theta = 90°$ if it is released from rest when $\theta = 0°$.

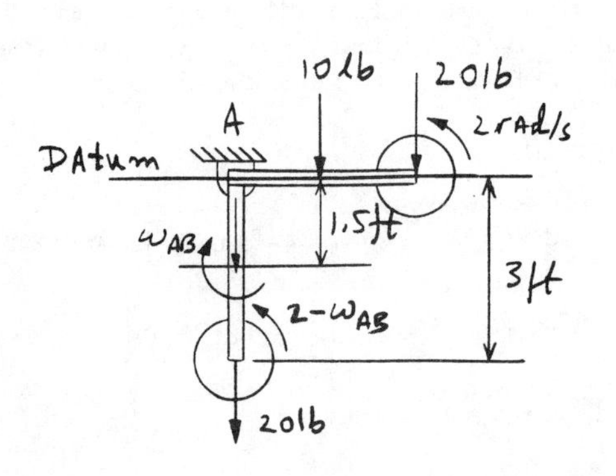

$$v_B = 3\omega_{AB}$$

Datum through A :

$$T_1 + V_1 = T_2 + V_2$$

$$\frac{1}{2}\left[\frac{1}{2}\left(\frac{20}{32.2}\right)(0.5)^2\right](2)^2 + 0 = \frac{1}{2}\left[\frac{1}{3}\left(\frac{10}{32.2}\right)(3)^2\right]\omega_{AB}^2 + \frac{1}{2}\left(\frac{20}{32.2}\right)(3\omega_{AB})^2$$

$$+\frac{1}{2}\left[\frac{1}{2}\left(\frac{20}{32.2}\right)(0.5)^2\right](2 - \omega_{AB})^2 - 1.5(10) - 3(20)$$

Solving for the positive root,

$\omega_{AB} = 4.79$ rad/s **Ans**

18-51. The uniform 150-lb stone (rectangular block) is being turned over on its side by pulling the vertical cable *slowly* upward until the stone begins to tip. If it then falls freely (**T** = **0**) from an essentially balanced at-rest position, determine the speed at which the corner A strikes the pad at B. The stone does not slip at its corner C as it falls.

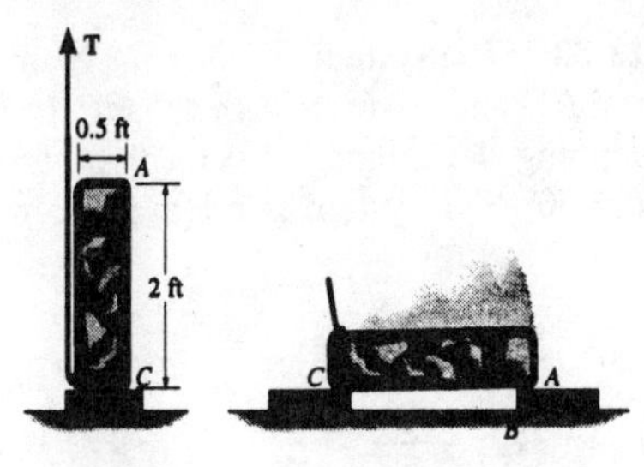

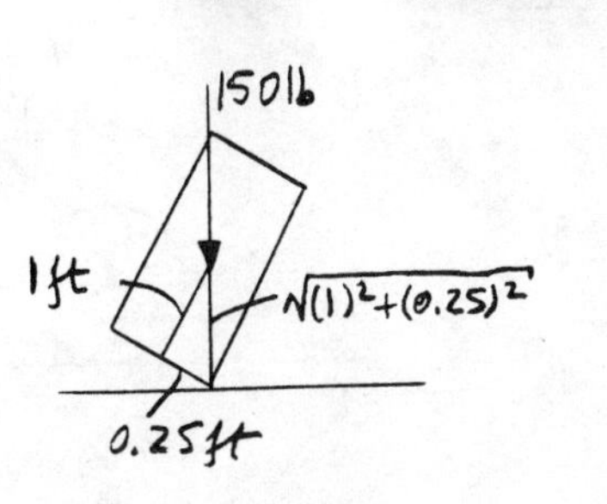

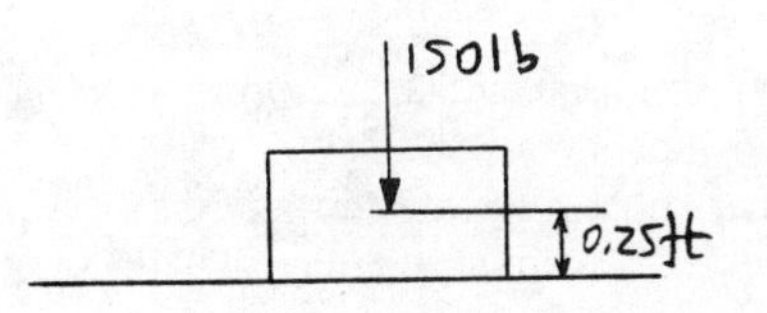

$$T_1 + V_1 = T_2 + V_2$$

$$0 + 1.0308(150) = \frac{1}{2}\left[\frac{1}{12}\left(\frac{150}{32.2}\right)(0.5^2 + 2^2)\right]\omega^2 + \frac{1}{2}\left(\frac{150}{32.2}\right)(1.0308\omega)^2 + (0.25)(150)$$

$\omega = 5.958$ rad/s

The IC is at C.

$v_A = 2\omega = 2(5.958) = 11.9$ ft/s **Ans**

***18-52.** The system consists of a 20-lb disk A, 4-lb slender rod BC, and a 1-lb smooth collar C. If the disk rolls without slipping, determine the velocity of the collar at the instant the rod becomes horizontal, i.e., $\theta = 0°$. The system is released from rest when $\theta = 45°$.

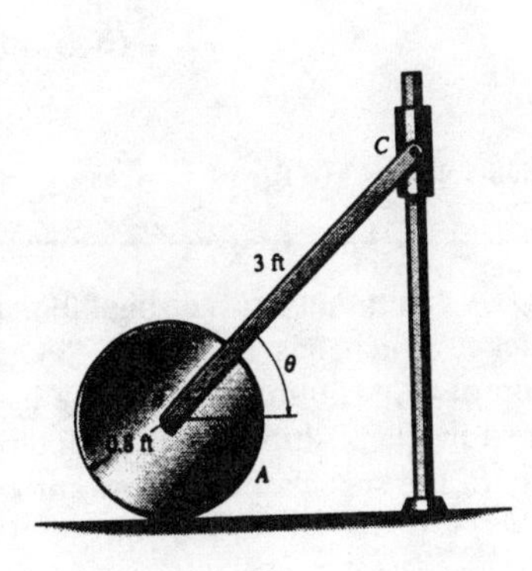

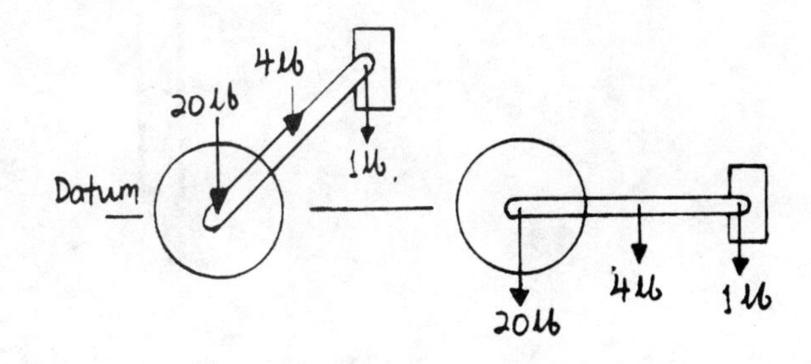

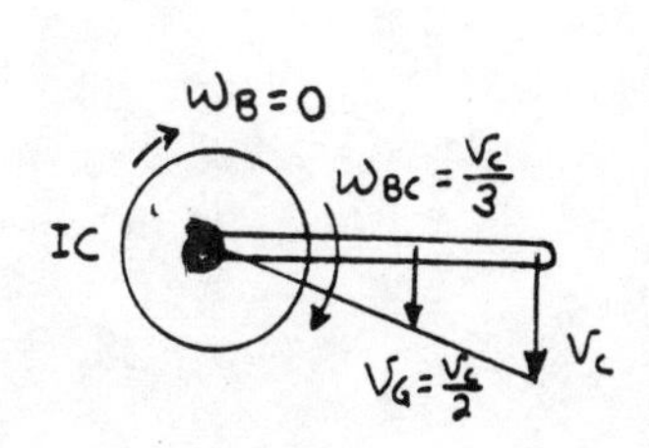

$$T_1 + V_1 = T_2 + V_2$$

$$0 + 4(1.5\sin 45°) + 1(3\sin 45°) = \frac{1}{2}\left[\frac{1}{3}\left(\frac{4}{32.2}\right)(3)^2\right]\left(\frac{v_C}{3}\right)^2 + \frac{1}{2}\left(\frac{1}{32.2}\right)(v_C)^2 + 0$$

$v_C = 13.3$ ft/s **Ans**

18-53. The system consists of a 20-lb disk A, 4-lb slender rod BC, and a 1-lb smooth collar C. If the disk rolls without slipping, determine the velocity of the collar at the instant $\theta = 30°$. The system is released from rest when $\theta = 45°$.

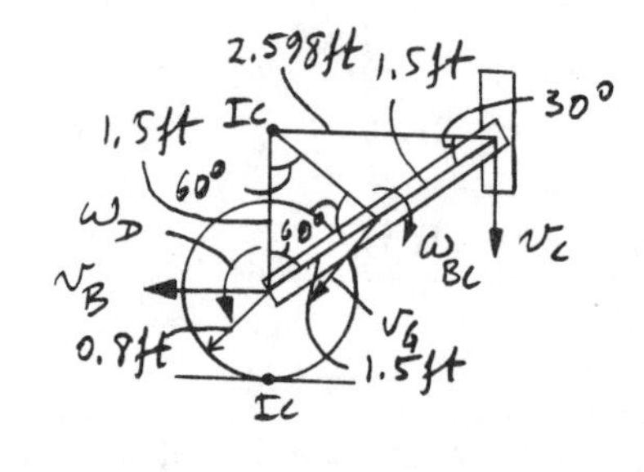

$v_B = 0.8\omega_D$

$\omega_{BC} = \frac{v_B}{1.5} = \frac{v_C}{2.598} = \frac{v_G}{1.5}$

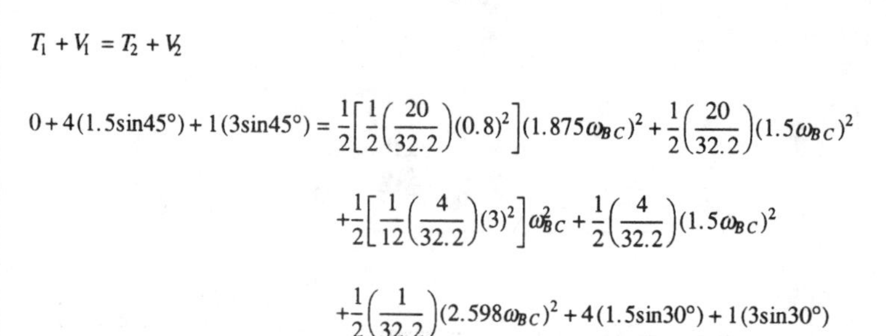

Thus,

$v_B = v_G = 1.5\omega_{BC}$

$v_C = 2.598\omega_{BC}$

$\omega_D = 1.875\omega_{BC}$

$T_1 + V_1 = T_2 + V_2$

$$0 + 4(1.5\sin 45°) + 1(3\sin 45°) = \frac{1}{2}\left[\frac{1}{2}\left(\frac{20}{32.2}\right)(0.8)^2\right](1.875\omega_{BC})^2 + \frac{1}{2}\left(\frac{20}{32.2}\right)(1.5\omega_{BC})^2$$

$$+\frac{1}{2}\left[\frac{1}{12}\left(\frac{4}{32.2}\right)(3)^2\right]\omega_{BC}^2 + \frac{1}{2}\left(\frac{4}{32.2}\right)(1.5\omega_{BC})^2$$

$$+\frac{1}{2}\left(\frac{1}{32.2}\right)(2.598\omega_{BC})^2 + 4(1.5\sin 30°) + 1(3\sin 30°)$$

$\omega_{BC} = 1.180$ rad/s

Thus

$v_C = 2.598(1.180) = 3.07$ ft/s **Ans**

18-54. A chain that has a negligible mass is draped over a sprocket which has a mass of 2 kg and a radius of gyration of $k_O = 50$ mm. If the 4-kg block A is released from rest in the position $s = 1$ m,. determine the angular velocity of the sprocket at the instant $s = 2$ m.

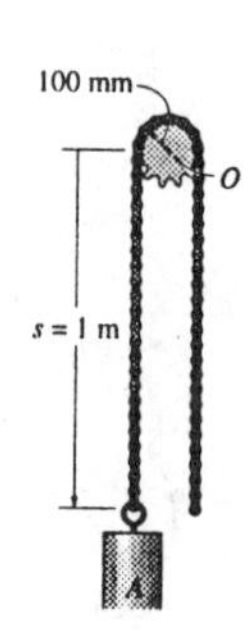

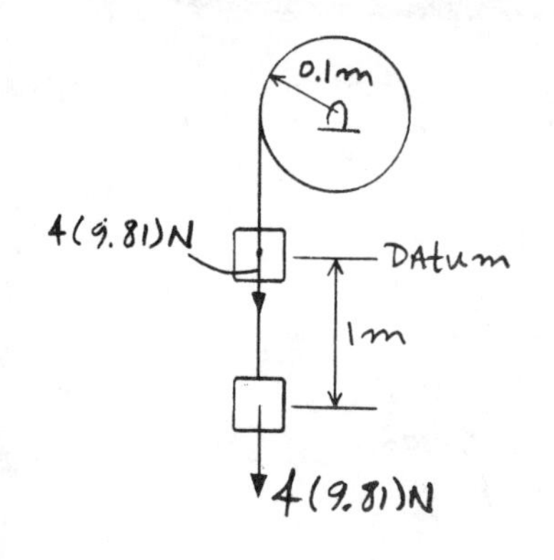

$T_1 + V_1 = T_2 + V_2$

$$0 + 0 + 0 = \frac{1}{2}(4)(0.1\omega)^2 + \frac{1}{2}\left[2(0.05)^2\right]\omega^2 - 4(9.81)(1)$$

$\omega = 41.8$ rad/s **Ans**

18-55. Solve Prob. 18-54 if the chain has a mass of 0.8 kg/m. For the calculation neglect the portion of the chain that wraps over the sprocket.

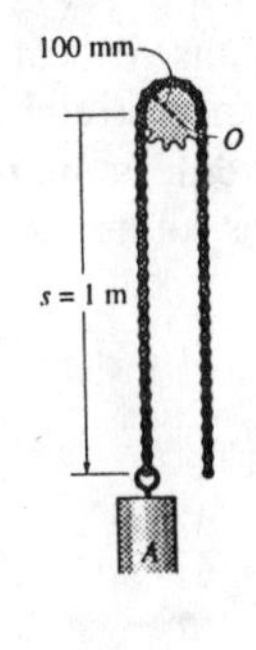

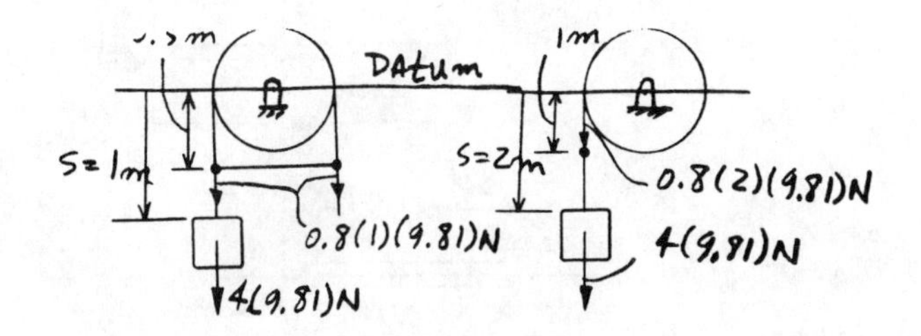

$T_1 + V_1 = T_2 + V_2$

$$0 - 4(9.81)(1) - 2\left[0.8(1)(9.81)(0.5)\right] = \frac{1}{2}(4)(0.1\omega)^2 + \frac{1}{2}\left[2(0.05)^2\right]\omega^2 + \frac{1}{2}(0.8)(2)(0.1\omega)^2$$
$$-4(9.81)(2) - 0.8(2)(9.81)(1)$$

$\omega = 39.3$ rad/s **Ans**

***18-56.** Pulley A has a weight of 30 lb and a centroidal radius of gyration $k_B = 0.6$ ft. Determine the speed of the 20-lb crate C at the instant $s = 10$ ft. Initially, the crate is released from rest when $s = 5$ ft. The pulley at P "rolls" downward on the cord without slipping. For the calculation, neglect the mass of this pulley and the cord as it unwinds from the inner and outer hubs of pulley A.

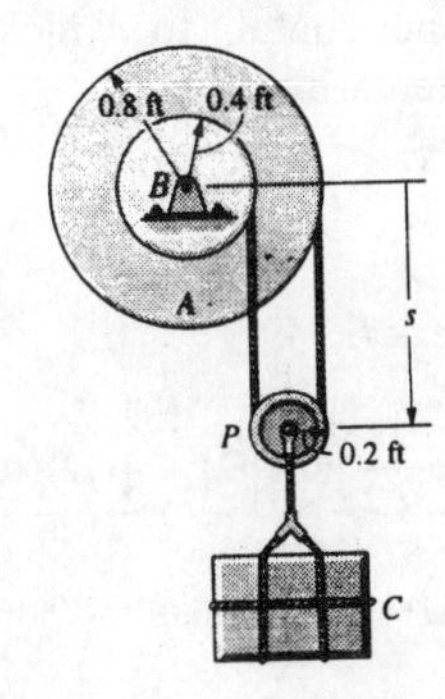

$$\frac{v_C}{0.6} = \frac{\omega(0.4)}{0.4}$$

$\omega = 1.667 v_C$

Datum through B.

$T_1 + V_1 = T_2 + V_2$

$$0 - 5(20) = \frac{1}{2}\left(\frac{20}{32.2}\right)v_C^2 + \frac{1}{2}\left[\left(\frac{30}{32.2}\right)(0.6)^2\right](1.667 v_C)^2 - 10(20)$$

$v_C = 11.3$ ft/s **Ans**

18-57. At the instant shown, the 50-lb bar is rotating downwards at 2 rad/s. The spring attached to its end always remains vertical due to the roller guide at C. If the spring has an unstretched length of 2 ft and a stiffness of $k = 6$ lb/ft, determine the angular velocity of the bar the instant it has rotated downward 30° below the horizontal.

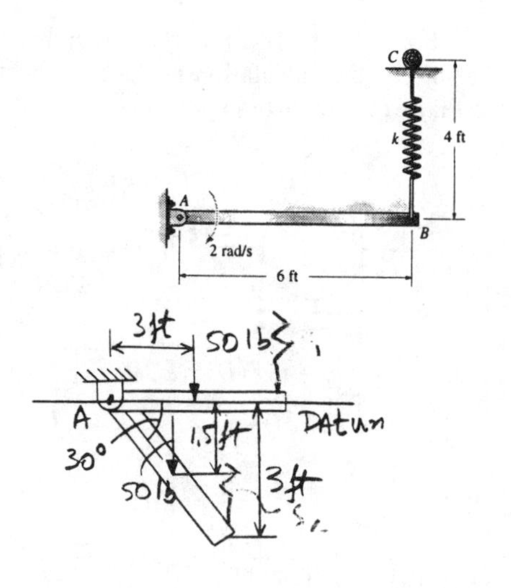

Datum through A.

$$T_1 + V_1 = T_2 + V_2$$

$$\frac{1}{2}\left[\frac{1}{3}\left(\frac{50}{32.2}\right)(6)^2\right](2)^2 + \frac{1}{2}(6)(4-2)^2 = \frac{1}{2}\left[\frac{1}{3}\left(\frac{50}{32.2}\right)(6)^2\right]\omega^2 + \frac{1}{2}(6)(7-2)^2 - 50(1.5)$$

$\omega = 2.30$ rad/s **Ans**

18-58. At the instant shown, the 50-lb bar is rotating downwards at 2 rad/s. The spring attached to its end always remains vertical due to the roller guide at C. If the spring has an unstretched length of 2 ft and a stiffness of $k = 12$ lb/ft, determine the angle θ, measured below the horizontal, to which the bar rotates before it momentarily stops.

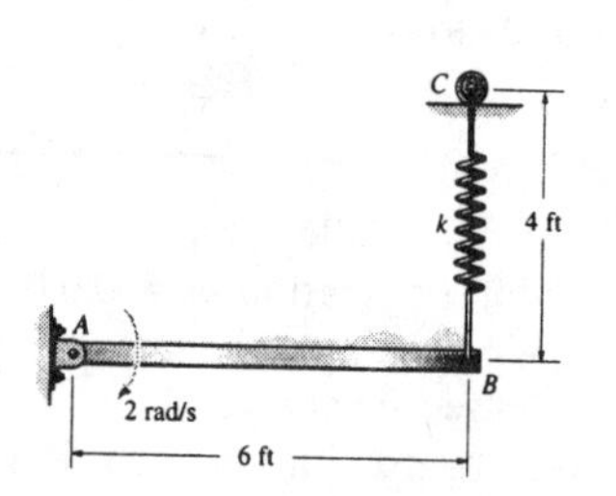

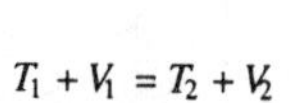

$$T_1 + V_1 = T_2 + V_2$$

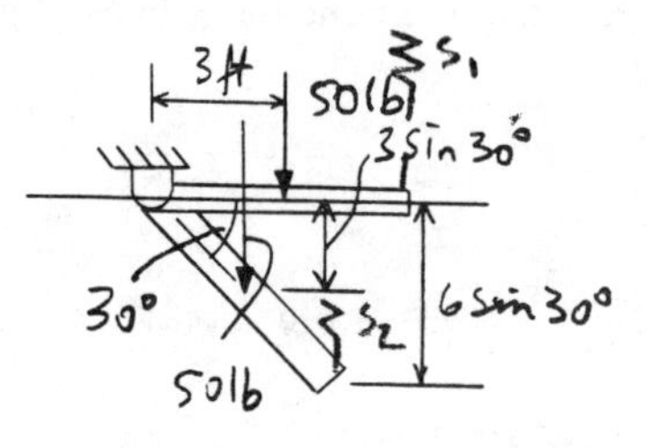

$$\frac{1}{2}\left[\frac{1}{3}\left(\frac{50}{32.2}\right)(6)^2\right](2)^2 + \frac{1}{2}(12)(4-2)^2 = 0 + \frac{1}{2}(12)(4 + 6\sin\theta - 2)^2 - 50(3\sin\theta)$$

$$61.2671 = 24(1 + 3\sin\theta)^2 - 150\sin\theta$$

$$37.2671 = -6\sin\theta + 216\sin^2\theta$$

Set $x = \sin\theta$, and solve the quadratic equation for the positive root:

$\sin\theta = 0.4295$

$\theta = 25.4°$ **Ans**

18-59. The uniform window shade AB has a total weight of 0.4 lb. When it is released, it winds up around the spring-loaded core O. Motion is caused by a spring within the core, which is coiled so that it exerts a torque $M = (0.3(10^{-3})\theta)$ lb·ft, where θ is in radians, on the core. If the shade is released from rest, determine the angular velocity of the core at the instant the shade is completely rolled up, i.e., after 12 revolutions. When this occurs, the spring becomes uncoiled and the radius of gyration of the shade about the axle at O is $k_O = 0.9$ in. *Note:* The elastic potential energy of the torsional spring is $V_e = \frac{1}{2}k\theta^2$, where $M = k\theta$ and $k = 0.3(10^{-3})$ lb·ft/rad.

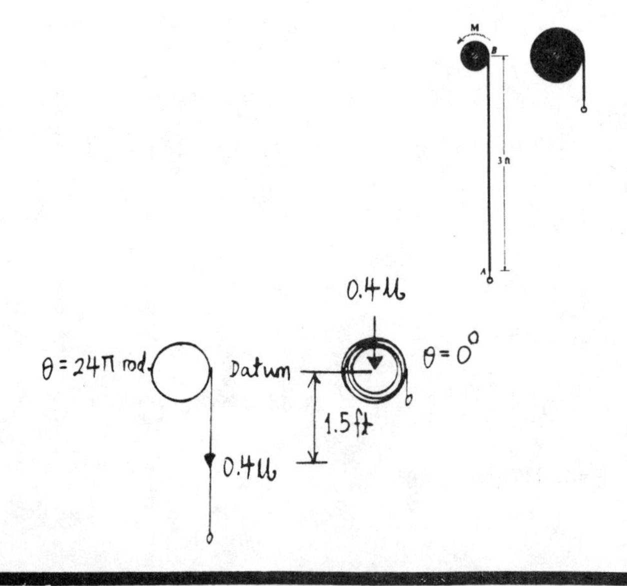

$$T_1 + V_1 = T_2 + V_2$$

$$0 - (0.4)(1.5) + \frac{1}{2}(0.3)\left(10^{-3}\right)(24\pi)^2 = \frac{1}{2}\left(\frac{0.4}{32.2}\right)\left(\frac{0.9}{12}\right)^2\omega^2$$

$\omega = 85.1$ rad/s **Ans**

***18-60.** The garage door CD has a mass of 50 kg and can be treated as a thin plate. If it is released from rest from the horizontal position shown, determine its angular velocity at the instant the two side bars AC become horizontal. Each of the two side springs has a stiffness of $k = 30$ N/m and is originally unstretched. Neglect the mass of the side bars AC.

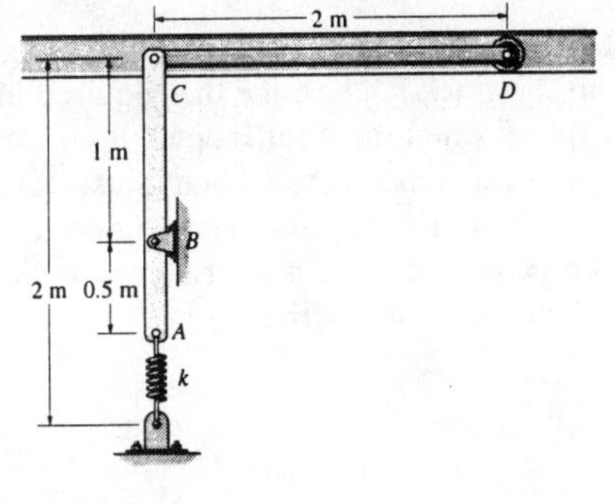

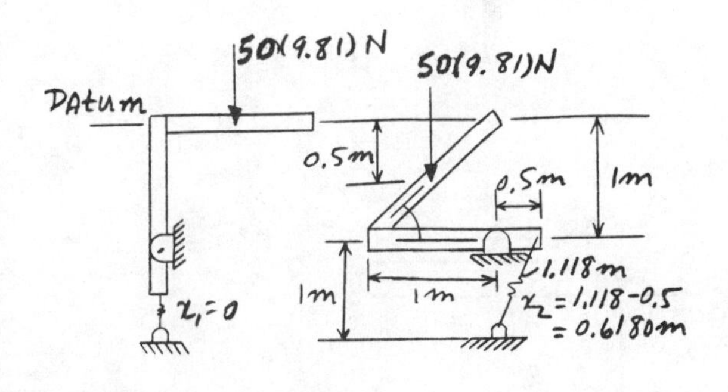

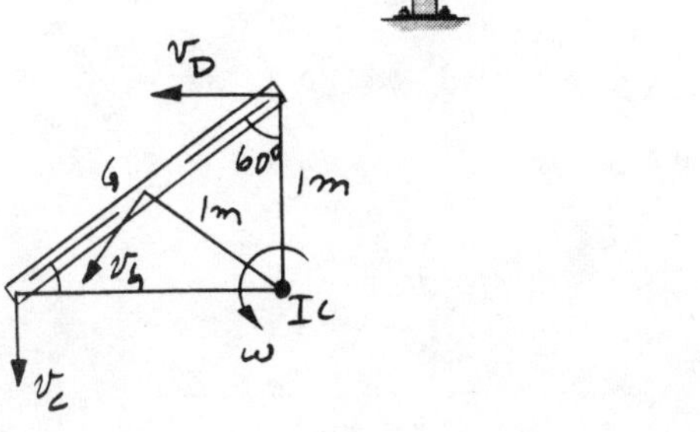

$v_G = (1)\omega$

$T_1 + V_1 = T_2 + V_2$

$$0 + 0 = \frac{1}{2}(50)\left[(1)\omega\right]^2 + \frac{1}{2}\left[\frac{1}{12}(50)(2)^2\right]\omega^2 - 50(9.81)(0.5) + 2\left[\frac{1}{2}(30)(0.6180)^2\right]$$

$\omega = 2.65$ rad/s **Ans**

18-61. The garage door CD has a mass of 50 kg and can be treated as a thin plate. Determine the required initial unstretched length of each of the two side springs when the door is open, so that when it falls freely it comes to rest when it reaches the fully closed position, i.e., when AC rotates 180°. Each of the two side springs has a stiffness of $k = 350$ N/m. Neglect the mass of the side bars AC.

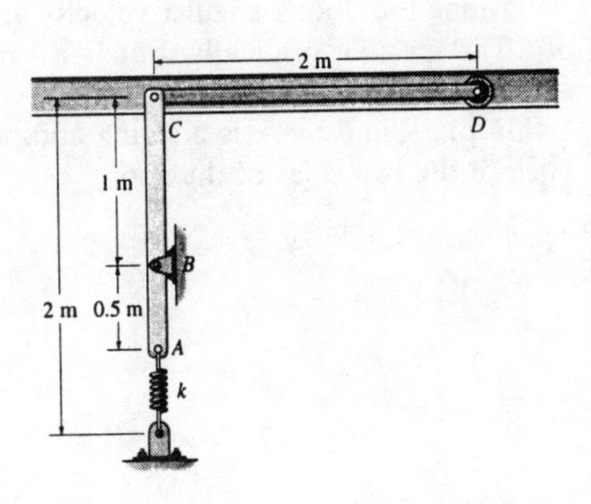

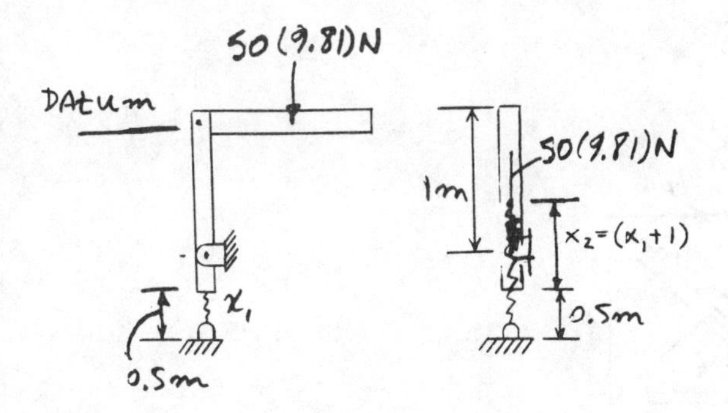

$T_1 + V_1 = T_2 + V_2$

$$0 + 2\left[\frac{1}{2}(350)(x_1)^2\right] = 0 + 2\left[\frac{1}{2}(350)(x_1 + 1)^2\right] - 50(9.81)(1)$$

$x_1 = 0.201$ m

Thus

$l_0 = 0.5\text{ m} - 0.201\text{ m} = 299$ mm **Ans**

18-62. The uniform 80-lb garage door is guided at its ends on the track. Determine the required initial stretch in the spring when the door is open, $\theta = 0°$, so that when it falls freely it comes to rest when it just reaches the fully closed position, $\theta = 90°$. Assume the door can be treated as a thin plate, and there is a spring and pulley system on each of the two sides of the door.

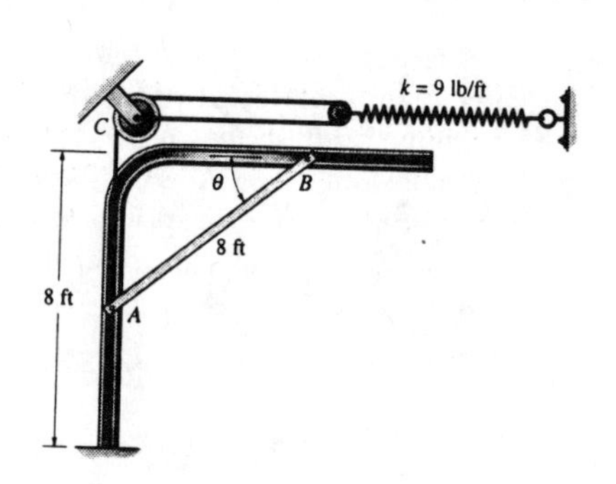

$s_A + 2s_s = l$

$\Delta s_A = -2\Delta s_s$

$8 \text{ ft} = -2\Delta s_s$

$\Delta s_s = -4 \text{ ft}$

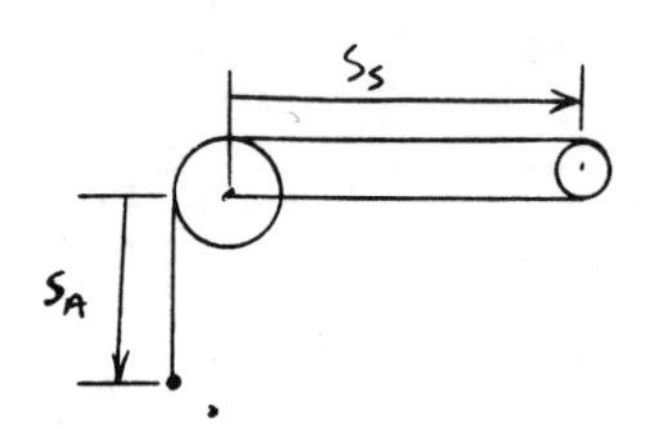

$T_1 + V_1 = T_2 + V_2$

$$0 + 2\left[\frac{1}{2}(9)s^2\right] = 0 - 80(4) + 2\left[\frac{1}{2}(9)(4+s)^2\right]$$

$9s^2 = -320 + 9\left(16 + 8s + s^2\right)$

$s = 2.44 \text{ ft}$ **Ans**

18-63. The uniform 80-lb garage door is guided at its ends on the track. If it is released from rest at $\theta = 0°$, determine the door's angular velocity at the instant $\theta = 30°$. The spring is originally stretched 1 ft when the door is held open, $\theta = 0°$. Assume the door can be treated as a thin plate, and there is a spring and pulley system on each of the two sides of the door.

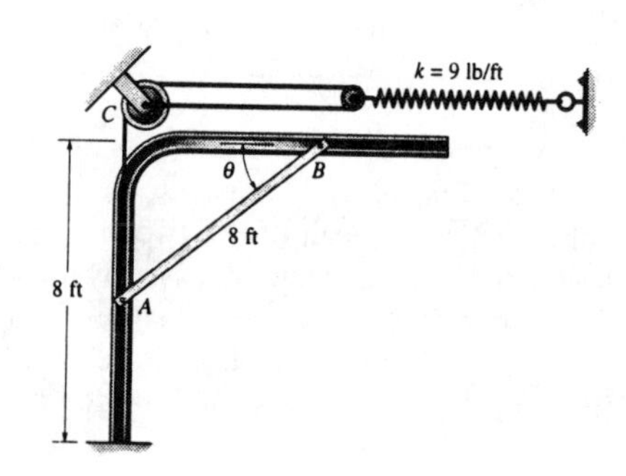

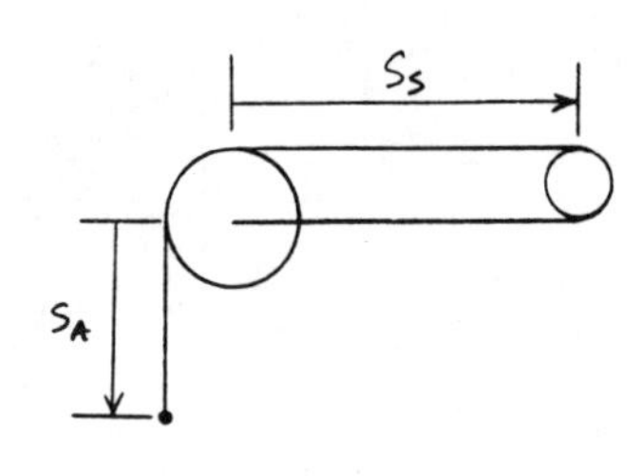

$v_G = 4\omega$

$s_A + 2s_s = l$

$\Delta s_A = -2\Delta s_s$

$4 \text{ ft} = -2\Delta s_s$

$\Delta s_s = -2 \text{ ft}$

$T_1 + V_1 = T_2 + V_2$

$$0 + 2\left[\frac{1}{2}(9)(1)^2\right] = \frac{1}{2}\left(\frac{80}{32.2}\right)(4\omega)^2 + \frac{1}{2}\left[\frac{1}{12}\left(\frac{80}{32.2}\right)(8)^2\right]\omega^2 - 80(4\sin 30°) + 2\left[\frac{1}{2}(9)(2+1)^2\right]$$

$\omega = 1.82 \text{ rad/s}$ **Ans**

***18-64.** The 500-g rod AB rests along the smooth inner surface of a hemispherical bowl. If the rod is released from rest from the position shown, determine its angular velocity at the instant it swings downward and becomes horizontal.

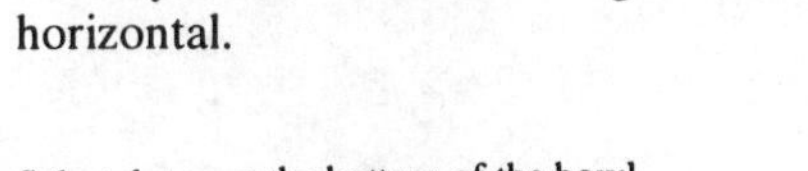

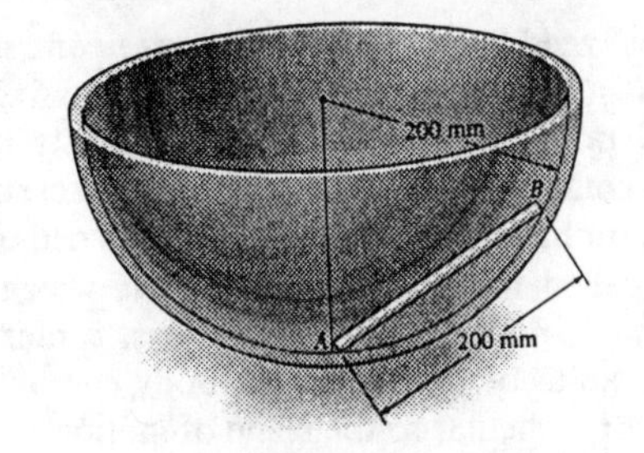

Select datum at the bottom of the bowl.

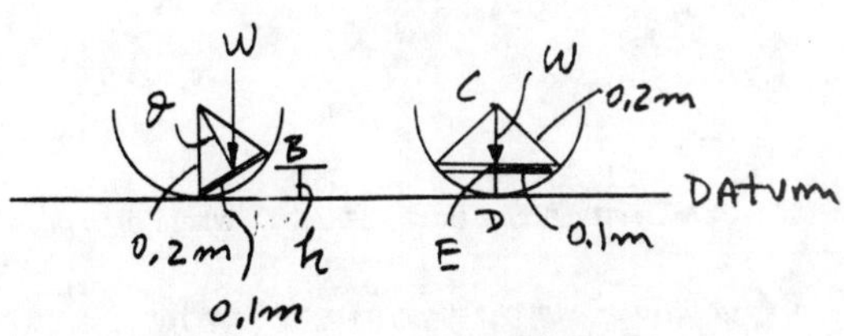

$$\theta = \sin^{-1}\left(\frac{0.1}{0.2}\right) = 30^\circ$$

$$h = 0.1\sin 30^\circ = 0.05$$

$$CE = \sqrt{(0.2)^2 - (0.1)^2} = 0.1732 \text{ m}$$

$$ED = 0.2 - 0.1732 = 0.02679$$

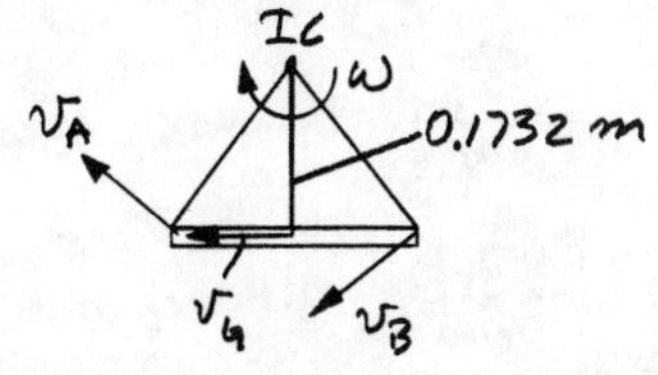

$$T_1 + V_1 = T_2 + V_2$$

$$0 + (0.5)(9.81)(0.05) = \frac{1}{2}\left[\frac{1}{12}(0.5)(0.2)^2\right]\omega_{AB}^2 + \frac{1}{2}(0.5)(v_G)^2 + (0.5)(9.81)(0.02679)$$

Since $v_G = 0.1732\omega_{AB}$

$\omega_{AB} = 3.70$ rad/s **Ans**

19-1. The rigid body (slab) has a mass m and is rotating with an angular velocity $\boldsymbol{\omega}$ about an axis passing through the fixed point O. Show that the momenta of all the particles composing the body can be represented by a single vector having a magnitude mv_G and acting through point P, called the *center of percussion*, which lies at a distance $r_{P/G} = k_G^2/r_{G/O}$ from the mass center G. Here k_G is the radius of gyration of the body, computed about an axis perpendicular to the plane of motion and passing through G.

$$H_O = (r_{G/O} + r_{P/G})mv_G = r_{G/O}(mv_G) + I_G\omega, \qquad \text{where } I_G = mk_G^2$$

$$r_{G/O}(mv_G) + r_{P/G}(mv_G) = r_{G/O}(mv_G) + \left(mk_G^2\right)\omega$$

$$r_{P/G} = \frac{k_G^2}{v_G/\omega} \qquad \text{However, } v_G = \omega r_{G/O} \text{ or } r_{G/O} = \frac{v_G}{\omega}$$

$$r_{P/G} = \frac{k_G^2}{r_{G/O}} \qquad \textbf{Q.E.D.}$$

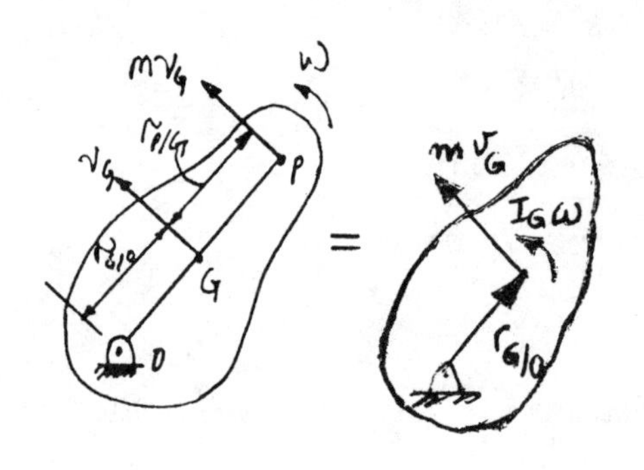

19-2. At a given instant, the body has a linear momentum $\mathbf{L} = m\mathbf{v}_G$ and an angular momentum $\mathbf{H}_G = I_G\boldsymbol{\omega}$ computed about its mass center. Show that the angular momentum of the body computed about the instantaneous center of zero velocity IC can be expressed as $\mathbf{H}_{IC} = I_{IC}\boldsymbol{\omega}$, where I_{IC} represents the body's moment of inertia computed about the instantaneous axis of zero velocity. As shown, the IC is located at a distance $r_{G/IC}$ away from the mass center G.

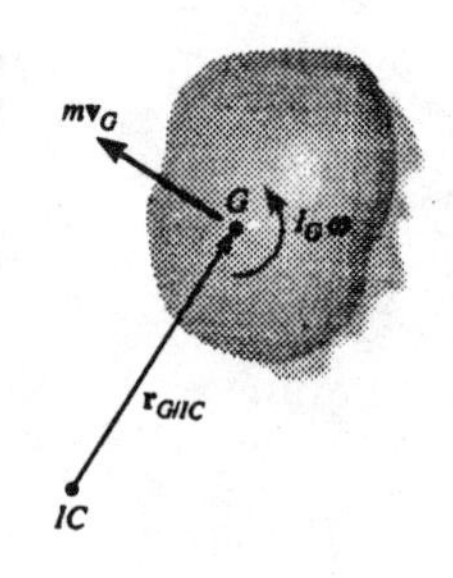

$$H_{IC} = r_{G/IC}(mv_G) + I_G\omega, \qquad \text{where } v_G = \omega r_{G/IC}$$

$$= r_{G/IC}(m\omega r_{G/IC}) + I_G\omega$$

$$= \left(I_G + mr_{G/IC}^2\right)\omega$$

$$= I_{IC}\omega \qquad \textbf{Q.E.D.}$$

19-3. Show that if a slab is rotating about a fixed axis perpendicular to the slab and passing through its mass center G, the angular momentum is the same when computed about any other point P on the slab.

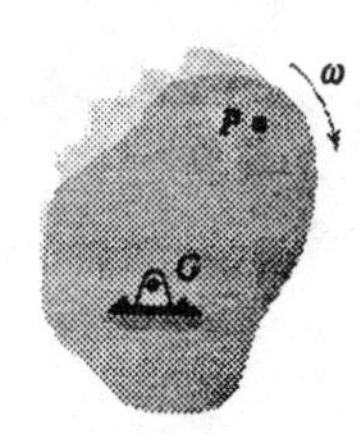

Since $v_G = 0$, the linear momentum $L = mv_G = 0$. Hence the angular momentum about any point P is

$$H_P = I_G\omega$$

Since ω is a free vector, so is $\mathbf{H}_P$. **Q.E.D.**

***19-4.** Gear A rotates along the inside of the circular gear rack R. If A has a weight of 4 lb and a radius of gyration of $k_B = 0.5$ ft, determine its angular momentum about point C when $\omega_{CB} = 30$ rad/s and (a) $\omega_R = 0$, (b) $\omega_R = 20$ rad/s.

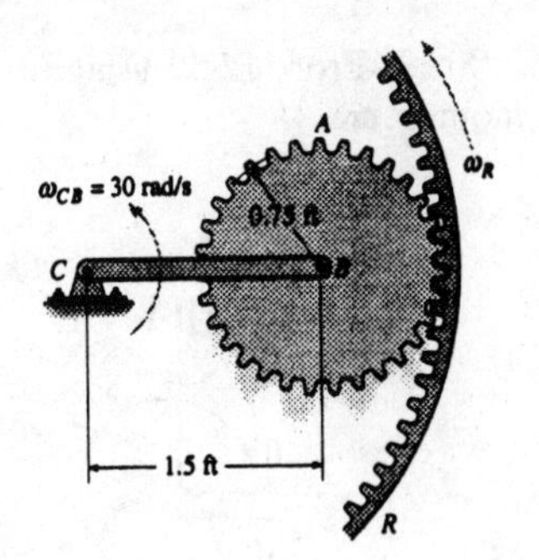

a)

$v_B = (1.5)(30) = 45$ ft/s

$$\omega_A = \frac{45}{0.75} = 60 \text{ rad/s}$$

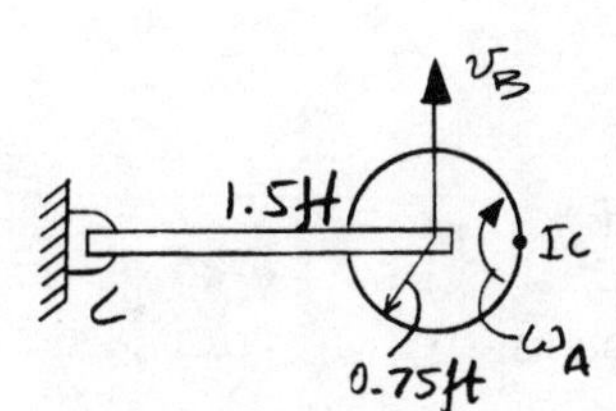

$$(\circlearrowleft +) \quad H_C = \left(\frac{4}{32.2}\right)(45)(1.5) - \left[\left(\frac{4}{32.2}\right)(0.5)^2\right](60)$$

$= 6.52$ slug $\cdot$ ft^2/s **Ans**

b)

$v_B = 45$ ft/s $v_P = 2.25(20) = 45$ ft/s

$v_B = 1.5(30) = 45$ ft/s

$\omega_A = 0$

$$(\circlearrowleft +) \quad H_C = \left(\frac{4}{32.2}\right)(45)(1.5)$$

$= 8.39$ slug $\cdot$ ft^2/s **Ans**

19-5. Solve Prob. 17-55 using the principle of impulse and momentum.

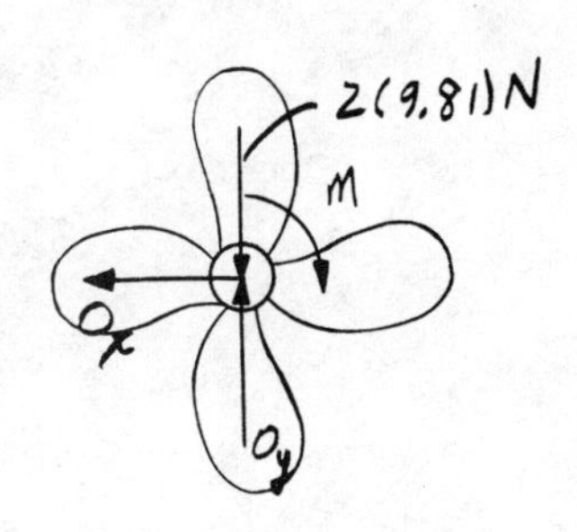

$$(\circlearrowleft +) \qquad (H_O)_1 + \Sigma\int M_O \, dt = (H_O)_2$$

$$0 + \int_0^4 3\left(1 - e^{-0.2t}\right) dt = (0.18)\,\omega$$

$$3\left(t + 5e^{-0.2t}\right)\Big|_0^4 = 0.18\omega$$

$\omega = 20.8$ rad/s **Ans**

19-6. Solve Prob. 17-59 using the principle of impulse and momentum.

30(9.81)N
O_y
2N·m
O_x
50N
0.025m

$$(\circlearrowleft +) \qquad (H_O)_1 + \Sigma \int M_O \, dt = (H_O)_2$$

$$30(0.15)^2(4) + 2(t) - 50(0.025)t = 30(0.15)^2(15)$$

$$t = 9.90 \text{ s} \qquad \textbf{Ans}$$

19-7. Solve Prob. 17-69 using the principle of impulse and momentum.

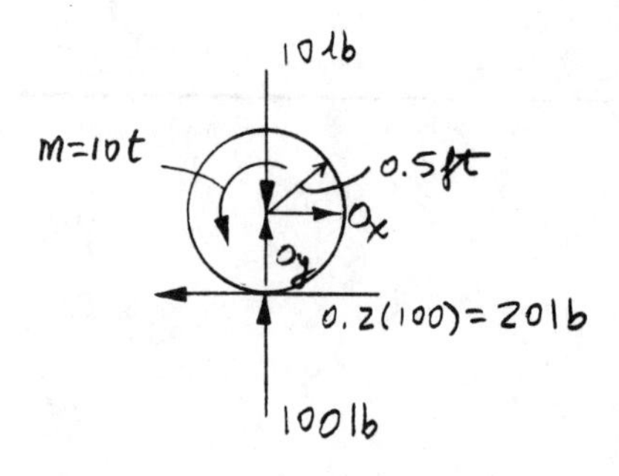

Time to start motion :

$F = 100(0.3) = 30$ lb

$$+\Sigma M_O = 0; \qquad 10t - 30(0.5) = 0$$

$$t = 1.5 \text{ s}$$

$$(\circlearrowleft +) \qquad (H_O)_1 + \Sigma \int M_O \, dt = (H_O)_2$$

$$0 + \int_{1.5}^{2} 10t \, dt - 20(2 - 1.5)(0.5) = \left[\frac{1}{2}\left(\frac{10}{32.2}\right)(0.5)^2\right]\omega$$

$$5t^2\Big|_{1.5}^{2} - 5 = 0.038820\omega$$

$$\omega = 96.6 \text{ rad/s} \qquad \textbf{Ans}$$

***19-8.** Solve Prob. 17-80 using the principle of impulse and momentum.

System :

$v_B = \omega(1.5)$

$(\curvearrowleft +) \qquad (H_A)_1 + \Sigma \int M_A \, dt = (H_A)_2$

$$0 + 5(1.5)(3) = \left[\left(\frac{180}{32.2}\right)(1.25)^2\right]\omega + \left(\frac{5}{32.2}\right)[\omega(1.5)](1.5)$$

$\omega = 2.48$ rad/s **Ans**

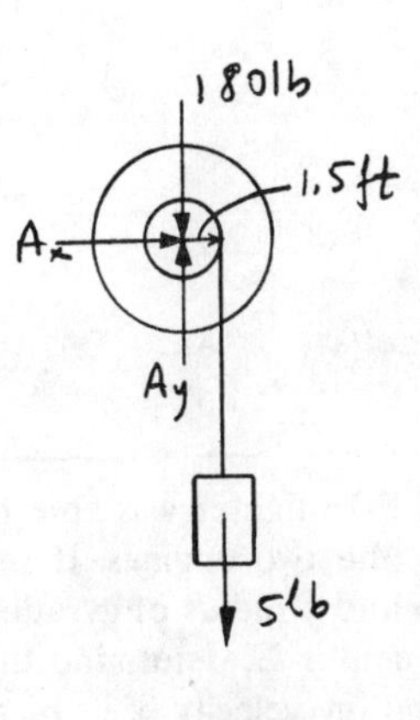

Block :

$v_B = \omega(1.5)$

$(+\downarrow) \qquad m(v_G)_1 + \Sigma \int F_y \, dt = m(v_G)_2$

$$0 + 5(3) - T(3) = \frac{5}{32.2}[\omega(1.5)] \qquad (1)$$

Spool :

$(\curvearrowleft +) \qquad (H_A)_1 + \Sigma \int M_A \, dt = (H_A)_2$

$$0 + T(1.5)(3) = \left[\left(\frac{180}{32.2}\right)(1.25)^2\right]\omega \qquad (2)$$

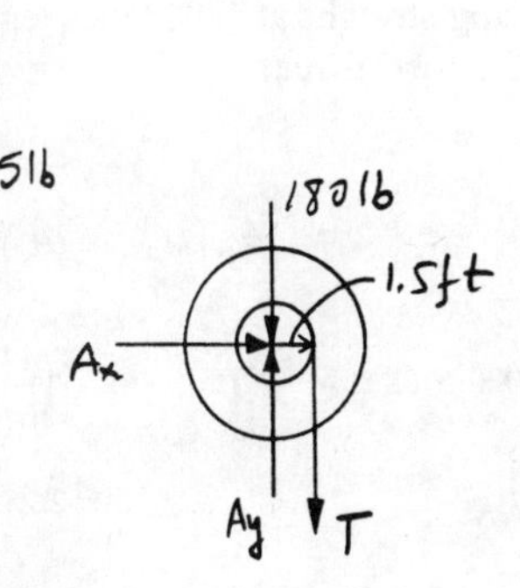

Solving Eqs. (1) and (2) :

$T = 4.81$ lb

$\omega = 2.48$ rad/s **Ans**

19-9. Solve Prob. 17-73 using the principle of impulse and momentum.

$(\overset{+}{\rightarrow}) \qquad m(v_{Gx})_1 + \Sigma \int F_x \, dt = m(v_{Gx})_2$

$$0 + T_{BC}\sin 30^\circ(t) - N_A t = 0$$

$$0.5T_{BC} = N_A$$

$(+\uparrow) \qquad m(v_{Gy})_1 + \Sigma \int F_y \, dt = m(v_{Gy})_2$

$$0 + T_{BC}\cos 30^\circ t - 20(9.81)t + 0.3N_A t = 0$$

$$0.86603T_{BC} + 0.3N_A = 196.2$$

Solving :

$T_{BC} = 193$ N **Ans**

$N_A = 96.553$ N

$(\curvearrowleft +) \qquad (H_B)_1 + \Sigma \int M_B \, dt = (H_B)_2$

$$\left[\frac{1}{2}(20)(0.15)^2\right](60) - 0.3(96.553)t(0.15) = 0$$

$t = 3.11$ s **Ans**

19-10. A flywheel has a mass of 60 kg and a radius of gyration of $k_G = 150$ mm about an axis of rotation passing through its mass center. If a motor supplies a clockwise torque having a magnitude of $M = (5t)$ N·m, where t is in seconds, determine the flywheel's angular velocity in $t = 3$ s. Initially the flywheel is rotating clockwise at $\omega_1 = 2$ rad/s.

(↻+) $\qquad (H_G)_1 + \Sigma\int M\,dt = (H_G)_2$

$$60(0.15)^2(2) + \int_0^3 5t\,dt = 60(0.15)^2\,\omega$$

$\omega = 18.7$ rad/s $\qquad$ **Ans**

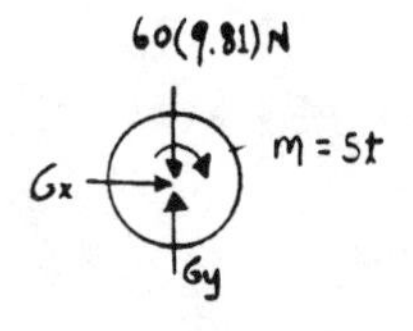

19-11. The pilot of a crippled F-15 fighter was able to control his plane by throttling the two engines. If the plane has a weight of 17 000 lb and a radius of gyration of $k_G = 4.7$ ft about the mass center G, determine the angular velocity of the plane and the velocity of its mass center G in $t = 5$ s if the thrust in each engine is altered to $T_1 = 5000$ lb and $T_2 = 800$ lb as shown. Originally the plane is flying straight at 1200 ft/s. Neglect the effects of drag and the loss of fuel.

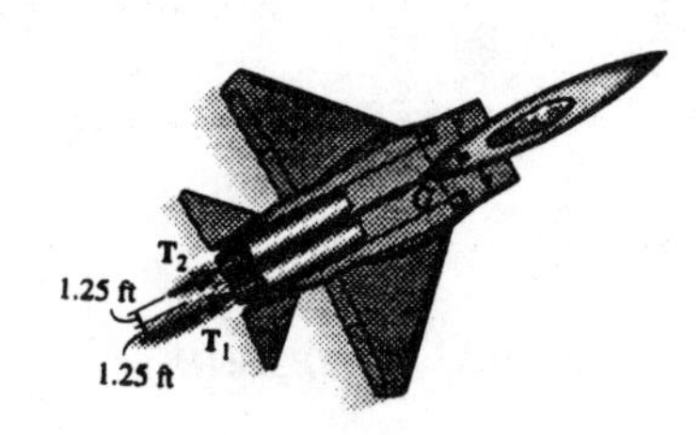

(↺+) $\qquad (H_G)_1 + \Sigma\int M_G\,dt = (H_G)_2$

$$0 + 5000(5)(1.25) - 800(5)(1.25) = \left[\left(\frac{17\,000}{32.2}\right)(4.7)^2\right]\omega$$

$\omega = 2.25$ rad/s $\qquad$ **Ans**

($\overset{+}{\rightarrow}$) $\qquad m(v_{Gx})_1 + \Sigma\int F_x\,dt = m(v_{Gx})_2$

$$\left(\frac{17\,000}{32.2}\right)(1200) + 5800(5) = \left(\frac{17\,000}{32.2}\right)(v_G)_2$$

$(v_G)_2 = 1.25(10^3)$ ft/s $\qquad$ **Ans**

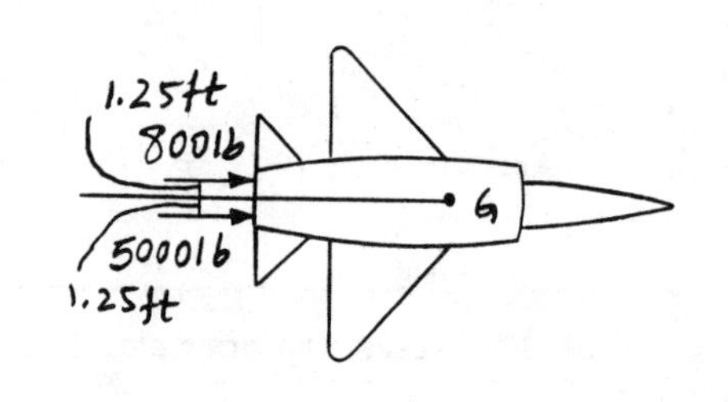

***19-12.** The disk has a weight of 10 lb and is pinned at its center O. If a vertical force of $P = 2$ lb is applied to the cord wrapped around its outer rim, determine the angular velocity of the disk in four seconds starting from rest. Neglect the mass of the cord.

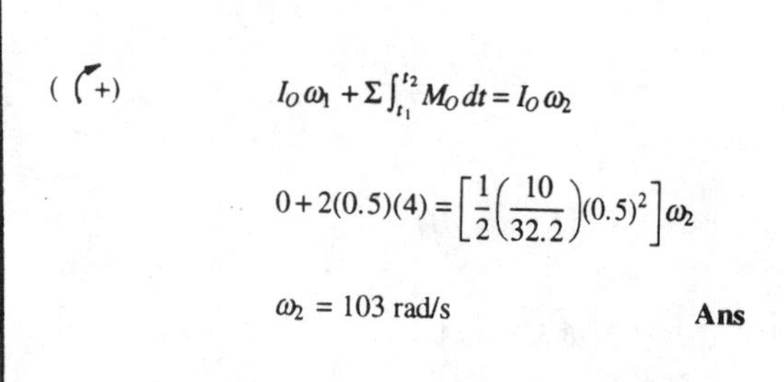

(↻+) $\qquad I_O\omega_1 + \Sigma\int_{t_1}^{t_2} M_O\,dt = I_O\omega_2$

$$0 + 2(0.5)(4) = \left[\frac{1}{2}\left(\frac{10}{32.2}\right)(0.5)^2\right]\omega_2$$

$\omega_2 = 103$ rad/s $\qquad$ **Ans**

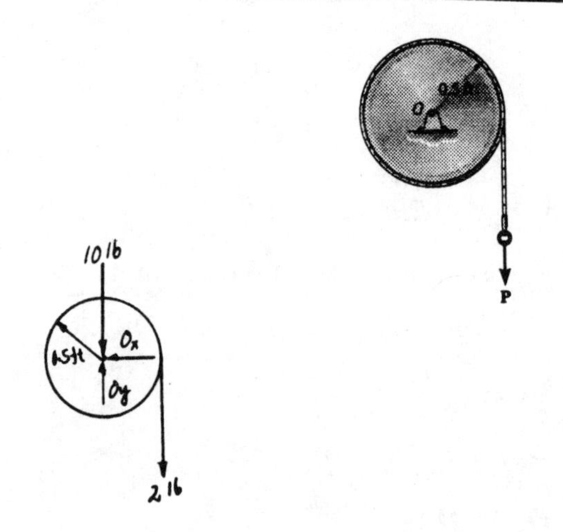

19-13. A wire of negligible mass is wrapped around the outer surface of the 2-kg disk. If the disk is released from rest, determine its angular velocity in 3 s.

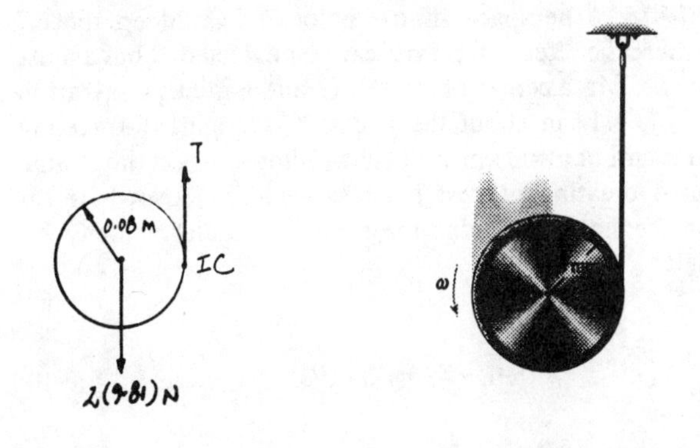

$$(\circlearrowleft +) \qquad I_{IC}\omega_1 + \Sigma\int_{t_1}^{t_2} M_{IC}\,dt = I_{IC}\omega_2$$

$$0 + 2(9.81)(0.08)(3) = \left[\frac{1}{2}(2)(0.08)^2 + 2(0.08)^2\right]\omega_2$$

$$\omega_2 = 245 \text{ rad/s} \qquad \textbf{Ans}$$

19-14. The 4-kg slender rod rests on a smooth floor. If it is kicked so as to receive a horizontal impulse $I = 8$ N·s at point A as shown, determine its angular velocity and the speed of its mass center.

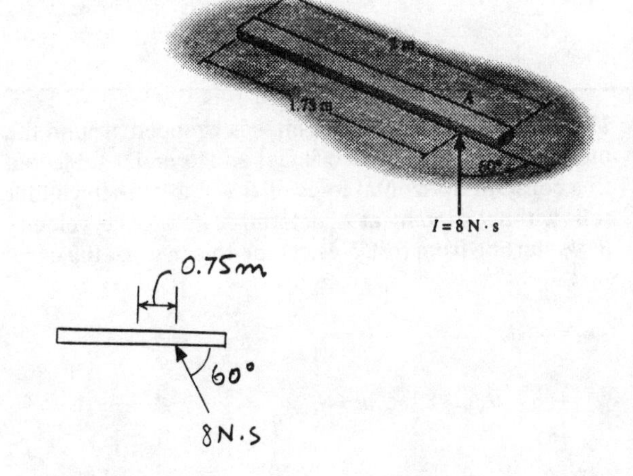

$$(\overset{+}{\leftarrow}) \qquad m(v_{Gx})_1 + \Sigma\int F_x\,dt = m(v_{Gx})_2$$

$$0 + 8\cos 60° = 4(v_G)_x$$

$$(v_G)_x = 1 \text{ m/s}$$

$$(+\uparrow) \qquad m(v_{Gy})_1 + \Sigma\int F_y\,dt = m(v_{Gy})_2$$

$$0 + 8\sin 60° = 4(v_G)_y$$

$$(v_G)_y = 1.732 \text{ m/s}$$

$$v_G = \sqrt{(1.732)^2 + (1)^2} = 2 \text{ m/s} \qquad \textbf{Ans}$$

$$(\circlearrowleft +) \qquad (H_G)_1 + \Sigma\int M_G\,dt = (H_G)_2$$

$$0 + 8\sin 60°(0.75) = \left[\frac{1}{12}(4)(2)^2\right]\omega$$

$$\omega = 3.90 \text{ rad/s} \qquad \textbf{Ans}$$

19-15. The impact wrench consists of a slender 1-kg rod AB which is 580 mm long, and cylindrical end weights at A and B that each have a diameter of 20 mm and a mass of 1 kg. This assembly is free to turn about the handle and socket, which are attached to the lug nut on the wheel of a car. If the rod AB is given an angular velocity of 4 rad/s and it strikes the bracket C on the handle without rebounding, determine the angular impulse imparted to the lug nut.

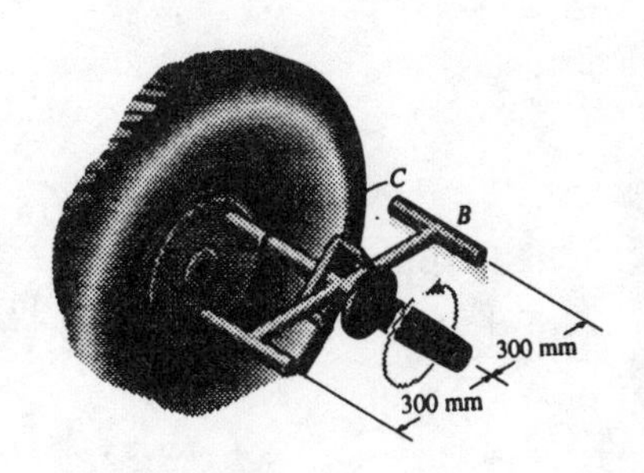

$$I_{axle} = \frac{1}{12}(1)(0.6 - 0.02)^2 + 2\left[\frac{1}{2}(1)(0.01)^2 + 1(0.3)^2\right] = 0.2081 \text{ kg}\cdot\text{m}^2$$

$$\int M dt = I_{axle}\,\omega = 0.2081(4) = 0.833 \text{ kg}\cdot\text{m}^2/\text{s} \qquad \textbf{Ans}$$

***19-16.** The space shuttle is located in "deep space," where the effects of gravity can be neglected. It has a mass of 120 Mg, a center of mass at G, and a radius of gyration $(k_G)_x = 14$ m about the x axis. It is originally traveling forward at $v = 3$ km/s when the pilot turns on the engine at A, creating a thrust $T = 600(1 - e^{-0.3t})$ kN, where t is in seconds. Determine the shuttle's angular velocity 2 s later.

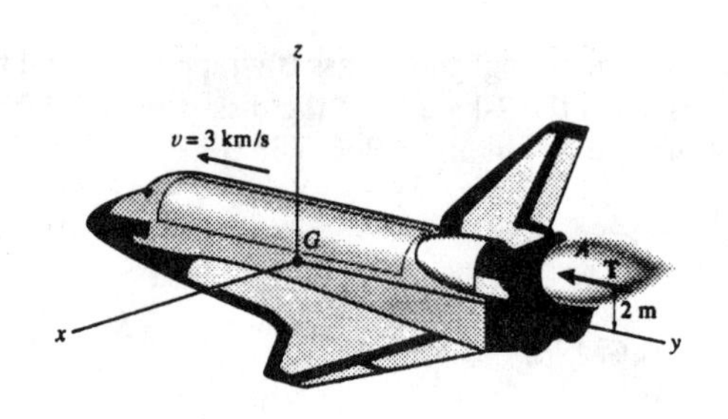

$$(\circlearrowleft +) \qquad (H_G)_1 + \Sigma\int M_G\,dt = (H_G)_2$$

$$0 + \int_0^2 600(10^3)(1 - e^{-0.3t})(2)\,dt = [120(10^3)(14)^2]\omega$$

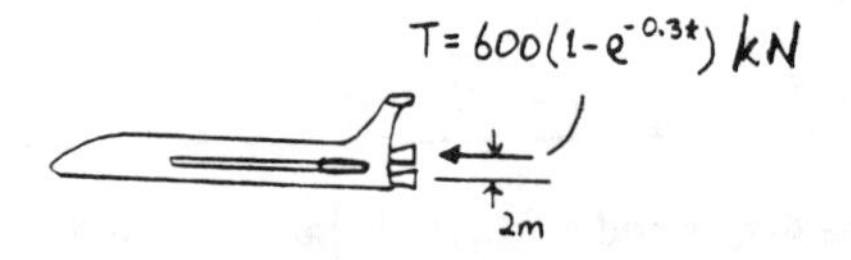

$$1200(10^3)\left[t + \frac{1}{0.3}e^{-0.3t}\right]_0^2 = 120(10^3)(14)^2\omega$$

$$\omega = 0.0253 \text{ rad/s} \qquad \textbf{Ans}$$

19-17. A cord of negligible mass is wrapped around the outer surface of the 50-lb cylinder and its end is subjected to a constant horizontal force of $P = 2$ lb. If the cylinder rolls without slipping at A, determine its angular velocity in 4 s starting from rest. Neglect the thickness of the cord.

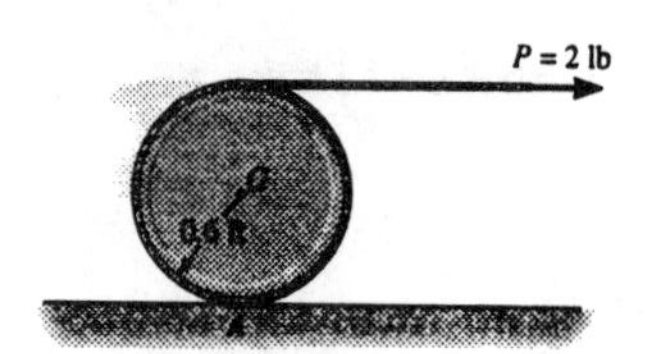

$v_G = 0.6\omega$

$$(\circlearrowleft +) \qquad (H_A)_1 + \Sigma\int M_A\,dt = (H_A)_2$$

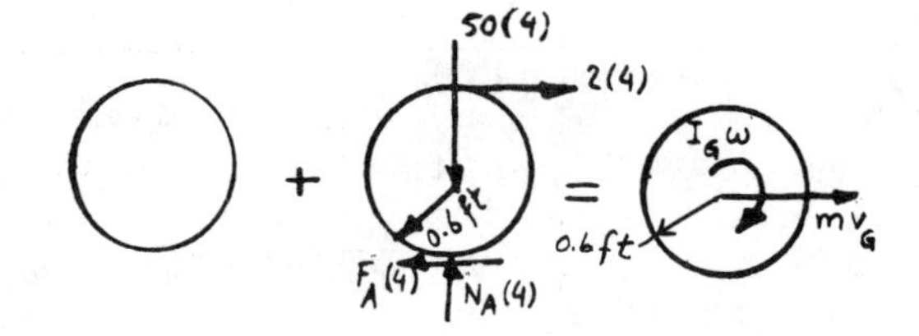

$$0 + 2(4)(1.2) = \left[\frac{1}{2}\left(\frac{50}{32.2}\right)(0.6)^2\right]\omega + \left(\frac{50}{32.2}\right)(0.6\omega)(0.6)$$

$$\omega = 11.4 \text{ rad/s} \qquad \textbf{Ans}$$

19-18. The two gears A and B have weights and radii of gyration of $W_A = 15$ lb, $k_A = 0.5$ ft and $W_B = 10$ lb, $k_B = 0.35$ ft, respectively. If a motor transmits a couple moment to gear B of $M = 2(1 - e^{-0.5t})$ lb·ft, where t is in seconds, determine the angular velocity of gear A in $t = 5$ s, starting from rest.

$\omega_A(0.8) = \omega_B(0.5)$

$\omega_B = 1.6\omega_A$

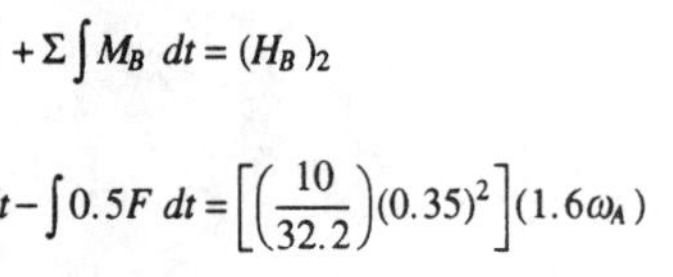

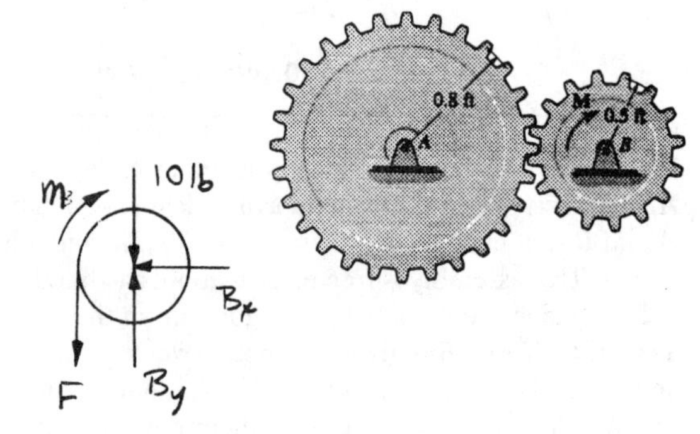

Gear B :

$$(\circlearrowleft +) \qquad (H_B)_1 + \Sigma\int M_B\,dt = (H_B)_2$$

$$0 + \int_0^5 2(1 - e^{-0.5t})\,dt - \int 0.5F\,dt = \left[\left(\frac{10}{32.2}\right)(0.35)^2\right](1.6\omega_A)$$

$$6.328 = 0.5\int F\,dt + 0.06087\omega_A \qquad (1)$$

Gear A :

$$(\circlearrowleft +) \qquad (H_A)_1 + \Sigma\int M_A\,dt = (H_A)_2$$

$$0 + \int 0.8F\,dt = \left[\left(\frac{15}{32.2}\right)(0.5)^2\right]\omega_A$$

$$0 = 0.8\int F\,dt - 0.1165\omega_A \qquad (2)$$

Eliminate $\int F\,dt$ between Eqs. (1) and (2), and solving for ω_A,

$$\omega_A = 47.3 \text{ rad/s} \qquad \textbf{Ans}$$

19-19. A motor transmits a torque of $M = 0.05$ N·m to the center of gear A. Determine the angular velocity of each of the three (equal) smaller gears in 2 s starting from rest. The smaller gears (B) are pinned at their centers, and the masses and centroidal radii of gyration of the gears are given in the figure.

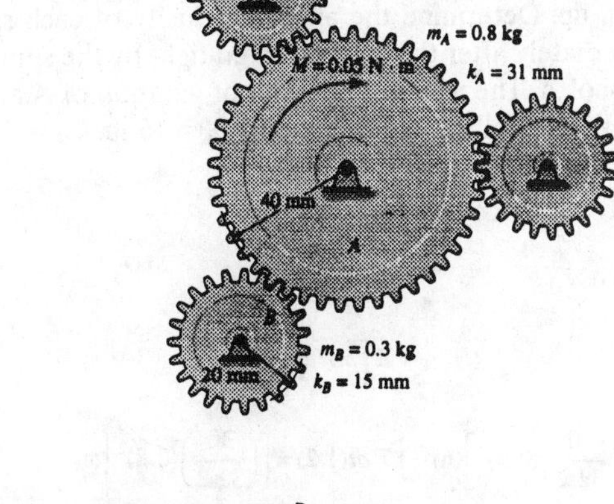

Gear A:

$$(\circlearrowleft +) \qquad (H_A)_1 + \Sigma \int M_A \, dt = (H_A)_2$$

$$0 - 3(F)(2)(0.04) + 0.05(2) = \left[0.8(0.031)^2\right]\omega_A$$

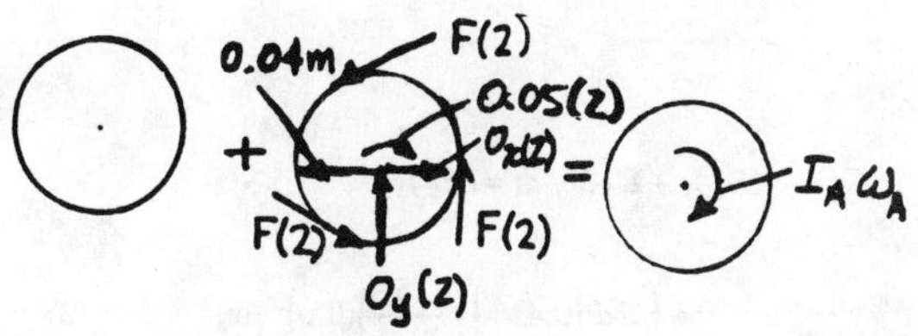

Gear B:

$$(\circlearrowright +) \qquad (H_B)_1 + \Sigma \int M_B \, dt = (H_B)_2$$

$$0 + F(2)(0.02) = \left[0.3(0.015)^2\right]\omega_B$$

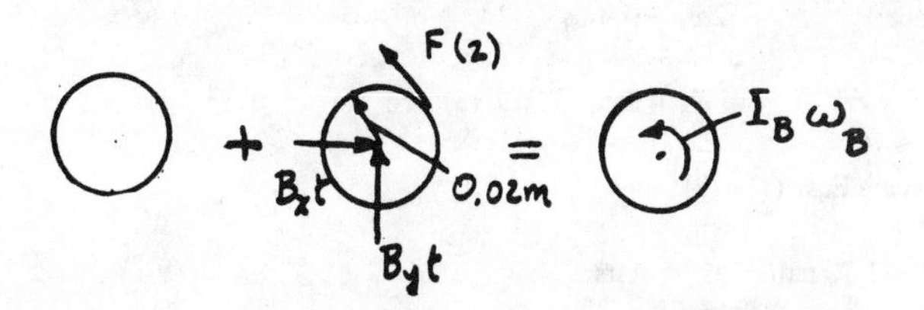

Since $0.04\omega_A = 0.02\omega_B$, or $\omega_B = 2\omega_A$, then solving,

$F = 0.214$ N

$\omega_A = 63.3$ rad/s

$\omega_B = 127$ rad/s **Ans**

***19-20.** The drum of mass m, radius r, and radius of gyration k_O rolls along an inclined plane for which the coefficient of static friction is μ_s. If the drum is released from rest, determine the maximum angle θ for the incline so that it rolls without slipping.

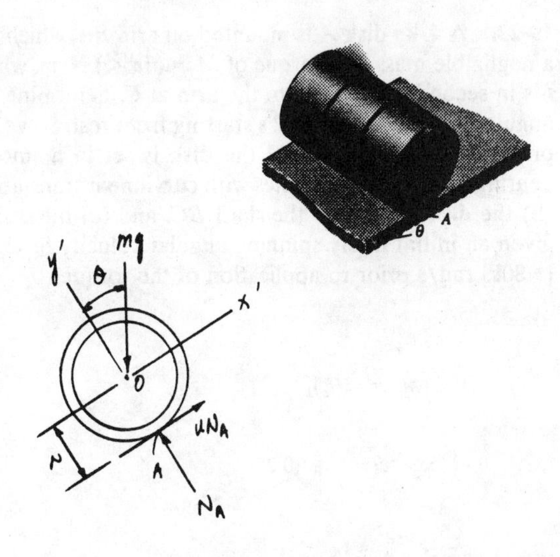

$$(+\nwarrow) \qquad m(v_{Oy'})_1 + \Sigma \int_{t_1}^{t_2} F_{y'} dt = m(v_{Oy'})_2$$

$$0 + N_A(t) - mg\cos\theta(t) = 0 \qquad N_A = mg\cos\theta$$

$$(+\swarrow) \qquad m(v_{Ox'})_1 + \Sigma \int_{t_1}^{t_2} F_{x'} dt = m(v_{Ox'})_2$$

$$0 + mg\sin\theta(t) - \mu_s mg\cos\theta(t) = mv_O \qquad (1)$$

Since no slipping occurs, $v_O = \omega r$. Hence Eq.(1) becomes

$$mg\sin\theta(t) - \mu_s mg\cos\theta(t) = m\omega r$$

$$t = \frac{m\omega r}{mg(\sin\theta - \mu_s\cos\theta)} \qquad (2)$$

$$(\circlearrowleft +) \qquad I_A\omega_1 + \Sigma \int_{t_1}^{t_2} M_A \, dt = I_A\omega_2$$

$$0 + mg\sin\theta(r)(t) = \left[mk_O^2 + mr^2\right]\omega$$

$$t = \frac{\left[mk_O^2 + mr^2\right]\omega}{mgr\sin\theta} \qquad (3)$$

Equating Eqs.(2) and (3)

$$\frac{m\omega r}{mg(\sin\theta - \mu_s\cos\theta)} = \frac{\left[mk_O^2 + mr^2\right]\omega}{mgr\sin\theta}$$

$$\theta = \tan^{-1}\left[\frac{\mu_s(k_O^2 + r^2)}{k_O^2}\right] \qquad \textbf{Ans}$$

19-21. Spool B is at rest and spool A is rotating at 6 rad/s when the slack in the cord connecting them is taken up. Determine the angular velocity of each spool immediately after the cord is jerked tight by the spinning of spool A. The weights and radii of gyration of A and B are $W_A = 30$ lb, $k_A = 0.8$ ft and $W_B = 15$ lb, $k_B = 0.6$ ft, respectively.

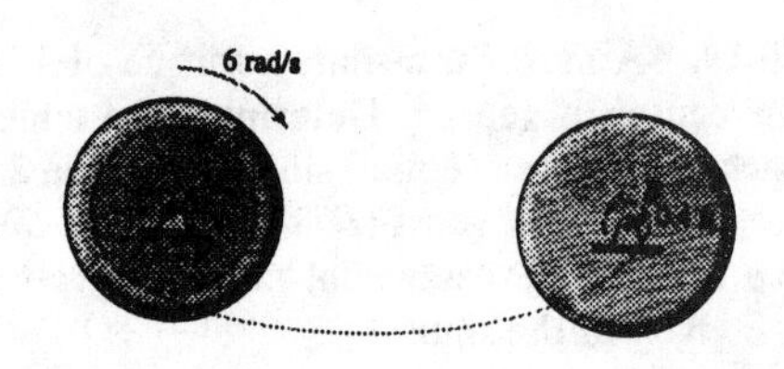

Spool A :

$$(\circlearrowleft+) \qquad (H_A)_1 + \Sigma\int M_A\, dt = (H_A)_2$$

$$\left[\left(\frac{30}{32.2}\right)(0.8)^2\right](6) - \int T\,dt(1.2) = \left[\left(\frac{30}{32.2}\right)(0.8)^2\right]\omega_A \qquad (1)$$

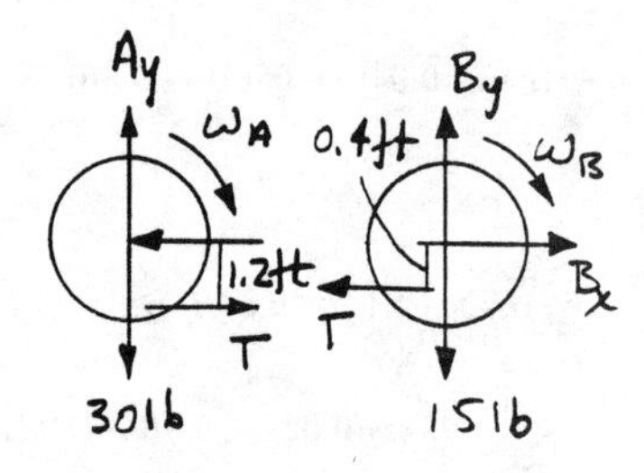

Spool B :

$$(\circlearrowleft+) \qquad (H_B)_1 + \Sigma\int M_B\, dt = (H_B)_2$$

$$0 + \int T\,dt(0.4) = \left[\left(\frac{15}{32.2}\right)(0.6)^2\right]\omega_B \qquad (2)$$

Since $\quad 1.2\omega_A = 0.4\omega_B$

$$\omega_B = 3\omega_A \qquad (3)$$

Solving Eqs. (1) – (3),

$\omega_A = 1.70$ rad/s **Ans**

$\omega_B = 5.10$ rad/s **Ans**

19-22. A 4-kg disk A is mounted on arm BC, which has a negligible mass. If a torque of $M = (5e^{0.5t})$ N·m, where t is in seconds, is applied to the arm at C, determine the angular velocity of BC in 2 s starting from rest. Solve the problem assuming that (a) the disk is set in a smooth bearing at B so that it rotates with curvilinear translation, (b) the disk is fixed to the shaft BC, and (c) the disk is given an initial freely spinning angular velocity of $\boldsymbol{\omega}_D = \{-80\mathbf{k}\}$ rad/s prior to application of the torque.

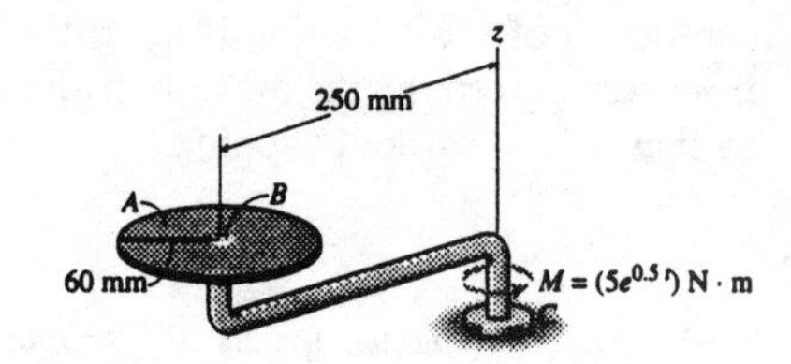

a)

$$(H_z)_1 + \Sigma\int M_z\, dt = (H_z)_2$$

$$0 + \int_0^2 5e^{0.5t}dt = 4(v_B)(0.25)$$

$$\frac{5}{0.5}e^{0.5t}\Big|_0^2 = v_B$$

$$v_B = 17.18 \text{ m/s}$$

Thus,

$$\omega_{BC} = \frac{17.18}{0.25} = 68.7 \text{ rad/s} \qquad \textbf{Ans}$$

b)

$$(H_z)_1 + \Sigma\int M_z\, dt = (H_z)_2$$

$$0 + \int_0^2 5e^{0.5t}dt = 4(v_B)(0.25) + \left[\frac{1}{2}(4)(0.06)^2\right]\omega_{BC}$$

Since $v_B = 0.25\omega_{BC}$, then

$\omega_{BC} = 66.8$ rad/s **Ans**

c)

$$(H_z)_1 + \Sigma\int M_z\, dt = (H_z)_2$$

$$-\left[\frac{1}{2}(4)(0.06)^2\right](80) + \int_0^2 5e^{0.5t}dt = 4(v_B)(0.25) - \left[\frac{1}{2}(4)(0.06)^2\right](80)$$

Since $v_B = 0.25\omega_{BC}$,

$\omega_{BC} = 68.7$ rad/s **Ans**

19-23. If the hoop has a weight W and radius r and is thrown onto a *rough surface* with a velocity $\mathbf{v}_G$ parallel to the surface, determine the amount of backspin, ω_0, it must be given so that it stops spinning at the same instant that its forward velocity is zero. It is not necessary to know the coefficient of kinetic friction at A for the calculation.

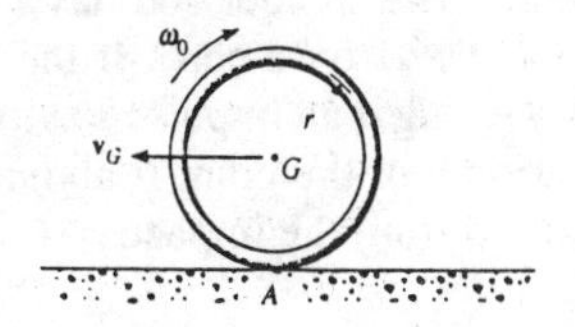

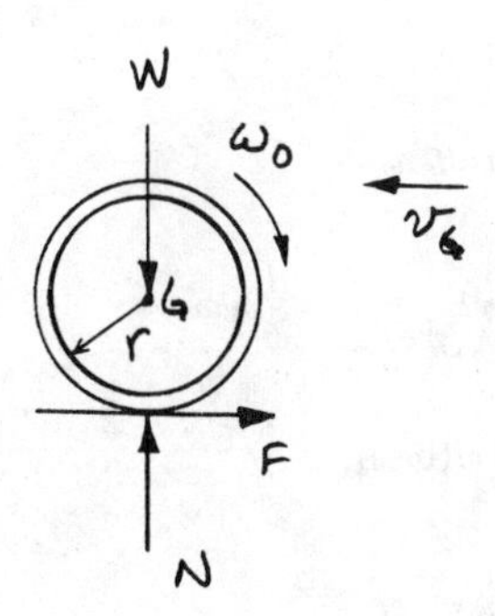

$(\overset{+}{\leftarrow})$ $\quad m(v_{Gx})_1 + \Sigma\int F_x\,dt = m(v_{Gx})_2$

$$\frac{W}{g}v_G - Ft = 0 \qquad (1)$$

$(\circlearrowleft +)$ $\quad (H_G)_1 + \Sigma\int M_G\,dt = (H_G)_2$

$$-\left(\frac{W}{g}r^2\right)\omega_0 + Ft(r) = 0 \qquad (2)$$

Eliminate Ft between Eqs. (1) and (2),

$$\omega_0 = \frac{v_G}{r} \qquad \textbf{Ans}$$

***19-24.** For safety reasons, the 20-kg supporting leg of a sign is designed to break away with negligible resistance at B when the leg is subjected to the impact of a car. Assuming that the leg is pin supported at A and approximates a thin rod, determine the impulse the car bumper exerts on it, if after the impact the leg appears to rotate upward to an angle of $\theta_{max} = 150°$.

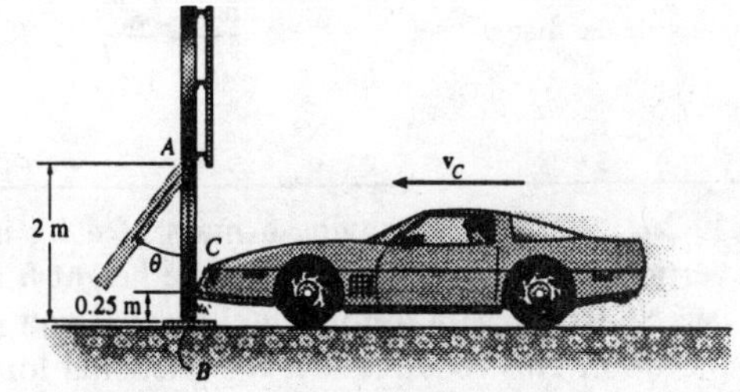

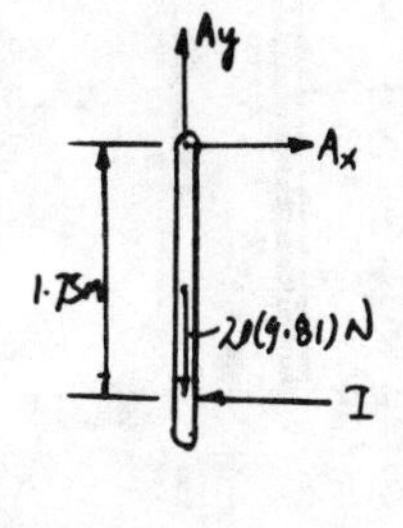

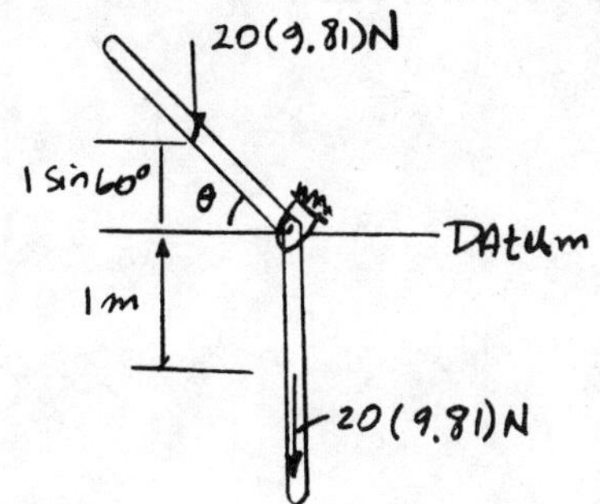

$(\circlearrowright +)$ $\quad I_A\,\omega_1 + \Sigma\int_{t_1}^{t_2} M_A\,dt = I_A\,\omega_2$

$$0 + I(1.75) = \left[\frac{1}{3}(20)(2)^2\right]\omega_2$$

$$\omega_2 = 0.065625I$$

$$T_2 + V_2 = T_3 + V_3$$

$$\frac{1}{2}\left[\frac{1}{3}(20)(2)^2\right](0.065625I)^2 + 20(9.81)(-1) = 0 + 20(9.81)(1\sin 60°)$$

$$I = 79.8\ \text{N}\cdot\text{s} \qquad \textbf{Ans}$$

19-25. The slender rod has a mass m and is suspended at its end A by a cord. If the rod receives a horizontal blow giving it an impulse $\mathbf{I}$ at its bottom B, determine the location y of the point P about which the rod appears to rotate during the impact.

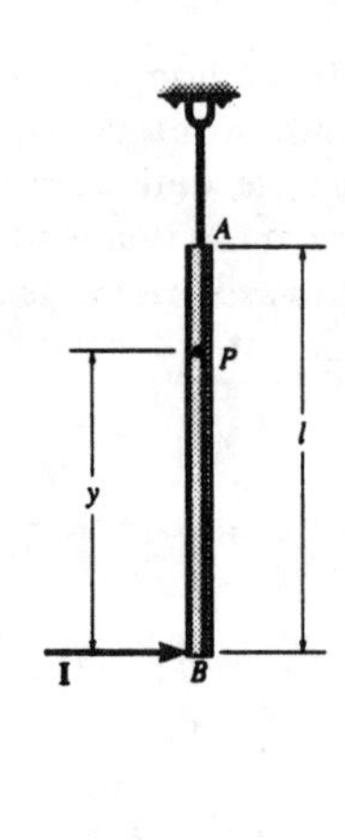

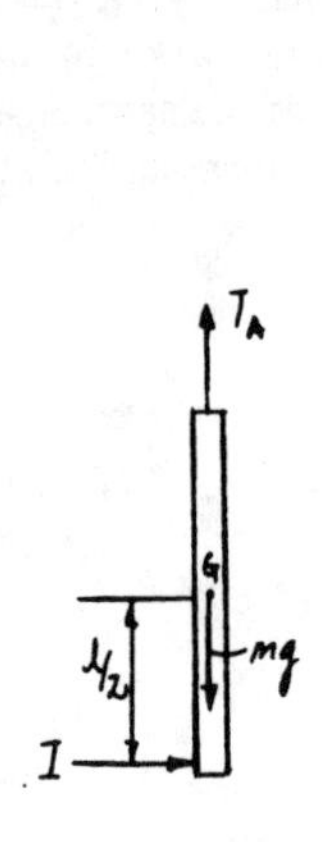

Principle of impulse and momentum :

$(\circlearrowleft +)$ $$I_G\omega_1 + \Sigma\int_{t_1}^{t_2} M_G dt = I_G\omega_2$$

$$0 + I\left(\frac{l}{2}\right) = \left[\frac{1}{12}ml^2\right]\omega \qquad I = \frac{1}{6}ml\omega$$

$(\overset{+}{\rightarrow})$ $$m(v_{Ax})_1 + \Sigma\int_{t_1}^{t_2} F_x dt = m(v_{Ax})_2$$

$$0 + \frac{1}{6}ml\omega = mv_G \qquad v_G = \frac{l}{6}\omega$$

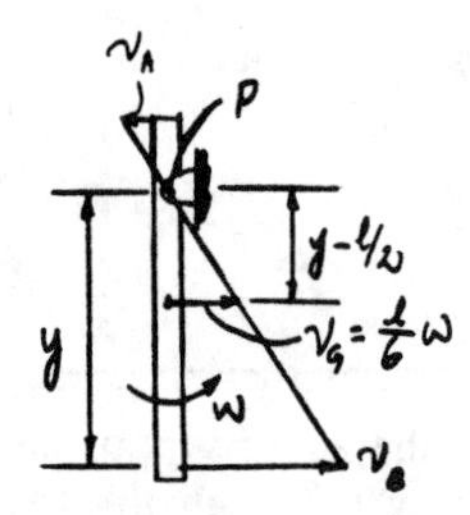

Kinematics : Point P is the IC.

$$v_B = \omega y$$

Using similar triangles $$\frac{\omega y}{y} = \frac{\frac{l}{6}\omega}{y - \frac{l}{2}} \qquad y = \frac{2}{3}l \qquad \textbf{Ans}$$

19-26. A thin rod having a mass of 4 kg is balanced vertically as shown. Determine the height h at which it can be struck with a horizontal force $\mathbf{F}$ and not slip on the floor. This requires that the frictional force at A be essentially zero.

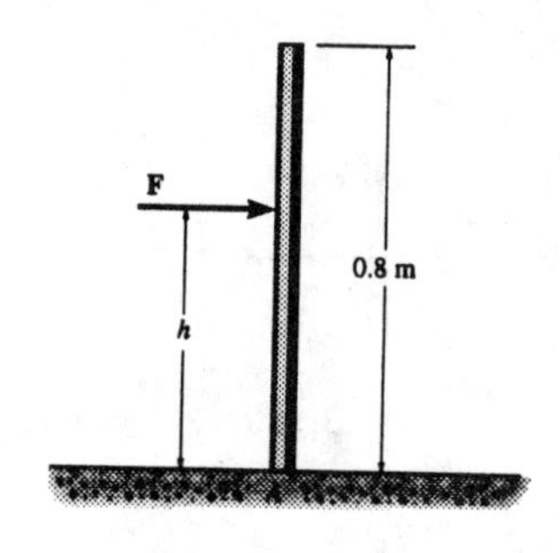

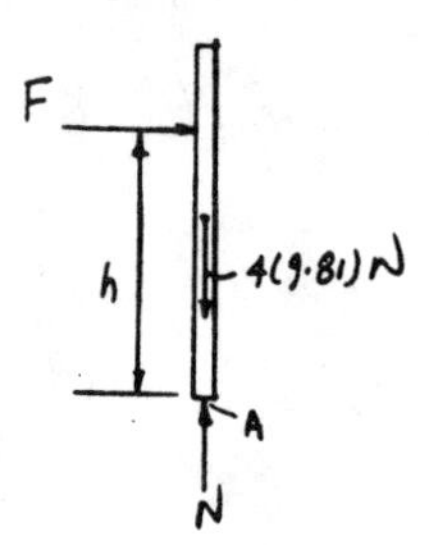

$(\overset{+}{\rightarrow})$ $$m(v_{Gx})_1 + \Sigma\int_{t_1}^{t_2} F_x dt = m(v_{Gx})_2$$

$$0 + F(t) = 4v_G \qquad (1)$$

$(\circlearrowleft +)$ $$I_A\omega_1 + \Sigma\int_{t_1}^{t_2} M_A dt = I_A\omega_2$$

$$0 + Fh(t) = \left[\frac{1}{3}(4)(0.8)^2\right]\omega \qquad (2)$$

However, $v_G = \omega(0.4)$ (3)

Susitute Eq.(3) into Eq. (1) $\qquad Ft = 1.6\omega \qquad (4)$

Divide Eq.(2) by Eq.(4) $$h = \frac{\frac{1}{3}(4)(0.8)^2}{1.6} = 0.533 \text{ m} \qquad \textbf{Ans}$$

19-27. Determine the height h of the bumper of the pool table, so that when the pool ball of mass m strikes it, no frictional force will be developed between the ball and the table at A. Assume the bumper exerts only a horizontal force on the ball.

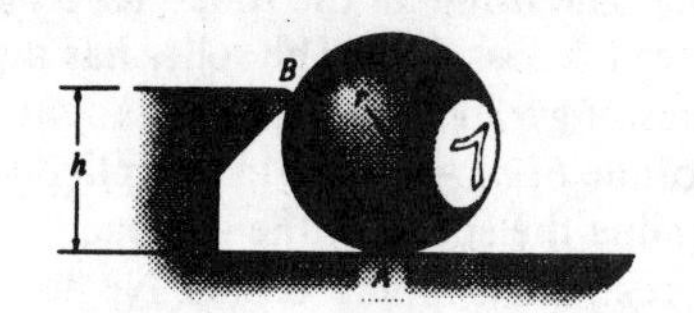

$(\overset{+}{\rightarrow})\qquad \Sigma\int_{t_1}^{t_2} F_x\,dt = m\Delta(v_{Ax})$

$$\int F\,dt = m\Delta v_G \qquad (1)$$

$(\curvearrowleft +)\qquad \Sigma\int_{t_1}^{t_2} M_G dt = I_G\Delta\omega_2$

$$(h-r)\int F\,dt = \left[\frac{2}{5}mr^2\right]\Delta\omega \qquad (2)$$

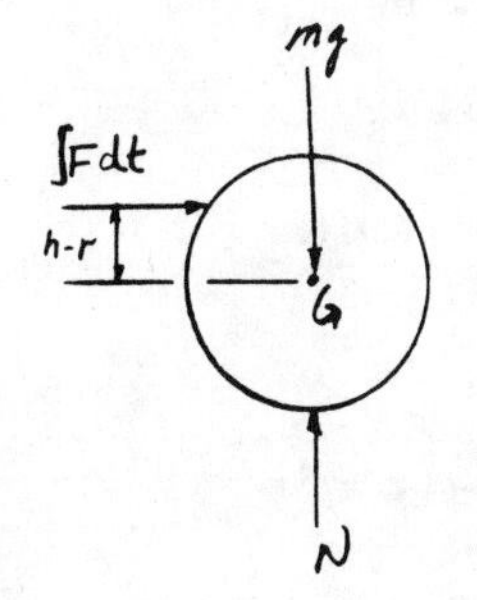

However, $\Delta v_G = \Delta\omega r \qquad (3)$

Substitute Eq.(3) into Eq. (1) yields $\quad \int F\,dt = m\Delta\omega r \qquad (4)$

Divide Eq.(2) by Eq.(4) yields $\quad h-r = \frac{2}{5}r \qquad h = \frac{7}{5}r$ **Ans**

***19-28.** If the ball has a weight W and radius r and is thrown onto a *rough surface* with a velocity $\mathbf{v}_0$ parallel to the surface, determine the amount of backspin, ω_0, it must be given so that it stops spinning at the same instant that its forward velocity is zero. It is not necessary to know the coefficient of friction at A for the calculation.

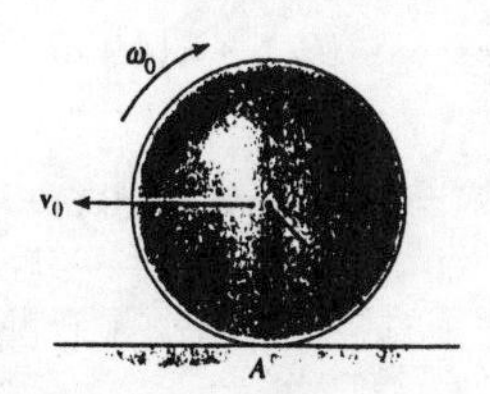

$(\overset{+}{\leftarrow})\qquad m(v_{Gx})_1 + \Sigma\int F_x\,dt = m(v_{Gx})_2$

$$\frac{W}{g}v_0 - Ft = 0 \qquad (1)$$

$(\curvearrowleft +)\qquad (H_G)_1 + \Sigma\int M_G\,dt = (H_G)_2$

$$-\frac{2}{5}\left(\frac{W}{g}r^2\right)\omega_0 + Ft(r) = 0 \qquad (2)$$

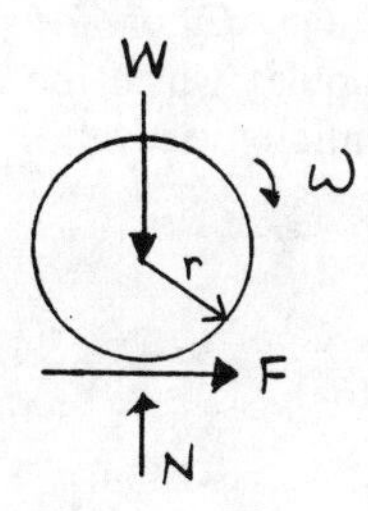

Eliminate Ft between Eqs. (1) and (2) :

$$\frac{2}{5}\left(\frac{W}{g}r^2\right)\omega_0 = \left[\frac{W}{g}\left(\frac{v_0}{t}\right)\right]t(r)$$

$$\omega_0 = 2.5\left(\frac{v_0}{r}\right) \qquad \textbf{Ans}$$

19-29. The frame of the roller has a mass of 5.5 Mg and a center of mass at G. The roller has a mass of 2 Mg and a radius of gyration about its mass center of $k_A = 0.45$ m. If a torque of $M = 600$ N·m is applied to the rear wheels, determine the speed of the compactor in $t = 4$ s, starting from rest. No slipping occurs. Neglect the mass of the driving wheels.

Driving Wheels : (mass is neglected)

$$\curvearrowleft + \Sigma M_D = 0; \qquad 600 - F_C(0.5) = 0$$

$$F_C = 1200 \text{ N}$$

Frame and driving wheels :

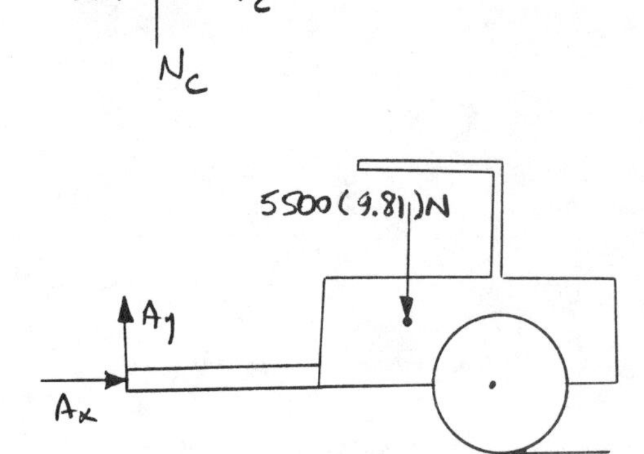

$$(\overset{+}{\leftarrow}) \qquad m(v_{Gx})_1 + \Sigma\int F_x\, dt = m(v_{Gx})_2$$

$$0 + 1200(4) - A_x(4) = 5500 v_G$$

$$A_x = 1200 - 1375 v_G \qquad (1)$$

Roller :

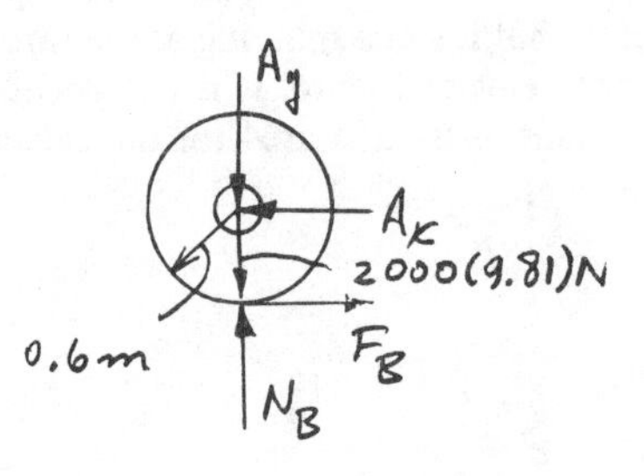

$$v_G = v_A = 0.6\omega$$

$$(\curvearrowleft +) \qquad (H_B)_1 + \Sigma\int M_B\, dt = (H_B)_2$$

$$0 + A_x(4)(0.6) = \left[2000(0.45)^2\right]\left(\frac{v_G}{0.6}\right) + \left[2000(v_G)\right](0.6)$$

$$A_x = 781.25 v_G \qquad (2)$$

Solving Eqs. (1) and (2) :

$A_x = 435$ N

$v_G = 0.557$ m/s **Ans**

19-30. The car strikes the side of a light pole, which is designed to break away from its base with negligible resistance. From a video taken of the collision it is observed that the pole was given an angular velocity of 60 rad/s when AC was vertical. The pole has a mass of 175 kg, a center of mass at G, and a radius of gyration about an axis perpendicular to the plane of the pole assembly and passing through G of $k_G = 2.25$ m. Determine the horizontal impulse which the car exerts on the pole while AC is essentially vertical.

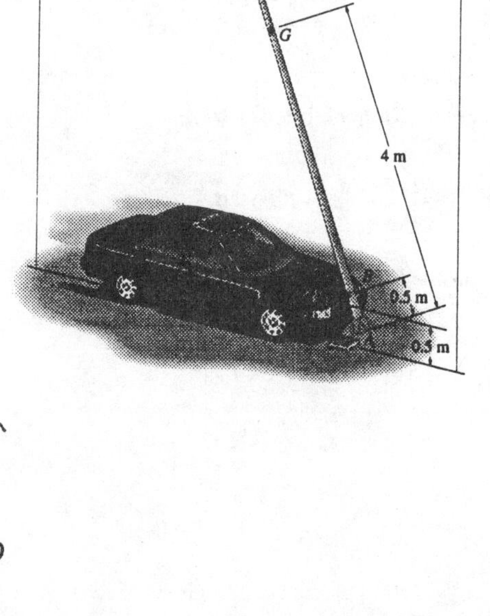

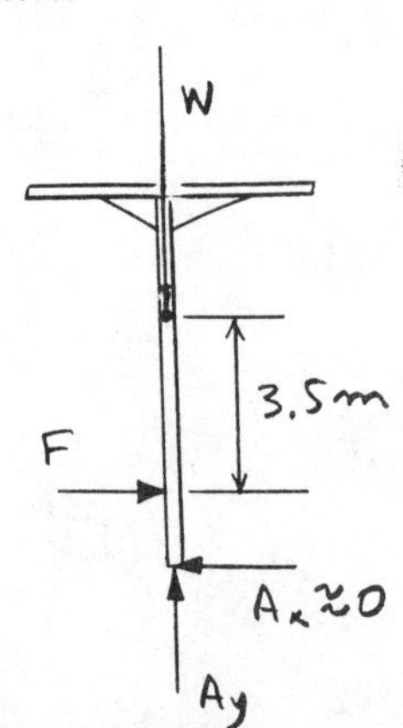

$$(\curvearrowleft +) \qquad (H_G)_1 + \Sigma\int M_G\, dt = (H_G)_2$$

$$0 + \left[\int F\, dt\right](3.5) = 175(2.25)^2(60)$$

$$\int F\, dt = 15.2 \text{ kN}\cdot\text{s} \qquad \textbf{Ans}$$

19-31. The double pulley consists of two wheels which are attached to one another and turn at the same rate. The pulley has a mass of 15 kg and a radius of gyration of $k_O = 110$ mm. If the block at A has a mass of 40 kg and the container at B has a mass of 85 kg, including its contents, determine the speed of the container when t = 3 s after it is released from rest.

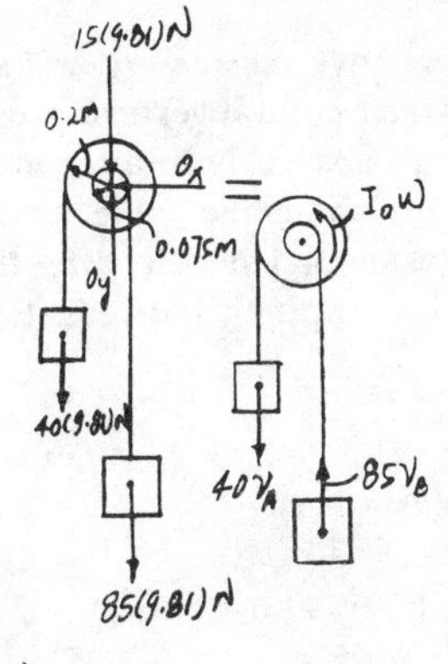

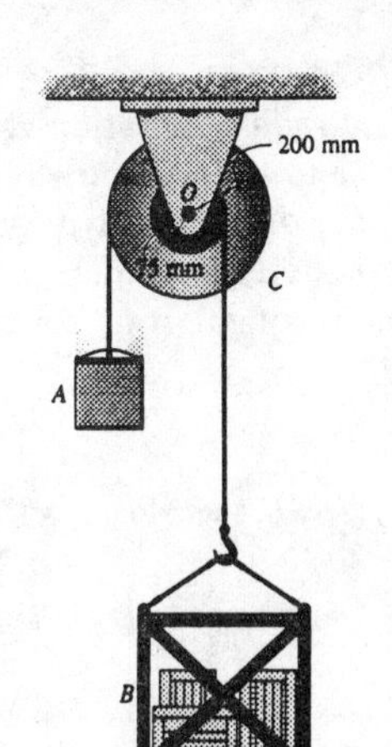

The angular velocity of the pulley can be related to the speed of container B by $\omega = \frac{v_B}{0.075} = 13.333 v_B$. Also the speed of block A $v_A = \omega(0.2) = 13.33 v_B(0.2) = 2.667 v_B$.

$$(\circlearrowleft +) \quad \left(\Sigma \text{Syst. Ang. Mom.}\right)_{O1} + (\Sigma \text{Syst. Ang. Imp.})_{O(1-2)} = \left(\Sigma \text{Syst. Ang. Mom.}\right)_{O2}$$

$$0 + 40(9.81)(0.2)(3) - 85(9.81)(0.075)(3)$$

$$= \left[15(0.110)^2\right](13.333 v_B) + 85 v_B(0.075) + 40(2.667 v_B)(0.2)$$

$$v_B = 1.59 \text{ m/s} \qquad \textbf{Ans}$$

***19-32.** The two rods each have a mass m and a length l, and lie on the smooth horizontal plane. If an impulse **I** is applied at an angle of 45° to one of the rods at mid length as shown, determine the angular velocity of each rod just after the impulse. The rods are pin connected at B.

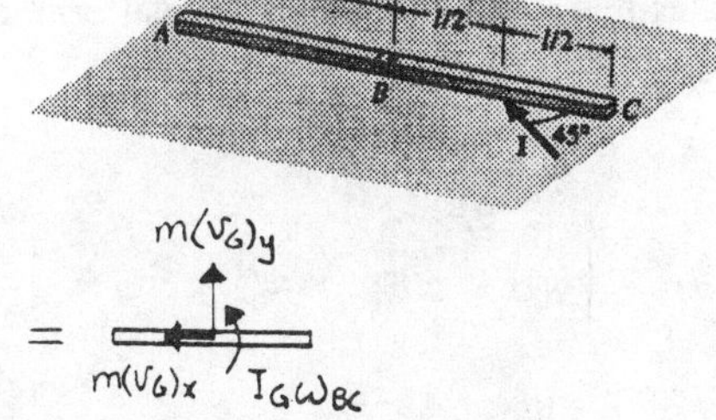

Bar BC:

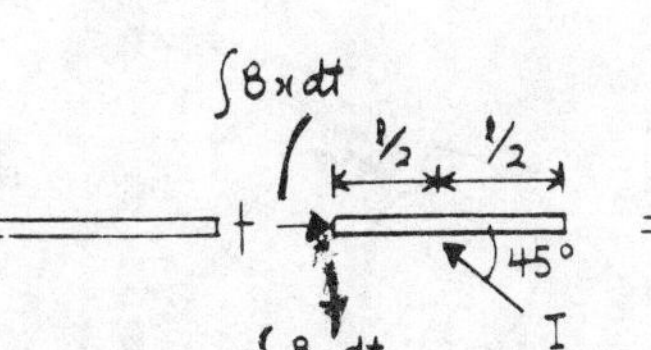

$$(\circlearrowleft +) \quad (H_G)_1 + \Sigma \int M_G \, dt = (H_G)_2$$

$$0 + \int B_y \, dt \left(\frac{l}{2}\right) = I_G \omega_{BC} \qquad (1)$$

$$(+\uparrow) \quad m(v_{Gy})_1 + \Sigma \int F_y \, dt = m(v_{Gy})_2$$

$$0 - \int B_y \, dt + I \sin 45° = m(v_G)_y \qquad (2)$$

Bar AB:

$$(\circlearrowleft +) \quad (H_{G'})_1 + \Sigma \int M_{G'} \, dt = (H_{G'})_2$$

$$0 + \int B_y \, dt \left(\frac{l}{2}\right) = I_G \omega_{AB} \qquad (3)$$

$$(+\uparrow) \quad m(v_{Gy})_1 + \Sigma \int F_y \, dt = m(v_{Gy})_2$$

$$0 + \int B_y \, dt = m(v_{G'})_y \qquad (4)$$

$$\mathbf{v}_B = \mathbf{v}_{G'} + \mathbf{v}_{B/G'} = \mathbf{v}_G + \mathbf{v}_{B/G}$$

$$(+\uparrow) \quad v_{By} = (v_{G'})_y + \omega_{AB}\left(\frac{l}{2}\right) = (v_G)_y - \omega_{BC}\left(\frac{l}{2}\right) \qquad (5)$$

Eliminate $\int B_y \, dt$ from Eqs. (1) and (2) and from Eqs. (3) and (4), and between Eqs. (1) and (3). This yields

$$m(v_{G'})_y \left(\frac{l}{2}\right) = I_G \omega_{AB}$$

$$I_G \omega_{BC} = \frac{l}{2}(I \sin 45° - m(v_G)_y)$$

$$\omega_{BC} = \omega_{AB}$$

Substituting into Eq. (5),

$$\frac{1}{m}\left(\frac{2}{l}\right) I_G \omega_{AB} + \omega_{AB}\left(\frac{l}{2}\right) = -\left[I_G\left(\frac{\omega_{AB}}{m}\right)\left(\frac{2}{l}\right)\right] + \frac{I}{m}\sin 45° - \omega_{AB}\left(\frac{l}{2}\right)$$

$$\left(\frac{4}{ml}\right) I_G \omega_{AB} + \omega_{AB} l = \frac{I}{m}\sin 45°$$

$$\left(\frac{4}{ml}\right)\left(\frac{1}{12} ml^2\right)\omega_{AB} + \omega_{AB} l = \frac{I}{m}\sin 45°$$

$$\frac{4}{3}\omega_{AB} l = \frac{I}{m}\sin 45°$$

$$\omega_{AB} = \omega_{BC} = \frac{3}{4\sqrt{2}}\left(\frac{I}{ml}\right) \qquad \textbf{Ans}$$

19-33. Two wheels A and B have masses m_A and m_B, and radii of gyration about their central vertical axes of k_A and k_B, respectively. If they are freely rotating in the same direction at ω_A and ω_B about the same vertical axis, determine their common angular velocity after they are brought into contact and slipping between them stops.

$(\Sigma \text{Syst. Ang. Mom.})_1 = (\Sigma \text{Syst. Ang. Mom.})_2$

$$(m_A k_A^2)\,\omega_A + (m_B k_B^2)\,\omega_B = (m_A k_A^2)\,\omega'_A + (m_B k_B^2)\,\omega'_B$$

Set $\omega'_A = \omega'_B = \omega$, then

$$\omega = \frac{m_A k_A^2 \omega_A + m_B k_B^2 \omega_B}{m_A k_A^2 + m_B k_B^2} \qquad \textbf{Ans}$$

19-34. A horizontal circular platform has a weight of 300 lb and a radius of gyration $k_z = 8$ ft about the z axis passing through its center O. The platform is free to rotate about the z axis and is initially at rest. A man having a weight of 150 lb begins to run along the edge in a circular path of radius 10 ft. If he has a speed of 4 ft/s and maintains this speed relative to the platform, determine the angular velocity of the platform. Neglect friction.

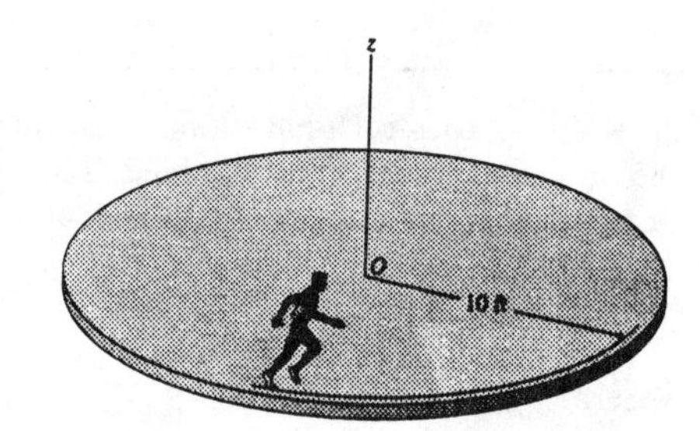

$\mathbf{v}_m = \mathbf{v}_p + \mathbf{v}_{m/p}$

$(\overset{+}{\rightarrow})\qquad v_m = -10\omega + 4$

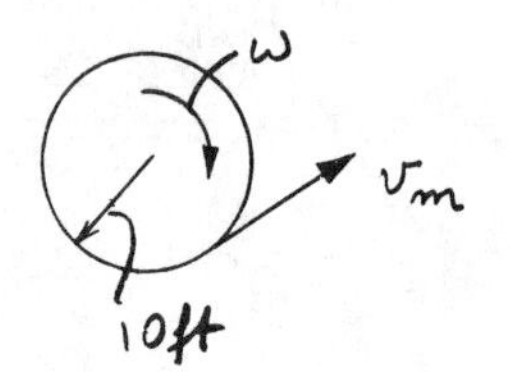

$(\circlearrowleft +)\qquad (H_z)_1 = (H_z)_2$

$$0 = -\left(\frac{300}{32.2}\right)(8)^2\,\omega + \left(\frac{150}{32.2}\right)(-10\omega + 4)(10)$$

$\omega = 0.175$ rad/s $\qquad$ **Ans**

19-35. A horizontal circular platform has a weight of 300 lb and a radius of gyration $k_z = 8$ ft about the z axis passing through its center O. The platform is free to rotate about the z axis and is initially at rest. A man having a weight of 150 lb throws a 15-lb block off the edge of the platform with a horizontal velocity of 5 ft/s, *measured relative to the platform.* Determine the angular velocity of the platform if the block is thrown (a) tangent to the platform, along the $+t$ axis, and (b) outward along a radial line, or $+n$ axis. Neglect the size of the man.

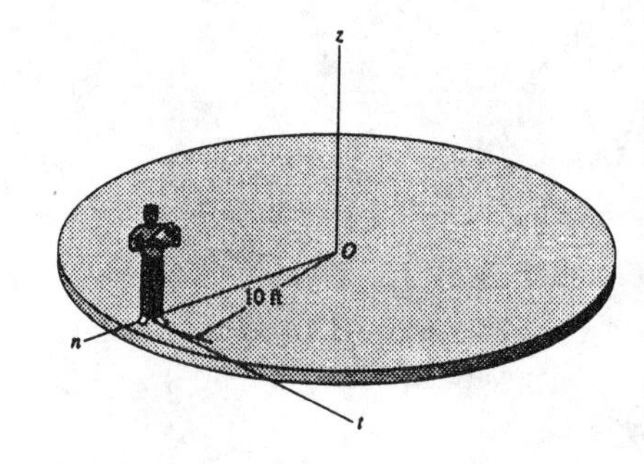

a)

$(H_z)_1 = (H_z)_2$

$$0 + 0 = \left(\frac{15}{32.2}\right)(v_b)(10) - \left(\frac{300}{32.2}\right)(8)^2\,\omega - \left(\frac{150}{32.2}\right)(10\omega)(10)$$

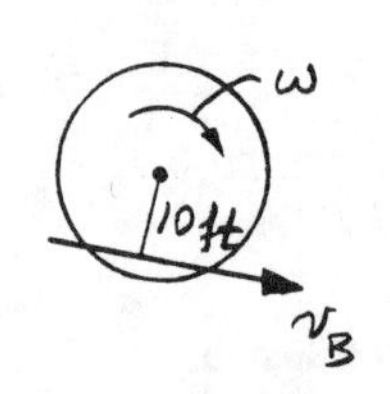

$v_b = 228\omega$

$\mathbf{v}_b = \mathbf{v}_m + \mathbf{v}_{b/m}$

$(\overset{+}{\rightarrow})\qquad v_b = -10\omega + 5$

$228\omega = -10\omega + 5$

$\omega = 0.0210$ rad/s $\qquad$ **Ans**

b)

$(H_z)_1 = (H_z)_2$

$$0 + 0 = 0 - \left(\frac{300}{32.2}\right)(8)^2\,\omega - \left(\frac{150}{32.2}\right)(10\omega)(10)$$

$\omega = 0$ $\qquad$ **Ans**

***19-36.** The Hubble Space Telescope is powered by two solar panels as shown. The body of the telescope has a mass of 11 Mg and radii of gyration $k_x = 1.64$ m and $k_y = 3.85$ m, whereas the solar panels can be considered as thin plates, each having a mass of 54 kg. Due to an internal drive, the panels are given an angular velocity of $\{0.6\mathbf{j}\}$ rad/s, measured relative to the telescope. Determine the angular velocity of the telescope due to the rotation of the panels. Prior to rotating the panels, the telescope was originally traveling at $\mathbf{v}_G = \{-400\mathbf{i} + 250\mathbf{j} + 175\mathbf{k}\}$ m/s. Neglect its orbital rotation.

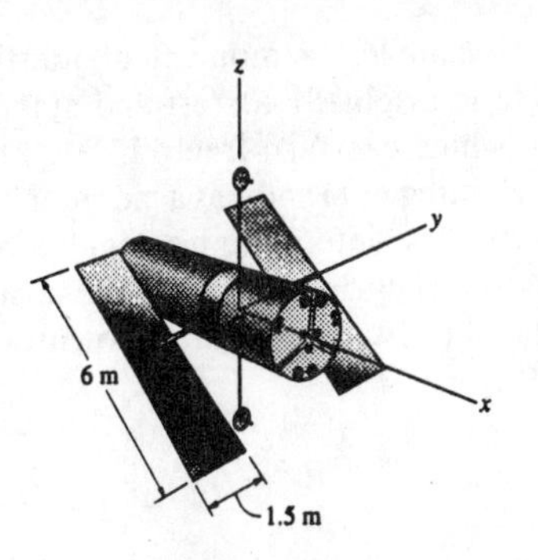

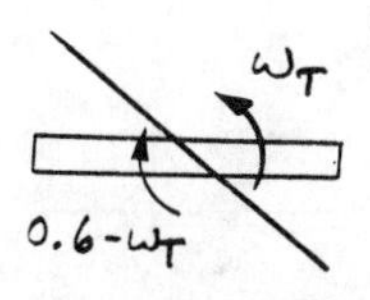

$$(H_y)_1 = (H_y)_2$$

$$0 = 2\left[\frac{1}{12}(54)(6)^2\right](0.6 - \omega_T) - \left[(11\,000)(3.85)^2\right]\omega_T$$

$\omega_T = 1.19\left(10^{-3}\right)$ rad/s **Ans**

19-37. The platform swing consists of a 200-lb flat plate suspended by four rods of negligible weight. When the swing is at rest, the 150-lb man jumps off the platform when his center of gravity G is 10 ft from the pin at A. This is done with a horizontal velocity of 5 ft/s, measured relative to the swing at the level of G. Determine the angular velocity he imparts to the swing just after jumping off.

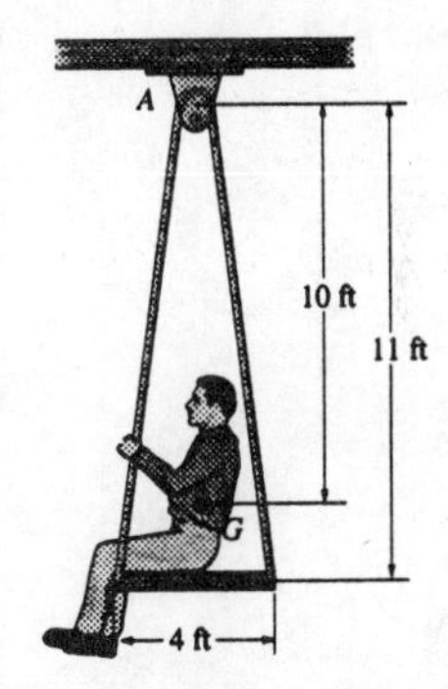

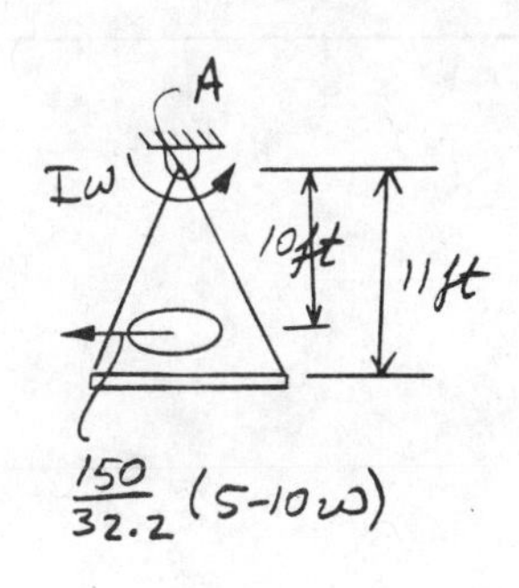

$$(\circlearrowleft +) \quad (H_A)_1 = (H_A)_2$$

$$0 + 0 = \left[\frac{1}{12}\left(\frac{200}{32.2}\right)(4)^2 + \frac{200}{32.2}(11)^2\right]\omega - \left[\left(\frac{150}{32.2}\right)(5 - 10\omega)\right](10)$$

$\omega = 0.190$ rad/s **Ans**

19-38. A man has a moment of inertia I_z about the z axis. He is originally at rest and standing on a small platform which can turn freely. If he is handed a wheel which is rotating at $\boldsymbol{\omega}$ and has a moment of inertia I about its spinning axis, determine his angular velocity if (a) he holds the wheel upright as shown, (b) turns the wheel out, $\theta = 90°$, and (c) turns the wheel downward, $\theta = 180°$.

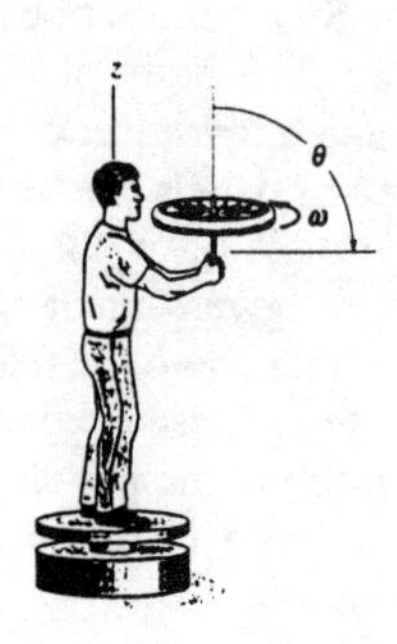

(a)

$$\Sigma(H_z)_1 = \Sigma(H_z)_2; \quad 0 + I\omega = I_z\omega_M + I\omega$$ **Ans**

(b)

$$\Sigma(H_z)_1 = \Sigma(H_z)_2; \quad 0 + I\omega = I_z\omega_M + 0 \quad \omega_M = \frac{I}{I_z}\omega$$ **Ans**

(c)

$$\Sigma(H_z)_1 = \Sigma(H_z)_2; \quad 0 + I\omega = I_z\omega_M - I\omega \quad \omega_M = \frac{2I}{I_z}\omega$$ **Ans**

19-39. A man has a moment of inertia I_z about the z axis. He is originally at rest and standing on a small platform which can turn freely. If he is handed a wheel when it is at *rest* and he starts it spinning with an angular velocity $\boldsymbol{\omega}$, determine his angular velocity if (a) he holds the wheel upright as shown, (b) turns the wheel out, $\theta = 90°$, and (c) turns the wheel downward, $\theta = 180°$.

(a)

$$\Sigma(H_z)_1 = \Sigma(H_z)_2; \quad 0 + 0 = I\omega - I_z\omega_M \quad \omega_M = \frac{I}{I_z}\omega$$ **Ans**

(b)

$$\Sigma(H_z)_1 = \Sigma(H_z)_2; \quad 0 + 0 = I_z\omega_M + 0 \quad \omega_M = 0$$ **Ans**

(c)

$$\Sigma(H_z)_1 = \Sigma(H_z)_2; \quad 0 + 0 = I_z\omega_M - I\omega \quad \omega_M = \frac{I}{I_z}\omega$$ **Ans**

***19-40.** A 7-g bullet having a velocity of 800 m/s is fired into the edge of the 5-kg disk as shown. Determine the angular velocity of the disk just after the bullet becomes embedded in it. Also, calculate how far θ the disk will swing until it momentarily stops. The disk is originally at rest.

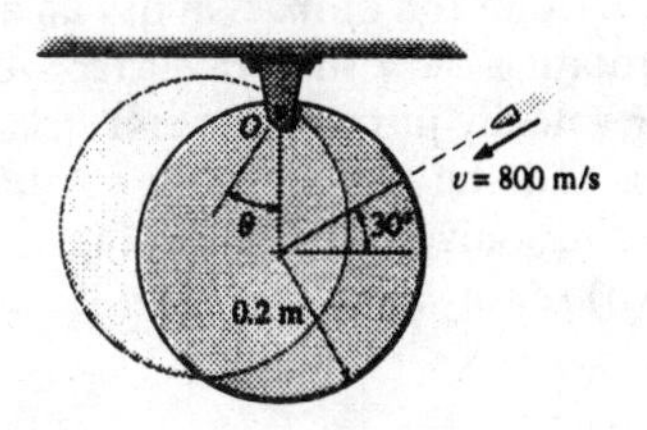

Disk and bullet :

$(\circlearrowleft +) \quad (H_O)_1 = (H_O)_2$

$$0.007(800)\cos 30°(0.2) = \left[\frac{3}{2}(5.007)(0.2)^2\right]\omega$$

$\omega = 3.229 = 3.23 \text{ rad/s}$ **Ans**

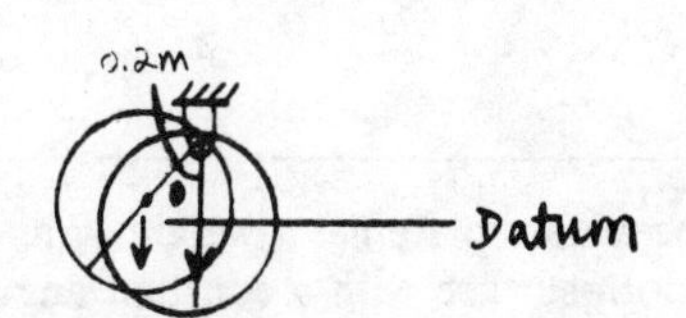

$$T_1 + V_1 = T_2 + V_2$$

$$\frac{1}{2}(5.007)\left[3.229(0.2)\right]^2 + \frac{1}{2}\left[\frac{1}{2}(5.007)(0.2)^2\right](3.229)^2 + 0 = 0 + 0.2(1-\cos\theta)(5.007)(9.81)$$

$\theta = 32.8°$ **Ans**

Note : This calculation is a close estimate because the center of mass for the disk – bullet is not actually known.

19-41. The pendulum consists of a slender 2-kg rod AB and 5-kg disk. It is released from rest without rotating. When it falls 0.3 m, the end A strikes the hook S, which provides a permanent connection. Determine the angular velocity of the pendulum after it has rotated 90°. Treat the pendulum's weight during impact as a nonimpulsive force.

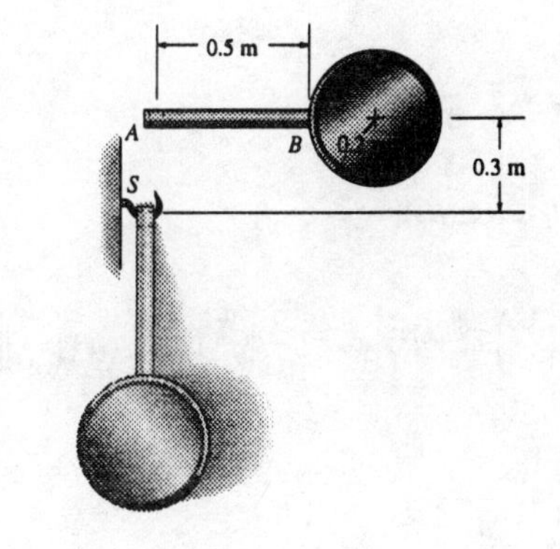

$$T_0 + V_0 = T_1 + V_1$$

$$0 + 2(9.81)(0.3) + 5(9.81)(0.3) = \frac{1}{2}(2)(v_G)_1^2 + \frac{1}{2}(5)(v_G)_1^2$$

$(v_G)_1 = 2.4261 \text{ m/s}$

$$\Sigma(H_S)_1 = \Sigma(H_S)_2$$

$$2(2.4261)(0.25) + 5(2.4261)(0.7) = \left[\frac{1}{12}(2)(0.5)^2 + 2(0.25)^2 + \frac{1}{2}(5)(0.2)^2 + 5(0.7)^2\right]\omega$$

$\omega = 3.572 \text{ rad/s}$

$$T_2 + V_2 = T_3 + V_3$$

$$\frac{1}{2}\left[\frac{1}{12}(2)(0.5)^2 + 2(0.25)^2 + \frac{1}{2}(5)(0.2)^2 + 5(0.7)^2\right](3.572)^2 + 0$$

$$= \frac{1}{2}\left[\frac{1}{12}(2)(0.5)^2 + 2(0.25)^2 + \frac{1}{2}(5)(0.2)^2 + 5(0.7)^2\right]\omega^2$$

$$+ 2(9.81)(-0.25) + 5(9.81)(-0.7)$$

$\omega = 6.45 \text{ rad/s}$ **Ans**

19-42. A thin rod of mass m has an angular velocity ω_0 while rotating on a smooth surface. Determine its new angular velocity just after its end strikes and hooks onto the peg and the rod starts to rotate about P without rebounding. Solve the problem (a) using the parameters given, (b) setting $m = 2$ kg, $\omega_0 = 4$ rad/s, $l = 1.5$ m.

(a)

$$\Sigma(H_P)_0 = \Sigma(H_P)_1$$

$$\left[\frac{1}{12}ml^2\right]\omega_0 = \left[\frac{1}{3}ml^2\right]\omega$$

$$\omega = \frac{1}{4}\omega_0 \qquad \textbf{Ans}$$

(b) From part (a) $\qquad \omega = \frac{1}{4}\omega_0 = \frac{1}{4}(4) = 1 \text{ rad/s} \qquad$ **Ans**

19-43. The square plate has a weight W and is rotating on the smooth surface with a constant angular velocity ω_0. Determine the new angular velocity of the plate just after its corner strikes the peg P and the plate starts to rotate about P without rebounding.

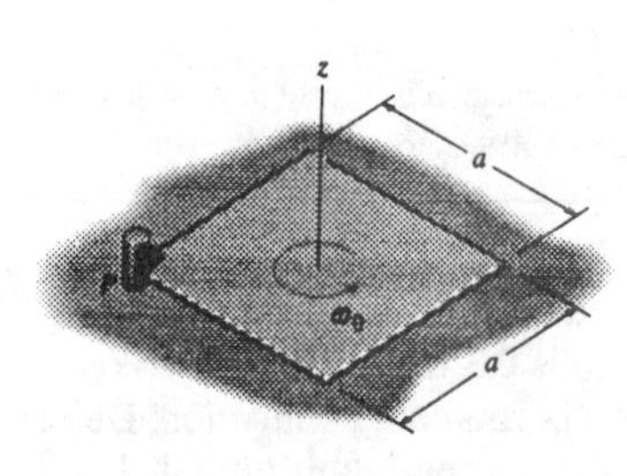

$$\Sigma(H_P)_0 = \Sigma(H_P)_1$$

$$\left[\frac{1}{12}\left(\frac{W}{g}\right)(a^2+a^2)\right]\omega_0 = \left[\frac{1}{12}\left(\frac{W}{g}\right)(a^2+a^2)+\left(\frac{W}{g}\right)\left(\frac{\sqrt{2}}{2}a\right)^2\right]\omega$$

$$\omega = \frac{1}{4}\omega_0 \qquad \textbf{Ans}$$

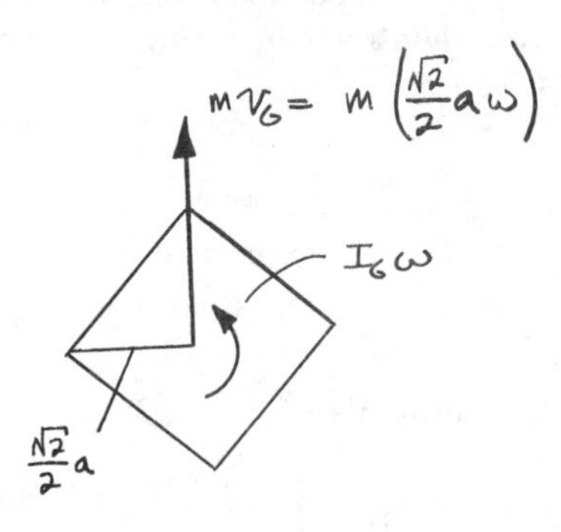

***19-44.** A ball having a mass of 8 kg and initial speed of $v_1 = 0.2$ m/s rolls over a 30-mm-long depression. Assuming that the ball rolls off the edges of contact, first A, then B, determine its final velocity $\mathbf{v}_2$ when it reaches the other side.

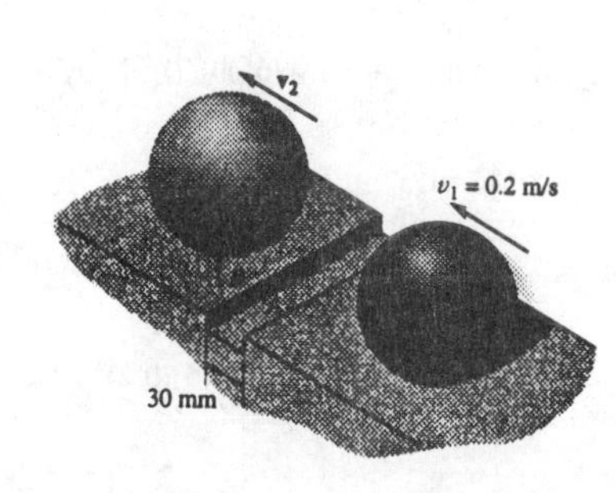

$$\omega_1 = \frac{0.2}{0.125} = 1.6 \text{ rad/s}, \quad \omega_2 = \frac{v_2}{0.125} = 8v_2$$

$$\theta = \sin^{-1}\left(\frac{15}{125}\right) = 6.8921°$$

$$h = 125 - 125\cos 6.8921° = 0.90326 \text{ mm}$$

$$T_1 + V_1 = T_2 + V_2$$

$$\frac{1}{2}(8)(0.2)^2 + \frac{1}{2}\left[\frac{2}{5}(8)(0.125)^2\right](1.6)^2 + 0 = -(0.90326)(10^{-3})8(9.81)$$

$$+\frac{1}{2}(8)\omega^2(0.125)^2 + \frac{1}{2}\left[\frac{2}{5}(8)(0.125)^2\right](\omega)^2$$

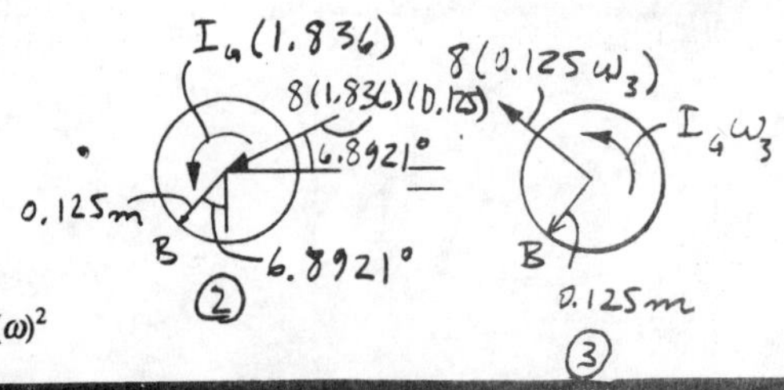

$\omega = 1.836$ rad/s

$$(H_B)_2 = (H_B)_3$$

$$\left[\frac{2}{5}(8)(0.125)^2\right](1.836) + 8(1.836)(0.125)\cos 6.892°(0.125\cos 6.892°) -$$

$$8(0.22948\sin 6.892°)(0.125\sin 6.892°) = \left[\frac{2}{5}(8)(0.125)^2\right]\omega_3 + 8(0.125)\omega_3(0.125)$$

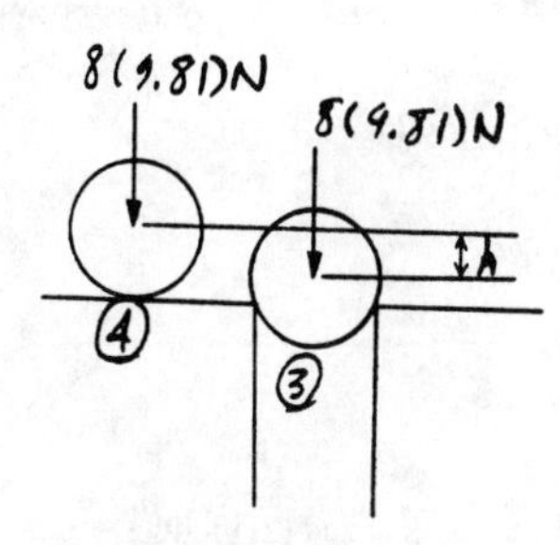

$\omega_3 = 1.7980$ rad/s

$$T_3 + V_3 = T_4 + V_4$$

$$\frac{1}{2}\left[\frac{2}{5}(8)(0.125)^2\right](1.7980)^2 + \frac{1}{2}(8)(1.7980)^2(0.125)^2 + 0 = 8(9.81)(0.90326(10^{-3})) +$$

$$\frac{1}{2}\left[\frac{2}{5}(8)(0.125)^2\right](\omega_4)^2 + \frac{1}{2}(8)(\omega_4)^2(0.125)^2$$

$\omega_4 = 1.56$ rad/s

So that $\quad v = 1.56(0.125) = 0.195$ m/s **Ans**

19-45. A thin ring having a mass of 15 kg strikes the 20-mm-high step. Determine the largest angular velocity ω_1 the ring can have so that it will not rebound off the step at A when it strikes it.

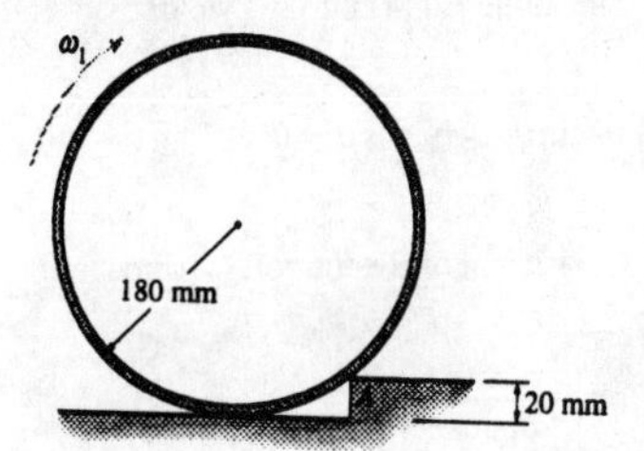

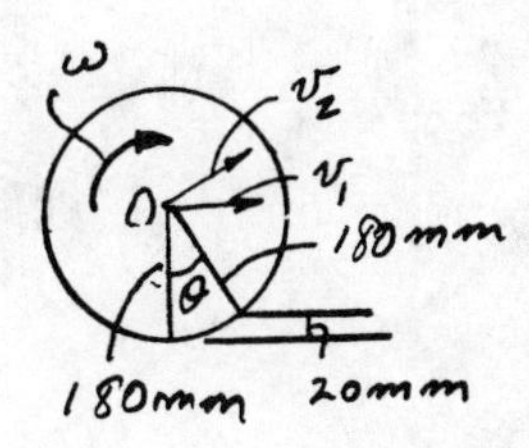

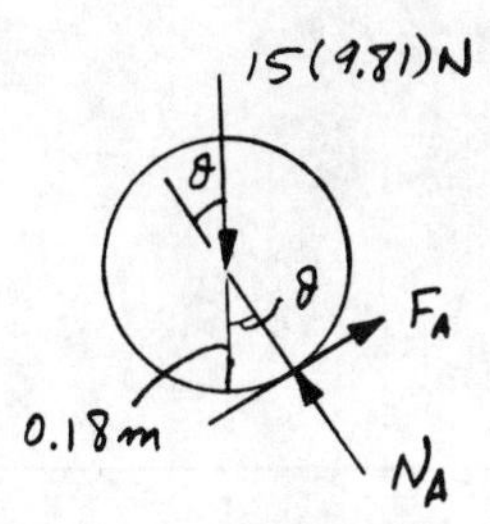

The weight is non – impulsive.

$(H_A)_1 = (H_A)_2$

$$15(\omega_1)(0.18)(0.18 - 0.02) + \left[15(0.18)^2\right](\omega_1) = \left[15(0.18)^2 + 15(0.18)^2\right]\omega_2$$

$\omega_2 = 0.9444\omega_1$

$+\searrow \Sigma F_n = m(a_G)_n; \quad (15)(9.81)\cos\theta - N_A = 15\omega_2^2(0.18)$

When hoop is about to rebound, $N_A \approx 0$. Also, $\cos\theta = \dfrac{160}{180}$, and so

$\omega_2 = 6.9602$ rad/s

$\omega_1 = \dfrac{6.9602}{0.9444} = 7.37$ rad/s **Ans**

19-46. The solid ball of mass m is dropped with a velocity $\mathbf{v}_1$ onto the edge of the rough step. If it rebounds horizontally off the step with a velocity $\mathbf{v}_2$, determine the angle θ at which contact occurs. Assume no slipping when the ball strikes the step. The coefficient of restitution is e.

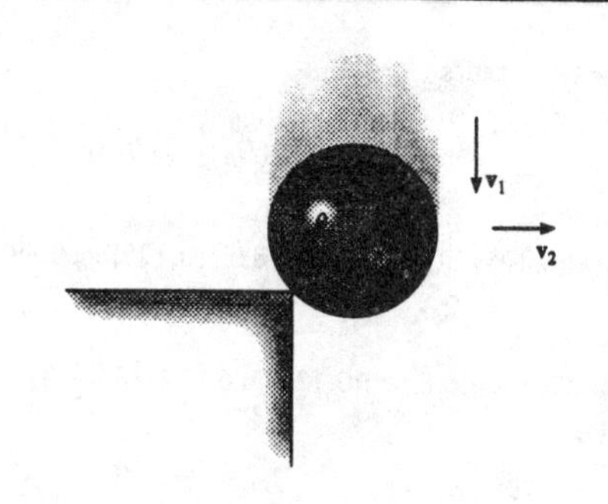

$$m(v_{Ax'})_1 + \Sigma\int F_{x'}dt = m(v_{Ax'})_2$$

$$mv_1 \sin\theta - F\Delta t = mv_2 \cos\theta \qquad (1)$$

$$I_G\omega_1 + \Sigma\int M_G dt = I_G\omega_2$$

$$0 + F\Delta t(r) = \left[\frac{2}{5}mr^2\right]\omega \qquad (2)$$

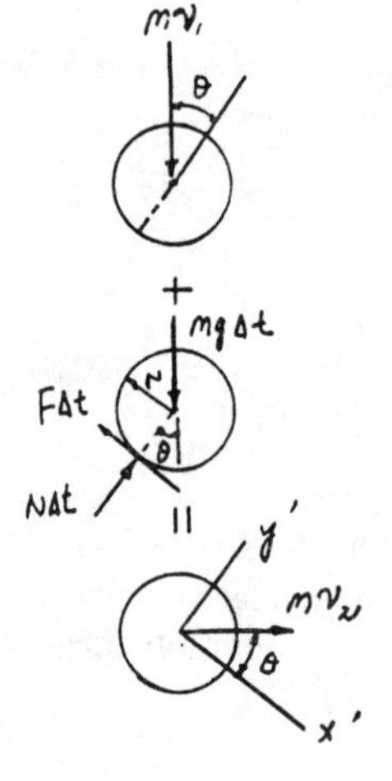

Eliminating $F\Delta t$ from Eqs.(1) and (2) yields :

$$v_1 \sin\theta - v_2 \cos\theta = \frac{2}{5}r\omega \qquad (3)$$

However, $\omega = \frac{v_2 \cos\theta}{r}$ (4)

Substitute Eq.(4) into (3) yields :

$$v_1 \sin\theta - \frac{7}{5}v_2 \cos\theta = 0 \qquad (5)$$

Coefficient of Restitution (y' direction) :

$$e = \frac{0-(-v_2 \sin\theta)}{v_1 \cos\theta - 0} \qquad v_1 = \frac{v_2}{e}\tan\theta \qquad (6)$$

Substitute Eq.(6) into (5) yields :

$$\left(\frac{v_2}{e}\tan\theta\right)\sin\theta - \frac{7}{5}v_2\cos\theta = 0$$

$$\tan^2\theta = \frac{7}{5}e \qquad \theta = \tan^{-1}\left[\sqrt{\frac{7}{5}e}\right] \qquad \textbf{Ans}$$

19-47. Tests of impact on the fixed crash dummy are conducted using the 300-lb ram that is released from rest at $\theta = 30°$, and allowed to fall and strike the dummy at $\theta = 90°$. If the coefficient of restitution between the dummy and the ram is $e = 0.4$, determine the angle θ to which the ram will rebound before momentarily coming to rest.

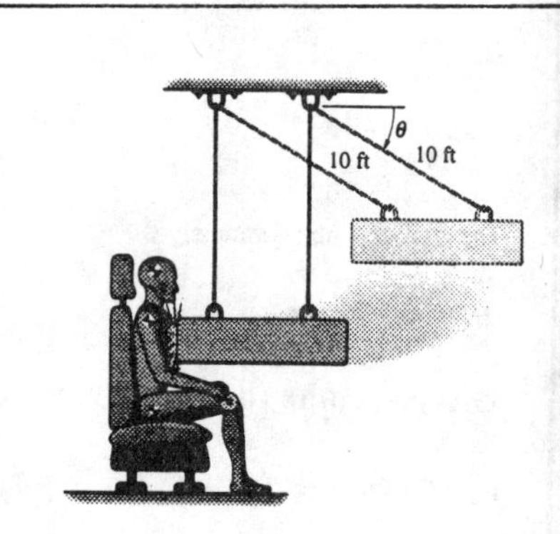

Datum through pin support at ceiling.

$$T_1 + V_1 = T_2 + V_2$$

$$0 - 300(10\sin 30°) = \frac{1}{2}\left(\frac{300}{32.2}\right)(v)^2 - 300(10)$$

$v = 17.944$ ft/s

$$(\overset{+}{\rightarrow}) \qquad e = 0.4 = \frac{v' - 0}{0-(-17.944)}$$

$v' = 7.178$ ft/s

$$T_2 + V_2 = T_3 + V_3$$

$$\frac{1}{2}\left(\frac{300}{32.2}\right)(7.178)^2 - 300(10) = 0 - 300(10\sin\theta)$$

$\theta = 66.9°$ **Ans**

***19-48.** The disk has a mass of 15 kg. If it is released from rest when $\theta = 30°$, determine the maximum angle θ of rebound after it collides with the wall. The coefficient of restitution between the disk and the wall is $e = 0.6$. When $\theta = 0°$, the disk hangs such that it just touches the wall. Neglect friction at the pin C.

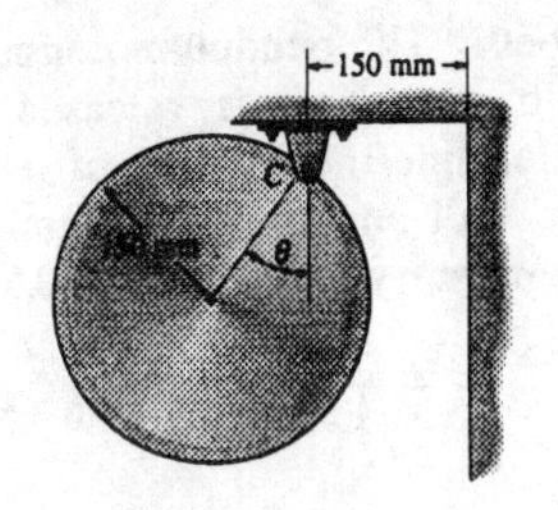

Datum at lower position of G.

$$T_1 + V_1 = T_2 + V_2$$

$$0 + (15)(9.81)(0.15)(1 - \cos 30°) = \frac{1}{2}\left[\frac{3}{2}(15)(0.15)^2\right]\omega^2 + 0$$

$\omega = 3.418$ rad/s

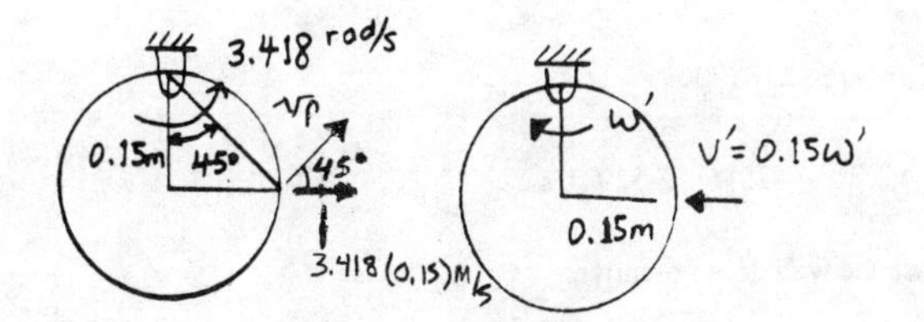

$$(\overset{+}{\rightarrow}) \qquad e = 0.6 = \frac{0 - (-0.15\omega')}{3.418(0.15) - 0}$$

$\omega' = 2.0508$ rad/s

$$T_2 + V_2 = T_3 + V_3$$

$$\frac{1}{2}\left[\frac{3}{2}(15)(0.15)^2\right](2.0508)^2 + 0 = 0 + 15(9.81)(0.15)(1 - \cos\theta)$$

$\theta = 17.9°$ **Ans**

19-49. The 6-lb slender rod AB is released from rest when it is in the *horizontal position* so that it begins to rotate clockwise. A 1-lb ball is thrown at the rod with a velocity $v = 50$ ft/s. The ball strikes the rod at C at the instant the rod is in the vertical position as shown. Determine the angular velocity of the rod just after the impact. Take $e = 0.7$.

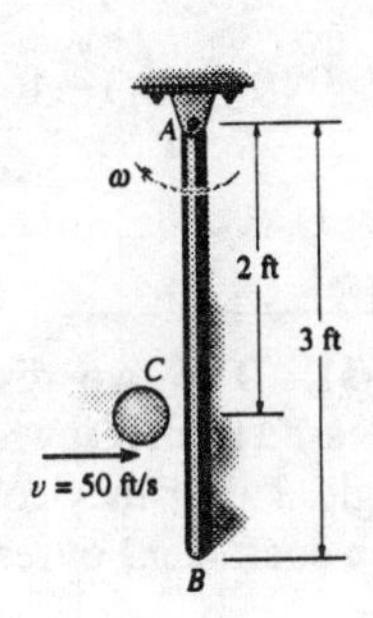

Datum at A:

$$T_1 + V_1 = T_2 + V_2$$

$$0 + 0 = \frac{1}{2}\left[\frac{1}{3}\left(\frac{6}{32.2}\right)(3)^2\right]\omega^2 - 6(1.5)$$

$\omega = 5.675$ rad/s

$$(\circlearrowleft +) \qquad (H_A)_1 = (H_A)_2$$

$$\frac{1}{32.2}(50)(2) - \left[\frac{1}{3}\left(\frac{6}{32.2}\right)(3)^2\right](5.675) = \frac{1}{32.2}(v_{bl})(2) + \left[\frac{1}{3}\left(\frac{6}{32.2}\right)(3)^2\right]\omega_2$$

$$(\overset{+}{\rightarrow}) \qquad e = 0.7 = \frac{v_C - v_{bl}}{50 - [-5.675(2)]}$$

Also,

$v_C = 2\omega_2$

Solving:

$\omega_2 = 3.81$ rad/s **Ans**

$v_{bl} = -35.3$ ft/s

$v_C = 7.61$ ft/s

19-50. The pendulum consists of a 10-lb solid ball and 4-lb rod. If it is released from rest when $\theta_0 = 0°$, determine the angle θ_1 of rebound after the ball strikes the wall and the pendulum swings up to the point of momentary rest. Take $e = 0.6$.

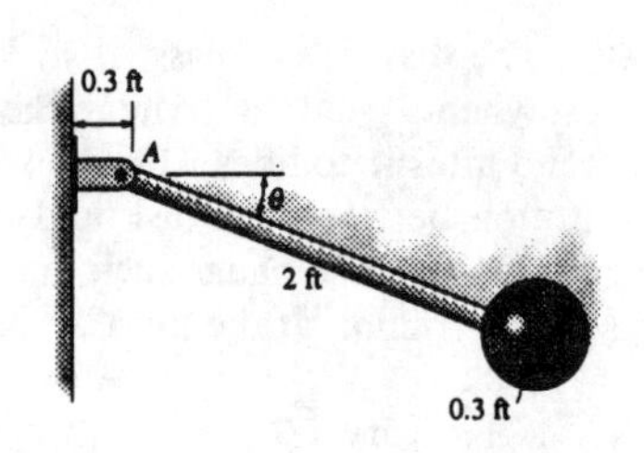

$$I_A = \frac{1}{3}\left(\frac{4}{32.2}\right)(2)^2 + \frac{2}{5}\left(\frac{10}{32.2}\right)(0.3)^2 + \left(\frac{10}{32.2}\right)(2.3)^2 = 1.8197 \text{ slug}\cdot\text{ft}^2$$

Just before impact :

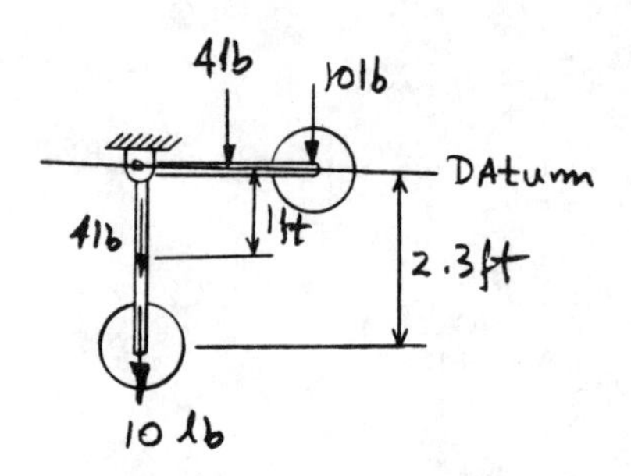

$$T_1 + V_1 = T_2 + V_2$$

$$0 + 0 = \frac{1}{2}(1.8197)\omega^2 - 4(1) - 10(2.3)$$

$$\omega = 5.4475 \text{ rad/s}$$

$$v_P = 2.3(5.4475) = 12.529 \text{ ft/s}$$

Since the wall does not move,

$$(\overset{+}{\rightarrow}) \qquad e = 0.6 = \frac{(v_{P'}) - 0}{0 - (-12.529)}$$

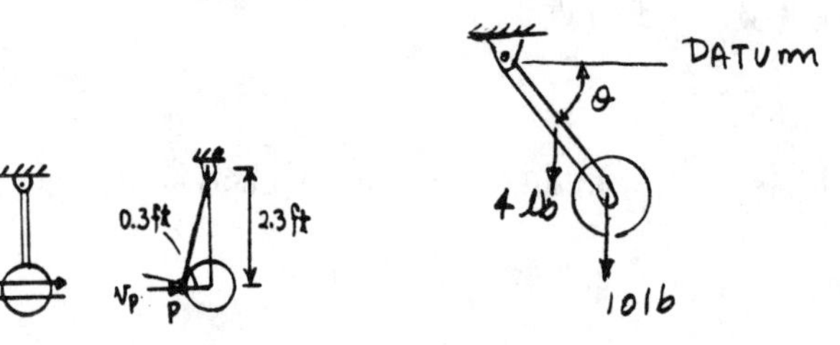

$$(v_{P'}) = 7.518 \text{ ft/s}$$

$$\omega' = \frac{7.518}{2.3} = 3.2685 \text{ rad/s}$$

$$T_3 + V_3 = T_4 + V_4$$

$$\frac{1}{2}(1.8197)(3.2685)^2 - 4(1) - 10(2.3) = 0 - 4(1)\sin\theta_1 - 10(2.3\sin\theta_1)$$

$$\theta_1 = 39.8° \qquad \textbf{Ans}$$

19-51. The two disks each weigh 10 lb. If they are released from rest when $\theta = 30°$, determine the maximum angle θ after they collide and rebound from each other. The coefficient of restitution is $e = 0.75$.

$$I_C = \frac{3}{2}\left(\frac{10}{32.2}\right)(1)^2 = 0.4658 \text{ slug}\cdot\text{ft}^2$$

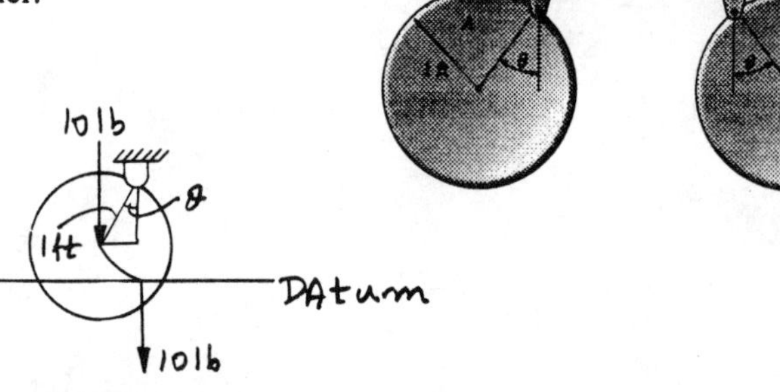

$$T_1 + V_1 = T_2 + V_2$$

$$0 + 10(1 - \cos 30°) = \frac{1}{2}(0.4658)\omega_1^2 + 0$$

$$\omega_1 = 2.398 \text{ rad/s}$$

Coefficient of restitution :

$$(\overset{+}{\rightarrow}) \qquad e = 0.75 = \frac{1\omega_2 - (-1\omega_2)}{2.398 - (-2.398)}$$

$$\omega_2 = 1.799 \text{ rad/s}$$

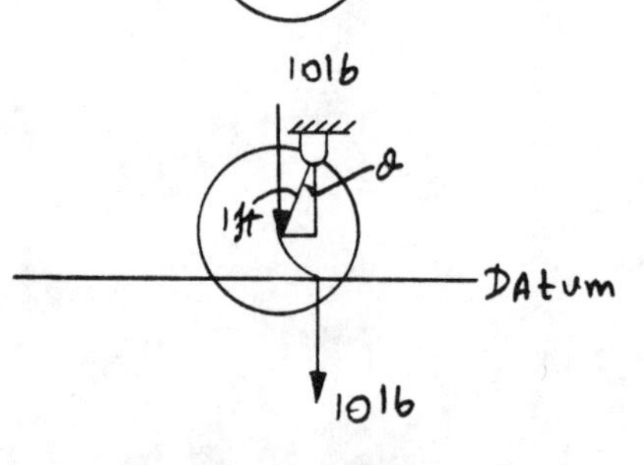

$$T_1 + V_1 = T_2 + V_2$$

$$\frac{1}{2}(0.4658)(1.799)^2 + 0 = 0 + 10(1 - \cos\theta)$$

$$\theta = 22.4° \qquad \textbf{Ans}$$

***19-52.** The pendulum consists of a 10-lb sphere and 4-lb rod. If it is released from rest when $\theta_0 = 90°$, determine the angle θ_1 of rebound after the sphere strikes the floor. Take $e = 0.8$.

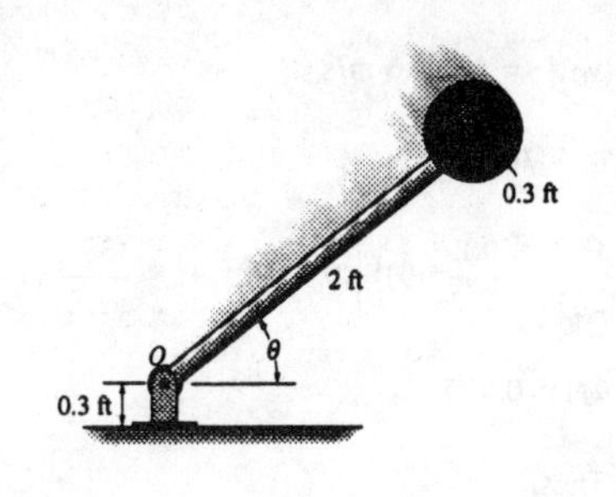

$$I_A = \frac{1}{3}\left(\frac{4}{32.2}\right)(2)^2 + \frac{2}{5}\left(\frac{10}{32.2}\right)(0.3)^2 + \left(\frac{10}{32.2}\right)(2.3)^2 = 1.8197 \text{ slug}\cdot\text{ ft}^2$$

Just before impact :

Datum through O.

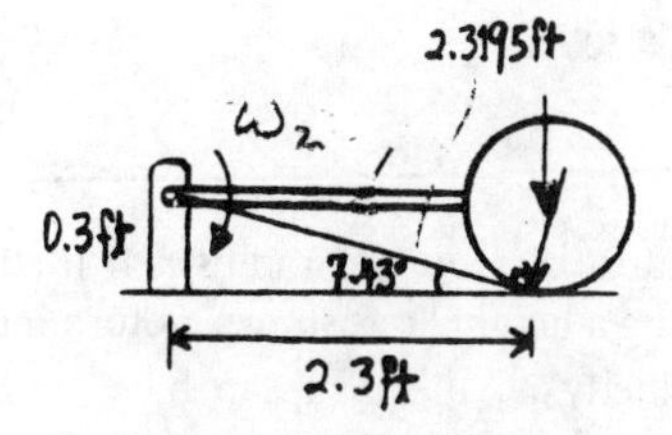

$$T_1 + V_1 = T_2 + V_2$$

$$0 + 4(1) + 10(2.3) = \frac{1}{2}(1.8197)\omega^2 + 0$$

$$\omega_2 = 5.4475 \text{ rad/s}$$

$$v = 2.3(5.4475) = 12.529 \text{ ft/s}$$

Since the floor does not move,

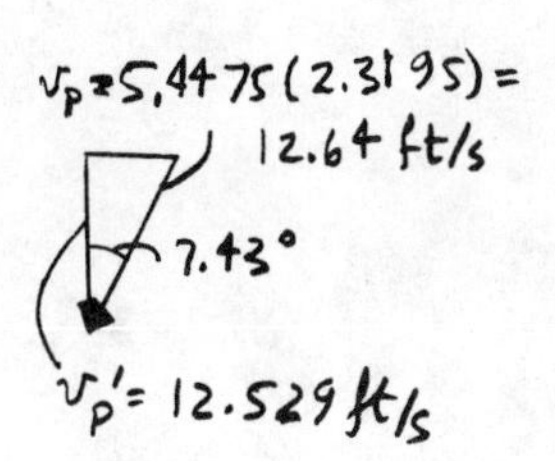

$$(+\uparrow) \qquad e = 0.8 = \frac{(v_{P'}) - 0}{0 - (-12.529)}$$

$$(v_{P'})_3 = 10.023 \text{ ft/s}$$

$$\omega_3 = \frac{10.023}{2.3} = 4.358 \text{ rad/s}$$

$$T_3 + V_3 = T_4 + V_4$$

$$\frac{1}{2}(1.8197)(4.358)^2 + 0 = 4(1\sin\theta_1) + 10(2.3\sin\theta_1)$$

$$\theta_1 = 39.8° \qquad \textbf{Ans}$$

19-53. The plank has a weight of 30 lb, center of gravity at G, and it rests on the two sawhorses at A and B. If the end D is raised 2 ft above the top of the sawhorses and is released from rest, determine how high end C will rise from the top of the sawhorses after the plank falls so that it rotates clockwise about A, strikes and pivots on the sawhorse at B, and rotates clockwise off the sawhorse at A.

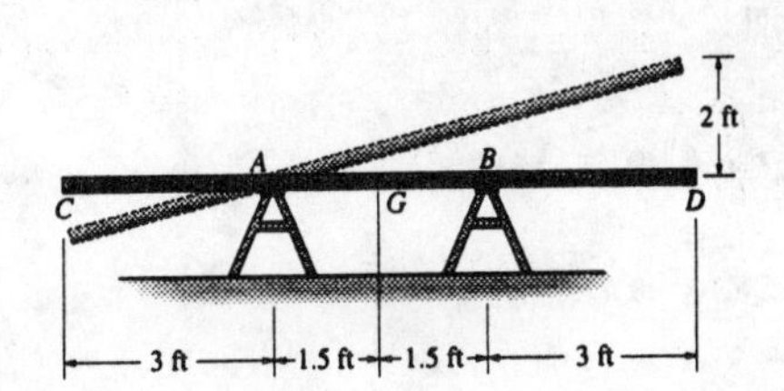

Establishing a datum through AB, the angular velocity of the plank just before striking B is

$$T_1 + V_1 = T_2 + V_2$$

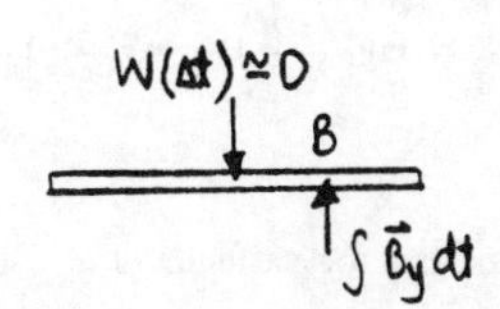

$$0 + 30\left[\frac{2}{6}(1.5)\right] = \frac{1}{2}\left[\frac{1}{12}\left(\frac{30}{32.2}\right)(9)^2 + \frac{30}{32.2}(1.5)^2\right](\omega_{CD})_2^2 + 0$$

$$(\omega_{CD})_2 = 1.8915 \text{ rad/s}$$

$$(v_G)_2 = 1.8915(1.5) = 2.837 \text{ m/s}$$

$$(\curvearrowleft +) \qquad (H_B)_2 = (H_B)_3$$

$$\left[\frac{1}{12}\left(\frac{30}{32.2}\right)(9)^2\right](1.8915) - \frac{30}{32.2}(2.837)(1.5) = \left[\frac{1}{12}\left(\frac{30}{32.2}\right)(9)^2\right](\omega_{AB})_3 + \frac{30}{32.2}(v_G)_3(1.5)$$

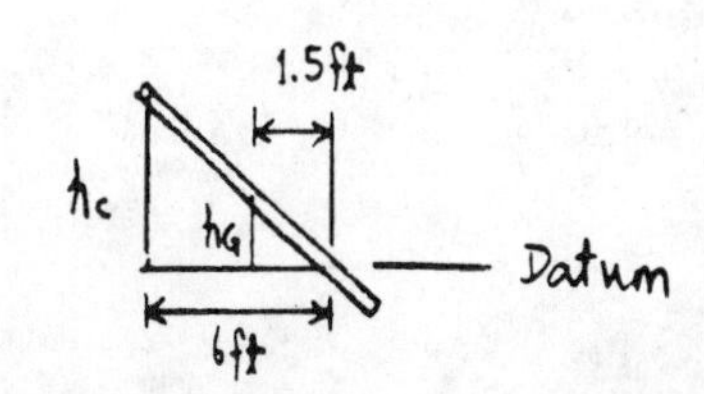

Since $(v_G)_3 = 1.5(\omega_{AB})_3$

$$(\omega_{AB})_3 = 0.9458 \text{ rad/s}$$

$(v_G)_3 = 1.4186 \text{ m/s}$

$T_3 + V_3 = T_4 + V_4$

$$\frac{1}{2}\left[\frac{1}{12}\left(\frac{30}{32.2}\right)(9)^2\right](0.9458)^2 + \frac{1}{2}\left(\frac{30}{32.2}\right)(1.4186)^2 + 0 = 0 + 30\,h_G$$

$h_G = 0.125$

Thus,

$h_C = \frac{6}{1.5}(0.125) = 0.500 \text{ ft}$ **Ans**

19-54. The disk has a mass m and radius r. If it strikes the rough step having a height $\frac{1}{8}r$ as shown, determine the largest angular velocity ω_1 the disk can have and not rebound off the step when it strikes it.

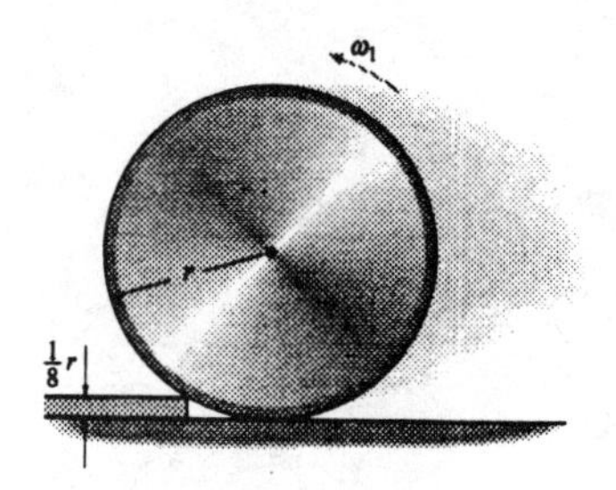

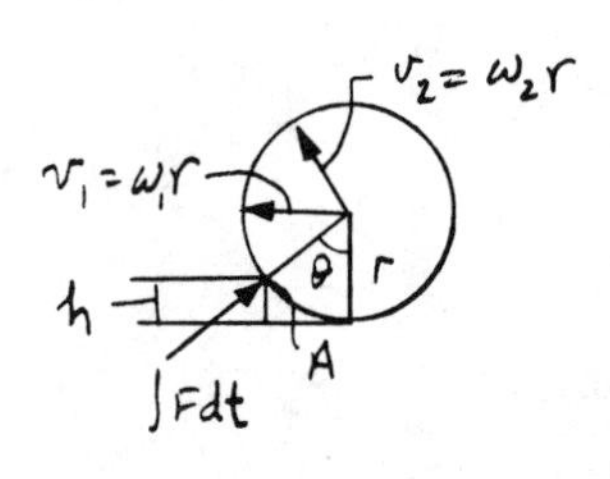

$$(H_A)_1 = \frac{1}{2}m\,r^2(\omega_1) + m(\omega_1 r)(r-h)$$

$$(H_A)_2 = \frac{1}{2}m\,r^2(\omega_2) + m(\omega_2 r)(r)$$

$$(H_A)_1 = (H_A)_2$$

$$\left[\frac{1}{2}m\,r^2 + m\,r\,(r-h)\right]\omega_1 = \frac{3}{2}\,m\,r^2\omega_2$$

$$\left(\frac{3}{2}r - h\right)\omega_1 = \frac{3}{2}r\,\omega_2$$

$$\nearrow\!+\Sigma F_n = m\,a_n; \quad W\cos\theta - F = m(\omega_2^2\,r)$$

$$F = mg\left(\frac{r-h}{r}\right) - m(\omega_2^2 r)$$

$$F = mg\left(\frac{r-h}{r}\right) - m\,r\left(\frac{2}{3}\right)^2\left(\frac{\frac{3}{2}r - h}{r}\right)^2\omega_1^2$$

Set $h = \frac{1}{8}r$; also note that for maximum ω_1 F will approach zero. Thus

$$mg\left(\frac{r - \frac{1}{8}r}{r}\right) - mr\left(\frac{2}{3}\right)^2\left(\frac{\frac{3}{2}r - \frac{r}{8}}{r}\right)^2\omega_1^2$$

$\omega_1 = 1.02\sqrt{\frac{g}{r}}$ **Ans**

19-55. The 15-lb rod AB is released from rest in the vertical position. If the coefficient of restitution between the floor and the cushion at B is $e = 0.7$, determine how high the end of the rod rebounds after impact with the floor.

$$T_1 + V_1 = T_2 + V_2$$

$$0 + 15(1) = \frac{1}{2}\left[\frac{1}{3}\left(\frac{15}{32.2}\right)(2)^2\right]\omega_2^2$$

$\omega_2 = 6.950$ rad/s   Hence $(v_B)_2 = 6.950(2) = 13.90$ rad/s

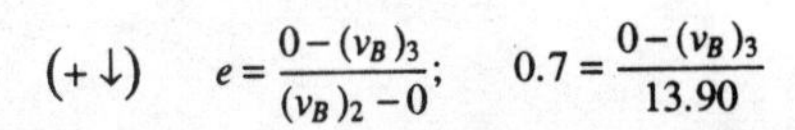

$$(+\downarrow) \qquad e = \frac{0 - (v_B)_3}{(v_B)_2 - 0}; \qquad 0.7 = \frac{0 - (v_B)_3}{13.90}$$

$$(v_B)_3 = -9.730 \text{ ft/s} = 9.730 \text{ ft/s} \uparrow$$

$$\omega_3 = \frac{(v_B)_3}{2} = \frac{9.730}{2} = 4.865 \text{ rad/s}$$

$$T_3 + V_3 = T_4 + V_4$$

$$\frac{1}{2}\left[\frac{1}{3}\left(\frac{15}{32.2}\right)(2)^2\right](4.865)^2 = 0 + 15(h_G)$$

$h_G = 0.490$ ft

$h_B = 2h_G = 0.980$ ft   **Ans**

***19-56.** A solid ball with a mass m is thrown on the ground such that at the instant of contact it has an angular velocity ω_1 and velocity components $(\mathbf{v}_G)_{x1}$ and $(\mathbf{v}_G)_{y1}$ as shown. If the ground is rough so no slipping occurs, determine the components of the velocity of its mass center just after impact. The coefficient of restitution is e.

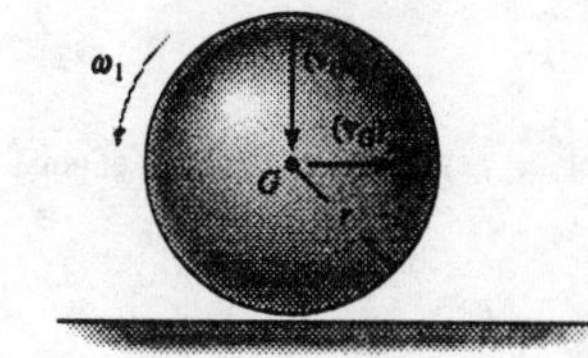

Coefficient of Restitution (y direction) :

$$(+\downarrow) \qquad e = \frac{0 - (v_G)_{y2}}{(v_G)_{y1} - 0} \qquad (v_G)_{y2} = -e(v_G)_{y1} = e(v_G)_{y1} \uparrow$$ **Ans**

Conservation of angular momentum about point on the ground :

$$(\circlearrowleft +) \qquad (H_A)_1 = (H_A)_2$$

$$-\frac{2}{5}mr^2\omega_1 + m(v_G)_{x1}r = \frac{2}{5}mr^2\omega_2 + m(v_G)_{x2}r$$

Since no slipping, $(v_G)_{x2} = \omega_2 r$ then,

$$\omega_2 = \frac{5\left((v_G)_{x1} - \frac{2}{5}\omega_1 r\right)}{7r}$$

Therefore

$$(v_G)_{x2} = \frac{5}{7}\left((v_G)_{x1} - \frac{2}{5}\omega_1 r\right)$$ **Ans**

R2-1. An automobile transmission consists of the planetary gear system shown. If the ring gear R is held fixed so that $\omega_R = 0$, and the shaft s, which is fixed to the sun gear S, is rotating at 20 rad/s, determine the angular velocity of each planet gear P and the angular velocity of the connecting rack D, which is free to turn about the center shaft s.

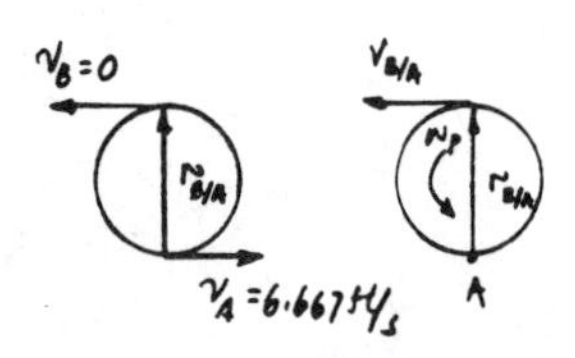

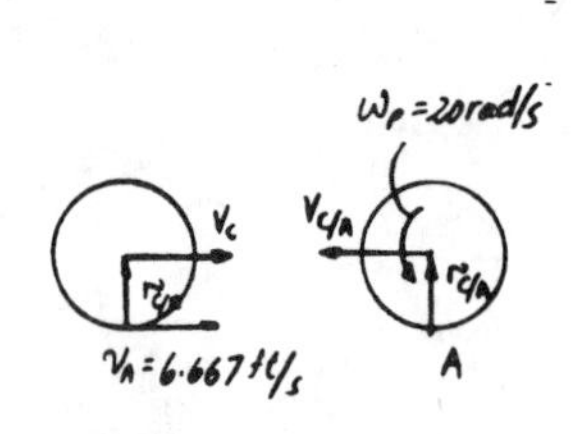

For planet gear P : The velocity of point A is $v_A = \omega_S r_S = 20\left(\frac{4}{12}\right) = 6.667$ ft/s.

$$\mathbf{v}_B = \mathbf{v}_A + \mathbf{v}_{B/A}$$

$$0 = \left[\underset{\rightarrow}{6.667}\right] + \left[\underset{\leftarrow}{\omega_P\left(\frac{4}{12}\right)}\right]$$

$(\overset{+}{\rightarrow})$ $\quad 0 = 6.667 - \omega_P\left(\frac{4}{12}\right) \qquad \omega_P = 20$ rad/s $\qquad$ **Ans**

For connecting rack D :

$$\mathbf{v}_C = \mathbf{v}_A + \mathbf{v}_{C/A}$$

$$\left[\underset{\rightarrow}{v_C}\right] = \left[\underset{\rightarrow}{6.667}\right] + \left[\underset{\leftarrow}{20\left(\frac{2}{12}\right)}\right]$$

$(\overset{+}{\rightarrow})$ $\quad v_C = 6.667 - 20\left(\frac{2}{12}\right) \qquad v_C = 3.333$ ft/s

The rack is rotating about a fixed axis (shaft s). Hence

$$v_C = \omega_D r_D$$

$3.333 = \omega_D\left(\frac{6}{12}\right) \qquad \omega_D = 6.67$ rad/s $\qquad$ **Ans**

R2-2. An automobile transmission consists of the planetary gear system shown. If the ring gear R is rotating at $\omega_R = 2$ rad/s, and the shaft s, which is fixed to the sun gear S, is rotating at 20 rad/s, determine the angular velocity of each planet gear P and the angular velocity of the connecting rack D, which is free to turn about the center shaft s.

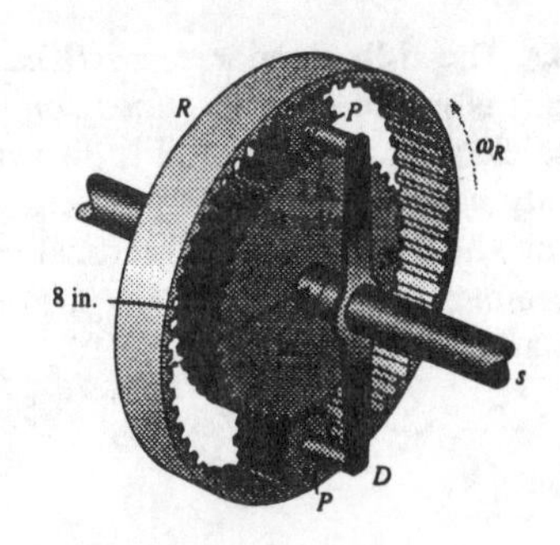

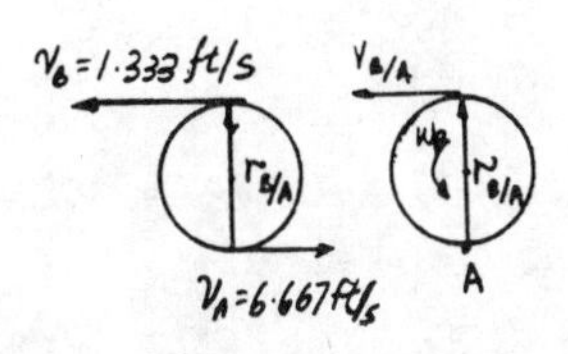

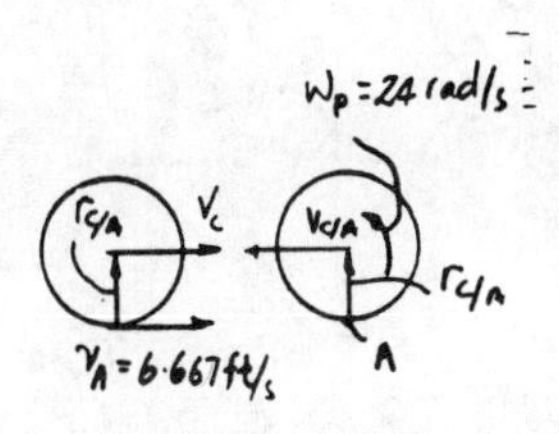

For planet gear P : The velocity of points A and B are $v_A = \omega_S r_S = 20\left(\frac{4}{12}\right)$ $= 6.667$ ft/s and $v_B = \omega_R r_R = 2\left(\frac{8}{12}\right) = 1.333$ ft/s.

$$\mathbf{v}_B = \mathbf{v}_A + \mathbf{v}_{B/A}$$

$$\left[\underset{\leftarrow}{1.333}\right] = \left[\underset{\rightarrow}{6.667}\right] + \left[\underset{\leftarrow}{\omega_P\left(\frac{4}{12}\right)}\right]$$

$(\overset{+}{\rightarrow})$ $\quad -1.333 = 6.667 - \omega_P\left(\frac{4}{12}\right) \qquad \omega_P = 24$ rad/s **Ans**

For connecting rack D :

$$\mathbf{v}_C = \mathbf{v}_A + \mathbf{v}_{C/A}$$

$$\left[\underset{\rightarrow}{v_C}\right] = \left[\underset{\rightarrow}{6.667}\right] + \left[\underset{\leftarrow}{24\left(\frac{2}{12}\right)}\right]$$

$(\overset{+}{\rightarrow})$ $\quad v_C = 6.667 - 24\left(\frac{2}{12}\right) \qquad v_C = 2.667$ ft/s

The rack is rotating about a fixed axis (shaft s). Hence

$$v_C = \omega_D r_D$$

$2.667 = \omega_D\left(\frac{6}{12}\right) \qquad \omega_D = 5.33$ rad/s **Ans**

R2-3. The 6-lb slender rod AB is released from rest when it is in the *horizontal position* so that it begins to rotate clockwise. A 1-lb ball is thrown at the rod with a velocity $v = 50$ ft/s. The ball strikes the rod at C at the instant the rod is in the vertical position as shown. Determine the angular velocity of the rod just after the impact. Take $e = 0.7$ and $d = 2$ ft.

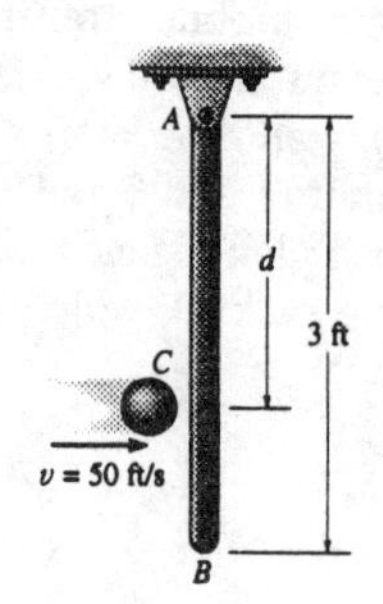

Datum at A:

$$T_1 + V_1 = T_2 + V_2$$

$$0 + 0 = \frac{1}{2}\left[\frac{1}{3}\left(\frac{6}{32.2}\right)(3)^2\right]\omega^2 - 6(1.5)$$

$$\omega = 5.675 \text{ rad/s}$$

$$(+ \quad (H_A)_1 = (H_A)_2$$

$$\frac{1}{32.2}(50)(2) - \left[\frac{1}{3}\left(\frac{6}{32.2}\right)(3)^2\right](5.675) = \left[\frac{1}{3}\left(\frac{6}{32.2}\right)(3)^2\right]\omega_2 + \frac{1}{32.2}(v_{BL})(2)$$

$$e = 0.7 = \frac{v_C - v_{BL}}{50 - [-5.675(2)]}$$

$$v_C = 2\omega_2$$

Solving,

$\omega_2 = 3.82$ rad/s **Ans**

$v_{BL} = -35.3$ ft/s

$v_C = 7.64$ ft/s

***R2-4.** The 6-lb slender rod AB is originally at rest, suspended in the vertical position. A 1-lb ball is thrown at the rod with a velocity $v = 50$ ft/s and strikes the rod at C. Determine the angular velocity of the rod just after the impact. Take $e = 0.7$ and $d = 2$ ft.

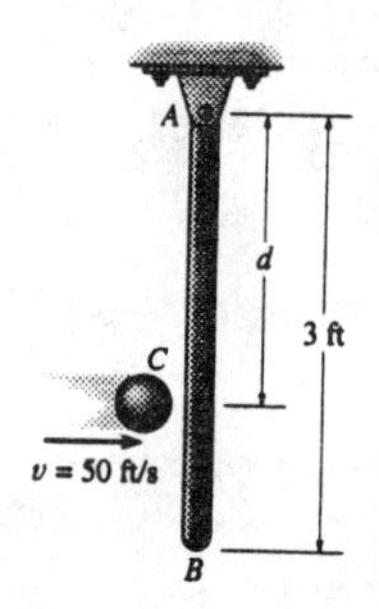

$$(+ \quad (H_A)_1 = (H_A)_2$$

$$\left(\frac{1}{32.2}\right)(50)(2) = \left[\frac{1}{3}\left(\frac{6}{32.2}\right)(3)^2\right]\omega_2 + \frac{1}{32.2}(v_{BL})(2)$$

$$e = 0.7 = \frac{v_C - v_{BL}}{50 - 0}$$

$$v_C = 2\omega_2$$

Thus,

$\omega_2 = 7.73$ rad/s **Ans**

$v_{BL} = -19.5$ ft/s

R2-5. The 6-lb slender rod is originally at rest, suspended in the vertical position. Determine the distance d where the 1-lb ball, traveling at $v = 50$ ft/s, should strike the rod so that it does not create a horizontal impulse at A. What is the rod's angular velocity just after the impact? Take $e = 0.5$.

Rod :

$$\circlearrowleft + \quad (H_G)_1 + \Sigma\int M_G\, dt = (H_G)_2$$

$$0 + \int F\, dt\,(d - 1.5) = \left(\frac{1}{12}(m)(1.5)^2\right)\omega$$

$$m(v_G)_1 + \Sigma\int F\, dt = m(v_G)_2$$

$$0 + \int F\, dt = m(1.5\omega)$$

Thus,

$$m(1.5\omega)(d - 1.5) = \frac{1}{12}(m)(1.5)^2\,\omega$$

$d = 2$ ft **Ans**

This is called the center of percussion. See Example 19 - 5.

$$\circlearrowleft + \quad (H_A)_1 = (H_A)_2$$

$$\frac{1}{32.2}(50)(2) = \left[\frac{1}{3}\left(\frac{6}{32.2}\right)(3)^2\right]\omega_2 + \frac{1}{32.2}(v_{BL})(2)$$

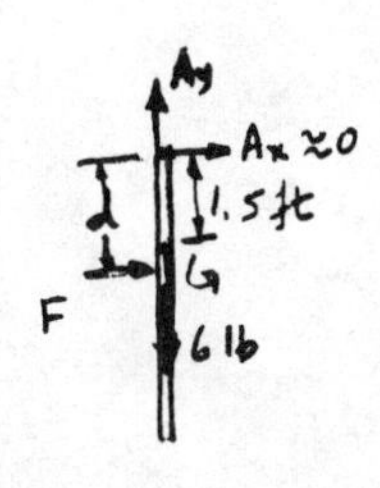

$$e = 0.5 = \frac{v_C - v_{BL}}{50 - 0}$$

$$v_C = 2\omega_2$$

Thus,

$\omega_2 = 6.82$ rad/s **Ans**

$v_{BL} = -11.4$ ft/s

R2-6. At a given instant, the wheel is rotating with the angular motions shown. Determine the acceleration of the collar at A at this instant.

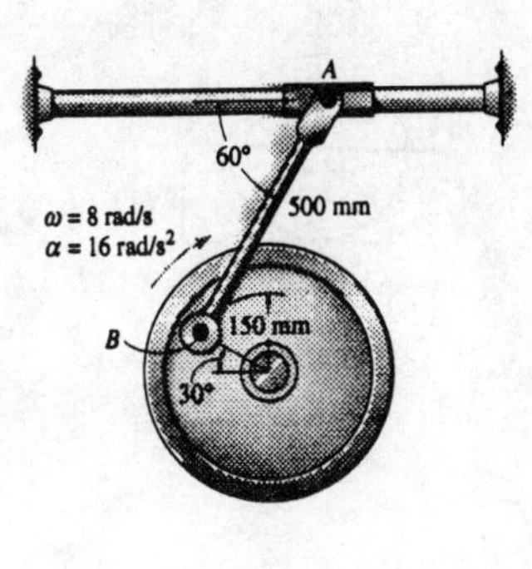

Using instantaneous center method.

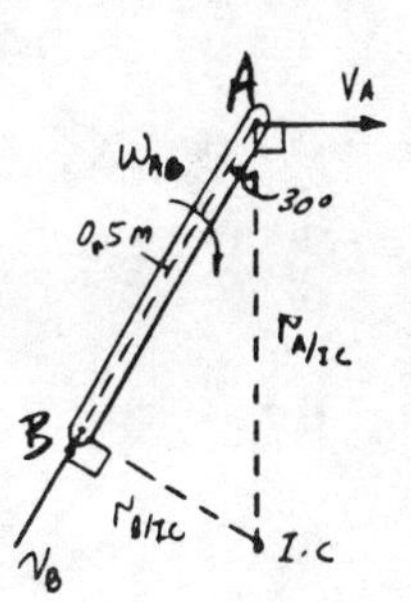

$$\omega_{AB} = \frac{v_B}{r_{B/IC}} = \frac{8(0.15)}{0.5\tan 30^\circ} = 4.157 \text{ rad/s}$$

$$\mathbf{a}_B = 16(0.15)\mathbf{i} - 8^2(0.15)\mathbf{j} = \{2.4\mathbf{i} - 9.6\mathbf{j}\} \text{ m/s}^2$$

$$\mathbf{a}_A = -a_A\cos 60^\circ\mathbf{i} + a_A\sin 60^\circ\mathbf{j} \qquad \boldsymbol{\alpha} = \alpha\mathbf{k} \qquad \mathbf{r}_{B/A} = \{-0.5\mathbf{i}\}\text{m}$$

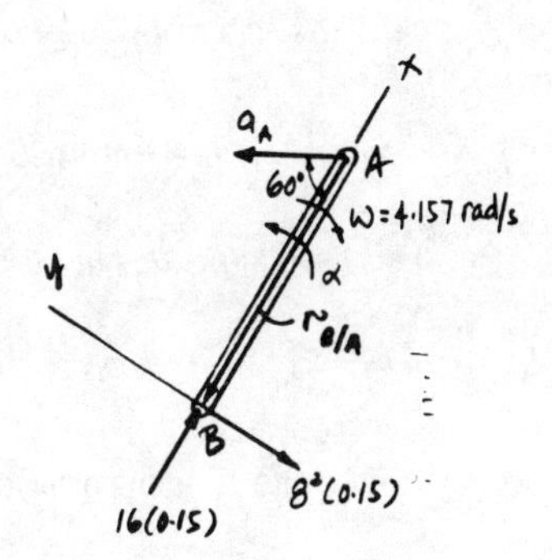

$$\mathbf{a}_B = \mathbf{a}_A + \boldsymbol{\alpha}\times\mathbf{r}_{B/A} - \omega^2\mathbf{r}_{B/A}$$

$$2.4\mathbf{i} - 9.6\mathbf{j} = (-a_A\cos 60^\circ\mathbf{i} + a_A\sin 60^\circ\mathbf{j}) + (\alpha\mathbf{k})\times(-0.5\mathbf{i}) - (4.157)^2(-0.5\mathbf{i})$$

$$2.4\mathbf{i} - 9.6\mathbf{j} = (-a_A\cos 60^\circ + 8.64)\mathbf{i} + (-0.5\alpha + a_A\sin 60^\circ)\mathbf{j}$$

Equating the **i** and **j** components yields :

$2.4 = -a_A\cos 60^\circ + 8.64$ $\qquad a_A = 12.5\text{ m/s}^2 \leftarrow$ **Ans**

$-9.6 = -0.5\alpha + (12.5)\sin 60^\circ$ $\qquad \alpha = 40.8\text{ rad/s}^2 \circlearrowright$

R2-7. The transmission gear fixed onto the frame of an electric train turns at a constant rate of $\omega_t = 30$ rad/s. This gear is in mesh with the gear that is fixed to the axle of the engine. Determine the velocity of the train, assuming the wheels do not slip on the track.

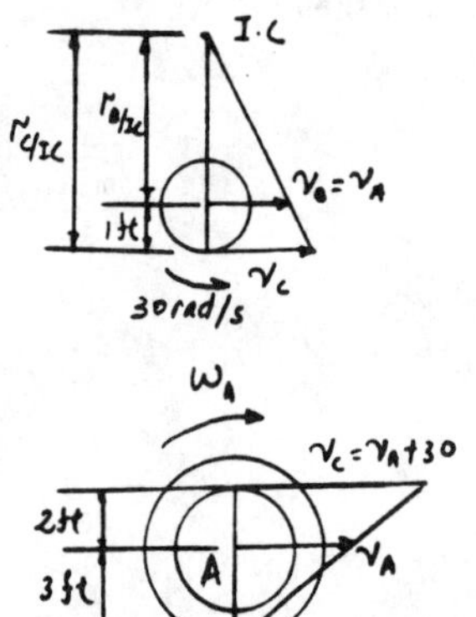

For the transmission gear

$$r_{B/IC} = \frac{v_A}{30} \qquad r_{C/IC} = r_{B/IC} + 1 = \frac{v_A}{30} + 1$$

$$v_C = \omega_t r_{C/IC} = 30\left(\frac{v_A}{30} + 1\right) = v_A + 30$$

For the wheel : Using similar triangles

$$\frac{v_A + 30}{5} = \frac{v_A}{3} \qquad v_A = 45 \text{ ft/s} \rightarrow \qquad \textbf{Ans}$$

***R2-8.** The 50-kg cylinder has an angular velocity of 30 rad/s when it is brought into contact with the surface at C. If the coefficient of kinetic friction is $\mu_k = 0.2$, determine how long it takes for the cylinder to stop spinning. What force is developed in the link AB during this time? The axis of the cylinder is connected to *two* symmetrical links. (Only AB is shown.) For the computation, neglect the weight of the links.

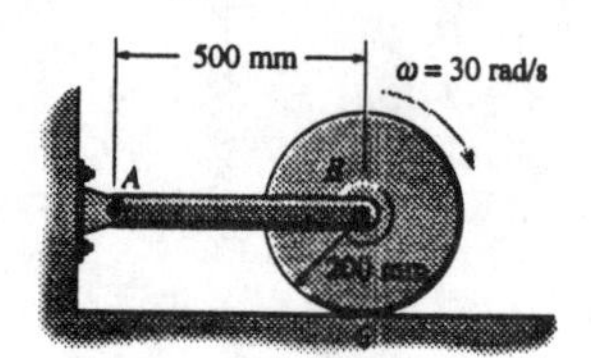

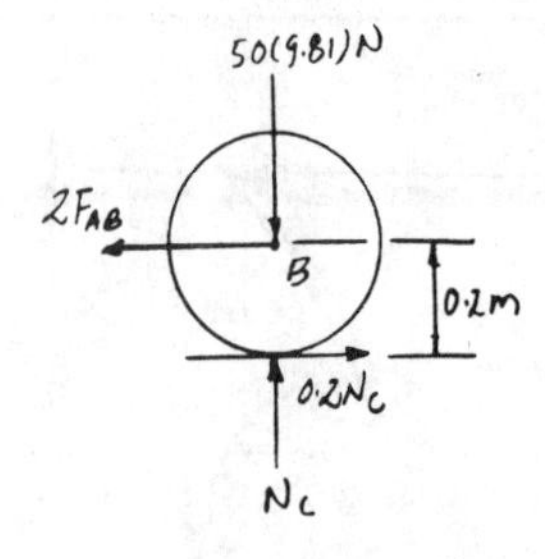

$$(+\uparrow) \qquad m(v_{Ay})_1 + \Sigma\int_{t_1}^{t_2} F_y\,dt = m(v_{Ay})_2$$

$$0 + N_C(t) - 50(9.81)(t) = 0 \qquad N_C = 490.5 \text{ N}$$

$$(\overset{+}{\rightarrow}) \qquad m(v_{Ax})_1 + \Sigma\int_{t_1}^{t_2} F_x\,dt = m(v_{Ax})_2$$

$$0 + 0.2(490.5)(t) - 2F_{AB}(t) = 0 \qquad F_{AB} = 49.0 \text{ N} \qquad \textbf{Ans}$$

$$(\circlearrowleft +) \qquad I_B\omega_1 + \Sigma\int_{t_1}^{t_2} M_B\,dt = I_B\omega_2$$

$$-\left[\frac{1}{2}(50)(0.2)^2\right](30) + 0.2(490.5)(0.2)(t) = 0$$

$$t = 1.53 \text{ s} \qquad \textbf{Ans}$$

R2-9. The gear rack has a mass of 6 kg, and the gears each have a mass of 4 kg and a radius of gyration of $k = 30$ mm. If the rack is originally moving downward at 2 m/s, when $s = 0$, determine the speed of the rack when $s = 600$ mm. The gears are free to turn about their centers, A and B.

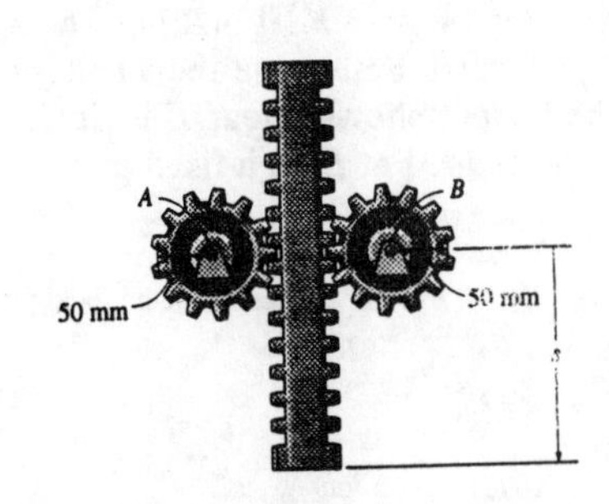

Originally, both gears rotate with an angular velocity of

$\omega_1 = \dfrac{2}{0.05} = 40$ rad/s. After the rack has traveled $s = 600$ mm, both gears rotate

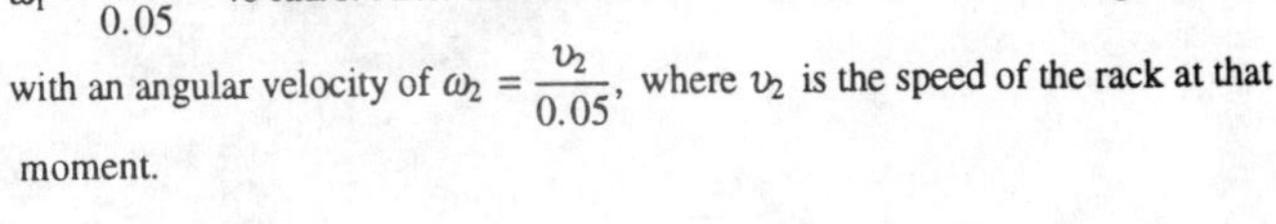

with an angular velocity of $\omega_2 = \dfrac{v_2}{0.05}$, where v_2 is the speed of the rack at that moment.

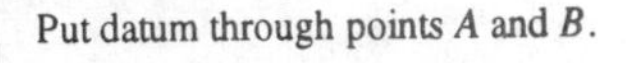

Put datum through points A and B.

$$T_1 + V_1 = T_2 + V_2$$

$$\frac{1}{2}(6)(2)^2 + 2\left\{\frac{1}{2}\left[4(0.03)^2\right](40)^2\right\} + 0 = \frac{1}{2}(6)v_2^2 + 2\left\{\frac{1}{2}\left[4(0.03)^2\right]\left(\frac{v_2}{0.05}\right)^2\right\} - 6(9.81)(0.6)$$

$v_2 = 3.46$ m/s **Ans**

R2-10. The gear has a mass of 2 kg and a radius of gyration $k_A = 0.15$ m. The connecting link (slender rod) and slider block at B have a mass of 4 kg and 1 kg, respectively. If the gear has an angular velocity $\omega = 8$ rad/s at the instant $\theta = 45°$, determine the gear's angular velocity when $\theta = 0°$.

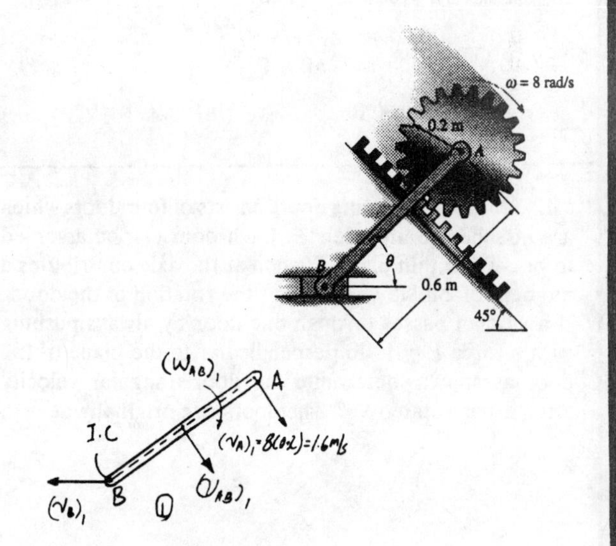

At position 1 :

$$(\omega_{AB})_1 = \frac{(v_A)_1}{r_{A/IC}} = \frac{1.6}{0.6} = 2.6667 \text{ rad/s} \qquad (v_B)_1 = 0$$

$$(v_{AB})_1 = (\omega_{AB})_1\, r_{G/IC} = 2.6667(0.3) = 0.8 \text{ m/s}$$

At position 2 :

$$(\omega_{AB})_2 = \frac{(v_A)_2}{r_{A/IC}} = \frac{\omega_2(0.2)}{\frac{0.6}{\cos 45°}} = 0.2357\omega_2$$

$$(v_B)_2 = (\omega_{AB})_2\, r_{B/IC} = 0.2357\omega_2(0.6) = 0.1414\omega_2$$

$$(v_{AB})_2 = (\omega_{AB})_2\, r_{G/IC} = 0.2357\omega_2(0.6708) = 0.1581\omega_2$$

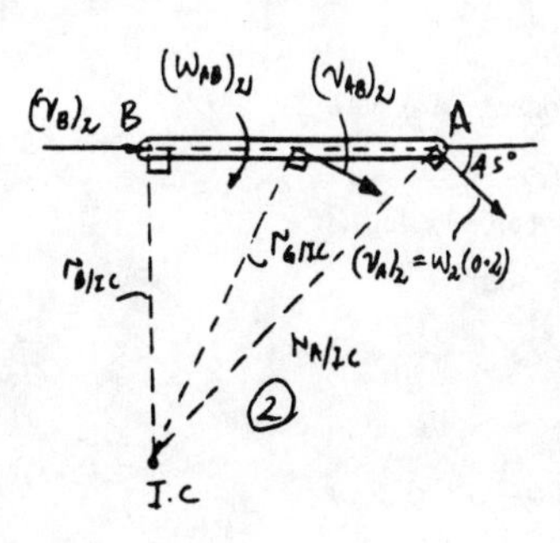

$$T_1 = \frac{1}{2}\left[(2)(0.15)^2\right](8)^2 + \frac{1}{2}(2)(1.6)^2 + \frac{1}{2}(4)(0.8)^2 + \frac{1}{2}\left[\frac{1}{12}(4)(0.6)^2\right](2.6667)^2$$

$$= 5.7067 \text{ J}$$

$$T_2 = \frac{1}{2}\left[(2)(0.15)^2\right](\omega_2)^2 + \frac{1}{2}(2)(0.2\omega_2)^2 + \frac{1}{2}(4)(0.1581\omega_2)^2$$

$$+ \frac{1}{2}\left[\frac{1}{12}(4)(0.6)^2\right](0.2357\omega_2)^2 + \frac{1}{2}(1)(0.1414\omega_2)^2$$

$$T_2 = 0.1258\omega_2^2$$

Put datum through bar in position 2.

$$V_1 = 2(9.81)(0.6\sin 45°) + 4(9.81)(0.3\sin 45°) = 16.6481 \text{ J} \qquad V_2 = 0$$

$$T_1 + V_1 = T_2 + V_2$$

$$5.7067 + 16.6481 = 0.1258\omega_2^2 + 0$$

$\omega_2 = 13.3$ rad/s **Ans**

R2-11. The rotation of link AB creates an oscillating movement of gear F. If AB has an angular velocity of $\omega_{AB} = 6$ rad/s, determine the angular velocity of gear F at the instant shown. Gear E is rigidly attached to arm CD and pinned at D to a fixed point.

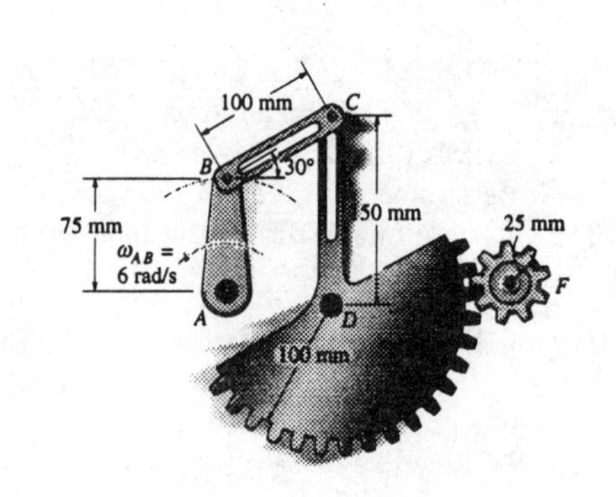

For link AB : Link AB rotates about the fixed point A. Hence

$$v_B = \omega_{AB} r_{AB}$$

$$= 6(0.075) = 0.45 \text{ m/s}$$

For link BC

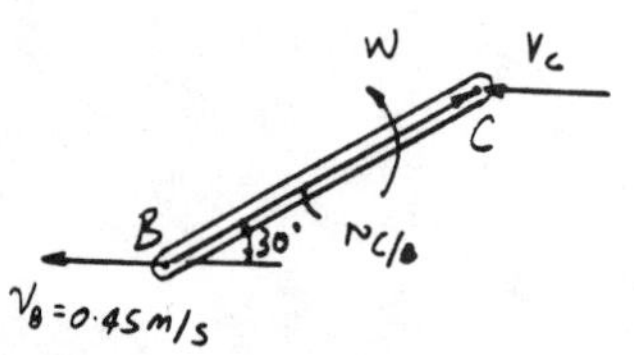

$\mathbf{v}_B = \{-0.45\mathbf{i}\}$ m/s $\quad \mathbf{v}_C = -v_C\mathbf{i} \quad \omega = \omega_{BC}\mathbf{k}$

$\mathbf{r}_{C/B} = \{0.1\cos 30°\mathbf{i} + 0.1\sin 30°\mathbf{j}\}$ m

$$\mathbf{v}_C = \mathbf{v}_B + \omega \times \mathbf{r}_{C/B}$$

$$-v_C\mathbf{i} = -0.45\mathbf{i} + (\omega_{BC}\mathbf{k}) \times (0.1\cos 30°\mathbf{i} + 0.1\sin 30°\mathbf{j})$$

$$-v_C\mathbf{i} = (-0.45 - 0.1\sin 30°\,\omega_{BC})\mathbf{i} + 0.1\cos 30°\,\omega_{BC}\mathbf{j}$$

Equating the **i** and **j** components yields :

$0 = 0.1\cos 30°\,\omega_{BC} \qquad \omega_{BC} = 0$

$-v_C = -0.45 - 0.1\sin 30°(0) \qquad v_C = 0.45$ m/s

Arm CD rotates about the fixed point D. Hence

$$v_C = \omega_{CD} r_{CD}$$

$0.45 = \omega_{CD}(0.15) \qquad \omega_{CD} = 3$ rad/s

$$\omega_F = \frac{r_E}{r_F}\omega_{CD} = \frac{0.1}{0.025}(3) = 12 \text{ rad/s} \quad \textbf{Ans}$$

***R2-12.** The revolving door consists of four doors which are attached to an axle AB. Each door can be assumed to be a 50-lb thin plate. Friction at the axle contributes a moment of 2 lb·ft which resists the rotation of the doors. If a woman passes through one door by always pushing with a force $P = 15$ lb perpendicular to the plane of the door as shown, determine the door's angular velocity after it has rotated 90°. The doors are originally at rest.

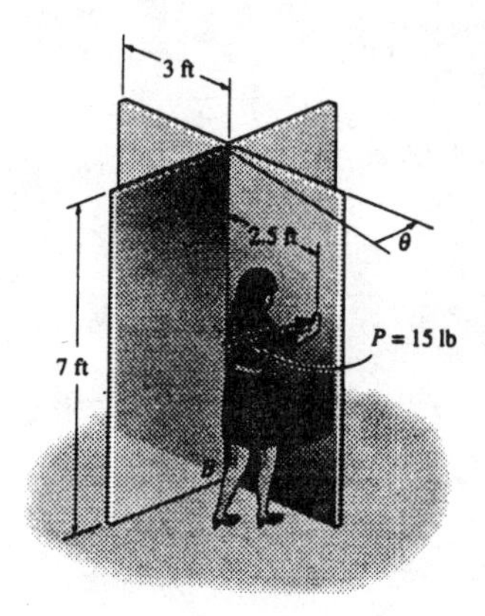

Moment of inertia of the door about axle AB :

$$I_{AB} = 2\left[\frac{1}{12}\left(\frac{100}{32.2}\right)(6)^2\right] = 18.6335 \text{ slug} \cdot \text{ft}^2$$

$$T_1 + \Sigma U_{1-2} = T_2$$

$$0 + \left\{15(2.5)\left(\frac{\pi}{2}\right) - 2\left(\frac{\pi}{2}\right)\right\} = \frac{1}{2}(18.6335)\omega^2$$

$\omega = 2.45$ rad/s $\quad$ **Ans**

R2-13. The 10-lb cylinder rests on the 20-lb dolly. If the system is released from rest, determine the angular velocity of the cylinder in 2 s. The cylinder does not slip on the dolly. Neglect the mass of the wheels on the dolly.

For the cylinder

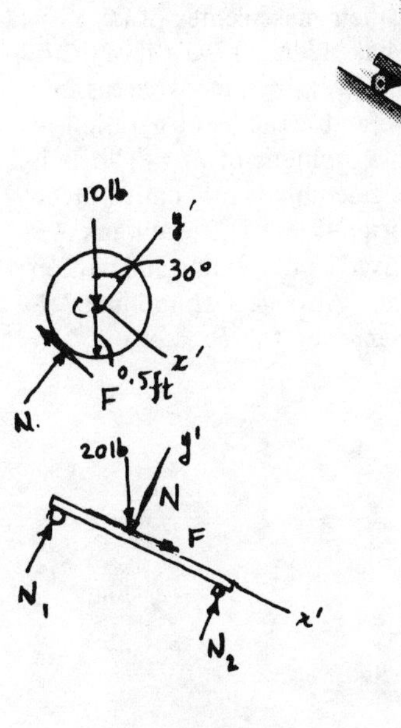

$(+\searrow)\ m(v_{Cx'})_1 + \Sigma\int_{t_1}^{t_2} F_{x'}dt = m(v_{Cx'})_2$

$$0 + 10\sin 30°(2) - F(2) = \left(\frac{10}{32.2}\right)v_C \qquad (1)$$

$(\circlearrowright +)\ I_C\omega_1 + \Sigma\int_{t_1}^{t_2} M_C dt = I_C\omega_2$

$$0 + F(0.5)(2) = \left[\frac{1}{2}\left(\frac{10}{32.2}\right)(0.5)^2\right]\omega \qquad (2)$$

For the dolly

$(+\searrow)\ m(v_{Dx'})_1 + \Sigma\int_{t_1}^{t_2} F_{x'}dt = m(v_{Dx'})_2$

$$0 + F(2) + 20\sin 30°(2) = \left(\frac{20}{32.2}\right)v_D \qquad (3)$$

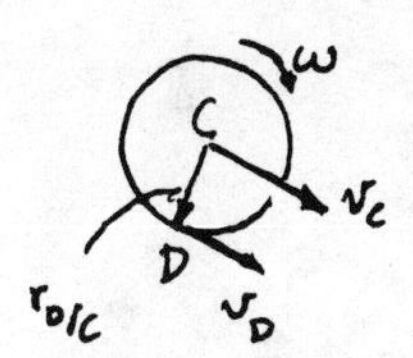

$(+\searrow)\ \mathbf{v}_D = \mathbf{v}_C + \mathbf{v}_{D/C}$

$$v_D = v_C - 0.5\omega \qquad (4)$$

Solving Eqs. (1) to (4) yields :

$\omega = 0$ **Ans**

$v_C = 32.2$ ft/s $\quad v_D = 32.2$ ft/s $\quad F = 0$

R2-14. Solve Prob. R2-13 if the coefficients of static and kinetic friction between the cylinder and the dolly are $\mu_s = 0.3$ and $\mu_k = 0.2$, respectively.

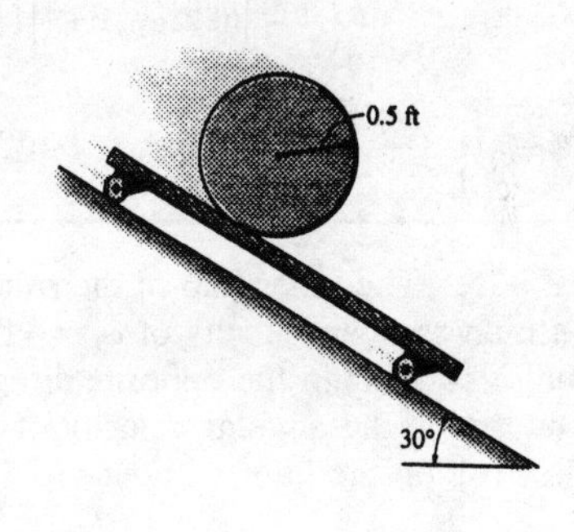

For the cylinder

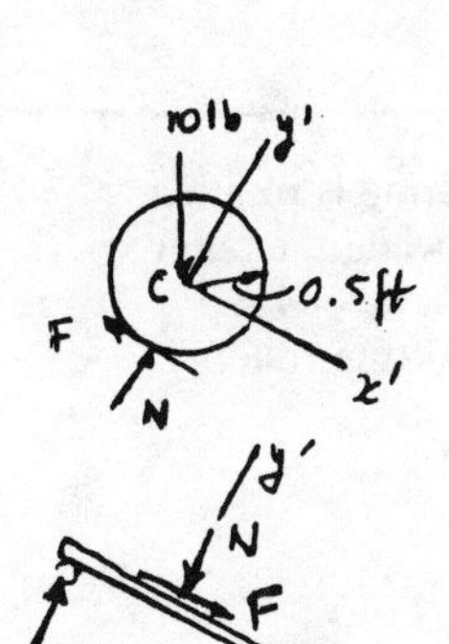

$(+\searrow)\ m(v_{Cx'})_1 + \Sigma\int_{t_1}^{t_2} F_{x'}dt = m(v_{Cx'})_2$

$$0 + 10\sin 30°(2) - F(2) = \left(\frac{10}{32.2}\right)v_C \qquad (1)$$

$(\circlearrowright +)\ I_C\omega_1 + \Sigma\int_{t_1}^{t_2} M_C dt = I_C\omega_2$

$$0 + F(0.5)(2) = \left[\frac{1}{2}\left(\frac{10}{32.2}\right)(0.5)^2\right]\omega \qquad (2)$$

For the dolly

$(+\searrow)\ m(v_{Dx'})_1 + \Sigma\int_{t_1}^{t_2} F_{x'}dt = m(v_{Dx'})_2$

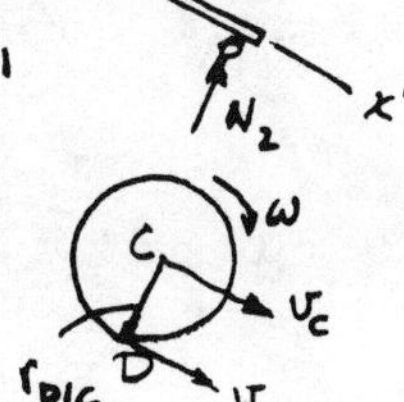

$$0 + F(2) + 20\sin 30°(2) = \left(\frac{20}{32.2}\right)v_D \qquad (3)$$

$(+\searrow)\ \mathbf{v}_D = \mathbf{v}_C + \mathbf{v}_{D/C}$

$$v_D = v_C - 0.5\omega \qquad (4)$$

Solving Eqs. (1) to (4) yields :

$\omega = 0$ **Ans**

$v_C = 32.2$ ft/s $\quad v_D = 32.2$ ft/s $\quad F = 0$

Note : No friction force develops.

R2-15. Gears C and H each have a weight of 0.4 lb and a radius of gyration about their mass center of $(k_H)_B = (k_C)_A = 2$ in. The uniform link AB has a weight of 0.2 lb and a radius of gyration of $(k_{AB})_A = 3$ in., whereas link DE has a weight of 0.15 lb and a radius of gyration of $(k_{DE})_B = 4.5$ in. If a couple moment of $M = 3$ lb·ft is applied to link AB and the assembly is originally at rest, determine the angular velocity of link DE when link AB has rotated 360°. Gear C is fixed from rotating and motion occurs in the horizontal plane. Also, gear H and link DE rotate together about the same shaft at B.

For link AB

$$v_B = \omega_{AB} r_{AB} = \omega_{AB}\left(\frac{6}{12}\right) = 0.5\omega_{AB}$$

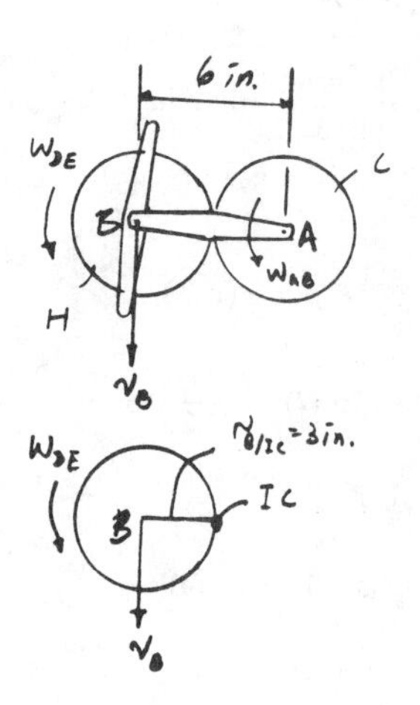

For gear H

$$\omega_{DE} = \frac{v_B}{r_{B/IC}} = \frac{0.5\omega_{AB}}{3/12} = 2\omega_{AB}$$

$$\omega_{AB} = \frac{1}{2}\omega_{DE}$$

$$v_B = \left(\frac{1}{2}\omega_{DE}\right)\frac{6}{12} = 0.25\omega_{DE}$$

Principle of work and energy for the system :

$$T_1 + \Sigma U_{1-2} = T_2$$

$$0 + 3(2\pi) = \frac{1}{2}\left[\left(\frac{0.2}{32.2}\right)\left(\frac{3}{12}\right)^2\right]\left(\frac{1}{2}\omega_{DE}\right)^2 + \frac{1}{2}\left[\left(\frac{0.4}{32.2}\right)\left(\frac{2}{12}\right)^2\right]\omega_{DE}^2$$

$$+ \frac{1}{2}\left(\frac{0.4}{32.2}\right)(0.25\omega_{DE})^2 + \frac{1}{2}\left[\left(\frac{0.15}{32.2}\right)\left(\frac{4.5}{12}\right)^2\right]\omega_{DE}^2 + \frac{1}{2}\left(\frac{0.15}{32.2}\right)(0$$

$\omega_{DE} = 132$ rad/s **Ans**

***R2-16.** The inner hub of the roller bearing is rotating with an angular velocity of $\omega_i = 6$ rad/s, while the outer hub is rotating in the opposite direction at $\omega_o = 4$ rad/s. Determine the angular velocity of each of the rollers if they roll on the hubs without slipping.

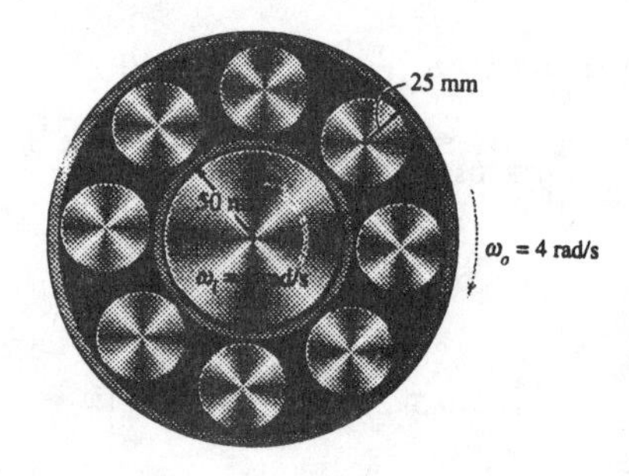

Since the hub does not slip, $v_A = \omega_i r_i = 6(0.05) = 0.3$ m/s and $v_B = \omega_O r_O = 4(0.1) = 0.4$ m/s.

$$\mathbf{v}_B = \mathbf{v}_A + \mathbf{v}_{B/A}$$

$$\left[\underset{\downarrow}{0.4}\right] = \left[\underset{\uparrow}{0.3}\right] + \left[\underset{\downarrow}{\omega(0.05)}\right]$$

$(+\downarrow)$ $\quad 0.4 = -0.3 + 0.05\omega \qquad \omega = 14$ rad/s **Ans**

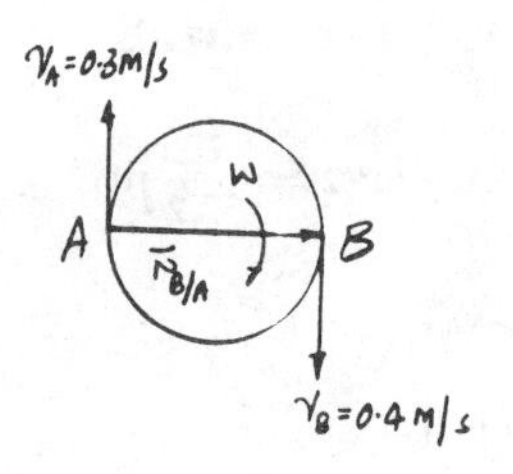

R2-17. The hoop (thin ring) has a mass of 5 kg and is released down the inclined plane such that it has a backspin $\omega = 8$ rad/s and its center has a velocity $v_G = 3$ m/s as shown. If the coefficient of kinetic friction between the hoop and the plane is $\mu_k = 0.6$, determine how long the hoop rolls before it stops slipping.

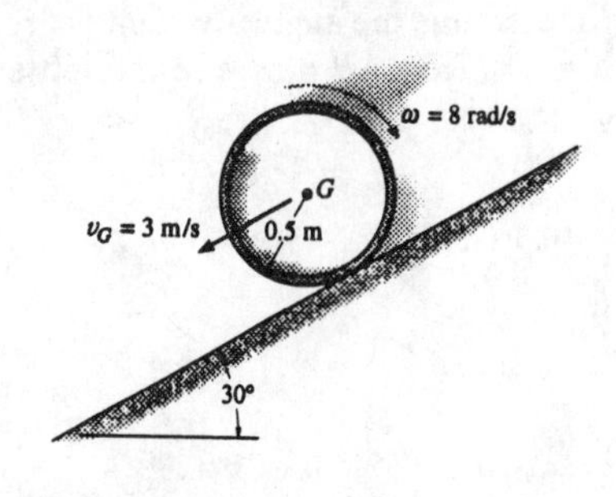

$(+\nwarrow)\qquad m v_{y1} + \Sigma\int F_y\,dt = m v_{y2}$

$$0 + N_h(t) - 5(9.81)t = 0$$

$$N_h = 42.479 \text{ N}$$

$$F_h = 0.6N_h = 0.6(42.479 \text{ N}) = 25.487 \text{ N}$$

$(\swarrow +)\qquad m v_{x1} + \Sigma\int F_x\,dt = m v_{x2}$

$$5(3) + 5(9.81)\sin 30^\circ(t) - 25.487t = 5v_G$$

$(\curvearrowleft +)\qquad (H_G)_1 + \Sigma\int M_G\,dt = (H_G)_2$

$$-5(0.5)^2(8) + 25.487(0.5)(t) = 5(0.5)^2\left(\frac{v_G}{0.5}\right)$$

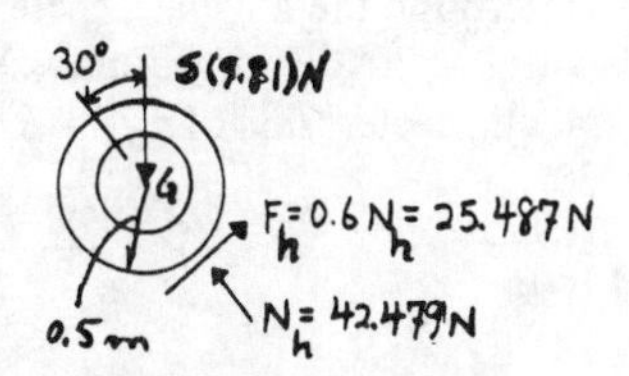

Solving,

$$v_G = 2.75 \text{ m/s}$$

$$t = 1.32 \text{ s} \qquad \textbf{Ans}$$

R2-18. The hoop (thin ring) has a mass of 5 kg and is released down the inclined plane such that it has a backspin $\omega = 8$ rad/s and its center has a velocity $v_G = 3$ m/s as shown. If the coefficient of kinetic friction between the hoop and the plane is $\mu_k = 0.6$, determine the hoop's angular velocity in 1 s.

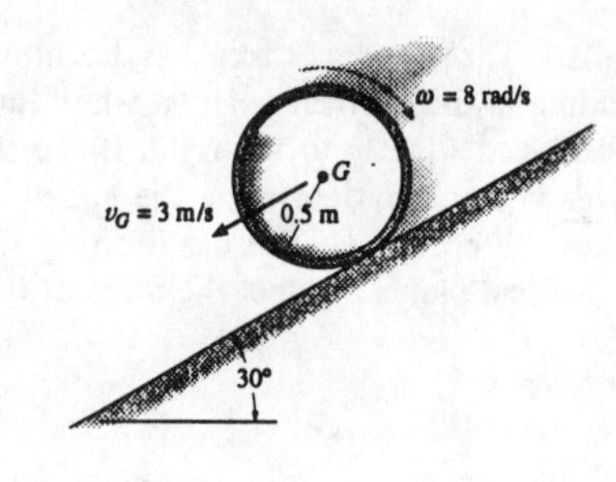

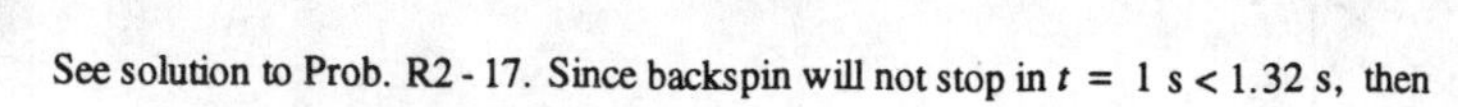
See solution to Prob. R2 - 17. Since backspin will not stop in $t = 1$ s < 1.32 s, then

$(+\nwarrow)\qquad m v_{y1} + \Sigma\int F_y\,dt = m v_{y2}$

$$0 + N_h(t) - 5(9.81)t = 0$$

$$N_h = 42.479 \text{ N}$$

$$F_h = 0.6N_h = 0.6(42.479 \text{ N}) = 25.487 \text{ N}$$

$(\curvearrowleft +)\qquad (H_G)_1 + \Sigma\int M\,dt = (H_G)_2$

$$-5(0.5)^2(8) + 25.487(0.5)(1) = 5(0.5)^2\omega$$

$$\omega = 2.19 \text{ rad/s} \curvearrowright \qquad \textbf{Ans}$$

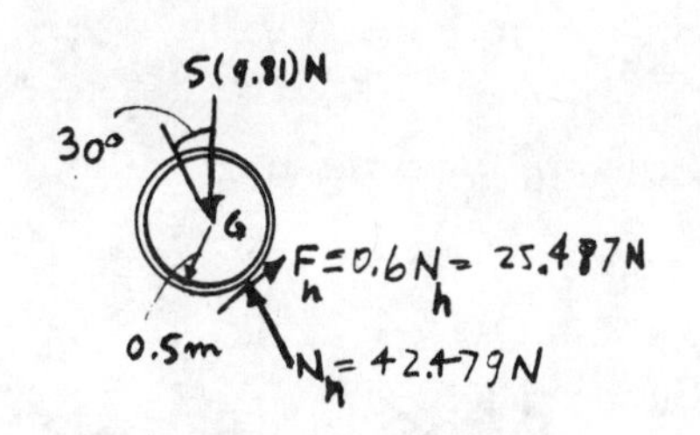

R2-19. Determine the angular velocity of rod CD at the instant $\theta = 30°$. Rod AB moves to the left at a constant rate $v_{AB} = 5$ m/s.

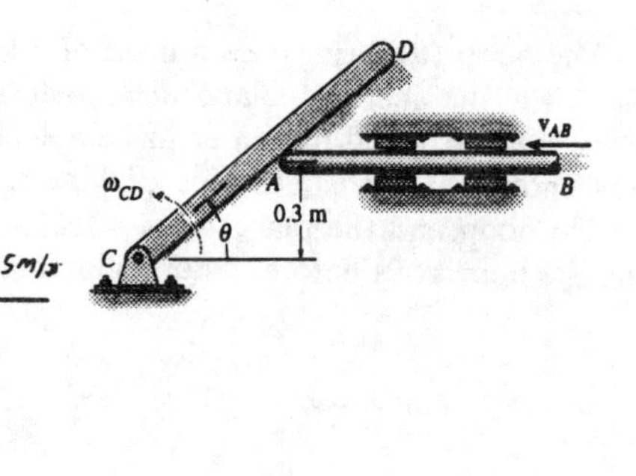

$$x = \frac{0.3}{\tan\theta} = 0.3\cot\theta$$

$$\dot{x} = v_{AB} = -0.3\csc^2\theta\,\dot{\theta}$$

Here $\dot{\theta} = \omega_{CD}$, $v_{AB} = -5$ m/s and $\theta = 30°$.

$-5 = -0.3\csc^2 30°(\omega_{CD})$ $\qquad \omega_{CD} = 4.17$ rad/s $\qquad$ **Ans**

***R2-20.** Determine the angular acceleration of rod CD at the instant $\theta = 30°$. Rod AB has zero velocity, i.e., $v_{AB} = 0$, and an acceleration of $a_{AB} = 2$ m/s^2 to the right when $\theta = 30°$.

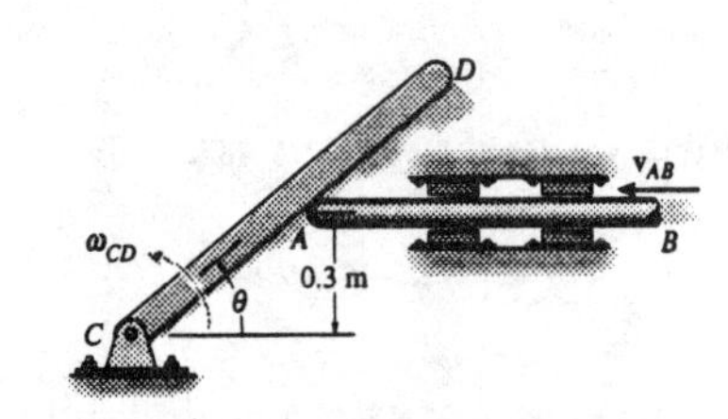

$$x = \frac{0.3}{\tan\theta} = 0.3\cot\theta$$

$$\dot{x} = v_{AB} = -0.3\csc^2\theta\,\dot{\theta}$$

$$\ddot{x} = a_{AB} = -0.3\left[\csc^2\theta\ddot{\theta} - 2\csc^2\theta\cot\theta\dot{\theta}^2\right] = 0.3\csc^2\theta\left(2\cot\theta\dot{\theta}^2 - \ddot{\theta}\right)$$

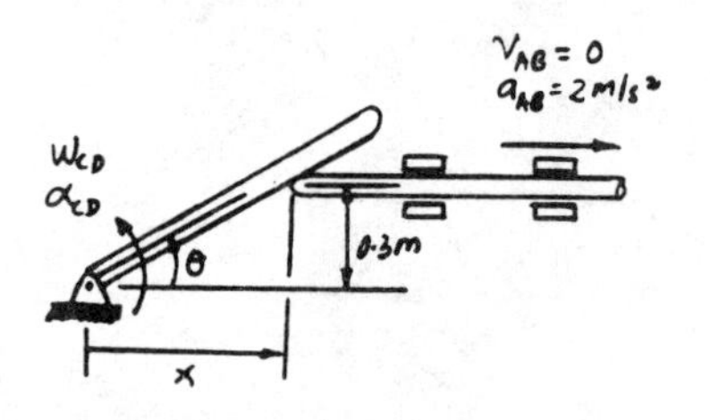

Here $\dot{\theta} = \omega_{CD}$, $v_{AB} = 0$, $a_{AB} = 2$ m/s^2, $\ddot{\theta} = \alpha_{CD}$ and $\theta = 30°$.

$0 = -0.3\csc^2 30°(\omega_{CD})$ $\qquad \omega_{CD} = 0$

$2 = 0.3\csc^2 30°\left[2\cot 30°(0)^2 - \alpha_{CD}\right]$ $\qquad \alpha_{CD} = -1.67$ rad/s^2 $\qquad$ **Ans**

R2-21. The slender 15-kg bar is initially at rest and standing in the vertical position when the bottom end A is displaced slightly to the right. If the track in which it moves is smooth, determine the speed at which end A strikes the corner D. The bar is constrained to move in the vertical plane. Neglect the mass of the cord BC.

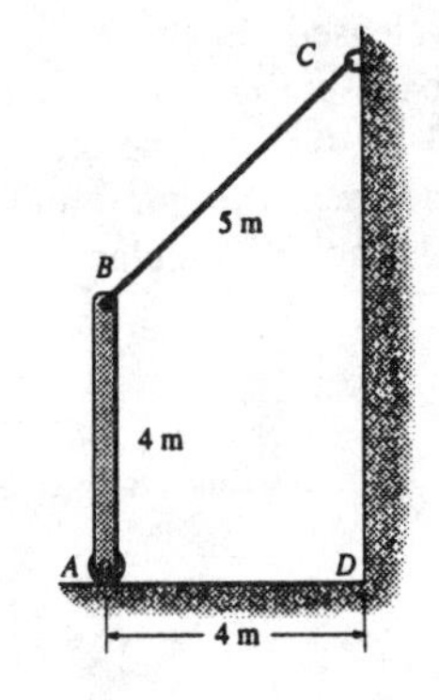

$$x^2 + y^2 = 5^2$$

$$x^2 + (7-y)^2 = 4^2$$

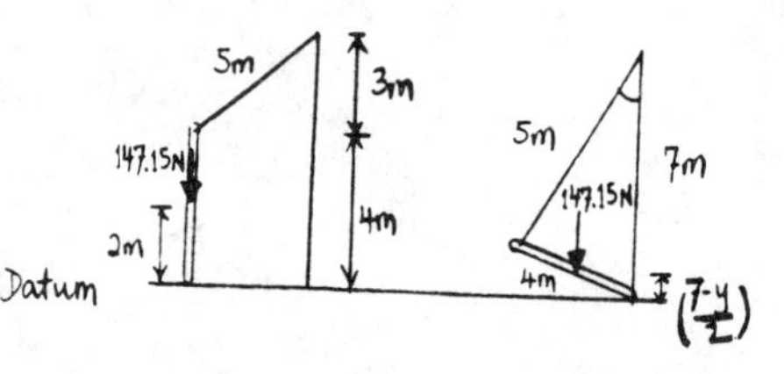

Thus, $\quad y = 4.1429$ m

$x = 2.7994$ m

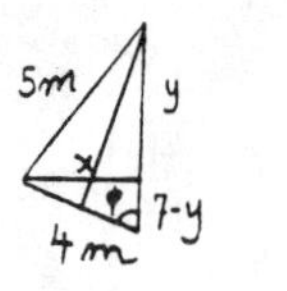

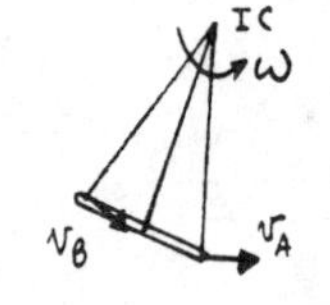

$$(5)^2 = (4)^2 + (7)^2 - 2(4)(7)\cos\phi$$

$$\phi = 44.42°$$

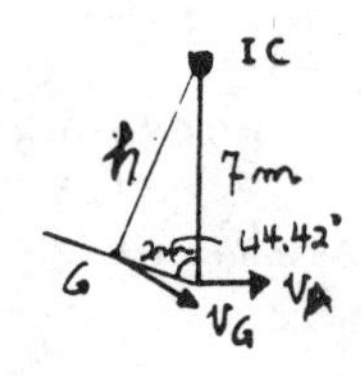

$$h^2 = (2)^2 + (7)^2 - 2(2)(7)\cos 44.42°$$

$$h = 5.745 \text{ m}$$

$$T_1 + V_1 = T_2 + V_2$$

$$0 + 147.15(2) = \frac{1}{2}\left[\frac{1}{12}(15)(4)^2\right]\omega^2 + \frac{1}{2}(15)(5.745\omega)^2 + 147.15\left(\frac{7-4.1429}{2}\right)$$

$\omega = 0.5714$ rad/s

$v_A = 0.5714(7) = 4.00$ m/s $\qquad$ **Ans**

R2-22. The four-wheeler has a weight of 335 lb and center of gravity at G_1, whereas the rider has a weight of 150 lb and center of gravity at G_2. If the engine can develop enough torque to cause the wheels to slip, determine the largest coefficient of static friction between the rear wheels and the ground so that the vehicle will accelerate without tipping over. What is this maximum acceleration? In order to *increase* the acceleration, should the rider crouch down or sit up straight from the position shown? Explain. The front wheels are free to roll. Neglect the mass of the wheels in the calculation.

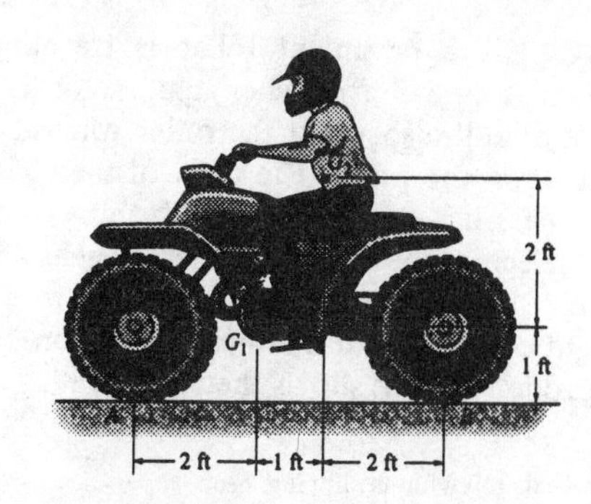

Equations of motion :

$$\circlearrowleft + \Sigma M_B = \Sigma (M_k)_B; \quad 335(3) + 150(2) = \frac{335}{32.2} a\,(1) + \frac{150}{32.2} a\,(3) \qquad (1)$$

$$a = 53.53 \text{ ft/s}^2 = 53.5 \text{ ft/s}^2 \qquad \textbf{Ans}$$

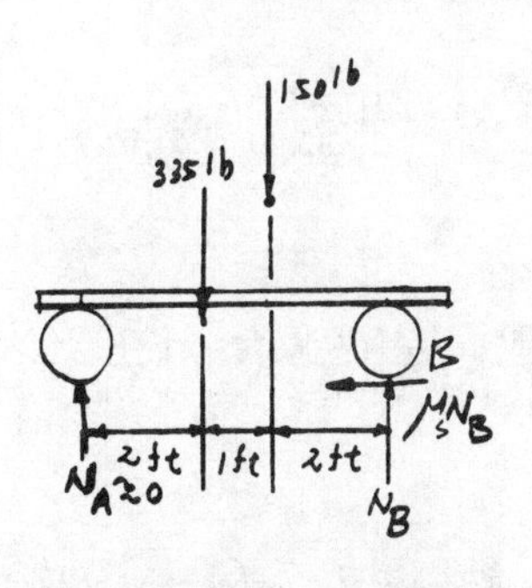

$$+\uparrow \Sigma F_y = m(a_G)_y; \quad N_B - 335 - 150 = 0 \quad N_B = 485 \text{ lb}$$

$$\overset{+}{\leftarrow} \Sigma F_x = m(a_G)_x; \quad \mu_s(485) = \frac{335}{32.2}(53.53) + \frac{150}{32.2}(53.53)$$

$$\mu_s = 1.66 \qquad \textbf{Ans}$$

From Eq. (1), in order to have a greater acceleration, the rider has to **crouch down** to lower his center of gravity G_2. This increases the moment of the rider's weight, and decreases the moment of ma_G.

R2-23. The four-wheeler has a weight of 335 lb and center of gravity at G_1, whereas the rider has a weight of 150 lb and center of gravity at G_2. If the coefficient of static friction between the rear wheels and the ground is $\mu_s = 0.3$, determine the greatest acceleration the vehicle can have. The front wheels are free to roll. Neglect the mass of the wheels in the calculation.

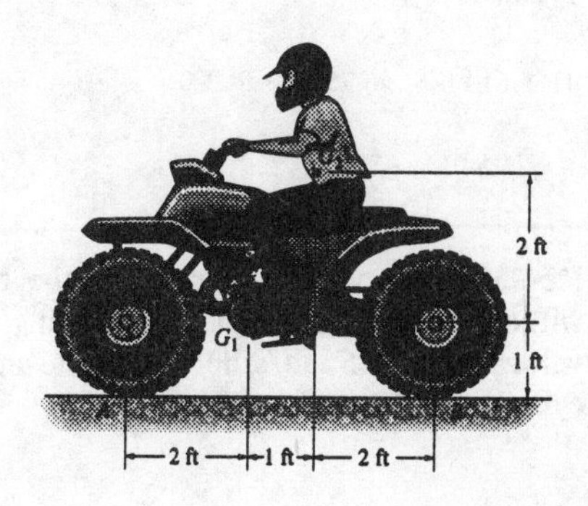

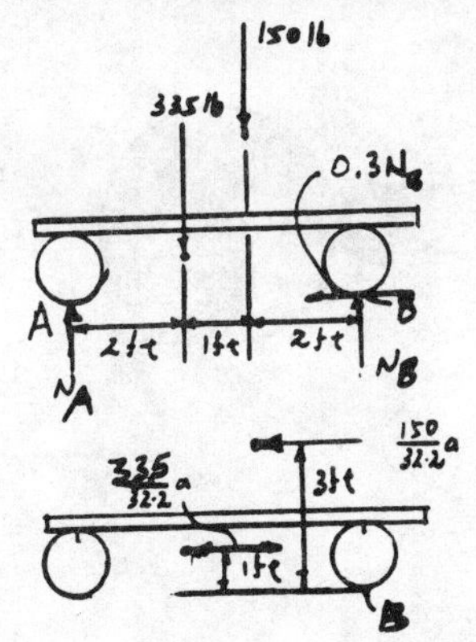

$$\circlearrowleft + \Sigma M_A = \Sigma (M_k)_A; \quad -335(2) - 150(3) + N_B(5) = \frac{335}{32.2} a\,(1) + \frac{150}{32.2} a\,(3) \qquad (1)$$

$$\overset{+}{\leftarrow} \Sigma F_x = m(a_G)_x; \quad 0.3N_B = \frac{335}{32.2} a + \frac{150}{32.2} a \qquad (2)$$

Solving Eqs. (1) and (2) yields :

$a = 4.94 \text{ ft/s}^2$ **Ans**

$N_B = 248 \text{ lb}$

***R2-24.** The pavement roller is traveling down the incline at $v_1 = 5$ ft/s when the motor is disengaged. Determine the speed of the roller when it has traveled 20 ft down the plane. The body of the roller, excluding the rollers, has a weight of 8000 lb and a center of gravity at G. Each of the two rear rollers weighs 400 lb and has a radius of gyration of $k_A = 3.3$ ft. The front roller has a weight of 800 lb and a radius of gyration of $k_B = 1.8$ ft. The rollers do not slip as they rotate.

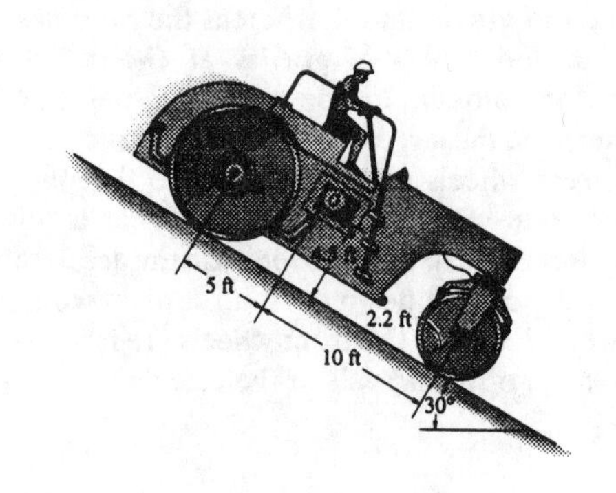

The wheels roll without slipping, hence $\omega = \dfrac{v_G}{r}$.

$$T_1 = \frac{1}{2}\left(\frac{8000+800+800}{32.2}\right)(5)^2 + \frac{1}{2}\left[\left(\frac{800}{32.2}\right)(3.3)^2\right]\left(\frac{5}{3.8}\right)^2 + \frac{1}{2}\left[\left(\frac{800}{32.2}\right)(1.8)^2\right]\left(\frac{5}{2.2}\right)^2$$

$$= 4168.81 \text{ ft} \cdot \text{lb}$$

$$T_2 = \frac{1}{2}\left(\frac{8000+800+800}{32.2}\right)v^2 + \frac{1}{2}\left[\left(\frac{800}{32.2}\right)(3.3)^2\right]\left(\frac{v}{3.8}\right)^2 + \frac{1}{2}\left[\left(\frac{800}{32.2}\right)(1.8)^2\right]\left(\frac{v}{2.2}\right)^2$$

$$= 166.753v^2$$

Put datum through the mass center of the wheels and body of the roller when it is in the initial position.

$$V_1 = 0$$

$$V_2 = -800(20\sin 30°) - 8000(20\sin 30°) - 800(20\sin 30°)$$

$$= -96000 \text{ ft} \cdot \text{lb}$$

$$T_1 + V_1 = T_2 + V_2$$

$$4168.81 + 0 = 166.753v^2 - 96000$$

$v = 24.5$ ft/s **Ans**

R2-25. The cylinder B rolls on the fixed cylinder A without slipping. If bar CD is rotating with an angular velocity $\omega_{CD} = 5$ rad/s, determine the angular velocity of cylinder B. Point C is a fixed point.

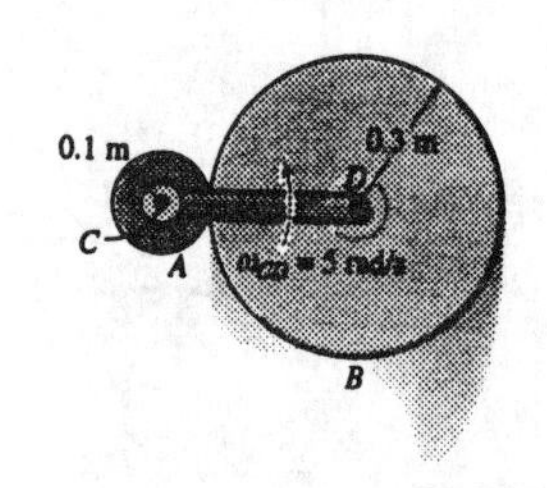

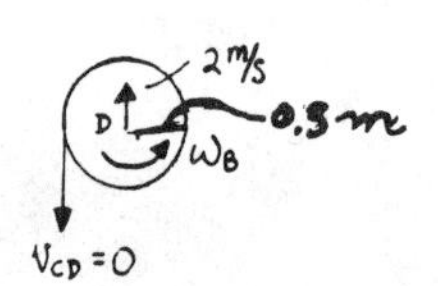

$$v_D = 5(0.4) = 2 \text{ m/s}$$

$\omega_B = \dfrac{2}{0.3} = 6.67$ rad/s **Ans**

R2-26. The disk has a mass M and a radius R. If a block of mass m is attached to the cord, determine the angular acceleration of the disk when the block is released from rest. Also, what is the distance the block falls from rest in the time t?

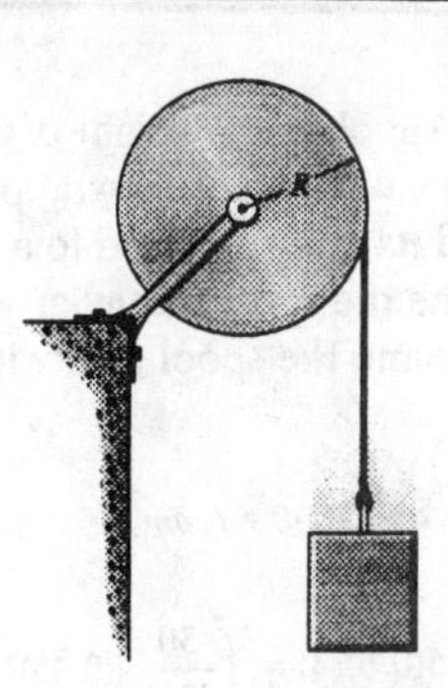

$$I_O = \frac{1}{2}MR^2$$

$$\circlearrowleft + \Sigma M_O = \Sigma (M_k)_O; \quad mgR = \frac{1}{2}MR^2(\alpha) + m(\alpha R)R$$

$$\alpha = \frac{2mg}{R(M+2m)} \qquad \textbf{Ans}$$

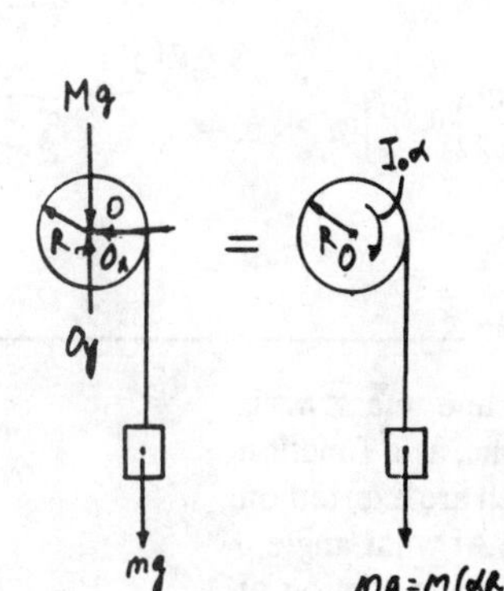

The displacement $h = R\theta$, hence $\quad \theta = \dfrac{h}{R}$

$$\theta = \theta_0 + \omega_0 t + \frac{1}{2}\alpha_c t^2$$

$$\frac{h}{R} = 0 + 0 + \frac{1}{2}\left(\frac{2mg}{R(M+2m)}\right)t^2$$

$$h = \frac{mg}{M+2m}t^2 \qquad \textbf{Ans}$$

R2-27. The tub of the mixer has a weight of 70 lb and a radius of gyration $k_G = 1.3$ ft about its center of gravity. If a constant torque $M = 60$ lb·ft is applied to the dumping wheel, determine the angular velocity of the tub when it has rotated $\theta = 90°$. Originally the tub is at rest when $\theta = 0°$.

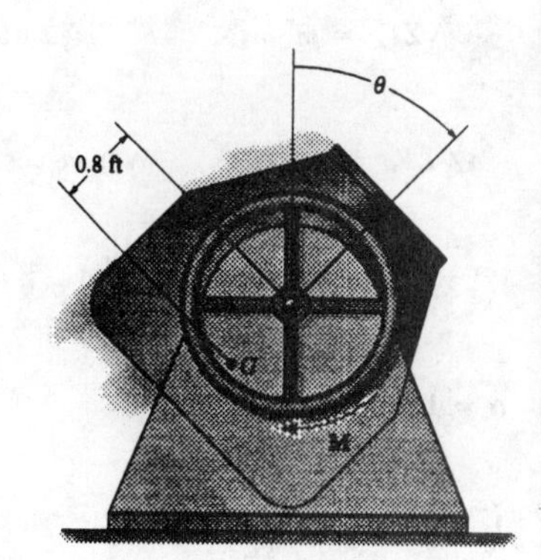

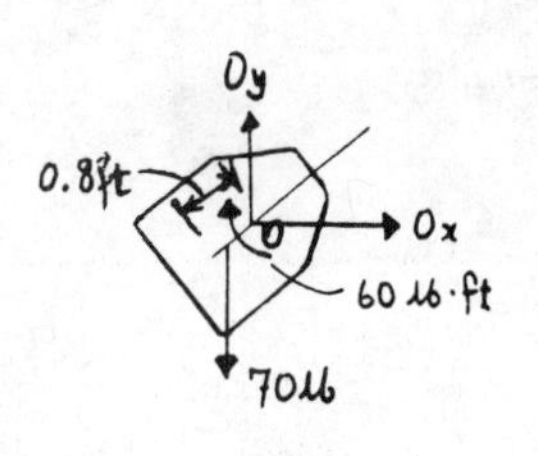

$$T_1 + \Sigma U_{1-2} = T_2$$

$$0 + 60\left(\frac{\pi}{2}\right) - 70(0.8) = \frac{1}{2}\left[\left(\frac{70}{32.2}\right)(1.3)^2\right](\omega)^2 + \frac{1}{2}\left[\frac{70}{32.2}\right](0.8\omega)^2$$

$\omega = 3.89$ rad/s $\qquad$ **Ans**

***R2-28.** Solve Prob. R2-27 if the applied torque is $M = (50\theta)$ lb·ft, where θ is in radians.

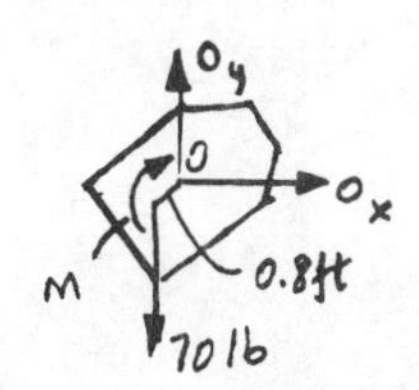

$$T_1 + \Sigma U_{1-2} = T_2$$

$$0 + \int_0^{\pi/2} 50\theta \, d\theta - 70(0.8) = \frac{1}{2}\left[\left(\frac{70}{32.2}\right)(1.3)^2\right]\omega^2 + \frac{1}{2}\left[\frac{70}{32.2}\right](0.8\omega)^2$$

$\omega = 1.50$ rad/s $\qquad$ **Ans**

R2-29. The spool has a weight of 30 lb and a radius of gyration $k_O = 0.45$ ft. A cord is wrapped around the spool's inner hub and its end subjected to a horizontal force $P = 5$ lb. Determine the spool's angular velocity in 4 s starting from rest. Assume the spool rolls without slipping.

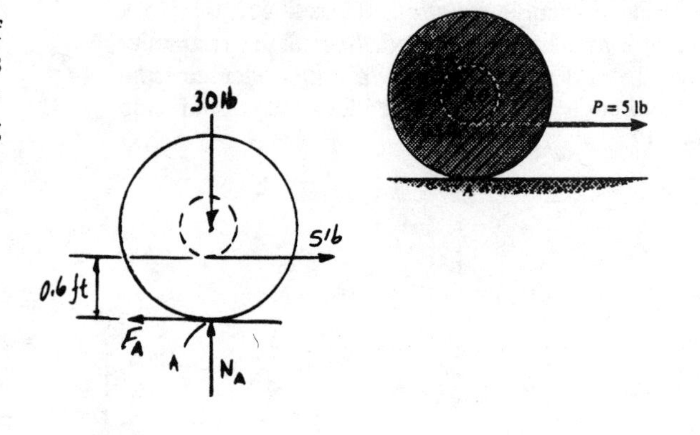

$$(+\circlearrowright) \quad I_A\,\omega_1 + \Sigma\int_{t_1}^{t_2} M_A\,dt = I_A\,\omega_2$$

$$0 + 5(0.6)(4) = \left[\left(\frac{30}{32.2}\right)(0.45)^2 + \left(\frac{30}{32.2}\right)(0.9)^2\right]\omega_2$$

$$\omega_2 = 12.7 \text{ rad/s} \qquad \textbf{Ans}$$

R2-30. The slender rod of length L and mass m is released from rest when $\theta = 0°$. Determine, as a function of θ, the normal and frictional forces which are exerted on the ledge at A as the rod falls downward. At what angle θ does it begin to slip if the coefficient of static friction at A is μ_s?

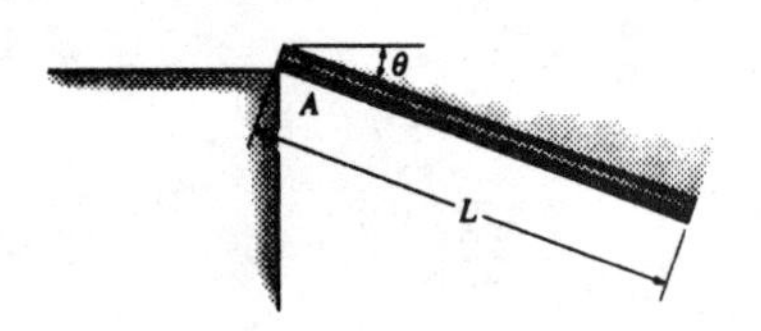

$$\nwarrow + \Sigma F_{x'} = m(a_G)_{x'}; \qquad F - mg\sin\theta = m\omega^2\left(\frac{L}{2}\right)$$

$$+\swarrow \Sigma F_{y'} = m(a_G)_{y'}; \qquad -N + mg\cos\theta = m\left(\frac{L}{2}\right)\alpha$$

$$\circlearrowright + \Sigma M_A = I_A\,\alpha; \qquad (mg)\left(\frac{L}{2}\right)\cos\theta = \frac{1}{3}mL^2\alpha$$

$$\alpha = 1.5\left(\frac{g}{L}\right)\cos\theta$$

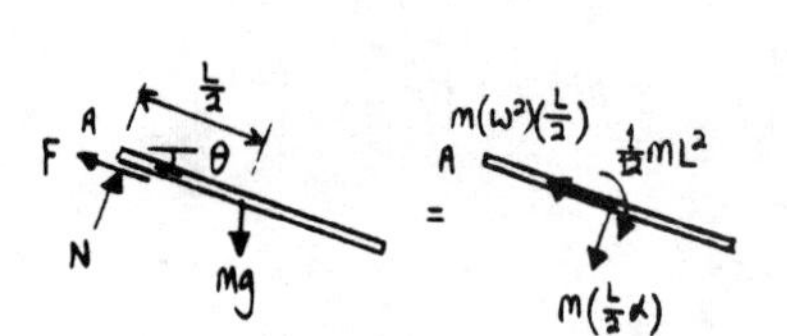

$$\int_0^{\omega}\omega\,d\omega = \int_0^{\theta} 1.5\left(\frac{g}{L}\right)\cos\theta\,d\theta$$

$$\omega^2 = 3\left(\frac{g}{L}\right)\sin\theta$$

$$F = mg\sin\theta + m\left(\frac{L}{2}\right)\left(3\frac{g}{L}\right)\sin\theta$$

$$F = 2.5\,mg\sin\theta \qquad \textbf{Ans}$$

$$N = mg\cos\theta - m\left(\frac{L}{2}\right)\left(1.5\frac{g}{L}\right)\cos\theta$$

$$N = 0.25\,mg\cos\theta \qquad \textbf{Ans}$$

Set $F = \mu_s N$

$$2.5mg\sin\theta = \mu_s(0.25mg\cos\theta)$$

$$\theta = \tan^{-1}(0.1\mu_s) \qquad \textbf{Ans}$$

R2-31. A sphere and cylinder are released from rest on the ramp at $t = 0$. If each has a mass m and a radius r, determine their angular velocities at time t. Assume no slipping occurs.

Principle of impulse and momentum : For the sphere

$(+\circlearrowleft)$ $\quad I_A\,\omega_1 + \Sigma\int_{t_1}^{t_2} M_A\,dt = I_A\,\omega_2$

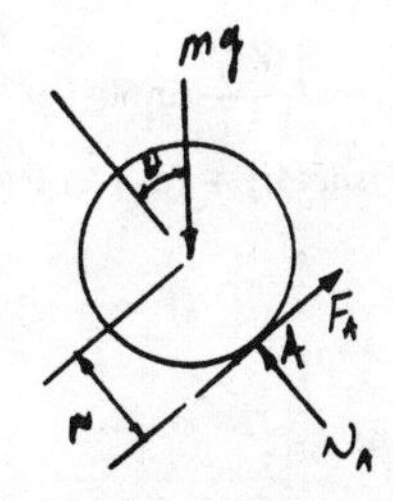

$$0 + mg\sin\theta(r)(t) = \left[\frac{2}{5}mr^2 + mr^2\right](\omega_S)_2$$

$$(\omega_S)_2 = \frac{5g\sin\theta}{7r}t \qquad \textbf{Ans}$$

Principle of impulse and momentum : For the cylinder

$(+\circlearrowleft)$ $\quad I_A\,\omega_1 + \Sigma\int_{t_1}^{t_2} M_A\,dt = I_A\,\omega_2$

$$0 + mg\sin\theta(r)(t) = \left[\frac{1}{2}mr^2 + mr^2\right](\omega_C)_2$$

$$(\omega_C)_2 = \frac{2g\sin\theta}{3r}t \qquad \textbf{Ans}$$

***R2-32.** At a given instant, link AB has an angular acceleration $\alpha_{AB} = 12\ \text{rad/s}^2$ and an angular velocity $\omega_{AB} = 4\ \text{rad/s}$. Determine the angular velocity and angular acceleration of link CD at this instant.

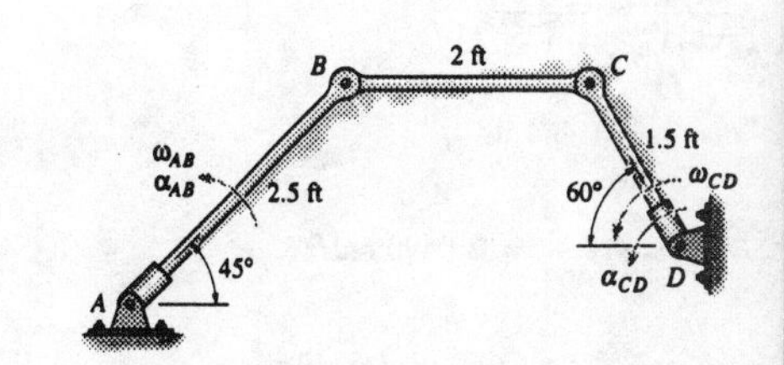

$\mathbf{v}_C = \mathbf{v}_B + \mathbf{v}_{C/B}$

$$\begin{bmatrix} v_C \\ 30^\circ\swarrow \end{bmatrix} = \begin{bmatrix} 10 \\ 45^\circ\nwarrow \end{bmatrix} + \begin{bmatrix} 2\omega_{BC} \\ \downarrow \end{bmatrix}$$

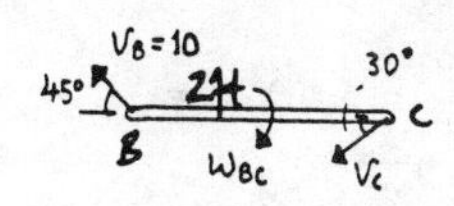

$(\overset{+}{\leftarrow})$ $\quad v_C\cos 30^\circ = 10\cos 45^\circ + 0$

$(+\downarrow)$ $\quad v_C\sin 30^\circ = -10\sin 45^\circ + 2\omega_{BC}$

$\omega_{BC} = 5.58\ \text{rad/s},$

$v_C = 8.16\ \text{ft/s}$

$$\omega_{CD} = \frac{8.16}{1.5} = 5.44\ \text{rad/s} \qquad \textbf{Ans}$$

$\mathbf{a}_C = \mathbf{a}_B + \mathbf{a}_{C/B}$

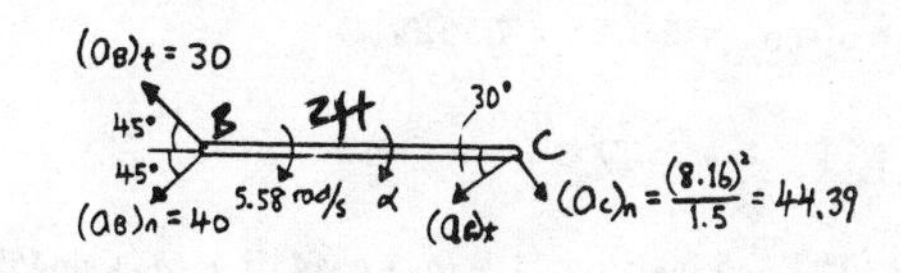

$$\begin{bmatrix} 44.44 \\ \searrow 60^\circ \end{bmatrix} + \begin{bmatrix} (a_C)_t \\ 30^\circ\swarrow \end{bmatrix} = \begin{bmatrix} 30 \\ 45^\circ\nwarrow \end{bmatrix} + \begin{bmatrix} 40 \\ 45^\circ\swarrow \end{bmatrix} + \begin{bmatrix} 2(5.58)^2 \\ \leftarrow \end{bmatrix} + \begin{bmatrix} 2\alpha_{BC} \\ \downarrow \end{bmatrix}$$

$(\overset{+}{\leftarrow})$ $\quad -44.44\cos 60^\circ + (a_C)_t\cos 30^\circ = 30\cos 45^\circ + 40\cos 45^\circ + 62.21$

$(+\downarrow)$ $\quad 44.44\sin 60^\circ + (a_C)_t\sin 30^\circ = -30\sin 45^\circ + 40\sin 45^\circ + 2\alpha_{BC}$

$(a_C)_t = 155\ \text{ft/s}^2, \qquad \alpha_{BC} = 54.4\ \text{rad/s}^2$

$$\alpha_{CD} = \frac{155}{1.5} = 103\ \text{rad/s}^2 \circlearrowright \qquad \textbf{Ans}$$

Also :

$10 = \omega(1.793); \qquad \omega = 5.577 \text{ rad/s}$

$v_C = (1.464)(5.577) = 8.164 \text{ ft/s}$

$\mathbf{a}_C = \mathbf{a}_B + \boldsymbol{\alpha}_{BC} \times \mathbf{r}_{C/B} - \omega^2 \mathbf{r}_{C/B}$

$$-(a_C)_t \cos 30°\mathbf{i} - (a_C)_t \sin 30°\mathbf{j} + \frac{(8.164)^2}{1.5}\cos 60°\mathbf{i} - \frac{(8.164)^2}{1.5}\sin 60°\mathbf{j} = -(4)^2(2.5)\cos 45°\mathbf{i}$$
$$-(4)^2(2.5)\cos 45°\mathbf{j} + (-12)(2.5)\cos 45°\mathbf{i} + (12)(2.5)\sin 45°\mathbf{j} + (\alpha_{BC}\mathbf{k}) \times (2\mathbf{i}) - (5.577)^2(2\mathbf{i})$$

$(a_C)_t = 155 \text{ ft/s}^2, \qquad \alpha_{BC} = 54.4 \text{ rad/s}^2$

$\alpha_{CD} = \dfrac{155}{1.5} = 103 \text{ rad/s}^2$ ↻ **Ans**

R2-33. At a given instant, link CD has an angular acceleration $\alpha_{CD} = 5 \text{ rad/s}^2$ and angular velocity $\omega_{CD} = 2$ rad/s. Determine the angular velocity and angular acceleration of link AB at this instant.

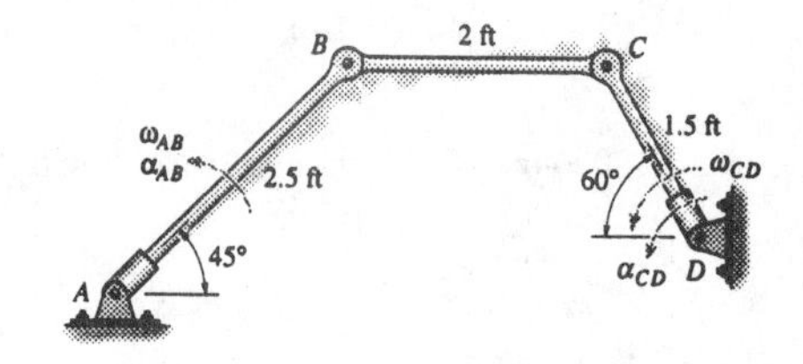

$\dfrac{r_{IC-C}}{\sin 45°} = \dfrac{2}{\sin 75°}$

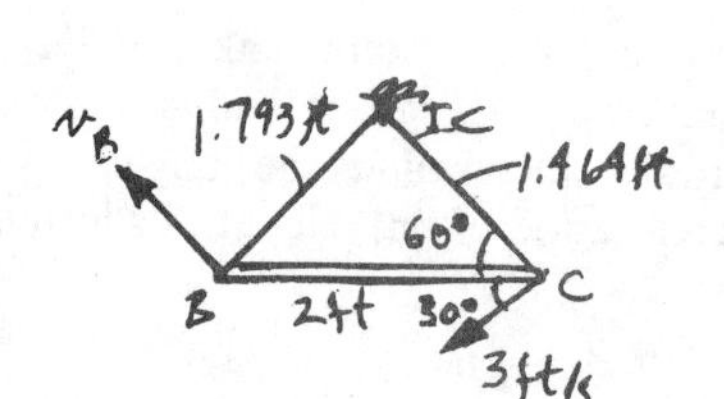

$r_{IC-C} = 1.464 \text{ ft}$

$\omega_{BC} = \dfrac{3}{1.464} = 2.0490 \text{ rad/s}$

$v_B = 2.0490(1.793) = 3.6742 \text{ ft/s}$

$\omega_{AB} = \dfrac{3.6742}{2.5} = 1.47 \text{ rad/s}$ ↻ **Ans**

$(a_B)_n = \dfrac{v_B^2}{r_{BA}} = \dfrac{(3.6742)^2}{2.5} = 5.4000 \text{ ft/s}^2$

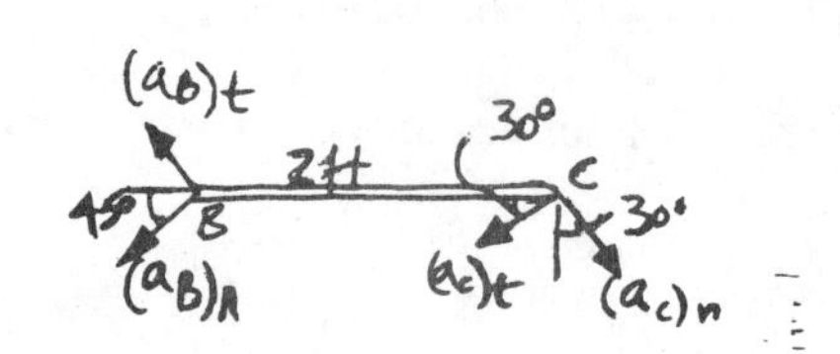

$(a_C)_n = \dfrac{v_C^2}{r_{CD}} = \dfrac{(3)^2}{1.5} = 6 \text{ ft/s}^2$

$(a_C)_t = \alpha_{CD}(r_{CD}) = 5(1.5) = 7.5 \text{ ft/s}^2$

$\mathbf{a}_B = \mathbf{a}_C + \boldsymbol{\alpha} \times \mathbf{r}_{B/C} - \omega^2 \mathbf{r}_{B/C}$

$$-5.400\cos 45°\mathbf{i} - 5.400\sin 45°\mathbf{j} - (a_B)_t \cos 45°\mathbf{i} + (a_B)_t \sin 45°\mathbf{j} = 6\sin 30°\mathbf{i} - 6\cos 30°\mathbf{j}$$
$$-7.5\cos 30°\mathbf{i} - 7.5\sin 30°\mathbf{j} + (\alpha\mathbf{k}) \times (-2\mathbf{i}) - (2.0490)^2(-2\mathbf{i})$$

$-3.818 - (a_B)_t(0.7071) = 3 - 6.495 + 8.3971$

$-3.818 + (a_B)_t(0.7071) = -5.1962 - 3.75 - 2\alpha$

$(a_B)_t = -12.332 \text{ ft/s}^2$

$\alpha = 1.80 \text{ rad/s}^2$

$\alpha_{AB} = \dfrac{12.332}{2.5} = 4.93 \text{ rad/s}^2$ ↻ **Ans**

R2-34. The spool and the wire wrapped around its core have a mass of 50 kg and a centroidal radius of gyration of $k_G = 235$ mm. If the coefficient of kinetic friction at the surface is $\mu_k = 0.15$, determine the angular acceleration of the spool after it is released from rest.

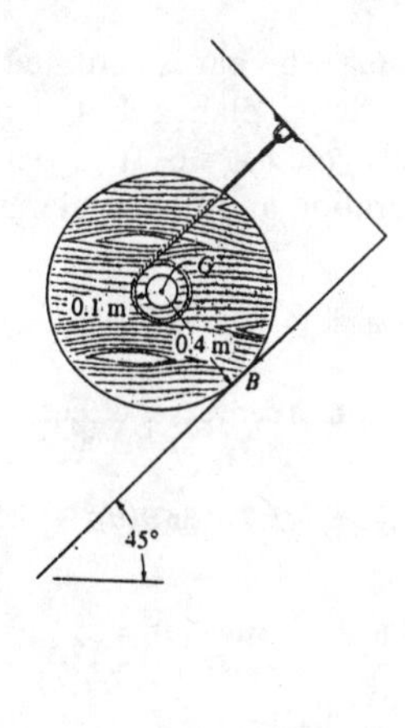

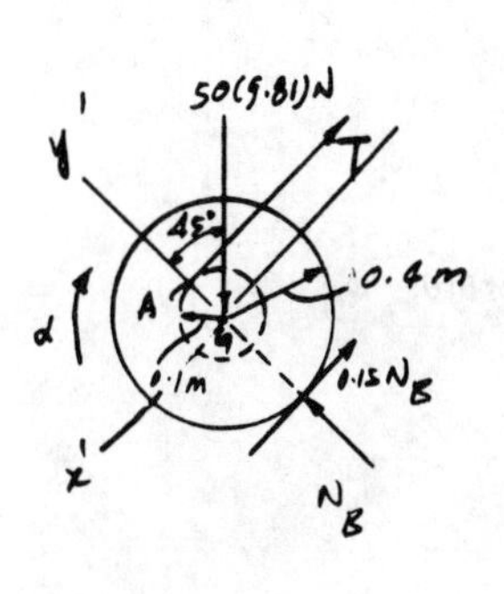

$I_G = mk_G^2 = 50(0.235)^2 = 2.76125 \text{ kg}\cdot\text{m}^2$

$+\swarrow \Sigma F_{x'} = m(a_G)_{x'};\quad 50(9.81)\sin 45° - T - 0.15N = 50a_G \quad (1)$

$+\nwarrow \Sigma F_{y'} = m(a_G)_{y'};\quad N_B - 50(9.81)\cos 45° = 0 \quad (2)$

$\circlearrowleft + \Sigma M_G = I_G\alpha;\quad T(0.1) - 0.15N_B(0.4) = 2.76125\alpha \quad (3)$

The spool does not slip at point A, therefore $a_G = 0.1\alpha$ (4)

Solving Eqs. (1) to (4) yields :

$N_B = 346.8$ N $\quad T = 281.5$ N $\quad a = 0.2659 \text{ m/s}^2$

$\alpha = 2.66 \text{ rad/s}^2$ ↻ **Ans**

R2-35. The bar is confined to move along the vertical and inclined planes. If the velocity of the roller at A is $v_A = 6$ ft/s when $\theta = 45°$, determine the bar's angular velocity and the velocity of B at this instant.

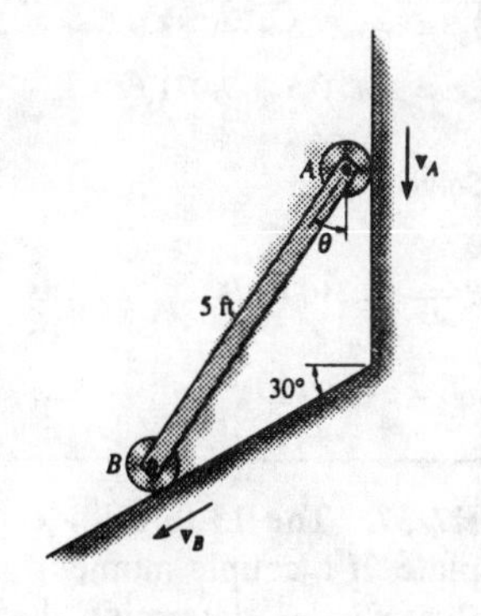

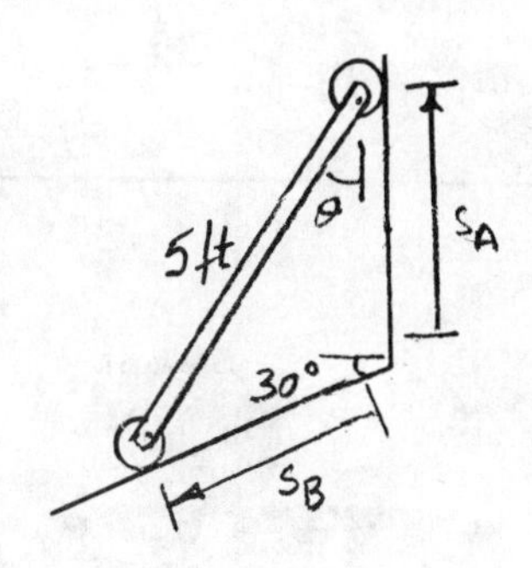

$s_B \cos 30° = 5 \sin\theta$

$s_B = 5.774 \sin\theta$

$\dot{s}_B = 5.774 \cos\theta\ \dot{\theta} \quad (1)$

$5\cos\theta = s_A + s_B \sin 30°$

$-5\sin\theta\ \dot{\theta} = \dot{s}_A + \dot{s}_B \sin 30° \quad (2)$

Combine Eqs. (1) and (2) :

$-5\sin\theta\ \dot{\theta} = -6 + 5.774\cos\theta(\dot{\theta})(\sin 30°)$

$-3.536\dot{\theta} = -6 + 2.041\dot{\theta}$

$\omega = \dot{\theta} = 1.08$ rad/s **Ans**

From Eq. (1) :

$v_B = \dot{s}_B = 5.774\cos 45°(1.076) = 4.39$ ft/s **Ans**

***R2-36.** The bar is confined to move along the vertical and inclined planes. If the roller at A has a constant velocity of $v_A = 6$ ft/s, determine the bar's angular acceleration and the acceleration of B when $\theta = 45°$.

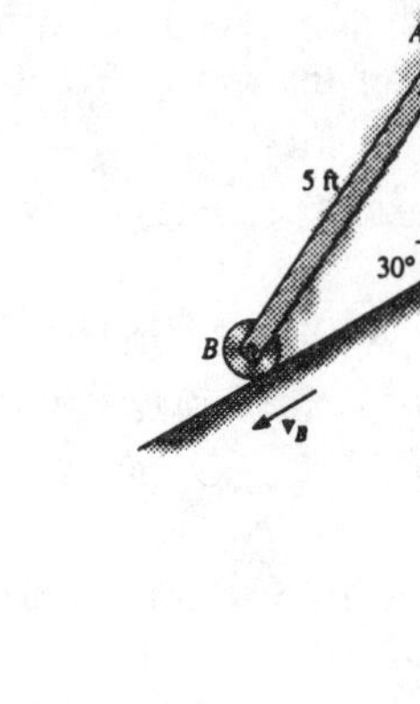

See solution to Prob. R2-35,

Taking the time derivatives of Eqs. (1) and (2) yields

$$a_B = \ddot{s}_B = -5.774\sin\theta(\dot{\theta})^2 + 5.774\cos\theta(\ddot{\theta})$$

$$-5\cos\theta\,\dot{\theta}^2 - 5\sin\theta(\ddot{\theta}) = \ddot{s}_A + \ddot{s}_B\sin 30°$$

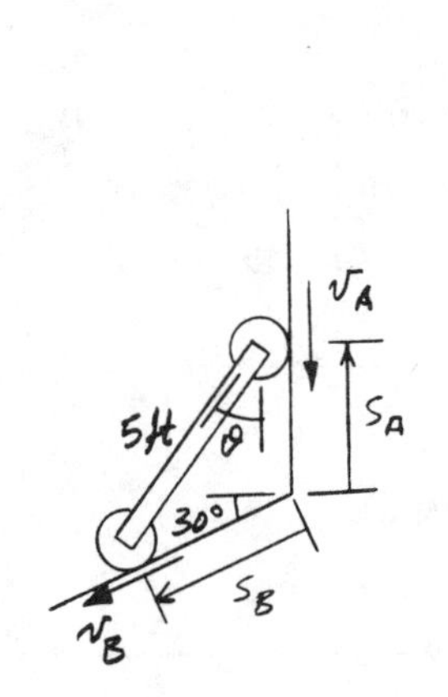

Substitute the data :

$$a_B = -5.774\sin 45°(1.076)^2 + 5.774\cos 45°(\ddot{\theta})$$

$$-5\cos 45°(1.076)^2 - 5\sin 45°(\ddot{\theta}) = 0 + a_B\sin 30°$$

$$a_B = -4.726 + 4.083\,\ddot{\theta}$$

$$a_B = -8.185 - 7.071\,\ddot{\theta}$$

Solving;

$\ddot{\theta} = -0.310$ rad/s^2 **Ans**

$a_B = -5.99$ ft/s^2 **Ans**

R2-37. The 15-lb cylinder is initially at rest on a 5-lb plate. If a couple moment of $M = 40$ lb·ft is applied to the cylinder, determine the angular acceleration of the cylinder and the time needed for the end B of the plate to travel 3 ft and strike the wall. Assume the cylinder does not slip on the plate, and neglect the mass of the rollers under the plate.

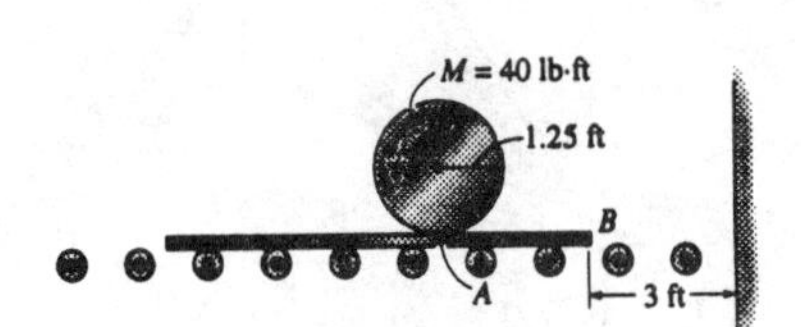

For the cylinder

$$\overset{+}{\leftarrow}\Sigma F_x = m(a_G)_x; \quad F_A = \left(\frac{15}{32.2}\right)a_G \quad (1)$$

$$\circlearrowleft +\Sigma M_G = I_G\alpha; \quad 40 - F_A(1.25) = \frac{1}{2}\left(\frac{15}{32.2}\right)(1.25)^2\alpha \quad (2)$$

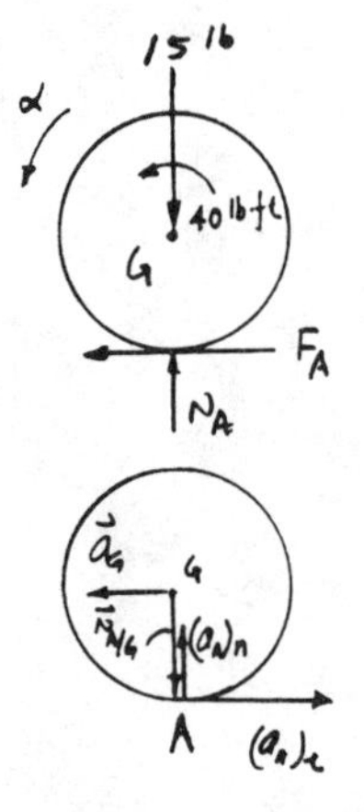

$$\mathbf{a}_A = \mathbf{a}_G + (\mathbf{a}_{A/G})_t + (\mathbf{a}_{A/G})_n$$

$$\left[\underset{\rightarrow}{(a_A)_t}\right] + \left[\underset{\uparrow}{(a_A)_n}\right] = \left[\underset{\leftarrow}{a_G}\right] + \left[\underset{\rightarrow}{1.25\alpha}\right] + \left[\underset{\uparrow}{(a_{A/G})_n}\right]$$

$$(\overset{+}{\rightarrow}) \quad (a_A)_t = -a_G + 1.25\alpha \quad (3)$$

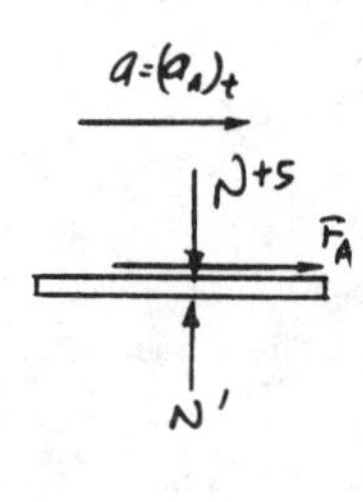

Since no slipping occurs at the point of contact, the plate will move with an acceleration of $a = (a_A)_t$.

$$\overset{+}{\rightarrow}\Sigma F_x = ma_x; \quad F_A = \left(\frac{5}{32.2}\right)(a_A)_t \quad (4)$$

Solving Eqs. (1) to (4) yields :

$a_G = 22.90$ ft/s^2 $\quad F_A = 10.67$ lb $\quad (a_A)_t = 68.69$ ft/s^2

$\alpha = 73.3$ rad/s^2 **Ans**

For the plate

$$(\overset{+}{\rightarrow}) \quad s = s_0 + v_0 t + \frac{1}{2}a_c t^2$$

$$3 = 0 + 0 + \frac{1}{2}(68.69)t^2 \quad t = 0.296 \text{ s}$$ **Ans**

R2-38. Each gear has a mass of 2 kg and a radius of gyration about its pinned mass centers A and B of k_g = 40 mm. Each link has a mass of 2 kg and a radius of gyration about its pinned ends A and B of k_l = 50 mm. If originally the spring is unstretched when the couple moment M = 20 N·m is applied to link AC, determine the angular velocities of the links at the instant link AC rotates $\theta = 45°$. Each gear and link is connected together and rotates about the fixed pins A and B.

Consider the system of both gears and the limks.
The spring stretches $s = 2(0.2\sin 45°) = 0.2828$ m.

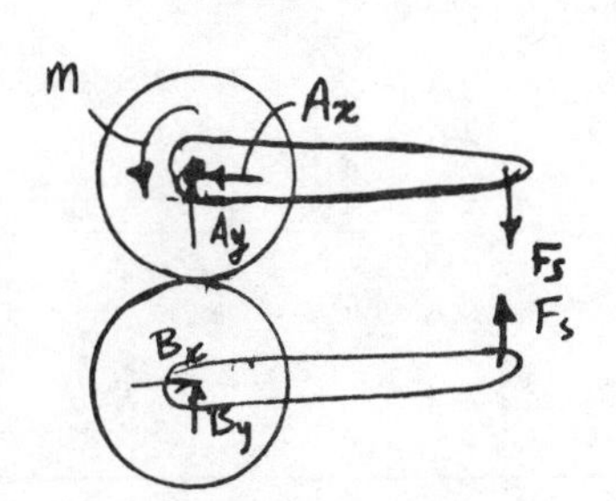

$$T_1 + \Sigma U_{1-2} = T_2$$

$$0+\left\{20\left(\frac{\pi}{4}\right)-\frac{1}{2}(200)(0.2828)^2\right\}=2\left\{\frac{1}{2}\left[(2)(0.05)^2+(2)(0.04)^2\right]\omega^2\right\}$$

$\omega = 30.7$ rad/s **Ans**

Note that work is done by the tangential force between the gears since each move. For the system, though, this force is equal but opposite and the work cancels.

R2-39. The 5-lb rod AB supports the 3-lb disk at its end. If the disk is given an angular velocity ω_D = 8 rad/s while the rod is held stationary and then released, determine the angular velocity of the rod after the disk has stopped spinning relative to the rod due to frictional resistance at the bearing A. Motion is in the *horizontal plane*. Neglect friction at the fixed bearing B.

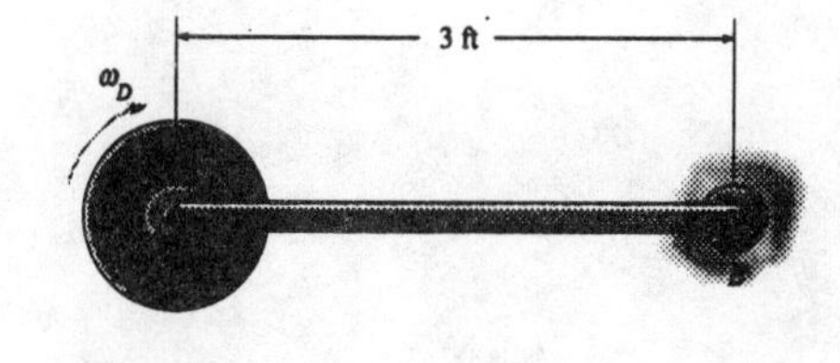

Conservation of momentum :

$$\circlearrowleft + \Sigma(H_B)_1 = \Sigma(H_B)_2$$

$$\left[\frac{1}{2}\left(\frac{3}{32.2}\right)(0.5)^2\right](8)+0=\left[\frac{1}{3}\left(\frac{5}{32.2}\right)(3)^2\right]\omega$$

$$+\left[\frac{1}{2}\left(\frac{3}{32.2}\right)(0.5)^2\right]\omega+\left(\frac{3}{32.2}\right)(3\omega)(3)$$

$\omega = 0.0708$ rad/s **Ans**

***R2-40.** If link AB is rotating at $\omega_{AB} = 3$ rad/s, determine the angular velocity of link CD at the instant shown.

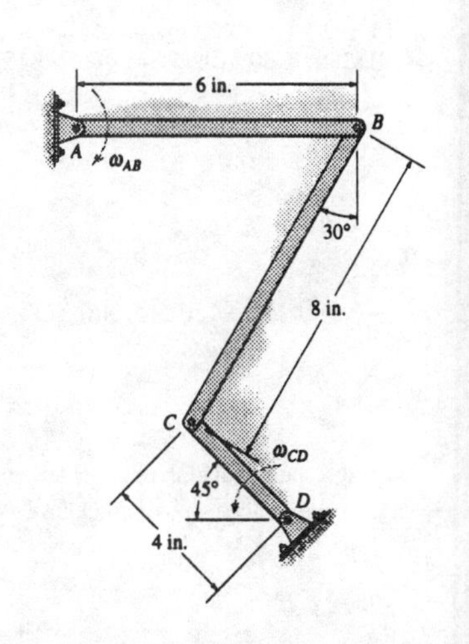

$$\mathbf{v}_B = \omega_{AB} \times \mathbf{r}_{B/A}$$

$$\mathbf{v}_C = \omega_{CD} \times \mathbf{r}_{C/D}$$

$$\mathbf{v}_C = \mathbf{v}_B + \omega_{BC} \times \mathbf{r}_{C/B}$$

$$(\omega_{CD}\mathbf{k})\times(-4\cos 45°\mathbf{i} + 4\sin 45°\mathbf{j}) = (-3\mathbf{k})\times(6\mathbf{i}) + (\omega_{BC}\mathbf{k})\times(-8\sin 30°\mathbf{i} - 8\cos 30°\mathbf{j})$$

$$-2.828\omega_{CD} = 0 + 6.928\omega_{BC}$$

$$-2.828\omega_{CD} = -18 - 4\omega_{BC}$$

Solving,

$\omega_{BC} = -1.65$ rad/s

$\omega_{CD} = 4.03$ rad/s **Ans**

R2-41. If link CD is rotating at $\omega_{CD} = 5$ rad/s, determine the angular velocity of link AB at the instant shown.

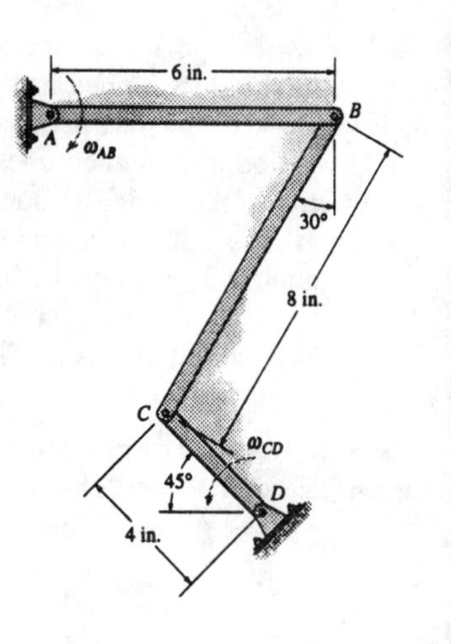

$$\mathbf{v}_B = \omega_{AB} \times \mathbf{r}_{B/A}$$

$$\mathbf{v}_C = \omega_{CD} \times \mathbf{r}_{C/D}$$

$$\mathbf{v}_B = \mathbf{v}_C + \omega_{BC} \times \mathbf{r}_{B/C}$$

$$(-\omega_{AB}\mathbf{k}) \times (6\mathbf{i}) = (5\mathbf{k}) \times (-4\cos 45^\circ\mathbf{i} + 4\sin 45^\circ\mathbf{j}) + (\omega_{BC}\mathbf{k}) \times (8\sin 30^\circ\mathbf{i} + 8\cos 30^\circ\mathbf{j})$$

$$0 = -14.142 - 6.9282\omega_{BC}$$

$$-6\omega_{AB} = -14.142 + 4\omega_{BC}$$

Solving,

$\omega_{AB} = 3.72$ rad/s **Ans**

$\omega_{BC} = -2.04$ rad/s

R2-42. The 15-kg disk is pinned at O and is initially at rest. If a 10-g bullet is fired into the disk with a velocity of 200 m/s, as shown, determine the maximum angle θ to which the disk swings. The bullet becomes embedded in the disk.

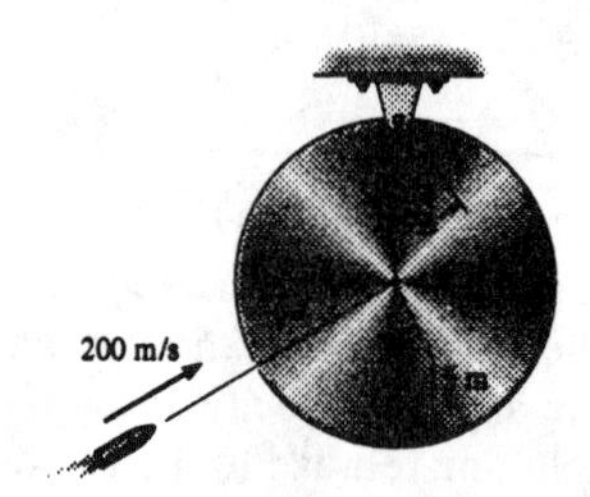

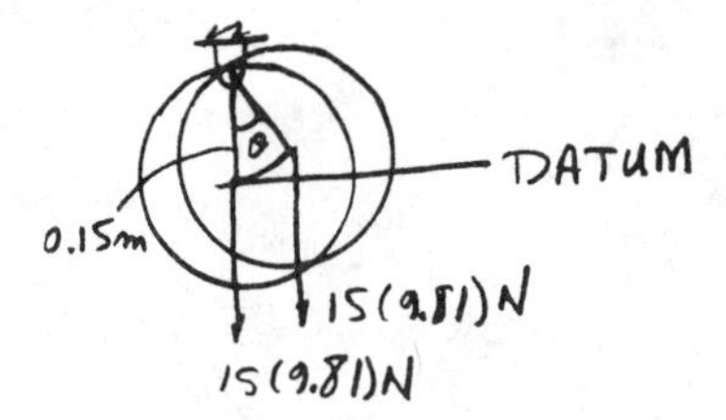

$$\circlearrowleft + (H_O)_1 = (H_O)_2$$

$$0.01(200\cos 30^\circ)(0.15) = \left[\frac{1}{2}(15)(0.15)^2 + 15(0.15)^2\right]\omega$$

$$\omega = 0.5132 \text{ rad/s}$$

$$T_1 + V_1 = T_2 + V_2$$

$$\frac{1}{2}\left[\frac{1}{2}(15)(0.15)^2 + 15(0.15)^2\right](0.5132)^2 + 0 = 0 + 15(9.81)(0.15)(1 - \cos\theta)$$

$\theta = 4.45^\circ$ **Ans**

Note that the calculation neglects the small mass of the bullet after it becomes embedded in the plate, since its position in the plate is not specified.

R2-43. The disk is rotating at a constant rate of 4 rad/s as it falls freely, its center G having an acceleration of $32.2\ \text{ft/s}^2$. Determine the accelerations of points A and B on the rim of the disk at the instant shown.

$$\mathbf{a}_A = \mathbf{a}_G + (\mathbf{a}_{A/G})_t + (\mathbf{a}_{A/G})_n$$

$$\left[\underset{\rightarrow}{(a_A)_x}\right] + \left[\underset{\uparrow}{(a_A)_y}\right] = \left[\underset{\downarrow}{32.2}\right] + 0 + \left[\underset{\downarrow}{(4)^2(1.5)}\right]$$

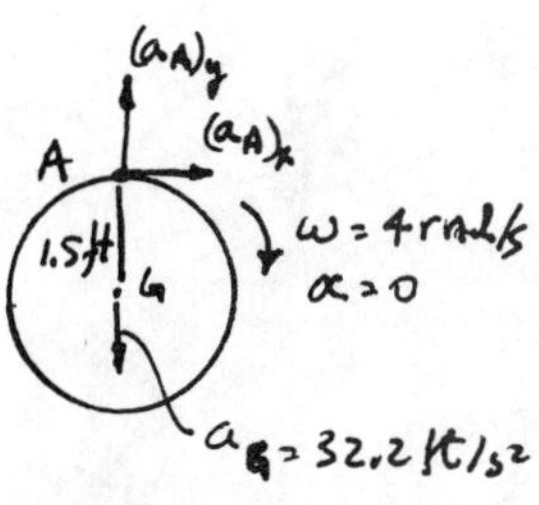

$(\overset{+}{\rightarrow})$ $\quad (a_A)_x = 0$

$(+\uparrow)$ $\quad (a_A)_y = -32.2 - (4)^2(1.5) = -56.2\ \text{ft/s}^2 = 56.2\ \text{ft/s}^2 \downarrow$

$$a_A = (a_A)_y = 56.2\ \text{ft/s}^2 \downarrow \qquad \textbf{Ans}$$

$$\mathbf{a}_B = \mathbf{a}_G + (\mathbf{a}_{B/G})_t + (\mathbf{a}_{B/G})_n$$

$$\left[\underset{\rightarrow}{(a_B)_x}\right] + \left[\underset{\uparrow}{(a_B)_y}\right] = \left[\underset{\downarrow}{32.2}\right] + 0 + \left[\underset{\leftarrow}{(4)^2(1.5)}\right]$$

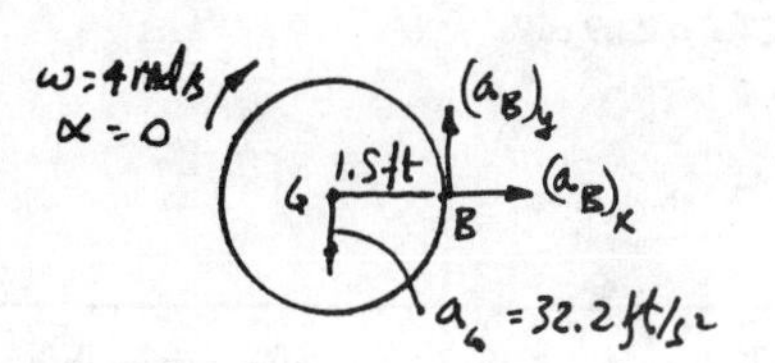

$(\overset{+}{\rightarrow})$ $\quad (a_B)_x = -(4)^2(1.5) = -24\ \text{ft/s}^2 = 24\ \text{ft/s}^2 \leftarrow$

$(+\uparrow)$ $\quad (a_B)_y = -32.2\ \text{ft/s}^2 = 32.2\ \text{ft/s}^2 \downarrow$

$$a_B = \sqrt{(a_B)_x^2 + (a_B)_y^2} = \sqrt{24^2 + 32.2^2} = 40.2\ \text{ft/s}^2 \qquad \textbf{Ans}$$

$$\theta = \tan^{-1}\left(\frac{(a_B)_y}{(a_B)_x}\right) = \tan^{-1}\left(\frac{32.2}{24}\right) = 53.3^\circ \qquad \textbf{Ans}$$

***R2-44.** The operation of "reverse" for a three-speed automotive transmission is illustrated schematically in the figure. If the crankshaft G is turning with an angular speed of 60 rad/s, determine the angular speed of the drive shaft H. Each of the gears rotates about a fixed axis. Note that gears A and B, C and D, E and F are in mesh. The radius of each of these gears is reported in the figure.

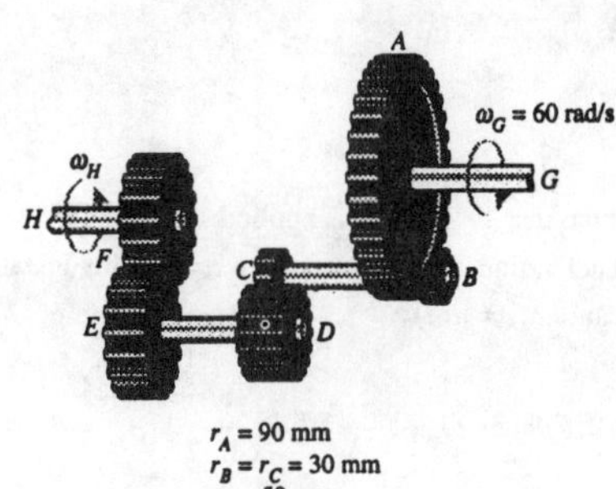

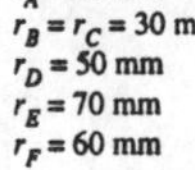

$$\omega_C = \omega_B = \frac{r_A}{r_B}\omega_G = \frac{90}{30}(60) = 180\ \text{rad/s}$$

$$\omega_E = \omega_D = \frac{r_C}{r_D}\omega_C = \frac{30}{50}(180) = 108\ \text{rad/s}$$

$$\omega_H = \frac{r_E}{r_F}\omega_E = \frac{70}{60}(108) = 126\ \text{rad/s} \qquad \textbf{Ans}$$

R2-45. Shown is the internal gearing of a "spinner" used for drilling wells. With constant angular acceleration, the motor M rotates the shaft S to 100 rev/min in $t = 2$ s starting from rest. Determine the angular acceleration of the drill-pipe connection D and the number of revolutions it makes during the 2-s startup.

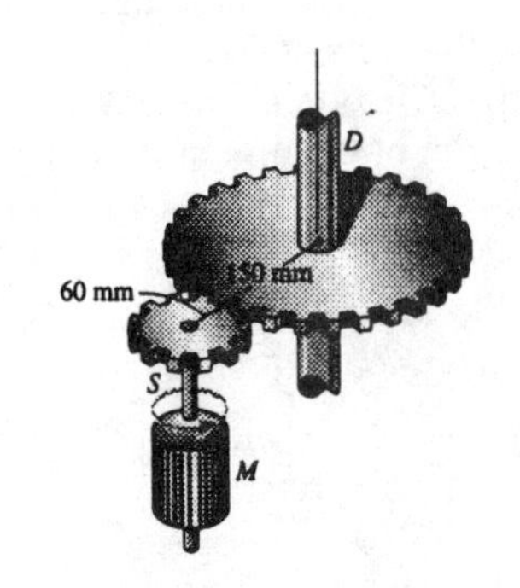

For shaft S:

$$\omega = \omega_0 + \alpha_c t$$

$$\frac{100(2\pi)}{60} = 0 + \alpha_S(2) \qquad \alpha_S = 5.236 \text{ rad/s}^2$$

$$\theta = \theta_0 + \omega_0 t + \frac{1}{2}\alpha_c t^2$$

$$\theta_S = 0 + 0 + \frac{1}{2}(5.236)(2)^2 = 10.472 \text{ rad}$$

For connection D:

$$\alpha_D = \frac{r_S}{r_D}\alpha_S = \frac{60}{150}(5.236) = 2.09 \text{ rad/s}^2 \qquad \textbf{Ans}$$

$$\theta_D = \frac{r_S}{r_D}\theta_S = \frac{60}{150}(10.472) = 4.19 \text{ rad} = 0.667 \text{ rev} \qquad \textbf{Ans}$$

R2-46. The plate weighs 40 lb and is supported by a roller at A. If a horizontal force of $F = 70$ lb is suddenly applied to the roller, determine the acceleration of the center of the roller at the instant the force is applied. The plate has a moment of inertia about its center of mass of $I_G = 0.414$ slug·ft². Neglect the weight and the size d of the roller.

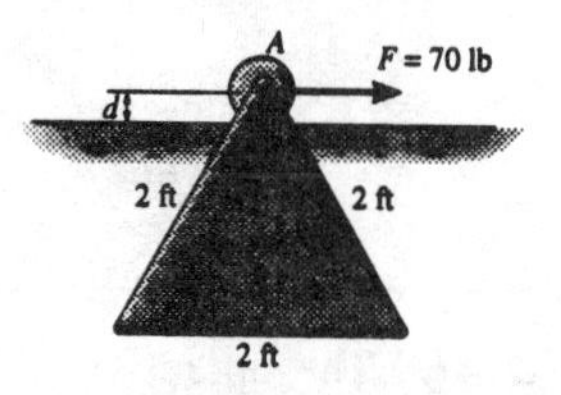

At the instant force **F** is applied, the normal component of acceleration of G is zero since ω=0 at this instant. As a result $a_G = (a_G)_t$.

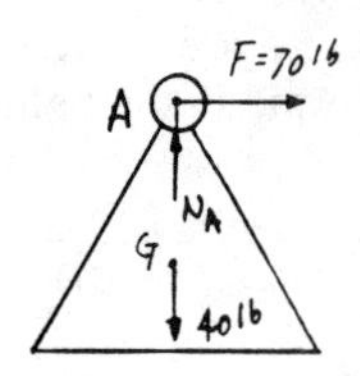

$$\overset{+}{\rightarrow}\Sigma F_x = m(a_G)_x; \qquad 70 = \left(\frac{40}{32.2}\right)a_G \qquad a_G = 56.35 \text{ ft/s}^2$$

$$\circlearrowleft + \Sigma M_A = \Sigma (M_k)_A; \qquad 0 = \left(\frac{40}{32.2}\right)(56.35)\left(\frac{2}{3}\right)(2\sin 60°) - 0.414\alpha$$

$$\alpha = 195.2 \text{ rad/s}^2$$

=

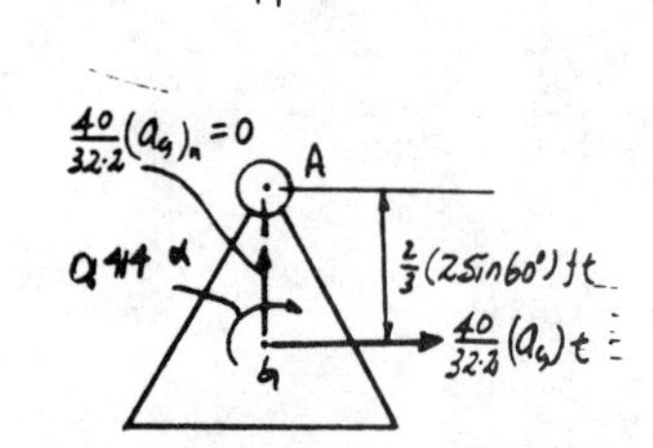

$$\mathbf{a}_A = \mathbf{a}_G + (\mathbf{a}_{A/G})_t + (\mathbf{a}_{A/G})_n$$

$$\left[\underset{\rightarrow}{a_A}\right] = \left[\underset{\rightarrow}{56.35}\right] + \left[\underset{\rightarrow}{195.2\left(\frac{2}{3}\right)(2\sin 60°)}\right] + [0]$$

$$(\overset{+}{\rightarrow}) \qquad a_A = 56.35 + 225.4 = 282 \text{ ft/s}^2 \qquad \textbf{Ans}$$

R2-47. The 15-kg cylinder is rotating with an angular velocity of $\omega = 40$ rad/s. If a force $F = 6$ N is applied to link AB, as shown, determine the time needed to stop the rotation. The coefficient of kinetic friction between AB and the cylinder is $\mu_k = 0.4$.

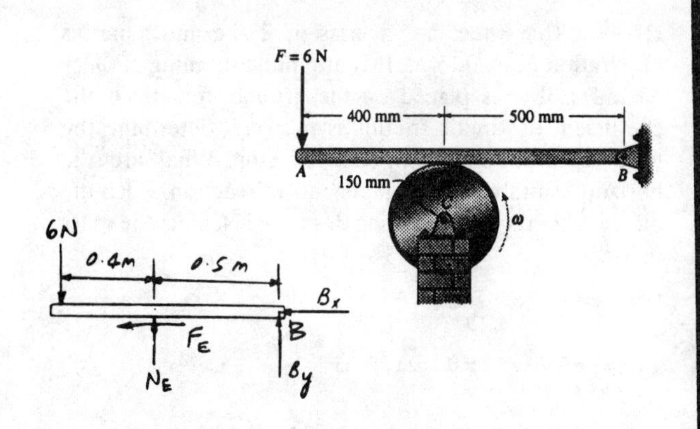

For link AB

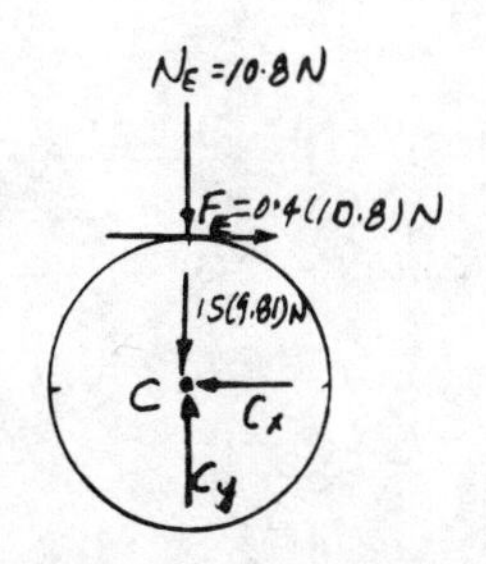

$\circlearrowleft + \Sigma M_B = 0; \quad 6(0.9) - N_E(0.5) = 0 \quad N_E = 10.8$ N

$I_C = \frac{1}{2}mr^2 = \frac{1}{2}(15)(0.15)^2 = 0.16875$ kg · m²

$\circlearrowleft + \Sigma M_C = I_C\alpha; \quad -0.4(10.8)(0.15) = 0.16875(\alpha) \quad \alpha = -3.84$ rad/s²

$\circlearrowleft + \ \omega = \omega_0 + \alpha t$

$0 = 40 + (-3.84)t$

$t = 10.4$ s **Ans**

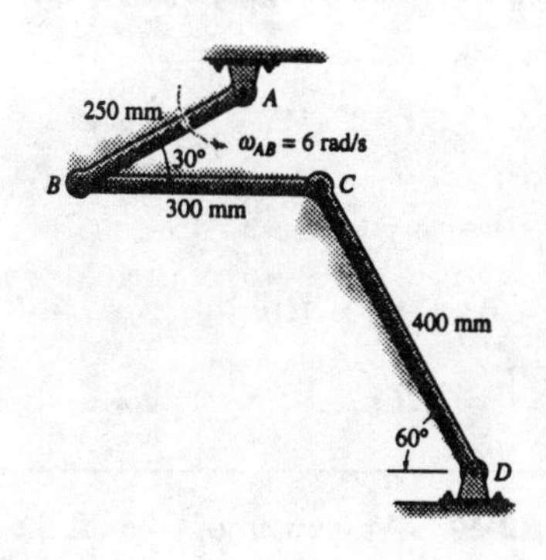

***R2-48.** If link AB is rotating at $\omega_{AB} = 6$ rad/s, determine the angular velocities of links BC and CD at the instant shown.

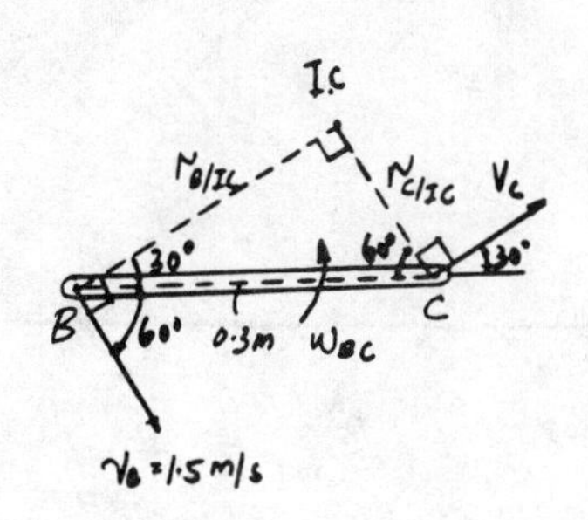

Link AB rotates about the fixed point A. Hence

$v_B = \omega_{AB} r_{AB} = 6(0.25) = 1.5$ m/s

For link BC

$r_{B/IC} = 0.3\cos 30° = 0.2598$ m $\quad r_{C/IC} = 0.3\cos 60° = 0.15$ m

$\omega_{BC} = \frac{v_B}{r_{B/IC}} = \frac{1.5}{0.2598} = 5.77$ rad/s **Ans**

$v_C = \omega_{BC} r_{C/IC} = 5.77(0.15) = 0.8660$ m/s

Link CD rotates about the fixed point D. Hence

$v_C = \omega_{CD} r_{CD}$

$0.8660 = \omega_{CD}(0.4) \quad \omega_{CD} = 2.17$ rad/s **Ans**

R2-49. The wheel has a mass of 25 kg and a radius of gyration $k_B = 0.15$ m. It is originally spinning at $\omega_1 = 40$ rad/s. If it is placed on the ground, for which the coefficient of kinetic friction is $\mu_k = 0.5$, determine the time required for the motion to stop. What are the horizontal and vertical components of reaction which the pin at A exerts on AB during this time? Neglect the mass of AB.

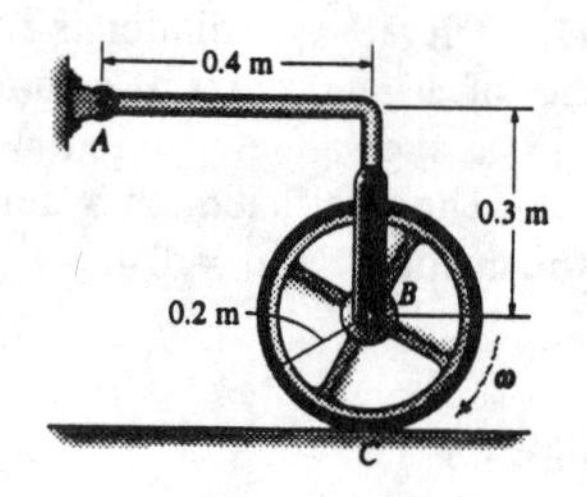

AB is a two – force member.

$\bar{I}_B = mk_B^2 = 25(0.15)^2 = 0.5625 \text{ kg}\cdot\text{m}^2$

$+\uparrow \Sigma F_y = m(a_G)_y;\quad (\tfrac{3}{5})F_{AB} + N_C - 25(9.81) = 0 \qquad (1)$

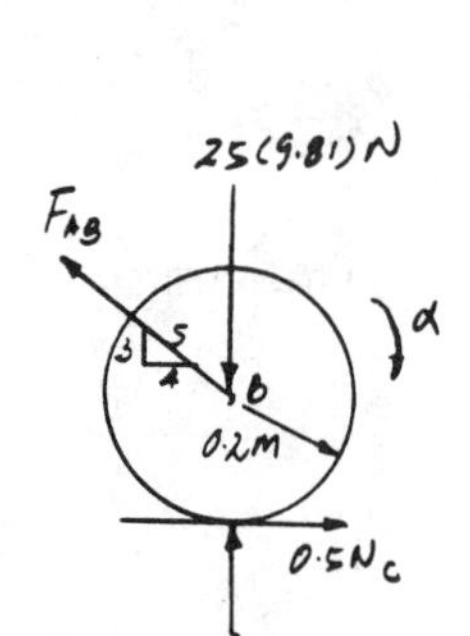

$\xrightarrow{+} \Sigma F_x = m(a_G)_x;\quad 0.5N_C - (\tfrac{4}{5})F_{AB} = 0 \qquad (2)$

$\circlearrowleft + \Sigma M_B = I_B\,\alpha;\quad 0.5N_C(0.2) = -0.5625(\alpha) \qquad (3)$

Solving Eqs. (1), (2) and (3) yields :

$F_{AB} = 111.48$ N $\quad N_C = 178.4$ N

$\alpha = -31.71$ rad/s^2

$A_x = \frac{4}{5}F_{AB} = 0.8(111.48) = 89.2 \text{ N} \leftarrow$ **Ans**

$A_y = \frac{3}{5}F_{AB} = 0.6(111.48) = 66.9 \text{ N} \uparrow$ **Ans**

$+\ \omega = \omega_0 + \alpha_c t$

$0 = 40 + (-31.71)t$

$t = 1.26$ s **Ans**

R2-50. At the *start* of take-off, the propeller on the 2-Mg plane exerts a horizontal thrust of 600 N on the plane. Determine the plane's acceleration and the vertical reactions at the nose wheel A and each of the *two* wing wheels B. Neglect the lifting force of the wings since the plane is originally at rest. The mass center is at G.

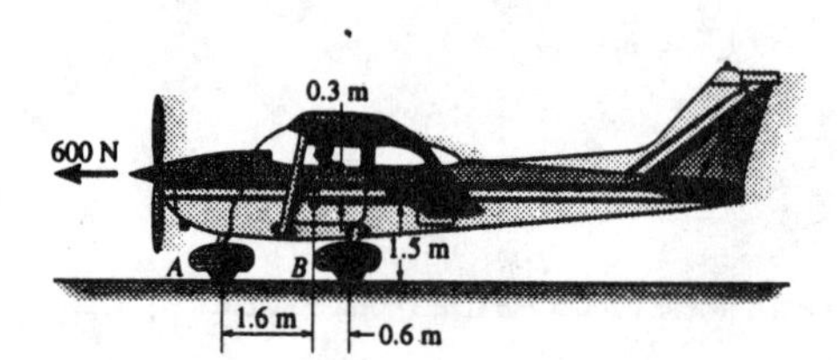

$\xleftarrow{+} \Sigma F_x = m(a_G)_x;\quad 600 = 2(10^3)a_G \quad a_G = 0.3 \text{ m/s}^2$ **Ans**

$+\uparrow \Sigma F_y = m(a_G)_y;\quad N_A + 2N_B - 2(10^3)(9.81) = 0 \qquad (1)$

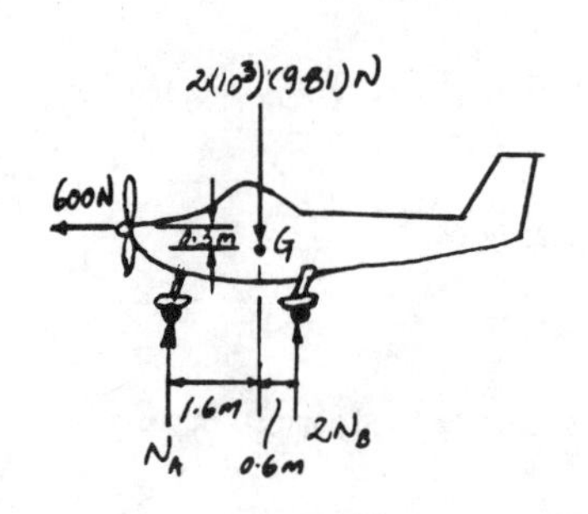

$\circlearrowleft + \Sigma M_G = 0;\quad 2N_B(0.6) + 600(0.3) - N_A(1.6) = 0 \qquad (2)$

Solving Eqs. (1) and (2) yields :

$N_B = 7094 \text{ N} = 7.09$ kN $\quad N_A = 5433 \text{ N} = 5.43$ kN **Ans**

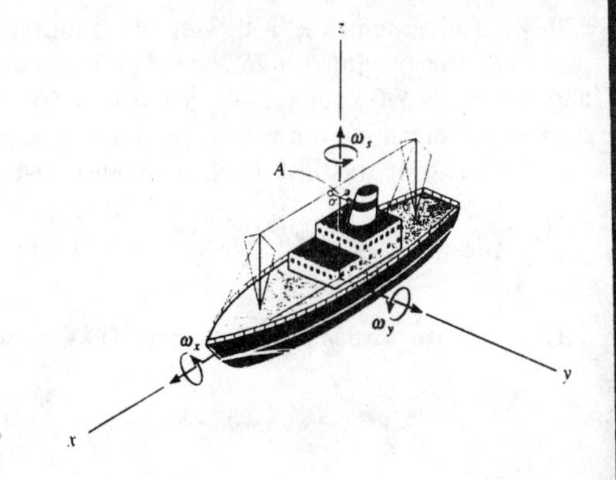

20-1. The anemometer located on the ship at A is spinning about its own axis at a rate ω_s, while the ship is rolling about the x axis at the rate ω_x and about the y axis at the rate ω_y. Determine the angular velocity and angular acceleration of the anemometer at the instant the ship is level as shown. Assume that the magnitudes of all components of angular velocity are constant and that the rolling motion caused by the sea is independent in the x and y directions.

$\omega = \omega_x \mathbf{i} + \omega_y \mathbf{j} + \omega_s \mathbf{k}$ **Ans**

Let $\Omega = \omega_x \mathbf{i} + \omega_y \mathbf{j}$.

Since ω_x and ω_y are independent of one another, they do not change their direction or magnitude. Thus,

$$\alpha = \dot{\omega} = \left(\dot{\omega}\right)_{xyz} + (\omega_x + \omega_y) \times \omega_s$$

$$\alpha = \mathbf{0} + (\omega_x \mathbf{i} + \omega_y \mathbf{j}) \times (\omega_s \mathbf{k})$$

$\alpha = \omega_y \omega_s \mathbf{i} - \omega_x \omega_s \mathbf{j}$ **Ans**

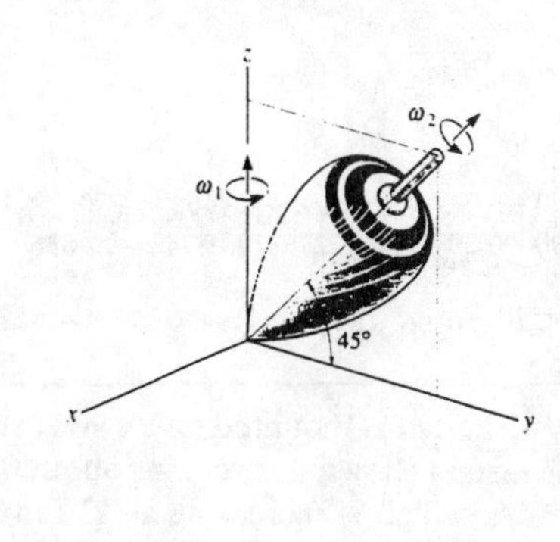

20-2. The motion of the top is such that at the instant shown it is rotating about the z axis at $\omega_1 = 0.6$ rad/s, while it is spinning at $\omega_2 = 8$ rad/s. Determine the angular velocity and angular acceleration of the top at this instant. Express the result as a Cartesian vector.

$$\omega = \omega_1 + \omega_2$$

$$\omega = 0.6\mathbf{k} + 8\cos 45^\circ \mathbf{j} + 8\sin 45^\circ \mathbf{k}$$

$\omega = \{5.66\mathbf{j} + 6.26\mathbf{k}\}$ rad/s **Ans**

$$\dot{\omega} = \dot{\omega}_1 + \dot{\omega}_2$$

Let x, y, z axes have angular velocity of $\Omega = \omega_1$, thus

$$\dot{\omega}_1 = 0$$

$$\dot{\omega}_2 = \left(\dot{\omega}_2\right)_{xyz} + (\omega_1 \times \omega_2) = \mathbf{0} + (0.6\mathbf{k}) \times (8\cos 45^\circ \mathbf{j} + 8\sin 45^\circ \mathbf{k}) = -3.394\mathbf{i}$$

$\alpha = \dot{\omega} = \{-3.39\mathbf{i}\}$ rad/s^2 **Ans**

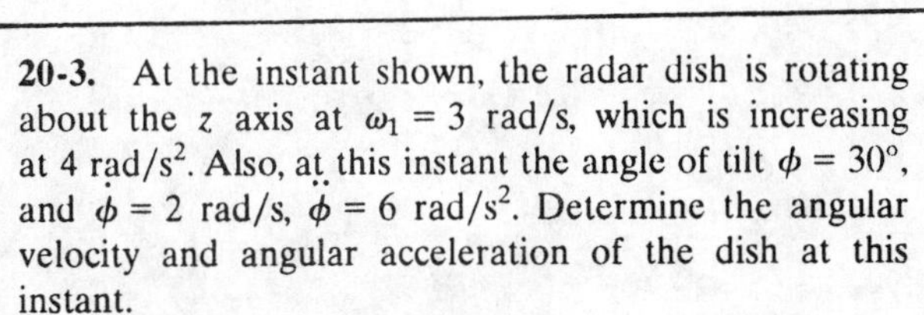

20-3. At the instant shown, the radar dish is rotating about the z axis at $\omega_1 = 3$ rad/s, which is increasing at 4 rad/s^2. Also, at this instant the angle of tilt $\phi = 30^\circ$, and $\dot{\phi} = 2$ rad/s, $\ddot{\phi} = 6$ rad/s^2. Determine the angular velocity and angular acceleration of the dish at this instant.

$\omega = \omega_1 + \dot{\phi} = 3\mathbf{k} + 2\mathbf{i} = \{2\mathbf{i} + 3\mathbf{k}\}$ rad/s **Ans**

For ω_2, $\Omega = \omega_1 = \{3\mathbf{k}\}$ rad/s.

$$(\dot{\omega}_2)_{XYZ} = (\dot{\omega}_2)_{xyz} + \Omega \times \omega_2$$

$$= (6\mathbf{i}) + (3\mathbf{k}) \times (2\mathbf{i})$$

$$= \{6\mathbf{i} + 6\mathbf{j}\} \text{ rad/s}^2$$

For ω_1, $\Omega = \mathbf{0}$.

$$(\dot{\omega}_1)_{XYZ} = (\dot{\omega}_1)_{xyz} + \Omega \times \omega_1 = (4\mathbf{k}) + \mathbf{0} = \{4\mathbf{k}\} \text{ rad/s}^2$$

$$\alpha = \dot{\omega} = (\dot{\omega}_1)_{XYZ} + (\dot{\omega}_2)_{XYZ}$$

$\alpha = 4\mathbf{k} + (6\mathbf{i} + 6\mathbf{j}) = \{6\mathbf{i} + 6\mathbf{j} + 4\mathbf{k}\}$ rad/s^2 **Ans**

***20-4.** The antenna is following the motion of a jet plane. At the instant $\theta = 25°$ and $\phi = 75°$, the constant angular rates of change are $\dot{\theta} = 0.4$ rad/s and $\dot{\phi} = 0.6$ rad/s. Determine the velocity and acceleration of the signal horn A at this instant. The distance OA is 0.8 m.

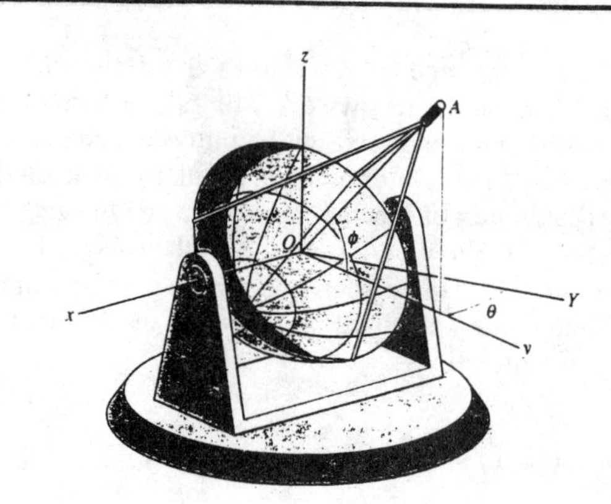

$$\omega = \dot{\phi}\mathbf{i} - \dot{\theta}\mathbf{k} = 0.6\mathbf{i} - 0.4\mathbf{k}$$

Let the x, y, z axes have an angular velocity of $\Omega = -\dot{\theta}\mathbf{k}$. Then

$$\dot{\omega} = \left(\dot{\omega}\right)_{xyz} + \left(-\dot{\theta}\mathbf{k}\right)\times\left(\dot{\phi}\mathbf{i} - \dot{\theta}\mathbf{k}\right) = \mathbf{0} + -\dot{\phi}\dot{\theta}\mathbf{j} = -(0.4)(0.6)\mathbf{j}$$

$$\alpha = \dot{\omega} = -0.24\mathbf{j}$$

Hence,

$$\mathbf{r}_A = 0.8\cos 75°\mathbf{j} + 0.8\sin 75°\mathbf{k} = 0.2071\mathbf{j} + 0.7727\mathbf{k}$$

$$\mathbf{v}_A = \omega\times\mathbf{r}_A = \begin{vmatrix} \mathbf{i} & \mathbf{j} & \mathbf{k} \\ 0.6 & 0 & -0.4 \\ 0 & 0.2071 & 0.7727 \end{vmatrix}$$

$$\mathbf{v}_A = \{0.0828\mathbf{i} - 0.464\mathbf{j} + 0.124\mathbf{k}\} \text{ m/s} \qquad \textbf{Ans}$$

$$\mathbf{a}_A = \alpha\times\mathbf{r}_A + \omega\times\mathbf{v}_A$$

$$\mathbf{a}_A = \begin{vmatrix} \mathbf{i} & \mathbf{j} & \mathbf{k} \\ 0 & -0.24 & 0 \\ 0 & 0.2071 & 0.7727 \end{vmatrix} + \begin{vmatrix} \mathbf{i} & \mathbf{j} & \mathbf{k} \\ 0.6 & 0 & -0.4 \\ 0.0828 & -0.4636 & 0.1242 \end{vmatrix}$$

$$\mathbf{a}_A = \{-0.371\mathbf{i} - 0.108\mathbf{j} - 0.278\mathbf{k}\} \text{ m/s}^2 \qquad \textbf{Ans}$$

20-5. The fan is mounted on a swivel support such that at the instant shown it is rotating about the z axis at $\omega_1 = 0.8$ rad/s, which is increasing at 12 rad/s^2. The blade is spinning at $\omega_2 = 16$ rad/s, which is decreasing at 2 rad/s^2. Determine the angular velocity and angular acceleration of the blade at this instant.

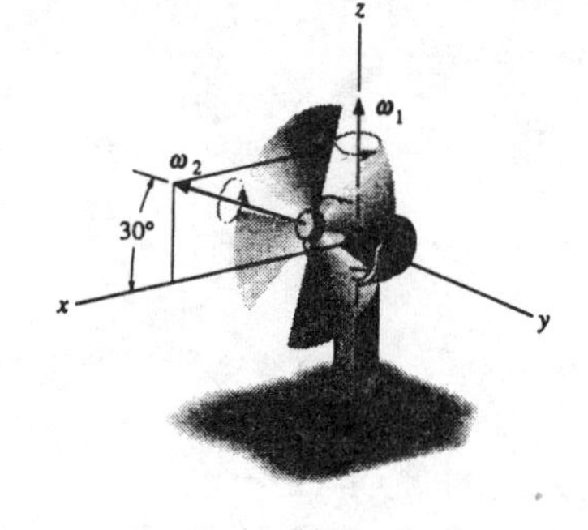

$$\omega = \omega_1 + \omega_2$$

$$= 0.8\mathbf{k} + (16\cos 30°\mathbf{i} + 16\sin 30°\mathbf{k})$$

$$= \{13.9\mathbf{i} + 8.80\mathbf{k}\} \text{ rad/s} \qquad \textbf{Ans}$$

For ω_2, $\Omega = \omega_1 = \{0.8\mathbf{k}\}$ rad/s.

$$(\dot{\omega}_2)_{XYZ} = (\dot{\omega}_2)_{xyz} + \Omega\times\omega_2$$

$$= (-2\cos 30°\mathbf{i} - 2\sin 30°\mathbf{k}) + (0.8\mathbf{k})\times(16\cos 30°\mathbf{i} + 16\sin 30°\mathbf{k})$$

$$= \{-1.7320\mathbf{i} + 11.0851\mathbf{j} - 1\mathbf{k}\} \text{ rad/s}^2$$

For ω_1, $\Omega = \mathbf{0}$.

$$(\dot{\omega}_1)_{XYZ} = (\dot{\omega}_1)_{xyz} + \Omega\times\omega_1$$

$$= (12\mathbf{k}) + \mathbf{0}$$

$$= \{12\mathbf{k}\} \text{ rad/s}^2$$

$$\alpha = \dot{\omega} = (\dot{\omega}_1)_{XYZ} + (\dot{\omega}_2)_{XYZ}$$

$$\alpha = 12\mathbf{k} + (-1.7320\mathbf{i} + 11.0851\mathbf{j} - 1\mathbf{k})$$

$$= \{-1.73\mathbf{i} + 11.1\mathbf{j} + 11.0\mathbf{k}\} \text{ rad/s}^2 \qquad \textbf{Ans}$$

20-6. Gear B is connected to the rotating shaft, while the plate gear A is fixed. If the shaft is turning at a constant rate of $\omega_z = 10$ rad/s about the z axis, determine the magnitudes of the angular velocity and the angular acceleration of gear B. Also, determine the magnitudes of the velocity and acceleration of point P.

$\omega_z = 10$ rad/s

$\omega_y = -10\tan 75.96° = -40$ rad/s

$\omega_x = 0$

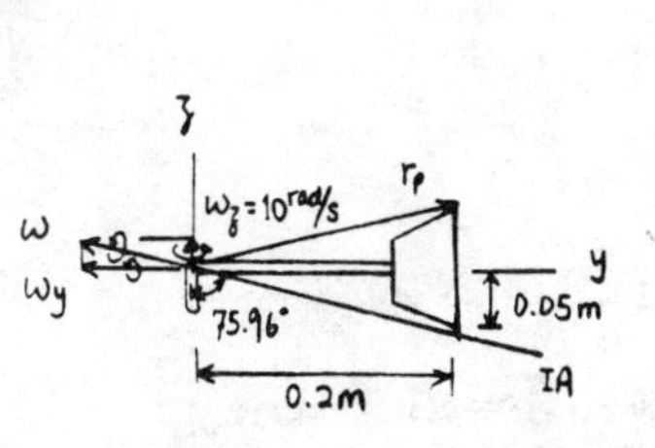

$\omega = \{-40\mathbf{j} + 10\mathbf{k}\}$ rad/s

$\omega = \sqrt{(-40)^2 + (10)^2} = 41.2$ rad/s **Ans**

$\mathbf{r}_P = \{0.2\mathbf{j} + 0.05\mathbf{k}\}$ m

$\mathbf{v}_P = \omega \times \mathbf{r}_P = (-40\mathbf{j} + 10\mathbf{k}) \times (0.2\mathbf{j} + 0.05\mathbf{k})$

$\mathbf{v}_P = \{-4\mathbf{i}\}$ m/s

$v_P = 4.00$ m/s **Ans**

Let $\Omega = \omega_z$,

$\dot{\omega} = \left(\dot{\omega}\right)_{xyz} + \Omega \times \omega$

$= \mathbf{0} + (10\mathbf{k}) \times (-40\mathbf{j} + 10\mathbf{k}) = \{400\mathbf{i}\}$ rad/s^2

$\alpha = \dot{\omega} = 400$ rad/s^2 **Ans**

$\mathbf{a}_P = \alpha \times \mathbf{r}_P + \omega \times \mathbf{v}_P = (400\mathbf{i}) \times (0.2\mathbf{j} + 0.05\mathbf{k}) + (-40\mathbf{j} + 10\mathbf{k}) \times (-4\mathbf{i})$

$= \{-60\mathbf{j} - 80\mathbf{k}\}$ m/s^2

$a_P = \sqrt{(-60)^2 + (-80)^2} = 100$ m/s^2 **Ans**

20-7. Gears A and B are fixed, while gears C and D are free to rotate on the shaft S. If the shaft is turning about the z axis at a constant rate of $\omega_1 = 4$ rad/s, determine the angular velocity and angular acceleration of gear C.

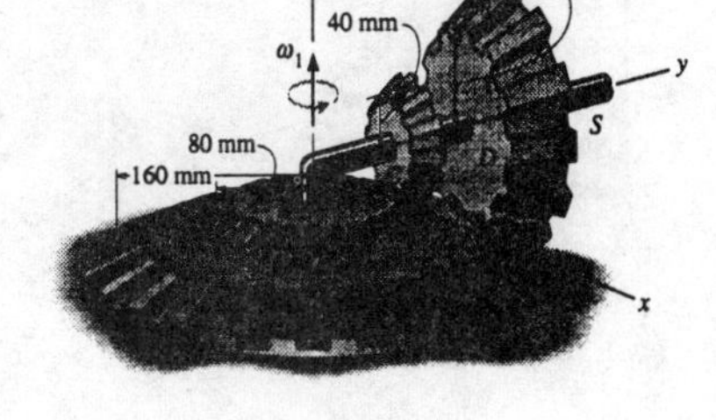

The resultant angular velocity $\omega = \omega_1 + \omega_2$ is always directed along the instantaneous axis of zero velocity IA.

$$\omega = \omega_1 + \omega_2$$

$$\frac{2}{\sqrt{5}}\omega\mathbf{j} - \frac{1}{\sqrt{5}}\omega\mathbf{k} = 4\mathbf{k} + \omega_2\mathbf{j}$$

Equating $\mathbf{j}$ and $\mathbf{k}$ components

$$-\frac{1}{\sqrt{5}}\omega = 4 \qquad \omega = -8.944 \text{ rad/s}$$

$$\omega_2 = \frac{2}{\sqrt{5}}(-8.944) = -8.0 \text{ rad/s}$$

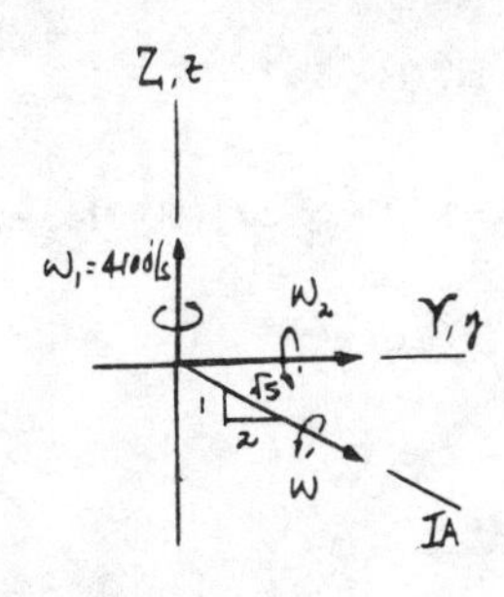

Hence $\omega = \frac{2}{\sqrt{5}}(-8.944)\mathbf{j} - \frac{1}{\sqrt{5}}(-8.944)\mathbf{k} = \{-8.0\mathbf{j} + 4.0\mathbf{k}\}$ rad/s **Ans**

For ω_2, $\Omega = \omega_1 = \{4\mathbf{k}\}$ rad/s.

$(\dot{\omega}_2)_{XYZ} = (\dot{\omega}_2)_{xyz} + \Omega \times \omega_2$

$= \mathbf{0} + (4\mathbf{k}) \times (-8\mathbf{j})$

$= \{32\mathbf{i}\}$ rad/s^2

For ω_1, $\Omega = \mathbf{0}$.

$(\dot{\omega}_1)_{XYZ} = (\dot{\omega}_1)_{xyz} + \Omega \times \omega_1 = \mathbf{0} + \mathbf{0} = \mathbf{0}$

$\alpha = \dot{\omega} = (\dot{\omega}_1)_{XYZ} + (\dot{\omega}_2)_{XYZ}$

$\alpha = \mathbf{0} + (32\mathbf{i}) = \{32\mathbf{i}\}$ rad/s^2 **Ans**

***20-8.** The propeller of an airplane is rotating at a constant speed $\omega_s \mathbf{i}$, while the plane is undergoing a turn at a constant rate ω_t. Determine the angular acceleration of the propeller if (a) the turn is horizontal, i.e., $\omega_t \mathbf{k}$, and (b) the turn is vertical, downward, i.e., $\omega_t \mathbf{j}$.

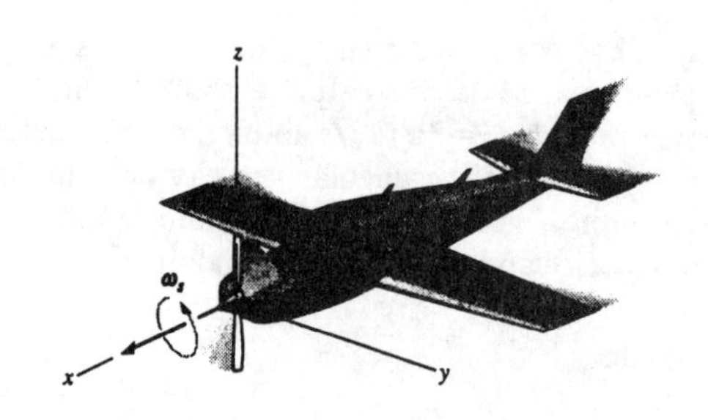

(a) For ω_s, $\Omega = \omega_t \mathbf{k}$.

$$(\dot{\omega}_s)_{XYZ} = (\dot{\omega}_s)_{xyz} + \Omega \times \omega_s$$

$$= 0 + (\omega_t \mathbf{k}) \times (\omega_s \mathbf{i}) = \omega_s \omega_t \mathbf{j}$$

For ω_t, $\Omega = 0$.

$$(\dot{\omega}_t)_{XYZ} = (\dot{\omega}_t)_{xyz} + \Omega \times \omega_t = 0 + 0 = 0$$

$$\alpha = \dot{\omega} = (\dot{\omega}_s)_{XYZ} + (\dot{\omega}_t)_{XYZ}$$

$$\alpha = \omega_s \omega_t \mathbf{j} + 0 = \omega_s \omega_t \mathbf{j}$$ **Ans**

(b) For ω_s, $\Omega = \omega_t \mathbf{j}$.

$$(\dot{\omega}_s)_{XYZ} = (\dot{\omega}_s)_{xyz} + \Omega \times \omega_s$$

$$= 0 + (\omega_t \mathbf{j}) \times (\omega_s \mathbf{i}) = -\omega_s \omega_t \mathbf{k}$$

For ω_t, $\Omega = 0$.

$$(\dot{\omega}_t)_{XYZ} = (\dot{\omega}_t)_{xyz} + \Omega \times \omega_t = 0 + 0 = 0$$

$$\alpha = \dot{\omega} = (\dot{\omega}_s)_{XYZ} + (\dot{\omega}_t)_{XYZ}$$

$$\alpha = -\omega_s \omega_t \mathbf{k} + 0 = -\omega_s \omega_t \mathbf{k}$$ **Ans**

20-9. The conical spool rolls on the plane without slipping. If the axle has an angular velocity of $\omega_1 = 3$ rad/s and an angular acceleration of $\alpha_1 = 2$ rad/s^2 at the instant shown, determine the angular velocity and angular acceleration of the spool at this instant.

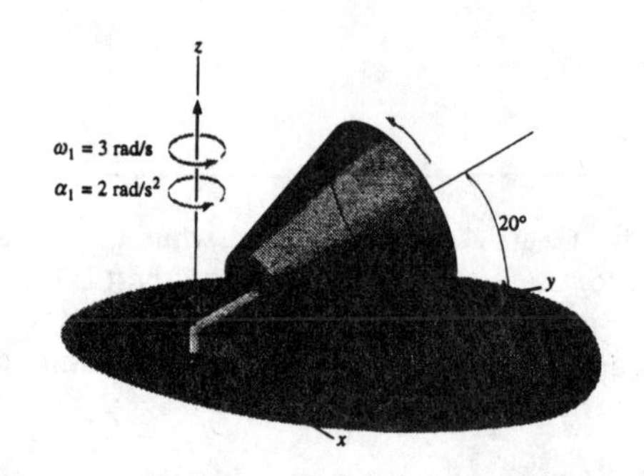

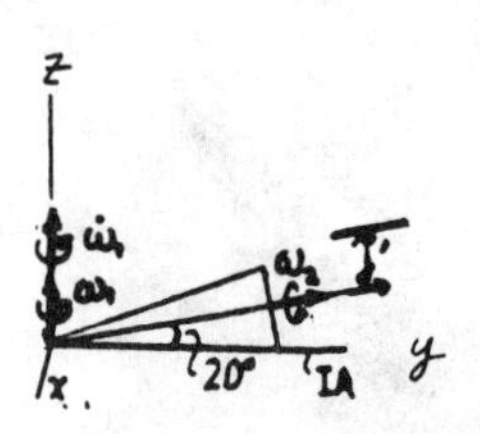

$$\omega_1 = 3 \text{ rad/s}$$

$$\omega_2 = -\frac{3}{\sin 20^\circ} = -8.7714 \text{ rad/s}$$

$$\omega = \omega_1 + \omega_2 = 3\mathbf{k} - 8.7714 \cos 20^\circ \mathbf{j} - 8.7714 \sin 20^\circ \mathbf{k}$$

$$= \{-8.24\mathbf{j}\} \text{ rad/s}$$ **Ans**

$$\left(\dot{\omega}_1\right)_{xyz} = 2 \text{ rad/s}^2$$

$$\left(\dot{\omega}_2\right)_{xyz} = -\frac{2}{\sin 20^\circ} = -5.8476 \text{ rad/s}^2$$

$$\alpha = \dot{\omega} = \left(\dot{\omega}_1\right)_{xyz} + \omega_1 \times \omega_1 + \left(\dot{\omega}_2\right)_{xyz} + \omega_1 \times \omega_2$$

$$= 2\mathbf{k} + 0 + (-5.8476 \cos 20^\circ \mathbf{j} - 5.8476 \sin 20^\circ \mathbf{k}) + (3\mathbf{k}) \times (-8.7714 \cos 20^\circ \mathbf{j} - 8.7714 \sin 20^\circ \mathbf{k})$$

$$\alpha = \{24.7\mathbf{i} - 5.49\mathbf{j}\} \text{ rad/s}^2$$ **Ans**

20-10. If the top gear B is rotating at a constant rate of ω, determine the angular velocity of gear A, which is free to turn about the shaft and rolls on the bottom fixed gear C.

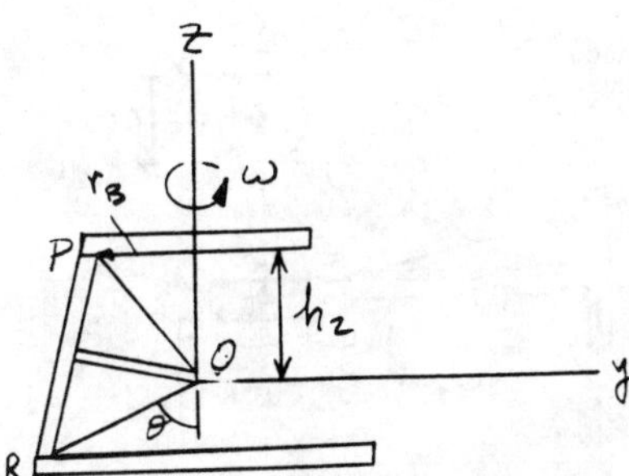

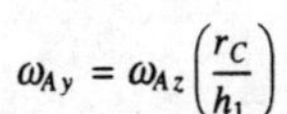

$$\mathbf{v}_P = \omega\mathbf{k}\times(-r_B\mathbf{j}) = \omega r_B\,\mathbf{i}$$

Also,

$$\mathbf{v}_P = \omega_A \times(-r_B\mathbf{j}+h_2\mathbf{k}) = \begin{vmatrix} \mathbf{i} & \mathbf{j} & \mathbf{k} \\ \omega_{Ax} & \omega_{Ay} & \omega_{Az} \\ 0 & -r_B & h_2 \end{vmatrix}$$

$$= (\omega_{Ay}h_2 + \omega_{Az}r_B)\mathbf{i} - (\omega_{Ax}h_2)\mathbf{j} - \omega_{Ax}r_B\,\mathbf{k}$$

Thus,

$$\omega r_B = \omega_{Ay}h_2 + \omega_{Az}r_B \qquad (1)$$

$$0 = \omega_{Ax}h_2$$

$$0 = \omega_{Ax}r_B$$

$$\omega_{Ax} = 0$$

$$\mathbf{v}_R = \mathbf{0} = \begin{vmatrix} \mathbf{i} & \mathbf{j} & \mathbf{k} \\ 0 & \omega_{Ay} & \omega_{Az} \\ 0 & -r_C & -h_1 \end{vmatrix} = (-\omega_{Ay}h_1 + \omega_{Az}r_C)\mathbf{i}$$

$$\omega_{Ay} = \omega_{Az}\left(\frac{r_C}{h_1}\right)$$

From Eq. (1)

$$\omega r_B = \omega_{Az}\left[\left(\frac{r_C h_2}{h_1}\right) + r_B\right]$$

$$\omega_{Az} = \frac{r_B h_1\,\omega}{r_C h_2 + r_B h_1}; \qquad \omega_{Ay} = \left(\frac{r_C}{h_1}\right)\left(\frac{r_B h_1\,\omega}{r_C h_2 + r_B h_1}\right)$$

$$\omega_A = \left(\frac{r_C}{h_1}\right)\left(\frac{r_B h_1\,\omega}{r_C h_2 + r_B h_1}\right)\mathbf{j} + \left(\frac{r_B h_1\,\omega}{r_C h_2 + r_B h_1}\right)\mathbf{k} \qquad \textbf{Ans}$$

20-11. The telescope is mounted on the frame F that allows it to be directed to any point in the sky. At the instant shown, the frame has an angular acceleration of $\alpha_{y'} = 0.2\ \text{rad/s}^2$ and an angular velocity of $\omega_{y'} = 0.3$ rad/s about the y' axis, and $\ddot{\theta} = 0.5\ \text{rad/s}^2$ while $\dot{\theta} = 0.4$ rad/s. Determine the velocity and acceleration of the observing capsule at C at this instant when $\theta = 30°$.

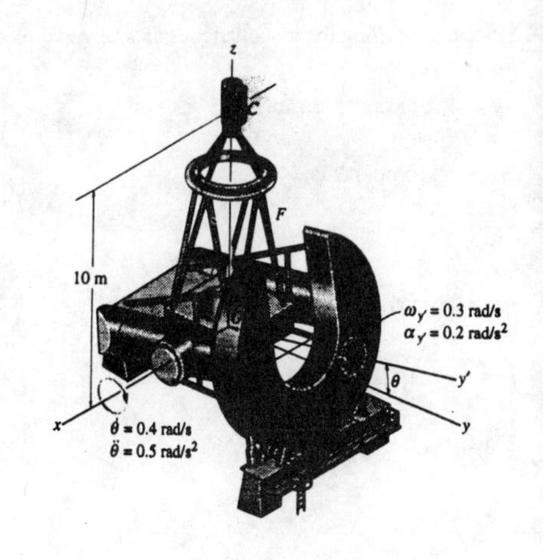

$$\omega = \omega_x + \omega_y + \omega_z = -0.4\mathbf{i} + 0.3\cos 30°\mathbf{j} + 0.3\sin 30°\mathbf{k}$$

$$\omega = \{-0.4\mathbf{i} + 0.2598\mathbf{j} + 0.15\mathbf{k}\}\ \text{rad/s}$$

$$\mathbf{v}_C = \omega\times\mathbf{r}_{OC} \begin{vmatrix} \mathbf{i} & \mathbf{j} & \mathbf{k} \\ -0.4 & 0.2598 & 0.15 \\ 0 & 0 & 10 \end{vmatrix} = 2.598\mathbf{i} + 4.00\mathbf{j}$$

$$\mathbf{v}_C = \{2.60\mathbf{i} + 4.00\mathbf{j}\}\ \text{m/s} \qquad \textbf{Ans}$$

$$\alpha = \dot{\omega} = \dot{\omega}_x + \dot{\omega}_{y'} = -0.5\mathbf{i} + (0.3\cos 30°\mathbf{j} + 0.3\sin 30°\mathbf{k})\times(-0.4\mathbf{i}) + (0.2\cos 30°\mathbf{j} + 0.2\sin 30°\mathbf{k})$$

$$= \{-0.5\mathbf{i} + 0.1132\mathbf{j} + 0.2039\mathbf{k}\}\ \text{rad/s}^2$$

$$\mathbf{a}_C = \alpha\times\mathbf{r}_{OC} + \omega\times(\omega\times\mathbf{r}_{OC}) = \begin{vmatrix} \mathbf{i} & \mathbf{j} & \mathbf{k} \\ -0.5 & 0.1132 & 0.2039 \\ 0 & 0 & 10 \end{vmatrix} + \begin{vmatrix} \mathbf{i} & \mathbf{j} & \mathbf{k} \\ -0.4 & 0.2598 & 0.15 \\ 2.598 & 4.00 & 0 \end{vmatrix}$$

$$= 1.132\mathbf{i} + 5\mathbf{j} - 0.6\mathbf{i} + 0.3897\mathbf{j} - 2.275\mathbf{k}$$

$$= \{0.532\mathbf{i} + 5.39\mathbf{j} - 2.28\mathbf{k}\}\ \text{m/s}^2 \qquad \textbf{Ans}$$

***20-12.** If the plate gears A and B are rotating with the angular velocities shown, determine the angular velocity of gear C about the shaft DE. What is the angular velocity of DE about the y axis?

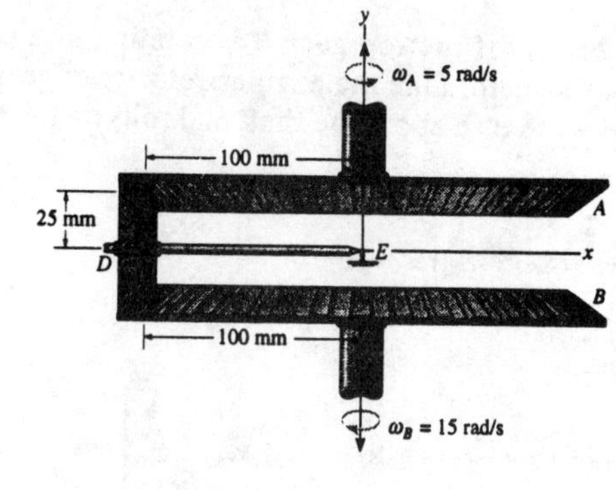

The speeds of points P and P', located at the top and bottom of gear C, are

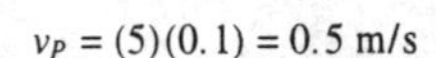

$v_P = (5)(0.1) = 0.5$ m/s

$v_{P'} = (15)(0.1) = 1.5$ m/s

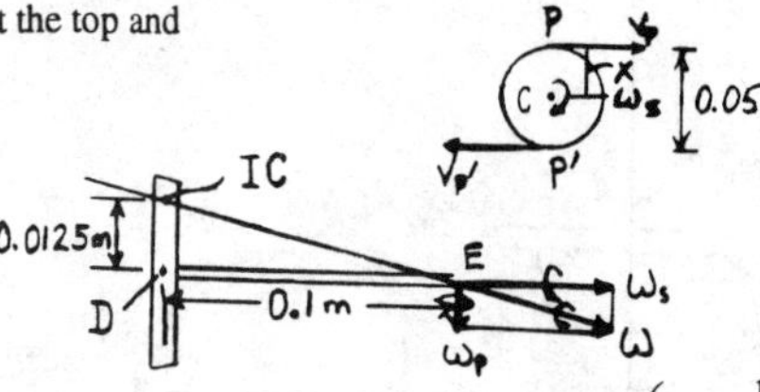

The IC is located as shown.

$$\frac{0.5}{x} = \frac{1.5}{(0.05-x)}; \qquad x = 0.0125 \text{ m}$$

$$\frac{\omega_s}{0.1} = \frac{\omega_p}{0.0125}; \qquad \omega_s = 8\omega_p$$

$$\boldsymbol{\omega} = \omega_s\mathbf{i} - \omega_p\mathbf{j} = \omega_s\mathbf{i} - \frac{1}{8}\omega_s\mathbf{j}$$

$$\mathbf{v} = \boldsymbol{\omega}\times\mathbf{r}$$

$$0.5\mathbf{k} = \left(\omega_s\mathbf{i} - \frac{1}{8}\omega_s\mathbf{j}\right)\times(-0.1\mathbf{i}+0.025\mathbf{j})$$

$$0.5\mathbf{k} = \begin{vmatrix} \mathbf{i} & \mathbf{j} & \mathbf{k} \\ \omega_s & -\frac{1}{8}\omega_s & 0 \\ -0.1 & 0.025 & 0 \end{vmatrix} = 0.0125\omega_s\mathbf{k}$$

$$\omega_s = \frac{0.5}{0.0125} = 40 \text{ rad/s}$$ **Ans** (Angular velocity of C about DE)

$$\omega_p = \frac{1}{8}(40) = 5 \text{ rad/s}$$ **Ans** (Angular velocity of DE about y axis)

20-13. The right circular cone rotates about the z axis at a constant rate of $\omega_1 = 4$ rad/s without slipping on the horizontal plane. Determine the magnitudes of the velocity and acceleration of points B and C.

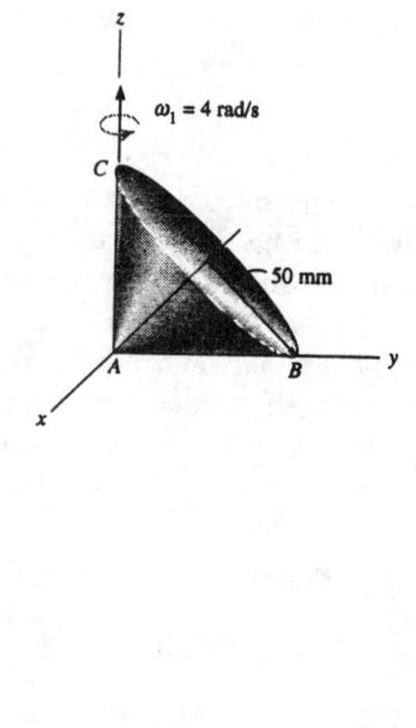

$$\boldsymbol{\omega} = \boldsymbol{\omega}_1 + \boldsymbol{\omega}_2$$

Since ω acts along the instantaneous axis of zero velocity

$$\omega\mathbf{j} = 4\mathbf{k} + \omega_2\cos45^\circ\mathbf{j} + \omega_2\sin45^\circ\mathbf{k}.$$

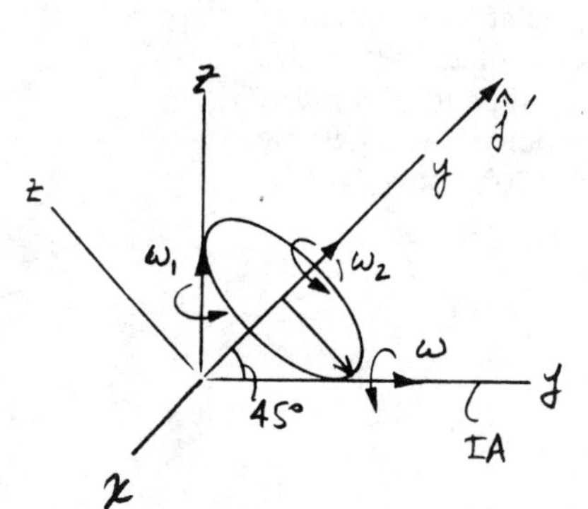

Equating components,

$$\omega = 0.707\omega_2$$

$$0 = 4 + 0.707\omega_2$$

$$\omega = -4 \text{ rad/s}, \qquad \omega_2 = -5.66 \text{ rad/s}$$

Thus,

$$\boldsymbol{\omega} = \{-4\mathbf{j}\} \text{ rad/s}$$

$$\boldsymbol{\Omega} = \boldsymbol{\omega}_1$$

$$\dot{\boldsymbol{\omega}} = \dot{\boldsymbol{\omega}}_1 + \dot{\boldsymbol{\omega}}_2 = \mathbf{0} + \boldsymbol{\omega}_1\times\boldsymbol{\omega}_2$$

$$= \mathbf{0} + (4\mathbf{k})\times(-5.66\cos45^\circ\mathbf{j} - 5.66\sin45^\circ\mathbf{k})$$

$$\boldsymbol{\alpha} = \dot{\boldsymbol{\omega}} = \{16\mathbf{i}\} \text{ rad/s}^2$$

$$\mathbf{v}_B = \boldsymbol{\omega}\times\mathbf{r}_B = (-4\mathbf{j})\times(0.1(0.707)\mathbf{j}) = \mathbf{0}$$

$v_B = 0$ **Ans**

$$\mathbf{v}_C = \boldsymbol{\omega}\times\mathbf{r}_C = (-4\mathbf{j})\times(0.1(0.707)\mathbf{k}) = \{-0.2828\mathbf{i}\} \text{ m/s}$$

$v_C = 0.283$ m/s **Ans**

$$\mathbf{a}_B = \boldsymbol{\alpha}\times\mathbf{r}_B + \boldsymbol{\omega}\times\mathbf{v}_B = 16\mathbf{i}\times(0.1)(0.707)\mathbf{j} + \mathbf{0}$$

$$\mathbf{a}_B = \{1.131\mathbf{k}\} \text{ m/s}$$

$a_B = 1.13$ m/s^2 **Ans**

$$\mathbf{a}_C = \boldsymbol{\alpha}\times\mathbf{r}_C + \boldsymbol{\omega}\times\mathbf{v}_C = 16\mathbf{i}\times(0.1)(0.707)\mathbf{k} + (-4\mathbf{j})\times(-0.2828\mathbf{i})$$

$$\mathbf{a}_C = \{-1.131\mathbf{j} - 1.131\mathbf{k}\} \text{ m/s}^2$$

$a_C = 1.60$ m/s^2 **Ans**

20-14. The tower crane is rotating about the z axis at a constant rate $\omega_1 = 0.25$ rad/s, while the boom OA is rotating downward at a constant rate $\omega_2 = 0.4$ rad/s. Determine the velocity and acceleration of point A located at the top of the boom at the instant shown.

$$\boldsymbol{\omega} = \boldsymbol{\omega}_1 + \boldsymbol{\omega}_2 = \{-0.4\mathbf{i} + 0.25\mathbf{k}\} \text{ rad/s}$$

$$\boldsymbol{\Omega} = \{0.25\mathbf{k}\} \text{ rad/s}$$

$$\dot{\boldsymbol{\omega}} = \left(\dot{\boldsymbol{\omega}}\right)_{xyz} + \boldsymbol{\Omega} \times \boldsymbol{\omega} = \mathbf{0} + (0.25\mathbf{k}) \times (-0.4\mathbf{i} + 0.25\mathbf{k}) = \{-0.1\mathbf{j}\} \text{ rad/s}^2$$

$$\mathbf{r}_A = 40\cos 30^\circ \mathbf{j} + 40\sin 30^\circ \mathbf{k} = \{34.64\mathbf{j} + 20\mathbf{k}\} \text{ ft}$$

$$\mathbf{v}_A = \boldsymbol{\omega} \times \mathbf{r}_A = (-0.4\mathbf{i} + 0.25\mathbf{k}) \times (34.64\mathbf{j} + 20\mathbf{k})$$

$$\mathbf{v}_A = \{-8.66\mathbf{i} + 8.00\mathbf{j} - 13.9\mathbf{k}\} \text{ ft/s}$$ **Ans**

$$\mathbf{a}_A = \boldsymbol{\alpha} \times \mathbf{r}_A + \boldsymbol{\omega} \times \mathbf{v}_A = (-0.1\mathbf{j}) \times (34.64\mathbf{j} + 20\mathbf{k}) + (-0.4\mathbf{i} + 0.25\mathbf{k}) \times (-8.66\mathbf{i} + 8.00\mathbf{j} - 13.9\mathbf{k})$$

$$\mathbf{a}_A = \{-4.00\mathbf{i} - 7.71\mathbf{j} - 3.20\mathbf{k}\} \text{ ft/s}^2$$ **Ans**

20-15. At the instant shown, the tower crane is rotating about the z axis with an angular velocity $\omega_1 = 0.25$ rad/s, which is increasing at 0.6 rad/s^2. The boom OA is rotating downward with an angular velocity $\omega_2 = 0.4$ rad/s, which is increasing at 0.8 rad/s^2. Determine the velocity and acceleration of point A located at the top of the boom at this instant.

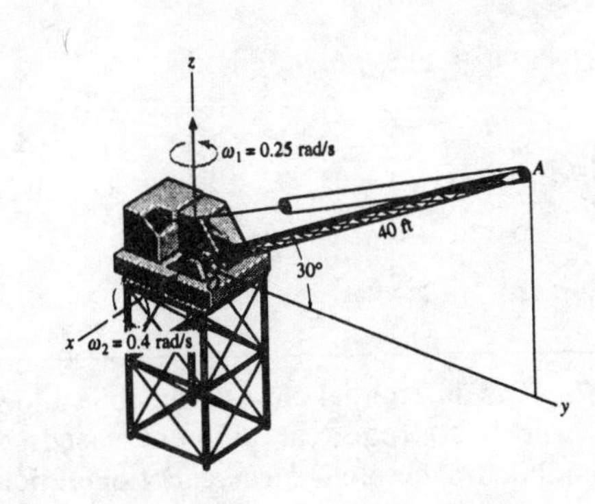

$$\boldsymbol{\omega} = \boldsymbol{\omega}_1 + \boldsymbol{\omega}_2 = \{-0.4\mathbf{i} + 0.25\mathbf{k}\} \text{ rad/s}$$

$$\boldsymbol{\Omega} = \{0.25\mathbf{k}\} \text{ rad/s}$$

$$\dot{\boldsymbol{\omega}} = \left(\dot{\boldsymbol{\omega}}\right)_{xyz} + \boldsymbol{\Omega} \times \boldsymbol{\omega} = (-0.8\mathbf{i} + 0.6\mathbf{k}) + (0.25\mathbf{k}) \times (-0.4\mathbf{i} + 0.25\mathbf{k})$$

$$= \{-0.8\mathbf{i} - 0.1\mathbf{j} + 0.6\mathbf{k}\} \text{ rad/s}^2$$

$$\mathbf{r}_A = 40\cos 30^\circ \mathbf{j} + 40\sin 30^\circ \mathbf{k} = \{34.64\mathbf{j} + 20\mathbf{k}\} \text{ ft}$$

$$\mathbf{v}_A = \boldsymbol{\omega} \times \mathbf{r}_A = (-0.4\mathbf{i} + 0.25\mathbf{k}) \times (34.64\mathbf{j} + 20\mathbf{k})$$

$$\mathbf{v}_A = \{-8.66\mathbf{i} + 8.00\mathbf{j} - 13.9\mathbf{k}\} \text{ ft/s}$$ **Ans**

$$\mathbf{a}_A = \boldsymbol{\alpha} \times \mathbf{r}_A + \boldsymbol{\omega} \times \mathbf{v}_A = (-0.8\mathbf{i} - 0.1\mathbf{j} + 0.6\mathbf{k}) \times (34.64\mathbf{j} + 20\mathbf{k}) + (-0.4\mathbf{i} + 0.25\mathbf{k}) \times (-8.66\mathbf{i} + 8.00\mathbf{j} - 13.9\mathbf{k})$$

$$\mathbf{a}_A = \{-24.8\mathbf{i} + 8.29\mathbf{j} - 30.9\mathbf{k}\} \text{ ft/s}^2$$ **Ans**

***20-16.** The construction boom OA is rotating about the z axis with a constant angular velocity of $\omega_1 = 0.15$ rad/s, while it is rotating downward with a constant angular velocity of $\omega_2 = 0.2$ rad/s. Determine the velocity and acceleration of point A located at the tip of the boom at the instant shown.

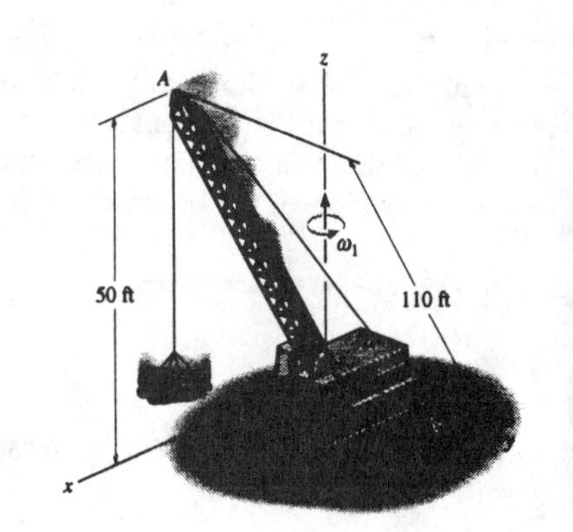

$$\omega = \omega_1 + \omega_2 = \{0.2\mathbf{j} + 0.15\mathbf{k}\}\ \text{rad/s}$$

$$\dot{\omega} = \dot{\omega}_1 + \dot{\omega}_2$$

Let the x, y, z axes rotate at $\Omega = \omega_1$, then

$$\alpha = \dot{\omega} = \left(\dot{\omega}\right)_{xyz} + \omega_1 \times \omega_2$$

$$\alpha = \mathbf{0} + 0.15\mathbf{k} \times 0.2\mathbf{j} = \{-0.03\mathbf{i}\}\ \text{rad/s}^2$$

$$\mathbf{r}_A = \left[\sqrt{(110)^2 - (50)^2}\right]\mathbf{i} + 50\mathbf{k} = \{97.98\mathbf{i} + 50\mathbf{k}\}\ \text{ft}$$

$$\mathbf{v}_A = \omega \times \mathbf{r}_A = \begin{vmatrix} \mathbf{i} & \mathbf{j} & \mathbf{k} \\ 0 & 0.2 & 0.15 \\ 97.98 & 0 & 50 \end{vmatrix}$$

$$\mathbf{v}_A = \{10\mathbf{i} + 14.7\mathbf{j} - 19.6\mathbf{k}\}\ \text{ft/s} \qquad \textbf{Ans}$$

$$\mathbf{a}_A = \alpha \times \mathbf{r}_A + \omega \times \mathbf{v}_A = \begin{vmatrix} \mathbf{i} & \mathbf{j} & \mathbf{k} \\ -0.03 & 0 & 0 \\ 97.98 & 0 & 50 \end{vmatrix} + \begin{vmatrix} \mathbf{i} & \mathbf{j} & \mathbf{k} \\ 0 & 0.2 & 0.15 \\ 10 & 14.7 & -19.6 \end{vmatrix}$$

$$\mathbf{a}_A = \{-6.12\mathbf{i} + 3\mathbf{j} - 2\mathbf{k}\}\ \text{ft/s}^2 \qquad \textbf{Ans}$$

20-17. The differential of an automobile allows the two rear wheels to rotate at different speeds when the automobile travels along a curve. For operation, the rear axles are attached to the wheels at one end and have beveled gears A and B on their other ends. The differential case D is placed over the left axle but can rotate about C independent of the axle. The case supports a pinion gear E on a shaft, which meshes with gears A and B. Finally, a ring gear G is *fixed* to the differential case so that the case rotates with the ring gear when the latter is driven by the drive pinion H. This gear, like the differential case, is free to rotate about the left wheel axle. If the drive pinion is turning at $\omega_H = 100$ rad/s and the pinion gear E is spinning about its shaft at $\omega_E = 30$ rad/s, determine the angular velocity, ω_A and ω_B, of each axle.

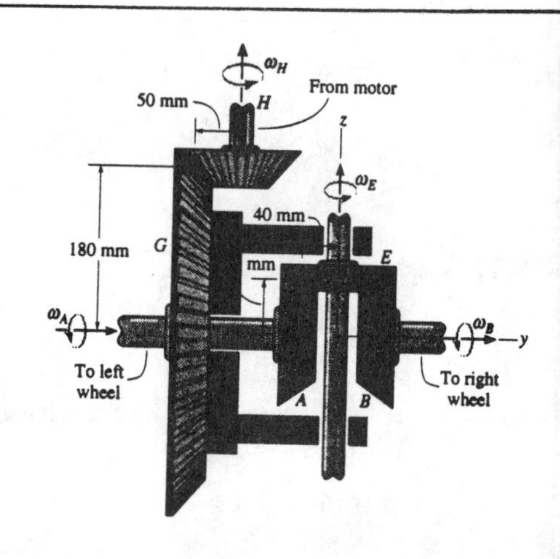

$$v_P = \omega_H r_H = 100(50) = 5000\ \text{mm/s}$$

$$\omega_G = \frac{5000}{180} = 27.78\ \text{rad/s}$$

Point O is a fixed point of rotation for gears A, E, and B.

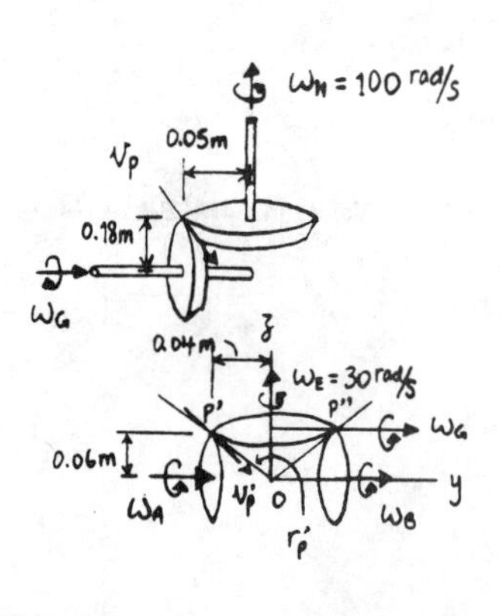

$$\Omega = \omega_G + \omega_E = \{27.78\mathbf{j} + 30\mathbf{k}\}\ \text{rad/s}$$

$$\mathbf{v}_{P'} = \Omega \times \mathbf{r}_{P'} = (27.78\mathbf{j} + 30\mathbf{k}) \times (-40\mathbf{j} + 60\mathbf{k}) = \{2866.7\mathbf{i}\}\ \text{mm/s}$$

$$\omega_A = \frac{2866.7}{60} = 47.8\ \text{rad/s} \qquad \textbf{Ans}$$

$$\mathbf{v}_{P''} = \Omega \times \mathbf{r}_{P''} = (27.78\mathbf{j} + 30\mathbf{k}) \times (40\mathbf{j} + 60\mathbf{k}) = \{466.7\mathbf{i}\}\ \text{mm/s}$$

$$\omega_B = \frac{466.2}{60} = 7.78\ \text{rad/s} \qquad \textbf{Ans}$$

20-18. The rod AB is attached to collars at its ends by ball-and-socket joints. If collar A has a velocity $v_A = 15$ ft/s at the instant shown, determine the velocity of collar B.

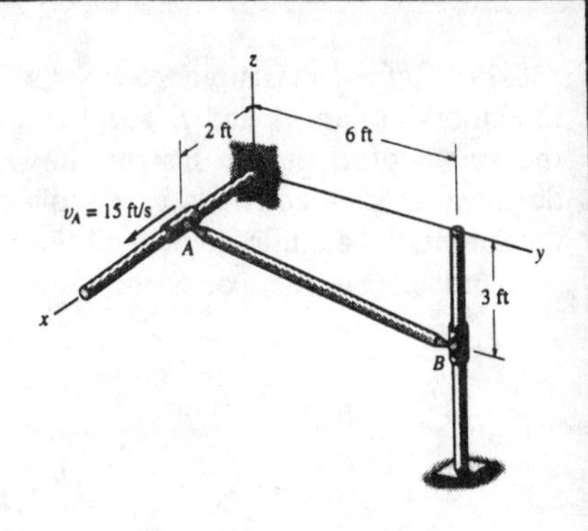

$\mathbf{v}_A = \{15\mathbf{i}\}$ ft/s $\quad \mathbf{v}_B = v_B\mathbf{k} \quad \omega_{AB} = \omega_x\mathbf{i} + \omega_y\mathbf{j} + \omega_z\mathbf{k}$

$\mathbf{r}_{B/A} = \{-2\mathbf{i} + 6\mathbf{j} - 3\mathbf{k}\}$ ft

$$\mathbf{v}_B = \mathbf{v}_A + \omega_{AB} \times \mathbf{r}_{B/A}$$

$$v_B\mathbf{k} = 15\mathbf{i} + \begin{vmatrix} \mathbf{i} & \mathbf{j} & \mathbf{k} \\ \omega_x & \omega_y & \omega_z \\ -2 & 6 & -3 \end{vmatrix}$$

Equating **i**, **j** and **k** components yields :

$$15 - 3\omega_y - 6\omega_z = 0 \qquad (1)$$

$$3\omega_x - 2\omega_z = 0 \qquad (2)$$

$$6\omega_x + 2\omega_y = v_B \qquad (3)$$

If ω_{AB} is perpendicular to the axis of the rod,

$$\omega_{AB} \cdot \mathbf{r}_{B/A} = (\omega_x\mathbf{i} + \omega_y\mathbf{j} + \omega_z\mathbf{k}) \cdot (-2\mathbf{i} + 6\mathbf{j} - 3\mathbf{k}) = 0$$

$$-2\omega_x + 6\omega_y - 3\omega_z = 0 \qquad (4)$$

Solving Eqs. (1) to (4) yields :

$\omega_x = 1.2245$ rad/s $\quad \omega_y = 1.3265$ rad/s $\quad \omega_z = 1.8367$ rad/s $\quad v_B = 10$ ft/s

Note: v_B can be obtained by solving Eqs. (1) - (3) without knowing the direction of ω.

Hence $\quad \omega_{AB} = \{1.2245\mathbf{i} + 1.3265\mathbf{j} + 1.8367\mathbf{k}\}$ rad/s

$\mathbf{v}_B = \{10\mathbf{k}\}$ ft/s $\quad$ **Ans**

20-19. The rod AB is attached to collars at its ends by ball-and-socket joints. If collar A has an acceleration of $a_A = 2$ ft/s^2 at the instant shown, determine the acceleration of collar B.

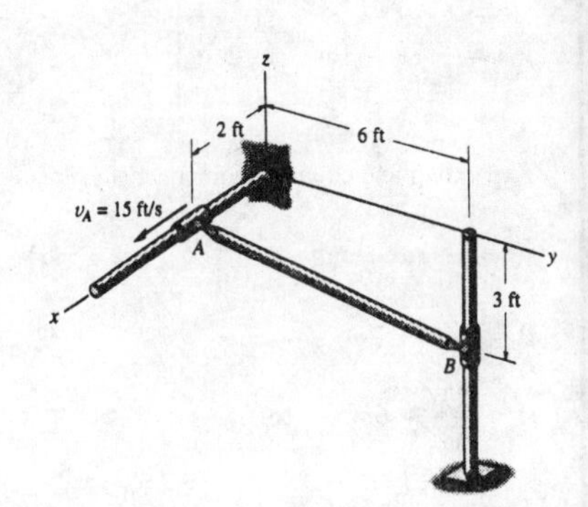

From Prob. 20 – 18

$\omega_{AB} = \{1.2245\mathbf{i} + 1.3265\mathbf{j} + 1.8367\mathbf{k}\}$ rad/s

$\mathbf{r}_{B/A} = \{-2\mathbf{i} + 6\mathbf{j} - 3\mathbf{k}\}$ ft

$\alpha_{AB} = \alpha_x\mathbf{i} + \alpha_y\mathbf{j} + \alpha_z\mathbf{k}$

$\mathbf{a}_A = \{2\mathbf{i}\}$ ft/s^2 $\qquad \mathbf{a}_B = a_B\mathbf{k}$

$$\mathbf{a}_B = \mathbf{a}_A + \alpha_{AB} \times \mathbf{r}_{B/A} + \omega_{AB} \times (\omega_{AB} \times \mathbf{r}_{B/A})$$

$$a_B\mathbf{k} = 2\mathbf{i} + (\alpha_x\mathbf{i} + \alpha_y\mathbf{j} + \alpha_z\mathbf{k}) \times (-2\mathbf{i} + 6\mathbf{j} - 3\mathbf{k})$$
$$+ (1.2245\mathbf{i} + 1.3265\mathbf{j} + 1.8367\mathbf{k})$$
$$\times [(1.2245\mathbf{i} + 1.3265\mathbf{j} + 1.8367\mathbf{k}) \times (-2\mathbf{i} + 6\mathbf{j} - 3\mathbf{k})]$$

If α_{AB} is perpendicular to the axis of the rod,

$$\alpha_{AB} \cdot \mathbf{r}_{B/A} = (\alpha_x\mathbf{i} + \alpha_y\mathbf{j} + \alpha_z\mathbf{k}) \cdot (-2\mathbf{i} + 6\mathbf{j} - 3\mathbf{k}) = 0$$

$$-2\alpha_x + 6\alpha_y - 3\alpha_z = 0 \qquad (4)$$

Equating **i**, **j** and **k** components yields :

$$15.2653 - 3\alpha_y - 6\alpha_z = 0 \qquad (1)$$

$$3\alpha_x - 2\alpha_z - 39.7955 = 0 \qquad (2)$$

$$6\alpha_x + 2\alpha_y + 19.8975 = a_B \qquad (3)$$

Solving Eqs.(1) to(4) yields :

$\alpha_x = 13.43$ rad/s^2 $\quad \alpha_y = 4.599$ rad/s^2 $\quad \alpha_z = 0.2449$ rad/s^2 $\quad a_B = 109.7$ ft/s^2

Note: a_B can be obtained by solving Eqs. (1) - (3) without knowing the direction of α

Hence $\quad \mathbf{a}_B = \{110\mathbf{k}\}$ ft/s^2 $\quad$ **Ans**

***20-20.** If the rod is attached with ball-and-socket joints to smooth collars A and B at its end points, determine the speed of B at the instant shown if A is moving downward at a constant speed of $v_A = 8$ ft/s. Also, determine the angular velocity of the rod if it is directed perpendicular to the axis of the rod.

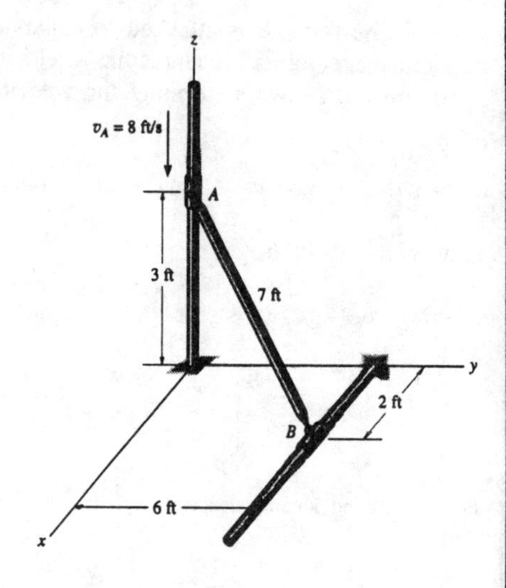

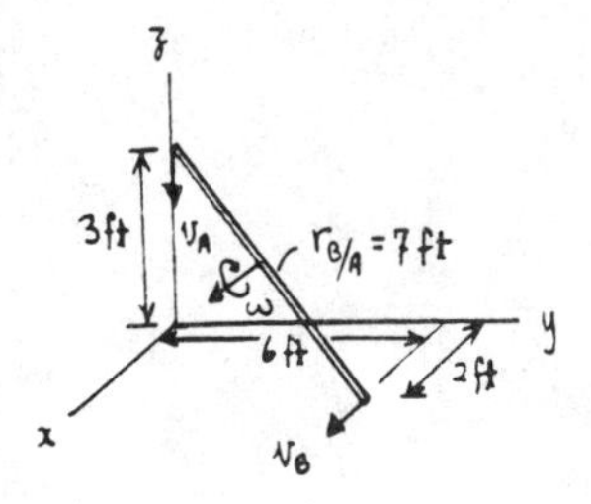

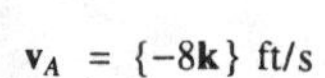

$\mathbf{v}_A = \{-8\mathbf{k}\}$ ft/s

$\mathbf{v}_B = v_B\,\mathbf{i}$

$\mathbf{r}_{B/A} = \{2\mathbf{i} + 6\mathbf{j} - 3\mathbf{k}\}$ ft

$\omega = \{\omega_x\mathbf{i} + \omega_y\mathbf{j} + \omega_z\mathbf{k}\}$ rad/s

$\mathbf{v}_B = \mathbf{v}_A + \omega \times \mathbf{r}_{B/A}$

$$v_B\,\mathbf{i} = -8\mathbf{k} + \begin{vmatrix} \mathbf{i} & \mathbf{j} & \mathbf{k} \\ \omega_x & \omega_y & \omega_z \\ 2 & 6 & -3 \end{vmatrix}$$

Expanding and equating components yields :

$v_B = -3\omega_y - 6\omega_z \qquad (1)$

$0 = 3\omega_x + 2\omega_z \qquad (2)$

$0 = -8 + 6\omega_x - 2\omega_y \qquad (3)$

Also, $\quad \omega \cdot \mathbf{r}_{B/A} = 0$

$(\omega_x\mathbf{i} + \omega_y\mathbf{j} + \omega_z\mathbf{k}) \cdot (2\mathbf{i} + 6\mathbf{j} - 3\mathbf{k}) = 0$

$2\omega_x + 6\omega_y - 3\omega_z = 0 \qquad (4)$

Solving Eqs. (1) – (4) yields

$\omega_x = 0.9796$ rad/s

$\omega_y = -1.061$ rad/s

$\omega_z = -1.469$ rad/s

$v_B = 12$ ft/s

$\omega = \{0.980\mathbf{i} - 1.06\mathbf{j} - 1.47\mathbf{k}\}$ rad/s **Ans**

$\mathbf{v}_B = \{12.0\mathbf{i}\}$ ft/s **Ans**

20-21. If the collar at A is moving downward with an acceleration $\mathbf{a}_A = \{-5\mathbf{k}\}$ ft/s^2, at the instant its speed is $v_A = 8$ ft/s, determine the acceleration of the collar at B at this instant.

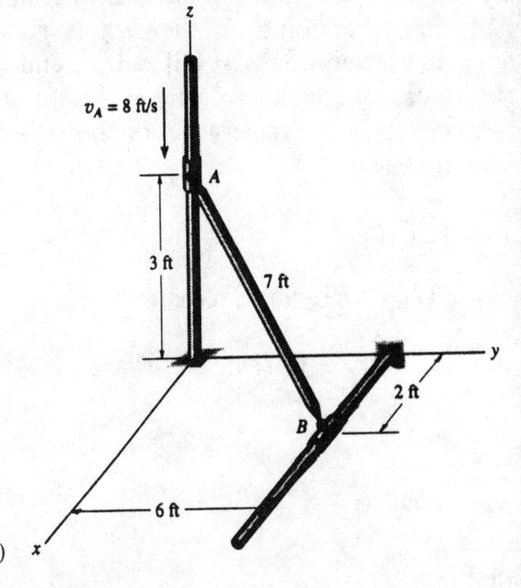

$\mathbf{a}_B = a_B\,\mathbf{i}, \qquad \mathbf{a}_A = -5\mathbf{k}$

From Prob. 20-20.

$\omega = 0.9796\mathbf{i} - 1.0612\mathbf{j} - 1.4694\mathbf{k}$

$\mathbf{a}_B = \mathbf{a}_A + \mathbf{a}_{B/A}$

$\mathbf{a}_{B/A} = \omega \times \mathbf{v}_{B/A} + \alpha \times \mathbf{r}_{B/A}$

$\mathbf{v}_{B/A} = \mathbf{v}_B - \mathbf{v}_A = 12\mathbf{i} + 8\mathbf{k}$

$\mathbf{a}_{B/A} = \{0.9796\mathbf{i} - 1.0612\mathbf{j} - 1.4694\mathbf{k}\} \times (12\mathbf{i} + 8\mathbf{k}) + \{\alpha_x\mathbf{i} + \alpha_y\mathbf{j} + \alpha_z\mathbf{k}\} \times (2\mathbf{i} + 6\mathbf{j} - 3\mathbf{k})$

$a_B\mathbf{i} = -5\mathbf{k} + \{0.9796\mathbf{i} - 1.0612\mathbf{j} - 1.4694\mathbf{k}\}\times(12\mathbf{i} + 8\mathbf{k})$
$\quad + \{(-3\alpha_y - 6\alpha_z)\mathbf{i} + (2\alpha_z + 3\alpha_x)\mathbf{j} + (6\alpha_x - 2\alpha_y)\mathbf{k}\}$

$3\alpha_y + 6\alpha_z + a_B = -8.4898$

$-3\alpha_x - 2\alpha_z = -25.4694$

$-6\alpha_x + 2\alpha_y = 7.7347$

Solving these equations

$a_B = -96.5$ ft/s^2

$\mathbf{a}_B = \{-96.5\mathbf{i}\}$ ft/s^2 **Ans**

20-22. Rod AB is attached to a disk and a collar by ball-and-socket joints. If the disk is rotating at a constant angular velocity $\boldsymbol{\omega} = \{2\mathbf{i}\}$ rad/s, determine the velocity and acceleration of the collar at A at the instant shown. Assume the angular velocity is directed perpendicular to the rod.

$\mathbf{v}_A = \mathbf{v}_B + \omega \times \mathbf{r}_{A/B}$

$$v_A\mathbf{i} = -(1)(2)\mathbf{j} + \begin{vmatrix} \mathbf{i} & \mathbf{j} & \mathbf{k} \\ \omega_x & \omega_y & \omega_z \\ 3 & -1 & -1 \end{vmatrix}$$

Expand and equate components:

$v_A = -\omega_y + \omega_z \qquad (1)$

$2 = \omega_x + 3\omega_z \qquad (2)$

$0 = -\omega_x - 3\omega_y \qquad (3)$

Also:

$\omega \cdot \mathbf{r}_{A/B} = 0$

$3\omega_x - \omega_y - \omega_z = 0 \qquad (4)$

Solving Eqs. (1)-(4):

$\omega_x = 0.1818$ rad/s

$\omega_y = -0.06061$ rad/s

$\omega_z = 0.6061$ rad/s

$v_A = 0.667$ ft/s

Thus $\mathbf{v}_A = \{0.667\mathbf{i}\}$ ft/s **Ans**

$\mathbf{v}_{A/B} = \mathbf{v}_A - \mathbf{v}_B$

$\quad = \{0.667\mathbf{i} + 2\mathbf{j}\}$ ft/s

$\mathbf{a}_A = \mathbf{a}_B + \alpha \times \mathbf{r}_{A/B} + \omega \times \mathbf{v}_{A/B}$

$$a_A\mathbf{i} = -4\mathbf{k} + \begin{vmatrix} \mathbf{i} & \mathbf{j} & \mathbf{k} \\ \alpha_x & \alpha_y & \alpha_z \\ 3 & -1 & -1 \end{vmatrix} + \begin{vmatrix} \mathbf{i} & \mathbf{j} & \mathbf{k} \\ 0.1818 & -0.06061 & 0.6061 \\ 0.667 & 2 & 0 \end{vmatrix}$$

Expand and equate components:

$a_A = -\alpha_y + \alpha_z - 1.212$

$0 = \alpha_x + 3\alpha_z + 0.404$

$0 = -\alpha_x - 3\alpha_y - 3.596$

Solve for a_A

$\mathbf{a}_A = \{-0.148\mathbf{i}\}$ ft/s^2 **Ans**

20-23. Rod AB is attached to a disk and a collar by ball-and-socket joints. If the disk is rotating with an angular acceleration $\boldsymbol{\alpha} = \{4\mathbf{i}\}$ rad/s^2, and at the instant shown has an angular velocity $\boldsymbol{\omega} = \{2\mathbf{i}\}$ rad/s, determine the velocity and acceleration of the collar at A at the instant shown.

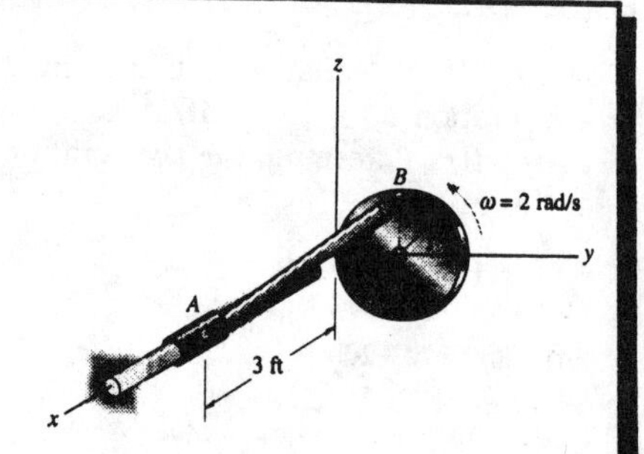

From Prob. 20-22,

$\Omega = 0.1818\mathbf{i} - 0.06061\mathbf{j} + 0.6061\mathbf{k}$

$\mathbf{v}_{A/B} = \mathbf{v}_A - \mathbf{v}_B = 0.66667\mathbf{i} + 2\mathbf{j}$

$(a_B)_t = (4)(1) = 4$

$(a_B)_n = (2)^2(1) = 4$

So that

$\mathbf{a}_B = -4\mathbf{j} - 4\mathbf{k}, \qquad \mathbf{a}_A = a_A\mathbf{i}$

$\mathbf{a}_A = \mathbf{a}_B + \mathbf{a}_{A/B}$

$\mathbf{a}_{A/B} = (0.1818\mathbf{i} - 0.06061\mathbf{j} + 0.6061\mathbf{k})\times(0.66667\mathbf{i} + 2\mathbf{j}) + (\alpha_x\mathbf{i} + \alpha_y\mathbf{j} + \alpha_z\mathbf{k}) \times (3\mathbf{i} - 1\mathbf{j} - 1\mathbf{k})$

$\mathbf{a}_{A/B} = (-1.2121 - \alpha_y + \alpha_z)\mathbf{i} + (0.4040 + 3\alpha_z + \alpha_x)\mathbf{j} + (0.4040 - \alpha_x - 3\alpha_y)\mathbf{k}$

$a_A\mathbf{i} = -4\mathbf{j} - 4\mathbf{k} + (-1.2121 - \alpha_y + \alpha_z)\mathbf{i} + (0.4040 + 3\alpha_z + \alpha_x)\mathbf{j} + (0.4040 - \alpha_x - 3\alpha_y)\mathbf{k}$

$a_A = -1.2121 - \alpha_y + \alpha_z$

$0 = -4 + (0.4040 + 3\alpha_z + \alpha_x)$

$0 = -4 + (0.4040 - \alpha_x - 3\alpha_y)$

Solving for a_A,

$a_A = 1.185 = 1.19$ ft/s^2

$\mathbf{a}_A = \{1.19\mathbf{i}\}$ ft/s^2 **Ans**

***20-24.** Disk A is rotating at a constant angular velocity of 10 rad/s. If rod BC is joined to the disk and a collar by ball-and-socket joints, determine the velocity of collar B at the instant shown. Also, what is the rod's angular velocity $\boldsymbol{\omega}_{BC}$ if it is directed perpendicular to the axis of the rod?

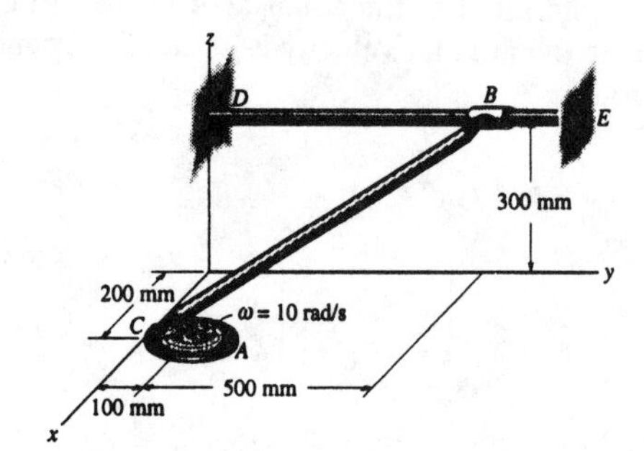

$\mathbf{v}_C = \{1\mathbf{i}\}$ m/s $\qquad \mathbf{v}_B = -v_B\mathbf{j} \qquad \boldsymbol{\omega}_{BC} = \omega_x\mathbf{i} + \omega_y\mathbf{j} + \omega_z\mathbf{k}$

$\mathbf{r}_{B/C} = \{-0.2\mathbf{i} + 0.6\mathbf{j} + 0.3\mathbf{k}\}$ m

$\mathbf{v}_B = \mathbf{v}_C + \boldsymbol{\omega}_{BC} \times \mathbf{r}_{B/C}$

$$-v_B\mathbf{j} = 1\mathbf{i} + \begin{vmatrix} \mathbf{i} & \mathbf{j} & \mathbf{k} \\ \omega_x & \omega_y & \omega_z \\ -0.2 & 0.6 & 0.3 \end{vmatrix}$$

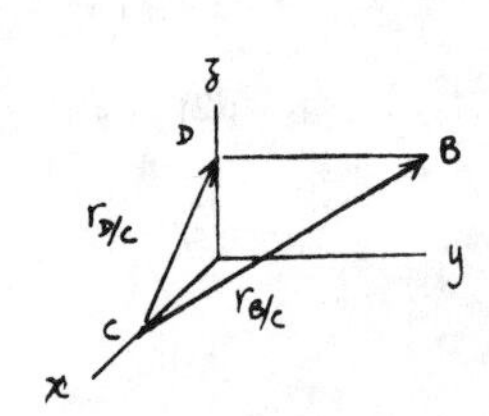

Equating **i**, **j** and **k** components

$1 + 0.3\omega_y - 0.6\omega_z = 0 \qquad (1)$

$0.3\omega_x + 0.2\omega_z = v_B \qquad (2)$

$0.6\omega_x + 0.2\omega_y = 0 \qquad (3)$

Since ω_{BC} is perpendicular to the axis of the rod,

$\boldsymbol{\omega}_{BC} \cdot \mathbf{r}_{B/C} = (\omega_x\mathbf{i} + \omega_y\mathbf{j} + \omega_z\mathbf{k}) \cdot (-0.2\mathbf{i} + 0.6\mathbf{j} + 0.3\mathbf{k}) = 0$

$-0.2\omega_x + 0.6\omega_y + 0.3\omega_z = 0 \qquad (4)$

Solving Eqs.(1) to (4) yields :

$\omega_x = 0.204$ rad/s $\qquad \omega_y = -0.612$ rad/s $\qquad \omega_z = 1.36$ rad/s $\qquad v_B = 0.333$ m/s

Then

$\boldsymbol{\omega}_{BC} = \{0.204\mathbf{i} - 0.612\mathbf{j} + 1.36\mathbf{k}\}$ rad/s **Ans**

$\mathbf{v}_B = \{-0.333\mathbf{j}\}$ m/s **Ans**

***20-25.** Solve Prob. 20-24 if the connection at B consists of a pin as shown in the figure below, rather than a ball-and-socket joint. *Hint:* The constraint allows rotation of the rod both along bar DE ($\mathbf{j}$ direction) and along the axis of the pin ($\mathbf{n}$ direction). Since there is no rotational component in the $\mathbf{u}$ direction, i.e., perpendicular to $\mathbf{n}$ and $\mathbf{j}$ where $\mathbf{u} = \mathbf{j} \times \mathbf{n}$, an additional equation for solution can be obtained from $\boldsymbol{\omega} \cdot \mathbf{u} = 0$. The vector $\mathbf{n}$ is in the same direction as $\mathbf{r}_{B/C} \times \mathbf{r}_{D/C}$.

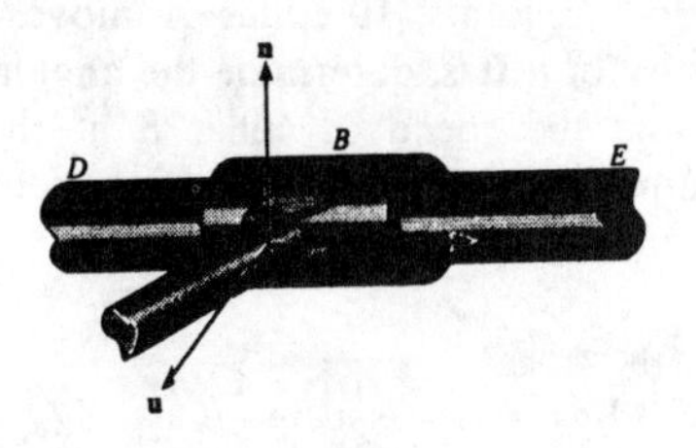

$\mathbf{v}_C = \{1\mathbf{i}\}$ m/s $\quad \mathbf{v}_B = -v_B\mathbf{j} \quad \boldsymbol{\omega}_{BC} = \omega_x\mathbf{i} + \omega_y\mathbf{j} + \omega_z\mathbf{k}$

$\mathbf{r}_{B/C} = \{-0.2\mathbf{i} + 0.6\mathbf{j} + 0.3\mathbf{k}\}$ m

$$\mathbf{v}_B = \mathbf{v}_C + \boldsymbol{\omega}_{BC} \times \mathbf{r}_{B/C}$$

$$-v_B\mathbf{j} = 1\mathbf{i} + \begin{vmatrix} \mathbf{i} & \mathbf{j} & \mathbf{k} \\ \omega_x & \omega_y & \omega_z \\ -0.2 & 0.6 & 0.3 \end{vmatrix}$$

Equating $\mathbf{i}$, $\mathbf{j}$ and $\mathbf{k}$ components

$$1 + 0.3\omega_y - 0.6\omega_z = 0 \qquad (1)$$

$$0.3\omega_x + 0.2\omega_z = v_B \qquad (2)$$

$$0.6\omega_x + 0.2\omega_y = 0 \qquad (3)$$

Aiso,

$\mathbf{r}_{B/C} = \{-0.2\mathbf{i} + 0.6\mathbf{j} + 0.3\mathbf{k}\}$ m

$\mathbf{r}_{D/C} = \{-0.2\mathbf{i} + 0.3\mathbf{k}\}$ m

$$\mathbf{r}_{B/C} \times \mathbf{r}_{D/C} = \begin{vmatrix} \mathbf{i} & \mathbf{j} & \mathbf{k} \\ -0.2 & 0.6 & 0.3 \\ -0.2 & 0 & 0.3 \end{vmatrix} = \{0.18\mathbf{i} + 0.12\mathbf{k}\} \text{ m}^2$$

$$\mathbf{n} = \frac{0.18\mathbf{i} + 0.12\mathbf{k}}{\sqrt{0.18^2 + 0.12^2}} = 0.8321\mathbf{i} + 0.5547\mathbf{k}$$

$$\mathbf{u} = \mathbf{j} \times \mathbf{n} = \mathbf{j} \times (0.8321\mathbf{i} + 0.5547\mathbf{k}) = 0.5547\mathbf{i} - 0.8321\mathbf{k}$$

$$\boldsymbol{\omega}_{BC} \cdot \mathbf{u} = (\omega_x\mathbf{i} + \omega_y\mathbf{j} + \omega_z\mathbf{k}) \cdot (0.5547\mathbf{i} - 0.8321\mathbf{k})$$

$$0.5547\omega_x - 0.8321\omega_z = 0 \qquad (4)$$

Solving Eqs.(1) to (4) yields :

$\omega_x = 0.769$ rad/s $\quad \omega_y = -2.31$ rad/s $\quad \omega_z = 0.513$ rad/s $\quad v_B = 0.333$ m/s

Then

$\boldsymbol{\omega}_{BC} = \{0.769\mathbf{i} - 2.31\mathbf{j} + 0.513\mathbf{k}\}$ rad/s **Ans**

$\mathbf{v}_B = \{-0.333\mathbf{j}\}$ m/s **Ans**

20-26. Rod AB is attached to the rotating arm using ball-and-socket joints. If AC is rotating with a constant angular velocity of 8 rad/s about the pin at C, determine the angular velocity of link BD at the instant shown.

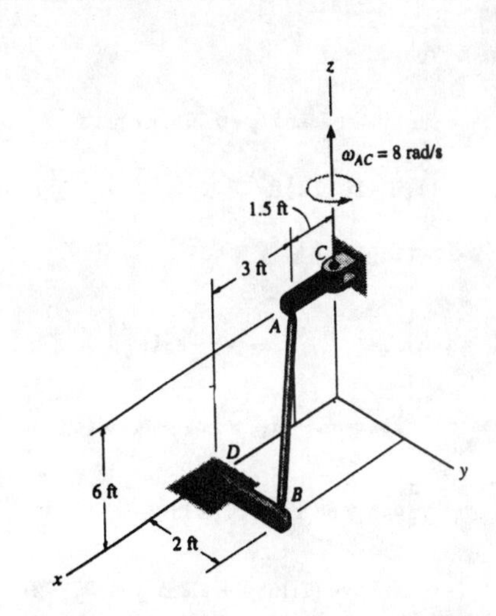

$\mathbf{v}_A = \{12\mathbf{j}\}$ ft/s $\quad \mathbf{v}_B = -v_B\mathbf{k} \quad \boldsymbol{\omega}_{AB} = \omega_x\mathbf{i} + \omega_y\mathbf{j} + \omega_z\mathbf{k}$

$\mathbf{r}_{B/A} = \{3\mathbf{i} + 2\mathbf{j} - 6\mathbf{k}\}$ ft

$$\mathbf{v}_B = \mathbf{v}_A + \boldsymbol{\omega}_{AB} \times \mathbf{r}_{B/A}$$

$$-v_B\mathbf{k} = 12\mathbf{j} + \begin{vmatrix} \mathbf{i} & \mathbf{j} & \mathbf{k} \\ \omega_x & \omega_y & \omega_z \\ 3 & 2 & -6 \end{vmatrix}$$

Equating $\mathbf{i}$, $\mathbf{j}$ and $\mathbf{k}$ components yields :

$$6\omega_y + 2\omega_z = 0 \qquad (1)$$

$$12 + 6\omega_x + 3\omega_z = 0 \qquad (2)$$

$$2\omega_x - 3\omega_y = -v_B \qquad (3)$$

If ω_{AB} is perpendicular to the axis of the rod,

$$\boldsymbol{\omega}_{AB} \cdot \mathbf{r}_{B/A} = (\omega_x\mathbf{i} + \omega_y\mathbf{j} + \omega_z\mathbf{k}) \cdot (3\mathbf{i} + 2\mathbf{j} - 6\mathbf{k}) = 0$$

$$3\omega_x + 2\omega_y - 6\omega_z = 0 \qquad (4)$$

Solving Eqs.(1) to(4) yields :

$\omega_x = -1.63$ rad/s $\quad \omega_y = 0.245$ rad/s $\quad \omega_z = -0.735$ rad/s $\quad v_B = 4$ ft/s $\downarrow$

Note: v_B can be obtained by solving Eqs. (1) - (3) without knowing the direction of ω.

For link BD

$\omega_{BD} = \dfrac{v_B}{r_{BD}} = \dfrac{4}{2} = 2$ rad/s $\quad$ then $\quad \boldsymbol{\omega} = \{-2\mathbf{i}\}$ rad/s  **Ans**

20-27. Rod AB is attached to collars at its ends by ball-and-socket joints. If collar A moves upward with a velocity of 8 ft/s, determine the angular velocity of the rod and the speed of collar B at the instant shown. Assume that the rod's angular velocity is directed perpendicular to the rod.

$\mathbf{v}_A = \{8\mathbf{k}\}$ ft/s $\quad \mathbf{v}_B = -\frac{3}{5}v_B\mathbf{i} + \frac{4}{5}v_B\mathbf{k} \quad \boldsymbol{\omega}_{AB} = \omega_x\mathbf{i} + \omega_y\mathbf{j} + \omega_z\mathbf{k}$

$\mathbf{r}_{B/A} = \{1.5\mathbf{i} - 2\mathbf{j} - 1\mathbf{k}\}$ ft

$$\mathbf{v}_B = \mathbf{v}_A + \boldsymbol{\omega}_{AB} \times \mathbf{r}_{B/A}$$

$$-\frac{3}{5}v_B\mathbf{i} + \frac{4}{5}v_B\mathbf{k} = 8\mathbf{k} + \begin{vmatrix} \mathbf{i} & \mathbf{j} & \mathbf{k} \\ \omega_x & \omega_y & \omega_z \\ 1.5 & -2 & -1 \end{vmatrix}$$

Equating $\mathbf{i}$, $\mathbf{j}$ and $\mathbf{k}$

$$-\omega_y + 2\omega_z = -\frac{3}{5}v_B \qquad (1)$$

$$\omega_x + 1.5\omega_z = 0 \qquad (2)$$

$$8 - 2\omega_x - 1.5\omega_y = \frac{4}{5}v_B \qquad (3)$$

Since ω_{AB} is perpendicular to the axis of the rod,

$$\boldsymbol{\omega}_{AB} \cdot \mathbf{r}_{B/A} = (\omega_x\mathbf{i} + \omega_y\mathbf{j} + \omega_z\mathbf{k}) \cdot (1.5\mathbf{i} - 2\mathbf{j} - 1\mathbf{k}) = 0$$

$$1.5\omega_x - 2\omega_y - \omega_z = 0 \qquad (4)$$

Solving Eqs.(1) to(4) yields :

$\omega_x = 1.1684$ rad/s $\quad \omega_y = 1.2657$ rad/s $\quad \omega_z = -0.7789$ rad/s

$v_B = 4.71$ ft/s **Ans**

Then $\quad \boldsymbol{\omega}_{AB} = \{1.17\mathbf{i} + 1.27\mathbf{j} - 0.779\mathbf{k}\}$ rad/s 

***20-28.** Rod AB is attached to collars at its ends by ball-and-socket joints. If collar A moves upward with an acceleration of $a_A = 4$ ft/s^2, determine the angular acceleration of rod AB and the magnitude of acceleration of collar B. Assume that the rod's angular acceleration is directed perpendicular to the rod.

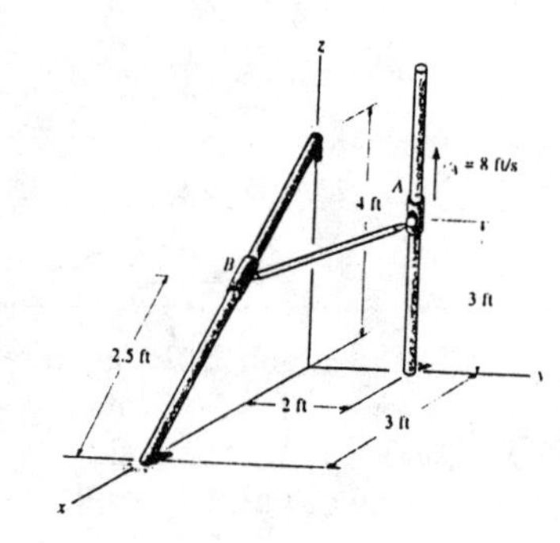

From Prob. 20 – 27

$\boldsymbol{\omega}_{AB} = \{1.1684\mathbf{i} + 1.2657\mathbf{j} - 0.7789\mathbf{k}\}$ rad/s

$\mathbf{r}_{B/A} = \{1.5\mathbf{i} - 2\mathbf{j} - 1\mathbf{k}\}$ ft

$\boldsymbol{\alpha}_{AB} = \alpha_x\mathbf{i} + \alpha_y\mathbf{j} + \alpha_z\mathbf{k}$

$\mathbf{a}_A = \{4\mathbf{k}\}$ ft/s^2 $\quad \mathbf{a}_B = -\frac{3}{5}a_B\mathbf{i} + \frac{4}{5}a_B\mathbf{k}$

$$\mathbf{a}_B = \mathbf{a}_A + \boldsymbol{\alpha}_{AB} \times \mathbf{r}_{B/A} + \boldsymbol{\omega}_{AB} \times (\boldsymbol{\omega}_{AB} \times \mathbf{r}_{B/A})$$

$$-\frac{3}{5}a_B\mathbf{i} + \frac{4}{5}a_B\mathbf{k} = 4\mathbf{k} + (\alpha_x\mathbf{i} + \alpha_y\mathbf{j} + \alpha_z\mathbf{k}) \times (1.5\mathbf{i} - 2\mathbf{j} - 1\mathbf{k})$$

$$+(1.1684\mathbf{i} + 1.2657\mathbf{j} - 0.7789\mathbf{k})$$

$$\times\left[(1.1684\mathbf{i} + 1.2657\mathbf{j} - 0.7789\mathbf{k}) \times (1.5\mathbf{i} - 2\mathbf{j} - 1\mathbf{k})\right]$$

Equating $\mathbf{i}$, $\mathbf{j}$ and $\mathbf{k}$ components

$$-\alpha_y + 2\alpha_z - 5.3607 = -\frac{3}{5}a_B \qquad (1)$$

$$\alpha_x + 1.5\alpha_z + 7.1479 = 0 \qquad (2)$$

$$7.5737 - 2\alpha_x - 1.5\alpha_y = \frac{4}{5}a_B \qquad (3)$$

Since α_{AB} is perpendicular to the axis of the rod,

$$\boldsymbol{\alpha}_{AB} \cdot \mathbf{r}_{B/A} = (\alpha_x\mathbf{i} + \alpha_y\mathbf{j} + \alpha_z\mathbf{k}) \cdot (1.5\mathbf{i} - 2\mathbf{j} - 1\mathbf{k}) = 0$$

$$1.5\alpha_x - 2\alpha_y - \alpha_z = 0 \qquad (4)$$

Solving Eqs.(1) to (4) yields :

$\alpha_x = -2.7794$ rad/s^2 $\quad \alpha_y = -0.6285$ rad/s^2 $\quad \alpha_z = -2.91213$ rad/s^2

$a_B = 17.6$ ft/s^2 **Ans**

Then $\quad \boldsymbol{\alpha}_{AB} = \{-2.78\mathbf{i} - 0.628\mathbf{j} - 2.91\mathbf{k}\}$ rad/s^2 **Ans**

20-29. The triangular plate ABC is supported at A by a ball-and-socket joint and at C by the x–z plane. The side AB lies in the x–y plane. At the instant $\theta = 60°$, $\dot{\theta} = 2$ rad/s and point C has the coordinates shown. Determine the angular velocity of the plate and the velocity of point C at this instant.

$\dot{\theta} = 2$ rad/s, 2.5 ft, 60°, $v_B = (2.5)(2) = 5$ ft/s

$$\mathbf{v}_B = -5\sin 60°\mathbf{i} + 5\cos 60°\mathbf{j}$$

$$= \{-4.33\mathbf{i} + 2.5\mathbf{j}\} \text{ ft/s}$$

$$\mathbf{v}_C = (v_C)_x\mathbf{i} + (v_C)_z\mathbf{k}$$

$$\mathbf{r}_{C/A} = \{3\mathbf{i} + 4\mathbf{k}\} \text{ ft}$$

$$\mathbf{r}_{B/A} = \{1.25\mathbf{i} + 2.165\mathbf{j}\} \text{ ft}$$

$$\mathbf{v}_B = \omega \times \mathbf{r}_{B/A}$$

$$-4.33\mathbf{i} + 2.5\mathbf{j} = \begin{vmatrix} \mathbf{i} & \mathbf{j} & \mathbf{k} \\ \omega_x & \omega_y & \omega_z \\ 1.25 & 2.165 & 0 \end{vmatrix}$$

$$-2.165\omega_z = -4.33; \quad \omega_z = 2 \text{ rad/s}$$

$$2.165\omega_x - 1.25\omega_y = 0; \quad \omega_y = 1.732\omega_x$$

$$\mathbf{v}_C = \omega \times \mathbf{r}_{C/A}$$

$$(v_C)_x\mathbf{i} + (v_C)_z\mathbf{k} = \begin{vmatrix} \mathbf{i} & \mathbf{j} & \mathbf{k} \\ \omega_x & \omega_y & 2 \\ 3 & 0 & 4 \end{vmatrix}$$

$$(v_C)_x = 4\omega_y$$

$$0 = 4\omega_x - 6; \quad \omega_x = 1.5 \text{ rad/s}$$

$$(v_C)_z = -3\omega_y$$

Solving,

$$\omega_y = 2.5981 \text{ rad/s}$$

$$(v_C)_x = 10.392 \text{ ft/s}$$

$$(v_C)_z = -7.7942 \text{ ft/s}$$

Thus,

$$\omega = \{1.50\mathbf{i} + 2.60\mathbf{j} + 2.00\mathbf{k}\} \text{ rad/s} \qquad \textbf{Ans}$$

$$\mathbf{v}_C = \{10.4\mathbf{i} - 7.79\mathbf{k}\} \text{ ft/s} \qquad \textbf{Ans}$$

20-30. The triangular plate ABC is supported at A by a ball-and-socket joint and at C by the x–z plane. The side AB lies in the x–y plane. At the instant $\theta = 60°$, $\dot{\theta} = 2$ rad/s, $\ddot{\theta} = 3$ rad/s^2, and point C has the coordinates shown. Determine the angular acceleration of the plate and the acceleration of point C at this instant.

From Prob. 20–29,

$$\omega = 1.5\mathbf{i} + 2.5981\mathbf{j} + 2\mathbf{k}$$

$$\mathbf{r}_{B/A} = 1.25\mathbf{i} + 2.165\mathbf{j}$$

$$\mathbf{v}_B = -4.33\mathbf{i} + 2.5\mathbf{j}$$

$$(a_B)_t = 3(2.5) = 7.5 \text{ ft/s}^2$$

$$(a_B)_n = (2)^2(2.5) = 10 \text{ ft/s}^2$$

$$\mathbf{a}_B = -7.5\sin 60°\mathbf{i} + 7.5\cos 60°\mathbf{j} - 10\cos 60°\mathbf{i} - 10\sin 60°\mathbf{j}$$

$$\mathbf{a}_B = -11.4952\mathbf{i} - 4.91025\mathbf{j}$$

$$\mathbf{a}_B = \alpha \times \mathbf{r}_{B/A} + \omega \times \mathbf{v}_{B/A}$$

$$-11.4952\mathbf{i} - 4.91025\mathbf{j} = \begin{vmatrix} \mathbf{i} & \mathbf{j} & \mathbf{k} \\ \alpha_x & \alpha_y & \alpha_z \\ 1.25 & 2.165 & 0 \end{vmatrix} + \begin{vmatrix} \mathbf{i} & \mathbf{j} & \mathbf{k} \\ 1.5 & 2.5981 & 2 \\ -4.33 & 2.5 & 0 \end{vmatrix}$$

$$-11.4952 = -2.165\alpha_z - 5$$

$$-4.91025 = 1.25\alpha_z - 8.66$$

$$\alpha_z = 3 \text{ rad/s}^2$$

$$0 = 2.165\alpha_x - 1.25\alpha_y + 15 \qquad (1)$$

$$\mathbf{a}_C = \alpha \times \mathbf{r}_{C/A} + \omega \times \mathbf{v}_{C/A}$$

$$\mathbf{v}_{C/A} = 10.39\mathbf{i} - 7.794\mathbf{k}$$

$$\mathbf{a}_C = (a_C)_x\mathbf{i} + (a_C)_z\mathbf{k} = \begin{vmatrix} \mathbf{i} & \mathbf{j} & \mathbf{k} \\ \alpha_x & \alpha_y & \alpha_z \\ 3 & 0 & 4 \end{vmatrix} + \begin{vmatrix} & & \\ & & \\ & & \end{vmatrix}$$

$$(a_C)_x = 4\alpha_y - 20.25 \qquad (2)$$

$$0 = 3\alpha_z - 4\alpha_x + 32.4760 \qquad (3)$$

$$(a_C)_z = -3\alpha_y - 27 \qquad (4)$$

Solving Eqs. (1)–(4),

$$\alpha_x = 10.369 \text{ rad/s}^2$$

$$\alpha_y = 29.96 \text{ rad/s}^2$$

$$(a_C)_x = 99.6 \text{ ft/s}^2$$

$$(a_C)_z = -117 \text{ ft/s}^2$$

$$\mathbf{a}_C = \{99.6\mathbf{i} - 117\mathbf{k}\} \text{ ft/s}^2 \qquad \textbf{Ans}$$

$$\alpha = \{10.4\mathbf{i} + 30.0\mathbf{j} + 3\mathbf{k}\} \text{ rad/s}^2 \qquad \textbf{Ans}$$

20-31. Solve Example 20–5 such that the x, y, z axes move with curvilinear translation, $\boldsymbol{\Omega} = \mathbf{0}$, in which case the collar appears to have both an angular velocity $\boldsymbol{\Omega}_{xyz} = \boldsymbol{\omega}_1 + \boldsymbol{\omega}_2$ and radial motion.

Relative to XYZ, let xyz have

$\boldsymbol{\Omega} = \mathbf{0} \qquad \dot{\boldsymbol{\Omega}} = \mathbf{0}$

$\mathbf{r}_B = \{-0.5\mathbf{k}\}$ m

$\mathbf{v}_B = \{2\mathbf{j}\}$ m/s

$\mathbf{a}_B = \{0.75\mathbf{j} + 8\mathbf{k}\}$ m/s^2

Relative to xyz, let $x'y'z'$ be coincident with xyz and be fixed to BD. Then

$\boldsymbol{\Omega}_{xyz} = \boldsymbol{\omega}_1 + \boldsymbol{\omega}_2 = \{4\mathbf{i} + 5\mathbf{k}\}$ rad/s $\qquad \dot{\boldsymbol{\Omega}}_{xyz} = \dot{\boldsymbol{\omega}}_1 + \dot{\boldsymbol{\omega}}_2 = \{1.5\mathbf{i} - 6\mathbf{k}\}$ rad/s^2

$(\mathbf{r}_{C/B})_{xyz} = \{0.2\mathbf{j}\}$ m

$$(\mathbf{v}_{C/B})_{xyz} = (\dot{\mathbf{r}}_{C/B})_{xyz} = (\dot{\mathbf{r}}_{C/B})_{x'y'z'} + (\boldsymbol{\omega}_1 + \boldsymbol{\omega}_2) \times (\mathbf{r}_{C/B})_{xyz}$$

$$= 3\mathbf{j} + (4\mathbf{i} + 5\mathbf{k}) \times (0.2\mathbf{j})$$

$$= \{-1\mathbf{i} + 3\mathbf{j} + 0.8\mathbf{k}\} \text{ m/s}$$

$$(\mathbf{a}_{C/B})_{xyz} = (\ddot{\mathbf{r}}_{C/B})_{xyz} = \left[(\ddot{\mathbf{r}}_{C/B})_{x'y'z'} + (\boldsymbol{\omega}_1 + \boldsymbol{\omega}_2) \times (\dot{\mathbf{r}}_{C/B})_{x'y'z'}\right]$$

$$+ \left[(\dot{\boldsymbol{\omega}}_1 + \dot{\boldsymbol{\omega}}_2) \times (\mathbf{r}_{C/B})_{xyz}\right] + \left[(\boldsymbol{\omega}_1 + \boldsymbol{\omega}_2) \times (\dot{\mathbf{r}}_{C/B})_{xyz}\right]$$

$$(\mathbf{a}_{C/B})_{xyz} = \left[2\mathbf{j} + (4\mathbf{i} + 5\mathbf{k}) \times 3\mathbf{j}\right] + \left[(1.5\mathbf{i} - 6\mathbf{k}) \times 0.2\mathbf{j}\right] + \left[(4\mathbf{i} + 5\mathbf{k}) \times (-1\mathbf{i} + 3\mathbf{j} + 0.8\mathbf{k})\right]$$

$$= \{-28.8\mathbf{i} - 6.2\mathbf{j} + 24.3\mathbf{k}\} \text{ m/s}^2$$

$$\mathbf{v}_C = \mathbf{v}_B + \boldsymbol{\Omega} \times \mathbf{r}_{C/B} + (\mathbf{v}_{C/B})_{xyz}$$

$$= 2\mathbf{j} + \mathbf{0} + (-1\mathbf{i} + 3\mathbf{j} + 0.8\mathbf{k})$$

$$= \{-1.00\mathbf{i} + 5.00\mathbf{j} + 0.800\mathbf{k}\} \text{ m/s}$$ **Ans**

$$\mathbf{a}_C = \mathbf{a}_B + \dot{\boldsymbol{\Omega}} \times \mathbf{r}_{C/B} + \boldsymbol{\Omega} \times (\boldsymbol{\Omega} \times \mathbf{r}_{C/B}) + 2\boldsymbol{\Omega} \times (\mathbf{v}_{C/B})_{xyz} + (\mathbf{a}_{C/B})_{xyz}$$

$$= (0.75\mathbf{j} + 8\mathbf{k}) + \mathbf{0} + \mathbf{0} + \mathbf{0} + (-28.8\mathbf{i} - 6.2\mathbf{j} + 24.3\mathbf{k})$$

$$= \{-28.8\mathbf{i} - 5.45\mathbf{j} + 32.3\mathbf{k}\} \text{ m/s}^2$$ **Ans**

***20-32.** Solve Example 20–5 by fixing x, y, z axes to rod BD so that $\mathbf{\Omega} = \boldsymbol{\omega}_1 + \boldsymbol{\omega}_2$. In this case the collar appears only to move radially outward along BD; hence $\mathbf{\Omega}_{xyz} = \mathbf{0}$.

Relative to XYZ, let $x'y'z'$ be coincident with XYZ and have $\Omega' = \omega_1$ and $\dot{\Omega}' = \dot{\omega}_1$.

$$\Omega = \omega_1 + \omega_2 = \{4\mathbf{i} + 5\mathbf{k}\} \text{ rad/s}$$

$$\dot{\Omega} = \dot{\omega}_1 + \dot{\omega}_2 = \left[\left(\dot{\omega}_1\right)_{x'y'z'} + \omega_1 \times \omega_1\right] + \left[\left(\dot{\omega}_2\right)_{x'y'z'} + \omega_1 \times \omega_2\right]$$

$$= (1.5\mathbf{i} + \mathbf{0}) + \left[-6\mathbf{k} + (4\mathbf{i}) \times (5\mathbf{k})\right] = \{1.5\mathbf{i} - 20\mathbf{j} - 6\mathbf{k}\} \text{ rad/s}^2$$

$$\mathbf{r}_B = \{-0.5\mathbf{k}\} \text{ m}$$

$$\mathbf{v}_B = \dot{\mathbf{r}}_B = \left(\dot{\mathbf{r}}_B\right)_{x'y'z'} + \omega_1 \times \mathbf{r}_B = \mathbf{0} + (4\mathbf{i}) \times (-0.5\mathbf{k}) = \{2\mathbf{j}\} \text{ m/s}$$

$$\mathbf{a}_B = \ddot{\mathbf{r}}_B = \left[\left(\ddot{\mathbf{r}}_B\right)_{x'y'z'} + \omega_1 \times \left(\dot{\mathbf{r}}_B\right)_{x'y'z'}\right] + \dot{\omega}_1 \times \mathbf{r}_B + \omega_1 \times \dot{\mathbf{r}}_B$$

$$= \mathbf{0} + \mathbf{0} + \left[(1.5\mathbf{i}) \times (-0.5\mathbf{k})\right] + (4\mathbf{i} \times 2\mathbf{j}) = \{0.75\mathbf{j} + 8\mathbf{k}\} \text{ m/s}^2$$

Relative to $x'y'z'$, let xyz have

$$\mathbf{\Omega}_{x'y'z'} = \mathbf{0}; \qquad \dot{\mathbf{\Omega}}_{x'y'z'} = \mathbf{0};$$

$$\left(r_{C/B}\right)_{xyz} = \{0.2\mathbf{j}\} \text{ m}$$

$$(\mathbf{v}_{C/B})_{xyz} = \{3\mathbf{j}\} \text{ m/s}$$

$$(\mathbf{a}_{C/B})_{xyz} = \{2\mathbf{j}\} \text{ m/s}^2$$

$$\mathbf{v}_C = \mathbf{v}_B + \mathbf{\Omega} \times \mathbf{r}_{C/B} + (\mathbf{v}_{C/B})_{xyz}$$

$$= 2\mathbf{j} + \left[(4\mathbf{i} + 5\mathbf{k}) \times (0.2\mathbf{j})\right] + 3\mathbf{j}$$

$$= \{-1\mathbf{i} + 5\mathbf{j} + 0.8\mathbf{k}\} \text{ m/s} \qquad \textbf{Ans}$$

$$\mathbf{a}_C = \mathbf{a}_B + \dot{\mathbf{\Omega}} \times \mathbf{r}_{C/B} + \mathbf{\Omega} \times (\mathbf{\Omega} \times \mathbf{r}_{C/B}) + 2\mathbf{\Omega} \times (\mathbf{v}_{C/B})_{xyz} + (\mathbf{a}_{C/B})_{xyz}$$

$$= (0.75\mathbf{j} + 8\mathbf{k}) + \left[(1.5\mathbf{i} - 20\mathbf{j} - 6\mathbf{k}) \times (0.2\mathbf{j})\right] + (4\mathbf{i} + 5\mathbf{k}) \times \left[(4\mathbf{i} + 5\mathbf{k}) \times (0.2\mathbf{j})\right] + 2\left[(4\mathbf{i} + 5\mathbf{k}) \times (3\mathbf{j})\right] + 2\mathbf{j}$$

$$\mathbf{a}_C = \{-28.8\mathbf{i} - 5.45\mathbf{j} + 32.3\mathbf{k}\} \text{ m/s}^2 \qquad \textbf{Ans}$$

20-33. At a given instant, the antenna has angular motion $\omega_1 = 3$ rad/s and $\dot{\omega}_1 = 2$ rad/s^2 about the z axis. At this same instant $\theta = 30°$, the angular motion about the x axis is $\omega_2 = 1.5$ rad/s and $\dot{\omega}_2 = 4$ rad/s^2. Determine the velocity and acceleration of the signal horn A at this instant. The distance from O to A is $d = 3$ ft.

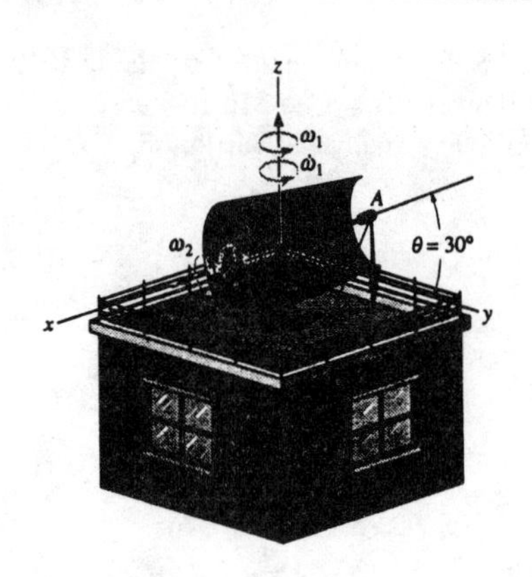

Relative to XYZ, let xyz have

$\Omega = \omega_1 = \{3\mathbf{k}\}$ rad/s $\quad \dot{\Omega} = \dot{\omega}_1 = \{2\mathbf{k}\}$ rad/s^2 (Ω does not change direction relative to XYZ.)

$\mathbf{r}_O = \mathbf{0}, \quad \mathbf{v}_O = \mathbf{0}, \quad \mathbf{a}_O = \mathbf{0}$

Relative to xyz, let $x'y'z'$ be coincident with xyz so that

$\Omega_{xyz} = \{1.5\mathbf{i}\}$ rad/s, $\quad \dot{\Omega}_{xyz} = \{4\mathbf{i}\}$ rad/s^2 (Ω_{xyz} does not change direction relative to xyz.)

$(\mathbf{r}_{A/O})_{xyz} = 3\cos 30°\mathbf{j} + 3\sin 30°\mathbf{k} = \{2.598\mathbf{j} + 1.5\mathbf{k}\}$ ft $\quad$ ($(\mathbf{r}_{A/O})_{xyz}$ changes direction relative to xyz)

$$(\mathbf{v}_{A/O})_{xyz} = \left(\dot{\mathbf{r}}_{A/O}\right)_{xyz} = \left(\dot{\mathbf{r}}_{A/O}\right)_{x'y'z'} + \Omega_{xyz} \times \mathbf{r}_{A/O} = \mathbf{0} + (1.5\mathbf{i}) \times (2.598\mathbf{j} + 1.5\mathbf{k})$$

$$= \{-2.25\mathbf{j} + 3.897\mathbf{k}\} \text{ ft/s}$$

$$(\mathbf{a}_{A/O})_{xyz} = \left(\ddot{\mathbf{r}}_{A/O}\right)_{xyz} = \left[\left(\ddot{\mathbf{r}}_{A/O}\right)_{x'y'z'} + \Omega_{xyz} \times \left(\dot{\mathbf{r}}_{A/O}\right)_{x'y'z'}\right] + \dot{\Omega}_{xyz} \times \left(\mathbf{r}_{A/O}\right)_{xyz} + \Omega_{xyz} \times \left(\dot{\mathbf{r}}_{A/O}\right)_{xyz}$$

$$= \mathbf{0} + \mathbf{0} + (4\mathbf{i}) \times (2.598\mathbf{j} + 1.5\mathbf{k}) + (1.5\mathbf{i}) \times (-2.25\mathbf{j} + 3.897\mathbf{k}) = \{-11.846\mathbf{j} + 7.017\mathbf{k}\} \text{ ft/s}^2$$

Thus,

$$\mathbf{v}_A = \mathbf{v}_O + \Omega \times \mathbf{r}_{A/O} + (\mathbf{v}_{A/O})_{xyz}$$

$$= \mathbf{0} + (3\mathbf{k}) \times (2.598\mathbf{j} + 1.5\mathbf{k}) + (-2.25\mathbf{j} + 3.897\mathbf{k})$$

$$\mathbf{v}_A = \{-7.79\mathbf{i} - 2.25\mathbf{j} + 3.90\mathbf{k}\} \text{ ft/s} \qquad \textbf{Ans}$$

$$\mathbf{a}_A = \mathbf{a}_O + \dot{\Omega} \times \mathbf{r}_{A/O} + \Omega \times \left(\Omega \times \mathbf{r}_{A/O}\right) + 2\Omega \times (\mathbf{v}_{A/O})_{xyz} + (\mathbf{a}_{A/O})_{xyz}$$

$$= \mathbf{0} + (2\mathbf{k}) \times (2.60\mathbf{j} + 1.5\mathbf{k}) + (3\mathbf{k}) \times \left[(3\mathbf{k}) \times (2.60\mathbf{j} + 1.5\mathbf{k})\right]$$

$$+ 2(3\mathbf{k}) \times (-2.25\mathbf{j} + 3.90\mathbf{k}) + (-11.85\mathbf{j} + 7.017\mathbf{k})$$

$$\mathbf{a}_A = \{8.30\mathbf{i} - 35.2\mathbf{j} + 7.02\mathbf{k}\} \text{ ft/s}^2 \qquad \textbf{Ans}$$

20-34. The boom AB of the crane is rotating about the z axis with an angular velocity $\omega_z = 0.75$ rad/s and an angular acceleration of $\dot{\omega}_z = 2$ rad/s^2. At the same instant, $\theta = 60°$ and the boom is rotating upward at a constant rate $\dot{\theta} = 0.5$ rad/s. Determine the velocity and acceleration of the tip B of the boom at this instant.

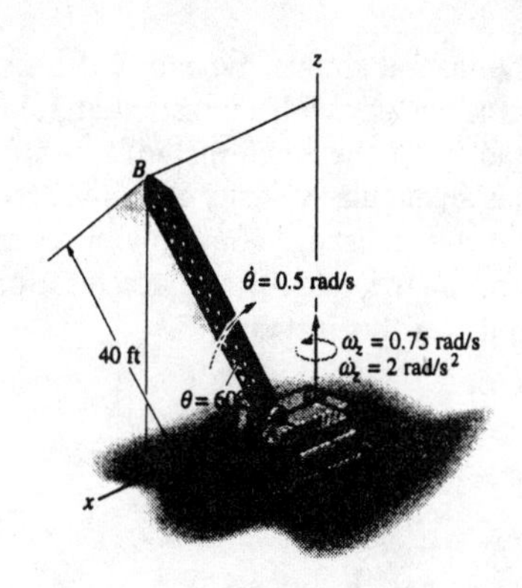

Relative to XYZ, let xyz have

$\Omega = \omega_z = \{0.75\mathbf{k}\}$ rad/s, $\quad \dot{\Omega} = \dot{\omega}_z = \{2\mathbf{k}\}$ rad/s^2 (Ω does not change direction relative to XYZ.)

$\mathbf{r}_A = (\mathbf{r}_A)_{xyz} = \{5\mathbf{i}\}$ ft ($\mathbf{r}_A$ changes direction relative to XYZ.)

$$\mathbf{v}_A = \dot{\mathbf{r}}_A = \left(\dot{\mathbf{r}}_A\right)_{xyz} + \Omega \times \mathbf{r}_A = \mathbf{0} + (0.75\mathbf{k}) \times (5\mathbf{i}) = \{3.75\mathbf{j}\} \text{ ft/s}$$

$$\mathbf{a}_A = \ddot{\mathbf{r}}_A = \left[\left(\ddot{\mathbf{r}}_A\right)_{xyz} + \Omega \times \left(\dot{\mathbf{r}}_A\right)_{xyz}\right] + \dot{\Omega} \times \mathbf{r}_A + \Omega \times \dot{\mathbf{r}}_A$$

$$= \mathbf{0} + \mathbf{0} + (2\mathbf{k}) \times (5\mathbf{i}) + (0.75\mathbf{k}) \times (3.75\mathbf{j}) = \{-2.8125\mathbf{i} + 10\mathbf{j}\} \text{ ft/s}^2$$

Relative to xyz, let $x'y'z'$ be located at A, be parallel to xyz and have

$\Omega_{xyz} = \{-0.5\mathbf{j}\}$ rad/s, $\quad \dot{\Omega}_{xyz} = \mathbf{0}$ (Ω_{xyz} does not change direction relative to xyz.)

$(\mathbf{r}_{B/A})_{xyz} = 40\cos 60°\mathbf{i} + 40\sin 60°\mathbf{k} = \{20\mathbf{i} + 34.64\mathbf{k}\}$ ft/s ($(\mathbf{r}_{B/A})_{xyz}$ changes direction relative to xyz.)

$$(\mathbf{v}_{B/A})_{xyz} = (\dot{\mathbf{r}}_{B/A})_{xyz} = \left(\dot{\mathbf{r}}_{B/A}\right)_{x'y'z'} + \Omega_{x'y'z'} \times (\mathbf{r}_{B/A})_{xyz} = \mathbf{0} + (-0.5\mathbf{j}) \times (20\mathbf{i} + 34.64\mathbf{k}) = \{-17.32\mathbf{i} + 10\mathbf{k}\} \text{ ft/s}$$

$$(\mathbf{a}_{B/A})_{xyz} = (\ddot{\mathbf{r}}_{B/A})_{xyz} = \left[\left(\ddot{\mathbf{r}}_{B/A}\right)_{x'y'z'} + \Omega_{B/O} \times \left(\dot{\mathbf{r}}_{B/A}\right)_{x'y'z'}\right] + \dot{\Omega}_{x'y'z'} \times (\mathbf{r}_{B/A})_{xyz} + \Omega_{x'y'z'} \times (\dot{\mathbf{r}}_{B/A})_{xyz}$$

$$= \mathbf{0} + \mathbf{0} + \mathbf{0} + (-0.5\mathbf{j}) \times (-17.32\mathbf{i} + 10\mathbf{k}) = \{-5\mathbf{i} - 8.66\mathbf{k}\} \text{ ft/s}^2$$

Thus,

$$\mathbf{v}_B = \mathbf{v}_A + \Omega \times \mathbf{r}_{B/A} + (\mathbf{v}_{B/A})_{xyz} = (3.75\mathbf{j}) + (0.75\mathbf{k}) \times (20\mathbf{i} + 34.64\mathbf{k}) + (-17.32\mathbf{i} + 10\mathbf{k})$$

$\mathbf{v}_B = \{-17.3\mathbf{i} + 18.8\mathbf{j} + 10.0\mathbf{k}\}$ ft/s **Ans**

$$\mathbf{a}_B = \mathbf{a}_A + \dot{\Omega} \times \mathbf{r}_{B/A} + \Omega \times \left(\Omega \times \mathbf{r}_{B/A}\right) + 2\Omega \times (\mathbf{v}_{B/A})_{xyz} + (\mathbf{a}_{B/A})_{xyz}$$

$$= (-2.8125\mathbf{i} + 10\mathbf{j}) + (2\mathbf{k}) \times (20\mathbf{i} + 34.64\mathbf{k}) + (0.75\mathbf{k}) \times \left[(0.75\mathbf{k}) \times (20\mathbf{i} + 34.64\mathbf{k})\right]$$

$$+ 2(0.75\mathbf{k}) \times (-17.32\mathbf{i} + 10\mathbf{k}) + (-5\mathbf{i} - 8.66\mathbf{k})$$

$\mathbf{a}_B = \{-19.1\mathbf{i} + 24.0\mathbf{j} - 8.66\mathbf{k}\}$ ft/s^2 **Ans**

20-35. At the instant shown, the cab of the excavator is rotating about the z axis with a constant angular velocity of $\omega_z = 0.3$ rad/s. At the same instant $\theta = 60°$, and the boom OBC has an angular velocity of $\dot{\theta} = 0.6$ rad/s, which is increasing at $\ddot{\theta} = 0.2$ rad/s^2, both measured relative to the cab. Determine the velocity and acceleration of point C on the grapple at this instant.

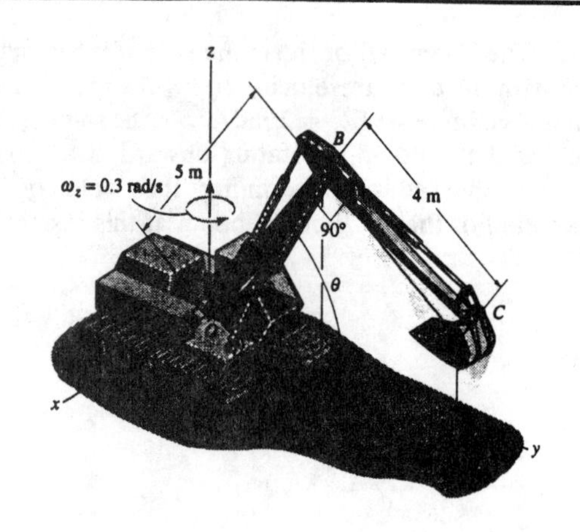

Relative to XYZ, let xyz have

$\Omega = \{0.3\mathbf{k}\}$ rad/s, $\quad \dot{\Omega} = \mathbf{0}$ (Ω does not change direction relative to XYZ.)

$\mathbf{r}_O = \mathbf{0}, \quad \mathbf{v}_O = \mathbf{0}, \quad \mathbf{a}_O = \mathbf{0}$

Relative to xyz, let $x'y'z'$ be coincident with xyz at O and have

$\Omega_{xyz} = \{0.6\mathbf{i}\}$ rad/s, $\quad \dot{\Omega}_{xyz} = \{0.2\mathbf{i}\}$ rad/s^2 (Ω_{xyz} does not change direction relative to XYZ.)

$$(\mathbf{r}_{C/O})_{xyz} = (5\cos 60° + 4\cos 30°)\mathbf{j} + (5\sin 60° - 4\sin 30°)\mathbf{k} = \{5.9641\mathbf{j} + 2.3301\mathbf{k}\} \text{ m}$$

($\mathbf{r}_{C/O})_{xyz}$ changes direction relative to XYZ.)

$$(\mathbf{v}_{C/O})_{xyz} = \left(\dot{\mathbf{r}}_{C/O}\right)_{xyz} = \left(\dot{\mathbf{r}}_{C/O}\right)_{x'y'z'} + \Omega_{xyz} \times \left(\mathbf{r}_{C/O}\right)_{xyz}$$

$$= \mathbf{0} + (0.6\mathbf{i}) \times (5.9641\mathbf{j} + 2.3301\mathbf{k}) = \{-1.3981\mathbf{j} + 3.5785\mathbf{k}\} \text{ m/s}$$

$$(\mathbf{a}_{C/O})_{xyz} = \left(\ddot{\mathbf{r}}_{C/O}\right)_{xyz} = \left[\left(\ddot{\mathbf{r}}_{C/O}\right)_{x'y'z'} + \Omega_{xyz} \times \left(\dot{\mathbf{r}}_{C/O}\right)_{x'y'z'}\right] + \dot{\Omega}_{xyz} \times \mathbf{r}_{C/O} + \Omega_{xyz} \times \dot{\mathbf{r}}_{C/O}$$

$$= \left[\mathbf{0} + \mathbf{0}\right] + (0.2\mathbf{i}) \times (5.9641\mathbf{j} + 2.3301\mathbf{k}) + (0.6\mathbf{i}) \times (-1.3981\mathbf{j} + 3.5785\mathbf{k})$$

$$= \{-2.61310\mathbf{j} + 0.35397\mathbf{k}\} \text{ m/s}^2$$

Thus,

$$\mathbf{v}_C = \mathbf{v}_O + \Omega \times \mathbf{r}_{C/O} + (\mathbf{v}_{C/O})_{xyz} = \mathbf{0} + (0.3\mathbf{k}) \times (5.9641\mathbf{j} + 2.3301\mathbf{k}) - 1.3981\mathbf{j} + 3.5785\mathbf{k}$$

$$= \{-1.79\mathbf{i} - 1.40\mathbf{j} + 3.58\mathbf{k}\} \text{ m/s} \qquad \textbf{Ans}$$

$$\mathbf{a}_C = \mathbf{a}_O + \dot{\Omega} \times \mathbf{r}_{C/O} + \Omega \times \left(\Omega \times \mathbf{r}_{C/O}\right) + 2\Omega \times (\mathbf{v}_{C/O})_{xyz} + (\mathbf{a}_{C/O})_{xyz}$$

$$= \mathbf{0} + \mathbf{0} + (0.3\mathbf{k}) \times \left[(0.3\mathbf{k}) \times (5.9641\mathbf{j} + 2.3301\mathbf{k})\right]$$

$$+ 2(0.3\mathbf{k}) \times (-1.3981\mathbf{j} + 3.5785\mathbf{k}) - 2.61310\mathbf{j} + 0.35397\mathbf{k}$$

$$= \{0.839\mathbf{i} - 3.15\mathbf{j} + 0.354\mathbf{k}\} \text{ m/s}^2 \qquad \textbf{Ans}$$

***20-36.** At the instant shown, the frame of the excavator is traveling forward in the y direction with a velocity of 2 m/s and an acceleration of 1 m/s², while the cab is rotating about the z axis with an angular velocity of $\omega_z = 0.3$ rad/s, which is increasing at $\alpha_z = 0.4$ rad/s². At the same instant $\theta = 60°$, and the boom OBC has an angular velocity of $\dot{\theta} = 0.6$ rad/s, which is increasing at $\ddot{\theta} = 0.2$ rad/s², both measured relative to the cab. Determine the velocity and acceleration of point C on the grapple at this instant.

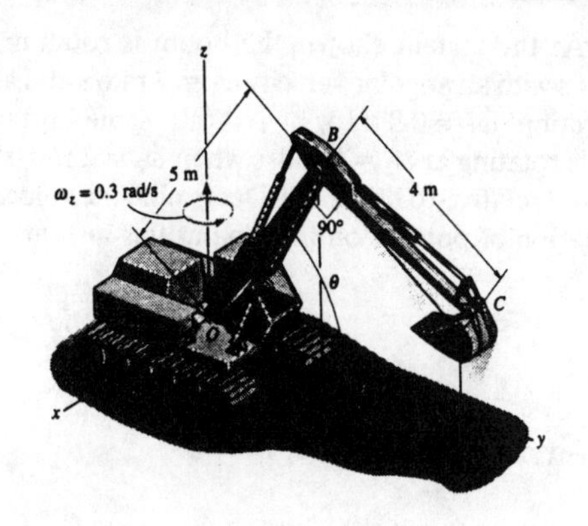

Relative to XYZ, let xyz have

$\mathbf{\Omega} = \{0.3\mathbf{k}\}$ rad/s, $\quad \dot{\mathbf{\Omega}} = \{0.4\mathbf{k}\}$ rad/s² ($\mathbf{\Omega}$ does not change direction relative to XYZ.)

$\mathbf{r}_O = \mathbf{0}$ ($\mathbf{r}_O$ does not change direction relative to XYZ.)

$\mathbf{v}_O = \{2\mathbf{j}\}$ m/s

$\mathbf{a}_O = \{1\mathbf{j}\}$ m/s²

Relative to xyz, let $x'y'z'$ have

$\mathbf{\Omega}_{xyz} = \{0.6\mathbf{i}\}$ rad/s, $\quad \dot{\mathbf{\Omega}}_{xyz} = \{0.2\mathbf{i}\}$ rad/s² ($\mathbf{\Omega}_{xyz}$ does not change direction relative to xyz.)

$(\mathbf{r}_{C/O})_{xyz} = (5\cos 60° + 4\cos 30°)\mathbf{j} + (5\sin 60° - 4\sin 30°)\mathbf{k} = \{5.9641\mathbf{j} + 2.3301\mathbf{k}\}$ m

($(\mathbf{r}_{C/O})_{xyz}$ changes direction relative to xyz.)

$$(\mathbf{v}_{C/O})_{xyz} = \left(\dot{\mathbf{r}}_{C/O}\right)_{xyz} = \left(\dot{\mathbf{r}}_{C/O}\right)_{x'y'z'} + \mathbf{\Omega}_{xyz} \times \left(\mathbf{r}_{C/O}\right)_{xyz}$$

$$= \mathbf{0} + (0.6\mathbf{i}) \times (5.9641\mathbf{j} + 2.3301\mathbf{k}) = \{-1.3981\mathbf{j} + 3.5785\mathbf{k}\} \text{ m/s}$$

$$(\mathbf{a}_{C/O})_{xyz} = \left(\ddot{\mathbf{r}}_{C/O}\right)_{xyz} = \left[\left(\ddot{\mathbf{r}}_{C/O}\right)_{x'y'z'} + \mathbf{\Omega}_{xyz} \times \left(\dot{\mathbf{r}}_{C/O}\right)_{x'y'z'}\right] + \dot{\mathbf{\Omega}}_{xyz} \times \left(\mathbf{r}_{C/O}\right)_{xyz} + \mathbf{\Omega}_{xyz} \times \left(\dot{\mathbf{r}}_{C/O}\right)_{xyz}$$

$$= [\mathbf{0} + \mathbf{0}] + (0.2\mathbf{i}) \times (5.9641\mathbf{j} + 2.3301\mathbf{k}) + (0.6\mathbf{i}) \times (-1.3981\mathbf{j} + 3.5785\mathbf{k})$$

$$= \{-2.61310\mathbf{j} + 0.35397\mathbf{k}\} \text{ m/s}^2$$

Thus,

$$\mathbf{v}_C = \mathbf{v}_O + \mathbf{\Omega} \times \mathbf{r}_{C/O} + (\mathbf{v}_{C/O})_{xyz} = 2\mathbf{j} + (0.3\mathbf{k}) \times (5.9641\mathbf{j} + 2.3301\mathbf{k}) - 1.3981\mathbf{j} + 3.5785\mathbf{k}$$

$$= \{-1.79\mathbf{i} + 0.602\mathbf{j} + 3.58\mathbf{k}\} \text{ m/s} \qquad \textbf{Ans}$$

$$\mathbf{a}_C = \mathbf{a}_O + \dot{\mathbf{\Omega}} \times \mathbf{r}_{C/O} + \mathbf{\Omega} \times \left(\mathbf{\Omega} \times \mathbf{r}_{C/O}\right) + 2\mathbf{\Omega} \times (\mathbf{v}_{C/O})_{xyz} + (\mathbf{a}_{C/O})_{xyz}$$

$$= 1\mathbf{j} + 0.4\mathbf{k} \times (5.9641\mathbf{j} + 2.3301\mathbf{k}) + (0.3\mathbf{k}) \times [(0.3\mathbf{k}) \times (5.9641\mathbf{j} + 2.3301\mathbf{k})]$$

$$+ 2(0.3\mathbf{k}) \times (-1.3981\mathbf{j} + 3.5785\mathbf{k}) - 2.61310\mathbf{j} + 0.35397\mathbf{k}$$

$$= \{-1.55\mathbf{i} - 2.15\mathbf{j} + 0.354\mathbf{k}\} \text{ m/s}^2 \qquad \textbf{Ans}$$

20-37. At the instant shown, the boom is rotating about the z axis with an angular velocity $\omega_1 = 2$ rad/s and angular acceleration $\dot{\omega}_1 = 0.8$ rad/s^2. At this same instant the swivel is rotating at $\omega_2 = 3$ rad/s when $\dot{\omega}_2 = 2$ rad/s^2, both measured relative to the boom. Determine the velocity and acceleration of point P on the pipe at this instant.

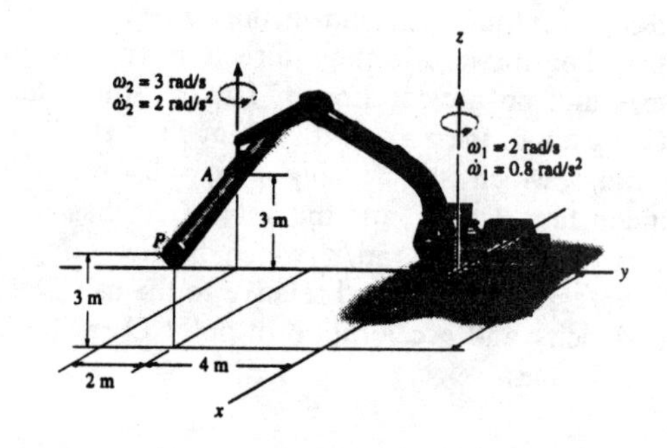

Relative to XYZ, let xyz have

$\Omega = \{2\mathbf{k}\}$ rad/s $\quad \dot{\Omega} = \{0.8\mathbf{k}\}$ rad/s^2 (Ω does not change direction relative to XYZ.)

$\mathbf{r}_A = \{-6\mathbf{j} + 3\mathbf{k}\}$ m ($\mathbf{r}_A$ changes direction relative to XYZ.)

$$\mathbf{v}_A = \dot{\mathbf{r}}_A = (\dot{\mathbf{r}}_A)_{xyz} + \Omega \times \mathbf{r}_A = 0 + (2\mathbf{k}) \times (-6\mathbf{j} + 3\mathbf{kj}) = \{12\mathbf{i}\} \text{ m/s}$$

$$\mathbf{a}_A = \ddot{\mathbf{r}}_A = \left[(\ddot{\mathbf{r}}_A)_{xyz} + \Omega \times (\dot{\mathbf{r}}_A)_{xyz}\right] + \dot{\Omega} \times \mathbf{r}_A + \Omega \times \dot{\mathbf{r}}_A$$

$$= 0 + 0 + (0.8\mathbf{k}) \times (-6\mathbf{j} + 3\mathbf{k}) + (2\mathbf{k}) \times (12\mathbf{i})$$

$$= \{4.8\mathbf{i} + 24\mathbf{j}\} \text{ m/s}^2$$

Relative to xyz, let $x'y'z'$ have the origin at A and

$\Omega_{xyz} = \{3\mathbf{k}\}$ rad/s $\quad \dot{\Omega}_{xyz} = \{2\mathbf{k}\}$ rad/s^2 (Ω_{xyz} does not change direction relative to xyz.)

$(\mathbf{r}_{P/A})_{xyz} = \{4\mathbf{i} + 2\mathbf{j}\}$ m ($(\mathbf{r}_{P/A})_{xyz}$ changes direction relative to xyz.)

$$(\mathbf{v}_{P/A})_{xyz} = (\dot{\mathbf{r}}_{P/A})_{xyz} = (\dot{\mathbf{r}}_{P/A})_{x'y'z'} + \Omega_{xyz} \times (\mathbf{r}_{P/A})_{xyz}$$

$$= 0 + (3\mathbf{k}) \times (4\mathbf{i} + 2\mathbf{j})$$

$$= \{-6\mathbf{i} + 12\mathbf{j}\} \text{ m/s}$$

$$(\mathbf{a}_{P/A})_{xyz} = (\ddot{\mathbf{r}}_{P/A})_{xyz} = \left[(\ddot{\mathbf{r}}_{P/A})_{x'y'z'} + \Omega_{xyz} \times (\dot{\mathbf{r}}_{P/A})_{x'y'z'}\right] + \left[\dot{\Omega}_{xyz} \times (\mathbf{r}_{P/A})_{xyz}\right] + \left[\Omega_{xyz} \times (\dot{\mathbf{r}}_{P/A})_{xyz}\right]$$

$$= 0 + 0 + \left[(2\mathbf{k}) \times (4\mathbf{i} + 2\mathbf{j})\right] + \left[(3\mathbf{k}) \times (-6\mathbf{i} + 12\mathbf{j})\right]$$

$$= \{-40\mathbf{i} - 10\mathbf{j}\} \text{ m/s}^2$$

Thus,

$$\mathbf{v}_P = \mathbf{v}_A + \Omega \times \mathbf{r}_{P/A} + (\mathbf{v}_{P/A})_{xyz}$$

$$= 12\mathbf{i} + \left[(2\mathbf{k}) \times (4\mathbf{i} + 2\mathbf{j})\right] + (-6\mathbf{i} + 12\mathbf{j})$$

$$= \{2\mathbf{i} + 20\mathbf{j}\} \text{ m/s} \qquad \textbf{Ans}$$

$$\mathbf{a}_P = \mathbf{a}_A + \dot{\Omega} \times \mathbf{r}_{P/A} + \Omega \times (\Omega \times \mathbf{r}_{P/A}) + 2\Omega \times (\mathbf{v}_{P/A})_{xyz} + (\mathbf{a}_{P/A})_{xyz}$$

$$= (4.8\mathbf{i} + 24\mathbf{j}) + \left[(0.8\mathbf{k}) \times (4\mathbf{i} + 2\mathbf{j})\right] + 2\mathbf{k} \times \left[2\mathbf{k} \times (4\mathbf{i} + 2\mathbf{j})\right]$$

$$+ \left[2(2\mathbf{k}) \times (-6\mathbf{i} + 12\mathbf{j})\right] + (-40\mathbf{i} - 10\mathbf{j})$$

$$= \{-101\mathbf{i} - 14.8\mathbf{j}\} \text{ m/s}^2 \qquad \textbf{Ans}$$

20-38. At the instant shown, the frame of the brush cutter is traveling forward in the x direction with a constant velocity of 1 m/s, and the cab is rotating about the vertical axis with a constant angular velocity of $\omega_1 = 0.5$ rad/s. At the same instant the boom AB has an angular velocity of $\dot{\theta} = 0.8$ rad/s in the direction shown. Determine the velocity and acceleration of point B at the connection to the mower at this instant.

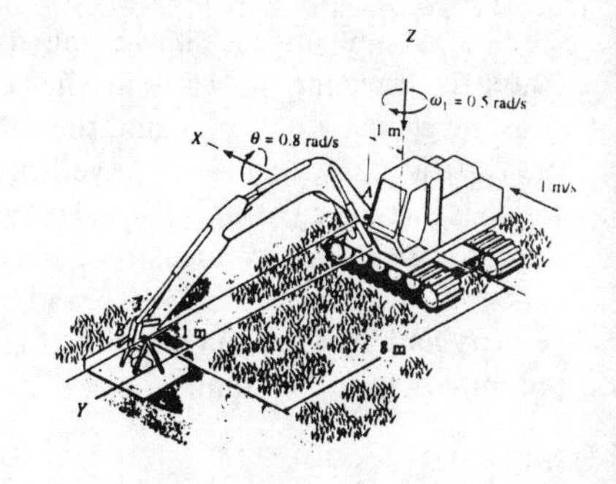

Let $x'y'z'$ with the origin O be coincident with XYZ, then $x'y'z'$ has translation of

$\mathbf{v}_O = \{1\mathbf{i}\}$ m/s
$\mathbf{a}_O = \mathbf{0}$

Relative to $x'y'z'$, let xyz have origin at A and

$\Omega = \{-0.5\mathbf{k}\}$ rad/s, $\quad \dot{\Omega} = \mathbf{0}$ (Ω does not change direction relative to $x'y'z'$.)

$\mathbf{r}_A = \{1\mathbf{i}\}$ m $\mathbf{r}_A$ changes direction relative to XYZ.)

$$\mathbf{v}_A = \dot{\mathbf{r}}_A = \left(\dot{\mathbf{r}}_A\right)_{x'y'z'} + \Omega \times \mathbf{r}_A = \mathbf{0} + (-0.5\mathbf{k}) \times (1\mathbf{j}) = \{-0.5\mathbf{j}\} \text{ m/s}$$

$$\mathbf{a}_A = \ddot{\mathbf{r}}_A = \left[\left(\ddot{\mathbf{r}}_A\right)_{x'y'z'} + \Omega \times \left(\dot{\mathbf{r}}_A\right)_{x'y'z'}\right] + \dot{\Omega} \times \mathbf{r}_A + \Omega \times \dot{\mathbf{r}}_A$$

$$= [\mathbf{0} + \mathbf{0}] + \mathbf{0} + (-0.5\mathbf{k}) \times (-0.5\mathbf{j}) = \{-0.25\mathbf{i}\} \text{ m/s}^2$$

Relative to xyz, let $x''y''z''$ have

$\Omega_{xyz} = \{0.8\mathbf{i}\}$ rad/s, $\quad \dot{\Omega}_{xyz} = \mathbf{0}$ (Ω_{xyz} does not change direction relative to xyz.)

$(\mathbf{r}_{B/A})_{xyz} = \{8\mathbf{j}\}$ m $\;((\mathbf{r}_{B/A})_{x''y''z''}$ changes direction relative to xyz.)

$$(\mathbf{v}_{B/A})_{xyz} = \left(\dot{\mathbf{r}}_{B/A}\right)_{xyz} = \left(\dot{\mathbf{r}}_{B/A}\right)_{x'y'z'} + \Omega_{xyz} \times \left(\mathbf{r}_{B/A}\right)_{xyz} = \mathbf{0} + (0.8\mathbf{i}) \times (8\mathbf{j}) = \{6.4\mathbf{k}\} \text{ m/s}$$

$$(\mathbf{a}_{B/A})_{xyz} = \left(\ddot{\mathbf{r}}_{B/A}\right)_{xyz} = \left[\left(\ddot{\mathbf{r}}_{B/A}\right)_{x''y''z''} + \Omega_{xyz} \times \left(\dot{\mathbf{r}}_{B/A}\right)_{x''y''z''}\right] + \dot{\Omega}_{xyz} \times \left(\mathbf{r}_{B/A}\right)_{xyz} + \Omega_{xyz} \times \left(\dot{\mathbf{r}}_{B/A}\right)_{xyz}$$

$$= \mathbf{0} + \mathbf{0} + \mathbf{0} + (0.8\mathbf{i}) \times (6.4\mathbf{k}) = \{-5.12\mathbf{j}\} \text{ m/s}^2$$

Thus,

$$\mathbf{v}_B = \mathbf{v}_A + \Omega \times \mathbf{r}_{B/A} + (\mathbf{v}_{B/A})_{xyz} = (-0.5\mathbf{j}) + (-0.5\mathbf{k}) \times (8\mathbf{j}) + (6.4\mathbf{k})$$

$$= \{4\mathbf{i} - 0.5\mathbf{j} + 6.4\mathbf{k}\} \text{ m/s}$$

$$\mathbf{a}_B = \mathbf{a}_A + \dot{\Omega} \times \mathbf{r}_{B/A} + \Omega \times \left(\Omega \times \mathbf{r}_{B/A}\right) + 2\Omega \times (\mathbf{v}_{B/A})_{xyz} + (\mathbf{a}_{B/A})_{xyz}$$

$$= (-0.25\mathbf{i}) + \mathbf{0} + (-0.5\mathbf{k}) \times [(-0.5\mathbf{k}) \times (8\mathbf{j})] + 2(-0.5\mathbf{k}) \times (6.4\mathbf{k}) - 5.12\mathbf{j}$$

$$= \{-0.25\mathbf{i} - 7.12\mathbf{j}\} \text{ m/s}^2$$

These results are measured from the $x'y'z'$ axes. Thus, relative to the XYZ axes

$\mathbf{v}_B = 1\mathbf{i} + (4\mathbf{i} - 0.5\mathbf{j} + 6.4\mathbf{k}) = \{5\mathbf{i} - 0.5\mathbf{j} + 6.4\mathbf{k}\}$ m/s **Ans**

$\mathbf{a}_B = \mathbf{0} + (-0.25\mathbf{i} - 7.12\mathbf{j}) = \{-0.25\mathbf{i} - 7.12\mathbf{j}\}$ m/s^2 **Ans**

20-39. At the instant shown, the frame of the brush cutter is traveling forward in the x direction with a constant velocity of 1 m/s, and the cab is rotating about the vertical axis with an angular velocity of $\omega_1 = 0.5$ rad/s, which is increasing at $\dot{\omega}_1 = 0.4$ rad/s^2. At the same instant the boom AB has an angular velocity of $\dot{\theta} = 0.8$ rad/s, which is increasing at $\ddot{\theta} = 0.9$ rad/s^2. Determine the velocity and acceleration of point B at the connection to the mower at this instant.

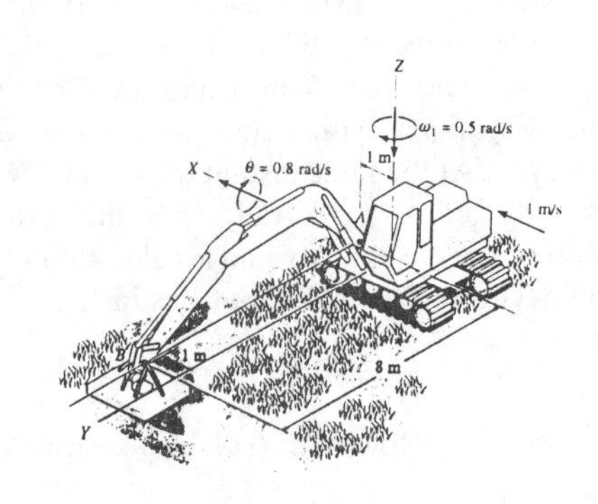

Let $x'y'z'$ with the origin O be coincident with XYZ, then $x'y'z'$ has translation of

$\mathbf{v}_O = \{1\mathbf{i}\}$ m/s

$\mathbf{a}_O = \mathbf{0}$

Relative to $x'y'z'$, let xyz have origin at A and

$\boldsymbol{\Omega} = \{-0.5\mathbf{k}\}$ rad/s, $\quad \dot{\boldsymbol{\Omega}} = \mathbf{0}$ (Ω does not change direction relative to $x'y'z'$.)

$\mathbf{r}_A = \{1\mathbf{i}\}$ m ($\mathbf{r}_A$ changes direction relative to $x'y'z'$.)

$$\mathbf{v}_A = \dot{\mathbf{r}}_A = \left(\dot{\mathbf{r}}_A\right)_{x'y'z'} + \boldsymbol{\Omega}\times\mathbf{r}_A = \mathbf{0} + (-0.5\mathbf{k})\times(1\mathbf{i}) = \{-0.5\mathbf{j}\} \text{ m/s}$$

$$\mathbf{a}_A = \ddot{\mathbf{r}}_A = \left[\left(\ddot{\mathbf{r}}_A\right)_{x'y'z'} + \boldsymbol{\Omega}\times\left(\dot{\mathbf{r}}_A\right)_{x'y'z'}\right] + \dot{\boldsymbol{\Omega}}\times\mathbf{r}_A + \boldsymbol{\Omega}\times\dot{\mathbf{r}}_A$$

$$= [\mathbf{0}+\mathbf{0}] + (-0.4\mathbf{k})\times(1\mathbf{i}) + (-0.5\mathbf{k})\times(-0.5\mathbf{j}) = \{-0.25\mathbf{i} - 0.4\mathbf{j}\} \text{ m/s}^2$$

Relative to xyz, let conicident $x'y''z''$ have

$\boldsymbol{\Omega}_{xyz} = \{0.8\mathbf{i}\}$ rad/s, $\quad \dot{\boldsymbol{\Omega}}_{xyz} = \{0.9\mathbf{i}\}$ rad/s^2 (Ω_{xyz} does not change direction relative to xyz.)

$(\mathbf{r}_{B/A})_{xyz} = \{8\mathbf{j}\}$ m ($(\mathbf{r}_{B/A})_{xyz}$ changes direction relative to xyz.)

$$(\mathbf{v}_{B/A})_{xyz} = \left(\dot{\mathbf{r}}_{B/A}\right)_{xyz} = \left(\dot{\mathbf{r}}_{B/A}\right)_{x''y''z''} + \boldsymbol{\Omega}_{xyz}\times\left(\mathbf{r}_{B/A}\right)_{xyz} = \mathbf{0} + (0.8\mathbf{i})\times(8\mathbf{j}) = \{6.4\mathbf{k}\} \text{ m/s}$$

$$(\mathbf{a}_{B/A})_{xyz} = \left(\ddot{\mathbf{r}}_{B/A}\right)_{xyz} = \left[\left(\ddot{\mathbf{r}}_{B/A}\right)_{x''y''z''} + \boldsymbol{\Omega}_{xyz}\times\left(\dot{\mathbf{r}}_{B/A}\right)_{x''y''z''}\right] + \dot{\boldsymbol{\Omega}}_{xyz}\times\left(\mathbf{r}_{B/A}\right)_{xyz} + \boldsymbol{\Omega}_{xyz}\times\left(\dot{\mathbf{r}}_{B/A}\right)_{xyz}$$

$$= \mathbf{0} + \mathbf{0} + (0.9\mathbf{i})\times(8\mathbf{j}) + (0.8\mathbf{i})\times(6.4\mathbf{k}) = \{-5.12\mathbf{j} + 7.2\mathbf{k}\} \text{ m/s}^2$$

Thus,

$$\mathbf{v}_B = \mathbf{v}_A + \boldsymbol{\Omega}\times\mathbf{r}_{B/A} + (\mathbf{v}_{B/A})_{xyz} = (1\mathbf{i} - 0.5\mathbf{j}) + (-0.5\mathbf{k})\times(8\mathbf{j}) + (6.4\mathbf{k})$$

$$= \{4\mathbf{i} - 0.5\mathbf{j} + 6.4\mathbf{k}\} \text{ m/s} \qquad \textbf{Ans}$$

$$\mathbf{a}_B = \mathbf{a}_A + \dot{\boldsymbol{\Omega}}\times\mathbf{r}_{B/A} + \boldsymbol{\Omega}\times\left(\boldsymbol{\Omega}\times\mathbf{r}_{B/A}\right) + 2\boldsymbol{\Omega}\times(\mathbf{v}_{B/A})_{xyz} + (\mathbf{a}_{B/A})_{xyz}$$

$$= (-0.25\mathbf{i} - 0.4\mathbf{j}) + (-0.4\mathbf{k})\times(8\mathbf{j}) + (-0.5\mathbf{k})\times\left[(-0.5\mathbf{k})\times(8\mathbf{j})\right] + 2(-0.5\mathbf{k})\times(6.4\mathbf{k}) - 5.12\mathbf{j} + 7.2\mathbf{k}$$

$$= \{2.95\mathbf{i} - 7.52\mathbf{j} + 7.20\mathbf{k}\} \text{ m/s}^2 \qquad \textbf{Ans}$$

These results are measured from the $x'y'z'$ axes. Thus, relative to the XYZ axes

$$\mathbf{v}_B = 1\mathbf{i} + (4\mathbf{i} - 0.5\mathbf{j} + 6.4\mathbf{k}) = \{5\mathbf{i} - 0.5\mathbf{j} + 6.4\mathbf{k}\} \text{ m/s} \qquad \textbf{Ans}$$

$$\mathbf{a}_B = \mathbf{0} + (2.95\mathbf{i} - 7.52\mathbf{j} + 7.20\mathbf{k}) = \{2.95\mathbf{i} - 7.52\mathbf{j} + 7.20\mathbf{k}\} \text{ m/s}^2 \qquad \textbf{Ans}$$

***20-40.** At the instant shown, the helicopter is moving upwards with a velocity $v_H = 4$ ft/s and has an acceleration $a_H = 2$ ft/s². At the same instant the frame H, *not* the horizontal blade, is rotating about a vertical axis with a constant angular velocity $\omega_H = 0.9$ rad/s. If the tail blade B is rotating with a constant angular velocity $\omega_{B/H} = 180$ rad/s, measured relative to H, determine the velocity and acceleration of point P, located on the tip of the blade, at the instant the blade is in the vertical position.

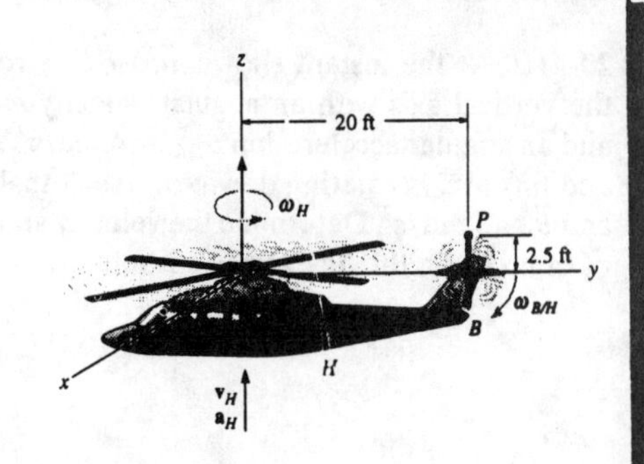

Relative to XYZ, let xyz have

$\Omega = \{0.9\mathbf{k}\}$ rad/s $\quad \dot{\Omega} = \mathbf{0}$ (Ω does not change direction relative to XYZ.)

$\mathbf{r}_B = \{20\mathbf{j}\}$ ft ($\mathbf{r}_B$ changes direction relative to XYZ.)

$$\mathbf{v}_B = \dot{\mathbf{r}}_B = (\dot{\mathbf{r}}_B)_{xyz} + \Omega \times \mathbf{r}_B = 4\mathbf{k} + (0.9\mathbf{k}) \times (20\mathbf{j}) = \{-18\mathbf{i} + 4\mathbf{k}\} \text{ ft/s}$$

$$\mathbf{a}_B = \ddot{\mathbf{r}}_B = \left[(\ddot{\mathbf{r}}_B)_{xyz} + \Omega \times \left(\dot{\mathbf{r}}_B\right)_{xyz}\right] + \dot{\Omega} \times \mathbf{r}_B + \Omega \times \dot{\mathbf{r}}_B$$

$$= \left[2\mathbf{k} + \mathbf{0}\right] + \mathbf{0} + \left[(0.9\mathbf{k}) \times (-18\mathbf{i} + 4\mathbf{k})\right]$$

$$= \{-16.2\mathbf{j} + 2\mathbf{k}\} \text{ ft/s}^2$$

Relative to xyz, let $x'y'z'$ have

$\Omega_{xyz} = \{-180\mathbf{i}\}$ rad/s $\quad \dot{\Omega}_{xyz} = \mathbf{0}$ (Ω_{xyz} does not change direction relative to xyz.)

$(\mathbf{r}_{P/B})_{xyz} = \{2.5\mathbf{k}\}$ ft ($(\mathbf{r}_{P/B})_{xyz}$ changes direction relative to xyz.)

$$(\mathbf{v}_{P/B})_{xyz} = (\dot{\mathbf{r}}_{P/B})_{xyz} = (\dot{\mathbf{r}}_{P/B})_{x'y'z'} + \Omega_{xyz} \times (\mathbf{r}_{P/B})_{xyz} = \mathbf{0} + (-180\mathbf{i}) \times (2.5\mathbf{k}) = \{450\mathbf{j}\} \text{ ft/s}$$

$$(\mathbf{a}_{P/B})_{xyz} = (\ddot{\mathbf{r}}_{P/B})_{xyz} = \left[(\ddot{\mathbf{r}}_{P/B})_{x'y'z'} + \Omega_{xyz} \times (\dot{\mathbf{r}}_{P/B})_{x'y'z'}\right] + \dot{\Omega}_{xyz} \times (\mathbf{r}_{P/B})_{xyz} + \Omega_{xyz} \times (\dot{\mathbf{r}}_{P/B})_{xyz}$$

$$(\mathbf{a}_{B/A})_{xyz} = \left[\mathbf{0} + \mathbf{0}\right] + \mathbf{0} + (-180\mathbf{i}) \times (450\mathbf{j}) = \{-81\,000\mathbf{k}\} \text{ ft/s}^2$$

Thus,

$$\mathbf{v}_P = \mathbf{v}_B + \Omega \times \mathbf{r}_{P/B} + (\mathbf{v}_{P/B})_{xyz}$$

$$= (-18\mathbf{i} + 4\mathbf{k}) + \left[(0.9\mathbf{k}) \times (2.5\mathbf{k})\right] + (450\mathbf{j})$$

$$= \{-18\mathbf{i} + 450\mathbf{j} + 4\mathbf{k}\} \text{ ft/s}$$ **Ans**

$$\mathbf{a}_P = \mathbf{a}_B + \dot{\Omega} \times \mathbf{r}_{P/B} + \Omega \times (\Omega \times \mathbf{r}_{P/B}) + 2\Omega \times (\mathbf{v}_{P/B})_{xyz} + (\mathbf{a}_{P/B})_{xyz}$$

$$= (-16.2\mathbf{j} + 2\mathbf{k}) + \mathbf{0} + (0.9\mathbf{k}) \times \left[(0.9\mathbf{k}) \times (2.5\mathbf{k})\right] + \left[2(0.9\mathbf{k}) \times (450\mathbf{j})\right] + (-81\,000\mathbf{k})$$

$$= \{-810\mathbf{i} - 16.2\mathbf{j} - 81\,000\mathbf{k}\} \text{ ft/s}^2$$ **Ans**

20-41. At the instant shown, rod BD is rotating about the vertical axis with an angular velocity $\omega_{BD} = 7$ rad/s and an angular acceleration $\alpha_{BD} = 4$ rad/s^2. Also, $\theta = 60°$ and link AC is rotating downward such that $\dot{\theta} = 2$ rad/s and $\ddot{\theta} = 3$ rad/s^2. Determine the velocity and acceleration of point A on the link at this instant.

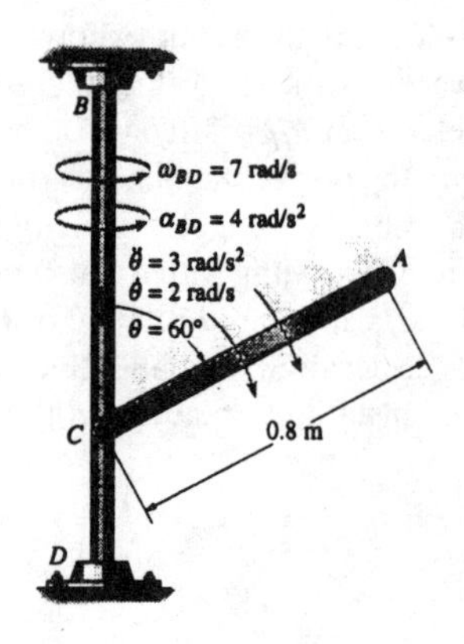

Relative to XYZ, with origin at C let coincident xyz have

$\Omega = \{7\mathbf{k}\}$ rad/s, $\dot{\Omega} = \{4\mathbf{k}\}$ rad/s^2 (Ω does not change direction relative to XYZ.)

$\mathbf{r}_C = \mathbf{0}$; $\mathbf{v}_C = \mathbf{0}$; $\mathbf{a}_C = \mathbf{0}$

Relative to xyz, let coincident $x'y'z'$ have

$\Omega_{xyz} = \{-2\mathbf{i}\}$ rad/s, $\dot{\Omega}_{xyz} = \{-3\mathbf{i}\}$ rad/s^2 (Ω_{xyz} does not change direction relative to xyz.)

$(\mathbf{r}_{A/C})_{xyz} = 0.8\sin 60°\mathbf{j} + 0.8\cos 60°\mathbf{k} = \{0.693\mathbf{j} + 0.4\mathbf{k}\}$ m/s ($(\mathbf{r}_{A/C})_{xyz}$ changes direction relative to xyz.)

$$(\mathbf{v}_{A/C})_{xyz} = \left(\dot{\mathbf{r}}_{A/C}\right)_{xyz} = \left(\dot{\mathbf{r}}_{A/C}\right)_{x'y'z'} + \Omega_{xyz} \times \left(\mathbf{r}_{A/C}\right)_{xyz} = \mathbf{0} + (-2\mathbf{i}) \times (0.693\mathbf{j} + 0.4\mathbf{k})$$

$$= \{0.8\mathbf{j} - 1.386\mathbf{k}\} \text{ m/s}$$

$$(\mathbf{a}_{A/C})_{xyz} = \left(\ddot{\mathbf{r}}_{A/C}\right)_{xyz} = \left[\left(\ddot{\mathbf{r}}_{A/C}\right)_{x'y'z'} + \Omega_{xyz} \times \left(\dot{\mathbf{r}}_{A/C}\right)_{x'y'z'}\right] + \dot{\Omega}_{xyz} \times \left(\mathbf{r}_{A/C}\right)_{xyz} + \Omega_{xyz} \times \left(\dot{\mathbf{r}}_{A/C}\right)_{xyz}$$

$$= \left[\mathbf{0} + \mathbf{0}\right] + (-3\mathbf{i}) \times (0.693\mathbf{j} + 0.4\mathbf{k}) + (-2\mathbf{i}) \times (0.8\mathbf{j} - 1.386\mathbf{k})$$

$$= -2.078\mathbf{k} + 1.2\mathbf{j} - 1.6\mathbf{k} - 2.771\mathbf{j} = \{-1.571\mathbf{j} - 3.678\mathbf{k}\} \text{ m/s}^2$$

Thus,

$$\mathbf{v}_A = \mathbf{v}_C + \Omega \times \mathbf{r}_{A/C} + (\mathbf{v}_{A/C})_{xyz} = \mathbf{0} + (7\mathbf{k}) \times (0.693\mathbf{j} + 0.4\mathbf{k}) + 0.8\mathbf{j} - 1.386\mathbf{k}$$

$$= \{-4.85\mathbf{i} + 0.800\mathbf{j} - 1.39\mathbf{k}\} \text{ m/s} \qquad \textbf{Ans}$$

$$\mathbf{a}_A = \mathbf{a}_C + \dot{\Omega} \times \mathbf{r}_{A/C} + \Omega \times \left(\Omega \times \mathbf{r}_{A/C}\right) + 2\Omega \times (\mathbf{v}_{A/C})_{xyz} + (\mathbf{a}_{A/C})_{xyz}$$

$$= \mathbf{0} + (4\mathbf{k}) \times (0.693\mathbf{j} + 0.4\mathbf{k}) + (7\mathbf{k}) \times \left[(7\mathbf{k}) \times (0.693\mathbf{j} + 0.4\mathbf{k})\right]$$

$$+ 2(7\mathbf{k}) \times (0.8\mathbf{j} - 1.386\mathbf{k}) - 1.572\mathbf{j} - 3.678\mathbf{k}$$

$$= \{-14.0\mathbf{i} - 35.5\mathbf{j} - 3.68\mathbf{k}\} \text{ m/s}^2 \qquad \textbf{Ans}$$

20-42. At a given instant, the rod has the angular motions shown, while the collar C is moving down *relative* to the rod with a velocity of 6 ft/s and an acceleration of 2 ft/s^2. Determine the collar's velocity and acceleration at this instant.

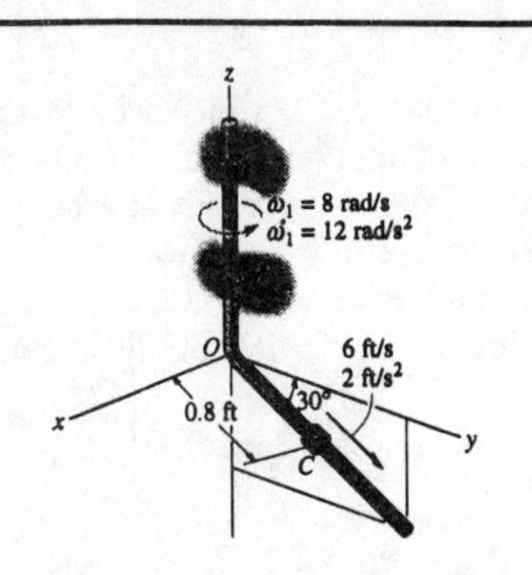

Relative to XYZ, let xyz have

$\Omega = \{8\mathbf{k}\}$ rad/s, $\quad \dot{\Omega} = \{12\mathbf{k}\}$ rad/s^2 (Ω does not change direction relative to XYZ.)

$\mathbf{r}_O = \mathbf{0}$

$\mathbf{v}_O = \mathbf{0}$

$\mathbf{a}_O = \mathbf{0}$

$$(\mathbf{r}_{C/O})_{xyz} = 0.8\cos 30°\mathbf{j} - 0.8\sin 30°\mathbf{k} = \{0.6928\mathbf{j} - 0.4\mathbf{k}\}\text{ ft}$$

$$(\mathbf{v}_{C/O})_{xyz} = (6\cos 30°\mathbf{j} - 6\sin 30°\mathbf{k})$$

$$= \{5.196\mathbf{j} - 3\mathbf{k}\}\text{ ft/s}$$

$$(\mathbf{a}_{C/O})_{xyz} = (2\cos 30°\mathbf{j} - 2\sin 30°\mathbf{k})$$

$$= \{1.732\mathbf{j} - 1\mathbf{k}\}\text{ ft/s}^2$$

Thus,

$$\mathbf{v}_C = \mathbf{v}_O + \Omega \times \mathbf{r}_{C/O} + (\mathbf{v}_{C/O})_{xyz}$$

$$= \mathbf{0} + (8\mathbf{k}) \times (0.6928\mathbf{j} - 0.4\mathbf{k}) + (5.196\mathbf{j} - 3\mathbf{k})$$

$$\mathbf{v}_C = \{-5.54\mathbf{i} + 5.20\mathbf{j} - 3.00\mathbf{k}\}\text{ ft/s} \qquad \textbf{Ans}$$

$$\mathbf{a}_C = \mathbf{a}_O + \dot{\Omega} \times \mathbf{r}_{C/O} + \Omega \times (\Omega \times \mathbf{r}_{C/O}) + 2\Omega \times (\mathbf{v}_{C/O})_{xyz} + (\mathbf{a}_{C/O})_{xyz}$$

$$= \mathbf{0} + (12\mathbf{k}) \times (0.6928\mathbf{j} - 0.4\mathbf{k}) + (8\mathbf{k}) \times [(8\mathbf{k}) \times (0.6928\mathbf{j} - 0.4\mathbf{k})]$$
$$+ 2(8\mathbf{k}) \times (5.196\mathbf{j} - 3\mathbf{k}) + (1.732\mathbf{j} - 1\mathbf{k})$$

$$\mathbf{a}_C = \{-91.5\mathbf{i} - 42.6\mathbf{j} - 1.00\mathbf{k}\}\text{ ft/s}^2 \qquad \textbf{Ans}$$

20-43. The particle P slides around the circular hoop with a constant angular velocity of $\dot{\theta} = 6$ rad/s, while the hoop rotates about the x axis at a constant rate of $\omega = 4$ rad/s. If at the instant shown the hoop is in the x–y plane and the angle $\theta = 45°$, determine the velocity and acceleration of the particle at this instant.

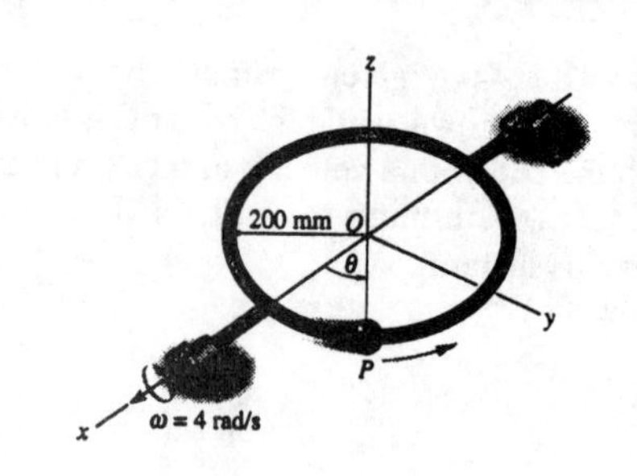

Relative to XYZ, let xyz have

$\Omega = \omega = \{4\mathbf{i}\}$ rad/s, $\quad \dot{\Omega} = \dot{\omega} = \mathbf{0}$ (Ω does not change direction relative to XYZ.)

$\mathbf{r}_O = \mathbf{0}; \quad \mathbf{v}_O = \mathbf{0}; \quad \mathbf{a}_O = \mathbf{0}$

Relative to xyz, let coincident $x'y'z'$ have

$\Omega_{xyz} = \{6\mathbf{k}\}$ rad/s, $\quad \dot{\Omega}_{xyz} = \mathbf{0}$ (Ω_{xyz} does not change direction relative to XYZ.)

$(\mathbf{r}_{P/O})_{xyz} = 0.2\cos 45°\mathbf{i} + 0.2\sin 45°\mathbf{j} = \{0.1414\mathbf{i} + 0.1414\mathbf{j}\}$ m $\quad$ ($(\mathbf{r}_{P/O})_{xyz}$ changes direction relative to XYZ.)

$$(\mathbf{v}_{P/O})_{xyz} = \left(\dot{\mathbf{r}}_{P/O}\right)_{xyz} = \left(\dot{\mathbf{r}}_{P/O}\right)_{x'y'z'} + \Omega_{xyz} \times \left(\mathbf{r}_{P/O}\right)_{xyz} = \mathbf{0} + (6\mathbf{k}) \times (0.1414\mathbf{i} + 0.1414\mathbf{j})$$

$$= \{-0.8485\mathbf{i} + 0.8485\mathbf{j}\} \text{ m/s}$$

$$(\mathbf{a}_{P/O})_{xyz} = \left(\ddot{\mathbf{r}}_{P/O}\right)_{xyz} = \left[\left(\ddot{\mathbf{r}}_{P/O}\right)_{x'y'z'} + \Omega_{xyz} \times \left(\dot{\mathbf{r}}_{P/O}\right)_{x'y'z'}\right] + \dot{\Omega} \times \left(\mathbf{r}_{P/O}\right)_{xyz} + \Omega \times \left(\dot{\mathbf{r}}_{P/O}\right)_{xyz}$$

$$= \left[\mathbf{0} + \mathbf{0}\right] + \mathbf{0} + (6\mathbf{k}) \times (-0.8485\mathbf{i} + 0.8485\mathbf{j}) = \{-5.0912\mathbf{i} - 5.0912\mathbf{j}\} \text{ m/s}^2$$

Thus,

$$\mathbf{v}_P = \mathbf{v}_O + \Omega \times \mathbf{r}_{P/O} + (\mathbf{v}_{P/O})_{xyz} = \mathbf{0} + (4\mathbf{i}) \times (0.1414\mathbf{i} + 0.1414\mathbf{j}) - 0.8485\mathbf{i} + 0.8485\mathbf{j}$$

$$= \{-0.849\mathbf{i} + 0.849\mathbf{j} + 0.566\mathbf{k}\} \text{ m/s} \qquad \textbf{Ans}$$

$$\mathbf{a}_P = \mathbf{a}_O + \dot{\Omega} \times \mathbf{r}_{P/O} + \Omega \times \left(\Omega \times \mathbf{r}_{P/O}\right) + 2\Omega \times (\mathbf{v}_{P/O})_{xyz} + (\mathbf{a}_{P/O})_{xyz}$$

$$= \mathbf{0} + \mathbf{0} + (4\mathbf{i}) \times \left[(4\mathbf{i}) \times (0.1414\mathbf{i} + 0.1414\mathbf{j})\right] + 2(4\mathbf{i}) \times (-0.8485\mathbf{i} + 0.8485\mathbf{j}) - 5.0912\mathbf{i} - 5.0912\mathbf{j}$$

$$= \{-5.09\mathbf{i} - 7.35\mathbf{j} + 6.79\mathbf{k}\} \text{ m/s}^2 \qquad \textbf{Ans}$$

***20-44.** At the given instant, the rod is spinning about the z axis with an angular velocity $\omega_1 = 3$ rad/s and angular acceleration $\dot{\omega}_1 = 4$ rad/s². At this same instant, the disk is spinning at $\omega_2 = 2$ rad/s when $\dot{\omega}_2 = 1$ rad/s², both measured *relative* to the rod. Determine the velocity and acceleration of point P on the disk at this instant.

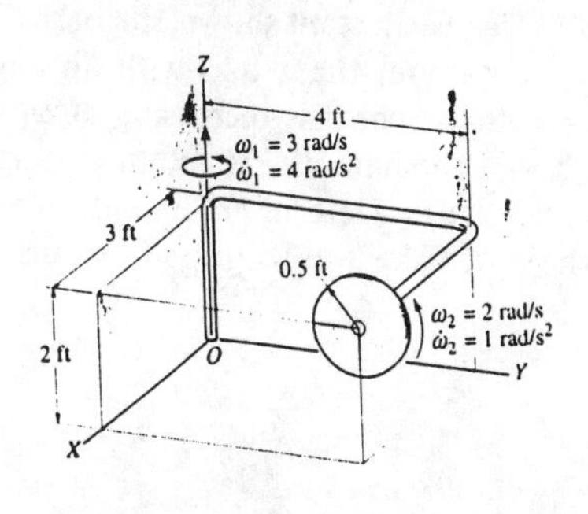

Relative to XYZ with origin at O, let xyz with origin at A have

$\Omega = \{3\mathbf{k}\}$ rad/s, $\dot{\Omega} = \{4\mathbf{k}\}$ rad/s² (Ω does not change direction relative to XYZ.)

$\mathbf{r}_A = \{3\mathbf{i} + 4\mathbf{j} + 2\mathbf{k}\}$ ft ($\mathbf{r}_A$ changes direction relative to XYZ.)

$$\mathbf{v}_A = \left(\dot{\mathbf{r}}_A\right)_{xyz} + \Omega \times \mathbf{r}_A = \mathbf{0} + (3\mathbf{k}) \times (3\mathbf{i} + 4\mathbf{j} + 2\mathbf{k}) = \{-12\mathbf{i} + 9\mathbf{j}\} \text{ ft/s}$$

$$\mathbf{a}_A = \left[\left(\ddot{\mathbf{r}}_A\right)_{xyz} + \Omega \times \left(\dot{\mathbf{r}}_A\right)_{xyz}\right] + \dot{\Omega} \times \mathbf{r}_A + \Omega \times \dot{\mathbf{r}}_A$$

$$= \mathbf{0} + \mathbf{0} + (4\mathbf{k}) \times (3\mathbf{i} + 4\mathbf{j} + 2\mathbf{k}) + (3\mathbf{k}) \times (-12\mathbf{i} + 9\mathbf{j}) = \{-43\mathbf{i} - 24\mathbf{j}\} \text{ ft/s}^2$$

Relative to xyz, let $x'y'z'$ have

$\Omega_{xyz} = \{2\mathbf{i}\}$ rad/s, $\dot{\Omega}_{xyz} = \{1\mathbf{i}\}$ rad/s² (Ω_{xyz} does not change direction relative to xyz.)

$(\mathbf{r}_{P/A})_{xyz} = \{-0.5\mathbf{j}\}$ ft ($(\mathbf{r}_{P/A})_{xyz}$ changes direction relative to xyz.)

$$(\mathbf{v}_{P/A})_{xyz} = \left(\dot{\mathbf{r}}_{P/A}\right)_{xyz} = \left(\dot{\mathbf{r}}_{P/A}\right)_{x'y'z'} + \Omega_{xyz} \times \mathbf{r}_{P/A}$$

$$= \mathbf{0} + (2\mathbf{i}) \times (-0.5\mathbf{j}) = \{-1\mathbf{k}\} \text{ ft/s}$$

$$(\mathbf{a}_{P/A})_{xyz} = \left(\ddot{\mathbf{r}}_{P/A}\right)_{xyz} = \left[\left(\ddot{\mathbf{r}}_{P/A}\right)_{x'y'z'} + \Omega_{xyz} \times \left(\dot{\mathbf{r}}_{P/A}\right)_{x'y'z'}\right] + \dot{\Omega}_{xyz} \times \left(\mathbf{r}_{P/A}\right)_{xyz} + \Omega_{xyz} \times \left(\dot{\mathbf{r}}_{P/A}\right)_{xyz}$$

$$= \mathbf{0} + \mathbf{0} + (1\mathbf{i}) \times (-0.5\mathbf{j}) + (2\mathbf{i}) \times (-1\mathbf{k}) = \{2\mathbf{j} - 0.5\mathbf{k}\} \text{ ft/s}^2$$

Thus,

$$\mathbf{v}_P = \mathbf{v}_A + \Omega \times \mathbf{r}_{P/A} + (\mathbf{v}_{P/A})_{xyz} = (-12\mathbf{i} + 9\mathbf{j}) + (3\mathbf{k}) \times (-0.5\mathbf{j}) + (-1\mathbf{k})$$

$$= \{-10.5\mathbf{i} + 9.00\mathbf{j} - 1.00\mathbf{k}\} \text{ ft/s} \qquad \textbf{Ans}$$

$$\mathbf{a}_P = \mathbf{a}_A + \dot{\Omega} \times \mathbf{r}_{P/A} + \Omega \times \left(\Omega \times \mathbf{r}_{P/A}\right) + 2\Omega \times (\mathbf{v}_{P/A})_{xyz} + (\mathbf{a}_{P/A})_{xyz}$$

$$= (-43\mathbf{i} - 24\mathbf{j}) + (4\mathbf{k}) \times (-0.5\mathbf{j}) + (3\mathbf{k}) \times \left[(3\mathbf{k}) \times (-0.5\mathbf{j})\right] + 2(3\mathbf{k}) \times (-1\mathbf{k}) + (2\mathbf{j} - 0.5\mathbf{k})$$

$$= \{-41.0\mathbf{i} - 17.5\mathbf{j} - 0.500\mathbf{k}\} \text{ ft/s}^2 \qquad \textbf{Ans}$$

20-45. At the instant shown, the base of the robotic arm is turning about the z axis with an angular velocity of $\omega_1 = 4$ rad/s, which is increasing at $\dot{\omega}_1 = 3$ rad/s^2. Also, the boom segment BC is rotating at a constant rate of $\omega_{BC} = 8$ rad/s. Determine the velocity and acceleration of the part C held in its grip at this instant.

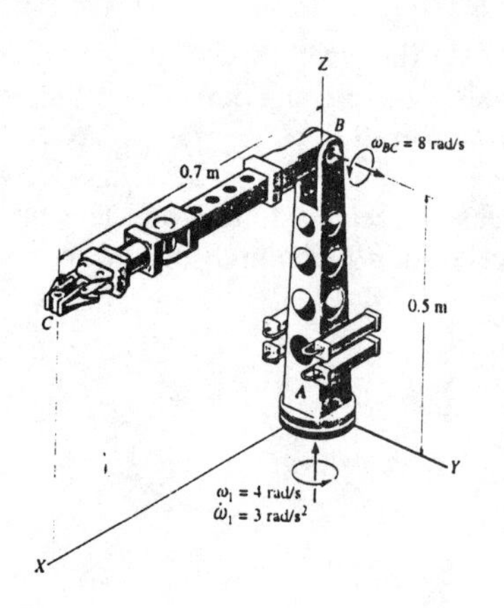

Relative to XYZ, let xyz have origin at B and have

$\Omega = \{4\mathbf{k}\}$ rad/s, $\quad \dot{\Omega} = \{3\mathbf{k}\}$ rad/s^2 (Ω does not change direction relative to XYZ.)

$\mathbf{r}_B = \{0.5\mathbf{k}\}$ m ($\mathbf{r}_B$ does not change direction relative to XYZ.)

$\mathbf{v}_B = 0$

$\mathbf{a}_B = 0$

Relative to xyz, let coincident $x'y'z'$ have origin at B and have

$\Omega_{xyz} = \{8\mathbf{j}\}$ rad/s, $\quad \dot{\Omega}_{xyz} = \mathbf{0}$ (Ω_{xyz} does not change direction relative to xyz.)

$(\mathbf{r}_{C/B})_{xyz} = \{0.7\mathbf{i}\}$ m ($(\mathbf{r}_{C/B})_{xyz}$ changes direction relative to xyz.)

$$(\mathbf{v}_{C/B})_{xyz} = \left(\dot{\mathbf{r}}_{C/B}\right)_{xyz} = \left(\dot{\mathbf{r}}_{C/B}\right)_{x'y'z'} + \Omega_{xyz} \times \left(\mathbf{r}_{C/B}\right)_{xyz} = \mathbf{0} + (8\mathbf{j}) \times (0.7\mathbf{i}) = \{-5.6\mathbf{k}\} \text{ m/s}$$

$$(\mathbf{a}_{C/B})_{xyz} = \left(\ddot{\mathbf{r}}_{C/B}\right)_{xyz} = \left[\left(\ddot{\mathbf{r}}_{C/B}\right)_{x'y'z'} + \Omega_{xyz} \times \left(\dot{\mathbf{r}}_{C/B}\right)_{x'y'z'}\right] + \dot{\Omega}_{xyz} \times \left(\mathbf{r}_{C/B}\right)_{xyz} + \Omega_{xyz} \times \left(\dot{\mathbf{r}}_{C/B}\right)_{xyz}$$

$$= \mathbf{0} + \mathbf{0} + \mathbf{0} + (8\mathbf{j}) \times (-5.6\mathbf{k}) = \{-44.8\mathbf{i}\} \text{ m/s}^2$$

Thus,

$$\mathbf{v}_C = \mathbf{v}_B + \Omega \times \mathbf{r}_{C/B} + (\mathbf{v}_{C/B})_{xyz} = \mathbf{0} + (4\mathbf{k}) \times (0.7\mathbf{i}) + (-5.6\mathbf{k})$$

$$= \{2.80\mathbf{j} - 5.60\mathbf{k}\} \text{ m/s} \qquad \textbf{Ans}$$

$$\mathbf{a}_C = \mathbf{a}_B + \dot{\Omega} \times \mathbf{r}_{C/B} + \Omega \times \left(\Omega \times \mathbf{r}_{C/B}\right) + 2\Omega \times (\mathbf{v}_{C/B})_{xyz} + (\mathbf{a}_{C/B})_{xyz}$$

$$= \mathbf{0} + (3\mathbf{k}) \times (0.7\mathbf{i}) + (4\mathbf{k}) \times \left[(4\mathbf{k}) \times (0.7\mathbf{i})\right]$$

$$+ 2(4\mathbf{k}) \times (-5.6\mathbf{k}) - 44.8\mathbf{i}$$

$$= \{-56\mathbf{i} + 2.1\mathbf{j}\} \text{ m/s}^2 \qquad \textbf{Ans}$$

20-46. At the instant shown, the base of the robotic arm is turning about the z axis with an angular velocity of $\omega_1 = 4$ rad/s, which is increasing at $\dot{\omega}_1 = 3$ rad/s^2. Also, the boom segment BC is rotating at $\omega_{BC} = 8$ rad/s, which is increasing at $\dot{\omega}_{BC} = 2$ rad/s^2. Determine the velocity and acceleration of the part C held in its grip at this instant.

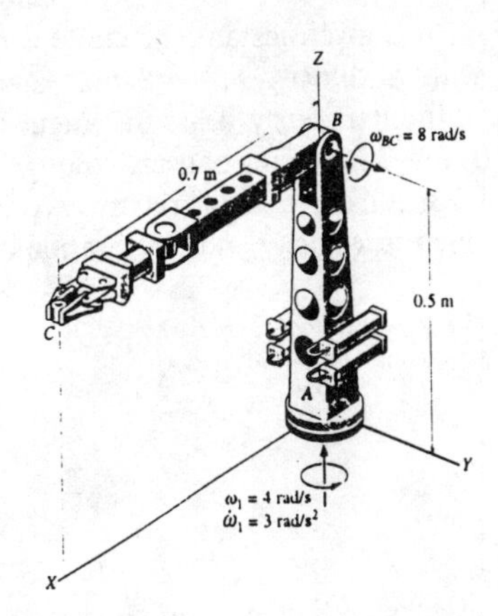

Relative to XYZ, let xyz with origin at B have

$\Omega = \{4\mathbf{k}\}$ rad/s, $\quad \dot{\Omega} = \{3\mathbf{k}\}$ rad/s^2 (Ω does not change direction relative to XYZ.)

$\mathbf{r}_B = \{0.5\mathbf{k}\}$ m ($\mathbf{r}_B$ does not change direction relative to XYZ.)

$\mathbf{v}_B = \mathbf{0}$

$\mathbf{a}_B = \mathbf{0}$

Relative to xyz, let coincident $x'y'z'$ have origin at B and have

$\Omega_{xyz} = \{8\mathbf{j}\}$ rad/s, $\quad \dot{\Omega}_{xyz} = \{2\mathbf{j}\}$ rad/s^2 (Ω does not change direction relative to xyz.)

$(\mathbf{r}_{C/B})_{xyz} = \{0.7\mathbf{i}\}$ m (Ω does not change direction relative to xyz.)

$$(\mathbf{v}_{C/B})_{xyz} = \left(\dot{\mathbf{r}}_{C/B}\right)_{xyz} = \left(\dot{\mathbf{r}}_{C/B}\right)_{x'y'z'} + \Omega_{xyz} \times \left(\mathbf{r}_{C/B}\right)_{xyz} = \mathbf{0} + (8\mathbf{j}) \times (0.7\mathbf{i}) = \{-5.6\mathbf{k}\} \text{ m/s}$$

$$(\mathbf{a}_{C/B})_{xyz} = \left(\ddot{\mathbf{r}}_{C/B}\right)_{xyz} = \left[\left(\ddot{\mathbf{r}}_{C/B}\right)_{x'y'z'} + \Omega_{xyz} \times \left(\dot{\mathbf{r}}_{C/B}\right)_{x'y'z'}\right] + \dot{\Omega}_{xyz} \times \left(\mathbf{r}_{C/B}\right)_{xyz} + \Omega_{xyz} \times \left(\dot{\mathbf{r}}_{C/B}\right)_{xyz}$$

$$= \mathbf{0} + \mathbf{0} + (2\mathbf{j}) \times (0.7\mathbf{i}) + (8\mathbf{j}) \times (-5.6\mathbf{k}) = \{-44.8\mathbf{i} - 1.40\mathbf{k}\} \text{ m/s}^2$$

Thus,

$$\mathbf{v}_C = \mathbf{v}_B + \Omega \times \mathbf{r}_{C/B} + (\mathbf{v}_{C/B})_{xyz} = \mathbf{0} + (4\mathbf{k}) \times (0.7\mathbf{i}) + (-5.6\mathbf{k})$$

$$= \{2.80\mathbf{j} - 5.60\mathbf{k}\} \text{ m/s} \qquad \textbf{Ans}$$

$$\mathbf{a}_C = \mathbf{a}_B + \dot{\Omega} \times \mathbf{r}_{C/B} + \Omega \times \left(\Omega \times \mathbf{r}_{C/B}\right) + 2\Omega \times (\mathbf{v}_{C/B})_{xyz} + (\mathbf{a}_{C/B})_{xyz}$$

$$= \mathbf{0} + (3\mathbf{k}) \times (0.7\mathbf{i}) + (4\mathbf{k}) \times \left[(4\mathbf{k}) \times (0.7\mathbf{i})\right]$$

$$+ 2(4\mathbf{k}) \times (-5.6\mathbf{k}) - 44.8\mathbf{i} - 1.40\mathbf{k}$$

$$= \{-56\mathbf{i} + 2.1\mathbf{j} - 1.40\mathbf{k}\} \text{ m/s}^2 \qquad \textbf{Ans}$$

20-47. At a given instant, the crane is moving along the track with a velocity $v_{CD} = 8$ m/s and acceleration of 9 m/s^2. Simultaneously, it has the angular motions shown. If the trolley T is moving outwards along the boom AB with a relative speed of 3 m/s and relative acceleration of 5 m/s^2, determine the velocity and acceleration of the trolley.

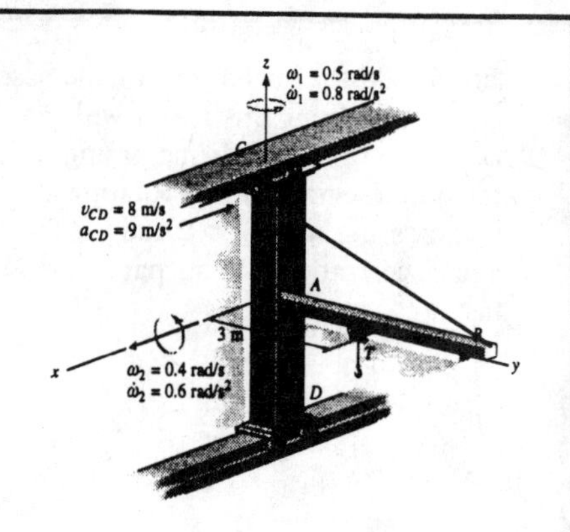

Relative to XYZ, let xyz have

$\mathbf{\Omega} = \{0.5\mathbf{k}\}$ rad/s, $\quad \dot{\mathbf{\Omega}} = \{0.8\mathbf{k}\}$ rad/s^2 (Ω does not change direction relative to XYZ.)

$\mathbf{r}_A = \mathbf{0}$ ($\mathbf{r}_A$ does not change direction relative to XYZ.)

$\mathbf{v}_A = \{-8\mathbf{i}\}$ m/s

$\mathbf{a}_A = \{-9\mathbf{i}\}$ m/s^2

Relative to xyz, let $x'y'z'$ have

$\mathbf{\Omega}_{xyz} = \{0.4\mathbf{i}\}$ rad/s, $\quad \dot{\mathbf{\Omega}}_{xyz} = \{0.6\mathbf{i}\}$ rad/s^2 (Ω_{xyz} does not change direction relative to xyz.)

$(\mathbf{r}_{T/A})_{xyz} = \{3\mathbf{j}\}$ m $\quad ((\mathbf{r}_{T/A})_{xyz}$ changes direction relative to xyz.)

$$(\mathbf{v}_{T/A})_{xyz} = \left(\dot{\mathbf{r}}_{T/A}\right)_{xyz} = \left(\dot{\mathbf{r}}_{T/A}\right)_{x'y'z'} + \mathbf{\Omega}_{xyz} \times \left(\mathbf{r}_{T/A}\right)_{xyz}$$

$$= (3\mathbf{j}) + (0.4\mathbf{i}) \times (3\mathbf{j}) = \{3\mathbf{j} + 1.2\mathbf{k}\} \text{ m/s}$$

$$(\mathbf{a}_{T/A})_{xyz} = \left(\ddot{\mathbf{r}}_{T/A}\right)_{xyz} = \left[\left(\ddot{\mathbf{r}}_{T/A}\right)_{x'y'z'} + \mathbf{\Omega}_{xyz} \times \left(\dot{\mathbf{r}}_{T/A}\right)_{x'y'z'}\right] + \dot{\mathbf{\Omega}}_{xyz} \times \left(\mathbf{r}_{T/A}\right)_{xyz} + \mathbf{\Omega}_{xyz} \times \left(\dot{\mathbf{r}}_{T/A}\right)_{xyz}$$

$$= (5\mathbf{j}) + (0.4\mathbf{i}) \times (3\mathbf{j}) + (0.6\mathbf{i}) \times (3\mathbf{j}) + (0.4\mathbf{i}) \times (3\mathbf{j} + 1.2\mathbf{k}) = \{4.52\mathbf{j} + 4.20\mathbf{k}\} \text{ m/s}^2$$

Thus,

$$\mathbf{v}_T = \mathbf{v}_A + \mathbf{\Omega} \times \mathbf{r}_{T/A} + (\mathbf{v}_{T/A})_{xyz} = (-8\mathbf{i}) + (0.5\mathbf{k}) \times (3\mathbf{j}) + 3\mathbf{j} + 1.2\mathbf{k}$$

$$= \{-9.50\mathbf{i} + 3\mathbf{j} + 1.20\mathbf{k}\} \text{ m/s} \qquad \textbf{Ans}$$

$$\mathbf{a}_T = \mathbf{a}_A + \dot{\mathbf{\Omega}} \times \mathbf{r}_{T/A} + \mathbf{\Omega} \times \left(\mathbf{\Omega} \times \mathbf{r}_{T/A}\right) + 2\mathbf{\Omega} \times (\mathbf{v}_{T/A})_{xyz} + (\mathbf{a}_{T/A})_{xyz}$$

$$= (-9\mathbf{i}) + (0.8\mathbf{k}) \times (3\mathbf{j}) + (0.5\mathbf{k}) \times \left[(0.5\mathbf{k}) \times (3\mathbf{j})\right] + 2(0.5\mathbf{k}) \times (3\mathbf{j} + 1.2\mathbf{k}) + (4.52\mathbf{j} + 4.20\mathbf{k})$$

$$= \{-14.4\mathbf{i} + 3.77\mathbf{j} + 4.20\mathbf{k}\} \text{ m/s}^2 \qquad \textbf{Ans}$$

***20-48.** The load is being lifted upward at a constant rate of 9 m/s relative to the crane boom *AB*. At the instant shown, the boom is rotating about the vertical axis at a constant rate of $\omega_1 = 4$ rad/s, and the trolley *T* is moving outward along the boom at a constant rate of $v_t = 2$ m/s. Furthermore, at this same instant the retractable arm supporting the load is vertical and is swinging in the y–z plane at an angular rate of 5 rad/s, with an increase in the rate of swing of 7 rad/s^2. Determine the velocity and acceleration of the center *G* of the load at this instant, i.e., when $s = 4$ m and $h = 3$ m.

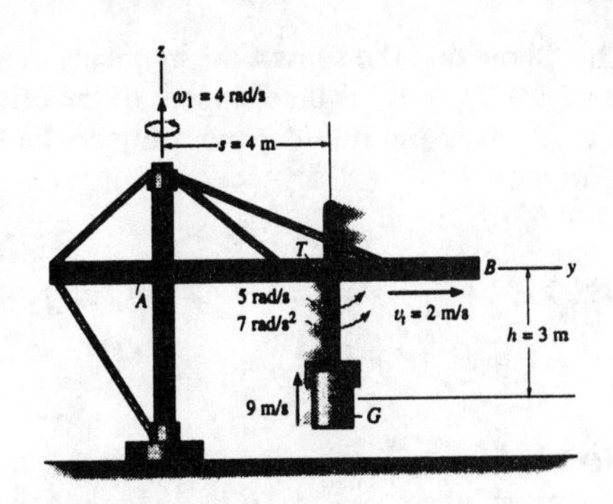

Relative to *XYZ*, with origin at *T* let *xyz* have

$\Omega = \omega_1 = 4\mathbf{k}, \quad \dot{\Omega} = \mathbf{0}$ (Ω does not change direction relative to *XYZ*.)

$\mathbf{r}_T = \{4\mathbf{j}\}$ m (Ω changes direction relative to *XYZ*.)

$$\mathbf{v}_T = \dot{\mathbf{r}}_T = \left(\dot{\mathbf{r}}_T\right)_{xyz} + \Omega \times \mathbf{r}_T = 2\mathbf{j} + (4\mathbf{k}) \times (4\mathbf{j}) = \{-16\mathbf{i} + 2\mathbf{j}\} \text{ m/s}$$

$$\mathbf{a}_T = \ddot{\mathbf{r}}_T = \left[\left(\ddot{\mathbf{r}}_T\right)_{xyz} + \Omega \times \left(\dot{\mathbf{r}}_T\right)_{xyz}\right] + \dot{\Omega} \times \mathbf{r}_T + \Omega \times \dot{\mathbf{r}}_T$$

$$= \left[\mathbf{0} + (4\mathbf{k}) \times (2\mathbf{j})\right] + \mathbf{0} + (4\mathbf{k}) \times (-16\mathbf{i} + 2\mathbf{j}) = \{-16\mathbf{i} - 64\mathbf{j}\} \text{ m/s}^2$$

Relative to *xyz*, let *x' y' z'* have

$\Omega_{xyz} = \{5\mathbf{i}\}$ rad/s, $\quad \dot{\Omega}_{xyz} = \{7\mathbf{i}\}$ rad/s^2 (Ω_{xyz} does not change direction relative to *xyz*.)

$(\mathbf{r}_{G/T})_{xyz} = \{-3\mathbf{k}\}$ m ($(\mathbf{r}_{G/T})_{xyz}$ changes direction relative to *xyz*.)

$$(\mathbf{v}_{G/T})_{xyz} = \left(\dot{\mathbf{r}}_{G/T}\right)_{xyz} = \left(\dot{\mathbf{r}}_{G/T}\right)_{x'y'z'} + \Omega_{xyz} \times \left(\mathbf{r}_{G/T}\right)_{xyz}$$

$$= (9\mathbf{k}) + (5\mathbf{i}) \times (-3\mathbf{k}) = \{15\mathbf{j} + 9\mathbf{k}\} \text{ m/s}$$

$$(\mathbf{a}_{G/T})_{xyz} = \left(\ddot{\mathbf{r}}_{G/T}\right)_{xyz} = \left[\left(\ddot{\mathbf{r}}_{G/T}\right)_{x'y'z'} + \Omega_{xyz} \times \left(\dot{\mathbf{r}}_{G/T}\right)_{x'y'z'}\right] + \dot{\Omega}_{xyz} \times \left(\mathbf{r}_{G/T}\right)_{xyz} + \Omega_{xyz} \times \left(\dot{\mathbf{r}}_{G/T}\right)_{xyz}$$

$$= \left[\mathbf{0} + (5\mathbf{i}) \times (9\mathbf{k})\right] + (7\mathbf{i}) \times (-3\mathbf{k}) + (5\mathbf{i}) \times (15\mathbf{j} + 9\mathbf{k}) = -69\mathbf{j} + 75\mathbf{k}$$

Thus,

$$\mathbf{v}_G = \mathbf{v}_T + \Omega \times \mathbf{r}_{G/T} + (\mathbf{v}_{G/T})_{xyz} = (-16\mathbf{i} + 2\mathbf{j}) + (4\mathbf{k}) \times (-3\mathbf{k}) + 15\mathbf{j} + 9\mathbf{k}$$

$$= \{-16\mathbf{i} + 17\mathbf{j} + 9\mathbf{k}\} \text{ m/s} \qquad \textbf{Ans}$$

$$\mathbf{a}_G = \mathbf{a}_T + \dot{\Omega} \times \mathbf{r}_{G/T} + \Omega \times \left(\Omega \times \mathbf{r}_{G/T}\right) + 2\Omega \times (\mathbf{v}_{G/T})_{xyz} + (\mathbf{a}_{G/T})_{xyz}$$

$$= (-16\mathbf{i} - 64\mathbf{j}) + \mathbf{0} + (4\mathbf{k}) \times \left[(4\mathbf{k}) \times (-3\mathbf{k})\right] + 2(4\mathbf{k}) \times (15\mathbf{j} + 9\mathbf{k}) - 69\mathbf{j} + 75\mathbf{k}$$

$$= \{-136\mathbf{i} - 133\mathbf{j} + 75\mathbf{k}\} \text{ m/s}^2 \qquad \textbf{Ans}$$

21-1. Show that the sum of the moments of inertia of a body, $I_{xx} + I_{yy} + I_{zz}$, is independent of the orientation of the x, y, z axes and thus depends only on the location of the origin.

$$I_{xx} + I_{yy} + I_{zz} = \int_m (y^2 + z^2)\,dm + \int_m (x^2 + z^2)\,dm + \int_m (x^2 + y^2)\,dm$$
$$= 2\int_m (x^2 + y^2 + z^2)\,dm$$

However, $x^2 + y^2 + z^2 = r^2$, where r is the distance from the origin O to dm. Since $|r|$ is constant, it does not depend on the orientation of the x, y, z axis. Consequently, $I_{xx} + I_{yy} + I_{zz}$ is also indepenent of the orientation of the x, y, z axis. **Q.E.D,**

21-2. Determine the moment of inertia of the cone with respect to a vertical $\bar{y}$ axis passing through the cone's center of mass. What is the moment of inertia about a parallel axis y' passing through the diameter of the base of the cone? The cone has a mass m.

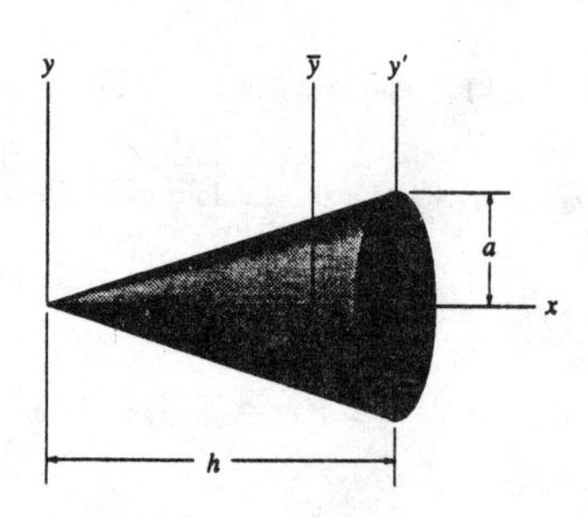

The mass of the differential element is $dm = \rho dV = \rho(\pi y^2)\,dx = \dfrac{\rho\pi a^2}{h^2}x^2\,dx$.

$$dI_y = \frac{1}{4}dm y^2 + dm x^2$$

$$= \frac{1}{4}\left[\frac{\rho\pi a^2}{h^2}x^2\,dx\right]\left(\frac{a}{h}x\right)^2 + \left(\frac{\rho\pi a^2}{h^2}x^2\right)x^2\,dx$$

$$= \frac{\rho\pi a^2}{4h^4}(4h^2 + a^2)x^4\,dx$$

$$I_y = \int dI_y = \frac{\rho\pi a^2}{4h^4}(4h^2 + a^2)\int_0^h x^4\,dx = \frac{\rho\pi a^2 h}{20}(4h^2 + a^2)$$

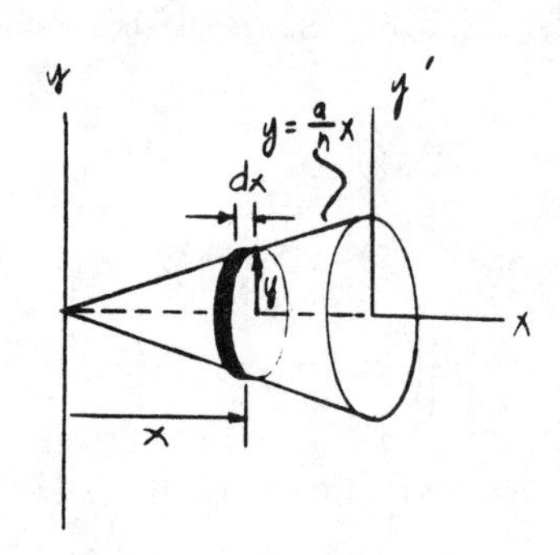

However, $$m = \int_m dm = \frac{\rho\pi a^2}{h^2}\int_0^h x^2\,dx = \frac{\rho\pi a^2 h}{3}$$

Hence, $$I_y = \frac{3m}{20}(4h^2 + a^2)$$

Using the parallel axis theorem :

$$I_y = I_{\bar{y}} + md^2$$

$$\frac{3m}{20}(4h^2 + a^2) = I_{\bar{y}} + m\left(\frac{3h}{4}\right)^2$$

$$I_{\bar{y}} = \frac{3m}{80}(h^2 + 4a^2) \qquad \textbf{Ans}$$

$$I_{y'} = I_{\bar{y}} + md^2$$

$$= \frac{3m}{80}(h^2 + 4a^2) + m\left(\frac{h}{4}\right)^2$$

$$= \frac{m}{20}(2h^2 + 3a^2) \qquad \textbf{Ans}$$

21-3. Determine the moments of inertia I_x and I_y of the paraboloid of revolution. The mass of the paraboloid is m.

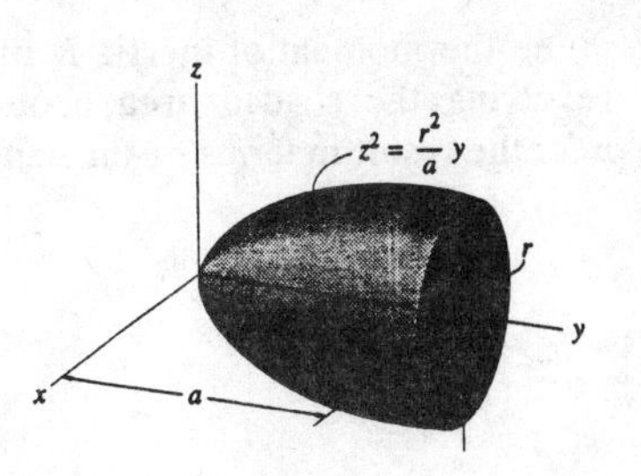

$$m = \rho\int_0^a \pi z^2\, dy = \rho\pi\int_0^a\left(\frac{r^2}{a}\right)y\, dy = \rho\pi\left(\frac{r^2}{2}\right)a$$

$$I_y = \int_m \frac{1}{2}dm\, z^2 = \frac{1}{2}\rho\pi\int_0^a z^4\, dy = \frac{1}{2}\rho\pi\left(\frac{r^4}{a^2}\right)\int_0^a y^2\, dy = \rho\pi\left(\frac{r^4}{6}\right)a$$

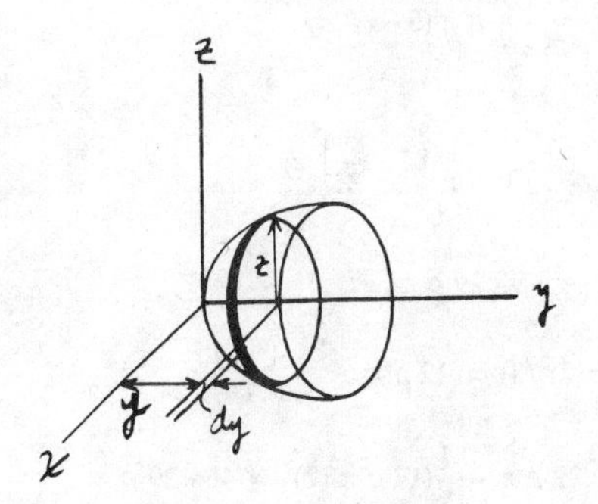

Thus,

$$I_y = \frac{1}{3}mr^2 \qquad \textbf{Ans}$$

$$I_x = \int_m\left(\frac{1}{4}dm\, z^2 + dm\, y^2\right) = \frac{1}{4}\rho\pi\int_0^a z^4\, dy + \rho\int_0^a \pi z^2 y^2\, dy$$

$$= \frac{1}{4}\rho\pi\left(\frac{r^4}{a^2}\right)\int_0^a y^2\, dy + \rho\pi\left(\frac{r^2}{a}\right)\int_0^a y^3 dy = \frac{\rho\pi r^4 a}{12} + \frac{\rho\pi r^2 a^3}{4} = \frac{1}{6}mr^2 + \frac{1}{2}ma^2$$

$$I_x = \frac{m}{6}(r^2 + 3a^2) \qquad \textbf{Ans}$$

***21-4.** Determine the product of inertia I_{xy} of the body formed by revolving the shaded area about the line $x = 5$ ft. Express the result in terms of the density of the material, ρ.

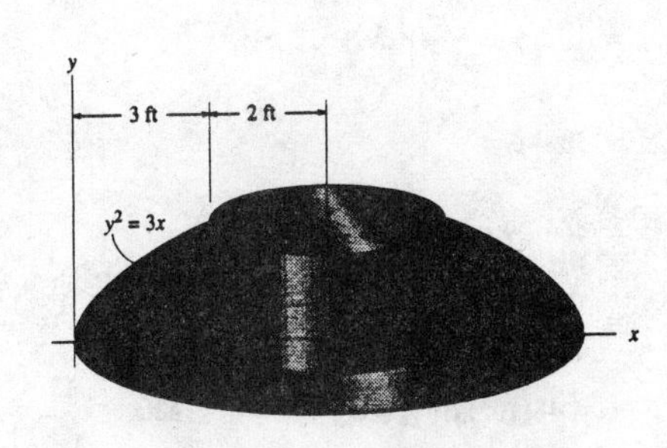

$$\int_0^3 dm = \rho 2\pi\int_0^3 (5-x)y\, dx = \rho 2\pi\int_0^3 (5-x)\sqrt{3x}\, dx = 38.4\rho\pi$$

$$\int_0^3 \tilde{y}\, dm = \rho 2\pi\int_0^3 \frac{y}{2}(5-x)y\, dx$$

$$= \rho\pi\int_0^3 (5-x)(3x)\, dx$$

$$= 40.5\rho\pi$$

Thus, $\bar{y} = \dfrac{40.5\rho\pi}{38.4\rho\pi} = 1.055$ ft

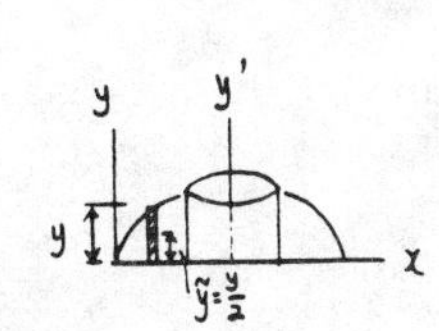

The solid is symmetric about y', thus

$$I_{xy'} = 0$$

$$I_{xy} = I_{xy'} + \bar{x}\bar{y}m$$

$$= 0 + 5(1.055)(38.4\rho\pi)$$

$$I_{xy} = 636\rho \qquad \textbf{Ans}$$

21-5. Determine the moment of inertia I_y of the body formed by revolving the shaded area about the line $x = 5$ ft. Express the result in terms of the density of the material, ρ.

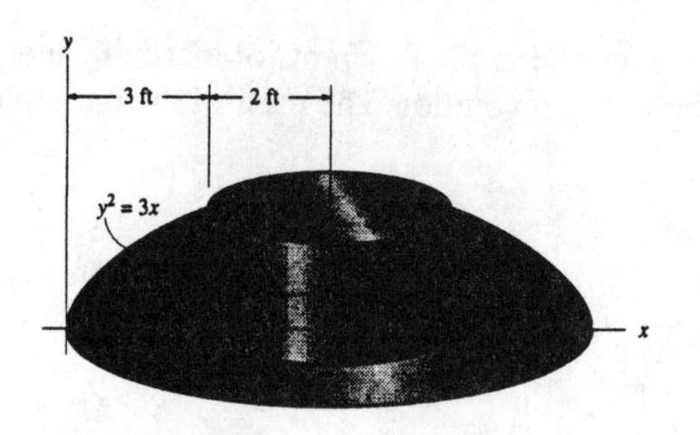

$$I_{y'} = \int_0^3 \frac{1}{2} dm\, r^2 - \frac{1}{2}(m')(2)^2$$

$$\int_0^3 \frac{1}{2} dm\, r^2 = \frac{1}{2}\int_0^3 \rho\, \pi\, (5-x)^4\, dy$$

$$= \frac{1}{2}\rho\, \pi \int_0^3 \left(5 - \frac{y^2}{3}\right)^4 dy$$

$$= 490.29\, \rho\, \pi$$

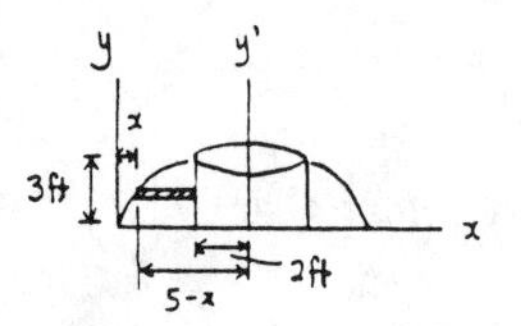

$$m' = \rho\, \pi\, (2)^2(3) = 12\, \rho\, \pi$$

$$I_{y'} = 490.29\, \rho\, \pi - \frac{1}{2}(12\, \rho\, \pi)(2)^2 = 466.29\, \rho\, \pi$$

Mass of body;

$$m = \int_0^3 \rho\, \pi\, (5-x)^2 dy - m'$$

$$= \int_0^3 \rho\, \pi\, (5 - \frac{y^2}{3})^2 dy - 12\, \rho\, \pi$$

$$= 38.4\, \rho\, \pi$$

$$I_y = 466.29\, \rho\, \pi + (38.4\, \rho\, \pi)(5)^2$$

$$= 1426.29\, \rho\, \pi$$

$$I_y = 4.48(10^3)\, \rho \qquad \textbf{Ans}$$

Also,

$$I_{y'} = \int_0^3 r^2 dm$$

$$= \int_0^3 (5-x)^2 \rho (2\pi)(5-x) y\, dx$$

$$= 2\, \rho\, \pi \int_0^3 (5-x)^3 (3x)^{1/2} dx$$

$$= 466.29\, \rho\, \pi$$

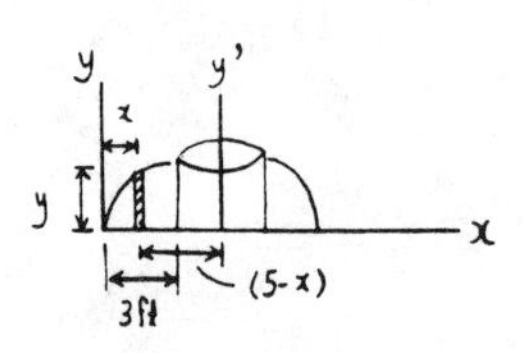

$$m = \int_0^3 dm$$

$$= 2\, \rho\, \pi \int_0^3 (5-x) y\, dx$$

$$= 2\, \rho\, \pi \int_0^3 (5-x)(3x)^{1/2} dx$$

$$= 38.4\, \rho\, \pi$$

$$I_y = 466.29\, \rho\, \pi + 38.4\, \rho\, \pi\, (5)^2 = 4.48(10^3)\rho \qquad \textbf{Ans}$$

21-6. Determine by direct integration the product of inertia I_{yz} for the homogeneous prism. The density of the material is ρ. Express the result in terms of the mass m of the prism.

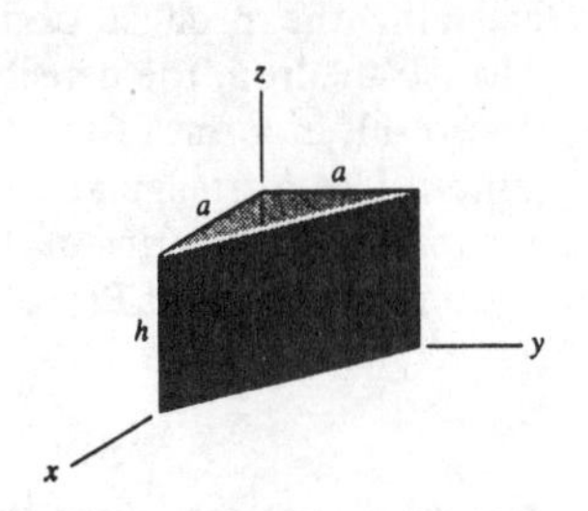

The mass of the differential element is $dm = \rho dV = \rho hxdy = \rho h(a-y)dy$.

$$m = \int_m dm = \rho h \int_0^a (a-y)dy = \frac{\rho a^2 h}{2}$$

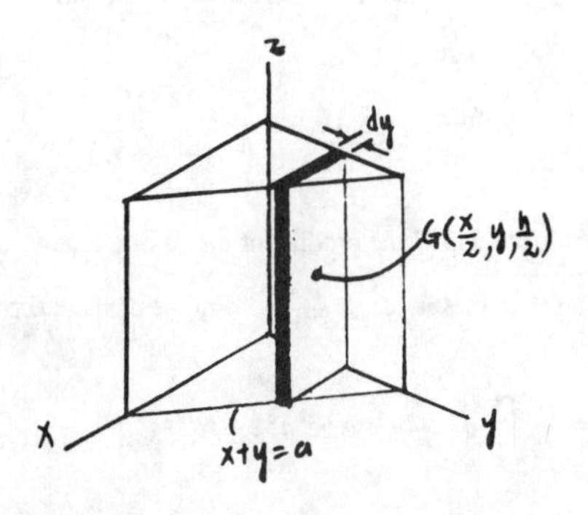

Using the parallel axis theorem :

$$dI_{yz} = (dI_{y'z'})_G + dm y_G z_G$$

$$= 0 + (\rho hxdy)(y)\left(\frac{h}{2}\right)$$

$$= \frac{\rho h^2}{2} xydy$$

$$= \frac{\rho h^2}{2}(ay - y^2)\, dy$$

$$I_{yz} = \frac{\rho h^2}{2}\int_0^a (ay - y^2)\, dy = \frac{\rho a^3 h^2}{12} = \frac{1}{6}\left(\frac{\rho a^2 h}{2}\right)(ah) = \frac{m}{6} ah \quad \textbf{Ans}$$

21-7. Determine by direct integration the product of inertia I_{xy} for the homogeneous prism. The density of the material is ρ. Express the result in terms of the mass m of the prism.

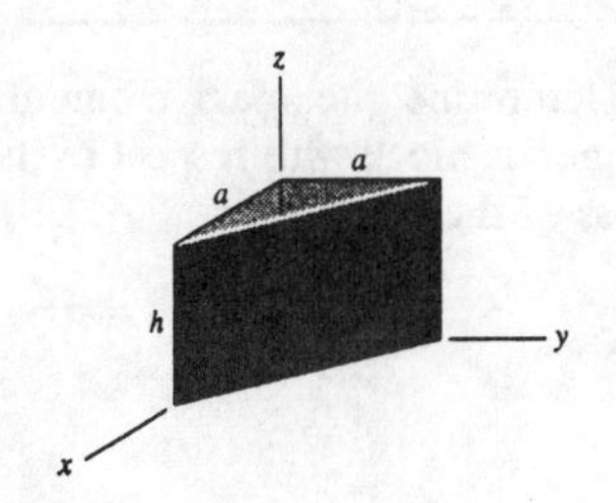

The mass of the differential element is $dm = \rho dV = \rho hxdy = \rho h(a-y)dy$.

$$m = \int_m dm = \rho h \int_0^a (a-y)dy = \frac{\rho a^2 h}{2}$$

Using the parallel axis theorem :

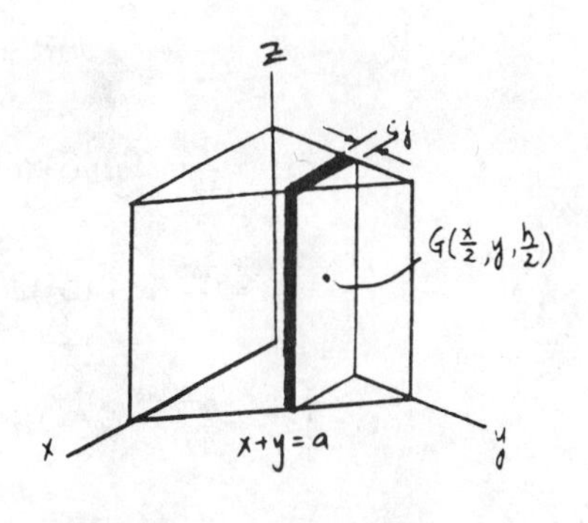

$$dI_{xy} = (dI_{x'y'})_G + dm x_G y_G$$

$$= 0 + (\rho hxdy)\left(\frac{x}{2}\right)(y)$$

$$= \frac{\rho h}{2} x^2 ydy$$

$$= \frac{\rho h}{2}(y^3 - 2ay^2 + a^2 y)\, dy$$

$$I_{xy} = \frac{\rho h}{2}\int_0^a (y^3 - 2ay^2 + a^2 y)\, dy$$

$$= \frac{\rho a^4 h}{24} = \frac{1}{12}\left(\frac{\rho a^2 h}{2}\right)a^2 = \frac{m}{12}a^2 \quad \textbf{Ans}$$

***21-8.** Determine the product of inertia I_{xy} for the homogeneous tetrahedron. The density of the material is ρ. Express the result in terms of the total mass m of the solid. *Suggestion:* Use a triangular element of thickness dz and then express dI_{xy} in terms of the size and mass of the element using the result of Prob. 21-7.

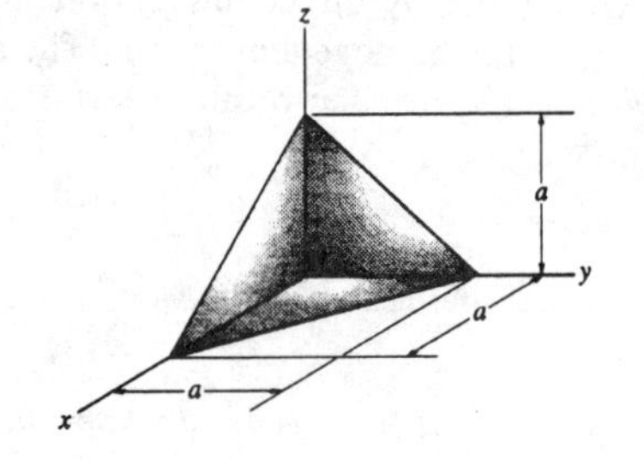

$$dm = \rho\, dV = \rho\left[\frac{1}{2}(a-z)(a-z)\right]dz = \frac{\rho}{2}(a-z)^2\,dz$$

$$m = \frac{\rho}{2}\int_0^a \left(a^2 - 2az + z^2\right)dz = \frac{\rho a^3}{6}$$

From Prob. 21 − 7 the product of inertia of a triangular prism with respect to the xz and yz planes is $I_{xy} = \dfrac{\rho a^4 h}{24}$. For the element above, $dI_{xy} = \dfrac{\rho dz}{24}(a-z)^4$. Hence,

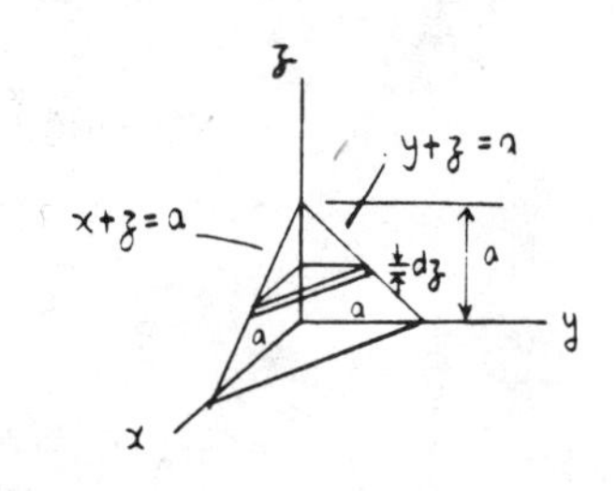

$$I_{xy} = \frac{\rho}{24}\int_0^a \left(a^4 - 4a^3 z + 6z^2 a^2 - 4az^3 + z^4\right)dz$$

$$I_{xy} = \frac{\rho a^5}{120}$$

or,

$$I_{xy} = \frac{ma^2}{20} \qquad \textbf{Ans}$$

21-9. Determine the mass moment of inertia of the homogeneous block with respect to its centroidal x' axis. The mass of the block is m.

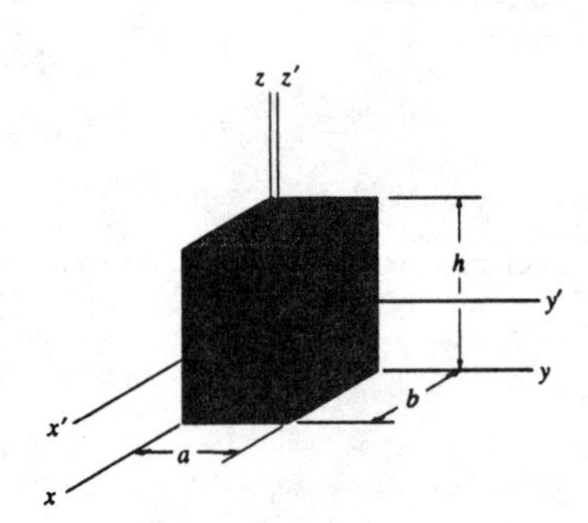

The mass of the differential element is $dm = \rho dV = \rho abdz$.

$$dI_{x'} = \frac{1}{12}dma^2 + dmz^2$$

$$= \frac{1}{12}(\rho abdz)a^2 + (\rho abdz)z^2$$

$$= \frac{\rho ab}{12}(a^2 + 12z^2)\,dz$$

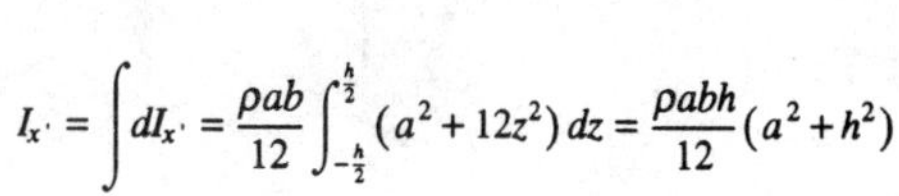

$$I_{x'} = \int dI_{x'} = \frac{\rho ab}{12}\int_{-\frac{h}{2}}^{\frac{h}{2}}(a^2 + 12z^2)\,dz = \frac{\rho abh}{12}(a^2 + h^2)$$

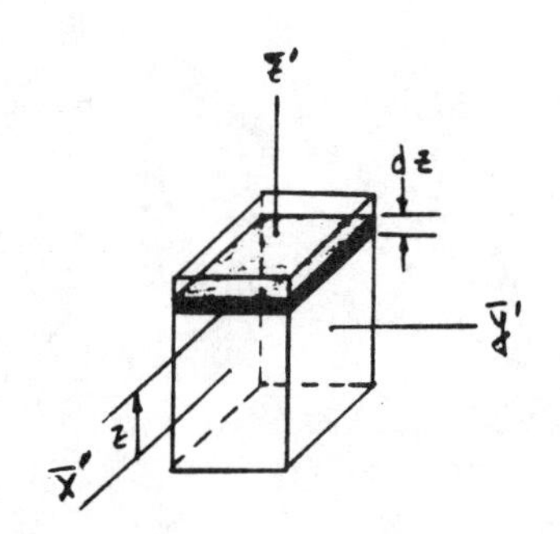

However,

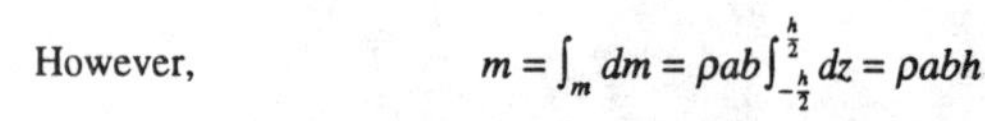

$$m = \int_m dm = \rho ab\int_{-\frac{h}{2}}^{\frac{h}{2}} dz = \rho abh$$

Hence,

$$I_{x'} = \frac{m}{12}(a^2 + h^2) \qquad \textbf{Ans}$$

21-10. Determine the moments of inertia for the homogeneous cylinder of mass m with respect to the x', y', z' axes.

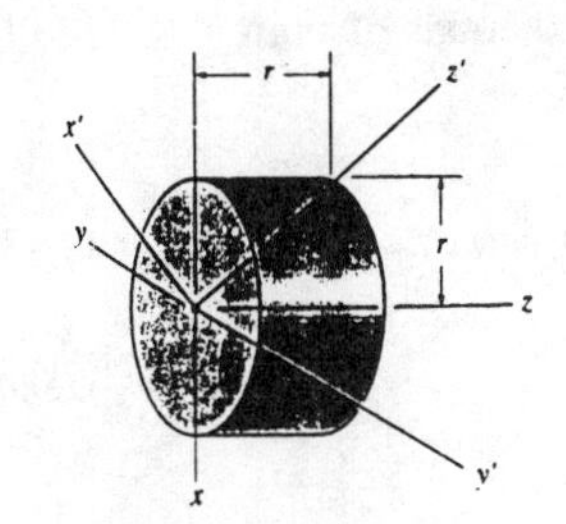

Due to symmetry $\quad I_{xy} = I_{yz} = I_{zx} = 0$

$$I_y = I_x = \frac{1}{12}m(3r^2 + r^2) + m\left(\frac{r}{2}\right)^2 = \frac{7mr^2}{12} \qquad I_z = \frac{1}{2}mr^2$$

For x',

$$u_x = \cos 135° = -\frac{1}{\sqrt{2}}, \qquad u_y = \cos 90° = 0, \qquad u_z = \cos 135° = -\frac{1}{\sqrt{2}}$$

$$I_{x'} = I_x u_x^2 + I_y u_y^2 + I_z u_z^2 - 2I_{xy}u_x u_y - 2I_{yz}u_y u_z - 2I_{zx}u_z u_x$$

$$= \frac{7mr^2}{12}\left(-\frac{1}{\sqrt{2}}\right)^2 + 0 + \frac{1}{2}mr^2\left(-\frac{1}{\sqrt{2}}\right)^2 - 0 - 0 - 0$$

$$= \frac{13}{24}mr^2 \qquad \textbf{Ans}$$

For y',

$$I_{y'} = I_y = \frac{7mr^2}{12} \qquad \textbf{Ans}$$

For z',

$$u_x = \cos 135° = -\frac{1}{\sqrt{2}}, \qquad u_y = \cos 90° = 0, \qquad u_z = \cos 45° = \frac{1}{\sqrt{2}}$$

$$I_{z'} = I_x u_x^2 + I_y u_y^2 + I_z u_z^2 - 2I_{xy}u_x u_y - 2I_{yz}u_y u_z - 2I_{zx}u_z u_x$$

$$= \frac{7mr^2}{12}\left(-\frac{1}{\sqrt{2}}\right)^2 + 0 + \frac{1}{2}mr^2\left(\frac{1}{\sqrt{2}}\right)^2 - 0 - 0 - 0$$

$$= \frac{13}{24}mr^2 \qquad \textbf{Ans}$$

21-11. Determine the moments of inertia about the x, y, z axes of the rod assembly. The rods have a mass of 0.75 kg/m.

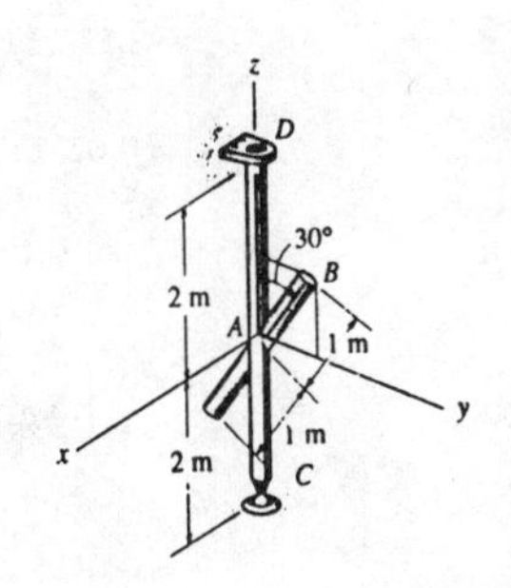

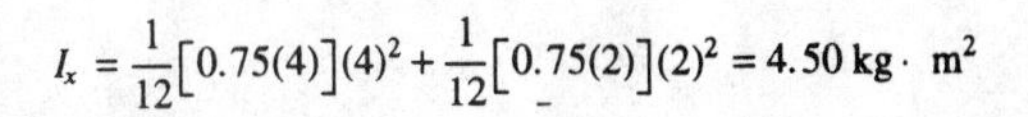

$$I_x = \frac{1}{12}[0.75(4)](4)^2 + \frac{1}{12}[0.75(2)](2)^2 = 4.50 \text{ kg}\cdot\text{m}^2 \qquad \textbf{Ans}$$

$$I_y = \frac{1}{12}[0.75(4)](4)^2 + \frac{1}{12}[0.75(2)](2\cos 30°)^2 = 4.38 \text{ kg}\cdot\text{m}^2 \qquad \textbf{Ans}$$

$$I_z = 0 + \frac{1}{12}[0.75(2)](2\sin 30°)^2 = 0.125 \text{ kg}\cdot\text{m}^2 \qquad \textbf{Ans}$$

***21-12.** Determine the moment of inertia of the disk about the axis of shaft AB. The disk has a mass of 15 kg.

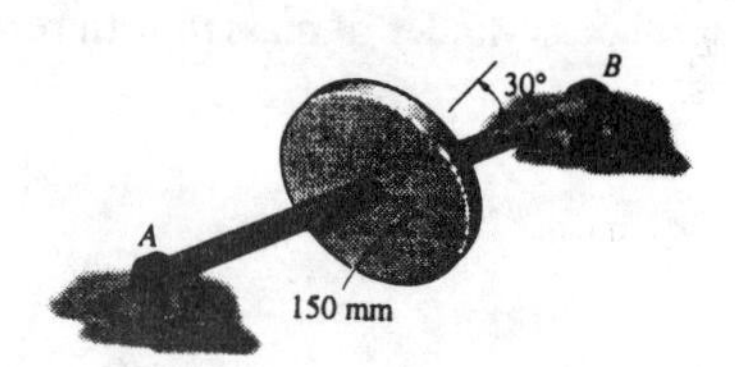

Due to symmetry $\quad I_{xy} = I_{yz} = I_{zx} = 0$

$$\bar{I}_x = \bar{I}_z = \frac{1}{4}(15)(0.15)^2 = 0.084375 \text{ kg.m}^2$$

$$I_y = \frac{1}{2}(15)(0.15)^2 = 0.16875 \text{ kg.m}^2$$

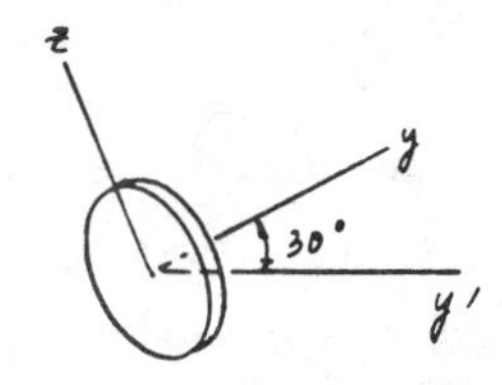

$$u_x = \cos 90° = 0, \qquad u_y = \cos 30° = 0.8660$$

$$u_z = \cos(30° + 90°) = -0.5$$

$$I_{y'} = I_x u_x^2 + I_y u_y^2 + I_z u_z^2 - 2I_{xy}u_x u_y - 2I_{yz}u_y u_z - 2I_{zx}u_z u_x$$

$$= 0 + 0.16875(0.8660)^2 + 0.084375(-0.5)^2 - 0 - 0 - 0$$

$$= 0.148 \text{ kg} \cdot \text{m}^2 \qquad \textbf{Ans}$$

21-13. The bent rod has a weight of 1.5 lb/ft. Locate the center of gravity $G(\bar{x}, \bar{y})$ and determine the principal moments of inertia $I_{x'}$, $I_{y'}$, and $I_{z'}$ of the rod with respect to the x', y', z' axes.

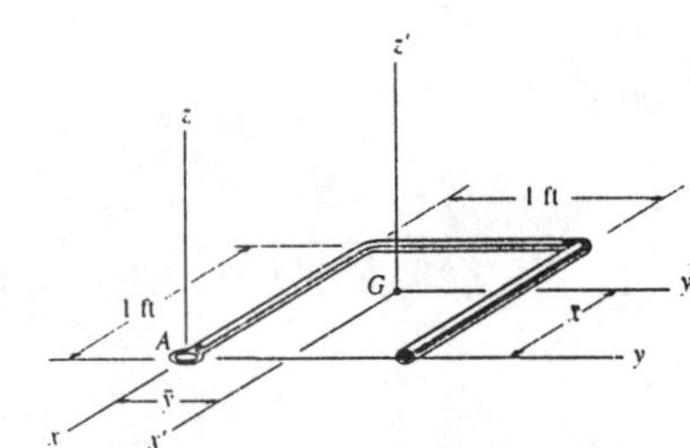

Due to symmetry $\quad \bar{y} = 0.5 \text{ ft} \qquad \textbf{Ans}$

$$\bar{x} = \frac{\Sigma \tilde{x} W}{\Sigma w} = \frac{(-1)(1.5)(1) + 2[(-0.5)(1.5)(1)]}{3[1.5(1)]} = -0.667 \text{ ft} \qquad \textbf{Ans}$$

$$I_{x'} = 2\left[\left(\frac{1.5}{32.2}\right)(0.5)^2\right] + \frac{1}{12}\left(\frac{1.5}{32.2}\right)(1)^2$$

$$= 0.0272 \text{ slug} \cdot \text{ft}^2 \qquad \textbf{Ans}$$

$$I_{y'} = 2\left[\frac{1}{12}\left(\frac{1.5}{32.2}\right)(1)^2 + \left(\frac{1.5}{32.2}\right)(0.667 - 0.5)^2\right]$$

$$+ \left(\frac{1.5}{32.2}\right)(1 - 0.667)^2$$

$$= 0.0155 \text{ slug} \cdot \text{ft}^2 \qquad \textbf{Ans}$$

$$I_{z'} = 2\left[\frac{1}{12}\left(\frac{1.5}{32.2}\right)(1)^2 + \left(\frac{1.5}{32.2}\right)(0.5^2 + 0.1667^2)\right]$$

$$+ \frac{1}{12}\left(\frac{1.5}{32.2}\right)(1)^2 + \left(\frac{1.5}{32.2}\right)(0.3333)^2$$

$$= 0.0427 \text{ slug} \cdot \text{ft}^2 \qquad \textbf{Ans}$$

21-14. Determine the moment of inertia of both the 1.5-kg rod and 4-kg disk about the z' axis.

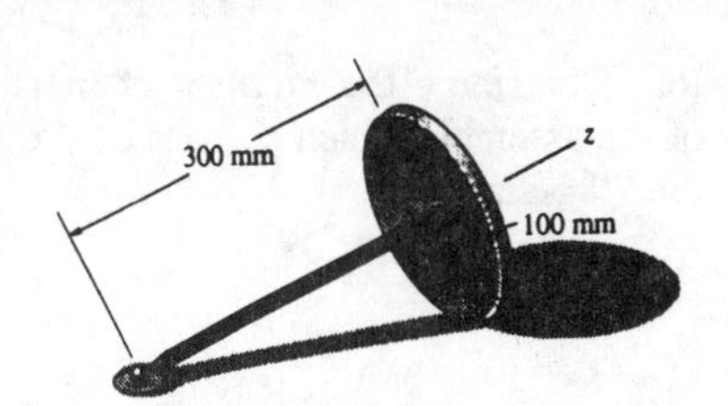

Due to symmetry $\quad I_{xy} = I_{yz} = I_{zx} = 0$

$$I_y = I_x = \left[\frac{1}{4}(4)(0.1)^2 + 4(0.3)^2\right] + \frac{1}{3}(1.5)(0.3)^2$$

$$= 0.415 \text{ kg} \cdot \text{m}^2$$

$$I_z = \frac{1}{2}(4)(0.1)^2 = 0.02 \text{ kg} \cdot \text{m}^2$$

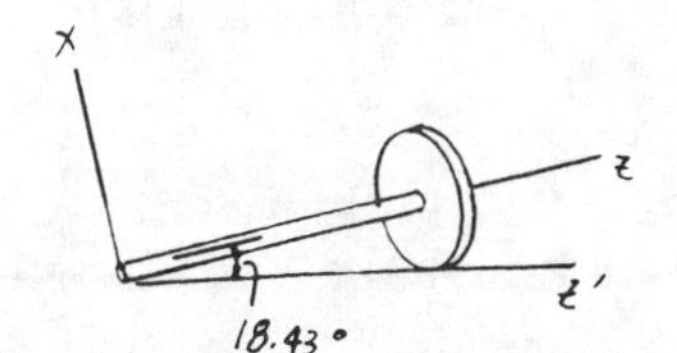

$u_z = \cos(18.43°) = 0.9487, \qquad u_y = \cos 90° = 0,$

$u_x = \cos(90° + 18.43°) = -0.3162$

$$I_{z'} = I_x u_x^2 + I_y u_y^2 + I_z u_z^2 - 2I_{xy}u_x u_y - 2I_{yz}u_y u_z - 2I_{zx}u_z u_x$$

$$= 0.415(-0.3162)^2 + 0 + 0.02(0.9487)^2 - 0 - 0 - 0$$

$$= 0.0595 \text{ kg} \cdot \text{m}^2 \qquad \textbf{Ans}$$

21-15. The top consists of a cone having a mass of 0.7 kg and a hemisphere of mass 0.2 kg. Determine the moment of inertia I_z when the top is in the position shown.

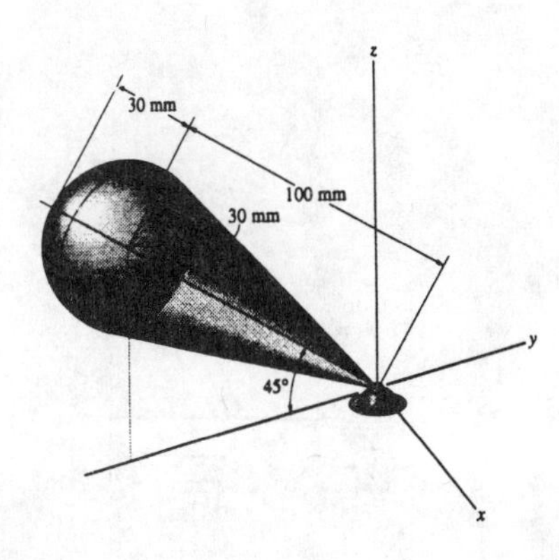

$$I_{x'} = I_{y'} = \frac{3}{80}(0.7)\left[(4)(0.03)^2 + (0.1)^2\right] + (0.7)\left[\frac{3}{4}(0.1)\right]^2$$

$$+\left(\frac{83}{320}\right)(0.2)(0.03)^2 + (0.2)\left[\frac{3}{8}(0.03) + (0.1)\right]^2 = 6.816\left(10^{-3}\right) \text{ kg} \cdot \text{m}^2$$

$$I_{z'} = \left(\frac{3}{10}\right)(0.7)(0.03)^2 + \left(\frac{2}{5}\right)(0.2)(0.03)^2$$

$$I_{z'} = 0.261\left(10^{-3}\right) \text{ kg} \cdot \text{m}^2$$

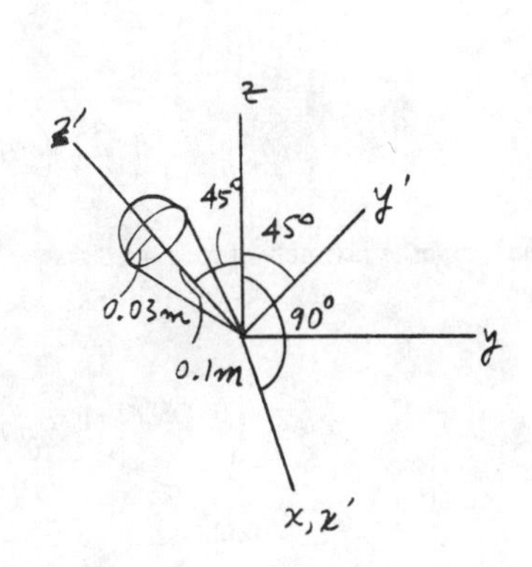

$u_{x'} = \cos 90° = 0, \qquad u_{y'} = \cos 45° = 0.7071, \qquad u_{z'} = \cos 45° = 0.7071$

$$I_z = I_{x'}u_{x'}^2 + I_{y'}u_{y'}^2 + I_{z'}u_{z'}^2 - 2I_{x'y'}u_{x'}u_{y'} - 2I_{y'z'}u_{y'}u_{z'} - 2I_{x'z'}u_{x'}u_{z'}$$

$$= 0 + 6.816\left(10^{-3}\right)(0.7071)^2 + (0.261)\left(10^{-3}\right)(0.7071)^2 - 0 - 0 - 0$$

$$I_z = 3.54\left(10^{-3}\right) \text{ kg} \cdot \text{m}^2 \qquad \textbf{Ans}$$

***21-16.** Determine the moment of inertia about the z axis of the assembly which consists of the 1.5-kg rod CD and the 7-kg disk.

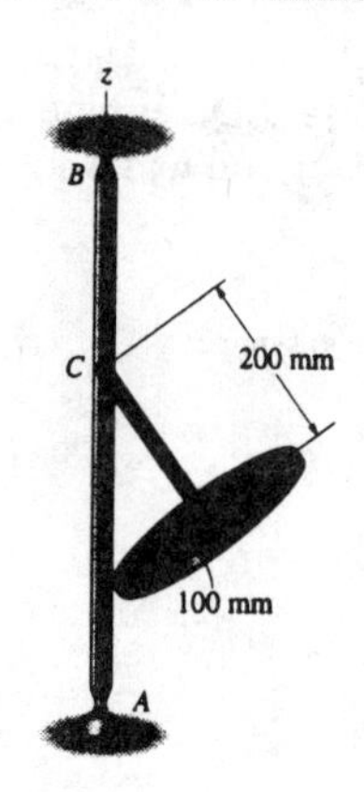

$$I_{xy} = I_{yz} = I_{zx} = 0$$

$$I_z = \frac{1}{2}(7)(0.1)^2 = 0.035 \text{ kg}\cdot\text{m}^2$$

$$I_x = I_y = \frac{1}{4}(7)(0.1)^2 + 7(0.2)^2 + \frac{1}{12}(1.5)(0.2)^2 + (1.5)(0.1)^2 = 0.3175 \text{ kg}\cdot\text{m}^2$$

$$\theta = \tan^{-1}\left(\frac{0.1}{0.2}\right) = 26.57°$$

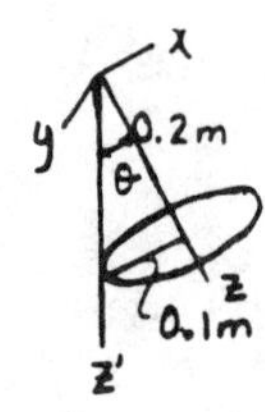

$u_x = \cos(90° + 26.57°) = -0.4472,\qquad u_y = \cos 90° = 0,\qquad u_z = \cos 26.57° = 0.8944$

$$I_{z'} = I_x u_x^2 + I_y u_y^2 + I_z u_z^2 - 2I_{xy} u_x u_y - 2I_{yz} u_y u_z - 2I_{zx} u_z u_x$$

$$= (0.3175)(-0.4472)^2 + (0.3175)(0) + (0.035)(0.8944)^2 - 0 - 0 - 0$$

$I_z = 0.0915 \text{ kg}\cdot\text{m}^2$ **Ans**

21-17. The thin plate has a weight of 5 lb and each of the four rods weighs 3 lb. Determine the moment of inertia of the assembly about the z axis.

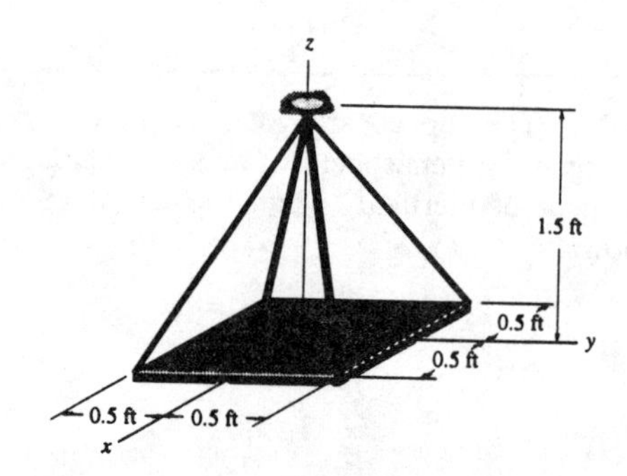

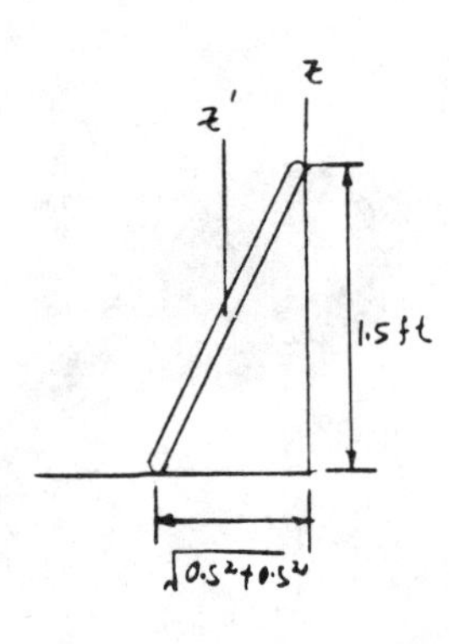

For the rod :

$$I_{z'} = \frac{1}{12}\left(\frac{3}{32.2}\right)\left(\sqrt{0.5^2 + 0.5^2}\right)^2 = 0.003882 \text{ slug}\cdot\text{ft}^2$$

For the composite assembly of rods and disks :

$$I_z = 4\left[0.003882 + \left(\frac{3}{32.2}\right)\left(\frac{\sqrt{0.5^2 + 0.5^2}}{2}\right)^2\right] + \frac{1}{12}\left(\frac{5}{32.2}\right)(1^2 + 1^2)$$

$= 0.0880 \text{ slug}\cdot\text{ft}^2$ **Ans**

21-18. Determine the moment of inertia of the rod-and-thin-ring assembly about the z axis. The rods and ring have a mass of 2 kg/m.

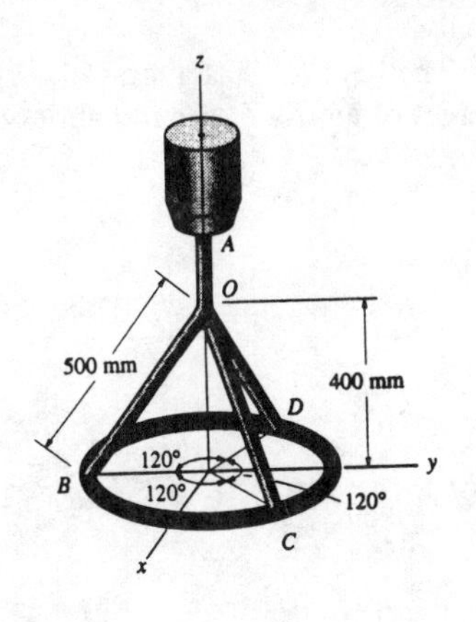

For the rod,

$$u_{x'} = 0.6, \quad u_{y'} = 0, \quad u_{z'} = 0.8$$

$$I_{x'} = I_{y'} = \frac{1}{3}[(0.5)(2)](0.5)^2 = 0.08333 \text{ kg}\cdot\text{m}^2$$

$$I_{z'} = 0$$

$$I_{x'y'} = I_{y'z'} = I_{x'z'} = 0$$

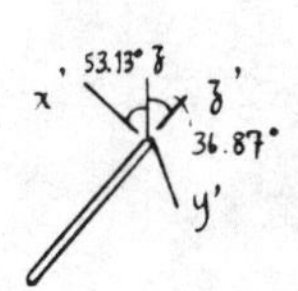

From Eq. 21 – 5,

$$I_z = 0.08333(0.6)^2 + 0 + 0 - 0 - 0 - 0$$

$$I_z = 0.03 \text{ kg}\cdot\text{m}^2$$

For the ring,

The radius is $r = 0.3$ m

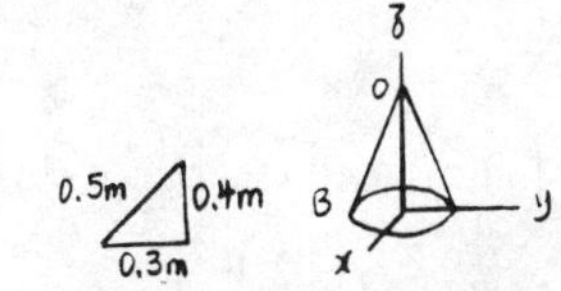

Thus,

$$I_z = mR^2 = [2(2\pi)(0.3)](0.3)^2 = 0.3393 \text{ kg}\cdot\text{m}^2$$

Thus the moment of inertia of the assembly is

$$I_z = 3(0.03) + 0.339 = 0.429 \text{ kg}\cdot\text{m}^2 \qquad \textbf{Ans}$$

21-19. Determine the moment of inertia of the rod-and-disk assembly about the x axis. The disks each have a weight of 12 lb. The two rods each have a weight of 4 lb, and their ends extend to the rims of the disks.

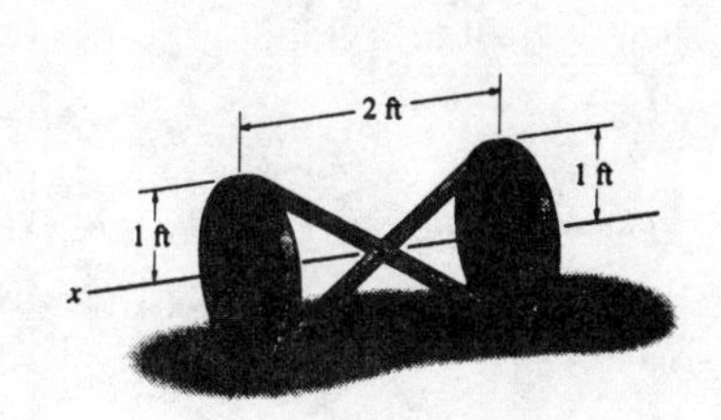

For a rod :

$$\theta = \tan^{-1}\left(\frac{1}{1}\right) = 45°$$

$$u_{x'} = \cos 90° = 0, \quad u_{y'} = \cos 45° = 0.7071, \quad u_{z'} = \cos(90° + 45°) = -0.7071$$

$$I_{x'} = I_{z'} = \left(\frac{1}{12}\right)\left(\frac{4}{32.2}\right)\left[(2)^2 + (2)^2\right] = 0.08282 \text{ slug}\cdot\text{ft}^2$$

$$I_{y'} = 0$$

$$I_{x'y'} = I_{y'z'} = I_{x'z'} = 0$$

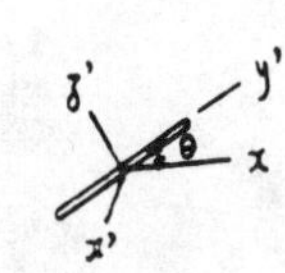

$$I_x = 0 + 0 + (0.08282)(-0.7071)^2 = 0.04141 \text{ slug}\cdot\text{ft}^2$$

For a disk :

$$I_x = \left(\frac{1}{2}\right)\left(\frac{12}{32.2}\right)(1)^2 = 0.1863 \text{ slug}\cdot\text{ft}^2$$

Thus,

$$I_x = 2(0.04141) + 2(0.1863) = 0.455 \text{ slug}\cdot\text{ft}^2 \qquad \textbf{Ans}$$

***21-20.** The bent rod has a mass of 4 kg/m. Determine the moment of inertia of the rod about the Oa axis.

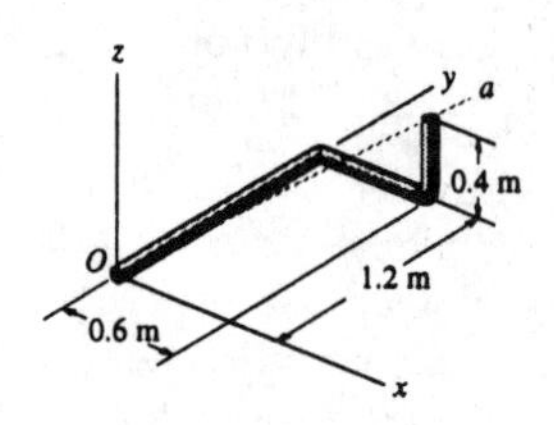

$$I_{xy} = [4(1.2)](0)(0.6) + [4(0.6)](0.3)(1.2) + [4(0.4)](0.6)(1.2) = 2.016 \text{ kg}\cdot\text{m}^2$$

$$I_{yz} = [4(1.2)](0.6)(0) + [4(0.6)](1.2)(0) + [4(0.4)](1.2)(0.2) = 0.384 \text{ kg}\cdot\text{m}^2$$

$$I_{zx} = [4(1.2)](0)(0) + [4(0.6)](0)(0.3) + [4(0.4)](0.2)(0.6) = 0.192 \text{ kg}\cdot\text{m}^2$$

$$I_x = \frac{1}{3}[4(1.2)](1.2)^2 + [4(0.6)](1.2)^2 + \left[\frac{1}{12}[4(0.4)](0.4)^2 + [4(0.4)](1.2^2 + 0.2^2)\right]$$

$$= 8.1493 \text{ kg}\cdot\text{m}^2$$

$$I_y = 0 + \frac{1}{3}[4(0.6)](0.6)^2 + \left[\frac{1}{12}[4(0.4)](0.4)^2 + [4(0.4)](0.6^2 + 0.2^2)\right]$$

$$= 0.9493 \text{ kg}\cdot\text{m}^2$$

$$I_z = \frac{1}{3}[4(1.2)](1.2)^2 + \left[\frac{1}{12}[4(0.6)](0.6)^2 + [4(0.6)](0.3^2 + 1.2^2)\right] + [4(0.4)](1.2^2 + 0.6^2)$$

$$= 8.9280 \text{ kg}\cdot\text{m}^2$$

$$\mathbf{u}_{Oa} = \frac{0.6\mathbf{i} + 1.2\mathbf{j} + 0.4\mathbf{k}}{\sqrt{0.6^2 + 1.2^2 + 0.4^2}} = \frac{3}{7}\mathbf{i} + \frac{6}{7}\mathbf{j} + \frac{2}{7}\mathbf{k}$$

$$I_{Oa} = I_x u_x^2 + I_y u_y^2 + I_z u_z^2 - 2I_{xy} u_x u_y - 2I_{yz} u_y u_z - 2I_{zx} u_z u_x$$

$$= 8.1493\left(\frac{3}{7}\right)^2 + 0.9493\left(\frac{6}{7}\right)^2 + 8.9280\left(\frac{2}{7}\right)^2 - 2(2.016)\left(\frac{3}{7}\right)\left(\frac{6}{7}\right)$$

$$- 2(0.384)\left(\frac{6}{7}\right)\left(\frac{2}{7}\right) - 2(0.192)\left(\frac{2}{7}\right)\left(\frac{3}{7}\right)$$

$$= 1.21 \text{ kg}\cdot\text{m}^2$$ **Ans**

21-21. If a body contains *no planes of symmetry,* the principal moments of inertia can be determined mathematically. To show how this is done, consider the rigid body which is spinning with an angular velocity $\boldsymbol{\omega}$, directed along one of its principal axes of inertia. If the principal moment of inertia about this axis is I, the angular momentum can be expressed as $\mathbf{H} = I\boldsymbol{\omega} = I\omega_x\mathbf{i} + I\omega_y\mathbf{j} + I\omega_z\mathbf{k}$. The components of $\mathbf{H}$ may also be expressed by Eqs. 21–10, where the inertia tensor is assumed to be known. Equate the $\mathbf{i}$, $\mathbf{j}$, and $\mathbf{k}$ components of both expressions for $\mathbf{H}$ and consider ω_x, ω_y, and ω_z to be unknown. The solution of these three equations is obtained provided the determinant of the coefficients is zero. Show that this determinant, when expanded, yields the cubic equation

$$I^3 - (I_{xx} + I_{yy} + I_{zz})I^2 + (I_{xx}I_{yy} + I_{yy}I_{zz} + I_{zz}I_{xx} - I_{xy}^2 - I_{yz}^2 - I_{zx}^2)I - (I_{xx}I_{yy}I_{zz} - 2I_{xy}I_{yz}I_{zx} - I_{xx}I_{yz}^2 - I_{yy}I_{zx}^2 - I_{zz}I_{xy}^2) = 0$$

The three positive roots of I, obtained from the solution of this equation, represent the principal moments of inertia I_x, I_y, and I_z.

$$\mathbf{H} = I\omega = I\omega_x\mathbf{i} + I\omega_y\mathbf{j} + I\omega_z\mathbf{k}$$

Equating the **i**, **j**, **k** components to the scalar equations (Eq. 21-10) yields

$$(I_{xx} - I)\omega_x - I_{xy}\omega_y - I_{xz}\omega_z = 0$$

$$-I_{yx}\omega_x + (I_{yy} - I)\omega_y - I_{yz}\omega_z = 0$$

$$-I_{zx}\omega_x - I_{zy}\omega_y + (I_{zz} - I)\omega_z = 0$$

Solution for ω_x, ω_y, and ω_z requires

$$\begin{vmatrix} (I_{xx} - I) & -I_{xy} & -I_{xz} \\ -I_{yx} & (I_{yy} - I) & -I_{yz} \\ -I_{zx} & -I_{zy} & (I_{zz} - I) \end{vmatrix} = 0$$

Expanding

$$I^3 - (I_{xx} + I_{yy} + I_{zz})I^2 + \left(I_{xx}I_{yy} + I_{yy}I_{zz} + I_{zz}I_{xx} - I_{xy}^2 - I_{yz}^2 - I_{zx}^2\right)I - \left(I_{xx}I_{yy}I_{zz} - 2I_{xy}I_{yz}I_{zx} - I_{xx}I_{yz}^2 - I_{yy}I_{zx}^2 - I_{zz}I_{xy}^2\right) = 0$$ **QED**

21-22. Show that if the angular momentum of a body is determined with respect to an arbitrary point A, then $\mathbf{H}_A$ can be expressed by Eq. 21–9. This requires substituting $\boldsymbol{\rho}_A = \boldsymbol{\rho}_G + \boldsymbol{\rho}_{G/A}$ into Eq. 21–6 and expanding, noting that $\int \boldsymbol{\rho}_G\, dm = \mathbf{0}$ by definition of the mass center and $\mathbf{v}_G = \mathbf{v}_A + \boldsymbol{\omega} \times \boldsymbol{\rho}_{G/A}$.

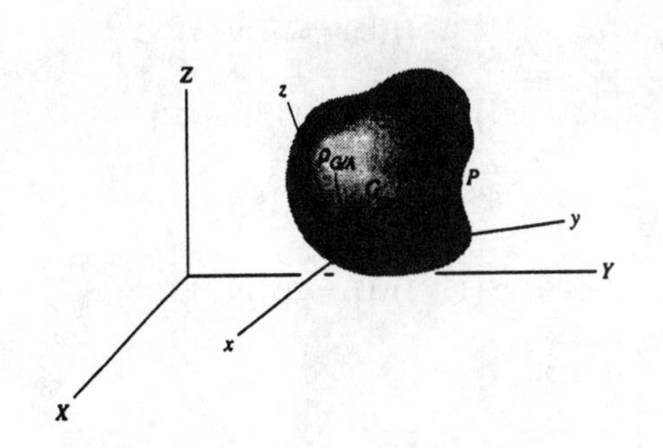

$$\mathbf{H}_A = \left(\int_m \rho_A\, dm\right) \times \mathbf{v}_A + \int_m \rho_A \times (\omega \times \rho_A)\, dm$$

$$= \left(\int_m (\rho_G + \rho_{G/A})\, dm\right) \times \mathbf{v}_A + \int_m (\rho_G + \rho_{G/A}) \times \left[\omega \times (\rho_G + \rho_{G/A})\right] dm$$

$$= \left(\int_m \rho_G dm\right) \times \mathbf{v}_A + (\rho_{G/A} \times \mathbf{v}_A)\int_m dm + \int_m \rho_G \times (\omega \times \rho_G)\, dm + \left(\int_m \rho_G dm\right) \times (\omega \times \rho_{G/A}) + \rho_{G/A} \times \left(\omega \times \int_m \rho_G dm\right) + \rho_{G/A} \times (\omega \times \rho_{G/A})\int_m dm$$

Since $\int_m \rho_G dm = 0$ and from Eq. 21-8 $\mathbf{H}_G = \int_m \rho_G \times (\omega \times \rho_G)\, dm$

$$\mathbf{H}_A = (\rho_{G/A} \times \mathbf{v}_A)m + \mathbf{H}_G + \rho_{G/A} \times (\omega \times \rho_{G/A})m$$

$$= \rho_{G/A} \times (\mathbf{v}_A + (\omega \times \rho_{G/A}))m + \mathbf{H}_G$$

$$= (\rho_{G/A} \times m\mathbf{v}_G) + \mathbf{H}_G$$ **Q.E.D.**

21-23. Gear A has a mass of 5 kg and a radius of gyration of $k_z = 75$ mm. Gears B and C each have a mass of 200 g and a radius of gyration about the axis of their connecting shaft of 15 mm. If the gears are in mesh and C has an angular velocity of $\omega_C = \{15\mathbf{j}\}$ rad/s, determine the total angular momentum for the system of three gears about point A.

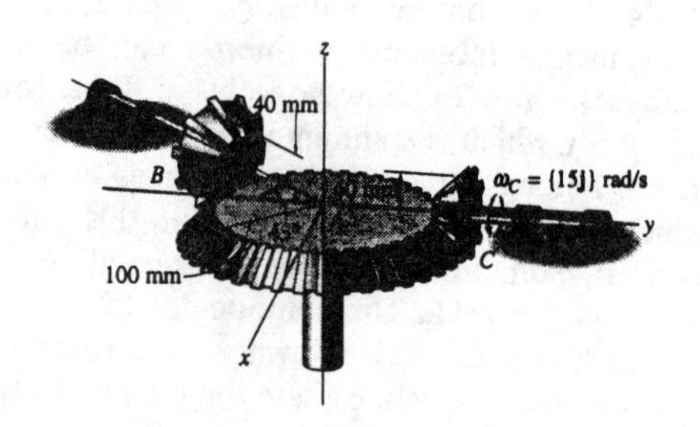

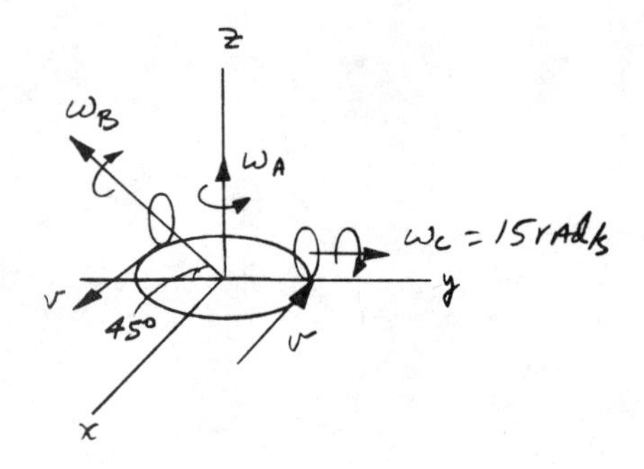

$$I_A = 5(0.075)^2 = 28.125(10^{-3})\ \text{kg}\cdot\text{m}^2$$

$$I_B = I_C = 0.2(0.015)^2 = 45(10^{-6})\ \text{kg}\cdot\text{m}^2$$

Kinematics :

$$\omega_C = \omega_B = 15\ \text{rad/s}$$

$$v = (0.04)(15) = 0.6\ \text{m/s}$$

$$\omega_A = \left(\frac{0.6}{0.1}\right) = 6\ \text{rad/s}$$

$$H_B = I_B\,\omega_B = (45(10^{-6}))(15) = 675(10^{-6})$$

$$\mathbf{H}_B = -675(10^{-6})\sin 45^\circ\mathbf{i} - 675(10^{-6})\cos 45^\circ\mathbf{j}$$

$$\mathbf{H}_B = -477.3(10^{-6})\mathbf{i} - 477.3(10^{-6})\mathbf{j}$$

$$H_C = I_C\,\omega_C = (45(10^{-6}))(15) = 675(10^{-6})$$

$$\mathbf{H}_C = 675(10^{-6})\mathbf{j}$$

$$H_A = I_A\,\omega_A = 28.125(10^{-3})(6) = 0.16875$$

$$\mathbf{H}_A = 0.16875\mathbf{k}$$

The total angular momentum is therefore,

$$\mathbf{H} = \mathbf{H}_B + \mathbf{H}_C + \mathbf{H}_A = \{-477(10^{-6})\mathbf{i} + 198(10^{-6})\mathbf{j} + 0.169\mathbf{k}\}\ \text{kg}\cdot\text{m}^2/\text{s} \qquad \textbf{Ans}$$

***21-24.** The 4-lb rod AB is attached to the disk and collar using ball-and-socket joints. If the disk has a constant angular velocity of 2 rad/s, determine the kinetic energy of the rod when it is in the position shown. Assume the angular velocity of the rod is directed perpendicular to the axis of the rod. *Hint:* The motion has been specified in Prob. 20-22.

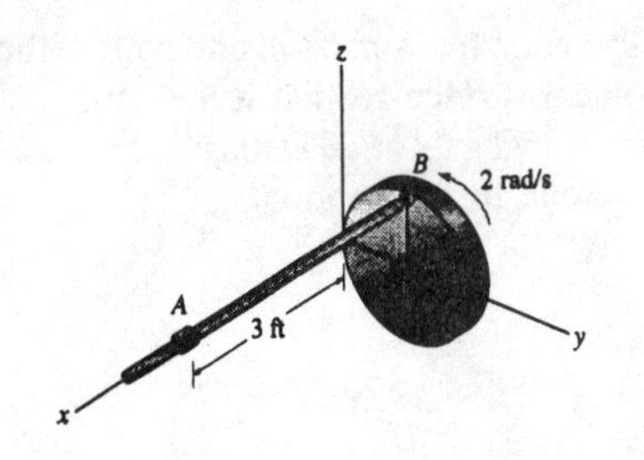

See Prob. 20 – 22.

ω is perpendicular to the rod.

$\omega_x = 0.1818$ rad/s, $\quad \omega_y = -0.06061$ rad/s, $\quad \omega_z = 0.6061$ rad/s

$\mathbf{v}_B = \{-2\mathbf{j}\}$ ft/s

$\mathbf{r}_{A/B} = \{3\mathbf{i} - 1\mathbf{j} - 1\mathbf{k}\}$ ft

$$\mathbf{v}_G = \mathbf{v}_B + \omega \times \frac{\mathbf{r}_{A/B}}{2}$$

$$\mathbf{v}_G = -2\mathbf{j} + \frac{1}{2}\begin{vmatrix} \mathbf{i} & \mathbf{j} & \mathbf{k} \\ 0.1818 & -0.06061 & 0.6061 \\ 3 & -1 & -1 \end{vmatrix}$$

$\mathbf{v}_G = \{0.333\mathbf{i} - 1\mathbf{j}\}$ ft/s

$v_G = \sqrt{(0.333)^2 + (-1)^2} = 1.054$ ft/s

$\omega = \sqrt{(0.1818)^2 + (-0.06061)^2 + (0.6061)^2} = 0.6356$ rad/s

$$T = \left(\frac{1}{2}\right)\left(\frac{4}{32.2}\right)(1.054)^2 + \left(\frac{1}{2}\right)\left[\frac{1}{12}\left(\frac{4}{32.2}\right)(3.3166)^2\right](0.6356)^2$$

$T = 0.0920$ ft · lb **Ans**

21-25. Determine the angular momentum of rod AB in Prob. 21-24 about its mass center at the instant shown. *Hint:* The motion has been specified in Prob. 20-22.

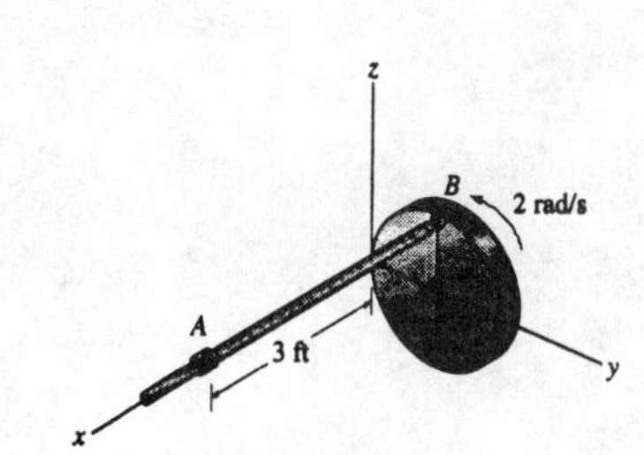

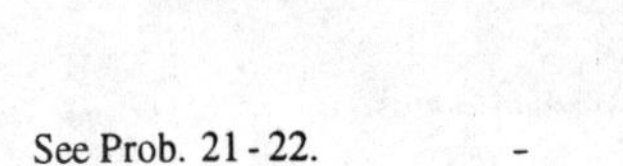

See Prob. 21 - 22.

$\omega_x = 0.1818$ rad/s, $\quad \omega_y = -0.06061$ rad/s, $\quad \omega_z = 0.6061$ rad/s

ω is perpendicular to the rod.

$r_{A/B} = \sqrt{(3)^2 + (-1)^2 + (-1)^2} = 3.3166$ ft

$I_G = \left(\frac{1}{12}\right)\left(\frac{4}{32.2}\right)(3.3166)^2 = 0.1139$ slug · ft^2

$\mathbf{H}_G = I_G\omega = 0.1139(0.1818\mathbf{i} - 0.06061\mathbf{j} + 0.6061\mathbf{k})$

$\mathbf{H}_G = \{0.0207\mathbf{i} - 0.00690\mathbf{j} + 0.0690\mathbf{k}\}$ slug · ft^2/s **Ans**

21-26. The cone has a mass m and rolls without slipping on the conical surface so that it has an angular velocity about the vertical axis of $\boldsymbol{\omega}$. Determine the kinetic energy of the cone due to this motion.

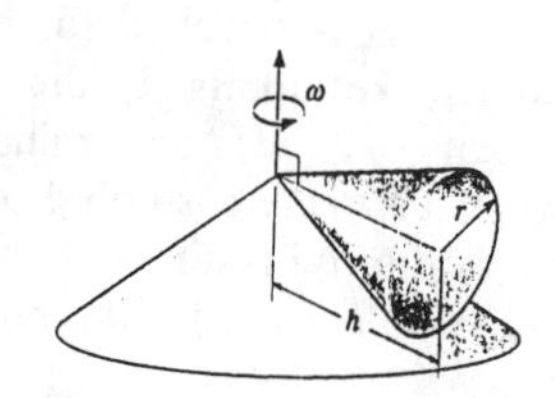

$$\frac{\omega_z}{r} = \frac{\omega_y}{h}$$

$$\omega_y = \left(\frac{h}{r}\right)\omega_z = \left(\frac{h}{r}\right)\omega$$

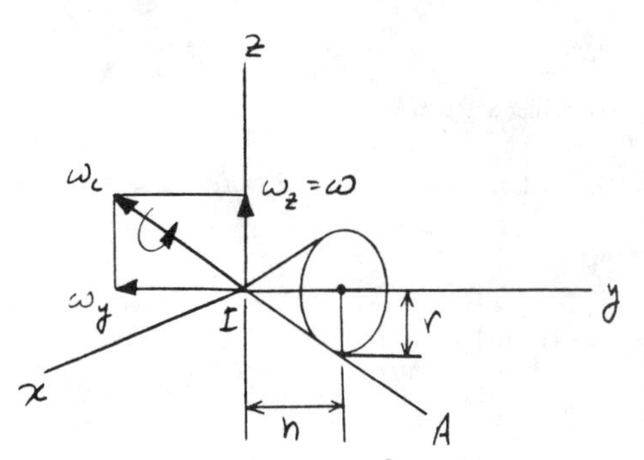

$$I_z = \left(\frac{3}{80}\right)m\left(4r^2+h^2\right) + m\left(\frac{3}{4}h\right)^2 = \left(\frac{3}{20}\right)mr^2 + \left(\frac{3}{5}\right)mh^2 = \left(\frac{3}{20}\right)m\left(r^2+4h^2\right)$$

$$T = \frac{1}{2}I_x\,\omega_x^2 + \frac{1}{2}I_y\,\omega_y^2 + \frac{1}{2}I_z\,\omega_z^2$$

$$= 0 + \frac{1}{2}\left(\frac{3}{10}mr^2\right)\left(\frac{h}{r}\omega\right)^2 + \frac{1}{2}\left[\left(\frac{3}{20}\right)m\left(r^2+4h^2\right)\right]\omega^2 = \frac{m\omega^2}{20}\left[3h^2 + \frac{3}{2}r^2 + 6h^2\right]$$

$$T = \frac{9mh^2}{20}\left[1+\frac{r^2}{6h^2}\right]\omega^2 \qquad \textbf{Ans}$$

21-27. A thin plate, having a mass of 4 kg, is suspended from one of its corners by a ball-and-socket joint O. If a stone strikes the plate perpendicular to its surface at an adjacent corner A with an impulse of $\mathbf{I}_s = \{-60\mathbf{i}\}$ N·s, determine the instantaneous axis of rotation for the plate and the impulse created at O.

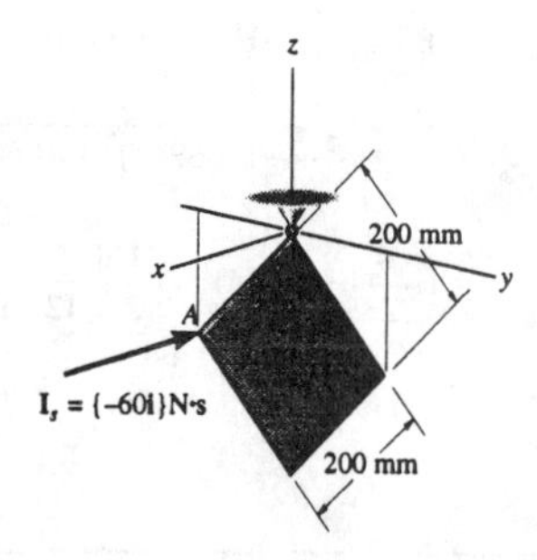

$$(\mathbf{H}_O)_1 + \Sigma\int \mathbf{M}_O\,dt = (\mathbf{H}_O)_2$$

$$0 + \mathbf{r}_{A/O}\times\mathbf{I}_S = (\mathbf{H}_O)_2$$

$$0 + (-0.2(0.7071)\mathbf{j} - 0.2(0.7071)\mathbf{k})\times(-60\mathbf{i}) = (I_O)_x\,\omega_x\mathbf{i} + (I_O)_y\,\omega_y\mathbf{j} + (I_O)_z\,\omega_z\mathbf{k}$$

Expand and equate components :

$$0 = (I_O)_x\,\omega_x \qquad (1)$$

$$8.4853 = (I_O)_y\,\omega_y \qquad (2)$$

$$-8.4853 = (I_O)_z\,\omega_z \qquad (3)$$

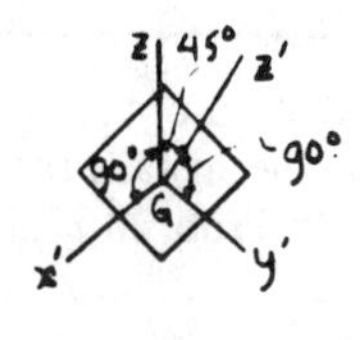

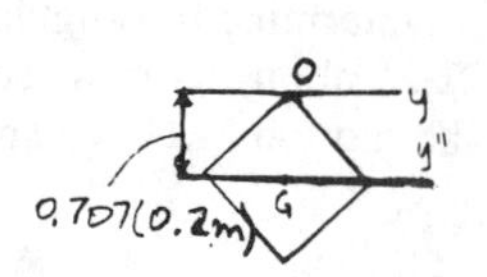

$$I_{x'y'} = 0, \quad I_{y'z'} = 0, \quad I_{x'z'} = 0$$

$$I_{y'} = \left(\frac{1}{12}\right)(4)(0.2)^2 = 0.01333, \qquad I_{z'} = \left(\frac{1}{12}\right)(4)(0.2)^2 = 0.01333$$

$$u_{x'} = \cos 90° = 0, \quad u_{y'} = \cos 135° = -0.7071, \quad u_{z'} = \cos 45° = 0.7071$$

$$(I_G)_z = I_{x'}u_{x'}^2 + I_{y'}u_{y'}^2 + I_{z'}u_{z'}^2 - 2I_{x'y'}u_{x'}u_{y'} - 2I_{y'z'}u_{y'}u_{z'} - 2I_{z'x'}u_{z'}u_{x'}$$

$$= 0 + (0.01333)(-0.7071)^2 + (0.01333)(0.7071)^2 - 0 - 0 - 0$$

$$(I_G)_z = (I_O)_z = 0.01333$$

For $(I_O)_y$, use the parallel axis theorem.

$$(I_O)_y = 0.01333 + 4\left[0.7071(0.2)\right]^2, \qquad (I_O)_y = 0.09333$$

Hence, from Eqs. (1) and (2): $\quad \omega_x = 0, \quad \omega_y = 90.914, \quad \omega_z = -636.340$

The instantaneous axis of rotation is thus,

$$\mathbf{u}_{IA} = \frac{90.914\mathbf{j} - 636.340\mathbf{k}}{\sqrt{(90.914)^2 + (-636.340)^2}} = 0.141\mathbf{j} - 0.990\mathbf{k} \qquad \textbf{Ans}$$

The velocity of G just after the plate is hit is

$$\mathbf{v}_G = \boldsymbol{\omega}\times\mathbf{r}_{G/O}$$

$$\mathbf{v}_G = (90.914\mathbf{j} - 636.340\mathbf{k})\times(-0.2(0.7071)\mathbf{k}) = -12.857\mathbf{i}$$

$$m(\mathbf{v}_G)_1 + \Sigma\int \mathbf{F}\,dt = m(\mathbf{v}_G)_2$$

$$0 - 60\mathbf{i} + \int \mathbf{F}_O\,dt = -4(12.857)\mathbf{i}$$

$$\int \mathbf{F}_O\,dt = \{8.57\mathbf{i}\}\ \text{N}\cdot\text{s} \qquad \textbf{Ans}$$

***21-28.** The 5-kg disk is connected to the 3-kg slender rod. If the assembly is attached to a ball-and-socket joint at A and the 5-N·m couple moment is applied, determine the angular velocity of the rod about the z axis after the assembly has made two revolutions about the z axis starting from rest. The disk rolls without slipping.

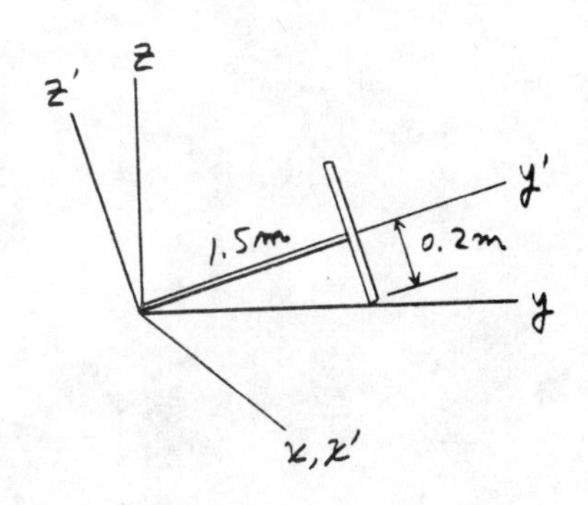

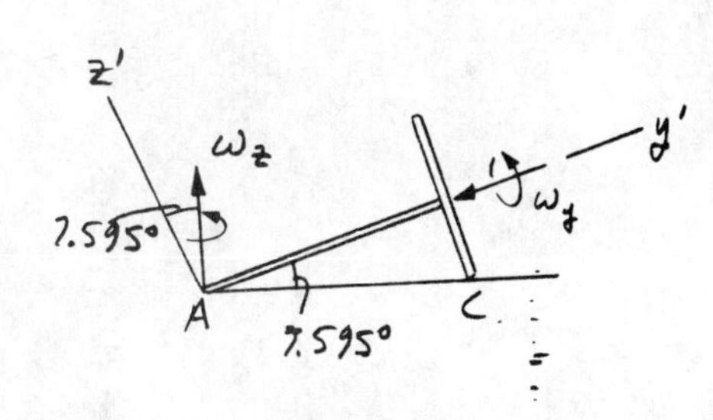

$$I_{x'} = I_{z'} = \frac{1}{4}(5)(0.2)^2 + 5(1.5)^2 + \frac{1}{3}(3)(1.5)^2 = 13.55$$

$$I_{y'} = \frac{1}{2}(5)(0.2)^2 = 0.100$$

$$\omega = -\omega_{y'}\mathbf{j}' + \omega_z\mathbf{k} = -\omega_{y'}\mathbf{j}' + \omega_z\sin 7.595°\mathbf{j}' + \omega_z\cos 7.595°\mathbf{k}'$$

$$= (0.13216\omega_z - \omega_{y'})\mathbf{j}' + 0.99123\omega_z\mathbf{k}'$$

Since points A and C have zero velocity,

$$\mathbf{v}_C = \mathbf{v}_A + \omega \times \mathbf{r}_{C/A}$$

$$\mathbf{0} = \mathbf{0} + \left[(0.13216\omega_z - \omega_{y'})\mathbf{j}' + 0.99123\omega_z\mathbf{k}'\right] \times (1.5\mathbf{j}' - 0.2\mathbf{k}')$$

$$0 = -1.48684\omega_z - 0.026433\omega_z + 0.2\omega_{y'}$$

$$\omega_{y'} = 7.5664\omega_z$$

Thus,

$$\omega = -7.4342\omega_z\mathbf{j}' + 0.99123\omega_z\mathbf{k}'$$

$$T_1 + \Sigma U_{1-2} = T_2$$

$$0 + 5(2\pi)(2) = 0 + \frac{1}{2}(0.100)(-7.4342\omega_z)^2 + \frac{1}{2}(13.55)(0.99123\omega_z)^2$$

$\omega_z = 2.58$ rad/s **Ans**

21-29. The 5-kg disk is connected to the 3-kg slender rod. If the assembly is attached to a ball-and-socket joint at A and the 5-N·m couple moment gives it an angular velocity about the z axis of $\omega_z = 2$ rad/s, determine the magnitude of the angular momentum of the assembly about A.

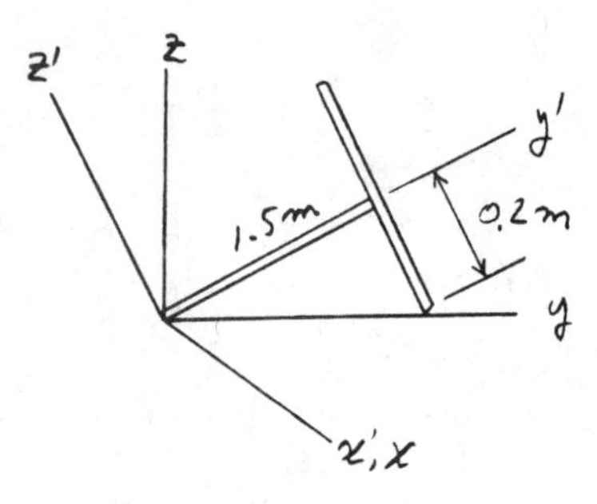

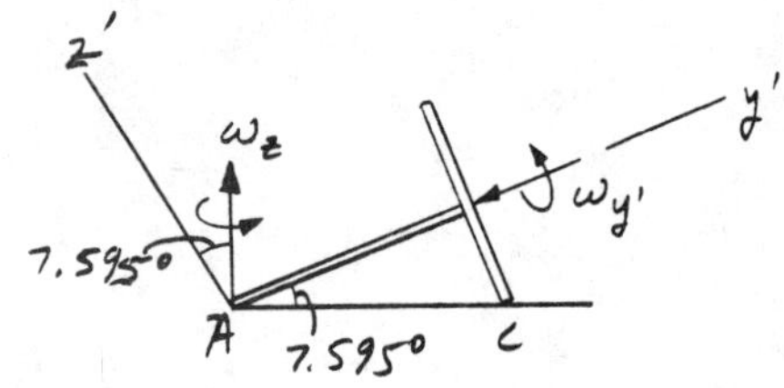

$$I_{x'} = I_{z'} = \frac{1}{4}(5)(0.2)^2 + 5(1.5)^2 + \frac{1}{3}(3)(1.5)^2 = 13.55$$

$$I_{y'} = \frac{1}{2}(5)(0.2)^2 = 0.100$$

$$\omega = -\omega_{y'}\mathbf{j}' + \omega_z\mathbf{k} = -\omega_{y'}\mathbf{j}' + \omega_z \sin 7.595°\mathbf{j}' + \omega_z \cos 7.595°\mathbf{k}'$$

$$= (0.13216\omega_z - \omega_{y'})\mathbf{j}' + 0.99123\omega_z\mathbf{k}'$$

Since points A and C have zero velocity,

$$\mathbf{v}_C = \mathbf{v}_A + \omega \times \mathbf{r}_{C/A}$$

$$\mathbf{0} = \mathbf{0} + \left[(0.13216\omega_z - \omega_{y'})\mathbf{j}' + 0.99123\omega_z\mathbf{k}'\right] \times (1.5\mathbf{j}' - 0.2\mathbf{k}')$$

$$0 = -1.48684\omega_z - 0.026433\omega_z + 0.2\omega_{y'}$$

$$\omega_{y'} = -7.5664\omega_z$$

Thus,

$$\omega = -7.4342\omega_z\mathbf{j}' + 0.99123\omega_z\mathbf{k}'$$

Since $\omega_z = 2$ rad/s

$$\omega = -14.868\mathbf{j}' + 1.9825\mathbf{k}'$$

So that,

$$\mathbf{H}_A = I_{x'}\omega_{x'}\mathbf{i}' + I_{y'}\omega_{y'}\mathbf{j} + I_{z'}\omega_{z'}\mathbf{k}' = \mathbf{0} + 0.100(-14.868)\mathbf{j}' + 13.55(1.9825)\mathbf{k}'$$

$$= -1.4868\mathbf{j}' + 26.862\mathbf{k}'$$

$$H_A = \sqrt{(-1.4868)^2 + (26.862)^2} = 26.9 \text{ kg}\cdot\text{m}^2/\text{s} \qquad \textbf{Ans}$$

21-30. Each of the two disks has a weight of 10 lb. The axle AB weighs 3 lb. If the assembly is rotating about the z axis at $\omega_z = 6$ rad/s, determine its angular momentum about the z axis and its kinetic energy. The disks roll without slipping.

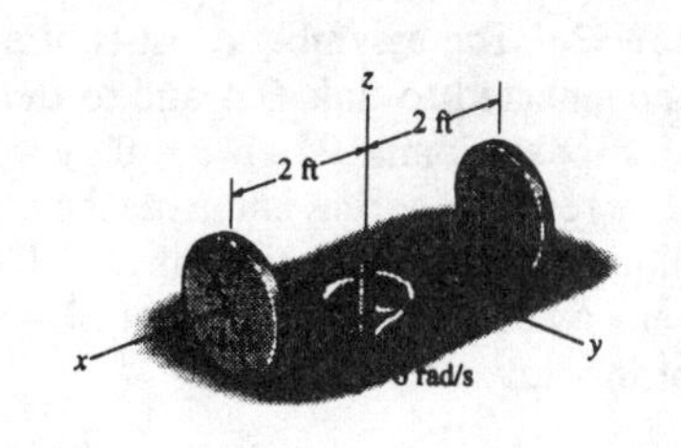

$$\frac{6}{\omega_x} = \frac{1}{2} \qquad \omega_x = 12 \text{ rad/s}$$

$$\omega_A = \{-12\mathbf{i}\} \text{ rad/s} \qquad \omega_B = \{12\mathbf{i}\} \text{ rad/s}$$

$$\mathbf{H}_z = \left[\frac{1}{2}\left(\frac{10}{32.2}\right)(1)^2\right](12\mathbf{i}) + \left[\frac{1}{2}\left(\frac{10}{32.2}\right)(1)^2\right](-12\mathbf{i})$$

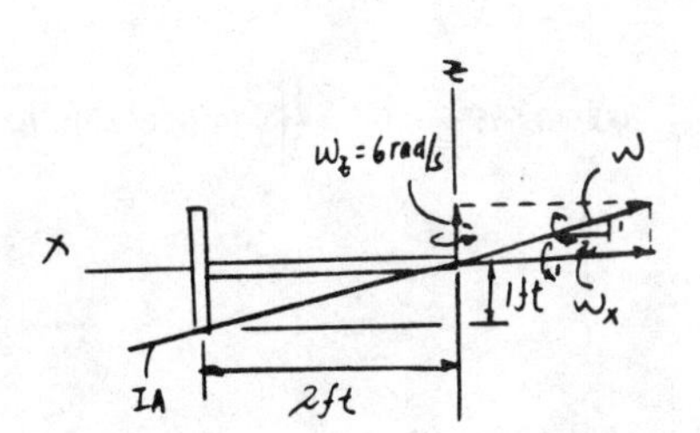

$$+\mathbf{0} + \left\{2\left[\frac{1}{4}\left(\frac{10}{32.2}\right)(1)^2 + \frac{10}{32.2}(2)^2\right](6) + \frac{1}{12}\left(\frac{3}{32.2}\right)(4)^2(6)\right\}\mathbf{k}$$

$$\mathbf{H}_z = \{16.6\mathbf{k}\} \text{ slug} \cdot \text{ft}^2/\text{s} \qquad \textbf{Ans}$$

$$T = \frac{1}{2}I_x\,\omega_x^2 + \frac{1}{2}I_y\,\omega_y^2 + \frac{1}{2}I_z\,\omega_z^2$$

$$= \frac{1}{2}\left[2\left(\frac{1}{2}\left(\frac{10}{32.2}\right)(1)^2\right)\right](12)^2 + 0$$

$$+\frac{1}{2}\left\{2\left[\frac{1}{4}\left(\frac{10}{32.2}\right)(1)^2 + \frac{10}{32.2}(2)^2\right] + \frac{1}{12}\left(\frac{3}{32.2}\right)(4)^2\right\}(6)^2$$

$$= 72.1 \text{ lb} \cdot \text{ft} \qquad \textbf{Ans}$$

21-31. The 2-kg thin disk is connected to the slender rod which is fixed to the ball-and-socket joint at A. If it is released from rest in the position shown, determine the spin of the disk about the rod when the disk reaches its lowest position. Neglect the mass of the rod. The disk rolls without slipping.

$$I_x = I_z = \frac{1}{4}(2)(0.1)^2 + 2(0.5)^2 = 0.505 \text{ kg} \cdot \text{m}^2$$

$$I_y = \frac{1}{2}(2)(0.1)^2 = 0.01 \text{ kg} \cdot \text{m}^2$$

$$\omega = \omega_y + \omega_{z'} = -\omega_y\,\mathbf{j} + \omega_{z'}\sin 11.31°\mathbf{j} + \omega_{z'}\cos 11.31°\mathbf{k}$$

$$= (0.19612\omega_{z'} - \omega_y)\mathbf{j} + (0.98058\omega_{z'})\mathbf{k}$$

Since $\mathbf{v}_A = \mathbf{v}_C = \mathbf{0}$, then

$$\mathbf{v}_C = \mathbf{v}_A + \omega \times \mathbf{r}_{C/A}$$

$$\mathbf{0} = \mathbf{0} + \left[(0.19612\omega_{z'} - \omega_y)\mathbf{j} + (0.98058\omega_{z'})\mathbf{k}\right] \times (0.5\mathbf{j} - 0.1\mathbf{k})$$

$$0 = -0.019612\omega_{z'} + 0.1\omega_y - 0.49029\omega_{z'}$$

$$\omega_{z'} = 0.19612\omega_y$$

Thus,

$$\omega = -0.96154\omega_y\,\mathbf{j} + 0.19231\omega_y\,\mathbf{k}$$

$$h_1 = 0.5\sin 41.31° = 0.3301 \text{ m}, \qquad h_2 = 0.5\sin 18.69° = 0.1602 \text{ m}$$

$$T_1 + V_1 = T_2 + V_2$$

$$0 + 2(9.81)(0.3301) = \left[0 + \frac{1}{2}(0.01)(-0.96154\omega_y)^2 + \frac{1}{2}(0.505)(0.19231\omega_y)^2\right] - 2(9.81)(0.1602)$$

$$\omega_y = 26.2 \text{ rad/s} \qquad \textbf{Ans}$$

***21-32.** The assembly consists of a 4-kg rod AB which is connected to link OA and to the collar at B by ball-and-socket joints. When $\theta = 0°$, $y = 600$ mm, the system is at rest, the spring is unstretched, and a 7-N·m couple moment is applied to the link at O. Determine the angular velocity of the link at the instant $\theta = 90°$. Neglect the mass of the link.

The length of AB $\quad AB = \sqrt{200^2 + 600^2 + 200^2} = 663.32 \text{ mm} = 0.6633 \text{ m}$

$$T_1 + \Sigma U_{1-2} = T_2$$

$$0 + 4(9.81)(0.1) + 7\left(\frac{\pi}{2}\right) - \frac{1}{2}(2000)(0.6633 - 0.6)^2 = \frac{1}{2}\left[\frac{1}{3}(4)(0.6633)^2\right]\omega_{AB}^2$$

$$\omega_{AB} = 6.098 \text{ rad/s}$$

Point B is the IC.

$$v_A = \omega_{AB} r_{A/IC} = 6.098(0.6633) = 4.045 \text{ m/s}$$

$$\omega_{OA} = \frac{v_A}{r_{OA}} = \frac{4.045}{0.2} = 20.2 \text{ rad/s} \qquad \textbf{Ans}$$

21-33. The 25-lb thin plate is suspended from a ball-and-socket joint at O. A 0.2-lb projectile is fired with a velocity of $\mathbf{v} = \{-300\mathbf{i} - 250\mathbf{j} + 300\mathbf{k}\}$ ft/s into the plate and becomes embedded in the plate at point A. Determine the angular velocity of the plate just after impact and the axis about which it begins to rotate.

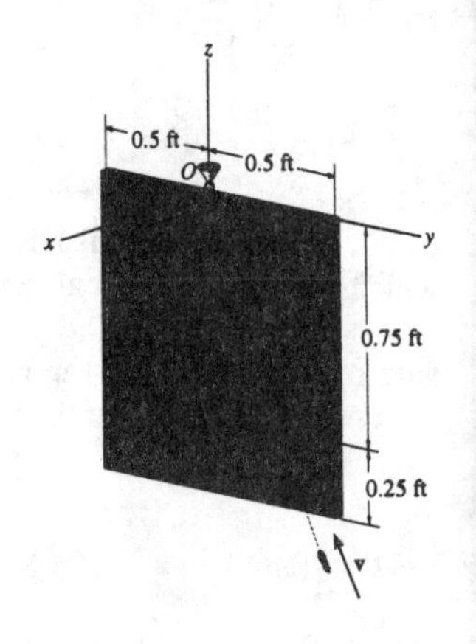

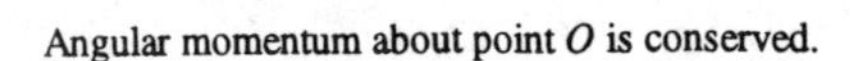

Angular momentum about point O is conserved.

$$(\mathbf{H}_O)_2 = (\mathbf{H}_O)_1 = \mathbf{r}_{OA} \times m_P \mathbf{v}_P$$

$$(\mathbf{H}_O)_1 = (0.25\mathbf{j} - 0.75\mathbf{k}) \times \left(\frac{0.2}{32.2}\right)(-300\mathbf{i} - 250\mathbf{j} + 300\mathbf{k}) = \{-0.6988\mathbf{i} + 1.3975\mathbf{j} + 0.4658\mathbf{k}\} \text{ lb}\cdot\text{ft}\cdot\text{s}$$

$$I_x = \left(\frac{1}{12}\right)\left(\frac{25}{32.2}\right)\left[(1)^2 + (1)^2\right] + \left(\frac{25}{32.2}\right)(0.5)^2 = 0.3235 \text{ slug}\cdot\text{ft}^2$$

$$I_y = \left(\frac{1}{12}\right)\left(\frac{25}{32.2}\right)(1)^2 + \left(\frac{25}{32.2}\right)(0.5)^2 = 0.2588 \text{ slug}\cdot\text{ft}^2$$

$$I_z = \left(\frac{1}{12}\right)\left(\frac{25}{32.2}\right)(1)^2 = 0.06470 \text{ slug}\cdot\text{ft}^2$$

$$(\mathbf{H}_O)_1 = (\mathbf{H}_O)_2$$

$$-0.6988\mathbf{i} + 1.3975\mathbf{j} + 0.4658\mathbf{k} = 0.3235\omega_x\mathbf{i} + 0.2588\omega_y\mathbf{j} + 0.06470\omega_z\mathbf{k}$$

$$\omega_x = \frac{-0.6988}{0.3235} = -2.160 \text{ rad/s}$$

$$\omega_y = \frac{1.3975}{0.2588} = 5.400 \text{ rad/s}$$

$$\omega_z = \frac{0.4658}{0.06470} = 7.200 \text{ rad/s}$$

$$\omega = \{-2.16\mathbf{i} + 5.40\mathbf{j} + 7.20\mathbf{k}\} \text{ rad/s} \qquad \textbf{Ans}$$

Axis of rotation line is along ω:

$$\mathbf{u}_A = \frac{-2.160\mathbf{i} + 5.400\mathbf{j} + 7.200\mathbf{k}}{\sqrt{(-2.160)^2 + (5.400)^2 + (7.200)^2}}$$

$$= -0.233\mathbf{i} + 0.583\mathbf{j} + 0.778\mathbf{k} \qquad \textbf{Ans}$$

21-34. Solve Prob. 21-33 if the projectile emerges from the plate with a velocity of 275 ft/s in the same direction.

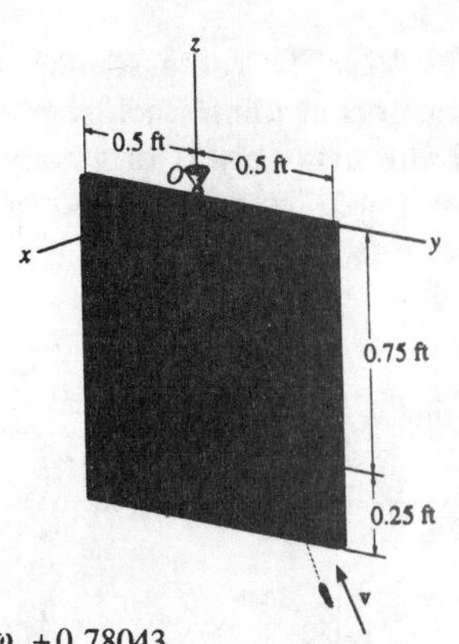

$$\mathbf{u}_v = \left(\frac{-300}{492.4}\right)\mathbf{i} - \left(\frac{250}{492.4}\right)\mathbf{j} + \left(\frac{300}{492.4}\right)\mathbf{k} = -0.6092\mathbf{i} - 0.5077\mathbf{j} + 0.6092\mathbf{k}$$

$$I_x = \left(\frac{1}{12}\right)\left(\frac{25}{32.2}\right)\left[(1)^2 + (1)^2\right] + \left(\frac{25}{32.2}\right)(0.5)^2 = 0.32350 \text{ slug}\cdot\text{ft}^2$$

$$I_y = \left(\frac{1}{12}\right)\left(\frac{25}{32.2}\right)(1)^2 + \left(\frac{25}{32.2}\right)(0.5)^2 = 0.25880 \text{ slug}\cdot\text{ft}^2$$

$$I_z = \left(\frac{1}{12}\right)\left(\frac{25}{32.2}\right)(1)^2 = 0.06470 \text{ slug}\cdot\text{ft}^2$$

$$\mathbf{H}_1 + \Sigma\int \mathbf{M}_O\, dt = \mathbf{H}_2$$

$$(0.25\mathbf{j} - 0.75\mathbf{k}) \times \left(\frac{0.2}{32.2}\right)(-300\mathbf{i} - 250\mathbf{j} + 300\mathbf{k}) + \mathbf{0} = 0.32350\omega_x\mathbf{i} + 0.25880\omega_y\mathbf{j} + 0.06470\omega_z\mathbf{k}$$

$$+(0.25\mathbf{j} - 0.75\mathbf{k}) \times \left(\frac{0.2}{32.2}\right)(275)(-0.6092\mathbf{i} - 0.5077\mathbf{j} + 0.6092\mathbf{k})$$

Expanding, the **i**, **j**, **k**, components are :

$$-0.6988 = 0.32350\omega_x - 0.390215$$

$$1.3975 = 0.25880\omega_y + 0.78043$$

$$0.4658 = 0.06470\omega_z + 0.26014$$

$$\omega_x = -0.9538, \quad \omega_y = 2.3844, \quad \omega_z = 3.179$$

$$\boldsymbol{\omega} = \{-0.954\mathbf{i} + 2.38\mathbf{j} - 3.18\mathbf{k}\} \text{ rad/s}$$ **Ans**

Axis of rotation is along $\boldsymbol{\omega}$:

$$\mathbf{u}_A = \frac{-0.954\mathbf{i} + 2.38\mathbf{j} + 3.18\mathbf{k}}{\sqrt{(-0.954)^2 + (2.38)^2 + (3.18)^2}}$$

$$\mathbf{u}_A = -0.233\mathbf{i} + 0.583\mathbf{j} + 0.778\mathbf{k}$$ **Ans**

21-35. The 4-lb rod AB is attached to the rod BC and collar A using ball-and-socket joints. If BC has a constant angular velocity of 2 rad/s, determine the kinetic energy of AB when it is in the position shown. Assume the angular velocity of AB is directed perpendicular to the axis of AB.

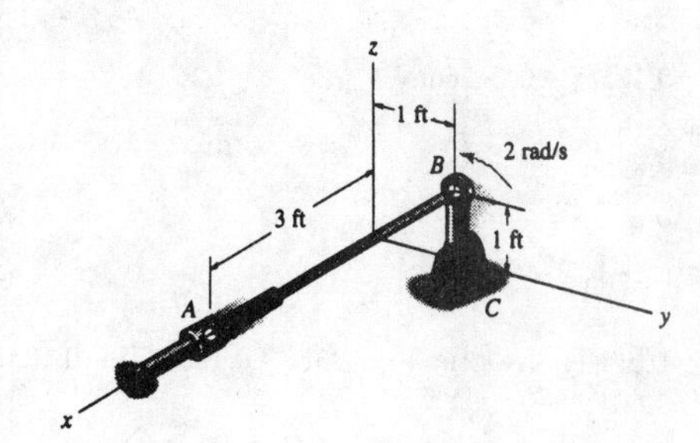

$\mathbf{v}_A = v_A\mathbf{i} \qquad \mathbf{v}_B = \{-2\mathbf{j}\}$ ft/s $\qquad \boldsymbol{\omega}_{AB} = \omega_x\mathbf{i} + \omega_y\mathbf{j} + \omega_z\mathbf{k}$

$\mathbf{r}_{B/A} = \{-3\mathbf{i} + 1\mathbf{j} + 1\mathbf{k}\}$ ft $\qquad \mathbf{r}_{G/B} = \{1.5\mathbf{i} - 0.5\mathbf{j} - 0.5\mathbf{k}\}$

$$\mathbf{v}_B = \mathbf{v}_A + \boldsymbol{\omega}_{AB} \times \mathbf{r}_{B/A}$$

$$-2\mathbf{j} = v_A\mathbf{i} + \begin{vmatrix} \mathbf{i} & \mathbf{j} & \mathbf{k} \\ \omega_x & \omega_y & \omega_z \\ -3 & 1 & 1 \end{vmatrix}$$

Equating **i**, **j** and **k** components

$$\omega_y - \omega_z + v_A = 0 \qquad (1)$$

$$\omega_x + 3\omega_z = 2 \qquad (2)$$

$$\omega_x + 3\omega_y = 0 \qquad (3)$$

Since $\boldsymbol{\omega}_{AB}$ is perpendicular to the axis of the rod,

$$\boldsymbol{\omega}_{AB}\cdot\mathbf{r}_{B/A} = (\omega_x\mathbf{i} + \omega_y\mathbf{j} + \omega_z\mathbf{k})\cdot(-3\mathbf{i} + 1\mathbf{j} + 1\mathbf{k}) = 0$$

$$-3\omega_x + 1\omega_y + 1\omega_z = 0 \qquad (4)$$

Solving Eqs.(1) to (4) yields :

$\omega_x = 0.1818$ rad/s $\qquad \omega_y = -0.06061$ rad/s $\qquad \omega_z = 0.6061$ rad/s

$v_A = 0.6667$ ft/s

Hence $\boldsymbol{\omega}_{AB} = \{0.1818\mathbf{i} - 0.06061\mathbf{j} + 0.6061\mathbf{k}\}$ rad/s $\qquad \mathbf{v}_A = \{0.6667\mathbf{i}\}$ ft/s

$$\mathbf{v}_G = \mathbf{v}_B + \boldsymbol{\omega}_{AB} \times \mathbf{r}_{G/B}$$

$$= -2\mathbf{j} + \begin{vmatrix} \mathbf{i} & \mathbf{j} & \mathbf{k} \\ 0.1818 & -0.06061 & 0.6061 \\ 1.5 & -0.5 & -0.5 \end{vmatrix}$$

$$= \{0.3333\mathbf{i} - 1.0\mathbf{j}\} \text{ ft/s}$$

$$\omega_{AB}^2 = 0.1818^2 + (-0.06061)^2 + 0.6061^2 = 0.4040$$

$$v_G^2 = (0.3333)^2 + (-1.0)^2 = 1.111$$

$$T = \frac{1}{2}mv_G^2 + \frac{1}{2}I_G\omega_{AB}^2$$

$$= \frac{1}{2}\left(\frac{4}{32.2}\right)(1.111) + \frac{1}{2}\left[\frac{1}{12}\left(\frac{4}{32.2}\right)\left(\sqrt{3^2 + 1^2 + 1^2}\right)^2\right](0.4040)$$

$$= 0.0920 \text{ ft}\cdot\text{lb}$$ **Ans**

***21-36.** The rod assembly is supported at G by a ball-and-socket joint. Each segment has a mass of 0.5 kg/m. If the assembly is originally at rest and an impulse of $\mathbf{I} = \{-8\mathbf{k}\}$ N·s is applied at D, determine the angular velocity of the assembly just after the impact.

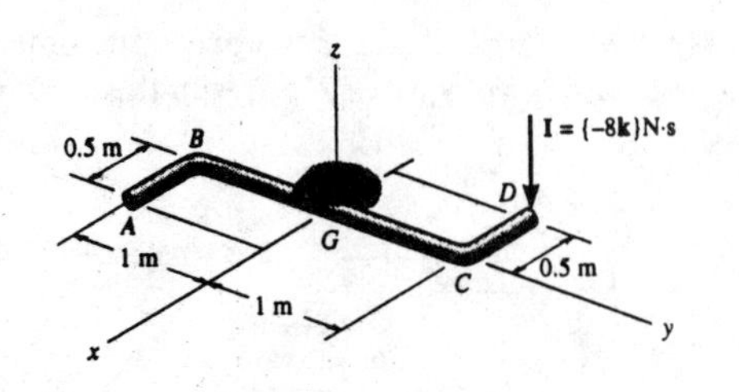

Moments and products of inertia :

$$I_{xx} = \frac{1}{12}\left[2(0.5)\right](2)^2 + 2\left[0.5(0.5)\right](1)^2 = 0.8333 \text{ kg}\cdot\text{ m}^2$$

$$I_{yy} = \frac{1}{12}\left[1(0.5)\right](1)^2 = 0.04166 \text{ kg}\cdot\text{ m}^2$$

$$I_{zz} = \frac{1}{12}\left[2(0.5)\right](2)^2 + 2\left[\frac{1}{12}\left[0.5(0.5)\right](0.5)^2 + \left[0.5(0.5)\right](1^2 + 0.25^2)\right]$$

$$= 0.875 \text{ kg}\cdot\text{ m}^2$$

$$I_{xy} = \left[0.5(0.5)\right](-0.25)(1) + \left[0.5(0.5)\right](0.25)(-1) = -0.125 \text{ kg.m}^2$$

$$I_{yz} = I_{xz} = 0$$

From Eq. 21-10

$$H_x = 0.8333\omega_x + 0.125\omega_y$$

$$H_y = 0.125\omega_x + 0.04166\omega_y$$

$$H_z = 0.875\omega_z$$

$$(\mathbf{H}_G)_1 + \Sigma\int_{t_1}^{t_2}\mathbf{M}_G dt = (\mathbf{H}_G)_2$$

$$0 + (-0.5\mathbf{i} + 1\mathbf{j}) \times (-8\mathbf{k}) = (0.8333\omega_x + 0.125\omega_y)\mathbf{i} + (0.125\omega_x + 0.04166\omega_y)\mathbf{j} + 0.875\omega_z\mathbf{k}$$

Equating **i**, **j** and **k** components

$$-8 = 0.8333\omega_x + 0.125\omega_y \qquad (1)$$

$$-4 = 0.125\omega_x + 0.04166\omega_y \qquad (2)$$

$$0 = 0.875\omega_z \qquad (3)$$

Solving Eqs.(1) to (3) yields :

$\omega_x = 8.73$ rad/s $\quad \omega_y = -122$ rad/s $\quad \omega_z = 0$

Then $\quad \omega = \{8.73\mathbf{i} - 122\mathbf{j}\}$ rad/s $\quad$ **Ans**

21-37. The 15-lb plate is subjected to a force $F = 8$ lb which is always directed perpendicular to the face of the plate. If the plate is originally at rest, determine its angular velocity after it has rotated one revolution (360°). The plate is supported by ball-and-socket joints at A and B.

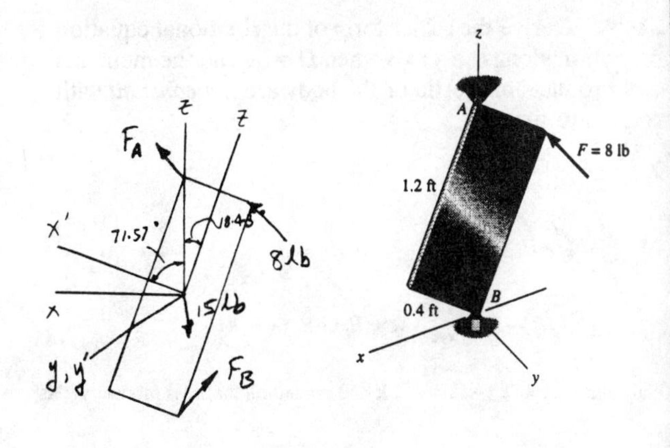

Due to symmetry $I_{x'y'} = I_{y'z'} = I_{z'x'} = 0$

$$I_{x'} = \frac{1}{12}\left(\frac{15}{32.2}\right)(1.2)^2 = 0.05590 \text{ slug}\cdot\text{ft}^2$$

$$I_{y'} = \frac{1}{12}\left(\frac{15}{32.2}\right)(1.2^2 + 0.4^2) = 0.06211 \text{ slug}\cdot\text{ft}^2$$

$$I_{z'} = \frac{1}{12}\left(\frac{15}{32.2}\right)(0.4)^2 = 0.006211 \text{ slug}\cdot\text{ft}^2$$

For z axis

$$u_{x'} = \cos 71.57° = 0.3162 \qquad u_{y'} = \cos 90° = 0$$

$$u_{z'} = \cos 18.43° = 0.9487$$

$$I_z = I_{x'}u_{x'}^2 + I_{y'}u_{y'}^2 + I_{z'}u_{z'}^2 - 2I_{x'y'}u_{x'}u_{y'} - 2I_{y'z'}u_{y'}u_{z'} - 2I_{z'x'}u_{z'}u_{x'}$$

$$= 0.05590(0.3162)^2 + 0 + 0.006211(0.9487)^2 - 0 - 0 - 0$$

$$= 0.01118 \text{ slug}\cdot\text{ft}^2$$

Principle of work and energy :

$$T_1 + \Sigma U_{1-2} = T_2$$

$$0 + 8(1.2\sin 18.43°)(2\pi) = \frac{1}{2}(0.01118)\omega^2$$

$\omega = 58.4$ rad/s **Ans**

21-38. The rod assembly has a mass of 2.5 kg/m and is rotating with a constant angular velocity of $\boldsymbol{\omega} = \{2\mathbf{k}\}$ rad/s when the looped end at C encounters a hook at S, which provides a permanent connection. Determine the angular velocity of the assembly immediately after impact.

$$I_z = \frac{1}{12}[1(2.5)](1)^2 = 0.2083 \text{ kg}\cdot\text{m}^2$$

$$I_y = \frac{1}{3}[0.5(2.5)](0.5)^2 + [1(2.5)](0.5)^2 = 0.7292 \text{ kg}\cdot\text{m}^2$$

Angular momentum is conserved about OC axis.

$$\mathbf{u}_{OC} = \frac{0.5\mathbf{j} - 0.5\mathbf{k}}{\sqrt{0.5^2 + (-0.5)^2}} = 0.7071\mathbf{j} - 0.7071\mathbf{k}$$

$$(\mathbf{H}_z)_1 = 0.2083(2\mathbf{k}) = \{0.4167\mathbf{k}\} \text{ kg}\cdot\text{m}^2/\text{s}$$

$$(H_{OC})_1 = (\mathbf{H}_z)_1\cdot\mathbf{u}_{OC} = (0.4167\mathbf{k})\cdot(0.7071\mathbf{j} - 0.7071\mathbf{k}) = -0.2946 \text{ kg}\cdot\text{m}^2/\text{s}$$

$$\omega = 0.7071\omega\mathbf{j} - 0.7071\omega\mathbf{k}$$

$$(\mathbf{H}_z)_2 = I_y\omega_y\mathbf{j} + I_z\omega_z\mathbf{k}$$

$$= 0.7292(0.7071\omega)\mathbf{j} + 0.2083(-0.7071\omega)\mathbf{k}$$

$$= 0.5156\omega\mathbf{j} - 0.1473\omega\mathbf{k}$$

$$(H_{OC})_2 = (\mathbf{H}_z)_2\cdot\mathbf{u}_{OC}$$

$$= (0.5156\omega\mathbf{j} - 0.1473\omega\mathbf{k})\cdot(0.7071\mathbf{j} - 0.7071\mathbf{k})$$

$$= 0.4688\omega$$

Requires,

$$(H_{OC})_1 = (H_{OC})_2$$

$$-0.2946 = 0.4688\omega$$

$$\omega = -0.6285 \text{ rad/s}$$

Hence

$$\omega = 0.7071(-0.6285)\mathbf{j} - 0.7071(-0.6285)\mathbf{k}$$

$$= \{-0.444\mathbf{j} + 0.444\mathbf{k}\} \text{ rad/s}$$

Ans 

21-39. Derive the scalar form of the rotational equation of motion along the x axis when $\boldsymbol{\Omega} \neq \boldsymbol{\omega}$ and the moments and products of inertia of the body are *not constant* with respect to time.

In general

$$\mathbf{M} = \frac{d}{dt}(H_x\mathbf{i}+H_y\mathbf{j}+H_z\mathbf{k})$$

$$= \left(\dot{H}_x\mathbf{i}+\dot{H}_y\mathbf{j}+\dot{H}_z\mathbf{k}\right)_{xyz} + \Omega\times(H_x\mathbf{i}+H_y\mathbf{j}+H_z\mathbf{k})$$

Substitute $\Omega = \Omega_x\mathbf{i}+\Omega_y\mathbf{j}+\Omega_z\mathbf{k}$ and expanding the cross product yields

$$\mathbf{M} = \left(\left(\dot{H}_x\right)_{xyz} - \Omega_z H_y + \Omega_y H_z\right)\mathbf{i} + \left(\left(\dot{H}_y\right)_{xyz} - \Omega_x H_z + \Omega_z H_x\right)\mathbf{j} + \left(\left(\dot{H}_z\right)_{xyz} - \Omega_y H_x + \Omega_x H_y\right)\mathbf{k}$$

Subsitute H_x , H_y and H_z using Eq. 21-10. For the **i** component

$$\Sigma M_x = \frac{d}{dt}(I_x\,\omega_x - I_{xy}\,\omega_y - I_{xz}\,\omega_z) - \Omega_z\,(I_y\,\omega_y - I_{yz}\,\omega_z - I_{yx}\,\omega_x) + \Omega_y\,(I_z\,\omega_z - I_{zx}\,\omega_x - I_{zy}\,\omega_y) \qquad \textbf{Ans}$$

One can obtain y and z components in a similar manner.

***21-40.** Derive the scalar form of the rotational equation of motion along the x axis when $\boldsymbol{\Omega} \neq \boldsymbol{\omega}$ and the moments and products of inertia of the body are *constant* with respect to time.

In general

$$\mathbf{M} = \frac{d}{dt}(H_x\mathbf{i}+H_y\mathbf{j}+H_z\mathbf{k})$$

$$= \left(\dot{H}_x\mathbf{i}+\dot{H}_y\mathbf{j}+\dot{H}_z\mathbf{k}\right)_{xyz} + \Omega\times(H_x\mathbf{i}+H_y\mathbf{j}+H_z\mathbf{k})$$

Substitute $\Omega = \Omega_x\mathbf{i}+\Omega_y\mathbf{j}+\Omega_z\mathbf{k}$ and expanding the cross product yields

$$\mathbf{M} = \left(\left(\dot{H}_x\right)_{xyz} - \Omega_z H_y + \Omega_y H_z\right)\mathbf{i} + \left(\left(\dot{H}_y\right)_{xyz} - \Omega_x H_z + \Omega_z H_x\right)\mathbf{j} + \left(\left(\dot{H}_z\right)_{xyz} - \Omega_y H_x + \Omega_x H_y\right)\mathbf{k}$$

Substitute H_x , H_y and H_z using Eq. 21-10. For the **i** component

$$\Sigma M_x = \frac{d}{dt}(I_x\,\omega_x - I_{xy}\,\omega_y - I_{xz}\,\omega_z) - \Omega_z\,(I_y\,\omega_y - I_{yz}\,\omega_z - I_{yx}\,\omega_x) + \Omega_y\,(I_z\,\omega_z - I_{zx}\,\omega_x - I_{zy}\,\omega_y)$$

For constant inertia, expanding the time derivative of the above equation yields

$$\Sigma M_x = (I_x\,\dot{\omega}_x - I_{xy}\,\dot{\omega}_y - I_{xz}\,\dot{\omega}_z) - \Omega_z\,(I_y\,\omega_y - I_{yz}\,\omega_z - I_{yx}\,\omega_x) + \Omega_y\,(I_z\,\omega_z - I_{zx}\,\omega_x - I_{zy}\,\omega_y) \qquad \textbf{Ans}$$

One can obtain y and z components in a similar manner.

21-41. Derive the Euler equations of motion for $\boldsymbol{\Omega} \neq \boldsymbol{\omega}$, i.e., Eqs. 21–26.

In general

$$\mathbf{M} = \frac{d}{dt}(H_x\mathbf{i} + H_y\mathbf{j} + H_z\mathbf{k})$$

$$= (\dot{H}_x\mathbf{i} + \dot{H}_y\mathbf{j} + \dot{H}_z\mathbf{k})_{xyz} + \Omega \times (H_x\mathbf{i} + H_y\mathbf{j} + H_z\mathbf{k})$$

Substitute $\Omega = \Omega_x\mathbf{i} + \Omega_y\mathbf{j} + \Omega_z\mathbf{k}$ and expanding the cross product yields

$$\mathbf{M} = \left((\dot{H}_x)_{xyz} - \Omega_z H_y + \Omega_y H_z\right)\mathbf{i} + \left((\dot{H}_y)_{xyz} - \Omega_x H_z + \Omega_z H_x\right)\mathbf{j} + \left((\dot{H}_z)_{xyz} - \Omega_y H_x + \Omega_x H_y\right)\mathbf{k}$$

Substitute H_x , H_y and H_z using Eq. 21-10. For the **i** component

$$\Sigma M_x = \frac{d}{dt}(I_x\omega_x - I_{xy}\omega_y - I_{xz}\omega_z) - \Omega_z(I_y\omega_y - I_{yz}\omega_z - I_{yx}\omega_x) + \Omega_y(I_z\omega_z - I_{zx}\omega_x - I_{zy}\omega_y)$$

Set $I_{xy} = I_{yz} = I_{zx} = 0$ and require I_x, I_y, I_z to be constant. This yields

$$\Sigma M_x = I_x\dot{\omega}_x - I_y\Omega_z\omega_y + I_z\Omega_y\omega_z \qquad \textbf{Ans}$$

One can obtain y and z components in a similar manner.

21-42. The 5-kg circular disk is mounted off center on a shaft which is supported by bearings at A and B. If the shaft is rotating at a constant rate of $\omega = 10$ rad/s, determine the vertical reactions at the bearings when the disk is in the position shown.

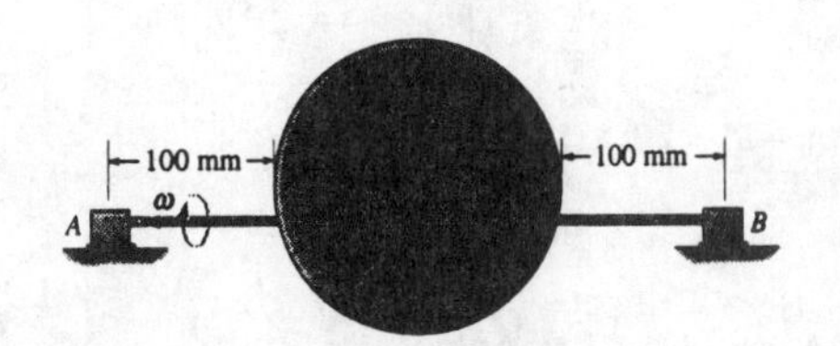

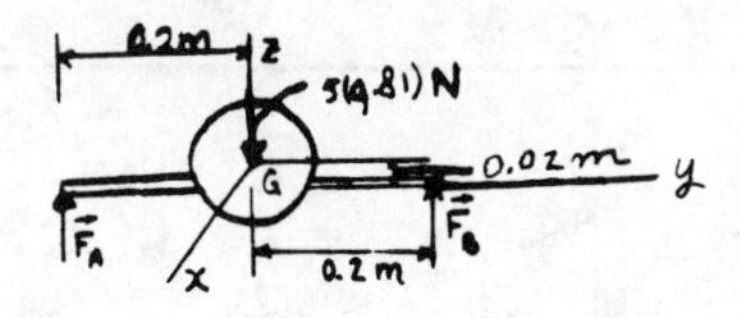

$\omega_x = 0, \quad \omega_y = -10$ rad/s, $\quad \omega_z = 0$

$\dot{\omega}_x = 0, \quad \dot{\omega}_y = 0, \quad \dot{\omega}_z = 0$

$$\circlearrowleft + \Sigma M_x = I_x\dot{\omega}_x - (I_y - I_z)\omega_y\omega_z$$

$$-(0.2)(F_A) + (0.2)(F_B) = 0$$

$F_A = F_B$

$+\downarrow \Sigma F_z = ma_z; \qquad F_A + F_B - 5(9.81) = -5(10)^2(0.02)$

$F_A = F_B = 19.5$ N $\qquad$ **Ans**

21-43. The uniform plate has a mass of $m = 2$ kg and is given a rotation of $\omega = 4$ rad/s about its bearings at A and B. If $a = 0.2$ m and $c = 0.3$ m, determine the vertical reactions at the instant shown. Use the x, y, z axes shown and note that $I_{zx} = -\left(\frac{mac}{12}\right)\left(\frac{c^2 - a^2}{c^2 + a^2}\right)$.

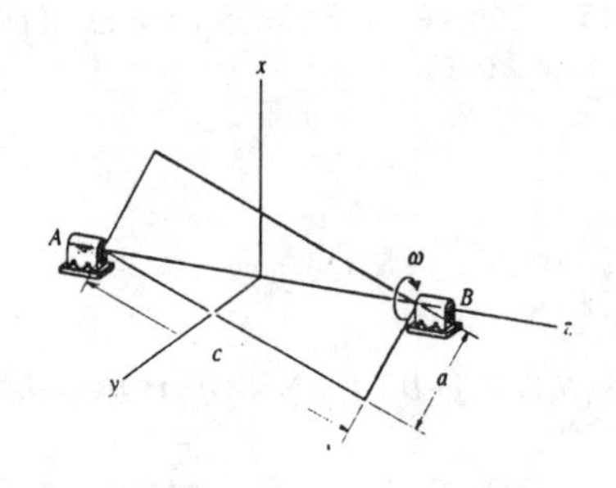

$\omega_x = 0, \quad \omega_y = 0, \quad \omega_z = -4$

$\dot{\omega}_x = 0, \quad \dot{\omega}_y = 0, \quad \dot{\omega}_z = 0$

$$\Sigma M_y = I_{yy}\dot{\omega}_y - (I_{zz} - I_{xx})\omega_z\omega_x - I_{yz}\left(\dot{\omega}_z - \omega_x\omega_y\right) - I_{zx}\left(\omega_z^2 - \omega_x^2\right) - I_{xy}\left(\dot{\omega}_z - \omega_y\omega_z\right)$$

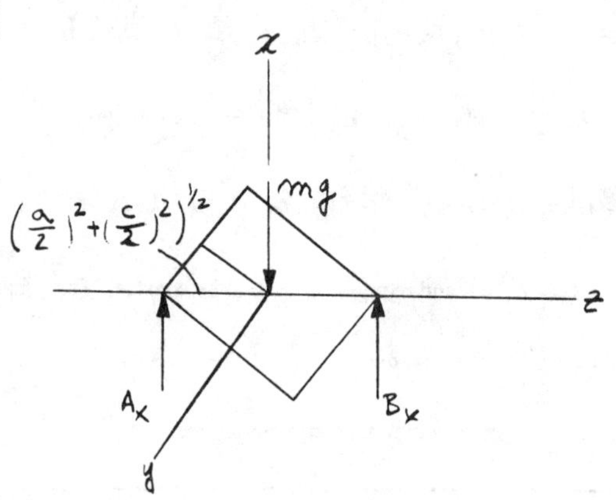

$$B_x\left[\left(\frac{a}{2}\right)^2 + \left(\frac{c}{2}\right)^2\right]^{\frac{1}{2}} - A_x\left[\left(\frac{a}{2}\right)^2 + \left(\frac{c}{2}\right)^2\right]^{\frac{1}{2}} = -I_{zx}(\omega)^2$$

$$B_x - A_x = \left(\frac{mac}{6}\right)\left(\frac{c^2 - a^2}{\left[a^2 + c^2\right]^{\frac{3}{2}}}\right)\omega^2$$

$\Sigma F_x = m(a_G)_x; \qquad A_x + B_x - mg = 0$

Substitute the data,

$$B_x - A_x = \frac{2(0.2)(0.3)}{6}\left[\frac{(0.3)^2 - (0.2)^2}{\left[(0.3)^2 + (0.2)^2\right]^{\frac{3}{2}}}\right](-4)^2 = 0.34135$$

$A_x + B_x = 2(9.81)$

Solving :

$A_x = 9.64$ N **Ans**

$B_x = 9.98$ N **Ans**

***21-44.** The 20-lb disk is mounted on the horizontal shaft AB such that its plane forms an angle of 10° with the vertical. If the shaft rotates with an angular velocity of 3 rad/s, determine the vertical reactions developed at the bearings when the disk is in the position shown.

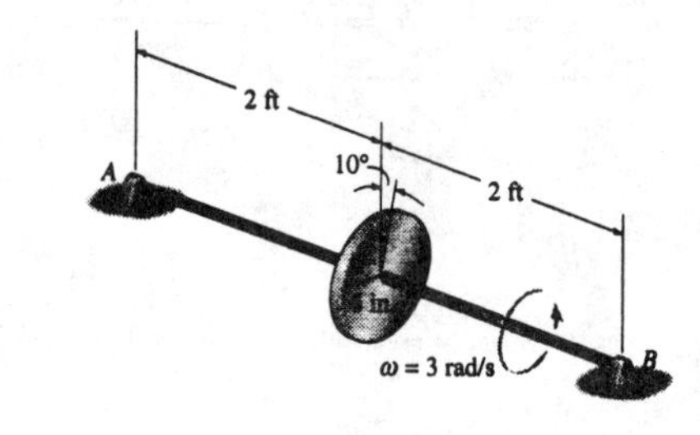

$$I_x = \frac{1}{2}\left(\frac{20}{32.2}\right)(0.5)^2 = 0.07764 \text{ slug} \cdot \text{ft}^2$$

$$I_y = I_z = \frac{1}{4}\left(\frac{20}{32.2}\right)(0.5)^2 = 0.03882 \text{ slug} \cdot \text{ft}^2$$

$\omega = 3\cos 10°\mathbf{i} + 3\sin 10°\mathbf{j} = \{2.9544\mathbf{i} + 0.5209\mathbf{j}\}$ rad/s

y
20 lb
10°
2 ft
2 ft
x
F_A
F_B

Applying the third of Eqs. 21-25 with $\omega_x = 2.9544$ rad/s $\omega_y = 0.5209$ rad/s $\dot{\omega}_z = 0$

$\Sigma M_z = I_z\dot{\omega}_z - (I_x - I_y)\omega_x\omega_y;$

$F_B(2) - F_A(2) = 0 - (0.07764 - 0.03882)(2.9544)(0.5209) \qquad (1)$

Also,

$\Sigma F_z = m(a_G)_z; \qquad F_A + F_B - 20 = 0 \qquad (2)$

Solving ,

$F_A = 10.0$ lb **Ans**

$F_B = 9.99$ lb **Ans**

21-45. The disk, having a mass of 3 kg, is mounted eccentrically on shaft AB. If the shaft is rotating at a constant rate of 9 rad/s, determine the reactions at the journal bearing supports when the disk is in the position shown.

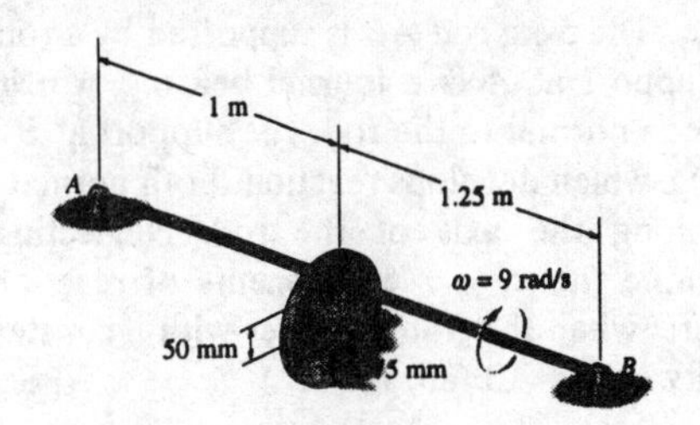

$\omega_x = 0, \quad \omega_y = -9, \quad \omega_z = 0$

$\Sigma M_x = I_x \dot{\omega}_x - (I_y - I_z)\,\omega_y\,\omega_z$

$B_z(1.25) - A_z(1) = 0 - 0$

$\Sigma M_z = I_z \dot{\omega}_z - (I_x - I_y)\,\omega_x\,\omega_y$

$A_x(1) - B_x(1.25) = 0 - 0$

$\Sigma F_x = ma_x; \quad A_x + B_x = 0$

$\Sigma F_z = ma_z; \quad A_z + B_z - 3(9.81) = 3(9)^2(0.05)$

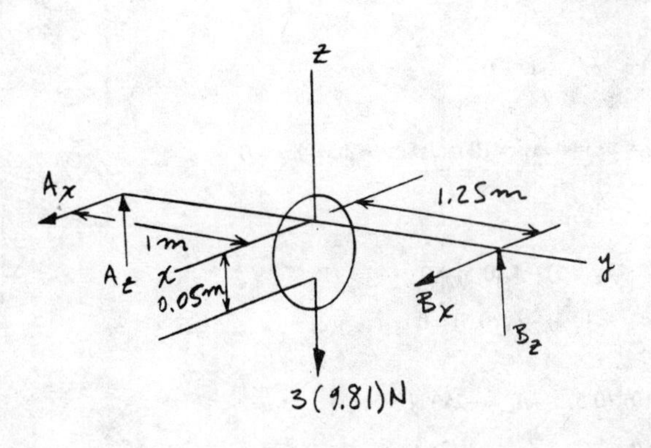

Solving,

$A_x = B_x = 0$ **Ans**

$A_z = 23.1$ N **Ans**

$B_z = 18.5$ N **Ans**

21-46. The conical pendulum consists of a bar of mass m and length L that is supported by the pin at its end A. If the pin is subjected to a rotation $\boldsymbol{\omega}$, determine the angle θ that the bar makes with the vertical as it rotates. Also, determine the components of reaction at the pin.

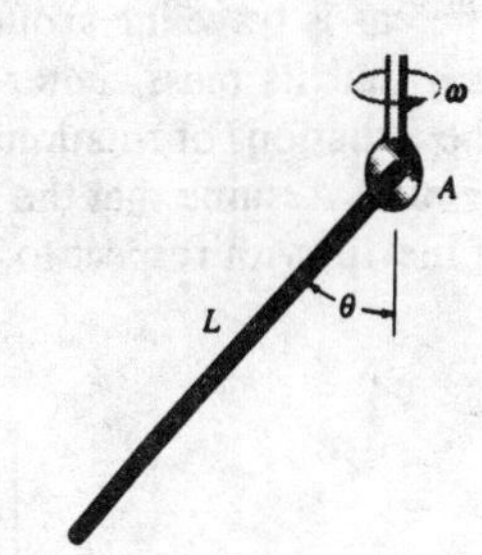

$I_x = I_z = \frac{1}{3}mL^2, \quad I_y = 0$

$\omega_x = 0, \quad \omega_y = -\omega\cos\theta, \quad \omega_z = \omega\sin\theta$

$\dot{\omega}_x = 0, \quad \dot{\omega}_y = 0, \quad \dot{\omega}_z = 0$

$\Sigma M_x = I_x \dot{\omega}_x - (I_y - I_z)\,\omega_y\,\omega_z$

$-mg\left(\frac{L}{2}\sin\theta\right) = 0 - \left(0 - \frac{1}{3}mL^2\right)(-\omega\cos\theta)(\omega\sin\theta)$

$\frac{g}{2} = \frac{1}{3}L\omega^2\cos\theta$

$\cos\theta = \dfrac{3g}{2L\omega^2}$

$\theta = \cos^{-1}\left(\dfrac{3g}{2L\omega^2}\right)$ **Ans**

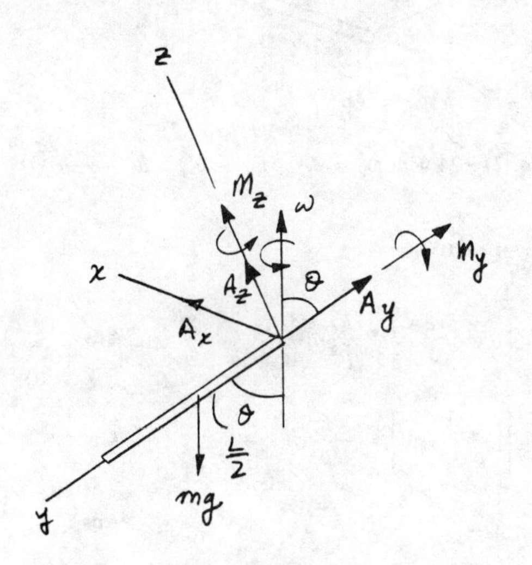

21-47. The 5-kg rod AB is supported by a rotating arm. The support at A is a journal bearing, which develops reactions normal to the rod. The support at B is a thrust bearing, which develops reactions both normal to the rod and along the axis of the rod. Neglecting friction, determine the x, y, z components of reaction at these supports when the frame rotates with a constant angular velocity of $\omega = 10$ rad/s.

$$I_y = I_z = \frac{1}{12}(5)(1)^2 = 0.4167 \text{ kg}\cdot\text{m}^2 \qquad I_x = 0$$

Applying Eq. 21-25 with $\omega_x = \omega_y = 0$ $\omega_z = 10$ rad/s $\dot{\omega}_x = \dot{\omega}_y = \dot{\omega}_z = 0$

$$\Sigma M_x = I_x\dot{\omega}_x - (I_y - I_z)\omega_y\omega_z; \qquad 0 = 0$$

$$\Sigma M_y = I_y\dot{\omega}_y - (I_z - I_x)\omega_z\omega_x; \qquad B_z(0.5) - A_z(0.5) = 0 \qquad (1)$$

$$\Sigma M_z = I_z\dot{\omega}_z - (I_x - I_y)\omega_x\omega_y; \qquad A_y(0.5) - B_y(0.5) = 0 \qquad (2)$$

Also,

$$\Sigma F_x = m(a_G)_x; \qquad B_x = -5(10)^2(0.5) \qquad B_x = -250\text{ N} \qquad \textbf{Ans}$$

$$\Sigma F_y = m(a_G)_y; \qquad A_y + B_y = 0 \qquad (3)$$

$$\Sigma F_z = m(a_G)_z; \qquad A_z + B_z - 5(9.81) = 0 \qquad (4)$$

Solving Eqs.(1) to (4) yields :

$$A_y = B_y = 0 \qquad A_z = B_z = 24.5\text{ N} \qquad \textbf{Ans}$$

***21-48.** The car is traveling around the curved road of radius ρ such that its mass center has a constant speed v_G. Write the equations of rotational motion with respect to the x, y, z axes. Assume that the car's six moments and products of inertia with respect to these axes are known.

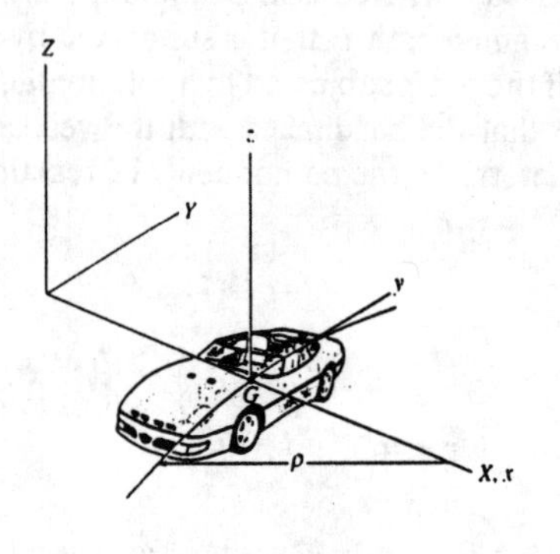

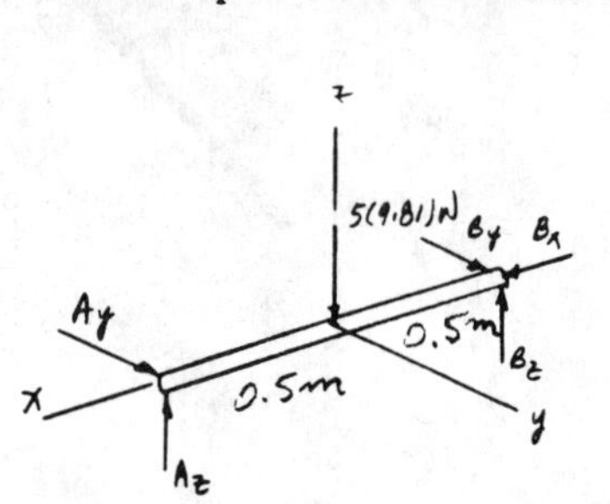

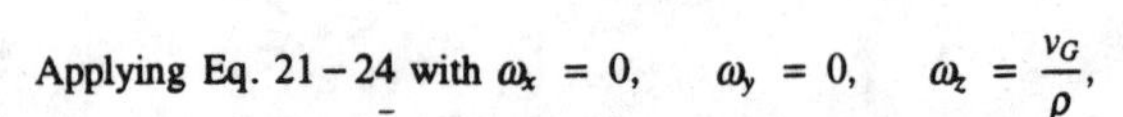

Applying Eq. 21 – 24 with $\omega_x = 0, \quad \omega_y = 0, \quad \omega_z = \dfrac{v_G}{\rho},$

$$\dot{\omega}_x = \dot{\omega}_y = \dot{\omega}_z = 0$$

$$\Sigma M_x = -I_{yz}\left[0 - \left(\frac{v_G}{\rho}\right)^2\right] = \frac{I_{yz}}{\rho^2}v_G^2 \qquad \textbf{Ans}$$

$$\Sigma M_y = -I_{zx}\left[\left(\frac{v_G}{\rho}\right)^2 - 0\right] = -\frac{I_{zx}}{\rho^2}v_G^2 \qquad \textbf{Ans}$$

$$\Sigma M_z = 0 \qquad \textbf{Ans}$$

Note : This result indicates the normal reactions of the tires on the ground are not all necessarily equal. Instead, they depend upon the speed of the car, radius of curvature, and the products of inertia I_{yz} and I_{zx}. (See Example 13 – 6)

21-49. The shaft is constructed from a rod which has a mass of 2 kg/m. Determine the components of reaction at the bearings A and B if at the instant shown the shaft is freely spinning and has an angular velocity of $\omega = 30$ rad/s. What is the angular acceleration of the shaft at this instant? Bearing A can support a component of force n the y direction, whereas bearing B cannot.

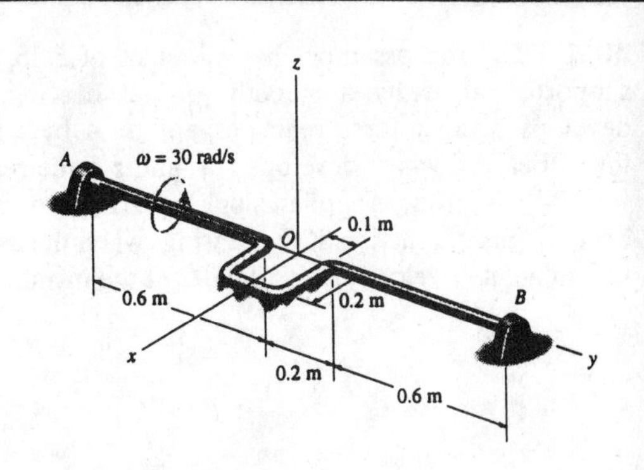

$$\Sigma W = [3(0.2) + 1.2](2)(9.81) = 35.316 \text{ N}$$

$$\Sigma \bar{x}W = 0[1.2(2)(9.81)] + 0.1[0.4(2)(9.81)] + 0.2[0.2(2)(9.81)] = 1.5696 \text{ N}\cdot\text{m}$$

$$\bar{x} = \frac{\Sigma \bar{x}W}{\Sigma W} = \frac{1.5696}{35.316} = 0.04444 \text{ m}$$

$$I_y = 2\left[\frac{1}{3}[0.2(2)](0.2)^2\right] + [0.2(2)](0.2)^2 = 0.02667 \text{ kg}\cdot\text{m}^2$$

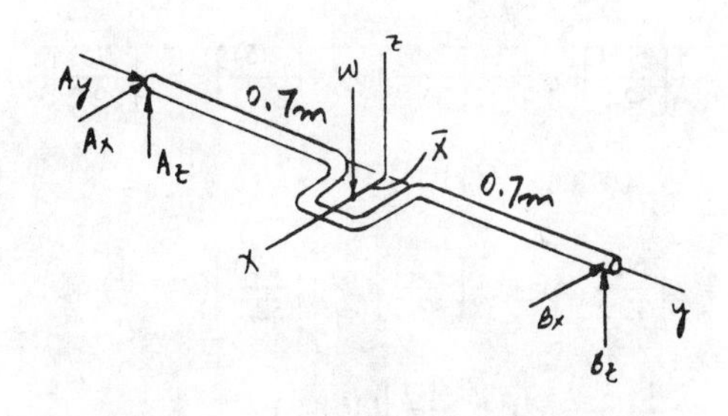

Applying Eq. 21-25 with $\omega_x = \omega_z = 0$ $\omega_y = 30$ rad/s $\dot{\omega}_x = \dot{\omega}_z = 0$

$$\Sigma M_x = I_x \dot{\omega}_x - (I_y - I_z)\omega_y \omega_z; \quad B_z(0.7) - A_z(0.7) = 0 \quad (1)$$

$$\Sigma M_y = I_y \dot{\omega}_y - (I_z - I_x)\omega_z \omega_x; \quad 35.316(0.04444) = 0.02667\dot{\omega}_y$$

$$\dot{\omega}_y = 58.9 \text{ rad/s}^2 \quad \textbf{Ans}$$

$$\Sigma M_z = I_z \dot{\omega}_z - (I_x - I_y)\omega_x \omega_y; \quad B_x(0.7) - A_x(0.7) = 0 \quad (2)$$

Also,

$$\Sigma F_x = m(a_G)_x; \quad -A_x - B_x = -1.8(2)(0.04444)(30)^2 \quad (3)$$

$$\Sigma F_y = m(a_G)_y; \quad A_y = 0 \quad \textbf{Ans}$$

$$\Sigma F_z = m(a_G)_z; \quad A_z + B_z - 35.316 = -1.8(2)(0.04444)(58.9) \quad (4)$$

Solving Eqs.(1) to (4) yields :

$A_x = B_x = 72.0$ N $\quad A_z = B_z = 12.9$ N $\quad$ **Ans**

21-50. A stone crusher consists of a large thin disk which is pin connected to a horizontal axle. If the axle is turning at a constant rate of 8 rad/s, determine the normal force which the disk exerts on the stones. Assume that the disk rolls without slipping and has a mass of 25 kg.

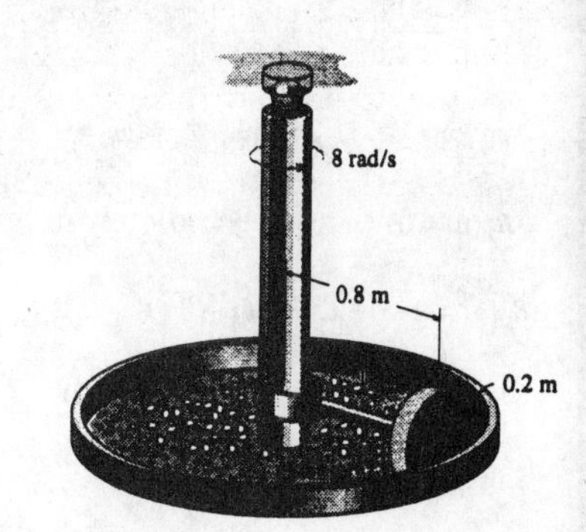

$$I_x = I_z = \frac{1}{4}(25)(0.2)^2 + 25(0.8)^2 = 16.25 \text{ kg}\cdot\text{m}^2$$

$$I_y = \frac{1}{2}(25)(0.2)^2 = 0.5 \text{ kg}\cdot\text{m}^2$$

$\omega = -\omega_y \mathbf{j} + \omega_z \mathbf{k}$, where $\omega_z = 8$ rad/s

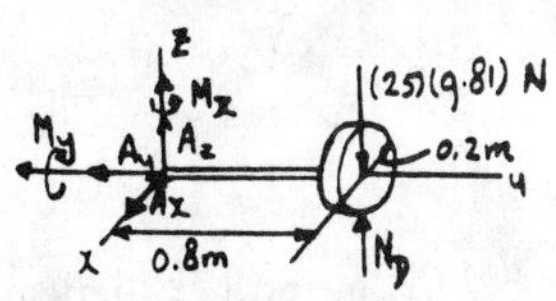

$$v = 0.8\omega_z = (0.8)(8) = 6.4 \text{ m/s}$$

$$\omega_y = -\frac{6.4}{0.2} = -32 \text{ rad/s}$$

Thus,

$$\omega = -32\mathbf{j} + 8\mathbf{k}$$

$$\dot{\omega} = \dot{\omega}_{xyz} + \Omega \times \omega = \mathbf{0} + (8\mathbf{k}) \times (-32\mathbf{j} + 8\mathbf{k}) = 256\mathbf{i}$$

$$\dot{\omega}_x = 256 \text{ rad/s}^2$$

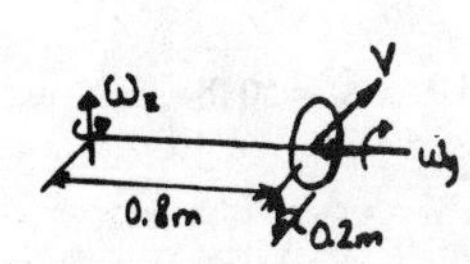

$$\Sigma M_x = I_x \dot{\omega}_x - (I_y - I_z)\omega_y \omega_z$$

$$N_D(0.8) - 25(9.81)(0.8) = (16.25)(256) - (0.5 - 16.25)(-32)(8)$$

$N_D = 405$ N $\quad$ **Ans**

21-51. The rod assembly has a weight of 5 lb/ft. It is supported at B by a smooth journal bearing, which develops x and y force reactions, and at A by a smooth thrust bearing, which develops x, y, and z force reactions. If a 50-lb·ft torque is applied along rod AB, determine the components of reaction at the bearings when the assembly has an angular velocity $\omega = 10$ rad/s at the instant shown.

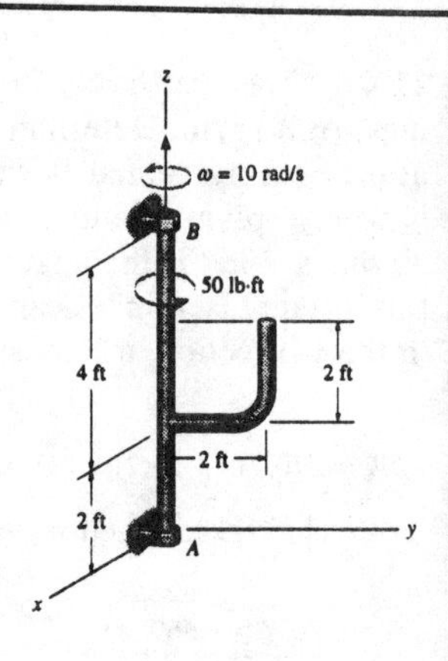

$$I_y = \frac{1}{3}\left[\frac{6(5)}{32.2}\right](6)^2 + \frac{1}{12}\left[\frac{2(5)}{32.2}\right](2)^2 + \left[\frac{2(5)}{32.2}\right](3)^2 + \left[\frac{2(5)}{32.2}\right](2)^2$$

$$= 15.3209 \text{ slug}\cdot\text{ft}^2$$

$$I_x = \frac{1}{3}\left[\frac{6(5)}{32.2}\right](6)^2 + \frac{1}{12}\left[\frac{2(5)}{32.2}\right](2)^2 + \left[\frac{2(5)}{32.2}\right](2^2+3^2)$$

$$+\frac{1}{12}\left[\frac{2(5)}{32.2}\right](2)^2 + \left[\frac{2(5)}{32.2}\right](1^2+2^2)$$

$$I_x = 16.9772 \text{ slug}\cdot\text{ft}^2$$

$$I_z = \frac{1}{3}\left[\frac{2(5)}{32.2}\right](2)^2 + \left[\frac{2(5)}{32.2}\right](2)^2 = 1.6563 \text{ slug}\cdot\text{ft}^2$$

$$I_{yz} = \left[\frac{2(5)}{32.2}\right](1)(2) + \left[\frac{2(5)}{32.2}\right](2)(3) = 2.4845 \text{ slug}\cdot\text{ft}^2 \qquad I_{xy} = I_{zx} = 0$$

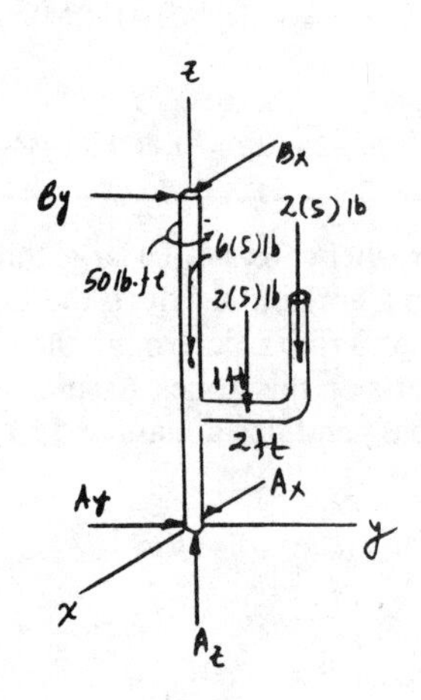

Applying Eq. 21-24 with $\omega_x = \omega_y = 0$ $\omega_z = 10$ rad/s $\dot{\omega}_x = \dot{\omega}_y = 0$

$-B_y(6) = 0 - 0 - 0 - 2.4845(0 - 10^2) - 0 \qquad B_y = -41.4$ lb **Ans**

$B_x(6) = 0 - 0 - 2.4845\dot{\omega}_z - 0 - 0 \qquad (1)$

$50 = 1.6563\dot{\omega}_z \qquad (2)$

Solving Eqs.(1) and (2) yields :

$\dot{\omega}_z = 30.19 \text{ rad/s}^2$

$B_x = -12.5$ lb **Ans**

$$\Sigma F_x = m(a_G)_x; \quad A_x + (-12.50) = -\left[\frac{2(5)}{32.2}\right](1)(30.19) - \left[\frac{2(5)}{32.2}\right](2)(30.19)$$

$A_x = -15.6$ lb **Ans**

$$\Sigma F_y = m(a_G)_y; \quad A_y + (-41.4) = -\left[\frac{2(5)}{32.2}\right](1)(10)^2 - \left[\frac{2(5)}{32.2}\right](2)(10)^2$$

$A_y = -51.8$ lb **Ans**

$\Sigma F_z = m(a_G)_z; \quad A_z - 2(5) - 2(5) - 6(5) = 0 \quad A_z = 50$ lb **Ans**

***21-52.** The 25-lb disk is *fixed* to rod BCD, which has negligible mass. Determine the torque T which must be applied to the vertical shaft so that the shaft has an angular acceleration of $\alpha = 6\ \text{rad/s}^2$.

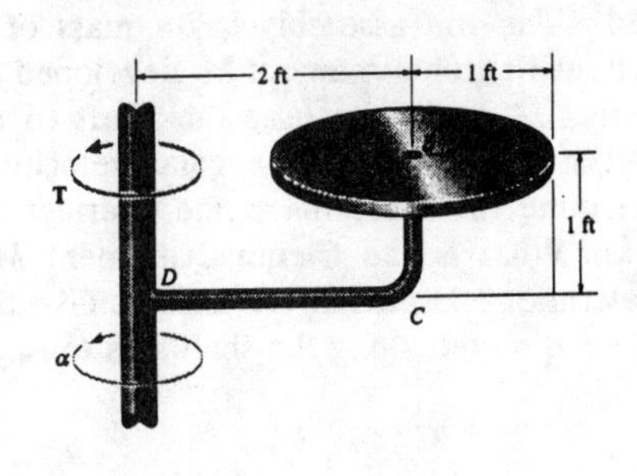

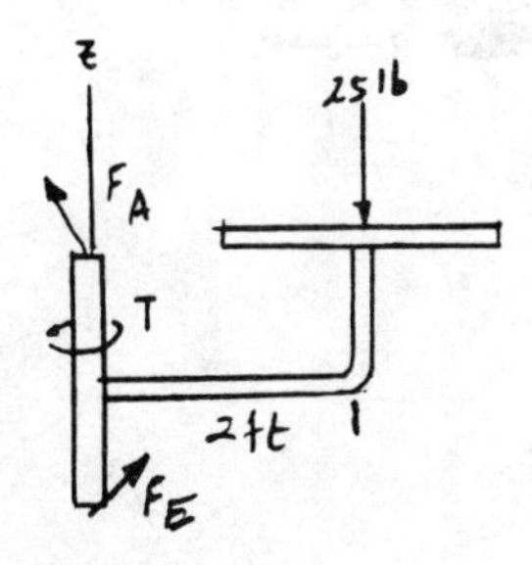

$$I_z = \frac{1}{2}\left(\frac{25}{32.2}\right)(1)^2 + \left(\frac{25}{32.2}\right)(2)^2 = 3.4938\ \text{slug}\cdot\text{ft}^2$$

Applying the third of Eq. 21-25 with $I_x = I_y$, $\omega_x = \omega_y = 0$, $\dot{\omega}_z = 6\ \text{rad/s}^2$

$\Sigma M_z = I_z\dot{\omega}_z - (I_x - I_y)\omega_x\omega_y;$ $\quad T = 3.4938(6) = 21.0\ \text{lb}\cdot\text{ft}$ **Ans**

21-53. Solve Prob. 21-52, assuming rod BCD has a weight of 2 lb/ft.

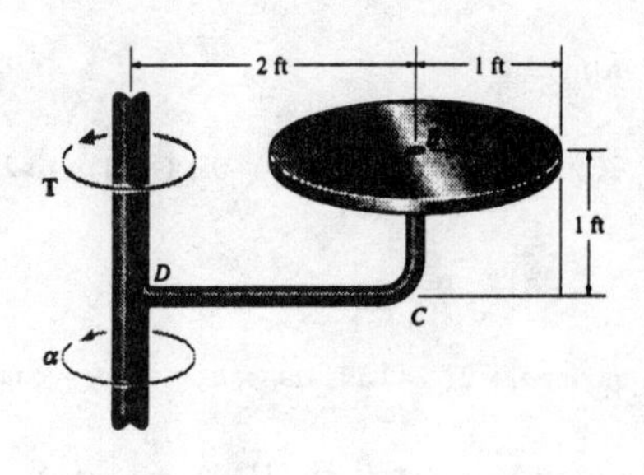

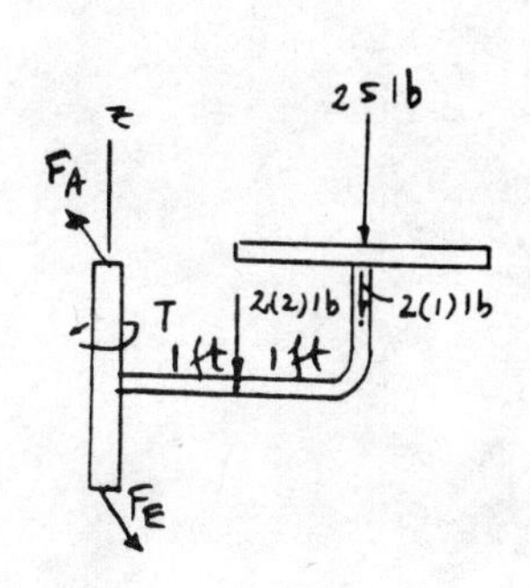

$$I_z = \frac{1}{2}\left(\frac{25}{32.2}\right)(1)^2 + \left(\frac{25}{32.2}\right)(2)^2 + \frac{1}{3}\left(\frac{2(2)}{32.2}\right)(2)^2 + \left(\frac{1(2)}{32.2}\right)(2)^2 = 3.9079\ \text{slug}\cdot\text{ft}^2$$

Applying the third of Eq. 21-25 with $I_x = I_y$, $\omega_x = \omega_y = 0$, $\dot{\omega}_z = 6\ \text{rad/s}^2$

$\Sigma M_z = I_z\dot{\omega}_z - (I_x - I_y)\omega_x\omega_y;$ $\quad T = 3.9079(6) = 23.4\ \text{lb}\cdot\text{ft}$ **Ans**

21-54. The rod assembly has a mass of 1.5 kg/m. A resultant frictional moment **M**, developed at the journal bearings A and B, causes the shaft to decelerate at 1 rad/s^2 when it has an angular velocity $\omega = 6$ rad/s. Determine the reactions at the bearings at the instant shown. What is the frictional moment M causing the deceleration? Establish the axes at G with $+y$ upward and $+z$ directed along the shaft axis toward A.

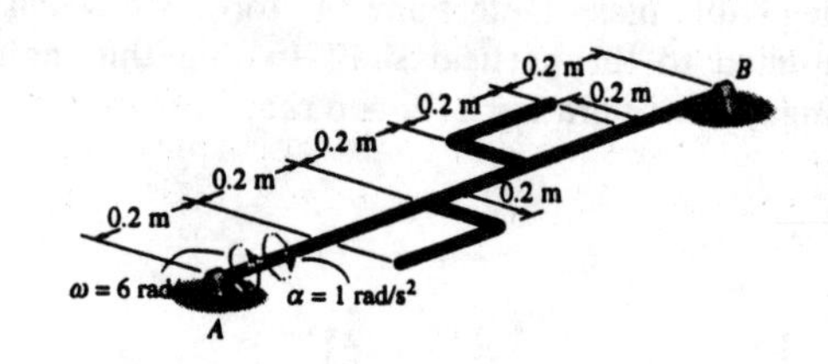

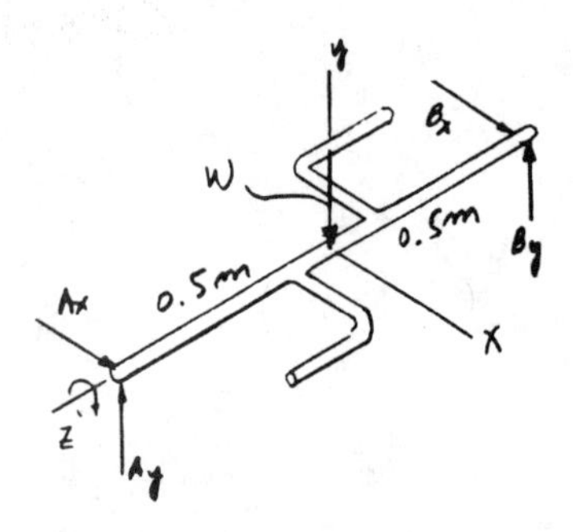

$\Sigma W = 1.8(1.5)(9.81) = 26.487$ N

$$I_z = 2\left[\frac{1}{3}[0.2(1.5)](0.2)^2 + [0.2(1.5)](0.2)^2\right] = 0.032 \text{ kg}\cdot\text{m}^2$$

$$I_{xy} = I_{yz} = 0$$

$$I_{zx} = 0.2(1.5)[(-0.2)(-0.2) + (-0.1)(-0.1) + (0.2)(0.2) + (0.1)(0.1)]$$

$$= 0.03 \text{ kg}\cdot\text{m}^2$$

Applying Eq. 21-24 with $\omega_x = \omega_y = 0$ $\omega_z = 6$ rad/s $\dot{\omega}_x = \dot{\omega}_y = 0$ $\dot{\omega}_z = -1$ rad/s^2

$B_y(0.5) - A_y(0.5) = -0.03(-1)$ (1)

$A_x(0.5) - B_x(0.5) = -0.03(6)^2$ (2)

$-M = 0.032(-1)$ $M = 0.0320$ N·m **Ans**

Also,

$\Sigma F_x = m(a_G)_x$; $A_x + B_x = 0$ (3)

$\Sigma F_y = m(a_G)_y$; $A_y + B_y - 26.487 = 0$ (4)

Solving Eqs.(1) to (4) yields :

$A_y = 13.2$ N $B_y = 13.3$ N $A_x = -1.08$ N $B_x = 1.08$ N **Ans**

21-55. A thin uniform plate having a mass of 0.4 kg is spinning with a constant angular velocity ω about its diagonal AB. If the person holding the corner of the plate at B releases his finger, the plate will fall downward on its side AC. Determine the necessary couple moment M which if applied to the plate would prevent this from happening.

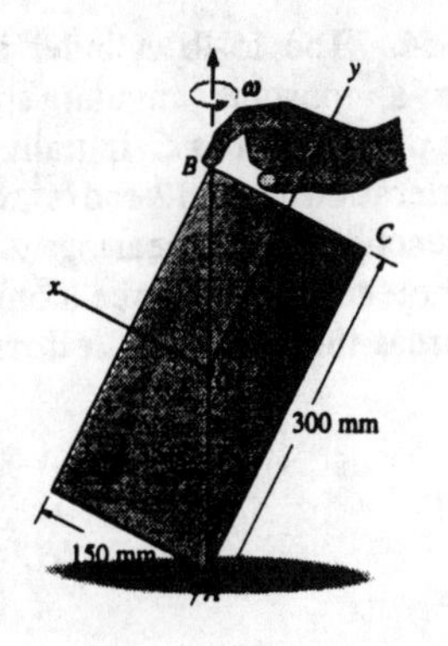

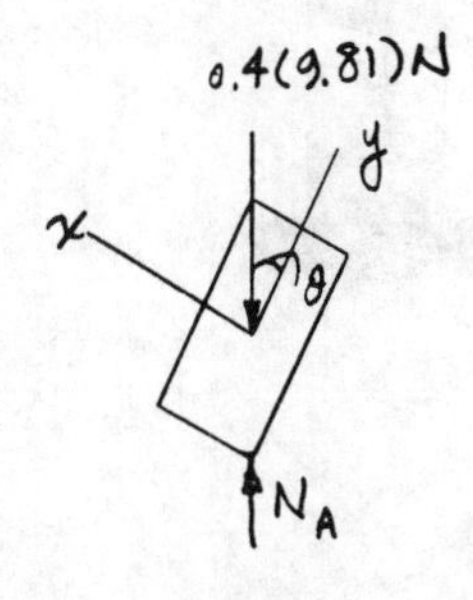

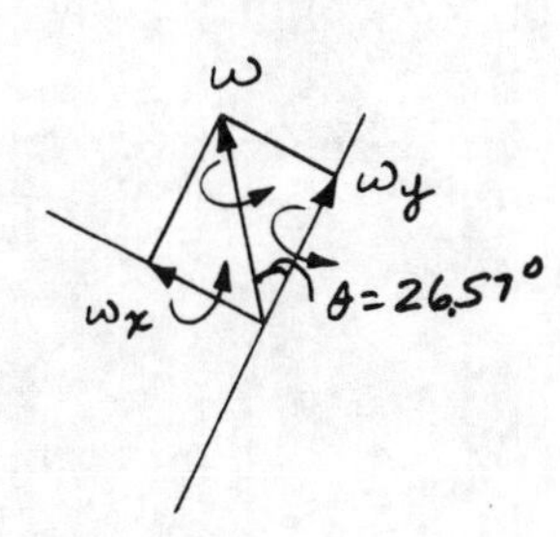

Using the principal axis shown,

$$I_x = \frac{1}{12}(0.4)(0.3)^2 = 3\left(10^{-3}\right)\ \text{kg}\cdot\text{m}^2$$

$$I_y = \frac{1}{12}(0.4)(0.15)^2 = 0.75\left(10^{-3}\right)\ \text{kg}\cdot\text{m}^2$$

$$I_z = \frac{1}{12}(0.4)\left[(0.3)^2 + (0.15)^2\right] = 3.75\left(10^{-3}\right)\ \text{kg}\cdot\text{m}^2$$

$$\theta = \tan^{-1}\left(\frac{75}{150}\right) = 26.57°$$

$$\omega_x = \omega\sin 26.57°, \qquad \dot{\omega}_x = 0$$

$$\omega_y = \omega\cos 26.57°, \qquad \dot{\omega}_y = 0$$

$$\omega_z = 0, \qquad \dot{\omega}_z = 0$$

$$\Sigma M_x = I_x\dot{\omega}_x - (I_y - I_z)\omega_y\omega_z$$

$$M_x = 0$$

$$\Sigma M_y = I_y\dot{\omega}_y - (I_z - I_x)\omega_z\omega_x$$

$$M_y = 0$$

$$\Sigma M_z = I_z\dot{\omega}_z - (I_x - I_y)\omega_x\omega_y$$

$$M_z = 0 - \left[3\left(10^{-3}\right) - 0.75\left(10^{-3}\right)\right]\omega^2\sin 26.57°\cos 26.57°$$

$$M_z = -0.9\left(10^{-3}\right)\omega^2\ \text{N}\cdot\text{m} = -0.9\omega^2\ \text{mN}\cdot\text{m} \qquad \textbf{Ans}$$

The couple acts outward, perpendicular to the face of the plate.

***21-56.** The 15-lb cylinder is rotating about shaft AB with a constant angular speed $\omega = 4$ rad/s. If the supporting shaft at C, initially at rest, is given an angular acceleration $\alpha_C = 12\ \text{rad/s}^2$, determine the components of reaction at the bearings A and B. The bearing at A cannot support a force component along the x axis, whereas the bearing at B does.

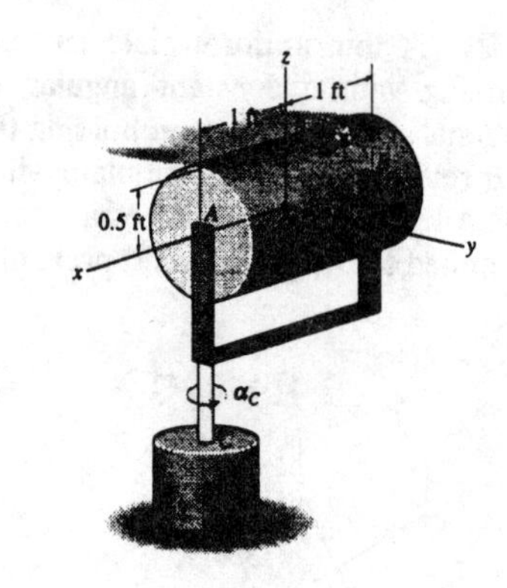

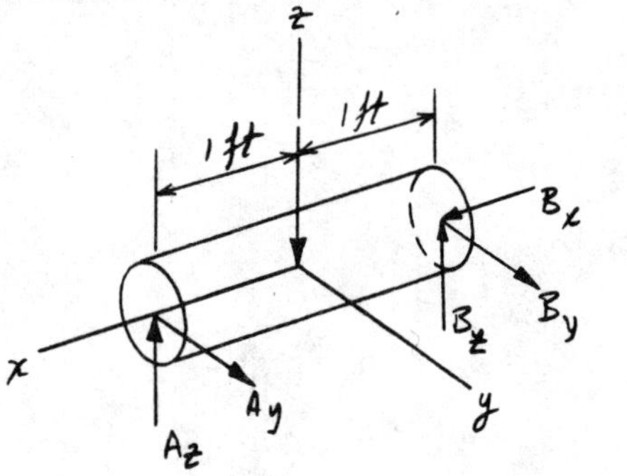

$\omega = \{-4\mathbf{i}\}$ rad/s

$\dot{\omega} = \dot{\omega}_{xyz} + \Omega \times \omega = 12\mathbf{k} + \mathbf{0} \times (-4\mathbf{i}) = \{12\mathbf{k}\}\ \text{rad/s}^2$

Hence

$\omega_x = -4, \qquad \omega_y = \omega_z = 0,$

$\dot{\omega}_x = \dot{\omega}_y = 0, \qquad \dot{\omega}_z = 12\ \text{rad/s}^2$

$\Sigma M_x = I_x \dot{\omega}_x - (I_y - I_z)\,\omega_y\,\omega_z, \qquad 0 = 0$

$\Sigma M_y = I_y \dot{\omega}_y - (I_z - I_x)\,\omega_z\,\omega_x, \qquad B_z(1) - A_z(1) = 0$

$\Sigma M_z = I_z \dot{\omega}_z - (I_x - I_y)\,\omega_x\,\omega_y, \qquad A_y(1) - B_y(1) = \left[\frac{1}{12}\left(\frac{15}{32.2}\right)\left(3(0.5)^2 + (2)^2\right)\right](12) - 0$

$\Sigma F_x = m(a_G)_x; \qquad B_x = 0$ **Ans**

$\Sigma F_y = m(a_G)_y; \qquad A_y + B_y = -\left(\frac{15}{32.2}\right)(1)(12)$

$\Sigma F_z = m(a_G)_z; \qquad A_z + B_z - 15 = 0$

Solving,

$A_y = -1.69$ lb **Ans**

$B_y = -3.90$ lb **Ans**

$A_z = B_z = 7.5$ lb **Ans**

21-57. The crankshaft is constructed from a rod which has a weight of 3 lb/ft. Determine its angular acceleration and the x and z components of reaction at bearings A and B at the instant the shaft is in the position shown.

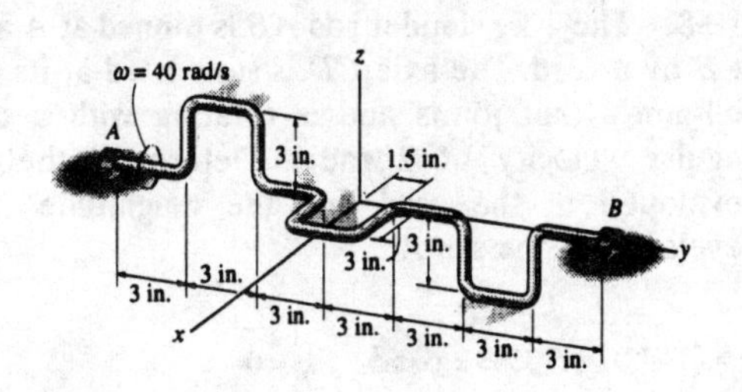

The x, y, z, axes are not prineipal axes.

$\omega_x = \omega_z = 0, \quad \omega_y = 40 \text{ rad/s}, \quad \dot{\omega}_x = \dot{\omega}_z = 0$

3 lb/ft = 0.25 lb/in.

$$I_{yz} = 2\left[\frac{0.25(3)}{32.2}\left(\frac{1.5}{12}\right)\left(-\frac{7.5}{12}\right) + \frac{0.25(3)}{32.2}\left(\frac{3}{12}\right)\left(-\frac{6}{12}\right) + \frac{0.25(3)}{32.2}\left(\frac{1.5}{12}\right)\left(-\frac{4.5}{12}\right)\right] = -0.011646 \text{ slug}\cdot\text{ft}^2$$

$I_{xy} = I_{zx} = 0$

$$I_{yy} = 6\left(\frac{1}{3}\right)\left(\frac{0.25(3)}{32.2}\right)\left(\frac{1}{4}\right)^2 + 3\left(\frac{0.25(3)}{32.2}\right)\left(\frac{1}{4}\right)^2 = 0.007279 \text{ slug}\cdot\text{ft}^2$$

$W = 13(3)(0.25) = 9.75$ lb

$$\bar{x} = \frac{2(3)(1.5) + 3(3)}{13(3)} = 0.4615 \text{ in.}$$

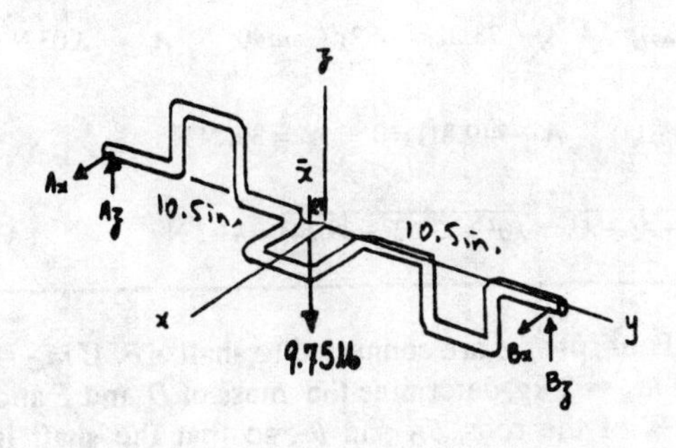

$\Sigma F_x = m(a_G)_x; \quad A_x + B_x = -\left(\frac{9.75}{32.2}\right)(40)^2\left(\frac{0.4615}{12}\right)$

$A_x + B_x = -18.634$

$\Sigma F_z = m(a_G)_z; \quad A_z + B_z - 9.75 = -\left(\frac{9.75}{32.2}\right)\left(\frac{0.4165}{12}\right)\dot{\omega}_y$

$A_z + B_z - 9.75 = -0.011646\dot{\omega}_y$

Equations 21 – 24 reduce to

$\Sigma M_x = -I_{yz}\omega_y^2; \quad -\frac{10.5}{12}(A_z) + \frac{10.5}{12}(B_z) = -(-0.011646)(40)^2$

$-A_z + B_z = 21.295$

$\Sigma M_y = I_{yy}\dot{\omega}_y; \quad \frac{0.4615}{12}(9.75) = 0.007279\dot{\omega}_y$

$\dot{\omega}_y = 51.52 \text{ rad/s}^2 = 51.5 \text{ rad/s}^2$ **Ans**

$\Sigma M_z = -I_{yz}\dot{\omega}_y; \quad A_x\left(\frac{10.5}{12}\right) - B_x\left(\frac{10.5}{12}\right) = -(-0.011646)(51.52)$

$A_x - B_x = 0.68571$

Solving,

$A_x = -8.97$ lb **Ans**

$B_x = -9.66$ lb **Ans**

$A_z = -6.07$ lb **Ans**

$B_z = 15.2$ lb **Ans**

21-58. The 4-kg slender rod AB is pinned at A and held at B by a cord. The axle CD is supported at its ends by ball-and-socket joints and is rotating with a constant angular velocity of 2 rad/s. Determine the tension developed in the cord and the magnitude of force developed at the pin A.

$I_z = \frac{1}{3}(4)(2)^2 = 5.3333 \text{ kg}\cdot\text{m}^2 \qquad I_y = 0$

Applying the third of Eq. 21-25 with

$\omega_y = 2\cos 40° = 1.5321$ rad/s

$\omega_z = 2\sin 40° = 1.2856$ rad/s

$\Sigma M_x = I_x\dot{\omega}_x - (I_y - I_z)\omega_y\omega_z;$

$T(2\cos 40°) - 4(9.81)(1\sin 40°) = 0 - (0 - 5.3333)(1.5321)(1.2856)$

$T = 23.3$ N **Ans**

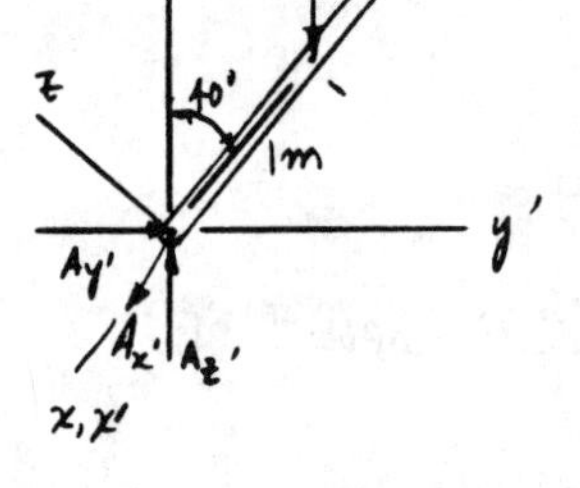

Also,

$\Sigma F_{x'} = m(a_G)_{x'}; \qquad A_{x'} = 0$

$\Sigma F_{y'} = m(a_G)_{y'}; \qquad A_{y'} - 23.32 = -4(2)^2(1\sin 40°) \qquad A_{y'} = 13.03$ N

$\Sigma F_{z'} = m(a_G)_{z'}; \qquad A_{z'} - 4(9.81) = 0 \qquad A_{z'} = 39.24$ N

$F_A = \sqrt{A_{x'}^2 + A_{y'}^2 + A_{z'}^2} = \sqrt{0^2 + 13.03^2 + 39.24^2} = 41.3$ N **Ans**

21-59. Four spheres are connected to shaft AB. If $m_C = 1$ kg and $m_E = 2$ kg, determine the mass of D and F and the angles of the rods, θ_D and θ_F, so that the shaft is dynamically balanced, that is, so that the bearings at A and B exert only vertical reactions on the shaft as it rotates. Neglect the mass of the rods.

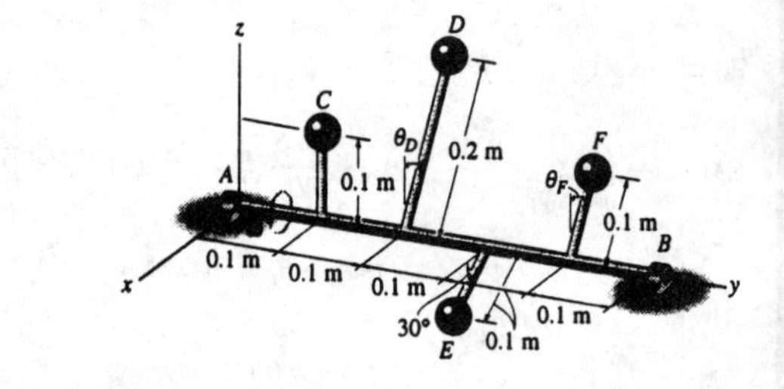

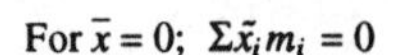

For $\bar{x} = 0$; $\Sigma\tilde{x}_i m_i = 0$

$(0.1\cos 30°)(2) - (0.1\sin\theta_F)m_F - (0.2\sin\theta_D)m_D = 0 \qquad (1)$

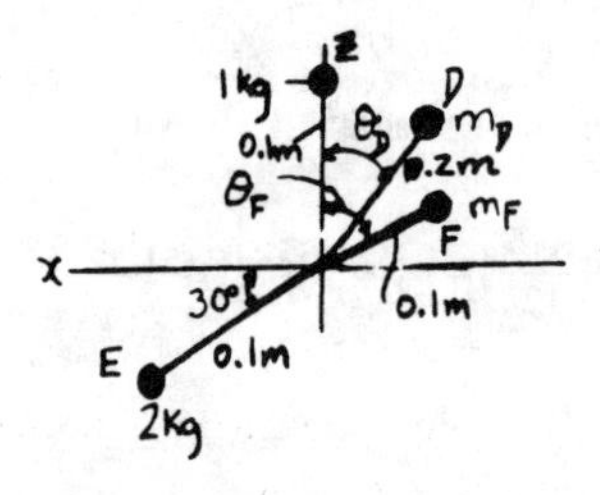

For $\bar{z} = 0$; $\Sigma\tilde{z}_i m_i = 0$

$(0.1)(1) - (0.1\sin 30°)(2) + (0.2\cos\theta_D)m_D + (0.1\cos\theta_F)m_F = 0 \qquad (2)$

For $I_{xy} = 0$; $\Sigma\tilde{x}_i\tilde{y}_i m_i = 0$

$-(0.2)(0.2\sin\theta_D)m_D + (0.3)(0.1\cos 30°)(2) - (0.4)(0.1\sin\theta_F)m_F = 0 \qquad (3)$

For $I_{zy} = 0$; $\Sigma\tilde{z}_i\tilde{y}_i m_i = 0$

$(0.1)(0.1)(1) + (0.2)(0.2\cos\theta_D)m_D - (0.3)(0.1\sin 30°)(2) + (0.1\cos\theta_F)(0.4)(m_F) = 0 \qquad (4)$

Solving,

$\theta_D = 139°$ **Ans**

$m_D = 0.661$ kg **Ans**

$\theta_F = 40.9°$ **Ans**

$m_F = 1.32$ kg **Ans**

***21-60.** Two uniform rods, each having a weight of 10 lb, are pin connected to the edge of a rotating disk. If the disk has a constant angular velocity $\omega_D = 4$ rad/s, determine the angle θ made by each rod during the motion, and the components of the force and moment developed at the pin *A*. *Suggestion:* Use the *x*, *y*, *z* axes oriented as shown.

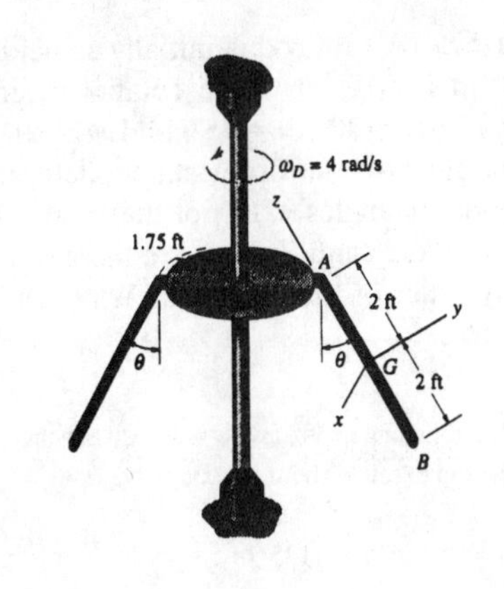

$I_y = \frac{1}{12}\left(\frac{10}{32.2}\right)(4)^2 = 0.4141$ slug · ft^2 $\quad I_z = 0$

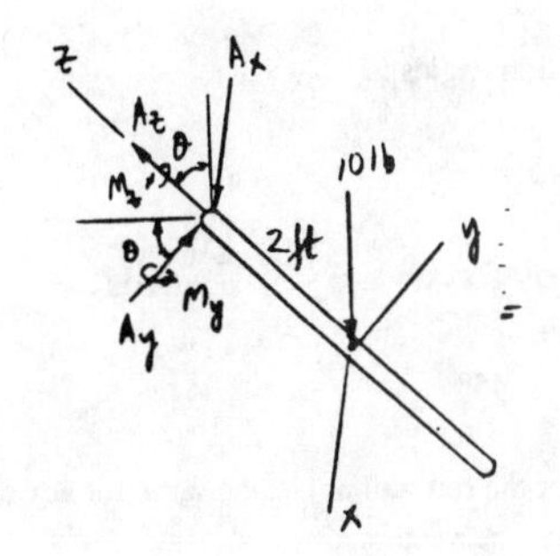

Applying Eq. 21-25 with

$\omega_y = 4\sin\theta \qquad \omega_z = 4\cos\theta \qquad \omega_x = 0$

$\dot{\omega}_x = \dot{\omega}_y = \dot{\omega}_z = 0$

$\Sigma M_x = I_x\dot{\omega}_x - (I_y - I_z)\omega_y\omega_z; \qquad -A_y(2) = 0 - (0.4141 - 0)(4\sin\theta)(4\cos\theta) \qquad (1)$

$\Sigma M_y = I_y\dot{\omega}_y - (I_z - I_x)\omega_z\omega_x; \qquad M_y + A_x(2) = 0 \qquad (2)$

$\Sigma M_z = I_z\dot{\omega}_z - (I_x - I_y)\omega_x\omega_y; \qquad M_z = 0$ **Ans**

Also,

$\Sigma F_x = m(a_G)_x; \qquad A_x = 0$ **Ans**

From Eq.(2) $\qquad M_y = 0$ **Ans**

$\Sigma F_y = m(a_G)_y; \qquad A_y - 10\sin\theta = -\left(\frac{10}{32.2}\right)(1.75 + 2\sin\theta)(4)^2\cos\theta \qquad (3)$

$\Sigma F_z = m(a_G)_z; \qquad A_z - 10\cos\theta = \left(\frac{10}{32.2}\right)(1.75 + 2\sin\theta)(4)^2\sin\theta \qquad (4)$

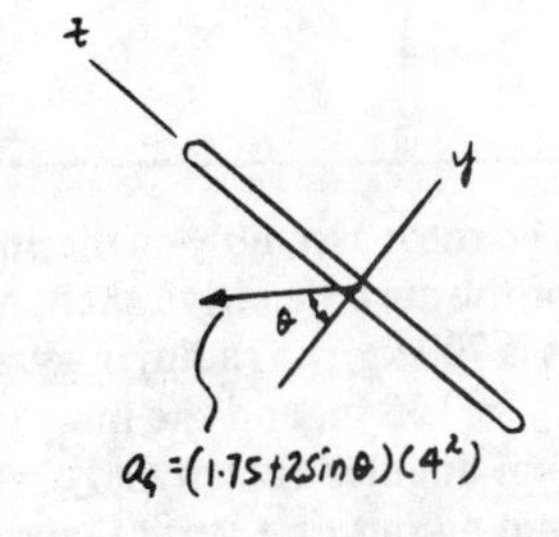

Solving Eqs.(1), (3) and (4) yields :

$\theta = 64.1°$ $\quad A_y = 1.30$ lb $\quad A_z = 20.2$ lb **Ans**

21-61. Show that the angular velocity of a body, in terms of Euler angles ϕ, θ, and ψ, may be expressed as $\boldsymbol{\omega} = (\dot{\phi}\sin\theta\sin\psi + \dot{\theta}\cos\psi)\mathbf{i} + (\dot{\phi}\sin\theta\cos\psi - \dot{\theta}\sin\psi)\mathbf{j} + (\dot{\phi}\cos\theta + \dot{\psi})\mathbf{k}$, where **i**, **j**, and **k** are directed along the *x*, *y*, *z* axes as shown in Fig. 21–15*d*.

From Fig. 21-15*b*, due to rotation ϕ, the *x*, *y*, *z* components of $\dot{\phi}$ are simply $\dot{\phi}$ along *z* axis.

From Fig. 21-15*c*, due to rotation θ, the *x*, *y*, *z* components of $\dot{\phi}$ and $\dot{\theta}$ are $\dot{\phi}\sin\theta$ in the *y* direction, $\dot{\phi}\cos\theta$ in the *z* direction, and $\dot{\theta}$ in the *x* direction.

Lastly, rotation ψ, Fig. 21-15*d*, produces the final components which yields

$\omega = (\dot{\phi}\sin\theta\sin\psi + \dot{\theta}\cos\psi)\mathbf{i} + (\dot{\phi}\sin\theta\cos\psi - \dot{\theta}\sin\psi)\mathbf{j} + (\dot{\phi}\cos\theta + \dot{\psi})\mathbf{k}$ **Q.E.D.**

21-62. A thin rod is initially coincident with the Z axis when it is given three rotations defined by the Euler angles $\phi = 30°$, $\theta = 45°$, and $\psi = 60°$. If these rotations are given in the order stated, determine the coordinate direction angles α, β, γ of the axis of the rod with respect to the X, Y, and Z axes. Are these directions the same for any order of the rotations? Why?

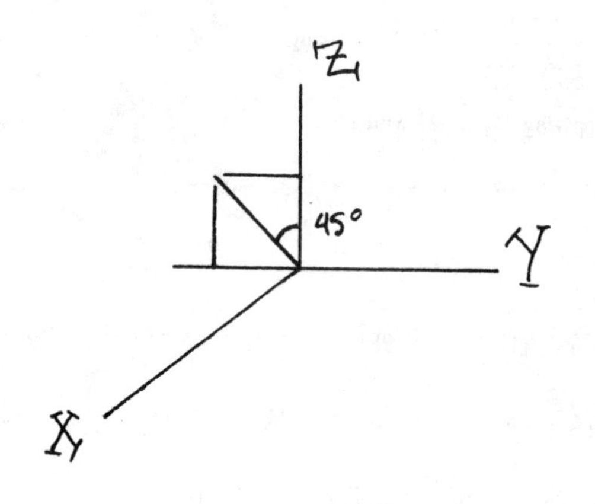

The rotations ϕ=30° and ψ= 60° do not change the orientation of the rod since it causes the rod to rotate about its axis only.

$$\mathbf{u} = \cos 90°\mathbf{i} + \cos 135°\mathbf{j} + \cos 45°\mathbf{k}$$

$$= -0.707\mathbf{j} + 0.707\mathbf{k}$$

Coordinate direction angles :

$\alpha = \cos^{-1}(0) = 90°$ **Ans**

$\beta = \cos^{-1}(-0.707) = 135°$ **Ans**

$\gamma = \cos^{-1}(0.707) = 45°$ **Ans**

The orientation of the rod will not be the same for any order of rotation because finite rotations are not vectors.

No **Ans**

21-63. The rotor assembly on the engine of a jet airplane consists of the turbine, drive shaft, and compressor. The total mass is 700 kg, the radius of gyration about the shaft axis is $k_{AB} = 0.35$ m, and the mass center is at G. If the rotor has an angular velocity $\omega_{AB} = 1000$ rad/s, and the plane is pulling out of a vertical curve while traveling at 250 m/s, determine the components of reaction at the bearings A and B due to the gyroscopic effect.

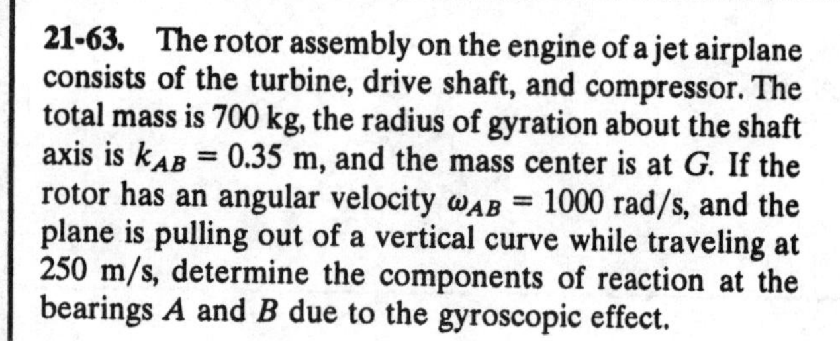

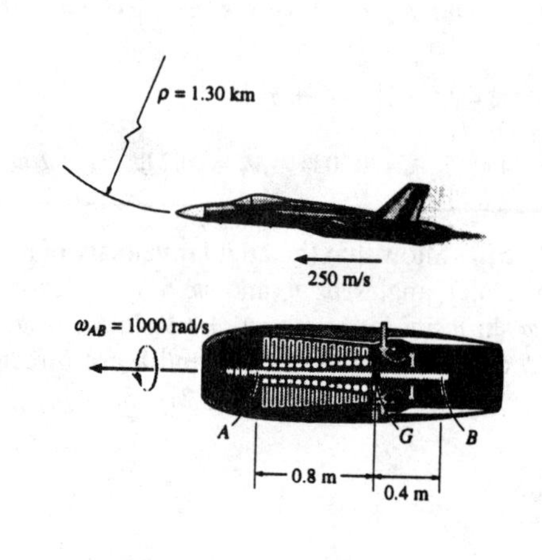

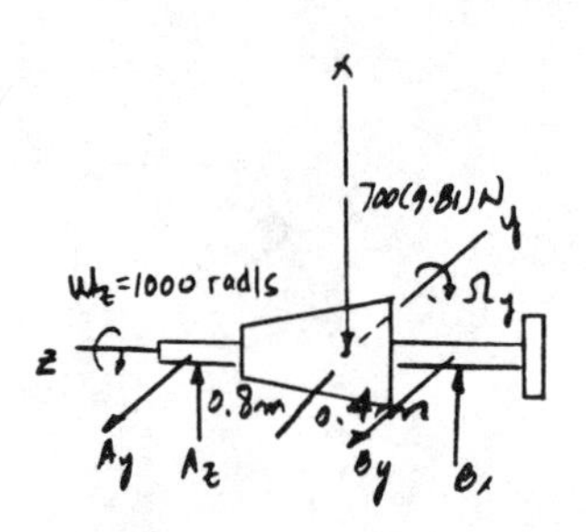

$$\omega_z = \omega_{AB} = 1000 \text{ rad/s} \qquad \Omega_y = \frac{250}{1300} = 0.1923 \text{ rad/s}$$

$$\Sigma M_x = I_z \Omega_y \omega_z ; \quad A_y(0.8) - B_y(0.4) = 700(0.35)^2(0.1923)(1000) \qquad (1)$$

$$\Sigma F_y = m(a_G)_y ; \quad A_y + B_y = 0 \qquad (2)$$

Solving Eqs.(1) and (2) yields :

$A_y = 13.7$ kN $\quad B_y = -13.7$ kN **Ans**

***21-64.** The 30-lb wheel rolls without slipping. If it has a radius of gyration $k_{AB} = 1.2$ ft about its axle AB, and the vertical drive shaft is turning at 8 rad/s, determine the normal reaction the wheel exerts on the ground at C.

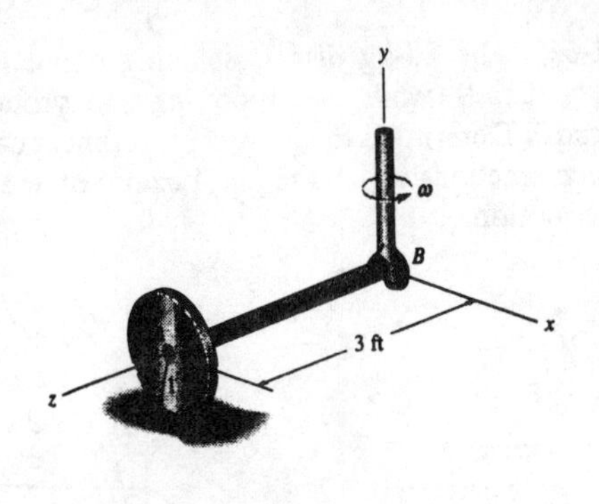

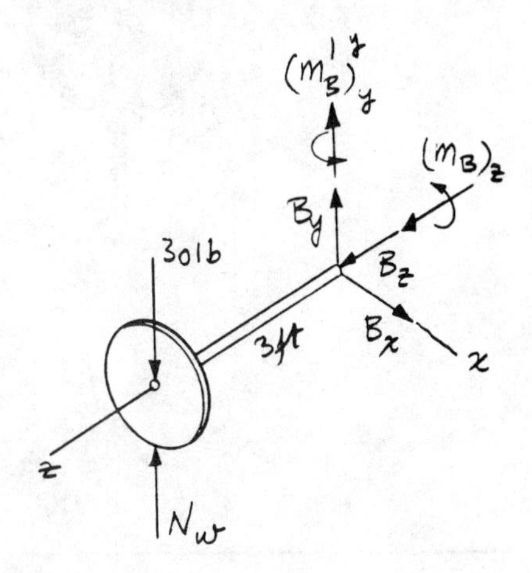

$\Omega_y = \omega = 8$ rad/s

$\omega_z = -\dfrac{3(8)}{1.8} = -13.33$ rad/s

$\Sigma M_x = I_z \Omega_y \omega_z;\qquad 30(3) - N_w(3) = \left[\left(\dfrac{30}{32.2}\right)(1.2)^2\right](8)(-13.33)$

$N_w = 77.7$ lb **Ans**

21-65. The 30-lb wheel rolls without slipping. If it has a radius of gyration $k_{AB} = 1.2$ ft about its axle AB, determine its angular velocity ω so that the normal reaction at C becomes 60 lb.

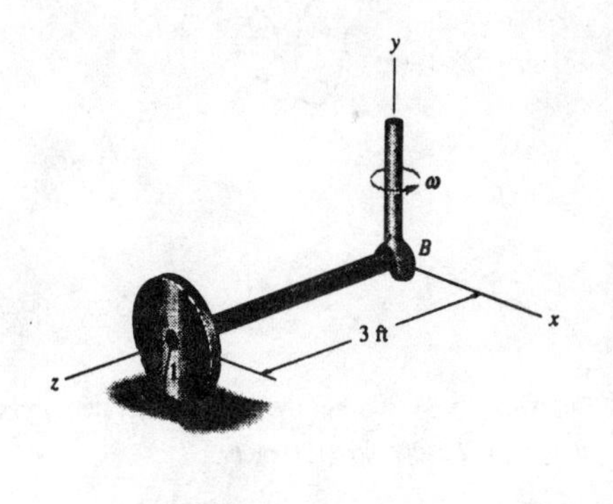

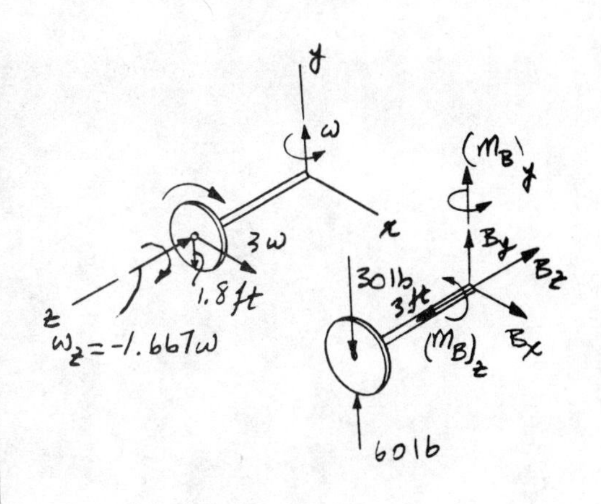

$\Sigma M_x = I_z \Omega_y \omega_z;\qquad 30(3) - 60(3) = \left[\dfrac{30}{32.2}(1.2)^2\right]\omega(-1.667\omega)$

$\omega = 6.34$ rad/s **Ans**

21-66. The 20-kg disk is spinning about its center at $\omega_s = 20$ rad/s while the supporting axle is rotating at $\omega_y = 6$ rad/s. Determine the gyroscopic moment caused by the force reactions which the pin A exerts on the disk due to the motion.

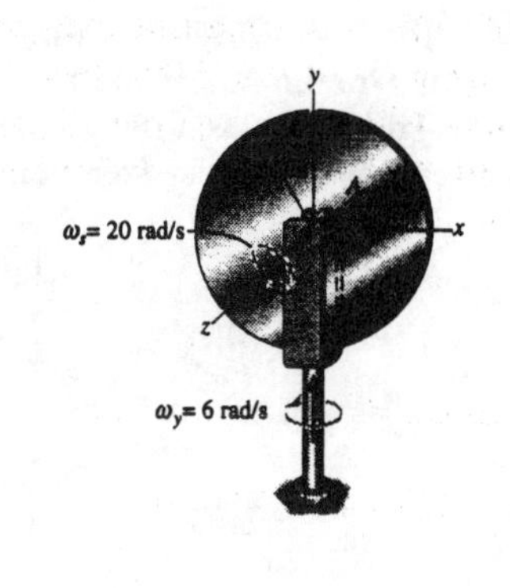

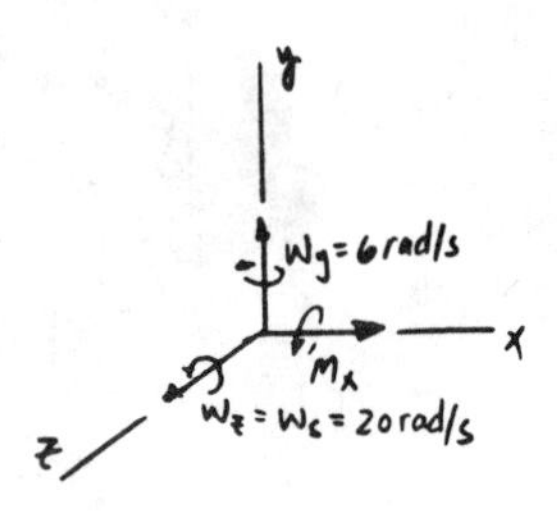

$$\Sigma M_x = I_z \Omega_y \omega_z; \quad M_x = \left[\frac{1}{2}(20)(0.15)^2\right](6)(20) = 27.0 \text{ N}\cdot\text{m} \quad \textbf{Ans}$$

21-67. The motor weighs 50 lb and has a radius of gyration of 0.2 ft about the z axis. The shaft of the motor is supported by bearings at A and B, and is turning at a constant rate of $\boldsymbol{\omega}_s = \{100\mathbf{k}\}$ rad/s, while the frame has an angular velocity of $\boldsymbol{\omega}_y = \{2\mathbf{j}\}$ rad/s. Determine the moment which the bearing forces at A and B exert on the shaft due to this motion.

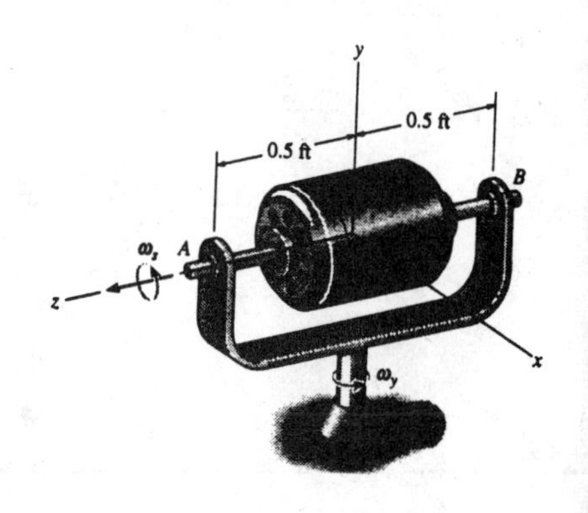

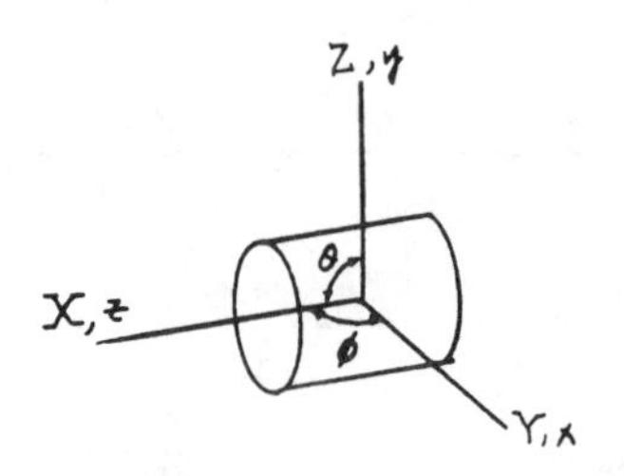

Applying Eq. 21-30 : For the coordinate system shown $\theta = 90°$ $\phi = 90°$
$\dot{\theta} = 0$ $\dot{\phi} = 2$ rad/s $\dot{\psi} = 100$ rad/s .

$$\Sigma M_x = -I\dot{\phi}^2 \sin\theta\cos\theta + I_z\dot{\phi}\sin\theta(\dot{\phi}\cos\theta + \dot{\psi}) \quad \text{reduces to}$$

$$\Sigma M_x = I_z\dot{\phi}\dot{\psi}; \quad M_x = \left[\left(\frac{50}{32.2}\right)(0.2)^2\right](2)(100) = 12.4 \text{ lb}\cdot\text{ft} \quad \textbf{Ans}$$

Since $\omega_x = 0$

$$\Sigma M_y = 0; \quad M_y = 0 \quad \textbf{Ans}$$

$$\Sigma M_z = 0; \quad M_z = 0 \quad \textbf{Ans}$$

***21-68.** **The homogeneous cone has a mass of 7 kg and a vertex angle of 90°. If the cone rolls on the horizontal surface without slipping, determine the greatest precessional speed ω_p it can have before the tip A starts to rise from the surface.**

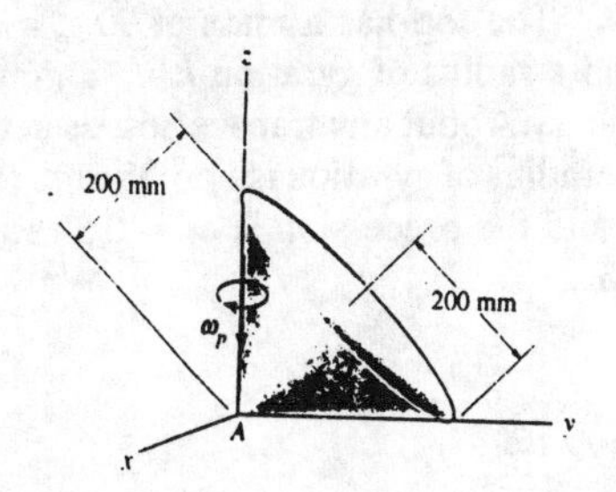

For the axis of symmetry, the speed of point P is

$$v_P = (0.2)\,\omega \sin 45° = 0.1414\omega$$

Also,

$$v_P = 0.2\omega_p \cos 45° = 0.1414\omega_p$$

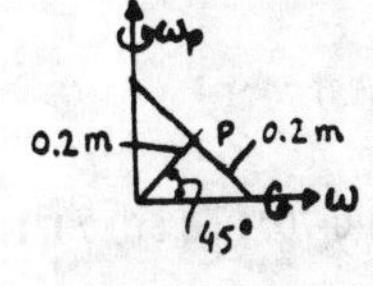

Thus, from these equations, $\omega = \omega_p$

Resolve ω into components along the principal axes of inertia.

$$\omega_y = \omega\cos 45°, \qquad \omega_z = \omega \sin 45°$$

Hence,

$$\omega_y = 0.7071\omega_p, \qquad \omega_z = 0.7071\omega_p$$

The angular momentum has components

$$H_x = 0$$

$$H_y = I_y\,\omega_y = \left[\frac{3}{80}(7)\left(4(0.2)^2 + (0.2)^2\right) + 7\left(\frac{3}{4}(0.2)\right)^2\right](0.7071\omega_p) = 0.1485\omega_p$$

$$H_z = I_z\,\omega_z = \left[\frac{3}{10}(7)(0.2)^2\right](0.7071\omega_p) = 0.05940\omega_p$$

H forms an angle ϕ with the z axis, where

$$\tan\phi = \frac{H_y}{H_z} = \frac{0.1485\omega_p}{0.05940\omega_p} \qquad \phi = 68.2°$$

so that

$$\gamma = 90° + 45° - 68.2° = 66.8°$$

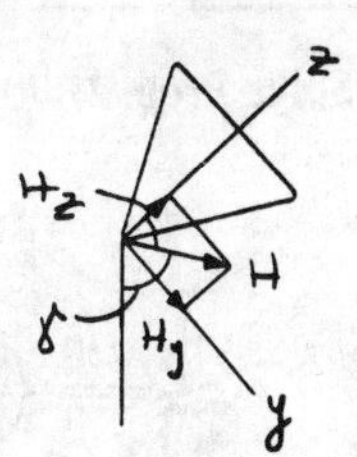

The magnitude of **H** is

$$H = \sqrt{(0.1485\omega_p)^2 + (0.05940\omega_p)^2} = 0.1599\omega_p$$

During the time dt **H** undergoes a rotation of $\omega_p\,dt$ about the vertical axis. Hence, in the x-direction,

$$\left|\frac{d\mathbf{H}}{dt}\right| = |\dot{\mathbf{H}}| = \omega_p H \sin\gamma = 0.1599\omega_p^2 \sin 66.8° = 0.147\omega_p^2$$

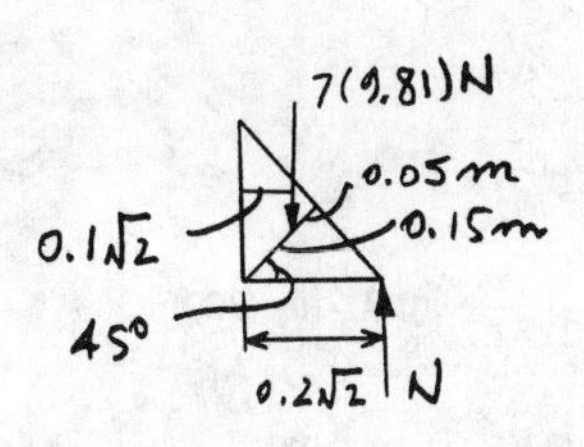

Since this equals ΣM_x, then

$$N(0.2\sqrt{2}) - (7)(9.81)(0.15\cos 45°) = 0.147\omega_p^2$$

When A lifts off the surface, $N = (7)(9.81)$. Thus,

$\omega_p = 9.09$ rad/s **Ans**

21-69. The top has a mass of 90 g, a center of mass at G, and a radius of gyration $k = 18$ mm about its axis of symmetry. About any transverse axis acting through point O the radius of gyration is $k_t = 35$ mm. If the top is pinned at O and the precession is $\omega_p = 0.5$ rad/s, determine the spin ω_s.

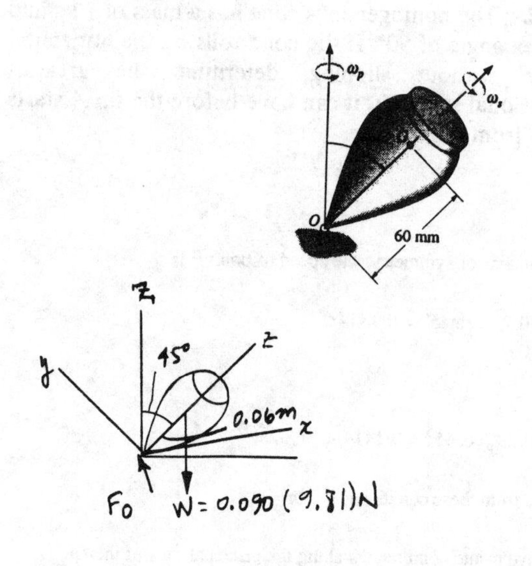

$\omega_p = 0.5$ rad/s

$$\Sigma M_x = -I\dot{\phi}^2\sin\theta\cos\theta + I_z\dot{\phi}\sin\theta\left(\dot{\phi}\cos\theta + \dot{\psi}\right)$$

$$0.090(9.81)(0.06)\sin 45° = -0.090(0.035)^2(0.5)^2(0.7071)^2$$
$$+0.090(0.018)^2(0.5)(0.7071)\left[0.5(0.7071) + \dot{\psi}\right]$$

$\omega_s = \dot{\psi} = 3.63\left(10^3\right)$ rad/s **Ans**

21-70. The top consists of a thin disk that has a weight of 8 lb and a radius of 0.3 ft. The rod has a negligible mass and a length of 0.5 ft. If the top is spinning with an angular velocity $\omega_s = 300$ rad/s, determine the steady-state precessional angular velocity ω_p of the rod when $\theta = 40°$.

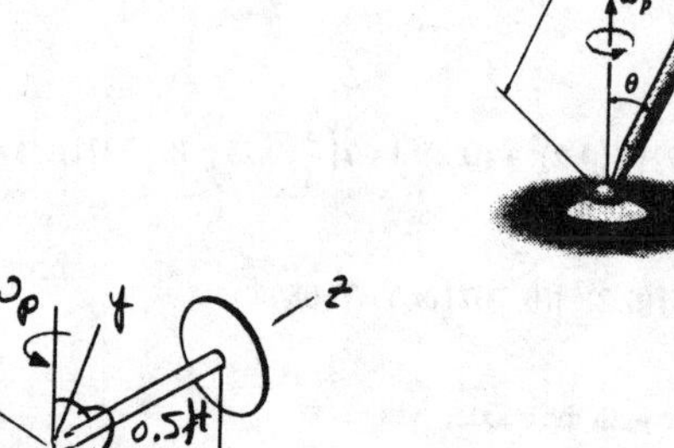

$$\Sigma M_x = -I\dot{\phi}^2\sin\theta\cos\theta + I_z\dot{\phi}\sin\theta\left(\dot{\phi}\cos\theta + \dot{\psi}\right)$$

$$8(0.5\sin 40°) = -\left[\frac{1}{4}\left(\frac{8}{32.2}\right)(0.3)^2 + \left(\frac{8}{32.2}\right)(0.5)^2\right]\omega_p^2\sin 40°\cos 40°$$
$$+\left[\frac{1}{2}\left(\frac{8}{32.2}\right)(0.3)^2\right]\omega_p\sin 40°\,(\omega_p\cos 40° + 300)$$

$$0.02783\omega_p^2 - 2.1559\omega_p + 2.571 = 0$$

$\omega_p = 1.21$ rad/s **Ans** (Low precession)

$\omega_p = 76.3$ rad/s **Ans** (High precession)

21-71. Solve Prob. 21-70 when $\theta = 90°$.

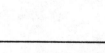

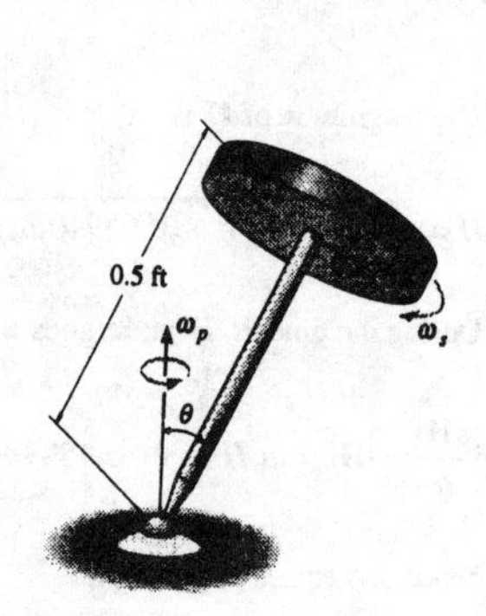

$$\Sigma M_x = I_z\Omega_y\omega_z$$

$$8(0.5) = \left[\frac{1}{2}\left(\frac{8}{32.2}\right)(0.3)^2\right]\omega_p\,(300)$$

$\omega_p = 1.19$ rad/s **Ans**

***21-72.** The top has a mass of 3 lb and can be considered as a solid cone. If it is observed to precess about the vertical axis at a constant rate of 5 rad/s, determine its spin.

$$I = \frac{3}{80}\left(\frac{3}{32.2}\right)\left[4\left(\frac{1.5}{12}\right)^2 + \left(\frac{6}{12}\right)^2\right] + \frac{3}{32.2}\left(\frac{4.5}{12}\right)^2 = 0.01419 \text{ slug}\cdot\text{ft}^2$$

$$I_z = \frac{3}{10}\left(\frac{3}{32.2}\right)\left(\frac{1.5}{12}\right)^2 = 0.43672\left(10^{-3}\right) \text{ slug}\cdot\text{ft}^2$$

$$\Sigma M_x = -I\dot{\phi}^2 \sin\theta\cos\theta + I_z\dot{\phi}\sin\theta\left(\dot{\phi}\cos\theta + \dot{\psi}\right)$$

$$(3)\left(\frac{4.5}{12}\right)(\sin 30^\circ) = -(0.01419)(5)^2\sin 30^\circ\cos 30^\circ + 0.43672\left(10^{-3}\right)(5)\sin 30^\circ\left(5\cos 30^\circ + \dot{\psi}\right)$$

$\dot{\psi} = 652$ rad/s **Ans**

21-73. The gyroscope consists of a uniform 450-g disk D which is attached to the axle AB of negligible mass. The supporting frame has a mass of 180 g and a center of mass at G. If the disk is rotating about the axle at $\omega_D = 90$ rad/s, determine the constant angular velocity ω_p at which the frame precesses about the pivot point O. The frame moves in the horizontal plane.

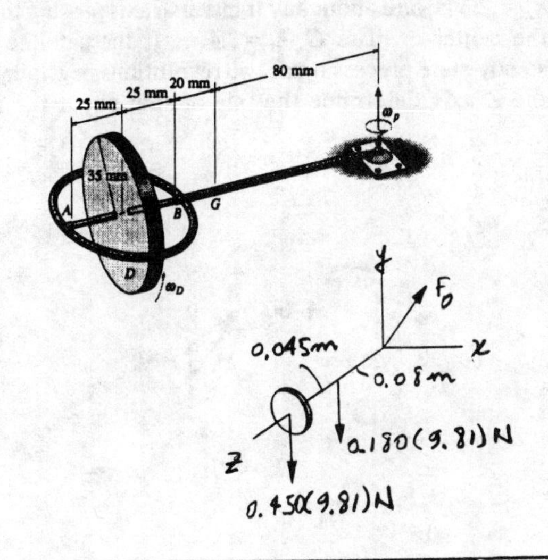

$$\Sigma M_x = I_z \Omega_y \omega_s$$

$$(0.450)(9.81)(0.125) + (0.180)(9.81)(0.080) = \frac{1}{2}(0.450)(0.035)^2\,\omega_p\,(90)$$

$\omega_p = 27.9$ rad/s **Ans**

21-74. The car is traveling at $v_C = 100$ km/h around the horizontal curve having a radius of 80 m. If each wheel has a mass of 16 kg, a radius of gyration $k_G = 300$ mm about its spinning axis, and a radius of 400 mm, determine the difference between the normal forces of the rear wheels, caused by the gyroscopic effect. The distance between the wheels is 1.30 m.

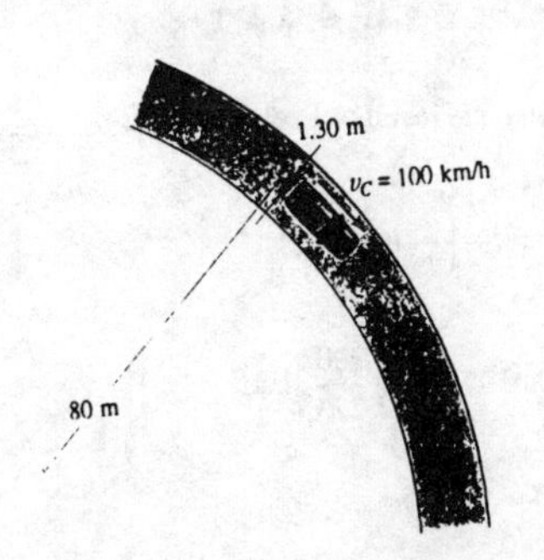

$$I = 2[16(0.3)^2] = 2.88 \text{ kg}\cdot\text{m}^2$$

$$\omega_s = \frac{100(1000)}{3600(0.4)} = 69.44 \text{ rad/s}$$

$$\omega_p = \frac{100(1000)}{80(3600)} = 0.347 \text{ rad/s}$$

$$M = I\,\omega_s\,\omega_p$$

$$\Delta F(1.30) = 2.88(69.44)(0.347)$$

$\Delta F = 53.4$ N **Ans**

21-75. The projectile shown is subjected to torque-free motion. The transverse and axial moments of inertia are I and I_z, respectively. If θ represents the angle between the precessional axis Z and the axis of symmetry z, and β is the angle between the angular velocity $\boldsymbol{\omega}$ and the z axis, show that β and θ are related by the equation $\tan\theta = (I/I_z)\tan\beta$.

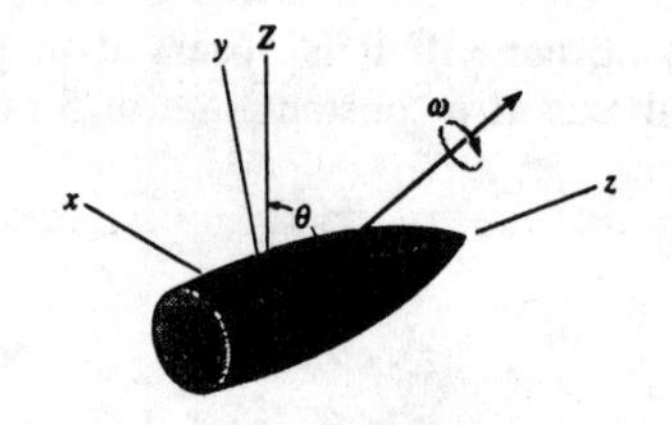

From Eq. 21-34 $\quad \omega_y = \dfrac{H_G \sin\theta}{I}$ and $\omega_z = \dfrac{H_G\cos\theta}{I_z}$ $\quad$ Hence $\dfrac{\omega_y}{\omega_z} = \dfrac{I_z}{I}\tan\theta$

However, $\omega_y = \omega\sin\beta$ and $\omega_z = \omega\cos\beta$

$$\frac{\omega_y}{\omega_z} = \tan\beta = \frac{I_z}{I}\tan\theta$$

$$\tan\theta = \frac{I}{I_z}\tan\beta \qquad \textbf{Q.E.D.}$$

***21-76.** The radius of gyration about an axis passing through the axis of symmetry of the 2.5-Mg satellite is $k_z = 2.3$ m, and about any transverse axis passing through the center of mass G, $k_t = 3.4$ m. If the satellite has a steady-state precession of two revolutions per hour about the Z axis, determine the rate of spin about the z axis.

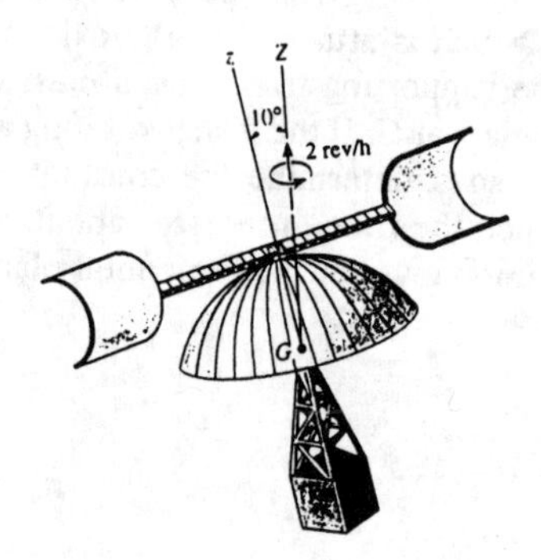

precession axis
10°
ω
ψ̇
2 rev/h
B
4.613°
ψ̇
ω
2 rev/h
5.387°

$$I_z = 2500(2.3)^2 = 13\,225 \text{ kg}\cdot\text{m}^2$$

$$I = 2500(3.4)^2 = 28\,900 \text{ kg}\cdot\text{m}^2$$

Use the result of Prob. 21 – 75.

$$\tan\theta = \left(\frac{I}{I_z}\right)\tan\beta$$

$$\tan 10^\circ = \left(\frac{28\,900}{13\,225}\right)\tan\beta$$

$$\beta = 4.613^\circ$$

From the law of sines,

$$\frac{\sin 5.387^\circ}{\dot\psi} = \frac{\sin 4.613^\circ}{2}$$

$$\dot\psi = 2.33 \text{ rev/h} \qquad \textbf{Ans}$$

21-77. The projectile has a mass of 0.9 kg and axial and transverse radii of gyration of $k_z = 20$ mm and $k_t = 25$ mm, respectively. If it is spinning at $\omega_s = 6$ rad/s when it leaves the barrel of a gun, determine its angular momentum. Precession occurs about the Z axis.

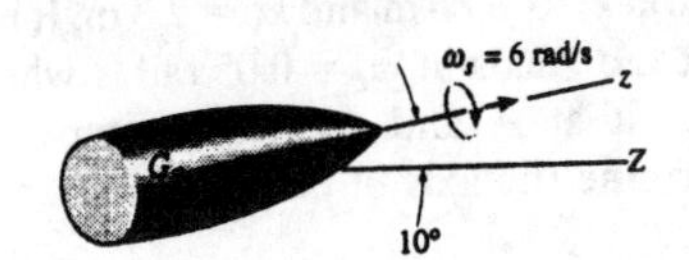

$$I = (0.9)(0.025)^2 = 0.5625\left(10^{-3}\right)\text{kg}\cdot\text{ m}^2$$

$$I_z = (0.9)(0.020)^2 = 0.360\left(10^{-3}\right)\text{kg}\cdot\text{ m}^2$$

$$\dot{\psi} = \frac{I - I_z}{I I_z} H_G \cos\theta$$

$$6 = \frac{0.5625(10^{-3}) - 0.360(10^{-3})}{(0.5625(10^{-3}))(0.360(10^{-3}))}(H_G \cos 10^\circ)$$

$$H_G = 6.09\left(10^{-3}\right)\text{ kg}\cdot\text{ m}^2/\text{s} \qquad \textbf{Ans}$$

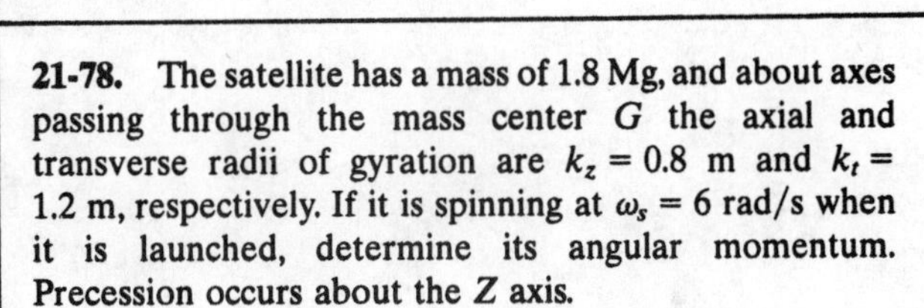

21-78. The satellite has a mass of 1.8 Mg, and about axes passing through the mass center G the axial and transverse radii of gyration are $k_z = 0.8$ m and $k_t = 1.2$ m, respectively. If it is spinning at $\omega_s = 6$ rad/s when it is launched, determine its angular momentum. Precession occurs about the Z axis.

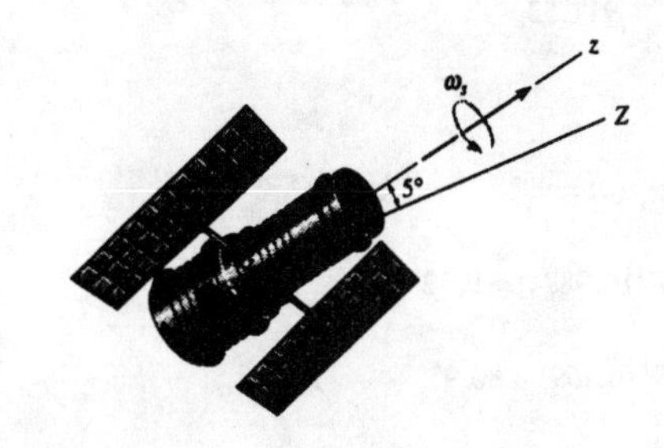

$$I = 1800(1.2)^2 = 2592\text{ kg}\cdot\text{ m}^2 \qquad I_z = 1800(0.8)^2 = 1152\text{ kg}\cdot\text{ m}^2$$

Applying the third of Eqs. 21-36 with $\theta = 5^\circ$ $\dot{\psi} = 6$ rad/s

$$\dot{\psi} = \frac{I - I_z}{I I_z} H_G \cos\theta$$

$$6 = \frac{2592 - 1152}{2592(1152)} H_G \cos 5^\circ$$

$$H_G = 12.5\text{ Mg}\cdot\text{ m}^2/\text{s} \qquad \textbf{Ans}$$

21-79. The radius of gyration about an axis passing through the axis of symmetry of the 1.6-Mg space capsule is $k_z = 1.2$ m, and about any transverse axis passing through the center of mass G, $k_t = 1.8$ m. If the capsule has a known steady-state precession of two revolutions per hour about the Z axis, determine the rate of spin about the z axis.

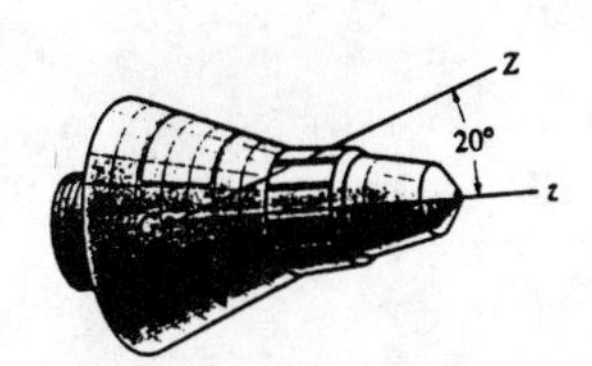

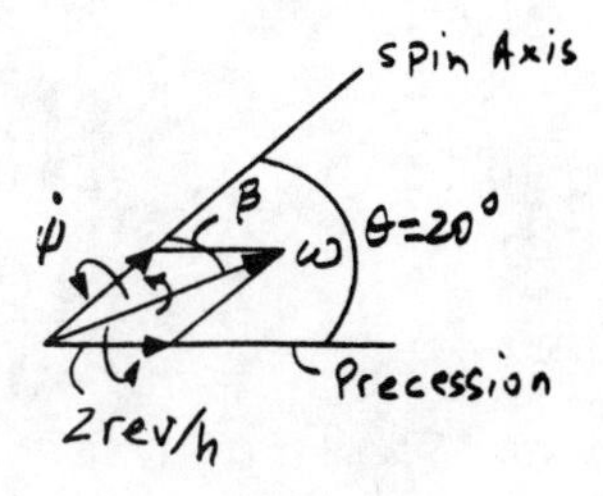

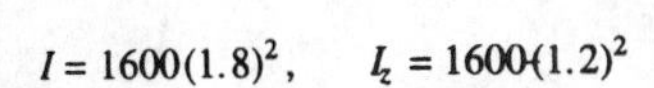

$$I = 1600(1.8)^2, \qquad I_z = 1600(1.2)^2$$

Use the result of Prob. 21 − 75.

$$\tan\theta = \left(\frac{I}{I_z}\right)\tan\beta$$

$$\tan 20^\circ = \left(\frac{1600(1.8)^2}{1600(1.2)^2}\right)\tan\beta$$

$$\beta = 9.189^\circ$$

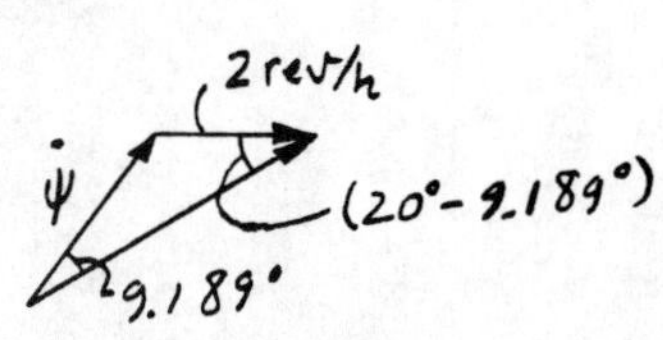

Using the law of sines :

$$\frac{\sin 9.189^\circ}{2} = \frac{\sin(20^\circ - 9.189^\circ)}{\dot{\psi}}$$

$$\dot{\psi} = 2.35\text{ rev/h} \qquad \textbf{Ans}$$

***21-80.** The rocket has a mass of 4 Mg and radii of gyration $k_z = 0.85$ m and $k_y = 2.3$ m. It is initially spinning about the z axis at $\omega_z = 0.05$ rad/s when a meteoroid M strikes it at A and creates an impulse $\mathbf{I} = \{300\mathbf{i}\}$ N·s. Determine the axis of precession after the impact.

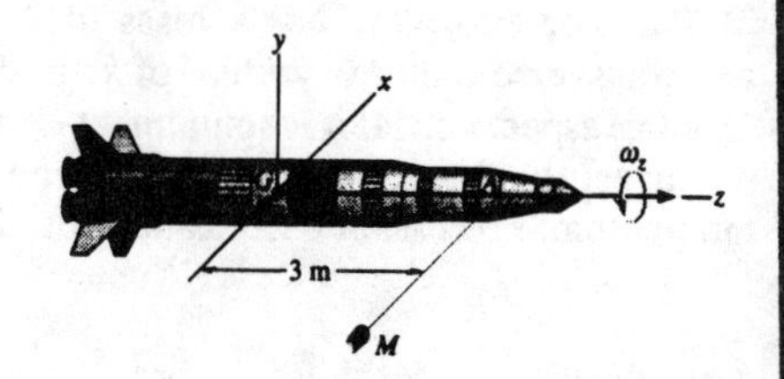

The impulse creates an angular momentum about the y axis of

$$H_y = 300(3) = 900 \text{ kg} \cdot \text{m}^2/\text{s}$$

Since

$$\omega_z = 0.05 \text{ rad/s},$$

then

$$\mathbf{H}_G = 900\mathbf{j} + \left[4000(0.85)^2\right](0.05)\mathbf{k} = 900\mathbf{j} + 144.5\mathbf{k}$$

The axis of precession is defined by $\mathbf{H}_G$.

$$\mathbf{u}_{H_G} = \frac{900\mathbf{j} + 144.5\mathbf{k}}{911.53} = 0.9874\mathbf{j} + 0.159\mathbf{k}$$

Thus,

$$\alpha = \cos^{-1}(0) = 90^\circ \qquad \textbf{Ans}$$

$$\beta = \cos^{-1}(0.9874) = 9.12^\circ \qquad \textbf{Ans}$$

$$\gamma = \cos^{-1}(0.159) = 80.9^\circ \qquad \textbf{Ans}$$

22-1. A spring is stretched 175 mm by an 8-kg block. If the block is displaced 100 mm downward from its equilibrium position and given a downward velocity of 1.50 m/s, determine the differential equation which describes the motion. Assume that positive displacement is downward. Also, determine the position of the block when $t = 0.22$ s.

$+\downarrow \Sigma F_y = ma_y;\quad mg - k(y + y_{st}) = m\ddot{y}\quad$ where $k\,y_{st} = mg$

$$\ddot{y} + \frac{k}{m}y = 0$$

Hence $\quad p = \sqrt{\frac{k}{m}}\quad$ Where $k = \frac{8(9.81)}{0.175} = 448.46$ N/m

$$= \sqrt{\frac{448.46}{8}} = 7.487$$

$\therefore\quad \ddot{y} + (7.487)^2 y = 0\qquad \ddot{y} + 56.1y = 0\qquad$ **Ans**

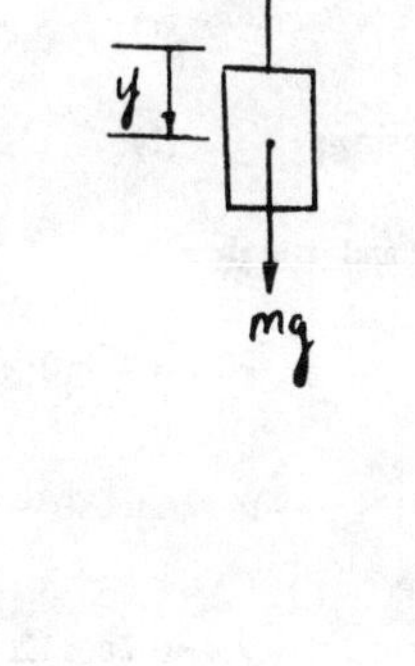

The solution of the above differential equation is of the form :

$y = A\sin pt + B\cos pt\qquad (1)$

$v = \dot{y} = Ap\cos pt - Bp\sin pt\qquad (2)$

At $t = 0$, $y = 0.1$ m and $v = v_0 = 1.50$ m/s

From Eq.(1) $\quad 0.1 = A\sin 0 + B\cos 0\qquad B = 0.1$ m

From Eq.(2) $\quad v_0 = Ap\cos 0 - 0\qquad A = \frac{v_0}{p} = \frac{1.50}{7.487} = 0.2003$ m

Hence $\quad y = 0.2003\sin 7.487t + 0.1\cos 7.487t$

At $t = 0.22$ s, $\quad y = 0.2003\sin[7.487(0.22)] + 0.1\cos[7.487(0.22)]$

$= 0.192$ m $\qquad$ **Ans**

22-2. When a 2-kg block is suspended from a spring, the spring is stretched a distance of 40 mm. Determine the natural frequency and the period of vibration for a 0.5-kg block attached to the same spring.

$$k = \frac{F}{x} = \frac{2(9.81)}{0.040} = 490.5 \text{ N/m}$$

$$p = \sqrt{\frac{k}{m}} = \sqrt{\frac{490.5}{0.5}} = 31.321$$

$f = \frac{p}{2\pi} = \frac{31.321}{2\pi} = 4.985$ Hz $\qquad$ **Ans**

$\tau = \frac{1}{f} = \frac{1}{4.985} = 0.201$ s $\qquad$ **Ans**

22-3. A block having a weight of 8 lb is suspended from a spring having a stiffness $k = 40$ lb/ft. If the block is pushed $y = 0.2$ ft upward from its equilibrium position and then released from rest, determine the equation which describes the motion. What are the amplitude and the natural frequency of the vibration? Assume that positive displacement is downward.

$+\downarrow \Sigma F_y = ma_y;\quad mg - k(y + y_{st}) = m\ddot{y}\quad$ where $k\,y_{st} = mg$

$$\ddot{y} + \frac{k}{m}y = 0$$

Hence $\quad p = \sqrt{\frac{k}{m}} = \sqrt{\frac{40}{8/32.2}} = 12.689$

$f = \frac{p}{2\pi} = \frac{12.689}{2\pi} = 2.02\text{ Hz}\quad$ **Ans**

$T = k(y + y_{st})$, y, mg

The solution of the above differential equation is of the form :

$y = A\sin pt + B\cos pt \quad (1)$

$v = \dot{y} = Ap\cos pt - Bp\sin pt \quad (2)$

At $t = 0,\ y = -0.2$ ft and $v = v_0 = 0$

From Eq.(1) $\quad -0.2 = A\sin 0° + B\cos 0° \quad B = -0.2$ ft

From Eq.(2) $\quad v_0 = Ap\cos 0° - 0 \quad A = \frac{v_0}{p} = \frac{0}{12.689} = 0$

Hence $\quad y = -0.2\cos 12.7t \quad$ **Ans**

Amplitude $\quad C = 0.2$ ft $\quad$ **Ans**

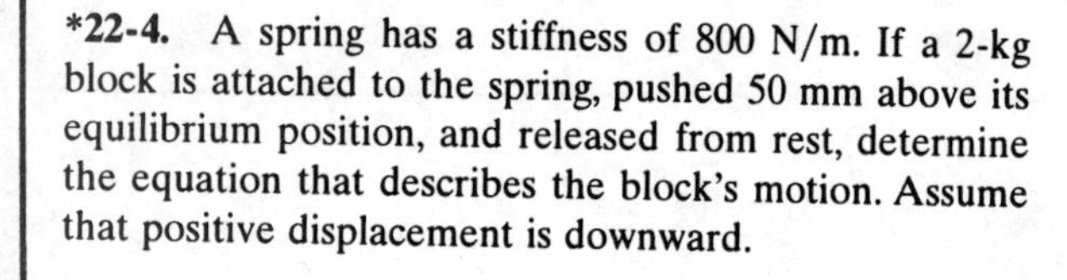

***22-4.** A spring has a stiffness of 800 N/m. If a 2-kg block is attached to the spring, pushed 50 mm above its equilibrium position, and released from rest, determine the equation that describes the block's motion. Assume that positive displacement is downward.

$p = \sqrt{\frac{k}{m}} = \sqrt{\frac{800}{2}} = 20$

$x = A\sin pt + B\cos pt$

$x = -0.05$ m when $t = 0$,

$-0.05 = 0 + B;\quad B = -0.05$

$v = Ap\cos pt - Bp\sin pt$

$v = 0$ when $t = 0$,

$0 = A(20) - 0;\quad A = 0$

Thus,

$x = -0.05\cos(20t)\quad$ **Ans**

22-5. A 2-kg block is suspended from a spring having a stiffness of 800 N/m. If the block is given an upward velocity of 2 m/s when it is displaced downward a distance of 150 mm from its equilibrium position, determine the equation which describes the motion. What is the amplitude of the motion? Assume that positive displacement is downward.

$$p = \sqrt{\frac{k}{m}} = \sqrt{\frac{800}{2}} = 20$$

$$x = A\sin pt + B\cos pt$$

$x = 0.150$ m when $t = 0$,

$$0.150 = 0 + B; \quad B = 0.150$$

$$v = Ap\cos pt - Bp\sin pt$$

$v = 2$ m/s when $t = 0$,

$$2 = A(20) - 0; \quad A = 0.1$$

Thus,

$$x = 0.1\sin(20t) + 0.150\cos(20t) \quad \textbf{Ans}$$

$$C = \sqrt{A^2 + B^2} = \sqrt{(0.1)^2 + (0.150)^2} = 0.180\text{ m} \quad \textbf{Ans}$$

22-6. A spring is stretched 200 mm by a 15-kg block. If the block is displaced 100 mm downward from its equilibrium position and given a downward velocity of 0.75 m/s, determine the equation which describes the motion. What is the phase angle? Assume that positive displacement is downward.

$$k = \frac{F}{x} = \frac{15(9.81)}{0.2} = 735.75\text{ N/m}$$

$$p = \sqrt{\frac{k}{m}} = \sqrt{\frac{735.75}{15}} = 7.00$$

$$x = A\sin pt + B\cos pt$$

$x = 0.1$ m when $t = 0$,

$$0.1 = 0 + B; \quad B = 0.1$$

$$v = Ap\cos pt - Bp\sin pt$$

$v = 0.75$ m/s when $t = 0$,

$$0.75 = A(7.00)$$

$$A = 0.107$$

$$x = 0.107\sin(7.00t) + 0.100\cos(7.00t) \quad \textbf{Ans}$$

$$\phi = \tan^{-1}\left(\frac{B}{A}\right) = \tan^{-1}\left(\frac{0.100}{0.107}\right) = 43.0^\circ \quad \textbf{Ans}$$

22-7. A 6-kg block is suspended from a spring having a stiffness of $k = 200$ N/m. If the block is given an upward velocity of 0.4 m/s when it is 75 mm above its equilibrium position, determine the equation which describes the motion and the maximum upward displacement of the block measured from the equilibrium position. Assume that positive displacement is downward.

$$p = \sqrt{\frac{k}{m}} = \sqrt{\frac{200}{6}} = 5.774$$

$$x = A\sin pt + B\cos pt$$

$x = -0.075$ m when $t = 0$,

$$-0.075 = 0 + B; \qquad B = -0.075$$

$$v = Ap\cos pt - Bp\sin pt$$

$v = -0.4$ m/s when $t = 0$,

$$-0.4 = A(5.774) - 0; \qquad A = -0.0693$$

Thus,

$$x = -0.0693\sin(5.77t) - 0.075\cos(5.77t) \qquad \textbf{Ans}$$

$$C = \sqrt{A^2 + B^2} = \sqrt{(-0.0693)^2 + (-0.075)^2} = 0.102 \text{ m} \qquad \textbf{Ans}$$

***22-8.** A 3-kg block is suspended from a spring having a stiffness of $k = 200$ N/m. If the block is pushed 50 mm upward from its equilibrium position and then released from rest, determine the equation that describes the motion. What are the amplitude and the natural frequency of the vibration? Assume that positive displacement is downward.

$$p = \sqrt{\frac{k}{m}} = \sqrt{\frac{200}{3}} = 8.165$$

$$f = \frac{p}{2\pi} = \frac{8.165}{2\pi} = 1.299 = 1.30 \text{ Hz} \qquad \textbf{Ans}$$

$$x = A\sin pt + B\cos pt$$

$x = -0.05$ m when $t = 0$,

$$-0.05 = 0 + B; \qquad B = -0.05$$

$$v = Ap\cos pt - Bp\sin pt$$

$v = 0$ when $t = 0$,

$$0 = A(8.165) - 0; \qquad A = 0$$

Hence,

$$x = -0.05\cos(8.16t) \qquad \textbf{Ans}$$

$$C = \sqrt{A^2 + B^2} = \sqrt{(0)^2 + (-0.05)^2} = 0.05 \text{ m} = 50 \text{ mm} \qquad \textbf{Ans}$$

22-9. A platform, having an unknown mass, is supported by *four* springs, each having the same stiffness k. When nothing is on the platform, the period of vertical vibration is measured as 2.35 s; whereas if a 3-kg block is supported on the platform, the period of vertical vibration is 5.23 s. Determine the mass of a block placed on the (empty) platform which causes the platform to vibrate vertically with a period of 5.62 s. What is the stiffness k of each of the springs?

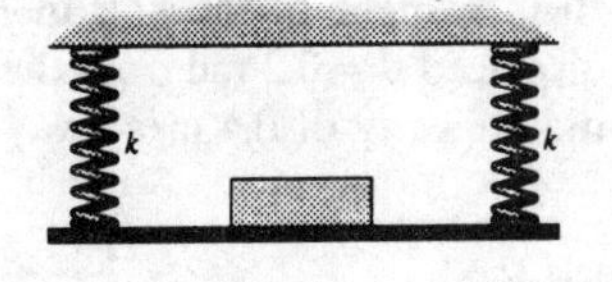

$+\downarrow \Sigma F_y = ma_y; \qquad m_T g - 4k(y + y_{st}) = m_T \ddot{y} \qquad$ Where $4k\, y_{st} = m_T g$

$$\ddot{y} + \frac{4k}{m_T} y = 0$$

Hence $\qquad p = \sqrt{\frac{4k}{m_T}}$

$$\tau = \frac{2\pi}{p} = 2\pi\sqrt{\frac{m_T}{4k}}$$

For empty platform $m_T = m_P$, where m_P is the mass of the platform.

$$2.35 = 2\pi\sqrt{\frac{m_P}{4k}} \qquad (1)$$

When 3-kg block is on the platform $m_T = m_P + 3$.

$$5.23 = 2\pi\sqrt{\frac{m_P + 3}{4k}} \qquad (2)$$

2k(y+yst) $\quad m_T g \quad$ 2k(y+yst) $\quad$ y

When an unknown mass is on the platform $m_T = m_P + m_B$.

$$5.62 = 2\pi\sqrt{\frac{m_P + m_B}{4k}} \qquad (3)$$

Solving Eqs.(1) to (3) yields :

$k = 1.36$ N/m $\qquad m_B = 3.58$ kg $\qquad$ **Ans**

$m_P = 0.7589$ kg

22-10. A pendulum has a 0.4-m-long cord and is given a tangential velocity of 0.2 m/s toward the vertical from a position $\theta = 0.3$ rad. Determine the equation which describes the angular motion.

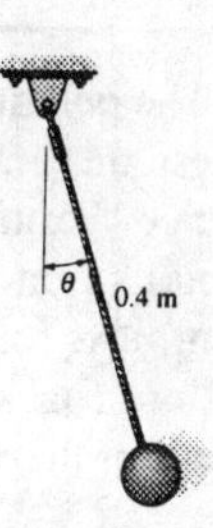

See Example 22 – 1.

$$p = \sqrt{\frac{g}{l}} = \sqrt{\frac{9.81}{0.4}} = 4.95$$

$\theta = A\sin pt + B\cos pt$

$\theta = 0.3$ rad when $t = 0$,

$0.3 = 0 + B; \qquad B = 0.3$

$\dot{\theta} = Ap\cos pt - Bp\sin pt$

Since $s = \theta l$, $v = \dot{\theta} l$. Hence,

$-0.2 = \dot{\theta}(0.4)$, $\dot{\theta} = -0.5$ when $t = 0$,

$-0.5 = A(4.95); \qquad A = -0.101$

Thus,

$\theta = -0.101\sin(4.95t) + 0.3\cos(4.95t) \qquad$ **Ans**

22-11. Determine to the nearest degree the maximum angular displacement of the bob in Prob. 22-10 if it is initially displaced $\theta = 0.2$ rad from the vertical and given a tangential velocity of 0.4 m/s away from the vertical.

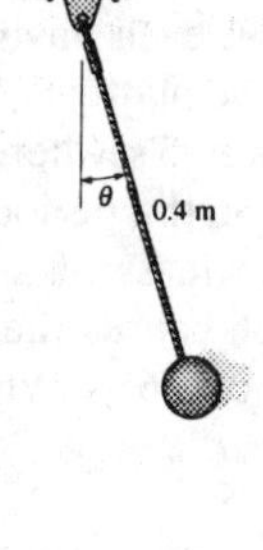

See Example 22 – 1.

$$p = \sqrt{\frac{g}{l}} = \sqrt{\frac{9.81}{0.4}} = 4.95$$

$$\theta = A\sin pt + B\cos pt$$

$\theta = 0.2$ rad when $t = 0$,

$$0.2 = 0 + B; \quad B = 0.2$$

$$\dot{\theta} = Ap\cos pt - Bp\sin pt$$

Since $s = \theta l$, $v = \dot{\theta} l$, then

$0.4 = \dot{\theta}(0.4)$; $\quad \dot{\theta} = 1$ rad/s when $t = 0$,

$$1 = A(4.95) - 0; \quad A = 0.202$$

Then,

$$\theta = 0.202\sin(4.95t) + 0.2\cos(4.95t)$$

$$C = \sqrt{A^2 + B^2} = \sqrt{(0.202)^2 + (0.2)^2} = 0.284 \text{ rad}$$

$$C = \left(\frac{180°}{\pi}\right)(0.284) = 16°$$ **Ans**

***22-12.** The pendulum of the Charpy impact machine is used to test the energy absorption of materials during impact. (See Example 19-5.) The pendulum weighs 46.7 lb, and its center of gravity is located at G. Through experiment, it is found that the pendulum will undergo 25 oscillations (back and forth) in 17 seconds. Determine the moment of inertia with respect to the pin at O.

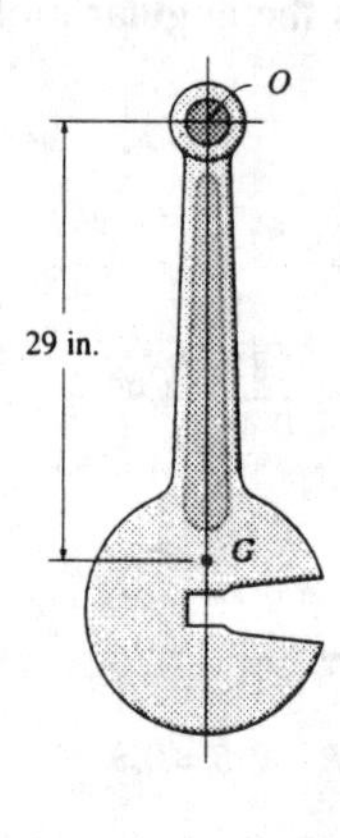

For small θ:

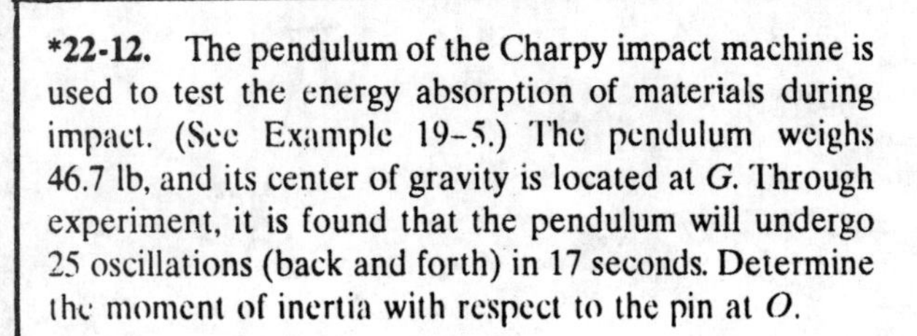

$$\curvearrowleft + \Sigma M_O = I_O\alpha; \quad -46.7\left(\frac{29}{12}\right)\theta = I_O\ddot{\theta}$$

$$\ddot{\theta} + \frac{112.858}{I_O}\theta = 0$$

$$p = \sqrt{\frac{112.858}{I_O}} = \frac{2\pi}{\tau}$$

$$\tau = \frac{17}{25} = 0.680 \text{ s}$$

Thus,

$$\left(\frac{2\pi}{0.680}\right)^2 = \frac{112.858}{I_O}$$

$$I_O = 1.32 \text{ slug} \cdot \text{ft}^2$$ **Ans**

22-13. The body of arbitrary shape has a mass m, mass center at G, and a radius of gyration about G of k_G. If it is displaced a slight amount θ from its equilibrium position and released, determine the natural period of vibration.

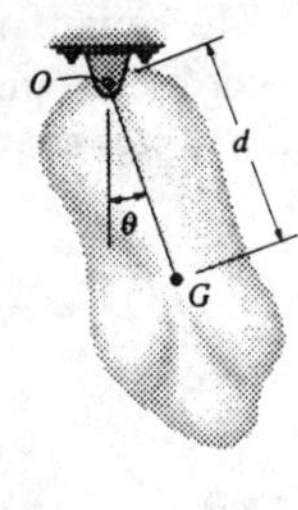

$$\circlearrowleft + \Sigma M_O = I_O \alpha; \qquad -mgd\sin\theta = \left[mk_G^2 + md^2\right]\ddot{\theta}$$

$$\ddot{\theta} + \frac{gd}{k_G^2 + d^2}\sin\theta = 0$$

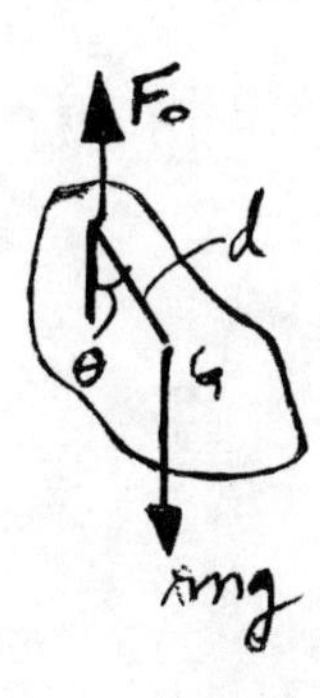

However, for small rotation $\sin\theta \approx \theta$. Hence

$$\ddot{\theta} + \frac{gd}{k_G^2 + d^2}\theta = 0$$

From the above differential equation, $p = \sqrt{\dfrac{gd}{k_G^2 + d^2}}$.

$$\tau = \frac{2\pi}{p} = \frac{2\pi}{\sqrt{\frac{gd}{k_G^2+d^2}}} = 2\pi\sqrt{\frac{k_G^2 + d^2}{gd}} \qquad \textbf{Ans}$$

22-14. The thin hoop of mass m is supported by a knife-edge. Determine the natural period of vibration for small amplitudes of swing.

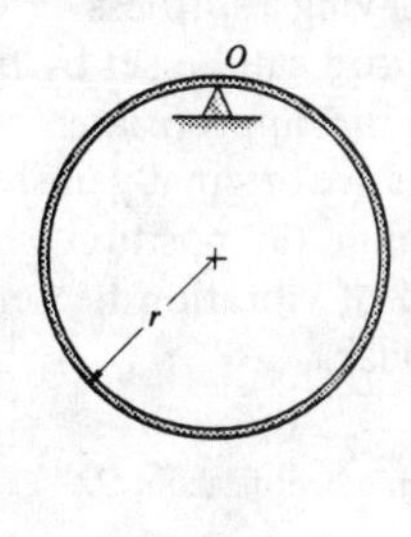

$$I_O = mr^2 + mr^2 = 2mr^2$$

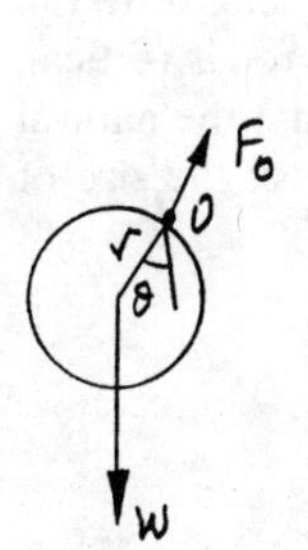

$$\circlearrowleft + \Sigma M_O = I_O \alpha; \qquad -mgr\theta = \left(2mr^2\right)\ddot{\theta}$$

$$\ddot{\theta} + \left(\frac{g}{2r}\right)\theta = 0$$

$$\tau = \frac{2\pi}{p} = 2\pi\sqrt{\frac{2r}{g}} \qquad \textbf{Ans}$$

22-15. The circular disk has a mass m and is pinned at O. Determine the natural period of vibration if it is displaced a small amount and released.

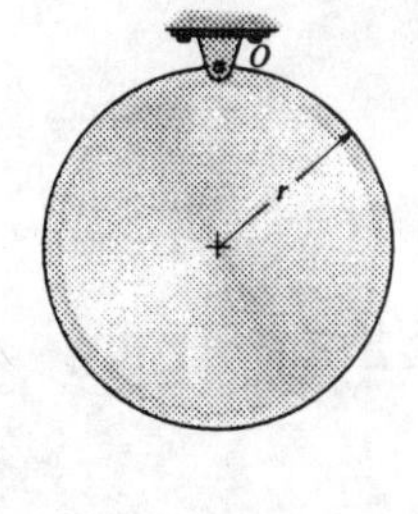

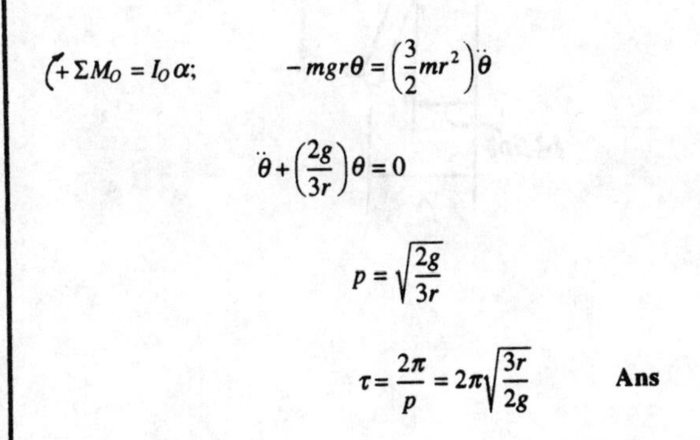

$$\circlearrowleft + \Sigma M_O = I_O \alpha; \qquad -mgr\theta = \left(\frac{3}{2}mr^2\right)\ddot{\theta}$$

$$\ddot{\theta} + \left(\frac{2g}{3r}\right)\theta = 0$$

$$p = \sqrt{\frac{2g}{3r}}$$

$$\tau = \frac{2\pi}{p} = 2\pi\sqrt{\frac{3r}{2g}} \qquad \textbf{Ans}$$

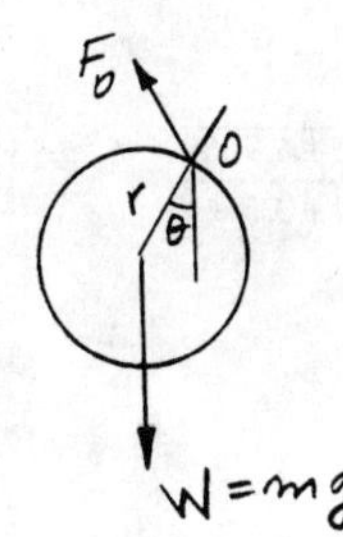

***22-16.** The square plate has a mass m and is suspended at its corner by the pin O. Determine the natural period of vibration if it is displaced a small amount and released.

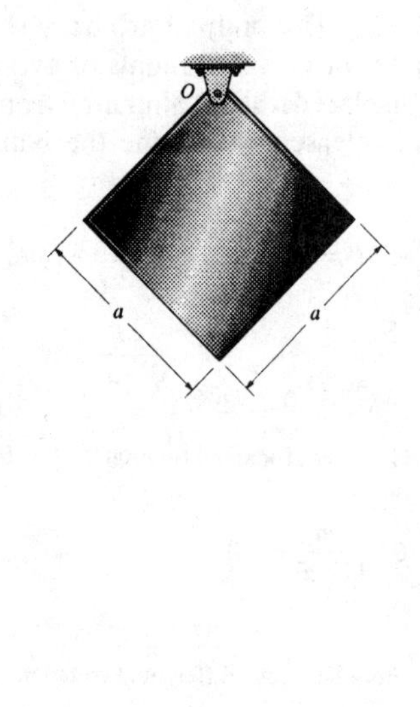

$$I_O = \frac{1}{12}m\left(a^2+a^2\right)+m\left(\frac{\sqrt{2}}{2}a\right)^2 = \frac{1}{6}ma^2+\frac{1}{2}ma^2 = \frac{2}{3}ma^2$$

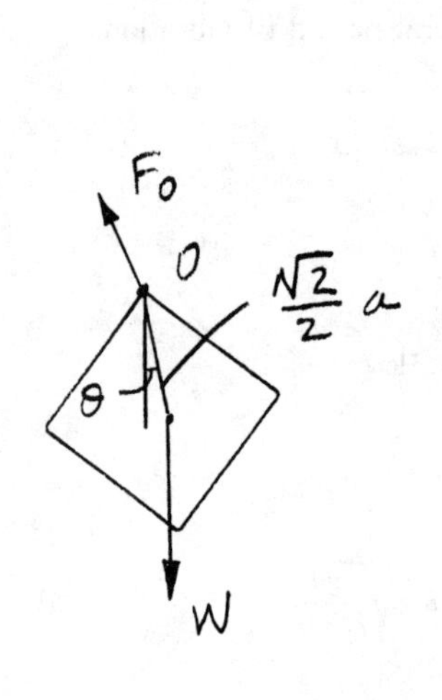

$\curvearrowleft + \Sigma M_O = I_O\alpha;$ $$-mg\left(\frac{\sqrt{2}}{2}a\right)\theta = \left(\frac{2}{3}ma^2\right)\ddot{\theta}$$

$$\ddot{\theta}+\left(\frac{3\sqrt{2}g}{4a}\right)\theta = 0$$

$$p = \sqrt{\frac{3\sqrt{2}g}{4a}}$$

$$\tau = \frac{2\pi}{p} = 6.10\sqrt{\frac{a}{g}} \qquad \textbf{Ans}$$

22-17. The slender rod has a mass of 0.2 kg and is supported at O by a pin and at its end A by two springs, each having a stiffness $k = 4$ N/m. The period of vibration of the rod can be set by fixing the 0.5-kg collar C to the rod at an appropriate location along its length. If the springs are originally unstretched when the rod is vertical, determine the position y of the collar so that the natural period of vibration becomes $\tau = 1$ s. Neglect the size of the collar.

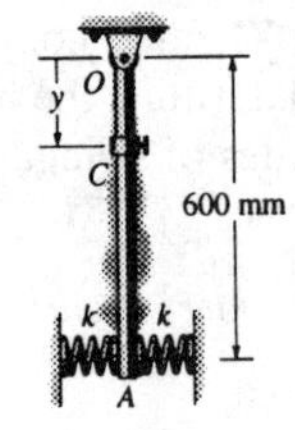

Moment of inertia about O:

$$I_O = \frac{1}{3}(0.2)(0.6)^2+0.5y^2 = 0.024+0.5y^2$$

Each spring force $F_s = kx = 4x$.

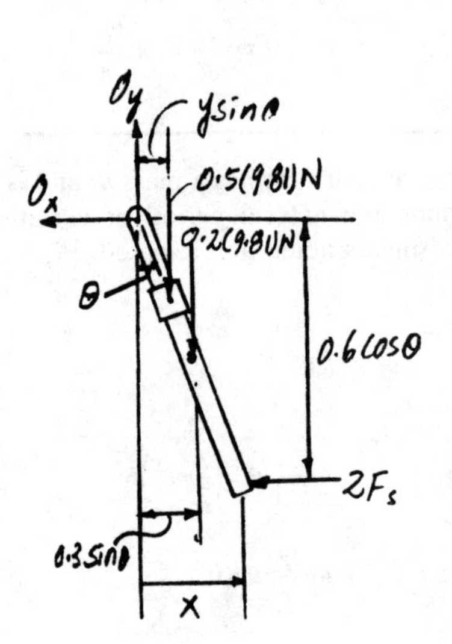

$\curvearrowleft + \Sigma M_O = I_O\alpha;$ $$-2(4x)(0.6\cos\theta)-0.2(9.81)(0.3\sin\theta)$$

$$-0.5(9.81)(y\sin\theta) = (0.024+0.5y^2)\ddot{\theta}$$

$$-4.8x\cos\theta-(0.5886+4.905y)\sin\theta = (0.024+0.5y^2)\ddot{\theta}$$

However, for small displacement $x = 0.6\theta$, $\sin\theta \approx \theta$ and $\cos\theta \approx 1$. Hence

$$\ddot{\theta}+\frac{3.4686+4.905y}{0.024+0.5y^2}\theta = 0$$

From the above differential equation, $p = \sqrt{\dfrac{3.4686+4.905y}{0.024+0.5y^2}}$.

$$\tau = \frac{2\pi}{p}$$

$$1 = \frac{2\pi}{\sqrt{\dfrac{3.4686+4.905y}{0.024+0.5y^2}}}$$

$$19.74y^2-4.905y-2.5211 = 0$$

$$y = 0.503 \text{ m} = 503 \text{ mm} \qquad \textbf{Ans}$$

22-18. Determine the *torsional stiffness* k of rod AB if the 4-kg thin rectangular plate has a natural period of vibration of $\tau = 0.3$ s as it oscillates around the axis of the rod. *Hint:* The torsional stiffness is defined from $M = k\theta$ and is measured in N·m/rad.

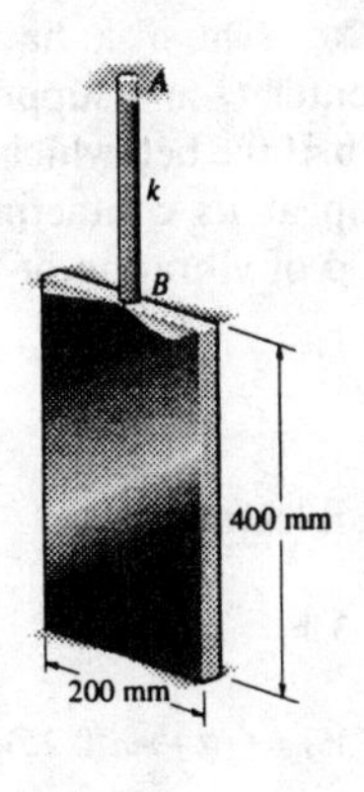

$\Sigma M_z = I_z \alpha; \qquad -k\theta = I_z \ddot{\theta}$

$I_z = \frac{1}{12}mb^2 = \frac{1}{12}(4)(0.2)^2 = 0.01333 \text{ kg} \cdot \text{m}^2$

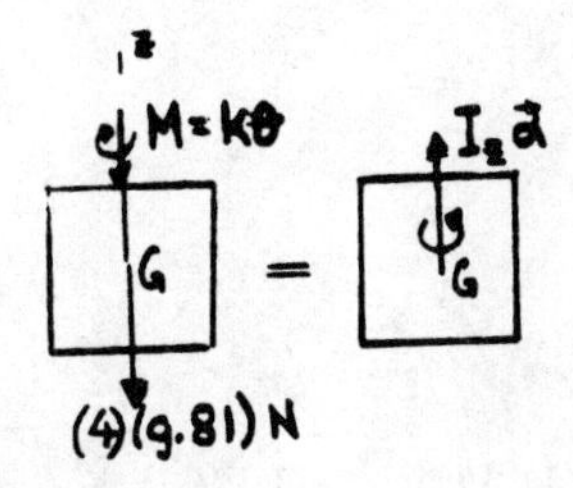

Hence,

$\ddot{\theta} + 75k\theta = 0$

$\tau = \frac{2\pi}{p}; \qquad 0.3 = \frac{2\pi}{\sqrt{75k}}$

$k = 5.85 \text{ N} \cdot \text{m/rad}$ **Ans**

22-19. If the lower end of the 30-kg slender rod is displaced a small amount and released from rest, determine the natural frequency of vibration. Each spring has a stiffness of $k = 500$ N/m and is unstretched when the rod is hanging vertically.

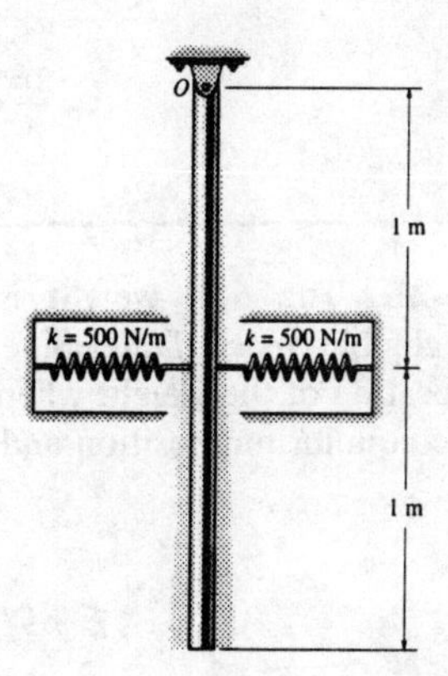

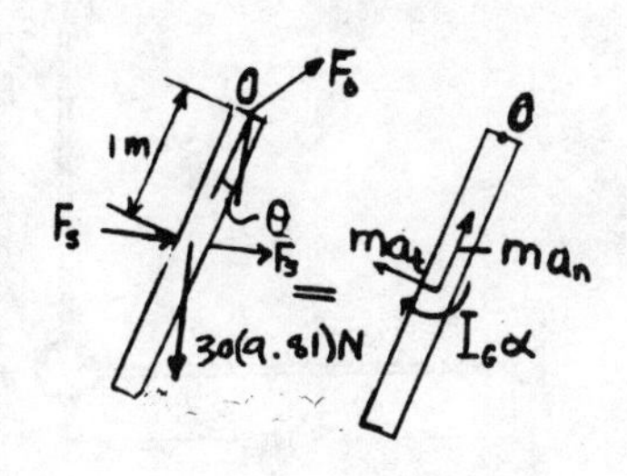

$\circlearrowleft + \Sigma M_O = I_O \alpha; \qquad -30(9.81)(1)(\sin\theta) - 2F_s(1)\cos\theta = \frac{1}{3}(30)(2)^2\ddot{\theta}$

For small θ, $\sin\theta \approx \theta$, $\cos\theta \approx 1$, also

$F_s = kx = (500)\left[(1)(\theta)\right] = 500\theta$

Thus,

$-294.3\theta - 1000\theta = 40\ddot{\theta}$

$\ddot{\theta} + 32.36\theta = 0$

$p = \sqrt{32.36} = 5.688$

$f = \frac{p}{2\pi} = \frac{5.688}{2\pi} = 0.905 \text{ Hz}$ **Ans**

***22-20.** The disk, having a weight of 15 lb, is pinned at its center O and supports the block A that has a weight of 3 lb. If the belt which passes over the disk is not allowed to slip at its contacting surface, determine the natural period of vibration of the system.

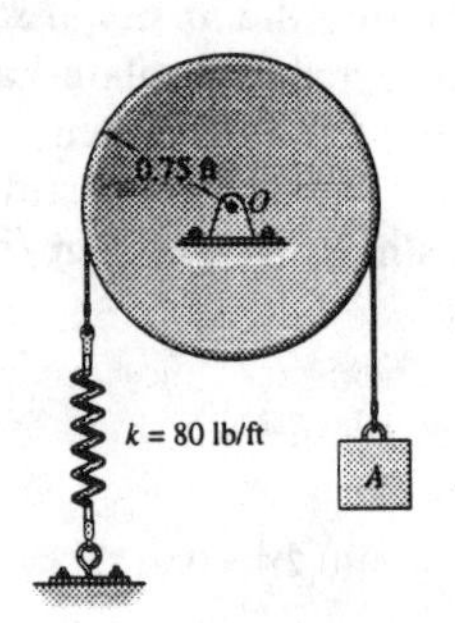

For equilibrium :

$T_{st} = 3$ lb

$\circlearrowleft + \Sigma M_O = I_O\alpha + ma(0.75)$

$a = 0.75\alpha$

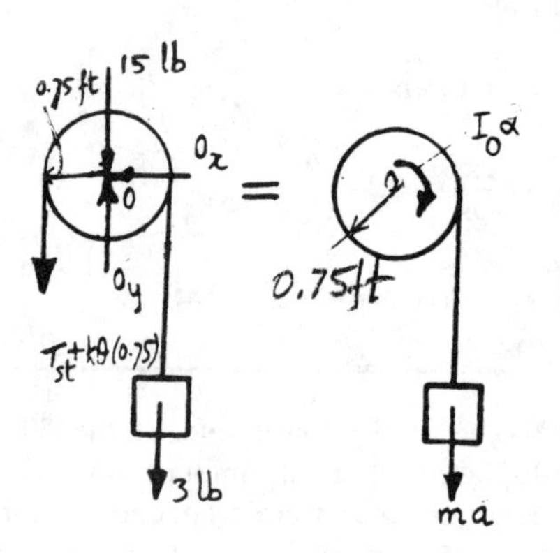

$$-T_{st}(0.75) - (80)(\theta)(0.75)(0.75) + (3)(0.75) = \left[\frac{1}{2}\left(\frac{15}{32.2}\right)(0.75)^2\right]\ddot{\theta} + \left(\frac{3}{32.2}\right)(0.75)\ddot{\theta}(0.75)$$

$$-2.25 - 45\theta + 2.25 = 0.131\ddot{\theta} + 0.05241\ddot{\theta}$$

$$\ddot{\theta} + 245.3\theta = 0$$

$$\tau = \frac{2\pi}{p} = \frac{2\pi}{\sqrt{245.3}} = 0.401 \text{ s} \qquad \textbf{Ans}$$

22-21. The 6-lb weight is attached to the rods of negligible mass. Determine the natural frequency of vibration of the weight when it is displaced slightly from the equilibrium position and released.

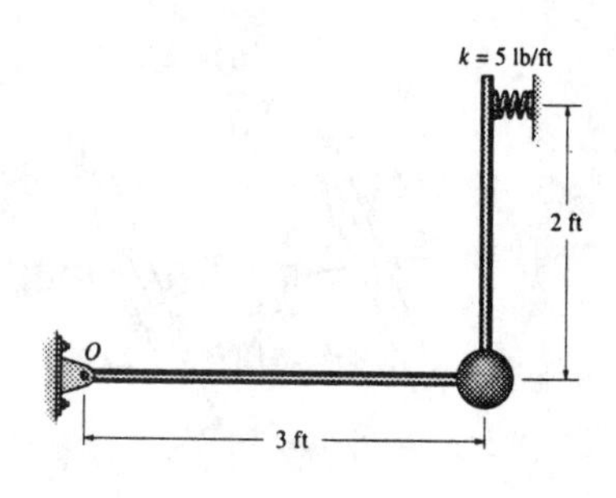

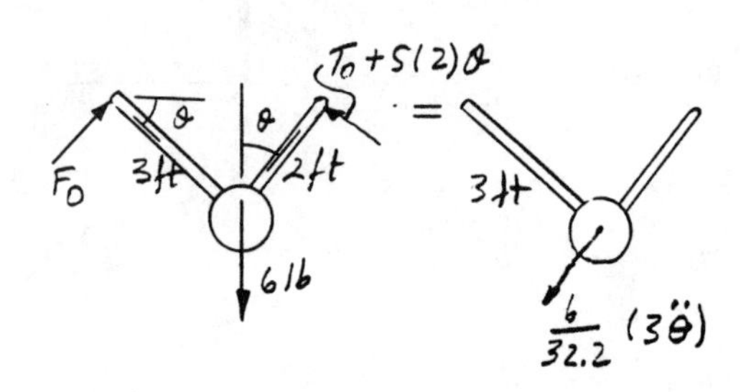

T_O is the equilibrium force.

$$T_O = \frac{6(3)}{2} = 9 \text{ lb}$$

Thus, for small θ,

$$\circlearrowleft + \Sigma M_O = I_O\alpha; \quad 6(3) - [9 + 5(2)\theta](2) = \left(\frac{6}{32.2}\right)(3\ddot{\theta})(3)$$

Thus,

$$\ddot{\theta} + 11.926\theta = 0$$

$$p = \sqrt{11.926} = 3.453 \text{ rad/s}$$

$$f = \frac{p}{2\pi} = \frac{3.453}{2\pi} = 0.550 \text{ Hz} \qquad \textbf{Ans}$$

22-22. The bell has a mass of 375 kg, a center of mass at G, and a radius of gyration about point D of $k_D = 0.4$ m. The tongue consists of a slender rod attached to the inside of the bell at C. If an 8-kg mass is attached to the end of the rod, determine the length l of the rod so that the bell will "ring silent," i.e., so that the natural period of vibration of the tongue is the same as that of the bell. For the calculation, neglect the small distance between C and D and neglect the mass of the rod.

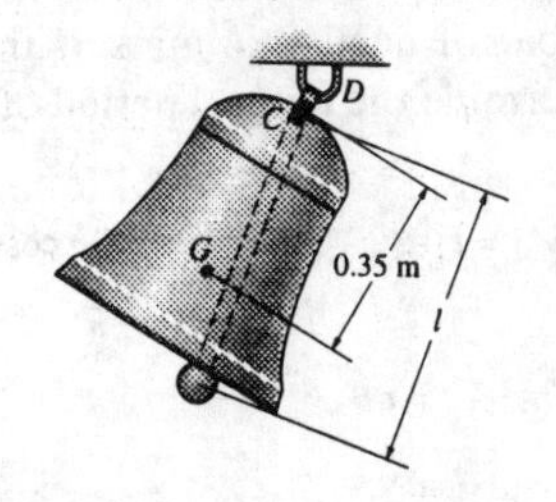

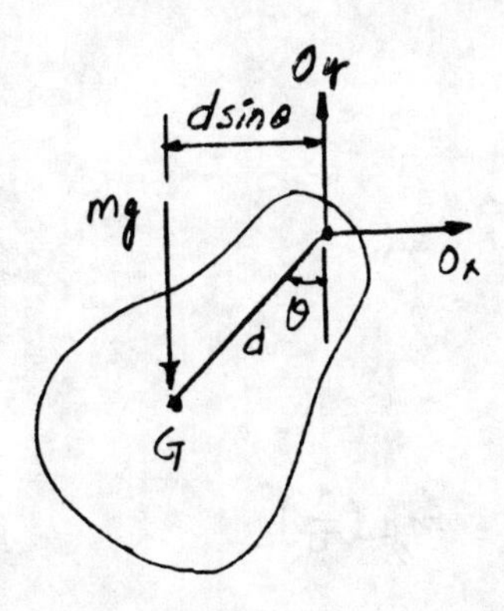

For an arbitrarily shaped body which rotates about a fixed point.

$\circlearrowleft + \Sigma M_O = I_O\alpha;\qquad mgd\sin\theta = -I_O\ddot{\theta}$

$$\ddot{\theta} + \frac{mgd}{I_O}\sin\theta = 0$$

However, for small rotation $\sin\theta \approx \theta$. Hence

$$\ddot{\theta} + \frac{mgd}{I_O}\theta = 0$$

From the above differential equation, $p = \sqrt{\dfrac{mgd}{I_O}}$.

$$\tau = \frac{2\pi}{p} = \frac{2\pi}{\sqrt{\dfrac{mgd}{I_O}}} = 2\pi\sqrt{\frac{I_O}{mgd}}$$

In order to have an equal period

$$\tau = 2\pi\sqrt{\frac{(I_O)_T}{m_T g d_T}} = 2\pi\sqrt{\frac{(I_O)_B}{m_B g d_B}}$$

$(I_O)_T$ = moment of inertia of tongue about O.

$(I_O)_B$ = moment of inertia of bell about O.

$$\frac{(I_O)_T}{m_T g d_T} = \frac{(I_O)_B}{m_B g d_B}$$

$$\frac{8(l^2)}{8gl} = \frac{375(0.4)^2}{375g(0.35)}$$

$$l = 0.457 \text{ m} \qquad \textbf{Ans}$$

22-23. The block has a mass m and is supported by a rigid bar of negligible mass. If the spring has a stiffness k, determine the natural period of vibration for the block.

$$+\Sigma M_O = I_O\alpha; \qquad mga\cos\theta - F_s b\cos\theta = m\left(\ddot{s}\right)a\cos\theta$$

$$s = a\theta, \qquad \ddot{s} = a\ddot{\theta}$$

$$\cos\theta \approx 1$$

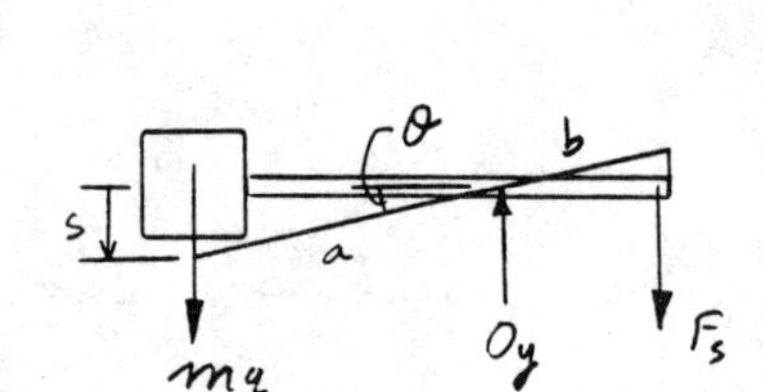

$$F_{eq} = mg\left(\frac{a}{b}\right) = kx_{eq}$$

$$x_{eq} = \frac{mga}{bk}$$

$$F_s = k(x_{eq} + b\theta)$$

Thus,

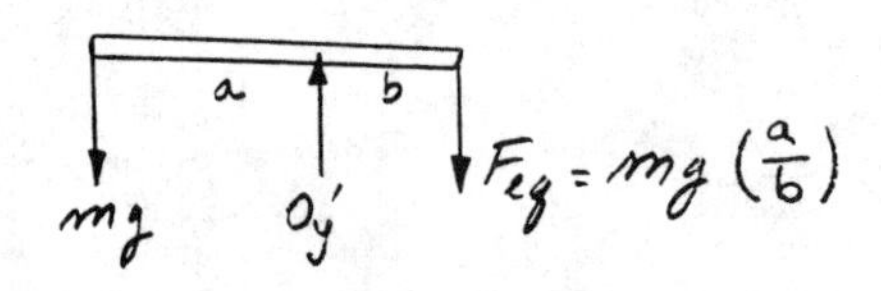

$$mga - k\left(\frac{mga}{bk} + b\theta\right)b = ma^2\ddot{\theta}$$

$$\ddot{\theta} + \frac{kb^2}{ma^2}\theta = 0$$

$$p = \left(\frac{b}{a}\right)\sqrt{\frac{k}{m}}$$

$$\tau = \frac{2\pi}{p} = 2\pi\left(\frac{a}{b}\right)\sqrt{\frac{m}{k}} \qquad \textbf{Ans}$$

***22-24.** The bar has a length l and mass m. It is supported at its ends by rollers of negligible mass. If it is given a small displacement and released, determine the natural frequency of vibration.

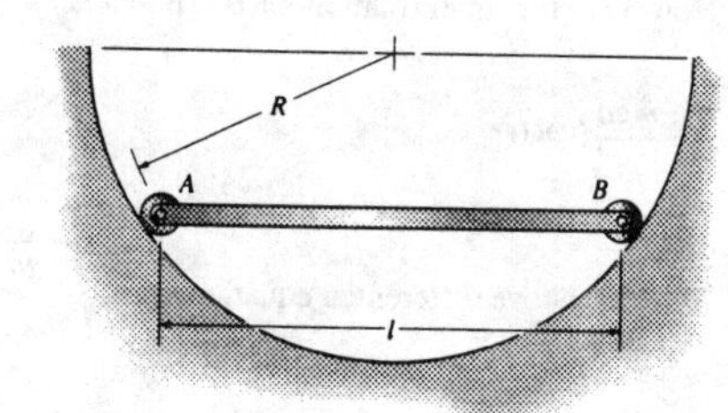

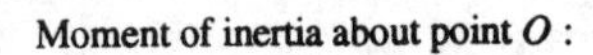

Moment of inertia about point O :

$$I_O = \frac{1}{12}ml^2 + m\left(\sqrt{R^2 - \frac{l^2}{4}}\right)^2 = m\left(R^2 - \frac{1}{6}l^2\right)$$

$$\circlearrowleft + \Sigma M_O = I_O\alpha; \qquad mg\left(\sqrt{R^2 - \frac{l^2}{4}}\right)\theta = -m\left(R^2 - \frac{1}{6}l^2\right)\ddot{\theta}$$

$$\ddot{\theta} + \frac{3g(4R^2 - l^2)^{\frac{1}{2}}}{6R^2 - l^2}\theta = 0$$

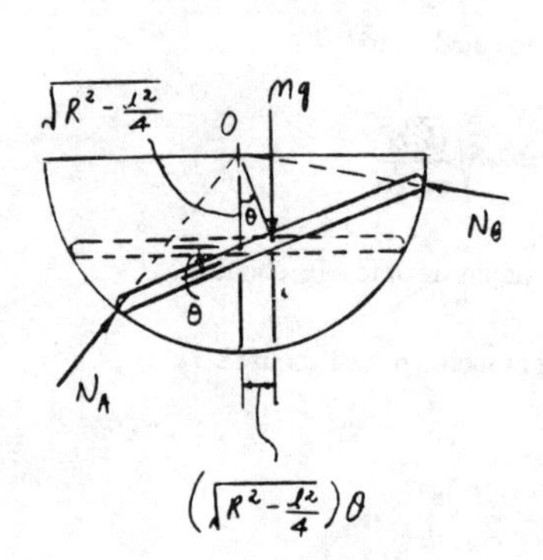

From the above differential equation, $p = \sqrt{\dfrac{3g(4R^2 - l^2)^{\frac{1}{2}}}{6R^2 - l^2}}$.

$$f = \frac{p}{2\pi} = \frac{1}{2\pi}\sqrt{\frac{3g(4R^2 - l^2)^{\frac{1}{2}}}{6R^2 - l^2}} \qquad \textbf{Ans}$$

22-25. Determine the natural frequency for small oscillations of the 10-lb sphere when the rod is displaced a slight distance and released. Neglect the size of the sphere and the mass of the rod. The spring has an unstretched length of 1 ft.

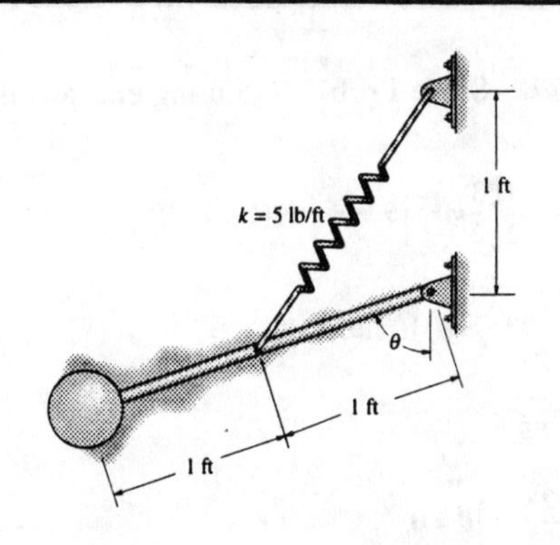

$$l' = \sqrt{l^2 + l^2 - 2l^2\cos(180^\circ - \theta)}$$

$$= l\sqrt{2\left[1-\cos(180^\circ-\theta)\right]} = l\sqrt{2(1+\cos\theta)} = 2l\sqrt{\frac{1+\cos\theta}{2}}$$

$$= 2l\cos\left(\frac{\theta}{2}\right)$$

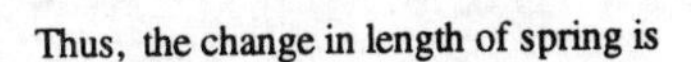

Thus, the change in length of spring is

$$\Delta l = l' - l = l\left(2\cos\left(\frac{\theta}{2}\right) - 1\right)$$

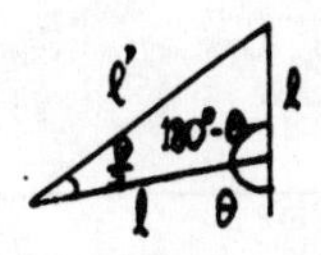

$$\circlearrowleft + \Sigma M_O = I_O\alpha; \qquad mg(2l\sin\theta) - \left(F_s \sin\frac{\theta}{2}\right)l = -m(2l)^2\ddot{\theta} \qquad (1)$$

$$F_s = kl\left(2\cos\left(\frac{\theta}{2}\right) - 1\right)$$

For small angle θ,

$$\sin\frac{\theta}{2} \approx \frac{\theta}{2}, \ \sin\theta \approx \theta, \ \cos\frac{\theta}{2} \approx 1$$

Thus, Eq. (1) reduces to

$$\ddot{\theta} + \frac{2mg - k\left(\frac{l}{2}\right)}{4ml}\theta = 0$$

$$f = \frac{p}{2\pi} = \frac{1}{2\pi}\sqrt{\frac{2mg - k\left(\frac{l}{2}\right)}{4ml}} = \frac{1}{2\pi}\sqrt{\frac{2(10) - 5(0.5)}{4\left(\frac{10}{32.2}\right)(1)}}$$

$f = 0.597$ Hz **Ans**

22-26. Solve Prob. 22-13 using energy methods.

$$T + V = \frac{1}{2}\left[mk_G^2 + md^2\right]\dot{\theta}^2 + mg(d)(1-\cos\theta)$$

$$\left(k_G^2 + d^2\right)\dot{\theta}\ddot{\theta} + gd(\sin\theta)\dot{\theta} = 0$$

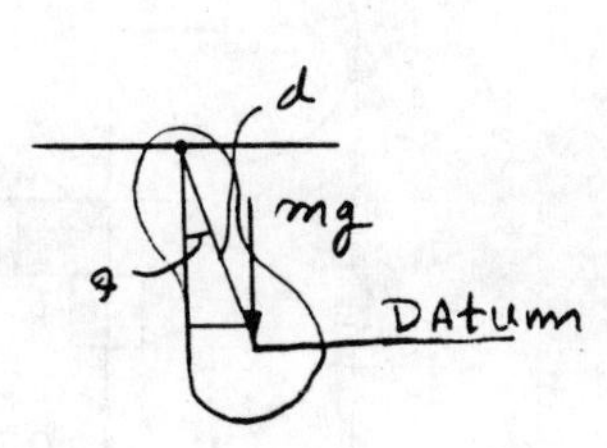

$$\sin\theta \approx \theta$$

$$\ddot{\theta} + \frac{gd}{\left(k_G^2 + d^2\right)}\theta = 0$$

$$\tau = \frac{2\pi}{p} = 2\pi\sqrt{\frac{\left(k_G^2 + d^2\right)}{gd}}$$ **Ans**

22-27. Solve Prob. 22-15 using energy methods.

$$T+V=\frac{1}{2}\left[\frac{3}{2}mr^2\right]\dot{\theta}^2+mg(r)(1-\cos\theta)$$

$$\frac{3}{2}mr^2\dot{\theta}\ddot{\theta}+mg(r)(\sin\theta)\dot{\theta}=0$$

$$\sin\theta\approx\theta$$

$$\ddot{\theta}+\left(\frac{2}{3}\right)\left(\frac{g}{r}\right)\theta=0$$

$$\tau=\frac{2\pi}{p}=2\pi\sqrt{\frac{3r}{2g}}$$ **Ans**

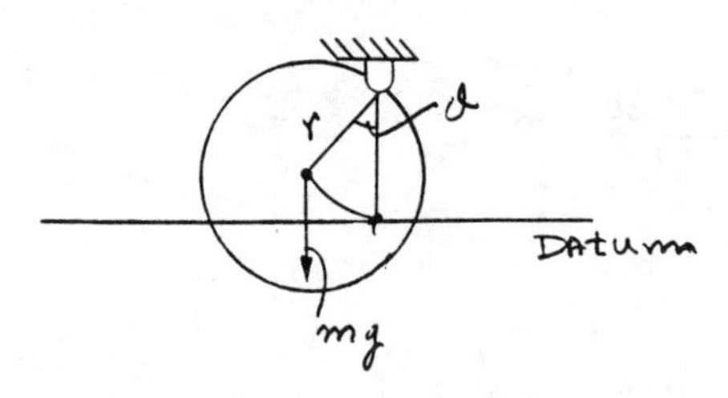

***22-28.** Solve Prob. 22-16 using energy methods.

$$T+V=\frac{1}{2}\left[\frac{1}{12}m\left(a^2+a^2\right)+m\left(\frac{a}{\sqrt{2}}\right)^2\right]\dot{\theta}^2+mg\left(\frac{a}{\sqrt{2}}\right)(1-\cos\theta)$$

$$\frac{2}{3}ma^2\dot{\theta}\ddot{\theta}+mg\left(\frac{a}{\sqrt{2}}\right)(\sin\theta)\dot{\theta}=0$$

$$\sin\theta\approx\theta$$

$$\ddot{\theta}+\frac{3g}{2\sqrt{2}a}\theta=0$$

$$\tau=\frac{2\pi}{p}=\frac{2\pi}{1.0299}\left(\sqrt{\frac{a}{g}}\right)=6.10\sqrt{\frac{a}{g}}$$ **Ans**

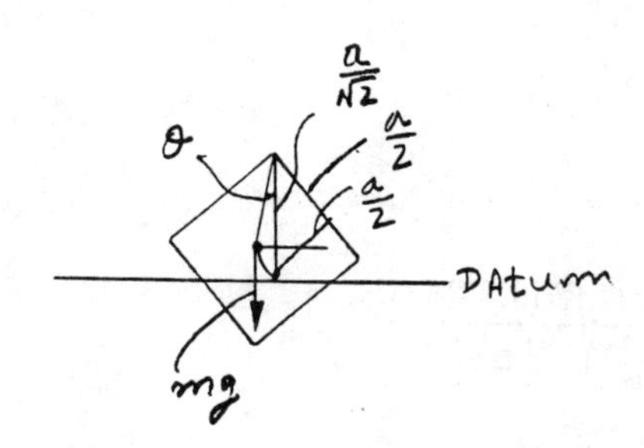

22-29. Solve Prob. 22-20 using energy methods.

$$s=0.75\theta,\qquad \dot{s}=0.75\dot{\theta}$$

$$T+V=\frac{1}{2}\left[\frac{1}{2}\left(\frac{15}{32.2}\right)(0.75)^2\right]\dot{\theta}^2+\frac{1}{2}\left(\frac{3}{32.2}\right)\left(0.75\dot{\theta}\right)^2$$

$$+\frac{1}{2}(80)(s_{eq}+0.75\theta)^2-3(0.75\theta)$$

$$0=0.1834\dot{\theta}\ddot{\theta}+80(s_{eq}+0.75\theta)\left(0.75\dot{\theta}\right)-2.25\dot{\theta}$$

$$F_{eq}=80s_{eq}=3$$

$$s_{eq}=0.0375\text{ ft}$$

Thus,

$$0.1834\ddot{\theta}+45\theta=0$$

$$\ddot{\theta}+245.3=0$$

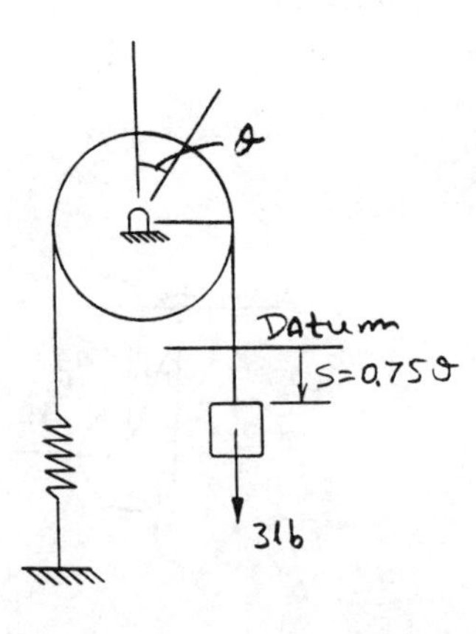

$$\tau=\frac{2\pi}{p}=\frac{2\pi}{\sqrt{245.3}}=0.401\text{ s}$$ **Ans**

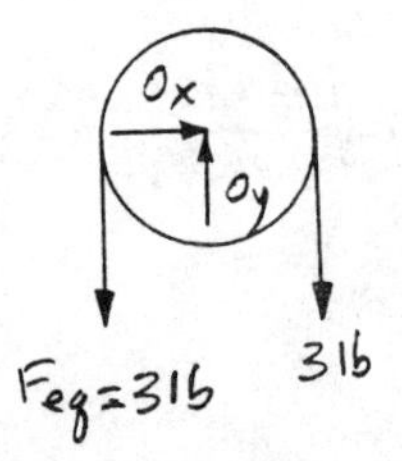

22-30. Determine the differential equation of motion of the block of mass m when it is displaced slightly and released. Motion occurs in the vertical plane. The springs are attached to the block.

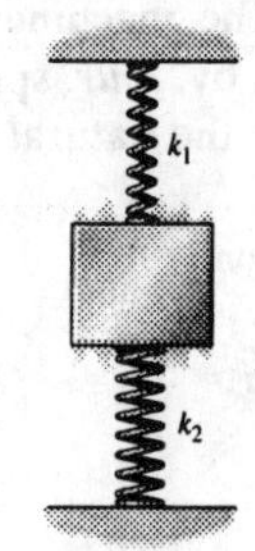

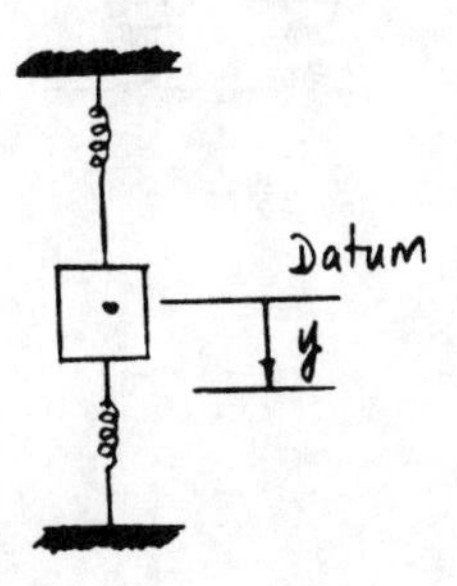

Energy equation :

$$T + V = \text{const.}$$

$$\frac{1}{2}m\dot{y}^2 + \frac{1}{2}k_1(y_{st} + y)^2 + \frac{1}{2}k_2(y_{st} + y)^2 - mgy = \text{const.}$$

Time derivative :

$$m\dot{y}\ddot{y} + k_1(y_{st_1} + y)\dot{y} + k_2(y_{st_2} + y)\dot{y} - mg\dot{y} = 0$$

$$\dot{y}\left[m\ddot{y} + (k_1 + k_2)y + k_1 y_{st_1} + k_2 y_{st_2} - mg\right] = 0 \qquad \text{Since } \dot{y} \neq 0$$

$$m\ddot{y} + (k_1 + k_2)y + k_1 y_{st_1} + k_2 y_{st_2} - mg = 0$$

From equilibrium $k_1 y_{st_1} + k_2 y_{st_2} = mg$. Hence

$$\ddot{y} + \frac{(k_1 + k_2)}{m}y = 0 \qquad \textbf{Ans}$$

22-31. Determine the differential equation of motion of the 3-kg block when it is displaced slightly and released. The surface is smooth and the springs are originally unstretched.

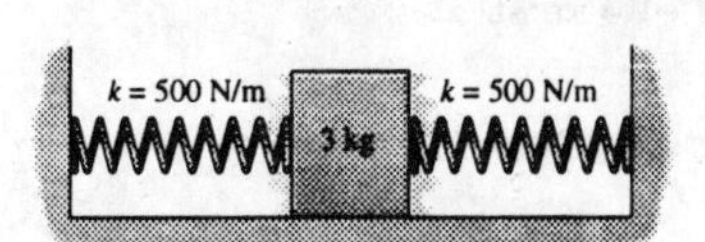

$$T + V = \text{const.}$$

$$T = \frac{1}{2}(3)\dot{x}^2$$

$$V = \frac{1}{2}(500)x^2 + \frac{1}{2}(500)x^2$$

$$T + V = 1.5\dot{x}^2 + 500x^2$$

$$1.5\left(2\dot{x}\right)\ddot{x} + 1000x\dot{x} = 0$$

$$3\ddot{x} + 1000x = 0$$

$$\ddot{x} + 333x = 0 \qquad \textbf{Ans}$$

***22-32.** The machine has a mass m and is uniformly supported by *four* springs, each having a stiffness k. Determine the natural period of vertical vibration.

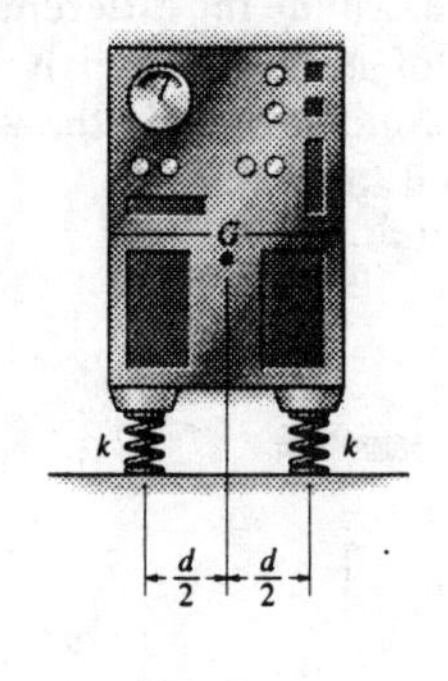

$T + V = \text{const.}$

$T = \frac{1}{2}m(\dot{y})^2$

$V = m\,g\,y + \frac{1}{2}(4k)(\Delta s - y)^2$

$T + V = \frac{1}{2}m(\dot{y})^2 + m\,g\,y + \frac{1}{2}(4k)(\Delta s - y)^2$

$m\,\dot{y}\,\ddot{y} + m\,g\,\dot{y} - 4k(\Delta s - y)\dot{y} = 0$

$m\,\ddot{y} + m\,g + 4ky - 4k\Delta s = 0$

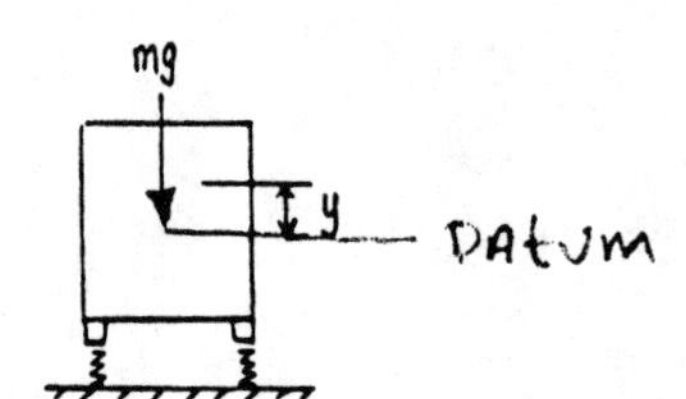

Since $\Delta s = \dfrac{mg}{4k}$

Then

$m\ddot{y} + 4ky = 0$

$\ddot{y} + \dfrac{4k}{m}y = 0$

$p = \sqrt{\dfrac{4k}{m}}$

$\tau = \dfrac{2\pi}{p} = \pi\sqrt{\dfrac{m}{k}}$ **Ans**

22-33. If the disk has a mass of 8 kg, determine the natural frequency of vibration. The springs are originally unstretched.

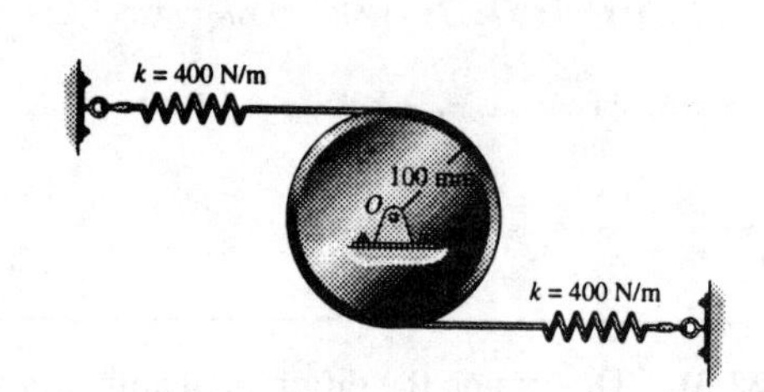

$I_O = \frac{1}{2}(8)(0.1)^2 = 0.04$

$T + V = \text{const.}$

$T = \frac{1}{2}(0.04)\dot{\theta}^2$

$V = 2\left[\frac{1}{2}(400)(0.1\theta)^2\right] = 4\theta^2$

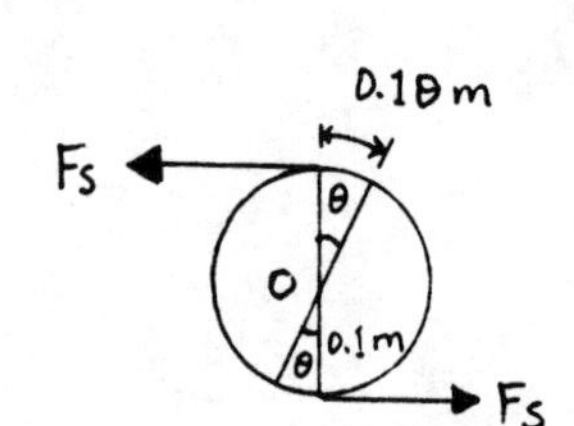

$T + V = 0.02\dot{\theta}^2 + 4\theta^2$

$0.04\dot{\theta}\,\ddot{\theta} + 4(2\theta)\dot{\theta} = 0$

$0.04\ddot{\theta} + 8\theta = 0$

$\ddot{\theta} + 200\theta = 0$

$f = \dfrac{p}{2\pi} = \dfrac{\sqrt{200}}{2\pi} = 2.25\text{ Hz}$ **Ans**

22-34. Determine the natural period of vibration of the pendulum. Consider the two rods to be slender, each having a weight of 8 lb/ft.

$$\bar{y} = \frac{1(8)(2) + 2(8)(2)}{8(2) + 8(2)} = 1.5 \text{ ft}$$

$$I_O = \frac{1}{32.2}\left[\frac{1}{12}(2)(8)(2)^2 + 2(8)(1)^2\right] + \frac{1}{32.2}\left[\frac{1}{12}(2)(8)(2)^2 + 2(8)(2)^2\right] = 2.8157 \text{ slug}\cdot\text{ft}^2$$

$$h = \bar{y}(1 - \cos\theta)$$

$$T + V = \text{const.}$$

$$T = \frac{1}{2}(2.8157)\left(\dot{\theta}\right)^2 = 1.4079\dot{\theta}^2$$

$$V = 8(4)(1.5)(1 - \cos\theta) = 48(1 - \cos\theta)$$

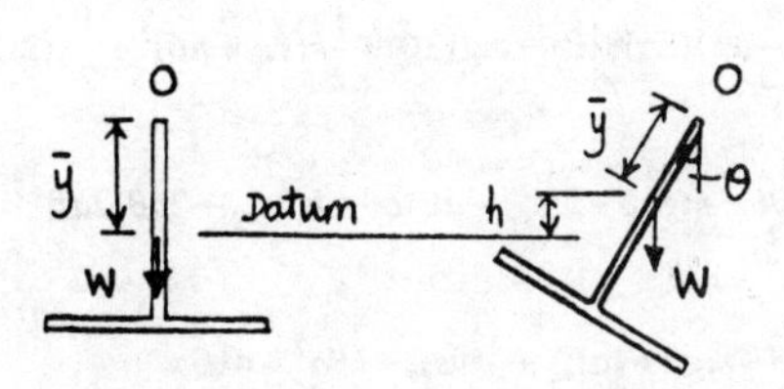

$$T + V = 1.4079\dot{\theta}^2 + 48(1 - \cos\theta)$$

$$1.4079\left(2\dot{\theta}\right)\ddot{\theta} + 48(\sin\theta)\dot{\theta} = 0$$

For small θ, $\sin\theta \approx \theta$, then

$$\ddot{\theta} + 17.047\theta = 0$$

$$\tau = \frac{2\pi}{p} = \frac{2\pi}{\sqrt{17.047}} = 1.52 \text{ s} \qquad \textbf{Ans}$$

22-35. Determine the natural period of vibration of the disk having a mass m and radius r. Assume the disk does not slip on the surface of contact as it oscillates.

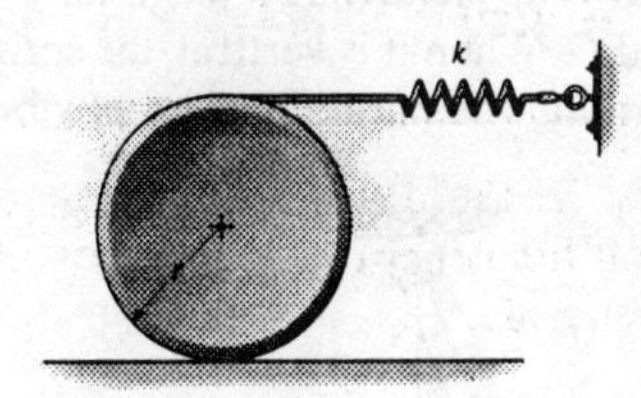

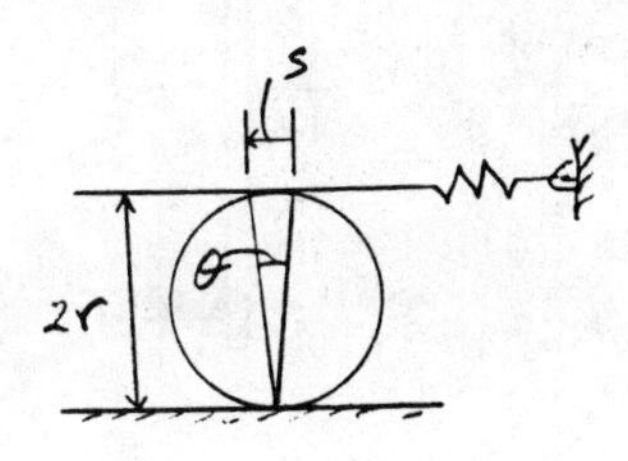

$$T + V = \text{const.}$$

$$s = (2r)\theta$$

$$T + V = \frac{1}{2}\left[\frac{1}{2}mr^2 + mr^2\right]\dot{\theta}^2 + \frac{1}{2}k(2r\,\theta)^2$$

$$0 = \frac{3}{2}mr^2\dot{\theta}\ddot{\theta} + 4kr^2\theta\dot{\theta}$$

$$\ddot{\theta} + \frac{8k}{3m}\theta = 0$$

$$\tau = \frac{2\pi}{p} = \frac{2\pi}{\sqrt{\frac{8k}{3m}}} = 3.85\sqrt{\frac{m}{k}} \qquad \textbf{Ans}$$

***22-36.** The bar has a mass m and is held in the horizontal position using the two springs. If the end of the bar is given a small displacement, determine the natural period of vibration.

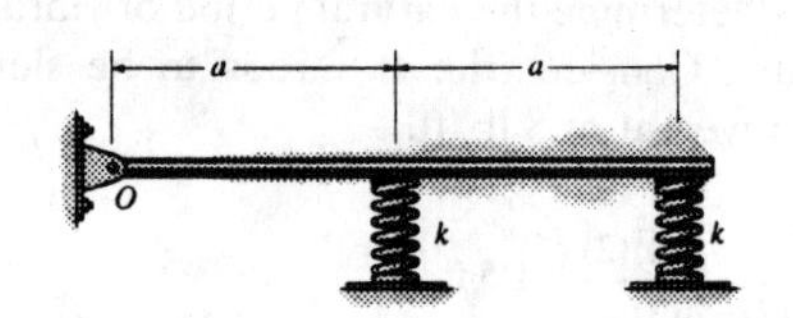

$$\circlearrowleft + \Sigma M_O = 0; \quad F_s a + F_s{}'(2a) - mg(a) = 0$$

$$F_s + 2F_s{}' = mg$$

$$ks_{eq} + 2(k)(2s_{eq}) = mg$$

$$s_{eq} = \frac{mg}{5k} \qquad (1)$$

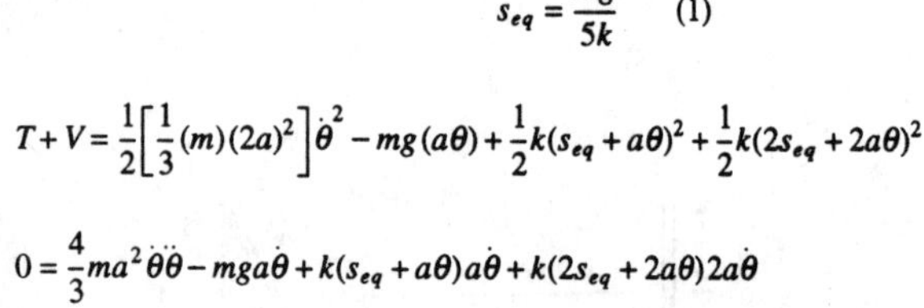

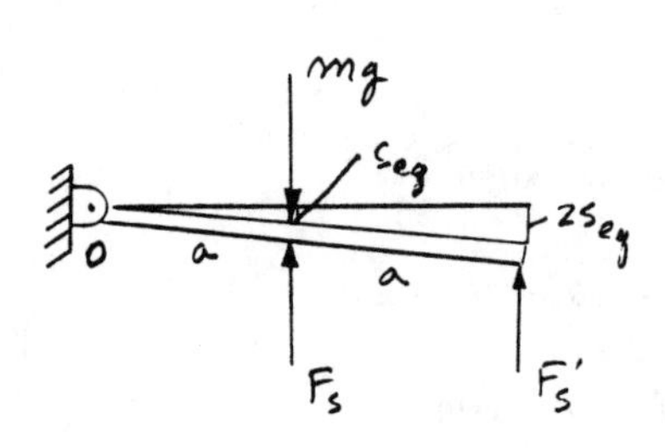

$$T + V = \frac{1}{2}\left[\frac{1}{3}(m)(2a)^2\right]\dot{\theta}^2 - mg(a\theta) + \frac{1}{2}k(s_{eq} + a\theta)^2 + \frac{1}{2}k(2s_{eq} + 2a\theta)^2$$

$$0 = \frac{4}{3}ma^2\dot{\theta}\ddot{\theta} - mga\dot{\theta} + k(s_{eq} + a\theta)a\dot{\theta} + k(2s_{eq} + 2a\theta)2a\dot{\theta}$$

$$0 = \frac{4}{3}ma^2\ddot{\theta} - mga + kas_{eq} + 4kas_{eq} + k\theta a^2 + 4k\theta a^2$$

Using Eq. (1) :

$$\ddot{\theta} + \left(\frac{15k}{4m}\right)\theta = 0$$

$$\tau = \frac{2\pi}{p} = 4\pi\sqrt{\frac{m}{15k}} \qquad \textbf{Ans}$$

22-37. The slender rod has a mass m and is pinned at its end O. When it is vertical, the springs are unstretched. Determine the natural period of vibration.

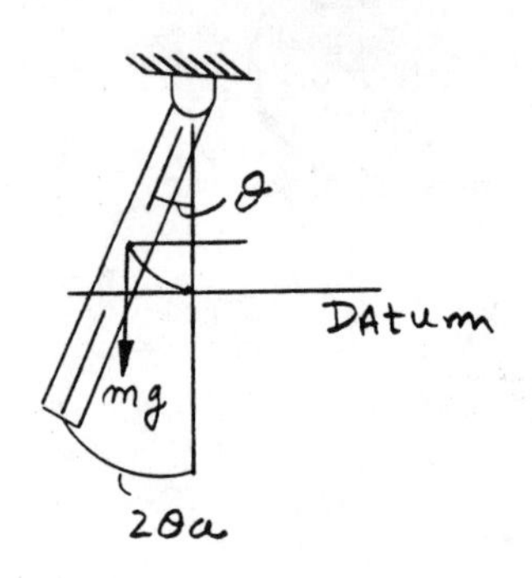

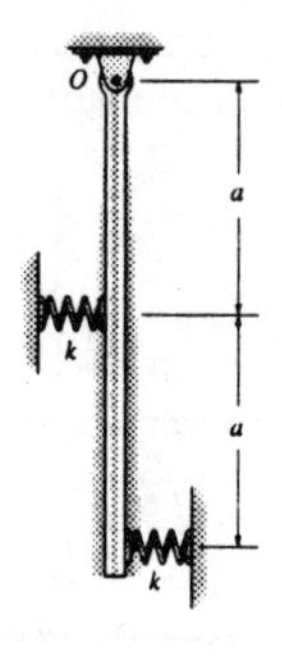

$$T + V = \frac{1}{2}\left[\frac{1}{3}m(2a)^2\right]\dot{\theta}^2 + \frac{1}{2}k(2\theta a)^2 + \frac{1}{2}k(\theta a)^2 + mga(1 - \cos\theta)$$

$$0 = \frac{4}{3}ma^2\dot{\theta}\ddot{\theta} + 4ka^2\theta\dot{\theta} + ka^2\theta\dot{\theta} + mga\sin\theta\dot{\theta}$$

$$\sin\theta \approx \theta$$

$$\frac{4}{3}ma^2\ddot{\theta} + 5ka^2\theta + mga\theta = 0$$

$$\ddot{\theta} + \left(\frac{15ka + 3mg}{4ma}\right)\theta = 0$$

$$\tau = \frac{2\pi}{p} = \frac{4\pi}{\sqrt{3}}\left(\frac{ma}{5ka + mg}\right)^{\frac{1}{2}} \qquad \textbf{Ans}$$

22-38. The 5-lb sphere is attached to a rod of negligible mass and rests in the horizontal position. Determine the natural frequency of vibration. Neglect the size of the sphere.

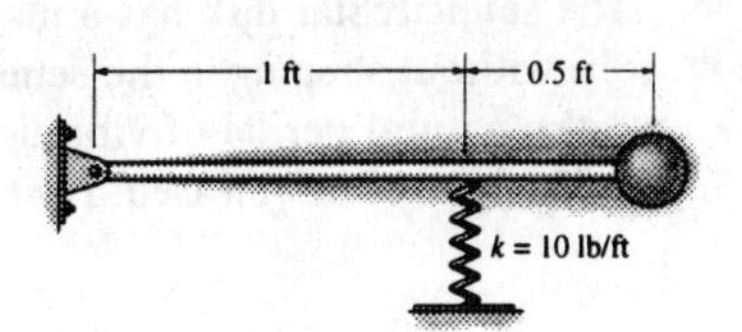

$E = T + V$

$= \frac{1}{2}\left[\left(\frac{5}{32.2}\right)(1.5)^2\right](\dot{\theta})^2 + \frac{1}{2}(10)(\delta_{st} + 1\sin\theta)^2 - (1.5\sin\theta)(5)$

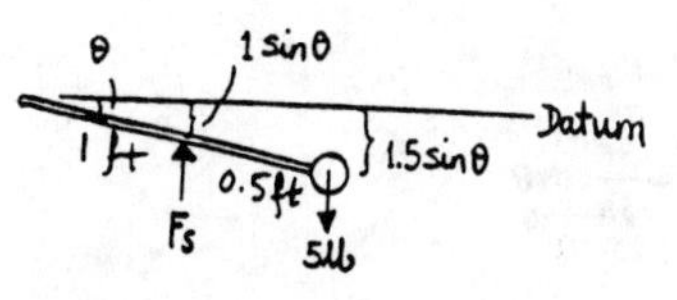

$\dot{E} = \dot{\theta}\left[10(\delta_{st} + \sin\theta)\cos\theta - 7.5\cos\theta + \frac{5}{32.2}(1.5)^2\ddot{\theta}\right] = 0$

By statics,

$F_s = \frac{5(1.5)}{1} = 7.5 \text{ lb}; \qquad \delta_{st} = \frac{7.5}{10} = 0.75$

For small θ, $\quad \sin\theta \approx \theta, \quad \cos\theta \approx 1$

Thus,

$7.5 + 10\theta - 7.5 + \frac{5}{32.2}(1.5)^2\ddot{\theta} = 0$

$\ddot{\theta} + 28.62\theta = 0$

$p = \sqrt{28.62} = 5.35 \text{ rad/s}$

$f = \frac{p}{2\pi} = \frac{5.35}{2\pi} = 0.851 \text{ Hz}$ **Ans**

22-39. Determine the natural period of vibration of the 10-lb semicircular disk. $I_O = \frac{1}{2}mr^2$.

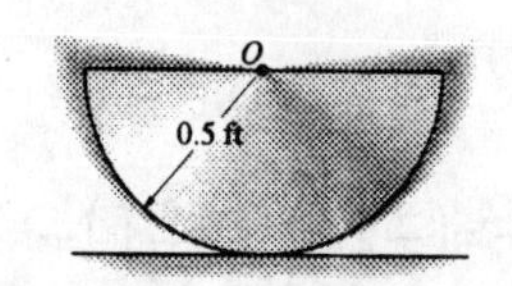

Datum at initial level of center of gravity of disk.

$\Delta = \bar{r}(1-\cos\theta)$

$E = T + V$

$= \frac{1}{2}I_{IC}(\dot{\theta})^2 + W\bar{r}(1-\cos\theta)$

$\dot{E} = \dot{\theta}(I_{IC}\ddot{\theta} + W\bar{r}\sin\theta) = 0$

For small θ, $\quad \sin\theta \approx \theta,$

$\ddot{\theta} + \frac{W\bar{r}}{I_{IC}}\theta = 0$

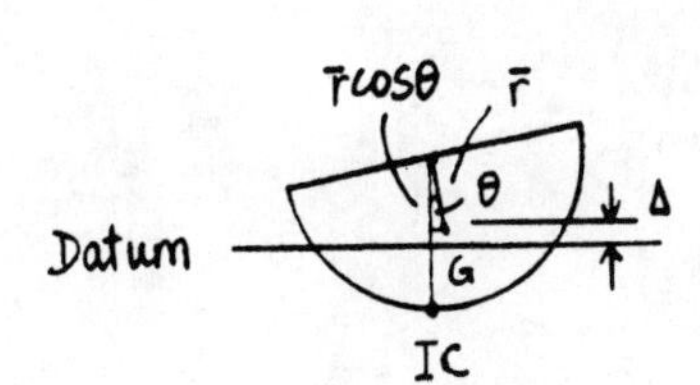

$\bar{r} = \frac{4(0.5)}{3\pi} = 0.212 \text{ ft}$

$I_O = I_G + m\bar{r}^2$

$\frac{1}{2}\left(\frac{10}{32.2}\right)(0.5)^2 = I_G + \frac{10}{32.2}(0.212)^2$

$I_G = 0.02483 \text{ slug}\cdot\text{ft}^2$

$I_{IC} = I_G + m(r-\bar{r})^2$

$= 0.02483 + \frac{10}{32.2}(0.5-0.212)^2$

$= 0.05056 \text{ slug}\cdot\text{ft}^2$

$\tau = \frac{2\pi}{p} = 2\pi\sqrt{\frac{I_{IC}}{W\bar{r}}} = 2\pi\sqrt{\frac{0.05056}{10(0.212)}}$

$\tau = 0.970 \text{ s}$ **Ans**

***22-40.** The semicircular disk has a mass m and radius r, and it rolls without slipping in the semicircular trough. Determine the natural period of vibration of the disk if it is displaced slightly and released. *Hint:* $I_O = \frac{1}{2}mr^2$.

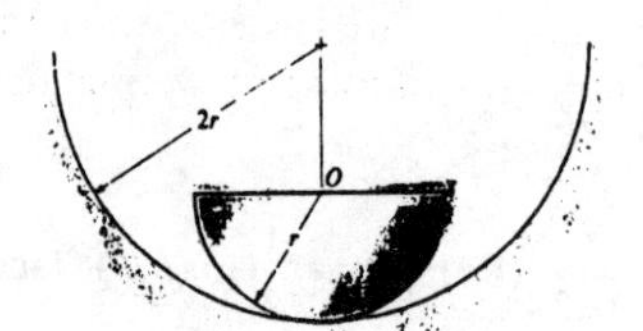

$AB = (2r - r)\cos\phi = r\cos\phi, \qquad BC = \dfrac{4r}{3\pi}\cos\theta$

$AC = r\cos\phi + \dfrac{4r}{3\pi}\cos\theta, \qquad DE = 2r\phi = r(\theta + \phi)$

$\phi = \theta$

$AC = r\left(1 + \dfrac{4}{3\pi}\right)\cos\theta$

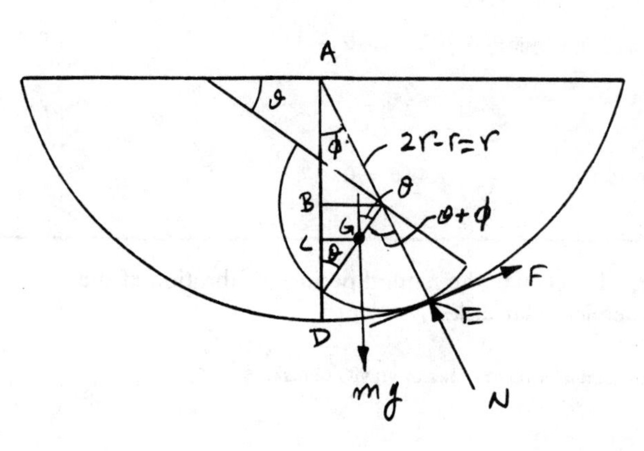

Thus, the change in elevation of G is,

$h = 2r - \left(r - \dfrac{4r}{3\pi}\right) - AC = r\left(1 + \dfrac{4}{3\pi}\right)(1 - \cos\theta)$

Since no slipping occurs,

$v_G = \dot{\theta}\left(r - \dfrac{4r}{3\pi}\right)$

$I_G = I_O - m\left(\dfrac{4r}{3\pi}\right)^2 = \left(\dfrac{1}{2} - \left(\dfrac{4}{3\pi}\right)^2\right)mr^2$

$T = \dfrac{1}{2}m\dot{\theta}^2 r^2\left(1 - \dfrac{4}{3\pi}\right)^2 + \dfrac{1}{2}\left(\dfrac{1}{2} - \left(\dfrac{4}{3\pi}\right)^2\right)mr^2\dot{\theta}^2 = \dfrac{1}{2}mr^2\left(\dfrac{3}{2} - \dfrac{8}{3\pi}\right)\dot{\theta}^2$

$T + V = \dfrac{1}{2}mr^2\left(\dfrac{3}{2} - \dfrac{8}{3\pi}\right)\dot{\theta}^2 + mgr\left(1 + \dfrac{4}{3\pi}\right)(1 - \cos\theta)$

$0 = mr^2\left(\dfrac{3}{2} - \dfrac{8}{3\pi}\right)\dot{\theta}\ddot{\theta} + mgr\left(1 + \dfrac{4}{3\pi}\right)\sin\theta\dot{\theta}$

$\sin\theta \approx \theta$

$$\ddot{\theta} + \frac{g\left(1 + \frac{4}{3\pi}\right)}{r\left(\frac{3}{2} - \frac{8}{3\pi}\right)}\theta = 0$$

$p = 1.479\sqrt{\dfrac{g}{r}}$

$\tau = \dfrac{2\pi}{p} = 4.25\sqrt{\dfrac{r}{g}}$ **Ans**

22-41. The block shown in Fig. 22–16 has a mass of 20 kg, and the spring has a stiffness $k = 600$ N/m. When the block is displaced and released, two successive amplitudes are measured as $x_1 = 150$ mm and $x_2 = 87$ mm. Determine the coefficient of viscous damping, c.

Assuming that the system is underdamped.

$$x_1 = De^{-\left(\frac{c}{2m}\right)t_1} \qquad (1)$$

$$x_2 = De^{-\left(\frac{c}{2m}\right)t_2} \qquad (2)$$

Divide Eq.(1) by Eq. (2) $\qquad \dfrac{x_1}{x_2} = \dfrac{e^{-\left(\frac{c}{2m}\right)t_1}}{e^{-\left(\frac{c}{2m}\right)t_2}}$

$$\ln\left(\frac{x_1}{x_2}\right) = \left(\frac{c}{2m}\right)(t_2 - t_1) \qquad (3)$$

However, $\qquad t_2 - t_1 = \tau_d = \dfrac{2\pi}{p_d} = \dfrac{2\pi}{p\sqrt{1-\left(\frac{c}{c_c}\right)^2}}$ $\qquad$ and $\qquad p = \dfrac{c_c}{2m}$

$$t_2 - t_1 = \frac{4m\pi}{c_c\sqrt{1-\left(\frac{c}{c_c}\right)^2}} \qquad (4)$$

Substitute Eq.(4) into Eq. (3) yields :

$$\ln\left(\frac{x_1}{x_2}\right) = \left(\frac{c}{2m}\right)\frac{4m\pi}{c_c\sqrt{1-\left(\frac{c}{c_c}\right)^2}}$$

$$\ln\left(\frac{x_1}{x_2}\right) = \frac{2\pi\left(\frac{c}{c_c}\right)}{\sqrt{1-\left(\frac{c}{c_c}\right)^2}} \qquad (5)$$

From Eq.(5)

$x_1 = 0.15$ m $\qquad x_2 = 0.087$ m $\qquad p = \sqrt{\dfrac{k}{m}} = \sqrt{\dfrac{600}{20}} = 5.477$ rad/s

$$c_c = 2mp = 2(20)(5.477) = 219.09 \text{ N}\cdot\text{s/m}$$

$$\ln\left(\frac{0.15}{0.087}\right) = \frac{2\pi\left(\frac{c}{219.09}\right)}{\sqrt{1-\left(\frac{c}{219.09}\right)^2}}$$

$c = 18.9$ N · s/m $\qquad$ **Ans**

Since $c < c_c$, the system is underdamped . Therefore, the assumption is OK!

22-42. If the block-and-spring model is subjected to the impressed force $F = F_0 \cos \omega t$, show that the differential equation of motion is $\ddot{x} + (k/m)x = (F_0/m) \cos \omega t$, where x is measured from the equilibrium position of the block. What is the general solution of this equation?

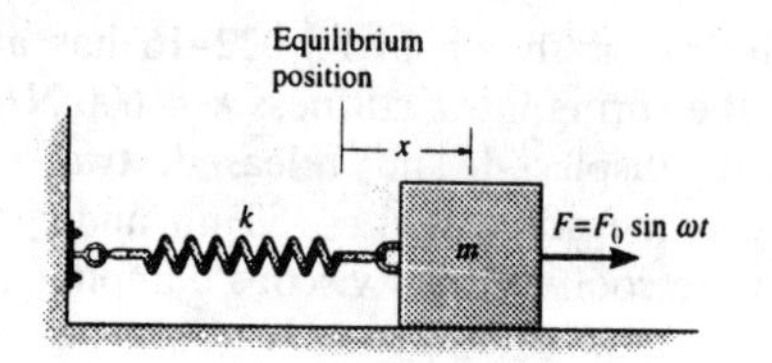

$\overset{+}{\rightarrow} \Sigma F_x = ma_x; \quad F_0 \cos \omega t - kx = m\ddot{x}$

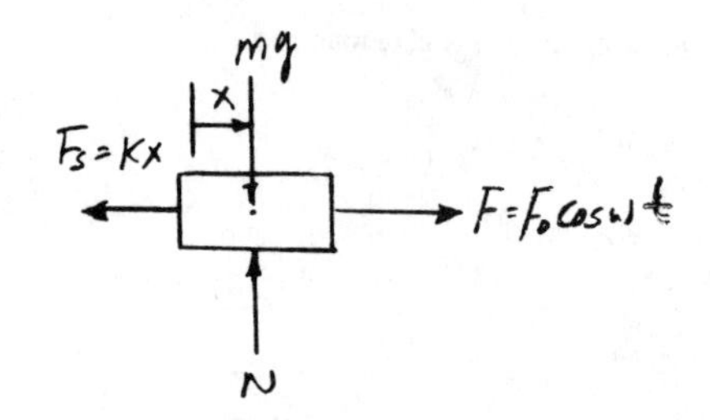

$$\ddot{x} + \frac{k}{m}x = \frac{F_0}{m}\cos \omega t \qquad \text{(Q.E.D.)}$$

$$\ddot{x} + p^2 x = \frac{F_0}{m}\cos \omega t \qquad \text{Where } p = \sqrt{\frac{k}{m}} \qquad (1)$$

The general solution of the above differential equation is of the form of $x = x_c + x_p$.

The complementary solution :

$x_c = A\sin pt + B\cos pt$

The particular solution :

$x_p = C\cos \omega t \qquad (2)$

$\ddot{x}_P = -C\omega^2 \cos \omega t \qquad (3)$

Substitute Eqs.(2) and (3) into(1) yields :

$$-C\omega^2 \cos \omega t + p^2 (C\cos \omega t) = \frac{F_0}{m}\cos \omega t$$

$$C = \frac{\frac{F_0}{m}}{p^2 - \omega^2} = \frac{F_0/k}{1 - \left(\frac{\omega}{p}\right)^2}$$

The general solution is therefore

$$x = A\sin pt + B\cos pt + \frac{F_0/k}{1 - \left(\frac{\omega}{p}\right)^2}\cos \omega t \qquad \textbf{Ans}$$

The constants A and B can be found from the initial conditions.

22-43. A 4-kg block is suspended from a spring that has a stiffness of $k = 600$ N/m. The block is drawn downward 50 mm from the equilibrium position and released from rest when $t = 0$. If the support moves with an impressed displacement of $\delta = (10 \sin 4t)$ mm, where t is in seconds, determine the equation that describes the vertical motion of the block. Assume positive displacement is downward.

$$p = \sqrt{\frac{k}{m}} = \sqrt{\frac{600}{4}} = 12.25$$

The general solution is defined by Eq. 22 – 23 with $k\delta_0$ substituted for F_0.

$$y = A\sin pt + B\cos pt + \left(\frac{\delta_0}{\left[1 - \left(\frac{\omega}{p}\right)^2\right]}\right)\sin \omega t$$

$\delta = (0.01\sin 4t)$ m, hence $\delta_0 = 0.01$, $\omega = 4$, so that

$y = A\sin 12.25t + B\cos 12.25t + 0.0112\sin 4t$

$y = 0.05$ when $t = 0$,

$0.05 = 0 + B + 0; \quad B = 0.05$ m

$y = A(12.25)\cos 12.25t - B(12.25)\sin 12.25t + 0.0112(4)\cos 4t$

$v = y = 0$ when $t = 0$,

$0 = A(12.25) - 0 + 0.0112(4); \quad A = -0.00366$ m

Expressing the result in mm, we have

$y = (-3.66\sin 12.25t + 50\cos 12.25t + 11.2\sin 4t)$ mm $\qquad$ **Ans**

***22-44.** If the block is subjected to the impressed force $F = F_0 \cos \omega t$, show that the differential equation of motion is $\ddot{y} + (k/m)y = (F_0/m) \cos \omega t$, where y is measured from the equilibrium position of the block. What is the general solution of this equation?

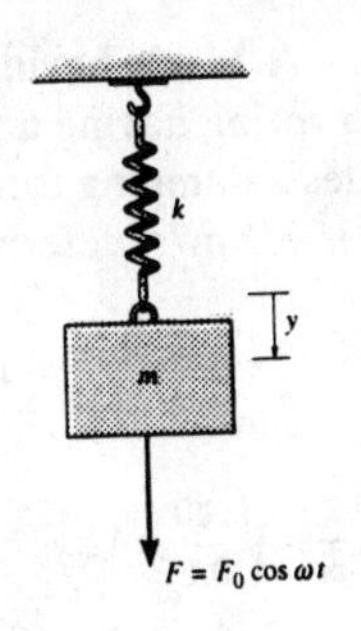

$+\downarrow \Sigma F_y = ma_y;\quad F_0 \cos \omega t + W - k\delta_{st} - ky = m\ddot{y}$

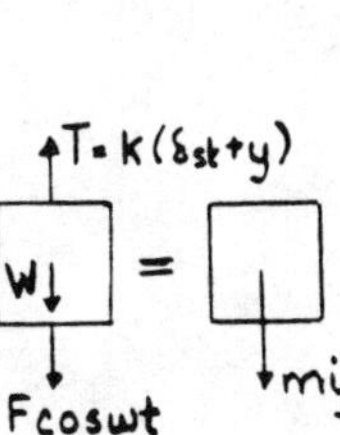
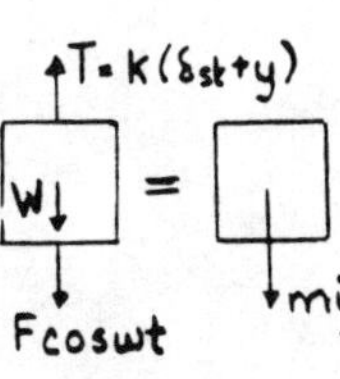

Since $W = k\delta_{st}$,

$$\ddot{y} + \left(\frac{k}{m}\right)y = \frac{F_0}{m}\cos \omega t \quad (1) \quad \textbf{Q.E.D.}$$

$y_c = A \sin py + B \cos py$ (complementary solution)

$y_p = C \cos \omega t$ (particular solution)

Substitute y_p into Eq. (1),

$$C\left(-\omega^2 + \frac{k}{m}\right)\cos \omega t = \frac{F_0}{m}\cos \omega t$$

$$C = \frac{\frac{F_0}{m}}{\left(\frac{k}{m} - \omega^2\right)}$$

$$y = y_c + y_p$$

$$y = A \sin pt + B \cos pt + \left(\frac{F_0}{(k - m\omega^2)}\right)\cos \omega t \quad \textbf{Ans}$$

22-45. The spring shown stretches 6 in. when it is loaded with a 50-lb weight. Determine the equation which describes the position of the weight as a function of time when the weight is pulled 4 in. below its equilibrium position and released from rest. The weight is subjected to the impressed force of $F = (-7 \sin 2t)$ lb, where t is in seconds.

$k(y + y_{st})$, y, mg, $F = F_0 \sin \omega t$; k, $F = -7 \sin 2t$

$+\uparrow \Sigma F_y = ma_y;\quad k(y_{st} + y) - mg - F_0 \sin \omega t = -m\ddot{y}$

$$m\ddot{y} + ky + ky_{st} - mg = F_0 \sin \omega t$$

However, from equilibrium $ky_{st} - mg = 0$, therefore

$$m\ddot{y} + ky = F_0 \sin \omega t$$

$$\ddot{y} + \frac{k}{m}y = \frac{F_0}{m}\sin \omega t \quad \text{where } p = \sqrt{\frac{k}{m}}$$

$$\ddot{y} + p^2 y = \frac{F_0}{m}\sin \omega t$$

From the text, the general solution of the above differential equation is

$$y = A \sin pt + B \cos pt + \frac{F_0/k}{1 - \left(\frac{\omega}{p}\right)^2}\sin \omega t$$

$$v = \dot{y} = Ap \cos pt - Bp \sin pt + \frac{(F_0/k)\omega}{1 - \left(\frac{\omega}{p}\right)^2}\cos \omega t$$

The initial condition when $t = 0$, $y = y_0$ and $v = v_0$.

$$y_0 = 0 + B + 0 \qquad B = y_0$$

$$v_0 = Ap - 0 + \frac{(F_0/k)\omega}{1 - \left(\frac{\omega}{p}\right)^2} \qquad A = \frac{v_0}{p} - \frac{(F_0/k)\omega}{p - \frac{\omega^2}{p}}$$

The solution is therefore

$$y = \left(\frac{v_0}{p} - \frac{(F_0/k)\omega}{p - \frac{\omega^2}{p}}\right)\sin pt + y_0 \cos pt + \frac{F_0/k}{1 - \left(\frac{\omega}{p}\right)^2}\sin \omega t$$

For this problem :

$$k = \frac{50}{6/12} = 100 \text{ lb/ft} \qquad p = \sqrt{\frac{k}{m}} = \sqrt{\frac{100}{50/32.2}} = 8.025$$

$$\frac{F_0/k}{1 - \left(\frac{\omega}{p}\right)^2} = \frac{-7/100}{1 - \left(\frac{2}{8.025}\right)^2} = -0.0746 \qquad y_0 = 0.333$$

$$\frac{v_0}{p} - \frac{(F_0/k)\omega}{p - \frac{\omega^2}{p}} = 0 - \frac{(-7/100)2}{8.025 - \frac{2^2}{8.025}} = 0.0186$$

$$y = (0.0186 \sin 8.02t + 0.333 \cos 8.02t - 0.0746 \sin 2t) \text{ ft} \quad \textbf{Ans}$$

22-46. A block having a mass of 0.8 kg is suspended from a spring having a stiffness of 120 N/m. If a dashpot provides a damping force of 2.5 N when the speed of the block is 0.2 m/s, determine the period of free vibration.

$$F = cv \qquad c = \frac{F}{v} = \frac{2.5}{0.2} = 12.5 \text{ N} \cdot \text{s/m}$$

$$p = \sqrt{\frac{k}{m}} = \sqrt{\frac{120}{0.8}} = 12.247 \text{ rad/s}$$

$$c_c = 2mp = 2(0.8)(12.247) = 19.60 \text{ N} \cdot \text{s/m}$$

$$p_d = p\sqrt{1-\left(\frac{c}{c_c}\right)^2} = 12.247\sqrt{1-\left(\frac{12.5}{19.6}\right)^2} = 9.432 \text{ rad/s}$$

$$\tau_d = \frac{2\pi}{p_d} = \frac{2\pi}{9.432} = 0.666 \text{ s} \qquad \textbf{Ans}$$

22-47. A 5-kg block is suspended from a spring having a stiffness of 300 N/m. If the block is acted upon by a vertical force $F = (7 \sin 8t)$ N, where t is in seconds, determine the equation which describes the motion of the block when it is pulled down 100 mm from the equilibrium position and released from rest at $t = 0$. Assume that positive displacement is downward.

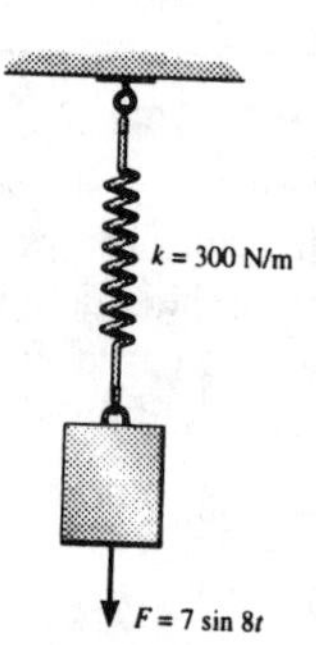

The general solution is defined by :

$$y = A\sin pt + B\cos pt + \left(\frac{\frac{F_0}{k}}{1-\left(\frac{\omega}{p}\right)^2}\right)\sin\omega t$$

Since

$$F = 7\sin 8t, \qquad F_0 = 7 \text{ N}, \qquad \omega = 8 \text{ rad/s}, \qquad k = 300 \text{ N/m}$$

$$p = \sqrt{\frac{k}{m}} = \sqrt{\frac{300}{5}} = 7.746 \text{ rad/s}$$

Thus,

$$y = A\sin 7.746t + B\cos 7.746t + \left(\frac{\frac{7}{300}}{1-\left(\frac{8}{7.746}\right)^2}\right)\sin 8t$$

$y = 0.1$ m when $t = 0$,

$$0.1 = 0 + B - 0; \qquad B = 0.1 \text{ m}$$

$$\dot{y} = A(7.746)\cos 7.746t - B(7.746)\sin 7.746t - (0.35)(8)\cos 8t$$

$v = \dot{y} = 0$ when $t = 0$,

$$\dot{y} = A(7.746) - 2.8 = 0; \qquad A = 0.361$$

Expressing the results in mm, we have

$$y = (361\sin 7.75t + 100\cos 7.75t - 350\sin 8t) \text{ mm} \qquad \textbf{Ans}$$

***22-48.** The electric motor has a mass of 50 kg and is supported by *four springs,* each spring having a stiffness of 100 N/m. If the motor turns a disk D which is mounted eccentrically, 20 mm from the disk's center, determine the angular rotation ω at which resonance occurs. Assume that the motor only vibrates in the vertical direction.

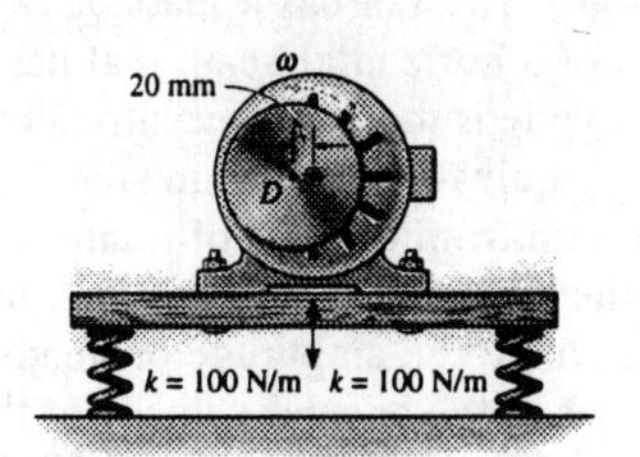

22-49. The fan has a mass of 25 kg and is fixed to the end of a horizontal beam that has a negligible mass. The fan blade is mounted eccentrically on the shaft such that it is equivalent to an unbalanced 3.5-kg mass located 100 mm from the axis of rotation. If the static deflection of the beam is 50 mm as a result of the weight of the fan, determine the angular speed of the fan at which resonance will occur. *Hint:* See the first part of Example 22–8.

$$k = \frac{F}{\Delta y} = \frac{25(9.81)}{0.05} = 4905 \text{ N/m}$$

$$p = \sqrt{\frac{k}{m}} = \sqrt{\frac{4905}{25}} = 14.01 \text{ rad/s}$$

Resonance occurs when

$$\omega = p = 14.0 \text{ rad/s} \qquad \textbf{Ans}$$

22-50. The fan has a mass of 25 kg and is fixed to the end of a horizontal beam that has a negligible mass. The fan blade is mounted eccentrically on the shaft such that it is equivalent to an unbalanced 3.5-kg mass located 100 mm from the axis of rotation. If the static deflection of the beam is 50 mm as a result of the weight of the fan, determine the amplitude of steady-state vibration of the fan when the angular velocity of the fan is 10 rad/s. *Hint:* See the first part of Example 22–8.

$$k = \frac{F}{\Delta y} = \frac{25(9.81)}{0.05} = 4905 \text{ N/m}$$

$$p = \sqrt{\frac{k}{m}} = \sqrt{\frac{4905}{25}} = 14.01 \text{ rad/s}$$

The force caused by the unbalanced rotor is

$$F_0 = mr\omega^2 = 3.5(0.1)(10)^2 = 35 \text{ N}$$

Using Eq. 22 – 22, the amplitude is

$$(x_p)_{max} = \left| \frac{\frac{F_0}{k}}{1 - (\frac{\omega}{p})^2} \right|$$

$$(x_p)_{max} = \left| \frac{\frac{35}{4905}}{1 - (\frac{10}{14.01})^2} \right| = 0.0146 \text{ m}$$

$(x_p)_{max} = 14.6$ mm **Ans**

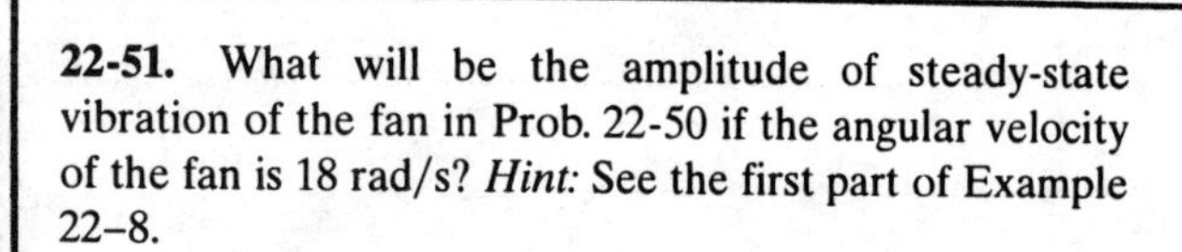

22-51. What will be the amplitude of steady-state vibration of the fan in Prob. 22-50 if the angular velocity of the fan is 18 rad/s? *Hint:* See the first part of Example 22–8.

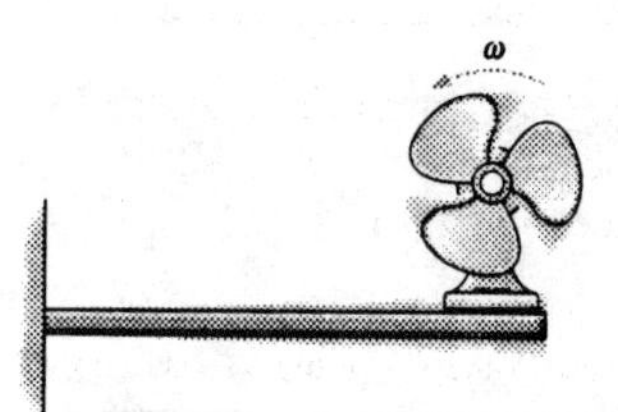

$$k = \frac{F}{\Delta y} = \frac{25(9.81)}{0.05} = 4905 \text{ N/m}$$

$$p = \sqrt{\frac{k}{m}} = \sqrt{\frac{4905}{25}} = 14.01 \text{ rad/s}$$

The force caused by the unbalanced rotor is

$$F_0 = mr\omega^2 = 3.5(0.1)(18)^2 = 113.4 \text{ N}$$

Using Eq. 22 – 22, the amplitude is

$$(x_p)_{max} = \left| \frac{\frac{F_0}{k}}{1 - \left(\frac{\omega}{p}\right)^2} \right|$$

$$(x_p)_{max} = \left| \frac{\frac{113.4}{4905}}{1 - \left(\frac{18}{14.01}\right)^2} \right| = 0.0355 \text{ m}$$

$(x_p)_{max} = 35.5$ mm **Ans**

***22-52.** The electric motor turns an eccentric flywheel which is equivalent to an unbalanced 0.25-lb weight located 10 in. from the axis of rotation. If the static deflection of the beam is 1 in. because of the weight of the motor, determine the angular velocity of the flywheel at which resonance will occur. The motor weighs 150 lb. Neglect the mass of the beam.

$$k = \frac{F}{\delta} = \frac{150}{1/12} = 1800 \text{ lb/ft} \qquad p = \sqrt{\frac{k}{m}} = \sqrt{\frac{1800}{150/32.2}} = 19.66$$

Resonance occurs when $\omega = p = 19.7$ rad/s **Ans**

22-53. What will be the amplitude of steady-state vibration of the motor in Prob. 22–52 if the angular velocity of the flywheel is 20 rad/s?

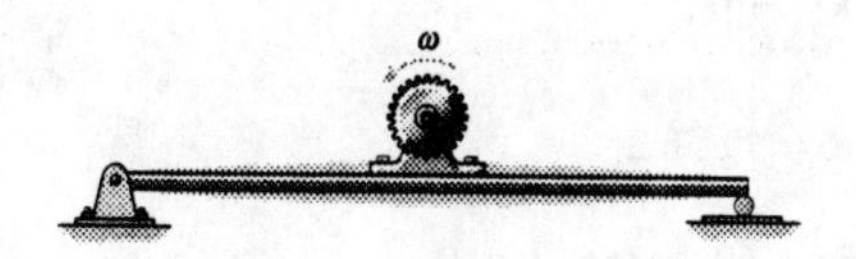

The constant value F_O of the periodic force is due to the centrifugal force of the unbalanced mass.

$$F_O = ma_n = mr\omega^2 = \left(\frac{0.25}{32.2}\right)\left(\frac{10}{12}\right)(20)^2 = 2.588 \text{ lb} \qquad \text{Hence } F = 2.588\sin 20t$$

$$k = \frac{F}{\delta} = \frac{150}{1/12} = 1800 \text{ lb/ft} \qquad p = \sqrt{\frac{k}{m}} = \sqrt{\frac{1800}{150/32.2}} = 19.657$$

From Eq. 22-21, the amplitude of the steady state motion is

$$C = \left|\frac{F_O/k}{1-\left(\frac{\omega}{p}\right)^2}\right| = \left|\frac{2.588/1800}{1-\left(\frac{20}{19.657}\right)^2}\right| = 0.04085 \text{ ft} = 0.490 \text{ in.} \qquad \textbf{Ans}$$

22-54. Determine the angular velocity of the flywheel in Prob. 22-52 which will produce an amplitude of vibration of 0.25 in.

The constant value F_O of the periodic force is due to the centrifugal force of the unbalanced mass.

$$F_O = ma_n = mr\omega^2 = \left(\frac{0.25}{32.2}\right)\left(\frac{10}{12}\right)\omega^2 = 0.006470\omega^2$$

$$F = 0.006470\omega^2 \sin \omega t$$

$$k = \frac{F}{\delta} = \frac{150}{1/12} = 1800 \text{ lb/ft} \qquad p = \sqrt{\frac{k}{m}} = \sqrt{\frac{1800}{150/32.2}} = 19.657$$

From Eq. 22.21, the amplitude of the steady state motion is

$$C = \left|\frac{F_O/k}{1-\left(\frac{\omega}{p}\right)^2}\right|$$

$$\frac{0.25}{12} = \left|\frac{0.006470\left(\frac{\omega^2}{1800}\right)}{1-\left(\frac{\omega}{19.657}\right)^2}\right|$$

$\omega = 19.0$ rad/s **Ans**

22-55. The engine is mounted on a foundation block which is spring-supported. Describe the steady-state vibration of the system if the block and engine have a total weight of 1500 lb and the engine, when running, creates an impressed force $F = (50 \sin 2t)$ lb, where t is in seconds. Assume that the system vibrates only in the vertical direction, with the positive displacement measured downward, and that the total stiffness of the springs can be represented as $k = 2000$ lb/ft.

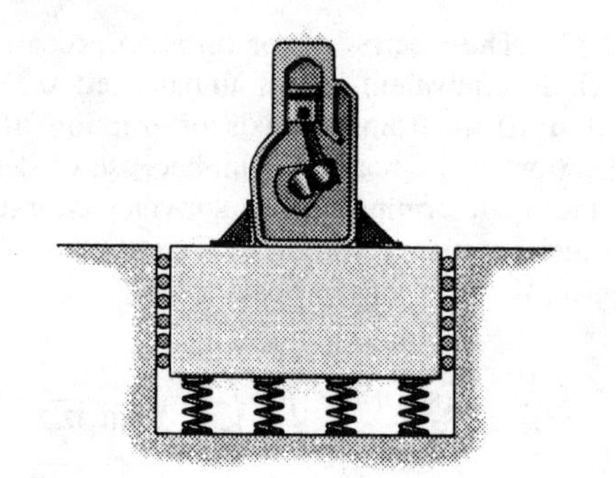

The steady - state vibration is defined by Eq. 22 - 22.

$$x_p = \frac{\frac{F_0}{k}}{1-\left(\frac{\omega}{p}\right)^2} \sin \omega t$$

Since $F = 50 \sin 2t$

Then $F_0 = 50$ lb, $\omega = 2$ rad/s

$k = 2000$ lb/ft

$$p = \sqrt{\frac{k}{m}} = \sqrt{\frac{2000}{\frac{1500}{32.2}}} = 6.55 \text{ rad/s}$$

$$\text{Hence, } x_p = \frac{\frac{50}{2000}}{1-\left(\frac{2}{6.55}\right)^2} \sin 2t$$

$x_p = (0.0276 \sin 2t)$ ft **Ans**

***22-56.** Determine the rotational speed ω of the engine in Prob. 22-55 which will cause resonance.

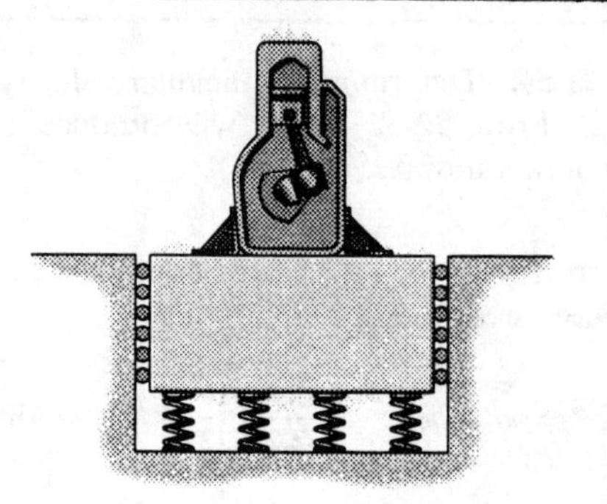

Resonance occurs when

$$\omega = p = \sqrt{\frac{k}{m}} = \sqrt{\frac{2000}{\frac{1500}{32.2}}} = 6.55 \text{ rad/s}$$ **Ans**

22-57. The block, having a weight of 12 lb, is immersed in a liquid such that the damping force acting on the block has a magnitude of $F = (0.7|v|)$ lb, where v is in ft/s. If the block is pulled down 0.62 ft and released from rest, determine the position of the block as a function of time. The spring has a stiffness of $k = 53$ lb/ft. Assume that positive displacement is downward.

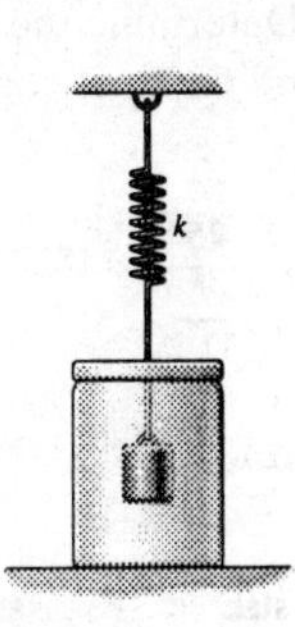

$c = 0.7$ lb · s/ft $\quad k = 53$ lb/ft $\quad m = \dfrac{12}{32.2} = 0.3727$ slug

$$p = \sqrt{\frac{k}{m}} = \sqrt{\frac{53}{0.3727}} = 11.925 \text{ rad/s}$$

$$c_c = 2mp = 2(0.3727)(11.925) = 8.889 \text{ lb}\cdot\text{s/ft}$$

Since $c < c_c$ the system is underdamped.

$$p_d = p\sqrt{1-\left(\frac{c}{c_c}\right)^2} = 11.925\sqrt{1-\left(\frac{0.7}{8.889}\right)^2} = 11.888 \text{ rad/s}$$

$$\frac{c}{2m} = \frac{0.7}{2(0.3727)} = 0.9392$$

From Eq.22-32 $\quad y = D\left[e^{-\left(\frac{c}{2m}\right)t}\sin(p_d t+\phi)\right]$

$$v = \dot{y} = D\left[e^{-\left(\frac{c}{2m}\right)t}p_d\cos(p_d t+\phi) + \left(-\frac{c}{2m}\right)e^{-\left(\frac{c}{2m}\right)t}\sin(p_d t+\phi)\right]$$

$$v = De^{-\left(\frac{c}{2m}\right)t}\left[p_d\cos(p_d t+\phi) - \frac{c}{2m}\sin(p_d t+\phi)\right]$$

Appling the initial condition at $t = 0$, $y = 0.62$ ft and $v = 0$.

$$0.62 = D\left[e^{-0}\sin(0+\phi)\right]$$

$$D\sin\phi = 0.62 \qquad (1)$$

$$0 = De^{-0}\left[11.888\cos(0+\phi) - 0.9392\sin(0+\phi)\right] \qquad \text{since } D \neq 0$$

$$11.888\cos\phi - 0.9392\sin\phi = 0 \qquad (2)$$

Solving Eqs.(1) and (2) yields :

$\phi = 85.5° = 1.49$ rad $\qquad D = 0.622$ ft

$$y = 0.622\left[e^{-0.939t}\sin(11.9t+1.49)\right] \qquad \textbf{Ans}$$

22-58. A 7-lb block is suspended from a spring having a stiffness of $k = 75$ lb/ft. The support to which the spring is attached is given simple harmonic motion which may be expressed as $\delta = (0.15 \sin 2t)$ ft, where t is in seconds. If the damping factor is $c/c_c = 0.8$, determine the phase angle ϕ of forced vibration.

$$p = \sqrt{\frac{k}{m}} = \sqrt{\frac{75}{\left(\frac{7}{32.2}\right)}} = 18.57$$

$$\delta = 0.15\sin 2t$$

$$\delta_0 = 0.15, \ \omega = 2$$

$$\phi' = \tan^{-1}\left(\frac{2\left(\frac{c}{c_c}\right)\left(\frac{\omega}{p}\right)}{1-\left(\frac{\omega}{p}\right)^2}\right) = \tan^{-1}\left(\frac{2(0.8)\left(\frac{2}{18.57}\right)}{1-\left(\frac{2}{18.57}\right)^2}\right)$$

$\phi' = 9.89°$ $\qquad$ **Ans**

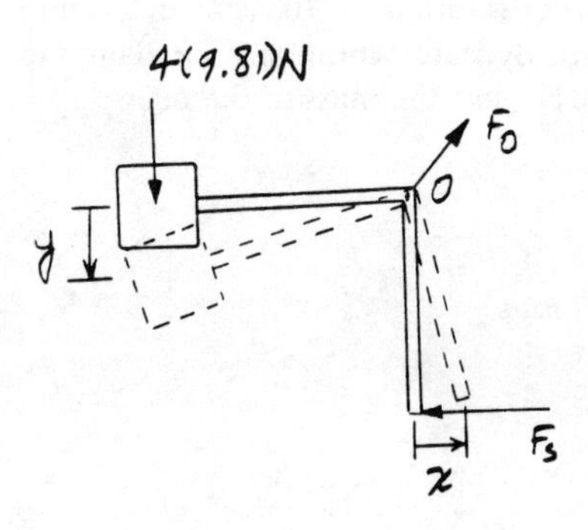

22-59. Determine the magnification factor of the block, spring, and dashpot combination in Prob. 22-58.

$$p = \sqrt{\frac{k}{m}} = \sqrt{\frac{75}{\left(\frac{7}{32.2}\right)}} = 18.57$$

$$\delta = 0.15\sin 2t$$

$$\delta_0 = 0.15, \ \omega = 2$$

$$\text{MF} = \frac{1}{\sqrt{\left[1-\left(\frac{\omega}{p}\right)^2\right]^2 + \left[2\left(\frac{c}{c_c}\right)\left(\frac{\omega}{p}\right)\right]^2}} = \frac{1}{\sqrt{\left[1-\left(\frac{2}{18.57}\right)^2\right]^2 + \left[2(0.8)\left(\frac{2}{18.57}\right)\right]^2}}$$

$\text{MF} = 0.997$ **Ans**

***22-60.** The 20-kg block is subjected to the action of the harmonic force $F = (90 \cos 6t)$ N, where t is in seconds. Write the equation which describes the steady-state motion.

$$F = 90\cos(6t), \quad F_0 = 90, \quad \omega = 6$$

$$p = \sqrt{\frac{k}{m}} = \sqrt{\frac{800}{20}} = 6.325$$

$$c_c = 2mp = (2)(20)(6.325) = 253.0$$

Here,

$$x = C'\cos(\omega t - \phi)$$

$$C' = \frac{\frac{F_0}{k}}{\sqrt{\left[1-\left(\frac{\omega}{p}\right)^2\right]^2 + \left[2\left(\frac{c}{c_c}\right)\left(\frac{\omega}{p}\right)\right]^2}} = \frac{\frac{90}{800}}{\sqrt{\left[1-\left(\frac{6}{6.32}\right)^2\right]^2 + \left[2\left(\frac{125}{253.0}\right)\left(\frac{6}{6.32}\right)\right]^2}}$$

$$C' = 0.119$$

$$\phi = \tan^{-1}\left(\frac{2\left(\frac{c}{c_c}\right)\left(\frac{\omega}{p}\right)}{1-\left(\frac{\omega}{p}\right)^2}\right) = \tan^{-1}\left(\frac{2\left(\frac{125}{253.0}\right)\left(\frac{6}{6.32}\right)}{1-\left(\frac{6}{6.32}\right)^2}\right) = 83.9^\circ$$

Thus,

$x = 0.119\cos(6t - 83.9^\circ)$ m **Ans**

22-61. The 200-lb electric motor is fastened to the midpoint of the simply supported beam. It is found that the beam deflects 2 in. when the motor is not running. The motor turns an eccentric flywheel which is equivalent to an unbalanced weight of 1 lb located 5 in. from the axis of rotation. If the motor is turning at 100 rpm, determine the amplitude of steady-state vibration. The damping factor is $c/c_c = 0.20$. Neglect the mass of the beam.

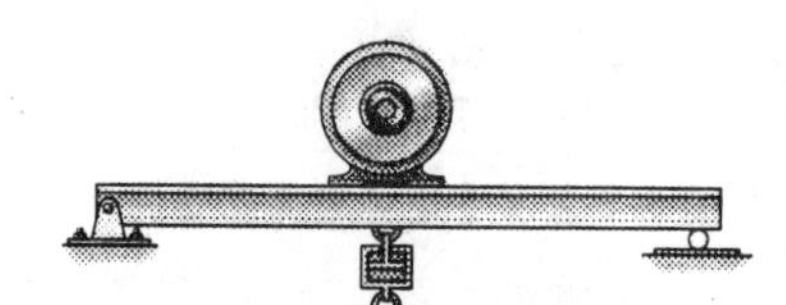

$$\delta = \frac{2}{12} = 0.167 \text{ ft}$$

$$\omega = 100\left(\frac{2\pi}{60}\right) = 10.47 \text{ rad/s}$$

$$k = \frac{200}{\frac{2}{12}} = 1200 \text{ lb/ft}$$

$$F_0 = mr\omega^2 = \left(\frac{1}{32.2}\right)\left(\frac{5}{12}\right)(10.47)^2 = 1.419 \text{ lb}$$

$$p = \sqrt{\frac{k}{m}} = \sqrt{\frac{1200}{\frac{200}{32.2}}} = 13.90 \text{ rad/s}$$

$$C' = \frac{\frac{F_0}{k}}{\sqrt{\left[1-\left(\frac{\omega}{p}\right)^2\right]^2 + \left[2\left(\frac{c}{c_c}\right)\left(\frac{\omega}{p}\right)\right]^2}}$$

$$= \frac{\frac{1.419}{1200}}{\sqrt{\left[1-\left(\frac{10.47}{13.90}\right)^2\right]^2 + \left[2(0.20)\left(\frac{10.47}{13.90}\right)\right]^2}}$$

$$= 0.00224 \text{ ft}$$

$C' = 0.0269$ in. **Ans**

22-62. A block having a mass of 7 kg is suspended from a spring that has a stiffness $k = 600$ N/m. If the block is given an upward velocity of 0.6 m/s from its equilibrium position at $t = 0$, determine its position as a function of time. Assume that positive displacement of the block is downward and that motion takes place in a medium which furnishes a damping force $F = (50|v|)$ N, where v is in m/s.

$c = 50$ N · s/m $\quad k = 600$ N/m $\quad m = 7$ kg

$$p = \sqrt{\frac{k}{m}} = \sqrt{\frac{600}{7}} = 9.258 \text{ rad/s}$$

$$c_c = 2mp = 2(7)(9.258) = 129.6 \text{ N} \cdot \text{s/m}$$

Since $c < c_c$ the system is underdamped.

$$p_d = p\sqrt{1-\left(\frac{c}{c_c}\right)^2} = 9.258\sqrt{1-\left(\frac{50}{129.6}\right)^2} = 8.542 \text{ rad/s}$$

$$\frac{c}{2m} = \frac{50}{2(7)} = 3.571$$

From Eq.22-32 $\quad y = D\left[e^{-\left(\frac{c}{2m}\right)t}\sin(p_d t+\phi)\right]$

$$v = \dot{y} = D\left[e^{-\left(\frac{c}{2m}\right)t}p_d\cos(p_d t+\phi)+\left(-\frac{c}{2m}\right)e^{-\left(\frac{c}{2m}\right)t}\sin(p_d t+\phi)\right]$$

$$v = De^{-\left(\frac{c}{2m}\right)t}\left[p_d\cos(p_d t+\phi)-\frac{c}{2m}\sin(p_d t+\phi)\right]$$

Applying the initial condition at $t = 0$, $y = 0$ and $v = -0.6$ m/s.

$$0 = D\left[e^{-0}\sin(0+\phi)\right] \quad \text{since } D \neq 0$$

$$\sin\phi = 0 \qquad \phi = 0°$$

$$-0.6 = De^{-0}\left[8.542\cos 0° - 0\right]$$

$$D = -0.0702 \text{ m}$$

$$y = \{-0.0702\left[e^{-3.57t}\sin(8.54t)\right]\} \text{ m} \qquad \textbf{Ans}$$

22-63. The damping factor, c/c_c, may be determined experimentally by measuring the successive amplitudes of vibrating motion of a system. If two of these maximum displacements can be approximated by x_1 and x_2, as shown in Fig. 22–17, show that the ratio $\ln x_1/x_2 = 2\pi(c/c_c)/\sqrt{1-(c/c_c)^2}$. The quantity $\ln x_1/x_2$ is called the *logarithmic decrement.*

Using Eq. 22 – 32,

$$x = D\left[e^{-\left(\frac{c}{2m}\right)t}\sin(p_d t + \phi)\right]$$

The maximum displacement is

$$x_{max} = De^{-\left(\frac{c}{2m}\right)t}$$

At $t = t_1$, and $t = t_2$

$$x_1 = De^{-\left(\frac{c}{2m}\right)t_1}$$

$$x_2 = De^{-\left(\frac{c}{2m}\right)t_2}$$

Hence,

$$\frac{x_1}{x_2} = \frac{De^{-\left(\frac{c}{2m}\right)t_1}}{De^{-\left(\frac{c}{2m}\right)t_2}} = e^{-\left(\frac{c}{2m}\right)(t_1-t_2)}$$

Since $p_d t_2 - p_d t_1 = 2\pi$

then $t_2 - t_1 = \dfrac{2\pi}{p_d}$

so that $\ln\left(\dfrac{x_1}{x_2}\right) = \dfrac{c\pi}{mp_d}$

Using Eq. 22 – 33, $c_c = 2mp$

$$p_d = p\sqrt{1-\left(\frac{c}{c_c}\right)^2} = \frac{c_c}{2m}\sqrt{1-\left(\frac{c}{c_c}\right)^2}$$

So that,

$$\ln\left(\frac{x_1}{x_2}\right) = \frac{2\pi\left(\frac{c}{c_c}\right)}{\sqrt{1-\left(\frac{c}{c_c}\right)^2}} \qquad \textbf{Q.E.D.}$$

***22-64.** The small block at A has a mass of 4 kg and is mounted on the bent rod having negligible mass. If the rotor at B causes a harmonic movement $\delta_B = (0.1 \cos 15t)$ m, where t is in seconds, determine the amplitude of vibration of the block.

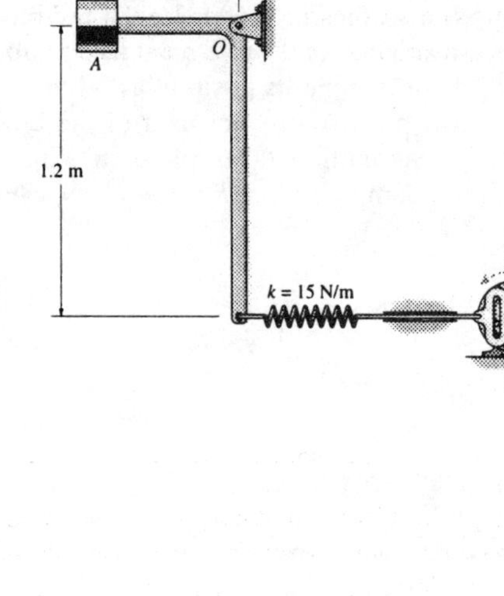

$$+\Sigma M_O = I_O\alpha; \quad 4(9.81)(0.6) - F_s(1.2) = 4(0.6)^2\ddot{\theta}$$

$$F_s = kx = 15(x + x_{st} - 0.1\cos 15t)$$

$$x_{st} = \frac{4(9.81)(0.6)}{1.2(15)}$$

Thus,

$$-15(x - 0.1\cos 15t)(1.2) = 4(0.6)^2\ddot{\theta}$$

$x = 1.2\theta$

$\ddot{\theta} + 15\theta = 1.25\cos 15t$

Set $x_p = C\cos 15t$

$-C(15)^2\cos 15t + 15(C\cos 15t) = 1.25\cos 15t$

$$C = \frac{1.25}{15 - (15)^2} = -0.00595 \text{ m}$$

Thus, amplitude is 5.95 mm **Ans**

22-65. Draw the electrical circuit that is equivalent to the mechanical system shown. Determine the differential equation which describes the charge q in the circuit.

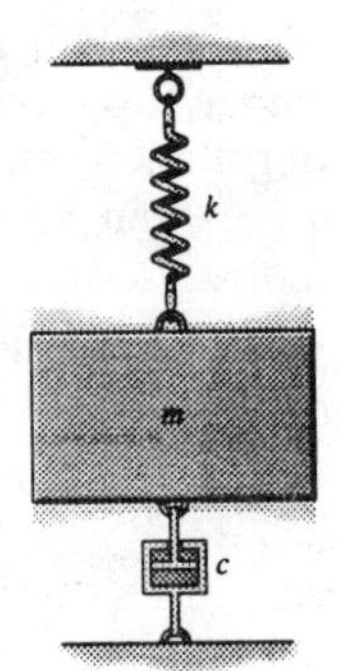

For the block

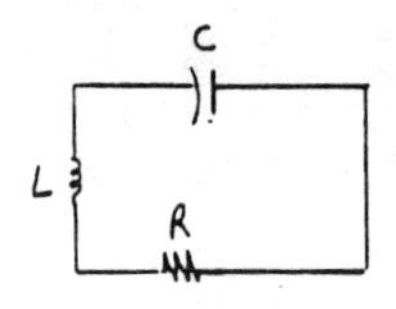

$m\ddot{y} + c\dot{y} + ky = 0$

Using Table 22 – 1

$L\ddot{q} + R\dot{q} + \frac{1}{C}q = 0$ **Ans**

22-66. Determine the mechanical analog for the electrical circuit. What differential equations describe the mechanical and electrical systems?

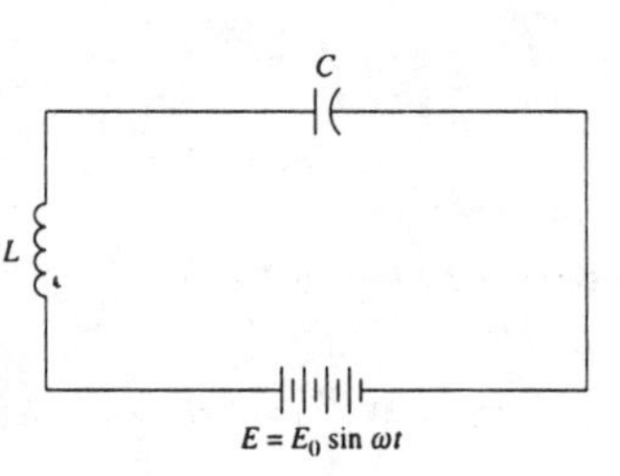

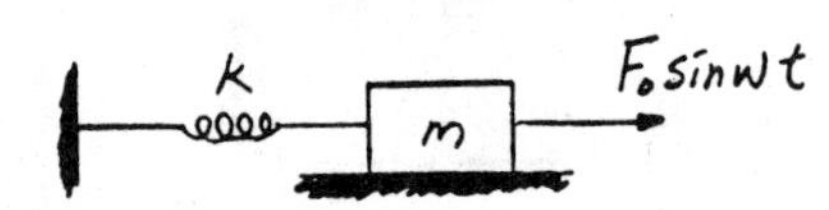

For the block:

$m\ddot{x} + kx = F_0 \sin\omega t$ **Ans**

Using Table 22 – 1

$L\ddot{q} + \frac{1}{C}q = E_0 \sin\omega t$ **Ans**

22-67. Draw the electrical circuit that is equivalent to the mechanical system shown. Determine the differential equation which describes the charge q in the circuit.

For the block,

$$m\ddot{x} + c\dot{x} + kx = F_0 \cos \omega t$$

Using Table 22–1,

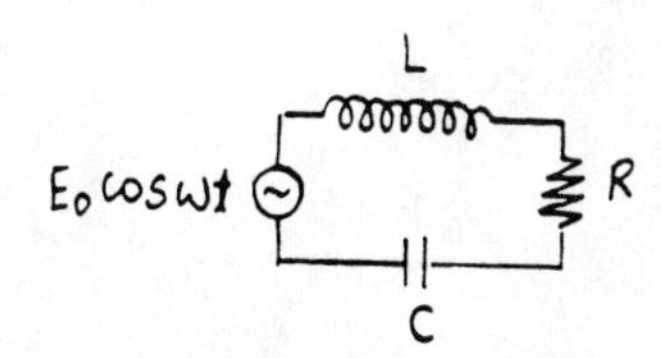

$$L\ddot{q} + R\dot{q} + (\frac{1}{C})q = E_0 \cos \omega t$$ **Ans**

***22-68.** Draw the electrical circuit that is equivalent to the mechanical system shown. What is the differential equation which describes the charge q in the circuit?

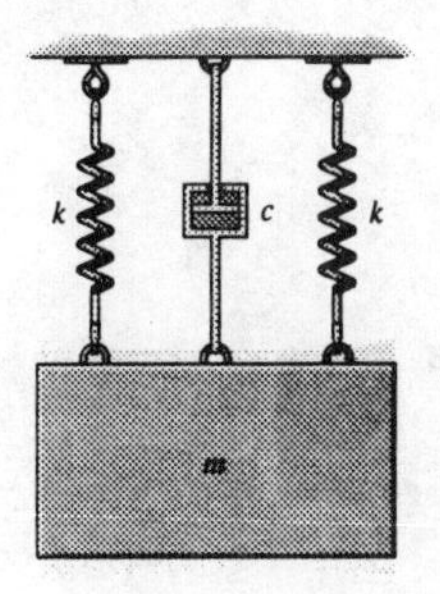

For the block,

$$m\ddot{x} + c\dot{x} + 2kx = 0$$

Using Table 22–1,

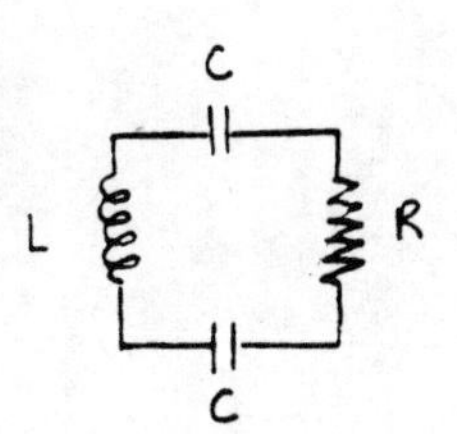

$$L\ddot{q} + R\dot{q} + (\frac{2}{C})q = 0$$ **Ans**

22-69. Determine the differential equation of motion for the damped vibratory system shown. What type of motion occurs?

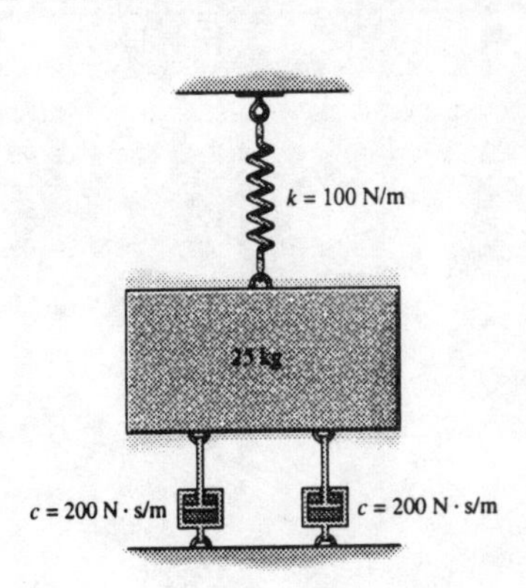

$+\downarrow \Sigma F_y = ma_y;\quad mg - k(y + y_{st}) - 2c\dot{y} = m\ddot{y}$

$$m\ddot{y} + ky + 2c\dot{y} + ky_{st} - mg = 0$$

Equilibrium $\quad ky_{st} - mg = 0$

$m\ddot{y} + 2c\dot{y} + ky = 0 \quad$ Here $m = 25$ kg $k = 100$ N/m

$c = 200$ N · s/m

$$25\ddot{y} + 400\dot{y} + 100y = 0 \qquad (1)$$

$$\ddot{y} + 16\dot{y} + 4y = 0$$ **Ans**

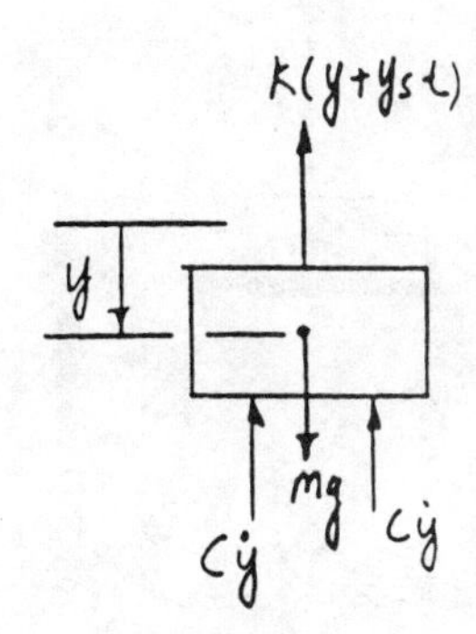

By comparing Eq.(1) to Eq. 22-27

$m = 25 \quad k = 100 \quad c = 400 \quad p = \sqrt{\frac{4}{1}} = 2$ rad/s

$c_c = 2mp = 2(25)(2) = 100$ N · s/m

Since $c > c_c$, the system will not vibrate. Therefore, it is **overdamped**. **Ans**